Perspectives in Mathematical Logic

Ω-Group:

R. O. Gandy H. Hermes A. Levy G. H. Müller
G. E. Sacks D. S. Scott

Ω-Bibliography of Mathematical Logic

Edited by Gert H. Müller

In Collaboration with Wolfgang Lenski

Volume II

Non-Classical Logics

Wolfgang Rautenberg (Editor)

Springer-Verlag Berlin Heidelberg New York
London Paris Tokyo

Gert H. Müller
Wolfgang Lenski
Mathematisches Institut, Universität Heidelberg
Im Neuenheimer Feld 288, D-6900 Heidelberg

Wolfgang Rautenberg
Mathematisches Institut, Freie Universität
Arnimallee 2, 1000 Berlin 33

The series *Perspectives in Mathematical Logic* is edited by the Ω-Group of the Heidelberger Akademie der Wissenschaften. The Group initially received a generous grant (1970–1973) from the Stiftung Volkswagenwerk and since 1974 its work has been incorporated into the general scientific program of the Heidelberger Akademie der Wissenschaften (Math. Naturwiss. Klasse).

ISBN 3-540-15521-X Springer-Verlag Berlin Heidelberg New York
ISBN 0-387-15521-X Springer-Verlag New York Berlin Heidelberg

Library of Congress Cataloging in Publication Data
[Omega]-bibliography of mathematical logic.
(Perspectives in mathematical logic)
Includes indexes.
Contents: v. 1. Classical logic / Wolfgang Rautenberg, ed. - v. 2. Non-classical logics / Wolfgang Rautenberg, ed. - v. 3. Model theory / Heinz-Dieter Ebbinghaus, ed. [etc.]
1. Locic, Symbolic and mathematical - Bibliography. I. Müller, G. H. (Gert Heinz), 1923 - II. Lenski, Wolfgang, 1952 -. III. Title: Bibliography of mathematical logic. IV. Series.
Z6654.M26047 1987 [QA9] 016.5113 86-31426

Printed in Germany

Media conversion, printing and bookbinding: Appl, Wemding
2141/3140-543210

Dedicated
to
Alonzo Church

whose bibliographic work for the
Journal of Symbolic Logic
was a milestone in the
development of modern logic.

Table of Contents

Preface			IX
Introduction			XV
User's Guide			XVII
Ω-Classification Scheme			XXV
Subject Index			1
Modal and tense logic		B45	3
Relevance and entailment		B46	38
Probability and inductive logic		B48	44
Many-valued logic		B50	53
Quantum logic		B51	81
Fuzzy logic		B52	87
Paraconsistent logic		B53	96
Intermediate and related logics		B55	100
Other logics		B60	107
Algorithmic and dynamic logic	Part of	B75	115
Proceedings	Part of	B97	119
Textbooks, surveys		B98	122
Author Index			133
Source Index			391
Journals			393
Series			414
Proceedings			417
Collection volumes			434
Publishers			442
Miscellaneous Indexes			453
External classifications			455
Alphabetization and alternative spellings of author names			465
International vehicle codes			467
Transliteration scheme for Cyrillic			469

Preface

Gert H. Müller

The growth of the number of publications in almost all scientific areas, as in the area of (mathematical) logic, is taken as a sign of our scientifically minded culture, but it also has a terrifying aspect. In addition, given the rapidly growing sophistication, specialization and hence subdivision of logic, researchers, students and teachers may have a hard time getting an overview of the existing literature, particularly if they do not have an extensive library available in their neighbourhood: they simply do not even know what to ask for! More specifically, if someone vaguely knows that something vaguely connected with his interests exists somewhere in the literature, he may not be able to find it even by searching through the publications scattered in the review journals. Answering this challenge was and is the central motivation for compiling this Bibliography.

The Bibliography comprises (presently) the following six volumes (listed with the corresponding Editors):

I.	Classical Logic	W. Rautenberg
II.	Non-classical Logics	W. Rautenberg
III.	Model Theory	H.-D. Ebbinghaus
IV.	Recursion Theory	P. G. Hinman
V.	Set Theory	A. R. Blass
VI.	Proof Theory; Constructive Mathematics	J. E. Kister; D. van Dalen & A. S. Troelstra.

Each volume is divided into four main parts:

1) The *Subject Index* is arranged in sections by topics, usually corresponding to sections in the classification scheme; each section is ordered chronologically by year, and within a given year the items are listed alphabetically by author with the titles of the publications and their full classifications added.

2) The *Author Index* is ordered alphabetically by author, and contains the full bibliographical data of each publication together with its review numbers in Mathematical Reviews (MR), Zentralblatt für Mathematik und ihre Grenzgebiete (Zbl), Journal of Symbolic Logic (JSL), and Jahrbuch über die Fortschritte der Mathematik (FdM). We much regret that we were not able to include reviews from Referativnyj Zhurnal Matematika in this edition.

3) The *Source Index* gives the full bibliographical data of each source (journals and books) for which only abbreviated forms are used in the Author Index.

4) The *Miscellaneous Indexes* contain various further indexes and tables to aid the reader in using the Bibliography.

For a more detailed technical description of the Bibliography see the *Table of Contents* and the *User's Guide*.

The uniform classification of all entries is a central feature of the Bibliography. The basic framework is the 03 section of the (1985 version of the) 1980 classification scheme of Mathematical Reviews and Zentralblatt für Mathematik und ihre Grenzgebiete. However, this has been modified in a number of ways. Indeed, the 1980 scheme was designed for the classification of works written after 1980, whereas the majority of entries in the Bibliography come before this date. In some areas this has made the classification of older works difficult, and we have tried to cope with this by adding a few new sections and altering slightly the interpretation of others. We have not designated the classifications assigned to a work as primary and secondary, because of the difficulty in doing so in many cases. Each volume contains the full annotated classification scheme together with a description of its general features. In their *introductions* the Editors discuss specifically their interpretations of the classification sections falling in their respective volumes.

The Subject Index is another central feature of the Bibliography. Reading through this Index gives a *historical perspective* for each classification section and provides a rather quick overview of the literature in it. By browsing through the entries of a specific area the reader may be rewarded by finding things (literature, subjects, questions) he was not aware of or had forgotten.

An obvious question now is the extent to which one can rely on the *completeness* and *correctness* of the Bibliography and on the *accuracy of the classifications*. We comment on each of these aspects separately.

In an effort to be as complete as possible, we consulted all sources available to us and decided in favour of inclusion in doubtful cases (so that certainly some papers with little bearing on mathematical logic are listed here and there). As the historical starting point for the Bibliography we chose the appearance of Frege's *Begriffsschrift* (1879). A certain restriction on scope stems from our decision to concentrate on mathematical logic and in particular on those areas defined by the titles of the six volumes. A major source of material was provided by the review journals mentioned above; we used them both to identify publications in the less known journals and to find review numbers and other bibliographical data of items found in other sources. We also made use of various lists of literature contained in books, survey articles, mimeographed notes, etc. Some especially valuable newer sources were:

W. Hodges: A Thousand Papers in Model Theory and Algebra

M. A. McRobbie, A. Barcan and P. B. Thistlewaite: Interpolation Theorems: A Bibliography

D. S. Scott and J. M. B. Moss †: A Bibliography of Books on Symbolic Logic, Foundations of Mathematics and Related Subjects

C. A. B. Peacocke and D. S. Scott: A Selective Bibliography of Philosophical Logic.

Various strategies and crosschecks were used to ensure the completeness of the bibliographical data and in particular of the reviews mentioned above. For each item listed in the Bibliography we tried to include any translations, reprintings in alternative sources and errata, and to give cross references for a work appearing in several parts.

On the whole this Bibliography was compiled and organized for use by the practising mathematician; there is no claim that the most rigorous standards of librarianship are met.

It is hard to say how successful our striving for completeness was. This is especially true for the most recent literature. No 1986 items were included. We checked all the main journals in logic, the reviews in MR, Zbl and JSL and Current Mathematical Publications for literature published up to the end of 1985, but undoubtedly some gaps remain.

As for correctness, in any ordinary book we can tolerate a number of printing errors because of our knowledge of the language and the context, but, when one organizes data connected by (abstract) pointers in a computer program, almost every typing error has far-reaching consequences. Various consistency tests were used to check the program and the input data. There are, however, many other sources for mistakes and errors.

For some items our references contained incomplete or ambiguous information. Although we tried to complete the bibliographical data, this was often difficult, particularly in cases where, for example, the source was obscure or the publisher was given only by location. Another source of errors lies in the identification of author names. An author may publish using abbreviations of his first, his second or both of his given names. This is generally not a problem for authors with uncommon surnames, but if the surname is, e.g., Smith or Brown the possibility of misidentification arises. We may have identified two different authors or failed to identify two or more different forms of an author name.

It is unavoidable in a project of this scope that there will be errors, particularly in the classification, so perhaps it is worthwhile to explain briefly the process by which the classification was done. Items entered before 1981 were originally classified according to a scheme unrelated to the current one. To begin the conversion to the 1980 scheme we used the computer to change old categories to their new versions wherever there was a well-defined correspondence. Then every entry was checked and if necessary reclassified by hand. From 1981 each new entry was classified shortly after being entered in the database. For the most part this was done on the basis of titles, reviews, and other information, but without consulting the works themselves. This was necessary to preserve the finiteness of the enterprise, but it has inevitably led to errors, certainly in some cases egregious ones. These were constantly being corrected during the final editing process, but many will remain.

Although the Editors have to some extent used different strategies in classifying the entries falling into their respective volumes, finally a reasonable degree of uniformity has arisen. The user is referred to the Editors' introductions for further details on the classifying procedure.

A special apology goes to the native speakers of languages with diacritical marks. Our central difficulty was to get the right spelling of names used in different forms in such a variety of sources. In addition, entering diacritical marks in a computer introduces yet another source of errors; so they have almost all been ignored (the User's Guide and the Miscellaneous Indexes contain details of those that have been transliterated). We appreciate that, although the absence of, for example, accents in the text of a French title may not create undue problems, the lack of diacritical marks in author names is particularly unfortunate. We hope that this omission will not be too misleading.

The future

By its nature a bibliography has lasting value to the extent it succeeds in "completing the past". But it should also serve for some years as an aid to current research. We have various plans to extend the scope of the present Bibliography by including new areas such as universal algebra, sheaves, philosophical logic (subdividing the present volumes I and II appropriately), and philosophy of mathematics. The present six volumes cover only approximately 80% of the data on our computer files.

The possibility of extending the classification scheme by developing a so-called thesaurus system was discussed on several occasions. Certainly this would be desirable; to some extent *Alonzo Church* tried to create such a system in connection with his bibliography in the Journal of Symbolic Logic. However there are difficult scientific problems connected with the creation of such a system and their solution requires much time and expertise.

Another way to extend the Bibliography which would perhaps better serve the purpose of providing an overview of certain special areas would be to commission a series of survey papers to appear from time to time as, say, an additional issue of the Journal of Symbolic Logic; each paper would include an annotated listing of the literature taken from the Bibliography.

There are plans to establish a bibliographical centre for Mathematical Logic and adjacent areas. A central function of such a centre would be to collect information on all new publications (including mimeographed notes, theses, etc.) as well as to correct errors and omissions in the current data. It is hoped that all logicians would provide information concerning their own publications as they appear. A continuation of the Bibliography together with supplements (to appear periodically) would be prepared at the centre. We also hope to make available an on-line system. From these activities and the flow of information from the individual logician to the centre and vice versa a "living Bibliography" would emerge. This would provide a way to determine the main trends in the progress (or decline) of specified directions of work. So a centre would exist at which it would be possible to gain some oversight of the rapidly developing field of mathematical logic.

Acknowledgements

Work on the Bibliography started at the same time as the Ω-Group came into being, early in 1969. To begin with, index cards were used for storing bibliographical data; it was *Horst Zeitler* and *Diana Schmidt* who convinced me that we are living in the 20th century and that the data should be computerized. They, together with *Ann Singleterry-Ferebee,* first brought the Bibliography to a workable computerized form at the end of the seventies. In this period I also had the help of *Ulrich Felgner* and *Klaus Gloede,* in particular in classifying the literature. At about this time, others contributed in many useful ways. In particular, important problems of principle were highlighted by a long list of intriguing questions from *Dana Scott:* "How do you classify this or that item . . .?" *Robert Harrison* worked faithfully collecting data for the Source Index.

The second period, beginning in the early eighties, was characterized by the programming necessary to manage the data. This was carried out by *Ulrich Burkhardt* (†) and *Werner Wolf* and finally by the *outstanding work* of *Rolf Bogus.* In this period we also changed the classification system and for this I had the continuous and intensive help of *Andreas Blass* and *Peter Hinman.* In addition, both of them, together with *Heinz-Dieter Ebbinghaus,* gave me much advice about organization and technical arrangements. Over the last four years the work of large groups of students has been essential for collecting reviews, entering corrections and new items into the computer, etc. Again and again I have been overwhelmed by their idealism and energy. Among them I wish to mention particularly the continuous help of *Elisabeth Wette* and *Ulrike Wieland.*

The Bibliography would not have reached publishable form without the work of my collaborator *Wolfgang Lenski* (in the second period). It would have been unthinkable for me to interfere anywhere in the process of the growth of the Bibliography without discussing the matter with him beforehand. He has accumulated a detailed knowledge of every aspect of the project and has devoted his talents for many years to the common enterprise.

My secretary, *Elfriede Ihrig,* has willingly assisted in the work of the Ω-Group and the Bibliography from the beginning, over many years, filled with ups and downs and with all kinds of tasks. She has always maintained her warmhearted balance.

To all I express my personal warm thanks!

The *Journal of Symbolic Logic* sent information concerning papers for which reviews were never published. We also acknowledge permission to use computer tapes with lists of literature covering certain periods of time from *Mathematical Reviews* and from *Zentralblatt für Mathematik und ihre Grenzgebiete.*

Yuzuru Kakuda and *Tosiyuki Tugue* collected and prepared the Japanese literature for us. *Petr Hajek* and *Gerd Wechsung* helped us with updating the bibliographical references of so many sources not available to us. *Mo Shaokui* corrected data on the Chinese literature and added items of which we were not aware.

The Editors filled many gaps, corrected all mistakes which came to their attention and undertook the burden of checking - and changing if necessary - the classification of the entries in their special areas. Here again I would like to mention *Rolf Bogus* and *Wolfgang Lenski* who organized the enormous exchange service for the transfer of literature among the Editors and the inputting of the many changes and corrections. *Andreas Blass* and *Peter Hinman* were also instrumental in this exchange; their preliminary classification of each item added to the Bibliography during the final editing process meant that the Editors had mainly to look at items inside their own areas. *Jane Kister* read through the whole Source Index correcting mistakes and suggesting valuable changes in it.

In collecting and organizing the data for the Bibliography we have received much help from various sources, and especially in letters from colleagues all over the world, containing information and suggestions. I apologize for being unable to answer them all individually, but all were read carefully.

We thank all those concerned.

As everybody can guess, the whole enterprise was indeed expensive. *Financial support was provided by the Heidelberger Akademie der Wissenschaften in the framework of the Ω-Group project.*

Special thanks go to the firm APPL, who transformed our computer tapes to the present printed form, and to the editorial and production staff of SPRINGER-Verlag for their continuous help, notably in the traditionally fine realization of the six volumes.

Finally, through working so many years on this project I have come to understand and appreciate more and more the immense work of Alonzo Church in building his Bibliography of Logic and its adjacent areas, together with a detailed classification, that is contained in so many volumes of the Journal of Symbolic Logic. Understanding comes from doing.

Introduction

Wolfgang Rautenberg

Non-classical logics are those conceptions of logic which either deviate from or go beyond the classical two-valued logic. For the particular case of intuitionistic logic, we refer to Volume VI.

The need for conceptions of logic other than the two-valued one arises primarily not from metamathematics but from the characteristics of certain disciplines, mainly philosophical ones, that differ in essential ways from mathematics. The formalisms of first-order logic are rather precisely adapted to the mathematical use of language. The methodology of other disciplines may well require a different use of language and thus a different framework of logical argumentation. The task therefore arises of analyzing such frameworks with the methods of mathematical logic.

The distinction between object-language and meta-language, mentioned in the introduction to Volume I, makes it possible to discuss in a precise mathematical fashion any conception of logic that has been implanted into the object-language. Often, variations in this conception do not even require variations in the syntax of the object-language but only in the rules that embody the intentions of the conceptions involved. The simplest example of this is a modal propositional language, having, in addition to classical propositional logic, one unary operator which can have very different interpretations, e.g., a modal one in the original sense of the word, a temporal one, a deductive one (provability in certain systems), etc.

Intensional (and therefore non-classical) conceptions of logic can be useful in connection with a semantic analysis of limited fragments of natural languages for the purpose of simplifying communication between humans and intelligent machines. For this, one would need an intensional language, understandable by humans and machines, and adapted to the sort of communication envisaged. This requires a precise linguistic and semantic analysis of corresponding conceptions of logic, an analysis that will be achieved primarily by using the highly developed techniques of mathematical logic.

For general remarks on the classification of entries in this volume, I refer to the introduction to Volume I. The borders with the primarily critical, philosophical, or linguistic literature are here particularly difficult to define, especially in view of the flood of literature on modal and intensional logic, on the so-called fuzzy logic, etc. The technical quality of a work is here, as in Volume I, not used as a criterion for its inclusion or exclusion; only the subject matter is relevant.

Historical and contentual descriptions of the individual subsections of this volume would in any case be incomplete, so we omit them and give only some essential information about our classification procedure and its differences from the Mathematical Reviews classification scheme.

For both contentual and practical reasons, several new sections have been added to the Mathematical Reviews system, namely B46 Relevance and entailment, B51 Quantum logic, and B53 Paraconsistent logic. In Mathematical Reviews, our B46 is part of B45 Modal logic.

I thank R. Routley and A. Urquhart for their help in separating the material in B46 from the rest of B45. A further subdivision of the unusually large subsection B45 would be desirable but must wait for a future edition. I thank P. Mittelstaedt for his help in separating B51 from G12; the intention was to extract from G12 the part that deals primarily with the problems of quantum logic itself rather than with the algebraic properties of those algebraic structures that are important in quantum logic. This distinction is analogous to that between classical propositional logic (B05) and the associated algebraic structures, Boolean algebras (G05).

It should be noted that the classification system puts into Section G the study of algebraic structures associated to various non-classical logics. Thus, the part of the literature on these logics that deals primarily with the corresponding algebraic structures is not included in this edition, although the database of the Ω-Bibliography contains a fairly complete listing of the literature in Section G.

I thank N.C.A. da Costa for his help in assembling the material for B53 (mostly from B60). My original goal of entirely replacing B60 with more specific classifications proved unattainable. Thus B60 still contains a quite heterogeneous body of literature, ranging from the entirely useful to the exotic.

Because the preparation of the database did not include a systematic search for material in the area of the logic of programming languages (B75), this edition contains from this area only works on algorithmic and dynamic logic. Only in these subfields of B75 is the Ω database sufficiently complete for publication.

Because Volumes I and II were split apart so late, it was impossible to split the material in B98 (textbooks and surveys). This is probably not a drawback, for the reader interested in the topics of Volume I will find in Volume II at least the textbook literature in the fields of Volume I and to some extent also of other fields of logic.

Acknowledgements

First of all I want to express my thanks to Gert Müller for his patience in the course of preparing this Bibliography. I also thank the other editors for valuable hints and suggestions. I am most obliged to my student Manfred Borzechowski. Not only is he very fast at finding a paper or a review in a large mathematical library, he also assisted in reading and classifying, and he selected the material for the part of B75 included in this edition of the Bibliography. Finally, I want to express my particular thanks to Andreas Blass who translated this introduction into English.

User's Guide

Wolfgang Lenski & Gert H. Müller

§1. Introduction

After some opening remarks, the organization of this Guide follows the main division of the volume: *Subject Index, Author Index, Source Index, Miscellaneous Indexes*. For each part we give first a general explanation followed by a more detailed description of typical entries in the index in question. The reader will probably find the User's Guide most helpful when he comes across an unclear entry in the Bibliography: he can then turn directly to the corresponding section in this Guide for an explanation of the abbreviations and conventions used.

§2. General remarks

The main languages of the Bibliography are English, French, German, Italian, and Spanish. For other languages translations (of titles, names of sources, etc.) are used - with some few exceptions in cases for which we had no translation. These translations were taken from various available sources or made by the Editors.

For practical reasons, all entries are in the Roman alphabet and diacritical marks have not been used. Thus, for languages other than English certain conventions have been adopted.

The *transliteration of Cyrillic* names and titles, the treatment of *diacritical marks* and the *alphabetization and alternative spelling* of author names are explained in detail in the Miscellaneous Indexes.

The *abbreviations* of *sources* were either taken from one of the various reviewing journals or invented by us. Although we had to abbreviate long titles, we hope that in most cases the abbreviation will suggest the full title in a sufficiently understandable way. How successful we were is left to the user to decide.

The *review numbers* given with the entries in the Author Index are from *Mathematical Reviews* (MR), *Zentralblatt für Mathematik und ihre Grenzgebiete* (Zbl), *Journal of Symbolic Logic* (JSL), and *Jahrbuch über die Fortschritte der Mathematik* (FdM). We made a serious attempt to include all reviews of any given item but we have doubts concerning our success. We also tried to avoid listing two reviews for a given item in those cases in which the second "review" simply points to the original review and does not give any additional information.

In case of *multipart publications* pointers are given to the other parts, as far as they are known, in the *Remarks* to the publication in question. It is not always the case that the different parts of a publication all have the same classifications. Thus it may happen that, for example, part I has a classification in this volume and part II does not. In this case the Remarks for part I indicate the author(s) and year of publication of part II. The user will need to consult the other volumes for further bibliographic information on part II.

The general way to search through the Bibliography is to use certain *pointers:* From the Subject Index to the Author Index the pointer is [Author, Year, Title]; from the Author Index to the Source Index the pointers are 5-digit codes; e.g. (**J** 1234) is a code which sends the user to the J-section of the Source Index.

A *word* of *caution*: In order to use the Bibliography for quotations in future publications it is necessary to use both the Author Index and the Source Index; it is not generally sufficient to quote just from the Author Index. For example, for the paper "AANDERAA, S.O. [1974] *On k-tape versus (k – 1)-tape real time computation*" a quotation of the source of this paper as given in the Author Index listing of this item, "Complexity of Computation; 1973 New York 75-96", would with high probability be misleading: one might try to find a book of this title published in New York in 1973 whereas in fact "1973 New York" denotes the date and place of the conference in the proceedings of which Aanderaa's paper appears. The volume was actually published in 1974. The source code (**P** 0761) for "Complexity of Computation; 1973 New York" should be used to find the full details of the source in the Source Index. The abbreviations of the sources may themselves be misleading without the corresponding additional details (e.g. country codes) given in the Source Index. For example, many abbreviations for conference proceedings do not include an abbreviation for "Proceedings of ...". Thus "Proceedings of the Third Brazilian Conference on Mathematical Logic" is abbreviated by "Brazil Conf Math Logic (3); 1979 Recife"; a reader without the Bibliography at hand might search in vain for the volume under "Brazil" in his library whereas in fact it might be found alphabetically under "Proceedings".

The Source Index includes, as far as they are known to us, *International Standard Serial Numbers* (ISSN) or *Book Numbers* (ISBN) and *Library of Congress* (LC) numbers. They may help in finding the source in question in libraries or bookstores.

To facilitate searches for works spanning two or more of the major subfields of logic, the first of the Miscellaneous Indexes lists the entries in the present volume that also occur in other volumes of the Bibliography.

Accidental occurrences of features not explained in the User's Guide are left as exercises to the user. HINT: Write to us (in any case), please.

§3. Subject Index

This is a listing of publication items ordered

first by the (special) *classification sections,*
then by the *year of appearance,*
and finally *alphabetically* by *author,*

showing the author, title and the codes of all classification sections which apply to the given publication.

• The *titles* are given in the main languages of the Bibliography; if the original title is in another language, this is indicated in parentheses, e.g. (Russian), but only a translation of the title is given. Information on summaries in languages other than the original is included.
• If a publication is by *multiple authors,* it occurs only *once,* under the alphabetically first name. (But see also the Author Index.)
• In order to get the full bibliographical data of a publication, use the author, year and title to find the item in the Author Index.
• The classification sections listed in each volume have been selected by the individual editors. Sections B96-F96 have been systematically omitted; for the collected works of an author refer to the Author Index.

§4. Author Index

This is a listing of publication items ordered

alphabetically by *author,* and for a given author
chronologically by the *year of appearance,* and therein
alphabetically by *title* of the item.

• The titles are given in the main languages as in the Subject Index.
• The *names* of the *authors* are written in the Roman alphabet using the Transliteration Table (see Miscellaneous Indexes) if necessary. There may be many versions of the name in use for a given author; (e.g. different combinations of the given name(s) or initials; different names used before and after marriage; different transliterations). The Miscellaneous Indexes include a table of different versions (known to us) and the corresponding form used in this Bibliography.
• Here publications with *multiple authors* are listed under *each* author but in the alphabetically later cases only the year is given and there is a pointer to the full entry given under the first author.
• The last entries for an author may contain a *reference to other name(s)* under which he/she also has publications in the Bibliography or *to other volumes* of the Bibliography where he/she has publications not mentioned in the present volume. A complete list of the author's papers contained in the six volumes is obtained by consulting the other volumes.
• In the following we explain the *individual entries* in more detail by giving an *idealized* example using fictitious names and sources showing all features that might occur; in a given case some features may not appear either because they do not apply or because our information is incomplete. The typefaces of the example and the order of its fields are as in the Author Index but, for *expository reasons only, here* we list all features on separate lines numbered by (1), (2), ...; we list explicitly those fields that begin a new line in the Author Index itself. (The foregoing description applies not only to the explanation of the Author Index treated here but also to the explanation of the Source Index later on.)

Example

(1) AUTHOR, K.J. & COMPANION, CECIL X. [1972]
(2) *On coding and decoding (Russian) (English and French summaries)*
(3) (**J** 9999) or (**S** 9998) or (**P** 9997) or (**C** 9996) or (**X** 9995)
(4) J Math 1*1-10
or Math Logic Series 1
or Logic Conf; 1999 London 3-10
or Math Publ xxv+200pp
(5) • ERR/ADD ibid 2*3-4 or (**J** 8888) Arch of Logic 2*3-4
(A new line begins here.)
(6) • LAST ED [1983] (**X** 9900) Logic Publ xx+100pp
(7) • REPR [1981] (**J** 9901) Math Logic J 2*3-8
(8) • TRANSL [1979] (**J** 9902) Math Transl 1*4-8
(A new line begins here.)
(9) ◇ B05 B20 C12 ◇
(10) • REV MR 99a:03001 Zbl 999#03001 JSL 99.321 FdM 99.123
(11) • REM This is an illustrative example
(12) • ID 12345

Explanations

(1) lists the authors followed by the year (in brackets) of publication of the item. Exceptionally a full given name (e.g. CECIL) is used to distinguish several authors with the same surname and initials.

(2) gives the title of the item followed, if the original is not in an official language of the bibliography, by the original language in parentheses and an indication of summaries in languages other than the original.

(3) is a pointer (or "source code") to the *Source Index;* there are five types: *Journal* (**J**), *Series* (**S**), *Proceedings Volume* (**P**), *Collection Volume* (**C**), and *Publisher* (**X**); one such code appears in (3). In order to find the full bibliographical data of the source use the pointer to locate the source in the Source Index; e.g. (**J** 9999) is given in the J-section of the Source Index.
Note: For a small number of items the source code is 0000, 1111, 2222 or 3333 (*not* preceded by J,S,P,C, or X). The code 0000, respectively 1111, indicates that the item is a thesis, respectively technical report. The code 2222, respectively 3333, is used for those cases in which the source, respectively publisher, is unknown. In each such case any further source information available is given in the Remarks (see line (11)).

(4) contains the abbreviation of the source indicated by the code in (3) followed by the paging as appropriate. Certain uniform features of the form of abbreviation used for proceedings and collection volumes should help the reader to recognise the volume. Abbreviations for proceedings (**P**) volumes end with an indication of the year and place of the corresponding conference, e.g. 1973 New York. Likewise, a name in parentheses, e.g. (Goedel), in an abbreviation of a collection (**C**) volume indicates the honorand to whom the volume is dedicated. A name followed by a colon, e.g. "Wang:", at the beginning of a collection volume abbreviation, indicates the author of all papers in the collection. The paging takes one of the following forms:

1*1-10 : Volume 1, pages 1-10 (for journals or series)
1/2*1-10 : Volume 1, Issue 2, pages 1-10 (for journals)
3-10 : pages 3-10 (for proceedings or collection volumes)
xx + 200pp: initial paging + paging of a book (following a publisher or series)

(5) The • here and later is intended to make the entries easier to read. It is used to separate different types of information. After the • is the bibliographical information for published errata or addenda to the item. The two ways ERR/ADD can be given correspond to the cases in which its source is the same as in (4) (indicated by "ibid") and that in which it is in a different source; in the latter case the entry is of the same form as in (3) and (4).

The remaining information is not strictly part of the bibliographical data but contains useful additions.

(6), (7), (8) list the most recent edition, reprintings and translations, respectively, given by source as in (3) and (4); note that (7) and/or (8) may contain several entries for one publication.

(9) The classification codes enclosed in ◇ always begin a new line. Note that the codes are given in alphabetical/numerical order; no distinction of primary and secondary classification is made. (The classifications often differ from those assigned to the item in MR or Zbl.)

(10) lists the reviews. Sometimes two reviews are given from one reviewing journal. This may happen, e.g., when an item and its erratum/addendum are reviewed separately or when two different editions of a book have independent reviews.

(11) contains additional information not appropriate for coding in one of the standard fields.

(12) Each entry ends with its *identification number.* It is not used elsewhere in the main body of this volume except occasionally in the Introduction and the Remarks of another item where it may be used to pinpoint an item not uniquely identified by author(s) and year. The identification number is used (together with author(s) and year) as a pointer in the External Classification Code Index. We ask that the identification number be used in any correspondence with the Editors concerning this publication, as the bibliographical data base is indexed by these numbers.

§5. Source Index

This index contains the bibliographical data of the sources of the publications listed in this volume. It is subdivided into the following parts.

J (*Journals*), **S** (*Series*), **C** (*Collection volumes*), **P** (*Proceedings*), **X** (*Publishers*).

• Each part is ordered by the 4-digit source code numbers. (There is no significance to the particular 4-digit number assigned to a given source other than as a way to find the entry in the source index. Numbers were assigned as the sources were entered into the data base and so the numbering does not correspond to alphabetical order or order of publication.) Each 4-digit number is used *only once* as a source code so that, e.g., 0007 is a source code for a journal and the number 0007 is not used as a code for a series, proceedings, collection volume or publisher.
• Titles are given in the original language, using the transliteration system (see Miscellaneous Indexes) where necessary, followed, if necessary, by a translation into one of the main languages in parentheses. Sometimes if the original title is unknown to us, we give only a translated title in parentheses. Sometimes a source, e.g., a journal, has more than one title (English, French, German); in this case all titles are given, separated by *. These measures were taken to ease the search in libraries. In order to explain the entries in the Source Index we again use idealized examples and apply the conventions described in §4 above.

Journals

Example of a journal entry:

(1) **J** 8888 Math Div • F
(A new line begins here)
(2) *Mathematica Diversa * Mathematiques Diverses*
(3) [1900ff] or [1905-1935] ISSN 0007-0882
(A new line begins here.)
(4) • CONT OF (**J** 8885) J Math Ser A
(5) • CONT AS (**J** 8887) J Math Ser C
(6) • TRANSL IN (**J** 9904) Math Transl
(7) • TRANSL OF (**J** 9905) Matemat
((4) - (7) may contain more than one entry)
(A new line begins here.)
(8) • REL PUBL (**J** 9903) Mathematica (Subseria)
(9) • REM This journal is a fiction

Explanations

(1) Source code and abbreviation of the journal as used in the Author Index followed by the *international vehicle code* of the country in which the journal is published. A list of these codes is included in the Miscellaneous Indexes.

(2) The form of title(s) (and translations) are explained above.

(3) [1900ff] indicates that this journal has appeared continuously since 1900; [1905–1935] indicates that the journal appeared from 1905 to 1935. The International Standard Serial Number (ISSN) is given whenever possible.

(4), (5) give the predecessors (continuation of) and successors (continued as) of the journal in (2). In some cases in (4) or (5) the source code may be missing; this means that there are no entries in the Author Index which refer to the continued source. (It is mentioned, however, for the convenience of the user.)

(6) lists the translation journals of (2) and (7) gives the journal of which (2) is a translation; the source code is shown only if the translation in question is used as a source in this Bibliography. (6) and (7) do not both occur in a single journal entry.

(8) lists further entries in the Bibliography related to this journal, e.g. a subseries of the journal.

(9) is intended for additional information of various kinds.

Series

It is often hard to determine what should and what should not be characterised as a series. Some serials that we have chosen to treat as series may elsewhere be considered to be journals. In other cases, in particular certain publication series of university mathematics departments, the series includes all publications of its publisher and so might reasonably be identified with the publisher. Despite these considerations, we have chosen to list series separately to accord with the form of quotation often used in the modern literature.

Example of a series entry:

(1) **S** 8999 Notae Log • NL
(A new line begins here)
(2) *Notae Logicae* * *Notas Logicas*
(3) [1900ff] or [1905–1935]
(4) • ED: EDITOR, A.A. & COEDITOR, B.B.
(5) • SER (**S** 8998) Notes in Phil
(6) • PUBL (**X** 9950) Logic Publ Co: Heidelberg
(7) • ALT PUBL (**X** 9951) Math Publ Inc: London
(A new line begins here.)
(8) • CONT OF (**S** 9975) Notes in Logic A
(9) • CONT AS (**S** 9901) Notes in Logic B
(10) • TRANSL IN (**S** 9902) Notes de Logique
(11) • TRANSL OF (**S** 9903) Logical Notes
(A new line begins here.)
(12) • ISSN 0011–11122 (or ISBN 0011–11123) LC-No 73–10000
(13) • REL PUBL (**S** 9900) Notae Logicae (Subseria)
(14) • REM The origins of this series are somewhat obscure

Explanations

Entries (1), (2), (3), (8)–(11), (13), and (14) correspond to (1), (2), (3), (4)–(7), (8), (9), respectively, of the *journal entry* described above.

(4) lists the editors of the series (given in the same form as in line (1) of the Author Index example).

(5) Occasionally a series is itself a subseries of another series or journal. This is indicated in (5) (with an **S** or **J** as appropriate).

(6) gives the publisher of (2). For those publishers not listed in the publisher section of the Source Index, an abbreviation is sometimes used if either the abbreviation is readily understandable or the full name is not known.

(7) Some sources are published by two or more publishers; ALT PUBL lists the alternative publisher(s).

(12) lists the ISSN (or ISBN) and the Library of Congress number.

Proceedings and Collection Volumes

Example of a proceedings or collection volume:

(1) **P** 9920 Atti Congr Mat; 1971 London, ON • CDN
or
C 9921 Atti Congr Mat • D

(A new line begins here.)

(2) [1972]

(3) *Atti del Congresso di Matematica * Actes du Congres de Mathematique*

(4) • ED: EDITOR, A.A. & COEDITOR, B.B.

(5) • SER (**S** 8999) Notes in Logic

(6) • PUBL (**X** 9950) Logic Publ Co: Heidelberg

(7) • ALT PUBL (**X** 9951) Math Publ Co: London

(A new line begins here.)

(8) • DAT&PL 1971 Aug;London, ON, CDN

(9) • ISBN 0-012-34567-X, LC-No 84-98765

(10) • REL PUBL (**P** 9947) Atti Congr Mat Vol Spez

(A new line begins here.)

(11) • TRANSL IN [1973] Conf de Logique Math (3); London, ON, CDN
• PUBL (**X** 9949) Livres: Paris

(12) • TRANSL OF [1971] Konf Math Logik (3); London, ON, CDN
• PUBL (**X** 9948) Buchverlag: Stuttgart

(A new line begins here.)

(13) • REM Not all the articles appear in the translation

Explanations

(1), (3), (4) - (7), (9), (11), (12), (13) correspond to (1), (2), (4)-(7), (12), (10), (11) and (14), respectively, of a *series entry*. In (11), (12) PUBL denotes the publisher of the translation or original, respectively.

(2) denotes the year of publication of the volume (and not, in the case of a proceedings, the year of the conference).

(8) is used for proceedings volumes to indicate the date (year and month) and place of the conference, given by the city, the state (for the USA and elsewhere) and the country using its code as defined above. Note in case of *Proceedings* (**P**) volumes in (1) the country code of the place of the conference is repeated for conformity reasons, whereas for *Collection* (**C**) volumes the country code in (1) refers to the location of the publisher as in the case of *Journals* and *Series*.

(10) lists further entries in the Bibliography related to this volume, e.g. another proceedings volume of the same conference or a journal of which the volume is a special issue.

Publisher

Example of a publisher entry:

(1) **X** 9950 *Logic Publishing Company* (Heidelberg, D & London, GB) ISBN 0-01

(2) • REL PUBL (**X** 9930) Editions Logiques: Paris, F

(3) • REM In London called Logic Publishing Corporation

Explanations

(1) lists the source code and full name of the publisher followed, in parenthesis, by the cities from which the publisher publishes and the ISBN. As in (8) of a **P** or **C** entry, codes are used for countries (see Miscellaneous Indexes).

(2) lists those publishers who have connections with the publisher listed in (1).

§6. Miscellaneous Indexes

This part contains the following indexes:

1. External classifications
2. Alphabetization and alternative spellings of author names
3. International vehicle codes
4. Transliteration scheme for Cyrillic

In each case a description of the contents and use are given in the corresponding introductory texts.

Ω-Classification Scheme

Andreas R. Blass
Peter G. Hinman

The classification scheme used for the Ω-Bibliography is a modified version of the section "03: Mathematical Logic and Foundations" of the 1985 Mathematics Subject Classification of *Mathematical Reviews* and *Zentralblatt für Mathematik und ihre Grenzgebiete*. For the sake of uniformity we have labeled all sections with a letter followed by a two-digit number; the prefix 03 is superfluous and therefore omitted. This decision has led to the creation of new sections to replace 03-01 through 03-06 (cf. X96-X98 and A10) and several sections with prefix other than 03 which have substantial logical content. Examples of the latter sort are B70 (to replace 94C10) and B75 (to replace the "logical part" of 68B10) (68Q55 and 68Q60 since 1985).

An important category of differences between the two schemes arises from the fact that whereas the MR/Zbl system is intended to classify works written after 1980, the majority of entries in the Ω-Bibliography were written before 1980. The subject matter of Mathematical Logic has, of course, changed immensely over the years, and today's categories are not always sufficient to distinguish properly important lines of earlier research. To deal with this problem we have added a few new sections (e.g. B22, B28, B65, C07, E07, and E47), renamed others (e.g. B35, C35, and E10), and altered slightly the interpretation of others (e.g. B25 and D65). To aid the reader in learning our conventions we have added descriptors to the section names. Topics preceded by a + (−) sign are specifically included (excluded) from a section. When this is in conflict with current MR/Zbl practice, this fact is also noted.

A

A05 Philosophical and critical

A10 History, Biography, Bibliography
MR uses 03-03 and 01A for history and biography
MR puts bibliography under specific fields.

B GENERAL LOGIC

B03 Syntax of logical languages

B05 Classical propositional logic and boolean functions
+ Axiomatizations of classical propositional logic
+ Boolean functions (machine manipulation is also in B35); MR puts these in G05 and in 06E30 and 94C10.
− Fragments of propositional logic: see B20
− Switching circuits: see B70; MR also uses 94C10

B10 Classical first-order logic
+ Many-sorted logic
+ Syntax and semantics up to the Completeness Theorem
− Model theory: see Cn, particularly C07
− Proof theory: see Fn

B15 Higher-order logic and type theory
+ Higher-order algebraic and other theories
– Higher-order model theory: see C85
– Set theory with classes: see E30 and E70
– Intuitionistic theory of types: see F35

B20 Fragments of classical logic
+ Fragments of propositional and of first-order logic
+ Fragments used in model theory, set theory, etc.
+ Syllogistic
– Classical propositional logic: see B05
– Weak axiomatizations without restrictions on formulas: see B55, B60, F50 ("Fragment" refers to reduced expressive power, not reduced deductive power; MR heading "Subsystems of classical logic" includes both)

B22 Abstract deductive systems
+ Consequence relations
MR uses B99

B25 Decidability of theories and sets of sentences
+ Decidability of satisfiability
+ Decidable Diophantine problems
– Decidable word problems: see D40
– Other decidability results: see subject of problem, e.g. D05, or D80; MR includes these results here.
– Undecidability results: see D35, D40, D80, etc.

B28 Classical foundations of number systems
+ Natural numbers, real numbers, ordinal numbers
+ Axiomatic foundations and set-theoretic foundations
MR uses B30

B30 Logical foundations of other classical theories; axiomatics
+ Axiomatic method
+ Geometry, probability, physics, etc.
+ Models for non-mathematical theories
– Foundations of parts of logic: see that part.
MR heading: "Foundations and axiomatics of classical theories" includes also B28

B35 Mechanization of proofs and logical operations
+ Theorem proving, proof checking by machine
+ Minimization algorithms for Boolean functions
+ Optimization of logical operations
MR sometimes uses 03-04 or 68G15 (68T15 since 1985)

B40 Combinatory logic and lambda-calculus
+ Models of lambda-calculus

B45 Modal and tense logic
+ Intensional logic; see also A05
+ Normative and deontic logic
+ Other non-truth-functional systems

B46 Relevance and entailment
+ Fragments
– Primarily modal logic
MR uses B45

B48 Probability and inductive logic
See also A05 and C90
+ Confirmation theory
– Foundations of probability: see B30; MR uses B48

B50 Many-valued logic
+ Matrix interpretations of propositional connectives unless used only as a tool for investigating classical propositional logic.
– Boolean valued set theory: see E40
– Probability logic: see B48 or C90

B51 Quantum logic
– Algebraic study of Quantum logic: see G12
MR uses only G12

B52 Fuzzy logic
+ Vagueness logic
– Papers demonstrating the fuzziness of the author's thought processes

B53 Paraconsistent logic
+ Discussive and dialectical logic
MR uses B60

B55 Intermediate and related logics
+ (Fragments of) propositional and predicate logics between intuitionistic or minimal and classical

B60 Other logics
– Intuitionistic logic: see F50 (MR uses B20)

B65 Logic of natural languages
– Computer languages: see B75
– Formal grammars unless applied to natural languages: see D05
– Natural language as a tool for the study of thought, reality, etc.: see A05
MR uses B65, B99, and 68Fn (68Sn since 1985)

B70 Logic in computer design; switching circuits
+ Hardware related to logic
MR uses 94Cn

B75 Logic of algorithmic and programming languages
+ Algorithmic and dynamic logic; MR uses B70 (formerly B45)
+ Logical analysis of programs
+ Logical aspects of database query languages and information retrieval
+ Semantics of programming languages related to logic
+ Software related to logic
– Specific algorithms: see subject of algorithm
MR uses B60, B70, 68Bn, 68Fn, and 68H05 (68Pn, 68Qn, and 68Tn since 1985)

B80 Other applications of logic
MR uses B99

B96 Collected works
+ Selected works
– Collections (almost) entirely in one subfield: see that subfield
MR uses 01A75, 03-03, and 03-06

B97 Proceedings
+ Collections of papers by various authors, even if they do not derive from any actual conference
– Proceedings (almost) entirely in one subfield: see that subfield
– Proceedings not concentrated in this field: see Source Index

MR uses 03-06

B98 Textbooks, surveys
MR uses 03-01 and 03-02

B99 None of the above or uncertain, but in this section

C MODEL THEORY

C05 Equational classes, universal algebra
+ Quasi-varieties, if the emphasis is algebraic
– Word problems: see D40

C07 Basic properties of first-order languages and structures
+ Completeness, compactness, Löwenheim-Skolem, and omitting types theorems for ordinary first-order logic; MR uses C50 for omitting types
+ General properties of first-order theories
+ Homomorphisms, automorphisms, and isomorphisms of first-order structures
– Analogues of these for stronger languages: see C55, C70, C75, etc.

C10 Quantifier elimination and related topics

C13 Finite structures
+ The spectrum problem
+ Probabilities of sentences being true in finite structures

C15 Denumerable structures

C20 Ultraproducts and related constructions
+ Applications of ultraproducts
+ Reduced products, limit ultrapowers, etc.
– General products: see C30

C25 Model-theoretic forcing
+ Existentially closed structures, model companions, etc.
– Model complete theories: see C35
– Set-theoretic forcing: see E35, E40

C30 Other model constructions
+ Contructions involving indiscernibles
+ Products, diagrams

C35 Categoricity and completeness of theories
+ Model completeness
– Gödel's completeness theorem: see C07
– Completeness of axiomatizations of other logics: see those logics, e.g., B45

C40 Interpolation, preservation, definability
+ Definability in classes of structures
– Definability in recursion theory: see appropriate Dn.
– Definability in set theory: see E15, E45, and E47

C45 Stability and related concepts
+ Rank, total transcendence (even before stability was defined)

C50 Models with special properties
+ Saturated, rigid, etc.

C52 Properties of classes of models

C55 Set-theoretic model theory
+ Cardinality and ordering of models
+ Generalized Löwenheim-Skolem results
– Applications of set theory to some part of model theory: see that part
– Models of set theory: see C62
– Original Löwenheim-Skolem theorem: see C07

C57 Recursion-theoretic model theory
+ Model theory of recursive, arithmetical, etc. structures, types, etc.
– Recursion theory without substantial model-theoretic content: see D45
MR uses D45

C60 Model-theoretic algebra
+ Applications of model theory to specific algebraic theories
– Applications of set theory to algebra: see E75
– Decidability questions for algebraic theories: see B25, D35, and D40
– Model theory of orderings: see C65
– Universal algebra: see C05

C62 Models of arithmetic and set theory
+ Admissible sets as models: see also C70 and D60
+ Nonstandard models of arithmetic, when model theory is emphasized
+ Omega-models of higher-order arithmetic
– Models introduced only for consistency results: see F25 and E35
– Nonstandard models of arithmetic, when non-standardness is emphasized: see H15 or H20
MR uses C62, C65, F30, or H15

C65 Models of other mathematical theories
+ Other applications of model theory outside logic
+ Theories of orderings
– Uses of models for purely foundational studies: see B30

C70 Logic on admissible sets
+ All sorts of "effective" infinitary logic

C75 Other infinitary logic
+ Infinitary logic even if not model theory, e.g., infinite terms in proof theory and infinitary definability in set theory

C80 Logic with extra quantifiers and operators
– Hilbert epsilon-theorems: see B10
– Modal or many-valued operators: see B45 or B50

C85 Second- and higher-order model theory
+ Weak second-order theories (quantification over finite sets)

C90 Nonclassical models
+ Boolean-valued models
+ Sheaf models
+ Kripke models (also in B45 or F50)
+ Probability models (often also in B48)
+ Topological models (unless the topological structure is condensed into a quantifier: see C80); MR uses C85
– Models of lambda calculus: see B40

C95 Abstract model theory
+ Lindström's theorem, delta-logics, etc.

C96 Collected works
+ Selected works
– Collections (almost) entirely in one subfield: see that subfield

MR uses 01A75, 03-03, and 03-06

C97 Proceedings
+ Collections of papers by various authors, even if they do not derive from any actual conference
– Proceedings (almost) entirely in one subfield: see that subfield
– Proceedings not concentrated in this field: see Source Index

MR uses 03-06

C98 Textbooks, surveys

MR uses 03-01 and 03-02

C99 None of the above or uncertain, but in this section

D RECURSION THEORY

D03 Thue and Post systems, etc.
+ Markov's normal algorithms

D05 Automata and formal grammars in connection with logical questions
+ Cellular automata
+ Finite automata
+ Generalized automata
+ Regular events
– Grammar of natural languages: see B65

MR uses 68 for most of these topics

D10 Turing machines and related notions
+ Potentially infinite automata
+ Probabilistic Turing machines

D15 Complexity of computation
+ Chaitin-Kolmogorov-Solomonoff complexity
+ Finer classification of decidable problems
+ Generalized complexity
+ Resource-bounded computability and reducibility
+ Speed-up theorems
– Complexity of derivations and proofs: see F20
– Complexity of specific non-logical problems (excluded from the Ω-Bibliography)
– Syntactic complexity, complexity of Boolean functions, etc.

MR uses also 68Q15

D20 Recursive functions and relations, subrecursive hierarchies

+ Computable functions of real numbers; MR uses D65 and F60
+ General theory of algorithms
+ Partial recursive functions
+ Primitive recursion

D25 Recursively enumerable sets and degrees

+ Finer classification of undecidable r.e. problems
+ Many-one, truth table, etc., degrees of r.e. sets
+ Sets whose theory is closely related to that of r.e. sets, e.g., productive sets: see also D50
− Generalizations of recursive enumerability: see D60 and D65
− Partial functions with r.e. graphs: see D20

D30 Other degrees; reducibilities

+ Degrees in generalized recursion and constructibility: see also D55, D60, D65, and E45
+ Jump operators
− Subrecursive reducibilities: see D15 and D20

D35 Undecidability and degrees of sets of sentences

+ Hilbert's tenth problem and extensions
+ Reduction classes of the predicate calculus (also in B20)
− Decidability results: see B25
− Halting problems, word problems, etc.: see D03, D05, D10, D30, D40, or D80

D40 Word problems, etc.

+ Conjugacy, isomorphism, and other algorithmic problems in algebra
+ Decidability and undecidability
+ Other algorithmic questions in classical algebra
− Problems concerning production systems or formal grammar: see D03 and D05
− Recursive functions on words: see D20

D45 Theory of numerations, effectively presented structures

+ Numberings of (partial) recursive functions
+ Numerations in the sense of Ershov
+ Recursive algebra, except when it is about recursive equivalence types: see D50
+ Recursive order types
− Classical recursive analysis: see F60
− Model theory of recursive structures: see C57
− Recursive arithmetic: see F30

D50 Recursive equivalence types of sets and structures, isols

+ Concepts traditionally associated with isols, e.g., regressiveness and immuneness

D55 Hierarchies

+ Arithmetical, Borel, analytical, projective, etc. hierarchies
− Descriptive Set Theory in which hierarchical questions are not central: see E15
− Hierarchies of definability in set theory: see E47
− Incidental use of hierarchies outside recursion theory
− Subrecursive hierarchies: see D15 and D20

D60 Recursion theory on ordinals, admissible sets, etc.
+ Beta-recursion on inadmissible ordinals
– Classification of ordinary recursive functions using ordinals: see D20
– Ordinal notations: see D45 and F15
– Other aspects of admissibility: see C62, C70, or E45

D65 Higher-type and set recursion
+ Primitive recursive set functions
– Functionals in Proof Theory: see F10
– Recursion on the hereditarily finite sets: see D20
– Recursion with all arguments and parameters of type $\leqslant 1$: see D20; MR includes this in D65 as long as there are type 1 arguments

D70 Inductive definability
+ Constructions equivalent to inductive definitions, e.g. set derivatives, game sentences, etc.
+ Recursion theory of inductive definitions and their duals
– Inductive definitions in proof theory: see F35 and F50
– Mechanics of inductive definitions: see B28, E20, or E30

D75 Abstract and axiomatic recursion theory
+ Algebras of (partial) recursive functions; MR uses D20
+ Recursion over general structures

D80 Applications
+ Decidability or undecidability results in areas outside logic and algebra
+ Effective versions of problems outside logic and algebra

D96 Collected works
+ Selected works
– Collections (almost) entirely in one subfield: see that subfield
MR uses 01A75, 03-03, and 03-06

D97 Proceedings
+ Collections of papers by various authors, even if they do not derive from any actual conference
– Proceedings (almost) entirely in one subfield: see that subfield
– Proceedings not concentrated in this field: see Source Index
MR uses 03-06

D98 Textbooks, surveys
MR uses 03-01 and 03-02

D99 None of the above or uncertain, but in this section

E SET THEORY

E05 Combinatorial set theory
+ Partition relations, ideals, ultrafilters, trees named after people; MR uses also 04A20
– Finite combinatorics (excluded from the Ω-Bibliography); MR uses 05Xn

E07 Relations and orderings
+ Relation algebras: see also G15; MR uses G15
– Theories about ordering: see C65
MR uses E20, 04A05, 04A20, or 06An

E10 Ordinal and cardinal numbers
+ Cardinal algebras, ordinal algebras
+ Dedekind finite cardinals
– Cardinal exponentiation and the (generalized) continuum hypothesis: see E50; MR sometimes uses 04A10
– Combinatorial aspects of cardinals and ordinals: see E05
– Large cardinals: see E55

E15 Descriptive set theory
+ Definability properties of sets (in the real line or similar spaces)
+ Effective descriptive set theory
– General topology, measure theory, etc.: see E75
MR sometimes uses 04A15
See also D55

E20 Other classical set theory
+ Set algebra

E25 Axiom of choice and related propositions
+ Weak axioms of choice and their negations
MR sometimes uses 04A25

E30 Axiomatics of classical set theory and its fragments
+ Zermelo-Fraenkel set theory and minor variants
+ Gödel-Bernays set theory (also in E70)
– Morse-Kelley set theory (a second order theory: see E70)
– New Foundations, etc.: see E70

E35 Consistency and independence results
+ Forcing used to prove consistency

E40 Other aspects of forcing and Boolean-valued models
+ Forcing in generalized recursion theory: see also D60 and D65
– Model theoretic forcing: see C25

E45 Constructibility, ordinal definability and related notions
+ Other inner models, e.g. the core model

E47 Other notions of set-theoretic definability
+ Lévy hierarchy, indescribability
– Formalization of branches of mathematics within set theory

E50 Continuum hypothesis and Martin's axiom
+ Cardinal exponentiation
+ Variants of Martin's axiom
MR sometimes uses 04A30

E55 Large cardinals
+ Effective (denumerable) analogues of large cardinals
+ Weakly inaccessible and larger cardinals
– Axioms of infinity provable in ZFC
– Large proof-theoretic ordinals: see F15

E60 Determinacy and related principles which contradict the axiom of choice
+ Infinite exponent partition relations
+ Projective determinacy, definable determinacy
+ Other uses of infinite games in set theory and logic
– Applications of games outside set theory and logic
– Weak axioms that merely contradict choice

E65 Other hypotheses and axioms
+ Reflection principles
+ Combinatorial principles

E70 Nonclassical and second-order set theories
+ Leśniewski's Ontology and Mereology; MR uses B60
+ Nonstandard theories, e.g. New Foundations, Ackermann
+ Set theories formulated in non-classical logic
+ Theory of real classes (Morse-Kelley, and Gödel-Bernays set theory); MR uses E30

E72 Fuzzy sets

E75 Applications
+ Independence from set theory of mathematical propositions (also in E35)
+ Results in other branches of mathematics obtained by set theoretic methods
– Set-theoretical foundations of mathematics: see B28 and B30

E96 Collected works
+ Selected works
– Collections (almost) entirely in one subfield: see that subfield

MR uses 01A75, 03-03, and 03-06

E97 Proceedings
+ Collections of papers by various authors, even if they do not derive from any actual conference
– Proceedings (almost) entirely in one subfield: see that subfield
– Proceedings not concentrated in this field: see Source Index

MR uses 03-06

E98 Textbooks, surveys

MR uses 03-01 and 03-02

E99 None of the above or uncertain, but in this section

F PROOF THEORY AND CONSTRUCTIVE MATHEMATICS

F05 Cut elimination and normal form theorems
+ Hilbert's epsilon symbol
– Cut elimination and normal form theorems for modal systems: see B45

F07 Structure of proofs
– Proof schemas used rather than studied: see B10, C07, etc.

F10 Functionals in proof theory
– Typed lambda-calculus: see B40

F15 Recursive ordinals and ordinal notations
+ Ordinal notations even if not proof theory
+ Transfinite progressions of theories (Turing, Feferman; also in F30)

F20 Complexity of proofs
– Complexity of non-proof-theoretic procedures: see D15
– Purely qualitative (rather than quantitative) properties of proofs: see F07

F25 Relative consistency and interpretations
- Consistency of systems of arithmetic: see F30 and F35
- Set theoretic consistency results: see E35

F30 First-order arithmetic and fragments
+ Gödel incompleteness theorems
+ Metamathematics of intuitionistic arithmetic
+ Provability logic; MR uses also B45 and F40
+ Provably recursive functions; MR uses also D20
+ Recursive arithmetic
- Model theory of arithmetic: see C62 and H15

F35 Second- and higher-order arithmetic and fragments
+ Metamathematics of intuitionistic analysis
+ Proof theory of systems of type theory
+ Proof theory of generalized inductive definitions
- Model theory : see C62

F40 Gödel numberings in proof theory
+ Any use of Gödel numbering of syntax
- Gödel numberings in recursion theory: see D20 and D45

F50 Metamathematics of constructive systems
+ Intuitionistic logic and subsystems; MR uses also B20
+ Model theoretic methods applied to constructive systems
+ Realizability
- Metamathematics of predicative systems: see F65

F55 Constructive and intuitionistic mathematics
+ Bishop school of constructivism
- Metamathematics: see F50

F60 Constructive recursive analysis
+ Classical recursive analysis
+ Soviet school of constructivism
- Metamathematics: see F50

F65 Other constructive mathematics
+ Constructive trends not covered by F55 or F60
+ Predicative mathematics
+ Metamathematics of predicative systems
- Other metamathematics: see F50

F96 Collected works
+ Selected works
- Collections (almost) entirely in one subfield: see that subfield

MR uses 01A75, 03-03, and 03-06

F97 Proceedings
+ Collections of papers by various authors, even if they do not derive from any actual conference
- Proceedings (almost) entirely in one subfield: see that subfield
- Proceedings not concentrated in this field: see Source Index

MR uses 03-06

F98 Textbooks, surveys
MR uses 03-01 and 03-02

F99 None of the above or uncertain, but in this section

G ALGEBRAIC LOGIC

G05 Boolean algebras
- \+ Boolean rings, etc.
- – Boolean functions : see B05; MR puts Boolean functions in G05, 06E30, and sometimes 94C10
- – Pseudo-Boolean algebras : see G10

G10 Lattices and related structures
- \+ Heyting algebras; MR uses also 06D20
- \+ Semilattices, continuous lattices; MR uses 06B35
- – Studies of "The lattice of..." where the lattice structure is not the main point

G12 Quantum logic
See also B51

G15 Cylindric and polyadic algebras, relation algebras

G20 Łukasiewicz and Post algebras
- \+ Lattices (or weaker structures) corresponding to many-valued logic

G25 Other algebras related to logic
- \+ Boolean algebras with provability and other operators
- \+ Implicative algebras, BCK algebras, etc.

G30 Categorical logic, topoi
- \+ Almost any connection between categories and logic, e.g. categories of models, logical foundations of category theory
- – Pure category theory (Excluded from the Ω-Bibliography); MR uses 18Xn

G96 Collected works
- \+ Selected works
- – Collections (almost) entirely in one subfield: see that subfield

MR uses 01A75, 03-03, and 03-06

G97 Proceedings
- \+ Collections of papers by various authors, even if they do not derive from any actual conference
- – Proceedings (almost) entirely in one subfield: see that subfield
- – Proceedings not concentrated in this field: see Source Index

MR uses 03-06

G98 Textbooks, surveys
MR uses 03-01 and 03-02

G99 None of the above or uncertain, but in this section

H NONSTANDARD MODELS

H05 Infinitesimal analysis in pure mathematics

H10 Other applications of infinitesimal analysis
- \+ Economics, physics, etc.

H15 Nonstandard models of arithmetic
- \+ Work emphasizing nonstandard methods
- – Work emphasizing model theory : see C62

H20 Other nonstandard models

H96 Collected works

+ Selected works
− Collections (almost) entirely in one subfield: see that subfield

MR uses 01A75, 03-03, and 03-06

H97 Proceedings

+ Collections of papers by various authors, even if they do not derive from any actual conference
− Proceedings (almost) entirely in one subfield: see that subfield
− Proceedings not concentrated in this field: see Source Index

MR uses 03-06

H98 Textbooks, surveys

MR uses 03-01 and 03-02

H99 None of the above or uncertain, but in this section

Subject Index

B45 Modal and tense logic

1918

LEWIS, C.I. *A survey of symbolic logic* ◇ B45 B46 B98 ◇

1920

LEWIS, C.I. *Strict implication – an emendation* ◇ A05 B45 B46 ◇

1924

MORGAN DE, A. *Formal logic; or, the calculus of inference, necessary and probable* ◇ A10 B45 B48 ◇

1930

BECKER, O. *Zur Logik der Modalitaeten* ◇ A05 B45 ◇

1931

GOEDEL, K. *Eine Interpretation des intuitionistischen Aussagenkalkuels* ◇ A05 B45 F50 ◇

1934

HUNTINGTON, E.V. *Independent postulates related to C.I.Lewis' theory of strict implication* ◇ B45 ◇

HUNTINGTON, E.V. *The relation between Lewis' strict implication and boolean algebra* ◇ B45 G05 ◇

1935

HUNTINGTON, E.V. *The mathematical structure of Lewis's theory of strict implication* ◇ B45 ◇

WARD, M. *A determination of all possible systems of strict implication* ◇ B45 ◇

1936

EMCH, A.F. *Implication and deducibility* ◇ B45 ◇

LEWIS, C.I. *Emch's calculus and strict implication* ◇ B45 B46 ◇

SMITH, HENRY BRADFORD *The algebra of propositions* ◇ B05 B45 ◇

SMITH, HENRY BRADFORD *The law of transitivity* ◇ B45 ◇

TANG, CAOZHEN *A paradox of Lewis's strict implication* ◇ B45 B46 ◇

1937

EMCH, A.F. *Deducibility with respect to necessary and impossible propositions* ◇ B45 ◇

FITCH, F.B. *Modal functions in two-valued logic* ◇ B45 ◇

SMITH, HENRY BRADFORD *Modal Logic – a revision* ◇ B45 ◇

1938

CHURCHMAN, C.W. *On finite and infinite modal systems* ◇ B45 ◇

FEYS, R. *Les logiques nouvelles des modalites* ◇ B45 ◇

MOISIL, G.C. *Sur la theorie classique de la modalite des jugements* ◇ B45 ◇

REACH, K. *The name relation and the logical antinomies* ◇ A05 B45 ◇

TANG, CAOZHEN *Algebraic postulates and a geometric interpretation for Lewis calculus of strict implication* ◇ B45 B46 G25 ◇

1939

FITCH, F.B. *Note on modal functions* ◇ B45 ◇

MENGER, K. *A logic of the doubtful. On optative and imperative logic* ◇ B45 ◇

VREDENDUIN, P.G.J. *A system of strict implication* ◇ B45 ◇

1940

DUGUNDJI, J. *Note on a property of matrices for Lewis and Langford's calculi of propositions* ◇ B45 ◇

MCKINSEY, J.C.C. *Proof that there are infinitely many modalities in Lewis's system* S_2 ◇ B45 ◇

WEYL, H. *The ghost of modality* ◇ A05 B45 ◇

1941

MCKINSEY, J.C.C. *A solution of the decision problem for the Lewis systems S2 and S4, with an application to topology* ◇ B25 B45 C65 ◇

MOISIL, G.C. *Remarques sur la logique modale du concept* ◇ B45 ◇

1942

MOISIL, G.C. *Logique modale* ◇ B45 ◇

1943

ALBAN, M.J. *Independence of the primitive symbols of Lewis's calculi of propositions* ◇ B45 ◇

BECKER, O. *Das formale System der ontologischen Modalitaeten* ◇ B45 ◇

QUINE, W.V.O. *Notes on existence and necessity* ◇ A05 B45 B65 ◇

1944

MCKINSEY, J.C.C. *On the number of complete extensions of the Lewis systems of sentential calculus* ◇ B45 ◇

1945

MCKINSEY, J.C.C. *On the syntactical construction of systems of modal logic* ◇ B45 ◇

1946

BARCAN, R.C. *A functional calculus of first order based on strict implication* ◇ B45 ◇

BARCAN, R.C. *The deduction theorem in a functional calculus of first order based on strict implication* ◇ B45 ◇

CARNAP, R. *Modalities and quantification* ◇ A05 B45 ◇

1947

CARNAP, R. *Meaning and necessity. A study in semantics and modal logic* ◇ A05 B45 ◇
QUINE, W.V.O. *The problem of interpreting modal logic* ◇ B45 ◇
STENIUS, E. *Natural implication and material implication* ◇ A05 B45 ◇

1948

FITCH, F.B. *Intuitionistic modal logic with quantifiers* ◇ B45 F50 ◇
MCKINSEY, J.C.C. & TARSKI, A. *Some theorems about the sentential calculi of Lewis and Heyting* ◇ B45 B55 ◇
SMULLYAN, A. *Modality and description* ◇ B45 ◇

1949

BERGMANN, G. *A syntactical characterization of S5* ◇ B45 ◇
BERGMANN, G. *The finite representations of S5* ◇ B45 ◇
DESTOUCHES, J.-L. *Intervention d'une logique de modalite dans une theorie physique* ◇ A05 B45 ◇
DRIESSCHE VAN DEN, R. *Sur le "de syllogismo hypothetico" de Boece* ◇ A05 A10 B45 ◇
FREUDENTHAL, H. *Sur le besoin d'une logique materielle* ◇ A05 B45 ◇
HALLDEN, S. *A reduction of the primitive symbols of the Lewis calculi* ◇ B45 ◇
HALLDEN, S. *An analogy in modal logic to the Lesniewski-Mihailescu theorem* ◇ B45 ◇
HALLDEN, S. *On the decision problem of Lewis' calculus S5* ◇ B25 B45 ◇
HALLDEN, S. *Results concerning the decision problem of Lewis's calculi S3 and S6* ◇ B25 B45 ◇
MCKINSEY, J.C.C. *Construction of systems of modal logic* ◇ B45 ◇
NOVIKOV, P.S. *On regularity classes (Russian)* ◇ B10 B45 ◇

1950

BARCAN MARCUS, R. *The elimination of contextually defined predicates in a modal system* ◇ B45 ◇
GOETLIND, E. *A System of postulates for Lewis's calculus* ◇ B45 ◇
HALLDEN, S. *Some results in modal logic (Swedish)* ◇ B45 ◇
MATSUMOTO, K. *Sur la structure concernant la logique moderne* ◇ A05 B45 F50 G05 ◇
MO, SHAOKUI *The deduction theorems and two new logical systems* ◇ B20 B45 ◇
SCHMIDT, H.A. *Systematische Basisreduktion der Modalitaeten bei Idempotenz der positiven Grundmodalitaeten* ◇ B45 ◇

1951

CHURCH, A. *A formulation of the logic of sense and denotation* ◇ B45 ◇
HALLDEN, S. *On the semantic non-completeness of certain Lewis calculi* ◇ B45 ◇
LEONARD, H.S. *Two-valued tables for modal functions* ◇ B45 ◇
RASIOWA, H. *Algebraic treatment of the functional calculi of Heyting and Lewis* ◇ B45 F50 G10 G25 ◇
SCROGGS, S.J. *Extensions of the Lewis system S5* ◇ B45 ◇
WRIGHT VON, G.H. *An essay in modal logic* ◇ B25 B45 ◇
WRIGHT VON, G.H. *Deontic logic* ◇ A05 B25 B45 ◇
YONEMITSU, N. *On systems of strict implication* ◇ B45 ◇

1952

BECKER, O. *Untersuchungen ueber den Modalkalkuel* ◇ B45 ◇
BLANCHE, R. *Quantity, modality, and other kindred systems of categories* ◇ B45 ◇
CURRY, H.B. *The elimination theorem when modality is present* ◇ B45 ◇
POIRIER, R. *Logique et modalite du point de vue organique* ◇ A05 B45 ◇
PRIOR, A.N. *Modality de dicto and modality de re* ◇ A05 B45 ◇
RIDDER, J. *Ueber modale Aussagenlogiken und ihren Zusammenhang mit Strukturen I,II* ◇ B45 ◇
ROSE, A. *Self dual primitives for modal logic* ◇ B45 ◇
VEATCH, H.B. *Intentional logic: a logic based on philosophical realism* ◇ A05 B45 ◇
WRIGHT VON, G.H. *Interpretations of modal logic* ◇ B45 ◇

1953

BEHMANN, H. *Die typenfreie Logik und die Modalitaet* ◇ A05 B45 ◇
FEYS, R. *A simplified proof of the reduction of all modalities to 42 in* S3 ◇ B45 ◇
FREUDENTHAL, H. *Context related interpretations in formal logic (Dutch)* ◇ A05 B45 ◇
LUKASIEWICZ, J. *A system of modal logic* ◇ B45 ◇
MCKINSEY, J.C.C. *Systems of modal logic which are not unreasonable in the sense of Hallden* ◇ B45 ◇
PRIOR, A.N. *On propositions neither necessary nor impossible* ◇ B45 ◇
QUINE, W.V.O. *Three grades of modal involvement* ◇ B45 ◇
QUINE, W.V.O. *Two theorems about truth-functions* ◇ B05 B45 ◇
RASIOWA, H. & SIKORSKI, R. *On satisfiability and decidability in non-classical functional calculi* ◇ B10 B25 B45 C90 F50 ◇
RIDDER, J. *Ueber modale Aussagenlogiken und ihren Zusammenhang mit Strukturen III,IV* ◇ B45 ◇
SIMONS, L. *New axiomatizations of S3 and S4* ◇ B45 ◇
SOBOCINSKI, B. *Note on a modal system of Feys-von Wright* ◇ B45 ◇
SUGIHARA, T. *The axiomatization of the aristotelian modal logic* ◇ B45 ◇
THOMAS, I. *Note on a modal system of Lukasiewicz* ◇ B45 ◇
WRIGHT VON, G.H. *A new system of modal logic* ◇ B45 ◇

1954

ANDERSON, A.R. *Improved decision procedures for Lewis's calculus* S4 *and von Wright's calculus* M ◇ B25 B45 ◇

ANDERSON, A.R. *On alternative formulations of a modal system of Feys - von Wright* ◇ B45 ◇
ANDERSON, A.R. *On the interpretation of a modal system of Lukasiewicz* ◇ B45 ◇
CHURCH, A. *Intensional isomorphism and identity of belief* ◇ A05 B45 ◇
DAVIS, CHANDLER *Modal operations, equivalence relations, and projective algebras* ◇ B45 ◇
ISHIMOTO, A. *A set of axioms of the modal propositional calculus equivalent to S3* ◇ B45 ◇
LORENZEN, P. *Zur Begruendung der Modallogik* ◇ B45 ◇
LUKASIEWICZ, J. *Arithmetic and modal logic* ◇ B45 ◇
LUKASIEWICZ, J. *On a controversial problem of Aristotle's modal syllogistic* ◇ A05 A10 B45 ◇
PRIOR, A.N. *The interpretation of two systems of modal logic* ◇ B45 ◇
REICHENBACH, H. *Nomological statements and admissible operations* ◇ B45 ◇
RIDDER, J. *Ueber modale Aussagenlogiken und ihren Zusammenhang mit Strukturen V, VI* ◇ B45 ◇
SCHMIDT, H.A. *Ein rein aussagenlogischer Zugang zu den Modalitaeten der strikten Logik* ◇ B45 ◇
YONEMITSU, N. *A decision method and a topological interpretation for systems of logical implication* ◇ B45 ◇
YONEMITSU, N. *A note on systems of logical implication* ◇ B45 ◇

1955

CHRISTIAN, C.C. *Die Elimination des Unendlichkeitstheorems in einem intensionalen Kalkuel* ◇ B45 ◇
MATSUMOTO, K. *Reduction theorem in Lewis' sentential calculi* ◇ B45 ◇
PRIOR, A.N. *Many-valued and modal systems: an intuitive approach* ◇ B45 B50 ◇
RASIOWA, H. & SIKORSKI, R. *An application of lattices to logic* ◇ B45 C90 G10 ◇
RASIOWA, H. & SIKORSKI, R. *On existential theorems in non-classical functional calculi* ◇ B45 C90 F50 G05 ◇
RIDDER, J. *Die Gentzenschen Schlussverfahren in modalen Aussagenlogiken I,II,III* ◇ B45 ◇
SANCHEZ-MAZAS, M. *Formalization of logic according to the perspective intensionality (Spanish) (French and English summaries)* ◇ B45 ◇
SUEKI, T. *Formalization of modal systems (Japanese) (English summary)* ◇ B45 ◇
YONEMITSU, N. *A note on modal systems, von Wright's M and Lewis's S1* ◇ B45 ◇

1956

ANDERSON, A.R. *Independent axiom schemata for S5* ◇ B45 ◇
BERGMANN, G. *The representations of S5* ◇ B45 ◇
ISHIMOTO, A. *A formulation of the modal propositional calculus equivalent to S4* ◇ B45 ◇
ISHIMOTO, A. *A note on the paper "A set of axioms of the modal propositional calculus equivalent to S3"* ◇ B45 ◇
KUBINSKI, T. *On a method of constructing modal logics (Polish)(Russian and English summaries)* ◇ B45 ◇
LEMMON, E.J. *Alternative postulate sets for Lewi's S5* ◇ B45 ◇
MEREDITH, D. *A correction to von Wright's decision procedure for the deontic system P* ◇ B25 B45 ◇
PRIOR, A.N. *Modality and quantification in S5* ◇ B45 ◇
QUINE, W.V.O. *Quantifiers and propositional attitudes* ◇ B45 ◇
SCHMIDT, H.A. *Das fundamentale Implikationssystem einer implikativen Modalitaetenstruktur mit idempotenter Moeglichkeit* ◇ B45 ◇
SIKORSKI, R. *A theorem on non-classical functional calculi* ◇ B45 F50 G10 ◇
WRIGHT VON, G.H. *A note on deontic logic and derived obligation* ◇ B45 ◇

1957

ALBRITTON, R. *Present truth and future contingencies* ◇ B45 ◇
ANDERSON, A.R. *Independent axiom schemata for von Wright's* M ◇ B45 ◇
EMDE, G. *Kriterien fuer die Herleitbarkeit in Modalitaetenstrukturen* ◇ B45 ◇
HALLDEN, S. *On the logic of "better"* ◇ A05 B45 ◇
HINTIKKA, K.J.J. *Modality as referential multiplicity* ◇ B45 ◇
HINTIKKA, K.J.J. *Quantifiers in deontic logic* ◇ B45 ◇
KANGER, S. *A note on quantification and modalities* ◇ B45 ◇
KANGER, S. *On the characterisation of modalities* ◇ B45 ◇
LEMMON, E.J. *New foundations for Lewis modal systems* ◇ B45 ◇
MATSUMOTO, K. & OHNISHI, M. & RIDDER, J. *Die Gentzenschen Schlussverfahren in modalen Aussagenlogiken I* ◇ B45 ◇
MATSUMOTO, K. & OHNISHI, M. *Gentzen method in modal calculi. I* ◇ B25 B45 ◇
MO, SHAOKUI *Modal systems with a finite number of modalities (Chinese) (English summary)* ◇ B45 ◇
PRIOR, A.N. *The necessary and the possible. The first of three talks on "The logic game."* ◇ A05 B45 ◇
PRIOR, A.N. *Time and modality. Being the John Locke lectures for 1955-6 delivered in the University of Oxford* ◇ A10 B45 ◇
SCHMIDT, H.A. *Die Gesamtheit der idempotenten implikativen Modalitaetenstrukturen* ◇ B45 ◇
SHEN, YOUDING *The basic calculus (Chinese) (English summary)* ◇ B45 ◇
WRIGHT VON, G.H. *A new system of modal logic* ◇ B45 ◇
WRIGHT VON, G.H. *On conditionals* ◇ A05 B45 ◇
YONEMITSU, N. *A note on modal systems II* ◇ B45 ◇

1958

ANDERSON, A.R. *Reduction of deontic logic to alethic modal logic* ◇ B45 ◇
BAYART, A. *Correction de la logique modale du premier et du second ordre S5* ◇ B45 ◇
BORKOWSKI, L. *On modal terms (Polish, Russian and English summaries)* ◇ B45 ◇

FREUDENTHAL, H. *Logique mathematique appliquee* ◇ A05 B05 B45 B70 ◇
GUILLAUME, M. *Calculs de consequences et tableaux d'epreuve pour les classes algebriques generales d'anneaux booleiens a operateurs* ◇ B45 C05 G05 ◇
GUILLAUME, M. *Rapports entre calculs propositionels modaux et topologie impliques par certaines extensions de la methode des tableaux semantiques. Systeme de Feys-von Wright* ◇ B45 ◇
GUILLAUME, M. *Rapports entre calculs propositionels modaux et topologie impliques par certaines extensions de la methode des tableaux semantiques. Systeme S4 de Lewis* ◇ B45 G05 ◇
LEMMON, E.J. *Quantifiers and modal operators* ◇ B45 ◇
MO, SHAOKUI *Fundamental finite modal systems (Chinese) (English summary)* ◇ B45 ◇
MYHILL, J.R. *Problems arising in the formalization of intensional logic* ◇ B45 ◇
PORTE, J. *Recherches sur les logiques modales* ◇ B45 ◇
PRIOR, A.N. *The syntax of time-distinctions* ◇ B45 ◇
RESCHER, N. *An axiom system for deontic logic* ◇ A05 B45 ◇
RIDDER, J. *Die Gentzenschen Schlussverfahren in modalen Aussagenlogiken II,III* ◇ B45 ◇
SCHMIDT, H.A. *Ueber einige neuere Untersuchungen zur Modalitaetenlogik* ◇ B45 ◇
SIKORSKI, R. *Some applications of interior mappings* ◇ B45 E75 F50 G05 G10 G25 ◇
TOERNEBOHM, H. *Notes on modal operators* ◇ B45 ◇

1959

BAYART, A. *Quasi-adequation de la logique modale du second ordre S5 et adequation de la logique modale du premier ordre S5* ◇ B45 ◇
CURRY, H.B. *The interpretation of formalized implication* ◇ B45 ◇
DAWSON, E.E. *A model for deontic logic* ◇ B45 ◇
DUMMETT, M. & LEMMON, E.J. *Modal logics between S4 and S5* ◇ B45 ◇
FENSTAD, J.E. *Notes on normative logic* ◇ B45 ◇
HAMBLIN, C.L. *The modal "probably"* ◇ A05 B45 ◇
ITO, MAKOTO *Modal logic and its application to sequential circuits* ◇ B45 ◇
KRIPKE, S.A. *A completeness theorem in modal logic* ◇ B45 ◇
LEMMON, E.J. *Symposium: Is there only one correct system of modal logic? I* ◇ A05 B45 ◇
MATSUMOTO, K. & OHNISHI, M. *Gentzen method in modal calculi II* ◇ B25 B45 ◇
MO, SHAOKUI *The modal system and implication system (Chinese) (English summary)* ◇ B45 ◇
PRIOR, A.N. *Notes on a group of new modal systems* ◇ B45 ◇
STAHL, G. *General considerations about modal sentences* ◇ B45 ◇

1960

BERG, J. *A note on deontic logic* ◇ B45 ◇
CASTANEDA, H.-N. *Obligation and modal logic* ◇ B45 ◇
HINTIKKA, K.J.J. *Aristotle's different possibilities* ◇ A05 A10 B45 ◇
LEMMON, E.J. *An extension algebra and the modal system T* ◇ B45 ◇
LEWIS, C.I. *A survey of symbolic logic (Reprinted edition)* ◇ B45 B98 ◇
MATSUMOTO, K. *Decision procedure for modal sentential calculus S3* ◇ B25 B45 ◇
YONEMITSU, N. *Systems of weak implications* ◇ B45 ◇

1961

FEYS, R. *Modeles a variables de differentes sortes pour les logiques modales M" ou S5* ◇ B45 ◇
GRZEGORCZYK, A. *Le traitement axiomatique de la notion de prolongement temporel (Polish) (French and Russian summaries)* ◇ B45 ◇
HINTIKKA, K.J.J. *Modality and quantification* ◇ B45 ◇
OHNISHI, M. *Gentzen decision procedures for Lewis's systems S2 and S3* ◇ B25 B45 ◇
OHNISHI, M. *Von Wright-Anderson's decision procedures for Lewis's systems* S2 *and* S3 ◇ B45 ◇
POLIFERNO, M.J. *Decision algorithms for some functional calculi with modality* ◇ B25 B45 ◇
PRIOR, A.N. *Some axiom-pairs for material and strict implication* ◇ B05 B45 ◇
RESCHER, N. *On the formalization of two modal theses* ◇ B45 ◇
SCHOCK, R. *Some definitions of subjunctive implication, of counterfactual implication, and of related concepts* ◇ A05 B45 ◇
SMILEY, T.J. *On Lukasiewicz's L-modal System* ◇ B45 ◇
SMULLYAN, R.M. *Extended canonical systems* ◇ B45 D03 D25 ◇

1962

AAQVIST, L. *A binary primitive in deontic logic* ◇ B45 ◇
ANGELL, R.B. *A propositional logic with subjunctive conditionals* ◇ B45 ◇
CATON, C.E. *A stipulation of logical truth in a modal propositional calculus* ◇ B45 ◇
DRAKE, F.R. *On Mckinsey's syntactical characterizations of systems of modal logic* ◇ B45 ◇
FISHER, M. *A system of deontic-alethic modal logic* ◇ B45 ◇
KRIPKE, S.A. *The undecidability of monadic modal quantification theory* ◇ B45 D35 ◇
MANSON, M. *Introduction a la semantica de los sistemas normativos* ◇ B45 ◇
MORGENSTERN, S. *Un algorithme du type de Zegalkin pour le probleme de la decidabilite en S2 (Romanian) (Russian and French summaries)* ◇ B45 ◇
NAKAMURA, A. *On the infinitely many-valued threshold logics and von Wright's System M* ◇ B45 ◇
NAKAMURA, A. *On the relation between modal logics and many-valued logic (Japanese)* ◇ B45 ◇
PRIOR, A.N. *Quantification and L-modality* ◇ B45 ◇
PRIOR, A.N. *Tense-logic and the continuity of time (Polish and Russian summaries)* ◇ B45 ◇
RESCHER, N. *Conditional permission in deontic logic* ◇ B45 ◇
RESCHER, N. *Modality conceived as a status* ◇ B45 ◇
RUBIN, J.E. *Bi-modal logic, double-closure algebras, and Hilbert space* ◇ B45 G20 ◇

SCHOCK, R. *A note on subjunctive and counterfactual implication* ◇ A05 B45 ◇
SIMONS, L. *A reduction in the number of independent axiom schemata for S4* ◇ B45 ◇
SOBOCINSKI, B. *A contribution to the axiomatization of Lewis' system S5* ◇ B45 ◇
SOBOCINSKI, B. *A note on the regular and irregular modal systems of Lewis* ◇ B45 ◇
SOBOCINSKI, B. *An axiom-system for* $\{K, N\}$ *-propositional calculus related to Simon's axiomatization of* S3 ◇ B45 ◇
SOBOCINSKI, B. *On the generalized Brouwerian axioms* ◇ B45 ◇
SUGIHARA, T. *The numbers of modalities in T supplemented by the axiom* CL^2pL^3p ◇ B45 ◇
THOMAS, I. *Finite limitations on Dummett's LC* ◇ B45 ◇
THOMAS, I. *Solutions of five modal problems of Sobocinski* ◇ B45 ◇

1963

AAQVIST, L. *Deontic logic based on a logic of "better"* ◇ A05 B45 ◇
AAQVIST, L. *Postulate sets and decision procedures for some systems of deontic logic* ◇ B45 ◇
BARCAN MARCUS, R. *Classes and attributes in extended modal systems* ◇ B45 ◇
BAUMER, W.H. *Discussion: Von Wright's paradoxes* ◇ A05 B45 ◇
CARNAP, R. *Donald Davidson on modalities and semantics* ◇ A05 B45 ◇
CARNAP, R. *John Myhill on modal logic and semantics* ◇ A05 B45 ◇
CARNAP, R. *Robert Feys on modalities* ◇ A05 B45 ◇
CATON, C.E. *A stipulation of a modal propositional calculus in terms of modalized truth values* ◇ B45 ◇
CHISHOLM, R.M. *Contrary-to-duty imperatives and deontic logic* ◇ B45 ◇
DAVIDSON, D. *The method of extension and intension* ◇ A05 B45 ◇
DONCHENKO, V.V. *Some questions connected with the decision problem for Ackermann's system of strict implication (Russian)* ◇ B45 ◇
FEYS, R. *Carnap on modalities* ◇ A05 B45 ◇
FITCH, F.B. *A logical analysis of some value concepts* ◇ B45 ◇
HACKING, I. *What is strict implication ?* ◇ A05 B45 ◇
HALLDEN, S. *A pragmatic approach to modal theory* ◇ B45 ◇
HINTIKKA, K.J.J. *The modes of modality* ◇ B45 ◇
KALINOWSKI, J. *La norme, l'action et la theorie des propositions normatives (Polish and Russian summaries)* ◇ B45 ◇
KRIPKE, S.A. *Semantical analysis of modal logic I. Normal modal propositional calculi* ◇ B45 G15 ◇
KRIPKE, S.A. *Semantical considerations on modal logic* ◇ B45 ◇
LEMMON, E.J. *A theory of attributes based on modal logic* ◇ B45 ◇
MATSUMOTO, K. & OHNISHI, M. *A system for strict implication* ◇ B45 ◇
MCCALL, S. *Aristotle's modal syllogisms* ◇ A10 B45 ◇
MONTAGUE, R. *Syntactical treatments of modality, with corollaries on reflexion principles and finite axiomatizability* ◇ B45 B65 C62 F30 ◇
MYHILL, J.R. *An alternative to the method of extension and intension* ◇ B45 ◇
RASIOWA, H. *On modal theories* ◇ B45 G05 G25 ◇
RESCHER, N. *A probabilistic approach to modal logic* ◇ B45 ◇
SMILEY, T.J. *Relative necessity* ◇ A05 B45 ◇
SOBOCINSKI, B. *A note on modal systems* ◇ B45 ◇
STENIUS, E. *The principles of a logic of normative systems* ◇ A05 B45 ◇
SWEET, A.M. *Toward a pragmatical explication of epistemic modalities* ◇ A05 B45 ◇
THOMAS, I. $S1^0$ *and Brouwerian axioms* ◇ B45 ◇
THOMAS, I. $S1^0$ *and generalized* S5*-axioms* ◇ B45 ◇
THOMAS, I. *A final note on* $S1^0$ *and the Brouwerian axioms* ◇ B45 ◇
TURQUETTE, A.R. *Modality, minimality, and many valuedness* ◇ A05 B45 B50 ◇
WRIGHT VON, G.H. *The logic of preference* ◇ B45 ◇
ZEMAN, J.J. *Bases for S4 and S4.2 without added axioms* ◇ B45 ◇

1964

AAQVIST, L. *Interpretations of deontic logic* ◇ A05 B45 ◇
AAQVIST, L. *On Dawson-models for deontic logic* ◇ B45 ◇
AAQVIST, L. *Results concerning some modal systems that contain S2* ◇ B45 ◇
BACK, K.W. *A nerve net system in modal logic* ◇ B45 ◇
BULL, R.A. *A note on the modal calculi S4.2 and S4.3* ◇ B45 ◇
BULL, R.A. *An axiomatization of Prior's modal calculus Q* ◇ B45 ◇
CANTY, J.T. *A natural deduction system for modal logic* ◇ B45 ◇
CASTANEDA, H.-N. *A note on S5* ◇ B45 ◇
FUJIMURA, T. *On T-principle (Japanese)* ◇ B45 ◇
GAO, HENGSHAN *On the simple completeness of the predicate calculus* $\mathscr{S}_\varepsilon^*$ *(Chinese)* ◇ B45 G05 ◇
GAO, HENGSHAN *Simple completeness of the predicate calculus* $\mathscr{S}_\varepsilon^*$ *(Chinese)* ◇ B45 G05 ◇
HINTIKKA, K.J.J. *The once and future sea fight: Aristotle's discussion of future contingents in De Interpretatione IX* ◇ A05 A10 B45 ◇
KALINOWSKI, G. *Obligation derivee et logique deontique relationnelle* ◇ A05 B45 ◇
MARTIN, R.M. *Toward a logic of intensions* ◇ A05 B45 ◇
MATSUMOTO, K. & OHNISHI, M. *A system for strict implication* ◇ B45 ◇
MEREDITH, C.A. & PRIOR, A.N. *Investigations into implicational S5* ◇ B45 ◇
NAKAMURA, A. *On the relation between modal logic and many-valued logic (Japanese)* ◇ B45 B50 ◇

NAKAMURA, A. *Truth-value stipulations for the von Wright system M′ and the Heyting system* ◇ B45 ◇
PIECZKOWSKI, A. *A set of axioms for the system Q_f of factorial implication* ◇ B45 ◇
PRIOR, A.N. *Axiomatisations of the modal calculus* Q ◇ B45 ◇
PRIOR, A.N. *K1, K2 and related modal systems* ◇ B45 ◇
PRIOR, A.N. *Two additions to positive implications* ◇ B20 B45 ◇
RESCHER, N. *A quantificational treatment of modality* ◇ B45 ◇
SOBOCINSKI, B. *A note on Prior's system in "the theory of deduction"* ◇ B45 ◇
SOBOCINSKI, B. *Family K of non-Lewis modal systems* ◇ B45 ◇
SOBOCINSKI, B. *Modal system S4.4* ◇ B45 ◇
SOBOCINSKI, B. *Remarks about axiomatization of certain modal systems* ◇ B45 ◇
SUGIHARA, T. *A study on modal logic (Japanese)* ◇ B45 ◇
THOMAS, I. *Decision procedures for* $S2^2$ *and* T^0 ◇ B45 ◇
THOMAS, I. *Modal systems in the neighbourhood of T* ◇ B45 ◇
THOMAS, I. *Ten modal models* ◇ B45 ◇
TIRNOVEANU, M. *Ueber die Modalitaet der Werte eines logischen nichtkonstruktiven Systems der Stufe I (rumaenisch)* ◇ B45 ◇
WISDOM, W.A. *Possibility-elimination in natural deduction* ◇ B45 ◇
WRIGHT VON, G.H. *A new system of deontic logic* ◇ B45 ◇

1965

BETH, E.W. & NIELAND, J.J.F. *Semantic construction of Lewis' systems* S4 *and* S5 ◇ B45 ◇
BULL, R.A. *A class of extensions of the modal system S4 with the finite model property* ◇ B45 ◇
BULL, R.A. *An algebraic study of diodorean modal systems* ◇ B45 ◇
BULL, R.A. *Some modal calculi based on IC* ◇ B45 ◇
CANTY, J.T. *A note on the axiomatization of Rubin's system (S)* ◇ B45 ◇
CANTY, J.T. *Systems classically axiomatized and properly contained in Lewis's S3* ◇ B45 ◇
CRESSWELL, M.J. *Another basis for S*4 ◇ B45 ◇
EHSAKIA, L.L. *A system and algebra with derivation (Russian) (Georgian summary)* ◇ B45 ◇
FEYS, R. *Modal logics* ◇ B45 ◇
FOLLESDAL, D. *Contributions to a discussion after Ruth Barcan Marcus: "modal logics"* ◇ B45 ◇
FOLLESDAL, D. *Quantification into causal contexts* ◇ A05 B45 ◇
GODDARD, L. *An augmented modal logic* ◇ B45 D35 ◇
ITO, MAKOTO *Boolean recursive functions and closure algebra* ◇ B25 B45 D20 ◇
KALINOWSKI, G. *Les themes actuels de la logique deontique. Esquisse de l'etat actuel des recherches dans la logique des normes (Polish and Russian summaries)* ◇ B45 ◇
KRIPKE, S.A. *Semantical analysis of modal logic II. Non-normal modal propositional calculi* ◇ B45 ◇
LEMMON, E.J. *Deontic logic and the logic of imperatives* ◇ B45 ◇
LEMMON, E.J. *Some results on finite axiomatizability in modal logic* ◇ B45 ◇
MASSEY, G.J. *Four simple systems of modal propositional logic* ◇ B45 ◇
MCCALL, S. *A modal logic with eight modalities* ◇ B45 ◇
MEREDITH, C.A. & PRIOR, A.N. *Modal logic with functorial variables and a contingent constant* ◇ B45 ◇
NAKAMURA, A. *On certain systems of modal logic* ◇ B45 ◇
RUZSA, I. *Axiomatischer Aufbau eines Systems der deontischen Logik* ◇ B45 ◇
SANCHEZ-MAZAS, M. *Sobre la estructura de la logica modal* ◇ B45 ◇
VAKARELOV, D. *A propositional calculus with functors for "likelyhood" and "doubt". I* ◇ B45 ◇
WRIGHT VON, G.H. *A correction to a new system of deontic logic* ◇ B45 ◇

1966

AAQVIST, L. *"Next" and "ought" alternative foundations for von Wright's tense logic, with an application to deontic logic* ◇ B45 ◇
BARCAN MARCUS, R. *Iterated deontic modalities* ◇ B45 ◇
BULL, R.A. *MIPC as the formalisation of an intuitionist concept of modality* ◇ B45 ◇
BULL, R.A. *That all normal extensions of S4.3 have the finite model property* ◇ B45 ◇
CANTY, J.T. & SCHARLE, T.W. *Note on the singularies of S5* ◇ B45 ◇
CLIFFORD, J.E. *Tense logic and the logic of change* ◇ A05 B45 ◇
CRESSWELL, M.J. *The completeness of SO.5* ◇ B45 ◇
FITCH, F.B. *Natural deduction rules for obligation* ◇ B45 F07 ◇
FOLLESDAL, D. *A model theoretic approach to causal logic* ◇ A05 B45 C90 ◇
GOBLE, L.F. *The iteration of deontic modalities* ◇ B45 ◇
HANSON, W.H. *On formalizing the distinction between logical and factual truth* ◇ A05 B45 ◇
HANSON, W.H. *On some alleged decision procedures for* S4 ◇ B25 B45 ◇
LEMMON, E.J. *A note on Hallden-incompleatness* ◇ B45 ◇
LEMMON, E.J. *Algebraic semantics for modal logics I,II* ◇ B45 ◇
LOEB, M.H. *Extensional interpretations of modal logics* ◇ B45 ◇
LUCE, DAVID RANDALL *A calculus of "before"* ◇ B45 ◇
MAKINSON, D. *On some completeness theorems in modal logic* ◇ B45 ◇
MAKINSON, D. *There are infinitely many diodorean modal functions* ◇ B45 ◇
MASHURYAN, A.S. *Distributivity axioms in the strict implication calculus of Ackermann (Russian) (Armenian and English summaries)* ◇ B45 ◇

MASSEY, G.J. *The theory of truth tabular connectives, both truth functional and modal* ◇ B45 ◇
MCCALL, S. *Connexive implication* ◇ A05 B45 ◇
MONTGOMERY, H. & ROUTLEY, R. *Contingency and non-contingency bases for normal modal logics* ◇ B45 ◇
PIECZKOWSKI, A. *The axiomatic system of the factorial implication (Polish and Russian summaries)* ◇ B45 ◇
POLLOCK, J.L. *Model theory and modal logic* ◇ B45 C90 ◇
PRIOR, A.N. *Postulates for tense-logic* ◇ B45 ◇
RESCHER, N. *On modal renderings of intuitionistic propositional logic* ◇ B45 F50 ◇
RESCHER, N. *On the logic of chronological propositions* ◇ B45 ◇
RESCHER, N. & ROBISON, J. *Temporally conditioned descriptions* ◇ B45 ◇
RESCHER, N. *The logic of commands* ◇ B45 ◇
SAITO, S. *On the completeness of the Leibnizian modal system with a restriction* ◇ B45 ◇
TIRNOVEANU, M. *Sur quelques proprietes d'un systeme logistique semi-constructif modal I* ◇ B45 ◇
VAKARELOV, D. *A propositional calculus with functors for "likelyhood" and "doubt". II* ◇ B45 ◇
WRIGHT VON, G.H. *"and then"* ◇ B45 B65 ◇
ZARNECKA-BIALY, E. *The intensional functions (Polish) (English summary)* ◇ B45 ◇

1967

AJDUKIEWICZ, K. *Intensional expressions* ◇ A05 B45 ◇
ANDERSON, A.R. & MOORE, O.K. *The formal analysis of normative concepts* ◇ B45 ◇
BARCAN MARCUS, R. *Essentialism and modal logic* ◇ B45 ◇
BARCAN MARCUS, R. *Modalities and intensional languages* ◇ B45 ◇
BULL, R.A. *On the extension of S4 with CLMpMLp* ◇ B45 ◇
BULL, R.A. *On three related extensions of S4* ◇ B45 ◇
COPI, I.M. & GOULD, JAMES A. *Deontic logic, introduction* ◇ B45 B98 ◇
COPI, I.M. & GOULD, JAMES A. *Modal logic, introduction* ◇ B45 ◇
CRESSWELL, M.J. *A Henkin completeness theorem for T* ◇ B45 ◇
CRESSWELL, M.J. *Alternative completeness theorems for modal systems* ◇ B45 ◇
CRESSWELL, M.J. *Note on a system of Aaqvist* ◇ B45 ◇
CRESSWELL, M.J. *Some further semantics for deontic logic* ◇ B45 ◇
CRESSWELL, M.J. *The interpretation of some Lewis systems of modal logic* ◇ B45 ◇
DANIELSSON, S. *Modal logic based on probability theory* ◇ B45 ◇
DRABBE, J. *Une propriete des matrices characteristiques des systemes S1, S2 et S3* ◇ B45 ◇
FITCH, F.B. *A complete and consistent modal set theory* ◇ B45 E70 ◇
FOLLESDAL, D. *Knowledge, identity and existence* ◇ A05 B45 ◇
GABBAY, D.M. & KASHER, A. *Improper definite descriptions* ◇ A05 B45 ◇
GARSON, J.W. & RESCHER, N. *A note on chronical logic* ◇ B45 ◇
GRZEGORCZYK, A. *Non-classical propositional calculi in relation to methodological patterns of scientific investigation (Polish) (Russian and English summaries)* ◇ A05 B45 F50 ◇
GRZEGORCZYK, A. *Some relational systems and the associated topological spaces* ◇ B45 C65 G25 ◇
HINTIKKA, K.J.J. *Existence and identity in epistemic contexts: A comment on Follesdal's paper* ◇ B45 ◇
HINTIKKA, K.J.J. *Individuals, possible worlds, and epistemic logic* ◇ B45 ◇
IVIN, A.A. *Some problems of the theory of deontic modalities (Russian)* ◇ B45 ◇
LEWIS, C.I. *The structure of system of strict implication* ◇ B45 B46 ◇
LUCKHARDT, H. *Von Wright's relative Modalitaeten* ◇ B45 ◇
MASSEY, G.J. *Binary connectives functionally complete by themselves in S5 modal logic* ◇ B45 ◇
MCCALL, S. *Connexive implication and the syllogism* ◇ B45 ◇
NAKAMURA, A. *A remark on the truth-value stipulation for the modal system M'* ◇ B45 ◇
OREVKOV, V.P. *The undecidability of a class of formulas containing just one single place predicate variable in modal calculus (Russian)* ◇ B45 D35 ◇
PARSONS, T. *Grades of essentialism in quantified modal logic* ◇ B45 ◇
PEKLO, B.T. *The logical inference from "it does not hold that x ought to be", to "it holds that x ought not to be"* ◇ B45 ◇
POLLOCK, J.L. *Basic modal logic* ◇ B45 ◇
POLLOCK, J.L. *The logic of logical necessity* ◇ B45 ◇
PRIOR, A.N. *Correspondence theory of truth* ◇ A05 B45 ◇
PRIOR, A.N. *Modal logic* ◇ A05 B45 ◇
PRIOR, A.N. *Stratified metric tense logic* ◇ B45 ◇
QUINE, W.V.O. *Reply to Prof. Marcus* ◇ B45 ◇
SEGERBERG, K. *Some modal logics based on a three-valued logic* ◇ B45 ◇
SHUKLA, A. *A note on the axiomatizations of certain modal systems* ◇ B45 ◇
SLININ, YA.A. *Theory of modalities in contemporary logic (Russian)* ◇ B45 ◇
SMIRNOV, G.A. *The proof of basic theorems in the theory of strong logical inference (Russian)* ◇ B45 ◇
SUSZKO, R. *An essay in the formal theory of extension and of intension (Polish and Russian summaries)* ◇ A05 B45 ◇
TAVANETS, P.V. (ED.) *Logical semantics and modal logic (Russian)* ◇ B45 ◇
THOMAS, I. *A theorem on S4.2 and S4.4* ◇ B45 ◇
THOMASON, R.H. *A decision procedure for Fitch's propositional calculus* ◇ B25 B45 ◇
VUCKOVIC, V. *A recursive model for the extended system A of B. Sobocinski* ◇ B45 F30 ◇
WRIGHT VON, G.H. *Deontic logic* ◇ B45 ◇

ZEMAN, J.J. *The deduction theorem in S4, S4.2 and S5* ◇ B45 ◇

1968

BARCAN MARCUS, R. *Modal logic* ◇ B45 ◇
BULL, R.A. *An algebraic study of tense logics with linear time* ◇ B45 ◇
BULL, R.A. *On possible worlds in propositional calculi* ◇ B45 ◇
CANTY, J.T. *On symbolizing singulary S5 functions* ◇ B45 ◇
COSTA DA, N.C.A. & DUBIKAJTIS, L. *Sur la logique discoursive de Jaskowski* ◇ B45 ◇
CRESSWELL, M.J. & HUGHES, G.E. *An introduction to modal logic* ◇ B45 B98 ◇
CRESSWELL, M.J. *Completeness without the Barcan formula* ◇ B45 ◇
CRESSWELL, M.J. *Some proofs of relative completeness in modal logic* ◇ B45 ◇
CRESSWELL, M.J. *The representation of intensional logic* ◇ B45 ◇
GARSON, J.W. & RESCHER, N. *Topological logic* ◇ B45 ◇
GODDARD, L. *Towards a logic of significance* ◇ A05 B45 ◇
GRZEGORCZYK, A. *Assertions depending on time and corresponding logical calculi* ◇ B45 ◇
HOOKER, C.A. & STOVE, D. *Relevance and the ravens* ◇ A05 B45 ◇
JOJA, A. *Essai sur la modalite du jugement* ◇ B45 ◇
LEBLANC, H. *On Meyer and Lambert's quantificational calculus FQ* ◇ B45 G15 ◇
LORENZEN, P. *On modal logic* ◇ B45 ◇
MARTIN, R.M. *Semantics, analytical truth and the new modalism* ◇ A05 B45 ◇
MASSEY, G.J. *Normal form generation of S5 functions via truth functions* ◇ B45 ◇
MINTS, G.E. *Cut-free calculi of the type S5 (Russian)* ◇ B45 ◇
MINTS, G.E. *Some calculi of modal logic (Russian)* ◇ B45 ◇
MONTGOMERY, D. & ROUTLEY, R. *Non-contigency axioms for S4 and S5* ◇ B45 ◇
MONTGOMERY, H. & ROUTLEY, R. *Modal reduction axioms in extension of SL* ◇ B45 ◇
MONTGOMERY, H. *The inadequacy of Kripke's semantical analysis of* D_2 *and* D_3 ◇ B45 ◇
NAKAMURA, A. *On an axiomatic system of modal logic* ◇ B45 ◇
PEKLO, B.T. *Deontic and alethic modalities* ◇ B45 ◇
PIECZKOWSKI, A. *The efficient implication (Polish and Russian summaries)* ◇ B45 ◇
PRIOR, A.N. *Modal logic and the logic of applicability* ◇ B45 ◇
PRIOR, A.N. *Papers on time and tense* ◇ B45 ◇
RENNIE, M.K. S3(5)=S3.5 ◇ B45 ◇
RENNIE, M.K. *A function which bounds truth-tabular calculations in S5* ◇ B25 B45 D15 ◇
RESCHER, N. *Chronological logic* ◇ B45 ◇
ROUTLEY, R. *Decision procedures and semantics for* C1, E1, *and* $S0.5^0$ ◇ B25 B45 ◇
ROUTLEY, R. *The decidability and semantical incompleteness of Lemmon's system S0.5* ◇ B25 B45 ◇
RUZSA, I. *Ein neues formales System der deontischen Logik* ◇ B45 ◇
SAITO, S. *On the Leibnizian modal system* ◇ A05 B45 ◇
SCHUETTE, K. *Vollstaendige Systeme modaler und intuitionistischer Logik* ◇ B45 C90 F50 F98 ◇
SEGERBERG, K. *Decidability of S4.1* ◇ B25 B45 ◇
SEGERBERG, K. *Decidability of four modal logics* ◇ B25 B45 ◇
STALNAKER, R.C. & THOMASON, R.H. *Abstraction in first order modal logic* ◇ B45 ◇
SUSZKO, R. *Non-Fregean logic and theories* ◇ A05 B45 ◇
THIELE, H. *Ueber einen sequentiellen Praedikatenkalkuel als Grundlage fuer den Aufbau problemorientierter algorithmischer Sprachen* ◇ B45 B75 ◇
THOMAS, I. *Replacement in some modal systems* ◇ B45 ◇
TREW, A. *Incompleteness of a logic of Routley's* ◇ A05 B45 ◇
VEATCH, H.B. *Intentional logic* ◇ B45 ◇
WALIGORSKI, S. *Implications in boolean algebras with a two-valued closure operator (Polish and Russian summaries)* ◇ B45 G05 G25 ◇
WEINGARTNER, P. *Modal logics with two kinds of necessity and possibility. In collaboration with Hans Knapp* ◇ B45 ◇
WRIGHT VON, G.H. *"Always"* ◇ A05 B45 ◇
WRIGHT VON, G.H. *An essay in deontic logic and the general theory of action. With a bibliography of deontic and imperative logic* ◇ A05 B45 ◇
ZARNECKA-BIALY, E. *A note on deduction theorem for Goedel's propositional calculus G4 (Polish and Russian summaries)* ◇ B45 ◇
ZEMAN, J.J. *Lemmon-style bases for the systems* $S1^o - S4^o$ ◇ B45 ◇
ZEMAN, J.J. *Some calculi with strong negation primitive* ◇ B45 F50 ◇
ZEMAN, J.J. *The propositional calculus MC and its modal analog* ◇ B45 ◇
ZEMAN, J.J. *The semisubstitutivity of strict implication* ◇ B45 ◇

1969

ALCHOURRON, C.E. *Logic of norms and logic of normative propositions* ◇ B45 ◇
BAR-HILLEL, Y. *Is modal logic different from the semantics of modal terms ?* ◇ B45 ◇
BERKA, K. *Remarks on the relations between the peripatetic and modern theories of modality* ◇ A10 B45 ◇
BUENO SANCHEZ, E. *Polyvalent modal logics (Spanish)* ◇ B45 ◇
BULL, R.A. *Note on a paper in tense logic* ◇ B45 ◇
BULL, R.A. *On a paper of Akira Nakamura* ◇ B45 ◇
BULL, R.A. *On modal logic with propositional quantifiers* ◇ B45 ◇
CHENDOV, B. *A system of dyadic (relative) modalities and its relation to the aristotelian syllogistic and some modern logical theories* ◇ A10 B45 ◇

COCCHIARELLA, N.B. *A completeness theorem in second order modal logic* ◇ A05 B45 ◇
CORCORAN, J. & WEAVER, G.E. *Logical consequences in modal logic: natural deduction in S5* ◇ B45 F07 ◇
CRESSWELL, M.J. *The conjunctive normal form for S3.5* ◇ B45 ◇
CRESSWELL, M.J. *The elimination of de re modalities* ◇ B45 ◇
FITTING, M. *Logics with several modal operators* ◇ B45 ◇
FOLLESDAL, D. *Quine on modality* ◇ A05 B45 ◇
GABBAY, D.M. *Montague type semantics for non-classical logic* ◇ B45 ◇
GOCHET, P. *Logique modale quantifiee et essentialisme* ◇ B45 ◇
HANSSON, B. *An analysis of some deontic logics* ◇ B45 ◇
HINTIKKA, K.J.J. *Models for modalities: selected essays* ◇ B45 ◇
JOJA, A. *La theorie de la modalite dans le de interpretatione* ◇ A10 B45 ◇
KRAUSKOPF, R. *Ein Entscheidungskalkuel fuer den Lewisschen Modalkalkuel S4* ◇ B45 ◇
LAMBERT, K. & LEBLANC, H. & MEYER, R.K. *A liberated version of S5* ◇ B45 ◇
LEBIEDIEWA, S. *The systems of modal calculus of names I (Polish and Russian summaries)* ◇ B45 ◇
LORENZEN, P. *Theophrastische Modallogik* ◇ A10 B45 ◇
MAKINSON, D. *A normal modal calculus between T and S4 without the finite model property* ◇ B45 ◇
MARTIN, R.M. *Belief, existence and meaning* ◇ A05 B45 ◇
MASSEY, G.J. *Sheffer functions for many-valued S5 modal logics* ◇ B45 B50 ◇
MCCALL, S. & NAT VANDER, A. *The system S9* ◇ B45 ◇
MCKEON, R. *Principles of modal logic, aristotelian and modern* ◇ A10 B45 ◇
MENNE, A. *Modalitaeten als Stufenfunktoren* ◇ B45 ◇
MINTS, G.E. *On the semantics of modal logic (Russian)* ◇ B45 ◇
MIURA, S. *Primitive logics with modalities and their characterization by evaluation* ◇ B45 ◇
MONTGOMERY, H. & ROUTLEY, R. *Modalities in a sequence of normal non-contigency modal systems* ◇ B45 ◇
MURRAY, W.J. *Les relations de raison et la modalite du jugement dans la logique d'Aristotle* ◇ A10 B45 ◇
POIRIER, R. *Sur les logiques de la modalite* ◇ B45 ◇
PRIOR, A.N. *On the calculus MCC* ◇ B45 ◇
QUINE, W.V.O. *Reply to Foellesdal* ◇ B45 ◇
RESCHER, N. *A contribution to modal logic* ◇ B45 ◇
RESCHER, N. *Appendix on the logic of determination and determinism* ◇ A05 B45 ◇
RESCHER, N. *Appendix I. A note on R-2* ◇ B45 ◇
RESCHER, N. *Appendix III. A summary of modal systems* ◇ A05 B45 ◇
RESCHER, N. *Assertion logic* ◇ A05 B45 ◇
RESCHER, N. *Deontic logic* ◇ A05 B45 ◇
RESCHER, N. *Epistemic modality: the problem of a logical theory of belief statements* ◇ A05 B45 ◇
RESCHER, N. *Nonstandard quantificational logic* ◇ A05 B45 C90 ◇
RESCHER, N. *The logic of preference* ◇ A05 B45 ◇
RESCHER, N. *Topics in philosophical logic* ◇ A05 B45 B98 C90 ◇
ROUTLEY, R. *Existence and identity in quantified modal logics* ◇ B45 ◇
SCHUMM, G.F. *On a modal system of D.C.Makinson and B.Sobocinski* ◇ B45 ◇
SCHUMM, G.F. *On some open questions of B.Sobocinski* ◇ B45 ◇
STINE, G.C. *Hintikka on quantification and belief* ◇ A05 B45 ◇
SUMNER, L.W. & WOODS, J. (EDS.) *Necessary truth: a book of readings* ◇ A05 B45 ◇
SURMA, S.J. *On the question of natural deduction in modal logic* ◇ B45 ◇
THOMASON, R.H. *Modal logic and metaphysics. The logical way of doing things* ◇ A05 B45 ◇
TIRNOVEANU, M. *Sur un systeme logistique deontique modal semi-constructif* ◇ B45 ◇
VIERU, S. *Sur les systemes de syllogistique apodictique* ◇ A10 B45 ◇
WEINGARTNER, P. *Ein Kalkuel fuer die Begriffe "beweisbar" und "entscheidbar"* ◇ A05 B45 ◇
WOODS, J. *Intensional relations (Polish and Russian summaries)* ◇ A05 B45 ◇
ZARNECKA-BIALY, E. *Undefinability of possibility and necessity functors in some poor purely implicational logical calculi (Polish) (English summary)* ◇ B45 ◇
ZEMAN, J.J. *Complete modalization in S4.4 and S4.04* ◇ B45 ◇
ZEMAN, J.J. *Decision procedures for* $S3^o$ *and* $S4^o$ ◇ B25 B45 ◇
ZEMAN, J.J. *Modal systems in which necessity is "factorable"* ◇ B45 ◇

1970

AAQVIST, L. *Knowing who* ◇ A05 B45 ◇
ANDERSON, A.R. *The logic of Hohfeldian propositions* ◇ A05 B45 ◇
BAYART, A. *On truth tables for M, B, S4 and S5* ◇ B45 ◇
BELNAP JR., N.D. *Conditional assertion and restricted quantification* ◇ B45 ◇
BULL, R.A. *An approach to tense logic* ◇ B45 ◇
CRESSWELL, M.J. *Classical intensional logics* ◇ B45 ◇
CRESSWELL, M.J. *Note on the interpretation of S0.5* ◇ B45 ◇
DUNN, J.M. *Comments on N.D.Belnap,jr.'s conditional assertion and restricted quantification* ◇ B45 ◇
DUTHIE, G.D. *Intensional propositional logic* ◇ B45 ◇
ESSLER, W.K. *Ueber Intensionen und Modalitaeten* ◇ B45 ◇
FINE, K. *Propositional quantifiers in modal logic* ◇ B45 ◇
FITTING, M. *An embedding of classical logic in S4* ◇ B45 ◇
FRAASSEN VAN, B.C. & LAMBERT, K. *Meaning relations, possible objects and possible worlds* ◇ B45 ◇
GABBAY, D.M. *Selective filtration in modal logic, I: Semantic tableaux method* ◇ B45 ◇

GOBLE, L.F. *Grades of modality* ◇ B45 ◇
GODDARD, L. *Nonsignificance* ◇ A05 B45 ◇
HANSSON, B. *Deontic logic and different levels of generality* ◇ B45 ◇
HERZBERGER, H.G. *Truth and modality in semantically closed languages* ◇ A05 B45 ◇
HINTIKKA, K.J.J. *Existential presuppositions and uniqueness presuppositions* ◇ A05 B45 ◇
HINTIKKA, K.J.J. *The semantics of modal notions and the indeterminancy of ontology* ◇ B45 ◇
IVIN, A.A. *Certain formulations of modal systems (Russian)* ◇ B45 ◇
IVIN, A.A. *Definitions of aletic and deontic modal functors in terms of material implication and constants (Russian)* ◇ B45 ◇
IVIN, A.A. *Temporal logic (Russian)* ◇ B45 ◇
JOJA, A. *The modality of judgments (Romanian)* ◇ B45 ◇
MAKINSON, D. *A generalisation of the concept of a relational model for modal logic* ◇ B45 ◇
MASSEY, G.J. *Binary closure-algebraic operations that are funtionally complete* ◇ B45 ◇
MONTAGUE, R. *Pragmatics and intensional logic* ◇ B45 B65 ◇
MONTAGUE, R. *Pragmatics and intensional logic* ◇ B45 ◇
NAKAMURA, A. *On the undecidability of monadic modal predicate logic* ◇ B45 D35 ◇
PURTILL, R.L. *Four-valued tables and modal logic* ◇ B45 ◇
QUINE, W.V.O. & ULLIAN, J.S. *The web of belief* ◇ A05 B45 ◇
RENNIE, M.K. *Models for multiply modal systems* ◇ B45 ◇
ROUTLEY, R. *Decision procedures and semantics for Feys' system* $S2^0$ *and surrounding systems* ◇ B25 B45 ◇
ROUTLEY, R. *Extensions of Makinsons completeness theorems in modal logic* ◇ B45 ◇
SCHINDLER, P. *Tense logic for discrete future time* ◇ A05 B45 ◇
SCHULTZ, KONRAD *Keine kontingenten Identitaeten in Lemmons modaler Mengenlehre* ◇ B45 E70 ◇
SCHULTZ, KONRAD *Modelle modaler Mengenlehren* ◇ B45 C62 C90 E35 E70 ◇
SCOTT, D.S. *Advice on modal logic* ◇ B45 ◇
SEGERBERG, K. *A remark concerning the Brouwer system (Swedish)* ◇ B45 ◇
SEGERBERG, K. *Kripke-type semantics for preference logic* ◇ B45 ◇
SEGERBERG, K. *Modal logics with linear alternative relations* ◇ B45 C90 ◇
SHUKLA, A. *Decision procedures for Lewis system S1 and related modal systems* ◇ B25 B45 ◇
SIDORENKO, E.A. *Certain variants of systems of logical inference (Russian)* ◇ B45 ◇
SIDORENKO, E.A. *Independence in systems of logical inference (Russian)* ◇ B45 ◇
SLININ, YA.A. *Iterated modalities in contemporary logic (Russian)* ◇ B45 ◇
SMIRNOV, G.A. *The forms of logical inference (Russian)* ◇ B45 ◇
SOBOCINSKI, B. *Certain extensions of modal system S4* ◇ B45 ◇
SOBOCINSKI, B. *Note on G.J.Massey's closure-algebraic operation* ◇ B45 G05 G25 ◇
SOBOCINSKI, B. *Note on Zeman's modal system S4.o4* ◇ B45 ◇
SURMA, S.J. *Concerning natural deduction in modal logic (Polish) (English summary)* ◇ B45 ◇
SUSZKO, R. *Non-Frege logic and theories that are based on it (Russian)* ◇ B45 ◇
THOMASON, R.H. *A Fitch-style formulation of conditional logic* ◇ B45 ◇
THOMASON, R.H. *Indeterminist time and truth-value gaps* ◇ A05 B45 ◇
THOMASON, R.H. *Some completeness results for modal predicate calculi* ◇ B45 ◇
WEINBERGER, O. *Normenlogik anwendbar im Recht* ◇ B45 ◇
WINNIE, J.A. *The completeness of Copi's system of natural deduction* ◇ B10 B45 ◇
WOODRUFF, P.W. *A note on JP'* ◇ B45 ◇
ZARNECKA-BIALY, E. *The deduction theorem for Goedel's propositional calculus G5 (Polish summary)* ◇ B45 ◇
ZINOV'EV, A.A. *Complex logic (Russian)* ◇ B45 ◇
ZINOV'EV, A.A. *Complex logic (a formal construction) (Russian)* ◇ A05 B45 ◇

1971

AAQVIST, L. *The completeness of some modal logics with circumstantials, subjunctive conditionals, transworld identity and dispositional predicates. A study in the prolegomena to the logic of science* ◇ A05 B45 ◇
ALCHOURRON, C.E. & BULYGIN, E. *Normative systems* ◇ B45 B98 ◇
BLOOM, S.L. *A completeness theorem for "theories of kind W" (Polish and Russian summaries)* ◇ B45 ◇
CAMPRI, P. *Un sistema di logica deontica al livello del calcolo die predicati* ◇ B45 ◇
CHELLAS, B.F. *Imperatives* ◇ B45 ◇
CRESSWELL, M.J. *The semantics for a logic of "better"* ◇ B45 ◇
D'AMORE, B. *Analisi ed ampliamenti della logica deontica proposizionale di von Wright* ◇ A05 B45 ◇
DUNN, J.M. & MEYER, R.K. *Algebraic completeness results for Dummett's LC and its extensions* ◇ B45 ◇
FINE, K. *The logics containing S4.3* ◇ B45 ◇
FOLLESDAL, D. & HILPINEN, R. *Deontic logic: an introduction* ◇ B45 ◇
GABBAY, D.M. *Decidability results in non-classical logic, III: Systems with statability operators* ◇ B25 B45 C90 ◇
GABBAY, D.M. *Montague type semantics for modal logics with propositional quantifiers* ◇ B45 ◇
GABBAY, D.M. *On decidable, finitely axiomatizable, modal and tense logics without the finite model property I, II* ◇ B25 B45 C90 ◇
GOBLE, L.F. *A system of modality* ◇ B45 ◇
HAMBLIN, C.L. *Instants and intervals* ◇ B45 ◇

HAMBLIN, C.L. *Starting and stopping* ◇ B45 ◇
HILPINEN, R. (ED.) *Deontic logic: introductory and systematic readings* ◇ B45 ◇
HINTIKKA, K.J.J. *Some main problems of deontic logic* ◇ B45 ◇
HUBIEN, H. *On two of Professor Rescher's modal theses* ◇ B45 ◇
IOAN, P. *The paradoxes of implication* ◇ A10 B45 ◇
KALINOWSKI, G. *Sur la syllogistique deontique de Zdzislaw Ziemba (Polish and Russian summaries)* ◇ B45 ◇
KAMP, H. *Formal properties of "now"* ◇ B45 ◇
KIELKOPF, C.F. *Provability as a deontic notion* ◇ B45 ◇
KIELKOPF, C.F. *Semantics for a utilitarian deontic logic* ◇ B45 ◇
LEWIS, D. *Completeness and decidability of three logics of counterfactual conditionals* ◇ B25 B45 ◇
LINSKY, L. *Reference and modality* ◇ A05 B45 ◇
MAKINSON, D. *Aspectos de la logica modal* ◇ B45 ◇
MAKINSON, D. *Some embedding theorems for modal logic* ◇ B45 ◇
MCLELLAND, J. *Epistemic logic and the paradox of the surprise examination* ◇ B45 ◇
MESKHI, V.YU. *To algebraic semantics of temporal logic (Russian)* ◇ B45 ◇
MORSCHER, E. *A matrix method for deontic logic* ◇ B45 ◇
PEKLO, B.T. *Analytical sentences and normative or evaluating logics* ◇ B45 ◇
PERZANOWSKI, J. & WRONSKI, A. *The deduction theorems for the system T of Feys-Wright (Polish summary)* ◇ B45 ◇
PRIOR, A.N. *Recent advances in tense logic* ◇ B45 ◇
PYATNITSYN, B.N. & SUBBOTIN, A.L. *De quelques approches de la classification des systemes logiques* ◇ B45 ◇
RENNIE, M.K. *Remark on Cresswell on S0.5* ◇ B45 ◇
RENNIE, M.K. *Semantics for* RK_t ◇ B45 ◇
RESCHER, N. & URQUHART, A.I.F. *Temporal logic* ◇ A05 B45 B98 ◇
ROUTLEY, R. *Conventionalist and contingency-oriented modal logics* ◇ B45 ◇
RUZSA, I. *Random models of logical systems II: Models of some modal logics* ◇ B45 C90 ◇
SCHOCK, R. *Quasi-connectives definable in concept theory* ◇ A05 B45 E70 ◇
SCHUMM, G.F. *Solutions to four modal problems of Soboconski* ◇ B45 ◇
SEGERBERG, K. *An essay in classical modal logic I,II,III* ◇ B45 ◇
SEGERBERG, K. *Some logics of commitment and obligation* ◇ B45 ◇
SESIC, B.V. *Foundations of the logic of change and development* ◇ B45 ◇
SHUKLA, A. *Finite model property for five modal calculi in the neighbourhood of S3* ◇ B45 ◇
SLUPECKI, J. *A generalization of modal logic (Polish and Russian summaries)* ◇ B45 ◇
SNYDER, D.P. *Modal logic and its applications* ◇ B45 ◇
SOBOCINSKI, B. *A new class of modal systems* ◇ B45 ◇
SOBOCINSKI, B. *A proper subsystem of S4.04* ◇ B45 ◇
SOBOCINSKI, B. *Concerning some extensions of S4* ◇ B45 ◇
SURMA, S.J. *Deduction theorems in modal systems constructed by Goedel's method (Polish summary)* ◇ B45 ◇
SUSZKO, R. *Identity connective and modality (Polish and Russian summaries)* ◇ B45 ◇
SUSZKO, R. & ZANDAROWSKA, W. *Lewi's system S4 and S5 and the identity connective (Polish) (Russian and English summaries)* ◇ B45 ◇
TIRNOVEANU, M. *On certain properties of the deonic logical system D1 (Romanian)* ◇ B45 ◇
TIRNOVEANU, M. *Sur quelques proprietes du systeme logistique modal deontique modal transfini* L'_D *(Romanian) (French summary)* ◇ B45 ◇
TOMBERLIN, J.E. *The sea battle tomorrow and fatalism* ◇ A05 B45 ◇
VETTER, H. *Deontic logic without deontic operators* ◇ B45 ◇
WASILEWSKA, A. *A formalization of the modal propositional S4 calculus (Polish and Russian summaries)* ◇ B45 ◇
WEINGARTNER, P. & WOLF, K. (EDS.) *Ernst Mally: Logische Schriften* ◇ A10 B45 ◇
ZEMAN, J.J. *A study of some systems in the neighbourhood of* S4.4 ◇ B45 ◇
ZIEMBA, Z. *Deontic syllogistics (Polish and Russian summaries)* ◇ B45 ◇

1972

AAQVIST, L. *Performatives and verifiability by the use of language. A study in the applied logic of indexicals and conditionals* ◇ A05 B45 ◇
ALCHOURRON, C.E. *The intuitive background of normative legal discourse and its formalization* ◇ A05 B45 ◇
BARNES, K.T. *Hintikka's argument for the need for quantifying into opaque contexts* ◇ A05 B45 ◇
BARTH, E.M. *Implication et necessite* ◇ A10 B45 ◇
BEATTY, H. *On evaluating deontic logics* ◇ B45 ◇
BELNAP JR., N.D. *Foreword to: A general interpreted modal calculus, by Aldo Bressan* ◇ B45 ◇
BLUM, A. *Isomorphisms between C1 and C2* ◇ B45 ◇
BRESSAN, A. *A general interpreted modal calculus* ◇ B45 ◇
BRYLL, G. & SLUPECKI, J. *The proof of T-decidability of Lewi's system S5* ◇ B45 ◇
CARNAP, R. *Bedeutung und Notwendigkeit: Eine Studie zur Semantik und modalen Logik* ◇ A05 B45 ◇
CHISHOLM, R.M. & KEIM, R.G. *A system of epistemic logic* ◇ B45 ◇
CHRISTIAN, C.C. *Modalkalkuel als formale Theorie und das Problem einer Praedikatoidenlogik* ◇ B45 ◇
CRESSWELL, M.J. *Intensional logics and logical truth* ◇ B45 ◇
CRESSWELL, M.J. *Second-order intensional logic* ◇ B15 B45 ◇
CRESSWELL, M.J. *S0.5 is alive and well* ◇ B45 ◇

Subject Index

CRESSWELL, M.J. *The completeness of S1 and some related systems* ◇ B45 ◇
EHSAKIA, L.L. *On the category of complete Kripke models (Russian)* ◇ B45 ◇
FACIONE, P.A. *A modal truth-tabular interpretation for necessary and sufficient conditions* ◇ B45 ◇
FINE, K. *In so many possible worlds* ◇ B45 ◇
FINE, K. *Logics containing S4 without the finite model property* ◇ B45 ◇
FITTING, M. *ε-calculus based axiom systems for some propositional modal logics* ◇ B45 ◇
FITTING, M. *An ε-calculus system for first order S4* ◇ B45 ◇
FITTING, M. *Non-classical logics and the independence results of set theory* ◇ B45 C90 E35 ◇
FITTING, M. *Tableau methods of proof for modal logics* ◇ B45 ◇
FRAASSEN VAN, B.C. *The logic of conditional obligation* ◇ B45 ◇
GABBAY, D.M. *A general filtration method for modal logics* ◇ B45 ◇
GABBAY, D.M. *Craig's interpolation theorem for modal logics* ◇ B45 C40 C90 ◇
GABBAY, D.M. *Tense systems with discrete moments of time I* ◇ B45 ◇
GARSON, J.W. *Two new interpretations of modality* ◇ B45 ◇
GRANDY, R.E. *A definition of truth for theories with intensional definite description operators* ◇ A05 B45 ◇
HAYES, S.E. *Extensions of* T^0 ◇ B45 ◇
HAZEN, A. *Semantics for S4.2* ◇ B45 ◇
HOCUTT, M.O. *Is epistemic logic possible?* ◇ A05 B45 ◇
JERVELL, H.R. *An Herbrand theorem for a modal logic* ◇ B45 ◇
KABZINSKI, J.K. *Some derivational formalizations of propositional modal logics (Polish) (English summary)* ◇ B45 ◇
KALINOWSKI, G. *Einfuehrung in die Normenlogik* ◇ A10 B45 ◇
KIELKOPF, C.F. *Kripke's axiomatization of S2* ◇ B45 ◇
LORENZEN, P. *Zur konstruktiven Deutung der semantischen Vollstaendigkeit klassischer Quantoren- und Modalkalkuele* ◇ B45 F50 ◇
MANOR, R. & RESCHER, N. *Modal elaborations of propositional logics* ◇ B45 ◇
MARTIN, N.M. *Systems of modal logic* ◇ B45 ◇
MASSEY, G.J. *The modal structure of the Prior-Rescher family of infinite product systems* ◇ B45 ◇
MIURA, S. *Probabilistic models of modal logics* ◇ B45 C90 ◇
MONTAGUE, R. *Pragmatics and intensial logic* ◇ B45 ◇
MORSCHER, E. & ZECHA, G. *Wozu deontische Logik?* ◇ B45 ◇
MOSS, B.P. *A picture of a Kripke model for S4* ◇ B45 ◇
MUNITZ, M.K. (ED.) *Identity and individuation* ◇ A05 B45 ◇
NAT VANDER, A. *Axiomatic, Sequenzen-Kalkul, and subordinate proof versions of S9* ◇ B45 ◇
PARKS, R.Z. *Classes and change* ◇ B45 ◇
PARKS, R.Z. *On formalizing Aristotle's theory of modal syllogisms* ◇ A05 A10 B45 ◇
PERZANOWSKI, J. *The first list of the deduction theorems characteristic for several modal calculi formalized after the manner of Lemmon* ◇ B45 ◇
PIETERS, J. *Et en meme temps* ◇ B45 ◇
PLEDGER, K.E. *Modalities of systems containing S3* ◇ B45 ◇
POSPESEL, H. *Scepticism and modal logic* ◇ A05 B45 ◇
PRUCNAL, T. *Structural completeness of Lewi's system S5 (Russian summary)* ◇ B45 ◇
SATRE, T.W. *Natural deduction rules for modal logics* ◇ B45 ◇
SATRE, T.W. *Natural deduction rules for* $S1^0 - S4^0$ ◇ B45 ◇
SEGERBERG, K. *Post completeness in modal logic* ◇ B45 C90 ◇
SHUKLA, A. *Consistent, independent, and distinct propositions* ◇ B45 ◇
SHUKLA, A. *The existence postulate and non-regular systems of modal logic* ◇ B45 ◇
SLATER, B.H. *The foundations of logic* ◇ A05 B45 ◇
SLININ, YA.A. *Die Modalitaetentheorie in der modernen Logik* ◇ B45 ◇
STAHL, G. *Temporal terms in functional systems* ◇ B45 ◇
SURMA, S.J. *The deduction theorems valid in certain fragments of the Lewis' system S2 and the system of Feys-von Wright* ◇ B45 ◇
SUSZKO, R. *A note on modal systems and SCI* ◇ B45 B60 ◇
THOMASON, S.K. *Noncompactness in propositional modal logic* ◇ B45 ◇
THOMASON, S.K. *Semantic analysis of tense logics* ◇ B45 ◇
TOMATIS, P. *Il teorema di eliminazione per una logica modale S4* ◇ B45 ◇
TURQUETTE, A.R. *Generalized modal sets* ◇ A10 B45 ◇
VANDAMME, F. *On negation* ◇ B45 ◇
WASILEWSKA, A. *The diagrams of formulas of the modal propositional* $S4^*$ *calculus (Polish and Russian summaries)* ◇ B45 ◇
WEINBERGER, O. *The concept of non-satisfaction and deontic logic* ◇ B45 ◇
WESSEL, H. (ED.) *Quantoren-Modalitaeten-Paradoxien* ◇ A05 B45 ◇
WOLNIEWICZ, B. *The notion of fact as a modal operator* ◇ B45 ◇
WRIGHT VON, G.H. *Some observations on modal logic and philosophical systems* ◇ A05 B45 ◇
WRONSKI, A. *Remarks on algebraic semantics for modal systems S0.9, S1 and S2* ◇ B45 ◇
ZEMAN, J.J. *Semantics for* S4.3.2 ◇ B45 ◇
ZEMAN, J.J. *S4.6 is S4.9* ◇ B45 ◇

1973

AAQVIST, L. *Modal logic with subjunctive conditional and dispositional predicates* ◇ B45 ◇
BAXTER, RODNEY J. *On some models for modal logic* ◇ B45 ◇

BEARD, R.W. *A natural deduction variant of Systems T, S4, S5, and the brouwerian system* ◇ B45 ◇
BEATTY, H. *On evaluating deontic logics. Comments on van Fraassen's paper* ◇ B45 ◇
BELNAP JR., N.D. *Restricted quantification and conditional assertion* ◇ A05 B45 ◇
BELNAP JR., N.D. *S-P interrogatives* ◇ A05 B45 ◇
BRESSAN, A. *Intensional descriptions and relative completeness in the general interpreted modal calculus* MC^{ν} ◇ B45 ◇
BRESSAN, A. *The interpreted type-free modal calculus* MC^{∞}*. I: The type free extensional calculus* MC^{∞} *involving individuals, and the interpreted language* ML^{∞} *on which* EC^{∞} *is based* ◇ B45 E70 ◇
BRESSAN, A. *The interpreted type-free modal calculus* MC^{∞}*. II. Foundations of* MC^{∞} ◇ B45 E70 ◇
BRYLL, G. & SLUPECKI, J. *A certain proof of the completeness of Lewis's system S5 (Polish) (English summary)* ◇ B45 ◇
BRYLL, G. & ROSIEK, M. *A note on the L-decidability of a certain modal logic of Lukasiewicz (Polish) (English summary)* ◇ B45 ◇
BRYLL, G. & SLUPECKI, J. *Proof of L-decidability of Lewis system S5 (Polish and Russian summaries)* ◇ B45 ◇
BYRD, M. *Knowledge and true belief in Hintikka's epistemic logic* ◇ B45 ◇
CHANG, C.C. *Modal model theory* ◇ B45 C90 ◇
CHERNIAVSKY, J.C. *The complexity of some non-classical logics* ◇ B45 D15 F20 F50 ◇
CHIDGEY, J.R. *A note on transitivity* ◇ B45 ◇
CHURCH, A. *Isomorfismo intensional e identidad de creencia (Spanish)* ◇ A05 B45 ◇
CHURCH, A. *Logica, analisis y intensionalidad (Spanish)* ◇ A05 B45 ◇
CHURCH, A. *Outline of a revised formulation of the logic of sense and denotation I* ◇ B45 ◇
CHURCH, A. *Postscript 1968* ◇ A05 B45 ◇
CHURCH, A. *Sobre el analisis del discurso indirecto propuesto por Scheffler* ◇ B45 B65 ◇
COLLIER, K.W. *Physical modalities and the system E* ◇ B45 ◇
CORCORAN, J. & WOOD, S. *Discussion: the switches "paradox"* ◇ B45 ◇
CRESSWELL, M.J. *Physical theories and possible worlds* ◇ B45 ◇
DAHN, B.I. *Generalized Kripke-models* ◇ B45 C90 ◇
DALLA CHIARA SCABIA, M.L. *Instanti e individui nelle logiche temporali* ◇ B45 ◇
DUNN, J.M. *A truth value semantics for modal logic* ◇ B45 ◇
EHSAKIA, L.L. *On some new results in the theory of modal and superintuitionistic systems (Russian)* ◇ B45 ◇
ENGLEBRETSEN, G. *Epistemic logic and mere belief* ◇ B45 ◇
FITCH, F.B. *A correlation between modal reduction principles and properties of relations* ◇ B45 ◇
FITTING, M. *A modal logic analog of Smullyan's fundamental theorem* ◇ B45 ◇
FITTING, M. *Model existence theorems for modal and intuitionistic logics* ◇ B45 F50 ◇
FRAASSEN VAN, B.C. *The logic of conditional obligation* ◇ B45 ◇
GABBAY, D.M. *A survey of decidability results for modal, tense and intermediate logics* ◇ B25 B45 B55 C98 F50 ◇
GABBAY, D.M. & MORAVCSIK, J.M.E. *Sameness and individuation* ◇ A05 B45 ◇
GAERDENFORS, P. & HANSSON, B. *A guide to intensional semantics* ◇ A05 B45 ◇
GAERDENFORS, P. *On the extensions of S5* ◇ B45 ◇
GAO, HENGSHAN *The decision problem of modal predicate calculi. I: On Slomson's reductions (Chinese)* ◇ B45 D35 ◇
GARDIES, J.-L. *Logique deontique et theorie generale des fonctions completives* ◇ B45 ◇
GARDIES, J.-L. *Vingt annees de recherches deontiques* ◇ A10 B45 ◇
GARSON, J.W. *Indefinite topological logic* ◇ B45 ◇
GARSON, J.W. *The completeness of an intensional logic: Definite topological logic* ◇ B45 ◇
GEACH, P.T. *On certain modal systems connected with S4.3* ◇ B45 ◇
GOBLE, L.F. *A new modal model* ◇ B45 C90 ◇
GOBLE, L.F. *A simplified semantics for modal logic* ◇ B45 ◇
GODDARD, L. & ROUTLEY, R. *The logic of significance and context* ◇ A05 B45 ◇
GOLDBLATT, R.I. *A model-theoretic study of some systems containing S3* ◇ B25 B45 C90 ◇
GOLDBLATT, R.I. *A new extension of S4* ◇ B45 ◇
GOLDBLATT, R.I. *Concerning the proper axiom for S4.04 and some related systems* ◇ B45 ◇
GREEF DE, J. *Professor Halberstadt on counterfactual conditionals and modality* ◇ A05 B45 ◇
HILPINEN, R. *On the semantics of personal directives* ◇ B45 ◇
HINTIKKA, K.J.J. *Surface semantics: definition and its motivation* ◇ A05 B10 B45 ◇
HUNG, CH'ENGWAN *On the axiomatic structures of intensional logic. III:* $B^+, N^+, T^+, E^+, S4^+, R^+, J^+$ ◇ B45 ◇
IVIN, A.A. *Modale Logik und die Theorie der Implikation (Russian)* ◇ A05 B45 ◇
JOHNSON WU, K. *A new approach to formalization of a logic of knowledge and belief* ◇ B45 ◇
KARAVAEV, E.F. *The tensed qualification of propositions and theory of logical inference (Russian)* ◇ B45 ◇
KEUTH, H. *Comments on Dr. Vetter's paper "Deontic logic without deontic operators"* ◇ B45 ◇
LEBLANC, H. *On dispensing with things and worlds* ◇ A05 B45 ◇
LEBLANC, H. *Semantic deviations* ◇ A05 B45 ◇
LEWIS, D. *Counterfactuals* ◇ A05 B45 ◇
LEWIS, D. *Counterfactuals and comparative possibility* ◇ B45 ◇
MAKINSON, D. *A warning about the choice of primitive operators in modal logic* ◇ B45 ◇

MARGOLIS, J. *Identity and the necessity operator* ◇ B45 ◇
MCLAUGHLIN, R.N. *On the logic of general conditionals* ◇ B45 ◇
MORGAN, C.G. *Systems of modal logic for impossible worlds* ◇ B45 ◇
MOTT, P.L. *On Chisholm's paradox* ◇ B45 ◇
NAT VANDER, A. & RESCHER, N. *On alternatives in epistemic logic* ◇ B45 ◇
PAHI, B. *Necessity and some non-modal propositional calculi* ◇ B45 ◇
PARKS, R.Z. & RESCHER, N. *Possible individuals, trans-world identity, and quantified modal logic* ◇ A05 B45 ◇
PERZANOWSKI, J. *The deduction theorems for the modal propositional calculi formalized after the manner of Lemmon I* ◇ B45 ◇
PURTILL, R.L. *Meinongian deontic logic* ◇ B45 ◇
RODRIGUEZ MARIN, J. *Deontic logic. Natural deduction and decision via semantic tables (spanish)* ◇ B45 ◇
RUZSA, I. *Prior-type modal logic I,II* ◇ B45 ◇
SAITO, S. *Modality and preference relation* ◇ B45 ◇
SANCHEZ-MAZAS, M. *Calculo de las normas* ◇ B45 ◇
SCOTT, D.S. *Background to formalization* ◇ A05 B45 ◇
SEGERBERG, K. *Franzen's proof of Bull's theorem* ◇ B45 ◇
SEGERBERG, K. *Two dimensional modal logic* ◇ B45 ◇
SMULLYAN, R.M. *A generalization of intuitionistic and modal logics* ◇ B45 F50 ◇
SOBOCINSKI, B. *A new axiomatization of modal system K1.2* ◇ B45 ◇
SOBOCINSKI, B. *Modal system S3 and the proper axioms of S4.02 and S4.04* ◇ B45 ◇
SOETEMAN, A. *Some remarks about two famous paradoxes of deontic logic* ◇ B45 ◇
SURMA, S.J. *The deduction theorems valid in certain fragments of the Lewi's system S2 and the system T of Feys-von Wright (Polish and Russian summaries)* ◇ B45 ◇
THOMAS, I. *Further extensions of* $S3^*$ ◇ B45 ◇
THOMAS, I. *Unusual feature of* $S3^*$ ◇ B45 ◇
THOMASON, R.H. *Philosophy and formal semantics* ◇ A05 B45 ◇
THOMASON, S.K. *A new representation of S5* ◇ B45 ◇
TICHY, P. *On "de dicto" modalities in quantified S5* ◇ B45 ◇
TIRNOVEANU, M. *On some properties of the transfinite assertoric-deontic logistic systems* $S_{ad}^{\alpha 210}, S_{ad}^{\alpha 2n0}$ *(Romanian)* ◇ B45 ◇
VANDERVEKEN, D.R. *La completude et la compacite de la pragmatique* ◇ B45 ◇
WEAVER, G.E. *Logical consequence in modal logic: alternative semantic systems for normal modal logics* ◇ B45 ◇
WEINBERGER, O. *Der Erlaubnisbegriff und der Aufbau der Normenlogik* ◇ B45 ◇
WEINBERGER, O. *Erwiderung auf A.Soetemans Kritik und seine Reformversuche der deontischen Logik* ◇ B45 ◇
WEINGARTNER, P. *A predicate calculus for intensional logic* ◇ A05 B45 ◇
WIREDU, J.E. *On the real logical structure of Lewis' "independent proof"* ◇ B45 ◇
WOODS, J. *Descriptions, essences and quantified modal logic* ◇ A05 B45 ◇
WOOLHOUSE, R.S. *Tensed modalities* ◇ B45 ◇
WRIGHT VON, G.H. *On the logic and epistemology of the causal relation* ◇ A05 B45 ◇
WRIGHT VON, G.H. *Remarks on the logic of predication* ◇ B45 ◇
ZACHOROWSKI, S. *Wajsberg's semantics for the system S5 of Lewis* ◇ B45 ◇
ZARNECKA-BIALY, E. *Modal functors and their definability in propositional calculus systems (Polish) (English summary)* ◇ B45 F50 ◇
ZEMAN, J.J. *Modal logic. The Lewis-modal system* ◇ B45 ◇

1974

AMBROSE, A. *Believing necessary propositions* ◇ A05 B45 ◇
ASHWORTH, E.J. *Language and logic in the post-medieval period* ◇ A05 A10 B45 ◇
BENTHEM VAN, J.F.A.K. *Some correspondance results in modal logic* ◇ B45 ◇
BERGSTROEM, L. *Hintikka on "prima facie" obligations* ◇ A05 B45 ◇
BLANCHE, R. *Modalite et temporalite* ◇ A05 B45 ◇
BRESSAN, A. *Comment's on Suppes' paper: the essential but implicit role of modal concepts in science* ◇ A05 B45 ◇
BRESSAN, A. *On the semantics for the language* ML^{ν} *based on a type system, and those for the type-free language* ML^{∞} ◇ B45 E70 ◇
BRESSAN, A. *On the usefulness of modal logic in axiomatizations of physics* ◇ B45 B80 ◇
BRESSAN, A. *The interpreted type-free modal calculus* MC^{∞} *III. Ordinals and cardinals in* MC^{∞} *(Italian summary)* ◇ B45 E70 ◇
BUDBAEVA, S.P. & PYATNITSYN, B.N. *The investigation and construction of pragmatic logics (Russian)* ◇ B45 ◇
CHELLAS, B.F. *Conditional obligation* ◇ A05 B45 ◇
CHURCH, A. *Outline of a revised formulation of the logic of sense and denotation II* ◇ B45 ◇
COCCHIARELLA, N.B. *La semantica della logica del tempo* ◇ A05 B45 ◇
COCCHIARELLA, N.B. *Logical atomism and modal logic* ◇ A05 B45 ◇
CORCORAN, J. & WEAVER, G.E. *Logical consequence in modal logic II: some semantic systems for S4* ◇ B45 ◇
CORNIDES, T. *Ordinale Deontik* ◇ B45 B98 ◇
DAHN, B.I. *A note on generalized Kripke-models* ◇ B45 C90 ◇
DAHN, B.I. *Contributions to the model theory for non-classical logics* ◇ B45 C90 ◇
DALEN VAN, D. *Variants of Rescher's semantics for preference logic and some completeness theorems* ◇ B45 ◇

DISHKANT, G.P. *Weak decidability of some modal extension of trigonometry* ◇ B45 ◇
EHSAKIA, L.L. *Topological Kripke models (Russian)* ◇ B45 B55 G05 G25 G30 ◇
ENGELER, E. *The logic of "can do"* ◇ B45 ◇
ENGLEBRETSEN, G. *A note on contrariety* ◇ B45 ◇
ERMOLAEVA, N.M. & MUCHNIK, A.A. *Modal extensions of logical calculi of Hao Wang type (Russian)* ◇ B45 ◇
FINE, K. *An ascending chain of S4 logics* ◇ B45 ◇
FINE, K. *An incomplete logic containing S4* ◇ B45 ◇
FINE, K. *Logics containing* K4 *I* ◇ B45 ◇
FOX, F. *A note on a consistency proof* ◇ B45 ◇
GABBAY, D.M. *A generalization of the concept of intensional semantics with applications* ◇ B45 ◇
GABBAY, D.M. *Tense logics and the tenses of English* ◇ B45 ◇
GARSON, J.W. *The substitution interpretation in topological logic* ◇ B45 ◇
GIRLE, R.A. *Possibility pre-supposition free logics* ◇ B45 ◇
GOBLE, L.F. *Gentzen systems for modal logic* ◇ B45 F50 ◇
GOLDBLATT, R.I. *A study of Z modal systems* ◇ B45 ◇
GOLDBLATT, R.I. *Metamathematics of modal logic* ◇ B45 ◇
HARPER, W.L. *A note on universal instantiation in the Stalnaker Thomason conditional logic and M type modal systems* ◇ B45 ◇
HERMAN, L. & PIZIAK, R. *Modal propositional logic on an orthomodular basis I* ◇ B45 ◇
HOEPELMAN, J.P. *Tense-logic and the semantics of the Russian aspects* ◇ B45 ◇
HUMBERSTONE, I.L. *Logic for saints and heroes* ◇ B45 ◇
HUNG, CH'ENGWAN *On the axiomatic structures of intensional logics. II: Gentzen type system GX, X=E, EQ* ◇ B45 ◇
IWANUS, B. *Concerning the so-called system BD of Aqvist* ◇ B45 ◇
JENNINGS, R.E. *A utilitarian semantics for deontic logic* ◇ B45 ◇
KARTTUNEN, L. *Presuppositions of compound sentences* ◇ B45 ◇
KIELKOPF, C.F. *K1 as a Dawson modeling of A.R.Anderson's sense of "ought"* ◇ B45 ◇
KNAPP, H.G. *Bemerkungen zu "Normative Wissenschaft ohne Normen -- aber mit Werten" von M. Bunge* ◇ B45 ◇
KULIKOWSKI, J.L. *An application of a tense-logic system to formal planning* ◇ B45 ◇
LACHLAN, A.H. *A note on Thomason's refined structures for tense logics* ◇ B45 ◇
LEWIS, D. *Intensional logics without iterative axioms* ◇ B45 ◇
LEWIS, D. *Semantic analyses for dyadic deontic logic* ◇ B45 ◇
MAKINSON, D. & SEGERBERG, K. *Post completeness and ultrafilters* ◇ B22 B45 G20 ◇
MAKSIMOVA, L.L. & RYBAKOV, V.V. *A lattice of normal modal logics (Russian)* ◇ B45 G10 ◇
MANOR, R. *A semantic analysis of conditional assertion* ◇ B45 ◇
MARSDEN, E.L. *Reducible implicative models* ◇ B20 B45 C05 G25 ◇
MATERNA, P. *Expressibility of propositions in* L_μ*-languages* ◇ B45 ◇
MCARTHUR, R.P. *Factuality and modality in the future tense* ◇ B45 ◇
MCARTHUR, R.P. *The Makinson completeness of tense logic* ◇ B45 ◇
MESKHI, V.YU. *Kripke semantics for modal systems including S4.3 (Russian)* ◇ B45 ◇
MINTS, G.E. *Lewis' systems and T (a survey 1965–1973) (Russian)* ◇ B45 ◇
MORGAN, C.G. *Liberated Brouwerian modal logic* ◇ B45 ◇
MORGAN, C.G. *Notes on garden of eden configurations of higher degree* ◇ B45 ◇
MORTIMER, M. *Some results in modal model theory* ◇ B45 C90 ◇
MULLOCK, P. *The permissiveness of powers* ◇ B45 ◇
MURUNGI, R.W. *On a nonthesis of classical modal logic* ◇ B45 ◇
PARKS, R.Z. *Semantics for contingent identity systems* ◇ B45 ◇
PARKS, R.Z. & SMITH, T. *The inadequacy of Hughes and Cresswells semantics for the CI systems* ◇ B45 ◇
PEKLO, B.T. *Sind die deontischen Funktoren distributiv?* ◇ B45 ◇
PLANTINGA, A. *The nature of necessity* ◇ B45 ◇
POERN, I. *Some basic concepts of action* ◇ A05 B45 ◇
SCHUMM, G.F. *K and Z* ◇ B45 ◇
SCHUMM, G.F. *S3.02 = S3.03* ◇ B45 ◇
SILVERSTEIN, H.S. *Von Wright's deontic logics* ◇ B45 ◇
SOBOCINSKI, B. *Concerning the proper axioms of S4.02* ◇ B45 ◇
STAHL, G. *Modal models corresponding to models* ◇ B45 ◇
STAHL, G. *Termes temporels dans des systemes fonctionnels* ◇ B45 ◇
SUGIHARA, T. *Temporal logic (Japanese)* ◇ B45 ◇
SUPPES, P. *The essential but implicit role of modal concepts in science* ◇ A05 B45 ◇
SUSZKO, R. *Some notions and theorems of McKinsey and Tarski and SCI* ◇ B45 C05 G05 ◇
SZEWCZAK, E. *Iterated modalities and the parallel between deontic and modal logics* ◇ A05 B45 ◇
THOMASON, S.K. *An incompleteness theorem in modal logic* ◇ B45 ◇
THOMASON, S.K. *Reduction of tense logic to modal logic, I* ◇ B45 ◇
TONDL, L. *Types of preference* ◇ B45 ◇
UCHII, S. *"Ought" and conditionals* ◇ B45 ◇
VOL'NEVICH, B. *The concept of a fact as a modal operator (Russian)* ◇ B45 ◇
WEINBERGER, O. *Contrary-to-fact and fact-transcendent conditionals* ◇ B45 ◇
WILSON, R.L. *Prenex normal form in the modal predicate logic PS*S and the Grosseteste algebra of sets GS*S* ◇ B45 ◇
WILSON, R.L. *The nine-valued modal logic ME*E* ◇ B45 ◇

WIREDU, J.E. *A remark on a certain consequence of connexive logic for Zermelo's set theory* ◇ B45 E30 E70 ◇
ZINOV'EV, A.A. *Certain systems of formal arithmetic (Russian)* ◇ B45 F30 ◇

1975

AAQVIST, L. & GUENTHNER, F. *Representability in QA of Hintikka intensional quantifiers and Keenan term quantifiers* ◇ B45 ◇
BACON, J. *Belief as relative knowledge* ◇ A05 B45 ◇
BALDWIN, THOMAS *The philosophical significance of intensional logic. II* ◇ A05 B45 ◇
BENTHEM VAN, J.F.A.K. *A note on modal formulae and relational properties* ◇ B45 ◇
BERNARDI, C. *The fixed-point theorem for diagonalizable algebras (the algebraization of the theories which express Theor. III)* ◇ B45 F30 G05 G25 ◇
BIELA, A. *Note on the structural incompleteness of some modal propositional calculi* ◇ B45 ◇
BLASZCZUK, J.J. & DZIOBIAK, W. *Modal systems related to* $S4_n$ *of Sobocinski* ◇ B45 ◇
BLASZCZUK, J.J. & DZIOBIAK, W. *Modal systems placed in the "triangle"* $S4 - T_1^* - T$ ◇ B45 ◇
BLASZCZUK, J.J. & DZIOBIAK, W. *Remarks on Perzanowski's modal system* ◇ B45 ◇
BOTEZATU, P. *Assertoric-problematic syllogistics (Romanian) (French summary)* ◇ B45 ◇
BOWEN, K.A. *Normal modal model theory* ◇ B45 C90 ◇
BROIDO, J. *Von Wright's principle of predication - some clarifications* ◇ B45 ◇
BYRD, M. *On incoherent quantification in languages without constants* ◇ B45 ◇
CASTANEDA, H.-N. *Thinking and doing. The philosophical foundations of institutions* ◇ A05 B45 ◇
CHAPMAN, T. *Prior's criticism of the Barcan formula* ◇ A05 B45 ◇
CHELLAS, B.F. *Basis conditional logic* ◇ A05 B45 ◇
CHELLAS, B.F. & MCKINNEY, A. *The completeness of monotonic modal logics* ◇ B45 ◇
CHISHOLM, R.M. *The intrinsic value in disjunctive states of affairs* ◇ A05 B45 ◇
CLIFFORD, J.E. *Tense and tense logic* ◇ B45 ◇
COCCHIARELLA, N.B. *Logical atomism, nominalism, and modal logic* ◇ A05 B45 ◇
COCCHIARELLA, N.B. *On the primary and secondary semantics of logical necessity* ◇ A05 B45 ◇
COFFA, J.A. *Fallacies of modality* ◇ B45 ◇
CRESSWELL, M.J. *Frames and models in modal logic* ◇ B45 ◇
CRESSWELL, M.J. *Hamblin on time* ◇ A05 B45 ◇
CRESSWELL, M.J. *Hyperintensional logic* ◇ B45 ◇
CRESSWELL, M.J. *Identity and intensional objects* ◇ B45 ◇
CRESSWELL, M.J. *Omnitemporal logic and tense logic* ◇ B45 ◇
CRESSWELL, M.J. & HUGHES, G.E. *Omnitemporal logic and converging time* ◇ B45 ◇
CZERMAK, J. *Ein Vollstaendigkeitsbeweis fuer die aussagenlogischen Modalitaetensysteme M, S4, BR und S5* ◇ B45 ◇
CZERMAK, J. *Embeddings of classical logic in S4* ◇ B45 ◇
CZERMAK, J. *Interpolation theorem for some modal logics* ◇ B45 ◇
DUNN, J.M. *Axiomatizing Belnap's conditional assertion* ◇ B45 ◇
FINE, K. *Normal forms in modal logic* ◇ B45 ◇
FINE, K. *Some connections between elementary and modal logic* ◇ B45 C07 C90 ◇
FINE, K. *Vagueness, truth and logic* ◇ A05 B45 ◇
FITTING, M. *A modal logic* ε*-calculus* ◇ B45 ◇
FREEMAN, J.B. *Compactness for RQ* ◇ B45 ◇
GABBAY, D.M. *A normal logic that is complete for neighborhood frames but not for Kripke frames* ◇ B45 ◇
GABBAY, D.M. *Decidability results in non-classical logics I* ◇ B25 B45 C90 F50 ◇
GABBAY, D.M. *Model theory for tense logics* ◇ B45 C20 C90 ◇
GAERDENFORS, P. & HANSSON, B. *Filtrations and the finite frame property in boolean semantics* ◇ B25 B45 C90 G05 G25 ◇
GALLIN, D. *Intensional and higher-order modal logic. With applications to Montague semantics* ◇ B45 C85 C90 E35 E50 E70 ◇
GARDIES, J.-L. *A propos de l'article de Georges Kalinowski "Sur un nouveau systeme de logique deontique"* ◇ A05 B45 ◇
GARDIES, J.-L. *Esquisse d'une theorie des predicats completifs* ◇ B45 ◇
GARDIES, J.-L. *La logique du temps* ◇ B45 ◇
GERSON, M.S. *An extension of S4 complete for the neighbourhood semantics but incomplete for the relational semantics* ◇ B45 ◇
GERSON, M.S. *The inadequacy of the neighbourhood semantics for modal logic* ◇ B45 ◇
GIBBARD, A. *Contingent identity* ◇ B45 ◇
GIRLE, R.A. $S1 \neq S0.9$ ◇ B45 ◇
GOLDBLATT, R.I. & THOMASON, S.K. *Axiomatic classes in propositional modal logic* ◇ B45 ◇
GOLDBLATT, R.I. *First order definability in modal logic* ◇ B45 C20 C90 ◇
GOLDBLATT, R.I. *Solution to a completeness problem of Lemmon and Scott* ◇ B45 ◇
GROENENDIJK, J. & STOKHOF, M. *Modality and conversational information* ◇ B45 B65 ◇
GUMANSKI, L. *Deontic logic without certain paradoxes* ◇ B45 ◇
HAMBLIN, C.L. *Cresswell's colleague TLM* ◇ A05 B45 ◇
HERZBERGER, H.G. *Canonical superlanguages* ◇ A05 B45 ◇
HINTIKKA, K.J.J. *Comment on Professor Bergstroem* ◇ B45 ◇
HINTIKKA, K.J.J. *Impossible possible worlds vindicated* ◇ A05 B45 B53 ◇
HINTIKKA, K.J.J. *The intentions of intentionality and other new models for modalities* ◇ A05 B45 ◇

HUGHES, G.E. *B(S4.3, S4) unveiled* ◇ B45 ◇
HUNG, CH'ENGWAN *Independence of connectives in E, EM, R, RM* ◇ B45 ◇
HUNG, CH'ENGWAN *On the axiomatic structures of intensional logics. I: E, EM, R, RM and Π′* ◇ B45 ◇
HUNG, CH'ENGWAN *The independence of the axiom schemata and the inference rules of E* ◇ B45 ◇
IWIN, A.A. *Grundlagen der Logik von Wertungen* ◇ A05 B45 ◇
JOHNSON WU, K. *On (C.KK*) and the KK-thesis* ◇ A05 B45 ◇
KESSLER, G. *A note on future branching time* ◇ B45 ◇
KIELKOPF, C.F. *Adjunction and paradoxical derivations* ◇ B45 ◇
KIELKOPF, C.F. *Kant's deontic logic* ◇ A05 B45 ◇
KOHN, R.V. *Generalization of a result of Hallden* ◇ B45 ◇
KORDIG, C.R. *Relativized deontic modalities* ◇ B45 ◇
LEWIS, D. *Counterfactuals and comparative possibility* ◇ B45 ◇
MAKSIMOVA, L.L. *Finite-level modal logics (Russian)* ◇ B45 ◇
MAKSIMOVA, L.L. *Pretabular extensions of Lewis* S4 *(Russian)* ◇ B45 ◇
MCARTHUR, R.P. S5 *with the* CBF ◇ B45 ◇
MCKAY, T.J. *Essentialism in quantified modal logic* ◇ A05 B45 ◇
MERRILL, G.H. *A free logic with intensions as possible values of terms* ◇ B45 ◇
MILLER, JAMES WILKIN *The logic of the synthetic a priori* ◇ B45 ◇
MORGAN, C.G. *Liberated versions of T, S4 and S5* ◇ B45 ◇
MORGAN, C.G. *Weak liberated versions of T and S4* ◇ B45 ◇
MULLOCK, P. *The Stone-Tammelo deontic logic* ◇ B45 ◇
NUTE, D.E. *Counterfactuals and the similarity of worlds* ◇ B45 ◇
ORENDUFF, J. *A partially truth-functional modal calculus* ◇ B45 ◇
PEKLO, B.T. *Erweiterte deontische Logik (EDL). Ein Versuch um weitere formale Ausbildung der deontischen Modallogik* ◇ B45 ◇
PERZANOWSKI, J. *On M-fragments and L-fragments of normal modal propositional logics* ◇ B45 ◇
PERZANOWSKI, J. *On homogenous fragments of normal modal propositional logics* ◇ B45 ◇
PIECZKOWSKI, A. *Causal implications of Jaskowski* ◇ B45 ◇
PLEDGER, K.E. *Some extensions of S3* ◇ B45 ◇
POTTINGER, G. *The* C_I *systems: an irenic theory of implications* ◇ B45 ◇
PURTILL, R.L. *Paradox-free deontic logics* ◇ B45 ◇
QUINE, W.V.O. *On the individuation of attributes* ◇ A05 B45 ◇
ROUTLEY, R. & ROUTLEY, V. *The role of inconsistent and incomplete theories in the logic of belief* ◇ B45 ◇
SAHLQVIST, H. *Completeness and correspondence in the first and second order semantics for modal logic* ◇ B45 ◇
SCHARLE, T.W. *Axiomatization of fragments of S5* ◇ B45 ◇
SCHUMM, G.F. *Disjunctive extensions of S4 and a conjecture of Goldblatt's* ◇ B45 ◇
SCHUMM, G.F. *Remark on a logic of preference* ◇ B45 ◇
SCHUMM, G.F. *Wajsberg normal forms for S5* ◇ B45 ◇
SEGERBERG, K. *That every extension of S4.3 is normal* ◇ B45 ◇
SIMMONS, H. *Topological aspects of suitable theories* ◇ B45 F30 G05 ◇
SOLOV'EV, A.E. *A certain approach to the description of control systems (Russian)* ◇ B45 ◇
SONOBE, O. *Bi-relational frameworks for minimal and intuitionistic logics* ◇ B45 C90 F50 ◇
SPOHN, W. *An analysis of Hansson's dyadic deontic logic* ◇ B45 ◇
STEPHAN, B.J. *Compactness and recursive enumerability in intensional logic* ◇ B45 D25 ◇
SUSZKO, R. *Abolition of the Fregean axiom* ◇ A05 B45 ◇
THOMASON, R.H. *Decidability in the logic of conditionals* ◇ B25 B45 ◇
THOMASON, S.K. *Categories of frames for modal logic* ◇ B45 C90 G30 ◇
THOMASON, S.K. *Reduction of second-order logic to modal logic* ◇ B45 ◇
THOMASON, S.K. *Reduction of tense logic to modal logic II* ◇ B45 ◇
THOMASON, S.K. *The logical consequence relation of propositional tense logic* ◇ B45 D20 D55 ◇
TICHY, P. *What do we talk about?* ◇ A05 B45 ◇
TIRNOVEANU, M. *On some syntactic properties of transfinite assertoric-deontic-modal logical systems of the form* $S^{\alpha 210}_{adm}, S^{\alpha 2n0}_{adm}$ *(Romanian) (French summary)* ◇ B45 ◇
VIERU, S. *Deontic logic, and the theory of modalities (Romanian) (French summary)* ◇ B45 ◇
WALTON, D.N. *Modal logic and agency* ◇ B45 ◇
WHITE, A.R. *Modal thinking* ◇ B45 B65 ◇
ZACHOROWSKI, S. *A proof of a conjecture of Roman Suszko* ◇ B45 B50 F50 ◇

1976

BEARD, J.B. *The modalities of* $KT4_nMG$ ◇ B45 ◇
BENTHEM VAN, J.F.A.K. *Modal formulas are either elementary or not ΣΔ-elementary* ◇ B45 ◇
BENTHEM VAN, J.F.A.K. *Modal reduction principles* ◇ B45 ◇
BERNARDI, C. *The uniqueness of the fixed-point in every diagonalizable algebra (The algebraization of the theories which express Theor. VIII.)* ◇ B45 F30 G05 G25 ◇
BIGELOW, J.C. *If-then meets the possible worlds* ◇ B45 ◇
BLASZCZUK, J.J. & DZIOBIAK, W. *An axiomatization of* M^n*-counterparts for some modal calculi* ◇ B45 ◇
BOOLOS, G. *On deciding the truth of certain statements involving the notion of consistency* ◇ B25 B45 F30 ◇
BROIDO, J. *Of the eliminability of de re modalities in some systems* ◇ B45 ◇

Subject Index

BYRD, M. *Single variable formulas in S4$_{\rightarrow}$* ◇ B45 ◇
CALUDE, C. *Une axiomatisation de la logique deontique a une infinite de valeurs* ◇ B45 ◇
CARROLL, M.J. *The logical modalities and the system S5* ◇ A05 B45 ◇
CASTANEDA, H.-N. *The twofold structure and the unity of practical thinking* ◇ A05 B45 ◇
CHISHOLM, R.M. *Knowledge and belief: "de dicto" and "de re"* ◇ A05 B45 ◇
CODE, A. *Aristotle's response to Quine's objections to modal logic* ◇ A05 B45 ◇
CZERMAK, J. *Distinct modalities are not equivalent in T* ◇ B45 ◇
CZERMAK, J. *Embeddings of classical logic in S4. Part II* ◇ B45 ◇
DAHN, B.I. *Neighbourhood semantics and generalized Kripke semantics* ◇ B45 ◇
DAVIDSON, B. & ELLIS, B. *Logic and strict coherence* ◇ B45 ◇
DAVIS, CHARLES C. & MCKIM, V.R. *Temporal modalities and the future* ◇ A05 B45 ◇
DZIOBIAK, W. *Classically axiomatizable modal propositional calculi containing the system T of Feys-von Wright* ◇ B45 ◇
DZIOBIAK, W. *Semantics of Kripke's style for some modal systems* ◇ B45 ◇
EHSAKIA, L.L. *On modal partners of superintuitionistic logics (Russian)* ◇ B45 B55 ◇
ERMOLAEVA, N.M. & MUCHNIK, A.A. *Modal logics, defined by endomorphisms of distributive lattices (Russian)* ◇ B45 G10 ◇
FRAASSEN VAN, B.C. *Probabilities of conditionals* ◇ A05 B45 B48 ◇
FRAASSEN VAN, B.C. *Report on conditionals* ◇ B45 ◇
FREEMAN, J.B. *Algebraic semantics for modal predicate logic* ◇ B45 G15 ◇
GABBAY, D.M. *Two-dimensional propositional tense logics* ◇ B45 ◇
GAO, HENGSHAN *The decision problem of modal predicate calculi. II: On Slomson's reductions (Chinese)* ◇ B45 C13 D35 ◇
GENRICH, H.J. & THIELER-MEVISSEN, G. *The calculus of facts* ◇ B45 D15 ◇
GEORGACARAKOS, G.N. *Semantics for S4.04, S4.4 and S4.3.2* ◇ B45 ◇
GERSON, M.S. *A neighbourhood frame for T with no equivalent relational frame* ◇ B45 ◇
GOLDBLATT, R.I. *Metamathematics of modal logic part I* ◇ B45 ◇
GOLDBLATT, R.I. *Metamathematics of modal logic part II* ◇ B45 ◇
HAACK, S. *Recent publications in logic* ◇ A05 B45 ◇
HARPER, W.L. *Ramsey test conditionals and iterated belief change (A response to Stalnaker.)* ◇ A05 B45 B48 ◇
HART, W.D. & MCGINN, C. *Knowledge and necessity* ◇ A05 A10 B45 ◇
HAZEN, A. *Expressive completeness in modal language* ◇ B45 ◇
HILPINEN, R. *Approximate truth and truthlikeness* ◇ A05 B45 ◇
HINTIKKA, K.J.J. *The semantics of questions and the questions of semantics: Case studies in the interrelations of logic, semantics, and syntax* ◇ A05 B45 B65 ◇
HUGHES, G.E. *Modal systems with no minimal proper extensions* ◇ B45 ◇
JOSHI, A.K. & WEISCHEDEL, R.M. *Some frills for modal tic-tac-toe: Semantics of predicate complement constructions* ◇ B45 ◇
KARAVAEV, E.F. *A note on iteration of tense-logical operators (Russian)* ◇ B45 ◇
KARAVAEV, E.F. *On entering of an operator of the present tense into tense-logic (Russian)* ◇ B45 ◇
KEENE, G.B. *The logical analysis of universal conditionals* ◇ B45 ◇
KULIKOWSKI, J.L. *Logical and algebraical models of the networks of activities* ◇ B45 ◇
KUTSCHERA VON, F. *Einfuehrung in die intensionale Semantik* ◇ A05 B45 ◇
KUTSCHERA VON, F. *Epistemic interpretation of conditionals* ◇ A05 B45 ◇
LEBLANC, H. & MCARTHUR, R.P. *A completeness result for quantificational tense logic* ◇ B45 ◇
LEHMANN, S.K. *An interpretation of "finite" modal first-order languages in classical second-order languages* ◇ B15 B45 ◇
LICHTBLAU, D.E. *Prior and the Barcan formula* ◇ A05 B45 ◇
MATERNA, P. *Pragmatic meaning and truth* ◇ A05 B45 ◇
MCARTHUR, R.P. *On proving linearity* ◇ B45 ◇
MCARTHUR, R.P. *Tense logic* ◇ B45 ◇
MCCALL, S. *Objective time flow* ◇ B45 ◇
MCLELLAND, J. *Epistemic logic with identifiers* ◇ B45 ◇
MESKHI, V.YU. *Critical normal modal systems containing K4 (Russian)* ◇ B45 ◇
MESKHI, V.YU. *Propositional diffusion logic (Russian)* ◇ B45 ◇
MEYER, R.K. & ROUTLEY, R. *Every sentential logic has a two-valued worlds semantics* ◇ B45 ◇
MEYER, R.K. *Metacompleteness* ◇ B45 F50 ◇
MEYER, R.K. *Negation disarmed* ◇ B45 ◇
MONTGOMERY, H. & ROUTLEY, R. *Algebraic semantics for $S2^0$ and necessitated extensions* ◇ B25 B45 ◇
MORTENSEN, C. *A sequence of normal modal systems with non-contingency bases* ◇ B45 ◇
OMODEO, E.G. *The elimination of descriptions from A. Bressan's modal language ML^{ν} on which the logical calculus MC^{ν} is based* ◇ B45 ◇
OMYLA, M. *Translatability in non-Fregean theories* ◇ B45 ◇
PARKS, R.Z. *Investigations into quantified modal logic. I* ◇ B45 ◇
PAULOS, J.A. *A model-theoretic semantics for modal logic* ◇ B45 C07 C90 ◇
PKHAKADZE, N.M. *On Gentzen's formulation of a bimodal system (Russian)* ◇ B45 ◇
POLLOCK, J.L. *Subjunctive reasoning* ◇ A05 B45 ◇
PRIEST, G. *Modality as a meta-concept* ◇ B45 ◇

RAUTENBERG, W. *Some properties of the hierarchy of modal logics (preliminary report)* ◇ B45 ◇
RODOS, V.B. *Semantics and modalities (Russian)* ◇ B45 ◇
RODRIGUEZ MARIN, J. *The reduction of deontic logic to modal logic* ◇ B45 ◇
RUZSA, I. *Semantics for von Wright's latest deontic logic* ◇ B45 ◇
RYBAKOV, V.V. *Hereditarily finitely axiomatizable extensions of logic S4 (Russian)* ◇ B45 ◇
SCHUMM, G.F. *Interpolation in S5 and some related systems* ◇ B45 ◇
SEGERBERG, K. *Discrete linear future time without axioms* ◇ B25 B45 C90 ◇
SEGERBERG, K. *The truth about some Post numbers* ◇ B45 ◇
SHUKLA, A. *Consistent, independent, and distinct propositions. II* ◇ B45 ◇
SLOGEDAL, S. *Unzugaengliche Welten* ◇ A05 B45 ◇
SOBOCINSKI, B. *Pledger lemma and the modal system* $S3^0$ ◇ B45 ◇
SOLOVAY, R.M. *Provability interpretations of modal logic* ◇ B45 E30 F30 ◇
TALLET, J. *The class of possibilities* ◇ B45 ◇
THOMASON, R.H. *Necessity, quotation, and truth: An indexical theory* ◇ A05 B45 ◇
ULRICH, D. *Generalization of a result of Pahi's* ◇ B45 ◇
ULRICH, D. *On a modal system of R.A.Bull's* ◇ B45 ◇
VANDERVEKEN, D.R. *A formal definition of the set of the logical connectors of pragmatics* ◇ A05 B45 ◇
WALTON, D.N. *Time and modality in the "can" of opportunity. With a comment by Keith Lehrer* ◇ B45 ◇
WASSERMAN, H.C. *A note of evaluation mappings* ◇ B05 B45 ◇
WASSERMAN, H.C. *An analysis of the counterfactual conditional* ◇ B25 B45 ◇
WHITE, M.J. *A suggestion regarding the semantical analysis of performatives* ◇ A05 B45 ◇
WIGGINS, D. *Identity, necessity and physicalism* ◇ A05 B45 ◇
WILSON, R.L. *On some modal logics related to the L-modal system* ◇ B45 ◇
WILSON, R.L. *Some remarks on metaphysics and the modal logics* F^*F ◇ A05 B45 ◇
WOJTYLAK, P. *Some generalisations of Makinson's theorem on structural completeness* ◇ B45 ◇
WRONSKI, A. *Remarks on Hallden completeness of modal and intermediate logics* ◇ B45 ◇
ZANDAROWSKA, W. *The modal system* S3 *and* SCI ◇ B45 ◇
ZIEMBINSKI, Z. *Practical logic (Polish)* ◇ A05 B45 B98 ◇

1977

AAQVIST, L. & GUENTHNER, F. (EDS.) *Tense logic* ◇ B45 ◇
ANDERSON, K.C. *A note on knowledge* ◇ A05 B45 ◇
ANGSTL, H. *On the logic of norms: a natural deontics* ◇ B45 ◇
BAILHACHE, P. *Semantiques pour des systemes deontiques integrant permission faible et permission forte* ◇ B45 ◇
BELL, J.M. & HUMBERSTONE, I.L. *Two systems of presupposition logic* ◇ B45 ◇
BENTHEM VAN, J.F.A.K. *Modal logic as second-order logic* ◇ B45 ◇
BENTHEM VAN, J.F.A.K. *Tense logic and standard logic* ◇ A05 B45 ◇
BLASZCZUK, J.J. & DZIOBIAK, W. *Modal logics connected with systems* $S4_n$ *of Sobocinski* ◇ B45 ◇
BLASZCZUK, J.J. *Remarks on* M^n*-counterparts of some normal calculi* ◇ B45 ◇
BLUM, A. *Two observations about S5* ◇ B45 ◇
BODE, J.R. *Knowledge and the evidential conditional* ◇ A05 B45 ◇
BOOLOS, G. *On deciding the provability of certain fixed point statements* ◇ B25 B45 F30 ◇
BOZZI, S. & MELONI, G.C. *Completezza proposizionale per l'operatore modale "accade localmente che" (English summary)* ◇ B45 ◇
BYRD, M. & ULLRICH, D. *The extensions of* $BAlt_3$ ◇ B45 ◇
CARRIER, L.S. *The irreducibility of knowledge* ◇ A05 B45 ◇
COCCHIARELLA, N.B. *Sortals, natural kinds and re-identification* ◇ A05 B45 ◇
CONTE, A.G. & HILPINEN, R. & WRIGHT VON, G.H. (EDS.) *Deontische Logik und Semantik* ◇ B45 ◇
COOPER, D.E. *Bivalence, determinism, and realism* ◇ A05 B45 ◇
COSTA DA, N.C.A. & KOTAS, J. *On some modal logical systems defined in connexion with Jaskowski's problem* ◇ B45 ◇
CRESSWELL, M.J. *Interval semantics and logical words* ◇ B45 B65 ◇
CROSSLEY, J.N. & HUMBERSTONE, L.L. *The logic of "actually"* ◇ B45 ◇
CUSMARIU, A. *"About belief de re."* ◇ B45 ◇
DAVIDSON, B. & JACKSON, F.C. & PARGETTER, R. *Modal trees for T and S5* ◇ B25 B45 ◇
DZIOBIAK, W. *On detachment-substitutional formalization in normal modal logics* ◇ B45 ◇
EHSAKIA, L.L. & MESKHI, V.YU. *Five critical modal systems* ◇ B45 ◇
ELLIS, B. & JACKSON, F.C. & PARGETTER, R. *An objection to possible-world semantics for counterfactual logics* ◇ A05 B45 ◇
EYTAN, M. *Logique modale propositionnelle: une vue cavaliere (English summary)* ◇ A10 B45 ◇
FINE, K. *Properties, propositions and sets* ◇ A05 B45 ◇
FISCHER SERVI, G. *On modal logic with an intuitionistic base* ◇ B45 F50 ◇
FITTING, M. *A tableau system for propositional S5* ◇ B45 ◇
GABBAY, D.M. *A tense system with split truth table* ◇ B45 ◇
GABBAY, D.M. *On some new intuitionistic propositional connectives I* ◇ B45 B55 F50 ◇
GEORGACARAKOS, G.N. *Abnormal worlds and the non-Lewis modal systems* ◇ B45 ◇

GEORGACARAKOS, G.N. *Additional extensions of S4* ◇ B45 ◇
GEORGACARAKOS, G.N. *Semantics for S4.03* ◇ B45 ◇
GOCHET, P. *Model theory and the pragmatics of indexicals* ◇ A05 B45 ◇
GOLDMAN, H.S. *David Lewis's semantics for deontic logic* ◇ B45 ◇
GUMANSKI, L. *An improvement of the deontic calculus DSC* ◇ B45 ◇
HOBBS, J.R. & ROSENSCHEIN, S.J. *Making computational sense of Montague's intensional logic* ◇ B45 ◇
HUGHES, G.E. *Omnitemporal logic and nodal time* ◇ B45 ◇
INWAGEN VAN, P. & MCKAY, T.J. *Counterfactuals with disjunctive antecedents* ◇ A05 B45 ◇
INWAGEN VAN, P. *Ontological arguments* ◇ A05 B45 ◇
ISARD, S.D. *A finitely axiomatizable undecidable extension of K* ◇ B45 D05 D35 ◇
KALINOWSKI, G. *Du prescriptif et du descriptif en logique deontique* ◇ B45 ◇
KOHN, R.V. *Some Post-complete extensions of S2 and S3* ◇ B45 ◇
LADNER, R.E. *The computational complexity of provability in systems of modal propositional logic* ◇ B45 D15 ◇
LEMMON, E.J. *An introduction to modal logic* ◇ A10 B45 B98 ◇
LEWIS, D. *Possible-world semantics for counterfactual logics: A rejoinder* ◇ A05 B45 ◇
LINSKY, L. *Names and descriptions* ◇ A05 B45 ◇
MAKOWSKY, J.A. & MARCJA, A. *Completeness theorems for modal model theory with the Montague-Chang semantics I* ◇ B45 C90 ◇
MCARTHUR, R.P. *Three-valued free tense logic* ◇ B45 ◇
MERRILL, G.H. *On an enduring non sequitur of Quine's* ◇ A05 B45 ◇
MIURA, S. & OHAMA, S. *A note on Thomason's representation of S5* ◇ B45 ◇
MIURA, S. & OHAMA, S. *Modally complete minimal extension of S5* ◇ B45 ◇
MURUNGI, R.W. *Necessitas consequentis in a singleton possible world* ◇ B45 ◇
ONO, H. *On some intuitionistic modal logics* ◇ B45 ◇
PIZZI, C. *Boethius' thesis and conditional logic* ◇ A05 A10 B45 ◇
PNUELI, A. *The temporal logic of programs* ◇ B45 B75 ◇
POPA, C.P. *An axiomatic system for a logical theory of liberty and constraint (Romanian)* ◇ B45 ◇
PRAKEL, J.M. *Some preliminary suggestions for the mirroring of non-metaphysical modalities in Lesniewski's ontology* ◇ A05 B45 ◇
PRIEST, G. *A refoundation of modal logic* ◇ B45 ◇
QUINE, W.V.O. *Intensions revisited* ◇ A05 B45 ◇
QUINN, P.L. *Improved foundations for a logic of intrinsic value* ◇ A05 B45 ◇
RAUTENBERG, W. *Der Verband der normalen verzweigten Modallogiken* ◇ B45 G10 ◇
RICE, L.C. *Von Wright, rationalism and modality* ◇ A05 B45 ◇
ROCKINGHAM GILL, R.R. *On Wandschneider's way out* ◇ A05 B45 E30 ◇
ROUTLEY, R. *Postscript: Some setbacks on the choice and descriptions adventure* ◇ A05 B45 ◇
ROUTLEY, R. *Welding semantics for weak strict modal logics into the general framework of modal logic semantics* ◇ B45 ◇
RUZSA, I. *Semantica per la logica deontica del primo ordine* ◇ B45 ◇
RYBAKOV, V.V. *Noncompact extensions of the logic S4 (Russian)* ◇ B45 C90 ◇
SALMON, W.C. *Laws, modalities and counterfactuals* ◇ B45 ◇
SATO, M. *A study of Kripke-type models for some modal logics by Gentzen's sequential method* ◇ B45 C90 ◇
SCALES, R. *A Russellian approach to truth* ◇ A05 B45 ◇
SHEKHTMAN, V.B. *A remark on: "Models for multiply modal systems" by M.K.Rennie* ◇ B45 ◇
SHREJDER, YU.A. & VILENKIN, N.YA. *Majorant spaces and "majority" quantifier (Russian) (English summary)* ◇ B45 C80 E20 ◇
SIEMENS JR., D.F. *Fitch-style rules for many modal logics* ◇ B45 ◇
SLAGHT, R.L. *Modal tree constructions* ◇ B45 ◇
SMOKLER, H. *Three grades of probabilistic involvement* ◇ A05 B45 B48 ◇
SONOBE, O. *Beth-type tableau proof systems for some intermediate and modal logics* ◇ B45 B55 ◇
SPADE, P.V. *General semantic closure* ◇ A05 B45 ◇
SUNDHOLM, G. *A completeness proof for an infinitary tense-logic* ◇ B45 C75 ◇
TAYLOR, B. *Tense and continuity* ◇ A05 B45 ◇
THOMASON, S.K. *Modal operators and functional completeness II* ◇ B45 ◇
TIRNOVEANU, M. *Some metalogical properties of transfinite assertorical, modal, deontic logistic systems of order zero (Romanian) (French summary)* ◇ B45 ◇
TSELISHCHEV, V.V. *Philosophical problems of the semantics of possible worlds (Russian)* ◇ A05 B45 ◇
WAND, M. *A characterization of weakest preconditions* ◇ B45 ◇
WILSON, R.L. *A note on metaphysics and the foundations of mathematics* ◇ A05 B45 ◇
WILSON, R.L. *The modal predicate logics* PF^*F ◇ B45 ◇
YAMAMOTO, A. *Algebraic proof of separability of certain modal logics* ◇ B45 ◇

1978

AL-HIBRI, A. *Deontic logic: a comprehensive appraisal and a new proposal* ◇ A05 B45 ◇
APRILE, G. *Calcolo inferenziale modale a due livelli* ◇ B45 ◇
AUSTIN, J. *Systemic causation* ◇ A05 B45 ◇
BAKER, J.R. *Essentialism and the modal semantics of K.J.J. Hintikka* ◇ B45 ◇
BAKER, J.R. *Some remarks on Quine's arguments against modal logic* ◇ A05 B45 ◇
BELLISSIMA, F. *On the modal logic corresponding to diagonalizable algebra theory* ◇ B45 G25 ◇
BENTHEM VAN, J.F.A.K. & BLOK, W.J. *Transitivity follows from Dummett's axiom* ◇ B45 ◇

BENTHEM VAN, J.F.A.K. *Two simple incomplete modal logics* ◇ B45 ◇
BLASZCZUK, J.J. *Normal modal logics and its extensions (Polish)* ◇ B45 ◇
BLASZCZUK, J.J. *Weakest normal calculi with respect to M^n-counterparts* ◇ B45 ◇
BLOK, W.J. *On the degree of incompleteness of modal logics* ◇ B45 G10 G25 ◇
BOWEN, K.A. *Model theory for modal logic. Kripke models for modal predicate calculi* ◇ B45 C90 ◇
BROWN, CHARLES D. *The ontological theorem* ◇ A05 B45 ◇
BURGESS, J.P. *The unreal future* ◇ A05 B45 ◇
BYRD, M. *On the addition of weakened L-reduction axioms to the Brouwer system* ◇ B45 ◇
BYRD, M. *The extensions of $BAlt_3$ - revisited* ◇ B45 ◇
CASTANEDA, H.-N. *Philosophical method and the theory of predication and identity* ◇ A05 B45 ◇
CHAPMAN, T. *A modal logic with temporal variables* ◇ A05 B45 ◇
CLARK, R. *Not every object of thought has being: a paradox in naive predication theory* ◇ A05 B45 ◇
CORISH, D. *Aristotle on temporal order: "Now," "Before," and "After"* ◇ A05 A10 B45 ◇
COX, A. *Hintikka and the interdefinability of obligation and forbiddance* ◇ B45 ◇
DANIELS, C.B. & FREEMAN, J.B. *Maximal propositions and the coherence theory of truth* ◇ A05 B45 ◇
DAVIES, M.K. *Weak necessity and truth theories* ◇ A05 B45 ◇
DOSEN, K. *A note on the law of identity and the converse parry property* ◇ B45 ◇
DZIOBIAK, W. *A note on incompleteness of modal logics with respect to neighbourhood semantics* ◇ B45 C90 ◇
EHSAKIA, L.L. *Semantic analysis of bimodal (temporal) systems (Russian)* ◇ B45 ◇
ELLIS, B. *A unified theory of conditionals* ◇ A05 B45 ◇
FINE, K. *Model theory for modal logic. part I: the de re/de dicto distinction. part II: the elimination of de re modality* ◇ A05 B45 C40 C90 ◇
FITTING, M. *Subformula results in some propositional modal logics* ◇ B45 ◇
GABBAY, D.M. *Tense logics with split truth tables* ◇ B45 ◇
GAERDENFORS, P. *On the interpretation of deontic logic* ◇ B45 ◇
GARSON, J.W. *Completeness of some quantified modal logics* ◇ B45 ◇
GEORGACARAKOS, G.N. *A modal system properly independent of both the Brouwerian system and S4* ◇ B45 ◇
GEORGACARAKOS, G.N. *A new family of modal systems* ◇ B45 ◇
GEORGACARAKOS, G.N. *Dawson modelling, deontic logic and moral necessity* ◇ A05 B45 ◇
GIRLE, R.A. *Logics for knowledge, possibility and existence* ◇ A05 B45 ◇
GOLDBLATT, R.I. *Arithmetical necessity, provability and intuitionistic logic* ◇ B45 F50 ◇
GRIGOLIYA, R.SH. & MESKHI, V.YU. *Critical bimodal systems (Russian)* ◇ B45 ◇
GUPTA, A. *Modal logic and truth* ◇ A05 B45 ◇
HART, W.D. & MCGINN, C. *On propositions* ◇ A05 B45 ◇
HAYASHI, T. *Disjunction property in McCarthy's propositional knowledge systems* ◇ B45 ◇
HAZEN, A. *The eliminability of the actuality operator in propositional modal logic* ◇ B45 ◇
HUNTER, D. & RICHTER, R. *Counterfactuals and Newcomb's paradox* ◇ A05 B45 B48 ◇
ISHIMOTO, A. *A bi-modal characterization of epistemic logic* ◇ B45 ◇
JENSEN, J.B. & LARSEN, P.F. & MACLELLAN, E.J. & SCHOTCH, P.K. *A note on three-valued modal logic* ◇ B45 ◇
KARAVAEV, E.F. *Combining calculi of modal logic as a need of methodology of knowledge* ◇ A05 B45 ◇
KESSLER, G. *Mathematics and modality* ◇ A05 B45 ◇
KRABBE, E.C.W. *Note on a completeness theorem in the theory of counterfactuals* ◇ A05 B45 ◇
KROLIKOSKI, S.J. *A proposed interpretation of Lukasiewicz's four-valued system of modal logic* ◇ B45 ◇
LENK, H. *Varieties of commitment: approaches to the symbolization of conditional obligations* ◇ A05 B45 ◇
LENZEN, W. *A rare accident* ◇ B45 ◇
LENZEN, W. *On some substitution instances of R1 and L1* ◇ B45 ◇
LENZEN, W. *Recent work in epistemic logic* ◇ A05 B45 ◇
LENZEN, W. *S4.1.4.=S4.1.2 and S4.021=S4.04* ◇ B45 ◇
LOPARIC, A. *The method of valuations in modal logic* ◇ B45 ◇
LORENZEN, P. *Praktische und theoretische Modalitaten* ◇ B45 ◇
MCKAY, T.J. *The principle of predication and the elimination of de re modalities* ◇ A05 B45 ◇
MESKHI, V.YU. *Algebraic analysis of modal fragments of tense systems (Russian)* ◇ B45 ◇
MILBERGER, M. *The minimal modal logic: A cautionary tale about primitives and definitions* ◇ A05 B45 ◇
MORI, S. *On S5 type modal logics based on intermediate logics with finitely many valued linear models* ◇ B45 B55 ◇
NUTE, D.E. *Algebraic semantics for conditional logics* ◇ B45 ◇
NUTE, D.E. *An incompleteness theorem for conditional logic* ◇ B45 ◇
PARSONS, T. *Nuclear and extranuclear properties, Meinong, and Leibniz* ◇ A05 B45 ◇
PIZZORNO, B. *Some logical aspects of the drama of Galileo* ◇ A10 B45 ◇
RABINOWICZ, W. *Utilitarianism and conflicting obligations* ◇ B45 ◇
ROUTLEY, R. *Constant domain semantics for quantified non-normal modal logics and for certain quantified quasi-entailment logics* ◇ B45 ◇

ROUTLEY, R. *Semantics for connexive logics. I* ◇ A05 B45 ◇
RUZSA, I. *A new approach to modal logic* ◇ B45 ◇
RYBAKOV, V.V. *A decidable noncompact extension of the logic S4 (Russian)* ◇ B25 B45 ◇
RYBAKOV, V.V. *Modal logics with LM-axioms (Russian)* ◇ B25 B45 ◇
SAMBIN, G. *Fixed points through the finite model property (The algebraization of the theories which express Theor. XI)* ◇ B45 F30 G25 ◇
SCHUMM, G.F. *An incomplete nonnormal extension of S3* ◇ B45 C90 ◇
SCHUMM, G.F. *Modalities in the extensions of B* ◇ B45 ◇
SCHUMM, G.F. *Putting $\mathfrak{K}$ in its place* ◇ B45 C90 ◇
SERGANT, M. *Reflexions sur un systeme de matrices a huit valeurs propose par N. D. Belnap* ◇ A05 B45 G10 ◇
SHEKHTMAN, V.B. *Two-dimensional modal logic (Russian)* ◇ B45 C90 ◇
SKYRMS, B. *An immaculate conception of modality* ◇ B45 ◇
SMORYNSKI, C.A. *Beth's theorem and self-referential sentences* ◇ B45 C40 F30 ◇
THOMASON, S.K. *Possible worlds and many truth values* ◇ B45 ◇
TICHY, P. *A new theory of subjunctive conditionals* ◇ A05 B45 ◇
TSELISHCHEV, V.V. *The concept of object in modal logic (Russian)* ◇ A05 B45 ◇
ULRICH, D. *Semantics for S4.1.2* ◇ B45 ◇
WILLIAMSON, J. *An ambiguity in modal logic* ◇ B45 ◇
YAMAMOTO, A. *Cut-elimination theorem for multi-modal logics* ◇ B45 ◇
ZEMAN, J.J. *Generalized normal logic* ◇ B45 G10 ◇

1979

AAQVIST, L. *A conjectured axiomatization of two-dimensional Reichenbachian tense logic* ◇ B45 ◇
AKIMOV, A.P. *On imbeddings of classical modal calculi in constructive calculi (Russian)* ◇ B45 F50 ◇
BACON, J. *The logical form of perception sentences* ◇ A05 B45 ◇
BAKER, L.R. & WALD, J.D. *Indexical reference and de re belief* ◇ A05 B45 ◇
BARNES, R.F. & GUMB, R.D. *The completeness of presupposition-free tense logic* ◇ B45 ◇
BEAZER, R. & MACNAB, D.S. *Modal extensions of Heyting algebras* ◇ B45 G10 G20 G25 ◇
BENTHEM VAN, J.F.A.K. *Canonical modal logics and ultrafilter extensions* ◇ B45 C90 G25 ◇
BENTHEM VAN, J.F.A.K. *Critical notice to: D. M. Gabbay's "Investigations in modal and tense logics with applications to problems in philosophy and linguistics"* ◇ A05 B45 ◇
BENTHEM VAN, J.F.A.K. *Minimal deontic logics* ◇ A05 B45 ◇
BENTHEM VAN, J.F.A.K. *Syntactic aspects of modal incompleteness theorems* ◇ B45 ◇
BENTON, R.A. *A descending chain of classical logics for which necessitation implies regularity* ◇ B45 ◇
BERMAN, F. *A completeness technique for D-axiomatizable semantics* ◇ B45 ◇
BLOK, W.J. *An axiomatization of the modal theory of the veiled recession frame* ◇ B45 ◇
BODE, J.R. *The possibility of a conditional logic* ◇ A05 B45 ◇
BOOLOS, G. *Reflection principles and iterated consistency assertions* ◇ B45 F30 ◇
BOOLOS, G. *The unprovability of consistency. An essay in modal logic* ◇ B25 B28 B45 B98 F30 F98 ◇
BRADLEY, R. & SWARTZ, N. *Possible worlds* ◇ A05 B45 ◇
BROWN, F.M. *A sequent calculus for modal quantificational logic* ◇ B45 ◇
BROY, M. & GNATZ, R. & WIRSING, M. *Semantics of nondeterministic and noncontinuous constructs* ◇ B45 B75 ◇
BUNDER, M.W. *Deduction theorems in significance logics* ◇ B45 ◇
BURGESS, J.P. *Logic and time* ◇ B45 ◇
CARROLL, M.J. *Reduction to first degree in quantificational S5* ◇ B45 ◇
CASANUEVA, S.P. *Investigaciones en logica modal. Desarollo y sistemas axiomaticos de Aristoteles a Clarence I.Lewis* ◇ A10 B45 ◇
CHAKRAVARTI, S.S. *A note on Kripke's distinction between rigid designators and non-rigid designators* ◇ A05 B45 ◇
CHELLAS, B.F. *Modalities in normal systems containing the S5 axiom* ◇ B45 ◇
CLEAVE, J.P. *The axiomatisation of theories of material necessity* ◇ B45 ◇
COLE, R. *Possibility matrices* ◇ A05 B45 G25 ◇
CRESSWELL, M.J. *B Seg has the finite model property* ◇ B45 ◇
CRESSWELL, M.J. *Interval semantics for some event expressions* ◇ B45 ◇
DEUTSCH, H. *The completeness of S* ◇ B45 ◇
EDELSTEIN, R. & SCHUMM, G.F. *Negation-free modal logics* ◇ B45 ◇
EHSAKIA, L.L. *On the variety of Grzegorczyk algebras (Russian)* ◇ B45 E70 G25 ◇
EHSAKIA, L.L. *To the theory of modal and superintuitionistic systems (Russian)* ◇ B45 B55 ◇
EPSTEIN, R.L. *Relatedness and implication* ◇ A05 B45 ◇
ERMOLAEVA, N.M. & MUCHNIK, A.A. *Pretabular temporal logic (Russian)* ◇ B45 ◇
FINE, K. *Failures of the interpolation lemma in quantified modal logic* ◇ B45 ◇
FINE, T.L. & WALLEY, P. *Varieties of modal (classificatory) and comparative probability* ◇ A05 B45 ◇
FISHER, M. *Rules as nests of beliefs and intentions* ◇ A05 B45 ◇
FONT I LLOVET, J.M. & VERDU I SOLANS, V. *Abstract logics, interior operators and modal logics S4 (Spanish, Catalan)* ◇ B45 ◇
FRAASSEN VAN, B.C. *Hidden variables and the modal interpretation of quantum theory* ◇ B45 G12 ◇
FULTON, J.A. *An intensional logic of predicates and predicate modifiers without modal operators* ◇ B45 ◇
GAO, HENGSHAN *Some remarks on Lewis modal system S5 (Chinese) (English summary)* ◇ B45 ◇

GARSON, J.W. *Free topological logic* ◇ A05 B45 ◇
GARSON, J.W. *The substitution interpretation and the expressive power of intensional logics* ◇ B45 ◇
GEORGESCU, G. *A representation theorem for tense polyadic algebras* ◇ B45 G15 ◇
GEORGESCU, G. *Modal polyadic algebras* ◇ B45 G15 ◇
GLADKICH, YU. *Singular terms, existence and truth: Some remarks on a first order logic of existence* ◇ A05 B45 ◇
GORGY, F.W. *The independence of the rule of syllogism in S_2* ◇ B45 F50 ◇
GUASPARI, D. & SOLOVAY, R.M. *Rosser sentences* ◇ B45 F30 ◇
GUENTHNER, F. & ROHRER, C. (EDS.) *Studies in formal semantics. Intensionality, temporality, negation. 2nd printing* ◇ A05 B45 ◇
GUMB, R.D. *Evolving theories* ◇ B45 ◇
HAREL, D. *Two results on process logic* ◇ B45 ◇
HEIJENOORT VAN, J. *Introduction a la semantique des logiques non-classiques* ◇ B45 C90 F50 ◇
HUMBERSTONE, L.L. *Interval semantics for tense logic: Some remarks* ◇ B45 ◇
INWAGEN VAN, P. *Laws and counterfactuals* ◇ A05 B45 ◇
KAPLAN, DAVID *On the logic of demonstratives* ◇ A05 B45 ◇
KLEIN, E. *On formalizing the referential/attributive distinction* ◇ A05 B45 ◇
KROLIKOSKI, S.J. *A deduction theorem for rejection theses in Lukasiewicz's system of modal logic* ◇ B45 ◇
KROLIKOSKI, S.J. *A second deduction theorem for rejection theses in Lukasiewicz's system of modal logic* ◇ B45 ◇
KUHN, STEVEN T. *The pragmatics of tense* ◇ B45 ◇
LARNER, A. *A matrix decision procedure for three modal logics* ◇ B45 ◇
LINSKY, L. *Believing and necessity* ◇ A05 B45 ◇
LOEWER, B.M. *Cotenability and counterfactual logics* ◇ A05 B45 ◇
MAKSIMOVA, L.L. *A classification of modal logics (Russian)* ◇ B45 ◇
MAKSIMOVA, L.L. *Interpolation theorems in modal logics and amalgamable varieties of topological boolean algebras (Russian)* ◇ B45 C05 C40 C90 G05 ◇
MANNA, Z. & PNUELI, A. *The modal logic of programs* ◇ B45 B75 ◇
MCCALL, S. *The strong future tense* ◇ B45 ◇
MCKINSEY, M. *Levels of obligation* ◇ A05 B45 ◇
MCLANE JR., H.E. *On the possibility of epistemic logic* ◇ A05 B45 ◇
MELLEMA, G. *An alternative semantics for knowledge* ◇ B45 ◇
MESKHI, V.YU. *Critical modal logics containing Brouwer's axioms (Russian)* ◇ B45 ◇
MICHAEL, E. *Some considerations in medieval tense logic* ◇ A05 A10 B45 ◇
MORGAN, C.G. *Local and global operators and many-valued modal logics* ◇ B45 ◇
MORGAN, C.G. *Note on a strong liberated modal logic and its relevance to possible world skepticism* ◇ B45 ◇
MULLOCK, P. *Logic and liberty* ◇ A05 B45 ◇
NAT VANDER, A. *Beyond non-normal possible worlds* ◇ B45 C90 ◇
NAT VANDER, A. *First-order indefinite and uniform neighbourhood semantics* ◇ B45 ◇
NIINILUOTO, I. & SAARINEN, E. (EDS.) *Intensional logic: theory and applications* ◇ B45 ◇
NISHIMURA, H. *Is the semantics of branching structures adequate for chronological modal logics?* ◇ B45 ◇
NISHIMURA, H. *Is the semantics of branching structures adequate for non-metric Ockhamist tense logics?* ◇ B45 ◇
NISHIMURA, H. *On the completeness of chronological logics with modal operators* ◇ B45 ◇
PARSONS, T. *Referring to nonexistent objects* ◇ A05 B45 ◇
PLA I CARRERA, J. *A formalization of the Lewis system S1 without rules of substitution* ◇ B45 G25 ◇
PORTE, J. *The Ω-system and the L-system of modal logic* ◇ B45 ◇
PRUCNAL, T. *On two problems of Harvey Friedman* ◇ B45 B55 ◇
QUINN, P.L. *Existence throughout an interval of time and existence at an instant of time* ◇ A05 B45 ◇
RABINOWICZ, W. *Reasonable beliefs* ◇ B45 ◇
RANTALA, V. *Possible worlds and formal semantics* ◇ A05 B45 ◇
RAUTENBERG, W. *Klassische und nichtklassische Aussagenlogik* ◇ B45 B50 B98 C90 ◇
RAUTENBERG, W. *More about the lattice of tense logics* ◇ B45 ◇
RESCHER, N. *Mondi possibili non-standard* ◇ B45 ◇
ROBINSON, O. *A modal natural deduction system for S4* ◇ B45 ◇
SAARINEN, E. *Backwards-looking operators in tense logic and in natural language* ◇ B45 B65 ◇
SCHUMM, G.F. *Modal logics with no minimal proper extensions* ◇ B45 ◇
SCHWARTZ, T. *Necessary truth as analyticity, and the eliminability of monadic de re formulas* ◇ A05 B45 ◇
SCHWIND, C.B. *Representing actions by state logic* ◇ B45 ◇
SEGERBERG, K. *Von Wright's temporal logic (Russian)* ◇ B45 ◇
SHREJDER, YU.A. *Majorant models for modal (deontic) logics (Russian)* ◇ B45 ◇
SMOKLER, H. *Single-case propensities, modality, and confirmation* ◇ A05 B45 ◇
SMOKLER, H. *The collapse of modal distinctions in probabilistic contexts* ◇ A05 B45 ◇
SMORYNSKI, C.A. *Calculating self-referential statements. I: Explicit calculations* ◇ B45 F30 G25 ◇
STAHL, G. *Le dominateur et ses problemes* ◇ A05 B45 ◇
STEINACKER, P. *Zur Semantik superschwacher modaler Kalkuele* ◇ B45 ◇
TOTH, K. *Modal logics with function symbols* ◇ B45 ◇
TSELISHCHEV, V.V. *Hintikka's possible worlds and rigid designators* ◇ A05 B45 ◇
URSINI, A. *A modal calculus analogous to K4W, based on intuitionistic propositional logic, I^0* ◇ B45 G25 ◇

Subject Index

URSINI, A. *Two remarks on models in modal logics (Italian summary)* ◇ B45 C90 ◇
VARDANYAN, V.A. *A formalization of the logic of decision making (Russian)* ◇ B45 ◇
WALTON, D.N. *Philosophical basis of relatedness logic* ◇ A05 B45 ◇
WALTON, D.N. *Relatedness in intensional action chains* ◇ A05 B45 ◇
WESSEL, H. *Ein System der strikten logischen Folgebeziehung* ◇ B45 ◇
WHITE, M.J. *An S5 Diodorean modal system* ◇ B45 ◇
WILLIAMSON, J. *S5 without modal axioms* ◇ B45 ◇
WIREDU, J.E. *On the necessity of S4* ◇ A05 B45 ◇
WRIGHT VON, G.H. *Diachronic and synchronic modalities* ◇ B45 ◇
ZEMAN, J.J. *Normal implications, bounded posets, and the existence of meets* ◇ B45 G10 ◇
ZEMAN, J.J. *Two basic pure-implicational systems* ◇ B45 ◇
ZIAI, H. *Modal propositions in Islamic philosophy* ◇ A05 A10 B45 ◇

1980

ANDERSON, E. & LACEY, H.M. *Spatial ontology and physical modalites* ◇ A05 B45 ◇
ARTEMOV, S.N. *Arithmetically complete modal theories (Russian)* ◇ B45 C90 F50 ◇
BACON, J. *Substance and first-order quantification over individual concepts* ◇ A05 B45 ◇
BAILHACHE, P. *Several possible systems of deontic weak and strong norms* ◇ A05 B45 ◇
BAYART, A. *Causalite,essai de definition* ◇ B45 ◇
BELLISSIMA, F. *Considerazioni semantiche sulle logiche modali predicative* ◇ B45 ◇
BENTHEM VAN, J.F.A.K. *Some kinds of modal completeness* ◇ B45 ◇
BLOK, W.J. *The lattice of modal logics: An algebraic investigation* ◇ B45 G10 ◇
BOOLOS, G. *On systems of modal logic with provability interpretations* ◇ B45 F30 ◇
BOOLOS, G. *Provability, truth, and modal logic* ◇ B45 ◇
BOOLOS, G. *Provability in arithmetic and a schema of Grzegorczyk* ◇ B45 ◇
BOYARSKAYA, Z.A. *A comparison of the expressive powers of propositional modal logics (Russian)* ◇ B45 ◇
BRYANT, J. *The logic of relative modality and the paradoxes of deontic logic* ◇ B45 ◇
BUNDER, M.W. *A note on quantified significance logics* ◇ B40 B45 ◇
BUNDER, M.W. *Significance and illative combinatory logics* ◇ B40 B45 ◇
BURGESS, J.P. *Decidability for branching time* ◇ B25 B45 C85 ◇
BYRD, M. *Eventual permanence* ◇ B45 ◇
CHELLAS, B.F. *Modal logic. An introduction* ◇ B45 B98 ◇
COOPER, D.E. *Referential occurrence* ◇ A05 B45 ◇
CRESSWELL, M.J. *Possible worlds* ◇ B45 ◇
CURRIE, G. & ODDIE, G. *Changing numbers* ◇ B45 ◇
DANIELS, C.B. & FREEMAN, J.B. *An analysis of the subjunctive conditional* ◇ A05 B45 ◇
DAVIES, M.K. & HUMBERSTONE, L.L. *Two notions of necessity* ◇ A05 B45 ◇
DIAZ, M.R. *Deductive completeness and conditionalization in systems of weak implication* ◇ B45 ◇
DISHKANT, G.P. *Set theory as modal logic* ◇ B45 ◇
ENOMOTO, H. & YONEZAKI, N. *Database system based on intensional logic* ◇ B45 ◇
FINE, K. *First-order modal theories II* ◇ B45 ◇
FONT I LLOVET, J.M. *Induction of interiors of order in abstract logics (Catalan) (English summary)* ◇ B45 ◇
GARSON, J.W. *The unaxiomatizability of a quantified intensional logic* ◇ B45 ◇
GODDARD, L. *Significance, necessity, and verification* ◇ A05 B45 ◇
GODFREY-SMITH, W. *Change and actuality* ◇ A05 B45 ◇
GOLDBLATT, R.I. *Diodorean modality in Minkowski spacetime* ◇ B45 ◇
GUMANSKI, L. *On deontic logic* ◇ A05 B45 ◇
GUPTA, A. *The logic of common nouns. An investigation in quantified modal logic* ◇ B45 ◇
HARKER, J.E. *A note on believing that one knows and Lehrer's proof that knowledge entails belief* ◇ A05 B45 ◇
HARNISH, R.M. *Searle and the logic of Moore's paradox* ◇ A10 B45 ◇
HART, A.M. *Toward a logic of doubt* ◇ B45 ◇
HAWRANEK, J. *A matrix adequate for S5 with MP and RN* ◇ B45 ◇
HINTIKKA, K.J.J. *Degrees and dimensions of intentionality* ◇ A05 B45 ◇
HUGHES, G.E. *Equivalence relations and S5* ◇ B45 ◇
HUGHES, G.E. *Modal logics between S4.2 and S4.3* ◇ B45 ◇
HUGHES, G.E. *Some extensions of the Brouwerian logic* ◇ B45 ◇
IVLEV, YU.V. *Intuitive semantics of modal logic (Russian)* ◇ B45 ◇
JENNINGS, R.E. & SCHOTCH, P.K. *Inference and necessity* ◇ A05 B45 ◇
JENNINGS, R.E. & JOHNSTON, D.K. & SCHOTCH, P.K. *Universal first-order definability in modal logic* ◇ B45 ◇
KUBINSKI, T. *On a realisation of the von Wright programme of construction of deontic theories* ◇ B45 ◇
KUHN, STEVEN T. *Quantifiers as modal operators* ◇ B45 ◇
KUZNETSOV, A.V. & MURAVITSKIJ, A.YU. *Provability as modality (Russian)* ◇ B45 ◇
LENZEN, W. *Beschraenkte und unbeschraenkte Reduktion von Konjunktionen von Modalitaeten in S4* ◇ B45 ◇
LENZEN, W. *Glauben, Wissen und Wahrscheinlichkeit. Systeme der epistemischen Logik* ◇ A05 B45 B48 ◇
LI, XIANG *The first-order theory of Kripke's semantical analysis of nonclassical logics (Chinese) (English summary)* ◇ B45 ◇
MACCHI, P. *Funzioni verita per* S_5 ◇ B45 ◇

MAKSIMOVA, L.L. *Interpolation theorems in modal logics. Sufficient conditions (Russian)* ◇ B45 G25 ◇
MATSUMOTO, K. *Foundation of modal logic (Japanese)* ◇ B45 ◇
MCMICHAEL, A. & ZALTA, E.N. *An alternative theory of nonexistent objects* ◇ A05 B45 ◇
MIKHAJLOVA, M. *Reduction of modalities in several intuitionistic modal logics* ◇ B45 ◇
MIRCHEVA, M. & VAKARELOV, D. *Modal Post algebras and many-valued modal logics* ◇ B45 G20 ◇
MIROLLI, M. *On the axiomatization of finite frames of the modal system GL (Italian summary)* ◇ B45 ◇
MONTAGNA, F. *Some modal logics with quantifiers (Italian summary)* ◇ B45 C80 C90 ◇
MONTAGNA, F. & PASINI, L. *The equational class corresponding to the modal logic K43W (Italian summary)* ◇ B45 C05 ◇
MORE, M.J. *Rigidity and scope* ◇ B45 ◇
MORI, S. *Axiomatization of S4 type modal logics based on finite linear intermediate logics* ◇ B45 B55 ◇
NAGAI, SHIGEO & OKUBO, T. *On the logic of a priori modalities* ◇ A05 B45 ◇
NAKAMURA, A. & ONO, H. *On the size of refutation Kripke models for some linear modal and tense logics* ◇ B45 D15 ◇
NISHIMURA, H. *A preservation theorem for tense logic* ◇ B45 C20 C40 C90 ◇
NISHIMURA, H. *A study of some tense logics by Gentzen's sequential method* ◇ B45 ◇
NISHIMURA, H. *Descriptively complete process logic* ◇ B45 B75 ◇
NISHIMURA, H. *Interval logics with applications to study of tense and aspect in English* ◇ B45 ◇
NISHIMURA, H. *Saturated and special models in modal model theory with applications to the modal and de re hierarchies* ◇ B45 C40 C50 C90 ◇
NUTE, D.E. *Topics in conditional logic* ◇ A05 B45 ◇
OMER, I.A. *Minimal law explanation* ◇ B45 ◇
OMODEO, E.G. *Three existence principles in a modal calculus without descriptions contained in A. Bressan's* MC^{ν} ◇ B45 ◇
PETERSON, T. *The law of time conversion: an algebraic property of time adverbs* ◇ B45 ◇
PLEDGER, K.E. *Location of some modal systems* ◇ B45 ◇
PORTE, J. *A research in modal logics* ◇ B45 ◇
PORTE, J. *Congruences in Lemmon's S0.5* ◇ B45 ◇
RABINOWICZ, W. *Some remarks about the family K of modal systems* ◇ B45 ◇
RAUTENBERG, W. *Splitting lattices of logics* ◇ B45 ◇
REHDER, W. *Thought-experiments and modal logics* ◇ B45 ◇
REINHARDT, W.N. *Necessity predicates and operators* ◇ A05 B45 ◇
REINHARDT, W.N. *Satisfaction definitions and axioms of infinity in a theory of properties with necessity operator* ◇ A05 B45 E47 E55 E70 ◇
ROEPER, P. *Intervals and tenses* ◇ A05 B30 B45 ◇
SAMBIN, G. & VALENTINI, S. *A modal sequent calculus for a fragment of arithmetic* ◇ B45 F05 F30 ◇
SAMBIN, G. *A simpler proof of Sahlqvist's theorem on completeness of modal logics* ◇ B45 ◇
SAMBIN, G. & VALENTINI, S. *Post completeness and free algebras* ◇ B45 ◇
SANCHEZ-MAZAS, M. *Essai de representation des systemes normatifs par des systemes d'equations* ◇ B45 ◇
SATO, M. *A cut-free Gentzen-type system for the modal logic S5* ◇ B45 ◇
SCHOCK, R. *A complete system of indexical logic* ◇ B45 ◇
SCHOCK, R. *A natural deduction system of indexical logic* ◇ B45 ◇
SCHWIND, C.B. *Natural language analysis by theorem proving methods: disambiguating pronouns in natural language texts* ◇ B35 B45 B65 ◇
SEGERBERG, K. *A note on the logic of elsewhere* ◇ B45 ◇
SEGERBERG, K. *Applying modal logic* ◇ A05 B45 ◇
SHEKHTMAN, V.B. *Topological models of propositional logics (Russian)* ◇ B45 C90 ◇
SIMI, G. *Algebre pre-modali. I: Su una classe di algebre atte all'introduzione di un operatore di possibilita. II: Sulla varieta delle algebre premodali* ◇ B45 B60 G25 ◇
SMORYNSKI, C.A. *Calculating self-referential statements* ◇ B45 F30 G25 ◇
SOKULER, Z.A. *Different interpretations of quantification (Russian)* ◇ B45 ◇
SOTIROV, V.KH. *Nonfinitely approximable intuitionistic modal logics (Russian)* ◇ B45 ◇
STEBBINS, S. *Necessity and natural language* ◇ A05 B45 B65 ◇
SWART DE, H.C.M. *Gentzen-type systems for C, K and several extensions of C and K; constructive completeness proofs and effective decision procedures for these systems* ◇ B25 B45 ◇
THOMASON, R.H. *A note on syntactical treatments of modality* ◇ A05 B45 ◇
THOMASON, S.K. *Independent propositional modal logics* ◇ B45 ◇
TICHY, P. *The logic of temporal discourse* ◇ A05 B45 ◇
TICHY, P. *The transiency of truth* ◇ B45 ◇
TITIEV, R.J. *On self-sustenance in systems of epistemic logic* ◇ A05 B45 ◇
TOTH, K. *Completeness in non-simple and stable modal logics* ◇ B45 ◇
VAKARELOV, D. *Simple examples of incomplete logics* ◇ B45 ◇
WHITE, M.J. *Necessity and unactualized possibilities in Aristotle* ◇ A05 B45 ◇

1981

AAQVIST, L. & HOEPELMAN, J.P. *Some theorems about a "tree" system of deontic tense logic* ◇ B45 ◇
ADAMS, R.M. *Actualism and thisness* ◇ A05 B45 ◇
ALCHOURRON, C.E. & MAKINSON, D. *Hierarchies of regulations and their logic* ◇ B45 ◇
ALCHOURRON, C.E. & BULYGIN, E. *The expressive conception of norms* ◇ B45 ◇
ALMOG, J. *Indexicals, demonstratives and the modality dynamics* ◇ B45 ◇

Subject Index

AUSTIN, D.F. *Plantinga on actualism and essences* ◇ A05 B45 ◇
BACON, J. *Purely physical modalities* ◇ B45 ◇
BAILHACHE, P. *Analytical deontic logic: Authorities and addressees* ◇ A05 B45 ◇
BARNES, R.F. *Interval temporal logic: a note* ◇ B45 ◇
BENCIVENGA, E. & WOODRUFF, P.W. *A new modal language with the λ operator* ◇ B45 ◇
BENTHEM VAN, J.F.A.K. *Tense logic, second-order logic and natural language* ◇ A05 B45 ◇
BERNERT, J. & BIELA, A. *On two different modal logics denoted by S9* ◇ B45 ◇
BERNINI, S. *A temporalization of modal semantics* ◇ A05 B45 ◇
BLUE, N. *A metalinguistic interpretation of counterfactual conditionals* ◇ A05 B45 ◇
BRESSAN, A. & MONTANARO, A. *Contributions to foundations of probability calculus on the basis of the modal logical calculus MC^{ν} or MC^{ν}_{*} II* ◇ B45 ◇
BRESSAN, A. *Extensions of the modal calculi MC^{ν} and MC^{∞}. Comparison of them with similar calculi endowed with different semantics. Application to probability theory* ◇ B45 ◇
BRESSAN, A. & ZANARDO, A. *General operators binding variables in the interpreted modal calculus MC* ◇ B45 ◇
BURGESS, J.P. *Quick completeness proofs for some logics of conditionals* ◇ B45 ◇
CASTANEDA, H.-N. *The paradoxes of deontic logic: the simplest solution to all of them in one fell swoop* ◇ B45 ◇
CHAPMAN, T. *Quantum logic and modality* ◇ B45 B51 ◇
DOSEN, K. *Minimal modal systems in which Heyting and classical logic can be embedded* ◇ B45 F50 ◇
DUBIKAJTIS, L. & MORAES DE, L. *On single operator for Lewis S5 modal logic* ◇ B45 ◇
DYWAN, Z. *The necessity of modal logic* S5 *is metalogical* ◇ B45 ◇
DZIOBIAK, W. *Nonexistence of a countable strongly adequate matrix semantics for neighbours of E* ◇ B45 ◇
EMDE BOAS VAN, P. & GROENENDIJK, J. & STOKHOF, M. *The Conway paradox: Its solution in an epistemic framework* ◇ A05 B45 ◇
FINE, K. *First-order modal theories. I. Sets* ◇ B45 E70 ◇
FINE, K. *Model theory for modal logic. III. Existence and predication* ◇ A05 B45 C90 ◇
FISCHER SERVI, G. *Completeness for nonnormal intuitionistic modal logics (Italian summary)* ◇ B45 ◇
FISCHER SERVI, G. *Remarks on Halmos' duality theory (Italian summary)* ◇ B45 G05 ◇
FISCHER SERVI, G. *Semantics for a class of intuitionistic modal calculi* ◇ B45 B55 F50 ◇
FISCHER SERVI, G. *Teoremi di completezza per calcoli bimodali (English summary)* ◇ B45 ◇
FORBES, G. *On the philosophical basis of essentialist theories* ◇ A05 B45 ◇
FRAASSEN VAN, B.C. *A modal interpretation of quantum mechanics* ◇ B45 ◇
FRAASSEN VAN, B.C. *Report on tense logic* ◇ B45 ◇
GABBAY, D.M. *An irreflexivity lemma with applications to axiomatizations of conditions on tense frames* ◇ B45 ◇
GABBAY, D.M. *Expressive functional completeness in tense logic* ◇ B45 ◇
GARDIES, J.-L. & MALINOWSKI, G. *D'un certain usage de la trivalence en logique deontique* ◇ B45 ◇
GERLA, G. & NOLA DI, A. *A three valued doxastic logic* ◇ B45 B50 ◇
GOLDBLATT, R.I. *Grothendieck topology as geometric modality* ◇ B25 B45 C90 F50 G30 ◇
GRZEGORCZYK, A. *Individualistic formal approach to deontic logic* ◇ B45 ◇
GUMANSKI, L. *The deontic calculus DSC_3* ◇ A05 B45 ◇
GUMB, R.D. *An explication of the notion of a consistent evolving theory* ◇ A05 B45 ◇
GUPTA, A. & THOMASON, R.H. *A theory of conditionals in the context of branching time* ◇ B45 ◇
HAJEK, P. *Decision problems of some statistically motivated monadic modal calculi* ◇ B25 B45 ◇
HAJEK, P. *On interpretability in theories containing arithmetics II* ◇ B28 B45 F25 F30 ◇
HILPINEN, R. (ED.) *New studies in deontic logic* ◇ B45 ◇
HINTIKKA, K.J.J. *Aristotle on the realization of possibilities in time* ◇ A10 B45 ◇
HINTIKKA, K.J.J. *Phenomenology vs.possible-worlds semantics: apparent and real differences* ◇ A05 B45 ◇
HINTIKKA, K.J.J. *Standard vs. nonstandard logic: higher order, modal, and first-order logics* ◇ B10 B15 B45 ◇
HOTTOIS, G. *Logique deontique et logique de l'action chez G. H. von Wright* ◇ A05 B45 ◇
HUGLY, P. & SAYWARD, C.W. *Completeness theorems for two propositional logics in which identity diverges from mutual entailment* ◇ A05 B45 ◇
HUMBERSTONE, I.L. *From worlds to possibilities* ◇ A05 B45 ◇
HUMBERSTONE, I.L. *Relative necessity revisited* ◇ A05 B45 ◇
HUSZCZA, J. *Foundations of Jean-Louis Gardies' deontic and modal algebra (Polish) (French summary)* ◇ B45 ◇
ISHMURATOV, A.T. *Logical theories of temporal contexts. Temporal logic (Russian)* ◇ B45 ◇
JENNINGS, R.E. *A note on the axiomatisation of Brouwersche modal logic* ◇ B45 ◇
JENNINGS, R.E. & SCHOTCH, P.K. *Epistemic logic, skepticism, and nonnormal modal logic* ◇ A05 B45 ◇
JENNINGS, R.E. & SCHOTCH, P.K. *Non-Kripkean deontic logic* ◇ B45 ◇
JENNINGS, R.E. & SCHOTCH, P.K. *Probabilistic considerations on modal semantics* ◇ A05 B45 ◇
JENNINGS, R.E. & SCHOTCH, P.K. *Some remarks on (weakly) weak modal logics* ◇ A05 B45 ◇

JENNINGS, R.E. & JOHNSTON, D.K. & SCHOTCH, P.K. *The n-adic first-order undefinability of the Geach formula* ◇ A05 B45 ◇
KALINOWSKI, G. *Sur les semantiques des mondes possibles pour les systemes de logique deontique* ◇ A05 B45 ◇
KAPETANOVIC, M. *A tableaux system in modal logic* ◇ B45 ◇
KAWAI, H. *On the completeness theorem for bimodal systems* LT_Ω *and* LET_Ω ◇ B45 ◇
KEARNS, J.T. *Modal semantics without possible worlds* ◇ B45 ◇
KNUUTTILA, S. (ED.) *Reforging the great chain of being* ◇ A05 B45 ◇
KORDIG, C.R. *A deontic argument for God's existence* ◇ A05 B45 ◇
KRATZER, A. *Partition and revision: the semantics of counterfactuals* ◇ A05 B45 ◇
LEIVANT, D. *On the proof theory of the modal logic for arithmetic provability* ◇ B45 F05 F30 ◇
LENZEN, W. *Doxastic logic and the Burge-Buridan-paradox* ◇ B45 ◇
LEWIS, D. *Counterfactuals and comparative possibility* ◇ B45 ◇
LEWIS, D. *Index, context, and content* ◇ B45 B65 ◇
LEWIS, D. *Ordering semantics and premise semantics for counterfactuals* ◇ A05 B45 ◇
MAKINSON, D. *Non-equivalent formulae in one variable in a strong omnitemporal modal logic* ◇ B45 ◇
MAKINSON, D. *Quantificational reefs in deontic waters* ◇ B45 ◇
MARTINO, A.A. *Deontic logic, computational linguistics and legal information systems. Vol.II* ◇ B45 ◇
MCGINN, C. *Modal reality* ◇ B45 ◇
MCGINN, C. *The mechanism of reference* ◇ B45 ◇
MCHEDLISHVILI, L.I. *Minimal temporal logic with a modalized temporal operator (Russian)* ◇ B45 ◇
MCHEDLISHVILI, L.I. *Modal interpretation of traditional relations between propositions (Russian)* ◇ B45 ◇
MILNER, R. *A modal characterisation of observable machine-behaviour* ◇ B45 ◇
MONTANARO, A. *Sui fondamenti del calcolo delle probabilita basato sui calcoli modali* MC^V *oppure* MC^V_* ◇ B45 ◇
NAGLE, M.C. *The decidability of normal K5 logics* ◇ B45 ◇
NEEDHAM, P. *Temporal intervals and temporal order* ◇ B45 ◇
NISHIMURA, H. *Model theory for tense logic: Saturated and special models with applications to the tense hierarchy* ◇ B45 C40 C50 C90 ◇
NISHIMURA, H. *The semantical characterization of de dicto in continuous modal model theory* ◇ B45 C40 C90 ◇
NUTE, D.E. *Introduction to the issue on the theory of conditionals* ◇ A05 B45 ◇
OKABE, M. *An interpretation of Aristotle's modal syllogism* ◇ A05 B45 ◇
PHILIPP, P. *Ein System der deontischen Logik II. Bedingte Normen* ◇ B45 ◇
POLLOCK, J.L. *A refined theory of counterfactuals* ◇ A05 B45 ◇
PORTE, J. *Notes on modal logics* ◇ B45 ◇
PORTE, J. *The deducibilities of S5* ◇ B45 ◇
PURICA, I.I. *The modal logic of events (Romanian)* ◇ B45 ◇
ROSENBERG, S. *On the modal version of the ontological argument* ◇ A05 B45 ◇
RUZSA, I. *An approach to intensional logic* ◇ B45 ◇
RUZSA, I. *Modal logic with descriptions* ◇ A05 B45 ◇
RYBAKOV, V.V. *Admissible rules for pretable modal logics (Russian)* ◇ B45 ◇
SCHUMM, G.F. *Bounded properties in modal logic* ◇ B45 ◇
SEGERBERG, K. *Action-games* ◇ B45 ◇
SHEKHTMAN, V.B. *Semantics of modal propositions (Russian)* ◇ B45 ◇
STAHL, G. *Divisions of intensional universes* ◇ B45 ◇
STALNAKER, R.C. *Indexical belief* ◇ A05 B45 ◇
STARK, W.R. *A logic of knowledge* ◇ B45 ◇
STEPIEN, T. *System S* ◇ B45 ◇
STERN, C.D. *Lewi's counterfactual analysis of causation* ◇ B45 ◇
SUSSMAN, A.N. *Counterfactuals and nontrivial de re modalities* ◇ B45 ◇
TEVZADZE, D.D. *Analysis of a difficulty in quantified modal logic (Russian)* ◇ B45 ◇
THOMASON, R.H. *Deontic logic as founded on tense logic* ◇ B45 ◇
THOMASON, R.H. *Deontic logic and the role of freedom in moral deliberation* ◇ B45 ◇
TOMBERLIN, J.E. *Contrary-to-duty imperatives and conditional obligation* ◇ B45 ◇
TORRANCE, S.B. *Prescriptivism and incompleteness* ◇ A05 B45 ◇
TURNER, R. *Counterfactuals without possible worlds* ◇ A05 B45 ◇
ULRICH, D. *RMLC: Solution to a problem left open by Lemmon* ◇ B45 ◇
ULRICH, D. *Strict implication in a sequence of extensions of S4* ◇ B45 ◇
UNTERHOLZNER, P. *Algebraic and relational semantics for tense logics* ◇ B45 ◇
URCHS, M.P. *Semantische Typen fuer klassische Modallogiken* ◇ B45 ◇
VAKARELOV, D. *Intuitionistic modal logics incompatible with the law of the excluded middle* ◇ B45 ◇
VALIEV, M.K. *The symmetrized von Wright temporal logic (Russian)* ◇ B45 ◇
VANDERVEKEN, D.R. *A strong completeness theorem for pragmatics* ◇ A05 B45 ◇
VARDANYAN, V.A. *The wise men:logic of knowledge and actions (Russian)* ◇ B45 ◇
VILLA, G. *Sul rapporto degli operatori deontici fra loro e sull'estensione alla logica deontica del teorema dello pseudo scoto* ◇ B45 ◇
VISSER, A. *A propositional logic with explicit fixed points* ◇ A05 B45 ◇
WAGNER DECEW, J. *Conditional obligation and counterfactuals* ◇ B45 ◇

WARMBROD, K. *An indexical theory of conditionals* ◇ B45 ◇
WARMBROD, K. *Counterfactuals and substitution of equivalent antecedents* ◇ A05 B45 ◇
WEYL, H. *Le fantome de la modalite (English summary)* ◇ A05 B45 ◇
WHITE, M.J. *On some ascending chains of Brouwerian modal logics* ◇ B45 ◇
WOLPER, P. *Temporal logic can be more expressive* ◇ B45 ◇
WRIGHT VON, G.H. *Explanation and understanding of action* ◇ A05 B45 ◇
WRIGHT VON, G.H. *On the logic of norms and actions* ◇ B45 ◇
WRIGHT VON, G.H. *Problems and prospects of deontic logic: a survey* ◇ B45 ◇
ZANARDO, A. *A completeness theorem for the general interpreted modal calculus* MC^{ν} *of A. Bressan* ◇ B45 ◇
ZANARDO, A. *Teorema di completezza per il calcolo modale interpretato* MC^{ν} ◇ B45 ◇

1982

AAQVIST, L. *How to handle the liar paradox in modal logic with sentential quantifiers and its own truth predicate* ◇ B45 ◇
AKIMOV, A.P. *Constructive variants of classical modal calculi (Russian)* ◇ B45 ◇
ALCHOURRON, C.E. *Normative order and derogation* ◇ B45 ◇
ANSHAKOV, O.M. & RYCHKOV, S.V. *On multivalued logical calculi (Russian)* ◇ B45 ◇
ARTEMOV, S.N. *Applications of modal logic in proof theory (Russian)* ◇ B45 F30 ◇
BEALER, G. *Quality and concept* ◇ A05 B45 ◇
BELNAP JR., N.D. *Display logic* ◇ B45 ◇
BENDOVA, K. & HAJEK, P. *A logical analysis of the truth-reaction paradox* ◇ B45 F30 ◇
BERMAN, F. *Compactness in models of propositional dynamic logic* ◇ B45 ◇
BERNARDO DI, G. *Antinomies and predicaments in normative systems* ◇ B45 ◇
BIAGIOLI, C. & NATALI, F.S. & SPINOSA, P.L. & TRIVISONNO, G. *Experiments on the "model" of Sanchez-Mazas* ◇ B45 ◇
BOCHVAR, D.A. *Some aspects of the investigations of reification paradoxes* ◇ B45 ◇
BONOTTO, C. *Synonymy for Bressan's modal calculus* MC^{ν}*. I,II* ◇ B45 ◇
BOOLOS, G. *On the nonexistence of certain normal forms in the logic of provability* ◇ B45 F30 ◇
BROADIE, A. *The logical syntax of deontic operators* ◇ B45 ◇
BROWN, M.A. *Generalized S2-like systems of propositional modal logic* ◇ B45 ◇
BURGESS, J.P. *Axioms for tense logic I: "Since" and "until". II: Time periods* ◇ B45 ◇
BUTTERFIELD, J. *Dummett on temporal operators* ◇ B45 ◇
BYSTROV, P.I. & SEREBRYANNIKOV, O.F. *Criteria for the quality of inferences in modal logic (Russian)* ◇ B45 ◇
CHANDLER, M. *The logic of "unless"* ◇ B45 ◇
CLARKE, E.M. & EMERSON, E.A. *Design and synthesis of synchronization skeletons using branching time temporal logic* ◇ B45 B75 ◇
CLARKE JR., E.M. & FRANCEZ, N. & GUREVICH, Y. & STULA, P.S. *Can message buffers be characterized in linear temporal logic?* ◇ B45 ◇
COPELAND, B.J. *A note on the Barcan formula and substitutional quantification* ◇ B45 ◇
CRESSWELL, M.J. *A canonical model for S2* ◇ B45 ◇
CRESSWELL, M.J. & HUGHES, G.E. *K1.1 is not canonical* ◇ B45 ◇
DARDZHANIYA, G.K. & PKHAKADZE, N.M. *On some universal modal theory (Russian) (English and Georgian summaries)* ◇ B45 ◇
DAVIDSON, B. *Modal trees for modal predicate logics* ◇ B45 ◇
DINI, G. *A modified version of an arithmetic model for legal informatics of Sanchez-Mazas* ◇ B45 ◇
ECK VAN, J.A. *A system of temporally relative modal and deontic predicate logic and its philosophical applications I,II* ◇ B45 ◇
FARINAS DEL CERRO, L. *A simple deduction method for modal logic* ◇ B35 B45 ◇
FARINAS DEL CERRO, L. *Prolegomena for programming in modal logic* ◇ B45 ◇
FARINAS DEL CERRO, L. *Valid reasoning: classical logic = analogical reasoning: nonclassical logic* ◇ A05 B45 ◇
FINE, K. *First-order modal theories III. Facts* ◇ B45 ◇
FONT I LLOVET, J.M. *An implication connective for classical and intuitionistic modal logics (Catalan)* ◇ B45 ◇
FONT I LLOVET, J.M. *Modal logic without primitive modal operators (Spanish) (English summary)* ◇ B45 ◇
FONT I LLOVET, J.M. *On the definition of necessity and possibility in non-classical modal logic (Catalan)* ◇ B45 ◇
FONT I LLOVET, J.M. *Some non-classical connectives for intuitionistic modal logic* ◇ B45 ◇
FRIEDMAN, JOEL I. *Kripkean necessity and the ontological argument* ◇ B45 ◇
GARGOV, G.K. & KIROV, K.A. *The logic of "strong box" in IGL is IS4Grz* ◇ B45 ◇
GERLA, G. *A note on the principle of predication* ◇ A05 B45 ◇
GIBBINS, P. *The strange modal logic of indeterminacy* ◇ B45 ◇
GUCCIONE, S. & TORTORA, R. *A semantical hierarchy for modal formulas* ◇ B45 ◇
GUDZHINSKAS, EH. *Syntactical proof of the elimination theorem for von Wright's temporal logic (Russian) (English and Lithuanian summaries)* ◇ B45 ◇
GUMB, R.D. & WEAVER, G.E. *First order properties of relations with the monotonic closure property* ◇ B45 E07 ◇
HAILPERN, B.T. *Verifying concurrent processes using temporal logic* ◇ B45 ◇

HAUSCHILD, K. *Model-theoretic properties of cause-and-effect structures* ◇ B45 C52 D35 ◇
HINTIKKA, K.J.J. *Is alethic modal logic possible?* ◇ B45 ◇
HINTIKKA, K.J.J. *Temporal discourse and semantical games* ◇ B45 ◇
HUGHES, G.E. *Some strong omnitemporal logics* ◇ B45 ◇
INHETVEEN, R. *Ein konstruktiver Weg zur Semantik der "moeglichen Welten"* ◇ B45 ◇
ISHMURATOV, A.T. *Logics with partial interpretations (Russian)* ◇ B45 ◇
JAGER, T. *An actualistic semantics for quantified modal logic* ◇ B45 ◇
KARAVAEV, E.F. *Temporal logic as a means of realization of the logical function of the category of time (Russian)* ◇ B45 ◇
KARPENKO, A.S. & SMIRNOV, V.A. (EDS.) *Modal and relevant logics (Russian)* ◇ B45 B97 ◇
KAWAI, H. *Eine Logik erster Stufe mit einem infinitaeren Zeitoperator* ◇ B45 C40 C75 C90 ◇
KIELKOPF, C.F. *A completeness proof for Porte's* S_a^0 *and* S_a ◇ B45 ◇
KUBINSKI, T. *Tense theories taking into account many change periods (Polish)* ◇ B45 ◇
LAMPORT, L. *Timesets. A new method for temporal reasoning about programs* ◇ B45 ◇
LEHMANN, D.J. & SHELAH, S. *Reasoning with time and chance* ◇ B45 B75 ◇
LEOPOLD-WILDBURGER, U. & SELTEN, R. *Subjunctive conditionals in decision and game theory* ◇ B45 ◇
LEWIS, D. *Logic for equivocators* ◇ B45 B53 ◇
LOUREIRO, I. *Axiomatisation et proprietes des algebres modales tetravalentes (English summary)* ◇ B45 G25 ◇
LOWE, E.J. *On the alleged necessity of true identity statements* ◇ A05 B45 ◇
LUNDBERG, B. *An axiomatization of events* ◇ B45 ◇
MAGARI, R. *Primi risultati sulla varieta di Boolos (English summary)* ◇ B45 F30 G25 ◇
MAKSIMOVA, L.L. *Absence of the interpolation property in modal companions of a Dummett logic (Russian)* ◇ B45 C40 C90 ◇
MAKSIMOVA, L.L. *Lyndon's interpolation theorem in modal logics (Russian)* ◇ B45 C40 C90 ◇
MARGHIDANU, D. *Five logical modalities of exclusion (Romanian) (French summary)* ◇ B45 ◇
MARKIN, V.I. *The neighborhood semantics for modalities de re (Russian)* ◇ B45 ◇
MATERNA, P. *Transparent intensional logic (Russian)* ◇ B45 ◇
MORGAN, C.G. *Simple probabilistic semantics for propositional K, T, B, S4, and S5* ◇ B45 C90 ◇
MORGAN, C.G. *There is a probabilistic semantics for every extension of classical sentence logic* ◇ B45 C90 ◇
MORSCHER, E. *Antinomies and incompatibilities within normative languages. Some semantic considerations* ◇ B45 ◇
MORTARI, C.A. *On a system of modal-temporal logic* ◇ B45 ◇
MOUTAFAKIS, N.J. *Rescher's logic of preference and linguistic analysis* ◇ B45 B65 ◇
NIINILUOTO, I. *Remarks on the logic of perception* ◇ B45 ◇
NOWAKOWSKA, M. *Knowledge and induced subjective temporal systems* ◇ B45 ◇
OHNISHI, M. *A new version of the Gentzen decision procedure for modal sentential calculus S5* ◇ B45 ◇
OMEL'YANCHIK, V.I. *Understanding and the semantics of classical and modal logic (Russian)* ◇ B45 ◇
ORLOWSKA, E. *Representation of temporal information* ◇ B45 ◇
ORLOWSKA, E. *Tense logic for nondeterministic time* ◇ B45 ◇
PARSONS, C. *Intensional logic in extensional language* ◇ A05 B45 ◇
POERN, I. *Meaning and intension* ◇ A05 B45 ◇
POPOV, V.A. *Gentzen-type calculi for some intuitionistic modal logics* ◇ B45 ◇
PORTE, J. *Boll-Reinhart modal logic* ◇ B45 ◇
PORTE, J. *S5 and the predicate calculus* ◇ B45 ◇
PRADE, H. *Modal semantics and fuzzy set theory* ◇ B45 E72 ◇
PRIEST, G. & ROUTLEY, R. *Lessons from Pseudo Scotus* ◇ A10 B45 ◇
QUEILLE, J.P. & SIFAKIS, J. *A temporal logic to deal with fairness in transition systems* ◇ B45 ◇
RANTALA, V. *Impossible worlds semantics and logical omniscience* ◇ B45 ◇
RANTALA, V. *Quantified modal logic: non-normal worlds and propositional attitudes* ◇ B45 C90 ◇
RATSA, M.F. *Functional completeness in modal logic (Russian)* ◇ B45 ◇
RATSA, M.F. *Nontabularity of the logic S4 with respect to functional completeness (Russian)* ◇ B45 ◇
REDDAM, J.P. *Van Fraassen on propositional attitudes* ◇ A05 B45 ◇
REHDER, W. *Von Wright's "and next" versus a sequential tense-logic* ◇ B45 ◇
RICHARD, M. *Tense, propositions, and meanings* ◇ A05 B45 ◇
ROPER, A. *Towards an eliminative reduction of possible worlds* ◇ B45 ◇
ROSADO HADDOCK, G.E. *Identity statements in the semantics of sense and reference* ◇ B45 ◇
RYBAKOV, V.V. *Bases of quasiidentities of finite modal algebras (Russian)* ◇ B45 ◇
RYBAKOV, V.V. *Completeness of modal logics of prefinite width (Russian)* ◇ B45 ◇
SAARINEN, E. *Propositional attitudes are not attitudes towards propositions* ◇ A05 B45 ◇
SAHLIN, N.-E. *On counterfactual probabilities and causation: a note* ◇ B45 ◇
SAMBIN, G. & VALENTINI, S. *The modal logic of provability. The sequential approach* ◇ B45 F05 F30 ◇
SAMCHENKO, V.N. *On Quine's paradox (Russian)* ◇ A05 B45 ◇
SANCHEZ-MAZAS, M. *Algebraic and arithmetical translations of normative systems and applications in legal informatics* ◇ B45 ◇

SCHIFFER, S. *Intention-based semantics* ◇ B45 ◇
SEGERBERG, K. *A deontic logic of action* ◇ B45 ◇
SEGERBERG, K. *The logic of deliberate action* ◇ B45 ◇
SEREBRYANNIKOV, O.F. *Gentzen's Hauptsatz for modal logic with quantifiers* ◇ B45 ◇
SIDORENKO, E.A. *On various notions of the conclusion from hypotheses (Russian)* ◇ B45 ◇
SMIRNOV, V.A. *The definition of modal operators by means of tense operators* ◇ B45 ◇
SMORYNSKI, C.A. *Commutativity and self-reference* ◇ B45 F30 ◇
SMORYNSKI, C.A. *The finite inseparability of the first-order theory of diagonalisable algebras* ◇ B45 C13 D35 F30 ◇
SPADE, P.V. *Three theories of obligations: Burley, Kilvington and Swyneshed on counterfactual reasoning* ◇ A10 B45 ◇
STENIUS, E. *Ross' paradox and well-formed codices* ◇ B45 ◇
THOMASON, S.K. *Undecidability of the completeness problem of modal logic* ◇ B45 ◇
TUOMELA, R. *Action generation* ◇ B45 ◇
VALENTINI, S. *Cut elimination in a modal sequent calculus for K (Italian summary)* ◇ B45 ◇
VOROBEJ, M. *Deontic accessibility* ◇ B45 ◇
WARMBROD, K. *A defense of the limit assumption* ◇ B45 ◇
WEINGARTNER, P. *Conditions of rationality for the concepts belief, knowledge, and assumption (French and German summaries)* ◇ A05 B45 ◇
WOLNIEWICZ, B. *A formal ontology of situations* ◇ B45 ◇
WRIGHT VON, G.H. *Diachronic and synchronic modalities* ◇ B45 ◇
WRIGHT VON, G.H. *Norms, truth and logic* ◇ A05 B45 ◇
YASHIN, A.D. *Intuitionistic predicate logic with the connective "tomorrow" (Russian) (English summary)* ◇ B45 ◇
ZEMACH, E.M. *Schematic objects and relative identity* ◇ B45 ◇

1983

AUSTIN, D.F. *Plantinga's theory of proper names* ◇ A05 B45 ◇
BARNETT, P.M. *Rational behavior in bargaining situations* ◇ B45 ◇
BELLISSIMA, F. & MIROLLI, M. *On the axiomatization of finite K-frames* ◇ B45 C13 C90 ◇
BELZER, M. & LOEWER, B.M. *Dyadic deontic detachment* ◇ B45 ◇
BEN-ARI, M. & MANNA, Z. & PNUELI, A. *The temporal logic of branching time* ◇ B45 ◇
BENTHEM VAN, J.F.A.K. *Determiners and logic* ◇ B45 ◇
BENTHEM VAN, J.F.A.K. *Five easy pieces* ◇ B45 ◇
BENTHEM VAN, J.F.A.K. & HUMBERSTONE, I.L. *Hallden-completeness by gluing of Kripke frames* ◇ B45 C30 C90 ◇
BENTHEM VAN, J.F.A.K. *Logical semantics as an empirical science* ◇ A05 B45 ◇
BENTHEM VAN, J.F.A.K. *The logic of time* ◇ B45 ◇
BLOK, W.J. & KOEHLER, P. *Algebraic semantics for quasiclassical modal logics* ◇ B45 C90 G25 ◇
BONEVAC, D. *Chellas on conditional obligation* ◇ B45 ◇
BORGA, M. *On some proof theoretical properties of the modal logic GL* ◇ B45 F05 F30 ◇
BUGAJSKI, S. *Languages of similarity* ◇ B45 ◇
CERLA, G. *Contralogical modalities (Italian)* ◇ B45 ◇
CHAGROV, A.V. *Polynomial finite approximability of modal and superintuitionistic logics (Russian)* ◇ B45 B55 ◇
CHELLAS, B.F. *$KG^{k,l,m,n}$ and the EFMP* ◇ B45 ◇
COPELAND, B.J. *Tense trees: a tree system for K_t* ◇ B45 ◇
CRESSWELL, M.J. *KM and the finite model property* ◇ B45 ◇
CRESSWELL, M.J. *The completeness of KW and K1.1* ◇ B45 ◇
CZERMAK, J. & WEINGARTNER, P. *Criteria for distinguishing logically necessary propositions* ◇ B45 B60 ◇
DALLA CHIARA SCABIA, M.L. *Physical implications in a Kripkian semantical approach to physical theories* ◇ B45 ◇
DRIVER, J. *Promises, obligations, and abilities* ◇ B45 ◇
DWYER, R. & MARTIN, E.P. & MEYER, R.K. *The fundamental S-theorem - a corollary* ◇ B45 ◇
DYWAN, Z. *The connective of necessity of modal logic S5 is metalogical* ◇ B45 ◇
DZIOBIAK, W. *Structural completeness of modal logics containing K4* ◇ B45 ◇
EMERSON, E.A. *Alternative semantics for temporal logics* ◇ B45 ◇
FARINAS DEL CERRO, L. & RAGGIO, A.R. *Some results in intuitionistic modal logic* ◇ B45 ◇
FINE, K. *The permutation principle in quantificational logic* ◇ B45 ◇
FITTING, M. *Proof methods for modal and intuitionistic logics* ◇ B45 B55 F07 F50 F98 ◇
FONT I LLOVET, J.M. *Intuitionistic implication in some non-classical modal logics of type S4* ◇ B45 ◇
FORBES, G. *Physicalism, instrumentalism and the semantics of modal logic* ◇ A05 B45 ◇
FORBES, G. *Thisness and vagueness* ◇ A05 B45 ◇
GARDIES, J.-L. *Logique de l'action (ou du changement) et logique deontique* ◇ B45 ◇
GEORGESCU, G. *Chang's modal operators in algebraic logic* ◇ B45 ◇
GERLA, G. *Model theory and modal logic (Italian)* ◇ B45 C25 C75 C90 ◇
GREEF DE, J. *La negation deontique dans "Directives and norms" d'Alf Ross* ◇ B45 ◇
GUMANSKI, L. *An extension of the deontic calculus DSC_1* ◇ B45 ◇
HAJEK, P. *Arithmetical interpretations of dynamic logic* ◇ B28 B45 ◇
HALPERN, J.Y. *Deterministic process logic is elementary* ◇ B45 ◇
HAUTAMAKI, A. *The logic of viewpoints* ◇ B45 ◇

HINTIKKA, K.J.J. *Situations, possible worlds, and attitudes* ◇ B45 ◇
HUMBERSTONE, I.L. *Inaccessible worlds* ◇ B45 ◇
KARAVAEV, E.F. *Foundations of tense-logic (Russian)* ◇ B45 ◇
KONDO, M. *Completeness theorem for some logic in terms of interval semantics* ◇ B45 ◇
KRAUT, R. *There are no de dicto attitudes* ◇ B45 ◇
LEHMANN, D.J. & SHELAH, S. *Reasoning with time and chance* ◇ B25 B45 ◇
LINDEMANS, J.F. *The reduction of deontic logic to the first-order predicate calculus* ◇ B45 ◇
LOPARIC, A. & MORTARI, C.A. *Valuation in temporal logic* ◇ B45 ◇
LOWE, E.J. *A simplification of the logic of conditionals* ◇ B45 ◇
LUE, QICI *Counterexample to Bowen's theorem and failures of interpolation theorem in modal logic (Chinese)* ◇ B45 C40 ◇
MACBEATH, M. *Communication and time reversal* ◇ B45 ◇
MAKINSON, D. *Individual actions are very seldom obligatory* ◇ B45 ◇
MCCALL, S. *If, since, and because: a study in conditional connection* ◇ B45 ◇
MCMICHAEL, A. *A new actualist modal semantics* ◇ B45 ◇
MESKHI, V.YU. *Critical modal logics containing the Brouwer axiom (Russian)* ◇ B45 ◇
MIURA, S. *Embedding of modal predicate systems into lower predicate calculus* ◇ B45 ◇
MIURA, S. *Embedding of modal predicate logics into lower predicate calculus I* ◇ B45 ◇
MONTAGNA, F. *Il teorema di Solovay e l'aritmetica di Heyting* ◇ B45 F30 ◇
MONTAGNA, F. *ZFC-models as Kripke-models* ◇ B45 C62 E30 F30 ◇
MOON, R.H. *Correction of the semantics for S4.03 and a note on literal disjunctive symmetry* ◇ B45 ◇
MUCHNIK, A.A. *Supplement of the translator to the paper "On alternation. I, II" (Russian)* ◇ B45 D05 D10 D15 F50 ◇
MURAVITSKIJ, A.YU. *Extensions of the provability logic (Russian)* ◇ B45 F30 ◇
NISHIMURA, H. *A cut-free sequential system for the propositional modal logic of finite chains* ◇ B45 B75 ◇
NISHIMURA, H. *Hauptsatz for higher-order modal logic* ◇ B45 ◇
PARIKH, R. *Propositional logics of programs: New directions* ◇ B45 ◇
PHILIPP, P. *Ein System der deontischen Logik III. Handlungen im Bereich von Normen* ◇ B45 ◇
PIZZI, C. *La definizione delle modalita fisiche nella logica dell'implicazione consequenziale* ◇ B45 ◇
POLLOCK, J.L. *A theory of direct inference I. A classical theory of direct inference* ◇ B45 ◇
PORTE, J. *Axiomatization and independence in S4 and in S5* ◇ B45 ◇
QUEILLE, J.P. & SIFAKIS, J. *Fairness and related properties in transition systems - a temporal logic to deal with fairness* ◇ B45 B75 ◇
RATSA, M.F. *Functional completeness in the modal logic S5 (Russian)* ◇ B45 ◇
RATSA, M.F. *Functional completeness in modal logics (Russian)* ◇ B45 ◇
RATSA, M.F. *Undecidability of the problem of functional expressibility in the modal logic S4 (Russian)* ◇ B45 ◇
RAUTENBERG, W. *Modal tableau calculi and interpolation* ◇ B45 C40 ◇
RICHARD, M. *Direct reference and ascriptions of belief* ◇ B45 ◇
ROUTLEY, R. & ROUTLEY, V. *Semantical foundations for value theory* ◇ B45 ◇
SCHWEIGERT, D. & SZYMANSKA, M. *On functionally complete modal algebras related to M and S4* ◇ B45 G25 ◇
SHEKHTMAN, V.B. *Denumerable approximability of superintuitionistic and modal logics (Russian)* ◇ B45 ◇
SHEKHTMAN, V.B. *Modal logics of domains on the real plane* ◇ B45 ◇
SHELTON, L.V. *A diachronic semantics for inexact reference* ◇ B45 ◇
SHUKLA, A. *Consistent, independent, and distinct propositions III: Modalities in S6* ◇ B45 ◇
SOLITRO, U. & VALENTINI, S. *The modal logic of consistency assertions of Peano arithmetic* ◇ B45 F30 ◇
SVEJDAR, V. *Modal analysis of generalized Rosser sentences* ◇ B45 F25 F30 ◇
SWART DE, H.C.M. *A Gentzen- or Beth-type system, a practical decision procedure and a constructive completeness proof for the counterfactual logics VC and VCS* ◇ B45 ◇
THIEFFINE, F. *Compatible complement in Piron's system and ordinary modal logic* ◇ B45 ◇
THOMASON, S.K. *Finite matrices for quasi-classical modal logics* ◇ B45 ◇
TODT, G. *Fuzzy logic and modal logic* ◇ B45 ◇
TSELKOV, V. *Intuitionistic modal logics contradicting the Rieger-Nishimura logics (Bulgarian summary)* ◇ B45 ◇
VALENTINI, S. *A "canonical" model for GL (Italian summary)* ◇ B45 F30 ◇
VALENTINI, S. *The modal logic of provability: cut-elimination* ◇ B45 F05 F30 ◇
VLACH, F. *On situation semantics for perception* ◇ B45 ◇
VUILLEMIN, J. *Le carre chrysippeen des modalites (English and German summaries)* ◇ B45 ◇
WIDERKER, D. *The extensionality argument* ◇ A05 B45 ◇
WILSON, M. *Why contigent identity is necessary* ◇ B45 ◇
WINSKEL, G. *A note on power domains and modality* ◇ B45 ◇
WOLPER, P. *Temporal logic can be more expressive* ◇ B45 B75 ◇
WRIGHT VON, G.H. *Norms of higher order* ◇ B45 ◇
ZANARDO, A. *On the equivalence between the calculi MC^{ν} and $EC^{\nu+1}$ of A. Bressan* ◇ B45 ◇

1984

ALCHOURRON, C.E. & BULYGIN, E. *Permission and permissive norms* ◇ B45 ◇

ALLEN, JAMES F. *Towards a general theory of action and time* ◇ B45 ◇

ARTEMOV, S.N. *On modal representations of extensions of Peano arithmetic* ◇ B45 F30 ◇

AVRON, A. *On modal systems having arithmetical interpretations* ◇ B45 F05 F30 ◇

BAILHACHE, P. *Quelques foncteurs faussement primitifs en logique deontique (trivalence et action)* ◇ B45 ◇

BALDWIN, THOMAS *E.J.Lowe on modalities de re: "On the alleged necessity of true identity statements"* ◇ A05 B45 ◇

BELLISSIMA, F. *On the relationship between one-point frames and degrees of unsatisfiability of modal formulas* ◇ B45 ◇

BENTHEM VAN, J.F.A.K. *Possible worlds semantics: a research program that cannot fail?* ◇ B45 ◇

BENTHEM VAN, J.F.A.K. *Tense logic and time* ◇ B45 ◇

BERNARDI, C. *A shorter proof of a recent result by R. Di Paola* ◇ B45 F30 ◇

BOOLOS, G. *The logic of provability* ◇ B45 F30 ◇

BOZIC, M. & DOSEN, K. *Models for normal intuitionistic modal logics* ◇ B45 C90 ◇

BULL, R.A. & SEGERBERG, K. *Basic modal logic* ◇ B45 ◇

BURGESS, J.P. *Beyond tense logic* ◇ B45 ◇

CAVALLI, A.R. & FARINAS DEL CERRO, L. *A decision method for linear temporal logic* ◇ B25 B45 ◇

CAVALLI, A.R. & FARINAS DEL CERRO, L. *Specification and verification of networks protocols using temporal logic* ◇ B45 ◇

CHRZ, T. *A system for knowledge representation based on intensional logic I,II* ◇ B45 ◇

CRESSWELL, M.J. & HUGHES, G.E. *A companion to modal logic* ◇ B45 ◇

CRESSWELL, M.J. *An incomplete decidable modal logic* ◇ B45 ◇

DARDZHANIYA, G.K. *A sequential variant of Grzegorczyk's modal system (Russian) (Grus. and English summaries)* ◇ B45 ◇

DOSEN, K. *Intuitionistic double negation as a necessity operator* ◇ B45 F50 ◇

DYWAN, Z. *On Lemmon's interpretations of the connective of necessity* ◇ B45 ◇

EMERSON, E.A. & SISTLA, A.P. *Deciding full branching time logic* ◇ B25 B45 D15 ◇

ENJALBERT, P. & MICHEL, M. *Many-sorted temporal logic for multiprocesses systems* ◇ B45 ◇

FARINAS DEL CERRO, L. *Un principe de resolution en logique modale* ◇ B45 ◇

FITTING, M. *A symmetric approach to axiomatizing quantifiers and modalities* ◇ B45 ◇

FITTING, M. *Linear reasoning in modal logic* ◇ B45 C40 ◇

FONT I LLOVET, J.M. *Implication and deduction in some intuitionistic modal logics* ◇ B45 ◇

FREUNDLICH, R. *Zur Begruendung einer formalen Normenlogik* ◇ B45 ◇

GARGOV, G.K. *A note on the provability logics of certain extensions of Heyting's arithmetic* ◇ B45 F30 ◇

GARRETT, B.J. *More on rigidity and scope* ◇ B45 ◇

GARSON, J.W. *Quantification in modal logic* ◇ B45 C90 C98 ◇

GERASIMOVA, I.A. *Neighborhood semantics for epistemic logic (Russian)* ◇ B45 ◇

GERLA, G. & VACCARO, V. *Modal logic and model theory* ◇ B45 C25 C75 C90 ◇

GOODMAN, NICOLAS D. *Epistemic arithmetic is a conservative extension of intuitionistic arithmetic* ◇ B45 F05 F30 F50 ◇

GRAF, S. *On Lamport's comparison between linear and branching time temporal logic* ◇ B45 ◇

GUMB, R.D. *An extended joint consistency theorem for a family of free modal logics with equality* ◇ B45 ◇

HILPINEN, R. *Disjunctive permissions and conditionals with disjunctive antecedents (Russian)* ◇ B45 ◇

HINTIKKA, K.J.J. *Are there nonexistent objects? Why not? But where are they?* ◇ A05 B45 ◇

HINTIKKA, K.J.J. *Is an aletic modal logic possible? (Russian)* ◇ B45 ◇

HODES, H.T. *Axioms for actuality* ◇ A05 B45 ◇

HODES, H.T. *On modal logics which enrich first-orders S5* ◇ B45 ◇

HODES, H.T. *Some theorems on the expressive limitations of modal languages* ◇ B45 ◇

HODES, H.T. *The modal theory of pure identity and some related decision problems* ◇ B25 B45 D35 ◇

HODES, H.T. *Well-behaved modal logics* ◇ B45 ◇

JENNINGS, R.E. & SCHOTCH, P.K. *The preservation of coherence* ◇ B45 ◇

KALINOWSKI, G. *Presuppositions, verite et normes* ◇ B45 ◇

KANE, R. *The modal ontological argument* ◇ A05 B45 ◇

KELLER, P. *Zum Verhaeltnis von Zeitlogik und Modallogik* ◇ B45 ◇

KIROV, K.A. *An intuitionistic analogue of the modal logic of provability in Peano's arithmetic* ◇ B45 F30 ◇

KLESHCHEV, A. *Frames (Russian)* ◇ B45 ◇

KONDO, M. *A note on a Brown's conjecture* ◇ B45 ◇

KONDO, M. *The completeness theorems for some intuitionistic epistemic logics in terms of interval semantics* ◇ B45 F50 ◇

KRAJICEK, J. *Modal set theory* ◇ B45 E35 E70 ◇

KRAWIETZ, W. & SCHELSKY, H. & SCHRAMM, A. & WINKLER, G. (EDS.) *Theorie der Normen* ◇ A05 B45 ◇

KROEGER, F. *A generalized nexttime operator in temporal logic* ◇ B45 ◇

KURBATOV, V.I. *Logical semantics of norms (Russian)* ◇ B45 ◇

LAVENDHOMME, R. & LUCAS, T. *Une interpretation modale de la logique intuitionniste (English summary)* ◇ B45 F50 ◇

LUE, QICI *Counterexample to Bowen's theorem and failures of interpolation theorem in modal logic (Chinese)* ◇ B45 ◇

MCCALL, S. *Counterfactuals based on real possible worlds* ◇ A05 B45 ◇

MELLIAR-SMITH, P.M. & SCHWARTZ, R.L. & VOGT, F.H. *An interval-based temporal logic* ◇ B45 ◇
MICHEL, M. *Algebre de machines et logique temporelle* ◇ B45 D05 ◇
MINTS, G.E. *Structural synthesis with independent subproblems and the modal logic S4 (Russian) (English and Estonian summaries)* ◇ B45 ◇
MONTAGNA, F. *The predicate modal logic of provability* ◇ B45 C90 F30 ◇
MORSCHER, E. *Sein-Sollen-Schluesse und wie Schluesse sein sollen* ◇ B45 ◇
OHAMA, S. *Conjunctive normal forms and weak modal logics without the axiom of necessity* ◇ B45 ◇
OHRSTROM, P. *Buridan on interval semantics for temporal logic* ◇ A10 B45 ◇
ORLOWSKA, E. & PAWLAK, Z. *Logical foundations of knowledge representation. I* ◇ B45 ◇
ORLOWSKA, E. *Modal logics in the theory of information systems* ◇ B45 ◇
PELLETIER, F.J. *The not-so-strange modal logic of indeterminacy* ◇ B45 ◇
PLAISTED, D.A. *A low level language for obtaining decision procedures for classes of temporal logics* ◇ B45 ◇
POERN, I. *Deontic detachment* ◇ B45 ◇
PORTE, J. *Lukasiewicz's L-modal system and classical refutability* ◇ B45 ◇
RADEV, S.R. *Interpolation in the infinitary propositional normal modal logic* KL_{ω_1} ◇ B45 C40 C75 ◇
RANTALA, V. *Facts and the choice of logical foundations* ◇ B45 ◇
RUZSA, I. *Intensional logic without intensional variables (Russian)* ◇ B45 ◇
RYBAKOV, V.V. *A criterion for admissibility of rules in the modal system S4 and the intuitionistic logic (Russian)* ◇ B45 ◇
RYBAKOV, V.V. *Admissible rules for logics containing S4.3 (Russian)* ◇ B45 ◇
RYBAKOV, V.V. *Decidability of the admissibility problem in layer-finite modal logics (Russian)* ◇ B25 B45 ◇
SAARINEN, E. *Does inference with objects of propositional purpose exist? (Russian)* ◇ B45 ◇
SCHLESINGER, G.N. *A theorem of epistemic logic* ◇ B45 ◇
SCHOTCH, P.K. *Remarks on the semantics of nonnormal modal logics* ◇ B45 ◇
SCHRAMM, A. *Norm-Folgern ohne Normenlogik* ◇ B45 ◇
SEREBRYANNIKOV, O.F. *Some generalizations of theorems about normal form in quantor modal logic (Russian)* ◇ B45 ◇
SHUM, A.A. *Varieties of algebraic systems and propositional calculi (Russian)* ◇ B45 C05 ◇
SKORDEV, D.G. *On a modal-type language for the predicate calculus* ◇ B45 ◇
SMIRNOV, V.A. *Logical systems with modal tense operators (Russian)* ◇ B45 ◇
SMOLENOV, KH. *Zeno's paradoxes and temporal becoming in dialectical atomism* ◇ A05 B45 ◇
SMORYNSKI, C.A. *Modal logic and self-reference* ◇ B45 F30 ◇
SOLT, K. *Deontic alternative worlds and the truth-value of "OA"* ◇ B45 ◇
SORENSEN, R.A. *Unique alternative guessing* ◇ B45 D20 ◇
SOTIROV, V.KH. *Modal theories with intuitionistic logic* ◇ B45 ◇
SPELLER, J. *Filling a gap in Professor von Kutschera's decision procedure for deontic logic* ◇ B45 ◇
STAHL, G. *Mythe et realite de la formule de Barcan* ◇ B45 ◇
STEINACKER, P. *Ueber eine intensionale Alternative bei C.I. Lewis* ◇ B45 ◇
SUCHON, W. *The deontic calculus* D_{KRZ} ◇ B45 ◇
SUHARA, K. *Another mode of metalinguistic speech: multi-modal logic on a new basis* ◇ B45 ◇
TAPSCOTT, B.L. *Correcting the tableau procedure for S4* ◇ B45 ◇
THOMASON, S.K. *On constructing instants from events* ◇ B45 ◇
TICHY, P. *Subjective conditionals: two parameters vs. three* ◇ B45 ◇
TRZESICKI, K. *Gentzen-style axiomatization of tense logic* ◇ B45 ◇
ULRICH, D. *Answer to a question suggested by Schumm* ◇ B45 ◇
URCHS, M.P. *Eine Familie von Kausalsystemen* ◇ B45 ◇
URCHS, M.P. *Kausallogik und Kripkesemantik* ◇ B45 ◇
VAKARELOV, D. *An application of the Rieger-Nishimura formulas to the intuitionistic modal logics* ◇ B45 ◇
VISSER, A. *The provability logic of recursively enumerable theories extending Peano arithmetic and arbitrary theories extending Peano arithmetic* ◇ B45 C62 F30 ◇
VOJSHVILLO, E.K. *Logical entailment and semantics of general state description (Russian)* ◇ B45 ◇
WANG, XUANYUAN *Mr.S and Mr.P Puzzle - a brief introduction to modal logic (Chinese)* ◇ B45 ◇
WHITE, M.J. *The necessity of the past and modal-tense logic incompleteness* ◇ B45 ◇
WRIGHT VON, G.H. *Bedingungsnormen - ein Pruefstein fuer die Normenlogik* ◇ B45 ◇
WRIGHTSON, G. *On some semantic tableau proof procedures for modal logic* ◇ B45 ◇
WROBLEWSKI, J. *Negation in law* ◇ B45 ◇
ZAKHAR'YASHCHEV, M.V. *Normal modal logics containing S4 (Russian)* ◇ B45 ◇
ZANARDO, A. *Individual concepts as propositional variables in* $ML^{\nu+1}$ ◇ B45 ◇
ZIEMBINSKI, Z. *Kinds of discordance of norms* ◇ A05 B45 ◇

1985

ARTEMOV, S.N. *Modal logics axiomatizing provability (Russian)* ◇ B45 F30 ◇
ARTEMOV, S.N. *Nonarithmeticity of truth predicate logics of provability (Russian)* ◇ B45 D35 D55 F30 ◇
BAILHACHE, P. *Logique deontique et temporelle (Polish summary)* ◇ B45 ◇

Subject Index

BARRINGER, H. & KUIPER, R. *Hierarchical development of concurrent systems in a temporal logic framework* ◇ B45 ◇
BELLISSIMA, F. *A test to determine distinct modalities in the extensions of* S4 ◇ B45 ◇
BOOLOS, G. & SAMBIN, G. *An incomplete system of modal logic* ◇ B45 ◇
BOOLOS, G. *1-consistency and the diamond* ◇ B45 F30 ◇
BORICIC, B.R. *On some subsystems of Dummett's LC* ◇ B45 ◇
BURGESS, J.P. & GUREVICH, Y. *The decision problem for linear temporal logic* ◇ B25 B45 ◇
CARMO, J. *The INFOLOG branching logic of events* ◇ B45 ◇
CARO DE, F. & FATTOROSI-BARNABA, M. *Grade modalities I* ◇ B45 ◇
CHANDRA, A.K. & HALPERN, J.Y. & MEYER, A.R. & PARIKH, R. *Equations between regular terms and application to process logic* ◇ B45 D05 ◇
CLARKE, E.M. & MISHRA, B. *Hierarchical verification of asynchronous circuits using temporal logic* ◇ B45 ◇
CLARKE, E.M. & SISTLA, A.P. *The complexity of propositional linear temporal logics* ◇ B45 ◇
CRESSWELL, M.J. *The decidable normal modal logics are not recursively enumerable* ◇ B25 B45 D25 D35 D80 ◇
DOSEN, K. *Models for stronger normal intuitionistic modal logics* ◇ B45 ◇
DOSEN, K. *Sequent-systems for modal logic* ◇ B45 ◇
DYWAN, Z. *On Lemmon's interpretations of the connective of necessity* ◇ B45 ◇
EMERSON, E.A. *Automata, tableaux, and temporal logics* ◇ B45 D05 ◇
EMERSON, E.A. & HALPERN, J.Y. *Decision procedures and expressiveness in the temporal logic of branching time* ◇ B45 ◇
FINE, K. *Logics containing K4. II* ◇ B45 ◇
FITTING, M. *A Kripke-Kleene semantics for logic programs* ◇ B45 ◇
FLAGG, R.C. *Epistemic set theory is a conservative extension of intuitionistic set theory* ◇ B45 E35 E70 F50 ◇
GARDIES, J.-L. *Comment articuler logique deontique et logique modale l'une sur l'autre? (Polish summary)* ◇ B45 ◇
GENSLER, H.J. *Ethical consistency principles* ◇ B45 ◇
GREEN, K. *Is a logic for belief sentences possible?* ◇ A05 B45 ◇
GUREVICH, Y. & SHELAH, S. *The decision problem for branching time logic* ◇ B25 B45 C65 D05 D25 ◇
GUREVICH, Y. & SHELAH, S. *To the decision problem for branching time logic* ◇ B45 ◇
HENDRY, H.E. & POKRIEFKA, M.L. *Carnapian extensions of S5* ◇ B45 ◇
HINTIKKA, K.J.J. & KULAS, J. *Anaphora and definite descriptions. Two applications of game-theoretical semantics* ◇ B45 ◇
IVLEV, YU.V. *Intensional semantics of modal logic (Russian)* ◇ B45 ◇
JANKOWSKI, A.W. *Disjunctions in closure spaces* ◇ B45 E75 ◇
JENNINGS, R.E. *Can there be a natural deontic logic?* ◇ B45 ◇
JOBE, E.K. *Explanation, causality, and counterfactuals* ◇ A05 B45 ◇
JONES, A.J.I. & POERN, I. *Ideality, subideality and deontic logic* ◇ A05 B45 ◇
KALINOWSKI, G. *Sur le fondement des normes et des enonces normatifs: a propos des idees de von Wright et de Castaneda* ◇ A05 B45 ◇
MALINOWSKI, J. *On the number of quasimodal algebras* ◇ B45 ◇
MONTAGNA, F. *First results on the modal predicate logic of provability (Italian)* ◇ B45 ◇
NAGLE, M.C. & THOMASON, S.K. *The extensions of the modal logic* k5 ◇ B45 ◇
NUTE, D.E. *Permission* ◇ B45 ◇
PEREZ LARAUDOGOITIA, Y. *Modal interpretation of quantum mechanics (Spanish) (English summary)* ◇ B45 B51 ◇
PFEIFFER, H. *Some system of predicate tense logic* ◇ B45 ◇
PLIUSKEVICIENE, A. *Generalized disjunction and existence properties for the logic of provability (Russian)* ◇ B35 B45 ◇
PLIUSKEVICIUS, R. *A conservative extension of the quantified modal logic S5 (Russian) (English and Lithuanian summaries)* ◇ B45 ◇
PNUELI, A. *In transition from global to modular temporal reasoning about programs* ◇ B35 B45 B75 ◇
PNUELI, A. *Linear and branching structures in the semantics and logics of reactive systems* ◇ B35 B45 B75 ◇
POTTINGER, G. *Intension, designation, and extension* ◇ B45 ◇
REIF, J.H. & SISTLA, A.P. *A multiprocess network logic with temporal and spatial modalities* ◇ B45 ◇
RYBAKOV, V.V. *A criterion for admissibility of deduction rules in modal and intuitionistic logic (Russian)* ◇ B45 B55 ◇
RYBAKOV, V.V. *Bases of admissible rules of the modal system Grz and intuitionistic logic (Russian)* ◇ B45 B55 ◇
RYBAKOV, V.V. *Bases of admisssible rules of the logics S4 and Int (Russian)* ◇ B45 ◇
SAWAMURA, H. *Axiomatization of computer-oriented modal logic and decision procedure* ◇ B25 B45 ◇
SCEDROV, A. *Extending Goedel's modal interpretation to type theory and set theory* ◇ B15 B45 E70 F35 F50 ◇
SEGERBERG, K. *Routines* ◇ B45 ◇
SHAPIRO, S. (ED.) *Intensional mathematics* ◇ B45 ◇
SISTLA, A.P. & VARDI, M.Y. & WOLPER, P. *The complementation problem for Buechi automata with applications to temporal logic (extended abstract)* ◇ B45 D05 ◇
SMORYNSKI, C.A. *Self-reference and modal logic* ◇ B45 F30 F98 ◇
SMULLYAN, R.M. *Modality and self-reference* ◇ B45 F30 ◇

SMULLYAN, R.M. *Some principles related to Loeb's theorem* ◇ B45 F30 ◇
STIRLING, C. *A complete modal proof system for a subset of SCCS* ◇ B45 B75 ◇
STIRLING, C. *A complete compositional modal proof system for a subset of CCS* ◇ B45 B75 ◇
TANG, TONGGAO *A completeness theorem of first order temporal logic with equality (Chinese)* ◇ B45 ◇
TANG, TONGGAO *Systems of first-order temporal logic with equality (Chinese) (English summary)* ◇ B45 ◇
THOMASON, R.H. *Note on tense and subjunctive conditionals* ◇ B45 ◇
ULRICH, D. *A descending chain of incomplete extensions of implicational S5* ◇ B45 ◇
URSINI, A. *Decision problems for classes of diagonalizable algebras* ◇ B25 B45 C05 D35 ◇
VENKATESH, G. *A decision method for temporal logic based on resolution* ◇ B35 B45 ◇
VINCENZI, A. *Alcune proprieta della logica di Chang* ◇ B45 C40 ◇
WOODWARD, D. *Modal logic without essences* ◇ B45 ◇
WRIGHT VON, G.H. *Probleme des Erklaerens und Verstehens von Handlungen (English summary)* ◇ A05 B45 ◇
YOKATA, S. *General charakterization results on intuitionistic modal propositional logics* ◇ B45 ◇
YOURGRAU, P. *On the logic of indeterminist time. With a comment by Paul Fitzgerald* ◇ B45 ◇

B46 Relevance and entailment

1918

LEWIS, C.I. *A survey of symbolic logic* ◇ B45 B46 B98 ◇

1920

LEWIS, C.I. *Strict implication – an emendation* ◇ A05 B45 B46 ◇

1936

LEWIS, C.I. *Emch's calculus and strict implication* ◇ B45 B46 ◇

TANG, CAOZHEN *A paradox of Lewis's strict implication* ◇ B45 B46 ◇

TANG, CAOZHEN *The theorem "p⥽q.=.pq=p" and Huntingtons's relation between Lewis's strict implication and boolean algebra* ◇ B46 B75 G05 ◇

1938

TANG, CAOZHEN *Algebraic postulates and a geometric interpretation for Lewis calculus of strict implication* ◇ B45 B46 G25 ◇

1939

PARRY, W.T. *Modalities in the survey system of strict implication* ◇ B46 ◇

1948

HALLDEN, S. *A note concerning the paradoxes of strict implication and Lewis's system S1* ◇ B46 ◇

HALLDEN, S. *A question concerning a logical calculus related to Lewis' system of strict implication, which is of special interest for the study of entailment* ◇ B46 ◇

1953

BARCAN MARCUS, R. *Strict implication, deducibility, and the deduction theorem* ◇ B46 ◇

1955

SUGIHARA, T. *Strict implication free from implicational paradoxes* ◇ A05 B46 ◇

1956

ACKERMANN, W. *Begruendung einer strengen Implikation* ◇ B46 ◇

1957

LEMMON, E.J. & MEREDITH, C.A. & MEREDITH, D. & PRIOR, A.N. & THOMAS, I. *Calculi of pure strict implication* ◇ B46 ◇

WRIGHT VON, G.H. *The concept of entailment* ◇ A05 B46 ◇

1958

ACKERMANN, W. *Ueber die Beziehung zwischen strikter und strenger Implikation* ◇ B46 ◇

1959

ANDERSON, A.R. & BELNAP JR., N.D. *Modalities in Ackermann's "rigorous implication"* ◇ B46 ◇

KOL'MAN, EH. *On material and formal implication (Russian)* ◇ B46 ◇

SMILEY, T.J. *Entailment and deducibility* ◇ A05 B46 ◇

WRIGHT VON, G.H. *A note on entailment* ◇ A05 B46 ◇

1960

ANDERSON, A.R. *Completeness theorems for the system E of entailment and EQ of entailment with quantification* ◇ B46 ◇

ANDERSON, A.R. & BELNAP JR., N.D. & WALLACE, J.R. *Independent axiom schemata for the pure theory of entailment* ◇ B46 ◇

APOSTEL, L. *Game theory and the interpretation of deontic logic* ◇ B46 ◇

BELNAP JR., N.D. *Entailment and relevance* ◇ B46 ◇

BELNAP JR., N.D. *EQ and the first order functional calculus* ◇ B46 ◇

MARTIN, N.M. *Deduction and strict implication* ◇ B46 ◇

1961

ANDERSON, A.R. & BELNAP JR., N.D. *Enthymemes* ◇ B46 ◇

1962

ANDERSON, A.R. & BELNAP JR., N.D. *Tautological entailments* ◇ A05 B46 ◇

ANDERSON, A.R. & BELNAP JR., N.D. *The pure calculus of entailment* ◇ B46 ◇

1963

ANDERSON, A.R. & BELNAP JR., N.D. *First degree entailments* ◇ B46 ◇

ANDERSON, A.R. *Some open problems concerning the system* E *of entailment* ◇ B46 ◇

ASHBY, R.W. *Entailment and modality* ◇ A05 B46 ◇

BELNAP JR., N.D. *An analysis of questions: Preliminary report* ◇ A05 B46 B60 ◇

PRIOR, A.N. *The theory of implication* ◇ A05 B46 ◇

1964

BACON, J. *Entailment and the modal fallacy* ◇ B46 ◇

MAKSIMOVA, L.L. *On a set of axioms for the system of rigorous implication (Russian)* ◇ B46 ◇

NELSON, JOHN O. *A question of entailment* ◇ A05 B46 ◇

1965

BELNAP JR., N.D. & WALLACE, J.R. *A decision procedure for the system* $E_{\bar{I}}$ *of entailment with negation* ◇ B46 ◇

HOCKNEY, D.J. & WILSON, K. *In defence of a relevance condition* ◇ A05 B46 ◇
MAKINSON, D. *An alternative characterisation of first-degree entailment* ◇ B46 ◇

1966

MAKSIMOVA, L.L. *Formal deduction in the system of rigorous implication (Russian)* ◇ B46 ◇
POLLOCK, J.L. *The paradoxes of strict implication* ◇ A05 B46 ◇
SNYDER, D.P. *Models for logical entailment* ◇ B46 ◇

1967

BELNAP JR., N.D. *Intensional models for first degree formulas* ◇ B46 ◇
ISEKI, K. *On the classical propositional calculus of A.R.Anderson and N.D.Belnap* ◇ B46 ◇
LEWIS, C.I. *The structure of system of strict implication* ◇ B45 B46 ◇
MAKSIMOVA, L.L. *On models of the calculus E (Russian)* ◇ B46 ◇
MAKSIMOVA, L.L. *Some problems of the Ackermann calculus (Russian)* ◇ B46 ◇

1968

MAKSIMOVA, L.L. *The calculus of strict implication (Russian)* ◇ B46 ◇
MEYER, R.K. *An undecidability result in the theory of relevant implication* ◇ B46 D35 ◇
MEYER, R.K. *Entailment and relevant implication* ◇ B46 ◇

1969

BAYART, A. *Pour une logique de l'"entailment"* ◇ B46 ◇
DUNN, J.M. & MEYER, R.K. *E, R and γ* ◇ B46 ◇
LEMMON, E.J. & MEREDITH, C.A. & MEREDITH, D. & PRIOR, A.N. & THOMAS, I. *Calculi of pure strict implication* ◇ B46 ◇
ROUTLEY, R. & ROUTLEY, V. *A fallacy of modality* ◇ B46 ◇

1970

DUNN, J.M. *Algebraic completeness results for R-mingle and its extensions* ◇ B46 ◇
DUNN, J.M. *Extensions of RM and LC* ◇ B46 ◇
MAKSIMOVA, L.L. *E-theories (Russian)* ◇ B46 ◇
MEYER, R.K. *R_i - the bounds of finitude* ◇ B46 ◇
MEYER, R.K. *E and S4* ◇ B46 ◇
PAULI, T. *Some different kinds of entailment relations (Swedish)* ◇ B46 ◇
SMIRNOV, V.A. *The so-called paradoxes of material implication, and systems of logic with a concept of strict implication (Russian)* ◇ B46 ◇
WILCOX, W.C. *On infinite matrices and the paradoxes of material implication* ◇ A05 A10 B46 ◇

1971

FILLMORE, C.J. *Entailment rules in a semantic theory* ◇ A05 B46 ◇
KNOX JR., J. *Material implication and "if...then"* ◇ A05 B46 ◇
MAKSIMOVA, L.L. *On interpretation and separation theorems for the calculi E and R (Russian)* ◇ B46 ◇
MEYER, R.K. *Entailment* ◇ B46 ◇
MEYER, R.K. *On coherence in modal logics* ◇ B46 ◇
TAMURA, S. *The implicational fragment of R-mingle* ◇ B46 ◇
URQUHART, A.I.F. *Completeness of weak implication* ◇ B46 ◇

1972

BUNGE, MARIO (ED.) *Proceedings of the First Symposium on Exact Philosophy* ◇ B46 B97 ◇
CHELLAS, B.F. *Notions of relevance* ◇ A05 B46 ◇
CHIDGEY, J.R. & PARKS, R.Z. *Necessity and ticket entailment* ◇ B46 ◇
DUNN, J.M. *A modification of Parry's analytic implication* ◇ B46 ◇
FOULKES, P. *The logic of "and" in the anatomy of proof: a new model for entailment* ◇ B46 ◇
GABBAY, D.M. *A general theory of the conditional in terms of a ternary operator* ◇ B46 ◇
GINSBERG, M. *The entailment-presupposition relationship* ◇ A05 B46 B65 ◇
KRON, A. *A note on E* ◇ B46 ◇
LEBLANC, H. (ED.) *Truth, syntax and modality* ◇ A05 B46 B97 ◇
MEYER, R.K. & ROUTLEY, R. *Algebraic analysis of entailment I* ◇ B46 ◇
MEYER, R.K. *On relevantly derivable disjunctions* ◇ B46 ◇
MEYER, R.K. & ROUTLEY, R. *The semantics of entailment II,III* ◇ B46 ◇
MINTS, G.E. *Cut-elimination theorem for relevant logics (Russian) (English summary)* ◇ B46 ◇
ROUTLEY, R. *A semantical analysis of implicational system I and of the first degree of entailment* ◇ B46 ◇
ROUTLEY, R. & ROUTLEY, V. *The semantics of first degree entailment* ◇ B46 ◇
ROUTLEY, R. *Vrendenduin's system of strict implication* ◇ B46 ◇
URQUHART, A.I.F. *Semantics for relevant logics* ◇ B46 ◇

1973

ANDERSON, A.R. *An intensional interpretation of truth-values* ◇ B46 ◇
BACON, J. *Kripke's deontic semantics again* ◇ B46 ◇
CHELLAS, B.F. *Notions of relevance.Comments on Leblanc's paper* ◇ A05 B46 ◇
GABBAY, D.M. *Applications of Scott's notion of consequence to the study of general binary intensional connectives and entailment* ◇ B22 B46 ◇
KANGER, S. *Entailment* ◇ B46 ◇
KRON, A. *Deduction theorems for relevant logics* ◇ B46 ◇
MAKSIMOVA, L.L. *A semantics for the calculus E of entailment* ◇ B46 ◇
MEYER, R.K. & ROUTLEY, R. *An undecidable relevant logic* ◇ B46 D35 D40 ◇
MEYER, R.K. & ROUTLEY, R. *Classical relevant logics I (Polish and Russian summaries)* ◇ B46 ◇
MEYER, R.K. *Conservative extension in relevant implication (Polish and Russian summaries)* ◇ B46 ◇
MEYER, R.K. *Intuitionism, entailment, negation* ◇ B46 F50 ◇

Meyer, R.K. & Routley, R. *The semantics of entailment I* ◇ B46 ◇
Urquhart, A.I.F. *A semantical theory of analytic implication* ◇ B46 ◇

1974

Cleave, J.P. *An account of entailment based on classical semantics* ◇ B46 ◇
Dunn, J.M. & Leblanc, H. & Meyer, R.K. *Completeness of relevant quantification theories* ◇ B46 ◇
Fine, K. *Models for entailment* ◇ B46 ◇
Meyer, R.K. & Routley, R. *Classical relevant logics II* ◇ B46 ◇
Meyer, R.K. & Routley, R. *E is a conservative extension of* $E_{\bar{I}}$ ◇ B46 ◇
Meyer, R.K. *Entailment is not strict implication* ◇ B46 ◇
Meyer, R.K. *New axiomatics for relevant logics. I* ◇ B46 ◇
Routley, R. *A rival account of logical consequence* ◇ A05 B46 ◇
Routley, R. *Semantical analyses of propositional systems of Fitch and Nelson* ◇ B25 B46 C90 F50 ◇
Sugihara, T. *Modalities in the entailment logic* ◇ B46 ◇

1975

Anderson, A.R. & Belnap Jr., N.D. *Entailment. The logic of relevance and necessity I* ◇ A05 B46 B98 ◇
Briskman, L. *Classical semantics and entailment* ◇ B46 ◇
Chidgey, J.R. *Independence* ◇ B46 ◇
Dunn, J.M. *Consecution formulation of positive R* ◇ B46 ◇
Dunn, J.M. *Extensions of RM* ◇ B46 ◇
Dunn, J.M. & Meyer, R.K. *The admissibility of* (γ) *in E* ◇ B46 ◇
Dunn, J.M. *The algebra of R* ◇ B46 G25 ◇
Fraassen van, B.C. *Facts and tautological entailments* ◇ A05 B46 ◇
Geach, P.T. *On entailment* ◇ B46 ◇
James, E.W. *On preserving entailment* ◇ B46 ◇
McCall, S. *Connexive implication* ◇ B46 ◇
Meyer, R.K. *Conservative extensions in R* ◇ B46 ◇
Meyer, R.K. *On conserving positive logics I* ◇ B46 ◇
Meyer, R.K. *On relevantly derivable disjunctions* ◇ B46 ◇
Meyer, R.K. *Relevance is not reducible to modality* ◇ B46 ◇
Meyer, R.K. *Sugihara is a characteristic matrix for RM* ◇ B46 ◇
Meyer, R.K. & Routley, R. *Towards a general semantical theory of implication and conditionals. I. Systems with normal conjunctions and disjunctions and aberrant and normal negations* ◇ B46 ◇
Parks, R.Z. *Indefinability of necessity in* $R_{\rightarrow}$ ◇ B46 ◇
Parks, R.Z. *Solution of* $T_{\rightarrow}$ ◇ B46 ◇
Patryas, W. *Concretizationand entailment (Comments on L. Nowak's article on "Idealizational laws and explanation".)* ◇ A05 B46 ◇
Ruzsa, I. *Two variants of the system of entailment* ◇ B46 ◇

1976

Almog, J. *Representation of the fuzziness of the concept of relevant implication within a multi-valued model theory* ◇ B46 E72 ◇
Almog, J. *The interrelations between fuzzy logics and the concept of relevant intailment* ◇ A05 B46 E72 ◇
Barker, J.A. *Presupposition and entailment* ◇ B46 ◇
Collier, K.W. *The story of NR and E* ◇ B46 ◇
Dunn, J.M. *A variation on the binary semantics for RM* ◇ B46 ◇
Dunn, J.M. *A Kripke-style semantics for R-mingle using a binary accessibility relation* ◇ B46 ◇
Dunn, J.M. *Intuitive semantics for first-degree entailments and coupled trees* ◇ B46 ◇
Dunn, J.M. *Quantification and RM* ◇ B46 ◇
Gabbay, D.M. *Investigations in modal and tense logics with applications to problems in philosophy and linguistics* ◇ A05 B25 B46 C90 ◇
Johnson, F.A. *A three-valued interpretation for a relevance logic* ◇ B46 ◇
Kalman, J.A. *Computer studies of* $T_{\rightarrow} - W - I$ ◇ B35 B46 ◇
Kron, A. *Deduction theorems for T, E and R reconsidered* ◇ B46 ◇
Lapara, N. *Semantics for a natural notion of entailment* ◇ B46 ◇
Lewy, C. *Meaning and modality* ◇ A05 B46 ◇
Meyer, R.K. *A general Gentzen system for implicational calculi* ◇ B46 ◇
Meyer, R.K. *Ackermann, Takeuti, und Schnitt:* γ *for higher-order relevant logic* ◇ B46 F05 F35 ◇
Meyer, R.K. *On* $\exists p(\exists q Aq \rightarrow Ap)$ ◇ B46 ◇
Meyer, R.K. *Relevant arithmetic* ◇ B46 F30 ◇
Meyer, R.K. *Two questions from Anderson and Belnap* ◇ B46 ◇
Parry, W.T. *Abstract: Entailment: Analytic implication vs. the entailment of Anderson & Belnap* ◇ B46 ◇
Parry, W.T. *Comparison of entailment theories* ◇ B46 ◇
Powers, L. *Latest on a logic problem* ◇ A05 B46 ◇
Powers, L. *On P-W* ◇ B46 ◇

1977

Belnap Jr., N.D. & McRobbie, M.A. *Relevant analytic tableaux* ◇ B46 ◇
Copeland, B.J. *The system RS. An amendation of E* ◇ B46 ◇
Deutsch, H. *Relevant analytic entailment* ◇ B46 ◇
Facione, P.A. *The entailment operator* ◇ B46 ◇
Johnson, F.A. *A natural deduction relevance logic* ◇ B46 ◇
Kielkopf, C.F. *Formal sentential entailment* ◇ B46 ◇
Meyer, R.K. & Routley, R. *Extensional reduction I* ◇ B46 ◇
Meyer, R.K. *First degree formulas in Curry's LD* ◇ B46 F50 ◇
Meyer, R.K. *On a purported implication* ◇ B46 ◇
Meyer, R.K. *S5--the poor man's connexive implication* ◇ B46 ◇
Meyer, R.K. & Routley, R. *Towards a general semantical theory of implication and conditionals. II.*

Improved negation theory and propositional identity ◇ B46 ◇
PARKS, R.Z. *The theory of limp implication* ◇ B46 ◇
POTTINGER, G. *A new classical relevant logic* ◇ B46 ◇
TOKARZ, M. *Deduction theorems for RM and its extensions* ◇ B46 G25 ◇
TOKARZ, M. *Degrees of maximality of three- and four-valued RM-extensions* ◇ B46 ◇

1978

BYRD, M. & HENRY, DENNIS *Sugihara's criterion and some structural parallels between* $E_{\rightarrow}$ *and* $S3_{\rightarrow}$ ◇ B46 ◇
GAERDENFORS, P. *On the logic of relevance* ◇ A05 B46 ◇
KIELKOPF, C.F. *A favorable mark for strict implication* ◇ A05 B46 ◇
KRON, A. *Decision procedures for two positive relevance logics* ◇ B25 B46 ◇

1979

ARRINGTON, R.L. *Criteria and entailment* ◇ A05 B46 ◇
BELNAP JR., N.D. & MCROBBIE, M.A. *Relevant analytic tableaux* ◇ B46 ◇
BUNDER, M.W. *A more relevant relevance logic* ◇ B46 ◇
BYRD, M. & HENRY, DENNIS *On defining necessity in terms of entailment* ◇ B46 ◇
CHIDGEY, J.R. *On the non-availability of Dawson-modeling into certain relevance alethic modal logics* ◇ B46 ◇
COPELAND, B.J. *On when a semantics is not a semantics: some reasons for disliking the Routley-Meyer semantics for relevance logic* ◇ A05 B46 ◇
DANIELS, C.B. & FREEMAN, J.B. *A second-order relevance logic with modality* ◇ B46 ◇
DUNN, J.M. *R-mingle and beneath* ◇ B46 ◇
DUNN, J.M. *Relevant Robinson's arithmetic* ◇ B46 ◇
HUNG, H.C. *Entailment and proof* ◇ A05 B46 ◇
MCROBBIE, M.A. & MEYER, R.K. *A note on the admissibility of cut in relevant tableau systems* ◇ B46 ◇
MCROBBIE, M.A. & MEYER, R.K. *Firesets and relevant implication* ◇ B46 ◇
MEYER, R.K. *Career induction stops here (and here = 2)* ◇ A05 B46 ◇
PABION, J.F. *Beth's tableaux for relevant logic* ◇ B46 ◇
POTTINGER, G. *A new classical relevance logic* ◇ B46 ◇
POTTINGER, G. *On analysing relevance constructively* ◇ B46 ◇
ROUTLEY, R. *Alternative semantics for quantified first degree relevant logic* ◇ B46 ◇
ROUTLEY, R. *Repairing proofs of Arrow's general impossibility theorem and enlarging the scope of the theorem* ◇ B46 ◇
SIDORENKO, E.A. *Some valid extensions of relevant systems (Russian)* ◇ B46 ◇
SMIRNOV, V.A. *Formal inference, deduction theorems and theories of implication (Russian)* ◇ B46 ◇
TOKARZ, M. *Deduction theorems for RM and its extensions* ◇ B46 G25 ◇
TOKARZ, M. *The existence of matrices strongly adequate for E, R and their fragments* ◇ B46 ◇
VOJSHVILLO, E.K. *Natural variants of certain systems of relevant logic (on the problem of explication of the notions of logical consequence and conditional relations) (Russian)* ◇ B46 ◇
WALTON, D.N. & WOODS, J. *Circular demonstration and von Wright-Geach entailment* ◇ B46 ◇

1980

BELNAP JR., N.D. & DUNN, J.M. & GUPTA, A. *A consecutive calculus for positive relevant implication with necessity* ◇ A05 B46 ◇
BUNDER, M.W. *Quantified relevance logic and generalised restricted generality* ◇ B46 ◇
CLARK, M. *The equivalence of tautological and "strict" entailment: Proof of an amended conjecture of Lewy's* ◇ B46 ◇
COPELAND, B.J. *The trouble Anderson and Belnap have with relevance* ◇ A05 B46 ◇
DALE, A.J. *Smiley's matrices and Dunn's semantics for tautological entailment* ◇ B46 ◇
DUNN, J.M. *A sieve for entailments* ◇ B46 ◇
GIAMBRONE, S. & MEYER, R.K. R_{+} *is contained in* T_{+} ◇ B46 ◇
HANSON, W.H. *First degree entailments and information* ◇ A05 B46 ◇
ISEMINGER, G. *Is relevance necessary for validity?* ◇ A05 B46 ◇
KRON, A. *Gentzen formulations of two positive relevance logics* ◇ B46 ◇
LOPARIC, A. & ROUTLEY, R. *Semantics for quantified relevant logics without replacement* ◇ B46 ◇
LOPTSON, P.J. *Q, entailment, and the Parry property* ◇ B46 ◇
MEYER, R.K. *Career induction for quantifiers* ◇ B46 ◇
MEYER, R.K. *Sentential constants in relevant implication* ◇ B46 ◇
PEACOCKE, C. *Causal modalities and realism* ◇ B46 B65 ◇
PRIEST, G. *Sense, entailment and modus ponens* ◇ A05 B46 ◇
ROUTLEY, R. *Problems and solutions in the semantics of quantified relevant logics. I* ◇ B46 ◇
SANCHEZ P., J. *Logical entailment and semantic intension of boolean formulas in relevant logic (Russian)* ◇ B46 ◇
TENNANT, N. *A proof-theoretic approach to entailment* ◇ B46 F05 ◇
TOKARZ, M. *Essays in matrix semantics of relevant logics* ◇ B46 G25 ◇
VOJSHVILLO, E.K. *Logical entailment and implication (semantic analysis) (Russian)* ◇ B46 ◇
VOJSHVILLO, E.K. *The concepts of intensional information and the intensional relation of logical entailment (intensional analysis) (Russian)* ◇ B46 ◇
WOJCICKI, R. *Entailment semantics for T* ◇ B46 ◇

1981

BELNAP JR., N.D. & DUNN, J.M. *Entailment and the disjunctive syllogism* ◇ B46 ◇

BELNAP JR., N.D. *Modal and relevance logics: 1977* ◇ B46 ◇

BURGESS, J.P. *Relevance: A fallacy?* ◇ A05 B46 ◇

CHARLWOOD, G. *An axiomatic version of positive semilattice relevance logic* ◇ B46 ◇

DIAZ, M.R. *Topics in the logic of relevance* ◇ A05 B46 ◇

DOSEN, K. *A reduction of classical propositional logic to the conjunction-negation fragment of an intuitionistic relevant logic* ◇ B46 B55 F50 ◇

DZIOBIAK, W. *On matrices characteristic of relevant logics* ◇ B46 ◇

FRAASSEN VAN, B.C. *Essences and laws of nature* ◇ A05 B46 ◇

GIAMBRONE, S. & MEYER, R.K. *Strict implication in T* ◇ B46 ◇

MCARTHUR, R.P. *Anderson's deontic logic and relevant implication* ◇ B46 ◇

MEYER, R.K. *Almost Skolem forms for relevant (and other) logics* ◇ B46 ◇

OVER, D. *Game theoretical semantics and entailment* ◇ B46 ◇

URCHS, M.P. *Kripke-style semantics for Jaskowski's system Q_f* ◇ B46 B53 ◇

1982

BRADY, R.T. *Completeness proofs for the systems RM3 and BN4* ◇ B46 ◇

BRADY, R.T. & MEYER, R.K. & PLUMWOOD, V. & ROUTLEY, R. *Relevant logics and their rivals. Part I* ◇ B46 ◇

DALE, A.J. *Material equivalence and tautological entailment* ◇ A05 B46 ◇

DUNN, J.M. *Anderson and Belnap, and Lewy on entailment* ◇ B46 ◇

EBERLE, R.A. *The relativization of truth to functions: its expressive power and ontic import* ◇ B46 ◇

GIAMBRONE, S. *Gentzen systems and decision procedures for relevant logics* ◇ B46 ◇

MARTIN, E.P. & MEYER, R.K. & ROUTLEY, R. & ROUTLEY, V. *On the philosophical bases of relevant logic semantics* ◇ A05 B46 ◇

MARTIN, E.P. & MEYER, R.K. *Solution to the P-W problem* ◇ B46 ◇

SIDORENKO, E.A. *On extensions of E* ◇ B46 ◇

VOJSHVILLO, E.K. *Semantics of generalized state descriptions* ◇ B46 ◇

1983

BURGESS, J.P. *Common sense and "relevance"* ◇ A05 B46 ◇

CHRISTENSEN, D. *Glymour on evidential relevance* ◇ B46 ◇

COPELAND, B.J. *A rejoinder to: "On the philosophical bases of relevant logic semantics"* ◇ A05 B46 ◇

COPELAND, B.J. *Pure semantics and applied semantics: a response to R.Routley, V.Routley, R.K.Meyer and E.P.Martin: "On the philosophical bases of relevant logic semantics" (J. Non-Classical Logic 1 (1982), no. 1, 71–105)* ◇ A05 B46 ◇

CZELAKOWSKI, J. *Some theorems on structural entailment relations* ◇ B46 ◇

DZIOBIAK, W. *There are $2^{\aleph_0}$ logics with the relevance principle between R and RM* ◇ B46 ◇

MEYER, R.K. *A note on R-matrices* ◇ B46 ◇

MEYER, R.K. & ROUTLEY, R. *Relevant logics and their semantics remain viable and undamaged by Lewis's equivocation charge* ◇ B46 ◇

MORTENSEN, C. *Relevance and verisimilitude* ◇ B46 ◇

PORTE, J. *Antitheses in systems of relevant implication* ◇ B46 ◇

READ, S. *A paradox of relevant implication* ◇ B46 ◇

READ, S. *Burgess on relevance: a fallacy indeed* ◇ B46 ◇

SIDORENKO, E.A. *Logical entailment and conditional propositions (Russian)* ◇ B46 ◇

SIDORENKO, E.A. *The strong proof from hypotheses and conditionals: some theorems of deduction for relevant systems* ◇ B46 ◇

SKVORTSOV, D.P. *The structure of extensions of the propositional fragment of the Ackermann logic (Russian)* ◇ B46 ◇

SZABO, M.E. *The continuous realizability of entailment* ◇ B46 ◇

URQUHART, A.I.F. *Relevant implication and projective geometry* ◇ B46 ◇

VOJSHVILLO, E.K. *A decision procedure for the system E (of entailment) I* ◇ B25 B46 ◇

1984

AVRON, A. *Relevant entailment – semantics and formal systems* ◇ B46 ◇

BRADY, R.T. *Natural deduction systems for some quantified relevant logics* ◇ B46 C90 ◇

BRADY, R.T. & GIAMBRONE, S. & MEYER, R.K. *Where gamma fails* ◇ B46 ◇

BURGESS, J.P. *Read on relevance: a rejoinder* ◇ B46 ◇

MEYER, R.K. & MORTENSEN, C. *Inconsistent models for relevant arithmetics* ◇ B46 ◇

OHLBACH, H.-J. & WRIGHTSON, G. *Solving a problem in relevance logic with an automated theorem prover* ◇ B35 B46 ◇

POPOV, V.M. *Semantical and syntactical analysis of an implicative-negative fragment of the system RM (Russian)* ◇ B46 ◇

ROUTLEY, R. *The American plan completed: Alternative classical-style semantics, without stars, for relevant and paraconsistent logics* ◇ B46 B53 ◇

SIDORENKO, E.A. *Relevance and amplification (Russian)* ◇ B46 ◇

SLANEY, J.K. *A metacompleteness theorem for contraction-free relevant logics* ◇ B46 ◇

SMOLENOV, KH. *Truthfulness and nontrivial contraction-free relevant logics* ◇ B46 ◇

URQUHART, A.I.F. *The undecidability of entailment and relevant implication* ◇ B46 D35 G10 ◇

1985

BATENS, D. & BENDEGEM VAN, J.P. *Relevant derivability and classical derivability in Fitch-style and axiomatic formulations of relevant logics* ◇ B46 ◇

DEUTSCH, H. *A note on the decidability of a strong relevant logic* ◇ B25 B46 ◇
DEUTSCH, H. *Relevance and conformity* ◇ B46 ◇
DEUTSCH, H. *Relevance and first-degree entailments* ◇ B46 ◇
GIAMBRONE, S. *TW_+ and RW_+ are decidable* ◇ B25 B46 ◇
GIAMBRONE, S. *On purported Gentzen formulations of two positive relevant logics* ◇ B46 F07 ◇
KRABBE, E.C.W. *Noncumulative dialectical models and formal dialectics* ◇ A05 B46 F99 ◇
KRON, A. *A constructive proof of a theorem in relevance logic* ◇ B46 ◇
MENDEZ, J.M. *Systems with the converse Ackermann property* ◇ B46 ◇
MEYER, R.K. *A farewell to entailment* ◇ B46 ◇
MEYER, R.K. & MORTENSEN, C. *Relevant quantum arithmetic* ◇ B46 F30 ◇
SLANEY, J.K. *3088 varieties: a solution to the Ackermann constant problem* ◇ B46 C05 ◇
SWART DE, H.C.M. *Gentzen-type or Beth-type systems,constructive completeness proofs and practical decision procedures (with special attention to relevance logic)* ◇ B46 ◇
WEINGARTNER, P. *A simple relevance criterion for natural language and its semantics* ◇ B46 ◇

Subject Index

B48 Probability and inductive logic

1883

PEIRCE, C.S. *A theory of probable inference. Note b. The logic of relatives* ◇ A05 B48 ◇

1924

MORGAN DE, A. *Formal logic; or, the calculus of inference, necessary and probable* ◇ A10 B45 B48 ◇

NICOD, J. *The logical problem of induction (French)* ◇ B48 ◇

1932

REICHENBACH, H. *Wahrscheinlichkeitslogik* ◇ A05 B30 B48 ◇

1934

ZAWIRSKI, Z. *Bedeutung der mehrwertigen Logik fuer die Erkenntnis und ihr Zusammenhang mit der Wahrscheinlichkeitsrechnung (Polnisch)* ◇ B48 ◇

1935

REICHENBACH, H. *Wahrscheinlichkeitslogik* ◇ A05 B48 ◇

1937

FEYS, R. *Les logiques nouvelles des modalites* ◇ B20 B48 ◇

1938

HEMPEL, C.G. *On the logical form of probability-statements* ◇ A05 B48 ◇

1939

BOREL, E. *Valeur pratique et philosophie des probabilites* ◇ A05 B48 ◇

EVANS, H.P. & KLEENE, S.C. *A postulational basis for probability* ◇ B30 B48 ◇

REICHENBACH, H. *Ueber die semantische und die Objekt-Auffassung von Wahrscheinlichkeitsausdruecken* ◇ B30 B48 ◇

1940

HOSIASSON-LINDENBAUM, J. *On confirmation* ◇ B48 ◇

1945

CARNAP, R. *On inductive logic* ◇ A05 B48 ◇

HELMER, O. & OPPENHEIM, P. *A syntactical definition of probability and of degree of confirmation* ◇ B48 ◇

HEMPEL, C.G. & OPPENHEIM, P. *A definition of "degree of confirmation"* ◇ B48 ◇

HEMPEL, C.G. *Studies in the logic of confirmation* ◇ A05 B48 ◇

1946

CARNAP, R. *Remarks on induction and truth* ◇ A05 B48 ◇

1947

CARNAP, R. *On the application of inductive logic* ◇ A05 B48 ◇

WILLIAMS, D.C. *The ground of induction* ◇ A05 B48 ◇

1948

POLYA, G. *On patterns of plausible inference* ◇ A05 B48 ◇

1950

CARNAP, R. *Logical foundations of probability* ◇ A05 B48 ◇

CARNAP, R. *Testability and meaning* ◇ A05 B48 ◇

1951

CARNAP, R. *The problem of relations in inductive logic* ◇ B48 ◇

MENGER, K. *Probabilistic theories of relations* ◇ B48 E07 ◇

WRIGHT VON, G.H. *A treatise on induction and probability* ◇ A05 B48 ◇

1952

CARNAP, R. *The continuum of inductive methods* ◇ A05 B48 ◇

1953

BAR-HILLEL, Y. *A note on comparative inductive logic* ◇ A05 B48 ◇

BRAITHWAITE, R.B. *Scientific explanation: A study of the function of theory, probability and law in science* ◇ A05 B48 ◇

CARNAP, R. *On the comparative concept of confirmation* ◇ A05 B48 ◇

DANTZIG VAN, D. *Carnap's foundation of probability theory* ◇ B48 ◇

JUHOS VON, B. *Wahrscheinlichkeitsschluesse als syntaktische Schlussformen* ◇ B48 ◇

1954

CHURCH, A. *On Carnap's analysis of statements of assertion and belief* ◇ A05 B48 ◇

DESTOUCHES, J.-L. *Allgemeine Theorie der Voraussagen* ◇ A05 B48 ◇

POPPER, K.R. *Degree of confirmation* ◇ A05 B48 ◇

1955

BUSTO DEL, E.H. *Probability and Carnap's inductive logic (Spanish)* ◇ B48 ◇

CHISHOLM, R.M. *Law statements and counterfactual inference* ◇ B48 B65 ◇

GRENIEWSKI, H. *Elements of the logic of induction (Russian)* ◇ A05 B48 ◇

KEMENY, J.G. *Fair bets and inductive probabilities* ◇ B48 ◇

LEHMAN, R.S. *On confirmation and rational betting* ◇ A05 B48 ◇
RUBIN, H. & SUPPES, P. *A note on two-place predicates and fitting sequences of measure functions* ◇ B48 ◇
SHIMONY, A. *Coherence and the axioms of confirmation* ◇ A05 B48 ◇

1956

CHISHOLM, R.M. *Sentences about believing* ◇ B48 ◇
NEUMANN VON, J. *Probabilistic logics and the synthesis of reliable organisms from unreliable components* ◇ B48 ◇

1957

CZERWINSKI, Z. *The problem of probabilistic justification of enumerative induction (Polish) (Russian and English summaries)* ◇ A05 B48 ◇
KOKOSZYNSKA, M. *On "good" and "bad" induction (Polish) (Russian and English summaries)* ◇ B48 ◇
LEBLANC, H. *On logically false evidence statements* ◇ A05 B48 ◇
LUSZCZEWSKA-ROMAHNOWA, S. *Induction and probability (Polish) (Russian and English summaries)* ◇ B48 ◇
WRIGHT VON, G.H. *The logical problem of induction (2nd edition)* ◇ A05 B48 ◇

1958

BARKER, S.F. *Induction and hypothesis, A study of the logic of confirmation* ◇ B48 ◇
CZERWINSKI, Z. *On the relation of statistical inference to traditional induction and deduction (Polish and Russian summaries)* ◇ B48 ◇
HARARY, F. *Note on Carnap's relational asymptotic relative frequencies* ◇ A05 B48 ◇
HERMES, H. *Zum Einfachheitsprinzip in der Wahrscheinlichkeitsrechnung* ◇ A05 B48 D20 D80 ◇
MARTIN, R.M. *A formalization of inductive logic* ◇ A05 B48 ◇

1959

LEBLANC, H. *On chances and estimated chances of being true* ◇ B48 ◇

1960

CZERWINSKI, Z. *Degree of confirmation and critical region* ◇ B48 ◇
CZERWINSKI, Z. *Enumerative induction and the theory of games (Polish and Russian summaries)* ◇ A05 B48 ◇
CZERWINSKI, Z. *On the concept of cause and Mill's method (Polish) (Russian and English summaries)* ◇ B48 ◇
FREUDENTHAL, H. *Models in applied probability* ◇ A05 B48 ◇
GAIFMAN, H. *Probability models and the completeness theorem* ◇ B48 C90 ◇
GIEDYMIN, J. *Confirmation, critical region and empirical content of hypotheses* ◇ B48 ◇
LEBLANC, H. *On a recent allotment of probabilities to open and closed sentences* ◇ B48 ◇
LEBLANC, H. *On requirements for conditional probability functions* ◇ B05 B30 B48 ◇
SPACEK, A. *Statistical estimation of provability in boolean logic* ◇ B48 G05 ◇
SZANIAWSKI, K. *A note on confirmation of statistical hypotheses* ◇ B48 ◇
SZANIAWSKI, K. *Some remarks concerning the criterion of rational decision making (Polish and Russian summaries)* ◇ A05 B48 ◇

1961

KYBURG JR., H.E. *Probability and the logic of rational belief* ◇ A05 B48 ◇
LEBLANC, H. *A new interpretation of c(h,e)* ◇ A05 B48 ◇
LEBLANC, H. *Probabilities as truth-value estimates* ◇ B48 ◇
SPACEK, A. *Statistical estimation of semantic provability* ◇ B48 G15 ◇
SZANIAWSKI, K. *A method of deciding between N statistical hypotheses (Polish and Russian summaries)* ◇ B48 ◇
SZANIAWSKI, K. *On some basic patterns of statistical inference (Polish and Russian summaries)* ◇ B48 ◇
ZIEMBA, Z. *Rational belief, probability and the justification of induction (Polish) (Russian and English summaries)* ◇ A05 B48 ◇

1962

ADAMS, E.W. *On rational betting systems* ◇ B48 ◇
CARNAP, R. *The aim of inductive logic* ◇ A05 B48 ◇
GOOD, I.J. *Subjective probability as the measure of a non-measurable set* ◇ B48 E75 ◇
KATZ, JERROLD J. *The problem of induction and its solution* ◇ B48 ◇
KRAMER, H.P. *A note on the self-consistency of definitions of generalization and inductive inference* ◇ B48 ◇
LEBLANC, H. *Statistical and inductive probabilities* ◇ A05 B48 ◇
MUROGA, S. *Majority logic and problems of probabilistic behaviour* ◇ B48 ◇
SZANIAWSKI, K. *On the justification of inductive rules of inference (Polish) (Russian and English summaries)* ◇ B48 ◇

1963

LEHRER, K. *Descriptive completeness and inductive methods* ◇ B48 ◇
LOS, J. *Semantic representation of the probability of formulas in formalized theories (Polish summary)* ◇ A05 B10 B48 B60 ◇
NAGEL, E. *Carnap's theory of induction* ◇ A05 B48 ◇
PUTNAM, H. *"Degree of confirmation" and inductive logic* ◇ A05 B48 ◇

1964

ADAMS, E.W. *On rational betting systems (continuation)* ◇ B48 ◇
GAIFMAN, H. *Concerning measures on first order calculi* ◇ B48 C90 ◇
MORTIMER, H. *Probabilistic definition as exemplified by a definition of genotype (Polish) (Russian and English summaries)* ◇ B48 ◇
TOERNEBOHM, H. *Information and confirmation* ◇ A05 B48 B60 ◇

1965

BERNAYS, P. & DOCKX, S. (EDS.) *Information and prediction in science* ◇ A05 B48 ◇
CURRY, H.B. *Remarks on inferential deduction* ◇ B48 ◇
HACKING, I. *Logic of statistical inference* ◇ A05 B48 ◇
HERMES, H. *Induktion und Wahrscheinlichkeit* ◇ A05 B48 ◇
HINTIKKA, K.J.J. *On a combined system of inductive logic* ◇ A05 B48 ◇
HINTIKKA, K.J.J. *Towards a theory of inductive generalization* ◇ A05 B48 ◇
JEFFREY, R.C. *The logic of decision* ◇ B48 B98 ◇
JONES, ROBERT M. *Discussion: the non-reducibility of Koopman's theorems of probability in Carnaps system for MC* ◇ A05 B48 ◇
KYBURG JR., H.E. *Probability, rationality, and a rule of detachment* ◇ A05 B48 ◇
SCHOCK, R. *On probability logics* ◇ B48 ◇

1966

ADAMS, E.W. *Probability and the logic of conditionals* ◇ B48 ◇
BLACK, M. *Notes on the "paradoxes of confirmation"* ◇ B48 ◇
HILPINEN, R. & HINTIKKA, K.J.J. *Knowledge, acceptance, and inductive logic* ◇ A05 B48 ◇
HINTIKKA, K.J.J. *A two-dimensional continuum of inductive methods* ◇ B48 ◇
HINTIKKA, K.J.J. & SUPPES, P. (EDS.) *Aspects of inductive logic* ◇ A05 B48 ◇
HINTIKKA, K.J.J. & PIETARINEN, J. *Semantic information and inductive logic* ◇ B48 ◇
KRAUSS, P.H. & SCOTT, D.S. *Assigning probabilities to logical formulas* ◇ B48 C75 C90 ◇
SKYRMS, B. *Choice and chance* ◇ A05 B48 B98 ◇
TUOMELA, R. *Inductive generalization in an ordered universe* ◇ A05 B48 ◇
WALK, K. *Simplicity, entropy and inductive logic* ◇ B48 ◇
WRIGHT VON, G.H. *The paradoxes of confirmation* ◇ A05 B48 ◇

1967

MORTIMER, H. *Consistency condition of two probabilistic postulates (Polish) (English and Russian summaries)* ◇ B48 ◇
PYATNITSYN, B.N. *The problem of the semantics of probabilistic and inductive logic (Russian)* ◇ A05 B48 ◇
RESCHER, N. *Semantic foundations for the logic of preference. With comments by A. R. Anderson and R. Ackermann* ◇ A05 B48 ◇
RUZAVIN, G.I. *The semantic conception of inductive logic (Russian)* ◇ A05 B48 ◇
SUSZKO, R. *A proposal concerning the formulation of the infinitistic axiom in the theory of logical probability* ◇ B48 C90 ◇
VETTER, H. *Wahrscheinlichkeit und logischer Spielraum* ◇ A05 B48 ◇

1968

BROAD, C.D. *Induction, probability and causation: Selected papers* ◇ A05 B48 ◇
CARNAP, R. *Inductive logic and inductive intuition* ◇ A05 B48 ◇
CHUDNOVSKY, D.V. *Probability on first order logics (Ukrainian) (Russian and English summaries)* ◇ B48 C90 ◇
FENSTAD, J.E. *The structure of logical probabilities* ◇ B48 ◇
HILPINEN, R. *Rules of acceptance and inductive logic* ◇ A05 B48 ◇
HINTIKKA, K.J.J. *Induction by enumeration and induction by elimination* ◇ A05 B48 ◇
LAKATOS, I. (ED.) *The problem of inductive logic* ◇ A05 B48 ◇
SAERNDAL, C.-E. *Some aspects of Carnap's theory of inductive inference* ◇ A05 B48 ◇
VENDLER, Z. *Adjectives and nominalizations* ◇ A05 B48 ◇

1969

BOTEZATU, P. *L'implication probable et l'implication certaine* ◇ B48 ◇
BURGESS, J.P. *Probability logic* ◇ B48 ◇
FENSTAD, J.E. *Logic and probability (Norwegian) (English summary)* ◇ B48 ◇
HACKING, I. *Linguistically invariant inductive logic* ◇ A05 B48 ◇
KUMAR, D. *Neutrality, contingency and undecidability* ◇ B48 ◇
MACKIE, J.L. *The relevance criterion of confirmation* ◇ B48 ◇
RESCHER, N. *Probability logic* ◇ B48 ◇
WATANABE, S. *Knowing and guessing: A quantitative study of inference and information* ◇ A05 B48 ◇

1970

CHUDNOVSKY, D.V. *Logical probability and conditional probability on Boolean algebras (Russian) (English summary)* ◇ B48 C90 G05 ◇
COHEN, L.J. *The implications of induction* ◇ A05 B48 B98 ◇
ESSLER, W.K. *Induktive Logik, Grundlagen und Voraussetzungen* ◇ B48 ◇
FISHBURN, P.C. *Utility theory for decision making* ◇ B48 ◇
HINTIKKA, K.J.J. & SUPPES, P. (EDS.) *Information and inference* ◇ A05 B48 ◇
KYBURG JR., H.E. *Probability and inductive logic* ◇ A05 B48 ◇
LEHRER, K. *Induction, reason and consistency* ◇ B48 ◇
LEVI, I. *Probability and evidence* ◇ A05 B48 ◇
MELTZER, B. *The semantics of induction and the possibility of complete systems of inductive inference* ◇ B48 ◇
MORTIMER, H. *Conditions for acceptance of probabilistic postulates (Polish and Russian summaries)* ◇ A05 B48 ◇
STALNAKER, R.C. *Probability and conditionals* ◇ A05 B48 ◇
SWAIN, M. (ED.) *Induction, acceptance, and rational belief* ◇ A05 B48 ◇

SWAIN, M. *The consistency of rational belief* ◇ A05 B48 ◇
VICKERS, J.M. *Probability and non-standard logics* ◇ B48 ◇
WRIGHT VON, G.H. *A note on confirmation theory and on the concept of evidence* ◇ A05 B48 ◇

1971

APOSTEL, L. *Logique inductive, modalites epistemologiques et logique de la preference* ◇ A05 B48 ◇
BATENS, D. *The paradoxes of confirmation* ◇ A05 B48 ◇
CARNAP, R. *A basic system of inductive logic part I* ◇ A05 B48 ◇
CARNAP, R. *Inductive logic and rational decisions* ◇ A05 B48 ◇
CARNAP, R. & JEFFREY, R.C. (EDS.) *Studies in inductive logic and probability I* ◇ B30 B48 ◇
CHIRKOV, M.K. *Ueber die Methode der wahrscheinlichkeitstheoretischen Logik (Russian)* ◇ B48 ◇
COHEN, L.J. *The inductive logic of progressive problem-shifts* ◇ A05 B48 ◇
GAIFMAN, H. *Applications of de Finetti's theorem to inductive logic* ◇ B48 ◇
HILPINEN, R. *Relational hypotheses and inductive inference* ◇ A05 B48 ◇
HILPINEN, R. & HINTIKKA, K.J.J. *Rules of acceptance, indices of lawlikeness, and singular inductive inference: reply to a critical discussion* ◇ A05 B48 ◇
HINTIKKA, K.J.J. *Inductive generalization and its problems: A comment on Kronthaler's comments* ◇ B48 ◇
HINTIKKA, K.J.J. *The "lottery paradox" and the concept of shared information* ◇ B48 ◇
HUMBURG, J. *The principle of instantial relevance* ◇ B48 ◇
KRONTHALER, E. *Bemerkungen zum zweidimensionalen Kontinuum induktiver Methoden von J.Hintikka* ◇ A05 B48 ◇
MICHALOS, A.C. *Hilpinen's rules of acceptance and inductive logic* ◇ A05 B48 ◇
NIINILUOTO, I. *Can we accept Lehrer's inductive rule?* ◇ A05 B48 ◇
SEGERBERG, K. *Qualitative probability in a modal setting* ◇ B48 ◇
VETTER, H. *Inductivism and falsificationism reconcilable* ◇ A05 B48 ◇

1972

BOUDOT, M. *Logique inductive et probabilite. Philosophie pour l'age de la science* ◇ A05 B48 ◇
CARNAP, R. *Notes on probability and induction* ◇ B48 ◇
FINETTI DE, B. *Probability, induction and statistics: The art of guessing* ◇ A05 B48 ◇
HART, W.D. *Probability as degree of possibility* ◇ B48 ◇
JEFFREY, R.C. *Carnap's inductive logic* ◇ B48 ◇
MASLOV, S.YU. & RUSAKOV, E.D. *Probabilistic canonical calculi (Russian) (English summary)* ◇ B48 D03 ◇
NIINILUOTO, I. *Inductive systematization: Definition and a critical survey* ◇ A05 B48 ◇
PIETARINEN, J. *Lawlikeness, analogy, and inductive logic* ◇ A05 B48 ◇
PYATNITSYN, B.N. *Ueber die Semantik der Wahrscheinlichkeitslogik und der induktiven Logik* ◇ A05 B48 ◇
RUZAVIN, G.I. *Die semantische Konzeption der induktiven Logik* ◇ A05 B48 ◇
WRIGHT VON, G.H. *The logic of preference reconsidered* ◇ B48 ◇

1973

AISA MOREU, D. *Alternatives for a logical foundation of inductive logic in R.Carnap (Spanish)* ◇ B48 ◇
BLUM, L. & BLUM, M. *Inductive inference: a recursion theoretic approach* ◇ B48 D15 D80 ◇
ELLIS, B. *The logic of subjective probability* ◇ A05 B48 ◇
GROFMAN, B. & HYMAN, G. *Probability and logic in belief systems* ◇ A05 B48 ◇
HILPINEN, R. *Carnap's new system of inductive logic* ◇ B48 ◇
KARTSAKLIS, A. *On the Carnap's C-T axiomatic system of space-time and the Basri's deductive theory of space and time* ◇ A05 B48 ◇
KUIPERS, T.A.F. *A generalization of Carnap's inductive logic* ◇ A05 B48 ◇
MORTIMER, H. *A rule of acceptance based on logical probability* ◇ A05 B48 ◇
NIINILUOTO, I. & TUOMELA, R. *Theoretical concepts and hypothetico-inductive inference* ◇ A05 B48 ◇
ROY, B.B. & SHUKLA, R. *Logical interpretation of probability calculus and derivation of many valued logic* ◇ B48 ◇
STEGMUELLER, W. *Probleme und Resultate der Wissenschaftstheorie und analytischen Philosophie. Band IV: Personelle und statistische Wahrscheinlichkeit. Erster Halbband: Personelle Wahrscheinlichkeit und rationale Entscheidung* ◇ A05 B48 ◇
STEGMUELLER, W. *Probleme und Resultate der Wissenschaftstheorie und analytischen Philosophie. Band IV: Personelle und statistische Wahrscheinlichkeit. Zweiter Halbband: Statistisches Schliessen. Statistische Begruendung. Statistische Analyse* ◇ A05 B48 ◇
SWINBURNE, R. *An introduction to confirmation theory* ◇ A05 B48 ◇
UCHII, S. *Inductive logic with causal modalities: a deterministic approach* ◇ A05 B48 ◇

1974

FINETTI DE, B. *Theory of probability. Vol.I* ◇ B48 ◇
GHOSE, A. *The logic of dynamic identity in "Meaning and structure of time"* ◇ A05 B48 ◇
HAVRANEK, T. *The computation of characteristic vectors of logical probabilistic expressions* ◇ B48 ◇
KARPOVICH, V.N. *Hypotheses introduced by reduction sentences (Russian)* ◇ A05 B48 B65 ◇
KOSTYUK, V.N. *The theory of inductive reasoning and its formalization (Russian)* ◇ A05 B48 ◇
KYBURG JR., H.E. *Propensities and probabilities* ◇ A05 B48 ◇
KYBURG JR., H.E. *The logical foundations of statistical inference* ◇ A05 B48 ◇

LAZEROWITZ, M. *Necessity and probability* ◇ A05 B48 ◇
MENGES, G. *Elements of an objective theory of inductive behaviour* ◇ A05 B48 ◇
POSPESEL, H. & WERNER, C.G. *Deductive inferences from particular to general* ◇ A05 B48 ◇
TONDL, L. *Semantic evaluation of prognostic statements and the base of probabilistic parameters* ◇ A05 B48 ◇

1975

ADAMS, E.W. *The logic of conditionals. An application of probability to deductive logic* ◇ A05 B48 B98 ◇
ANDERSON JR., R.M. & MAXWELL, G. (EDS.) *Induction, probability, and confirmation* ◇ A05 B48 ◇
BLUM, L. & BLUM, M. *Toward a mathematical theory of inductive inference* ◇ B48 D10 D15 ◇
FINETTI DE, B. *Theory of probability. Vol.II* ◇ B48 ◇
GAERDENFORS, P. *Qualitative probability as an intensional logic* ◇ B48 ◇
GAERDENFORS, P. *Some basic theorems of qualitative probability* ◇ B48 ◇
HARPER, W.L. & HOOKER, C.A. (EDS.) *Foundations of probability theory, statistical inference, and statistical theories of science I,II,III* ◇ B48 ◇
HOWSON, C. *The rule of succession, inductive logic and probability logic* ◇ B48 ◇
KOSTYUK, V.N. *The theory of inductive reasoning (Russian)* ◇ A05 B48 ◇
KRON, A. & MILOVANOVIC, V. *Preference and choice* ◇ A05 B48 ◇
LINDNER, R. *On the theory of inference operators* ◇ B48 F99 ◇
POGOSYAN, EH.M. *Comparative characteristics of matching inductors (Russian) (Armenian summary)* ◇ B48 ◇
POGOSYAN, EH.M. *Inductive inference with feedback (Russian) (Armenian summary)* ◇ B48 D15 ◇
VERLOREN VAN THEMAAT, W.A. *The confirmation of sentences by instances with different truth-values of its atoms* ◇ A05 B48 ◇
VINCENT, R.H. *Selective confirmation and the ravens* ◇ A05 B48 ◇

1976

ADAMS, E.W. *Prior probabilities and counterfactual conditionals* ◇ A05 B48 ◇
BIGELOW, J.C. *Possible worlds foundations for probability* ◇ B48 C90 ◇
BLUM, A. *A logic of belief* ◇ B48 ◇
COHEN, L.J. *A conspectus of the neo-classical theory of induction* ◇ A05 B48 ◇
DALE, A.I. *Probability logic and 𝔉-coherence* ◇ B48 ◇
FOULIS, D.J. & RANDALL, C.H. *A mathematical setting for inductive reasoning* ◇ B48 ◇
FRAASSEN VAN, B.C. *Probabilities of conditionals* ◇ A05 B45 B48 ◇
FRAASSEN VAN, B.C. *Representation of conditional probabilities* ◇ B48 ◇
GILES, R. *A logic for subjective belief* ◇ B48 ◇
HAILPERIN, T. *Boole's logic and probability* ◇ A10 B48 G05 ◇
HARPER, W.L. *Ramsey test conditionals and iterated belief change (A response to Stalnaker.)* ◇ A05 B45 B48 ◇
HARPER, W.L. & MAY, S. *Toward an optimization procedure for applying minimum change principles in probability kinematics* ◇ B48 ◇
HINTIKKA, K.J.J. & NIINILUOTO, I. *An axiomatic foundation for the logic of inductive generalization* ◇ A05 B48 ◇
HINTIKKA, K.J.J. *Information, causality, and the logic of perception* ◇ A05 B48 ◇
HOWSON, C. *The development of logical probability* ◇ A05 B30 B48 ◇
KOFLER, E. & MENGES, G. *Cognitive decisions under partial information* ◇ B48 ◇
KYBURG JR., H.E. *Local and global induction* ◇ A05 B48 ◇
LEHMANN, S.K. *A first-order logic of knowledge and belief with identity. I,II* ◇ A05 B48 ◇
LEVI, I. *Acceptance revisited* ◇ A05 B48 ◇
MICHALSKI, R.S. *Learning by inductive inference* ◇ B48 ◇
MILLER, DAVID *On probability measures for deductive systems I* ◇ B48 ◇
MINICOZZI, E. *Some natural properties of strong-identification in inductive inference* ◇ B48 ◇
TUOMELA, R. *Confirmation, explanation, and the paradoxes of transitivity* ◇ A05 B48 ◇
VICKERS, J.M. *Belief and probability* ◇ A05 B48 ◇

1977

ADAMS, E.W. *A note on comparing probabilistic and modal logics of conditionals* ◇ A05 B48 ◇
BIGELOW, J.C. *Semantics of probability* ◇ A05 B48 ◇
FERENCZI, M. *On valid assertions - in probability logic* ◇ B25 B48 C90 ◇
GORAN, V.P. *Two aspects of the problem of interpreting necessity and randomness (Russian)* ◇ A05 B48 ◇
KUGEL, P. *Induction, pure and simple* ◇ B48 ◇
NIINILUOTO, I. *On a k-dimensional System of inductive logic* ◇ B48 ◇
SALMON, W.C. (ED.) *Hans Reichenbach, logical empiricist. I,II,III,IV,V* ◇ B48 ◇
SMOKLER, H. *Semantical questions in Carnap's inductive logic* ◇ A05 B48 ◇
SMOKLER, H. *Three grades of probabilistic involvement* ◇ A05 B45 B48 ◇
WERNER, C.G. *Frequencies and beliefs* ◇ A05 B48 ◇

1978

FIELD, H.H. *A note on Jeffrey conditionalization* ◇ B48 ◇
GAERDENFORS, P. *Conditionals and changes of belief* ◇ B48 ◇
GAINES, B.R. *Fuzzy and probability uncertainty logics* ◇ B48 B52 ◇
GEORGESCU, G. *Extension of probabilities defined on polyadic algebras* ◇ B48 G15 ◇
GIBBARD, A. & HARPER, W.L. *Counterfactuals and two kinds of expected utility* ◇ B48 ◇
HOOKER, C.A. & LEACH, J.J. *Theoretical foundations* ◇ B48 ◇

HOOVER, D.N. *Probability logic* ◇ B48 C80 C90 ◇
HUNTER, D. & RICHTER, R. *Counterfactuals and Newcomb's paradox* ◇ A05 B45 B48 ◇
JEFFREY, R.C. *Axiomatizing the logic of decision* ◇ B48 ◇
KRAMOSIL, I. & SINDELAR, J. *Statistical deducibility testing with stochastic parameters* ◇ B25 B48 ◇
KUIPERS, T.A.F. *On the generalization of the continuum of inductive methods to universal hypotheses* ◇ A05 B48 ◇
KUIPERS, T.A.F. *Studies in inductive probability and rational expectation* ◇ B48 ◇
LEVI, I. *Irrelevance* ◇ B48 ◇
MILLER, DAVID *On probability measures for deductive systems II,III* ◇ B48 ◇
MONDADORI, M. *The probability of "laws of nature": a case for Hintikka's inductive logic* ◇ B48 ◇
TENNANT, N. *Entailment and proofs* ◇ B48 ◇

1979

BENDALL, K. *Belief-theoretic formal semantics for first-order logic and probability* ◇ A05 B48 ◇
BOON, L. *Repeated tests and repeated testing: how to corroborate low level hypotheses* ◇ A05 B48 ◇
COX, R.T. *Of inference and inquiry, an essay in inductive logic* ◇ B48 ◇
FARRELL, R.J. *Material implication, confirmation, and counterfactuals* ◇ B48 ◇
FRAASSEN VAN, B.C. & LEBLANC, H. *On Carnap and Popper probability functions* ◇ A05 B48 ◇
HUERTAS, J.A. & MARTINEZ RECIO, A. *Inductive logic and artificial intelligence boundaries (Spanish) (English summary)* ◇ B48 ◇
LARGEAULT, J. *Chance, probabilities and induction (French)* ◇ A05 B48 ◇
LEBLANC, H. *Probabilistic semantics for first-order logic* ◇ B48 ◇
MERRILL, G.H. *Confirmation and prediction* ◇ B48 ◇
MOHR, H.P. *Wahl einer induktiven Methode mit entscheidungslogischen Mitteln* ◇ A05 B48 ◇
NIINILUOTO, I. *Truthlikeness in first-order languages* ◇ A05 B48 ◇
PAULOS, J.A. *A model-theoretic account of confirmation* ◇ A05 B48 ◇
SHAJDA, J. *On probability of first order formulas in a given model* ◇ B48 ◇
VITYAEV, E.E. *Detection of regularities expressed by universal formulas (Russian)* ◇ A05 B48 ◇

1980

ARMENDT, B. *Is there a Dutch Book argument for probability kinematics?* ◇ B48 ◇
CARNAP, R. *A basic system of inductive logic. II* ◇ B48 ◇
CHUAQUI, R.B. *Foundations of statistical methods using a semantical definition of probability* ◇ B48 C90 ◇
DISHKANT, G.P. *Three propositional calculi of probability* ◇ B48 ◇
DOMOTOR, Z. *Probability kinematics and representation of belief change* ◇ B48 ◇
DUESBERG, K.J. *Zur Kritik einer neuen Definition von "Wahrscheinlichkeit"* ◇ B48 ◇
FENSTAD, J.E. *The structure of probabilities defined on first-order languages* ◇ B48 ◇
FRAASSEN VAN, B.C. *Rational belief and probability kinematics* ◇ A05 B48 ◇
FRENCH, S. *Updating of belief in the light of someone else's opinion* ◇ B48 ◇
GACS, P. *Exact expressions for some randomness tests* ◇ B48 ◇
GARBER, D. *Discussion: Field and Jeffrey conditionalization* ◇ B48 ◇
HALLDEN, S. *The foundations of decision logic* ◇ B48 ◇
HINTIKKA, K.J.J. *Aristotelian induction* ◇ A05 B48 ◇
HOOVER, D.N. *A note on regularity* ◇ B48 ◇
JEFFREY, R.C. (ED.) *Studies in inductive logic and probability II* ◇ B48 ◇
KAYSER, D. *Vers une modelisation du raisonnement "approximatif"* ◇ B48 ◇
KHARSHING, L. *Criteria for acceptance of scientific hypotheses (Russian)* ◇ B48 ◇
KOSTYUK, V.N. *Explanation, prediction and comprehension (Russian)* ◇ A05 B48 ◇
KUIPERS, T.A.F. *A survey of inductive systems* ◇ B48 ◇
LANE, D.A. *Fisher, Jeffreys, and the nature of probability* ◇ A05 B48 ◇
LEBEDEV, S.A. *The role of induction in the functioning of modern scientific knowledge (Russian) (English Summary)* ◇ A05 B48 ◇
LENZEN, W. *Glauben, Wissen und Wahrscheinlichkeit. Systeme der epistemischen Logik* ◇ A05 B45 B48 ◇
MICHALSKI, R.S. *Detection of conceptual patterns through inductive inference* ◇ B48 ◇
PYATNITSYN, B.N. & TYAGNIBEDINA, O.S. *The role of the concepts of "probability" and "degree of certainty" in B. Bolzano's logic (Russian)* ◇ A05 B48 ◇
ROZEBOOM, W.W. *Nicod's criterion: subtler than you think* ◇ B48 ◇
SCOZZAFAVA, R. *Bayesian inference and inductive "logic"* ◇ B48 ◇
SINDELAR, J. *Statistical theory of logical derivability* ◇ B48 ◇
SPOHN, W. *Stochastic independence, causal independence, and shieldability* ◇ B48 ◇

1981

ADAMS, E.W. *Transmissible improbabilities and marginal essentialness of premises in inferences involving indicative conditionals* ◇ A05 B48 ◇
BRESSAN, A. & MONTANARO, A. *Contributions to foundations of probability calculus on the basis of the modal logical calculus MC^{ν} or MC^{ν}_{*} I* ◇ B48 ◇
COHEN, L.J. *Inductive logic, 1945–1977* ◇ B48 ◇
COSTANTINI, D. & GEYMONAT, L. *For an objectivist notion of probability (Italian) (English and French summaries)* ◇ B48 ◇
COSTANTINI, D. *Inductive logic and inductive statistics* ◇ A05 B48 ◇
DIETTERICH, T.G. & MICHALSKI, R.S. *Inductive learning of structural descriptions: evaluation criteria and comparative review of selected methods* ◇ B48 ◇

DOMOTOR, Z. *Higher order probabilities* ◇ B48 ◇
DOMOTOR, Z. *Probabilistic and causal dependence structures* ◇ B48 ◇
FETZER, J.H. *Probability and explanation* ◇ B48 ◇
FRAASSEN VAN, B.C. *A temporal framework for conditionals and chance* ◇ B48 ◇
FRAASSEN VAN, B.C. *Probabilistic semantics objectified: I. Postulated and logics. II. Implication in probabilistic model sets* ◇ B48 ◇
GIBBARD, A. *Two recent theories of conditionals* ◇ B48 ◇
GOOD, I.J. *The weight of evidence provided by uncertain testimony or from an uncertain event* ◇ B48 ◇
HANNA, J.F. *Single case propensities and the explanation of particular events* ◇ B48 ◇
HARPER, W.L. *A sketch of some recent developments in the theory of conditionals* ◇ B48 B98 ◇
HARPER, W.L. & PEARCE, G. & STALNAKER, R.C. (EDS.) *Ifs* ◇ B48 ◇
HAVRANEK, T. *Formal systems for mechanized statistical inference* ◇ B48 ◇
HINTIKKA, K.J.J. *Intuitions and philosophical method* ◇ A05 B48 ◇
LEBLANC, H. *What price substitutivity? A note on probability theory* ◇ B48 ◇
LESSI, O. *Corrispondenza tra funzioni di conferma e distribuzioni di probilita condizionate e identificazione di una classe di funzioni rappresentative (English and French summaries)* ◇ B48 ◇
LEVI, I. *Direct inference and confirmational conditionalization* ◇ B48 ◇
LEWIS, D. *Probabilities of conditionals and conditional probabilities* ◇ B48 ◇
MCGEE, V. *Finite matrices and the logic of conditionals* ◇ B48 ◇
NARENS, L. *A general theory of ratio scalability with remarks about the measurement-theoretic concept of meaningfulness* ◇ B48 ◇
NOVIKOV, V.F. & VITYAEV, E.E. *Paradoxicality of methods of induction (Russian)* ◇ B48 ◇
OTTE, R. *A critique of Suppes' theory of probabilistic causality* ◇ A05 B48 ◇
PAULOS, J.A. *Probabilistic, truth-value, and standard semantics and the primacy of predicate logic* ◇ A05 B10 B48 ◇
POLLOCK, J.L. *Indicative conditionals and conditional probability* ◇ B48 ◇
ROSENKRANTZ, R.D. *Foundations and applications of inductive probability* ◇ B48 ◇
SCHWARTZ, R.J. *Approximate truth and confirmation* ◇ B48 ◇
SHAFER, G. *Constructive probability* ◇ B48 ◇
SHIRLEY, E.S. *An unnoticed flaw in Barker and Achinstein's solution to Goodman's new riddle of induction* ◇ B48 ◇
STALNAKER, R.C. *A defense of conditional excluded middle* ◇ B48 ◇
STALNAKER, R.C. *A theory of conditionals* ◇ B48 ◇
STALNAKER, R.C. *Indicative conditionals* ◇ B48 ◇
STALNAKER, R.C. *Probability and conditionals* ◇ B48 ◇
SUPPES, P. *Logique du probable. Demarche bayesienne et rationalite (French)* ◇ B48 ◇
TUOMELA, R. *Inductive explanation* ◇ A05 B48 ◇

1982

APPIAH, A. *Conversation and conditionals* ◇ A05 B48 ◇
BANDLER, W. & KOHOUT, L.J. *Axioms for conditional inference: probabilistic and possibilistic* ◇ B48 ◇
BARZDINS, J. *Some rules of inductive inference and their application (Russian)* ◇ B48 ◇
BENDALL, K. *A "definitive" probabilistic semantics for first-order logic* ◇ B48 ◇
COSTANTINI, D. & MONARI, P. *Inferenza predittiva e "fatti"* ◇ B48 ◇
EELLS, E. *Rational decision and causality* ◇ B48 ◇
FINETTI DE, B. *Probability: the different views and terminologies in a critical analysis* ◇ B48 ◇
FISHBURN, P.C. *The foundations of expected utility* ◇ B48 ◇
FREJVALD, R.V. & KINBER, E.B. & WIEHAGEN, R. *Inductive inference and computable one-one numberings* ◇ B48 D20 D45 ◇
FRENCH, S. *On the axiomatisation of subjective probabilities* ◇ B48 ◇
FU, K.S. & ISHIZUKA, M. & YAO, J.T.P. *Inference procedures under uncertainty for the problem-reduction method* ◇ B48 ◇
GAERDENFORS, P. & SAHLIN, N.-E. *Unreliable probabilities, risk taking, and decision making* ◇ B48 ◇
HORWICH, P. *Probability and evidence* ◇ B48 ◇
JACKSON, F.C. & PARGETTER, R. *Physical probability as a propensity* ◇ A05 B48 ◇
JOSHI, V.M. *A counter example of Hacking against the long run rule* ◇ B48 ◇
KATZ, M. *Metric similarities in the logic of approximation* ◇ B48 ◇
LEVI, I. *Ignorance, probability and rational choice* ◇ B48 ◇
LEVY, E. *Causal-relevance explanation: Salmon's theory and its relation to Reichenbach* ◇ A05 B48 ◇
LIDDELL, G.F. *A logic for propositions with indefinite truth values* ◇ B48 ◇
MEIXNER, J. *Are statistical explanations really explanatory?* ◇ B48 ◇
MILLER, DAVID *Impartial truth* ◇ A05 B48 ◇
PLATO VON, J. *The significance of the ergodic decomposition of stationary measures for the interpretation of probability* ◇ B48 ◇
RADU, J.B. (ED.) *Henry E.Kyburg, Jr.& Isaac Levi* ◇ A05 B48 ◇
REHDER, W. *Conditions for probabilities of conditionals to be conditional probabilities* ◇ B48 ◇
SEIDENFELD, T. *Paradoxes of conglomerability and fiducial inference* ◇ B48 ◇
STEMMER, N. *A solution to the lottery paradox* ◇ B48 ◇
WILL, U. *Eine pragmatische Rechtfertigung der Induktion* ◇ A05 B48 ◇
WILLIAMS, A.G.P. *Applicable inductive logic* ◇ B48 ◇
WOLNIEWICZ, B. *A closure system for elementary situations* ◇ B48 ◇

1983

BAIRD, D. *The Fisher/Pearson chi-squared controversy: a turning point for inductive inference* ◇ B48 ◇
BOTUSHAROV, O.I. *A recursion theory characterization of inductive inference with additional information classes* ◇ B48 D20 ◇
BRENY, H. *A propos du terme "probabilite"* ◇ B48 ◇
BRESSAN, A. & MONTANARO, A. *Contributions to foundations of probability calculus on the basis of the modal logic calculus* MC^{ν} *or* MC^{ν}_{*} *III: An analysis of the notions of random variables and probability spaces, based on modal logic* ◇ B48 ◇
CASE, J. & SMITH, C.H. *Comparison of identification criteria for machine inductive inference* ◇ B35 B48 D20 ◇
COOKE, R.M. *A result in Renyi's conditional probability theory with application to subjective probability* ◇ B48 ◇
COSTANTINI, D. *Il contributo di R. Carnap alla metodologia statistica* ◇ A10 B48 ◇
COSTANTINI, D. & MONARI, P. *Inferenze statistiche finitarie* ◇ B48 ◇
CZOGALA, E. *An approach to probabilistic L-valued logic* ◇ B48 ◇
DUBOIS, DIDIER & PRADE, H. *Unfair coins and necessity measures towards a possibilistic interpretation of histograms* ◇ B48 E72 ◇
FALMAGNE, J.-C. *A random utility model for a belief function* ◇ B48 ◇
FETZER, J.H. *Probability and objectivity in deterministic and indeterministic situations* ◇ B48 ◇
FOTR, J. *Subjective probabilities and methods for constructing them (Czech) (English summary)* ◇ B48 ◇
FRAASSEN VAN, B.C. *Gentlemen's wagers: relevant logic and probability* ◇ B48 ◇
FRAASSEN VAN, B.C. *Shafer on conditional probability* ◇ B48 ◇
GLYMOUR, C. *Discussion: revisions of bootstrap testing* ◇ B48 ◇
HAJEK, P. & HAVRANEK, T. *Logic, statistics and computers* ◇ B48 ◇
HALPERN, M. *Inductive inference in finite algebraic structures* ◇ B48 C13 ◇
HARSANYI, J.C. *Bayesian decision theory, subjective and objective probabilities, and acceptance of empirical hypotheses* ◇ B48 ◇
HAUSSER, R.R. *Vagueness and truth (German summary)* ◇ B48 ◇
JEFFREYS, H. *Theory of probability* ◇ B48 ◇
LEBLANC, H. & MORGAN, C.G. *Probabilistic semantics for intuitionistic logic* ◇ B48 F50 ◇
LEBLANC, H. & MORGAN, C.G. *Probability theory, intuitionism, semantics, and the Dutch book argument* ◇ B48 C90 F50 ◇
LEBLANC, H. *Probability functions and their assumption sets - the singulary case* ◇ B48 ◇
LEHRER, K. *Sellars on induction reconsidered* ◇ B48 ◇
LENZEN, W. *On the representation of classificatory value structures* ◇ B48 ◇
MARTIROSYAN, A.A. *On the influence of the consistency of hypotheses with initial information on the complexity of inductive inference (Russian) (Armenian summary)* ◇ B48 ◇
MAYO, D.G. *An objective theory of statistical testing* ◇ B48 ◇
MICHALSKI, R.S. *A theory and methodology of inductive learning* ◇ B48 ◇
NIINILUOTO, I. *Inductive logic as a methodological research programme* ◇ B48 ◇
SHAFER, G. *A subjective interpretation of conditional probability* ◇ B48 ◇
SWOYER, C. *Belief and predication* ◇ B48 ◇
WEIRICH, P. *Conditional probabilities and probabilities given knowledge of a condition* ◇ B48 ◇
YOURGRAU, P. *Knowledge and relevant alternatives* ◇ B48 ◇
ZEUGMANN, T. *On the synthesis of fastest programs in inductive inference* ◇ B48 ◇

1984

APPIAH, A. *Generalising the probabilistic semantics of conditionals* ◇ B48 ◇
BIBEL, W. *First-order reasoning about knowledge and belief* ◇ A05 B48 ◇
CZOGALA, E. *An introduction to probabilistic L-valued logic* ◇ B48 ◇
EDIDIN, A. *Inductive reasoning and the uniformity of nature* ◇ A05 B48 ◇
EELLS, E. *Metatickles and the dynamics of deliberation* ◇ B48 ◇
ESSLER, W.K. *Induktion und induktive Logik I,II* ◇ B48 ◇
FORGE, J. *Theoretical functions, theory and evidence* ◇ A05 B48 ◇
FREELING, A. *Possibilities versus fuzzy probabilities - two alternative decision aids* ◇ B48 ◇
FREJVALD, R.V. & KINBER, E.B. & WIEHAGEN, R. *On the power of probabilistic strategies in inductive inference* ◇ B48 D20 ◇
GAERDENFORS, P. *The dynamics of belief as a basis for logic* ◇ B48 ◇
GAINES, B.R. *Fundamentals of decision: Probabilistic, possibilistic and other forms of uncertainty in decision analysis* ◇ B48 ◇
HAILPERIN, T. *Probability logic* ◇ B25 B48 C90 ◇
JANTKE, K.P. *The main proof theoretic problems in inductive inference* ◇ B48 ◇
LEBLANC, H. & MORGAN, C.G. *Probability functions and their assumption sets - the binary case* ◇ B48 ◇
MORGAN, C.G. *Weak conditional comparative probability as a formal semantic theory* ◇ B48 ◇
ORLOWSKA, E. *Semantical analysis of inductive reasoning* ◇ B48 ◇
POLLOCK, J.L. *Foundations for direct inference* ◇ B48 ◇
PYATNITSYN, B.V. *Modality and probability - means to express the notion of uncertainty (Russian)* ◇ B48 ◇
WAGNER, C.G. *Aggregating subjective probabilities: some limitative theorems* ◇ B48 ◇

1985

AMER, M.A. *Classification of Boolean algebras of logic and probabilities defined on them by classical models* ◇ B48 G05 ◇

AMER, M.A. *Extension of relatively σ-additive probabilities on Boolean algebras of logic* ◇ B48 ◇

ANGLUIN, D. & SMITH, C.H. *Inductive inference: theory and methods* ◇ B48 ◇

BOTUSHAROV, O.I. *On inductive inference with additional information (Bulgarian summary)* ◇ B48 D20 ◇

CHUAQUI, R.B. *How to decide between different methods of statistical inference* ◇ B48 ◇

CHUAQUI, R.B. *Models for probability* ◇ B48 ◇

COSTANTINI, D. *Finitary inductive inferences (Italian)* ◇ B48 ◇

CUTLAND, N.J. *Partially observed stochastic controls based on a cumulative digital read out of the observations* ◇ B48 ◇

DOMOTOR, Z. *Probability kinematics, conditionals,and entropy principles* ◇ B48 ◇

FAJARDO, S. *Completeness theorems for the general theory of stochastic processes* ◇ B48 C65 C90 ◇

FAJARDO, S. *Probability logic with conditional expectation* ◇ B48 C40 C90 ◇

GAERDENFORS, P. *Propositional logic based on the dynamics of belief* ◇ B48 ◇

HARSANYI, J.C. *Acceptance of empirical statements: a Bayesian theory without cognitive utilities* ◇ B48 ◇

KOPPEL, M. *Towards a new theory of confirmation* ◇ B48 ◇

LEBLANC, H. *Unary probabilistic semantics* ◇ B48 ◇

LEVI, I. *Imprecision and indeterminacy in probability judgement* ◇ B48 ◇

ORLOWSKA, E. *Relative induction (Russian summary)* ◇ B48 ◇

ORLOWSKA, E. *Semantics of nondeterministic possible worlds* ◇ B48 B60 ◇

PITOWSKY, I. *On the status of statistical inferences* ◇ B48 ◇

POPPER, K.R. *The nonexistence of probabilistic inductive support* ◇ B48 ◇

SEN, A. *Rationality and uncertainty* ◇ A05 B48 ◇

SHINOHARA, T. *Inductive inference from negtive data* ◇ B48 ◇

WILLMOTT, R. *A probabilistic interpreation of a case of inference involving fuzzy quantification* ◇ B48 ◇

YANASE, M.M. *Fuzziness and probability* ◇ A05 B48 B52 E72 ◇

B50 Many-valued logic

1920

LUKASIEWICZ, J. *On three-valued logic (Polish)* ◇ A05 B50 ◇

1921

LUKASIEWICZ, J. *Two-valued logic (Polish)* ◇ A05 B05 B50 ◇

POST, E.L. *Introduction to a general theory of elementary propositions* ◇ B05 B20 B50 ◇

1930

LUKASIEWICZ, J. *Philosophische Bemerkungen zu mehrwertigen Systemen des Aussagenkalkuels* ◇ A05 B50 ◇

1931

WAJSBERG, M. *Axiomatization of the three-valued propositional calculus (Polish) (German summary)* ◇ B50 ◇

1935

ITO, MINORU *On the generation of all functions in the finitely many-valued algebra of logic (Japanese)* ◇ B50 ◇

LUKASIEWICZ, J. *Zur vollen dreiwertigen Aussagenlogik* ◇ B50 ◇

TARSKI, A. *Wahrscheinlichkeitslehre und mehrwertige Logik* ◇ B50 ◇

WEBB, D.L. *Generation of any n-valued logic by one binary operation* ◇ B50 ◇

ZAWIRSKI, Z. *Ueber das Verhaeltnis der mehrwertigen Logik zur Wahrscheinlichkeitsrechnung* ◇ B50 ◇

1936

SCHMIERER, Z. *On characteristic functions in many-valued systems of logic (Polish)* ◇ B50 ◇

SLUPECKI, J. *Der volle dreiwertige Aussagenkalkuel* ◇ B50 ◇

SOBOCINSKI, B. *Axiomatization of certain many-valued systems of the theory of deduction* ◇ B50 ◇

WAJSBERG, M. *Untersuchung ueber Unabhaengigkeitsbeweise nach der Matrizenmethode* ◇ B50 ◇

WEBB, D.L. *Definition of Post's generalized negative and maximum in terms of one binary operation* ◇ B50 ◇

1937

HAILPERIN, T. *Foundations of probability in mathematical logic* ◇ B50 ◇

HEMPEL, C.G. *A purely topological form of non-aristotelian logic* ◇ B50 ◇

HEMPEL, C.G. *Ein System verallgemeinerter Negationen* ◇ B50 ◇

WEBB, D.L. *The algebra of n-valued logic* ◇ B50 ◇

1938

BOCHVAR, D.A. *On a three-valued logical calculus and its application to the analysis of contradictions (Russian)* ◇ A05 B50 ◇

FRINK, O. *New algebras of logic* ◇ B50 F50 G05 ◇

VAIDYANATHASWAMY, R. *Quasi-boolean algebras and many-valued logics* ◇ B50 ◇

ZICH, O.V. *Sentential calculus with complex values (Czech)* ◇ B50 ◇

1939

ROSSER, J.B. *The introduction of quantification into a three-valued logic* ◇ B50 ◇

SLUPECKI, J. *A criterion of fullness of many-valued systems of propositional logic (Polish) (German summary)* ◇ B50 ◇

SLUPECKI, J. *Proof of the axiomatizability of full many-valued systems of propositional calculus (Polish) (German summary)* ◇ B50 ◇

1940

MCKINSEY, J.C.C. *A correction to Lewis and Langford's symbolic logic* ◇ B50 ◇

1941

MALISOFF, W.M. *Meanings in multi-valued logics* ◇ A05 B50 ◇

MOISIL, G.C. *Contribution a l'etude des logiques non-chrysippiennes I. Un nouveau systeme d'axiomes pour les algebres Lukasiewicziennes tetra-valentes* ◇ A05 A10 B50 G20 ◇

MOISIL, G.C. *Recherches sur la theorie des chaines* ◇ B50 E07 G10 ◇

MOISIL, G.C. *Sur la structure algebrique de la logique de M. Bochvar* ◇ B50 G15 G25 ◇

MOISIL, G.C. *Sur les anneaux de caracteristique 2 ou 3 et leurs applications* ◇ B50 ◇

MOISIL, G.C. *Sur les theories deductives a logique non-chrysippienne* ◇ B50 ◇

ROSSER, J.B. *On the many-valued logics* ◇ B50 ◇

1942

MOISIL, G.C. *Contributions a l'etude des logiques non-chrysippiennes II,III: Anneaux engendres par les algebres Lukasiewicziennes tetravalentes centrees* ◇ B50 ◇

1943

BOCHVAR, D.A. *On the consistency of a three-valued logical calculus (Russian) (English summary)* ◇ B50 ◇

1945

MOISIL, G.C. *Contributions a l'etude des logiques non-chrysippiennes IV. Sur la logique de M. Becker* ◇ A05 A10 B50 ◇

ROSSER, J.B. & TURQUETTE, A.R. *Axiom schemes for m-valued propositional calculi* ◇ B50 ◇

1946

SLUPECKI, J. *The complete three-valued propositional calculus (Polish)* ◇ B50 ◇

1947

RASIOWA, H. *Sur certaines matrices logiques* ◇ B50 ◇

1948

KALICKI, J. *On Tarski's matrix method* ◇ B50 ◇
LINKE, P.F. *Die mehrwertigen Logiken und das Wahrheitsproblem* ◇ B50 ◇
LOS, J. *Many-valued logics and the formalization of intensional functions (Polish)* ◇ B50 ◇
RIDDER, J. *Ueber mehrwertige Aussagenkalkuele und mehrwertige engere Praedikatenkalkuele* ◇ B50 ◇
ROSSER, J.B. & TURQUETTE, A.R. *Axiom schemes for m-valued functional calculi of first order. Part I. Definition of axiom schemes and proof of plausibility* ◇ B50 ◇

1949

DIENES, P. *On an implication function in many-valued systems of logic* ◇ B50 ◇
DIENES, P. *On ternary logic* ◇ B50 ◇
HU, SHIHUA *M-valued sub-system of (m + n)-valued propositional calculus* ◇ B50 ◇
LOS, J. *On logical matrices (Polish)* ◇ B22 B50 ◇
MARTIN, G. *Ueber ein zweiwertiges Modell einer vierwertigen Logik* ◇ A05 B50 ◇
RIDDER, J. *Sur quelques logiques multivalentes* ◇ B50 ◇
ROSSER, J.B. & TURQUETTE, A.R. *A note on the deductive completeness of m-valued propositional calculi* ◇ B50 ◇

1950

FOSTER, A.L. *On n-ality theories in rings and their logical algebras, including triality principle and their three valued logics* ◇ B50 G25 ◇
HU, SHIHUA *Construction of an $\aleph_0$-valued propositional calculus (Chinese)* ◇ B50 ◇
KALICKI, J. *A test for the existence of tautologies according to many-valued truth-tables* ◇ B50 ◇
MARTIN, N.M. *Some analogues of the Sheffer Stroke function in n-valued logic* ◇ B50 ◇
SUGIHARA, T. *Many-valued logic (Japanese)* ◇ B50 ◇
WAGNER, K. *Zum Repraesentantenproblem der Logik fuer Aussagenfunktionen mit beliebig endlich vielen Wahrheitswerten* ◇ B50 ◇

1951

CHEN, QIANGYE & HU, SHIHUA *A note on the four-valued propositional calculus and the four colour problem (Chinese summary)* ◇ B50 ◇
FOSTER, A.L. *p^k-rings and ring-logics* ◇ B50 ◇
FOSTER, A.L. *Ring-logics and p-rings* ◇ B50 ◇
GOETLIND, E. *A Lesniewski-Mihailescu-theorem for m-valued propositional calculi* ◇ B50 ◇
HENLE, P. *n-valued boolean algebra* ◇ B50 G25 ◇
MARTIN, N.M. *A note on Sheffer functions in n-valued logic* ◇ B50 ◇
MCNAUGHTON, R. *A theorem about infinite-valued sentential logic* ◇ B50 ◇
ROSE, A. *A lattice-theoretic characterization of the $\aleph_0$-valued propositional calculi* ◇ B50 G10 ◇
ROSE, A. *A new proof of a theorem of Dienes* ◇ B50 ◇
ROSE, A. *An axiom system for the three-valued logic* ◇ B50 ◇
ROSE, A. *Axiom systems for three-valued logic* ◇ B50 ◇
ROSE, A. *Conditioned disjunction as a primitive connective for the m-valued propositional calculus* ◇ B50 ◇
ROSE, A. *Remarques sur les notions d'independance et de noncontradiction* ◇ A05 B50 ◇
ROSE, A. *Systems of logic whose truth-values form lattices* ◇ B50 G10 ◇
ROSE, A. *The degree of completeness of some Lukasiewicz-Tarski propositional calculi* ◇ B20 B50 ◇
ROSSER, J.B. & TURQUETTE, A.R. *Axiom schemes for m-valued functional calculi of first order. Part II. Deductive completeness* ◇ B50 ◇
SUGIHARA, T. *Many-valued logical characteristics of Brouwerian logic (Japanese)* ◇ B50 G10 ◇

1952

GOETLIND, E. *Some Sheffer functions in n-valued logic* ◇ B50 ◇
MENNE, A. *Zu den triadischen bivalenten Aussagefunktoren* ◇ B50 ◇
ROSE, A. *A formalisation of Post's m-valued propositional calculus* ◇ B50 ◇
ROSE, A. *An extension on the calculus of non-contradiction* ◇ B50 ◇
ROSE, A. *Eight-valued geometry* ◇ B50 ◇
ROSE, A. *Extensions of some theorems of Schmidt and McKinsey. I* ◇ B50 G05 ◇
ROSE, A. *Le degre de saturation du calcul propositionnel implicatif a trois valeurs de Sobocinski* ◇ B50 ◇
ROSE, A. *Some generalized Sheffer functions* ◇ B50 ◇
ROSE, A. *The degree of completeness of m-valued Lukasiewicz propositional calculus* ◇ B50 ◇
ROSSER, J.B. & TURQUETTE, A.R. *Many-valued logics* ◇ A05 A10 B50 ◇
SOBOCINSKI, B. *Axiomatization of a partial system of three-value calculus of propositions* ◇ B50 ◇
SUEKI, T. *The formulization of two-valued and n-valued systems I (Japanese) (English summary)* ◇ B50 ◇
SUGIHARA, T. *Negation in many-valued logic* ◇ B50 ◇
SWIFT, J.D. *Algebraic properties of n-valued propositional calculi* ◇ B50 ◇

1953

ASSER, G. *Die endlichwertigen Lukasiewiczschen Aussagenkalkuele* ◇ B50 ◇
CHURCH, A. *Non-normal truth-tables for the propositional calculus* ◇ B50 ◇
CHURCH, A. *Tablas de verdad no normales para el calculo proposicional* ◇ B50 ◇
FREY, G. *Bemerkungen zum Problem der mehrwertigen Logiken* ◇ B50 ◇
MARTIN, N.M. *On completeness of decision element sets* ◇ B50 ◇

PRIOR, A.N. *Three-valued logic and future contingents* ◇ B50 ◇
ROSE, A. *A formalization of an $\aleph_0$-valued propositional calculus* ◇ B50 ◇
ROSE, A. *A formalization of Sobocinski's three-valued implicational propositional calculus* ◇ B50 ◇
ROSE, A. *Fragments of the m-valued propositional calculus* ◇ B50 ◇
ROSE, A. *Some self-dual primitive functions for propositional calculi* ◇ B50 ◇
ROSE, A. *The m-valued calculus of non-contradiction* ◇ B50 ◇
ROSE, A. *The degree of completeness of the $\aleph_0$-valued Lukasiewicz propositional calculus* ◇ B50 ◇
SUEKI, T. *The formulization of two-valued and n-valued systems II (Japanese) (English summary)* ◇ B50 ◇
VACCARINO, G. *La logiche polivalenti e non aristoteliche* ◇ B50 ◇

1954

KALICKI, J. *On equivalent truth-tables of many-valued logics* ◇ B50 ◇
MARTIN, N.M. *The Sheffer functions of 3-valued logic* ◇ B50 ◇
MO, SHAOKUI *Logical paradoxes for many-valued systems* ◇ A05 B50 B53 ◇
REICHENBACH, H. *Les fondements logiques de la theorie des quanta. Utilisation d'une logique a trois valeurs* ◇ B50 B51 G12 ◇
ROHLEDER, H. *Der dreiwertige Aussagenkalkuel der theoretischen Logik und seine Anwendung zur Beschreibung von Schaltungen, die aus Elementen mit zwei stabilen Zustaenden bestehen* ◇ B50 ◇
ROSE, A. *A Goedel theorem for an infinite-valued erweiterter Aussagenkalkuel* ◇ B50 ◇
ROSE, A. *Sur les fonctions definissables dans une logique a un nombre infini de valeurs* ◇ B50 ◇
SUBLET, J. *Essai de formalisation complete du raisonnement mathematique sur la base de trois operations* ◇ B50 E30 ◇
SUEKI, T. *The formulization of two-valued and n-valued systems III (Japanese) (English summary)* ◇ B50 ◇
TAKEKUMA, R. *On a nine-valued propositional calculus* ◇ B50 ◇
TURQUETTE, A.R. *Many-valued logics and systems of strict implication* ◇ A05 B50 ◇
YABLONSKIJ, S.V. *On functional completeness in a three-valued calculus (Russian)* ◇ B50 ◇
YONEMITSU, N. *Note on the completeness of m-valued propositional calculus* ◇ B50 ◇

1955

BADILLO BARALLAT, M.C. *Esquemas representativos di sistemas regidos por una logica polivalente (English summary)* ◇ B50 ◇
BADILLO BARALLAT, M.C. *Fundamentos en relacion con logica simbolica polivalente* ◇ B50 ◇
CHRESTENSON, H.E. *A class of generalized Walsh functions* ◇ B50 ◇
GUENTHER, G. *Dreiwertige Logik und die Heisenbergsche Unbestimmtheitsrelation* ◇ B50 ◇
HU, SHIHUA *Die endlichwertigen und funktionell vollstaendigen Subsysteme eines $\aleph_0$-wertigen Aussagenkalkuels (Chinese) (German summary)* ◇ B50 ◇
ITO, MAKOTO *On the lattice of n-valued functions (theory of n-valued logics) I,II* ◇ B50 ◇
KURIHARA, T. *On the representation of finitely many-valued logics by electric circuits (Japanese)* ◇ B50 ◇
METZE, G. *An application of multi-valued logic systems to circuits* ◇ B50 B70 ◇
PRIOR, A.N. *Curry's paradox and 3-valued logic* ◇ B50 ◇
PRIOR, A.N. *Many-valued and modal systems: an intuitive approach* ◇ B45 B50 ◇
ROSE, A. *A Goedel theorem for an infinite-valued erweiterter Aussagenkalkuel* ◇ B50 F30 ◇
ROSE, A. *Le degre de saturation du calcul propositionnel implicatif a m valeurs de Lukasiewicz* ◇ B50 ◇
YASUURA, K. *On the representation of many-valued propositional logics by relay circuits (Japanese)* ◇ B50 B70 ◇

1956

GRENIEWSKI, M. *Utilisation des logiques trivalentes dans la theorie des mecanismes automatiques. I. Realisation des functions fondamentales par des circuits (Roumanian) (Russian and French summaries)* ◇ B50 ◇
ITO, MAKOTO *General solution of 3-valued logical equations with m-variables* ◇ B50 ◇
ITO, MAKOTO *On the general solution of the n-valued function-lattice (logical) equation in several variables (Japanese)* ◇ B50 ◇
ITO, MINORU *On the general solution of the n-valued function-lattice (logical) equation in one variable (Japanese)* ◇ B50 ◇
MOISIL, G.C. *Utilisation des logiques trivalentes dans la theorie des mecanismes automatiques. II. Equation caracteristique d'un relai-polarise (Roumanian) (Russian and French summaries)* ◇ B50 ◇
ROSE, A. *An alternative formalisation of Sobocinski's three-valued implicational propositional calculus* ◇ B50 ◇
ROSE, A. *Formalisation du calcul propositionnel implicatif a $\aleph_0$ valeurs de Lukasiewicz* ◇ B50 ◇
ROSE, A. *Formalisation du calcul propositionnel implicatif a m valeurs de Lukasiewicz* ◇ B50 ◇
ROSE, A. *Some formalisations of $\aleph_0$-valued propositional calculi* ◇ B50 ◇
SUGIHARA, T. *Four-valued propositional calculus with one designated truth value* ◇ B50 ◇

1957

BADILLO BARALLAT, M.C. *Logica trivalente en la automatizacion de los circuitos* ◇ B50 ◇
BARKER, C.C.H. *Some calculations in logic* ◇ B50 ◇
BUTLER, J.W. *On complete and independent sets of truth functions in multi-valued logics* ◇ B50 ◇
CHANG, C.C. *Algebraization of infinitely many-valued logics* ◇ B50 G20 G25 ◇

DREBEN, B. *Relation of m-valued quantificational logic to 2-valued quantificational logic* ◇ B10 B50 ◇
EVANS, T. & HARDY, L. *Sheffer stroke functions in many-valued logics* ◇ B50 ◇
GAZALE, M.J. *Multi-valued switching functions (and a number- theoretical approach to their synthesis)* ◇ B50 ◇
GRENIEWSKI, H. 2^{n+1} *logical values* ◇ B50 ◇
MO, SHAOKUI *Axiomatization of many-valued logical systems (Chinese)* ◇ B50 ◇
MO, SHAOKUI *The axiomatization of the finite many valued matrix systems (Chinese)* ◇ B50 ◇
PRIOR, A.N. *Many-valued logics. The last of three talks of "The logic game."* ◇ B50 ◇
PUTNAM, H. *Three-valued logic* ◇ B50 ◇
ROSSER, J.B. *Many-valued logics* ◇ B50 ◇
SKOLEM, T.A. *Bemerkungen zum Komprehensionsaxiom* ◇ B50 E70 ◇
SKOLEM, T.A. *Mengenlehre gegruendet auf einer Logik mit unendlich vielen Wahrheitswerten* ◇ B50 E70 ◇
SUSZKO, R. *A formal theory of the logical values I (Polish) (Russian and English summaries)* ◇ B50 ◇

1958

BADILLO BARALLAT, M.C. *Aplicacion de la logica polivalente a la teoria de numeros* ◇ B50 ◇
CHANG, C.C. *Algebraic analysis of many valued logics* ◇ B50 G20 ◇
CHANG, C.C. *Proof of an axiom of Lukasiewicz* ◇ B50 ◇
EVANS, T. & SCHWARTZ, P.B. *On Slupecki T-functions* ◇ B50 ◇
GOODSTEIN, R.L. *Models of propositional calculi in recursive arithmetic* ◇ B05 B50 F30 F50 ◇
MARGARIS, A. *A problem of Rosser and Turquette* ◇ B50 ◇
MEREDITH, C.A. *The dependence of an axiom of Lukasiewicz* ◇ B50 ◇
ROSE, A. & ROSSER, J.B. *Fragments of many-valued statement calculi* ◇ B50 ◇
ROSE, A. *Many-valued logical machines* ◇ B50 ◇
ROSE, A. *Sur les definitions de l'implication et de la negation dans certains systemes de logique dont les valeurs forment des treillis* ◇ B50 G10 ◇
SUGIHARA, T. *A three-valued logic with meaning-operator* ◇ B50 ◇
THIELE, H. *Theorie der endlichwertigen Lukasiewiczschen Praedikatenkalkuele der ersten Stufe* ◇ B50 ◇
TURQUETTE, A.R. *Simplified axioms for many-valued quantification theory* ◇ B50 ◇
YABLONSKIJ, S.V. *Functional constructions in a k-valued logic (Russian)* ◇ B50 ◇
YABLONSKIJ, S.V. *On limit logics (Russian)* ◇ B50 ◇

1959

CHANG, C.C. *A new proof of the completeness of the Lukasiewicz axioms* ◇ B50 ◇
GAVRILOV, G.P. *Certain conditions for completeness in countable-valued logic (Russian)* ◇ B50 ◇
GAZALE, M.J. *Les structures de communication a m valeurs et les calculatrices numeriques* ◇ B50 ◇
KRICHEVSKIJ, R.E. *Realization of functions by superpositions (Russian)* ◇ B50 B70 ◇
LEVI, I. *Putnam's three truth values* ◇ B50 ◇
MUCHNIK, A.A. & YANOV, YU.I. *Existence of k-valued closed classes without a finite basis (Russian)* ◇ B30 B50 ◇
SALOMAA, A. *On many-valued systems of logic* ◇ B50 ◇
SUPPES, P. *Measurement,empirical meaningfulness and threevalued logic* ◇ A05 B50 ◇
YABLONSKIJ, S.V. *Functional constructions in many-valued logics (Russian)* ◇ B50 ◇
YABLONSKIJ, S.V. *Functional constructions in many-valued logic (Russian)* ◇ B50 ◇
YABLONSKIJ, S.V. *Some properties of enumerable closed classes from* $P_{\aleph_0}$ *(Russian)* ◇ B50 ◇

1960

CHAUVIN, A. *Sur les modeles du calcul* K_0 *de Bochvar, avec ou sans egalite, et l'interpretation des paradoxes de la logique dans la theorie des ensembles elementaires arithmetiques* ◇ B50 D55 ◇
MARTYNYUK, V.V. *Untersuchungen einiger Klassen von Funktionen in der mehrwertigen Logik (Russisch)* ◇ B50 ◇
MOISIL, G.C. *On the logic of Bochvar (Roumanian)* ◇ B50 ◇
RESCHER, N. *Many-sorted quantification* ◇ B50 ◇
ROSE, A. *An extension of a theorem of Margaris* ◇ B50 ◇
ROSE, A. *Nouvelle formalisation du calcul propositionnel bivalent dont les foncteurs primitifs forment un ensemble qui constitue son propre dual* ◇ B50 ◇
ROSE, A. *Sur les schemas d'axiomes pour les calculs propositionnels a m-valeurs ayant des valeurs sur designees* ◇ B50 ◇
ROSE, A. *Sur un ensemble independent de foncteurs primitifs pour le calcul propositionnel, lequel constitue son propre dual* ◇ B50 ◇
ROSSER, J.B. *Axiomatization of infinite valued logics* ◇ B50 ◇
RUTLEDGE, J.D. *On the definition of an infinitely-many valued predicate calculus* ◇ B50 G15 ◇
SALOMAA, A. *A theorem concerning the composition of functions of several variables ranging over a finite set* ◇ B50 ◇
SALOMAA, A. *On the composition of functions of several variables ranging over a finite set* ◇ B50 ◇
SKOLEM, T.A. *A set theory based on a certain three-valued logic* ◇ B50 E70 ◇
ZINOV'EV, A.A. *Philosophical problems of many-valued logic (Russian)* ◇ A05 B50 ◇

1961

BADILLO BARALLAT, M.C. *Many-valued logics of systems (Spanish)* ◇ B50 ◇
BOCHVAR, D.A. *On a three-valued calculus and its application to analysis of the paradoxes of the classical extended functional calculus (Romanian)* ◇ B50 ◇
COHN, M. *Canonical forms of functions in p-valued logics* ◇ B50 ◇

FISHER, M. *A three-valued calculus for deontic logic* ◇ B50 ◇
MIHAILESCU, E.G. *Sur la methode de J.Lukasiewicz* ◇ B50 ◇
MLEZIVA, M. *On the axiomatization of three-valued propositional logic (Czech) (Russian and English summaries)* ◇ B50 ◇
MOISIL, G.C. *Les logiques a plusieurs valeurs et l'automatique* ◇ B50 B70 ◇
MOISIL, G.C. *The predicate calculus in three-valued logic* ◇ B50 ◇
MOSTOWSKI, ANDRZEJ *An example of a non-axiomatizable many valued logic* ◇ B50 ◇
MOSTOWSKI, ANDRZEJ *Axiomatizability of some many valued predicate calculi* ◇ B50 ◇
ROSE, A. *Self-dual binary and ternary connectives for m-valued propositional calculi* ◇ B50 ◇
ROSE, A. *Sur certains calculs propositionnels a m-valeurs ayant un seul foncteur primitif lequel constitue son propre dual* ◇ B50 ◇
SIERPINSKI, W. *Sur un probleme de la logique a n valeurs* ◇ B50 ◇
SOBOCINSKI, B. *A note concerning the many-valued propositional calculi* ◇ B50 ◇
TIRNOVEANU, M. *Sur un domaine logique transfini (Romanian) (Russian and French summaries)* ◇ B50 ◇
TURQUETTE, A.R. *Solution to a problem of Rose and Rosser* ◇ B50 ◇
WANG, HAO *The calculus of partial predicates and its extension to set theory I* ◇ B50 E70 ◇
ZINOV'EV, A.A. *A method of describing the truth-functions of the n-valued propositional calculus (Russian) (Polish and English summaries)* ◇ B50 ◇

1962

CHANG, C.C. & KEISLER, H.J. *Model theories with truth values in a uniform space* ◇ B50 C20 C50 C90 ◇
CLAY, R.E. *A simple proof of functional completeness in many-valued logics based on Lukasiewicz's C and N* ◇ B50 ◇
CLAY, R.E. *Note on Slupecki T-functions* ◇ B50 ◇
FADINI, A. *Il calcolo delle classi in una logica a tre valori di verita* ◇ B50 E70 ◇
FOXLEY, E. *The determination of all Sheffer functions in 3-valued logic, using a logical computer* ◇ B35 B50 ◇
GNIDENKO, V.M. *Ermittlung der Ordnung der praevollstaendigen Klassen von Funktionen der dreiwertigen Logik (Russisch)* ◇ B50 ◇
JOBE, W.H. *Functional completeness and canonical forms in many-valued logics* ◇ B50 ◇
MOISIL, G.C. *Sur la logique a trois valeurs de Lukasiewicz* ◇ B50 ◇
MOSTOWSKI, ANDRZEJ *L'espace des modeles d'une theorie formalisee et quelques-unes de ses applications* ◇ B50 C07 C40 C57 C80 C85 C90 D45 ◇
NAKAMURA, A. *On the many-valued logics and an axiomatic system for them (Japanese)* ◇ B50 ◇
REICHBACH, J. *On the connection of the first-order functional calculus with many-valued propositional calculi* ◇ B10 B50 ◇
ROSE, A. *A simplified self m-al set of primitive functors for the m-valued propositional calculus* ◇ B50 ◇
ROSE, A. *An alternative generalisation of the concept of duality* ◇ B50 ◇
ROSE, A. *Extensions of some theorems of Anderson and Belnap* ◇ B50 ◇
ROSE, A. *Sur les applications de la logique polyvalente a la construction des machines Turing* ◇ B50 D10 ◇
ROSE, A. *Sur un ensemble de foncteurs primitifs pour le calcul propositionnel a m valeurs lequel constitue son propre m-al* ◇ B50 ◇
ROSE, A. *Sur un ensemble complet de functeurs primitifs independants pour le calcul propositionnel trivalent lequel constitue son propre trial* ◇ B50 ◇
SALOMAA, A. *On the number of simple bases of the set of functions over a finite domain* ◇ B50 ◇
SALOMAA, A. *Some completeness criteria for sets of functions over a finite domain* ◇ B50 ◇
SCARPELLINI, B. *Die Nichtaxiomatisierbarkeit des unendlichwertigen Praedikatenkalkuels von Lukasiewicz* ◇ B50 D35 ◇
TIRNOVEANU, M. *Sur les extensions des types P et Q de la logique $\mathfrak{L}_{\Omega^s}$ (Romanian) (Russian and French summaries)* ◇ B50 ◇
WILHELMY, A. *Bemerkungen zur Semantik quantifizierter mehrwertiger logistischer Systeme* ◇ B50 ◇

1963

BELLUCE, L.P. & CHANG, C.C. *A weak completeness theorem for infinite valued first-order logic* ◇ B50 ◇
CHANG, C.C. *Logic with positive and negative truth values* ◇ B50 ◇
CHANG, C.C. *The axiom of comprehension in infinite valued logic* ◇ B50 E70 ◇
CLAY, R.E. *A standard form for Lukasiewicz many-valued logics* ◇ B50 ◇
GNIDENKO, V.M. *Determinarea ordinelor claselor aproape complete in logica trivalenta (Romanian)* ◇ B50 ◇
HAY, L. *Axiomatization of the infinitive-valued predicate calculus* ◇ B50 ◇
IVAS'KIV, YU.L. & RABINOWICZ, Z.L. *A class of canonical forms for the representation of 3-valued functions* ◇ B50 ◇
KIRIN, V.G. *On the polynomial representation of operators in the n-valued propositional calculus* ◇ B50 ◇
KLOSS, B.M. & NECHIPORUK, EH.I. *Ueber eine Klassifikation der Funktionen der mehrwertigen Logik (Russisch)* ◇ B50 ◇
KRNIC, L. *A note on enumeration of bases of algebra of logic generated by functions of one and two variables (Russian)* ◇ B50 ◇
MO, SHAOKUI *The axiomatization of m-valued matrix systems (Chinese)* ◇ B50 ◇
MO, SHAOKUI *Two complete n-valued matrix-systems (Chinese)* ◇ B50 ◇
MOISIL, G.C. *Les logiques non-chrysippiennes et leurs applications* ◇ B50 G20 ◇
MOSTOWSKI, ANDRZEJ *The Hilbert ε function in many-valued logics* ◇ B50 ◇
NAKAMURA, A. *A note on truth-value functions in the infinitely many-valued logics* ◇ B50 ◇

NAKAMURA, A. *On a simple axiomatic system of the infinitely many-valued logic based on* (∧ , →) ◇ B50 ◇
NAKAMURA, A. *On an axiomatic system of the infinitely many-valued threshold predicate calculi* ◇ B50 ◇
REICHBACH, J. *About connection of the first-order functional calculus with many valued propositional calculi* ◇ B10 B50 ◇
SALOMAA, A. *On basic groups for the set of functions over a finite domain* ◇ B50 ◇
SALOMAA, A. *Some analogues of Sheffer functions in infinite-valued logics* ◇ B50 ◇
SALOMAA, A. *Some completeness criteria for sets of functions over a finite domain II* ◇ B50 ◇
SKOLEM, T.A. *Studies on the axiom of comprehension* ◇ A05 B50 E30 E70 ◇
TEH, H.H. *On 3-valued sentential calculus. An axiomatic approach* ◇ B50 ◇
TURQUETTE, A.R. *Independent axioms for infinite-valued logic* ◇ B50 ◇
TURQUETTE, A.R. *Modality, minimality, and many valuedness* ◇ A05 B45 B50 ◇
VISHIN, V.V. *Identical transformations in four-valued logic (Russian)* ◇ B50 ◇
YABLONSKIJ, S.V. *On superpositions of functions in* P_k *(Russian)* ◇ B50 ◇

1964

BELLUCE, L.P. *Further results on infinite valued predicate logic* ◇ B50 ◇
FENSTAD, J.E. *On the consistency of the axiom of comprehension in the Lukasiewicz infinite valued logic* ◇ B50 E70 ◇
GOLUNKOV, YU.V. *On the equivalence of formulae realizing transformations of functions of a k-valued logic (Russian)* ◇ B50 ◇
MENGER, K. *Superassociative systems and logical functors* ◇ B50 C05 ◇
MOISIL, G.C. *Sur les logiques de Lukasiewicz a un nombre fini de valeurs* ◇ B50 ◇
NAKAMURA, A. *On the relation between modal logic and many-valued logic (Japanese)* ◇ B45 B50 ◇
POGORZELSKI, W.A. *The deduction theorem for Lukasiewicz many-valued propositional calculi (Polish and Russian summmaries)* ◇ B50 ◇
REICHBACH, J. *A note about connection of the first-order functional calculus with many valued propositional calculi* ◇ B10 B50 ◇
RESCHER, N. *Quantifiers in many-valued logic* ◇ B50 ◇
ROSE, A. *Formalisations de calculus propositionnels polyvalents a foncteurs variables* ◇ B50 ◇
ROSE, A. *Note sur la formalisation de calculs propositionnels polyvalents a foncteurs variables* ◇ B50 ◇
ROSE, A. *Sur la saturation de l'axiome trouve par Lukasiewicz pour le calcul c-0 a foncteurs variables* ◇ B50 ◇
SADOWSKI, W. *A proof of axiomatizability of certain n-valued sentential calculi (Polish) (Russian and English summaries)* ◇ B50 ◇
SALOMAA, A. *On infinitely generated sets of operations in finite algebras* ◇ B50 C05 ◇
SCHOCK, R. *On denumerable many-valued logics* ◇ B50 ◇
SCHOCK, R. *On finitely many-valued logics* ◇ B50 ◇
SCHOFIELD, P. *On a correspondence between many-valued and two-valued logics* ◇ B50 ◇
SHESTAKOV, V.I. *On the relationship between certain three-valued logical calculi (Russian)* ◇ B50 ◇
SIOSON, F.M. *Further axiomatizations of the Lukasiewicz three-valued calculus* ◇ B50 ◇
SOSINSKIJ, L.M. *On the representation of functions by irreversible superpositions in three-valued logic (Russian)* ◇ B50 ◇
WANG, SHIQIANG & WU, WANGMING *On the classification of complemented lattices according to their tautologies. I (Chinese)(English summary)* ◇ B50 ◇
YABLONSKIJ, S.V. & ZAKHAROVA, E.YU. *On some properties of essential functions from* P_k *(Russian)* ◇ B50 E20 ◇

1965

AJZENBERG, N.N. *Representation of a sum* mod *m in a class of normal forms of functions of m-valued logic (Russian)* ◇ B50 ◇
BALUKA, E. *On verification of the expressions of many-valued sentential calculi I (Polish and Russian summaries)* ◇ B50 ◇
CHANG, C.C. *Infinite valued logic as a basis for set theory* ◇ B50 E70 ◇
CHIMEV, K.N. *Some systems of functions of* P_k *(Bulgarian) (Russian and French summaries)* ◇ B50 ◇
GAVRILOV, G.P. *On functional completeness in countable-valued logic (Russian)* ◇ B50 ◇
ITURRIOZ, L. *Axiomas para il calculo proposicional trivalente de Lukasiewicz* ◇ B50 ◇
KAKLUGIN, B.A. *Decomposition of three-valued logical functions (Russian)* ◇ B50 ◇
KLAUA, D. *Ueber einen Ansatz zur mehrwertigen Mengenlehre* ◇ B50 E70 ◇
LIN, FENGBU *On the congruenziation of k-valued logics (Chinese)* ◇ B50 ◇
LUO, ZHUKAI *On the maximal closed set in 4-valued logic (Chinese)* ◇ B50 ◇
MURSKIJ, V.L. *The existence in three-valued logic of a closed class with finite basis, not having a finite complete system of identities (Russian)* ◇ B22 B50 ◇
NAKAMURA, A. *On the infinitely many-valued double-threshold logic* ◇ B50 ◇
NICOLAU, E. *Synthese einer Klasse von Automaten, die Operatoren in der n-wertigen Logik modellieren mit gegebenen Zeitfunktionen (Rumaenisch)* ◇ B50 ◇
OVSIEVICH, B.L. *Some properties of symmetric functions of the three-valued logic (Russian)* ◇ B50 ◇
RASIOWA, H. *On non-classical calculi of classes* ◇ B20 B50 E70 G10 G25 ◇
RATSA, M.F. *Class of logic functions corresponding to the first matrix of Jaskowski (Russian)* ◇ B50 ◇
REICHBACH, J. *On the connection of the first-order functional calculus with* $\aleph_0$ *propositional calculus* ◇ B10 B50 ◇

RESCHER, N. *An intuitive interpretation of systems of four-valued logic* ◇ B50 ◇
ROMANKEVICH, A.M. *Minimization methods for functions of many-valued logic (Russian)* ◇ B50 ◇
ROSE, A. *A formalisation of Post's m-valued propositional calculus with variable functors* ◇ B50 ◇
ROSE, A. *Formalisations of certain propositional calculi with partially variable functors* ◇ B50 ◇
ROSENBERG, I.G. *La structure des fonctions de plusieurs variables sur un ensemble fini* ◇ B50 ◇
SALIJ, V.N. *Binary $\mathscr{L}$-relations (Russian)* ◇ B50 E07 E70 G10 G15 ◇
SALOMAA, A. *On some algebraic notions in the theory of truth-functions* ◇ B50 ◇
SCHOCK, R. *Some theorems on the relative strenghts of many-valued logic* ◇ B50 ◇

1966

ADZHI-VELI, YA.KH. *An application of the Blake-Poreckij method to the minimization of the functions of a many-valued logic (Russian) (Uzbek summary)* ◇ B50 ◇
BRYLL, G. *Ein neuer Beweis fuer die Vollstaendigkeit eines vielwertigen Systems des Aussagenkalkuels (Polnisch) (Deutsche und russische Zusammenfassung)* ◇ B50 ◇
BRYLL, G. *Ueber die halbvollen Systeme des Aussagenkalkuels (Polnisch) (Deutsche und russische Zusammenfassung)* ◇ B50 ◇
CARACCIOLO, A. *M-valued logics and m-ary selection functions* ◇ B50 ◇
CHANG, C.C. & KEISLER, H.J. *Continuous model theory* ◇ B50 C05 C20 C50 C90 C98 ◇
CHIMEV, K.N. *The essential dependence of the functions from P_k (Bulgarian) (Russian and English summaries)* ◇ B50 ◇
CIGNOLI, R. *Boolean multiplicative closures I,II* ◇ B50 G05 G10 G20 ◇
DAVIES, R.O. *Two theorems on essential variables* ◇ B50 ◇
FISCH, M.H. & TURQUETTE, A.R. *Peirce's triadic logic* ◇ A10 B50 ◇
FREJVALD, R.V. *A completeness criterion for partial functions of logic and many valued logic algebras (Russian)* ◇ B50 G15 ◇
GODDARD, L. *Predicates, relations and categories* ◇ A05 B10 B50 ◇
GOLUNKOV, YU.V. *Ueber die Klasse der durch Formeln realisierten Abbildungen des P_2^n auf sich selbst (Russian)* ◇ B50 ◇
ILKKA, S. *A new arithmetization for finitely many-valued propositional calculi* ◇ B50 ◇
ISEKI, K. & TANAKA, S. *A pentavalued logic and its algebraic theory* ◇ B50 ◇
ISEKI, K. *Some three valued logics and its algebraic representations* ◇ B50 ◇
KIRIN, V.G. *Gentzen's method for the many-valued propositional calculi* ◇ B50 ◇
KOGA, Y. & MINE, H. *A property of three-valued functions of one variable* ◇ B50 ◇
MAGARI, R. *Sull'introduzione dei connettivi in un calcolo generale. calcoli generali. III* ◇ B50 ◇
MAL'TSEV, A.I. *Iterative algebras and Post varieties (Russian)* ◇ B50 C05 G20 ◇
MCCALL, S. & MEYER, R.K. *Pure three-valued Lukasiewiczian implication* ◇ B50 ◇
MENGER, K. & WHITLOCK, H.I. *Two theorems on the generation of systems of functions* ◇ B50 ◇
MEYER, R.K. *Pure denumerable Lukasiewiczian implication* ◇ B50 ◇
MOISIL, G.C. *Application des algebres de Lukasiewicz a l'etude des schemas a contacts et relais. I (Polish)* ◇ B50 B70 ◇
POGORZELSKI, W.A. *A note on models of two-valued implicational-negational logic in many-valued Lukasiewicz's logics (Polish)* ◇ B50 ◇
PRUCNAL, T. *A proof of completeness of the three-valued C-N sentential calculus of Lukasiewicz (Polish and Russian summaries)* ◇ B50 ◇
RATSA, M.F. *A criterion of functional completeness in logic corresponding to the first Jaskowski matrix (Russian)* ◇ B50 B55 ◇
ROSE, A. *A formalisation of the m-valued Lukasiewicz implicational propositional calculus with variable functors* ◇ B50 ◇
ROSE, A. *A formalisation of an infinite-valued propositional calculus in which the truth values are lattice ordered* ◇ B50 ◇
ROSE, A. *An algebraic model of the m-valued propositional calculus with variable functors* ◇ B50 G20 ◇
ROSE, A. *Formalisation d'un calcul propositionnel a un nombre infini de valeurs lesquelles forment un treillis* ◇ B50 G10 ◇
ROSE, A. *Formalisation de certains calculs propositionnels polyvalents a foncteurs variables au cas ou certaines valeurs sont sur designees* ◇ B50 ◇
ROSENBERG, I.G. *Zu einigen Fragen der Superposition von Funktionen mehrerer Veraenderlicher* ◇ B50 C05 ◇
ROUTLEY, R. *On a signification theory* ◇ B50 ◇
SCHOFIELD, P. *Complete subsets of mappings over a finite domain* ◇ B50 ◇
STOJAKOVIC, M. *Matrizeninversion in der algebraischen Theorie der Wahrheitsfunktionen der mehrwertigen Logik* ◇ B50 ◇
WHEELER, R.F. *Complete connectives for the 3-valued propositional calculus* ◇ B50 ◇
ZAKHAROVA, E.YU. *On a sufficient condition for completeness in P_k (Russian)* ◇ B50 ◇

1967

ABDUGALIEV, U.A. *Normal forms of k-valued logic (Russian)* ◇ B50 ◇
ABIAN, A. & LAMACCHIA, S.E. *Examples of generalized Sheffer functions* ◇ B50 ◇
ABIAN, A. & LAMACCHIA, S.E. *Some generalized Sheffer functions* ◇ B50 ◇
ACKERMANN, R.J. *An introduction to many-valued logics* ◇ B50 B98 ◇
ACKERMANN, R.J. *Introduction to many-valued logics* ◇ B50 B98 ◇

BAJRAMOV, R.A. *A series of precomplete classes in k-valued logic (Russian)* ◇ B50 ◇
BAJRAMOV, R.A. *On a question of functional completeness in many-valued logic (Russian)* ◇ B50 ◇
BAJRAMOV, R.A. *Sheffer functions in many-valued logic (Russian)* ◇ B50 ◇
BRYLL, G. & PRUCNAL, T. & SLUPECKI, J. *Some remarks on three-valued logic of J. Lukasiewicz (Polish and Russian summaries)* ◇ B50 ◇
BURLE, G.A. *Classes of k-valued logic which contain all functions of a single variable (Russian)* ◇ B50 ◇
CHIMEV, K.N. *A certain kind of dependence of certain functions of P_k on their arguments (Bulgarian) (Russian and French summaries)* ◇ B50 ◇
CHIMEV, K.N. *The dependence of the functions of P_k on their arguments (Bulgarian) (Russian and French summaries)* ◇ B50 ◇
COPI, I.M. & GOULD, JAMES A. *Many valued logic, introduction* ◇ B50 ◇
GORELIK, A.G. *Geometric interpretation of a three-valued algebra (Russian)* ◇ B50 ◇
GRAHAM, R.L. *On n-valued functionally complete truth functions* ◇ B50 ◇
JORGENSEN, C.K. *The extensional classification of categorical propositions and trivalent logic* ◇ B50 ◇
MOISIL, G.C. *Sur le calcul des propositions de type superieur dans la logique de Lukasiewicz a plusieurs valeurs* ◇ B50 ◇
NIELAND, J.J.F. *Beth's tableau-method* ◇ B05 B25 B50 F07 ◇
O'CARROLL, M.J. *A three-valued, non-levelled logic consistent for all self-reference* ◇ B50 ◇
PRUCNAL, T. *A proof of the axiomatizability of Lukasiewicz's three-valued implicational propositional calculus (Polish) (Russian and English summaries)* ◇ B50 ◇
REICHBACH, J. *Boolean many valued calculi with quantifiers and a homogeny proof of Goedel-Skolem-Loewenheim-Herbrand's theorems* ◇ B50 ◇
ROSE, A. *A formalisation of the $\aleph_0$-valued Lukasiewicz implicational propositional calculus with variable functors* ◇ B50 ◇
ROSE, A. *A formalisation of the m-valued Lukasiewicz propositional calculus with super-designated truth-values* ◇ B50 ◇
ROSE, A. *Formalisation of the $\aleph_0$-valued Lukasiewicz propositional calculus with variable functors* ◇ B50 ◇
ROSE, A. *Nouveau calcul de non-contradiction* ◇ B50 ◇
ROSE, A. *Nouveau calcul de non-contradiction (methode alternative)* ◇ B50 ◇
ROSSER, J.B. & TURQUETTE, A.R. *Many valued logics* ◇ A05 B50 ◇
ROUSSEAU, G. *Completeness in finite algebras with a single operation* ◇ B50 C05 ◇
ROUSSEAU, G. *Sequents in many valued logic I* ◇ B50 ◇
RVACHEV, V.L. & SHKLYAROV, L.I. *Some functionally closed classes of functions in k-valued logic (Russian)* ◇ B50 ◇
SINGER, F.R. *Some Sheffer functions for m-valued logics* ◇ B50 ◇
SLUPECKI, J. *The full three-valued propositional calculus* ◇ B50 ◇
TAKAGI, M. *On an extension of two-valued logic. Extended boolean function and its application (Japanese summary)* ◇ B50 ◇
TAUTS, A. *The definition of truth values as formulae (Russian) (Estonian and German summaries)* ◇ B50 ◇
THOMAS, I. *Three-valued propositional fragments with classical implication* ◇ B50 ◇
TOLSTOVA, YU.N. *Modelling l-valued logic in k-valued logic k>1 (Russian)* ◇ B50 ◇
TURQUETTE, A.R. *Peirce's phi and psi operators for triadic logic* ◇ A05 A10 B50 ◇
VUCKOVIC, V. *Recursive models for three-valued propositional calculi with classical implication* ◇ B50 C57 C90 F30 ◇
ZAKHAROVA, E.YU. *Criteria of completeness for systems of functions from P_k (Russian)* ◇ B50 ◇

1968

AJZENBERG, N.N. *The representation of functions of k-valued logic by polynomials mod k (Russian)* ◇ B50 ◇
BAJRAMOV, R.A. *Letter to the editors (Russian)* ◇ B50 ◇
BAJRAMOV, R.A. *Predicate stabilizers and Sheffer functions in a finite-valued logic (Russian)* ◇ B50 ◇
CALL, R.L. *A minimal canonically complete M-valued proper logic for each M* ◇ B50 ◇
CARVALLO, M. *Logique a trois valeurs. Logique a seuils* ◇ B50 ◇
CARVALLO, M. *Sur la resolution des equations de Post a v valeurs* ◇ B50 G20 ◇
CAZACU, C. *Sur les logiques m-valentes* ◇ B50 ◇
CHIMEV, K.N. *Separable pairs and the representability of the functions of P_k (Bulgarian) (Russian and French summaries)* ◇ B50 ◇
CHIMEV, K.N. & KUTIRKOV, G.A. *The first order variables of the functions of many-valued logic (Bulgarian) (Russian and French summaries)* ◇ B50 ◇
CHIMEV, K.N. *The strongly essential variables of the functions of P_k (Bulgarian) (Russian and French summaries)* ◇ B50 ◇
CHIMEV, K.N. *The various kinds of dependence of the functions of many-valued logic on their arguments (Bulgarian) (Russian and French summaries)* ◇ B50 ◇
ENTRESS, G. *Bemerkungen ueber eine Logik fuer kontinuierliche Groessen* ◇ B50 G05 ◇
FREJVALD, R.V. *Completeness up to coding of systems of functions of k-valued logic and the complexity of its determination (Russian)* ◇ B50 D15 ◇
GIVONE, D.D. & SNELSIRE, R.W. *Final report on the design of multiple-valued logic systems* ◇ B50 ◇
KIRIN, V.G. *Post algebras as semantic bases of some many-valued logics* ◇ B50 G20 ◇
MIURA, S. & NAGATA, S. *Certain methods for generating series of logics* ◇ B50 ◇

MOISIL, G.C. *Sur la logique des propositions a modeles partiellement ordonnes* ◇ B50 ◇
OHYA, T. *Many valued logics extended to simple type theory* ◇ B50 ◇
ONO, K. *A remark on Peirce's rule in many-valued logics* ◇ B50 ◇
PETCU, A. *The definition of the trivalent Lukasiewicz algebras by three equations* ◇ B50 ◇
POPOVICI, N. *On the consensus theory for the trivalent argument functions* ◇ B50 ◇
POSPELOV, D.A. & TOSIC, Z. *Polynomdarstellungen in mehrwertigen Logiken (Russisch)* ◇ B50 ◇
PRUCNAL, T. *A definability criterion for the functions in Lukasiewicz's matrices (Polish) (Russian and English summaries)* ◇ B50 ◇
RESCHER, N. *Conspectus of recent work in many-valued logic* ◇ B50 ◇
ROSE, A. *Binary generators for the m-valued and $\aleph_0$-valued Lukasiewicz propositional calculi* ◇ B50 ◇
ROSE, A. *Formalisations of some $\aleph_0$-valued Lukasiewicz propositional calculi* ◇ B50 ◇
ROSENBERG, J. *Functional completeness in one variable* ◇ B50 ◇
RVACHEV, V.L. & SHKLYAROV, L.I. *R_k-functions (Ukrainian) (Russian and English summaries)* ◇ B50 ◇
RVACHEV, V.L. & SHKLYAROV, L.I. *Predicate description of domains of complex form (Russian)* ◇ B50 ◇
STOICHITA, R. *Der Wahrheitsbegriff. Beziehung der zweiwertigen zur mehrwertigen Logik* ◇ B50 ◇
TAKAHASHI, MOTO-O *Many-valued logics of extended Gentzen style. I* ◇ B50 ◇
TAKAMATSU, T. *On foundation of three-valued logic (Japanese)* ◇ B50 ◇
TAUTS, A. *Logic as a classification of formulae (Russian) (Estonian and German summaries)* ◇ B50 ◇
THOMAS, I. *The rule of Peirce* ◇ B50 ◇
TURQUETTE, A.R. *Dualizable quasi-strokes for m-state automata* ◇ B50 D05 ◇
TURQUETTE, A.R. *The Pascal triangle of Post sets for Lukasiewicz's many-valued logics* ◇ B50 ◇
VARLET, J.C. *Algebres de Lukasiewicz trivalentes* ◇ B50 ◇
ZINOV'EV, A.A. *An outline of many-valued logic (Russian)* ◇ A05 B50 ◇
ZOLL, E.J. *Logic: A programmed text for 2-valued and 3-valued logics* ◇ B50 B98 ◇

1969

ABDUGALIEV, U.A. *On the question of the representation of functions of a k-valued logic by a normal form I (Russian)* ◇ B50 ◇
ALEKSEEV, V.B. *The simple basis of a k-valued logic (Russian)* ◇ B50 ◇
BAJRAMOV, R.A. *Criteria of fundamentality for permutation groups and for transformation semigroups (Russian)* ◇ B50 ◇
BAJRAMOV, R.A. *The predicative characterizability of subalgebras of many-valued logic (Russian)* ◇ B50 ◇
BOERNER, W. *Eine Klasse von Darstellungen der Funktionen der k-wertigen Logik* ◇ B50 ◇
DUNDERDALE, H. *Current-mode circuits for ternary-logic realisation* ◇ B50 B70 ◇
EDMONDS, E.A. *Independence of Rose's axioms for m-valued implication* ◇ B50 ◇
FINN, V.K. *The precompleteness of a class of functions that corresponds to the three-valued logic of J.Lukasiewicz (Russian)* ◇ B50 ◇
GAVRILOV, G.P. *On the power of a set of limit logics with a finite base (Russian)* ◇ B50 ◇
HANSEL, G. & LEVY, G. *Construction d'un schema fonctionnel en logique a plusieurs valeurs* ◇ B50 ◇
HARRISON, M.A. *Sur la classification des fonctions logiques a plusieurs valeurs* ◇ B50 ◇
HENDRY, H.E. & MASSEY, G.J. *On the concepts of Sheffer functions* ◇ B50 ◇
IVAS'KIV, YU.L. & POSPELOV, D.A. & TOSIC, Z. *Representations in many-valued logics (Russian) (English summary insert)* ◇ B50 ◇
JANIN, P. *Bases de l'algebrisation de la logique trivalente* ◇ B50 ◇
JORGOVIC, M.M. *A remark concerning a base for many-valued propositional algebra (Serbo-Croatian) (English summary)* ◇ B50 ◇
KASAHARA, Y. & KITAHASHI, T. & NOMURA, H. & TEZUKA, Y. *The characterizing parameters of ternary logical functions and their applications to threshold logical functions* ◇ B50 ◇
KHASIN, L.S. *The realization of monotone symmetric functions by formulae in the base* $\vee$, $\wedge$, $\neg$ *(Russian)* ◇ B05 B50 ◇
KUDRYAVTSEV, V.B. & YABLONSKIJ, S.V. & ZAKHAROVA, E.YU. *Precomplete classes in k-valued logics (Russian)* ◇ B50 ◇
LEVY, G. *Formes disjonctives minimales dans une logique a plusieurs valeurs* ◇ B50 ◇
MASSEY, G.J. *Sheffer functions for many-valued S5 modal logics* ◇ B45 B50 ◇
NECHIPORUK, EH.I. *The topological principles of autocorrection (Russian)* ◇ B50 ◇
PORAT, D.I. *Three-valued digital systems* ◇ B50 ◇
RASIOWA, H. *A theorem on the existence of prime filters in Post algebras and the completeness theorem for some many-valued predicate calculi* ◇ B50 G20 ◇
RATSA, M.F. *On a class of ternary logic functions corresponding to the first Jaskowski matrix (Russian)* ◇ B50 ◇
RESCHER, N. *Many-valued logic* ◇ B50 ◇
ROMANKEVICH, A.M. *Minimization of multi-valued functions in a system that includes all one-place functions* ◇ B50 ◇
ROSE, A. *Nouvelle formalisation du calcul propositionnel a $\aleph_0$ valeurs de Lukasiewicz dans le cas ou il y a des foncteurs variables* ◇ B50 ◇
ROSE, A. *Some many-valued propositional calculi without single generators* ◇ B50 ◇
ROSENBERG, I.G. *Ueber die Verschiedenheit maximaler Klassen in P_K* ◇ B50 ◇

ROSENBERG, J. *The application of ternary semi-groups to the study of n-valued Sheffer functions* ◇ B50 ◇
ROUSSEAU, G. *Logical systems with finitely many truth-values (Russian summary)* ◇ B50 G20 ◇
RUZSA, I. *Random models of logical systems I: Models of valuing logics* ◇ B50 C90 ◇
SADE, A. *Fonctions propositionnelles monadiques dans la logique trivalente* ◇ B50 ◇
SADE, A. *Sur le premier systeme de Lukasiewicz* ◇ B50 ◇
SCHOFIELD, P. *Independent conditions for completeness of finite algebra with a single generator* ◇ B50 ◇
TAUTS, A. *Nonregular first order predicate calculus (Russian)* ◇ B50 ◇
TURQUETTE, A.R. *Generalizable Kleene logics* ◇ B50 ◇
TURQUETTE, A.R. *Peirce's complete systems of triadic logic* ◇ A05 A10 B50 ◇
VARSHAVSKIJ, V.I. *A majority operation in a multi-valued logic (Russian)* ◇ B50 ◇
WASZKIEWICZ, J. & WEGLORZ, B. *On products of structures for generalized logics (Polish and Russian summaries)* ◇ B50 C20 C30 C90 ◇

1970

ABDUGALIEV, U.A. *On the question of the representation of functions of a k-valued logic by a normal form II (Russian)* ◇ B50 ◇
BELNAP JR., N.D. & MCCALL, S. *Every functionally complete m-valued logic has a post-complete axiomatization* ◇ B50 ◇
BERTOLINI, F. *Kripke models and many-valued logics II* ◇ B50 ◇
BLACKWELL, P. *The equivalence problem for Post logic functions* ◇ B50 ◇
BLOKHINA, G.N. *The predicative description of Post classes (Russian)* ◇ B50 ◇
BRUSENTSOV, N.P. & SHAUMAN, A.M. (EDS.) *Computing techniques and questions of cybernetics No.7 (Russian)* ◇ B50 B97 ◇
BUENO SANCHEZ, E. *A certain class of many-valued modal systems (Russian)* ◇ B50 ◇
CHIMEV, K.N. *A certain class of functions (Bulgarian) (Russian and French summaries)* ◇ B50 ◇
CHIMEV, K.N. & KUTIRKOV, G.A. *Certain discrete functions (Bulgarian) (Russian and French summaries)* ◇ B50 ◇
CHIMEV, K.N. *Certain functions of many-valued logic (Bulgarian) (Russian and French summaries)* ◇ B50 ◇
CHIMEV, K.N. *Separable pairs for a certain class of functions (Bulgarian) (Russian and French summaries)* ◇ B50 ◇
CHIMEV, K.N. *The dependence of the function of k-valued logic on their arguments (Bulgarian) (Russian and French summaries)* ◇ B50 ◇
CIRULIS, J. *Quasilinear functions and precomplete classes (Russian) (Latvian and English summaries)* ◇ B50 ◇
FINN, V.K. *Certain characteristic truth-tables of classical logic and of the three-valued logic of J.Lukasiewicz (Russian)* ◇ B50 ◇
FREJVALD, R.V. *Coding of finite sets and a criterion of completeness up to coding in three-valued logic (Russian)* ◇ B50 ◇
KARPOVSKY, M.G. & MOSKALEV, EH.S. *Realization of partially-defined boolean functions by expansion in orthogonal series (Russian)* ◇ B50 ◇
KASAHARA, Y. & KITAHASHI, T. & NOMURA, H. & TEZUKA, Y. *Monotonicity of ternary logical functions and its applications to the analysis of ternary threshold functions (Japanese)* ◇ B50 ◇
KLEVACHEV, V.I. *Certain systems that are complete in P_k (Russian) (English summary)* ◇ B50 ◇
KUDRYAVTSEV, V.B. *The coverings of precomplete classes of k-valued logic (Russian)* ◇ B50 ◇
LEE, E.S. & SMITH, KENNETH C. & VRANESIC, Z.G. *A many-valued algebra for switching systems* ◇ B50 B70 ◇
LEVY, G. *Sur la partition de certains treillis en spheres concentriques* ◇ B50 ◇
MINE, H. & SHIMADA, R. *Permutations and lattice operations in many-valued logic (Japanese)* ◇ B50 ◇
MONTEIRO, L.F.T. *Les algebres de Heyting et de Lukasiewicz trivalentes* ◇ B50 G10 G20 ◇
MUZIO, J.C. *A decision process for 3-valued Sheffer functions I* ◇ B50 ◇
RASIOWA, H. *Ultraproducts of m-valued models and a generalization of the Loewenheim-Skolem-Goedel-Mal'cev theorem for theories based on m-valued logics (Russian summary)* ◇ B50 C20 C90 ◇
ROCKINGHAM GILL, R.R. *The Craig-Lyndon interpolation theorem in 3-valued logic* ◇ B50 C40 ◇
ROSENBERG, I.G. *Complete sets for finite algebras* ◇ B50 ◇
ROSENBERG, I.G. *Maximal clones on algebras A and A^r* ◇ B50 ◇
ROSENBERG, I.G. *Ueber die funktionale Vollstaendigkeit in den mehrwertigen Logiken. Struktur der Funktionen von mehreren Veraenderlichen auf endlichen Mengen* ◇ B50 ◇
ROUSSEAU, G. *Sequents in many valued logic II* ◇ B50 ◇
SADE, A. *Sur les axiomes de Goetlind* ◇ B50 ◇
SCHOTT, H.F. *Generalizability of the propositional and predicate calculi to infinite valued calculi* ◇ B50 ◇
SKYRMS, B. *Notes on quantification and self reference* ◇ A05 B50 ◇
SKYRMS, B. *Return of the liar: three-valued logic and the concept of truth* ◇ A05 B50 ◇
SOLODUKHIN, YU.N. *A concept of function-analogue and a criterion for functional incompleteness of systems of many-valued logic (Russian)* ◇ B50 ◇
SURMA, S.J. *The doctoral dissertation of Emil L. Post (Polish) (English summary)* ◇ A10 B50 ◇
TAHARA, M. & TANAKA, S. *Functional completeness and polypheks in 3-valued logic (Japanese)* ◇ B50 ◇
TAKAHASHI, MOTO-O *Continuous λ-ε logics* ◇ B50 ◇
TAKAHASHI, MOTO-O *Many-valued logics of extended Gentzen style II* ◇ B50 ◇
TALLET, J. *On the symmetry of many-valued logical systems* ◇ B50 ◇
TAUTS, A. *Nonfixed formulae (Russian) (Estonian and German summaries)* ◇ B50 ◇

TOLSTOVA, YU.N. *Letter to the editors (Russian)* ◇ B50 ◇

TOSIC, Z. *Polynomial representations of m-valued logical functions (Russian)* ◇ B50 ◇

TURQUETTE, A.R. *Pascal triangles of Post sets* ◇ A10 B50 ◇

VAKARELOV, D. *The concept of dogma and some systems of many-valued logic (Russian) (English summery)* ◇ A05 B50 ◇

VESSEL', KH.A. *Topological logic (Russian)* ◇ B50 ◇

VILLA, G. *Calcolo proposizionale trivalente dialettico* ◇ A05 B50 ◇

ZINOV'EV, A.A. *Minimal tautologies and contradictions in finitely many-valued logic (Russian)* ◇ B50 ◇

1971

ACKERMANN, R.J. *Matrix satisfiability and axiomatization* ◇ B50 ◇

AJZENBERG, N.N. & SEMION, I.V. & TSITKIN, A.I. *Polynomial representations of logical functions (Russian)* ◇ B50 G05 ◇

ALEKSEEV, V.B. *Simple bases and simple functions in k-valued logic (Russian) (English summary)* ◇ B50 ◇

ALEKSEEV, V.B. *The number of simple bases in k-valued logic (Russian)* ◇ B50 ◇

ALLEN, C.M. & ZAVISCA, E. *An approach to multiple-valued sequential circuit synthesis* ◇ B50 B70 ◇

ALOISIO, P. *Alcune osservazioni sulla definibilata di operatori finiti e sulla decidibilita algoritmica della completezza funzionale e deduttiva delle logiche a piu valori (English summary)* ◇ B50 ◇

APPLEBEE, R.C. & PAHI, B. *Some results on generalized truth-tables* ◇ B50 ◇

BAJRAMOV, R.A. *The classes of multiplace functions on finite sets (Russian, Azerbaijani and English summaries)* ◇ B50 ◇

BENDOVA, K. & HAJEK, P. & RENC, Z. *The GUHA method and the three-valued logic (Czech summary)* ◇ B50 ◇

BERTOLINI, F. *Kripke models and manyvalued logics* ◇ B50 F50 ◇

BLOKHINA, G.N. & KUDRYAVTSEV, V.B. *On basic groups in k-valued logics (Russian)* ◇ B50 ◇

BOICESCU, V. *Sur les systemes deductifs dans la logique Θ-valente* ◇ B50 G20 ◇

BRADY, R.T. *The consistency of the axioms of abstraction and extensionality in a three-valued logic* ◇ B50 E35 E70 ◇

CHEUNG, P.T. & SU, S.Y.H. *Algebraic properties of multi-valued symmetric switching functions* ◇ B50 ◇

CHEUNG, P.T. & SU, S.Y.H. *Computer-orientated algorithm for minimizing multi-valued switching functions* ◇ B50 B70 ◇

CHIMEV, K.N. *Certain properties of functions (Bulgarian) (Russian and French summaries)* ◇ B50 ◇

CHIMEV, K.N. *Separable pairs of functions (Bulgarian) (Russian and French summaries)* ◇ B50 ◇

DUMITRIU, A. *Logica polivalenta* ◇ B50 ◇

FINN, V.K. *Axiomatization of some 3-valued propositional calculi and of their algebras (Russian)* ◇ B50 ◇

FINN, V.K. *Ueber die Axiomatisierung einiger dreiwertiger Logiken (Russisch)* ◇ B50 ◇

FUJITA, Y. & KITAHASHI, T. & TANAKA, K. *Homomorphisms of logical functions and their applications* ◇ B50 ◇

GARDIES, J.-L. *Les particularites du systeme propositionnel trivalent de Lukasiewicz s'expliquent-elles par le conflit de deux exigences ?* ◇ B50 ◇

GEORGESCU, G. *Algebres de Lukasiewicz monadiques et polyadiques* ◇ B50 G15 ◇

GEORGESCU, G. *Les θ-ultrafiltres validants d'algebre de Lindenbaum-Tarski d'un calcul des predicats a θ valeurs* ◇ B50 G25 ◇

GEORGESCU, G. *The θ-valued Lukasiewicz algebras, I,II,III* ◇ B50 G20 ◇

GEORGESCU, G. *Theoreme de completude en logique θ-valente* ◇ B50 ◇

IVAS'KIV, YU.L. *Principles of multivalued implementation schemes (Russian)* ◇ B50 ◇

JANCZEWSKI, L.J. & SMITH, KENNETH C. & VRANESIC, Z.G. *Circuit implementations of multivalued logic* ◇ B50 ◇

KITAHASHI, T. & TANAKA, K. *An orthogonal expansion of multiple-valued logical functions* ◇ B50 ◇

KOLESNIKOV, M.A. & SHEJNBERGAS, I.M. *Sheffer functions in four valued logic (Russian)* ◇ B50 ◇

KOTS, V.S. *Synthesis of schemes of k-valued logic (Russian)* ◇ B50 ◇

KUDRYAVTSEV, V.B. *On S-systems of functions of k-valued logic (Russian)* ◇ B50 ◇

KUDRYAVTSEV, V.B. *Ueber Eigenschaften der S-Systeme von Funktionen der k-wertigen Logik (Russisch)* ◇ B50 ◇

LEE, E.T. & LEE, S.C. *Multivalued symmetric functions* ◇ B50 ◇

LIEBLER, M. & ROESSER, R. *Multiple-real-valued Walsh functions* ◇ B50 ◇

MUZIO, J.C. *A decision process for 3-valued Sheffer functions II* ◇ B50 ◇

MUZIO, J.C. *On the t-closure condition of Martin* ◇ B50 ◇

NUTTER, R.S. & SWARTWOUT, R.E. *A ternary logic minimization technique* ◇ B50 B70 ◇

PERKOWSKA, E. *Herbrand theorem for theories based on m-valued logics (Russian summary)* ◇ B50 ◇

PREPARATA, F.P. & YEH, R.T. *On a theory of continuosly valued logic* ◇ B50 ◇

ROMOV, B.A. *The uniprimitive foundations of the maximal subalgebras of Post algebras (Russian) (English summary)* ◇ B50 D03 G20 ◇

ROSE, A. *Note sur la formalisation des calculus propositionnels polyvalents* ◇ B50 ◇

ROSE, A. *Sur certains foncteurs a m arguments des calculus propositionnels a m valeurs* ◇ B50 ◇

RUZSA, I. *Random models of logical systems III: Models of quantified logics* ◇ B50 C90 ◇

RVACHEV, V.L. *A certain extension of the concept of r-function (Russian) (English summary)* ◇ B50 ◇

SCHMIDT, J. *A semigroup in function algebra* ◇ B50 C05 E20 ◇

SEESKIN, K.R. *Many-valued logic and future contingencies* ◇ A05 B50 ◇
SEPER, K. *A decision problem concerning autoduality in k-valued logic* ◇ B50 ◇
SETLUR, R.V. *Duality in finite many-valued logic* ◇ B50 ◇
SHOESMITH, D.J. & SMILEY, T.J. *Deducibility and many-valuedness* ◇ B50 ◇
SLUPECKI, J. *Proof of the axiomatizability of full many-valued systems of calculus of propositions* ◇ B50 ◇
SOKOLOV, V.A. *Remarks on the class of partial functions of countably-valued logic (Russian)* ◇ B50 ◇
TAKAOKA, T. *A construction method of fail-safe systems for many-valued logic* ◇ B50 ◇
THOMAS, I. *A proof of a theorem of Lukasiewicz* ◇ B50 ◇
WASZKIEWICZ, J. *The notions of isomorphism and identity for many-valued relational structures (Polish and Russian summaries)* ◇ B50 C90 ◇
ZAKHAROVA, E.YU. *A remark on the maximal order of a simple basis in P_k (Russian)* ◇ B50 ◇

1972

BAJRAMOV, R.A. *Deduction of criteria for basicness of permutation groups and transformation semigroups from the Rosenberg theorem (Russian) (English summary)* ◇ B50 G15 ◇
BECCHIO, D. *Nouvelle demonstration de la completude du systeme de Wajsberg axiomatisant la logique trivalente Lukasiewicz* ◇ B50 ◇
BELL, N. & PAGE, E.W. & THOMASON, M.G. *Extension of the Boolean difference concept to multi-valued logic systems* ◇ B50 ◇
BLANCHE, R. *Sur la trivalence* ◇ B50 ◇
BOBROV, A.E. & ZHURKIN, V.A. *Minimierung von Funktionen einer mehrwertigen Logik in einem System, das die zyklische Negation enthaelt (Russisch)* ◇ B50 ◇
BRADY, R.T. *The relative consistency of the class axioms of abstraction and extensionality and the axioms of NBG in a three-valued logic* ◇ B50 E35 E70 ◇
BUENO SANCHEZ, E. & RUIZ SHULCLOPER, J. *Construction of many-valued logics. Method of multiplication of matrices (Spanish)* ◇ B50 ◇
CHIMEV, K.N. *On second order variables (Bulgarian) (Russian and French summaries)* ◇ B50 ◇
CHIMEV, K.N. *The representability of certain functions (Bulgarian) (Russian and French summaries)* ◇ B50 ◇
COOPER, G.A. & PATT, Y.N. *Toward a characterization of logically complete switching functions in k-valued logic* ◇ B50 ◇
D'AMORE, B. & MATTEUZZI, M.L.M. *Sulla definizione di una implicazione nelle logiche polivalenti che soddisfi alla transitivita a che rispetti la struttura di reticoli di Boole (English summary)* ◇ B50 G05 G20 ◇
DEMETROVICS, J. *On cardinal numbers of sets of precomplete classes in limit logics (Russian) (English summary)* ◇ B50 ◇
DOERBAND, W. *Ein Vorschlag zur systematischen Bezeichnung aller n-wertigen und k-stelligen booleschen Funktionen und ein Algorithmus, nach dem auf dieser Grundlage die Funktionswerte dieser Funktionen berechnet werden koennen* ◇ B50 ◇
FUJITA, Y. & KITAHASHI, T. & TANAKA, K. *A consideration on compositions of many-valued logical functions* ◇ B50 ◇
FUJITA, Y. & KITAHASHI, T. & TANAKA, K. *An extension of the theorem of isomorphism concerning the compositions of many-valued switching functions* ◇ B50 ◇
GORELIK, A.G. *An indexed three-valued calculus and its geometric interpretation (Russian)* ◇ B50 ◇
HAMACHER, V.C. & VRANESIC, Z.G. *Ternary logic in parallel multipliers* ◇ B50 B70 ◇
HASEGAWA, T. *Many-valued logic and its applications (Japanese)* ◇ B50 ◇
KAEMMERER, W. *Ein Algorithmus zur Berechnung der Werte einer durch ihre charakteristische Zahl gegebenen n-wertigen und k-stelligen booleschen Funktion* ◇ B50 ◇
KORNEJCHUK, V.I. & TARASENKO, V.P. & TESLENKO, A.K. *Synthesis of combination schemes in a certain class of multivalued algebras (Russian)* ◇ B50 ◇
LEIGH, V. *Self-dual sets of unary and binary connectives for the 3-valued propositional calculus* ◇ B50 ◇
LEVIN, V.A. *Infinite-valued logic and transient processes in finite automata (Russian)* ◇ B50 D05 ◇
LOADER, J. *Universal decision elements in m-valued logic* ◇ B50 ◇
MADUCH, M. *A definability criterion for the operations in finite Lukasiewicz's implicational matrices* ◇ B50 ◇
MAYET, R. *Relations entre les anneaux Booleens, les anneaux monadiques et les algebres trivalentes de Lukasiewicz* ◇ B50 G05 G15 G20 ◇
MEYER, R.K. & PARKS, R.Z. *Independent axioms for the implicational fragment of Sobocinski's three-valued logic* ◇ B50 ◇
MOISIL, G.C. *Exemples de propositions a plusieurs valeurs* ◇ B50 ◇
MOISIL, G.C. *La logique des concepts nuances* ◇ B50 ◇
MOISIL, G.C. *Les axiomes des algebres de Lukasiewicz n-valentes* ◇ B50 ◇
MOISIL, G.C. *Les ensembles flous et la logique a trois valeurs* ◇ B50 E70 E72 ◇
MOISIL, G.C. *Sur les algebres de Lukasiewicz θ-valentes* ◇ B50 ◇
MONTEIRO, L.F.T. *Algebre du calcul propositionel trivalent de Heyting* ◇ B50 G10 ◇
MOTOHASHI, N. *Some investigations on many valued logics* ◇ B50 ◇
MUKAIDONO, M. *On the B-ternary logical function – a ternary logic considering ambiguity (Japanese)* ◇ B50 ◇
NGUYEN CAT HO *Generalized algebras of Post and their applications to many-valued logics with infinitely long formulas* ◇ B50 C75 G20 ◇
ORSIC, M. *Multiple-valued universal logic modules* ◇ B50 ◇
PAHI, B. *Maximal full matrices* ◇ B50 ◇
PARKS, R.Z. *A note on R-Mingle and Sobocinski's three-valued logic* ◇ B50 ◇
PARKS, R.Z. *Dependence of some axioms of Rose* ◇ B50 ◇

PREPARATA, F.P. & YEH, R.T. *Continuously valued logic* ◇ B50 G05 ◇
RASIOWA, H. *The Craig interpolation theorem for m-valued predicate calculi (Russian summary)* ◇ B50 C40 C90 ◇
ROSE, A. *Locally full $\aleph_0$-valued propositional calculi* ◇ B50 ◇
ROWE, I. *The generation of Gaussian-distributed pseudorandum noise sequences from multiple-valued logic* ◇ B50 ◇
SALONI, Z. *Gentzen rules for the m-valued logic* ◇ B50 ◇
SARRIS, A.A. & SU, S.Y.H. *The relationship between multivalued swiching algebra and boolean algebra under different definitions of complement* ◇ B50 B70 G05 ◇
SINTONEN, L. *On the realization of functions in n-valued logic* ◇ B50 B70 ◇
SLUPECKI, J. *A criterion of fullness of many-valued systems of propositional logic* ◇ B50 ◇
SUCHON, W. *On non equivalence of two definitions of the algebras of Lukasiewicz* ◇ B50 G20 ◇
TARASOV, V.V. *Certain properties of the essential function of k-valued logic (Russian)* ◇ B50 ◇
TARASOV, V.V. *Letter to the editors (Diskret. Analiz 20 (1972), 66 - 78) (Russian)* ◇ B50 ◇
TOKARZ, M. *On invariant systems and structural completeness of Lukasiewicz's logics* ◇ B50 ◇
TOKARZ, M. *On structural completeness of Lukasiewicz's logics (Polish and Russian summaries)* ◇ B50 ◇
TOSIC, Z. *Analytical representation of m-valued logical functions over the ring of integers modulo m* ◇ B50 ◇
VAKARELOV, D. *Notes on the semantics of three-valued Lukasiewicz logic (Russian)* ◇ B50 G20 ◇
VASILENKO, YU.A. *Multivalued structures (Russian)* ◇ B50 ◇
WOJCICKI, R. *Some results concerning many valued Lukasiewicz's calculi* ◇ B50 ◇
WOJCICKI, R. *The degrees of completeness of the finitely-valued propositional calculi (Polish summary)* ◇ B50 ◇
YABLONSKIJ, S.V. & ZAKHAROVA, E.YU. *Certain properties of nondegenerate superpositions in P_k (Russian)* ◇ B50 ◇
ZAKHAROVA, E.YU. *The realization of functions of the P_k by formulae ($k \geqslant 3$) (Russian)* ◇ B50 ◇
ZAVISCA, E. *Multiple-valued combinational function synthesis using a single gate type* ◇ B50 ◇

1973

BLOKHINA, G.N. & BUROSCH, G. & KUDRYAVTSEV, V.B. *Das Problem der Vollstaendigkeit fuer boolesche Funktionen ueber zwei Dualmengen mit nichtleerem Durchschnitt I* ◇ B50 ◇
BOICESCU, V. *On θ-valent logics (Romanian)* ◇ B50 ◇
BRADY, R.T. & ROUTLEY, R. *Don't care was made to care* ◇ B50 ◇
BRYLL, G. & SLUPECKI, J. *Syntactic proof of the L-decidability of classical logic and Lukasiewicz's three-valued logic (Polish) (English summary)* ◇ B25 B50 ◇
CABANES, A. & PICHAT, E. *Minimal representation of a function of several finite variables - applications to data bank* ◇ B50 ◇
CHIMEV, K.N. *On certain classes of functions (Bulgarian) (Russian and French summaries)* ◇ B50 ◇
CHIMEV, K.N. *Subfunctions,and the representability of certain functions of k-valued logic (Bulgarian) (Russian and French summaries)* ◇ B50 ◇
CHIMEV, K.N. *Subfunctions and strongly essential variables of functions (Bulgarian) (Russian and French summaries)* ◇ B50 ◇
CHIMEV, K.N. *The structural properties of functions in relation to their separable pairs (Bulgarian) (Russian and French summaries)* ◇ B50 ◇
DOERBAND, W. *Ein Algorithmus zur Berechnung aller Funktionswerte aller n-wertigen und k-stelligen booleschen Funktionen* ◇ B50 ◇
DUCASSE, E.G. & METZE, G. *Hazard-free realizations of Boolean functions using Post functions* ◇ B50 ◇
ELLOZY, H.A. *On the generation of all the one variable functions in m-valued logic* ◇ B50 ◇
ENGLEBRETSEN, G. *Suggested truth-tables for a three-valued sentential logic* ◇ B50 ◇
FORT, M. *Calcul des predicats trivalent* ◇ B50 ◇
FORT, M. *Deduction en logique trivalente* ◇ B50 ◇
GERASIMOV, V.A. *A completeness theorem for Bochvar's predicate logic (Russian)* ◇ B50 ◇
GNIAZDOWSKI, A. *A note on the strong adequacy of a certain matrix for the logic W (Polish) (English summary)* ◇ B50 ◇
GNIAZDOWSKI, A. *Nonexistence of finite full systems of rejected axioms for certain sentential logics (Polish) (English summary)* ◇ B50 ◇
GORLOV, V.V. *Congruences on closed Post classes (Russian)* ◇ B50 D03 ◇
GRIGOLIYA, R.SH. *Algebraic analysis of n-valued Lukasiewicz-Tarski logical systems (Russian) (Gregorian summary)* ◇ B50 G20 ◇
GROSJEAN, P.V. *La logique sur le corps de rupture des paradoxes* ◇ B50 ◇
IRVING, T.A. & NAGLE JR., H.T. *An approach to multi-valued sequential logic* ◇ B50 ◇
JANCZEWSKI, L.J. *Geometrical approach to multivalued logic function synthesis* ◇ B50 ◇
KLAUA, D. *Zur Arithmetik mehrwertiger Zahlen* ◇ B50 ◇
KRNIC, L. *Cardinals of bases in the 3-valued logic I (Serbo-Croatian summary)* ◇ B50 ◇
KUDRYAVTSEV, V.B. *The properties of S-sytems of functions of k-valued logic (Russian) (German and English summary)* ◇ B50 ◇
MADUCH, M. & PIROG-RZEPECKA, K. *An adequate matrix for a certain sentential calculus (Polish) (English summary)* ◇ B50 ◇
MIYAMA, T. *Another proof and an extension of Gill's interpolation theorem* ◇ B50 ◇
MOKLYAK, N.G. & POPOV, V.A. & SKIBENKO, I.T. *The number of types of systems of k-valued logical functions (Russian) (English summary)* ◇ B50 ◇
MUZIO, J.C. *Some large classes of n-valued Sheffer functions* ◇ B50 ◇

MUZIO, J.C. *The cosubstitution condition* ◇ B50 ◇
MUZIO, J.C. *The necessity of a condition of Shoenfield* ◇ B50 ◇
NECHIPORUK, EH.I. *The topological principles of autocorrection. II (Russian)* ◇ B50 ◇
NGUYEN CAT HO *Generalized Post algebras and their application to some infinitary many-valued logics* ◇ B50 C75 G20 ◇
ORSIC, M. *Incremental curve generation using mutiple-valued logic* ◇ B50 ◇
PATT, Y.N. *Nonlinearity: A new necessary condition for logical completeness in k-valued logic* ◇ B50 ◇
PIROG-RZEPECKA, K. & SLUPECKI, J. *An extension of the algebra of sets (Polish and Russian summaries)* ◇ B50 E70 ◇
POESCHEL, R. *Postsche Algebren von Funktionen ueber einer Familie endlicher Mengen* ◇ B50 G20 ◇
PORUS, V.N. *Ueber eine analytische Darstellungsweise mehrwertiger Matrizen (Russian)* ◇ B50 ◇
PRUCNAL, T. *Proof of structural completeness of a certain class of implicative propositional calculi (Polish and Russian summaries)* ◇ B50 ◇
RASIOWA, H. *Formalized ω^+ valued algorithmic systems (Russian summary)* ◇ B50 B75 D15 ◇
RASIOWA, H. *On generalized Post algebras of order ω^+ and ω^+ valued predicate calculi (Russian summary)* ◇ B50 G20 ◇
RINE, D.C. *A proposed multi-valued extension to ALGOL 68* ◇ B50 ◇
RINE, D.C. *Basic concepts of multiple-discrete valued logic and decision theory* ◇ B50 ◇
ROMOV, B.A. *The solving algorithm for the functional completeness problem in class of vector multiple-valued schemes (Russian)* ◇ B50 ◇
ROSE, A. *Note sur "l'erweiterter Aussagenkalkuel" ayant un seul quantificateur primitif* ◇ B50 ◇
ROSE, A. *Sur certains "erweiterter Aussagenkalkuel" dont le seul quantificateur primitif constitue son dual* ◇ B50 ◇
ROSENBERG, I.G. *Strongly rigid relations* ◇ B50 C05 C50 ◇
ROSENBERG, I.G. *The number of maximal closed classes in the set of functions over a finite domain* ◇ B50 C05 ◇
SALONI, Z. *The sequent Gentzen system for m-valued logic* ◇ B50 ◇
SAMOFALOV, K.G. *Synthese mehrwertiger Schaltungen in einer Klasse redundanter vollstaendiger Systeme (Ukrainian) (English and Russian summaries)* ◇ B50 B70 ◇
SANMARTIN ESPLUGUES, J. *Syllogistics, many-valued logic and model theory (Spanish)* ◇ A10 B20 B50 C90 ◇
SCHUH, E. *Many-valued logics and the Lewis paradoxes* ◇ A05 B50 ◇
SUCHON, W. *Application of the theory of logical matrices in the independence proofs* ◇ B50 ◇
SUCHON, W. *On defining Moisil's functors in n-valued propositional logic* ◇ B50 ◇
SURMA, S.J. *A note on indirect deduction theorems valid in Lukasiewicz's finitely-valued propositional calculi (Polish) (English and Russian summaries)* ◇ B50 ◇
SZCZECH, W. *Axiomatizability of finite matrices* ◇ B50 ◇
THELLIEZ, S. *Introduction a l'etude des structures ternaires de commutation* ◇ B50 B70 ◇
TOKARZ, M. *On mutual non-reconstructability of the Lukasiewicz calculi and their dual counterparts* ◇ B50 ◇
TSICHRITZIS, D.C. *Approximate logic* ◇ B50 ◇
URQUHART, A.I.F. *An interpretation of many-valued logic* ◇ B50 ◇
VISHNEVSKIJ, A.P. *Decomposition and minimization of multilogical functions in a single submultilogic (Russian)* ◇ B50 ◇
WOJCICKI, R. *On matrix representations of consequence operations of Lukasiewicz's calculi* ◇ B50 ◇
WOODRUFF, P.W. *On compactness in many-valued logic. I* ◇ B50 ◇
ZOLOTOVA, T.M. *Some representations of continuous-logic functions, conserving values of arguments and their complements (Russian)* ◇ B50 ◇

1974

AGRACHEV, A.A. *Superpositions of continuous functions (Russian)* ◇ B50 ◇
ALEKSEEV, V.B. *On the number of the monotone k-valued functions (Russian)* ◇ B50 ◇
BLOKHINA, G.N. & KUDRYAVTSEV, V.B. *A criterion for the basis property of groups in k-valued logic (Russian)* ◇ B50 ◇
BLYUMIN, S.L. *The geometric aspect of multivalued threshold logic (Russian) (English summary insert)* ◇ B50 ◇
BOBROV, A.E. & KOSTYANKO, N.F. & UGAROV, B.N. *The Y-complexity of systems of functions of many-valued logic (Russian)* ◇ B50 ◇
BOCHVAR, D.A. & FINN, V.K. *Quasilogical functions (Russian)* ◇ B50 ◇
BODENDIEK, R. *Ueber das Repraesentantenproblem in der k-wertigen Logik* ◇ B50 ◇
CAIN, R.G. & HONG, SEJUNE & OSTAPKO, D.L. *A practical approach to two-level minimization of multivalued logic* ◇ B50 ◇
CHUANG, H.Y.H. & DAS, S. *On the design of reliable and fail-safe digital systems using multiple-valued logic* ◇ B50 ◇
COLLIER, K.W. *A result of extending Bochvar's 3-valued logic* ◇ B50 ◇
DAHN, B.I. *Kripke-style semantics for some many-valued propositional calculi* ◇ B50 ◇
DAHN, B.I. *Meta-mathematics of some many-valued calculi* ◇ B50 C40 C90 ◇
DEMETROVICS, J. *On the properties of the minimal limit logic Q (Russian)* ◇ B50 ◇
DEMETROVICS, J. *The number of pairewise nonisomorphic limit logics (Russian)* ◇ B50 ◇
DUCASSE, E.G. & SU, S.Y.H. *Application of a multi-valued gate to the design of fault-tolerant digital systems* ◇ B50 B70 ◇
EDMONDS, E.A. *Model formation – an application of m-valued logic* ◇ B50 ◇
EPSTEIN, G. & FRIEDER, G. & RINE, D.C. *A summary of the development of multiple-valued logic as related to computer science* ◇ B50 ◇

EPSTEIN, G. & HORN, A. *Propositional calculi based on subresiduation* ◇ B50 F50 G10 ◇
EPSTEIN, G. & FRIEDER, G. & RINE, D.C. *The development of multiple-valued logic as related to computer science* ◇ B50 ◇
ETIEMBLE, D. & ISRAEL, M. *A new concept for ternary logic elements* ◇ B50 ◇
EVENDEN, J. *Generalized logic* ◇ B50 ◇
FINN, V.K. *A criterion of functional completeness for* $\mathfrak{B}^3$ ◇ B50 ◇
FINN, V.K. *A test for functional completeness for* $\mathfrak{B}^3$ *(Russian)* ◇ B50 ◇
FINN, V.K. *Axiomatization of some three-valued propositional calculi and their algebras (Russian)* ◇ B50 G20 ◇
GAVRILOV, G.P. *The power of the set of classes of finite height in countable-valued logic (Russian)* ◇ B50 ◇
GILES, R. *A non-classical logic for physics* ◇ B50 ◇
GOLDBERG, H. & LEBLANC, H. & WEAVER, G.E. *A strong completeness theorem for 3-valued logic* ◇ B50 ◇
GRANT, J. *A non-truth-functional 3-valued logic* ◇ B50 ◇
HARNAU, W. *Die vertauschbaren Funktionen der k-wertigen Logik und ein Basisproblem* ◇ B50 ◇
HARNAU, W. *Die Definition von Vertauschbarkeitsmengen in der k-wertigen Logik und das Maximalitaetsproblem* ◇ B50 ◇
HARNAU, W. *Einige numerische Invarianten der teilweise geordneten Menge* Φ_k *von Vertauschbarkeitsmengen der k-wertigen Logik* ◇ B50 ◇
HUBER, O. *An axiomatic system for multidimensional preference* ◇ B50 ◇
JORDAN, I.B. & MOUFTAH, H.T. *Integrated circuits for ternary logic* ◇ B50 ◇
KANG, A.N.C. & SHEN, S.N.T. *On the Boolean functions over 4-element Boolean algebra* ◇ B50 ◇
KERGALL, E. *Monomes premiers d'une fonction de plusieurs variables sur une algebre a p valeurs. Trois methodes de determination* ◇ B50 ◇
KODANDAPANI, K.L. & SETLUR, R.V. *Multi-valued algebraic generalizations of Reed-Muller canonical forms* ◇ B50 ◇
KOHOUT, L.J. *The Pinkava many-valued complete logic systems and their applications in the design of many-valued switching circuits* ◇ B50 ◇
KOLPAKOV, V.I. *Subalgebras of monotone functions in Post algebras (Russian)* ◇ B50 G20 ◇
KRIVINE, J.-L. *Langages a valeurs reelles et applications* ◇ B50 C40 C60 C90 ◇
LAM, C.I. & VRANESIC, Z.G. *Multiple-valued pseudorandom number generators* ◇ B50 ◇
LEONT'EV, V.K. *A theorem on monotone functions (Russian)* ◇ B50 ◇
LEVIN, V.A. *Equations in infinite-valued logic, and transition processes in finite automata (Russian)* ◇ B50 D05 ◇
LOADER, J. *A method for finding formulae corresponding to first order universal decision elements in m-valued logic* ◇ B50 B70 ◇
MAKSIMOVA, L.L. & VAKARELOV, D. *Semantics for* ω^+*-valued predicate calculi* ◇ B50 ◇
MALINOWSKI, G. & SPASOWSKI, M. *Dual counterparts of Lukasiewicz's sentential calculi* ◇ B50 ◇
MALINOWSKI, G. *S-algebras and the degrees of maximality of three and four valued logics of Lukasiewicz* ◇ B50 ◇
MARQUETTY, A. *Calculs propositionnels a p negations (aspect functionnel)* ◇ B50 ◇
MARQUETTY, A. *Calculs propositionnels a p negations (aspect ensembliste)* ◇ B50 ◇
MICHALSKI, R.S. *Variable-valued logic* ◇ B50 ◇
MIYAMA, T. *The interpolation theorem and Beth's theorem in many-valued logics* ◇ B50 C40 ◇
MORGAN, C.G. *A theory of equality for a class of many-valued predicate calculi* ◇ B50 ◇
MOTIL, J.M. *A multiple-valued digital "saturation" system* ◇ B50 ◇
NAGLE JR., H.T. & SHIVA, S.G. *On multiple-valued memory elements* ◇ B50 ◇
NEPEJVODA, N.N. *The language* Δ *with a weak three-valued logic (Russian)* ◇ B50 ◇
NUTTER JR., R.S. & RINE, D.C. & SWARTWOUT, R.E. *Equivalence and transformations for Post multivalued algebras* ◇ B50 ◇
POESCHEL, R. *Die funktionale Vollstaendigkeit von Funktionenklassen ueber einer Familie endlicher Mengen* ◇ B50 ◇
POPOV, V.A. & SKIBENKO, I.T. *Ueber die Komplementaritaet der Aequivalenzklassen k-wertiger logischer Funktionen (Russian)* ◇ B50 ◇
PRADHAN, D.K. *A multivalued switch algebra based on finite fields* ◇ B50 ◇
RASIOWA, H. *Post algebras as a semantic foundation of m-valued logics* ◇ B50 G20 ◇
REICHBACH, J. *About decidability, generalized models and main theorems of many valued predicate calculi* ◇ B25 B50 C90 ◇
REICHBACH, J. *Partitions in decidability and generalized models with many valued codes* ◇ B50 ◇
RINE, D.C. *Conditional and Post algebra expressions* ◇ B50 G20 ◇
RINE, D.C. *General cost factors for multiple-valued logic design of arithmetic units* ◇ B50 B70 ◇
RINE, D.C. *Multple-valued logic and computer sciences in the 20th century* ◇ B50 ◇
RINE, D.C. *There is more to boolean algebra than you would have thought* ◇ B50 G05 ◇
ROSE, A. *Sur certains ensembles de foncteurs connectifs primtifs pour des extensions du calcul propositionnel a* $\aleph_0$ *valeurs de Lukasiewicz lesquels constituent leurs propres duals* ◇ B50 ◇
ROSE, A. *Sur un calcul propositionnel a* $\aleph_0$ *valeurs ayant la disjonction conditionee comme foncteur primitif* ◇ B50 ◇
ROSE, A. *Sur un foncteur a cinq arguments d'un calcul propositionnel a* $\aleph_0$*-valeurs lequel constitue son propre dual* ◇ B50 ◇

ROSENBERG, I.G. *Completeness, closed classes and relations in multiple-valued logics* ◇ B50 ◇
ROSENBERG, I.G. *Some maximal closed classes of operations on infinite sets* ◇ B50 C05 ◇
ROUTLEY, R. & WOLF, R.G. *No rational sentential logic has a finite characteristic matrix* ◇ B50 ◇
SALOMAA, A. *Some remarks concerning many-valued propositional logics* ◇ B50 ◇
SCOTT, D.S. *Completeness and axiomatizability in many-valued logic* ◇ B50 ◇
SEESKIN, K.R. *Some remarks on truth and bivalence* ◇ B50 ◇
SESELJA, B. & VOJVODIC, G. *Implication in three-valued logic as an exponential function (Serbo-Croatian) (English summary)* ◇ B50 ◇
SLUPECKI, J. *Some remarks on the many-valued logics of Jan Lukasiewicz (Russian)* ◇ A05 B50 ◇
SMITH, KENNETH C. & VRANESIC, Z.G. *Engineering aspects of multi-valued logic systems* ◇ B50 ◇
SMITH III, W.R. *Minimization of multivalued functions* ◇ B50 ◇
STRAZDINS, I.E. *The number of self-dual types in finite-valued logics (Russian)* ◇ B50 ◇
SUCHON, W. *Definition des foncteurs modaux de Moisil dans le calcul n-valent des propositions de Lukasiewicz avec implication et negation* ◇ B50 ◇
SUCHON, W. *La methode de Smullyan de construire le calcul n-valent de Lukasiewicz avec implication et negation* ◇ B50 ◇
SURMA, S.J. *A method of the construction of finite Lukasiewiczian algebras and its application to a Gentzen-style characterization of finite logics* ◇ B50 G20 ◇
TOKARZ, M. *A method of axiomatization of Lukasiewicz logics* ◇ B50 ◇
TOKARZ, M. *A new proof of "topographic" theorem on Lukasiewicz logics* ◇ B50 ◇
TOKARZ, M. *Invariant systems of Lukasiewicz calculi* ◇ B50 ◇
TVERDOKHLEBOVA, N.N. *A number-theoretic method for the recognition of functions of k-valued logic (Russian)* ◇ B50 ◇
WESSELKAMPER, T.C. *A note on UDE's in an n-valued logic* ◇ B50 ◇
WESSELKAMPER, T.C. *Some completeness results for abelian semigroups and groups* ◇ B50 ◇
WHITE, B. *A note on natural deduction in many-valued logic* ◇ B50 ◇
WOJCICKI, R. *Note on deducibility and many-valuedness* ◇ B50 ◇
WOJCICKI, R. *The logics stronger than Lukasiewicz's three valued sentential calculus - The notion of degree of maximality versus the notion of degree of completeness* ◇ B50 ◇
WOODRUFF, P.W. *A modal interpretation of three-valued logic* ◇ B50 ◇
YABLONSKIJ, S.V. *The structure of the upper neighborhood for predicate-describable classes in P_k (Russian)* ◇ B50 ◇
ZYGMUNT, J. *A note on direct products and ultraproducts of logical matrices* ◇ B50 C05 C20 ◇

1975

ARON, E. & BARAK, A. *On the efficiency of ternary algorithms for multiplication and division* ◇ B50 B75 D20 ◇
BELLMAN, R.E. *Local logics* ◇ B50 ◇
BRUIJN DE, N.G. *Exact finite models for minimal propositional calculus over a finite alphabet* ◇ B50 ◇
CHEUNG, P.T. & PURVIS, D.M. *A computer-oriented heuristic minimization algorithm for multiple-output multi-valued switching functions* ◇ B50 ◇
CHIMEV, K.N. *Die Invarianz gewisser Eigenschaften von Funktionen bei Identifizierung ihrer Argumente (Bulgarian) (Russian and English summaries)* ◇ B50 ◇
CHIMEV, K.N. *Ueber separierte Teilmengen und stark wesentliche Variablen von Funktionen (Bulgarian) (Russian and English summaries)* ◇ B50 ◇
DEMETROVICS, J. *Certain homomorphisms and relations for limit logics (Russian)* ◇ B50 ◇
DEMETROVICS, J. *On homomorphism of limit logics (Hungarian) (Russian summary)* ◇ B50 ◇
DEMETROVICS, J. *On the principal diagonal of Sheffer functions* ◇ B50 ◇
DESCHAMPS, J.-P. & THAYSE, A. *Representation of discrete functions* ◇ B50 ◇
DORDEVIC, L.N. *Structural syntheses of sequential machines in three-valued logic* ◇ B50 ◇
DUCASSE, E.G. *Reducibility of Post functions* ◇ B50 ◇
DUJMOVIC, J.J. *Extended continuous logic and the theory of complex criteria* ◇ B50 ◇
DURAND, D. *Nouvelle demonstration de la completude fonctionelle du systeme $\{C, N, T\}$ de Rosser-Turquette, en calcul propositionnel m-valent (English summary)* ◇ B50 ◇
ELLOZY, H.A. & PATT, Y.N. *The linearity property and functional completeness in M-valued logic* ◇ B50 ◇
GAINES, B.R. & KOHOUT, L.J. *Possible automata* ◇ B50 D05 ◇
GALE, S. *Conjectures on many-valued logic, regions, and criteria for conflict resolution* ◇ B50 ◇
GILES, R. *Lukasiewicz logic and fuzzy set theory* ◇ B50 E72 ◇
GORLOV, V.V. *Kongruenzen auf abgeschlossenen Postschen Klassen* ◇ B50 ◇
GRIGOLIYA, R.SH. *On the algebras corresponding to the n-valued Lukasiewicz-Tarski logical systems* ◇ B50 G20 ◇
HARNAU, W. *Ueber Kettenlaengen der teilweise geordneten Menge Φ_k der Vertauschbarkeitsmengen der k-wertigen Logik* ◇ B50 ◇
HARNAU, W. *Vertauschbarkeitsmengen in mehrwertigen Logiken* ◇ B50 ◇
HAVRANEK, T. *Statistical quantifiers in observational calculi* ◇ B50 C80 C90 ◇
HERZBERGER, H.G. *Supervaluations in two dimensions* ◇ B50 ◇
HICKIN, K.K. & PLOTKIN, J.M. *Compactness and ρ-valued logics* ◇ B50 C75 C90 ◇

HUANG, H.K. & LEDLEY, R.S. *Multivalued logic design and Postian matrices* ◇ B50 B70 ◇
HUGHES, C.E. & SINGLETARY, W.E. *Triadic partial implicational propositional calculi* ◇ B50 ◇
JAHN, K.-U. *Eine auf den Intervall-Zahlen fussende 3-wertige lineare Algebra* ◇ B50 ◇
JAHN, K.-U. *Intervall-wertige Mengen* ◇ B50 E70 ◇
JAHN, K.-U. *Mehrwertige Mengenlehre und Klassifikationsprozesse* ◇ B50 E70 ◇
KATERINOCHKINA, N.N. *Location of the maximal upper zero of a monotonic logic-algebra function (Russian)* ◇ B50 ◇
KODANDAPANI, K.L. & SETLUR, R.V. *Reed-Muller canonical forms in multivalued logic* ◇ B50 ◇
KUDRYAVTSEV, V.B. *Ueber einige allgemeine Eigenschaften des Funktionalsystems* P_Σ ◇ B50 ◇
LABUNETS, V.G. & SITNIKOV, O.P. *Harmonic analysis of boolean functions and k-ary logic functions on finite fields (Russian)* ◇ B50 ◇
LAU, D. *Praevollstaendige Klassen von* $P_{(k,l)}$ ◇ B50 C05 ◇
LEBLANC, H. *A Henkin-type completeness proof for 3-valued logic with quantifiers* ◇ B50 ◇
LEVIN, V.A. *Equations in infinite-valued logic with deviating arguments (Russian)* ◇ B50 ◇
LEVIN, V.A. *Equations of infinite-valued logic that contain all logical operations (Russian)* ◇ B50 ◇
LEVIN, V.A. *Equations of infinite-valued logic with deviating arguments and with all possible logical operations (Russian)* ◇ B50 ◇
LOADER, J. *An alternative concept of the universal decision element in m-valued logic* ◇ B50 ◇
LOADER, J. *Second order and higher order universal decision elements in m-valued logic* ◇ B50 ◇
MAKSIMOVA, L.L. *The tautologies of* ω^+ *-valued logic (Russian)* ◇ B50 ◇
MALINOWSKI, G. *Matrix representation for the dual counterparts of Lukasiewicz n-valued sentential calculi and the problem of their degrees of maximality* ◇ B50 ◇
MARTIN, J.N. *A syntactic characterization of Kleene's strong connectives with two designated values* ◇ B50 ◇
MASSEY, G.J. *Concerning an alleged Sheffer function* ◇ B50 ◇
McCAWLEY, J.D. *Truth functionality and natural deduction* ◇ B50 ◇
MORAGA, C. & NAZARALA, J. *Bilineal separability of ternary functions* ◇ B50 ◇
MORAGA, C. *Polynomial separation of ternary functions* ◇ B50 ◇
MORGAN, C.G. *Similarity as a theory of graded equality for a class of many-valued predicate calculi* ◇ B50 ◇
MUZIO, J.C. *Ternary two-place functions that are complete with constants* ◇ B50 ◇
MUZIO, J.C. *The size of some large classes of n-valued Sheffer functions* ◇ B50 ◇
NAGATA, M. & NAKANISHI, M. & NISHIMURA, T. *Implementation of Lukasiewicz's, Kleene's and McCarthy's 3-valued logics* ◇ B50 ◇
NISHIMURA, T. & OHYA, T. *The formal system for various 3-valued logics. I* ◇ B50 ◇
PINKAVA, V. *Some further properties of the* π *-logics* ◇ B50 ◇
PLANAVERGNE, G. *Deduction et trivalence* ◇ B50 ◇
POESCHEL, R. *The number of maximal classes of functions closed with respect to superposition over a finite family of finite sets (Russian) (English summary)* ◇ B50 ◇
POESCHEL, R. *Zur Charakterisierung spezieller superpositionsabgeschlossener Funktionenklassen, die der Durchschnitt einer absteigenden Kette sind, durch invariante Relationen (English and Russian summaries)* ◇ B50 E07 ◇
PREVIALE, F. *Tavole semantiche per sistemi astratti di logica estensionale* ◇ B50 F50 ◇
RASIOWA, H. *Mixed-value predicate calculi* ◇ B50 ◇
RAUCH, J. *Ein Beitrag zu der GUHA Methode in der dreiwertigen Logik* ◇ B50 ◇
RINE, D.C. *Multiple-valued logic in programming and flowcharting* ◇ B50 ◇
ROHLEDER, H. *Ein vollstaendiger Ableitungsbegriff fuer die Aequivalenz in einem funktionell unvollstaendigen dreiwertigen Aussagenkalkuel* ◇ B50 G10 ◇
ROSE, A. *Sur un ensemble de calculs propositionnels localement sature a* $\aleph_0$ *valeurs ayant un seul foncteur primitif binaire (English summary)* ◇ B50 ◇
ROSENBERG, I.G. *Functional completeness in heterogeneous multiple-valued logics* ◇ B50 C05 ◇
ROSENBERG, I.G. *Polynomial functions over finite rings (Serbo-Croatian summary)* ◇ B50 ◇
ROSENBERG, I.G. *Special types of universal algebras preserving a relation* ◇ B50 C05 ◇
SAELI, D. *Problemi di decisione per algebre connesse a logiche a piu valori (English summary)* ◇ B25 B50 C60 G20 ◇
SCOTT, D.S. *Combinators and classes* ◇ B40 B50 E70 ◇
SPASOWSKI, M. *The degrees of completeness of dual counterparts of Lukasiewicz sentential calculi* ◇ B50 ◇
STRAZDINS, I.E. *Selfdual transformations in many-valued logics (Russian) (English summary insert)* ◇ B50 ◇
SUCHON, W. *Definition of Moisil functors in n-valued implicative-negative Lukasiewicz propositional calculus (Polish with English translation)* ◇ B50 ◇
SUCHON, W. *Matrix Lukasiewicz algebras* ◇ B50 ◇
SURMA, S.J. *A certain method for constructing Lukasiewicz algebras, and its application to the Gentzenisation of Lukasiewicz logics (Polish with English translation)* ◇ B50 ◇
SUSZKO, R. *Remarks on Lukasiewicz's three-valued logic* ◇ B50 ◇
TARASOV, V.V. *A test for the completeness of not everywhere defined functions of the algebra of logic (Russian)* ◇ B50 G20 ◇
TOKARZ, M. *Functions definable in Sugihara algebras and their fragments I* ◇ B50 G25 ◇
ULRICH, D. *Finitely-many-valued logics with infinitely-many-valued extensions: Two examples* ◇ B50 ◇

WESSELKAMPER, T.C. *A sole sufficient operator* ◇ B50 ◇
WESSELKAMPER, T.C. *Weak completeness and abelian semigroups* ◇ B50 ◇
WOJCICKI, R. *A theorem on the finiteness of the degree of maximality of the n-valued Lukasiewicz logic* ◇ B50 ◇
WOLF, R.G. *A critical survey of many-valued logics, 1966 - 1974* ◇ B50 ◇
ZACHOROWSKI, S. *A proof of a conjecture of Roman Suszko* ◇ B45 B50 F50 ◇
ZACHOROWSKI, S. *Consistency of the scheme of comprehension with the Lukasiewicz's logic (Polish and English)* ◇ B50 E35 E70 ◇
ZINOV'EV, A.A. *Logik und Sprache der Physik* ◇ A05 B50 ◇

1976

ALEKSEEV, V.B. *Deciphering of some classes of monotonic many-valued functions (Russian)* ◇ B50 ◇
AVRAMENKO, M.B. & VASILENKO, YU.A. *Erkennung der Symmetrie von Funktionen der mehrwertigen Logik (Russian)* ◇ B50 ◇
BAGYINSZKI, J. & DEMETROVICS, J. *The structure of linear classes in prime-valued logics* ◇ B50 ◇
BEAZER, R. *Products of Post algebras with applications* ◇ B50 G20 ◇
BECCHIO, D. *Calcul des sequents et deduction naturelle pour la logique trivalente de Lukasiewicz* ◇ B50 ◇
BIRNBAUM, L. *Algebre et logique tripolaire* ◇ B50 ◇
BOCHVAR, D.A. *On a general theory of logical matrices with a continuum of valences (Russian)* ◇ B50 ◇
BOCHVAR, D.A. & FINN, V.K. *Some supplements to articles on multivalued logics (Russian)* ◇ B50 ◇
BOUSSUET, G. *Random-valued switching algebra* ◇ B50 B70 ◇
BRADY, R.T. *Significance logics* ◇ B50 ◇
BRUCE, BERTRAM *A logic for unknown outcomes* ◇ B50 ◇
CHEUNG, P.T. & SU, S.Y.H. *Cubical representation for computer-aided processing of multiple-valued switching functions* ◇ B50 B70 ◇
CHEUNG, P.T. *Identification of different functional properties of multiple-valued switching functions* ◇ B50 ◇
CHIMEV, K.N. *Representability of some classes of functions of many-valued logic (Bulgarian) (Russian and French summaries)* ◇ B50 G20 G25 ◇
CLEAVE, J.P. *Comment: does many valued logic have any use* ◇ A05 B50 ◇
CLEAVE, J.P. *Quasi-boolean algebras, empirical continuity and 3-valued logic* ◇ B50 G10 ◇
DASSOW, J. *Kleene-Mengen und trennende Mengen* ◇ B50 D05 ◇
DEMETROVICS, J. *The comparison of limit logics when finitely valued logics are modelled in them (Russian) (English summary)* ◇ B50 ◇
DEMETROVICS, J. *The M maximal limit-logic (Hungarian) (English summary)* ◇ B50 ◇
DUGGAN, R.W. & SMITH, W.R. *Generation of value-consistent multi-valued prime implicants* ◇ B50 ◇
EPSTEIN, G. *Decisive Venn diagrams* ◇ B50 ◇
EPSTEIN, G. & HORN, A. *Logics which are characterized by subresiduated lattices* ◇ B50 F50 G10 ◇
GABBAY, D.M. *On Kreisel's notion of validity in Post systems* ◇ B50 D03 F50 ◇
GAINES, B.R. & HAACK, S. & KOTAHASHI, T. & POPPELBAUM, W.J. & RINE, D.C. & SMITH, KENNETH C. *Applications of multiple-valued logic* ◇ B50 ◇
GAVRILOV, G.P. *Pre-complete classes of partial countably-valued logic that contain all functions of one variable (Russian)* ◇ B50 ◇
GIRARD, J.-Y. *Three-valued logic and cut-elimination: The actual meaning of Takeuti's conjecture* ◇ B50 C85 C90 F05 F35 F50 ◇
GRIGOLIYA, R.SH. *The lattice of all finitely-approximable extensions of a countably-valued Lukasiewicz-logic (Russian)* ◇ B50 G20 ◇
GRISHIN, V.N. *On the algebraic semantics of a logic without reduction (Russian)* ◇ B50 G20 ◇
HARNAU, W. *Die teilweise geordnete Menge Φ_k der Vertauschbarkeitsmengen der k-wertigen Logik* ◇ B50 ◇
HARNAU, W. *Eine Verallgemeinerung der Vertauschbarkeit in der k-wertigen Logik* ◇ B50 ◇
HENNO, J. *On the completeness of idempotent functions on infinite sets* ◇ B50 ◇
KALMAN, J.A. *Axiomatizations of logics with values in groups* ◇ B50 ◇
KANEYORI, S. & KITAHASHI, T. & TANAKA, K. *On multi-valued perfect codes* ◇ B50 ◇
KUZ'MENKO, V.M. & KUZ'MIN, I.V. & SAMOJLENKO, N.I. *Determination of the minimal disjunctive normal form of logical functions of k-valued logic (Russian)* ◇ B50 ◇
LANTSOV, A.L. & RAKOV, M.A. *Synthesis of many-valued functions in a modular algebra with a composite modulus (Russian)* ◇ B50 ◇
LARSEN, J. *A multi-step formation of variable valued logic hypotheses* ◇ B50 ◇
LOADER, J. *The binary representation of m-valued logic with applications to universal decision elements* ◇ B50 ◇
MANGANI, P. *Alcune questioni di teoria dei modelli per linguaggi predicativi generali (English summary)* ◇ B50 C20 C90 ◇
MARTIN, J.N. *Classical indeterminacy, many-valued logic, and supervaluations* ◇ B50 ◇
MARTIN, N.M. *Direct analogues of the Sheffer stroke in m-valued logic* ◇ B50 ◇
MEREDITH, D. *A calculus of matrical descriptors* ◇ B50 ◇
MILLER, D.MICHAEL & MUZIO, J.C. *A ternary universal decision element* ◇ B50 ◇
MIYAMA, T. *Eliminability of descriptive definitions in many-valued logics* ◇ B50 ◇
MORGAN, C.G. *A resolution principle for a class of many-valued logics* ◇ B50 ◇
MORGAN, C.G. *Many-valued propositional intuitionism* ◇ B50 F50 ◇
MUZIO, J.C. *Binary functions that are complete with constants over* $\{0,1,2\}$ ◇ B50 ◇

MUZIO, J.C. *Concerning completeness and abelian semigroups* ◇ B50 ◇
NURLYBAEV, A.N. *A majorant local algorithm of index 1 for the construction of a sum of minimal normal forms of a function of k-valued logic (Russian) (Kazakh summary)* ◇ B50 ◇
NURLYBAEV, A.N. *Normal forms of k-valued logic (Russian)* ◇ B50 ◇
NURLYBAEV, A.N. *Simplification of normal forms of k-valued logic (Russian)* ◇ B50 ◇
OKOL'NISHNIKOVA, E.A.A. *The distribution of types of functions of q-valued logic according to cardinality (Russian)* ◇ B50 ◇
ORLOWSKA, E. *The Gentzen style axiomatization of* ω^+*-valued logic* ◇ B50 ◇
PIROG-RZEPECKA, K. *The problem of completeness of a certain sentential calculus system in different meanings of the term "complete" (Polish) (English summary)* ◇ B50 ◇
RINE, D.C. *A survey of multiple-valued algorithmic logics: from a practical point of view* ◇ B50 B75 ◇
ROSENBERG, I.G. *Some algebraic and combinatorial aspects of multiple-valued circuits* ◇ B50 ◇
ROSENBERG, I.G. *The set of maximal closed classes of operations on an infinite set A has cardinality* $2^{2^{|A|}}$ ◇ B50 C05 E20 ◇
RVACHEV, V.L. & SHKLYAROV, L.I. & TONITSA, V.S. *The closing functions of three-valued logic (Russian) (English summary)* ◇ B50 ◇
SCHWARTZ, DIETRICH *Das Homomorphietheorem fuer MV-Algebren endlicher Ordnung* ◇ B50 ◇
SCOTT, D.S. *Does many-valued logic have any use?* ◇ B50 ◇
SHAH, H. & SMITH, W.R. *Multiple-output multi-valued prime implicants* ◇ B50 ◇
SKYRMS, B. *Definitions of semantical reference and self-reference* ◇ A05 B50 ◇
SMILEY, T.J. *Does many-valued logic have any use?* ◇ A05 B50 ◇
SPADE, P.V. *An alternative to Brian Skyrms' approach to the liar* ◇ A05 B50 ◇
STOIDE, M. *On the completeness theorem for some many-valued predicate calculi with countable conjunctions and disjunctions* ◇ B50 ◇
THAYSE, A. *Difference operators and extended truth vectors for discrete functions* ◇ B50 ◇
THELLIEZ, S. *M-valued sequential cellular networks* ◇ B50 ◇
THOMAS, I. *Axiom sets equivalent to syllogism and Peirce* ◇ B50 ◇
TOKARZ, M. *Functions definable in Sugihara algebras and their fragments II* ◇ B50 G25 ◇
TURQUETTE, A.R. *Minimal axioms for Peirce's triadic logic* ◇ A10 B50 ◇
VILFAN, B. *Lower bounds for the size of expressions for certain functions in d-ary logic* ◇ B50 ◇
WESSELKAMPER, T.C. *No abelian semigroup operation is complete* ◇ B50 ◇
WILLIAMSON, J. *The complete axiomatization of any many-valued propositional logic* ◇ B50 ◇
WOJTYLAK, P. *A new proof of structural completeness of Lukasiewicz's logics* ◇ B50 ◇
WOJTYLAK, P. *On structural completeness of the infinite-valued Lukasiewicz's propositional calculus* ◇ B50 ◇
ZAKRZEWSKA, T. *Some remarks on three-valued implicative sentential calculi* ◇ B50 ◇

1977

AJZENBERG, N.N. & IVAS'KIV, YU.L. *Many-valued threshold logic (Russian)* ◇ B50 ◇
BECCHIO, D. & PABION, J.F. *Gentzen's techniques in the three-valued logic of Lukasiewicz* ◇ B50 ◇
BELNAP JR., N.D. *A useful four-valued logic* ◇ B50 ◇
BOCHVAR, D.A. & FINN, V.K. *A note to the article "Some supplements to articles on multivalued logics" (Russian) (English and German summaries)* ◇ B50 ◇
BOSSUET, G. *phi -algebras: some concepts* ◇ B50 G25 ◇
BUTLER, J.T. *Fanout-free networks of multivalued gates* ◇ B50 ◇
DANIL'CHENKO, A.F. *Parametric expressibility of functions of three-valued logic (Russian)* ◇ B50 ◇
DUNN, J.M. & EPSTEIN, G. (EDS.) *Modern uses of multiple-valued logic. Invited papers from the fifth international symposium on multiple-valued logic held at Indiana University, Bloomington* ◇ B50 B97 ◇
ENGLISH, W.R. *Canonical representation of nonlinear finite memory functionals in a Galois field* ◇ B50 ◇
GILMORE, P.C. *Defining and computing many-valued functions* ◇ B50 ◇
GORLOV, V.V. *On closed classes of k-valued logic, all superclasses of which have only trivial congruences (Russian)* ◇ B50 ◇
GORLOV, V.V. *The closed classes of k-valued logic, all of whose congruences are all trivial (Russian)* ◇ B50 C05 G20 ◇
GOTTWALD, S. *Untersuchungen zur mehrwertigen Mengenlehre. III* ◇ B50 E70 ◇
GRIGOLIYA, R.SH. *Algebraic analysis of Lukasiewicz-Tarski's n-valued logical systems* ◇ B50 G20 ◇
HAJEK, P. *Generalized quantifiers and finite sets* ◇ B25 B50 C13 C40 C80 ◇
HAWRANEK, J. & TOKARZ, M. *Matrices for predicate logics (Polish)* ◇ B50 ◇
HERZBERGER, H.G. *Tertium without plenum* ◇ B50 ◇
HINO, K. *On Yablonskii theorem concerning functional completeness of k-valued logic* ◇ B50 ◇
HOSOI, T. *On the axiomatic method and the algebraic method for dealing with propositional logics II* ◇ B50 ◇
ITURRIOZ, L. *An axiom system for three-valued Lukasiewicz propositional calculus* ◇ B50 ◇
ITURRIOZ, L. *Lukasiewicz and symmetrical Heyting algebras* ◇ B50 G10 G20 ◇
ITURRIOZ, L. *Two characteristic properties of monadic three-valued Lukasiewicz algebras* ◇ B50 G15 ◇
IVAS'KIV, YU.L. *Elementare Automaten mit einem Speicher mit k-wertigem strukturellem Alphabet (Russisch)* ◇ B50 ◇

Jensen, J.B. & Larsen, P.F. & Schotch, P.K. *A strong completeness theorem for Lukasiewicz finitely many-valued logics* ◇ B50 ◇
Kabulov, A.V. *Minimization of correcting functions in three-valued logic (Russian)* ◇ B50 ◇
Kapetanovic, M. *On a class of sentential functions* ◇ B50 ◇
Katerinochkina, N.N. *Search for a maximal upper zero for a class of monotone functions in k-valued logic (Russian)* ◇ B50 ◇
Keisler, H.J. *Hyperfinite model theory* ◇ B50 C65 C75 C80 C90 H20 ◇
Klukowski, J. *Operators in multi-valued logic* ◇ B50 G20 ◇
Krzystek, P.S. & Zachorowski, S. *Lukasiewicz logics have not the interpolation property* ◇ B50 C40 ◇
Lau, D. *Kongruenzen auf gewissen Teilklassen von* $P_{k,l}$ ◇ B50 ◇
Leblanc, H. *A strong completeness theorem for 3-valued logic. part II* ◇ B50 ◇
Loader, J. *Second order and higher order universal decision elements in m-valued logic* ◇ B50 ◇
Malinowski, G. *Classical characterization of n-valued Lukasiewicz calculi* ◇ B50 G20 ◇
Malinowski, G. *Degrees of maximality of Lukasiewicz-like sentential calculi* ◇ B50 ◇
Malinowski, G. *S-algebras for n-valued sentential calculi of Lukasiewicz. The degrees of maximality of some Lukasiewicz's logics* ◇ B50 ◇
Malinowski, G. & Wojcicki, R. (eds.) *Selected papers on Lukasiewicz sentential caculi* ◇ B50 G20 ◇
Martin, J.N. *An axiomatization of Herzberger's 2-dimensional presuppositional semantics* ◇ B50 ◇
Martin, J.N. *Many-valued logic, classical logic and the horizontal* ◇ B50 ◇
Miller, D.Michael *A canonical representation for many-valued symmetric functions* ◇ B50 ◇
Miller, David *On distance from the truth as a true distance (short version)* ◇ A05 B50 G05 G10 ◇
Moraga, C. *Ternary spectral logic* ◇ B50 ◇
Muzio, J.C. *The class structure of complete binary operations on* $\{1,2,3\}$ ◇ B50 G25 ◇
Muzio, J.C. *Three-valued unary generators based on Mouftah's primitives* ◇ B50 ◇
Nikitina, I.G. *Intersection of set of all precomplete classes in countable-valued logic (Russian)* ◇ B50 G25 ◇
Nishimura, T. & Ohya, T. *The formal system for various 3-valued logics. II* ◇ B50 ◇
Nurlybaev, A.N. *A class of functions of three-valued logic (Russian)* ◇ B50 ◇
Nurlybaev, A.N. *A local algorithm of index 1 for constructing the sum of dead-end disjunctive normal forms for functions of k-valued logic (Russian)* ◇ B50 ◇
Orlowska, E. *The Herbrand theorem for* ω^+ *-valued logic* ◇ B50 ◇
Patt, Y.N. *Independent necessary conditions for functional completeness in m-valued logic* ◇ B50 ◇
Poeschel, R. *Funktionale Vollstaendigkeit in mehrwertigen Logiken* ◇ B50 ◇
Popov, V.A. & Skibenko, I.T. *Self-complementary types of the algebra of multivalued functions for self-dual transformations (Russian) (English summary)* ◇ B50 ◇
Primenko, Eh.A. *On the number of types of invertible transformations in multivalued logics (Russian) (English summary)* ◇ B50 ◇
Rine, D.C. *A note on multi-valued interrogation logic of associative memories* ◇ B50 ◇
Rine, D.C. (ed.) *Computer science and multiple-valued logic. Theory and applications* ◇ B50 B97 ◇
Rosenberg, I.G. *On closed classes, basic sets and groups* ◇ B50 ◇
Serfati, M. *Une methode de resolution des equations postiennes a partir d'une solution particuliere* ◇ B50 G20 ◇
Suchon, W. *Smullyan's method of contructing Llukasiewicz's n-valued implicational- negational sentential calculus* ◇ B50 ◇
Takahashi, Moto-o *Many-valued logics and their algebras (Japanese)* ◇ B50 G20 ◇
Tkachev, G.A. *Complexity of realization of one sequence of functions of k-valued logic (Russian)* ◇ B50 B70 D15 ◇
Tokarz, M. *A remark on maximal matrix consequences* ◇ B50 ◇
Tokarz, M. *Degrees of completeness of Lukasiewicz logics* ◇ B50 ◇
Tokarz, M. *On structural completeness of Lukasiewicz logics* ◇ B50 ◇
Vakarelov, D. *Lattices related to Post algebras and their applications to some logical systems* ◇ B50 G10 G15 G20 ◇
Wojcicki, R. *On matrix representations of consequence operations of Lukasiewicz's sentencial calculi* ◇ B50 ◇
Wojcik, A.S. *Digital system design using 4-valued logic devices* ◇ B50 ◇
Wolf, R.G. *A survey of many-valued logic (1966–1974)* ◇ B50 ◇
Zygmunt, J. *Research project on strongly finite sentential calculi* ◇ B50 ◇

1978

Bagyinszki, J. *The lattice of closed classes of linear functions over a finite ring of square-free order* ◇ B50 ◇
Bajramov, R.A. *On sub-algebras, maximal in* P_k, *with a given* σ_k*-component and on subalgebras, maximal in* P_k, *with ideal* $\Omega_{k,k-1}$*-component (Russian)* ◇ B50 G20 ◇
Becchio, D. *Logique trivalente de Lukasiewicz* ◇ B50 G10 G20 ◇
Bordat, J. *Resolution des equations de la logique a p valeurs* ◇ B50 G10 G20 ◇
Canak, B. & Gilezan, K. *Quelques formes generales des fonctions pseudo-booleennes* ◇ B50 ◇
Ceccherini, P.V. & D'Andrea, R. *Nuove operazioni primitive in logiche polivalenti* ◇ B50 ◇
Chilin, V.I. *Continuous valuations on logics (Russian)* ◇ B50 E65 G05 ◇

CRIST, S.C. *A tri-state logic family* ◇ B50 ◇
CURRENT, K.W. & MOW, D.A. *Applications of multi-valued logic in LSI digital signal processing circuits* ◇ B50 B70 ◇
CZERMAK, J. *The reducibility of a logical space with many-valued dimensions* ◇ B50 ◇
DANIL'CHENKO, A.F. *Parametrically closed classes of functions of three-valued logic (Russian)* ◇ B50 G25 ◇
DAVIO, M. & DESCHAMPS, J.-P. *Addition in signed digit number systems* ◇ B50 B70 ◇
DEMETROVICS, J. *On the main diagonal of Sheffer functions* ◇ B50 ◇
EPSTEIN, G. *A summary of investigations into three and four valued logics* ◇ B50 ◇
FILIPOIU, A. *Analytic tableaux for θ-valued propositional logic* ◇ B50 G20 ◇
FRICKE, J. *Decomposition of multiple-valued logic functions* ◇ B50 ◇
GARDOS, E. *On the monotone classes of the M limit-logic* ◇ B50 ◇
GORELIK, A.G. *Geometric interpretations of indexed three-valued calculi (Russian)* ◇ B50 ◇
GRIFFIN, N. *Supervaluations and Tarski* ◇ A05 B50 ◇
HART, A.M. & HENDRY, H.E. *Some observations on a method of McKinsey* ◇ B50 F50 ◇
HAVRANEK, T. *Enumeration calculi and rank methods* ◇ B50 D10 ◇
HIKITA, T. *Completeness criterion for functions with delay defined over a domain of three elements* ◇ B50 G20 G25 ◇
KABULOV, A.V. *On functions correcting collections of incorrect algorithms (Russian)* ◇ B50 ◇
KLAUA, D. *Partielle boolesche Algebren in der Intervallmathematik* ◇ B50 G05 ◇
KNOEBEL, A. *Further conclusions on functional completeness* ◇ B50 ◇
KOHOUT, L.J. *Analysis of computing protection structures by means of multi-valued logic systems* ◇ B50 ◇
KOMORI, Y. *Super-Lukasiewicz implicational logics* ◇ B50 ◇
KOMORI, Y. *The separation theorem of the $\aleph_0$-valued Lukasiewicz propositional logic* ◇ B50 ◇
KRAMOSIL, I. *A boolean-valued probability theory* ◇ B50 G05 ◇
LAU, D. *Bestimmung der Ordnung maximaler Klassen von Funktionen der k-wertigen Logik* ◇ B50 G20 ◇
LAU, D. *Ueber die Anzahl von abgeschlossenen Mengen linearer Funktionen der n-wertigen Logik* ◇ B50 G25 ◇
LESCHINE, T.M. *Propositional logic for topology-like matrices: A calculus with restricted substitution* ◇ B50 G10 ◇
MADUCH, M. *On three-valued implicative systems* ◇ B50 ◇
MALINOWSKI, G. & ZYGMUNT, J. *Results in general theory of matrices for sentential calculi with applications to the Lukasiewicz logics* ◇ B50 ◇
MARTIN, L. & REISCHER, C. & ROSENBERG, I.G. *Completeness problems for switching circuits constructed from delayed gates* ◇ B50 ◇
MEREDITH, D. *Are there many-valued Scotan logics?* ◇ B50 ◇
MILLER, JOHN R. *Use of an infinite-valued propositional calculus in a document retrieval system* ◇ B50 ◇
MIYAMA, T. *Note on functional completeness in many valued logics* ◇ B50 ◇
MORAGA, C. *Complex spectral logic* ◇ B50 ◇
NAKANISHI, M. *Gentzen type formal system for many-valued propositional logic* ◇ B50 ◇
NEGRU, I.S. *On the identities on the structure of Post classes (Russian)* ◇ B50 ◇
ORLOV, YU.F. *Group theory approach to logic -- the wave logic* ◇ B50 ◇
ORLOWSKA, E. *Resolution system for ω^+-valued logic* ◇ B50 ◇
PINKAVA, V. *On a class of functionally complete multi-valued logical calculi* ◇ B50 ◇
PIOCHI, B. *Matrici adeguate per calcoli generali predicativi (English summary)* ◇ B50 ◇
ROSE, A. *A three-valued model for set theory* ◇ B50 E35 E70 ◇
ROSE, A. *Formalisations of further $\aleph_0$-valued Lukasiewicz propositional calculi* ◇ B50 ◇
ROSENBERG, I.G. *On generating large classes of Sheffer functions* ◇ B50 C05 ◇
ROSENBERG, I.G. *Subalgebra systems of direct powers* ◇ B50 C05 ◇
SASAO, T. *An application of multiple-valued logic to a design of programmable logic arrays* ◇ B50 ◇
STEWARD, M.H. *The arithmetic properties of certain number systems* ◇ B50 ◇
STOCK, W.G. *Zur Bestimmung der Negation in der Handlungslogik* ◇ B50 ◇
SZENDREI, A. *On closed sets of linear operations over a finite set of square-free cardinality* ◇ B50 ◇
TETRAULT, G.E. *Deviant, many-valued modellings of paradox: Outlines of indefiniteness as the formal source of exact logic* ◇ A05 B50 ◇
TURQUETTE, A.R. *Alternative axioms for Peirce's triadic logic* ◇ B50 ◇
UZIEMBLO, A.O. & ZURAWSKA, L. *Definability of functions on a finite set by means of compositions of addition and multiplication modulo some prime numbers* ◇ B50 ◇
VOJVODIC, G. *Some theorems for model theory of mixed-valued predicate calculi* ◇ B50 C20 C35 C90 ◇
WEAVER, G.E. *Compactness theorems for finitely-many-valued sentential logics* ◇ B50 ◇
WINKER, S.K. & WOS, L. *Automated generation of models and counter-examples and its application to open questions in ternary Boolean algebra* ◇ B50 ◇
WOJTYLAK, P. *On structural completeness of many-valued logics* ◇ B50 ◇
WOZNIAKOWSKA, B. *Algebraic proof of the separation theorem for the infinite-valued logic of Lukasiewicz* ◇ B50 ◇
WOZNIAKOWSKA, B. *The representation theorem for the algebras determined by the fragments of infinite-valued logic of Lukasiewicz* ◇ B20 B50 G20 G25 ◇

1979

Adamson, A. & Giles, R. *A game-based formal system for* L_∞ ◇ B25 B50 D15 ◇

Armstrong, J.R. & Pomper, G. *An efficient multivalued minimization algorithm* ◇ B50 ◇

Balabaev, V.M. & Gerasimov, I.N. *A numerical method for the minimization of ternary logic functions (Russian)* ◇ B50 ◇

Barton, S. *The functional completeness of Post's m-valued propositional calculus* ◇ B50 G20 ◇

Bochvar, D.A. & Shokurov, V.V. *The behaviour of the plausibility functional for "tertium non datur" in some sequences of logical matrices (Russian)* ◇ B50 ◇

Bondarev, V.M. *A decomposition theorem in predicate algebra (Russian)* ◇ B50 ◇

Bunder, M.W. *Λ-elimination in illative combinatory logic* ◇ B40 B50 ◇

Butler, J.T. & Ginzer, J.A. *Multiple-valued logic: 1974 - 1978 -- survey and analysis* ◇ B50 ◇

Byrd, M. *A formal interpretation of Lukasiewicz' logics* ◇ B50 ◇

Byrd, M. *On Lukasiewiczian logics with several designated values* ◇ B50 ◇

Colmerauer, A. *Un sous-ensemble interessant du francais* ◇ B50 B65 ◇

Coy, W. & Moraga, C. *Description and detection of faults in multiple-valued logic circuits* ◇ B50 B70 ◇

Current, K.W. *Simultaneous analog-to-quaternary conversion* ◇ B50 ◇

Davies, R.O. *On n-valued Sheffer functions* ◇ B50 G20 ◇

Demetrovics, J. & Hannak, L. *On the cardinality of self-dual closed classes in k-valued logics* ◇ B50 ◇

Demetrovics, J. & Hannak, L. *The cardinality of closed sets in pre-complete classes in k-valued logics* ◇ B50 G20 ◇

Derus, K.H. & Hansen, J.C. *Propositions with multiple dispositions and multiple truth values* ◇ B50 ◇

Duffy, M.J. *Modal interpretations of three valued logic. I,II* ◇ B50 ◇

Dunn, J.M. *A theorem in 3-valued model theory with connections to number theory, type theory, and relevant logic* ◇ B50 ◇

Ermolaeva, N.M. & Muchnik, A.A. *Functionally closed 4-valued extensions of the Boolean algebra and corresponding logics (Russian)* ◇ B50 G25 ◇

Finn, V.K. & Grigoliya, R.Sh. *Bochvar algebras and corresponding propositional calculi (Russian)* ◇ B50 G25 ◇

Franksen, O.I. *Group representations of finite polyvalent logic. A case study using APL notation* ◇ B50 B75 ◇

Furtek, F.C. *Doing without values* ◇ B50 ◇

Gilezan, K. & Rudeanu, S. *Interpolation formulas over finite sets* ◇ B50 ◇

Gorlov, V.V. *A sufficient condition for closed classes of k-valued logic to have only trivial congruences (Russian)* ◇ B50 ◇

Gottwald, S. *Eine Anwendungsvariante der mehrwertigen Logik* ◇ B50 E72 ◇

Henno, J. *The depth of monotone functions in multivalued logic* ◇ B50 ◇

Henno, J. *The depth of functions in m-valued logic* ◇ B50 ◇

Hinde, C.J. *Generalisation of Bochvar's 3-valued logic* ◇ B50 G25 ◇

Hirose, K. & Nagaoka, K. & Umezu, M. *Many-valued logic with truth values from finite lattices* ◇ B50 G10 ◇

Jennings, R.E. & Schotch, P.K. *Multiple-valued consistency* ◇ A05 B50 ◇

Kabulov, A.V. *The principle of the independence of coding for the continuation of functions of k-valued logic (Russian)* ◇ B50 G10 ◇

Kabzinski, J.K. *On equivalential fragment of the three-valued logic of Lukasiewicz* ◇ B50 G20 ◇

Kaluzhnin, L.A. & Poeschel, R. *Funktionen- und Relationenalgebren. Ein Kapitel der diskreten Mathematik* ◇ B50 E07 G15 G20 G98 ◇

Kearns, J.T. *The strong completeness of a system for Kleene's three-valued logic* ◇ B50 ◇

Kerntopf, P. *On finding roots on finite-valued functions* ◇ B50 ◇

Klaua, D. *Partielle boolesche Algebren in der Intervallmathematik. II* ◇ B50 G05 ◇

Krolikoski, S.J. *Did many-valued logic begin with a mistake?* ◇ A05 B50 ◇

Krolikoski, S.J. *Lukasiewicz's twin possibility functors* ◇ B50 ◇

Kubin, W. *Eine Axiomatisierung der mehrwertigen Logiken von Goedel* ◇ B50 ◇

Kuzyurin, N.N. *On the approximate search for the maximal upper zero for monotone functions of k-valued logics (Russian)* ◇ B50 ◇

Lau, D. *Automorphismen auf den maximalen Klassen der k-wertigen Logik* ◇ B50 G25 ◇

Malinowski, G. *S-algebras for n-valued Lukasiewicz propositional calculi (Russian)* ◇ B50 ◇

Marchenkov, S.S. *On closed classes of self-dual functions in a many-valued logic (Russian)* ◇ B50 G25 ◇

Markova, V.P. *Harmonic analysis of functions of many-valued logic (Russian)* ◇ B50 ◇

Miller, D.Michael & Muzio, J.C. *On the minimization of many-valued functions* ◇ B50 ◇

Miyama, T. *On the study of many-valued logics. I (Japanese)* ◇ B50 G20 ◇

Perfil'eva, I.G. *The construction of a hypercontinuum of precomplete classes of a countably many-valued logic (Russian)* ◇ B50 ◇

Peters, S. *A truth-conditional formulation of Karttunen's account of presupposition* ◇ A05 B50 ◇

Pinkava, V. *On some manipulative properties of the π-algebras* ◇ B50 G25 ◇

Popov, V.A. & Skibenko, I.T. *Combinatorial approach to the classification of k-valued logical functions (Russian)* ◇ B50 ◇

Presic, S.B. *A completeness theorem for one class of propositional calculi* ◇ B50 ◇

Priest, G. *The logic of paradox* ◇ A05 B50 ◇

Quesada, J.D. *Referential presuppositions and three-valued logic (Spanish)* ◇ A05 B50 ◇

RAUTENBERG, W. *Klassische und nichtklassische Aussagenlogik* ◇ B45 B50 B98 C90 ◇
ROBITASHVILI, N.G. *Semantics of languages containing not everywhere defined expressions* ◇ B50 ◇
ROSE, A. *Many-valued propositional calculi without constants* ◇ B50 ◇
RVACHEV, V.L. & SHKLYAROV, L.I. & TONITSA, V.S. *A set of closing prime functions of three-valued logic (Ukrainian) (English and Russian summaries)* ◇ B50 ◇
SASAO, T. & TERADA, H. *Multiple-valued logic and the design of programmable logic arrays with decoders* ◇ B50 ◇
SHOJLEV, KH. & TENEV, V. *Polynomial forms of many-valued logical functions. Minimization of polynomial form* ◇ B50 ◇
SHUM, A.A. *Propositional semantic systems (Russian)* ◇ B50 G25 ◇
THAYSE, A. *Integer expansions of discrete functions and their use in optimization problems* ◇ B50 ◇
TOKARZ, M. *Theorems derivable from the McNaughton criterion (Russian)* ◇ B50 ◇
TOKMEN, V.H. *Some properties of the spectra of ternary logic functions* ◇ B50 ◇
VARELA GARCIA, F.J. *The extended calculus of indications interpreted as a three-valued logic* ◇ A05 B50 ◇
VOJVODIC, G. *On the II ε-theorem for mixed-valued predicate calculi* ◇ B50 ◇
VRANESIC, Z.G. *Multivalued signaling in daisy chain bus control* ◇ B50 ◇
WASILEWSKA, A. *A constructive proof of Craig's interpolation lemma for m-valued logic* ◇ B50 C40 ◇
WESSELKAMPER, T.C. *The algebraic representation of partial functions* ◇ B50 G25 ◇
WHITE, R.B. *The consistency of the axiom of comprehension in the infinite-valued predicate logic of Lukasiewicz* ◇ B50 E35 E70 F05 ◇
YABLONSKIJ, S.V. *Introduction to discrete mathematics (Russian)* ◇ B50 ◇
ZASLAVSKIJ, I.D. *The realization of three-valued logical functions by means of recursive and Turing operators (Russian)* ◇ B50 D10 D20 ◇
ZHANG, HONGYU *A simplification for ternary functions (Chinese) (English summary)* ◇ B50 ◇
ZYGMUNT, J. *Entailment relations and matrices. I* ◇ B50 ◇

1980

ABDUGALIEV, U.A. *Irredundant disjunctive normal forms of k-valued logic functions (Russian)* ◇ B50 ◇
ALBERT, P. *The notion of consequence in many-valued logic* ◇ B50 ◇
ALEKSEEV, V.B. *Semi-simple bases of k-valued logic (Russian)* ◇ B50 ◇
ANSHAKOV, O.M. & FINN, V.K. & GRIGOLIYA, R.SH. & ZABEZHAJLO, M.I. *Many-valued logics as fragments of formalized semantics (Russian)* ◇ B50 ◇
ANSHAKOV, O.M. *On a certain constructivization of D. A. Bochvar's propositional logic (Russian)* ◇ B50 ◇
BAGYINSZKI, J. & DEMETROVICS, J. *The structure of the maximal linear classes in prime-valued logics* ◇ B50 G10 G25 ◇
BERMAN, J. *Algebraic properties of k-valued logics* ◇ B50 ◇
BRADY, R.T. *A theory of classes and individuals based on a 3-valued significance logic* ◇ B50 E35 E70 ◇
BRADY, R.T. *Significance range theory* ◇ B50 E70 ◇
BRADY, R.T. *Two remarks on "The logic of significance and context"* ◇ A05 B50 ◇
CANTINI, A. *A note on three-valued logic and Tarski theorem on truth definitions* ◇ B50 C40 F30 ◇
CERNY, E. *Characteristic functions in multivalued logic systems (French and German summaries)* ◇ B50 ◇
CIGNOLI, R. *Some algebraic aspects of many-valued logics* ◇ B50 G20 ◇
CLEAVE, J.P. *Some remarks on the interpretation of 3-valued logics* ◇ B50 ◇
COSGROVE, R.J. *A three-valued free logic for presuppositional languages* ◇ A05 B50 ◇
COSTA DA, N.C.A. & KOTAS, J. *Some problems on logical matrices and valorizations* ◇ B50 ◇
DEMETROVICS, J. & HANNAK, L. & MARCHENKOV, S.S. *On closed classes of self-dual functions in P_3 (Russian)* ◇ B50 G10 G25 ◇
DEMETROVICS, J. & HANNAK, L. & MARCHENKOV, S.S. *Some remarks on the structure of P_3* ◇ B50 G10 G25 ◇
EPSTEIN, G. & MILLER, D.MICHAEL & MUZIO, J.C. *Selecting don't-care sets for symmetric many-valued functions: a pictorial approach using matrices* ◇ B50 ◇
GILEZAN, K. *Interpolation formulas over finite sets* ◇ B50 ◇
GOLUNKOV, YU.V. *A criterion for the completeness of systems of operations in operator algorithms realizing functions of k-valued logics (Russian)* ◇ B50 ◇
HENDRY, H.E. *Functional completeness and non-Lukasiewiczian truth functions* ◇ B50 ◇
KABZINSKI, J.K. *Complete many-valued systems of propositonal logic* ◇ B50 ◇
KAPETANOVIC, M. *On a many-valued sentential calculus* ◇ B50 ◇
KATZ, M. *Inexact geometry* ◇ B50 ◇
KROLIKOSKI, S.J. *On substitution for variable one-place functors* ◇ B50 ◇
KRZYSTEK, P.S. *Equivalential fragment of the infinite valued logic of Lukasiewicz and the intermediate logics* ◇ B50 ◇
KUDRYAVTSEV, V.B. *Functional systems (Russian)* ◇ B50 C05 D05 ◇
LUO, ZHUKAI *The determination of all maximal closed sets of functions in multivalued logic (Chinese)* ◇ B50 ◇
LYASHENKO, L.N. & VASHCHENKO, V.P. *Multiple functional decomposition in K-valued logic (Russian)* ◇ B50 ◇
MARTIN, J.N. *The emergent nature of coherence and the possibility of truth-functional logic* ◇ B50 ◇
MILLER, D.MICHAEL & MUZIO, J.C. *A class of two-place three-valued unary generators* ◇ B50 ◇
MIYAMA, T. *On the study of many-valued logics. II (Japanese)* ◇ B50 G20 ◇
MORAGA, C. *Language inference using multiple-valued logic* ◇ B50 ◇

MUKAIDONO, M. *Some kinds of functional completeness of ternary logic functions* ◇ B50 ◇
MUZIO, J.C. & ROSENBERG, I.G. *Large classes of functionally complete operations. I* ◇ B50 ◇
NECHAEV, A.A. *Criterion for completeness of systems of functions of p^n-valued logic containing operations of addition and multiplication modulo p^n (Russian)* ◇ B50 ◇
PINKAVA, V. *On potential tautologies in k-valued calculi and their assessment by means of normal forms* ◇ B50 ◇
PIOCHI, B. *Nota su matrici adeguate per calcoli generali predicativi* ◇ B50 ◇
PRZYMUSINSKA, H. *Craig interpolation theorem and Hanf number for ν-valued infinitary predicate calculi (Russian summary)* ◇ B50 C40 C55 C75 G20 ◇
PRZYMUSINSKA, H. *Gentzen-type semantics for ν-valued infinitary predicate calculi (Russian summary)* ◇ B50 C75 G20 ◇
ROBBEL, G. *Many-valued logics based on L-algebras or on P-algebras* ◇ B50 G10 G25 ◇
ROMOV, B.A. *Maximal subalgebras of algebras of partial multivalued logic functions (Russian)* ◇ B50 G20 G25 ◇
SZENDREI, A. *On closed classes of quasilinear functions* ◇ B50 G10 ◇
TAKANO, M. *Valid sequents in many-valued logics* ◇ B50 ◇
TKACHEV, G.A. *On the influence of a basis on the behaviour of the Shanon function (Russian)* ◇ B50 ◇
TOKMEN, V.H. *Disjoint decomposability of multivalued functions by spectral means* ◇ B50 ◇
VOJVODIC, G. *The Craig interpolation theorem for mixed-valued predicate calculi (Serbo-Croatian summary)* ◇ B50 C40 C90 ◇
WANG, SHIQIANG & WENG, JIAFENG *Normal forms in lattice-valued predicate calculi (Chinese) (English summary)* ◇ B50 ◇
WANG, SHIQIANG & WU, WANGMING *On a classification problem for complemented lattices (Chinese)* ◇ B50 G10 ◇
WHITE, R.B. *Natural deduction in the Lukasiewicz logics* ◇ B50 ◇
WILLMOTT, R. *On the transitivity of implication and equivalence in some many-valued logics* ◇ B50 ◇
WOJCICKI, R. *More about referential matrices* ◇ B50 ◇
YABLONSKIJ, S.V. *On some results in the theory of function systems (Russian)* ◇ B50 D20 G25 ◇
YAGER, R.R. *Generalized "and/or" operators for multivalued and fuzzy logic* ◇ B50 B97 ◇
ZHANG, JINWEN *A unified treatment of fuzzy set theory and boolean-valued set theory - fuzzy set structures and normal fuzzy set structures* ◇ B50 E40 E72 ◇

1981

ALSINA, C. & TRILLAS, E. *On Sheffer stroke* ◇ B50 ◇
ANSHAKOV, O.M. *The logic B_3 and Boolean algebras (Russian)* ◇ B50 ◇
ATLAS, J.D. *Is "not" logical?* ◇ A05 B50 ◇
BAJRAMOV, R.A. *Some new results in the theory of function algebras of finite-valued logics* ◇ B50 G20 ◇
BAKHMUTOVA, I.V. *A variant of the theory of prediction for the case of four-valued logic (Russian)* ◇ B50 ◇
BENDOVA, K. *On the relation of three-valued logic to modal logic* ◇ B50 ◇
BERGMANN, MERRIE *'Only','even', and clefts in two-dimensional logic* ◇ B50 ◇
BOROWIK, P. *On Gentzen axiomatization of the reducts of multivalued logic (Polish) (English summary)* ◇ B50 ◇
BRINK, C. *An eight-valued logic* ◇ B50 ◇
COY, W. *A common approach to the description, implementation and test generation of multivalued functions* ◇ B50 ◇
CSAKANY, B. & ROSENBERG, I.G. (EDS.) *Finite algebra and multiple-valued logic (Colloquium held in Szeged, August 27-31, 1979)* ◇ B50 ◇
DANIL'CHENKO, A.F. *On parametrical expressibility of the functions of k-valued logic* ◇ B50 ◇
DEMETROVICS, J. & HANNAK, L. & MARCHENKOV, S.S. *On closed classes of selfdual functions in P_3* ◇ B50 ◇
DOMINGO, X. & TRILLAS, E. & VALVERDE, L. *Pushing Lukasiewicz-Tarski implication a little further* ◇ B50 ◇
ESTEVA, F. *On the form of negations in posets* ◇ B50 ◇
GARDOS, E. *On the monotone classes of maximal limit-logic M* ◇ B50 ◇
GERLA, G. & NOLA DI, A. *A three valued doxastic logic* ◇ B45 B50 ◇
GUCCIONE, S. & TORTORA, R. & VACCARO, V. *Deduction theorems in Lukasiewicz propositional calculi* ◇ B50 ◇
HENNO, J. *The depth of functions in many-sorted logic* ◇ B50 G25 ◇
HIKITA, T. *Completeness properties of k-valued functions with delays inclusions among closed spectra* ◇ B50 ◇
HIKITA, T. *On completeness for k-valued functions with delay* ◇ B50 ◇
KABULOV, A.V. & SHESTEROVA, N.A. *A representation of functions of the k-valued logic by formulas (Russian)* ◇ B50 ◇
KABZINSKI, J.K. *Many-valued logic* ◇ B50 ◇
KARAKHANYAN, L.M. & SAPOZHENKO, A.A. *On the negative effects of eliminating unessential variables (Russian)* ◇ B50 ◇
KARPOVSKY, M.G. *Spectral methods for decomposition, design, and testing of multiple-valued logical networks* ◇ B50 ◇
KATERINOCHKINA, N.N. *Location of the maximal upper zero for a certain class of monotone functions of k-valued logic (Russian)* ◇ B50 ◇
KATZ, M. *Lukasiewicz logic and the foundations of measurement* ◇ B50 ◇
KATZ, M. *Two systems of many-valued logic for science* ◇ B50 ◇
KOLETSOS, G. *Notational and logical completness in three-valued logic* ◇ B50 ◇
KUDRYAVTSEV, V.B. *Function systems (Russian)* ◇ B50 E20 ◇
KUDRYAVTSEV, V.B. *On completeness for function systems* ◇ B50 ◇

LAU, D. *Congruences on closed sets of k-valued logic* ◇ B50 ◇
LEVIN, V.A. *The infinite-valued logic and the problem of making group decisions (Russian)* ◇ B50 ◇
LOWESMITH, B.J. *Three-valued commutative pseudo-Sheffer functions* ◇ B50 ◇
MACHIDA, H. *On closed sets of three-valued monotone logical functions* ◇ B50 ◇
MALINOWSKI, G. & MICHALCZYK, M. *Interpolation properties for a class of many-valued propositional calculi* ◇ B50 C40 ◇
MARCHENKOV, S.S. *Cardinality of the set of precomplete classes in certain classes of functions of countably valued logic (Russian)* ◇ B50 ◇
MARCHINI, C. *Modelli di Kripke e modelli algebrici* $S_{\pi\iota}$ *(English summary)* ◇ B50 ◇
MILLER, D.MICHAEL *Spectral symmetry tests* ◇ B50 ◇
MILLER, D.MICHAEL *The fanout-free realization of multiple-valued functions* ◇ B50 ◇
MINE, H. & YAMAMOTO, Y. *Testing and realization of three-valued majority functions* ◇ B50 ◇
MIYAKAWA, M. *Enumerations of bases of three-valued logical functions* ◇ B50 ◇
MOSHCHENSKIJ, V.A. *Special completeness of some languages of k-valued logics (Russian)* ◇ B50 ◇
MOTOHASHI, N. & SHIRAI, K. *On a generalisation of theorems of Skolem and Mints (Japanese)* ◇ B50 ◇
NOZAKI, A. *Completeness criteria for a set of delayed functions with or without nonuniform compositions* ◇ B50 ◇
PINKAVA, V. *On Sheffer functions in k-valued logical calculi* ◇ B50 ◇
RAGAZ, M. *Arithmetische Klassifikation von Formelmengen der unendlichwertigen Logik* ◇ B50 ◇
RINE, D.C. *Picture processing using multiple-valued logic* ◇ B50 ◇
ROSE, A. *Extensions de quelques theoremes de McNaughton (English summary)* ◇ B50 ◇
ROSE, A. *Many-valued logics* ◇ B50 ◇
ROSENBERG, I.G. *Large classes of functionally complete groupoids II* ◇ B50 ◇
SASAO, T. *Multiple-valued decomposition of generalized boolean functions and the complexity of programmable logic arrays* ◇ B50 B70 ◇
SHESTEROVA, N.A. *Some metric properties of functions of the k-valued logic (Russian)* ◇ B50 ◇
SMITH, KENNETH C. *The prospects for multivalued logic: a technology and applications view* ◇ B50 B70 ◇
SOKOLOV, N.A. *The search for the maximum upper zero for a class of monotonic functions of finite-valued logic (Russian)* ◇ B50 ◇
STRAZDINS, I.E. *On fundamental transformation groups in the algebra of logic* ◇ B05 B50 ◇
SURMA, S.J. *On closure operators with the deduction property* ◇ B50 ◇
SZENDREI, A. *Clones of linear operations on finite sets* ◇ B50 C05 ◇
TURQUETTE, A.R. *Quantification for Peirce's preferred system of triadic logic* ◇ B50 ◇
YASUGI, M. *Definitive valuation of set theory* ◇ B50 E70 ◇
ZLATOS, P. *Two-levelled logic and model theory* ◇ B50 C90 G25 ◇

1982

AIHARA, R. & AKO, T. & AMAMITSU, S. & DOI, T. *Synchronous problems of digital networks in terms of multiple stability problems (Japanese)* ◇ B50 ◇
ANSHAKOV, O.M. & FINN, V.K. & GRIGOLIYA, R.SH. & ZABEZHAJLO, M.I. *Many-valud logics as fragments of formalized semantics* ◇ B50 ◇
ANSHAKOV, O.M. & RYCHKOV, S.V. *Many-valued logical calculi (Russian)* ◇ B50 ◇
BAGYINSZKI, J. & DEMETROVICS, J. *The lattice of linear classes in prime-valued logics* ◇ B50 ◇
BAGYINSZKI, J. *The solution of the Hosszu equation over finite fields (Russian and Hungarian summaries)* ◇ B50 ◇
BAJRAMOV, R.A. *On independent sets of functions in finite-valued logics* ◇ B50 ◇
BATENS, D. *A bridge between two-valued and many-valued semantic systems: n-tuple semantics* ◇ B50 ◇
CAI, QINGSHENG & HUANG, KECHENG & TANG, CHESHAN *Four-valued logic and a formulation of the CKP (Chinese)* ◇ B50 ◇
CHAKRABORTY, B.C. *Algebraic representations of n-valued logics, n being prime* ◇ B50 ◇
CHIMEV, K.N. *On a class of functions (Russian)* ◇ B50 ◇
CIGNOLI, R. *Proper n-valued Lukasiewicz algebras as S-algebras of Lukasiewicz n-valued propositional calculi* ◇ B50 ◇
DEMETROVICS, J. & HANNAK, L. *How to construct a large set of nonequivalent functionally complete algebras (Russian and Hungarian summaries)* ◇ B50 C05 ◇
DEMETROVICS, J. & HANNAK, L. & RONYAI, L. *On functional completeness of prime-element algebras (Russian and Hungarian summaries)* ◇ B50 C05 ◇
DENECKE, K. *Preprimal algebras* ◇ B50 C05 C13 ◇
DOMBI, J. *Basic concepts for a theory of evaluation: the aggregative operator* ◇ B50 ◇
DOROSLOVACKI, R. & TOSIC, R. *A combinatorial identity and its applications* ◇ B50 ◇
EPSTEIN, G. & LIU, YOWEI *Positive multiple-valued switching functions – an extension of Dedekind's problem* ◇ B50 B70 ◇
FANG, ZHENGXIAN *Ternary algebra and fault diagnosis in ternary logic (Chinese) (English summary)* ◇ B50 ◇
FUKAYAMA, T. *On the state of research in many-valued logic (Japanese)* ◇ B50 ◇
GARDOS, E. *Generalization of the self-dualism in the limit-logic M (German and Hungarian summaries)* ◇ B50 ◇
GEORGESCU, G. *On Moisil's work in many-valued logics* ◇ A10 B50 ◇
GORLOV, V.V. & LAU, D. *On a family of M-classes of k-valued logic and congruences on submaximal classes of* P_3 *(Russian) (English and German summaries)* ◇ B50 ◇
GOTO, M. & KAO, S. & NINOMIYA, T. *Improvement of the axioms of Wajsberg's three-valued logic, using a specific four-valued logic (Japanese)* ◇ B50 ◇

GUCCIONE, S. & TORTORA, R. *Deducibility in many-valued logics* ◇ B50 ◇
HASEGAWA, T. & OKURA, Y. & SHIMADA, R. *Symmetric ternary scaled cyclic AN-codes (Japanese)* ◇ B50 ◇
HENNO, J. *Equivalent sets of functions of multiple-valued logic* ◇ B50 ◇
HENNO, J. *On equivalent sets of functions* ◇ B50 ◇
HERZBERGER, H.G. *The algebra of supervaluations* ◇ B50 ◇
HIKITA, T. *Completeness of functions with delay (Japanese)* ◇ B50 ◇
IMANISHI, S. & MURANAKA, N. *Applications of the clause expansion method to functions in three-valued logic (Japanese)* ◇ B50 ◇
IMANISHI, S. & MURANAKA, N. *Three-valued ordered circuits* ◇ B50 ◇
ISHIKAWA, K. & SASAO, T. & TERADA, H. *On simplified algorithms for equations in many-valued logic (Japanese)* ◇ B50 ◇
JONES, JOHN *A formalisation of an* $\aleph_0$*-valued propositional calculus with variable functors* ◇ B50 ◇
KABULOV, A.V. *Synthesis of bases of complete systems of logical functions (Russian)* ◇ B50 ◇
KABULOV, A.V. *The method of invariant extension of partially defined functions of k-valued logic (Russian)* ◇ B50 ◇
KAGA, Y. & MURAKI, M. *Properties of many-valued logical functions and their structure (Japanese)* ◇ B50 ◇
KARAKHANYAN, L.M. *Metric comparison of minimal (in various senses) disjunctive normal forms for partial functions of logic algebra (Russian)* ◇ B50 ◇
KARPENKO, A.S. *Aristotle, Lukasiewicz and factor-semantics* ◇ A05 A10 B50 ◇
KARPENKO, A.S. *Matrix logic for prime numbers and its generalization (Russian)* ◇ B50 ◇
KATZ, M. *Measures of proximity and dominance* ◇ B50 ◇
KNOEBEL, A. *The finite interpolation property for small sets of classical polynomials* ◇ B50 C05 ◇
KUDRYAVTSEV, V.B. *Truth functional systems (Russian)* ◇ B50 B75 ◇
LAU, D. *Die maximalen Klassen von* $Pol_k(0)$ ◇ B50 ◇
LAU, D. *Submaximale Klasse von* P_3 *(English and Russian summaries)* ◇ B50 ◇
LINDSTROEM, T.L. *A Loeb-measure approach to theorems by Prohorov, Sazonov and Gross* ◇ B50 H05 ◇
LOU, ZHUKAI *The problem of functional completeness in many-valued logics (Chinese) (English summary)* ◇ B50 ◇
MACHIDA, H. *On the lattice of closed sets of functions of three-valued logic (Japanese)* ◇ B50 ◇
MACHIDA, H. *Toward a classification of minimal closed sets in 3-valued logic* ◇ B50 ◇
MALAI, V.P. *Chain precomplete classes of functions of many-valued logic (Russian)* ◇ B50 ◇
MIYAKAWA, M. *Enumeration of bases of a submaximal set of three-valued logical functions* ◇ B50 ◇
MIYAMA, T. *On the study of many-valued logics III (Japanese)* ◇ B50 ◇
MUZIO, J.C. & ROSENBERG, I.G. *Large classes of functionally complete groupoids. I* ◇ B50 ◇
NAKAMURA, A. *On a three-valued logic connected with incomplete information data bases* ◇ B50 ◇
OVCHINNIKOV, S.V. *Social choice and Lukasiewicz logic* ◇ B50 ◇
PRADE, H. *Possibility sets, fuzzy sets and their relation to Lukasiewicz logic* ◇ B50 E70 ◇
SALES, T. *A multivalued Boolean logic (Catalan)* ◇ B50 ◇
SZENDREI, A. *Algebras of prime cardinality with a cyclic automorphism* ◇ B50 C05 ◇
TARASOV, V.V. *Realization of not everywhere defined functions of the algebra of logic (Russian)* ◇ B50 ◇
THOMAS, G.G. *On permutographs* ◇ B50 ◇
WEGENER, I. *Best possible asymptotic bounds on the depth of monotone functions in multivalued logic* ◇ B50 ◇
WOJCICKI, R. *Referential matrix semantics for propositional calculi* ◇ B22 B50 ◇
XU, XIAOSHU *The difference of the maximal closed sets in* P_k *(Chinese) (English summary)* ◇ B50 ◇
YASUGI, M. *Continuous valuation and logic* ◇ B50 ◇

1983

ANSHAKOV, O.M. *Some constructivizations of the propositional logics of D.A. Bochvar and S. Hallden (Russian)* ◇ B50 ◇
BLAU, U. *Three-valued analysis of precise, vague and presupposing quantifiers* ◇ B50 ◇
CHEREPOV, A.N. *Description of the structure of closed classes in* P_k *containing the class of polynomials (Russian)* ◇ B50 ◇
DEMETROVICS, J. & HANNAK, L. *Cardinality of sets of closed classes contained in precomplete classes in* P_k *(Russian) (English summary)* ◇ B50 ◇
DENEV, I.D. & GYUDZHENOV, I.D. *Separable subsets of arguments of functions from* P_k *(Russian) (English summary)* ◇ B50 ◇
DONG, PING *The optimization of GMC over GF(p)* ◇ B50 ◇
EPSTEIN, G. *The underlying ground for hypothetical propositions* ◇ B50 ◇
FEJZULLAEV, R.B. *Algebras of polymorphisms of* n*-ary models (Russian) (English and Azerbaijani summaries)* ◇ B50 ◇
GNIAZDOWSKI, A. *Two sentential logics determined by a four-element quasi-Boolean algebra (Polish) (English summary)* ◇ B50 ◇
GORLOV, V.V. & LAU, D. *Congruences on closed sets of self-dual functions in many-valued logics and on closed sets of linear functions in prime-valued logics* ◇ B50 ◇
GOTO, M. & KAO, S. & NINOMIYA, T. *Synthesis of axiom systems for the three-valued predicate logic by means of the special four-valued logic* ◇ B50 ◇
GRUENWALD, N. *Bestimmung saemtlicher abgeschlossener Mengen aus* $P_{3,2}$*, deren Projektion* F_8^n *ist* ◇ B50 ◇
GRUENWALD, N. *Bestimmung aller abgeschlossenen*

Mengen aus $P_{3,2}$, deren Projektion F_8^2 ist, mit Hilfe von Relationen ◇ B50 ◇
GU, WEINAN & ZHENG, QILUN *Lattice symmetric ternary logic system and its simplification (Chinese)* ◇ B50 ◇
GUCCIONE, S. & TERMINI, S. & TORTORA, R. *In the labyrinth of many valued logics* ◇ B50 ◇
HENDRY, H.E. *Minimally incomplete sets of Lukasiewiczian truth functions* ◇ B50 ◇
JHA, J.S. & LAL, R.N. *n-valued logic of contacts* ◇ B50 ◇
JONES, JOHN *A formalization of an m-valued propositional calculus with variable functors* ◇ B50 ◇
JONES, JOHN *Implication and iterated implication* ◇ B50 ◇
KARPENKO, A.S. *Factor semantics for n-valued logics* ◇ B50 ◇
KATZ, M. *Quotient algebras for logics of imprecision* ◇ B50 ◇
KATZ, M. *Towards a formal multi-valued utility theory* ◇ B50 ◇
LINN, R. *Kompositionsvollstaendigkeit von Algebren mit (m,n)-stelligen Operationen* ◇ B50 ◇
LOU, ZHUKAI *Galois theory of total and partial functions in many-valued logics (Chinese) (English summary)* ◇ B50 ◇
MARCHENKOV, S.S. *On closed classes of self-dual functions of many-valued logic II (Russian)* ◇ B50 ◇
NAKAMURA, A. *Three-valued logic and its application to the query language of incomplete information* ◇ B50 ◇
PAEPPINGHAUS, P. & WIRSING, M. *Nondeterministic three-valued logic: isotonic and guarded truth functions* ◇ B50 B75 ◇
PENA, L. *(Quasi)transitive algebras* ◇ B50 ◇
PERFIL'EVA, I.G. *Representation of functions in $P_{\aleph_0}$ by superpositions of one-place functions and addition (Russian)* ◇ B50 ◇
RAGAZ, M. *Die Unentscheidbarkeit der einstelligen unendlichwertigen Praedikatenlogik* ◇ B50 D35 D55 ◇
ROSE, A. *Completeness of sets of three-valued Sheffer functions* ◇ B50 ◇
SARABIA ALVAREZ-UDE, E.J. *On metalogic and three-valued logic (Spanish)* ◇ B50 ◇
SARNAVSKIJ, N.G. *Recognition of an ordering predicate as a function of three-valued logic (Russian)* ◇ B50 ◇
SHESTAKOV, V.I. *Application of three-valued logic for analysis of relations between physical values (Russian)* ◇ B50 ◇
STOJMENOVIC, I. & TOSIC, R. *Enumeration of monotone symmetric functions of three-valued logic (Russian) (Serbo-Croatian summary)* ◇ B50 ◇
STOJMENOVIC, I. *Enumeration of symmetric functions of precomplete classes of three-valued logic (Russian) (Serbo-Croatian summary)* ◇ B50 ◇
TAJMANOV, V.A. *On cartesian powers of P_2 (Russian)* ◇ B50 ◇
TAJMANOV, V.A. *On function systems in k-valued logic with closure operations of program type (Russian)* ◇ B50 ◇
UGOL'NIKOV, A.B. *Realization of functions from closed classes by schemes of functional elements in a complete basis* ◇ B50 ◇
ULRICH, D. *Models of three-valued calculi in implicational S5* ◇ B50 ◇

1984

ALISEJCHIK, P.A. *On maximal length of a base in P_3* ◇ B50 ◇
ANSHAKOV, O.M. & RYCHKOV, S.V. *Axiomatization of finite-valued logical calculi (Russian)* ◇ B50 C90 ◇
ANSHAKOV, O.M. & RYCHKOV, S.V. *On a way of the formalization and classification of many valued logics (Russian)* ◇ B50 C90 ◇
BOCHVAR, D.A. *On the consistency of a three-valued logical calculus* ◇ B50 ◇
DEMETROVICS, J. & MAL'TSEV, I.A. *Essentially minimal TC-clones on a three-element set (Russian) (English and Hungarian summaries)* ◇ B50 C05 ◇
DENISOVA, R.A. *A class of functions of k-valued logic (Russian)* ◇ B50 ◇
DENISOVA, R.A. *A method of synthesis of irredundant representative collections for k-valued tables (Russian)* ◇ B50 ◇
DROZDOVA, G.D. & GRIGOR'EV, V.V. *Combinatorial enumeration in the set of many-valued logic functions (Russian)* ◇ B50 ◇
GAVRILOV, G.P. *Inductive representations of Boolean functions and the finite generation of the Post classes (Russian)* ◇ B50 ◇
GOTO, M. & KAO, S. & OKADA, Y. *A formulation of a system of specific four-valued logic axioms for the analysis of multi-valued logic axioms (Japanese) (English summary)* ◇ B50 ◇
GOTTWALD, S. *T-Normen und φ-Operatoren als Wahrheitswertfunktion mehrwertiger Junktoren* ◇ B50 ◇
HANNAK, L. *On the construction of many-valued logics (Hungarian)* ◇ B50 ◇
HANNAK, L. *On the structure of many-valued logic (Hungarian)* ◇ B50 ◇
JONES, JOHN *Some propositional calculi with constant and variable functors* ◇ B50 ◇
KABZINSKI, J.K. *An axiomatization of the equivalential fragment of the three-valued logic of Lukasievicz* ◇ B50 ◇
KARPENKO, A.S. *Factor-semantics for n-valued logic (Russian)* ◇ B50 ◇
KATERINOCHKINA, N.N. *Some relations for subsets of layers of an n-dimensional k-valued lattice (Russian) (English summary)* ◇ B50 G10 ◇
KATZ, M. *Controlled-error theories of proximity and dominance* ◇ B50 ◇
KOVALEV, M.M. & MILANOV, P. *Monotone functions of many-valued logic and supermatroids (Russian)* ◇ B50 ◇
KUDRYAVTSEV, V.B. *Concerning functional systems (Russian)* ◇ B50 ◇
LAU, D. *Die maximalen Klassen von $Pol_k\{(x,x+1 \bmod k) \mid x\in E_k\}$* ◇ B50 ◇

LAU, D. *Unterhalbgruppen von (P_3^1,∗)* ◇ B50 ◇
LI, WENGUANG & WU, XUNWEI *Three-valued unary functions and their CMOS realization (Chinese) (English summary)* ◇ B50 ◇
LOWESMITH, B.J. & ROSE, A. *A generalisation of Slupecki's criterion for functional completeness* ◇ B50 ◇
LUO, ZHUKAI *Maximal closed sets in the set of partial many-valued logic functions (Chinese)* ◇ B50 ◇
LUO, ZHUKAI *The classification of normal relations in many-valued logics (Chinese) (English summary)* ◇ B50 ◇
LUO, ZHUKAI *The completeness theory of partial many-valued logic functions (Chinese)* ◇ B50 ◇
MARCHENKOV, S.S. *Closed classes in P_k containing homogeneous functions (Russian) (English summary)* ◇ B50 ◇
MARTIN, J.N. *Epistemic semantics for classical and intuitionistic logic* ◇ B50 F50 ◇
MILICIC, M.I. *Galois correspondences for closed classes of functions with delay I, II (Russian)* ◇ B50 ◇
MIODUSZEWSKA, E. *Referential presupposition within Ulrich Blau's three-valued logic system* ◇ B50 ◇
MUNDICI, D. *Abstract model theory of many-valued logics and K-theory of certain C^*-algebras* ◇ B50 C90 C95 ◇
MYCROFT, A. *Logic programs and many-valued logic* ◇ B50 ◇
PETKOV, P.P. *The uniqueness of syntactical analysis for some calculi that are similar to Post's calculus. A generalization of the cut elimination theorem for classical propositional calculus (Russian)* ◇ B50 F05 ◇
ROSE, A. *Generalised functional completeness of sets of m-valued Sheffer functions* ◇ B50 ◇
SCHULTZ, KONRAD *A generalization of Lindenbaum's theorem for predicate calculi* ◇ B50 ◇
SHAPIRO, D.G. & TONG, R.M. *An experiment with multiple-valued logics in an expert system* ◇ B50 ◇
SHEVCHENKO, V.N. *On some functions on many-valued logics related to integer programming (Russian)* ◇ B50 ◇
STOJMENOVIC, I. *Classification of P_3 and the enumeration of bases of P_3* ◇ B50 ◇
STOJMENOVIC, I. *Enumeration of the bases of three-valued monotone logical functions* ◇ B50 ◇
STREHLE, P. *Zur Notwendigkeit und zum Aufbau mehrdimensionaler mehrwertiger Logiken* ◇ B50 ◇
SUCHON, W. *An elementary method of determining the degree of completeness of n-valued Lukasiewicz propositional calculus* ◇ B50 ◇
VISSER, A. *Four valued semantics and the Liar* ◇ B50 ◇
VUKOVIC, A. *On the bases of the three-valued logic (Serbo-Croatian summary)* ◇ B50 ◇
VUKOVIC, A. *On three-values logic function types* ◇ B50 ◇
VUKOVIC, A. *Three-valued logic function inside the basic type ‹012›* ◇ B50 ◇
XU, ZHIWEI *Multivalued logic and fuzzy logic - their relationship, minimization, and application to fault diagnosis* ◇ B50 B52 ◇

1985

ALEKSEEV, V.B. & EMEL'YANOV, N.R. *A method for constructing fast algorithms in k-valued logic (Russian)* ◇ B50 ◇
AVSARKISYAN, G.S. *Polynomial forms of partial functions of k-valued logic (Russian) (English summary)* ◇ B50 ◇
BUNDER, M.W. & MEYER, R.K. *A result for combinators, BCK logics and BCK algebras* ◇ B50 ◇
DAVIES, R.O. & ROSENBERG, I.G. *Precomplete classes of operations on an uncountable set* ◇ B50 ◇
D'OTTAVIANO, I.M.L. *The model extension theorems for Π_3-theories* ◇ B50 C07 C52 ◇
DYWAN, Z. *On a certain method of producing logical matrices* ◇ B50 ◇
EMEL'YANOV, N.R. *Complexity of the problem of expressibility in many-valued logics (Russian)* ◇ B50 ◇
GOTTWALD, S. *A generalized Lukasiewicz-style identity logic* ◇ B50 ◇
GRIGOLIYA, R.SH. *Formulas of m variables in Lukasiewicz calculi (Russian) (Polish summary)* ◇ B50 ◇
HROMKOVIC, J. *On the number of monotonic functions from two-valued logic to k-valued logic* ◇ B50 ◇
JONES, JOHN *Simplified axiom schemes for implication and iterated implication* ◇ B50 ◇
JONES, JOHN *The rule of procedure Re in Lukasiewicz's many-valued propositional calculi* ◇ B50 ◇
KOPPELBERG, S. *Booleschwertige Logik* ◇ B50 C90 C98 E35 E45 E50 E65 G05 ◇
MIYAKAWA, M. *A note to the classification and base enumeration of three-valued logical functions* ◇ B50 ◇
RASIOWA, H. *Topological representations of Post algebras of order ω^+ and open theories based on ω^+-valued Post logic* ◇ B50 C90 G20 ◇
REISCHER, C. & SIMOVICI, D.A. *Iteration properties of transformations of finite sets with application to multivalued logic* ◇ B50 ◇
ROMOV, B.A. *A local analoque of Slupecki's theorem for infinitely many valued logic (Russian) (English summary)* ◇ B50 ◇
SHTRAKOV, S.V. *On some transformations on k-valued logic (Bulgarian summary)* ◇ B50 ◇
SOKHATSKIJ, F.N. *A generalization of two theorems of Belousov for strongly dependent functions of k-valued logic (Russian)* ◇ B50 ◇
WEISPFENNING, V. *Quantifier elimination for distributive lattices and measure algebras* ◇ B25 B50 C10 C65 G05 G10 ◇
XIAO, XIAN & ZHENG, YUXIN & ZHU, WUJIA *Finite-valued of infinite-valued logical paradoxes (Chinese)* ◇ B50 ◇

B51 Quantum logic

1936

BIRKHOFF, GARRETT & NEUMANN VON, J. *The logic of quantum mechanics* ◇ B51 ◇

1945

BENGY-PUYVALLEE DE, R. *Sur les regles de composabilite dans les logiques de la complementarite de Mme. Destouches-Fevrier* ◇ B51 ◇

DESTOUCHES-FEVRIER, P. *Logique adaptee aux theories quantiques* ◇ B51 G12 ◇

1948

REICHENBACH, H. *Philosophic foundations of quantum mechanics* ◇ A05 B51 G12 ◇

1949

BENGY-PUYVALLEE DE, R. *Sur la relation de composabilite dans les logiques de complementarite* ◇ B51 G12 ◇

DESTOUCHES-FEVRIER, P. *Logique et theories physiques* ◇ A05 B51 ◇

1950

JORDAN, P. *Zur Quanten-Logik* ◇ B51 G12 ◇

1951

REICHENBACH, H. *Ueber die erkenntnistheoretische Problemlage und den Gebrauch einer dreiwertigen Logik in der Quantenmechanik* ◇ A05 B51 G12 ◇

1953

REICHENBACH, H. *Les fondements logiques de la mecanique des quanta* ◇ B51 G12 ◇

1954

DESTOUCHES-FEVRIER, P. *La logique des propositions experimentales* ◇ B51 G12 ◇

REICHENBACH, H. *Les fondements logiques de la theorie des quanta. Utilisation d'une logique a trois valeurs* ◇ B50 B51 G12 ◇

1956

DESTOUCHES, J.-L. *Ueber den Aussagenkalkuel der Experimentalaussagen* ◇ B51 G12 ◇

1958

FEYERABEND, P. *Reichenbach's interpretation of quantum-mechanics* ◇ B51 ◇

1959

JORDAN, P. *Quantenlogik und das kommutative Gesetz* ◇ A05 B51 G12 ◇

1961

BEDAU, H. & OPPENHEIM, P. *Complementarity in quantum mechanics: A logical analysis* ◇ A05 B51 G12 ◇

1962

JORDAN, P. *Bemerkungen zur Quantenlogik* ◇ B51 G12 ◇

1963

KOTAS, J. *Axiom for Birkhoff-v.Neumann quantum logic* ◇ B51 G12 ◇

1964

KUNSEMUELLER, H. *Zur Axiomatik der Quantenlogik* ◇ B51 G12 ◇

1965

KOCHEN, S. & SPECKER, E. *Logical structures arising in quantum theory* ◇ B51 G12 ◇

SCHLESSINGER, N. & ZIERLER, N. *Boolean embeddings of orthomodular sets and quantum logic* ◇ A05 B51 G05 G10 G12 ◇

1967

KOTAS, J. *An axiom system for the modular logic (Polish and Russian summaries)* ◇ B51 G12 ◇

1968

JAUCH, J.M. *Foundation of quantum theory* ◇ B51 G12 ◇

1969

FINCH, P.D. *On the structure of quantum logic* ◇ B51 G12 ◇

GUDDER, S.P. *On the quantum logic approach to quantum mechanics* ◇ A05 B51 G12 ◇

LEININGER, C.W. *Concerning some proposals for quantum logic* ◇ B51 G12 ◇

1970

FOULIS, D.J. & RANDALL, C.H. *An approach to empirical logic* ◇ B51 G12 ◇

GUDDER, S.P. *Axiomatic quantum mechanics and generalized probability theory* ◇ B51 G12 ◇

HEELAN, P.A. *Quantum logic and classical logic: their respective roles* ◇ B51 ◇

JAUCH, J.M. & PIRON, C. *What is "quantum-logic"?* ◇ B51 ◇

MITTELSTAEDT, P. *Quantenlogische Interpretation orthokomplementaerer quasimodularer Verbaende* ◇ B51 G10 G12 ◇

1971

BLASI, A.A. *Difficulties of a quantum mechanics over finite and p-adic fields* ◇ B51 ◇

GREECHIE, R.J. *Combinatorial quantum logic* ◇ B51 G12 ◇

GREECHIE, R.J. & GUDDER, S.P. *Is a quantum logic a logic?* ◇ B51 ◇

JADCZYK, A.Z. *Quantum logic and indefinite metric spaces* ◇ B51 G12 ◇
MACZYNSKI, M.J. *Boolean properties of observables in axiomatic quantum mechanics* ◇ B51 G05 G12 ◇

1972

DISHKANT, G.P. *Semantics of the minimal logic of quantum mechanics (Polish and Russian summaries)* ◇ B51 ◇
FRIEDMAN, M. & GLYMOUR, C. *If quanta had logic* ◇ A05 B51 G12 ◇
GARDNER, MICHAEL R. *Two deviant logics for quantum theory: Bohr and Reichenbach* ◇ B51 G12 ◇
JEFFCOTT, B. *The center of an orthologic* ◇ B51 G12 G25 ◇
MITTELSTAEDT, P. *On the interpretation of the lattice of subspaces of the Hlbert space as a propositional calculus* ◇ B51 G12 ◇

1973

BENIOFF, P.A. *On definitions of validity applied to quantum theories* ◇ B51 G12 ◇
CLARK, I.D. *An axiomatisation of quantum logic* ◇ B51 ◇
DELIYANNIS, P.C. *Vector space models of abstract quantum logics* ◇ B51 ◇
FRAASSEN VAN, B.C. *Semantic analysis of quantum logic* ◇ A05 B51 G12 ◇
GREECHIE, R.J. & GUDDER, S.P. *Quantum logics* ◇ B51 G12 ◇

1974

DISHKANT, G.P. *The first order predicate calculus based on the logic of quantum mechanics* ◇ B51 ◇
FOULIS, D.J. & RANDALL, C.H. *Empirical logic and quantum mechanics* ◇ B51 G12 ◇
FRAASSEN VAN, B.C. *Hidden variables in conditional logic* ◇ B51 G12 ◇
FRAASSEN VAN, B.C. *The labyrinth of quantum logics* ◇ B51 ◇
GOLDBLATT, R.I. *Semantic analysis of orthologic* ◇ B51 G10 G12 ◇
GREECHIE, R.J. *Some results from the combinatorial approach to quantum logic* ◇ B51 G12 ◇
HARDEGREE, G.M. *The conditional in quantum logic* ◇ B51 G12 ◇
HEELAN, P.A. *Quantum logic and classical logic: their respective roles* ◇ B51 ◇
JEFFCOTT, B. *A note on commutivity in the composite product of orthologics* ◇ B51 ◇
KALMBACH, G. *Orthomodular logic* ◇ B51 G10 G12 ◇
PUTNAM, H. *How to think quantum-logically* ◇ A05 B51 G12 ◇
SCHEIBE, E. *Popper and quantum logic* ◇ A05 B51 G12 ◇
TARSKI, J. *Quantum field theory; an unusual discipline* ◇ B51 G12 ◇

1975

BELTRAMETTI, E.G. & CASSINELLI, G. *Ideal, first-kind measurements in a proposition-state structure* ◇ B51 ◇
FRAASSEN VAN, B.C. *Incomplete assertion and Belnap connectives* ◇ B51 G12 ◇
GREECHIE, R.J. *On three dimensional quantum proposition systems* ◇ B51 G12 ◇
HARDEGREE, G.M. *Quasi-implicative lattices and the logic of quantum mechanics* ◇ B51 G10 G12 ◇
HARDEGREE, G.M. *Stalnaker conditionals and quantum logic* ◇ B51 G12 ◇
HOOKER, C.A. (ED.) *The logico-algebraic approach to quantum mechanics. Vol. I: Historical evolution* ◇ A10 B51 G12 ◇

1976

DELIYANNIS, P.C. *Conditioning of states* ◇ B51 ◇
DELIYANNIS, P.C. *Superposition of states and the structure of quantum logics* ◇ B51 ◇
DISHKANT, G.P. *Logic of quantum mechanics* ◇ A05 B51 ◇
DOMOTOR, Z. *The probability structure of quantum-mechanical systems* ◇ B51 G12 ◇
FINCH, P.D. *Quantum mechanical physical quantities as random variables* ◇ B51 ◇
FINKELSTEIN, D. *Classical and quantum probability and set theory* ◇ A05 B51 E70 G12 ◇
FOULIS, D.J. & RANDALL, C.H. *Empirical logic and quantum mechanics* ◇ B51 G12 ◇
FRAASSEN VAN, B.C. & HOOKER, C.A. *A semantic analysis of Niels Bohr's philosophy of quantum theory* ◇ A05 B51 G12 ◇
GREECHIE, R.J. *Some results from the combinatorial approach to quantum logic* ◇ B51 G12 ◇
HARDEGREE, G.M. *The conditional in quantum logic* ◇ B51 G12 ◇
MITTELSTAEDT, P. *On the applicability of the probability concept to quantum theory* ◇ B51 G12 ◇
MITTELSTAEDT, P. *Quantum logic* ◇ B51 G12 ◇
PUTNAM, H. *How to think quantum-logically* ◇ A05 B51 G12 ◇
SASIADA, E. *A notion of entropy which does not increase* ◇ B51 ◇
STACHOW, E.-W. *Completeness of quantum logic* ◇ B51 G12 ◇

1977

BELTRAMETTI, E.G. & CASSINELLI, G. *On state transformations induced by yes-no experiments, in the context of quantum logic* ◇ B51 ◇
BUB, J. *Von Neumann's projection postulate as a probability conditionalization rule in quantum mechanics* ◇ A05 B51 ◇
CEGLA, W. & JADCZYK, A.Z. *Causal logic of Minkowski space* ◇ B51 G12 ◇
DALLA CHIARA SCABIA, M.L. *Logical selfreference, set theoretical paradoxes and the measurement problem in quantum mechanics* ◇ A05 B51 E30 G12 ◇
DALLA CHIARA SCABIA, M.L. *Quantum logic and physical modalities* ◇ B51 ◇
DENECKE, H.-M. *Quantum logic of quantifiers* ◇ B51 ◇
DISHKANT, G.P. *Imbedding of the quantum logic in the modal system of Brouwer* ◇ B51 ◇
DISHKANT, G.P. *The connective "becoming" and the paradox of electron diffraction* ◇ B51 ◇

DRIESCHNER, M. *Is (quantum) logic empirical?* ◇ A05 B51 ◇
FINKELSTEIN, D. *The Leibniz project* ◇ B51 G12 G25 ◇
GUZ, W. *Axioms for nonrelativistic quantum mechanics* ◇ B51 ◇
HARDEGREE, G.M. *Reichenbach and the logic of quantum mechanics* ◇ A05 B51 G12 ◇
JONES, ROGER *Causal anomalies and the completeness of quantum theory* ◇ A05 B51 G12 ◇
KAEGI-ROMANO, U. *Quantum logic and generalized probability theory* ◇ B51 G12 ◇
LAHTI, P.J. *Complementary logic in the description of physical systems* ◇ B51 G12 ◇
MITTELSTAEDT, P. (ED.) *Special issue: Symposium on Quantum Logic* ◇ B51 G12 ◇
MITTELSTAEDT, P. *Time dependent propositions and quantum logic* ◇ A05 B51 G12 ◇
PIRON, C. *On the logic of quantum logic* ◇ A05 B51 G12 ◇
PULMANNOVA, S. *Symmetries in quantum logics* ◇ B51 G12 ◇
STACHOW, E.-W. *How does quantum logic correspond to physical reality?* ◇ A05 B51 G12 ◇

1978

AERTS, D. & DAUBECHIES, I. *About the structure-preserving maps of a quantum mechanical propositional system* ◇ B51 ◇
ALMOG, J. *Perhaps (?), new logical foundations are needed for quantum mechanics* ◇ A05 B51 G12 ◇
BEK, R. *Discourse on one way in which a quantum-mechanics language on the classical logical base can be built up* ◇ B51 ◇
BUGAJSKI, S. *Probability implication in the logics of classical and quantum mechanics* ◇ B51 ◇
COOK, T.A. *The geometry of generalized quantum logics* ◇ B51 ◇
CORBETT, J. *Probability in quantum mechanics* ◇ B51 ◇
DISHKANT, G.P. *An extension of the Lukasiewicz logic to the modal logic of quantum mechanics* ◇ B51 ◇
ERWIN, E. *Quantum logic and the status of classical logic* ◇ B51 G12 ◇
FAY, G. & TOEROES, R. *Quantum logic (Hungarian)* ◇ B51 ◇
FRIEDMAN, M. & PUTNAM, H. *Quantum logic, conditional probability, and interference* ◇ A05 B51 G12 ◇
GUZ, W. *On the lattice structure of quantum logics* ◇ B51 G12 ◇
GUZ, W. *On the simultaneous verifiability of yes-no measurements* ◇ B51 G12 ◇
IVERT, P.A. & SJOEDIN, T. *On the impossibility of a finite propositional lattice for quantum mechanics* ◇ B51 G12 ◇
LUNGARZO, C.A. *Superposition of states in quantum logic from a set theoretical point of view* ◇ B51 E75 G12 ◇
MAJEWSKI, M. *On some matrix of the Birkhoff and v. Neumann quantum logic* ◇ B51 G12 ◇
MITTELSTAEDT, P. *Quantum logic* ◇ B51 G10 G12 ◇
MITTELSTAEDT, P. & STACHOW, E.-W. *The principle of excluded middle in quantum logic* ◇ B51 G12 ◇
PULMANNOVA, S. *Joint distributions of observables on quantum logics* ◇ B51 G12 ◇
STACHOW, E.-W. *Quantum logical calculi and lattice structures* ◇ B51 G12 ◇

1979

AERTS, D. & DAUBECHIES, I. *A characterization of subsystems in physics* ◇ B51 G12 ◇
AERTS, D. & DAUBECHIES, I. *A mathematical condition for a sublattice of a propositional system to represent a physical subsystem, with a physical interpretation* ◇ B51 G12 ◇
BARONE, F. & GALDI, G.P. *On the question of atomicity and determinism in boolean systems* ◇ B51 G12 ◇
BELTRAMETTI, E.G. & CASSINELLI, G. *Properties of states in quantum logic* ◇ B51 ◇
BIGELOW, J.C. *Quantum probability in logical space* ◇ B51 G12 ◇
BUB, J. *Some reflections on quantum logic and Schroedinger's cat* ◇ A05 B51 ◇
BUB, J. *The measurement problem of quantum mechanics* ◇ B51 ◇
CASSINELLI, G. & TRUINI, P. *Toward a generalized probability theory: Conditional probabilities* ◇ B51 G12 ◇
DALLA CHIARA SCABIA, M.L. & TORALDO DI FRANCIA, G.G. *Formal analysis of physical theories* ◇ A05 B30 B51 ◇
DEMOPOULOS, W. *Boolean representations of physical magnitudes and locality* ◇ B51 G12 ◇
DRECHSEL, P. *Traditional logic and quantum logic* ◇ B51 ◇
DVURECHENSKIJ, A. *Laws of large numbers and the central limit theorems on a logic* ◇ B51 ◇
FOULIS, D.J. & RANDALL, C.H. *The operational approach to quantum mechanics* ◇ B51 G12 ◇
GEORGACARAKOS, G.N. *Orthomodularity and relevance* ◇ B51 G12 ◇
GILES, R. *The concept of a proposition in classical and quantum physics* ◇ A05 B51 G12 ◇
GUDDER, S.P. *A survey of axiomatic quantum mechanics* ◇ B51 G12 ◇
GUZ, W. *An improved formulation of axioms for quantum mechanics* ◇ B51 ◇
HARDEGREE, G.M. *The conditional in abstract and concrete quantum logic* ◇ B51 G12 ◇
HOOKER, C.A. (ED.) *Physical theory as logico-operational structure* ◇ A05 B51 G12 ◇
HOOKER, C.A. (ED.) *The logico-algebraic approach to quantum mechanics. Vol.II* ◇ B51 G12 ◇
LAHTI, P.J. *On the expectation value of an observable in quantum logic* ◇ B51 G12 ◇
MITTELSTAEDT, P. *Quantum logic* ◇ B51 G12 ◇
MITTELSTAEDT, P. *The modal logic of quantum logic* ◇ B51 G12 ◇
PULMANNOVA, S. *Superposition principle and sectors in quantum logics* ◇ B51 ◇
STACHOW, E.-W. *An operational approach to quantum probability* ◇ B51 G12 ◇

STACHOW, E.-W. *Completeness of quantum logic* ◇ B51 G12 ◇
STACHOW, E.-W. *Quantum logical calculi and lattice structures* ◇ B51 G12 ◇
STOUT, L.N. *Laminations, or how to build a quantum-logic-valued model of set theory* ◇ B51 E70 G12 G30 ◇
ZEMAN, J.J. *Quantum logic with implication* ◇ B51 G12 ◇

1980

BURGHARDT, F.J. *Modal quantum logic and its dialogic foundation* ◇ B51 G12 ◇
DVURECHENSKIJ, A. *On a sum of observables in a logic* ◇ B51 ◇
DVURECHENSKIJ, A. & RIECHAN, B. *On the individual ergodic theorem on a logic* ◇ B51 ◇
DVURECHENSKIJ, A. & PULMANNOVA, S. *On the sum of observables in a logic* ◇ B51 ◇
FRESCURA, F.A.M. & HILEY, B.J. *The implicate order, algebras, and the spinor* ◇ B51 G12 ◇
GAROLA, C. *Propositions and orthocomplementation in quantum logic* ◇ B51 G12 ◇
GUZ, W. *Conditional probability in quantum axiomatics* ◇ B51 G12 ◇
GUZ, W. *Event-phase-space structure: An alternative to quantum logic* ◇ B51 G12 ◇
LAHTI, P.J. *Characterization of quantum logics* ◇ B51 G12 ◇
MATVEJCHUK, M.S. *A theorem on the states of quantum logics (Russian) (English summary)* ◇ B51 ◇
NISHIMURA, H. *Sequential method in quantum logic* ◇ B51 G12 ◇
PULMANNOVA, S. *Semiobservables on quantum logics* ◇ B51 G12 ◇
REHDER, W. *Quantum logic of sequential events and their objectivistic probabilities* ◇ B51 G12 ◇
REHDER, W. *When do projections commute?* ◇ B51 G12 ◇
STACHOW, E.-W. *Logical foundation of quantum mechanics* ◇ B51 G12 ◇

1981

ABBATI, M.C. & MANIA, A. *The quantum logical and the operational description for physical systems* ◇ B51 ◇
AERTS, D. *Description of compound physical systems and logical interaction of physical systems* ◇ B51 G12 ◇
BANAI, M. *Propositional systems in field theories and lattice-valued quantum logic* ◇ B51 ◇
BANAI, M. *Propositional systems in local field theories* ◇ B51 G12 ◇
BARONE, F. *On implemented state automorphisms with the logico-algebraic approach to deterministic mechanics* ◇ B51 ◇
BELTRAMETTI, E.G. & FRAASSEN VAN, B.C. (EDS.) *Current issues in quantum logic* ◇ B51 B97 ◇
BELTRAMETTI, E.G. & CASSINELLI, G. *Problems of the proposition-state structure of quantum mechanics* ◇ B51 ◇
BELTRAMETTI, E.G. & CASSINELLI, G. *The logic of quantum mechanics* ◇ B51 ◇
BERNINI, S. *Quantum logic as an extension of classical logic* ◇ B51 ◇
BRIDGES, D.S. *Towards a constructive foundation for quantum mechanics* ◇ B51 G12 ◇
BUB, J. *What does quantum logic explain?* ◇ B51 G12 ◇
BUGAJSKI, S. *The inner language of operational quantum mechanics* ◇ B51 ◇
CEGLA, W. *Causal logic of Minkowski space* ◇ B51 ◇
CHAPMAN, T. *Quantum logic and modality* ◇ B45 B51 ◇
CHENTSOV, N.N. & MOROZOVA, E.A. *Noncommutative quantum logic (finite-dimensional theory) (Russian) (English summary)* ◇ B51 ◇
COOKE, R.M. & HILGEVOORD, J. *A new approach to equivalence in quantum logic* ◇ B51 G12 ◇
CORSI, G. *Deduzione naturale e logica quantistica* ◇ B51 ◇
CZELAKOWSKI, J. *Partial boolean algebras in a broader sense as a semantics for quantum logic* ◇ B51 G12 ◇
CZELAKOWSKI, J. *Partial referential matrices for quantum logics* ◇ B51 ◇
DALLA CHIARA SCABIA, M.L. *Is there a logic of empirical sciences?* ◇ A05 B51 G12 ◇
DALLA CHIARA SCABIA, M.L. *Logical foundations of quantum mechanics* ◇ B51 ◇
DALLA CHIARA SCABIA, M.L. *Some metalogical pathologies of quantum logic* ◇ B51 ◇
DORLING, J. *How to rewrite a stochastic dynamical theory so as to generate a measurement paradox* ◇ B51 G12 ◇
DVURECHENSKIJ, A. *On extension properties for observables (Russian summary)* ◇ B51 ◇
DVURECHENSKIJ, A. & PULMANNOVA, S. *Random measures on a logic* ◇ B51 ◇
FINKELSTEIN, D. *Quantum sets, assemblies and plexi* ◇ B51 E70 ◇
FOULIS, D.J. & RANDALL, C.H. *Empirical logic and tensor products* ◇ B51 ◇
FOULIS, D.J. & RANDALL, C.H. *What are quantum logics and what ought they to be?* ◇ B51 ◇
FRAASSEN VAN, B.C. *Assumptions and interpretations of quantum logic* ◇ B51 ◇
FRAZER, P.J. & HARDEGREE, G.M. *Charting the labyrinth of quantum logics: a progress report* ◇ B51 ◇
GIBBINS, P. *A note on quantum logic and the uncertainty principle* ◇ B51 ◇
GREECHIE, R.J. *A nonstandard quantum logic with a strong set of states* ◇ B51 G12 ◇
GUCCIONE, S. *Quantum logic and the two-slit experiment* ◇ B51 ◇
GUDDER, S.P. *Comparison of the quantum logic, convexity, and algebraic approaches to quantum mechanics* ◇ B51 G12 ◇
GUDDER, S.P. *Measure and integration in quantum set theory* ◇ B51 G12 ◇
GUDDER, S.P. *Representations of Baer *-semigroups and quantum logics in Hilbert space* ◇ B51 G12 ◇
GUZ, W. *Projection postulate and superposition principle in non-lattice quantum logics* ◇ B51 G12 ◇

HARDEGREE, G.M. *An axiom system for orthomodular quantum logic* ◇ B51 G12 ◇
HARDEGREE, G.M. *Some problems and methods in formal quantum logic* ◇ B51 ◇
HELLMAN, GEOFFREY *Quantum logic and the projection postulate* ◇ B51 ◇
HUGHES, R.J.G. *Realism and quantum logic* ◇ A05 B51 G12 ◇
KALMBACH, G. *Omologic as a Hilbert type calculus* ◇ B51 ◇
KRON, A. & MARIC, Z. & VUJOSEVIC, S.T. *Entailment and quantum logic* ◇ B51 ◇
MACZYNSKI, M.J. *Commutativity and generalized transition probability in quantum logic* ◇ B51 ◇
MARLOW, A.R. *Space-time structure from quantum logic* ◇ B51 G12 ◇
MATVEJCHUK, M.S. *A theorem on the states of quantum logics. II (Russian) (English summary)* ◇ B51 G12 ◇
MIELNIK, B. *Motion and form* ◇ B51 G12 ◇
MITTELSTAEDT, P. *Classification of different areas of work afferent to quantum logic* ◇ B51 ◇
MITTELSTAEDT, P. *The dialogic approach to modalities in the language of quantum physics* ◇ B51 ◇
NEUMANN, H. (ED.) *Interpretations and foundations of quantum theory* ◇ B51 G12 ◇
ORLOV, YU.F. *A quantum model of doubt* ◇ B51 ◇
PULMANNOVA, S. *On the observables on quantum logics* ◇ B51 ◇
RUETTIMANN, G.T. *Detectable properties and spectral quantum logics* ◇ B51 G12 ◇
STACHOW, E.-W. *Der quantenlogische Wahrscheinlichkeitskalkuel* ◇ B51 ◇
STACHOW, E.-W. *Sequential quantum logic* ◇ B51 ◇
TAKEUTI, G. *Logic and set theory* ◇ B51 B98 E40 E70 F50 ◇
WEIZSAECKER VON, C.F. *In welchem Sinne ist die Quantenlogik eine zeitliche Logik?* ◇ B51 ◇
WESSELS, L. *The "EPR" argument: a post-mortem* ◇ B51 ◇
ZECCA, A. *Products of logics* ◇ B51 ◇

1982

AERTS, D. *Description of many separated physical entities without the paradoxes encountered in quantum mechanics* ◇ B51 ◇
BELL, J.L. & HALLETT, M. *Logic, quantum logic and empiricism* ◇ B51 ◇
BORN, R. *Kausalitaet und Quantenlogik* ◇ B51 G12 ◇
BRABEC, J. & PTAK, P. *On compatibility in quantum logics* ◇ B51 G12 ◇
BUB, J. *Quantum logic, conditional probability, and interference* ◇ B51 ◇
BUGAJSKI, S. *What is quantum logic?* ◇ B51 G12 ◇
CUTLAND, N.J. & GIBBINS, P. *A regular sequent calculus for quantum logic in which* ∧ *and* ∨ *are dual* ◇ B51 G12 ◇
EBERHARD, P.H. *Constraints of determinism and of Bell's inequalities are not equivalent* ◇ B51 ◇
FINE, A.I. *Comments on the significance of Bell's theorem* ◇ B51 ◇
FRAASSEN VAN, B.C. *The Charybdis of realism: epistemological implications of Bell's inequality* ◇ A05 B51 ◇
GREECHIE, R.J. & GUDDER, S.P. & RUETTIMANN, G.T. *Measurements, Hilbert space and quantum logics* ◇ B51 ◇
GUDDER, S.P. *Hilbertian interpretations of manuals* ◇ B51 G12 ◇
HELLMAN, GEOFFREY *Einstein and Bell: strengthening the case for microphysical randomness* ◇ B51 ◇
JAMMER, M. *A note on Peter Gibbins':"A note on quantum logic and the uncertainty principle"* ◇ B51 ◇
KATZ, M. *The logic of approximation in quantum theory* ◇ B51 ◇
PITOWSKY, I. *Substitution and truth in quantum logic* ◇ B51 ◇
PULMANNOVA, S. *Individual ergodic theorem on a logic (Russian summary)* ◇ B51 ◇
STAIRS, A. *Discussion: quantum logic and the Lueders rule* ◇ B51 ◇
STAPP, H.P. *Bell's theorem as a nonlocality property of quantum theory* ◇ B51 ◇
SUTHERLAND, R.I. *On Kochen and Specker's impossibility proof (Italian and Russian summaries)* ◇ B51 G12 ◇

1983

BAEZ, J.C. *Recursivity in quantum mechanics* ◇ B51 D80 ◇
BEAVER, O.R. *Regularity and decomposability of finitely additive functions on a quantum logic* ◇ B51 ◇
BELL, J.L. *Orthologic, forcing, and the manifestation of attributes* ◇ B51 C90 ◇
BUGAJSKI, S. *Semantics in Banach spaces* ◇ B51 C65 C90 ◇
DORNINGER, D. & LAENGER, H. & MACZYNSKI, M.J. *Zur Darstellung von Observablen auf σ-stetigen Quantenlogiken* ◇ B51 G12 ◇
DRUMMOND, P.D. *Violations of Bell's inequality in cooperative states* ◇ B51 ◇
FOULIS, D.J. & RANDALL, C.H. *Properties and operational propositions in quantum mechanics* ◇ B51 G12 ◇
GARDEN, R.W. *Modern logic and quantum mechanics* ◇ B51 G12 ◇
GIBBINS, P. *Quantum logic and ensembles* ◇ B51 G12 ◇
HOLDSWORTH, D.G. & HOOKER, C.A. *A critical survey of quantum logic* ◇ B51 ◇
MAJEWSKI, M. *About the intermediate logics between modular logic and the classical propositional calculus* ◇ B51 ◇
MATVEJCHUK, M.S. *A theorem on states on quantum logics. States in Jordan algebras (Russian)* ◇ B51 G12 ◇
MERMIN, N.D. *Pair distributions and conditional independence: some hints about the structure of strange quantum correlations* ◇ B51 G12 ◇
MITTELSTAEDT, P. & STACHOW, E.-W. *Analysis of the Einstein-Podolsky-Rosen experiment by relativistic quantum logic* ◇ B51 G12 ◇
MITTELSTAEDT, P. *Relativistic quantum logic* ◇ B51 ◇

MITTELSTAEDT, P. *Wahrheit, Wirklichkeit und Logik in der Sprache der Physik* ◇ B51 ◇
MOLDAUER, P.A. *Comment on : "Bell's theorem as a nonlocality property of quantum theory" by H.P. Stapp* ◇ B51 ◇
MORGAN, C.G. *Probabilistic semantics for orthologic and quantum logic* ◇ B51 ◇
MOROZ, B.Z. *Reflections on quantum logic* ◇ B51 ◇
NEUBRUNN, T. & PULMANNOVA, S. *On compatability in quantum logics (Russian and Slovak summaries)* ◇ B51 G12 ◇
PTAK, P. & ROGALEWICZ, V. *Regularly full logics and the uniqueness problem for observables* ◇ B51 ◇
PULMANNOVA, S. *Coupling of quantum logics* ◇ B51 G12 ◇
STAIRS, A. *On the logic of pairs of quantum systems* ◇ B51 ◇
STAIRS, A. *Quantum logic, realism, and value definiteness* ◇ B51 G12 ◇

1984

BAZHANOV, V.A. *Logic of quantum mechanics and the problem of its decidability (Russian)* ◇ B51 D35 ◇
BELNAP JR., N.D. & MCROBBIE, M.A. & NUEL JR., D. *Proof tableau formulations of some first-order relevant orthologics* ◇ B51 ◇
BRODY, T.A. *On quantum logic* ◇ B51 G12 ◇
BURGHARDT, F.J. *Modalities and quantum mechanics* ◇ B51 ◇
CATTANEO, G. & NISTICO, G. *Orthogonality and orthocomplementations in the axiomatic approach to quantum mechanics: Remarks about some critiques* ◇ B51 G12 ◇
CIRULIS, J. *Variations on the theme of quantum logic (Russian)* ◇ B51 ◇
DANIEL, W. *The entropy of observables on quantum logic* ◇ B51 G12 ◇
DELIYANNIS, P.C. *Geometrical models for quantum logics with conditioning* ◇ B51 ◇
DVURECHENSKIJ, A. & PULMANNOVA, S. *Connection between joint distribution and compatibility* ◇ B51 ◇
GODOWSKI, R. & GREECHIE, R.J. *Some equations related to states on orthomodular lattices* ◇ B51 ◇
GOLDBLATT, R.I. *Orthomodularity is not elementary* ◇ B51 C65 G10 ◇
HARTHONG, J. *Etudes sur la mecanique quantique (French)* ◇ B51 ◇
HERBUT, F. *On a possible empirical meaning of meets and joins for quantum propositions* ◇ B51 ◇
MUKHERJEE, M.K. *A generalized characterization theorem for quantum logics* ◇ B51 ◇
PTAK, P. *On centers and state spaces of logics* ◇ B51 ◇
ROGALEWICZ, V. *On the uniqueness problem for quite full logics (French summary)* ◇ B51 ◇
TAKEUTI, G. *Quantum logic and quantization* ◇ B51 C90 E40 E75 ◇

1985

AKEN VAN, J. *Analysis of quantum probability theory I* ◇ B51 ◇
COOK, T.A. & RUETTIMANN, G.T. *Symmetries on quantum logics* ◇ B51 G12 ◇
DUBOIS, DIDIER & PRADE, H. *A review of fuzzy set aggregation connectives* ◇ B51 E72 ◇
DVURECHENSKIJ, A. & PULMANNOVA, S. *Uncertainty principle and joint distributions of observables* ◇ B51 ◇
GAROLA, C. *Embedding of posets into lattices in quantum logic* ◇ B51 ◇
HARDEGREE, G.M. & LOCK, P.F. *Connections among quantum logics I: Quantum propositional logics. II: Quantum event logics* ◇ B51 ◇
HUGHES, R.J.G. *Semantic alternatives in partial boolean quantum logic* ◇ B51 ◇
MITTELSTAEDT, P. & STACHOW, E.-W. *Recent developments in quantum logic* ◇ B51 ◇
OVCHINNIKOV, P.G. *Finitely additive functions on extensions of quantum logics (Russian)* ◇ B51 ◇
PEREZ LARAUDOGOITIA, Y. *Modal interpretation of quantum mechanics (Spanish) (English summary)* ◇ B45 B51 ◇
PTAK, P. & WRIGHT, J.D.M. *On the concreteness of quantum logics* ◇ B51 ◇

B52 Fuzzy logic

1956

BADARAU, D. *On indeterminacy in particular judgements (Romanian)* ◇ B52 ◇

1965

ZADEH, L.A. *Fuzzy sets* ◇ B52 E72 ◇

1967

KUMAR, D. *Logic and inexact predicates* ◇ B52 ◇

1968

ZADEH, L.A. *Probability measures of fuzzy events* ◇ B52 ◇

1969

FU, K.S. & WEE, W.G. *A formulation of fuzzy automata and its application as a model of learning systems* ◇ B52 ◇

GOGUEN, J.A. *The logic of inexact concepts* ◇ A05 B52 ◇

MARINOS, P. *Fuzzy logic and its application to switching systems* ◇ B52 ◇

TSICHRITZIS, D.C. *Fuzzy computability* ◇ B52 D75 ◇

1970

BELLMAN, R.E. & ZADEH, L.A. *Decision-making in a fuzzy environment* ◇ B52 ◇

CLEAVE, J.P. *The notion of validity in logical systems with inexact predicates* ◇ B52 ◇

1971

BROWN, J.G. *A note on fuzzy sets* ◇ B52 E72 ◇

CHANG, C.L. & LEE, R.C.T. *Some properties of fuzzy logic* ◇ B52 ◇

KUMAR, D. *Vagueness and subjunctivity* ◇ B52 ◇

TIEN, P.S. *The analysis and synthesis of fuzzy switching circuits* ◇ B52 ◇

ZADEH, L.A. *Quantitative fuzzy semantics* ◇ B52 E72 ◇

1972

CHANG, S.S.L. & ZADEH, L.A. *On fuzzy mapping and control* ◇ B52 ◇

LEE, R.C.T. *Fuzzy logic and the resolution principle* ◇ B35 B52 ◇

1973

BELLMAN, R.E. & GIERTZ, M. *On the analytic formalism of the theory of fuzzy sets* ◇ B52 E72 ◇

KANDEL, A. *Comment on an algorithm that generates fuzzy prime implicants by Lee and Chang* ◇ B52 ◇

KANDEL, A. *On the analysis of fuzzy logic* ◇ B52 ◇

KAUFMANN, A. *Introduction a la theorie des sous-ensembles flous a l'usage des ingenieurs. Tome I: Elements theoriques de base* ◇ B52 E70 E72 E98 ◇

LAKOFF, G. *Hedges: A study in meaning criteria and the logic of fuzzy concepts* ◇ B52 ◇

1974

CHEN, C.S. & SIY, P. *Fuzzy logic for handwritten numeral character recognition* ◇ B52 ◇

CLEAVE, J.P. *The notion of logical consequences in the logic of inexact predicates* ◇ B52 ◇

KANDEL, A. *Fuzzy represantation, CNF minimization and their application to fuzzy transmission structures* ◇ B52 ◇

KANDEL, A. *On the minimization of incompletely specified fuzzy functions* ◇ B52 ◇

KLING, R. *Fuzzy-PLANNER: Reasoning with inexact concepts in a procedural problem-solving language* ◇ B52 ◇

MENGES, G. & SKALA, H.J. *On the problem of vagueness in the social sciences* ◇ A05 B52 ◇

NEGOITA, C.V. & RALESCU, D.A. *Applications of fuzzy sets to systems analysis (Romanian)* ◇ B52 E70 E72 ◇

ZADEH, L.A. *Fuzzy logic and its application to approximate reasoning* ◇ B52 ◇

1975

ARBIB, M.A. & MANES, E.G. *Fuzzy machines in a category* ◇ B52 D05 G30 ◇

CARLSTROM, I.F. *Truth and entailment for a vague quantifier* ◇ B52 C80 ◇

COOLS, M. & DUBOIS, T. & KAUFMANN, A. *Exercices avec solutions sur la theorie des sous-ensembles flous* ◇ B52 E72 E98 ◇

FRAASSEN VAN, B.C. *Comments: Lakoff's fuzzy propositional logic* ◇ A05 B52 ◇

LAKOFF, G. *Hedges: a study in meaning criteria and the logic of fuzzy concepts* ◇ A05 B52 B65 ◇

MUKAIDONO, M. *On some properties of fuzzy logic* ◇ B52 ◇

NEGOITA, C.V. & RALESCU, D.A. *Representation theorems for fuzzy concepts* ◇ B52 E72 ◇

SCHOTCH, P.K. *Fuzzy modal logic* ◇ B52 ◇

SHIMURA, M. *An approach to pattern recognition and associative memories using fuzzy logic* ◇ B52 ◇

SRINI, V.P. *Realization of fuzzy forms* ◇ B52 ◇

ZADEH, L.A. *Calculus of fuzzy restrictions* ◇ B52 ◇

ZADEH, L.A. *Fuzzy logic and approximate reasoning* ◇ B52 ◇

ZADEH, L.A. *The concept of a linguistic variable and its application to approximate reasoning. I,II,III* ◇ B52 B65 ◇

1976

GAINES, B.R. *Foundations of fuzzy reasoning* ◇ B52 E72 ◇
GAINES, B.R. *Fuzzy reasoning and the logic of uncertainty* ◇ B52 ◇
GRATTAN-GUINNESS, I. *Fuzzy membership mapped onto intervals and many-valued quantities* ◇ B52 E72 ◇
KANDEL, A. *Block decomposition of imprecise models* ◇ B52 ◇
KANDEL, A. *Fuzzy maps and their applications in the simplification of fuzzy switching functions* ◇ B52 ◇
KANDEL, A. *On the decomposition of fuzzy functions* ◇ B52 ◇
MACHINA, K.F. *Truth, belief, and vagueness* ◇ A05 B52 ◇
MAMDANI, E.H. *Application of fuzzy logic to approximate reasoning using linguistic synthesis* ◇ B52 ◇
NEGOITA, C.V. & RALESCU, D.A. *Comment on a "Comment on an algorithm that generates fuzzy prime implicants by Lee and Chang"* ◇ B52 ◇
PEARL, J. *On the complexity of deriving imprecise assertions* ◇ B52 ◇
PINKAVA, V. *"Fuzzification" of binary and finite multivalued logical calculi* ◇ B52 ◇
SANCHEZ, E. *Resolution of composite fuzzy relation equations* ◇ B52 E72 ◇
SANTOS, E.S. *Fuzzy and probabilistic programs* ◇ B52 B75 D10 ◇
SCHWEDE, G.W. *N-variable fuzzy maps with application to disjunctive decomposition of fuzzy switching functions* ◇ B52 ◇
SKALA, H.J. *On the problem of imprecision* ◇ B52 ◇
ZADEH, L.A. *A fuzzy-algorithmic approach to the definition of complex or imprecise concepts* ◇ B52 E72 ◇
ZADEH, L.A. *Semantic influence from fuzzy premises* ◇ B52 ◇

1977

BELLMAN, R.E. & ZADEH, L.A. *Local and fuzzy logics* ◇ B52 ◇
BENHALCEN, D. & HIRSCH, G. & LAMOTTE, M. *Codage et minimisation des fonctions floues dans une algebre floue* ◇ B52 ◇
EYTAN, M. *Semantique preordonnee des ensembles flous* ◇ B52 E72 ◇
EZHKOVA, I.V. & POSPELOV, D.A. *Decision making with fuzzy premises. I: Universel scale (Russian)* ◇ B52 ◇
GAINES, B.R. & KOHOUT, L.J. *The fuzzy decade: A bibliography of fuzzy systems and closely related topics* ◇ A10 B52 E72 ◇
GOTTWALD, S. *Theoretische Betrachtungen ueber Fuzzy-Logik* ◇ B52 ◇
HAJNAL, M. & KOCZY, L.T. *A new attempt to axiomatize fuzzy algebra with an application example* ◇ B52 E72 ◇
HAJNAL, M. & KOCZY, L.T. *A new fuzzy calculus and its application as a pattern recognition technique* ◇ B52 E72 G25 ◇
KANDEL, A. & RICKMAN, S.M. *Column table approach for the minimization of fuzzy functions* ◇ B52 ◇
KANDEL, A. *Comments on comments by Lee - Author's reply* ◇ B52 ◇
LEE, E.T. *Comments on two theorems by Kandel* ◇ B52 ◇
MACVICAR-WHELAN, P.J. *Fuzzy and multivalued logic* ◇ B52 ◇
MESKHI, S. *Fuzzy propositional logic. Algebraic approach* ◇ B52 G10 G25 ◇
MORGAN, C.G. & PELLETIER, F.J. *Some notes concerning fuzzy logics* ◇ B52 ◇
MUKAIDONO, M. *On some properties of a quantization in fuzzy logic* ◇ B52 ◇
PAPKE, W. & POSPELOV, D.A. & VAGIN, V.N. *Application of fuzzy logic in control systems* ◇ B52 ◇
PAVELKA, J. *On L-fuzzy semantic* ◇ B52 ◇
THAYSE, A. *Static-hazard detection in switching circuits by prime-implicant examination in fuzzy functions* ◇ B52 B70 ◇

1978

ALBERT, P. *The algebra of fuzzy logic* ◇ B52 E72 G25 ◇
BUTNARIU, D. *Fuzzy games: A description of the concept* ◇ B52 ◇
CZAJA-POSPIECH, D. & CZOGALA, E. & PEDRYCZ, W. *Fuzzy-control as a mathematical formalisation of a heuristic approach to the control of complex processes* ◇ B52 ◇
EZHKOVA, I.V. & POSPELOV, D.A. *Decision making with fuzzy premises. II: The inference schemes (Russian)* ◇ B52 ◇
FADINI, A. *Indecisive sets and their algebra* ◇ B52 E70 ◇
FICHTE, J. *Ueber den Wahrheitswertbreich von fuzzy Aussagenlogiken sowie Bemerkungen zu deren Axiomatisierung* ◇ B52 ◇
FUKAMI, S. & MIZUMOTO, M. & TANAKA, K. *On fuzzy reasoning* ◇ B52 ◇
GAINES, B.R. *Fuzzy and probability uncertainty logics* ◇ B48 B52 ◇
GAINES, B.R. *General fuzzy logics* ◇ B52 ◇
HAMACHER, H.C. *Ueber logische Verknuepfungen unscharfer Aussagen und deren zugehoerige Bewertungsfunktion* ◇ B52 ◇
KANDEL, A. *On the compactification and enumeration of distinct fuzzy switching functions* ◇ B52 ◇
KICKERT, W.J.M. & MAMDANI, E.H. *Analysis of a fuzzy logic controller* ◇ B52 ◇
SKALA, H.J. *On many-valued logics, fuzzy sets, fuzzy logics and their applications* ◇ B52 E72 ◇
STICKEL, M.E. *Fuzzy four-valued logic for inconsistency and uncertainty* ◇ B52 ◇
TONG, R.M. *Synthesis of fuzzy models for industrial processes - some recent results* ◇ B52 ◇
WECHLER, W. *The concept of fuzziness in automata and language theory* ◇ B52 D05 E72 ◇
WU, XUEMOU *Studies and applications of pansystems analysis Ia: A brief survey of pansystems analysis (Chinese)* ◇ B52 E72 ◇
ZADEH, L.A. *PRUF - a meaning representation language for natural languages* ◇ B52 ◇

1979

ALVES, E.H. & ARRUDA, A.I. *A semantical study of some systems of vagueness logic* ◇ B52 ◇

ALVES, E.H. & ARRUDA, A.I. *Some remarks on the logic of vagueness* ◇ B52 ◇

BALDWIN, J.F. & PILSWORTH, B.W. *A model of fuzzy reasoning through multi-valued logic and set theory* ◇ B52 E72 ◇

BALDWIN, J.F. *A new approach to approximate reasoning using a fuzzy logic* ◇ B52 ◇

BALDWIN, J.F. & GUILD, N.C.F. *Fuzlog: A computer program for fuzzy reasoning* ◇ B52 ◇

BALDWIN, J.F. *Fuzzy logic and approximate reasoning for mixed input arguments* ◇ B52 ◇

BALDWIN, J.F. *Fuzzy logic and fuzzy reasoning* ◇ B52 ◇

BALDWIN, J.F. *Fuzzy logic and its application to fuzzy reasoning* ◇ B52 ◇

BALDWIN, J.F. & PILSWORTH, B.W. *Fuzzy truth definition of possibility measure for decision classification* ◇ B52 E72 ◇

BALDWIN, J.F. & GUILD, N.C.F. *On the satisfaction of a fuzzy relation by a set of inputs* ◇ B52 E72 ◇

BANDLER, W. & KOHOUT, L.J. *Application of fuzzy logics to computer protection structures* ◇ B52 ◇

DISHKANT, G.P. *Formalization of a soft conjunction in the theory of approximate decision-making (Russian)* ◇ B52 ◇

DISHKANT, G.P. *The Lotfi Zadeh logic of taking approximate decisions as a game (Russian)* ◇ B52 ◇

DUBOIS, DIDIER & PRADE, H. *Operations in a fuzzy-valued logic* ◇ B52 E72 ◇

ESHRAGH, F. & MAMDANI, E.H. *A general approach to linguistic approximation* ◇ B52 ◇

FUKAMI, S. & MIZUMOTO, M. & TANAKA, K. *Some methods of fuzzy reasoning* ◇ B52 ◇

GILES, R. *A formal system for fuzzy reasoning* ◇ A05 B52 ◇

GOTTWALD, S. *Eine Anwendung des Formalismus der mehrwertigen Logik* ◇ B52 E72 G20 ◇

GRATTAN-GUINNESS, I. *Forays into the meta-theory of fuzzy set theory* ◇ B52 E72 ◇

HAACK, S. *Do we need "fuzzy logic"?* ◇ A05 B52 ◇

HAMACHER, H.C. *Ueber das Zadeh'sche Konzept der Fuzzy Sets und dessen Verhaeltnis zu den Wahrscheinlichkeitsmodellen von Koopman und Kolmogoroff* ◇ A05 B52 E72 ◇

HIROTA, K. & IIJIMA, T. *Logical basis in probabilistic set theory -- probabilistic expression of ambiguity and subjectivity* ◇ B52 E72 ◇

HISDAL, E. *Possibilistically dependent variables and a general theory of fuzzy sets* ◇ B52 E72 ◇

KANDEL, A. *Evaluating procedures and fuzzy logic* ◇ B52 ◇

KANDEL, A. *On the theory of fuzzy switching mechanisms (FSM's)* ◇ B52 ◇

KOCZY, L.T. *Some questions of b-algebras of fuzzy objects of type N* ◇ B52 E72 G25 ◇

MACVICAR-WHELAN, P.J. *Fuzzy-logic: An alternative approach* ◇ B52 ◇

MAMDANI, E.H. & SEMBI, B.S. *On the nature of implication in fuzzy logic* ◇ B52 ◇

MAMDANI, E.H. (ED.) *Special issue on fuzzy reasoning. Papers presented at a Workshop held at Queen Mary College, London, September 1978* ◇ B52 ◇

MUKAIDONO, M. *A necessary and sufficient condition for fuzzy logic functions* ◇ B52 ◇

NALIMOV, V.V. *The probability distribution function as a method of defining fuzzy sets. Metatheoretic sketches (Russian)* ◇ B52 E72 ◇

OSIS, YA.YA. & SHAJTSANE, V.A. *Determination of diagnostic parameters of the checking of good conditions for a complex fuzzy system with a continuous process of functioning (Russian)* ◇ B52 ◇

PAVELKA, J. *On fuzzy logic. I: Many-valued rules of inference. III: Semantical completeness of some many-valued propositional calculi* ◇ B52 ◇

PAVELKA, J. *On fuzzy logic. II. Enriched residuated lattices and semantics of propositional calculi* ◇ B52 G25 ◇

TRILLAS, E. *Note on negation functions for fuzzy sets (Spanish)* ◇ B52 E72 ◇

TRILLAS, E. *On negation functions in the theory of fuzzy sets (Spanish)* ◇ B52 E72 ◇

TSUKAMOTO, Y. *An approach to fuzzy reasoning method* ◇ B52 ◇

WENSTOP, F. *Exploring linguistic consequences of assertions in social sciences* ◇ B52 ◇

WU, XUEMOU *Studies and applications of pansystems analysis Ib: A brief survey of pansystems analysis (Chinese)* ◇ B52 E72 ◇

YAGER, R.R. *A note on fuzziness in a standard uncertainty logic* ◇ B52 ◇

ZADEH, L.A. *A theory of approximate reasoning* ◇ B52 ◇

ZHANG, JINWEN *Fuzzy logic and its formal system* ◇ B52 ◇

1980

ALMOG, J. *Modal counterfactual logic. Semantical considerations on modal counterfactual logic with corollaries on decidability, completeness, and consistency questions* ◇ B52 ◇

ALSINA, C. & TRILLAS, E. & VALVERDE, L. *On nondistributive logical connectives for the theory of fuzzy sets (Spanish) (English summary)* ◇ B52 E72 ◇

ARIGONI, A.O. *Mathematical developments arising from "semantical implication" and the evaluation of membership characteristic functions* ◇ B52 G10 ◇

ARONSON, A.R. & JACOBS, B.E. & MINKER, J. *A note on fuzzy deduction* ◇ B52 ◇

BALDWIN, J.F. & PILSWORTH, B.W. *Axiomatic approach to implication for approximate reasoning with fuzzy logic* ◇ B52 ◇

BALDWIN, J.F. & GUILD, N.C.F. *Feasible algorithms for approximate reasoning using fuzzy logic* ◇ B52 ◇

BALDWIN, J.F. & GUILD, N.C.F. *The resolution of two paradoxes by approximate reasoning using a fuzzy logic* ◇ A05 B52 ◇

BANDLER, W. & KOHOUT, L.J. *Fuzzy power sets and fuzzy implication operators* ◇ B52 E72 ◇

BHAT, K.V.S. *Comments on: "Simplification of fuzzy switching functions" by Thomas P. Neff and Abraham Kandel* ◇ B52 B70 ◇

BLISHUN, A.F. *Truth of fuzzy set similarity (Russian and English)* ◇ B52 ◇
CATTANEO, G. *Fuzzy events and fuzzy logics in classical information systems* ◇ B52 G12 ◇
CAYROL, M. & FARRENY, H. & PRADE, H. *Fuzzy reasoning based on multivalent logics in the framework of production-rules systems* ◇ B52 ◇
CHEN, HUAITING & WU, XUEMOU *Transition logic, fuzzy logic and pansystem logic I (Chinese) (English summary)* ◇ B52 ◇
CHEN, XINGGOU & WANG, PEIZHUANG *An improvement and strengthening of the Negoita-Ralescu representation theorem (Chinese) (English summary)* ◇ B52 ◇
CZAJA-POSPIECH, D. *On some rules of fuzzy logics (Polish) (English and Russian summaries)* ◇ B52 ◇
CZAJA-POSPIECH, D. & CZOGALA, E. & PEDRYCZ, W. *The multiple implications and compositional rules of inference in multivalued logics (fuzzylogics) (Polish) (English and Russian summaries)* ◇ B52 ◇
DOMINGO, X. & ESTEVA, F. *On weak negation functions for the theory of fuzzy sets (Catalan)* ◇ B52 E72 ◇
EDMONDS, E.A. *Lattice fuzzy logics* ◇ B52 E72 ◇
EFSTATHIOU, J. & TONG, R.M. *Rule-based decomposition of fuzzy relational models* ◇ B52 ◇
FUKAMI, S. & MIZUMOTO, M. & TANAKA, K. *Some considerations on fuzzy conditional inference* ◇ B52 ◇
GILES, R. *A computer program for fuzzy reasoning* ◇ B52 ◇
GOL'DMAN, R.S. *Theory of fuzzy sets (Russian) (English summary)* ◇ B52 E72 ◇
GOTTWALD, S. *Fuzzy propositional logics* ◇ B52 ◇
KAMPE DE FERIET, J. *Une interpretation des mesures de plausibilite et de credibilite au sens de G. Shafer et de la fonction d'appartenance definissant un ensemble flou de L. Zadeh* ◇ B52 E72 ◇
KANDEL, A. *On the control and evaluation of uncertain processes* ◇ B52 ◇
KANDEL, A. *On the theory of fuzzy logics* ◇ B52 ◇
LI, BANGRONG & LIU, GUOHAN *Some properties of fuzzy space and fuzzy logic (Chinese) (English summary)* ◇ B52 E72 ◇
LI, CHULIN & WU, XUEMOU & YU, HONGZU *A binary comparison average method for fuzzy pattern recognition (Chinese)* ◇ B52 ◇
MAMDANI, E.H. & SEMBI, B.S. *Process control using fuzzy logic* ◇ B52 ◇
NAKAMURA, A. & ONO, H. *Decidability results on a query language for data bases with incomplete information* ◇ B25 B52 B75 ◇
PEDRYCZ, W. *On the use of a fuzzy Lukasiewicz logic for fuzzy control* ◇ B52 ◇
RIERA, T. & TRILLAS, E. *On a special kind of variables in fuzzy environment* ◇ B52 ◇
SANFORD, D.H. *Notes on logics of vagueness and some applications* ◇ B52 ◇
SCHEFE, P. *On foundations of reasoning with uncertain facts and vague concepts* ◇ B52 ◇
SCOTT, L.L. *Necessary and sufficient conditions for the values of a function of fuzzy variables to lie in a specified subinterval of (0, 1)* ◇ B52 ◇
SHI, NIANDONG & YUE, YUJUN *Notes on the W_1L calculus (Chinese) (English summary)* ◇ B52 ◇
SPIGLER, R. *The structure of the relations "much better" and "much worse" in fuzzy calculus* ◇ B52 E72 ◇
WILLMOTT, R. *Two fuzzier implication operators in the theory of fuzzy power sets* ◇ B52 E72 ◇
WU, XUEMOU & ZHU, YUANZHAO *Pansystem analysis of dynamic yinyang logic and fuzziness I (Chinese) (English summary)* ◇ B52 ◇
WU, XUEMOU *Studies and applications of pansystems analysis IV: Pansystem observocontrollability, pansystem logic and fuzziness (Chinese) (English summary)* ◇ B52 E72 ◇
YAGER, R.R. *A foundation for a theory of possibility* ◇ B52 ◇
YAGER, R.R. *An approach to inference in approximate reasoning* ◇ B52 ◇
YAGER, R.R. *Fuzzy thinking as quick and efficient* ◇ B52 E72 ◇
YAGER, R.R. *On a general class of fuzzy connectives* ◇ B52 ◇
YAGER, R.R. *On choosing between fuzzy subsets* ◇ B52 E72 ◇
YAGER, R.R. *Some observations on probabilistic qualification in approximate reasoning* ◇ B52 ◇
YANASE, M.M. *Hidden realism. I* ◇ A05 B52 ◇
ZADEH, L.A. *Inference in fuzzy logic* ◇ B52 ◇

1981

ANSHAKOV, O.M. & FINN, V.K. *So-called fuzzy logics and one-implication calculi (Russian)* ◇ B52 ◇
AZORIN POCH, F. *Uncertainty and fuzziness in primary data and estimation procedures* ◇ B52 ◇
BALDWIN, J.F. *Fuzzy logic knowledge bases and automated fuzzy reasoning* ◇ B52 ◇
BALDWIN, J.F. & GUILD, N.C.F. *Fuzzy reasoning applied to control models* ◇ B52 ◇
BALDWIN, J.F. & PILSWORTH, B.W. *Fuzzy reasoning with probability* ◇ B52 ◇
BEN, D.M. & CLARK, C.M. & KANDEL, A. *On the enumeration of distinct fuzzy switching functions* ◇ B52 ◇
BESSLICH, P.W. *Incompletely specified fuzzy switching function minimization: an in-place transform approach* ◇ B52 ◇
BHAT, K.V.S. *Minimization of disjunctive normal forms of fuzzy logic functions* ◇ B52 ◇
BOUCHON, B. *Information transmitted by a system of fuzzy events* ◇ B52 ◇
CHEN, GUOQUAN & LEE, S.C. & PAN, WEIJIE *D-fuzzy logic* ◇ B52 ◇
CLARK, C.M. & KANDEL, A. *New results in the enumeration of minimized fuzzy-valued switching functions* ◇ B52 ◇
CZOGALA, E. & PEDRYCZ, W. *Identification and control in fuzzy systems* ◇ B52 ◇
CZOGALA, E. & PEDRYCZ, W. *Some problems concerning the construction of algorithms of decision-making in fuzzy systems* ◇ B52 ◇
DISHKANT, G.P. *About membership functions estimation* ◇ B52 E72 ◇

DOMINGO, X. & ESTEVA, F. & TRILLAS, E. *Weak and strong negation functions for fuzzy set theory* ◇ B52 E72 ◇
FOX, J. *Towards a reconciliation of fuzzy logic and standard logic* ◇ B52 ◇
GAINES, B.R. & MAMDANI, E.H. (EDS.) *Fuzzy reasoning and its applications* ◇ B52 B98 ◇
HISDAL, E. *The IF THEN ELSE statement and interval-valued fuzzy sets of higher type* ◇ A05 B52 B75 E72 ◇
HOEHLE, U. *A mathematical theory of uncertainty (fuzzy experiments and their realizations)* ◇ B52 ◇
HUANG, ZHENDE & TONG, ZHENGXIANG *Fuzzy probability and possibility spaces I (Chinese) (English summary)* ◇ B52 ◇
KANDEL, A. & MARGOLIS, I.B. *Absolutely dispensable prime implicants of fuzzy switching functions* ◇ B52 ◇
KLEMENT, E.P. & LOWEN, R. & SCHWYHLA, W. *Fuzzy probability measures* ◇ B52 E72 ◇
KOROVIN, S.YA. & MELIKHOV, A.N. & TSELYKH, A.N. *Some principles of program realization for fuzzy control algorithms (Russian)* ◇ B52 ◇
LIU, GUOHAN *Some transforming relations among the pansystem logic spaces with soft algebra shadow system (Chinese) (English summary)* ◇ B52 ◇
MCCAWLEY, J.D. *Fuzzy logic and restricted quantifiers* ◇ B52 ◇
MIZUMOTO, M. *Fuzzy reasonings with "if...then...else..."* ◇ B52 ◇
MIZUMOTO, M. *Note on the arithmetic rule by Zadeh for fuzzy conditional inference* ◇ B52 ◇
MIZUMOTO, M. *Various fuzzy reasoning methods: the case of "If...then..." (Japanese)* ◇ B52 ◇
NOWAKOWSKA, M. *A new theory of time: generation of time from fuzzy temporal relations* ◇ B52 ◇
PEACOCKE, C. *Are vague predicates incoherent?* ◇ A05 B52 ◇
PENA, L. *Prenexation, comparatives, and non-Archimedean infinite-valued fuzzy logic* ◇ B52 ◇
RADECKI, T. *Outline of a fuzzy logic approach to information retrieval* ◇ B52 ◇
SHU, YONGCHANG & WANG, YIZHI *Fuzzy languages and fuzzy grammars (Chinese)* ◇ B52 D05 ◇
SMETS, P. *The degree of belief in a fuzzy event* ◇ B52 ◇
TRILLAS, E. & VALVERDE, L. *On some functionally expressible implications for fuzzy set theory* ◇ B52 ◇
WU, XUEMOU *Fuzziness, reliability and pansystem V: Analytic pansystem logic and its applications to the analysis of fuzzy sets (Chinese) (English summary)* ◇ B52 ◇
WU, XUEMOU *Pansystems analysis: some new investigations of logic, observocontrollability and fuzziness* ◇ B52 E72 ◇
WU, XUEMOU *Studies and applications of pansystems analysis III: Pansystems analysis - a new exploration of interdisciplinary investigation (Chinese) (English summary)* ◇ B52 E72 ◇
YAGER, R.R. *Quantified propositions in a linguistic logic* ◇ B52 ◇
ZHANG, JINWEN *System logic and fuzzy set theory (Chinese) (English summary)* ◇ B52 E72 ◇
ZHANG, MINGYI *Some projection theorems in pansystem logic conservations (Chinese) (English summary)* ◇ B52 ◇
ZHUKOVIN, V.E. *Models and procedures in decision-making (Russian)* ◇ B52 ◇
ZHUKOVIN, V.E. *On the relationship between multicriteria and fuzzy representations of decision making problems (Russian)* ◇ B52 ◇

1982

ADAM, ADOLF & BERAN, H. *On the problem of the application of fuzzy-logic concepts* ◇ B52 ◇
ARIGONI, A.O. *Transformational-generative grammar for description of formal properties* ◇ B52 ◇
ASAI, K. & TANAKA, H. & TSUKIYAMA, T. *A fuzzy system model based on the logical structure* ◇ B52 ◇
ASAI, K. & TANAKA, H. & TSUKIYAMA, T. *Logical-systems identification by fuzzy input-output data (Japanese) (English summary)* ◇ B52 ◇
BALDWIN, J.F. *An automated fuzzy reasoning algorithm* ◇ B52 ◇
BELLMAN, R.E. & ESOGBUE, A.O. *Constributions to fuzzy dynamic programming* ◇ B52 B75 ◇
BHAT, K.V.S. *Further remarks on Neff-Kandel algorithm for minimization of fuzzy conjunctive forms: "Simplification of fuzzy switching functions" (Internat. J. Comput. Information Sci. 6/1*55–70, 1977)* ◇ B52 ◇
BHAT, K.V.S. *On the notion of fuzzy consensus* ◇ B52 ◇
BLANCHARD, N. *Fuzzy-lip functions - the fuzzy-lip category* ◇ B52 ◇
BONISSONE, P.P. *A fuzzy sets based linguistic approach: theory and applications* ◇ B52 ◇
BORISOV, A.N. & LEVCHENKOV, A.S. *Methods of interactive evaluation of decisions (Russian)* ◇ B52 ◇
BOTTA, O. & DELORME, M. *An f-sets universe* $\tilde{V}$ ◇ B52 E72 ◇
CAO, ZHIJIANG *Structure of solution sets of fuzzy linear equations* ◇ B52 ◇
CAO, ZHIJIANG *The eigen fuzzy sets of a fuzzy matrix* ◇ B52 ◇
CAVALLO, R.E. & KLIR, G.J. *Decision making in reconstructability analysis* ◇ B52 ◇
CHATTERJI, B.N. *Character recognition using fuzzy similarity relations* ◇ B52 ◇
CHEN, GUOQUAN & LEE, S.C. & PAN, WEIJIE *D-fuzzy logic (Chinese summary)* ◇ B52 ◇
CHEN, YONGYI *An approach to fuzzy operators I (Chinese) (English summary)* ◇ B52 ◇
CHEN, YONGYI *An approach to fuzzy operators* ◇ B52 ◇
CHENG, WEIMIN & WU, WANGMING & XU, WENLI *An algorithm to solve the max-min composite fuzzy relational equations* ◇ B52 ◇
CHENG, ZHAOZHONG & LIU, DSOSU *Some problem in fuzzy probability* ◇ B52 ◇
CLARK, C.M. & KANDEL, A. *The enumeration of distinct fuzzy-valued switching functions* ◇ B52 B70 ◇
CZOGALA, E. & GOTTWALD, S. & PEDRYCZ, W. *Aspects for the evaluation of decision situations* ◇ B52 ◇
CZOGALA, E. & GOTTWALD, S. & PEDRYCZ, W. *Quality indices of fuzzy sets for decision making in fuzzy environments* ◇ B52 ◇

DAVIS, H.W. & WINSLOW, L.E. *Computational power in query languages* ◇ B52 ◇
DENG, JULONG & PENG, GUOZHONG *The digital simulation and computer-aided design of a fuzzy integrated system (Chinese) (English summary)* ◇ B52 ◇
DIMITROV, V. *Creative decision-making through fuzzy catastrophes* ◇ B52 ◇
DOMPERE, K.K. *The theory of fuzzy decisions* ◇ B52 ◇
DUBOIS, DIDIER & PRADE, H. *The use of fuzzy numbers in decision analysis* ◇ B52 ◇
DUBOIS, DIDIER & PRADE, H. *Towards fuzzy differential calculus I: Integration of fuzzy mappings. II:Integration on fuzzy intervals. III: Differentiation* ◇ B52 ◇
DUTA, D.L. & FABIAN, C. *The cutting stock problem with several objective functions and in fuzzy conditions* ◇ B52 ◇
EFSTATHIOU, J. & TONG, R.M. *A critical assessment of truth functional modification and its use in approximate reasoning* ◇ B52 ◇
ENTA, Y. *Fuzzy decision theory* ◇ B52 ◇
ESCHBACH, M. *The logic of fuzzy Bayesian inference.* ◇ B52 ◇
GAINES, B.R. *Fuzzy and probability uncertainty logics (Chinese)* ◇ B52 ◇
GILES, R. *Foundations for a theory of possibility* ◇ B52 ◇
GILES, R. *Semantics for fuzzy reasoning* ◇ B52 ◇
HISDAL, E. *A fuzzy IF THEN ELSE relation with guaranteed correct inference* ◇ B52 ◇
HOEHLE, U. *A mathematical theory of uncertainty* ◇ B52 ◇
IWAI, S. & NAKAMURA, K. *A representation of analogical inference by fuzzy sets and its application to information retrieval system* ◇ B52 E72 ◇
IWAI, S. & NAKAMURA, K. *Topological fuzzy sets as a quantitative description of analogical inference and its application to question-answering systems for information retrieval* ◇ B52 E72 ◇
KABBARA, G. *New utilization of fuzzy optimization method* ◇ B52 ◇
KACPRZYK, J. *Multistage decision processes in a fuzzy environment: a survey* ◇ B52 ◇
KONG, YOUKUN & ZHANG, WENGQIAN *A fuzzy control model for weather forecasting (Chinese) (English summary)* ◇ B52 ◇
LE, HUILING & ZHANG, WENXIU *Extended fuzzy operators and fuzzy truth-valued possibility (Chinese) (English summary)* ◇ B52 ◇
LI, BIQIANG & WANG, PEIZHUANG *Fuzzy logistic reasoning of second order (Chinese) (English summary)* ◇ B52 ◇
LINDLEY, D.V. *Scoring rules and the inevitability of probability (French summary)* ◇ B52 ◇
LIU, YUNFENG *An axiomatic structure of fuzzy set theory and fuzzy logic (Chinese) (English summary)* ◇ B52 E72 ◇
LOPES, L.L. & ODEN, G.C. *On the internal structure of fuzzy subjective categories* ◇ B52 ◇
LOU, SHIBO & TONG, ZHENGXIANG *LSG powers of an L-fuzzy matrix* ◇ B52 ◇
MANES, E.G. *A class of fuzzy theories* ◇ B52 ◇
MIZUMOTO, M. & SEKITA, Y. *A fuzzified relevance tree approach for solving the complex planning* ◇ B52 ◇
MIZUMOTO, M. & ZIMMERMANN, H.-J. *Comparison of fuzzy reasoning methods* ◇ B52 ◇
MIZUMOTO, M. *Fuzzy conditional inference under max-⊙ composition* ◇ B52 ◇
MIZUMOTO, M. *Fuzzy inference using max-$\wedge$ composition in the compositional rule of inference* ◇ B52 ◇
MIZUMOTO, M. *Fuzzy reasoning with a fuzzy conditional proposition "if...then...else..."* ◇ B52 ◇
MIZUMOTO, M. *Fuzzy reasoning with various fuzzy premises (Japanese)* ◇ B52 ◇
MUKAIDONO, M. *Fuzzy inference of resolution style* ◇ B52 ◇
MUKAIDONO, M. *New canonical forms of fuzzy switching functions* ◇ B52 ◇
MUKAIDONO, M. *New canonical forms and their applications to enumerating fuzzy switching functions* ◇ B52 ◇
MUKAIDONO, M. *New canonical forms for and their application to fuzzy switching functions* ◇ B52 ◇
ORLOVSKY, S.A. *Effective alternatives for multiple fuzzy preference relations* ◇ B52 ◇
ORLOWSKA, E. *Logic of vague concepts* ◇ B52 ◇
ORLOWSKA, E. *Logic of vague concepts. Applications of rough sets* ◇ B52 ◇
ORLOWSKA, E. *Semantics of vague concepts. Applications of rough sets* ◇ B52 ◇
OVCHINNIKOV, S.V. *On fuzzy relational systems* ◇ B52 ◇
PEDRYCZ, W. *Some aspects of fuzzy decision-making* ◇ B52 ◇
PRASAD, J. & RAO, S.V.L.N. *Definition of kriging in terms of fuzzy logic* ◇ B52 ◇
PULTR, A. *Fuzzyness and fuzzy equality* ◇ B52 E72 ◇
RALESCU, D.A. *Fuzzy logic and statistical estimation* ◇ B52 ◇
RENNA E SOUZA DE, C. & SILVA DE, O. *Changing the fuzzy rule of detachment* ◇ B52 ◇
RIERA, T. & TRILLAS, E. *From measures of fuzzyness to Booleanity control* ◇ B52 ◇
RUSPINI, E.H. *Possibilistic data structures for the representation of uncertainty* ◇ B52 ◇
SKALA, H.J. *Modelling vagueness* ◇ B52 ◇
SMETS, P. *Probability of a fuzzy event: an axiomatic approach* ◇ B52 ◇
SMETS, P. *Subjective probability and fuzzy measures* ◇ B52 ◇
SUSTAL, J. *On the uncertainty of fuzzy classifications* ◇ B52 ◇
TAKEDA, E. *Interactive identification of fuzzy outranking relations in a multicriteria decision problem* ◇ B52 ◇
TANG, XUZHANG *The existence of fuzziness and its meaning (Chinese) (English summary)* ◇ A05 B52 ◇
TRILLAS, E. & VALVERDE, L. *A few remarks on some lattice-type properties of fuzzy connectives* ◇ B52 ◇
TRILLAS, E. *An essay on indistinguishability relations (Catalan)* ◇ B52 ◇
TUERKSEN, I.B. & YAO, D.D.W. *Bounds for fuzzy inference* ◇ B52 ◇

VALVERDE, L. *Notes on a study of the "complements" in fuzzy logic (Catalan) (English summary)* ◇ B52 ◇
WANG, JINCHENG *An informal discussion of fuzzy mathematics (Chinese)* ◇ B52 E72 ◇
WANG, PEIZHUANG *Fuzzy contactability and fuzzy variables* ◇ B52 ◇
WIERZCHON, S.T. *Applications of fuzzy decision-making theory to coping with ill-defined problems* ◇ B52 ◇
YAGER, R.R. (ED.) *Fuzzy set and possibility theory. Recent developments* ◇ B52 ◇
YAGER, R.R. *Generalized probabilities of fuzzy events from fuzzy belief structures* ◇ B52 ◇
YAGER, R.R. *Some procedures for selecting fuzzy set-theoretic operators* ◇ B52 ◇
YU, LIANSHENG *Research into the theory of fuzzy clustering: principle of the maximum element of a matrix (Chinese) (English summary)* ◇ B52 E72 ◇
ZHANG, WENXIU & ZHAO, QINPING *Possibility degree theories (Chinese)* ◇ B52 ◇
ZHAO, ZHEN *A new map approach for the minimization of fuzzy logic functions (Chinese) (English summary)* ◇ B52 ◇

1983

ALSINA, C. & TRILLAS, E. & VALVERDE, L. *On some logical connectives for fuzzy sets theory* ◇ B52 ◇
AMBROSIO, R. & MARTINI, G.B. *Resolution of max-min equations with fuzzy symbols* ◇ B52 ◇
ASAI, K. & IZUMI, K. & TANAKA, H. *Duality of fuzzy systems based on the intuitionistic logical system LJ (Japanese)* ◇ B52 ◇
BALLMER, T.T. & PINKAL, M. (EDS.) *Approaching vagueness* ◇ B52 ◇
BANDLER, W. & KOHOUT, L.J. *Probabilistic vs. fuzzy production rules in expert systems* ◇ B52 ◇
BOECH, P. *"Vagueness" is context-dependence. A solution to the sorites paradox* ◇ B52 ◇
BRENNENSTUHL, W. *A door's closing-a contribution to the vagueness semantics of tense and aspect* ◇ B52 ◇
BRILL JR., E.D. & CHANG, SHOOUYUH & HOPKINS, L.D. *Modeling to generate alternatives: a fuzzy approach* ◇ B52 ◇
BRUDZY, K. & KISZKA, J.B. *The reproducibility property of fuzzy control systems* ◇ B52 B75 ◇
CARLSSON, C. *An approach to handling fuzzy problem structures* ◇ B52 ◇
CARLSSON, C. *Some features of a methodology for dealing with multiple- criteria problems* ◇ B52 ◇
CATTANEO, G. *Canonical embedding of an abstract quantum logic into the partial Baer*-ring of complex fuzzy events* ◇ B52 ◇
CHEN, YONGYI & WANG, PEIZHUANG *Optimum fuzzy implication and direct method of approximate reasoning* ◇ B52 ◇
CHEN, ZHAOKUAN *Fixed points for a class of fuzzy mappings (Chinese)* ◇ B52 ◇
CZOGALA, E. & GOTTWALD, S. & PEDRYCZ, W. *Logical connectives of probabilistic sets* ◇ B52 E72 ◇
DOMBI, J. & VAS, Z. *Basic theoretical treatment of fuzzy connectives* ◇ B52 ◇
EIKMEYER, H.-J. & RIESER, H. *A formal theory of context dependence and context change* ◇ B52 ◇
EZAWA, Y. *Homomorphic linguistic hedges and reasonable fuzzy inferences* ◇ B52 ◇
FENG, YINGJUN *A method using fuzzy mathematics to solve the vectormaximum problem* ◇ B52 ◇
FRANCIONI, J.M. & KANDEL, A. *Decomposable fuzzy-valued switching functions* ◇ B52 ◇
HINDE, C.J. *Inference of fyzzy relational tableaux from fuzzy exemplifications* ◇ B52 ◇
HIROTA, K. & PEDRYCZ, W. *Analysis and synthesis of fuzzy systems by the use of probabilistic interpretation of histograms* ◇ B52 ◇
HONG, HENGLING *Method of fuzzy statistics and applications* ◇ B52 ◇
KACPRZYK, J. *Multistage decision-making under fuzziness. Theory and applications* ◇ B52 ◇
KANDEL, A. & LUGINBUHL, D.R. *A comparison of fuzzy switching functions and multiple-valued switching functions* ◇ B52 ◇
LAARHOVEN VAN, P.J.M. & PEDRYCZ, W. *A fuzzy extension of Saaty's priority theory* ◇ B52 ◇
LIU, XUHUA *Minimization of fuzzy logic formulas (Chinese) (English summary)* ◇ B52 ◇
MENGES, G. *Unscharfe Konzepte im Operations Research* ◇ B52 ◇
MIZUMOTO, M. *Fuzzy inferences with various fuzzy inputs (Chinese summary)* ◇ B52 ◇
MIZUMOTO, M. *Fuzzy reasoning with various fuzzy inputs (Japanese)* ◇ B52 ◇
MIZUMOTO, M. *Fuzzy reasoning under new compositonal rules of inference* ◇ B52 ◇
NATVIG, B. *Possibility versus probability* ◇ B52 ◇
NOVAK, VILEM *Basic operations with fuzzy sets from the point of fuzzy logic* ◇ B52 E72 ◇
PARIKH, R. *The problem of vague predicates* ◇ B52 ◇
PAUN, G. *An impossibility theorem for indicators aggregation* ◇ B52 ◇
PEDRYCZ, W. *Fuzzy relational equations with generalized connectives and their applications* ◇ B52 ◇
PEDRYCZ, W. *Some applicational aspects of fuzzy relational equations in system analysis* ◇ B52 ◇
PINKAL, M. *Towards a semantics of precisation* ◇ B52 ◇
PRADE, H. *Data bases with fuzzy information and approximate reasoning in expert systems* ◇ B52 B75 ◇
RAMIK, J. *Extension principle and fuzzy-mathematical programming* ◇ B52 ◇
SHAJDA, J. *An application of fuzzy logic and approximate reasoning* ◇ B52 ◇
SONG, DAHE *A mathematical model and analysis of the algorithm of a fuzzy controller (Chinese) (English summary)* ◇ B52 ◇
SUGENO, M. & TANG, DUOYUAN & TERANO, T. *Linguistic models of processes and "optimal" control (Chinese) (English summary)* ◇ B52 ◇
SUGENO, M. & TAKAGI, T. *Multi-dimensional fuzzy reasoning* ◇ B52 ◇
SULTAN, L.H. *Fuzzy logic implementation in the realization of fuzzy algorithms* ◇ B52 ◇

TAUTU, P. *Belief functions and analogous fuzzy measures (with applications to Bayes' theorem)* ◇ B52 ◇
TOGAI, M. & WANG, P.P. *A study of fuzzy relations and their inverse problem* ◇ B52 E72 ◇
TUERKSEN, I.B. *Inference regions for fuzzy propositions* ◇ B52 ◇
WEBER, S. *A general concept of fuzzy connectives, negations and implications based on t-norms and t-conorms* ◇ B52 ◇
WILDGEN, W. *Modelling vagueness in catastrophe-theoretic semantics* ◇ B52 ◇
XIAO, XIAN *Fuzzy expressions and the property of order preserving (Chinese) (English summary)* ◇ B52 ◇
YAGER, R.R. *Entropy and specificity in a mathematical theory of evidence* ◇ B52 ◇
YAGER, R.R. *On the implication operator in fuzzy logic* ◇ B52 ◇
YAGER, R.R. *Presupposition in binary and fuzzy logics* ◇ B52 ◇
YAGER, R.R. *Some relationships between possibility, truth and certainty* ◇ B52 E72 ◇
ZADEH, L.A. *A computational approach to fuzzy quantifiers in natural languages* ◇ B52 ◇
ZADEH, L.A. *The role of fuzzy logic in the management of uncertainty in expert systems* ◇ B52 ◇
ZHANG, JINWEN *Fuzzy set structure with strong implication* ◇ B52 ◇
ZHAO, ZHEN *A graphical algorithm for decomposing fuzzy logic functions (Chinese) (English summary)* ◇ B52 ◇

1984

ALMUKDAD, A. & NELSON, D. *Constructible falsity and inexact predicates* ◇ B52 F50 ◇
ASAI, K. & IZUMI, K. & TANAKA, H. *Fuzzy logical construction of linguistic modal hedges (Japanese)* ◇ B52 ◇
ATANASSOV, K. & STOEVA, S. *Intuitionistic L-fuzzy sets* ◇ B52 E72 ◇
BALDWIN, J.F. & ZHOU, S.Q. *A fuzzy relational inference language* ◇ B52 ◇
BECVAR, J. *Notes on vagueness and mathematics* ◇ B52 ◇
BERMAN, J. & MUKAIDONO, M. *Enumerating fuzzy switching functions and free Kleene algebras* ◇ B52 ◇
BOURRELLY, L. & CHOURAQUI, E. & RICARD, M. *Formalisation of an approximate reasoning: the analogical reasoning* ◇ B52 ◇
CERNY, M. & NEKOLA, J. & NOVAK, VILEM *Fuzzy sets - perspectives, problems and applications (Czech)* ◇ B52 E72 ◇
DUBOIS, DIDIER & PRADE, H. *Fuzzy logics and the generalized modus ponens revisted* ◇ B52 ◇
DYCKHOFF, H. & PEDRYCZ, W. *Generalized means as a model of compensative connectives* ◇ B52 E72 ◇
EZAWA, Y. & MIZUMOTO, M. *Linguistic hedges and reasonable fuzzy inferences* ◇ B52 ◇
GAINES, B.R. & SHAW, M.L.G. *Deriving the constructs underlying decision* ◇ B52 ◇
GLAS DE, M. *Invariance and stability of fuzzy systems* ◇ B52 ◇
GOTTWALD, S. *Criteria for non-interactivity of fuzzy logic controller rules* ◇ B52 ◇
HATANO, Y. & TSUKAMOTO, Y. *Fuzzy statement formation by means of linguistic measures* ◇ B52 ◇
HE, ZHONGXIONG & XIAO, XIAN & YUAN, XIANGWAN & ZHU, WUJIA *Some opinions on fuzzy mathematics and its foundations (Chinese)* ◇ A05 B52 E72 ◇
HOEHLE, U. *Fuzzy plausability measures* ◇ B52 ◇
HOEHLE, U. & KLEMENT, E.P. *Plausibility measures - a general framework for possibility and fuzzy probability measures* ◇ B52 ◇
KANDEL, A. & THUM, M. *On the complexity of growth of the number of distinct fuzzy switching functions* ◇ B52 ◇
KUTSCHERA VON, F. *Eine Logik vager Saetze* ◇ B52 ◇
LE, HUILING & ZHANG, WENXIU *Fuzzy truth possibility degrees (Chinese) (English summary)* ◇ B52 ◇
LE, HUILING & ZHANG, WENXIU *The extended fuzzy operators and the fuzzy truth-valued possibility degree* ◇ B52 ◇
LI, TAIHANG *Degree of abstraction of concepts and longitudinal reasoning* ◇ B52 ◇
LI, TAIHANG *Fuzzy longitudinal reasoning and selfgeneration of membership degrees (Chinese) (English summary)* ◇ B52 ◇
MIZUMOTO, M. *Fuzzy reasoning with various fuzzy inputs* ◇ B52 ◇
NEKOLA, J. & NOVAK, VILEM *Basic operations with fuzzy sets from the point of view of fuzzy logic* ◇ B52 E72 ◇
ORLOWSKA, E. & WIERZCHON, S.T. *Mechanical reasoning in fuzzy logics* ◇ B52 ◇
PENA, L. *A neo-Fregean (onto)logical fuzzy framework* ◇ B52 ◇
PENA, L. *Identity, fuzziness and noncontradiction* ◇ B52 ◇
RALESCU, A.L. & RALESCU, D.A. *Probability and fuzziness* ◇ B52 ◇
SCHMUCKER, K.J. *Fuzzy sets, natural language computations, and risk analysis, Foreword by Lotfi A. Zadeh* ◇ B52 B98 ◇
SHAJDA, J. *On some basic problems of propositional fuzzy logic* ◇ B52 ◇
SMITHSON, M. *Multivariate analysis using "and" and "or"* ◇ B52 ◇
TERMINI, S. *Aspects of vagueness and some epistemological problems related to their formalization* ◇ B52 ◇
TUERKSEN, I.B. *Measurement of fuzziness: an interpretation of the axioms of measurement* ◇ B52 ◇
TUERKSEN, I.B. & YAO, D.D.W. *Representations of connectives in fuzzy reasoning: the view through normal forms* ◇ B52 ◇
VITEK, M. *Fuzzy information and fuzzy time* ◇ B52 ◇
XIAO, XIAN & ZHU, WUJIA *Foundations of classical mathematics and fuzzy mathematics (Chinese) (English summary)* ◇ B52 ◇
XIAO, XIAN *The improvement in bounds on the number of fuzzy expressions (Chinese)* ◇ B52 ◇
XU, ZHIWEI *Multivalued logic and fuzzy logic - their relationship, minimization, and application to fault diagnosis* ◇ B50 B52 ◇

ZADEH, L.A. *A theory of commonsense knowledge* ◇ B52 ◇

ZADEH, L.A. *Coping with the imprecision of the real world* ◇ B52 ◇

ZADEH, L.A. *Fuzzy probabilities* ◇ B52 ◇

ZHAO, ZHEN *A discussion on graphical minimization and decomposition of multivariate fuzzy logic functions (Chinese) (English summary)* ◇ B52 ◇

1985

BANDEMER, H. *Evaluating explicit functional relationships from fuzzy observations* ◇ B52 ◇

BUCKLEY, J.J. *Ranking alternatives using fuzzy numbers* ◇ B52 ◇

CALUWE DE, R.M.M. & KERRE, E.E. & ZENNER, R.B.R.C. *A new approach to information retrieval systems using fuzzy expressions* ◇ B52 ◇

CHEN, YONGYI & WANG, PEIZHUANG *Optimum fuzzy implication and a direct method of approximate reasoning (Chinese) (English summary)* ◇ B52 ◇

DUBOIS, DIDIER & PRADE, H. *Fuzzy cardinality and the modeling of imprecise quantification* ◇ B52 E72 ◇

FINKELSTEIN, D. & RODRIGUEZ, E. *Application of quantum set theory to quantum time-space* ◇ B52 E70 ◇

FLOWERS, P.L. & KANDEL, A. *Possibilistic search trees* ◇ B52 ◇

GERLA, G. *Pavelka's fuzzy logic and free L-subsemigroups* ◇ B52 ◇

KISZKA, J.B. & KOCHANSKA, M.E. & SLIWINSKA, D.S. *The influence of some fuzzy implication operators on the accuracy of a fuzzy model. I,II* ◇ B52 ◇

LIAO, QUN *Solutions of fuzzy equations and the inverse problem of fuzzy reasoning based on $\bar{t}$ (Chinese) (English summary)* ◇ B52 E72 ◇

MIZUMOTO, M. *Fuzzy reasoning under new compositional rules of inference* ◇ B52 ◇

ORLOWSKA, E. *Semantics of vague concepts* ◇ B52 ◇

PEDRYCZ, W. *Applications for fuzzy relational equations for methods of reasoning in presence of fuzzy data* ◇ B52 ◇

TOGAI, M. *A fuzzy inverse relation based on Goedelian logic and its applications* ◇ B52 ◇

YAGER, R.R. *Aggregation evidence using quantified statements* ◇ B52 B75 ◇

YAGER, R.R. *Knowledge trees in complex knowledge bases* ◇ B52 ◇

YAGER, R.R. *Reasoning with fuzzy quantified statements I* ◇ B52 ◇

YAGER, R.R. *Strong truth and rules of inference in fuzzy logic and approximate reasoning* ◇ B52 ◇

YANASE, M.M. *Fuzziness and probability* ◇ A05 B48 B52 E72 ◇

ZADEH, L.A. *Syllogistic reasoning in fuzzy logic and its applications to usuality and reasoning with dispositions* ◇ B52 ◇

ZIMMERMANN, H.-J. *Fuzzy set theory – and its applications* ◇ B52 E72 ◇

B53 Paraconsistent logic

1940

POPPER, K.R. *What is dialectic?* ◇ A05 B53 ◇

1942

JEFFREYS, H. *Does a contradiction entail every proposition?* ◇ B53 ◇

1948

JASKOWSKI, S. *Un calcul des propositions pour les systemes deductifs contradictoires (Polish) (French summary)* ◇ B53 ◇

1949

JASKOWSKI, S. *Sur la conjonction discursive dans le calcul des propositions pour les systemes contradictoires (Polish) (French summary)* ◇ B53 ◇

1951

POPOV, P.S. *Dialectic and the subject matter of formal logic (Russian)* ◇ A05 B53 ◇

1954

MO, SHAOKUI *Logical paradoxes for many-valued systems* ◇ A05 B50 B53 ◇

1958

COSTA DA, N.C.A. *Note on the concept of contradiction (Portuguese) (English summary)* ◇ A05 B53 ◇

1959

COSTA DA, N.C.A. *Observations on the concept of existence in mathematics (Portuguese) (English summary)* ◇ A05 B53 ◇

1960

JOHNSTONE, H.W. *The law of contradiction* ◇ B53 ◇

1962

GORSKIJ, D.P. *Application of dialectical logic to the study of thought processes (Russian)* ◇ A05 B53 ◇

KEDROV, B.M. *Operating with scientific concepts in dialectical and formal logic (Russian)* ◇ A05 B53 ◇

LORENZEN, P. *Das Problem einer Formalisierung der Hegelschen Logik* ◇ A05 B53 ◇

1963

ARRUDA, A.I. & COSTA DA, N.C.A. *The paradox of Curry-Moh Shaw-Kwei (Portuguese)* ◇ B53 ◇

COSTA DA, N.C.A. *Calculs propositionnels pour les systemes formels inconsistants* ◇ B53 ◇

COSTA DA, N.C.A. *The present situation in set theory (Portuguese)* ◇ B53 E98 ◇

1964

ARRUDA, A.I. & COSTA DA, N.C.A. *Sur une hierarchie de systemes formels* ◇ B53 ◇

COSTA DA, N.C.A. *Calculs des predicats pour les systemes formels inconsistants* ◇ B53 ◇

COSTA DA, N.C.A. *Calculs de predicats avec egalite pour les systemes formels inconsistants* ◇ B53 ◇

COSTA DA, N.C.A. *Calculs des descriptions pour les systemes formels inconsistants* ◇ B53 ◇

COSTA DA, N.C.A. & GUILLAUME, M. *Sur les calculs* $\mathscr{C}_n$ ◇ B53 ◇

COSTA DA, N.C.A. *Sur un systeme inconsistant de theorie des ensembles* ◇ B53 E70 ◇

1965

ASENJO, F.G. *Dialectic logic* ◇ B53 ◇

BALLESTREM, K.G. *Dialectical logic* ◇ B53 ◇

COSTA DA, N.C.A. & GUILLAUME, M. *Negations composees et loi de Peirce dans les systemes* $\mathscr{C}_n$ ◇ B53 ◇

COSTA DA, N.C.A. *Sur les systemes formels* $\mathscr{C}_i, \mathscr{C}_i^*, \mathscr{C}_i^=, \mathscr{D}_i$ *et* NF_i ◇ B53 ◇

1966

ASENJO, F.G. *A calculus of antinomies* ◇ B53 ◇

KOSOK, M. *The formalization of Hegel's dialectical logic* ◇ A05 B53 ◇

TSERETELI, S.B. *The concept of dialectical logic* ◇ A05 B53 ◇

VASSAILS, G. *Elements de formalisation d'une logique dialectique* ◇ B53 ◇

1967

ARRUDA, A.I. *Sur certaines hierarchies de calculus propositionnels* ◇ B53 ◇

COSTA DA, N.C.A. *Two formal systems of set theory* ◇ B53 E30 E55 E70 ◇

COSTA DA, N.C.A. *Une nouvelle hierarchie de theories inconsistantes* ◇ B53 E70 ◇

GAUTHIER, Y. *Logique hegelienne et formalisation* ◇ B53 ◇

VASSAILS, G. *Les loqiques metriques divalentes* ◇ B53 ◇

VASSAILS, G. *Note sur l'antinomie* ◇ A05 B53 ◇

1968

ARRUDA, A.I. *Sur certaines hierarchies de calculs propositionnels* ◇ B53 ◇

GOTESKY, R. *The uses of inconsistency* ◇ B53 ◇

RAGGIO, A.R. *Propositional sequence-calculi for inconsistent systems* ◇ B53 ◇

1969

ARRUDA, A.I. *Sur certaines hierarchies de calculus de predicats* ◇ B53 ◇

ARRUDA, A.I. *Sur certaines algebres de classes non classiques* ◇ B53 G15 ◇

COSTA DA, N.C.A. & SETTE, A.-M. *Les algebres $\mathscr{C}_{\omega}$* ◇ B53 G15 ◇
DUBARLE, D. *Sur une formalisation de la logique hegelienne* ◇ B53 ◇
JASKOWSKI, S. *Propositional calculus for contradictory deductive systems (French summary)* ◇ B53 ◇
WRIGHT VON, G.H. *Time, change and contradiction* ◇ A05 B53 ◇

1970

ARRUDA, A.I. & COSTA DA, N.C.A. *Sur le schema de la separation* ◇ B53 E70 ◇
ARRUDA, A.I. *Sur le systeme NF_{ω}* ◇ B53 E70 ◇
ARRUDA, A.I. *Sur les systemes NF_i de da Costa* ◇ B53 E70 ◇
COSTA DA, N.C.A. *On the underlying logic of two systems of set theory* ◇ B53 E70 ◇
COSTA DA, N.C.A. & D'OTTAVIANO, I.M.L. *Sur un probleme de Jaskowski* ◇ B53 ◇
MANOR, R. & RESCHER, N. *On inference from inconsistent premisses* ◇ A05 B53 ◇
SPISANI, F. *Fondamenti di logica produttiva. Logica produttiva e logica formale* ◇ A05 B53 ◇
SPISANI, F. *Fondamenti di logica produttiva. La differenza dell'identico e le logiche dogmatiche* ◇ B53 ◇

1971

ARRUDA, A.I. *La mathematique classique dans NF_{ω}* ◇ B53 E30 E35 E70 ◇
ASENJO, F.G. *Logica dialectica* ◇ A05 B53 ◇
COSTA DA, N.C.A. *Remarques sur le systeme NF_1* ◇ B53 E70 ◇
KOTAS, J. *On the algebra of classes of formulae of Jaskowski's discussive system (Polish and Russian summaries)* ◇ B53 G15 ◇
SPISANI, F. *Fondamenti di logica produttiva. Annibale alle porte* ◇ A05 B53 ◇

1972

ASENJO, F.G. *On dialectic logic* ◇ A05 B53 ◇
COSTA DA, N.C.A. *On the theory of inconsistent formal systems* ◇ B53 ◇
DOZ, A. & DUBARLE, D. *Logique et Dialectique* ◇ A05 B53 ◇
SPISANI, F. *Foundations of productive logic (Italian, English)* ◇ A05 B53 ◇
SPISANI, F. *Outlines of productive logic (Italian and English)* ◇ B53 ◇

1973

KOTAS, J. *The axiomatization of S.Jaskowski's discussive systems* ◇ B53 ◇

1974

ARRUDA, A.I. & COSTA DA, N.C.A. *Le schema de la separation dans les calculs $\mathscr{J}_n$* ◇ B53 E70 ◇
COSTA DA, N.C.A. *On the theory of inconsistent formal systems* ◇ B53 E35 E70 ◇
COSTA DA, N.C.A. *Remarques sur les calculs $\mathscr{C}_n$, $\mathscr{C}^*_n$, $\mathscr{C}^=_n$ et $\mathscr{D}_n$* ◇ B25 B53 ◇
GRANT, J. *Incomplete models* ◇ B53 C90 ◇
KOTAS, J. *On quantity of logical values in the discussive D_2 system and in modular logic* ◇ B53 G12 ◇
KOTAS, J. *The axiomatization of S.Jaskowski's discussive system* ◇ B53 ◇
MAGARI, R. *Sur certe teorie non enumerabili. Sulle limitazioni dei sistemi formali I* ◇ B53 B60 F30 ◇

1975

ALLINSON, R.E. *When is a self contradictory proposition true: A sortie into Marxist logic* ◇ A05 B53 ◇
ARRUDA, A.I. *Le schema de la separation dans les systemes NF_n (English summary)* ◇ B53 E70 ◇
ARRUDA, A.I. *Remarques sur les systemes $\mathscr{C}_n$ (English summary)* ◇ B53 ◇
ASENJO, F.G. & TAMBURINO, J. *Logic of antinomies* ◇ B53 E70 ◇
COSTA DA, N.C.A. *Remarks on Jaskowski's discussive logic* ◇ B53 ◇
FURMANOWSKI, T. *Remarks on discussive propositional calculus* ◇ B53 ◇
GRANT, J. *Inconsistent and incomplete logics* ◇ B22 B53 ◇
HINTIKKA, K.J.J. *Impossible possible worlds vindicated* ◇ A05 B45 B53 ◇
KOTAS, J. *Discussive sentential calculus of Jaskowski* ◇ B53 ◇
SPISANI, F. *Outlines of productive logic (Italian and English)* ◇ B53 ◇
STANCIU, L. & STANCOVICI, V. *The structure of space and time in Franco Spisani's productive logic (Italian and English)* ◇ B53 ◇
WALD, H. *Introduction to dialectical logic* ◇ A05 B53 ◇

1976

ALVES, E.H. & COSTA DA, N.C.A. *Une semantique pour le calcul $\mathscr{C}_1$ (English summary)* ◇ B53 ◇
MEYER, R.K. & ROUTLEY, R. *Dialectical logic, classical logic and the consistency of the world* ◇ A05 B53 ◇
SPISANI, F. *Outlines of productive logic (Italian)* ◇ A05 B53 ◇

1977

ALVES, E.H. & COSTA DA, N.C.A. *A semantical analysis of the calculi $\mathscr{C}_n$* ◇ B53 ◇
ARRUDA, A.I. *On the imaginary logic of N.A. Vasil'ev* ◇ B53 ◇
ARRUDA, A.I. & COSTA DA, N.C.A. *Une semantique pour le calcul $\mathscr{C}^=_1$* ◇ B53 E70 ◇
COSTA DA, N.C.A. & DUBIKAJTIS, L. *On Jaskowski's discursive logic* ◇ B53 ◇
FIDEL, M.M. *The decidability of the calculi $\mathscr{C}_n$* ◇ B25 B53 G25 ◇
GRANT, J. *Incompleteness and inconsistency in propositional logic* ◇ B53 ◇
KRAMOSIL, I. *A classification of inconsistent theories* ◇ B53 ◇
LOPES DOS SANTOS, L.H. *Some remarks on discussive logic* ◇ B53 ◇

1978

ALMEIDA MOURA DE, J.E. & ALVES, E.H. *On some higher-order paraconsistent predicate calculi* ◇ B53 ◇

ALVES, E.H. *On decidability of a system of dialectical propositional logic* ◇ B25 B53 ◇

GRANT, J. *Classifications for inconsistent theories* ◇ B53 C07 ◇

LOPARIC, A. & ROUTLEY, R. *Semantical analysis of Arruda da Costa P systems and adjacent non-replacement relevant systems* ◇ B53 ◇

LOPES DOS SANTOS, L. *Discussive versions of the modal calculi T, B, S4 and S5* ◇ B53 ◇

1979

APOSTEL, L. *Logica e dialettica in Hegel* ◇ A05 B53 ◇

APOSTEL, L. *Logique et dialectique* ◇ B53 ◇

BENTHEM VAN, J.F.A.K. *What is dialectical logic?* ◇ A05 B53 ◇

BUNDER, M.W. *Paraconsistent combinatory logic* ◇ B53 ◇

BUNDER, M.W. *Some definitions of negation leading to paraconsistent logics* ◇ B53 ◇

COSTA DA, N.C.A. & KOTAS, J. *A new formulation of discussive logic* ◇ B53 ◇

FOLEY, R. *Justified inconsistent beliefs* ◇ B53 ◇

MACKENZIE, J.D. *Question-begging in non-cumulative systems* ◇ A05 B53 ◇

MARCONI, D. *La formalizzazione della dialettica* ◇ A05 B53 ◇

PETROV, S. *Hegel's thesis of contradictory truths* ◇ A05 B53 ◇

ROUTLEY, R. *Dialectical logic, semantics, and metamathematics* ◇ B53 ◇

1980

ALVES, E.H. & LOPARIC, A. *The semantics of the systems* C_n *of da Costa* ◇ B53 ◇

ARRUDA, A.I. *A survey of paraconsistent logic* ◇ A05 B53 B98 E70 ◇

ARRUDA, A.I. *The paradox of Russell in the systems* NF_n ◇ B53 E70 ◇

BATENS, D. *Paraconsistent extensional propositional logics* ◇ A05 B53 ◇

BRANDOM, R.B. & RESCHER, N. *The logic of inconsistency* ◇ B53 ◇

BUNDER, M.W. *A new hierarchy of paraconsistent logics* ◇ B53 ◇

COSTA DA, N.C.A. & WOLF, R.G. *Studies in paraconsistent logic. I* ◇ B53 ◇

DOYLE, J. & MCDERMOTT, D. *Non-monotonic logic. I* ◇ B53 ◇

DUBIKAJTIS, L. & DUDEK, E. & KONIOR, J. *On axiomatics of Jaskowski's discussive propositional calculus* ◇ B53 ◇

GAGNON, L.S. *Three theories of dialectic* ◇ A05 B53 ◇

GUCCIONE, S. *A semantics for a dialectical logic* ◇ B53 ◇

MARCONI, D. *A decision method for the calculus* C_1 ◇ A05 B25 B53 ◇

PENA, L. *The philosophical relevance of a contradictorial system of logic: Ap* ◇ A05 B53 ◇

PETERSEN, U. *Die logischen Grundlagen der Dialektik* ◇ A05 B53 ◇

PINTER, C. *The logic of inherent ambiguity* ◇ B53 ◇

ZHAO, ZONGGUAN *Discussion of the problem of logical contradiction and dialectical contradiction* ◇ B53 ◇

1981

ACHTELIK, G. & DUBIKAJTIS, L. & DUDEK, E. & KONIOR, J. *On independence of axioms of Jaskowski's discussive propositional calculus* ◇ B53 ◇

ALVES, E.H. & COSTA DA, N.C.A. *Relations between paraconsistent logic and many-valued logic* ◇ B53 ◇

GOODMAN, NICOLAS D. *The logic of contradiction* ◇ B53 E70 ◇

GRISHIN, V.N. *Predicate and set-theoretic calculi based on a logic without contractions (Russian)* ◇ B53 B60 E30 E70 F07 ◇

HAVAS, K.G. *Some remarks on an attempt at formalizing dialectical logic* ◇ B53 ◇

URCHS, M.P. *Kripke-style semantics for Jaskowski's system* Q_f ◇ B46 B53 ◇

WARAGAI, T. *The ontological law of contradiction and its logical structure* ◇ A10 B53 ◇

WILLIAMS, J.N. *Inconsistency and contradiction* ◇ B53 ◇

1982

ALVES, E.H. & COSTA DA, N.C.A. *On a paraconsistent predicate calculus* ◇ B53 ◇

ARRUDA, A.I. & COSTA DA, N.C.A. *Remarks on da Costa's paraconsistent set theories* ◇ B53 E70 ◇

BARTH, E.M. *A normative-pragmatical foundation of the rules of some systems of formal*$_3$ *dialectics* ◇ B53 ◇

COSTA DA, N.C.A. *The philosophical import of paraconsistent logic* ◇ B53 ◇

GUCCIONE, S. & TAMBURRINI, G. *The labyrinth of dialectical logics* ◇ B53 ◇

GUPTA, A. *Truth and paradox* ◇ A05 B53 ◇

KOMORI, Y. & ONO, H. *Logics without the contradiction rule* ◇ B20 B53 ◇

KRABBE, E.C.W. *Formal dialectics as immanent criticism of philosophical systems* ◇ A05 B53 ◇

LEWIS, D. *Logic for equivocators* ◇ B45 B53 ◇

MASON, R.O. & MITROFF, I.I. *On the structure of dialectical reasoning in the social and policy sciences* ◇ B53 ◇

PRIEST, G. *To be and not to be: dialectical tense logic* ◇ B53 ◇

WEDIN, M.V. *Atistotle on the range of the principle of noncontradiction* ◇ A10 B53 ◇

WILLIAMS, J.N. *Believing the self-contradictory* ◇ A05 B53 ◇

1983

BUNDER, M.W. *On Arruda and Da Costa's logics* J_1 *to* J_5 ◇ B53 ◇

GRANA, N. *Logica paraconsistente: Una introduzione* ◇ B53 ◇

MASON, R.O. & MITROFF, I.I. & QUINTON, H. *Beyound contradiction and consistency: a design for a dialectical policy system* ◇ B53 ◇

PETERSEN, U. *What do the antinomies teach us? Elements of a dialectical approach towards the contradictions of comprehension* ◇ B53 E30 E70 ◇
ROZONOEHR, L.I. *Proving contradictions in formal theories. I, II (Russian)* ◇ B53 ◇
SMOLENOV, KH. *Paraconsistency, paracompleteness and intentional contradictions* ◇ B53 ◇

1984

ALCANTARA DE, L.P. & CARNIELLI, W.A. *Paraconsistent logics* ◇ B53 ◇
ALVES, E.H. *Paraconsistent logic and model theory* ◇ B53 C40 C52 C90 ◇
ARRUDA, A.I. *N.A. Vasil'ev: a forerunner of paraconsistent logic* ◇ B53 ◇
ARRUDA, A.I. & COSTA DA, N.C.A. *On the relevant systems P and P^* and some related systems* ◇ B53 ◇
BLASZCZUK, J.J. *Some paraconsistent sentential calculi* ◇ B53 ◇
BRADY, R.T. *Depth relevance of some paraconsistent logics* ◇ B53 ◇
BUNDER, M.W. *Some definitions of negation leading to paraconsistent logics* ◇ B53 ◇
COSTA DA, N.C.A. & LOPARIC, A. *Paraconsistency, paracompleteness and valuations* ◇ B53 ◇
DEUTSCH, H. *Paraconsistent analytic implication* ◇ B53 ◇
PRIEST, G. & ROUTLEY, R. *Introduction: Paraconsistent logics* ◇ B53 ◇
PRIEST, G. *Semantic closure* ◇ B53 ◇
ROUTLEY, R. *The American plan completed: Alternative classical-style semantics, without stars, for relevant and paraconsistent logics* ◇ B46 B53 ◇
TENNANT, N. *Perfect validity, entailment ond paraconsistency* ◇ B53 ◇

1985

ANDREEV, I.D. *Dialectical logic (Russian)* ◇ B53 ◇
D'OTTAVIANO, I.M.L. & LOPEZ-ESCOBAR, E.G.K. *The conditional and paraconsistent logics* ◇ B53 ◇
KRZYSZTOF, R. (ED.) *Logics and models of concurrent systems* ◇ B53 ◇

B55 Intermediate and related logics

1927

Levy, P. *Logique classique, logique Brouwerienne et logique mixte* ◇ A05 B55 F50 ◇

1936

Jaskowski, S. *Recherches sur le systeme de la logique intuitionniste* ◇ B55 F50 ◇

1948

Destouches-Fevrier, P. *Logique de l'intuitionisme sans negation et logique de l'intuitionisme positif* ◇ B55 F50 ◇

McKinsey, J.C.C. & Tarski, A. *Some theorems about the sentential calculi of Lewis and Heyting* ◇ B45 B55 ◇

1955

Umezawa, T. *Ueber die Zwischensysteme der Aussagenlogik* ◇ B55 ◇

1957

Kreisel, G. & Putnam, H. *Eine Unableitbarkeitsbeweismethode fuer den intuitionistischen Aussagenkalkuel* ◇ B55 F50 ◇

1959

Dummett, M. *A propositional calculus with denumerable matrix* ◇ B55 ◇

Umezawa, T. *On intermediate many-valued logics* ◇ B55 ◇

Umezawa, T. *On intermediate propositional logics* ◇ B55 ◇

Umezawa, T. *On logics intermediate between intuitionistic and classical predicate logic* ◇ B55 ◇

Umezawa, T. *On some properties of intermediate logics* ◇ B55 ◇

1960

Umezawa, T. *On an application of intermediate logics* ◇ B55 ◇

1962

Bull, R.A. *The implicational fragment of Dummett's LC* ◇ B55 ◇

Monteiro, A. *Linearisation de la logique positive de Hilbert-Bernays* ◇ B55 ◇

Nishimura, I. *On a classification of intermediate propositional calculi (Japanese) (English summary)* ◇ B55 ◇

Resnik, M.D. *A decision procedure for positive implication* ◇ B55 ◇

1963

Yankov, V.A. *Realizable formulas of propositional logic (Russian)* ◇ B55 F50 ◇

Yankov, V.A. *Some superconstructive propositional calculi (Russian)* ◇ B55 ◇

Yankov, V.A. *The relationship between deducibility in the intuitionistic propositional calculus and finite implicational structures (Russian)* ◇ B55 F50 G25 ◇

1964

Bull, R.A. *Some results for implicational calculi* ◇ B55 ◇

Vorob'ev, N.N. *A constructive calculus of statements with strong negation (Russian)* ◇ B55 F50 ◇

1965

Bull, R.A. *A modal extension of intuitionist logic* ◇ B55 ◇

Kripke, S.A. *Semantical analysis of intuitionistic logic I* ◇ B55 C90 F50 ◇

Stralberg, A.H. *On the reduction of classical logic. An extension of some theorems of Glivenko and Goedel* ◇ B55 F30 F50 ◇

Troelstra, A.S. *On intermediate propositional logics* ◇ B55 ◇

Varpakhovskij, F.L. *The nonrealizability of a disjunction of nonrealizable formulas of propositional logic (Russian)* ◇ B55 F50 ◇

1966

Diego, A. *Sur les algebres de Hilbert* ◇ B55 G25 ◇

Hosoi, T. *Algebraic proof of the seperation theorem on Dummett's LC* ◇ B55 ◇

Hosoi, T. *On the separation theorem of intermediate propositional calculi* ◇ B55 ◇

Hosoi, T. *The axiomatization of the intermediate propositional systems S_n of Goedel* ◇ B55 ◇

Hosoi, T. *The separable axiomatization of the intermediate propositional systems S_n of Goedel* ◇ B55 ◇

Nagata, S. *A series of successive modifications of Peirce's rule* ◇ B55 ◇

Parsons, C. *A propositional calculus intermediate between the minimal calculus and the classical* ◇ B55 ◇

Ratsa, M.F. *A criterion of functional completeness in logic corresponding to the first Jaskowski matrix (Russian)* ◇ B50 B55 ◇

Standley, G.B. *Testing singly quantified tautologies* ◇ B55 ◇

1967

Hosoi, T. *A criterion for the separable axiomatization of Goedel's S_n* ◇ B55 ◇

Hosoi, T. *On intermediate logics* ◇ B55 ◇

Hosoi, T. *On the axiomatic method and the algebraic method for dealing with propositional logics* ◇ B55 ◇

McKay, C.G. *A note on the Jaskowski sequence* ◇ B55 F50 ◇
McKay, C.G. *On finite logics* ◇ B55 ◇
McKay, C.G. *Some completeness results for intermediate propositional logics* ◇ B55 ◇
Mints, G.E. *Embedding operations related to S.Kripke's semantics (Russian)* ◇ B55 ◇
Yankov, V.A. *Finite validity of formulas of a special form (Russian)* ◇ B55 C13 F50 ◇

1968

Malmnaes, P.E. & Prawitz, D. *A survey of some connections between classical, intuitionistic and minimal logic* ◇ B55 F05 F25 F50 ◇
McKay, C.G. *The decidability of certain intermediate propositional logics* ◇ B25 B55 ◇
McKay, C.G. *The non-separability of a certain finite extension of Heyting's propositional logic* ◇ B55 ◇
Ono, K. *On tabooistic treatment of proposition logics* ◇ B55 ◇
Segerberg, K. *Propositional logics related to Heyting and Johansson* ◇ B55 F50 ◇
Yankov, V.A. *Constructing a sequence of strongly independent superintuitionistic propositional calculi (Russian)* ◇ B55 ◇
Yankov, V.A. *On the extension of the intuitionist propositional calculus to the classical calculus, and the minimal calculus to the intuitionist calculus* ◇ B55 F50 ◇
Yankov, V.A. *The calculus of the weak "law of excluded middle" (Russian)* ◇ B55 ◇
Yankov, V.A. *Three sequences of formulas with two variables in positive propositional logic (Russian)* ◇ B55 ◇

1969

Anderson, J.G. *A note on finite intermediate logics* ◇ B55 ◇
Anderson, J.G. *An application of Kripke's completeness theorem for intuitionism to superconstructive propositional calculi* ◇ B55 C90 ◇
Golota, Ya.Ya. *Nets of marks and deducibility in the intuitionistic propositional calculus (Russian)* ◇ B55 B75 F50 ◇
Horn, A. *Logic with truth values in a linearly ordered Heyting algebra* ◇ B55 ◇
Hosoi, T. *On intermediate logics II* ◇ B55 ◇
Levin, L.A. *Some syntactic theorems on the calculus of finite problems of Ju.T.Medvedev (Russian)* ◇ B55 F50 ◇
Ruzsa, I. *Random models of some quasi-intuitionistic logics* ◇ B55 C90 ◇
Thomason, R.H. *A semantical study of constructible falsity* ◇ B55 ◇
Yankov, V.A. *Conjunctively indecomposable formulas in propositional calculi (Russian)* ◇ B55 F50 G10 ◇

1970

Gabbay, D.M. *The decidability of the Kreisel-Putnam system* ◇ B25 B55 F50 ◇
Gerchiu, V.Ya. & Kuznetsov, A.V. *On finitely axiomatizable superintuitionistic logics (Russian)* ◇ B55 ◇
Gerchiu, V.Ya. *Superindependent superintuitionistic logics (Russian)* ◇ B55 ◇
Gerchiu, V.Ya. & Kuznetsov, A.V. *Superintuitionistic logic and finite approximability (Russian)* ◇ B55 ◇
Hosoi, T. & Ono, H. *The intermediate logics on the second slice* ◇ B55 ◇
Hung, Ch'engwan *Remarks on intermediate logics* ◇ B55 ◇
Ono, H. *Kripke models and intermediate logics* ◇ B55 ◇
Ratsa, M.F. *Functional completeness in certain logics that are intermediate between the classical and intuitionistic ones (Russian)* ◇ B55 ◇
Rousseau, G. *The separation theorem for fragments of the intuitionistic propositional calculus* ◇ B55 F50 ◇
Semenenko, M.I. *Properties of some subsystems of classical and intuitionistic propositional calculi (Russian)* ◇ B55 F50 ◇
Wojcicki, R. *On reconstructability of the classical propositional logic in intuitionistic logic (Russian summary)* ◇ B55 F50 ◇

1971

Anderson, J.G. *The degree of completeness of Dummett's LC and Thomas's LC_n* ◇ B55 ◇
Belding, W.R. *Intuitionistic negation* ◇ B55 F55 ◇
Bowen, K.A. *An extension of the intuitionistic propositional calculus* ◇ B55 F50 ◇
Gabbay, D.M. *Semantic proof of the Craig interpolation theorem for intuitionistic logic and extensions I,II* ◇ B55 C40 C90 F50 ◇
Goernemann, S. *A logic stronger than intuitionism* ◇ B55 C90 F50 ◇
Klemke, D. *Ein Henkin-Beweis fuer die Vollstaendigkeit eines Kalkuels relativ zur Grzegorczyk-Semantik* ◇ B55 C90 F50 ◇
Kuznetsov, A.V. *The functional expressibility in superintuitionistic logics (Russian)* ◇ B55 ◇
McKay, C.G. *A class of decidable intermediate propositional logics* ◇ B25 B55 ◇
Ono, H. *On the finite model property for Kripke models* ◇ B55 ◇
Rootselaar van, B. *A class of models for intermediate logics* ◇ B55 G10 ◇
Wronski, A. *Axiomatization of the implicational Goedel's matrices by Kalmar's method (Polish summary)* ◇ B55 F50 ◇
Wronski, A. *Axiomatization of the implicational matrix $\Gamma((\mathscr{L}_2)^2)$ of Jaskowski's family $(\mathscr{L}_1)_{\Pi\Gamma}$ (Polish summary)* ◇ B55 ◇

1972

Anderson, J.G. *Superconstructive propositional calculi with extra axiom schemes containing one variable* ◇ B55 ◇
Gabbay, D.M. *Application of trees to intermediate logics* ◇ B55 C90 ◇
Gabbay, D.M. *Model theory for intuitionistic logic* ◇ B55 C20 C90 F50 ◇
Gerchiu, V.Ya. *The finite approximability of superintuitionistic logics (Russian)* ◇ B55 ◇

Hosoi, T. & Ono, H. *Axiomatization of models for intermediate logics constructed with boolean models by piling up* ◇ B55 G05 G25 ◇

Maksimova, L.L. *Pretabular superintuitionistic logics (Russian)* ◇ B55 ◇

Mints, G.E. *Derivability of addmissible rules (Russian) (English summary)* ◇ B55 F50 ◇

Ono, H. *Some results on the intermediate logics* ◇ B55 ◇

Prucnal, T. *On the structural completeness of some pure implicational propositional calculi (Polish and Russian summaries)* ◇ B55 ◇

Selman, A.L. *Completeness of calculi for axiomatically defined classes of algebras* ◇ B55 C05 C07 ◇

Wronski, A. *An algorithm for finding finite axiomatization of finite intermediate logics (Polish summary)* ◇ B55 ◇

Wronski, A. *Intermediate logics and the disjunction property. I,II* ◇ B55 ◇

1973

Dzik, W. & Wronski, A. *Structural completeness of Goedel's and Dummett's propositional calculi (Polish and Russian summaries)* ◇ B55 ◇

Gabbay, D.M. *A survey of decidability results for modal, tense and intermediate logics* ◇ B25 B45 B55 C98 F50 ◇

Hosoi, T. & Ono, H. *Intermediate propositional logics (a survey)* ◇ B55 ◇

Lorenz, K. *Rules versus theorems. A new approach for mediation between intuitionistic and two-valued logic* ◇ A05 B55 F50 ◇

Nagai, Satoru *On a semantics for non-classical logics* ◇ B55 ◇

Nagashima, T. *An intermediate predicate logic* ◇ B55 ◇

Ono, H. *A study of intermediate predicate logics* ◇ B55 ◇

Ono, H. *Incompleteness of semantics for intermediate predicate logics. I.Kripke's semantics* ◇ B55 ◇

Sato, M. *Characterization of pseudo-boolean models by boolean models and its applications to intermediate logics* ◇ B55 ◇

Smorynski, C.A. *Investigations of intuitionistic formal systems by means of Kripke models* ◇ B55 C90 F50 ◇

Surma, S.J. *Jaskowski's matrix criterion for the intuitionistic propositional calculus* ◇ B55 F50 ◇

Tax, R.E. *On the intuitionistic equivalent calculus* ◇ B55 ◇

Wronski, A. *Axiomatization of the implicational Goedel's matrices by Kalmar's method* ◇ B55 ◇

Wronski, A. *Intermediate logics and the disjunction property* ◇ B55 ◇

Wronski, A. *On the degree of completeness of positive logic* ◇ B55 ◇

Wronski, A. *Remarks on intermediate logics with axioms containing only one variable* ◇ B55 ◇

1974

Anderson, J.G. *A note on finite intermediate logics* ◇ B55 ◇

Bielak, P. *A note on weak functional completeness* ◇ B55 ◇

Czermak, J. *Matrix calculi SS1M and SS1I compared with axiomatic systems* ◇ B55 ◇

Ehsakia, L.L. *Topological Kripke models (Russian)* ◇ B45 B55 G05 G25 G30 ◇

Gabbay, D.M. & Jongh de, D.H.J. *A sequence of decidable finitely axiomatizable intermediate logics with the disjunction property* ◇ B22 B25 B55 ◇

Georgieva, N.V. *An extension of the Kreisel-Putnam System* ◇ B55 ◇

Goldblatt, R.I. *Decidability of some extensions of J* ◇ B25 B55 ◇

Graczynska, E. & Wronski, A. *Constructing denumerable matrices strongly adequate for prefinite logics* ◇ B55 G10 ◇

Hosoi, T. *On intermediate logics III* ◇ B55 ◇

Nagai, Satoru & Ono, H. *Applications of general Kripke models to intermediate logics* ◇ B55 ◇

Prucnal, T. *Interpretations of classical implicational sentential calculus in nonclassical implicational calculi* ◇ B55 ◇

Radu, E. *On a certain theorem of J.Porte (Romanian) (French summary)* ◇ B55 ◇

Rauszer, C. *A formalization of the propositional calculus of H-B logic* ◇ B55 F50 G10 ◇

Segerberg, K. *Proof of a conjecture of McKay* ◇ B55 F50 ◇

Surma, S.J. & Wronski, A. & Zachorowski, S. *On Jaskowski-type semantics for the intuitionistic propositional logic* ◇ B55 F50 ◇

Tokarz, M. *Binary functions definable in implicational Goedel algebras* ◇ B55 ◇

Wronski, A. *On cardinalities of matrices strongly adequate for the intuitionistic propositional logic* ◇ B55 F50 ◇

Wronski, A. *On equivalential fragments of some intermediate logics* ◇ B55 ◇

Wronski, A. *Remarks on intermediate logics with axioms containing only one variable* ◇ B55 ◇

1975

Dahn, B.I. *Eine Anwendung des Basistheorems in der nichtklassischen Logik* ◇ B55 C40 C90 ◇

Gargov, G.K. *Intuitionistic analysis, and intermediate logics (Russian)* ◇ B55 F50 ◇

Georgieva, N.V. *On the decidability of p′, p′′ and P′′* ◇ B55 ◇

Hosoi, T. *On the implicational fragment logics* ◇ B55 ◇

Komori, Y. *The finite model property of the intermediate propositional logics on finite slices* ◇ B55 ◇

Kuznetsov, A.V. *On superintuitionistic logics* ◇ B25 B55 ◇

Kuznetsov, A.V. *Superintuitionistic logics (Russian)* ◇ B55 ◇

Pahi, B. *Jankow-theorems for some implicational calculi* ◇ B55 ◇

Porebska, M. & Wronski, A. *A characterization of fragments of the intuitionistic propositional logic* ◇ B55 F50 ◇

Prucnal, T. *Structural completeness and the disjunction property of intermediate logics* ◇ B55 ◇

SONOBE, O. *A Gentzen-type formulation of some intermediate propositional logics* ◇ B55 ◇

SURMA, S.J. & WRONSKI, A. & ZACHOROWSKI, S. *On Jaskowski-type semantics for the intuitionistic propositional logic* ◇ B55 F50 ◇

TAUTS, A. *Search for deduction by means of a semantic model (Russian) (Estonian and German summaries)* ◇ B55 C75 C90 F50 ◇

WRONSKI, A. *On equivalential fragments of some intermediate logics* ◇ B55 ◇

1976

BAKER, K.A. *Equational axioms for classes of Heyting algebras* ◇ B55 C05 G05 G10 ◇

EHSAKIA, L.L. *On modal partners of superintuitionistic logics (Russian)* ◇ B45 B55 ◇

FISCHER SERVI, G. *Un'algebrizzazione del calcolo intuizionista monadico (English summary)* ◇ B55 F50 G10 ◇

GABBAY, D.M. *Completeness properties of Heyting's predicate calculus with respect to RE models* ◇ B55 F50 ◇

HOSOI, T. *Axiomatizability of fragments of intermediate finite logics* ◇ B55 ◇

HOSOI, T. *Non-separable intermediate propositional logics* ◇ B55 ◇

KHOMICH, V.I. *A separability theorem for superintuitionistic propositional calculi (Russian)* ◇ B55 ◇

KHOMICH, V.I. *A separability theorem for a certain pre-truth table superintuitionistic propositional logic (Russian)* ◇ B55 ◇

PRUCNAL, T. *Structural completeness of Medvedev's propositional calculus* ◇ B55 ◇

RAUSZER, C. *On the strong semantical completeness of any extension of the intuitionistic predicate calculus* ◇ B55 F50 ◇

SONOBE, O. *Tableau methods of proof for LPn and LQn* ◇ B55 ◇

1977

BERNHARDT, K. *Abschliessbarkeit von Kripke-Semantiken* ◇ B55 C90 ◇

BESSONOV, A.V. *New operations in intuitionistic calculus (Russian)* ◇ B55 F50 ◇

DARDZHANIYA, G.K. *Intuitionistic system without contraction* ◇ B55 F50 ◇

GABBAY, D.M. *Craig interpolation theorem for intuitionistic logic and extensions part III* ◇ B55 C40 C90 F50 ◇

GABBAY, D.M. *On some new intuitionistic propositional connectives I* ◇ B45 B55 F50 ◇

HOSOI, T. & SHUNDO, S. *A finer classification of intermediate propositional logics* ◇ B55 ◇

MAKSIMOVA, L.L. *Craig's interpolation theorem and amalgamable varieties (Russian)* ◇ B22 B55 C05 C40 C90 G10 ◇

MAKSIMOVA, L.L. *Craig's theorem in superintuitionistic logics and amalgamable varieties of pseudo-boolean algebras (Russian)* ◇ B55 C05 C40 C90 G05 G10 ◇

NAGASHIMA, T. *A theorem on an intermediate predicate logic* ◇ B55 ◇

POPOV, S.V. *Undecidable interval arithmetic (Russian) (English Summary)* ◇ B55 D35 ◇

RAUSZER, C. *Applications of Kripke models to Heyting-Brouwer logic* ◇ B55 C90 F50 ◇

RAUSZER, C. *Model theory for an extension of intuitionistic logic* ◇ B55 C90 F50 ◇

SHANIN, N.A. *On the quantifier of limiting realizability* ◇ B55 C57 C80 F50 ◇

SHATALOV, V.V. *Topological models for intuitionistic logic and the spectrum of refutions of a formula in the Fitch logic (Russian)* ◇ B55 ◇

SHEKHTMAN, V.B. *On incomplete propositional logics (Russian)* ◇ B55 ◇

SHUNDO, S. *Some results concerning the new classification of intermediate propositional logics* ◇ B55 ◇

SOBOLEV, S.K. *On finite approximability of superintuitionistic logics (Russian)* ◇ B55 ◇

SOBOLEV, S.K. *On finite-dimensional superintuitionistic logics (Russian)* ◇ B55 D05 F50 G10 G25 ◇

SOBOLEV, S.K. *The intuitionistic propositional calculus with quantifiers (Russian)* ◇ B55 C80 C90 D35 F50 ◇

SONOBE, O. *Beth-type tableau proof systems for some intermediate and modal logics* ◇ B45 B55 ◇

TSITKIN, A.I. *On admissible rules of intuitionistic propositional logic (Russian)* ◇ B55 F50 ◇

VAKARELOV, D. *Notes on $\mathscr{N}$-lattices and constructive logic with strong negation* ◇ B55 F50 G10 ◇

YASHIN, A.L. *On an extension of intuitionistic logic and its interpretations (Russian)* ◇ B55 C90 ◇

ZACHOROWSKI, S. *Intermediate logics without the interpolation property* ◇ B55 C40 ◇

1978

ARRUDA, A.I. *Some remarks on Griss' logic of negationless intuitionistic mathematics* ◇ B55 F50 F55 G25 ◇

DRUGUSH, YA.M. *A class of logics without disjunction property (Russian)* ◇ B55 F50 ◇

FIDEL, M.M. *An algebraic study of a propositional system of Nelson* ◇ B55 F50 G10 G25 ◇

FISCHER SERVI, G. *Semantics for a class of intuitionistic modal calculi* ◇ B55 G25 ◇

FISCHER SERVI, G. *The finite model property for MIPQ and some consequences* ◇ B55 F50 G10 ◇

HOSOI, T. *Axiomatization of certain intermediate logics* ◇ B55 ◇

HOSOI, T. *The delta operation on the intermediate logics and the modal logics* ◇ B55 ◇

KOMORI, Y. *Logics without Craig's interpolation property* ◇ B55 C40 ◇

MORI, S. *On S5 type modal logics based on intermediate logics with finitely many valued linear models* ◇ B45 B55 ◇

ROUTLEY, R. *An inadequacy in Kripke-semantics for intuitionistic quantificational logic* ◇ B55 C90 F50 ◇

SHEKHTMAN, V.B. *An undecidable superintuitionistic propositional calculus (Russian)* ◇ B55 D35 ◇

SHEKHTMAN, V.B. *Rieger-Nishimura lattices (Russian)* ◇ B55 C90 G10 G25 ◇

SHUNDO, S. *A procedure for classifying the intermediate propositional logics and its application to the finite slices* ◇ B55 ◇
SMORYNSKI, C.A. *The axiomatization problem for fragments* ◇ B55 F50 ◇
TRAGESSER, R.S. & ZUCKER, J.I. *The adequacy problem for inferential logic* ◇ A05 B55 F50 ◇
TSITKIN, A.I. *On structurally complete superintuitionistic logics (Russian)* ◇ B55 G10 ◇
WENDEL, N. *The inconsistency of Bernini's very strong intuitionistic theory* ◇ B55 F35 F50 ◇
ZACHOROWSKI, S. *Dummett's LC has the interpolation property* ◇ B55 C40 C90 ◇
ZACHOROWSKI, S. *Remarks on interpolation property for intermediate logics* ◇ B55 ◇
ZASLAVSKIJ, I.D. *Symmetric constructive logic (Russian)* ◇ B55 F05 F30 F50 ◇

1979

CAPINSKA, E. *On standard consequence operations in the implicationless language* ◇ B55 ◇
DARDZHANIYA, G.K. *On the complexity of countermodels for intuitionistic propositional calculus (Russian) (English summary)* ◇ B55 F50 ◇
EHSAKIA, L.L. *To the theory of modal and superintuitionistic systems (Russian)* ◇ B45 B55 ◇
GIMON, V.V. *Quantifier-free formulas in A. A. Markov's ramified semantics (Russian)* ◇ B55 F50 ◇
HOSOI, T. *Evaluation of Peirce's axiom on intermediate Kripke models and its application* ◇ B55 ◇
JOHNSTONE, P.T. *Conditions related to de Morgan's law* ◇ B55 C65 C90 F50 G30 ◇
KHOMICH, V.I. *Separability of superintuitionistic propositional calculi (Russian)* ◇ B20 B55 ◇
KIRK, R.E. *Some classes of Kripke frames characteristic for the intuitionistic logic* ◇ B55 C90 F50 ◇
MAKSIMOVA, L.L. *Interpolation properties of superintuitionistic logics* ◇ B55 ◇
MAKSIMOVA, L.L. & SHEKHTMAN, V.B. & SKVORTSOV, D.P. *The impossibility of a finite axiomatization of Medvedev's logic of finitary problems (Russian)* ◇ B55 ◇
MEDVEDEV, YU.T. *Transformations of information and calculi that describe them: types of information and their possible transformations (Russian)* ◇ B55 ◇
PRUCNAL, T. *On two problems of Harvey Friedman* ◇ B45 B55 ◇
SHEKHTMAN, V.B. *Topological semantics of superintuitionistic logics (Russian)* ◇ B55 C90 ◇
SKVORTSOV, D.P. *Logic of infinite problems and Kripke models on atomic semilattices of sets (Russian)* ◇ B55 C90 F50 G20 ◇
SKVORTSOV, D.P. *On some propositional logics connected with the concept of Yu. T. Medvedev's types of information (Russian)* ◇ B55 ◇
SKVORTSOV, D.P. *Realizability and finite validity of propositional formulas with restructions on the occurrences of implication (Russian)* ◇ B55 ◇

1980

BOZZI, S. & MELONI, G.C. *Representation of Heyting algebras with covering and propositional intuitionistic logic with local operator* ◇ B25 B55 C90 F50 G10 ◇
DARSOW, W. & KITTEL, P.M. *On intermediate consequence relations* ◇ B55 ◇
DZIOBIAK, W. *The degrees of maximality of the intuitionistic propositional logic and of some of its fragments* ◇ B55 ◇
FIDEL, M.M. *An algebraic study of logic with constructible negation* ◇ B25 B55 G25 ◇
HAZEN, A. *Comments on the logic of constructible falsity (strong negation)* ◇ B55 ◇
JANICKA-ZUK, I. *Finite axiomatization for some intermediate logics* ◇ B55 ◇
JONGH DE, D.H.J. *A class of intuitionistic connectives* ◇ B55 C75 C90 F50 ◇
KABZINSKI, J.K. *What is the equivalence connective?* ◇ B55 C05 G25 ◇
KALICKI, C. *Infinitary propositional intuitionistic logic* ◇ B55 C75 F50 ◇
KHOMICH, V.I. *On the problem of separability for superintuitionistic propositional logics (Russian)* ◇ B55 ◇
KIRK, R.E. *A characterization of the classes of finite tree frames which are adequate for the intuitionistic logic* ◇ B55 C90 F50 ◇
MILLER, D.MICHAEL *Spectral addition techniques for multiple-valued functions* ◇ B55 ◇
MORI, S. *Axiomatization of S4 type modal logics based on finite linear intermediate logics* ◇ B45 B55 ◇
RAUSZER, C. *An algebraic and Kripke-style approach to a certain extension of intuitionistic logic* ◇ B55 C20 C90 F50 G25 ◇
SKVORTSOV, D.P. *On the connection of finitary general validity of certain propositional formulas with derivability in the Kreisel-Putnam logic (Russian)* ◇ B55 F50 ◇
UMEZAWA, T. *Cut elimination in intuitionistic and some intermediate predicate logics* ◇ B55 F05 F50 ◇
WRONSKI, A. *On reducts of intermediate logics* ◇ B55 ◇

1981

BORICIC, B.R. *Equational reformulations of intuitionistic propositional calculus and classical first-order predicate calculus* ◇ B55 F50 ◇
BURGESS, J.P. *The completeness of intuitionistic propositional calculus for its intended interpretation* ◇ B55 F50 ◇
CAPINSKA, E. *On intermediate logics which can be axiomatized by means of implicationless formulas* ◇ B55 ◇
CHAGROV, A.V. *Superintuitionist fragments of nonnormal modal logics (Russian)* ◇ B55 ◇
DOSEN, K. *A reduction of classical propositional logic to the conjunction-negation fragment of an intuitionistic relevant logic* ◇ B46 B55 F50 ◇
DZIOBIAK, W. *Strong completeness with respect to finite Kripke models* ◇ B55 ◇

DZIOBIAK, W. *The degrees of maximality of the intuitionistic propositional logic and of some of its fragments* ◇ B55 ◇

FISCHER SERVI, G. *Semantics for a class of intuitionistic modal calculi* ◇ B45 B55 F50 ◇

GABBAY, D.M. *Semantical investigations in Heyting's intuitionistic logic* ◇ B55 F50 F98 ◇

HENDRY, H.E. *Does IPC have a binary indigenous Sheffer function?* ◇ B55 ◇

KABZINSKI, J.K. & POREBSKA, M. & WRONSKI, A. *On the* $\{\leftrightarrow, \sim\}$ *-reduct of the intuitionistic consequence operation* ◇ B55 ◇

KOMORI, Y. *Super-Lukasiewicz proposotional logics* ◇ B55 ◇

LOPEZ-ESCOBAR, E.G.K. *On the interpolation theorem for the logic of constant domains* ◇ B55 C40 C90 F50 ◇

LOPEZ-ESCOBAR, E.G.K. *Variations on a system of Gentzen* ◇ B55 F50 ◇

MURAVITSKIJ, A.YU. *Finite approximability of the calculus* I^{Δ} *and the nonmodelability of some of its extensions (Russian)* ◇ B55 ◇

MURAVITSKIJ, A.YU. *Strong equivalence on an intuitionistic Kripke model and assertorically equivolumetric logics (Russian)* ◇ B55 C90 F50 ◇

POPOV, S.V. *Nondecidable intermediate calculus (Russian)* ◇ B55 D35 ◇

RENARDEL DE LAVALETTE, G.R. *The interpolation theorem in fragments of logics* ◇ B20 B55 C40 F50 ◇

SHUM, A.A. *Finite axiomatization of the logic of infinite problems in an extended language (Russian)* ◇ B55 ◇

SZATKOWSKI, M. *On fragments of Medvedev's logic* ◇ B55 ◇

TROELSTRA, A.S. *On a second order propositional operator in intuitionistic logic* ◇ B55 F35 F50 ◇

ZAKHAR'YASHCHEV, M.V. *Some classes of intermediate logics (Russian) (English summary)* ◇ B55 ◇

1982

DRUGUSH, YA.M. *Forest superintuitionistic logics (Russian)* ◇ B55 ◇

DRUGUSH, YA.M. *Union of logics modeled by finite trees* ◇ B55 ◇

DZIOBIAK, W. *On finite approximability of* ψ *-intermediate logics* ◇ B55 ◇

GABBAY, D.M. *Intuitionistic basis for non-monotonic logic* ◇ B55 C90 F50 ◇

GEORGACARAKOS, G.N. *The semantics of minimal intuitionism* ◇ B55 F50 ◇

KABZINSKI, J.K. *Basic properties of the equivalence* ◇ B55 G25 ◇

LOPEZ-ESCOBAR, E.G.K. *A natural deduction system for some intermediate logics* ◇ B55 ◇

LOPEZ-ESCOBAR, E.G.K. *Implicational logics in natural deduction systems* ◇ B55 ◇

MAKSIMOVA, L.L. *Interpolation properties of superintuitionistic, positive and modal logics* ◇ B55 C40 C90 ◇

ONO, H. & RAUSZER, C. *On an algebraic and Kripke semantics for intermediate logics* ◇ B55 ◇

SHEKHTMAN, V.B. *Undecidable propositional calculi (Russian)* ◇ B22 B55 D35 ◇

1983

BIELAK, P. *Criteria of definability for positive fragments of some intermediate logics (Polish) (English summary)* ◇ B55 ◇

BORICIC, B.R. *A decision procedure for certain disjunction-free intermediate propositional calculi* ◇ B25 B55 ◇

BOZIC, M. & DOSEN, K. *Axiomatizations of intuitionistic double negation* ◇ B55 F50 ◇

CASARI, E. *Intermediate logics* ◇ B55 ◇

CASARI, E. & MINARI, P. *Negation-free intermediate predicate logics (Italian summary)* ◇ B55 ◇

CHAGROV, A.V. *Polynomial finite approximability of modal and superintuitionistic logics (Russian)* ◇ B45 B55 ◇

DZIOBIAK, W. *Cardinalities of proper ideals in some lattices of strengthenings of the intuitionistic propositional logic* ◇ B55 ◇

FITTING, M. *Proof methods for modal and intuitionistic logics* ◇ B45 B55 F07 F50 F98 ◇

HAWRANEK, J. *A characterization of prime theories in Johannsson's minimal logic* ◇ B55 ◇

KOMORI, Y. *Some results on the super-intuitionistic predicate logics* ◇ B55 ◇

LOPEZ-ESCOBAR, E.G.K. *A second paper on the interpolation theorem for the logic of constant domains* ◇ B55 C40 C90 F05 F50 ◇

MEREDITH, D. *Separating minimal, intuitionist and classical logic* ◇ B55 ◇

MIGLIOLI, P.A. & MOSCATO, U. & ORNAGHI, M. & USBERTI, G. *Alcuni calcoli intermedi costruttivi* ◇ B55 F65 ◇

MINARI, P. *Completeness theorems for some intermediate predicate calculi* ◇ B55 C90 ◇

MURAVITSKIJ, A.YU. *Comparison of the topological and relational semantics of superintuitionistic logics (Russian)* ◇ B55 C90 ◇

NAKAMURA, T. *Disjunction property for some intermediate predicate logics* ◇ B55 ◇

NEPEJVODA, N.N. *Notes concerning constructive implicative logics (Russian)* ◇ B55 F50 ◇

ONO, H. *Model extension theorem and Craig's interpolation theorem for intermediate predicate logics* ◇ B55 C20 C40 C90 ◇

PRUCNAL, T. *Structural completeness of some fragments of intermediate logics* ◇ B55 ◇

SHUM, A.A. *A pseudointuitionistic propositional calculus (Russian)* ◇ B55 ◇

SZOSTEK, B. *A certain classification of the Heyting-Brouwer intermediate propositional logics* ◇ B55 ◇

ZAKHAR'YASHCHEV, M.V. *On intermediate logics (Russian)* ◇ B55 C90 ◇

1984

BORICIC, B.R. *A note on some intermediate propositional calculi* ◇ B55 ◇

BOZIC, M. *Positive logic with double negation* ◇ B55 ◇

DOSEN, K. *Negative modal operators in intuitionistic logic* ◇ B55 F50 ◇
DRUGUSH, YA.M. *Finite approximability of forest superintuitionistic logics (Russian)* ◇ B25 B55 ◇
HAWRANEK, J. *On the degree of matrix complexity of Johansson's minimal logic* ◇ B55 ◇
MARDAEV, S.I. *Number of prelocally table superintuitionistic propositional logics (Russian)* ◇ B55 ◇
MURAVITSKIJ, A.YU. *A result on the completeness of superintuitionistic logics* ◇ B55 C90 ◇
MURAVITSKIJ, A.YU. *Superintuitionistic logics approximable by algebras with the descending chain condition (Russian)* ◇ B55 ◇
SCEDROV, A. *On some nonclassical extensions of second-order intuitionistic propositional calculus* ◇ B55 F50 ◇
SIKIC, Z. *Multiple forms of Gentzen's rules and some intermediate forms* ◇ B55 F07 ◇
SZATKOWSKI, M. *Concerning Kripke semantics for intermediate predicate logics* ◇ B55 C90 ◇

1985

BALLARD, D. *Independence in higher-order subclassical logic* ◇ B55 ◇
BOZIC, M. *Semantics for some intermediate logics* ◇ B55 C90 ◇
DALE, A.J. *A completeness property of negationless intuitionist propositional logics* ◇ B55 ◇
DOSEN, K. *An intuitionistic Sheffer function* ◇ B55 F50 ◇
GHILARDI, S. & MELONI, G.C. *Completeness for intermediate logics (Italian)* ◇ B55 ◇
GHILARDI, S. & MELONI, G.C. & PIERSANTINI, C. *Some results on intermediate propositional logics (Italian)* ◇ B55 ◇
GORANKO, V.F. *Propositional logics with strong negation and the Craig interpolation theorem* ◇ B55 C40 ◇
GORANKO, V.F. *The Craig interpolation theorem for propositional logics with strong negation* ◇ B55 C40 ◇
LOPEZ-ESCOBAR, E.G.K. *Proof functional connectives* ◇ B55 F07 F30 ◇
MCKAY, C.G. *A consistent propositional logic without any finite models* ◇ B55 ◇
POREBSKA, M. *Interpolation for fragments of intermediate logics* ◇ B55 C40 ◇
PRUCNAL, T. *Structural completeness of purely implicational intermediate logics* ◇ B55 ◇
RAUSZER, C. *Formalizations of certain intermediate logics. I* ◇ B55 C40 C90 ◇
RAUTENBERG, W. *A note on implicational intermediate consequences* ◇ B55 ◇
RYBAKOV, V.V. *A criterion for admissibility of deduction rules in modal and intuitionistic logic (Russian)* ◇ B45 B55 ◇
RYBAKOV, V.V. *Bases of admissible rules of the modal system Grz and intuitionistic logic (Russian)* ◇ B45 B55 ◇
SKVORTSOV, D.P. *A superintuitionistic propositional calculus (Russian)* ◇ B55 ◇
TAMURA, S. *Decision procedure for pseudocomplemented lattices* ◇ B55 ◇
VAKARELOV, D. *An application of Rieger-Nishimura formulas to the intuitionistic modal logics* ◇ B55 ◇
YASHIN, A.D. *Semantic characterization of intuitionistic logical connectives (Russian)* ◇ B55 ◇

B60 Other logics

1912

VASIL'EV, N.A. *Imaginary (non-Aristotelian) logic (Russian)* ◇ B60 ◇

1932

LEWIS, C.I. *Alternative systems of logic* ◇ B60 ◇

1936

SPERANTIA, E. *Remarques sur les propositions interrogatives. Projet d'une "logique du probleme"* ◇ B60 ◇

1938

CHWISTEK, L.B. *Remarques critiques concernant la notion de la variable dans le systeme de la semantique rationnelle (Polish) (French summary)* ◇ A05 B60 E70 ◇

1939

KLEENE, S.C. *On the term "analytic" in logical syntax* ◇ A05 B60 ◇

SOBOCINSKI, B. *Aus den Untersuchungen zur Protothetik (Polish)* ◇ B60 ◇

1943

HEMPEL, C.G. *A purely syntactical definition of confirmation* ◇ B60 ◇

MCCULLOCH, W.S. & PITTS, W. *A logical calculus of the ideas immanent in nervous activity* ◇ B60 D05 ◇

1947

LIEBER, L.R. *Mits, wits, and logic* ◇ B60 B98 ◇

1949

HALLDEN, S. *The logic of nonsense* ◇ B60 ◇

SOBOCINSKI, B. *An investigation of protothetic* ◇ B60 ◇

1951

LUKASIEWICZ, J. *On variable functors of propositional arguments* ◇ B60 ◇

MARTIN, R.M. & WOODGER, J.-H. *Toward an inscriptional semantics* ◇ A05 B60 ◇

MEREDITH, C.A. *On a extended system of the propositional calculus* ◇ B60 ◇

MILLER, JAMES WILKIN *The logic of terms* ◇ B60 ◇

1953

PRIOR, A.N. *Negative quantifiers* ◇ B60 ◇

SLUPECKI, J. *St. Lesniewski's protothetics (Polish and Russian summaries)* ◇ B60 ◇

1954

QUINE, W.V.O. *Quantification and the empty domain* ◇ B10 B60 ◇

1955

LEJEWSKI, C. *A contribution to Lesniewski's mereology* ◇ B60 E70 ◇

SLUPECKI, J. *St. Lesniewski's calculus of names (Polish and Russian summaries)* ◇ B60 ◇

1956

AJDUKIEWICZ, K. *Conditional sentence and material implication (Polish, English) (Russian summary)* ◇ B60 ◇

KRASZEWSKI, Z. *Logic of extensional relations (Polish, Russian) (English summary)* ◇ B60 ◇

LORENZEN, P. *Protologik. Ein Beitrag zum Begruendungsproblem der Logik* ◇ B60 ◇

MOCH, F. *La logique des attitudes* ◇ A05 B60 ◇

STAHL, G. *La logica de las preguntas* ◇ B60 ◇

TUTUGAN, F. *On the existence of certain valid syllogistic modi other than those of classical logic (Romanian)* ◇ B60 ◇

1957

MUELLER, GERT H. *Zur operativen Begruendung von Logik und Mathematik* ◇ B30 B60 ◇

1958

GOODSTEIN, R.L. *On the formalisation of indirect discourse* ◇ B60 B65 ◇

1959

LAMOUCHE, A. *Logique de la simplicite* ◇ A05 B60 ◇

1960

MENDELSON, E. *A semantic proof of the eliminability of descriptions* ◇ B60 ◇

1961

HARRAH, D. *A logic of questions and answers* ◇ B60 ◇

RESCHER, N. *Non-deductive rules of inference and problems in the analysis of inductive reasoning* ◇ B60 ◇

STAHL, G. *Preguntas y premisas* ◇ B60 ◇

ZIEMBINSKI, Z. *The propositional character of enacted rules (Polish) (Russian and English summaries)* ◇ B60 ◇

1962

AAQVIST, L. *Reflections on the logic of nonsense* ◇ A05 B60 ◇

GORN, S. *An axiomatic approach to prefix languages* ◇ B60 ◇

RESCHER, N. *Quasi-truth-functional systems of propositional logic* ◇ B60 ◇

1963

BELNAP JR., N.D. *An analysis of questions: Preliminary report* ◇ A05 B46 B60 ◇

LOS, J. *Semantic representation of the probability of formulas in formalized theories (Polish summary)* ◇ A05 B10 B48 B60 ◇

ROSE, A. *A formalisation of the propositional calculus corresponding to Wang's calculus of partial predicates* ◇ B60 ◇

STAHL, G. *Un developpement de la logique des questions* ◇ B60 ◇

TIRNOVEANU, M. *Probabilistic methods in nonconstructive logical systems (Romanian) (Russian and French summaries)* ◇ B60 ◇

1964

BLACK, F. *A deductive question-answering system* ◇ B60 ◇

PIECZKOWSKI, A. *On the systems Q and Q_f* ◇ B60 ◇

SOBOCINSKI, B. *On the propositional system A of Vuchkovich and its extension I,II* ◇ B20 B60 ◇

SUGIHARA, T. *Sequential truth-function* ◇ B60 ◇

TOERNEBOHM, H. *Information and confirmation* ◇ A05 B48 B60 ◇

1965

CRESSWELL, M.J. *On the logic of incomplete answers* ◇ B60 ◇

CRESSWELL, M.J. *The logic of interrogatives* ◇ B60 ◇

HERMES, H. *Eine Termlogik mit Auswahloperator* ◇ B10 B60 ◇

MO, SHAOKUI & SHEN, BAIYING *Semi-complete quasi basic system (Chinese)* ◇ B60 ◇

SEGERBERG, K. *A contribution to nonsense-logic* ◇ B60 ◇

1966

FRAASSEN VAN, B.C. *The completeness of free logic* ◇ B60 G15 ◇

KUBINSKI, T. *A review of some problems of the logic of questions (Polish) (Russian and English summaries)* ◇ A05 B60 ◇

MAGARI, R. *Calcoli generali e spazi V_α* ◇ B22 B60 ◇

MUELLER, GERT H. *Zur Einfuehrung der Logik* ◇ A05 B60 ◇

PIROG-RZEPECKA, K. *A propositional calculus in which expressions are loosing their sense (Polish) (Russian and English summaries)* ◇ B60 ◇

1967

BOTEZATU, P. *La logique extensionnaliste et la double negation* ◇ B60 ◇

EBERLE, R.A. *Some complete calculi of individuals* ◇ A05 B60 ◇

FRAASSEN VAN, B.C. & LAMBERT, K. *On free description theory* ◇ B60 G15 ◇

LAMBERT, K. & SCHARLE, T.W. *A translation theorem for two systems of free logic* ◇ B20 B60 F25 ◇

MAGARI, R. *Sun una questione riguardante le chiusure di Moore* ◇ B60 ◇

ZEMAN, J.J. *A system of implicit quantification* ◇ B60 ◇

1968

FRAASSEN VAN, B.C. *A topological proof of the Loewenheim-Skolem, compactness, and strong completeness theorems for free logic* ◇ B60 C07 ◇

GOE, G. *Modifications of Quine's ML and inclusive quantification systems* ◇ B60 E70 ◇

KEARNS, J.T. *The logical concept of existence* ◇ A05 B60 ◇

LAMBERT, K. & MEYER, R.K. *Universally free logic and standard quantification theory* ◇ B60 G15 ◇

LORENZ, K. *Dialogspiele als semantische Grundlage von Logikkalkuelen I,II* ◇ B60 F50 ◇

PIROG-RZEPECKA, K. *The proof of the non-existence of a finite matrix adequate for the sentential calculus in which expressions become meaningless (Polish and Russian summaries)* ◇ B60 ◇

PLIUSKEVICIUS, R. *Kanger's version of the predicate calculus with symbols for not everywhere defined functions (Russian)* ◇ B10 B60 ◇

TANAKA, S. *On axioms of ontology* ◇ B60 ◇

TANAKA, S. *On theorems of ontology* ◇ B60 ◇

1969

APOSTEL, L. *A proposal in the analysis of questions* ◇ B60 ◇

CANTY, J.T. *Lesniewski's terminological explanations as recursive concepts* ◇ B60 ◇

EBBINGHAUS, H.-D. *Ueber eine Praedikatenlogik mit partiell definierten Praedikaten und Funktionen* ◇ B60 C90 ◇

RAO, A.P. *A note on universally free first order quantification theory* ◇ B60 ◇

RESCHER, N. *The logic of existence* ◇ B60 ◇

STAHL, G. *The effectivity of questions* ◇ B60 ◇

SWEET, A.M. *The pragmatics of monadic quantification* ◇ A05 B60 ◇

WIELADEK, R. *A propositional calculus without the law of extensionality (Polish) (English and Russian summaries)* ◇ B60 ◇

1970

BOBROVA, L.A. *The completeness of systems of degenerate inference and quasiinference (Russian)* ◇ B60 ◇

CANTY, J.T. & KUENG, G. *Substitutional quantification and Lesniewskian quantifiers* ◇ A05 B60 C80 E70 ◇

CHAVCHANIDZE, V.V. *Logik von Interessen (Phasenlogik auf der Basis von Logik-Informationsfunktionen.) (Russian) (English summary)* ◇ B60 ◇

DORROUGH, D. *A logical calculus of analogy involving functions of order 2* ◇ A05 B60 ◇

EBERLE, R.A. *Nominalistic systems* ◇ A05 B60 ◇

FERREIRA BARBOSA, J.E. *On the operational logic of junctors (Portuguese and English)* ◇ B60 ◇

PARSONS, C. *Axiomatization of Aaqvist's CS-logics* ◇ B60 ◇

SOMMERS, F. *The calculus of terms* ◇ A05 B60 ◇

TANAKA, S. *On axiom systems of ontology. I* ◇ B60 ◇

TIRNOVEANU, M. *Sur quelques notions de proto-logistique semi-constructive* ◇ B60 F99 ◇

TREW, A. *Nonstandard theories of quantifications and identity* ◇ B20 B60 ◇

1971

AAQVIST, L. *Revised foundations for imperative-epistemic and interrogative logic* ◇ A05 B60 ◇
BLOOM, S.L. & SUSZKO, R. *Semantics for the sentential calculus with identity (Polish and Russian summaries)* ◇ B60 G05 ◇
CANTY, J.T. *Elementary logic without referential quantification* ◇ B60 ◇
KUBINSKI, T. *Introduction to the logical theory of questions* ◇ B60 ◇
MARKWALD, W. *Praedikatenlogik mit partiell definierten Funktionen I* ◇ B10 B60 ◇
MOISIL, G.C. *Sur le mode ponendo ponens (Romanian summary)* ◇ B60 ◇
PETKOV, P.P. *On the question of the possibility of introducing disjunction on the lower levels of A.A. Markov's hierarchy of mathematical logic (Russian)* ◇ B60 F50 ◇
PIECZKOWSKI, A. *On the definitive implication (Polish and Russian summaries)* ◇ B60 ◇
SUSZKO, R. *Quasi-completeness in non-Fregean logic (Polish and Russian summaries)* ◇ B60 C07 ◇
TANAKA, S. *On axiom systems of ontology. II* ◇ B60 ◇
TOLSTOVA, YU.N. *A weakening of intuitionistic logic (Russian) (English summary)* ◇ B60 F50 ◇
TRAN VAN TOAN *Le Strichkalkuel de B. von Freytag-Loeringhoff et la recherche des premisses sous-entendues* ◇ A05 B60 ◇

1972

BLOOM, S.L. & SUSZKO, R. *Investigations into the sentential calculus with identity* ◇ B60 G05 ◇
CLAY, R.E. *Note on induvtive finiteness in mereology* ◇ B60 ◇
LORENZEN, P. *Dialogkalkuele* ◇ B60 F50 ◇
MATERNA, P. *Intensional semantics of vague constants. An application of Tichy's concept of semantics* ◇ A05 B60 ◇
OMYLA, M. & SUSZKO, R. *Definitions in theories of kind W* ◇ B60 ◇
OMYLA, M. & SUSZKO, R. *Descriptions in theories of kind W* ◇ B60 ◇
ROSE, A. *Sur certains calculus propositionnels sans constantes ou tautologies ordinaires* ◇ B60 ◇
ROSE, A. *Tautologies sans variables propositionnelles* ◇ B60 ◇
SUSZKO, R. *A note on modal systems and SCI* ◇ B45 B60 ◇
SUSZKO, R. *A note on adequate models for non-Fregean sentential calculi* ◇ B60 ◇
THOMASON, R.H. *A semantic theory of sortal incorrectness* ◇ A05 B60 ◇
WESSEL, H. *Probleme topologischer Logiken* ◇ A05 B60 ◇

1973

CHURCH, A. *Proposiciones y oraciones (Spanish)* ◇ B60 ◇
GROVER, D.L. *Propositional quantification and quotation contexts* ◇ A05 B60 ◇
IBRAGIMOV, S.G. *Consequence logic of Paul Lorenzen (Russian)* ◇ B60 F50 ◇
IWANUS, B. *On Lesniewski's elementary ontology (Polish and Russian summaries)* ◇ B25 B60 E70 ◇
MADUCH, M. *Axiomatization of the logic W (Polish) (English summary)* ◇ B60 ◇
PIROG-RZEPECKA, K. *Post normal forms for the system W (Polish) (English summary)* ◇ B60 ◇
ROEHLE, F. *Der Informationsgehalt der Geraden. Ansaetze zu einer kommunikationstheoretischen Grundlegung der Mathematik* ◇ A05 B60 ◇
STENLUND, S. *The logic of description and existence* ◇ A05 B60 ◇
SUSZKO, R. *Adequate models for the non-fregean sentential calculus (SCI)* ◇ B60 ◇
WOJCICKI, R. *Set theoretic representations of empirical phenomena* ◇ A05 B60 ◇
WOODRUFF, P.W. *On constructive nonsense logic* ◇ B60 ◇

1974

AAQVIST, L. *A new approach to the logical theory of actions and causality* ◇ B60 ◇
CZERMAK, J. *A logical calculus with descriptions* ◇ A05 B60 ◇
EHRENFEUCHT, A. *Logic without iterations* ◇ B60 F65 ◇
GNANI, G. *Insiemi dialettici generalizzati* ◇ B60 D25 D55 ◇
GODDARD, L. & MEYER, R.K. & ROUTLEY, R. *Choice and descriptions in enriched intensional languages I* ◇ A05 B60 ◇
GRISHIN, V.N. *A nonstandard logic, and its application to set theory (Russian)* ◇ B60 E70 F07 ◇
HAACK, S. *Deviant logic* ◇ A05 B60 ◇
HINTIKKA, K.J.J. *Questions about questions (Russian)* ◇ A05 B60 B65 ◇
LEJEWSKI, C. *A system of logic for bicategorial ontology* ◇ B60 E70 ◇
LEMPICKI, A. *A new look at optical logic* ◇ B60 ◇
LINDNER, R. & WAGNER, K. *The axiomatisation of a certain sequential propositional calculus (Russian)* ◇ B60 B70 D05 D10 ◇
LIU, SHICHAO *A theory of first order logic with description* ◇ B60 ◇
MAGARI, R. *Sur certe teorie non enumerabili. Sulle limitazioni dei sistemi formali I* ◇ B53 B60 F30 ◇
MARKWALD, W. *Praedikatenlogik mit partiell definierten Funktionen. II* ◇ B10 B60 ◇
MICHAELS, A. *A uniform proof procedure for SCI tautologies* ◇ B60 ◇
MICHAELS, A. & SUSZKO, R. *EN-logic* ◇ B60 ◇
QUINE, W.V.O. *Truth and disquotation* ◇ B60 ◇
TAUTS, A. *A formal deduction of tautological formulas in pseudo-boolean algebras (Russian) (Estonian and German summaries)* ◇ B60 C90 F50 G05 G10 ◇
WATANABE, S. *Logic of the empirical world. With reference to identity theory and reductionism* ◇ A05 B60 ◇

1975

CLAY, R.E. *Single axioms for atomistic and atomless mereology* ◇ B60 E70 ◇
GAINES, B.R. & KOHOUT, L.J. *The logic of protection* ◇ B60 ◇

HAJEK, P. *Observationsfunktorenkalkuele und die Logik der automatisierten Forschung* ◇ B60 ◇
HAJEK, P. *On logics of discovery* ◇ B60 ◇
HARRAH, D. *A system for erotetic sentences* ◇ B60 ◇
HORWICH, P. *A formalization of "nothing"* ◇ A05 B60 ◇
JEROSLOW, R. *Experimental logics and Δ_2^0-theories* ◇ B60 D55 ◇
KALININ, V.V. *Dimension functions on Boolean logics I,II (Russian)* ◇ B60 ◇
LAMBROS, C.H. *Notes toward an axiomatization of recklessness* ◇ B60 ◇
MARTIC, B. *A certain nonclassical system of logic (Serbo-Croatian) (English summary)* ◇ B60 ◇
MOUTAFAKIS, N.J. *Imperatives and their logics* ◇ B60 ◇
PACKARD, D.J. *A preference logic minimally complete for expected utility maximization* ◇ B60 ◇
RAO, A.P. *A note on universally free description theory* ◇ B60 ◇
ROGAVA, M.G. *Cut elimination in SCI* ◇ B60 F05 ◇
ROSE, A. *A note on the existence of tautologies without constants* ◇ B60 ◇
VANDERVEKEN, D.R. *An extension of Lesniewski-Curry's formal theory of syntactical categories adequate for the categorially open functors* ◇ B40 B60 ◇
YABLON, P. *A generalized propositional calculus* ◇ B60 D35 ◇

1976

BALLMER, T.T. *Prolegomena to a logic of changing the context by operations and commands (A case study of objects of intention.)* ◇ A05 B60 ◇
BELNAP JR., N.D. & STEEL JR., T.B. *The logic of questions and answers* ◇ B60 B65 ◇
BENCIVENGA, E. *Set theory and free logic* ◇ A05 B60 E70 ◇
BINKLEY, R. *The logic of action* ◇ A05 B60 ◇
CEGLA, W. & JADCZYK, A.Z. *Logics generated by causality structures. Covariant representations of the Galilean logic* ◇ B60 ◇
CLARK, R. *Old foundations for a logic of perception* ◇ B60 ◇
COSTA DA, N.C.A. & PINTER, C. *α-logic and infinitary languages* ◇ B60 C75 ◇
DAVIS, CHARLES C. *A note on the axiom of choice in Lesniewski's ontology* ◇ B60 E25 E70 ◇
DESCLES, J.-P. *Operations sur des operateurs types (English summary)* ◇ B60 ◇
EFIMOV, E.I. *Purposeful system behaviour logic (Russian) (English summary)* ◇ B60 ◇
GRANDY, R.E. *Anadic logic and English* ◇ B60 B65 G15 ◇
GRANDY, R.E. *On the relation between free description theories and standard quantification theory* ◇ B60 C20 ◇
HAJEK, P. *Observationsfunktorenkalkuele und die Logik der automatisierten Forschung* ◇ B60 C13 C80 C90 ◇
HERMES, H. *Dialog games* ◇ A05 B60 F50 ◇
KIELAK, W. *ENE-logic* ◇ B60 ◇
MARTIC, B. *On the completeness of the non-classical system A'* ◇ B60 ◇
MICHAELS, A. & SUSZKO, R. *Sentential calculus of identity and negation* ◇ B60 ◇
MORAIS, R. *Projective logic* ◇ B60 C75 G05 ◇
NILSON, P.S. *Productive logic and the principle of self-contradiction* ◇ A05 B60 ◇
PIROG-RZEPECKA, K. *A theory of deduction systems strongly adequate in relation to the sentential calculus W (Polish) (English summary)* ◇ B60 ◇
PONCELET, G. *Sur la logique productive (P.L)* ◇ A05 B60 ◇
ROSE, A. *A note on the existence of tautologies in certain propositional calculi without propositional variables* ◇ B60 ◇
SAMOKHVALOV, K.F. *The impossibility theorem for universal theory of prediction* ◇ A05 B60 ◇
SESIC, B.V. *Productive logic and foundations of a logic of variable predicates (LVP)* ◇ B60 ◇
WETTE, E. *On the formalization of productive logic* ◇ B60 D20 ◇

1977

ASENJO, F.G. *Lesniewski's work and nonclassical set theories* ◇ A05 A10 B60 E70 ◇
BENCIVENGA, E. *Are arithmetical truths analytic? New results in free set theory* ◇ A05 B60 E70 ◇
CASEY, M.R. *Productive logic and congruency tables* ◇ B60 ◇
HAJEK, P. *Experimental logics and Π_3^0 theories* ◇ B60 C62 D55 F30 ◇
KONONOV, B.P. *The distinction relation: duality semantics (Russian) (English and German summaries)* ◇ B60 ◇
KROEGER, F. *LAR: A logic of algorithmic reasoning* ◇ B60 B75 ◇
KUDLEK, M. & NOVOTNY, M. *Reducing operators for normed general formal systems* ◇ B60 ◇
LEJEWSKI, C. *Systems of Lesniewski's ontology with the functor of weak inclusion as the only primitive term* ◇ B60 ◇
LIPSKI JR., W. *On the logic of incomplete information* ◇ B60 ◇
MORAIS, R. *Projective logics and projective boolean algebras* ◇ B60 C75 E15 G05 G25 ◇
NOLAN, P.C. *A semantics model for imperatives* ◇ B60 ◇
OMYLA, M. *Barcan formulas in SCI with quantifiers* ◇ B60 ◇
RAO, A.P. *A more natural alternative to Mostowski's (MFL)* ◇ B60 C07 ◇
RICKEY, V.F. *A survey of Lesniewski's logic* ◇ A05 A10 B60 E70 ◇
ROUTLEY, R. *Choice and descriptions in enriched intensional languages. II,III* ◇ A05 B60 ◇
SARLET, H.J. *Hintikka's free logic is not free* ◇ B60 ◇
SPISANI, F. *Implicazione, endometria, universo del discorso. Implication, endometry, universe of discourse* ◇ A05 B60 ◇
VANDAMME, R.W. *Productive logic and propositional paradoxes* ◇ A05 B60 ◇

1978

BENCIVENGA, E. *A semantics for a weak free logic* ◇ B60 ◇

DONNADIEU, M.-R. & RAMBAUD, C. *Theories conditionnelles (English summary)* ◇ B60 ◇

GARDIES, J.-L. *Formalization of that-clauses* ◇ A05 B60 B65 ◇

GOAD, C.A. *Monadic infinitary propositional logic: a special operator* ◇ B25 B60 C75 F50 ◇

HUMBERSTONE, I.L. *Two merits of the circumstantial operator language for conditional logics* ◇ A05 B60 ◇

KONONOV, B.P. *Eine stationaere Logik zur Entscheidungsfindung (Russian) (German and English summaries)* ◇ B60 ◇

KRABBE, E.C.W. *The adequacy of material dialogue-games* ◇ A05 B60 C07 ◇

LORENZ, K. & LORENZEN, P. *Dialogische Logik* ◇ A05 B60 F99 ◇

MERRETT, T. *QT logic: Simpler and more expressive than predicate calculus* ◇ B60 B75 ◇

WELSH JR., P.J. *Primitivity in mereology I,II* ◇ A05 B60 E70 ◇

1979

BUNCH, B.L. *Presupposition: An alternative approach* ◇ B60 ◇

CHRISTENSEN, N.E. *Mathematical logic and mathematical reasoning* ◇ A05 B60 ◇

CIRULIS, J. *Protothetics without typical ambiguity of expressions (Russian)* ◇ B60 ◇

CIRULIS, J. *The traditional means of inference in protothetics (Russian)* ◇ B60 ◇

GRIFFIN, N. & ROUTLEY, R. *Towards a logic of relative identity* ◇ A05 B60 ◇

GUMB, R.D. *An extended joint consistency theorem for free logic with equality* ◇ B60 C40 ◇

HOOGEWIJS, A. *On a formalization of the non-definedness notion* ◇ B60 C90 ◇

JOHNSON WU, K. *Natural deduction rules for free logic* ◇ B60 ◇

KUBINSKI, T. *Completeness theorems for a class of elementary action theories (Polish)* ◇ B60 ◇

KUBINSKI, T. *On semantics of change logics (Polish)* ◇ B60 ◇

LEHMANN, S.K. *A general propositional logic of conditionals* ◇ B60 ◇

MCCARTHY, J. *First order theories of individual concepts and propositions* ◇ A05 B60 ◇

MELONI, G.C. & RADAELLI, G. *Calcolo naturale per categorie con residui* ◇ B60 F07 G30 ◇

OMYLA, M. *Propositional quantifiers in non-Fregean theories* ◇ B60 ◇

PETERSON, P.L. *On the logic of "few", "many", and "most"* ◇ B60 C80 ◇

PRANK, R.K. *Expressibility in the elementary theory of recursive sets with realizability logic (Russian) (English summary)* ◇ B60 D20 F50 ◇

ROGAVA, M.G. *A new decision procedure for SCI (Russian)* ◇ B25 B60 F05 ◇

STELZNER, W. *Funktorenvariable, Funktionenvariable und nichtklassische Logik* ◇ A05 B60 ◇

WUTTICH, K. *Fragen des Psychologismus in der epistemischen Logik* ◇ A05 B60 ◇

ZEMAN, J.J. *Normal, Sasaki, and classical implications* ◇ B60 ◇

1980

ACZEL, P. & FEFERMAN, S. *Consistency of the unrestricted abstraction principle using an intensional equivalence operator* ◇ B60 E70 ◇

BENCIVENGA, E. *A weak free logic with the existence sign* ◇ B60 ◇

BENCIVENGA, E. *Free semantics for definite descriptions* ◇ B60 ◇

BIRNBAUM, L. *N-polar class logic. Class calculus, judgements, syllogisms* ◇ B60 ◇

BOBROW, D.G. (ED.) *Special issue on nonmonotonic logic* ◇ B60 ◇

CARVALHO DE, R.L. & VELOSO, P.A.S. *Towards a logic of limited perception* ◇ A05 B60 ◇

CURD, M.V. *The logic of discovery: an analysis of three approaches* ◇ A05 B60 ◇

KOHOUT, L.J. & PINKAVA, V. *The algebraic structure of the Spencer Brown and Varela calculi* ◇ B60 G25 ◇

KUBINSKI, T. *An outline of the logical theory of questions* ◇ A05 B60 ◇

KUTSCHERA VON, F. *Grundbegriffe der Handlungslogik* ◇ B60 ◇

LEHMANN, S.K. *Slightly nonstandard logic* ◇ B60 ◇

MCCARTHY, J. *Circumscription -- a form of nonmonotonic reasoning* ◇ B60 ◇

PRANK, R.K. *Semantics of realizability for a language with variables for recursively enumerable sets (Russian)* ◇ B60 D25 F35 F50 ◇

RIJEN VAN, J. *On criticizing deviant logics* ◇ B60 ◇

SIMI, G. *Algebre pre-modali. I: Su una classe di algebre atte all'introduzione di un operatore di possibilita. II: Sulla varieta delle algebre premodali* ◇ B45 B60 G25 ◇

SLATER, B.H. *Logic without tears* ◇ B60 ◇

1981

AAQVIST, L. *Predicate calculi with adjectives and nouns* ◇ A05 B60 ◇

ALPS, R.A. & NEVELN, R.C. *A predicate logic based on indefinite description and two notions of identity* ◇ B60 ◇

ASANIDZE, G.Z. *A dialogue justification of logic (Russian)* ◇ B60 F50 F65 ◇

BARWISE, J. & COOPER, R. *Generalized quantifiers and natural language* ◇ A05 B60 C80 ◇

BENCIVENGA, E. *Free semantics* ◇ B60 ◇

BERGMANN, MERRIE *Presupposition and two-dimensional logic* ◇ A05 B60 ◇

CHANKVETADZE, O.E. *The addition rule for implications and its applications (Russian) (English and Georgian summaries)* ◇ B60 ◇

CLARKE, B.L. *A calculus of individuals based on "connection"* ◇ B60 G25 ◇

DALLA CHIARA SCABIA, M.L. *Some metalogical pathologies of non distributive logics* ◇ B60 ◇

GRISHIN, V.N. *Predicate and set-theoretic calculi based on a logic without contractions (Russian)* ◇ B53 B60 E30 E70 F07 ◇
HINTIKKA, K.J.J. *On the logic of an interrogative model of scientific inquiry* ◇ B60 ◇
KANTOR, P.B. *The logic of weighted queries* ◇ B60 ◇
KUBINSKI, T. *Theorems on some classes of change theories with quantifiers* ◇ B60 ◇
MA, XIWEN *Formalization of the logical problem of "to know" (Chinese)* ◇ B60 ◇
MARCONI, D. *Types of non-Scotian logic* ◇ B60 ◇
OUELLET, R. *Inclusive first-order logic* ◇ B60 ◇
PRANK, R.K. *Expressibility in the elementary theory of recursively enumerable sets with realizability logic (Russian)* ◇ B60 D25 F50 ◇
SILVESTRINI, D. *Alcune considerazioni sul metodo delle supervalutazioni e una semantica a supervalutazioni per il calcolo predicativo classico* ◇ B35 B60 ◇
SIMONS, P.M. *A note on Lesniewski and free logic* ◇ B60 ◇
STALNAKER, R.C. *Logical semiotic* ◇ B60 ◇
TIURYN, J. *A survey of the logic of effective definitions* ◇ B60 B75 ◇
TIURYN, J. *Logic of effective definitions* ◇ B60 C55 C75 ◇
TOSIC, R. *Types of bases for a modification of the algebra of logic (Russian) (Serbo-Croatian summary)* ◇ B60 ◇
WEINGARTNER, P. *A new theory of intensions* ◇ A05 B60 ◇

1982

AJZERMAN, M.A. & MALISHEVSKIJ, A.V. *Problems of logical justification in general choice theory. Levels and criteria of classical choice rationality (Russian) (English summary)* ◇ B60 ◇
AJZERMAN, M.A. & MALISHEVSKIJ, A.V. *Problems of logical justification in general choice theory. Examples of the analysis of the rationality of choice mechanisms (Russian) (English summary)* ◇ B60 ◇
ALCHOURRON, C.E. & MAKINSON, D. *On the logic of theory change: contraction functions and their associated revision functions* ◇ B60 ◇
ALSINA, C. *Functional equations and negations of distribution functions (Catalan)* ◇ B60 ◇
APOSTEL, L. *Towards a general theory of argumentation* ◇ B60 ◇
BENNETT, J.H. *Psychology and semantics: comments on Schiffer's "Intention-based semantics"* ◇ B60 ◇
BERGMANN, MERRIE *Expressibility in two-dimensional languages for presupposition* ◇ B60 ◇
BERKLING, K.J. & FEHR, E. *A consistent extension of the lambda-calculus as a base for functional programming languages* ◇ B40 B60 B75 ◇
BESSONOV, A.V. *Logical semantics and ontology (Russian)* ◇ B60 ◇
BLOOM, S.L. *A note on the logic of signed equations* ◇ B60 C05 C07 ◇
CIRULIS, J. *Axioms of protothetics with a strengthened capacity rule (Russian)* ◇ B15 B60 ◇
CRESSWELL, M.J. *Urn models: A classical exposition* ◇ B60 ◇
DULIN, S.K. *Introduction to dissonant logic (Russian) (English summary)* ◇ B60 ◇
DUN VAN, F. *On the philosophy of argument and the logic of common morality* ◇ B60 ◇
ENGEL, P. & NEF, F. *Quelques remarques sur la logique des phrases d'action* ◇ B60 B65 ◇
FRAISSE, R. *La zerologie, une recherche aux frontieres de la logique et de l'art; application a la logique des relations de base vide* ◇ B60 C07 ◇
GUENTHER, A. *A set of concepts for the study of dialogic argumentation* ◇ B60 ◇
HALKOWSKA, K. *On some connections between conditional algebras and the systems of nonsense-logics* ◇ B60 ◇
HARRAH, D. *What should we teach about questions?* ◇ B60 ◇
HINTIKKA, K.J.J. *Game-theoretical semantics: Insights and prospects* ◇ A05 B60 C80 ◇
HINTIKKA, K.J.J. & LEOPOLD-WILDBURGER, U. *Regeln und Nutzen von Dialogspielen* ◇ A05 B60 ◇
HUBBELING, H.G. *A modern reconstruction of the cosmological argument* ◇ A05 B60 ◇
KRAJEWSKI, S. *On relatedness logic of Richard L. Epstein* ◇ B60 ◇
KUBINSKI, T. *On Diodorean theories* ◇ B60 ◇
LORENZ, K. *On the criteria for the choice of the rules of dialogic logic* ◇ A05 B60 ◇
LORENZEN, P. *Die dialogische Begruendung von Logikkalkuelen* ◇ A05 B60 ◇
MALACHOWSKI, A.R. *Carney's proposal* ◇ B60 ◇
MALINOWSKI, G. & MICHALCZYK, M. *That SCI has the interpolation property* ◇ B60 C40 ◇
MCDERMOTT, D. *Nonmonotonic logic II: nonmonotonic modal theories* ◇ B60 ◇
NEPEJVODA, N.N. *Constructive logics (Russian)* ◇ B60 C90 C95 F50 ◇
ODDIE, G. & TICHY, P. *The logic of ability, freedom and responsibility* ◇ B60 ◇
ORLOV, YU.F. *The wave logic of consciousness: a hypothesis* ◇ B60 ◇
PACKARD, D.J. *Cyclical preference logic* ◇ B60 ◇
PAWLOWSKY, V. & TRILLAS, E. *On rejection connectives* ◇ B60 ◇
POSPELOV, D.A. & VAROSYAN, S.O. *Nommetric spatial logic (Russian)* ◇ B60 ◇
POSY, C.J. *A free IPC is a natural logic: strong completeness for some intuitionistic free logics* ◇ B60 ◇
ROGAVA, M.G. *Nonlenghtening applications of the identity rules in the calculus SCI (Russian) (English and Georgian summaries)* ◇ B60 ◇
ROGAVA, M.G. *Proof of the elimination theorem for the calculus SCI (Russian) (English and Georgian summaries)* ◇ B60 ◇
SMIRNOV, V.A. *An immersion of St. Lesniewski's elementary ontology into second-order one-place predicate calculus (Russian)* ◇ B60 ◇
STAHL, G. *Analyse et applications des suites de questions* ◇ B60 ◇
THOMPSON, B. *Syllogisms using "few", "many", and "most"* ◇ B60 C80 ◇

TSKHADADZE, O.S. *Equivalence of sequential variants of the calculus SCI (Russian) (English and Georgian summaries)* ◇ B60 ◇

WALTON, D.N. & WOODS, J. *Question-begging and cumulativeness in dialectical games* ◇ A05 B60 ◇

1983

BORICIC, B.R. *One of the possible formal descriptions of deducibility* ◇ B60 ◇

CZERMAK, J. & WEINGARTNER, P. *Criteria for distinguishing logically necessary propositions* ◇ B45 B60 ◇

GEORGIEVA, N.V. *Some properties of conditional expressions (Russian)* ◇ B60 ◇

HOOGEWIJS, A. *A partial predicate calculus in a two-valued logic* ◇ B60 C90 ◇

KUBINSKI, T. *Some classes of elementary action theories (Polish) (English summary)* ◇ B60 ◇

KUENG, G. *The difficulty with the well-formedness of ontoligical statements* ◇ A05 B60 ◇

MARGHIDANU, D. *An exclusion logic for the complementary of a theory (Roumanian) (English summary)* ◇ B60 ◇

MULLIGAN, K. & SMITH, BARRY *Framework for formal ontology* ◇ B60 ◇

OSIPOV, G.S. & ZHOLONDZ', V.Y. *Pseudophysical logics in semiotic models* ◇ B60 ◇

SIMONS, P.M. *Class, mass and mereology* ◇ A05 B60 ◇

SPISANI, F. *Teoria generale dei numeri relativi. Vol.I (Italian and English)* ◇ B60 ◇

TAUTS, A. *The possibility of constructing a countermodel for an indeterminate formula (Russian) (German and Estonian summaries)* ◇ B60 ◇

VAROSYAN, S.O. *Space logic for robots (Russian) (English summary)* ◇ B60 ◇

WATANABE, S. *Theory of propensity. A new foundation of logic* ◇ B60 ◇

WILSON, N.L. *The transitivity of implication in tree logic* ◇ B60 ◇

ZINOV'EV, A.A. *Nonstandard logic and its applications* ◇ B60 ◇

1984

ALTAEV, N.K. *Principles of a physical logic (Russian)* ◇ B60 ◇

BATTEN, L.M. & WALTON, D.N. *Games, graphs and circular arguments* ◇ B60 ◇

BATTICELLI, N.A. *Integral logic* ◇ B60 ◇

BENCIVENGA, E. *A possibility-free logic of descriptions* ◇ B60 ◇

BENTHEM VAN, J.F.A.K. *Foundations of conditional logic* ◇ B60 B65 ◇

BLUTNER, R. *Eine Erlaeuterung zu Freges Begriff der "Erlaeuterung"; defaults und ihre Logik* ◇ A10 B60 ◇

BRYLL, G. & HALKOWSKA, K. *Slupecki's fragmentary systems* ◇ B60 ◇

EBERLE, R.A. *Logic with a relative truth predicate and "that"-terms* ◇ B60 ◇

LEMAN, M. *On the logic and criteriology of causality* ◇ B60 ◇

MANGANI, P. *General calculi "with types" and generalized logics (Italian) (English summary)* ◇ B60 ◇

MIGLIOLI, P.A. & USBERTI, G. *A logic of knowing (Italian)* ◇ A05 B60 ◇

NAVARA, M. *Two-valued states on a concrete logic and the additivity problem (Russian summary)* ◇ B60 ◇

ORLOWSKA, E. *Logic of nondeterministic information* ◇ B60 B75 ◇

PARKER, J.H. *Social logics and moral logics* ◇ B60 ◇

PESCHEL, K. *Argumentationsfelder als Grundlage einer Argumentationslogik* ◇ B60 ◇

POSPELOV, D.A. *Pseudophysical logics (Russian)* ◇ B60 ◇

SEGERBERG, K. *A topological logic of action* ◇ B60 ◇

SIMONS, P.M. *A Brentanian basis for Lesniewskian logic* ◇ B60 ◇

WASILEWSKA, A. *DFC-algorithms for Suszko logic SCI and one-to-one Gentzen type formalizations* ◇ B60 ◇

WOJCICKI, R. *R. Suszko's situational semantics* ◇ B60 ◇

WOLNIEWICZ, B. *A topology for logical space* ◇ B60 ◇

WOODRUFF, P.W. *Paradox, truth and logic. I. Paradox and truth* ◇ A05 B60 ◇

1985

ALCHOURRON, C.E. & GAERDENFORS, P. & MAKINSON, D. *On the logic of theory change: partial meet contraction and revision functions* ◇ B60 ◇

ARNBORG, S. & TIDEN, E. *Unification problems with one-side distributivity* ◇ B60 ◇

BENCIVENGA, E. *Strong completeness of a pure free logic* ◇ B60 ◇

BOSSU, G. & SIEGEL, P. *Saturation, nonmonotonic reasoning and the closed-world assumption* ◇ B60 ◇

CIRULIS, J. *A system of logic permitting vacuous and plural terms (Russian)* ◇ B60 ◇

GABBAY, D.M. *Theoretical foundations for nonmonotonic reasoning in expert systems* ◇ B60 ◇

GAERDENFORS, P. *Epistemic importance and the logic of theory change* ◇ A05 B60 ◇

HARRAH, D. *A logic of message and reply* ◇ B60 ◇

KANDA, A. *Numeration models of λ-calculus* ◇ B40 B60 C57 D45 ◇

LEBLANC, A. *Axioms for mereology* ◇ B60 ◇

LEBLANC, A. *Investigations in protothetic* ◇ B60 ◇

LEBLANC, A. *New axioms for mereology* ◇ B60 ◇

LEHRER, K. & WAGNER, C.G. *Intransitive indifference: the semi-order problem* ◇ B60 ◇

LEWIS, H.A. *Substitutional quantification and nonstandard quantifiers* ◇ B60 ◇

MAKINSON, D. *How to give it up: a survey of some formal aspects of the logic of theory change* ◇ B60 ◇

MALINOWSKI, G. *Non-fregean logic and other formalizations of propositional identity* ◇ B60 ◇

MALINOWSKI, G. *Notes on sentential logic with identity* ◇ B10 B60 ◇

MCKEE, T.A. *Generalized equivalence: A pattern of mathematical expression* ◇ B60 ◇

MOORE, R.C. *Semantical considerations on nonmonotonic logic* ◇ B60 ◇

MORAWIEC, A. & PIROG-RZEPECKA, K. *New results in logic of formulas which lose sense* ◇ B60 ◇

MUSKARDIN, V. *Logic of guaranty* ◇ B60 ◇
ORLOWSKA, E. *A logic of indiscernibility relations* ◇ B60 ◇
ORLOWSKA, E. *Logic of indiscernibility relations* ◇ B60 ◇
ORLOWSKA, E. *Logic of indiscernibility relations (Russian summary)* ◇ B60 ◇
ORLOWSKA, E. *Logic of nondeterministic information* ◇ B60 ◇
ORLOWSKA, E. *Semantics of nondeterministic possible worlds* ◇ B48 B60 ◇
PARIKH, R. & RAMANUJAM, R. *Distributed processes and the logic of knowledge* ◇ B60 ◇
PETERSON, P.L. *Higher quantity syllogisms* ◇ B60 ◇
RAUSZER, C. *An equivalence between indiscernibility relations in information systems and a fragment of intuitionistic logic* ◇ B60 B75 ◇
RAUSZER, C. *An equivalence between theory of functional dependencies and a fragment of intuitionistic logic (Russian summary)* ◇ B60 B75 ◇
SEPER, K. *Contra-intuitionist logic and symmetric Skolem algebras* ◇ B60 ◇
SIMONS, P.M. *Lesniewski's logic and its relation to classical and free logics* ◇ B60 ◇
STEPIEN, T. *Logic based on atomic entailment* ◇ B22 B60 ◇
TAKANO, M. *A semantical investigation into Lesniewski's axiom and his ontology* ◇ B60 E70 ◇
TALJA, J. *On the logic of omissions* ◇ B60 ◇
VINCENZI, A. *Effective logic II. Expressive power (Italian)* ◇ B60 ◇
VINCENZI, A. *Logica effettiva* L_e ◇ B60 ◇
WALTON, D.N. *New directions in the logic of dialogue* ◇ B60 ◇

B75 Algorithmic and dynamic logic

1971

KRECZMAR, A. *The set of all tautologies of algorithmic logic is hyperarithmetical* ◇ B75 D35 D55 ◇

MIRKOWSKA, G. *On formalized systems of algorithmic logic (Russian summary)* ◇ B75 ◇

1972

KRECZMAR, A. *Degree of recursive unsolvability of algorithmic logic (Russian summary)* ◇ B75 D30 D35 ◇

PERKOWSKA, E. *On algorithmic m-valued logics (Russian summary)* ◇ B75 ◇

PERKOWSKA, E. *On algorithmic m-valued logic* ◇ B75 ◇

1973

RASIOWA, H. *Formalized ω^+ valued algorithmic systems (Russian summary)* ◇ B50 B75 D15 ◇

RASIOWA, H. *On a logical structure of mix-valued programs and the ω^+ valued algorithmic logic (Russian summary)* ◇ B75 ◇

1974

KRECZMAR, A. *Effectivity problems of algorithmic logic* ◇ B75 ◇

MIRKOWSKA, G. *Herbrands theorem in the algorithmic logic* ◇ B75 ◇

RASIOWA, H. *A simplified formalization of ω^+ valued algorithmic logic. A formalized theory of programs* ◇ B75 ◇

RASIOWA, H. *Extended ω^+ valued algorithmic logic. A formalized theory of programs with recursive procedures* ◇ B75 ◇

RASIOWA, H. *On ω^+-valued algorithmic logic and related problems* ◇ B75 D05 ◇

1975

ENGELER, E. *Algorithmic logic* ◇ B25 B75 D15 ◇

RASIOWA, H. *Many-valued algorithmic logic* ◇ B75 G20 ◇

1976

MIRKOWSKA, G. & SALWICKI, A. *A complete characterization of algorithmic properties of block-structured programs with procedures* ◇ B75 ◇

RINE, D.C. *A note regarding multiple-valued algorithmic logics as a tool to investigate programs* ◇ B75 ◇

RINE, D.C. *A survey of multiple-valued algorithmic logics: from a practical point of view* ◇ B50 B75 ◇

1977

BANACHOWSKI, L. & KRECZMAR, L. & MIRKOWSKA, G. & RASIOWA, H. & SALWICKI, A. *An introduction to algorithmic logic* ◇ B75 ◇

BANACHOWSKI, L. *Investigations of properties of programs by means of the extended algorithmic logic. I,II* ◇ B75 ◇

KRECZMAR, A. *Effectivity problems of algorithmic logic* ◇ B75 ◇

MIRKOWSKA, G. *Algorithmic logic and its application in the theory of programs I,II* ◇ B40 B75 ◇

RASIOWA, H. *Algorithmic logic* ◇ B75 G20 ◇

RASIOWA, H. *Completeness theorem for extended algorithmic logic* ◇ B75 ◇

1978

DANKO, W. *Algorithmic properties of programs with tables* ◇ B75 ◇

EMDE BOAS VAN, P. *The connection between modal and algorithmic logic* ◇ B75 ◇

HAREL, D. *Arithmetical completeness in logics of programs* ◇ B75 ◇

HAREL, D. & PRATT, V.R. *Nondeterminism in logics of programs* ◇ B75 ◇

MIRKOWSKA, G. & ORLOWSKA, E. *An elimination of iteration quantifiers in a certain class of algorithmic formulas* ◇ B75 ◇

PARIKH, R. *The completeness of propositional dynamic logic* ◇ B75 ◇

PRATT, V.R. *A practical decision method for propositional dynamic logic* ◇ B75 ◇

1979

CHLEBUS, B.S. *On decidability of propositional algorithmic logic (Polish) (English summary)* ◇ B25 B75 D10 D15 ◇

DANKO, W. *Definability in algorithmic logic* ◇ B75 D05 D20 ◇

FISCHER, MICHAEL J. & LADNER, R.E. *Propositional dynamic logic of regular programs* ◇ B75 D15 ◇

HAREL, D. *First-order dynamic logic* ◇ B75 ◇

HAREL, D. *Recursion in logics of programs* ◇ B75 ◇

MIRKOWSKA, G. *On the propositional algorithmic logic* ◇ B75 ◇

NISHIMURA, H. *Sequential method in propositional dynamic logic* ◇ B75 ◇

PRATT, V.R. *Dynamic logic* ◇ B75 ◇

PRATT, V.R. *Models of program logics* ◇ B75 ◇

RASIOWA, H. *Algorithmic logic and its extensions, a survey* ◇ B75 ◇

RASIOWA, H. *Algorithmic logic. Multiple-valued extensions* ◇ B75 ◇

RASIOWA, H. *Logic of complex algorithms* ◇ B75 ◇

SALWICKI, A. *Algorithmic logics of programs* ◇ B75 ◇

TRAKHTENBROT, B.A. *Completeness of algorithmic logic (Russian)* ◇ B75 ◇

1980

BOWEN, K.A. *Interpolation in loop-free logic* ◇ B75 C40 ◇

BRIM, L. *Algebraic semantics for propositional dynamic logic* ◇ B75 ◇

ERIMBETOV, M.M. *Model-theoretic properties of languages of algorithmic logics (Russian)* ◇ B75 ◇

HAJEK, P. & KALASEK, P. & KURKA, P. *On dynamic logic (Czech)* ◇ B75 ◇

HAREL, D. *Proving the correctness of regular deterministic programs: a unifying survey using dynamic logic* ◇ B75 ◇

KOZEN, D. *A representation theorem for models of *-free PDL* ◇ B75 G25 ◇

MAKOWSKY, J.A. *Measuring the expressive power of dynamic logics: an application of abstract model theory* ◇ B75 C95 ◇

MIRKOWSKA, G. *Algorithmic logic with nondeterministic programs* ◇ B75 ◇

MIRKOWSKA, G. *Model existence theorem in algorithmic logic with nondeterministic programs* ◇ B75 ◇

NEMETI, I. *Some constructions of cylindric algebra theory applied to dynamic algebras of programs* ◇ B75 G15 ◇

ORLOWSKA, E. *Program logic with quantifiable propositional variables* ◇ B75 ◇

PRATT, V.R. *Application of modal logic to programming* ◇ B75 ◇

RASIOWA, H. *Completeness in classical logic of complex algorithms* ◇ B75 ◇

REITERMAN, J. & TRNKOVA, V. *Dynamic algebras which are not Kripke structures* ◇ B75 C90 G25 ◇

VALIEV, M.K. *Decision complexity of variants of propositional dynamic logic* ◇ B25 B75 D15 ◇

1981

BEN-ARI, M. & HALPERN, J.Y. & PNUELI, A. *Finite models for deterministic propositional dynamic logic* ◇ B75 ◇

BERMAN, F. & PATERSON, M.S. *Propositional dynamic logic is weaker without tests* ◇ B75 ◇

CHLEBUS, B.S. *On the computational complexity of satisfiability in propositional logic programs* ◇ B75 D15 ◇

ENGELER, E. (ED.) *Logic of programs* ◇ B75 ◇

HAJEK, P. & KURKA, P. *A second-order dynamic logic with array assignments* ◇ B75 C85 ◇

HAJEK, P. *Making dynamic logic first-order* ◇ B75 ◇

HAREL, D. & PNUELI, A. & STAVI, J. *Propositional dynamic logic of context-free programs* ◇ B75 ◇

KOZEN, D. & PARIKH, R. *An elementary proof of the completeness of PDL* ◇ B75 ◇

KOZEN, D. *On the duality of dynamic algebras and Kripke models* ◇ B75 C90 ◇

LEIVANT, D. *Proof theoretic methodology for propositional dynamic logic* ◇ B75 F99 ◇

MEYER, A.R. & PARIKH, R. *Definability in dynamic logic* ◇ B75 ◇

MEYER, A.R. & MIRKOWSKA, G. & STREETT, R.S. *The deducibility problem in propositional dynamic logic* ◇ B75 D25 D35 ◇

MIRKOWSKA, G. *Algorithmic logic with nondeterministic programs* ◇ B75 ◇

MIRKOWSKA, G. *PAL-propositional algorithmic logic* ◇ B75 ◇

NEMETI, I. *Dynamic algebras of programs* ◇ B75 ◇

NISHIMURA, H. *Arithmetical completeness in first-order dynamic logic for concurrent programs* ◇ B75 ◇

PARIKH, R. *Propositional dynamic logics of programs: A survey* ◇ B75 ◇

PLIUSKEVICIUS, R. *Symmetric dynamic logics (Russian) (English summary)* ◇ B75 ◇

RASIOWA, H. *On logic of complex algorithms* ◇ B75 ◇

REITERMAN, J. & TRNKOVA, V. *On representation of dynamic algebras with reversion* ◇ B75 G25 ◇

SAIN, I. *First order dynamic logic with decidable proofs and workable model theory* ◇ B75 ◇

STREETT, R.S. *Propositional dynamic logic of looping and converse* ◇ B75 ◇

TIURYN, J. *Unbounded program memory adds to the expressive power of first-order dynamic logic* ◇ B75 ◇

1982

AKYLBEKOVA, E.A. *On the completeness and solvability of some program logics (Russian)* ◇ B75 ◇

BEN-ARI, M. & HALPERN, J.Y. & PNUELI, A. *Deterministic propositional dynamic logic: finite models, complexity, and completeness* ◇ B75 ◇

BERGSTRA, J.A. & MEYER, J.-J.C. *On the elimination of iteration quantifiers in a fragment of algorithmic logic* ◇ B75 ◇

BERMAN, F. *Semantics of looping programs in propositional dynamic logic* ◇ B75 ◇

BERMAN, P. & HALPERN, J.Y. & TIURYN, J. *On the power of nondeterminism in dynamic logic* ◇ B75 ◇

CHLEBUS, B.S. *Completeness proofs for some logics of programs* ◇ B75 ◇

CHLEBUS, B.S. *On the computational complexity of satisfiability in propositional logics of programs* ◇ B75 D15 ◇

CHLEBUS, B.S. *On the decidability of propositional algorithmic logic* ◇ B25 B75 ◇

DALEY, R.P. & MANDERS, K.L. *The complexity of the validity problem for dynamic logic* ◇ B75 D15 ◇

FENG, YULIN *A basis for programming logic and a strong termination theorem (Chinese) (English summary)* ◇ B75 ◇

HAREL, D. & PNUELI, A. & STAVI, J. *Further results on propositional dynamic logic of nonregular programs* ◇ B75 ◇

HAREL, D. & SHERMAN, R. *Looping versus repeating in dynamic logic* ◇ B75 ◇

KOZEN, D. *On induction vs. *-continuity* ◇ B75 ◇

MEYER, A.R. & TIURYN, J. *A note on equivalences among logics of programms* ◇ B75 ◇

MEYER, A.R. & WINKLMANN, K. *Expressing program looping in regular dynamic logic* ◇ B75 ◇

MIRKOWSKA, G. *The representation theorem for algorithmic algebras* ◇ B75 ◇

NEMETI, I. *Nonstandard dynamic logic* ◇ B75 ◇

NISHIMURA, H. *Propositional dynamic logic for concurrent programs* ◇ B75 ◇

NISHIMURA, H. *Semantical analysis of constructiove PDL* ◇ B75 ◇
PLIUSKEVICIUS, R. *Symmetrization of a regular and inverse propositional dynamic logic (Russian) (English and Lithuanian summaries)* ◇ B75 ◇
PLIUSKEVICIUS, R. *Two theorems for symmetric propositional dynamic logic (Russian) (English and Lithuanian summaries)* ◇ B75 ◇
PRATT, V.R. *Dynamic logic* ◇ B75 ◇
PRATT, V.R. *Using graphs to understand PDL* ◇ B75 ◇
RASIOWA, H. *Lectures on infinitary logic and logics of programs* ◇ B75 C75 C98 ◇
SEGERBERG, K. *"After" and "during" in dynamic logic* ◇ B75 ◇
SEGERBERG, K. *A completeness theorem in the modal logic of programs* ◇ B75 C90 ◇
STREETT, R.S. *Propositional dynamic logic of looping and converse is elementarily decidable* ◇ B75 ◇

1983

BERGHAMMER, R. & SCHMIDT, GUNTHER *A relational view on gotos and dynamic logic (German summary)* ◇ B75 ◇
BERMAN, F. *Nonstandard models in propositional dynamic logic* ◇ B75 ◇
BERMAN, P. *Deterministic dynamic logic of recursive programs is weaker than dynamic logic* ◇ B75 ◇
BIELA, A. *Introduction to algorithmic logic. Part I (Polish)* ◇ B75 ◇
BROOKES, S.D. & ROUNDS, W.C. *Behavioural equivalence relations induced by programming logics* ◇ B75 ◇
DANKO, W. *Interpretability of algorithmic theories* ◇ B25 B75 C35 F25 ◇
ENJALBERT, P. *Algebraic semantics and program logics: algorithmic logic for program trees* ◇ B75 ◇
ERSHOV, YU.L. *Dynamic logic over admissible sets (Russian)* ◇ B75 C70 ◇
HALPERN, J.Y. & REIF, J.H. *The propositioal dynamic logic of deterministic, well-structured programs* ◇ B75 ◇
HAREL, D. & SHERMAN, R. *Propositional dynamic logic of flowcharts* ◇ B75 ◇
HAREL, D. & PNUELI, A. & STAVI, J. *Propositional dynamic logic of nonregular programs* ◇ B75 ◇
MEYER, A.R. & TIURYN, J. *A note on equivalence among logic of programs* ◇ B75 ◇
MIRKOWSKA, G. *Multimodal logic as a base of algorithmic logic* ◇ B75 ◇
MIRKOWSKA, G. *On the propositional algorithmic theory of arithmetic* ◇ B75 ◇
ORLOWSKA, E. *On some extensions of dynamic logic* ◇ B75 ◇
PETERMANN, U. *On algorithmic logic with partial operations* ◇ B75 ◇
PLIUSKEVICIUS, R. *Some disjunction properties for propositional classical and constructive dynamic logic (Russian) (English and Lithuanian summaries)* ◇ B75 F07 ◇
RASIOWA, H. *Logic LNCA of nondeterministic complex algorithms* ◇ B75 ◇
SAIN, I. *Total correctness in nonstandard dynamic logic* ◇ B75 H10 ◇
STOLBOUSHKIN, A.P. & TAJTSLIN, M.A. *Deterministic dynamic logic is strictly weaker than dynamic logic* ◇ B75 ◇
STOLBOUSHKIN, A.P. *Regular dynamic logic is not interpretable in deterministic context-free dynamic logic* ◇ B75 ◇
STOLBOUSHKIN, A.P. & TAJTSLIN, M.A. *The comparison of the expressive power of first-order dynamic logics* ◇ B75 ◇
URZYCZYN, P. *Deterministic context-free dynamic logic is more expressive than deterministic dynamic logic of regular programs* ◇ B75 ◇
VAKARELOV, D. *Filtration theorem for dynamic algebras with tests and inverse operator* ◇ B75 G25 ◇

1984

BIELA, A. *The program-substitution in algorithmic logic and algorithmic logic with non-deterministic programs* ◇ B75 ◇
DANECKI, R. *Propositional dynamic logic with strong loop predicate* ◇ B75 ◇
DANKO, W. *Algorithmic logic with stacks and its model-theoretic properties* ◇ B75 ◇
FELDMAN, Y.A. & HAREL, D. *A probabilistic dynamic logic* ◇ B75 ◇
HABASINSKI, Z. *Process logics: two decidability results* ◇ B75 ◇
HAREL, D. & PNUELI, A. & SHERMAN, R. *Is the interesting part of process logic uninteresting?: A translation from PL to PDL* ◇ B75 ◇
HAREL, D. & PELEG, D. *On static logics, dynamic logics, and complexity classes* ◇ B75 D15 ◇
KOREN, T. & PNUELI, A. *There exist decidable context free propositional dynamic logics* ◇ B75 ◇
MEYER, A.R. & TIURYN, J. *Equivalences among logics of programs* ◇ B75 ◇
MIRKOWSKA, G. *On a certain property not expressible in PAL* ◇ B75 ◇
PARIKH, R. *Logics of knowledge, games and dynamic logic* ◇ B75 ◇
PLIUSKEVICIUS, R. *On sequential form of functional dynamic logics* ◇ B75 ◇
REITERMAN, J. & TRNKOVA, V. *From dynamic algebras to test algebras* ◇ B75 ◇
SAIN, I. *Structured nonstandard dynamic logic* ◇ B75 ◇
SCHMITT, P.H. *Diamond formulas: A fragment of dynamic logic with recursively enumerable validity problem* ◇ B75 ◇
STOLBOUSHKIN, A.P. *The expressive power of certain dynamic logics (Russian)* ◇ B75 ◇

1985

DANECKI, R. *Nondeterministic propositional dynamic logic with intersection is decidable* ◇ B75 ◇
HAREL, D. & PELEG, D. *More on looping vs. repeating in dynamic logic* ◇ B75 ◇
KFOURY, A.J. *Definability by deterministic and nondeterministic programs (with applications to first-order dynamic logic)* ◇ B75 ◇
LEONE, M. *On temporal invariance of menu systems* ◇ B05 B75 C90 ◇

MAKOWSKY, J.A. & TIOMKIN, M.L. *Propositional dynamic logic with local assignments* ◇ B75 ◇

PASSY, S. & TINCHEV, T. *Quantifiers in combinatory PDL: completeness, definability, incompleteness* ◇ B75 ◇

RADEV, S.R. *Extensions of PDL and consequence relations* ◇ B75 ◇

SAKALAUSKAITE, J. *Completeness theorem for some propositional dynamic logics with infinite repetition (Russian) (English and Lithuanian summaries)* ◇ B75 ◇

STREETT, R.S. *Fixpoints and program looping reductions from the propositional mu-calculus into propositional dynamic logics of looping* ◇ B75 ◇

TINCHEV, T. & VAKARELOV, D. *Propositional dynamic logics with counters and stacks* ◇ B75 ◇

B97 Proceedings

1962

NAGEL, E. & SUPPES, P. & TARSKI, A. (EDS.) *Logic, methodology and philosophy of science* ◇ B97 ◇

1965

CROSSLEY, J.N. & DUMMETT, M. (EDS.) *Formal systems and recursive functions* ◇ B97 D20 D97 ◇

KALMAR, L. (ED.) *Colloque sur les fondements des mathematiques, les machines mathematiques, et leurs applications* ◇ B97 D05 D10 D97 ◇

TYMIENIECKA, A.-T. (ED.) *Contributions to logic and methodology, in honor of I.M.Bochenski* ◇ A05 B97 ◇

1967

COPI, I.M. & GOULD, JAMES A. (EDS.) *Contemporary readings in logical theory* ◇ A05 A10 B97 ◇

RESCHER, N. (ED.) *The logic of decision and action* ◇ A05 B97 ◇

SCHWARTZ, J.T. (ED.) *Mathematical aspects of computer science* ◇ B35 B65 B75 B97 D10 D80 D97 ◇

1968

LOEB, M.H. (ED.) *Proceedings of the summer school in logic, Leeds, 1967* ◇ B97 C97 F97 ◇

NARSKIJ, I.S. (ED.) *Problems in logic and epistemology (Russian)* ◇ A05 B97 ◇

OREVKOV, V.P. (ED.) *The calculi of symbolic logic I (Russian)* ◇ B97 F97 ◇

1970

BRUSENTSOV, N.P. & SHAUMAN, A.M. (EDS.) *Computing techniques and questions of cybernetics No.7 (Russian)* ◇ B50 B97 ◇

PAULI, T. (ED.) *Logic and value* ◇ A05 B97 ◇

TAVANETS, P.V. (ED.) *Nonclassical logic (Russian)* ◇ B97 ◇

TAVANETS, P.V. (ED.) *Studies in systems of logic (Russian)* ◇ B97 ◇

1971

ENGELER, E. (ED.) *Symposium on semantics of algorithmic languages* ◇ B75 B97 D05 ◇

MOISIL, G.C. (ED.) *Logique automatique, informatique* ◇ B75 B97 ◇

1972

BUNGE, MARIO (ED.) *Proceedings of the First Symposium on Exact Philosophy* ◇ B46 B97 ◇

DAVIDSON, D. & HARMAN, G. (EDS.) *Semantics of natural language* ◇ B97 ◇

HINTIKKA, K.J.J. (ED.) *Collection of articles in honor of Rudolf Carnap* ◇ B96 B97 ◇

HODGES, W. (ED.) *Conference in Mathematical Logic -- London '70* ◇ B97 C97 ◇

LEBLANC, H. (ED.) *Truth, syntax and modality* ◇ A05 B46 B97 ◇

1974

ADDISON, J.W. & CHANG, C.C. & CRAIG, W. & HENKIN, L. & SCOTT, D.S. & VAUGHT, R.L. (EDS.) *Proceedings of the Tarski Symposium* ◇ B97 C97 ◇

BARZDINS, J. (ED.) *Theory of algorithms and programs I (Russian)* ◇ B97 D97 ◇

BOCHVAR, D.A. (ED.) *Studies in formalized languages and nonclassical logics (Russian)* ◇ B97 ◇

SMIRNOV, V.A. & TAVANETS, P.V. (EDS.) *Philosphy and logic (Russian)* ◇ A05 B97 ◇

WEINGARTNER, P. (ED.) *Proceedings of a Colloquium on Logic and Ontology* ◇ A05 B97 ◇

1975

BARZDINS, J. (ED.) *Theory of algorithms and programs II (Russian)* ◇ B97 D97 ◇

BOGDAN, R.J. & NIINILUOTO, I. (EDS.) *Logic, language, and probability. A selection of papers contributed to sections IV, VI and XI of logic, methodology and philosophy of science* ◇ B97 ◇

HENKIN, L. & JOJA, A. & MOISIL, G.C. & SUPPES, P. (EDS.) *Logic, methodology and philosophy of science, IV. Proceedings of the fourth international congress for logic, methodology and philosophy of science* ◇ B97 ◇

KANGER, S. (ED.) *Proceedings of the third scandinavian logic symposium* ◇ B97 ◇

KOTAS, J. (ED.) *Stanislaw Jaskowski's achievements in mathematical logic* ◇ A10 B97 ◇

MUELLER, GERT H. & OBERSCHELP, A. & POTTHOFF, K. (EDS.) *ISILC Logic Conference* ◇ B97 ◇

ROSE, H.E. & SHEPHERDSON, J.C. (EDS.) *Proceedings of the Logic Colloquium (Bristol, July, 1973)* ◇ B97 ◇

1976

BOCHVAR, D.A. & GRISHIN, V.N. (EDS.) *Studies in set theory and nonclassical logics. Work collection (Russian)* ◇ B97 E97 ◇

BOGDAN, R.J. (ED.) *Local induction* ◇ A05 B97 ◇

KASHER, A. (ED.) *Language in focus: foundations, methods and systems. Essays in memory of Yehoshua Bar-Hillel* ◇ B97 ◇

KUZNETSOV, A.V. (ED.) *Fourth All-Union Conference on Mathematical Logic (Russian)* ◇ B97 ◇

MAREK, W. & SREBRNY, M. & ZARACH, A. (EDS.) *Set theory and hierarchy theory. A memorial tribute to Andrzej Mostowski* ◇ B97 D55 D97 E97 ◇

1977

ARRUDA, A.I. & CHUAQUI, R.B. & COSTA DA, N.C.A. (EDS.) *Nonclassical logics, model theory and computability. Proceedings of the Third Latin-American Symposium on Mathematical Logic* ◇ B97 C97 ◇

BONOMI, A. & MANGIONE, C. (EDS.) *Recent philosophical logic in Italy* ◇ A05 B97 ◇

BUTTS, R.E. & HINTIKKA, K.J.J. (EDS.) *Proceedings of the fifth international congress of logic, methodology and philosophy of science. I,II,III,IV* ◇ A05 A10 B97 ◇

CZERMAK, J. & MORSCHER, E. & WEINGARTNER, P. (EDS.) *Problems in logic and ontology* ◇ A05 B97 ◇

DUNN, J.M. & EPSTEIN, G. (EDS.) *Modern uses of multiple-valued logic. Invited papers from the fifth international symposium on multiple-valued logic held at Indiana University, Bloomington* ◇ B50 B97 ◇

RINE, D.C. (ED.) *Computer science and multiple-valued logic. Theory and applications* ◇ B50 B97 ◇

SURMA, S.J. (ED.) *On Lesniewski's systems. Proceedings of XXII conference on the history of logic* ◇ A10 B15 B97 ◇

TUOMELA, R. (ED.) *Dispositions* ◇ A05 B97 ◇

1978

ARRUDA, A.I. & CHUAQUI, R.B. & COSTA DA, N.C.A. (EDS.) *Mathematical logic. Proceedings of the first Brazilian conference on mathematical logic* ◇ B97 ◇

BIRYUKOV, B.V. & SPIRKIN, A.G. (EDS.) *Cybernetics and logic. The growth of cybernetical ideas and the development of computing machinery in their mathematical-logical aspects (Russian) (English summary)* ◇ B97 ◇

MACINTYRE, A. & PACHOLSKI, L. & PARIS, J.B. (EDS.) *Logic colloquium '77. Proceedings of the colloquium held in Wroclaw, august 1977* ◇ B97 C97 ◇

NIINILUOTO, I. & TUOMELA, R. (EDS.) *The logic and epistemology of scientific change* ◇ A05 B97 ◇

1979

BORISOV, YU.F. (ED.) *Methodological problems of mathematics (Russian)* ◇ B97 ◇

ERSHOV, A.P. & KNUTH, D.E. (EDS.) *Algorithms in modern mathematics and their applications I,II* ◇ B75 B97 D97 ◇

HILPINEN, R. & HINTIKKA, M.P. & NIINILUOTO, I. & SAARINEN, E. (EDS.) *Essays in honour of Jaakko Hintikka. On the occasion of his fiftieth birthday on January 12, 1979* ◇ B97 ◇

HINTIKKA, K.J.J. & NIINILUOTO, I. & SAARINEN, E. (EDS.) *Essays on mathematical and philosophical logic* ◇ A05 B97 ◇

JENSEN, F.V. & MAYOH, B.H. & MOELLER, K.K. (EDS.) *Proceedings from 5th Scandinavian Logic Symposium, Aalborg, 17-19, January 1979* ◇ B97 ◇

KHOMICH, V.I. & MARKOV, A.A. (EDS.) *Studies in the theory of algorithms and mathematical logic (Russian)* ◇ B97 D97 ◇

MIKHAJLOV, A.I. (ED.) *Studies in nonclassical logics and set theory (Russian)* ◇ B97 ◇

MORSCHER, E. & WEINGARTNER, P. (EDS.) *Ontology and logic* ◇ A05 B97 ◇

SMIRNOV, V.A. (ED.) *Logical inference (Russian)* ◇ B97 ◇

1980

ARRUDA, A.I. & CHUAQUI, R.B. & COSTA DA, N.C.A. (EDS.) *Mathematical logic in Latin America. Proceedings of the IV Latin American symposium on mathematical logic* ◇ B97 ◇

ARRUDA, A.I. & COSTA DA, N.C.A. & SETTE, A.-M. (EDS.) *Proceedings of the third Brazilian Conference on Mathematical logic (UFPe - Recife, December 17-22, 1979)* ◇ B97 ◇

BARWISE, J. & KEISLER, H.J. & KUNEN, K. (EDS.) *The Kleene Symposium* ◇ B97 ◇

BORILLO, M. (ED.) *Representation of knowledge and reason in the human sciences* ◇ B97 ◇

STARCHENKO, A.A. (ED.) *Studies in logical methodology (Russian)* ◇ A05 B97 ◇

YAGER, R.R. *Generalized "and/or" operators for multivalued and fuzzy logic* ◇ B50 B97 ◇

1981

AGAZZI, E. (ED.) *Modern logic - a survey. Historical, philosophical, and mathematical aspects of modern logic and its applications* ◇ A05 A10 B97 ◇

BELTRAMETTI, E.G. & FRAASSEN VAN, B.C. (EDS.) *Current issues in quantum logic* ◇ B51 B97 ◇

DOEMOELKI, B. & GERGELY, T. (EDS.) *Mathematical logic in computer science (Colloquium held in Salgotarjan, Hungary, September 10-15, 1978)* ◇ B97 ◇

MIKELADZE, Z.N. (ED.) *Studies in logic and semantics (Russian)* ◇ B97 ◇

TAJTSLIN, M.A. (ED.) *Investigations in theoretical programming (Russian)* ◇ B97 ◇

WEINKE, K. (ED.) *Logik, Ethik und Sprache. Festschrift fuer Rudolf Freundlich* ◇ B97 ◇

1982

BARTH, E.M. & MARTENS, J.L. (EDS.) *Argumentation* ◇ B97 ◇

CHUPAKHIN, I.YA. & PLOTNIKOV, A.M. (EDS.) *Logic and philosophical categories (Russian)* ◇ A05 B97 ◇

COHEN, L.J. & LOS, J. & PFEIFFER, H. & PODEWSKI, K.-P. (EDS.) *Logic, methodology and philosophy of science VI* ◇ B97 D98 ◇

CRESSWELL, M.J. & GOLDBLATT, R.I. & MEYERHOFF CRESSWELL, M. & SEGERBERG, K. *Symbolic logic. Proceedings of the 1981 Annual Conference of the Australian Association of Symbolic Logic held in Wellington, New Zealand, from 2-5 July 1981* ◇ B97 ◇

ENGELER, E. & LAEUCHLI, H. & STRASSEN, V. (EDS.) *Logic and algorithmic. An International Symposium held in Honnour of Ernst Specker (Zuerich, February 5-11, 1980)* ◇ B97 ◇

HORECKY, J. *COLING 82. Proceedings of the Ninth International Conference on Computational Linguistics, Prague, July 5-10, 1982* ◇ B97 ◇

KARPENKO, A.S. & SMIRNOV, V.A. (EDS.) *Modal and relevant logics (Russian)* ◇ B45 B97 ◇

SOBOLEV, S.L. (ED.) *Mathematical logic and the theory of algorithms (Russian)* ◇ B97 C97 D97 ◇

1983

BERNARDI, C. (ED.) *Atti degli incontri di logica matematica* ◇ B97 C97 ◇

CHONG, C.T. & WICKS, M.J. (EDS.) *Southeast Asian conference on logic* ◇ B97 ◇

COHEN, R.S. & WARTOFSKY, M.W. (EDS.) *Language, logic and method* ◇ B97 ◇

MIKHAJLOV, A.I. (ED.) *Studies in nonclassical logics and formal systems (Russian)* ◇ B97 ◇

NIINILUOTO, I. & ZYGMUNT, J. (EDS.) *Proceedings of the Finnish-Polish-Soviet logic conference* ◇ B97 ◇

1984

ERSHOV, YU.L. & LAVROV, I.A. & PAVILENIS, R.I. & PETROV, V.V. & SMIRNOV, V.A. *Logic, the foundations of mathematics and linguistics (Russian) (English summary)* ◇ B97 ◇

LANDMAN, F. & VELTMAN, F. (EDS.) *Varieties of formal semantics* ◇ B97 ◇

1985

ALCANTARA DE, L.P. (ED.) *Mathematical logic and formal systems. A collection of papers in honor of Professor Newton C.A. da Costa* ◇ B97 ◇

BLOOM, S.L. & WAGNER, E.G. *Many-sorted theories and their algebras with some applications to data types* ◇ B75 B97 ◇

HARRINGTON, L.A. & MORLEY, M.D. & SCEDROV, A. & SIMPSON, S.G. *Harvey Friedman's research on the foundations of mathematics* ◇ B97 C97 D97 E97 F97 ◇

JOUANNAUD, J.-P. (ED.) *Rewriting techniques and applications* ◇ B75 B97 ◇

PRISCO DI, C.A. (ED.) *Methods in mathematical logic. Proceedings of the 6th Latin American Symposium on Mathematical Logic held in Caracas, Venezuela, Aug. 1-6, 1983* ◇ B97 C97 ◇

SAINT-DIZIER, P. (ED.) *Natural language understanding and logic programming* ◇ B65 B75 B97 ◇

B98 Textbooks, surveys

1880

JEVONS, W.S. *Studies in deductive logic: a manual for students* ◇ B98 ◇

1881

VENN, J. *Symbolic logic* ◇ A10 B20 B98 ◇

1884

KEYNES, J.N. *Studies and exercises in formal logic, including a generalisation of logical processes in their application to complex inferences* ◇ B98 ◇

1887

DODGSON, C.L. *The game of logic* ◇ B98 ◇

1889

BAIN, A. *Logic: deductive and inductive* ◇ A05 B98 ◇

1890

SCHROEDER, F.W.K.E. *Vorlesungen ueber die Algebra der Logik (exakte Logik) Vol.1* ◇ A05 A10 B20 B98 G98 ◇

1891

SCHROEDER, F.W.K.E. *Vorlesungen ueber die Algebra der Logik (exakte Logik) Vol.2* ◇ A05 A10 B20 B98 G05 G15 G98 ◇

1894

BURALI-FORTI, C. *Logica matematica* ◇ B98 ◇

1895

DODGSON, C.L. *Symbolic logic. Part I. Elementary* ◇ B20 B98 ◇

SCHROEDER, F.W.K.E. *Vorlesungen ueber die Algebra der Logik (exakte Logik) Vol.3 Algebra und Logik der Relative* ◇ A05 A10 B20 B98 G98 ◇

SIGWART, C. *Logic. Vol.II: logical methods* ◇ B98 ◇

1896

SCHUBERT, H. *Mathematical essays and recreations* ◇ B98 ◇

1902

POINCARE, H. *Du role de l'intuition et de la logique en mathematiques* ◇ A05 B98 ◇

1904

COUTURAT, L. *L'algebre de la logique* ◇ B98 G05 ◇

NATORP, P. *Logik. Grundlegung und logischer Aufbau der Mathematik und mathematischen Naturwissenschaft* ◇ B98 ◇

SIGWART, C. *Logik Vol.1* ◇ B98 ◇

1906

MACCOLL, H. *Symbolic logic and its applications* ◇ B05 B98 ◇

1911

BOSANQUET, B. *Logic. Vol. II* ◇ B98 ◇

WHITEHEAD, A.N. *Introduction to mathematics* ◇ B98 ◇

1912

PADOA, A. *La logique deductive dans sa derniere phase de developpement* ◇ A10 B98 ◇

1918

LEWIS, C.I. *A survey of symbolic logic* ◇ B45 B46 B98 ◇

RIEBER, C.H. *Footnotes to formal logic* ◇ A05 B98 ◇

1926

RAMSEY, F.P. *The foundations of mathematics* ◇ A05 B98 ◇

1927

BEHMANN, H. *Mathematik und Logik* ◇ A05 B98 ◇

1928

HILBERT, D. & ACKERMANN, W. *Grundzuege der theoretischen Logik* ◇ B25 B98 D35 ◇

SKOLEM, T.A. *Ueber die mathematische Logik* ◇ A05 B25 B98 C07 C10 C35 ◇

1929

CARNAP, R. *Abriss der Logistik, mit besonderer Beruecksichtigung der Relationstheorie und ihrer Anwendungen* ◇ A05 B98 ◇

KOTARBINSKI, T. *Elements of epistemology, formal logic, and methodology of the sciences (Polish)* ◇ A05 B98 ◇

LUKASIEWICZ, J. *Elements of mathematical logic (Polish)* ◇ B98 ◇

1930

STEBBING, L.S. *A modern introduction to logic* ◇ B98 ◇

1931

EATON, R.M. *General logic: an introductory survey* ◇ B98 ◇

JEFFREYS, H. *Scientific inference* ◇ A05 B98 ◇

JORGENSEN, J. *A treatise of formal logic: its evolution and main branches, with its relations to mathematics and philosophy. I,II,III* ◇ A05 A10 B98 ◇

1932

BURKAMP, W. *Logik* ◇ B98 ◇

LANGFORD, C.H. & LEWIS, C.I. *Symbolic logic* ◇ B25 B98 C10 C98 ◇

WHITEHEAD, A.N. & RUSSELL, B. *Einfuehrung in die mathematische Logik* ◇ A05 B98 ◇

1933

CHAPMAN, F.M. & HENLE, P. *The fundamentals of logic* ◇ B98 ◇

MACE, C.A. *The principles of logic: an introductory survey* ◇ B98 ◇

1934

COHEN, M.R. & NAGEL, E. *An introduction to logic and scientific method* ◇ B98 ◇

HILBERT, D. & BERNAYS, P. *Grundlagen der Mathematik I* ◇ A05 B98 F05 F30 F98 ◇

QUINE, W.V.O. *A system of logistic* ◇ A05 B15 B98 ◇

ZHU, YANJUN *A summary on mathematical logic (Chinese)* ◇ B98 ◇

1936

CHURCH, A. *A bibliography of symbolic logic* ◇ A10 B98 ◇

CHURCH, A. *Mathematical logic (mimeographed notes)* ◇ B40 B98 D20 ◇

TARSKI, A. *On mathematical logic and the deductive method (Polish)* ◇ B98 ◇

USHENKO, A.P. *The theory of logic* ◇ B98 ◇

WAISMANN, F. *Einfuehrung in das mathematische Denken. Die Begriffsbildung der modernen Mathematik* ◇ A05 B98 ◇

ZHU, YANJUN *Introduction to mathematical logic* ◇ B98 ◇

1937

BOULIGAND, G. *Structure des theories. Problemes infinis* ◇ A05 B98 ◇

GONSETH, F. *Qu'est-ce que la logique* ◇ B98 ◇

LANGER, S.K. *An introduction to symbolic logic* ◇ B98 ◇

LUKASIEWICZ, J. *In defense of logistic (Polish)* ◇ A05 B98 ◇

TARSKI, A. *Einfuehrung in die mathematische Logik und in die Methodologie der Mathematik* ◇ A05 B30 B98 ◇

1938

BOCHENSKI, I.M. *Nove lezioni di logica simbolica* ◇ A05 B98 ◇

1939

BAYLIS, C.A. & BENNETT, A.A. *Formal logic: A modern introduction* ◇ B98 ◇

CARNAP, R. *Foundations of logic and mathematics* ◇ A05 B98 ◇

FEYS, R. *Principes de logistique, premier volume* ◇ A05 B98 ◇

HILBERT, D. & BERNAYS, P. *Grundlagen der Mathematik II* ◇ A05 B98 F05 F15 F30 F40 F98 ◇

REICHENBACH, H. *Introduction a la logistique* ◇ B98 ◇

1940

BETH, E.W. *Einfuehrung in die Philosophie der Mathematik (Dutch)* ◇ A05 B98 ◇

CHURCH, A. *Elementary topics in mathematical logic* ◇ B98 ◇

CHURCHMAN, C.W. *Elements of logic and formal science* ◇ B98 ◇

QUINE, W.V.O. *Mathematical logic* ◇ A05 B98 E70 ◇

1941

LUKASIEWICZ, J. *Die Logik und das Grundlagenproblem* ◇ A05 B98 ◇

QUINE, W.V.O. *Elementary logic* ◇ B98 ◇

USHENKO, A.P. *The problems of logic* ◇ A05 B98 ◇

1942

BETH, E.W. *Summulae logicales. Supplement to formal logic (Dutch)* ◇ A05 B98 ◇

BOLL, M. *Elements de logique scientifique* ◇ A05 B98 ◇

COOLEY, J.C. *A primer of formal logic* ◇ B98 ◇

1943

CARNAP, R. *Formalization of logic* ◇ A05 B98 ◇

STEBBING, L.S. *A modern elementary logic* ◇ B98 ◇

1944

CHURCH, A. *Introduction to mathematical logic. Part I* ◇ B98 ◇

FEYS, R. *Logistic. Formal logic I. General survey. Logic of propositions and classes (Dutch)* ◇ B98 ◇

GOEDEL, K. *Russell's mathematical logic* ◇ A10 B98 ◇

QUINE, W.V.O. *O sentido da nova logica* ◇ B98 ◇

1945

DAVAL, R. & GUILBAUD, G.T. *Le raisonnement mathematique* ◇ A05 B98 ◇

SERRUS, C. *Traite de logique* ◇ A05 B98 ◇

1946

QUINE, W.V.O. *A short course in logic* ◇ B98 ◇

1947

BOOLE, G. *The mathematical analysis of logic, being an essay toward a calculus of deductive reasoning* ◇ A05 A10 B98 G05 ◇

GOEDEL, K. *What is Cantor's continuum problem?* ◇ A05 B98 E50 E98 ◇

LIEBER, L.R. *Mits, wits, and logic* ◇ B60 B98 ◇

REICHENBACH, H. *Elements of symbolic logic* ◇ A05 B98 ◇

1948

AMBROSE, A. & LAZEROWITZ, M. *Fundamentals of symbolic logic* ◇ B98 ◇

LORENZEN, P. *Einfuehrung in die Logik* ◇ B98 ◇

MOSTOWSKI, ANDRZEJ *Logique mathematique. Cours donne a l'universite (Polish)* ◇ B98 ◇

QUINE, W.V.O. *Theory of deduction. Parts I-IV* ◇ B98 ◇

SCHOLZ, H. *Vorlesungen ueber Grundzuege der mathematischen Logik I* ◇ B98 ◇

1949

BLACK, M. *Language and philosophy: studies in method* ◇ A05 B98 ◇

BOCHENSKI, I.M. *Precis de logique mathematique* ◇ B98 ◇

BOURBAKI, N. *Foundations of mathematics for the working mathematician* ◇ A05 B98 ◇

REICHENBACH, H. *The theory of probability* ◇ A05 B98 ◇

SCHOLZ, H. *Vorlesungen ueber Grundzuege der mathematischen Logik II* ◇ B98 ◇

1950

BETH, E.W. *Les fondements logiques des mathematiques* ◇ A05 B98 ◇
MARC-WOGAU, K. *Modern logic. An elementary text-book (Swedish)* ◇ B98 ◇
QUINE, W.V.O. *Methods of logic* ◇ B98 ◇
ROSENBAUM, I. *Introduction to mathematical logic and its applications* ◇ B98 ◇
ROSENBLOOM, P.C. *The elements of mathematical logic* ◇ B98 ◇
SCHMIDT, H.A. *Mathematische Grundlagenforschung* ◇ B98 ◇

1951

BECKER, O. *Einfuehrung in die Logistik, vorzueglich in den Modalkalkuel* ◇ B98 ◇
FREYTAG-LOERINGHOFF BARON VON, B. *Logik I. Das System der reinen Logik und ihr Verhaeltnis zur Logistik* ◇ A05 B98 ◇
GOODSTEIN, R.L. *Constructive formalism: Essays on the foundations of mathematics* ◇ A05 B98 F30 F60 F98 ◇
HENLE, P. & KALLEN, H.M. & LANGER, S.K. (EDS.)*Structure, method, and meaning. Essays in honor of Henry M. Scheffer* ◇ B98 ◇
SESMAT, A. *Logique. vol.II: les raisonnements, la logistique* ◇ A05 B98 ◇
STENIUS, E. *Modern logic (Swedish)* ◇ B98 ◇
WAISMANN, F. *Introduction to mathematical thinking. The formation of concepts in modern mathematics* ◇ B98 ◇

1952

BERKELEY, E.C. *Symbolic logic. Twenty problems and solutions* ◇ B98 ◇
BOOLE, G. *Studies in logic and probability* ◇ A05 B98 ◇
CHURCH, A. *Brief bibliography of formal logic* ◇ A10 B98 ◇
CURRY, H.B. *Lecons de logique algebrique* ◇ B98 G15 G98 ◇
FITCH, F.B. *Symbolic logic. An introduction* ◇ B40 B98 ◇
HERMES, H. & SCHOLZ, H. *Mathematische Logik* ◇ B98 ◇
KLEENE, S.C. *Introduction to metamathematics* ◇ A05 B98 D98 F30 F50 F98 ◇
SKOLEM, T.A. *Consideraciones sobre los fundamentos de la matematica I* ◇ B98 F30 F50 ◇
STRAWSON, P.F. *Introduction to logical theory* ◇ A05 B98 ◇
WILDER, R.L. *Introduction to the foundations of mathematics* ◇ A05 B98 ◇

1953

BASSON, A.H. & O'CONNOR, D.J. *Introduction to symbolic logic* ◇ B98 ◇
COPI, I.M. *Introduction to logic* ◇ A10 B98 ◇
ROSSER, J.B. *Logic for mathematicians* ◇ B15 B98 E70 E98 ◇
SKOLEM, T.A. *Consideraciones sobre los fundamentos de la matematica II* ◇ B98 F30 F50 ◇
STABLER, E.R. *An introduction to mathematical thought* ◇ A05 B98 ◇

1954

ALBRECHT, W. *Die Logik der Logistik* ◇ B98 ◇
BOCHENSKI, I.M. & MENNE, A. *Grundriss der Logistik* ◇ B98 ◇
CARNAP, R. *Einfuehrung in die symbolische Logik mit besonderer Beruecksichtigung ihrer Anwendungen* ◇ A05 B30 B98 ◇
COPI, I.M. *Symbolic logic* ◇ B98 ◇
CURRY, H.B. *Remarks on the definition and nature of mathematics* ◇ A05 B98 ◇
DUERR, K. *Lehrbuch der Logistik* ◇ B98 ◇
JOHNSTONE JR., H.W. *Elementary deductive logic* ◇ B98 ◇
JUHOS VON, B. *Elemente der neuen Logik* ◇ B98 ◇
KREISEL, G. *Applications of mathematical logic to various branches of mathematics* ◇ B98 C65 ◇
KREYCHE, R.J. *Logic for undergraduates* ◇ B98 ◇
SKOLEM, T.A. *Results in investigations in the foundations (Norwegian)* ◇ A05 B98 D03 F55 ◇

1955

BETH, E.W. *Die Stellung der Logik im Gebaeude der heutigen Wissenschaft* ◇ B98 ◇
BLANCHE, R. *L'axiomatique* ◇ B98 ◇
GRENIEWSKI, H. *Elements of formal logic (Polish)* ◇ B98 ◇
GRZEGORCZYK, A. & JASKOWSKI, S. & LOS, J. & MAZUR, S. & MOSTOWSKI, ANDRZEJ & RASIOWA, H. & SIKORSKI, R. *The present state of investigations on the foundations of mathematics (Polish)* ◇ A10 B98 E98 ◇
LEBLANC, H. *An introduction to deductive logic* ◇ B98 ◇
LORENZEN, P. *Einfuehrung in die operative Logik und Mathematik* ◇ A05 B98 F50 F65 F98 ◇
MILLER, JAMES WILKIN *Exercises in introductory symbolic logic* ◇ B98 ◇
MOSAHEB, G.-H. *Introduction to formal logic (Iranian)* ◇ B98 ◇
PRIOR, A.N. *Formal logic* ◇ A05 A10 B98 E30 ◇
WANG, HAO *On formalization* ◇ A05 B98 ◇

1956

BETH, E.W. *L'existence en mathematiques* ◇ A05 B98 F55 ◇
BOCHENSKI, I.M. *Formale Logik* ◇ A10 B98 ◇
KEMENY, J.G. *Semantics as a branch of logic* ◇ B98 ◇
STAHL, G. *Introduccion a la logica simbolica* ◇ B98 ◇

1957

BLANCHE, R. *Introduction a la logique contemporaine* ◇ A05 B98 F50 ◇
BLYTH, J.W. *A modern introduction to logic* ◇ B98 ◇
BRENNAN, J.G. *A handbook of logic* ◇ B98 ◇
CHAUVINEAU, J. *La logique moderne* ◇ B98 ◇
CURRY, H.B. *Combinatory logic* ◇ B40 B98 ◇
DUBARLE, H.D. *Initiation a la logique* ◇ B98 ◇
GOODSTEIN, R.L. *Mathematical logic* ◇ B98 ◇
HERMES, H. *Einfuehrung in die mathematische Logik. Klassische Praedikatenlogik* ◇ B10 B98 ◇

HU, SHIHUA *Mathematical logic, its fundamental properties and scientific significance (Chinese)* ◇ B98 ◇
LADRIERE, J. *Les limitations internes des formalismes. Etude sur la signification du theoreme de Goedel et des theoremes apparentes dans la theorie des fondements des mathematiques* ◇ A05 B98 F30 F98 ◇
NIDDITCH, P.H. *Introductory formal logic of mathematics* ◇ B98 ◇
PASQUINELLI, A. *Introduzione alla logica simbolica* ◇ B98 ◇
SUPPES, P. *Introduction to logic* ◇ B98 ◇

1958

AJDUKIEWICZ, K. *Abriss der Logik* ◇ B98 ◇
BAR-HILLEL, Y. & DALEN VAN, D. & FRAENKEL, A.A. & LEVY, A. *Foundations of set theory* ◇ A05 B98 E30 E70 E98 F98 ◇
BERNAYS, P. & FRAENKEL, A.A. *Axiomatic set theory* ◇ B98 E98 ◇
BOOLE, G. *An investigation of the laws of thought, on which are founded the mathematical theories of logic and probabilities* ◇ A05 B98 G05 ◇
CHRISTIAN, R.R. *Introduction to logic and sets* ◇ B98 E98 ◇
CULBERTSON, J.T. *Mathematics and logic for digital devices* ◇ B98 G05 ◇
CURRY, H.B. & FEYS, R. *Combinatory logic. Volume I* ◇ B40 B98 ◇
HEYTING, A. *Intuitionism in mathematics* ◇ A05 B98 F50 F55 F98 ◇
MILLER, JAMES WILKIN *Logic workbook* ◇ B98 ◇
ROBINSON, A. *Outline of an introduction to mathematical logic I,II,III* ◇ B98 C07 G05 ◇
WANG, HAO *Eighty years of foundational studies (German and French summaries)* ◇ A05 A10 B98 ◇

1959

BETH, E.W. *The foundation of mathematics. A study in the philosophy of science* ◇ A05 B98 ◇
BOCHENSKI, I.M. *A precis of mathematical logic* ◇ B98 ◇
DAVIS, MARTIN D. *Lecture notes on mathematical logic* ◇ B98 ◇
EXNER, R.M. & ROSSKOPF, M.F. *Logic in elementary mathematics* ◇ B98 ◇
FRAENKEL, A.A. *Mengenlehre und Logik* ◇ B98 E98 ◇
NOVIKOV, P.S. *Elements of mathematical logic (Russian)* ◇ B98 ◇
POPOV, A.I. *Introduction to mathematical logic (Russian)* ◇ B98 ◇
ROBINSON, A. *Outline of an introduction to mathematical logic IV* ◇ B10 B98 C07 C98 G05 ◇
SCHIPPER, E.W. & SCHUH, E. *A first course in modern logic* ◇ B98 ◇

1960

CASARI, E. *Lineamenti di logica matematica* ◇ B98 ◇
HALBERSTADT, W.H. *Introduction to modern logic: an elementary textbook of symbolic logic* ◇ B98 ◇
KLEENE, S.C. *Mathematical logic: constructive and non-constructive operations* ◇ B98 D98 ◇
LEWIS, C.I. *A survey of symbolic logic (Reprinted edition)* ◇ B45 B98 ◇
LU, YANGCI & LU, ZHONGWAN & TANG, ZHISONG & WAN, ZHEXIAN *Mathematical logic and its application (Chinese)* ◇ B98 ◇
PRIJATELJ, N. *Introduction to mathematical logic (Slovenian)* ◇ B98 ◇
RUBY, L. *Logic: An introduction* ◇ B98 ◇
SCHMIDT, H.A. *Mathematische Gesetze der Logik. I. Vorlesungen ueber Aussagenlogik* ◇ B05 B98 F50 F98 ◇
SCHUETTE, K. *Beweistheorie* ◇ B98 F05 F15 F30 F35 F98 ◇
SUGIHARA, T. *Modern logic (Japanese)* ◇ B98 ◇

1961

ALLEN, L.E. & BROOKS, R.B.S. & DICKOFF, J.W. & JAMES, P.A. *The ALL project (accelerated learning of logic)* ◇ B98 ◇
AMBROSE, A. & LAZEROWITZ, M. *Logic: the theory of formal inference* ◇ B98 ◇
FREUDENTHAL, H. *Exact logic (Dutch)* ◇ A05 B98 ◇
GRZEGORCZYK, A. *An outline of mathematical logic (Polish)* ◇ B98 ◇
GRZEGORCZYK, A. *Logika popularna* ◇ B98 ◇
LEE, H.N. *Symbolic logic* ◇ B98 ◇
LUKASIEWICZ, J. *Logistic and philosophy (Polish)* ◇ A05 B98 ◇
MUNOZ, V. *De la axiomatica a los sistemas formales* ◇ B30 B98 ◇
SCHOLZ, H. & HASENJAEGER, G. *Grundzuege der mathematischen Logik* ◇ B98 ◇
SMULLYAN, R.M. *Theory of formal systems* ◇ B98 D03 D05 D20 D25 D98 ◇
STOLL, R.R. *Sets, logic and axiomatic theories* ◇ B98 E98 ◇

1962

ANDERSON, J.M. & JOHNSTONE JR., H.W. *Natural deduction. The logical basis of axiom systems* ◇ B98 F07 F98 ◇
ARNOLD, B.H. *Logic and boolean algebra* ◇ B05 B98 G05 ◇
BADAWI, A.R. *Formale und mathematische Logik (Arabic)* ◇ B98 ◇
BETH, E.W. *Formal methods. An introduction to symbolic logic and to the study of effective operations in arithmetic and logic* ◇ B28 B98 F30 F50 F98 ◇
CLARK, R. & WELSH JR., P.J. *Introduction to logic* ◇ B98 ◇
DOPP, J. *Logiques construites par une methode de deduction naturelle* ◇ B98 F05 F07 F50 F98 ◇
HASENJAEGER, G. *Einfuehrung in die Grundbegriffe und Probleme der modernen Logik* ◇ A05 A10 B98 ◇
KAESBAUER, M. & KUTSCHERA VON, F. (EDS.) *Logik und Logikkalkuel* ◇ B98 ◇
LORENZEN, P. *Metamathematik* ◇ A05 B98 C60 D20 D35 D98 F98 ◇
MITCHELL, D. *An introduction to logic* ◇ B98 ◇
POGORZELSKI, W.A. & SLUPECKI, J. *On mathematical proof (Polish)* ◇ B98 ◇

SCHUETTE, K. *Lecture notes in mathematical logic. Vol.I* ◇ B98 F15 ◇
SMULLYAN, A. *Fundamentals of logic* ◇ B98 ◇
ZINOV'EV, A.A. *Propositional logic and theory of deduction (Russian)* ◇ B98 ◇

1963

AJZERMAN, M.A. & GUSEV, L.A. & ROZONOEHR, L.I. & SMIRNOVA, I.M. & TAL', A.A. *Logic, automata and algorithms (Russian)* ◇ B98 D05 D10 D98 ◇
BLYTH, J.W. & JACOBSON JR., J.H. *Class logic. A programmed text* ◇ B98 ◇
BORKOWSKI, L. & SLUPECKI, J. *Elements of mathematical logic and set theory (Polish)* ◇ A05 B98 E98 ◇
BRAUN, E.L. *Digital computer design: logic, circuitry, and synthesis* ◇ B70 B98 ◇
CURRY, H.B. *Foundations of mathematical logic* ◇ B98 ◇
HARBECK, G. *Einfuehrung in die formale Logik* ◇ B98 ◇
HILTON, A.M. *Logic, computing machines, and automation* ◇ A05 B98 D05 ◇
KHARIN, N.N. *Mathematical logic and the theory of sets (Russian)* ◇ A05 B98 ◇
KNEEBONE, G.T. *Mathematical logic and the foundations of mathematics: an introduction* ◇ A05 A10 B98 ◇
MO, SHAOKUI *Notes on mathematical logic (Chinese)* ◇ B98 ◇
MOURANT, J.A. *Formal logic: an introductory text book* ◇ B98 ◇
QUINE, W.V.O. *Set theory and its logic* ◇ A10 B98 E30 E98 ◇
RASIOWA, H. & SIKORSKI, R. *The mathematics of metamathematics* ◇ B98 C98 F50 F98 G05 G10 G98 ◇
ROBINSON, A. *Introduction to model theory and to the metamathematics of algebra* ◇ B98 C60 C98 H15 ◇
SALMON, W.C. *Logic* ◇ B98 ◇
SCHUETTE, K. *Lecture notes in mathematical logic. Vol.II* ◇ B98 F15 ◇
SMULLYAN, R.M. *First order logic* ◇ B98 F07 F98 ◇
STOLL, R.R. *Set theory and logic* ◇ B98 E98 ◇
SULLIVAN, D.J. *Fundamentals of logic* ◇ B98 ◇
WANG, HAO *The predicate calculus* ◇ B98 ◇

1964

AGAZZI, E. *La logica simbolica* ◇ A05 B98 ◇
ANGELL, R.B. *Reasoning and logic* ◇ B98 ◇
BAR-HILLEL, Y. *Language and information: selected essays on their theory and application* ◇ A05 B65 B98 ◇
BLACK, M. *Critical thinking: an introduction to logic and scientific method* ◇ A05 B98 ◇
BROWNE, S.S.S. *Fundamentals of deductive logic* ◇ B98 ◇
CARNEY, J.D. & SCHEER, R.K. *Fundamentals of logic* ◇ B98 ◇
DEVIDE, V. *Mathematical logic. Part I (The classical logic of propositions) (Croatian) (English summary)* ◇ B05 B98 ◇
DINKINES, F. *Elementary concepts of modern mathematics. Part 1. Elementary theory of sets. Part 2. Introduction to mathematical logic* ◇ B98 E98 ◇
FISK, M. *A modern formal logic* ◇ B98 ◇
HILL, S. & SUPPES, P. *First course in mathematical logic* ◇ B98 ◇
KALISH, D. & MONTAGUE, R. *Logic. Techniques of formal reasoning* ◇ B98 ◇
KEENE, G.B. *First-order functional calculus* ◇ B98 ◇
KLAUS, G. *Moderne Logik. Abriss der formalen Logik* ◇ B98 ◇
LIGHTSTONE, A.H. *The axiomatic method: an introduction to mathematical logic* ◇ B98 ◇
MANGIONE, C. *Elementi di logica matematica* ◇ B98 ◇
MENDELSON, E. *Introduction to mathematical logic* ◇ B98 ◇
RESCHER, N. *Introduction to logic* ◇ A05 B98 ◇
STAHL, G. *Elementos de la metalogica y metamatematica* ◇ B98 ◇

1965

AAQVIST, L. *A new approach to the logical theory of interrogatives. Part I. Analysis* ◇ A05 B98 ◇
AJDUKIEWICZ, K. *Pragmatic logic (Polish)* ◇ A05 B98 ◇
BARKER, S.F. *The elements of logic* ◇ B98 ◇
BARLINGAY, S.S. *A modern introduction to Indian logic* ◇ A10 B98 ◇
BINFORD, F. *Solution to the exercises in: "First course in mathematical logic"* ◇ B98 ◇
BROWN, P.L. & STUBERMAN, W.E. *Elementary modern logic* ◇ B98 ◇
CHOMSKY, N. *Aspects of the theory of syntax* ◇ B98 ◇
DICKOFF, J.W. & JAMES, P.A. *Symbolic logic and language: a programmed text* ◇ B98 ◇
EVES, H.W. *An introduction to the foundations and fundamental concepts of mathematics* ◇ B98 ◇
HU, SZETIEN *Threshold logic* ◇ B70 B98 ◇
HUGHES, G.E. & LONDEY, D.G. *The elements of formal logic* ◇ B98 ◇
JEFFREY, R.C. *The logic of decision* ◇ B48 B98 ◇
KREISEL, G. *Mathematical logic* ◇ B98 F10 F35 F50 F98 ◇
KRETZMANN, N. *Elements of formal logic* ◇ B98 ◇
LEMMON, E.J. *Beginning logic* ◇ B98 ◇
LUCHINS, A.S. & LUCHINS, E.H. *Logical foundation of mathematics for behavioral scientists* ◇ B98 ◇
MATES, B. *Elementary logic* ◇ A10 B98 ◇
MO, SHAOKUI *Introduction to mathematical logic (Chinese)* ◇ B98 ◇
MOSTOWSKI, ANDRZEJ *Thirty years of foundational studies. Lectures on the development of mathematical logic and the studies of the foundations of mathematics in 1930-1964* ◇ A10 B98 C98 D98 E98 F98 ◇
PRAWITZ, D. *Natural deduction. A proof-theoretical study* ◇ B15 B98 F05 F07 F50 F98 ◇

1966

ACKERMANN, R.J. *Non-deductive inference* ◇ B98 ◇
CAPALDI, N. *Introduction to deductive logic* ◇ B98 ◇
CORNMAN, J.W. *Metaphysics, reference and language* ◇ A05 B98 ◇
DUTTON, J.D. (ED.) *Logics: an introduction with exercises* ◇ B98 ◇

ESSLER, W.K. *Einfuehrung in die Logik* ◇ B98 ◇
GRASSMANN, G. *Die Formenlehre oder Mathematik. 1.die Groessenlehre. 2.die Begriffslehre der Logik. 3.die Bindelehre oder Kombinationslehre. 4.die Zahlenlehre oder Arithmetik. 5.die Aussenlehre oder Ausdehnungslehre (reprinted)* ◇ B98 ◇
HACKSTAFF, L.H. *Systems of formal logic* ◇ B98 ◇
HASSE, M. *Grundbegriffe der Mengenlehre und Logik* ◇ B98 E98 ◇
KORFHAGE, R. *Logic and algorithms with applications to the computer and information sciences* ◇ B98 D98 ◇
KUPPERMAN, J. & MCGRADE, A.S. *Fundamentals of logic* ◇ B98 ◇
LEBLANC, H. *Techniques of deductive inference* ◇ B98 ◇
MAEHARA, S. *Introduction to mathematical logic (Japanese)* ◇ B98 ◇
MENNE, A. *Einfuehrung in die Logik* ◇ B98 ◇
ROBINSON, A. *Non-standard analysis* ◇ A10 B98 H05 H10 H15 H98 ◇
SKYRMS, B. *Choice and chance* ◇ A05 B48 B98 ◇
THOMAS, N.L. *Modern logic. An introduction* ◇ B98 ◇

1967

ACKERMANN, R.J. *An introduction to many-valued logics* ◇ B50 B98 ◇
ACKERMANN, R.J. *Introduction to many-valued logics* ◇ B50 B98 ◇
BLUMBERG, A.E. *Modern logic* ◇ A05 B98 ◇
CANGELOSI, V.E. *Compound statements and mathematical logic* ◇ B98 ◇
COATES, C.L. & LEWIS II, P.M. *Threshold logic* ◇ B70 B98 ◇
COPI, I.M. & GOULD, JAMES A. *Deontic logic, introduction* ◇ B45 B98 ◇
CURRY, H.B. *Combinatory logic* ◇ A05 B40 B98 ◇
DOPP, J. *Notions de logique formelle* ◇ B98 ◇
FRAISSE, R. *Cours de logique mathematique. Tome I. Relation, formule logique, compacite, completude* ◇ B98 C98 ◇
HAMBLIN, C.L. *Elementary formal logic* ◇ B98 ◇
JEFFREY, R.C. *Formal logic: Its scope and limits* ◇ B98 ◇
KENELLY, J.W. *Informal logic* ◇ B98 ◇
KILMISTER, C.W. *Language, logic and mathematics* ◇ A05 B98 D35 ◇
KLEENE, S.C. *Mathematical logic* ◇ B98 D20 D25 D55 D98 F30 F98 ◇
KREISEL, G. & KRIVINE, J.-L. *Elements de logique mathematique. Theorie des modeles* ◇ B98 C98 ◇
KUTSCHERA VON, F. *Elementare Logik* ◇ B98 ◇
LEONARD, H.S. *Principles of reasoning: an introduction to logic, methodology, and the theory of signs* ◇ A05 B98 ◇
MARGARIS, A. *First order mathematical logic* ◇ B98 ◇
NEIDORF, R. *Deductive forms: An elementary logic* ◇ B98 ◇
PONASSE, D. *Logique mathematique. Elements de base: calcul propositionnel, calcul des predicats* ◇ B98 ◇
RIEGER, L. *Algebraic methods of mathematical logic* ◇ A05 B98 G05 G15 ◇
ROURE, M.-L. *Elements de logique contemporaine* ◇ B98 ◇
SCHOCK, R. *Logik* ◇ B98 ◇
SHOENFIELD, J.R. *Mathematical logic* ◇ B98 C98 D98 E98 F98 ◇
SMIRNOVA, E.D. & TAVANETS, P.V. *Semantics in logic (Russian)* ◇ A05 B98 ◇
TERRELL, D.B. *Logic: A modern introduction to deductive reasoning* ◇ B98 ◇

1968

CHOMSKY, N. *Language and mind* ◇ A05 B65 B98 ◇
CRESSWELL, M.J. & HUGHES, G.E. *An introduction to modal logic* ◇ B45 B98 ◇
EHLERS, F. *Logic by way of set theory* ◇ B98 E98 ◇
HATCHER, W.S. *Foundations of mathematics* ◇ B98 E30 E70 E98 G30 ◇
HERMES, H. *Methodik der Mathematik und Logik* ◇ A05 A10 B98 ◇
ISEMINGER, G. *An introduction to deductive logic* ◇ B98 ◇
KAC, M. & ULAM, S.M. *Mathematics and logic: Retrospect and prospects* ◇ A05 A10 B98 ◇
KILGORE, W.J. *An introductory logic* ◇ B98 ◇
KRUGER, A.N. & MANICAS, P.T. *Essentials of logic* ◇ B98 ◇
PENZOV, YU.E. *Elements of mathematical logic and set theory (Russian)* ◇ B98 E98 ◇
ROOTSELAAR VAN, B. & STAAL, J.F. (EDS.) *Logic, methodology and philosophy of science III* ◇ B98 ◇
SCHAGRIN, M.L. *The language of logic. A programmed text* ◇ B98 ◇
SUMMERS, G.J. *New puzzles in logical deduction* ◇ B98 ◇
ZOLL, E.J. *Logic: A programmed text for 2-valued and 3-valued logics* ◇ B50 B98 ◇

1969

CHELLAS, B.F. *The logical form of imperatives* ◇ A05 B98 ◇
DODGE, C.W. *Sets,logic and numbers* ◇ B98 E98 ◇
ENNIS, R.H. *Ordinary logic* ◇ B98 ◇
FEYS, R. & FITCH, F.B. *Dictionary of symbols of mathematical logic* ◇ B98 ◇
FITTING, M. *Intuitionistic logic, model theory, and forcing* ◇ B98 C90 C98 E25 E35 E45 E50 F50 F98 ◇
FOULIS, D.J. *Fundamental concepts of mathematics* ◇ B98 ◇
GRIZE, J.-B. *Logique moderne. fasc.I: Logique des propositions et des predicats, deduction naturelle* ◇ B98 ◇
HIRSCHFELD, J. & MACHOVER, M. *Lectures on non-standard analysis* ◇ B98 H05 ◇
KAHANE, H. *Logic and philosophy. A modern introduction* ◇ B98 ◇
KEARNS, J.T. *Deductive logic: a programmed introduction* ◇ B98 ◇
KLEENE, S.C. *The new logic* ◇ B98 ◇
MICHALOS, A.C. *Principles of logic* ◇ B98 ◇
MIHAILESCU, E.G. *Mathematical logic. Elements of propositional and predicate calculus (Romanian)* ◇ B98 ◇
NAESS, A. *Introduction to logic and scientific method* ◇ B98 ◇

POLLOCK, J.L. *Introduction to symbolic logic* ◇ B98 ◇
RESCHER, N. *Topics in philosophical logic* ◇ A05 B45 B98 C90 ◇
ROBBIN, J.W. *Mathematical logic: a first course* ◇ B98 ◇
ROBISON, G.B. *An introduction to mathematical logic* ◇ B98 ◇
SHENG, C.L. *Threshold logic* ◇ B75 B98 ◇

1970

ACKERMANN, R.J. *Modern deductive logic* ◇ B98 ◇
AVENOSO, F.J. & CHEIFETZ, P.M. *Logic and set theory* ◇ B98 E98 ◇
BEACH, J. *Introduction to logic* ◇ B98 ◇
BETH, E.W. *Aspects of modern logic* ◇ B98 ◇
BITTINGER, M.L. *Logic and proof* ◇ B98 ◇
CARNEY, J.D. *Introduction to symbolic logic* ◇ B98 ◇
CASEY, H. & CLARK, M. *Logic: a practical approach* ◇ B98 ◇
COHEN, L.J. *The implications of induction* ◇ A05 B48 B98 ◇
DELONG, H. *A profile of mathematical logic* ◇ A05 A10 B98 ◇
DETLOVS, V.K. *Elements of mathematical logic and set theory (Latvian)* ◇ B98 E98 ◇
FIORENTINI, M. & MARRUCCELLI, A. *Complementi di matematiche moderne; Logica matematicea, teorie degli insiemi, strutture algebriche* ◇ B98 E98 ◇
LAVROV, I.A. *Logic and algorithms (Russian)* ◇ B98 D98 ◇
LAVROV, I.A. & MAKSIMOVA, L.L. *Problems in logic (Russian)* ◇ B98 E98 ◇
LORENZEN, P. *Logica formal* ◇ B98 ◇
MASSEY, G.J. *Understanding symbolic logic* ◇ A05 B98 ◇
MATSUMOTO, K. *Mathematical logic (Japanese)* ◇ B98 ◇
MINTS, G.E. (ED.) *Mathematical logic (Russian)* ◇ A10 B98 ◇
MITCHELL, D. *An introduction to logic* ◇ B98 ◇
MOSTOWSKI, A.WLODZIMIERZ & PAWLAK, Z. *Logik fuer Ingenieure (Polnisch)* ◇ B98 ◇
OBERSCHELP, A. *Mengenlehre und Logik* ◇ B98 E98 ◇
PANCZAKIEWICZ, M. *Mathematical logic (Polish)* ◇ B98 ◇
RENYI, A. *Foundations of probability* ◇ B98 ◇
RESNIK, M.D. *Elementary logic* ◇ B98 ◇
STOLYAR, A.A. *Introduction to elementary mathematical logic* ◇ B98 ◇
TAJTSLIN, M.A. *Model theory (Russian)* ◇ B98 C98 ◇
THOMASON, R.H. *Symbolic logic. An introduction* ◇ B98 ◇

1971

ALCHOURRON, C.E. & BULYGIN, E. *Normative systems* ◇ B45 B98 ◇
ALTHAM, J.E.J. *The logic of plurality* ◇ A05 B98 C80 ◇
BIZAM, G. & HERCZEG, J. *Play and logic (Hungarian)* ◇ B98 ◇
BREITKOPF, A. & KUTSCHERA VON, F. *Einfuehrung in die moderne Logik* ◇ B98 ◇
COMBES, M. *Fondements des mathematiques* ◇ A05 A10 B98 E98 ◇
FRAASSEN VAN, B.C. *Formal semantics and logic* ◇ B98 ◇
GEORGESCU, G. *Praedikatenkalkuel I* ◇ B98 ◇
GLUSHKOV, V.M. *Einfuehrung in die technische Kybernetik Band 1, 2* ◇ B98 ◇
GRIZE, J.-B. *Logique moderne. fasc.II: Logique des propositions et des predicats, tables de verite et axiomatisation* ◇ B98 ◇
HANNA, S.C. & SABER, J.C. *Sets and logic* ◇ B98 E98 ◇
HUNTER, G. *Metalogic: an introduction to the metatheory of standard firstorder logic* ◇ B10 B98 ◇
KNEEBONE, G.T. *Mathematical logic in relation to ordinary mathematics* ◇ A05 B98 ◇
ONICESCU, O. *Principes de logique et de philosophie mathematique* ◇ A05 B98 ◇
POSPESEL, H. *Arguments: deductive logic exercises* ◇ B98 ◇
PURTILL, R.L. *Logic for philosophers* ◇ B98 ◇
RASIOWA, H. *Introduction to modern mathematics (Polish)* ◇ B30 B98 C05 E98 ◇
RESCHER, N. & URQUHART, A.I.F. *Temporal logic* ◇ A05 B45 B98 ◇
ROGERS, R. *Mathematical logic and formalized theories. A survey of basic concepts and results* ◇ B98 C98 ◇
SABBAGH, G. *Logique mathematique I. Generalites* ◇ B98 ◇
STANDLEY, G.B. *New methods in symbolic logic* ◇ B98 ◇
YASUHARA, A. *Recursive function theory and logic* ◇ B98 D10 D20 D98 ◇

1972

ASH, C.J. & BRICKHILL, C.J. & CROSSLEY, J.N. & STILLWELL, J.C. & WILLIAMS, N.H. *What is mathematical logic ?* ◇ B98 ◇
BALLARD, K.E. *Study guide for Copi: Introduction to logic. A self-instructional supplement* ◇ B98 ◇
BERNARDO DI, G. *Introduzione alla logica dei sistemi normativi* ◇ B98 ◇
BRODSKIJ, I.N. *Elementare Einfuehrung in die Logik. 2.ueberarb. Aufl (Russian)* ◇ B98 ◇
BULL, R.A. *Mathematical logic* ◇ B98 ◇
CHOMSKY, N. *Studies on semantics in generative grammars* ◇ B65 B98 D05 ◇
DALEN VAN, D. *Logik und formale Theorien (Niederlaendisch)* ◇ B98 ◇
EMDE, H. & REYERSBACH, W. & STROMBACH, W. *Mathematische Logik. Ihre Grundprobleme in Theorie und Anwendung* ◇ B98 G05 G10 ◇
ENDERTON, H.B. *A mathematical introduction to logic* ◇ B98 C98 F30 ◇
FRAISSE, R. *Cours de logique mathematique. Tome 2: Theorie des modeles* ◇ B98 C98 ◇
GRADSHTEJN, I.S. *Direct and inverse theorems (Russian)* ◇ B98 ◇
GUTTENPLAN, S.D. & TAMNY, M. *Logic: a comprehensive introduction* ◇ A05 B98 ◇
HACKING, I. *A concise introduction to logic* ◇ B98 ◇
HERINGER, H.J. *Formale Logik und Grammatik* ◇ B65 B98 D05 ◇

HINDLEY, J.R. & LERCHER, B. & SELDIN, J.P. *Introduction to combinatory logic* ◇ B40 B98 F98 ◇
KLOETZER, G. & RAUTENBERG, W. *Im Grenzbereich Algebra, Logik, Maschinen. 10 Jahre Forschungsarbeit der Nowosibirsker Schule A.I. Malcev's (eine Studie)* ◇ A10 B98 C05 D05 D98 ◇
LEBLANC, H. & WISDOM, W.A. *Deductive logic* ◇ B98 ◇
MAREK, W. & ONYSZKIEWICZ, J. *Elements of logic and foundations of mathematics in problems (Polish)* ◇ B98 C98 D98 E98 ◇
MARKWALD, W. *Einfuehrung in die formale Logik und Metamathematik* ◇ B98 ◇
OTEPANOV, V.I. & VERKHOZIN, O.M. *Problems in logic (Russian)* ◇ B98 E98 ◇
PORTE, J. *La logique mathematique et le calcul mecanique* ◇ B98 D98 ◇
PURTILL, R.L. *Logical thinking* ◇ B98 ◇
RASIOWA, H. *Introduction to set theory and mathematical logic (Bulgarian)* ◇ B98 C05 E98 ◇
RAUTENBERG, W. *Zur praktischen und theoretischen Wirksamkeit der mathematischen Logik* ◇ B98 ◇
SEREBRYANNIKOV, O.F. *Heuristic principles and logical calculi* ◇ B98 ◇
SMIRNOV, V.A. *Formal derivation and logical calculi (Russian)* ◇ B98 C07 C40 C98 F05 F07 F98 ◇
STEEN, S.W.P. *Mathematical logic, with special reference to the natural numbers* ◇ B98 F30 F98 ◇

1973

BAGINSKI, M. & GLAEWE, W. & GOLL, P. & LIST, G. & LOESCHAU, G. & MERTENS, A. & SCHWANITZ, G. & WALTER, M. *Einfuehrung in die mathematische Logik, Einfuehrung in die Mengenlehre, Aufbau der Zahlenbereiche* ◇ B05 B28 B98 E98 ◇
BRODY, B.A. *Logic, theoretical and applied* ◇ B98 ◇
BYERLY, H.C. *A primer of logic* ◇ B98 ◇
CARNAP, R. *Grundlagen der Logik und Mathematik* ◇ A05 B98 ◇
CHANG, C.L. & LEE, R.C.T. *Symbolic logic and mechanical theorem proving* ◇ B35 B98 ◇
CRESSWELL, M.J. *Logics and languages* ◇ A05 B98 ◇
EARLE, J.N.F. *Logic* ◇ B98 ◇
ERSHOV, YU.L. & PALYUTIN, E.A. & TAJTSLIN, M.A. *Mathematical logic (Russian)* ◇ B98 C98 ◇
GEORGESCU, G. *The theory of categories and mathematical logic (Romanian)* ◇ B98 ◇
GIRLING, B. & MORING, H. *Logic and logic design* ◇ B70 B98 ◇
MAKINSON, D. *Topics in modern logic* ◇ B98 ◇
MOSHCHENSKIJ, V.A. *Lectures on mathematical logic (Russian)* ◇ B98 D10 D20 ◇
NIINILUOTO, I. *Mathematical logic (Finnish)* ◇ B98 ◇
PONASSE, D. *Mathematical logic* ◇ B98 ◇
RUTZ, P. *Zweiwertige und mehrwertige Logik - ein Beitrag zur Geschichte und Einheit der Logik* ◇ A10 B98 ◇
SEIFFERT, H. *Einfuehrung in die Logik. Logische Propaedeutik und formale Logik* ◇ B98 ◇
STAHL, G. *Elementos de metamatematica* ◇ B98 ◇

1974

ASSER, G. *Mathematische Logik und Grundlagen der Mathematik* ◇ B98 ◇
BOOLOS, G. & JEFFREY, R.C. *Computability and logic* ◇ B98 D98 ◇
CHENIQUE, F. *Comprendre la logique moderne. Tome I: Classes, propositions et predicats* ◇ B98 E98 ◇
CHENIQUE, F. *Comprendre la logique moderne. Tome II: Logiques non classiques, relations et structures* ◇ B98 ◇
COOPER, W.S. *Set theory and syntactic description* ◇ B98 E75 ◇
CORNIDES, T. *Ordinale Deontik* ◇ B45 B98 ◇
DAVIS, MARTIN D. *Computability* ◇ B98 D98 ◇
DETLOVS, V.K. *Mathematical logic (Latvian)* ◇ B98 ◇
FITCH, F.B. *Elements of combinatory logic* ◇ B40 B98 ◇
GLADKIJ, A.V. *Lecture notes on mathematical logic and set theory (Russian)* ◇ B98 E98 ◇
HALLERBERG, A.E. *Logic in mathematics: An elementary approach* ◇ B98 ◇
HALLERBERG, A.E. *Mathematical proof: An elementary approach* ◇ B98 ◇
JACOBSON, A. & KLEMKE, E.D. & ZABECH, F. (EDS.) *Readings in semantics* ◇ A05 B98 ◇
OBERSCHELP, A. *Elementare Logik und Mengenlehre I* ◇ B98 E98 ◇
PETERS, F.E. *Einfuehrung in mathematische Methoden der Informatik* ◇ B70 B98 D80 D98 ◇
POGORZELSKI, W.A. & PRUCNAL, T. *Introduction to mathematical logic. Part I. Elements of the algebra of propositional logic (Polish)* ◇ B98 G05 ◇
POSPESEL, H. *Introduction to logic. Propositional logic* ◇ B05 B98 ◇
RASIOWA, H. *An algebraic approach to non-classical logics* ◇ B98 F50 G05 G10 G20 G25 G98 ◇
SCHENK, G. *Die Logik* ◇ A10 B98 ◇
STYAZHKIN, N.I. *Logic with the elements of mathematical logic (Russian)* ◇ B98 ◇

1975

ADAMS, E.W. *The logic of conditionals. An application of probability to deductive logic* ◇ A05 B48 B98 ◇
ANDERSON, A.R. & BELNAP JR., N.D. *Entailment. The logic of relevance and necessity I* ◇ A05 B46 B98 ◇
ANDERSON, A.R. & BARCAN MARCUS, R. & MARTIN, R.M. (EDS.) *The logical enterprise* ◇ A05 A10 B98 ◇
BARNES, D.W. & MACK, J.M. *An algebraic introduction to mathematical logic* ◇ B98 G98 ◇
BARWISE, J. *Admissible sets and structures. An approach to definability theory* ◇ B98 C40 C70 C98 D60 D98 E30 E98 ◇
BISWAS, N.N. *Introduction to logic and switching theory* ◇ B98 ◇
BODDENBERG, E. *Logik. I* ◇ B98 ◇
CIRULIS, J. *Lectures on mathematical logic and set theory. Part I: Mathematical logic (Russian)* ◇ B98 ◇
CROSSLEY, J.N. *What is mathematical logic ?* ◇ B98 F30 ◇

EDEL'MAN, S.L. *Mathematical logic (Russian)* ◇ B98 ◇

FRIEDMAN, H.M. *One hundred and two problems in mathematical logic* ◇ B98 C98 D98 E98 F98 ◇

HEGENBERG, L. *Logica: Simbolizacao e deducao* ◇ B98 ◇

KREISEL, G. *Was hat die Logik in den letzten 25 Jahren fur die Mathematik geleistet?* ◇ A05 B98 C75 D40 ◇

LAVROV, I.A. & MAKSIMOVA, L.L. *Problems in set theory, mathematical logic and the theory of algorithms (Russian)* ◇ B98 D98 E98 ◇

MANASTER, A.B. *Completeness, compactness, and undecidability: an introduction to mathematical logic* ◇ B98 C07 D35 D98 ◇

MOSS, J.M.B. & SCOTT, D.S. *Bibliography of logic books* ◇ A10 B98 ◇

RESCHER, N. *A theory of possibility* ◇ A05 B98 ◇

WESSEL, H. & ZINOV'EV, A.A. *Logische Sprachregeln. Eine Einfuehrung in die Logik* ◇ A05 B98 ◇

1976

BLUMBERG, A.E. *Logic: A first course* ◇ B98 ◇

BORKOWSKI, L. *Formale Logik. Logische Systeme. Einfuehrung in die Metalogik* ◇ A05 B98 ◇

CHRISTIAN, C.C. *Die Bedeutung der Mengentheorie als Grundlagenwissenschaft* ◇ A05 B98 E10 E30 ◇

CHUAQUI, R.B. *Introduccion a la metamatematica y sus aplicaciones* ◇ B98 ◇

COURVOISIER, M. & LAGASSE, J. & RICHARD, J. *Logique combinatoire. 3e ed* ◇ B40 B98 ◇

HALKOWSKA, K. & PIROG-RZEPECKA, K. & SLUPECKI, J. *Mathematical logic (Polish)* ◇ B98 C07 C98 E98 ◇

LEBLANC, H. *Truth-value semantics* ◇ B98 ◇

MALATESTA, M. *Logistica I: Introduzione. La logica degli enunciati* ◇ B98 ◇

MONK, J.D. *Mathematical logic* ◇ B98 ◇

MOSTERIN, J. *Logica de primer order* ◇ B98 ◇

PABION, J.F. *Logique mathematique* ◇ B98 ◇

POSPESEL, H. *Introduction to logic. Predicate logic* ◇ B10 B98 ◇

RICHTER, M.M. *Mathematische Logik I* ◇ B98 ◇

TAPSCOTT, B.L. *Elementary applied symbolic logic* ◇ B98 ◇

ZIEMBINSKI, Z. *Practical logic (Polish)* ◇ A05 B45 B98 ◇

1977

BARWISE, J. (ED.) *Handbook of mathematical logic* ◇ B98 C98 D98 E98 F98 H98 ◇

BELL, J.L. & MACHOVER, M. *A course in mathematical logic* ◇ B98 ◇

BERGMANN, E. & NOLL, H. *Mathematische Logik mit Informatik-Anwendungen* ◇ A05 A10 B98 ◇

BURKS, A.W. *Chance, cause, reason. An inquiry into the nature of scientific evidence* ◇ A05 B98 ◇

GEORGE, F.H. *Precision, language and logic* ◇ A05 B98 ◇

GLADKIJ, A.V. *The language of mathematical logic* ◇ B98 ◇

GRANDY, R.E. *Advanced logic for applications* ◇ B98 ◇

HODGES, W. *Logic* ◇ B98 ◇

KAAZ, M.A. *Elemente der mathematischen Logik fuer den Gebrauch in Physik und Technik* ◇ B98 ◇

KARPOV, V.G. & MOSHCHENSKIJ, V.A. *Mathematical logic and discrete mathematics (Russian)* ◇ B98 E98 ◇

KREMPA, J. & MAZBIC-KULMA, B. *Elements of logic, set theory and algebra (Elementy logiki, teorii mnogosci i algebry) (Polish)* ◇ B98 E98 ◇

LEMMON, E.J. *An introduction to modal logic* ◇ A10 B45 B98 ◇

MANIN, YU.I. *A course in mathematical logic* ◇ B98 C07 D98 E35 E50 F30 G12 ◇

SCHREIBER, P. *Grundlagen der Mathematik* ◇ B98 ◇

STILLWELL, J.C. *Concise survey of mathematical logic* ◇ B98 C98 D10 D35 ◇

VASILACHE, S. *Ensembles, structures, categories, faisceaux* ◇ B98 E98 G30 ◇

1978

BAXANDALL, P.R. & BROWN, W.S. & ROSE, G.S.C. & WATSON, F.R. *Proof in mathematics ("If", "then" and "perhaps"). A collection of material illustrating the nature and variety of the idea of proof in mathematics* ◇ B98 ◇

COOPER, W.S. *Foundations of logico-linguistics: A unified theory of information, language, and logic* ◇ A05 B98 ◇

DENNING, P.J. & DENNIS, J.B. & QUALITZ, J.E. *Machines, languages, and computation* ◇ B98 D05 D10 D98 ◇

DILLER, J. *Klassische Praedikatenlogik* ◇ B98 ◇

DURNEV, V.G. *Elements of set theory and mathematical logic (Russian)* ◇ B98 E98 ◇

EBBINGHAUS, H.-D. & FLUM, J. & THOMAS, WOLFGANG *Einfuehrung in die mathematische Logik* ◇ B98 C95 C98 ◇

FACIONE, P.A. & SCHERER, D. *Logic and logical thinking: a modular approach* ◇ A05 B98 ◇

GAUTHIER, Y. *Methodes et concepts de la logique formelle* ◇ A05 B98 ◇

HALKOWSKA, K. & PIROG-RZEPECKA, K. & SLUPECKI, J. *Logic and set theory (Polish)* ◇ B98 E98 ◇

HAMILTON, A.G. *Logic for mathematicians* ◇ B98 ◇

KHROMOJ, YA.V. *A collection of exercises and problems in mathematical logic (Ukrainian)* ◇ B98 ◇

KONDAKOV, N.I. *Woerterbuch der Logik* ◇ A05 A10 B98 ◇

LEMMON, E.J. *Beginning logic* ◇ B98 ◇

LIGHTSTONE, A.H. *Mathematical logic. An introduction to model theory. Edited by H. B. Enderton* ◇ B98 C98 H05 ◇

LOLLI, G. *Lezioni di logica matematica* ◇ B98 ◇

OBERSCHELP, A. *Elementare Logik und Mengenlehre II* ◇ B98 E98 ◇

RICHTER, M.M. *Logikkalkuele* ◇ B35 B98 F05 G10 ◇

RUTKOWSKI, A. *Elements of mathematical logic (Polish)* ◇ B98 ◇

SMULLYAN, R.M. *What is the name of this book? The riddle of Dracula and other logical puzzles* ◇ B98 ◇

TAKEUTI, G. *Two applications of logic to mathematics* ◇ B98 C90 E40 E75 F05 F30 F35 G12 ◇

TENNANT, N. *Natural logic* ◇ B98 ◇

1979

APT, K.R. *Ten years of Hoare's logic, a survey* ◇ B98 ◇
BOOLOS, G. *The unprovability of consistency. An essay in modal logic* ◇ B25 B28 B45 B98 F30 F98 ◇
BOYER, R.S. & MOORE, J.S. *A computational logic* ◇ B98 D70 ◇
CARDWELL, C.E. *Arguments and inference. An introduction to symbolic logic* ◇ B98 ◇
EDWARDS, R.E. *A formal background to mathematics. Ia, Ib: Logic, sets and numbers* ◇ B98 E98 ◇
ERSHOV, YU.L. & PALYUTIN, E.A. *Mathematical logic (Russian)* ◇ B98 ◇
GOLDBLATT, R.I. *Topoi. The categorial analysis of logic* ◇ B98 C98 E98 F35 F50 F98 G30 ◇
GOMEZ CALDERON, J. *Logica simbolica* ◇ B98 ◇
HOFSTADTER, D.R. *Goedel, Escher, Bach: an eternal golden braid* ◇ A05 B98 D99 ◇
MALITZ, J. *Introduction to mathematical logic. Set theory, computable functions, model theory* ◇ B98 C98 D98 ◇
MANIN, YU.I. *Provable and unprovable (Russian)* ◇ A05 B98 E35 E50 F30 F98 G12 ◇
MOSTOWSKI, ANDRZEJ *An excerpt from the book Mathematical logic* ◇ B98 ◇
PURTILL, R.L. *Logic. Argument, refutation, and proof* ◇ B98 ◇
RAUTENBERG, W. *Klassische und nichtklassische Aussagenlogik* ◇ B45 B50 B98 C90 ◇
ROBINSON, JOHN ALAN *Logic: form and function. The mechanization of deductive reasoning* ◇ A10 B10 B35 B98 ◇
THIELE, R. *Mathematische Beweise* ◇ B98 ◇
WOJCIECHOWSKA, A. *Elements of logic and set theory* ◇ B98 E98 ◇
WU, YUNZENG *Two problems of modern mathematical logic and its philosophic significance (Chinese)* ◇ A05 B98 ◇

1980

ARRUDA, A.I. *A survey of paraconsistent logic* ◇ A05 B53 B98 E70 ◇
BARTNICK, J. *Predicate logic without bound variables* ◇ B98 ◇
BAUDISCH, A. & SEESE, D.G. & TUSCHIK, H.-P. & WEESE, M. *Decidability and generalized quantifiers* ◇ B25 B98 C10 C60 C80 C98 ◇
BERGMANN, MERRIE & MOOR, J. & NELSON, JACK *The logic book* ◇ B98 ◇
CHELLAS, B.F. *Modal logic. An introduction* ◇ B45 B98 ◇
DALEN VAN, D. *Logic and structure* ◇ B98 C98 ◇
LYNCH, E.P. *Applied symbolic logic* ◇ B98 ◇
MO, SHAOKUI *Introduction to mathematical logic (Chinese)* ◇ B98 ◇
PUTNAM, H. *Models and reality* ◇ A05 B98 C99 ◇
SKORUBSKIJ, V.I. *Arithmetical and logical foundations of digital machines. Textbook (Russian)* ◇ B70 B98 ◇

1981

ASSER, G. *Einfuehrung in die mathematische Logik. Teil III: Praedikatenlogik hoeherer Stufe* ◇ B15 B98 ◇
BARENDREGT, H.P. *The lambda calculus, its syntax and semantics* ◇ B40 B98 ◇
BECKMAN, F.S. *Mathematical foundations of programming* ◇ B98 ◇
BOEHME, G. *Einstieg in die mathematische Logik* ◇ B98 ◇
GAINES, B.R. & MAMDANI, E.H. (EDS.) *Fuzzy reasoning and its applications* ◇ B52 B98 ◇
GLADKIJ, M. *Mathematical logic and mathematical linguistics (Russian)* ◇ B98 ◇
GOZE, M. & LUTZ, R. *Nonstandard analysis. A pratical guide with applications* ◇ B98 H05 ◇
HARPER, W.L. *A sketch of some recent developments in the theory of conditionals* ◇ B48 B98 ◇
HU, SHIHUA & LU, ZHONGWAN *Foundation of mathematical logic I (Chinese)* ◇ B98 ◇
KOSOVSKIJ, N.K. *Elements of mathematical logic and its applications to the theory of subrecursive algorithms (Russian)* ◇ B98 D20 D98 F30 F60 F98 ◇
MCCAWLEY, J.D. *Everything that linguists have always wanted to know about logic, but were ashamed to ask* ◇ B98 ◇
NUTE, D.E. *Essential formal semantics* ◇ B98 ◇
POGORZELSKI, W.A. *Classical calculus of quantifiers. Outline of the theory (Polish)* ◇ B98 ◇
TAKEUTI, G. *Logic and set theory* ◇ B51 B98 E40 E70 F50 ◇
WANG, HAO *Popular lectures on mathematical logic (Chinese)* ◇ A05 B98 C98 E98 F30 F98 ◇

1982

BITTINGER, M.L. *Logic, proof, and sets* ◇ B98 ◇
FISHER, A. *Formal number theory and computability* ◇ B98 D20 F30 F98 ◇
HATCHER, W.S. *The logical foundations of mathematics* ◇ B98 ◇
HU, SHIHUA & LU, ZHONGWAN *Foundation of mathematical logic II (Chinese)* ◇ B98 ◇
PRIJATELJ, N. *Foundations of mathematical logic I (Slovenian)* ◇ B98 ◇
SCOTT, D.S. *Lectures on a mathematical theory of computation* ◇ B40 B98 D75 ◇
TSELISHCHEV, V.V. (ED.) *Problems of logic and methodology of science (Russian)* ◇ B98 ◇
WANG, XIANJUN *Introduction to mathematical logic (Chinese)* ◇ B98 ◇
XU, LIZHI & YUAN, XIANGWAN & ZHENG, YUXIN & ZHU, WUJIA *Antinomies and the foundational problem of mathematics II (Chinese) (English summary)* ◇ A05 B98 ◇
ZAHN, P. *Ein argumentativer Weg zur Logik* ◇ B98 ◇

1983

BACHMANN, H. *Der Weg der mathematischen Grundlagenforschung* ◇ A10 B98 ◇
EICHHORN, H. *Conceptual and conventional definitions in the mathematical sciences* ◇ A05 B98 ◇
ENGELER, E. *Metamathematik der Elementarmathematik* ◇ B98 ◇
GABBAY, D.M. & GUENTHNER, F. (EDS.) *Handbook of philosophical logic Vol. I: Elements of classical logic* ◇ B98 C98 ◇

GLUBRECHT, J.-M. & OBERSCHELP, A. & TODT, G. *Klassenlogik* ◇ B10 B98 E30 ◇
GOE, G. *Lezioni di logica* ◇ B98 ◇
SMULLYAN, R.M. *Dame oder Tiger? Logische Denkspiele und eine mathematische Novelle ueber Goedels grosse Entdeckung* ◇ B98 F99 ◇
TAMAS, G. (ED.) *Studien zur Logik* ◇ B98 ◇

1984

ATIYAH, M. & HOARE, C.A.R. & SHEPHERDSON, J.C. (EDS.) *Mathematical logic and programming languages* ◇ B98 ◇
BENCIVENGA, E. *Il primo libro di logica. Introduzione ai metodi della logica contemporanea* ◇ B98 ◇
GABBAY, D.M. & GUENTHNER, F. (EDS.) *Handbook of philosophical logic Vol II. Extensions of classical logic* ◇ B98 C90 C98 ◇
HAAS, G. *Konstruktive Einfuehrung in die formale Logik* ◇ B98 F98 ◇
MARKOV, A.A. *Elements of mathematical logic (Russian)* ◇ B98 ◇
MO, SHAOKUI & SHEN, BAIYING & XU, YONGSHEN *Mathematical logic (Chinese)* ◇ B98 ◇
SCHMUCKER, K.J. *Fuzzy sets, natural language computations, and risk analysis, Foreword by Lotfi A. Zadeh* ◇ B52 B98 ◇
WOJCICKI, R. *Lectures on propositional calculi* ◇ B98 ◇

1985

DALLA CHIARA SCABIA, M.L. (ED.) *Present state of the problem of the foundations of mathematics* ◇ B98 ◇
DRAKE, F.R. *How recent work in mathematical logic relates to the foundations of mathematics* ◇ A05 B98 E98 ◇
GINDIKIN, S.G. *Algebraic logic* ◇ B98 ◇
KREISEL, G. *Mathematical logic: tool and object lesson for science* ◇ A05 B98 ◇
ROBERT, A. *Analyse non standard* ◇ B98 H05 ◇

Author Index

AAQVIST, L. [1962] *A binary primitive in deontic logic* (**J** 0079) Logique & Anal, NS 5*90-97
◇ B45 ◇ REV MR 27 # 4738 JSL 36.519 • ID 00434

AAQVIST, L. [1962] *Reflections on the logic of nonsense* (**J** 0105) Theoria (Lund) 28*138-157
◇ A05 B60 ◇ REV JSL 33.134 • ID 03835

AAQVIST, L. [1963] *Deontic logic based on a logic of "better"* (**J** 0096) Acta Philos Fenn 16*285-290
◇ A05 B45 ◇ REV MR 31 # 38 JSL 31.278 • ID 00436

AAQVIST, L. [1963] *Postulate sets and decision procedures for some systems of deontic logic* (**J** 0105) Theoria (Lund) 29*154-175
◇ B45 ◇ REV JSL 31.278 • ID 00435

AAQVIST, L. [1964] *Interpretations of deontic logic* (**J** 0094) Mind 73*246-253
◇ A05 B45 ◇ REV JSL 31.120 JSL 33.137 • ID 00437

AAQVIST, L. [1964] *On Dawson-models for deontic logic* (**J** 0079) Logique & Anal, NS 7*14-21
◇ B45 ◇ REV MR 31 # 3319 JSL 31.666 • ID 00438

AAQVIST, L. [1964] *Results concerning some modal systems that contain S2* (**J** 0036) J Symb Logic 29*79-87
◇ B45 ◇ REV MR 30 # 3839 JSL 34.648 • ID 00439

AAQVIST, L. [1965] *A new approach to the logical theory of interrogatives. Part I. Analysis* (**X** 0882) Univ Filos Foeren: Uppsala iv + 174pp
◇ A05 B98 ◇ REV JSL 32.403 • ID 16716

AAQVIST, L. [1966] *"Next" and "ought" alternative foundations for von Wright's tense logic, with an application to deontic logic* (**J** 0079) Logique & Anal, NS 9*231-251
◇ B45 ◇ REV MR 37 # 6163 • ID 00440

AAQVIST, L. [1970] *Knowing who* (**C** 0735) Logic & Value (Dahlquist) 227-234
◇ A05 B45 ◇ REV MR 50 # 4245 • ID 17201

AAQVIST, L. [1971] *Revised foundations for imperative-epistemic and interrogative logic* (**J** 0105) Theoria (Lund) 37*33-73
◇ A05 B60 ◇ REV Zbl 341 # 02017 • ID 29740

AAQVIST, L. [1971] *The completeness of some modal logics with circumstantials, subjunctive conditionals, transworld identity and dispositional predicates. A study in the prolegomena to the logic of science* (**X** 0882) Univ Filos Foeren: Uppsala ii + 90pp
◇ A05 B45 ◇ REV MR 49 # 2262 Zbl 318 # 02022 • ID 23604

AAQVIST, L. [1972] *Performatives and verifiability by the use of language. A study in the applied logic of indexicals and conditionals* (**X** 0882) Univ Filos Foeren: Uppsala iii + 53pp
◇ A05 B45 ◇ REV MR 50 # 1843 • ID 17144

AAQVIST, L. [1973] *Modal logic with subjunctive conditional and dispositional predicates* (**J** 0122) J Philos Logic 2*1-76
◇ B45 ◇ REV MR 54 # 4924 Zbl 259 # 02015 • ID 00601

AAQVIST, L. [1974] *A new approach to the logical theory of actions and causality* (**C** 1936) Log Th & Semant Anal (Kanger) 73-91
◇ B60 ◇ REV Zbl 301 # 02027 • ID 60174

AAQVIST, L. & GUENTHNER, F. [1975] *Representability in QA of Hintikka intensional quantifiers and Keenan term quantifiers* (**J** 1517) Theor Linguist 2*21-44
◇ B45 ◇ REV MR 58 # 21461 Zbl 356 # 02015 • ID 50287

AAQVIST, L. & GUENTHNER, F. (EDS.) [1977] *Tense logic* (**J** 0079) Logique & Anal, NS 20*357-517
◇ B45 ◇ REV MR 58 # 16084 • ID 70086

AAQVIST, L. [1979] *A conjectured axiomatization of two-dimensional Reichenbachian tense logic* (**J** 0122) J Philos Logic 8*1-45
◇ B45 ◇ REV MR 80d:03014 Zbl 407 # 03024 • ID 56190

AAQVIST, L. [1981] *Predicate calculi with adjectives and nouns* (**J** 0122) J Philos Logic 10*1-26
◇ A05 B60 ◇ REV MR 82b:03060 Zbl 473 # 03007 • ID 55338

AAQVIST, L. & HOEPELMAN, J.P. [1981] *Some theorems about a "tree" system of deontic tense logic* (**C** 3731) New Studies Deontic Log 187-221
◇ B45 ◇ REV MR 83j:03028 • ID 34863

AAQVIST, L. [1982] *How to handle the liar paradox in modal logic with sentential quantifiers and its own truth predicate* (**J** 1517) Theor Linguist 9*111-129
◇ B45 ◇ REV MR 83k:03022 Zbl 487 # 03009 • ID 36162

ABBATI, M.C. & MANIA, A. [1981] *The quantum logical and the operational description for physical systems* (**P** 3820) Curr Iss in Quantum Log;1979 Erice 119-127
◇ B51 ◇ REV MR 84j:03118 Zbl 537 # 03044 • ID 34708

ABBATI, M.C. see Vol. III for further entries

ABDUGALIEV, U.A. [1967] *Normal forms of k-valued logic (Russian)* (**J** 0040) Kibernetika, Akad Nauk Ukr SSR 1967/1*16-20
• TRANSL [1967] (**J** 0021) Cybernetics 3/1*13-16
◇ B50 ◇ REV MR 44 # 1540 Zbl 204.2 • ID 00022

ABDUGALIEV, U.A. [1969] *On the question of the representation of functions of a k-valued logic by a normal form I (Russian)* (**J** 0071) Met Diskr Analiz (Novosibirsk) 15*3-24
◇ B50 ◇ REV MR 44 # 2578a Zbl 211.11 • REM Part II 1970 • ID 00023

ABDUGALIEV, U.A. [1970] *On the question of the representation of functions of a k-valued logic by a normal form II (Russian)* (**J** 0071) Met Diskr Analiz (Novosibirsk) 16*3-15
◇ B50 ◇ REV MR 44 # 2578b Zbl 215.321 • REM Part I 1969 • ID 28014

ABDUGALIEV, U.A. [1980] *Irredundant disjunctive normal forms of k-valued logic functions (Russian)* (**J** 0403) Izv Akad Nauk Kazak SSR, Ser Fiz-Mat 1980/1*53-55
◇ B50 ◇ REV MR 82a:94110 Zbl 437 # 94009 • ID 69030

ABDUGALIEV, U.A. see Vol. I for further entries

ABIAN, A. & LAMACCHIA, S.E. [1967] *Examples of generalized Sheffer functions* (**J** 0001) Acta Math Acad Sci Hung 18*189-190
◇ B50 ◇ REV MR 38 # 986 Zbl 208.24 • ID 00041

ABIAN, A. & LAMACCHIA, S.E. [1967] *Some generalized Sheffer functions* (**J** 0009) Arch Math Logik Grundlagenforsch 10*6-7
◇ B50 ◇ REV MR 35 # 5294 Zbl 162.311 • ID 00079

ABIAN, A. also published under the name ABIAN, S.

ABIAN, A. see Vol. I, III, IV, V for further entries

ACHTELIK, G. & DUBIKAJTIS, L. & DUDEK, E. & KONIOR, J. [1981] *On independence of axioms of Jaskowski's discussive propositional calculus* (**J** 0302) Rep Math Logic, Krakow & Katowice 11*3-11
◇ B53 ◇ REV MR 82g:03042 Zbl 459 # 03015 • ID 54455

ACKERMANN, R.J. [1966] *Non-deductive inference* (**X** 0866) Routledge & Kegan Paul: Henley on Thames v+130pp
◇ B98 ◇ ID 25397

ACKERMANN, R.J. [1967] *An introduction to many-valued logics* (**X** 0813) Dover: New York 50pp
◇ B50 B98 ◇ ID 00099

ACKERMANN, R.J. [1967] *Introduction to many-valued logics* (**X** 0866) Routledge & Kegan Paul: Henley on Thames v+90pp
◇ B50 B98 ◇ ID 25260

ACKERMANN, R.J. [1970] *Modern deductive logic* (**X** 0878) Doubleday: London ix+261pp
◇ B98 ◇ REV MR 50 # 12628 • ID 04128

ACKERMANN, R.J. [1971] *Matrix satisfiability and axiomatization* (**J** 0047) Notre Dame J Formal Log 12*309-321
◇ B50 ◇ REV MR 45 # 8515 Zbl 184.280 • ID 00100

ACKERMANN, W. [1928] see HILBERT, D.

ACKERMANN, W. [1956] *Begruendung einer strengen Implikation* (**J** 0036) J Symb Logic 21*113-128
◇ B46 ◇ REV MR 18.270 Zbl 72.1 JSL 22.327 • ID 00128

ACKERMANN, W. [1958] *Ueber die Beziehung zwischen strikter und strenger Implikation* (**J** 0076) Dialectica 12*213-222
◇ B46 ◇ REV MR 21 # 3323 Zbl 99.5 JSL 25.349 • ID 00130

ACKERMANN, W. see Vol. I, III, IV, V, VI for further entries

ACZEL, P. & FEFERMAN, S. [1980] *Consistency of the unrestricted abstraction principle using an intensional equivalence operator* (**C** 3050) Essays Combin Log, Lambda Calc & Formalism (Curry) 67-98
◇ B60 E70 ◇ REV MR 82b:03037 Zbl 469 # 03006 • ID 70302

ACZEL, P. see Vol. III, IV, V, VI for further entries

ADAM, ADOLF & BERAN, H. [1982] *On the problem of the application of fuzzy-logic concepts* (**C** 3433) Prog in Cybern & Syst Res, Vol 6 143-150
◇ B52 ◇ REV MR 83d:03030 Zbl 474 # 03008 • ID 55412

ADAMS, E.W. [1962] *On rational betting systems* (**J** 0009) Arch Math Logik Grundlagenforsch 6*7-29
◇ B48 ◇ REV MR 25 # 3794 • ID 00159

ADAMS, E.W. [1964] *On rational betting systems (continuation)* (**J** 0009) Arch Math Logik Grundlagenforsch 6*112-128
◇ B48 ◇ REV MR 30 # 4581 • ID 00160

ADAMS, E.W. [1966] *Probability and the logic of conditionals* (**C** 1107) Aspects Inductive Log 265-316
◇ B48 ◇ REV Zbl 202.300 JSL 39.609 • ID 27710

ADAMS, E.W. [1975] *The logic of conditionals. An application of probability to deductive logic* (**S** 3307) Synth Libr 86*xiii+155pp
◇ A05 B48 B98 ◇ REV MR 58 # 5043 Zbl 324 # 02002 • ID 60060

ADAMS, E.W. [1976] *Prior probabilities and counterfactual conditionals* (**P** 2411) Found Probab Th, Stat Inf & Stat Th Sci;1973 London ON 1*1-21
◇ A05 B48 ◇ REV Zbl 322 # 02029 • ID 60059

ADAMS, E.W. [1977] *A note on comparing probabilistic and modal logics of conditionals* (**J** 0105) Theoria (Lund) 43*186-194
◇ A05 B48 ◇ REV MR 80a:03020 Zbl 389 # 03007 • ID 52296

ADAMS, E.W. [1981] *Transmissible improbabilities and marginal essentialness of premises in inferences involving indicative conditionals* (**J** 0122) J Philos Logic 10*149-177
◇ A05 B48 ◇ REV MR 82j:03025 Zbl 473 # 03012 • ID 55343

ADAMS, E.W. see Vol. III for further entries

ADAMS, R.M. [1981] *Actualism and thisness* (**J** 0154) Synthese 49*3-41
◇ A05 B45 ◇ REV MR 82i:03005 • ID 70315

ADAMSON, A. & GILES, R. [1979] *A game-based formal system for L_∞* (**J** 0063) Studia Logica 38*49-73
◇ B25 B50 D15 ◇ REV MR 80m:03049 Zbl 417 # 03008 • ID 53245

ADAMSON, A. see Vol. III, V for further entries

ADDISON, J.W. & CHANG, C.C. & CRAIG, W. & HENKIN, L. & SCOTT, D.S. & VAUGHT, R.L. (EDS.) [1974] *Proceedings of the Tarski Symposium* (**S** 3304) Proc Symp Pure Math 25*xxi+498pp
◇ B97 C97 ◇ REV MR 50 # 1829 • REM Corr. ed. 1979; xx+498pp. Contains bibliography of A.Tarski with a supplement in the corr. ed. • ID 70206

ADDISON, J.W. see Vol. I, III, IV, V for further entries

ADZHI-VELI, YA.KH. [1966] *An application of the Blake-Poreckij method to the minimization of the functions of a many-valued logic (Russian) (Uzbek summary)* (**J** 0024) Dokl Akad Nauk Uzb SSR 1966/1*10-12
◇ B50 ◇ REV MR 42 # 4379 • ID 00198

AERTS, D. & DAUBECHIES, I. [1978] *About the structure-preserving maps of a quantum mechanical propositional system* (**J** 2146) Helv Phys Acta 51*637-660
◇ B51 ◇ REV MR 81h:81010 • ID 80515

AERTS, D. & DAUBECHIES, I. [1979] *A characterization of subsystems in physics* (**J** 2777) Lett Math Phys 3*11-17
◇ B51 G12 ◇ REV MR 80f:81009a Zbl 451 # 03025 • ID 54040

Aerts, D. & Daubechies, I. [1979] *A mathematical condition for a sublattice of a propositional system to represent a physical subsystem, with a physical interpretation* (**J** 2777) Lett Math Phys 3*19-27
◇ B51 G12 ◇ rev MR 80f:81009b Zbl 451 # 03026
• id 54041

Aerts, D. [1981] *Description of compound physical systems and logical interaction of physical systems* (**P** 3820) Curr Iss in Quantum Log;1979 Erice 381-403
◇ B51 G12 ◇ rev MR 84m:81031 Zbl 537 # 03044
• id 44879

Aerts, D. [1982] *Description of many separated physical entities without the paradoxes encountered in quantum mechanics* (**J** 1678) Found Phys 12*1131-1170
◇ B51 ◇ rev MR 84b:81008 • id 38963

Agazzi, E. [1964] *La logica simbolica* (**X** 1348) Scuola: Brescia 396pp
◇ A05 B98 ◇ rev Zbl 166.248 JSL 39.327 • id 16707

Agazzi, E. (ed.) [1981] *Modern logic - a survey. Historical, philosophical, and mathematical aspects of modern logic and its applications* (**S** 3307) Synth Libr 149*viii + 475pp
◇ A05 A10 B97 ◇ rev MR 82f:03002 Zbl 464 # 03001
• id 54591

Agazzi, E. see Vol. I, VI for further entries

Agrachev, A.A. [1974] *Superpositions of continuous functions (Russian)* (**J** 0087) Mat Zametki (Akad Nauk SSSR) 16*517-522
• transl [1974] (**J** 1044) Math Notes, Acad Sci USSR 16*897-900
◇ B50 ◇ rev MR 51 # 10550 Zbl 307 # 26013 • id 17509

Aihara, R. & Ako, T. & Amamitsu, S. & Doi, T. [1982] *Synchronous problems of digital networks in terms of multiple stability problems (Japanese)* (**P** 4081) Many-Val Log & Appl;1982 Kyoto 245-259
◇ B50 ◇ id 47636

Aisa Moreu, D. [1973] *Alternatives for a logical foundation of inductive logic in R.Carnap (Spanish)* (**J** 0162) Teorema (Valencia) 3*523-541
◇ B48 ◇ rev MR 50 # 4241 • id 03821

Ajdukiewicz, K. [1956] *Conditional sentence and material implication (Polish, English) (Russian summary)* (**J** 0063) Studia Logica 4*117-134,135-153
◇ B60 ◇ id 45795

Ajdukiewicz, K. [1958] *Abriss der Logik* (**X** 1230) Aufbau: Berlin 204pp
◇ B98 ◇ rev JSL 19.235 • id 22390

Ajdukiewicz, K. [1965] *Pragmatic logic (Polish)* (**X** 1034) PWN: Warsaw
• transl [1974] (**S** 3307) Synth Libr 62*xv + 460pp
◇ A05 B98 ◇ rev Zbl 307 # 02005 • id 60083

Ajdukiewicz, K. [1967] *Intensional expressions* (**J** 0063) Studia Logica 20*63-86
◇ A05 B45 ◇ rev MR 35 # 6532 Zbl 301 # 02007
• id 00234

Ajdukiewicz, K. see Vol. I for further entries

Ajzenberg, N.N. [1965] *Representation of a sum* mod *m in a class of normal forms of functions of m-valued logic (Russian)* (**J** 0040) Kibernetika, Akad Nauk Ukr SSR 1965/4*101-102
◇ B50 ◇ rev MR 34 # 42 Zbl 223 # 02012 • id 41904

Ajzenberg, N.N. [1968] *The representation of functions of k-valued logic by polynomials mod k (Russian)* (**J** 0040) Kibernetika, Akad Nauk Ukr SSR 1968/2*102
• transl [1968] (**J** 0021) Cybernetics 4/2*83
◇ B50 ◇ rev MR 45 # 6588 • id 00212

Ajzenberg, N.N. & Semion, I.V. & Tsitkin, A.I. [1971] *Polynomial representations of logical functions (Russian)* (**J** 0474) Avtom Vychis Tekh, Akad Nauk Latv SSR 1971/2*6-13
• transl [1971] (**J** 2666) Autom Control Comput Sci 5/2*5-11
◇ B50 G05 ◇ rev MR 46 # 10498 Zbl 218 # 02014
• id 26273

Ajzenberg, N.N. & Ivas'kiv, Yu.L. [1977] *Many-valued threshold logic (Russian)* (**X** 2199) Naukova Dumka: Kiev 147pp
◇ B50 ◇ rev MR 58 # 26554 Zbl 405 # 94020 • id 80525

Ajzerman, M.A. & Gusev, L.A. & Rozonoehr, L.I. & Smirnova, I.M. & Tal', A.A. [1963] *Logic, automata and algorithms (Russian)* (**X** 3709) Izdat Fiz-Mat Lit: Moskva 556pp
• transl [1971] (**X** 0801) Academic Pr: New York xii + 433pp [1967] (**X** 0814) Oldenbourg: Muenchen x + 431pp (German) [1971] (**X** 1226) Academia: Prague 408pp (Czech)
◇ B98 D05 D10 D98 ◇ rev MR 29 # 5690 MR 35 # 1411 MR 43 # 3044 MR 48 # 1818 Zbl 131.8 Zbl 216.7 JSL 31.109 JSL 37.625 • id 00220

Ajzerman, M.A. & Malishevskij, A.V. [1982] *Problems of logical justification in general choice theory. Levels and criteria of classical choice rationality (Russian) (English summary)* (**X** 0670) Inst Problem Upravlen: Moskva 80pp
◇ B60 ◇ rev MR 84i:90019a • id 40219

Ajzerman, M.A. & Malishevskij, A.V. [1982] *Problems of logical justification in general choice theory. Examples of the analysis of the rationality of choice mechanisms (Russian) (English summary)* (**X** 0670) Inst Problem Upravlen: Moskva 72pp
◇ B60 ◇ rev MR 84i:90019b • id 40221

Ajzerman, M.A. see Vol. IV for further entries

Aken van, J. [1985] *Analysis of quantum probability theory I* (**J** 0122) J Philos Logic 14*267-296
◇ B51 ◇ id 42660

Akimov, A.P. [1979] *On imbeddings of classical modal calculi in constructive calculi (Russian)* (**J** 0023) Dokl Akad Nauk SSSR 247*777-778
• transl [1979] (**J** 0062) Sov Math, Dokl 20*781-782
◇ B45 F50 ◇ rev MR 81d:03015 Zbl 426 # 03028
• id 70347

Akimov, A.P. [1982] *Constructive variants of classical modal calculi (Russian)* (**C** 3832) Avtom Issl Mat 85-91
◇ B45 ◇ rev MR 85i:03039 Zbl 531 # 03008 • id 37673

Ako, T. [1982] see Aihara, R.

Akylbekova, E.A. [1982] *On the completeness and solvability of some program logics (Russian)* (**S** 2874) Vopr Kibern, Akad Nauk SSSR 75*143-164
◇ B75 ◇ rev Zbl 519 # 03020 • id 37537

Author Index

AL-HIBRI, A. [1978] *Deontic logic: a comprehensive appraisal and a new proposal* (**X** 1946) Univ Pr Amer: Lanham x+179pp
◇ A05 B45 ◇ REV Zbl 452 # 03018 • ID 54082

ALBAN, M.J. [1943] *Independence of the primitive symbols of Lewis's calculi of propositions* (**J** 0036) J Symb Logic 8*25-26
◇ B45 ◇ REV MR 4.182 JSL 8.56 • ID 00251

ALBERT, P. [1978] *The algebra of fuzzy logic* (**J** 2720) Fuzzy Sets Syst 1*203-230
◇ B52 E72 G25 ◇ REV MR 80a:03079 Zbl 407 # 03031 • ID 56197

ALBERT, P. [1980] *The notion of consequence in many-valued logic* (**J** 2293) Comp Linguist & Comp Lang 14*165-181
◇ B50 ◇ REV MR 84d:03028 Zbl 469 # 03013 • ID 55141

ALBRECHT, W. [1954] *Die Logik der Logistik* (**X** 1258) Duncker & Humblot: Berlin 60pp
◇ B98 ◇ ID 22391

ALBRITTON, R. [1957] *Present truth and future contingencies* (**J** 0101) Phil Rev 66*29-46
◇ B45 ◇ REV JSL 33.483 • ID 43181

ALCANTARA DE, L.P. & CARNIELLI, W.A. [1984] *Paraconsistent logics* (**J** 0063) Studia Logica 43*79-88
◇ B53 ◇ ID 42362

ALCANTARA DE, L.P. (ED.) [1985] *Mathematical logic and formal systems. A collection of papers in honor of Professor Newton C.A. da Costa* (**S** 3310) Lect Notes Pure Appl Math 94*xiv+297pp
◇ B97 ◇ REV MR 86e:03001 Zbl 563 # 00002 • ID 47463

ALCANTARA DE, L.P. see Vol. I, III, V for further entries

ALCHOURRON, C.E. [1969] *Logic of norms and logic of normative propositions* (**J** 0079) Logique & Anal, NS 12*242-268
◇ B45 ◇ REV MR 45 # 30 Zbl 197.276 JSL 37.628 • ID 00258

ALCHOURRON, C.E. & BULYGIN, E. [1971] *Normative systems* (**X** 0902) Springer: Wien xviii+208pp
◇ B45 B98 ◇ REV Zbl 231 # 02006 JSL 38.326 • ID 15012

ALCHOURRON, C.E. [1972] *The intuitive background of normative legal discourse and its formalization* (**J** 0122) J Philos Logic 1*447-463
◇ A05 B45 ◇ REV MR 58 # 21459 Zbl 248 # 02027 • ID 60092

ALCHOURRON, C.E. & MAKINSON, D. [1981] *Hierarchies of regulations and their logic* (**C** 3731) New Studies Deontic Log 125-148
◇ B45 ◇ REV MR 83a:03014 • ID 35058

ALCHOURRON, C.E. & BULYGIN, E. [1981] *The expressive conception of norms* (**C** 3731) New Studies Deontic Log 95-124
◇ B45 ◇ REV MR 83d:03016 • ID 35170

ALCHOURRON, C.E. [1982] *Normative order and derogation* (**P** 4041) Log, Inform, Law;1981 Firenze 2*51-63
◇ B45 ◇ REV MR 85j:03016 • ID 45209

ALCHOURRON, C.E. & MAKINSON, D. [1982] *On the logic of theory change: contraction functions and their associated revision functions* (**J** 0105) Theoria (Lund) 48*14-37
◇ B60 ◇ REV MR 85g:03023 Zbl 503 # 03005 Zbl 525 # 03001 • ID 36919

ALCHOURRON, C.E. & BULYGIN, E. [1984] *Permission and permissive norms* (**C** 4065) Theorie der Normen 349-371
◇ B45 ◇ REV MR 85k:03015 • ID 44534

ALCHOURRON, C.E. & GAERDENFORS, P. & MAKINSON, D. [1985] *On the logic of theory change: partial meet contraction and revision functions* (**J** 0036) J Symb Logic 50*510-530
◇ B60 ◇ ID 42562

ALEKSEEV, V.B. [1969] *The simple basis of a k-valued logic (Russian)* (**J** 0087) Mat Zametki (Akad Nauk SSSR) 5*471-482
• TRANSL [1969] (**J** 1044) Math Notes, Acad Sci USSR 5*282-287
◇ B50 ◇ REV MR 40 # 2522 Zbl 182.4 • ID 00263

ALEKSEEV, V.B. [1971] *Simple bases and simple functions in k-valued logic (Russian) (English summary)* (**J** 0040) Kibernetika, Akad Nauk Ukr SSR 1971/5*137-141
◇ B50 ◇ REV MR 46 # 8811 Zbl 253 # 02012 • ID 00264

ALEKSEEV, V.B. [1971] *The number of simple bases in k-valued logic (Russian)* (**J** 0071) Met Diskr Analiz (Novosibirsk) 19*3-10
◇ B50 ◇ REV MR 51 # 2863 Zbl 252 # 02008 • ID 17326

ALEKSEEV, V.B. [1974] *On the number of the monotone k-valued functions (Russian)* (**J** 0052) Probl Kibern 28*5-24,278
• ERR/ADD ibid 29*248
◇ B50 ◇ REV MR 51 # 3005 Zbl 293 # 02014 • ID 60096

ALEKSEEV, V.B. [1976] *Deciphering of some classes of monotonic many-valued functions (Russian)* (**J** 0199) Zh Vychisl Mat i Mat Fiz 16*189-198
• TRANSL [1977] (**J** 1049) USSR Comput Math & Math Phys 16/1*180-189
◇ B50 ◇ REV MR 53 # 15008 Zbl 337 # 94025 • ID 60099

ALEKSEEV, V.B. [1980] *Semi-simple bases of k-valued logic (Russian)* (**J** 0087) Mat Zametki (Akad Nauk SSSR) 28*407-422,478
• TRANSL [1980] (**J** 1044) Math Notes, Acad Sci USSR 28*672-680
◇ B50 ◇ REV MR 82c:03027 Zbl 453 # 03021 • ID 54151

ALEKSEEV, V.B. & EMEL'YANOV, N.R. [1985] *A method for constructing fast algorithms in k-valued logic (Russian)* (**J** 0087) Mat Zametki (Akad Nauk SSSR) 38*148-156,171
◇ B50 ◇ ID 48594

ALISEJCHIK, P.A. [1984] *On maximal length of a base in* P_3 (**S** 3382) Sem-ber, Humboldt-Univ Berlin, Sekt Math 56*1-3
◇ B50 ◇ REV Zbl 548 # 03008 • ID 43191

ALLEN, C.M. & ZAVISCA, E. [1971] *An approach to multiple-valued sequential circuit synthesis* (**P** 2007) Symp Th & Appl of Multi-Val Log Design;1971 Buffalo
◇ B50 B70 ◇ ID 42003

ALLEN, C.M. see Vol. I for further entries

ALLEN, JAMES F. [1984] *Towards a general theory of action and time* (**J** 0503) Artif Intell 23*123-154
◇ B45 ◇ REV Zbl 567 # 68025 • ID 49216

ALLEN, L.E. & BROOKS, R.B.S. & DICKOFF, J.W. & JAMES, P.A. [1961] *The ALL project (accelerated learning of logic)* (**J** 0005) Amer Math Mon 68*497-500
◇ B98 ◇ REV JSL 35.484 • ID 00270

ALLEN, L.E. see Vol. I for further entries

ALLINSON, R.E. [1975] *When is a self contradictory proposition true: A sortie into Marxist logic* (J 2026) J West Virginia Phil Soc 9*18-19
◇ A05 B53 ◇ ID 42796

ALMEIDA MOURA DE, J.E. & ALVES, E.H. [1978] *On some higher-order paraconsistent predicate calculi* (P 1800) Brazil Conf Math Log (1);1977 Campinas 1-8
◇ B53 ◇ REV MR 81h:03054 Zbl 385 # 03010 • ID 52118

ALMOG, J. [1976] *Representation of the fuzziness of the concept of relevant implication within a multi-valued model theory* (J 1893) Relevance Logic Newslett 1*100-118
◇ B46 E72 ◇ REV Zbl 358 # 02015 • ID 50469

ALMOG, J. [1976] *The interrelations between fuzzy logics and the concept of relevant intailment* (J 1893) Relevance Logic Newslett 1*39-55
◇ A05 B46 E72 ◇ REV Zbl 358 # 02014 • ID 50468

ALMOG, J. [1978] *Perhaps (?), new logical foundations are needed for quantum mechanics* (J 0079) Logique & Anal, NS 21*253-277 • ERR/ADD ibid 21/84*369
◇ A05 B51 G12 ◇ REV MR 80d:03023 MR 81e:03020 Zbl 394 # 03010 • ID 52487

ALMOG, J. [1980] *Modal counterfactual logic. Semantical considerations on modal counterfactual logic with corollaries on decidability, completeness, and consistency questions* (J 0047) Notre Dame J Formal Log 21*467-479
◇ B52 ◇ REV MR 81e:03010 Zbl 349 # 02013 • ID 53772

ALMOG, J. [1981] *Indexicals, demonstratives and the modality dynamics* (J 0079) Logique & Anal, NS 24*331-349
◇ B45 ◇ REV MR 83k:03021 • ID 36161

ALMUKDAD, A. & NELSON, D. [1984] *Constructible falsity and inexact predicates* (J 0036) J Symb Logic 49*231-233
◇ B52 F50 ◇ REV MR 86c:03020 Zbl 575 # 03016 • ID 42414

ALOISIO, P. [1971] *Alcune osservazioni sulla definibilata di operatori finiti e sulla decidibilita algoritmica della completezza funzionale e deduttiva delle logiche a piu valori (English summary)* (J 0088) Ann Univ Ferrara, NS, Sez 7 16*157-165
◇ B50 ◇ REV MR 46 # 3267 Zbl 238 # 02015 • ID 00282

ALOISIO, P. see Vol. IV for further entries

ALPS, R.A. & NEVELN, R.C. [1981] *A predicate logic based on indefinite description and two notions of identity* (J 0047) Notre Dame J Formal Log 22*251-263
◇ B60 ◇ REV MR 82f:03020 Zbl 438 # 03031 • ID 55546

ALSINA, C. & TRILLAS, E. & VALVERDE, L. [1980] *On nondistributive logical connectives for the theory of fuzzy sets (Spanish) (English summary)* (P 3512) Jorn Mat Luso-Espanol (7);1980 St Feliu de Guixois 1*69-72
◇ B52 E72 ◇ ID 44653

ALSINA, C. & TRILLAS, E. [1981] *On Sheffer stroke* (P 3705) Int Symp Multi-Val Log (11);1981 Oklahoma City & Norman 244-245
◇ B50 ◇ REV MR 83m:94033 Zbl 553 # 03039 • ID 43356

ALSINA, C. [1982] *Functional equations and negations of distribution functions (Catalan)* (P 3870) Congr Catala de Log Mat (1);1982 Barcelona 63-65
◇ B60 ◇ REV MR 84i:03003 Zbl 513 # 39009 • ID 47581

ALSINA, C. & TRILLAS, E. & VALVERDE, L. [1983] *On some logical connectives for fuzzy sets theory* (J 0034) J Math Anal & Appl 93*15-26
◇ B52 ◇ REV MR 84h:03052 Zbl 522 # 03012 • ID 37401

ALSINA, C. see Vol. V for further entries

ALTAEV, N.K. [1984] *Principles of a physical logic (Russian)* (X 2235) VINITI: Moskva 7471-84
◇ B60 ◇ ID 46863

ALTHAM, J.E.J. [1971] *The logic of plurality* (X 0816) Methuen: London & New York ix + 84pp
◇ A05 B98 C80 ◇ REV MR 47 # 12 Zbl 276 # 02008 • ID 00285

ALTHAM, J.E.J. see Vol. III for further entries

ALVES, E.H. & COSTA DA, N.C.A. [1976] *Une semantique pour le calcul $\mathscr{C}_1$ (English summary)* (J 2313) C R Acad Sci, Paris, Ser A-B 283*A729-A731
◇ B53 ◇ REV MR 54 # 12490 Zbl 344 # 02016 • ID 31722

ALVES, E.H. & COSTA DA, N.C.A. [1977] *A semantical analysis of the calculi $\mathscr{C}_n$* (J 0047) Notre Dame J Formal Log 18*621-630
◇ B53 ◇ REV MR 58 # 16180 Zbl 349 # 02022 • ID 31719

ALVES, E.H. [1978] *On decidability of a system of dialectical propositional logic* (J 0387) Bull Sect Logic, Pol Acad Sci 7*179-184
◇ B25 B53 ◇ REV MR 80b:03028 Zbl 412 # 03012 • ID 52943

ALVES, E.H. [1978] see ALMEIDA MOURA DE, J.E.

ALVES, E.H. & ARRUDA, A.I. [1979] *A semantical study of some systems of vagueness logic* (J 0387) Bull Sect Logic, Pol Acad Sci 8*139-144
◇ B52 ◇ REV MR 81c:03016b Zbl 423 # 03024 • ID 53536

ALVES, E.H. & ARRUDA, A.I. [1979] *Some remarks on the logic of vagueness* (J 0387) Bull Sect Logic, Pol Acad Sci 8*133-138
◇ B52 ◇ REV MR 81c:03016a Zbl 423 # 03023 • ID 53535

ALVES, E.H. & LOPARIC, A. [1980] *The semantics of the systems C_n of da Costa* (P 3006) Brazil Conf Math Log (3);1979 Recife 161-172
◇ B53 ◇ REV MR 83a:03027 Zbl 448 # 03015 • ID 56615

ALVES, E.H. & COSTA DA, N.C.A. [1981] *Relations between paraconsistent logic and many-valued logic* (J 0387) Bull Sect Logic, Pol Acad Sci 10*185-191
◇ B53 ◇ REV MR 84f:03024 Zbl 494 # 03019 • ID 42830

ALVES, E.H. & COSTA DA, N.C.A. [1982] *On a paraconsistent predicate calculus* (P 3860) Coll Papers to Farah on Retirement;1981 Sao Paulo 83-90
◇ B53 ◇ REV MR 85b:03036 Zbl 522 # 03010 • ID 37779

ALVES, E.H. [1984] *Paraconsistent logic and model theory* (J 0063) Studia Logica 43*17-32
◇ B53 C40 C52 C90 ◇ ID 39893

AMAMITSU, S. [1982] see AIHARA, R.

AMBROSE, A. & LAZEROWITZ, M. [1948] *Fundamentals of symbolic logic* (X 0818) Holt Rinehart & Winston: New York ix + 310pp
◇ B98 ◇ REV JSL 14.191 JSL 40.607 • ID 14474

AMBROSE, A. & LAZEROWITZ, M. [1961] *Logic: the theory of formal inference* (**X** 0818) Holt Rinehart & Winston: New York vi+78pp
◇ B98 ◇ REV Zbl 98.241 JSL 28.169 • ID 07900

AMBROSE, A. [1974] *Believing necessary propositions* (**J** 0094) Mind 83*286-290
◇ A05 B45 ◇ REV MR 58 # 21399b • ID 70384

AMBROSE, A. see Vol. VI for further entries

AMBROSIO, R. & MARTINI, G.B. [1983] *Resolution of max-min equations with fuzzy symbols* (**J** 3919) BUSEFAL 16*61-68
◇ B52 ◇ REV Zbl 566 # 94025 • ID 48755

AMBROSIO, R. see Vol. V for further entries

AMER, M.A. [1985] *Classification of Boolean algebras of logic and probabilities defined on them by classical models* (**J** 0068) Z Math Logik Grundlagen Math 31*509-515
◇ B48 G05 ◇ REV Zbl 561 # 03032 • ID 47801

AMER, M.A. [1985] *Extension of relatively σ-additive probabilities on Boolean algebras of logic* (**J** 0036) J Symb Logic 50*589-596
◇ B48 ◇ ID 47365

AMER, M.A. see Vol. I, III, V, VI for further entries

ANDERSON, A.R. [1954] *Improved decision procedures for Lewis's calculus* S4 *and von Wright's calculus* M (**J** 0036) J Symb Logic 19*201-214 • ERR/ADD ibid 20*150
◇ B25 B45 ◇ REV MR 16.103 Zbl 57.8 JSL 20.302 • ID 00309

ANDERSON, A.R. [1954] *On alternative formulations of a modal system of Feys - von Wright* (**J** 0093) J Comp Syst 1*211-212
◇ B45 ◇ REV MR 16.554 Zbl 56.248 JSL 20.302 • ID 00310

ANDERSON, A.R. [1954] *On the interpretation of a modal system of Lukasiewicz* (**J** 0093) J Comp Syst 1*209-210
◇ B45 ◇ REV MR 16.554 Zbl 58.246 JSL 25.293 • ID 00311

ANDERSON, A.R. [1956] *Independent axiom schemata for S5* (**J** 0036) J Symb Logic 21*255-256
◇ B45 ◇ REV MR 18.552 Zbl 74.247 JSL 22.327 • ID 00313

ANDERSON, A.R. [1957] *Independent axiom schemata for von Wright's* M (**J** 0036) J Symb Logic 22*241-244
◇ B45 ◇ REV MR 24 # A1211 Zbl 81.11 JSL 23.346 • ID 00314

ANDERSON, A.R. [1958] *Reduction of deontic logic to alethic modal logic* (**J** 0094) Mind 67*100-103
◇ B45 ◇ REV JSL 24.177 • ID 00315

ANDERSON, A.R. & BELNAP JR., N.D. [1959] *Modalities in Ackermann's "rigorous implication"* (**J** 0036) J Symb Logic 24*107-111
◇ B46 ◇ REV MR 22 # 6700 Zbl 90.9 JSL 34.120 • ID 00317

ANDERSON, A.R. [1960] *Completeness theorems for the system E of entailment and EQ of entailment with quantification* (**J** 0068) Z Math Logik Grundlagen Math 6*201-216
◇ B46 ◇ REV Zbl 95.8 JSL 36.520 • ID 00319

ANDERSON, A.R. & BELNAP JR., N.D. & WALLACE, J.R. [1960] *Independent axiom schemata for the pure theory of entailment* (**J** 0068) Z Math Logik Grundlagen Math 6*93-95
◇ B46 ◇ REV Zbl 95.7 JSL 38.327 • ID 00320

ANDERSON, A.R. & BELNAP JR., N.D. [1961] *Enthymemes* (**J** 0301) J Phil 58*713-723
◇ B46 ◇ REV JSL 27.115 • ID 43018

ANDERSON, A.R. & BELNAP JR., N.D. [1962] *Tautological entailments* (**J** 0095) Philos Stud 13*9-24
◇ A05 B46 ◇ REV JSL 33.608 • ID 00327

ANDERSON, A.R. & BELNAP JR., N.D. [1962] *The pure calculus of entailment* (**J** 0036) J Symb Logic 27*19-52
◇ B46 ◇ REV MR 27 # 4739 Zbl 113.4 JSL 30.240 • ID 00328

ANDERSON, A.R. & BELNAP JR., N.D. [1963] *First degree entailments* (**J** 0043) Math Ann 149*302-319
◇ B46 ◇ REV MR 27 # 3506 Zbl 113.4 JSL 36.520 • ID 00330

ANDERSON, A.R. [1963] *Some open problems concerning the system* E *of entailment* (**J** 0096) Acta Philos Fenn 16*7-18
◇ B46 ◇ REV MR 28 # 2968 Zbl 119.11 • ID 00329

ANDERSON, A.R. & MOORE, O.K. [1967] *The formal analysis of normative concepts* (**C** 0721) Contemp Readings in Log Th 319-331
◇ B45 ◇ REV JSL 24.177 • ID 17145

ANDERSON, A.R. [1970] *The logic of Hohfeldian propositions* (**J** 0079) Logique & Anal, NS 13*231-263
◇ A05 B45 ◇ REV Zbl 198.15 • ID 16292

ANDERSON, A.R. [1973] *An intensional interpretation of truth-values* (**P** 0580) Int Congr Log, Meth & Phil of Sci (4,Sel Pap);1971 Bucharest 3-28
◇ B46 ◇ REV MR 58 # 21460 Zbl 284 # 02004 • ID 30513

ANDERSON, A.R. & BELNAP JR., N.D. [1975] *Entailment. The logic of relevance and necessity I* (**X** 0857) Princeton Univ Pr: Princeton xxxii+543pp
◇ A05 B46 B98 ◇ REV MR 53 # 10542 Zbl 323 # 02030 JSL 42.311 • ID 23056

ANDERSON, A.R. & BARCAN MARCUS, R. & MARTIN, R.M. (EDS.) [1975] *The logical enterprise* (**X** 0875) Yale Univ Pr: New Haven x+261pp
◇ A05 A10 B98 ◇ REV Zbl 344 # 00006 • ID 48638

ANDERSON, A.R. see Vol. I, VI for further entries

ANDERSON, E. & LACEY, H.M. [1980] *Spatial ontology and physical modalites* (**J** 0095) Philos Stud 38*261-285
◇ A05 B45 ◇ REV MR 83i:03008 • ID 35486

ANDERSON, J.G. [1969] *A note on finite intermediate logics* (**P** 4314) Canad Math Congr;1969 Montreal
◇ B55 ◇ REV Zbl 236 # 02024 • ID 41528

ANDERSON, J.G. [1969] *An application of Kripke's completeness theorem for intuitionism to superconstructive propositional calculi* (**J** 0068) Z Math Logik Grundlagen Math 15*259-288
◇ B55 C90 ◇ REV MR 40 # 1261 Zbl 216.288 • ID 00344

ANDERSON, J.G. [1971] *The degree of completeness of Dummett's* LC *and Thomas's* LC_n (**J** 3172) J London Math Soc, Ser 2 3*558-560
◇ B55 ◇ REV MR 44 # 1539 Zbl 214.8 • ID 00345

ANDERSON, J.G. [1972] *Superconstructive propositional calculi with extra axiom schemes containing one variable* (**J** 0068) Z Math Logik Grundlagen Math 18*113-130
◇ B55 ◇ REV MR 46 # 8815 Zbl 246 # 02020 • ID 00346

ANDERSON, J.G. [1974] *A note on finite intermediate logics* (J 0047) Notre Dame J Formal Log 15*149-155
◇ B55 ◇ REV MR 50#4254 Zbl 236#02024 • ID 00347

ANDERSON, J.G. see Vol. I for further entries

ANDERSON, J.M. & JOHNSTONE JR., H.W. [1962] *Natural deduction. The logical basis of axiom systems* (X 0821) Wadsworth Publ: Belmont xii+418pp
◇ B98 F07 F98 ◇ REV MR 25#4986 JSL 29.93 • ID 00341

ANDERSON, K.C. [1977] *A note on knowledge* (J 0094) Mind 86*249-251
◇ A05 B45 ◇ REV MR 57#9456 • ID 70411

ANDERSON JR., R.M. & MAXWELL, G. (EDS.) [1975] *Induction, probability, and confirmation* (X 1307) Univ Minnesota Pr: Minneapolis vii+551pp
◇ A05 B48 ◇ REV MR 57#15935 • ID 70095

ANDREEV, I.D. [1985] *Dialectical logic (Russian)* (X 3407) "Vysshaya Shkola": Moskva 368pp
◇ B53 ◇ ID 49593

ANGELL, R.B. [1962] *A propositional logic with subjunctive conditionals* (J 0036) J Symb Logic 27*327-343
◇ B45 ◇ REV MR 30#1029 Zbl 121.10 JSL 35.464 • ID 00373

ANGELL, R.B. [1964] *Reasoning and logic* (X 1228) Appleton-Century-Crofts: New York xiv+625pp
◇ B98 ◇ REV JSL 31.674 • ID 22456

ANGELL, R.B. see Vol. I for further entries

ANGLUIN, D. & SMITH, C.H. [1985] *Inductive inference: theory and methods* (J 2331) ACM Comp Surveys 15*237-269
◇ B48 ◇ ID 47260

ANGLUIN, D. see Vol. IV for further entries

ANGSTL, H. [1977] *On the logic of norms: a natural deontics* (J 1022) Ratio (Oxford) 19*58-67
◇ B45 ◇ REV MR 56#11752 • ID 70439

ANGSTL, H. see Vol. I for further entries

ANSHAKOV, O.M. & FINN, V.K. & GRIGOLIYA, R.SH. & ZABEZHAJLO, M.I. [1980] *Many-valued logics as fragments of formalized semantics (Russian)* (S 2582) Semiotika & Inf, Akad Nauk SSSR 15*27-60
◇ B50 ◇ REV MR 82j:03027 Zbl 463#03011 • ID 54551

ANSHAKOV, O.M. [1980] *On a certain constructivization of D. A. Bochvar's propositional logic (Russian)* (S 2582) Semiotika & Inf, Akad Nauk SSSR 15*61-73
◇ B50 ◇ REV MR 82m:03033 Zbl 459#03012 • ID 54452

ANSHAKOV, O.M. & FINN, V.K. [1981] *So-called fuzzy logics and one-implication calculi (Russian)* (S 2582) Semiotika & Inf, Akad Nauk SSSR 17*71-89
◇ B52 ◇ REV MR 84m:03036 Zbl 485#03009 • ID 35740

ANSHAKOV, O.M. [1981] *The logic B_3 and Boolean algebras (Russian)* (S 2582) Semiotika & Inf, Akad Nauk SSSR 16*8-13
◇ B50 ◇ REV MR 83d:03026 Zbl 536#03010 • ID 35175

ANSHAKOV, O.M. & FINN, V.K. & GRIGOLIYA, R.SH. & ZABEZHAJLO, M.I. [1982] *Many-valud logics as fragments of formalized semantics* (P 3800) Intens Log: Th & Appl;1979 Moskva 239-272
◇ B50 ◇ REV MR 86f:03040 Zbl 521#03011 • ID 37066

ANSHAKOV, O.M. & RYCHKOV, S.V. [1982] *Many-valued logical calculi (Russian)* (S 2582) Semiotika & Inf, Akad Nauk SSSR 19*90-117
◇ B50 ◇ REV MR 84e:03031 Zbl 501#03012 • ID 34370

ANSHAKOV, O.M. & RYCHKOV, S.V. [1982] *On multivalued logical calculi (Russian)* (J 0023) Dokl Akad Nauk SSSR 264*267-270
• TRANSL [1982] (J 0062) Sov Math, Dokl 25*599-602
◇ B45 ◇ REV MR 83j:03040 Zbl 507#03011 • ID 35345

ANSHAKOV, O.M. [1983] *Some constructivizations of the propositional logics of D.A. Bochvar and S. Hallden (Russian)* (C 3807) Issl Neklass Log & Formal Sist 335-359
◇ B50 ◇ REV MR 85e:03023 • ID 40595

ANSHAKOV, O.M. & RYCHKOV, S.V. [1984] *Axiomatization of finite-valued logical calculi (Russian)* (J 0142) Mat Sb, Akad Nauk SSSR, NS 123(165)*477-495
• TRANSL [1984] (J 0349) Math of USSR, Sbor 51*473-491
◇ B50 C90 ◇ REV MR 85i:03065 Zbl 551#03015 Zbl 566#03010 • ID 44158

ANSHAKOV, O.M. & RYCHKOV, S.V. [1984] *On a way of the formalization and classification of many valued logics (Russian)* (S 2582) Semiotika & Inf, Akad Nauk SSSR 23*78-106,156
◇ B50 C90 ◇ REV Zbl 562#03012 • ID 47425

APOSTEL, L. [1960] *Game theory and the interpretation of deontic logic* (J 0079) Logique & Anal, NS 3*70-90
◇ B46 ◇ REV JSL 30.242 • ID 00404

APOSTEL, L. [1969] *A proposal in the analysis of questions* (J 0079) Logique & Anal, NS 12*376-381
◇ B60 ◇ REV MR 43#6051 Zbl 195.299 • ID 48011

APOSTEL, L. [1971] *Logique inductive, modalites epistemologiques et logique de la preference* (J 2076) Rev Int Philos 25*78-100
◇ A05 B48 ◇ REV MR 58#27195 • ID 70445

APOSTEL, L. [1979] *Logica e dialettica in Hegel* (C 2041) Formal della Dialettica 85-113
◇ A05 B53 ◇ ID 42797

APOSTEL, L. [1979] *Logique et dialectique* (X 2301) Commun & Cognition: Ghent
◇ B53 ◇ ID 42798

APOSTEL, L. [1982] *Towards a general theory of argumentation* (P 3754) Argumentation;1978 Groningen 93-122
◇ B60 ◇ ID 45902

APPIAH, A. [1982] *Conversation and conditionals* (J 2811) Phil Quart (St Andrews) 32*327-338
◇ A05 B48 ◇ REV MR 85d:03003 • ID 40921

APPIAH, A. [1984] *Generalising the probabilistic semantics of conditionals* (J 0122) J Philos Logic 13*351-372
◇ B48 ◇ REV MR 86c:03001 • ID 42646

APPLEBEE, R.C. & PAHI, B. [1971] *Some results on generalized truth-tables* (J 0047) Notre Dame J Formal Log 12*435-440
◇ B50 ◇ REV MR 45#6591 Zbl 223#02009 JSL 38.521 • ID 00422

APPLEBEE, R.C. see Vol. I, IV for further entries

APRILE, G. [1978] *Calcolo inferenziale modale a due livelli* (J 0104) Atti Accad Sci Lett Arti Palermo, Ser 4/I 37*179-188
◇ B45 ◇ REV MR 82m:03042b Zbl 449#03015 • ID 56682

APRILE, G. see Vol. I, IV, V for further entries

APT, K.R. [1979] *Ten years of Hoare's logic, a survey* (**P** 2615) Scand Logic Symp (5);1979 Aalborg 1-44
◇ B98 ◇ REV MR 82m:68018 Zbl 426 # 68004 • REM Part I. Part II 1983 • ID 53676

APT, K.R. see Vol. I, III, IV, V, VI for further entries

ARBIB, M.A. & MANES, E.G. [1975] *Fuzzy machines in a category* (**J** 0016) Bull Austral Math Soc 13*169-210
◇ B52 D05 G30 ◇ REV MR 53 # 10889 Zbl 318 # 18008 • ID 23098

ARBIB, M.A. see Vol. IV, V, VI for further entries

ARIGONI, A.O. [1980] *Mathematical developments arising from "semantical implication" and the evaluation of membership characteristic functions* (**J** 2720) Fuzzy Sets Syst 4*167-183
◇ B52 G10 ◇ REV MR 81j:03083 Zbl 443 # 03013 • ID 56419

ARIGONI, A.O. [1982] *Transformational-generative grammar for description of formal properties* (**J** 2720) Fuzzy Sets Syst 8*311-322
◇ B52 ◇ REV MR 84h:68061 Zbl 491 # 68082 • ID 39519

ARIGONI, A.O. see Vol. V for further entries

ARMENDT, B. [1980] *Is there a Dutch Book argument for probability kinematics?* (**J** 0153) Phil of Sci (East Lansing) 47*583-588
◇ B48 ◇ REV MR 82m:03024 • ID 70485

ARMSTRONG, J.R. & POMPER, G. [1979] *An efficient multivalued minimization algorithm* (**P** 3003) Int Symp Multi-Val Log (9);1979 Bath 300-303
◇ B50 ◇ ID 35964

ARNBORG, S. & TIDEN, E. [1985] *Unification problems with one-side distributivity* (**P** 4244) Rewriting Techn & Appl (1);1985 Dijon 398-406
◇ B60 ◇ ID 49756

ARNOLD, B.H. [1962] *Logic and boolean algebra* (**X** 0819) Prentice Hall: Englewood Cliffs viii+144pp
◇ B05 B98 G05 ◇ REV MR 27 # 65 Zbl 121.27 JSL 29.95 • ID 23577

ARON, E. & BARAK, A. [1975] *On the efficiency of ternary algorithms for multiplication and division* (**P** 1805) Int Symp Multi-Val Log (5,Proc);1975 Bloomington 331-343
◇ B50 B75 D20 ◇ REV MR 58 # 8466 • ID 35833

ARONSON, A.R. & JACOBS, B.E. & MINKER, J. [1980] *A note on fuzzy deduction* (**J** 0037) ACM J 27*595-603
◇ B52 ◇ REV MR 82g:03038 Zbl 464 # 03024 • ID 54614

ARRINGTON, R.L. [1979] *Criteria and entailment* (**J** 1022) Ratio (Oxford) 21*62-72
◇ A05 B46 ◇ REV MR 81a:03015 • ID 70488

ARRUDA, A.I. & COSTA DA, N.C.A. [1963] *The paradox of Curry-Moh Shaw-Kwei (Portuguese)* (**J** 0108) Bol Soc Mat Sao Paulo 18*82-89
◇ B53 ◇ REV MR 35 # 5291 Zbl 168.1 • ID 00479

ARRUDA, A.I. & COSTA DA, N.C.A. [1964] *Sur une hierarchie de systemes formels* (**J** 0109) C R Acad Sci, Paris 259*2943-2945
◇ B53 ◇ REV MR 31 # 3335 Zbl 134.15 • ID 00483

ARRUDA, A.I. [1967] *Sur certaines hierarchies de calculus propositionnels* (**J** 2313) C R Acad Sci, Paris, Ser A-B 265*A641-A644
◇ B53 ◇ REV MR 36 # 4959 Zbl 153.4 • REM Part I. Parts II,III 1968 • ID 31612

ARRUDA, A.I. [1968] *Sur certaines hierarchies de calculs propositionnels* (**J** 2313) C R Acad Sci, Paris, Ser A-B 266*A37-A39,A897-A900
◇ B53 ◇ REV MR 38 # 46 MR 38 # 47 Zbl 157.14 • REM Parts II,III. Part I 1967 • ID 00487

ARRUDA, A.I. [1969] *Sur certaines hierarchies de calculus de predicats* (**J** 2313) C R Acad Sci, Paris, Ser A-B 268*A629-A632
◇ B53 ◇ REV MR 39 # 5313 Zbl 175.260 • ID 00489

ARRUDA, A.I. [1969] *Sur certaines algebres de classes non classiques* (**J** 2313) C R Acad Sci, Paris, Ser A-B 268*A677-A680
◇ B53 G15 ◇ REV MR 39 # 5314 Zbl 175.260 • ID 00488

ARRUDA, A.I. & COSTA DA, N.C.A. [1970] *Sur le schema de la separation* (**J** 0111) Nagoya Math J 38*71-84
◇ B53 E70 ◇ REV MR 41 # 6681 Zbl 199.297 • ID 00490

ARRUDA, A.I. [1970] *Sur le systeme* NF_{ω} (**J** 2313) C R Acad Sci, Paris, Ser A-B 270*A1137-A1139
◇ B53 E70 ◇ REV MR 41 # 6680 Zbl 193.294 • ID 00493

ARRUDA, A.I. [1970] *Sur les systemes* NF_i *de da Costa* (**J** 2313) C R Acad Sci, Paris, Ser A-B 270*A1081-A1084
◇ B53 E70 ◇ REV MR 41 # 6679 Zbl 193.294 • ID 00492

ARRUDA, A.I. [1971] *La mathematique classique dans* NF_{ω} (**J** 2313) C R Acad Sci, Paris, Ser A-B 272*A1152-A1153
◇ B53 E30 E35 E70 ◇ REV MR 44 # 58 Zbl 214.12 • ID 00494

ARRUDA, A.I. & COSTA DA, N.C.A. [1974] *Le schema de la separation dans les calculs* $\mathcal{J}_n$ (**J** 0352) Math Jap 19*183-186
◇ B53 E70 ◇ REV MR 53 # 101 Zbl 311 # 02070 • ID 16566

ARRUDA, A.I. [1975] *Le schema de la separation dans les systemes* NF_n *(English summary)* (**J** 2313) C R Acad Sci, Paris, Ser A-B 280*A1341-A1343
◇ B53 E70 ◇ REV MR 57 # 9542 Zbl 307 # 02047 • ID 27257

ARRUDA, A.I. [1975] *Remarques sur les systemes* $\mathcal{C}_n$ *(English summary)* (**J** 2313) C R Acad Sci, Paris, Ser A-B 280*A1253-A1256
◇ B53 ◇ REV MR 57 # 12169 Zbl 304 # 02011 • ID 60186

ARRUDA, A.I. & CHUAQUI, R.B. & COSTA DA, N.C.A. (EDS.) [1977] *Nonclassical logics, model theory and computability. Proceedings of the Third Latin-American Symposium on Mathematical Logic* (**X** 0809) North Holland: Amsterdam xviii+307pp
◇ B97 C97 ◇ REV MR 55 # 12445 Zbl 348 # 00003 • ID 16589

ARRUDA, A.I. [1977] *On the imaginary logic of N.A. Vasil'ev* (**P** 1076) Latin Amer Symp Math Log (3);1976 Campinas 1976*3-24
◇ B53 ◇ REV MR 58 # 21462 Zbl 359 # 02015 • ID 50563

Arruda, A.I. & Costa da, N.C.A. [1977] *Une semantique pour le calcul* $\mathscr{C}_1^=$ (J 2313) C R Acad Sci, Paris, Ser A-B 284*A279-A282
◇ B53 E70 ◇ rev MR 55 # 2553 Zbl 351 # 02020 • id 31721

Arruda, A.I. & Chuaqui, R.B. & Costa da, N.C.A. (eds.) [1978] *Mathematical logic. Proceedings of the first Brazilian conference on mathematical logic* (X 1684) Dekker: New York xii + 303pp
◇ B97 ◇ rev MR 58 # 16082 Zbl 377 # 00004 • id 32251

Arruda, A.I. [1978] *Some remarks on Griss' logic of negationless intuitionistic mathematics* (P 1800) Brazil Conf Math Log (1);1977 Campinas 9-29
◇ B55 F50 F55 G25 ◇ rev MR 80b:03095 Zbl 383 # 03014 • id 31617

Arruda, A.I. [1979] see Alves, E.H.

Arruda, A.I. [1980] *A survey of paraconsistent logic* (P 2958) Latin Amer Symp Math Log (4);1978 Santiago 1-41
◇ A05 B53 B98 E70 ◇ rev MR 81i:03033 Zbl 426 # 03031 • id 53627

Arruda, A.I. & Chuaqui, R.B. & Costa da, N.C.A. (eds.) [1980] *Mathematical logic in Latin America. Proceedings of the IV Latin American symposium on mathematical logic* (S 3303) Stud Logic Found Math 99*xii + 392pp
◇ B97 ◇ rev MR 81d:03002 Zbl 419 # 00003 • id 53342

Arruda, A.I. & Costa da, N.C.A. & Sette, A.-M. (eds.) [1980] *Proceedings of the third Brazilian Conference on Mathematical logic (UFPe - Recife, December 17-22, 1979)* (X 2836) Soc Brasil Log: Sao Paulo vi + 336pp
◇ B97 ◇ rev MR 81m:03005 Zbl 436 # 00009 • id 55839

Arruda, A.I. [1980] *The paradox of Russell in the systems* NF_n (P 3006) Brazil Conf Math Log (3);1979 Recife 1-12
◇ B53 E70 ◇ rev MR 82j:03068 Zbl 451 # 03021 • id 54036

Arruda, A.I. & Costa da, N.C.A. [1982] *Remarks on da Costa's paraconsistent set theories* (P 2047) Latin Amer Symp Math Log (5);
◇ B53 E70 ◇ id 42801

Arruda, A.I. [1984] *N.A. Vasil'ev: a forerunner of paraconsistent logic* (J 0455) Phil Naturalis 21*472-491
◇ B53 ◇ id 47915

Arruda, A.I. & Costa da, N.C.A. [1984] *On the relevant systems* P *and* P^* *and some related systems* (J 0063) Studia Logica 43*33-50
◇ B53 ◇ id 42357

Arruda, A.I. see Vol. I, V for further entries

Artemov, S.N. [1980] *Arithmetically complete modal theories (Russian)* (S 2582) Semiotika & Inf, Akad Nauk SSSR 14*115-133
◇ B45 C90 F50 ◇ rev MR 82a:03056 Zbl 463 # 03006 • id 54546

Artemov, S.N. [1982] *Applications of modal logic in proof theory (Russian)* (S 2874) Vopr Kibern, Akad Nauk SSSR 75*3-22
◇ B45 F30 ◇ rev Zbl 499 # 03046 • id 38128

Artemov, S.N. [1984] *On modal representations of extensions of Peano arithmetic* (J 2128) C R Math Acad Sci, Soc Roy Canada 6*129-132
◇ B45 F30 ◇ rev Zbl 555 # 03010 • id 45832

Artemov, S.N. [1985] *Modal logics axiomatizing provability (Russian)* (J 0216) Izv Akad Nauk SSSR, Ser Mat 49*1123-1154,1342
◇ B45 F30 ◇ id 49322

Artemov, S.N. [1985] *Nonarithmeticity of truth predicate logics of provability (Russian)* (J 0023) Dokl Akad Nauk SSSR 284*270-271
• transl [1985] (J 0062) Sov Math, Dokl 32*403-405
◇ B45 D35 D55 F30 ◇ id 48972

Asai, K. & Tanaka, H. & Tsukiyama, T. [1982] *A fuzzy system model based on the logical structure* (P 4051) Fuzzy Set & Possibility Th;1980 Acapulco 257-274
◇ B52 ◇ rev MR 84b:03004 • id 46330

Asai, K. & Tanaka, H. & Tsukiyama, T. [1982] *Logical-systems identification by fuzzy input-output data (Japanese) (English summary)* (J 2844) Systems and Control 26*319-326
◇ B52 ◇ rev MR 83k:93031 • id 40317

Asai, K. & Izumi, K. & Tanaka, H. [1983] *Duality of fuzzy systems based on the intuitionistic logical system LJ (Japanese)* (J 0979) Denshi Tsushin Gakkai Ronbunshi, Sect A-D A-D/10*1114-1121
• transl [1983] (J 0464) Syst-Comp-Controls 14/5*45-52
◇ B52 ◇ rev MR 86g:94069 • id 45462

Asai, K. & Izumi, K. & Tanaka, H. [1984] *Fuzzy logical construction of linguistic modal hedges (Japanese)* (J 0979) Denshi Tsushin Gakkai Ronbunshi, Sect A-D A-D67/6*655-660
• transl [1984] (J 0464) Syst-Comp-Controls 15/6*19-26
◇ B52 ◇ rev MR 86d:03026 • id 44660

Asanidze, G.Z. [1981] *A dialogue justification of logic (Russian)* (C 3810) Log-Semat Issl (Tbilisi) 36-53
◇ B60 F50 F65 ◇ rev MR 84j:03061 • id 34653

Asenjo, F.G. [1965] *Dialectic logic* (J 0079) Logique & Anal, NS 8*321-326
◇ B53 ◇ rev Zbl 143.5 • id 30995

Asenjo, F.G. [1966] *A calculus of antinomies* (J 0047) Notre Dame J Formal Log 7*103-105
◇ B53 ◇ rev MR 34 # 5658 Zbl 145.5 • id 00516

Asenjo, F.G. [1971] *Logica dialectica* (J 0162) Teorema (Valencia) 1*7-13
◇ A05 B53 ◇ id 42802

Asenjo, F.G. [1972] *On dialectic logic* (J 0162) Teorema (Valencia) 1972/2*133-134
◇ A05 B53 ◇ id 42803

Asenjo, F.G. & Tamburino, J. [1975] *Logic of antinomies* (J 0047) Notre Dame J Formal Log 16*17-44
◇ B53 E70 ◇ rev MR 50 # 9545 Zbl 246 # 02023 • id 00521

Asenjo, F.G. [1977] *Lesniewski's work and nonclassical set theories* (J 0063) Studia Logica 36*249-255
◇ A05 A10 B60 E70 ◇ rev MR 80a:03069 Zbl 387 # 03020 • id 52236

Asenjo, F.G. see Vol. I, V for further entries

Ash, C.J. & Brickhill, C.J. & Crossley, J.N. & Stillwell, J.C. & Williams, N.H. [1972] *What is mathematical logic ?* (X 0894) Oxford Univ Pr: Oxford ix + 82pp
◇ B98 ◇ rev MR 54 # 2411 Zbl 251 # 02001 JSL 40.241 • id 17019

Ash, C.J. see Vol. III, IV, V for further entries

Ashby, R.W. [1963] *Entailment and modality* (**J** 0113) Proc Aristotelian Soc 63*203-216
◇ A05 B46 ◇ REV JSL 38.668 • ID 00524

Ashworth, E.J. [1974] *Language and logic in the post-medieval period* (**X** 0835) Reidel: Dordrecht xi+304pp
◇ A05 A10 B45 ◇ REV Zbl 306#02004 • ID 60204

Ashworth, E.J. see Vol. I for further entries

Asser, G. [1953] *Die endlichwertigen Lukasiewiczschen Aussagenkalkuele* (**P** 0626) Ber Math-Tagung Berlin;1953 Berlin 15-18
◇ B50 ◇ REV MR 16.782 Zbl 52.10 JSL 21.87 • ID 00539

Asser, G. [1974] *Mathematische Logik und Grundlagen der Mathematik* (**C** 2318) Entwicklung Math in DDR 3-17
◇ B98 ◇ REV Zbl 311#02006 • ID 29592

Asser, G. [1981] *Einfuehrung in die mathematische Logik. Teil III: Praedikatenlogik hoeherer Stufe* (**X** 1079) Teubner: Leipzig 164pp
• REPR [1981] (**X** 1054) Harri Deutsch: Frankfurt 164pp
◇ B15 B98 ◇ REV MR 84k:03032 Zbl 471#03001 Zbl 471#03002 • ID 55197

Asser, G. see Vol. I, III, IV, V for further entries

Atanassov, K. & Stoeva, S. [1984] *Intuitionistic L-fuzzy sets* (**P** 4008) Cybern & Systems Research (7);1984 Wien 539-540
◇ B52 E72 ◇ REV Zbl 547#03030 • ID 43217

Atiyah, M. & Hoare, C.A.R. & Shepherdson, J.C. (eds.) [1984] *Mathematical logic and programming languages* (**J** 0354) Phil Trans Roy Soc London, Ser A 312*343-518
• REPR [1985] (**X** 0819) Prentice Hall: Englewood Cliffs 184pp
◇ B98 ◇ REV MR 85k:68005 • ID 47111

Atlas, J.D. [1981] *Is "not" logical?* (**P** 3705) Int Symp Multi-Val Log (11);1981 Oklahoma City & Norman 124-128
◇ A05 B50 ◇ REV MR 83m:94033 • ID 45932

Austin, D.F. [1981] *Plantinga on actualism and essences* (**J** 0095) Philos Stud 39*35-42
◇ A05 B45 ◇ REV MR 82i:03025 • ID 70545

Austin, D.F. [1983] *Plantinga's theory of proper names* (**J** 0047) Notre Dame J Formal Log 24*115-132
◇ A05 B45 ◇ REV MR 84d:03003 Zbl 464#03009 • ID 54599

Austin, J. [1978] *Systemic causation* (**J** 3273) Stud Hist Philos Sci 9*83-97
◇ A05 B45 ◇ REV MR 58#21370 Zbl 395#03003 • ID 52554

Avenoso, F.J. & Cheifetz, P.M. [1970] *Logic and set theory* (**X** 0821) Wadsworth Publ: Belmont 196pp
◇ B98 E98 ◇ ID 22631

Avramenko, M.B. & Vasilenko, Yu.A. [1976] *Erkennung der Symmetrie von Funktionen der mehrwertigen Logik (Russian)* (**J** 0474) Avtom Vychis Tekh, Akad Nauk Latv SSR 1976/2*22-23
◇ B50 ◇ REV Zbl 323#02027 • ID 30411

Avron, A. [1984] *On modal systems having arithmetical interpretations* (**J** 0036) J Symb Logic 49*935-942
◇ B45 F05 F30 ◇ REV MR 86g:03028 • ID 42415

Avron, A. [1984] *Relevant entailment - semantics and formal systems* (**J** 0036) J Symb Logic 49*334-342
◇ B46 ◇ REV MR 85k:03010 • ID 42417

Avsarkisyan, G.S. [1985] *Polynomial forms of partial functions of k-valued logic (Russian) (English summary)* (**J** 0040) Kibernetika, Akad Nauk Ukr SSR 1985/4*32-36,46,134
◇ B50 ◇ ID 49394

Avsarkisyan, G.S. see Vol. I for further entries

Azorin Poch, F. [1981] *Uncertainty and fuzziness in primary data and estimation procedures* (**P** 4050) Appl Syst & Cybern;1980 Acapulco 2734-2739
◇ B52 ◇ REV MR 84b:00009 • ID 46296

Bachmann, H. [1983] *Der Weg der mathematischen Grundlagenforschung* (**X** 2865) Lang: Frankfurt 240pp
◇ A10 B98 ◇ REV MR 85a:01002 • ID 38832

Bachmann, H. see Vol. V, VI for further entries

Back, K.W. [1964] *A nerve net system in modal logic* (**J** 0079) Logique & Anal, NS 7*22-31
◇ B45 ◇ REV MR 31#1182 Zbl 115.351 • ID 00614

Bacon, J. [1964] *Entailment and the modal fallacy* (**J** 0125) Rev of Metaphysics 18*566-571
◇ B46 ◇ REV JSL 38.668 • ID 00615

Bacon, J. [1973] *Kripke's deontic semantics again* (**J** 0047) Notre Dame J Formal Log 14*581-582
◇ B46 ◇ REV MR 49#2263 Zbl 254#02022 • ID 03848

Bacon, J. [1975] *Belief as relative knowledge* (**C** 1856) Log Enterprise 189-210
◇ A05 B45 ◇ REV Zbl 385#03016 • ID 52124

Bacon, J. [1979] *The logical form of perception sentences* (**J** 0154) Synthese 41*271-308
◇ A05 B45 ◇ REV MR 81c:03003 • ID 70564

Bacon, J. [1980] *Substance and first-order quantification over individual concepts* (**J** 0036) J Symb Logic 45*193-203
◇ A05 B45 ◇ REV MR 81h:03033 Zbl 432#03014 • ID 70563

Bacon, J. [1981] *Purely physical modalities* (**J** 0105) Theoria (Lund) 47*134-141
◇ B45 ◇ REV MR 84f:03013 Zbl 492#03006 • ID 34431

Bacon, J. see Vol. I for further entries

Badarau, D. [1956] *On indeterminacy in particular judgements (Romanian)* (**S** 1613) Probl Logic (Bucharest) 1956*237-313
◇ B52 ◇ ID 47812

Badawi, A.R. [1962] *Formale und mathematische Logik (Arabic)* (**X** 3333) Unknown Publisher: See Remarks 330pp
◇ B98 ◇ REM 4th ed. 1976, Kuwait • ID 43354

Badillo Barallat, M.C. [1955] *Esquemas representativos di sistemas regidos por una logica polivalente (English summary)* (**J** 4441) Rev Calc Autom Cibern 4/9*54-62
◇ B50 ◇ REV MR 17.1013 JSL 25.184 • ID 48773

Badillo Barallat, M.C. [1955] *Fundamentos en relacion con logica simbolica polivalente* (**J** 4443) Gac Mat (Madrid) 7*7-13
◇ B50 ◇ REV MR 17.447 Zbl 64.12 JSL 27.112 • ID 04230

BADILLO BARALLAT, M.C. [1957] *Logica trivalente en la automatizacion de los circuitos* (J 4441) Rev Calc Autom Cibern 6*1-7
◇ B50 ◇ REV JSL 27.112 • ID 48779

BADILLO BARALLAT, M.C. [1958] *Aplicacion de la logica polivalente a la teoria de numeros* (J 4443) Gac Mat (Madrid) 10*76-81
◇ B50 ◇ REV Zbl 80.242 JSL 27.112 • ID 48781

BADILLO BARALLAT, M.C. [1961] *Many-valued logics of systems (Spanish)* (J 4443) Gac Mat (Madrid) 13*185-195
◇ B50 ◇ REV Zbl 156.7 • ID 41879

BAEZ, J.C. [1983] *Recursivity in quantum mechanics* (J 0064) Trans Amer Math Soc 280*339-350
◇ B51 D80 ◇ REV MR 84i:81010 Zbl 547 # 03041 • ID 40205

BAGINSKI, M. & GLAEWE, W. & GOLL, P. & LIST, G. & LOESCHAU, G. & MERTENS, A. & SCHWANITZ, G. & WALTER, M. [1973] *Einfuehrung in die mathematische Logik, Einfuehrung in die Mengenlehre, Aufbau der Zahlenbereiche* (X 1036) Volk & Wissen: Berlin 327pp
◇ B05 B28 B98 E98 ◇ REV MR 50 # 12630 • ID 75585

BAGYINSZKI, J. & DEMETROVICS, J. [1976] *The structure of linear classes in prime-valued logics* (J 2774) Koezlem MTA Szam & Autom: Kutat Intez 16*25-52
◇ B50 ◇ REV MR 57 # 9478 • ID 70586

BAGYINSZKI, J. [1978] *The lattice of closed classes of linear functions over a finite ring of square-free order* (X 3151) Karl Marx Univ Dpt Math: Budapest 21pp
◇ B50 ◇ REV MR 83c:06011 Zbl 419 # 06004 • ID 39029

BAGYINSZKI, J. & DEMETROVICS, J. [1980] *The structure of the maximal linear classes in prime-valued logics* (J 2128) C R Math Acad Sci, Soc Roy Canada 2*209-213
◇ B50 G10 G25 ◇ REV MR 81m:03027 Zbl 468 # 03041 • ID 55106

BAGYINSZKI, J. & DEMETROVICS, J. [1982] *The lattice of linear classes in prime-valued logics* (P 3787) Discr Math;1977 Warsaw 105-123
◇ B50 ◇ REV MR 84g:03103 Zbl 538 # 03020 • ID 34207

BAGYINSZKI, J. [1982] *The solution of the Hosszu equation over finite fields (Russian and Hungarian summaries)* (J 2774) Koezlem MTA Szam & Autom: Kutat Intez 25*25-33
◇ B50 ◇ REV MR 84d:39008 • ID 39683

BAILHACHE, P. [1977] *Semantiques pour des systemes deontiques integrant permission faible et permission forte* (J 0079) Logique & Anal, NS 20*286-316
◇ B45 ◇ REV MR 58 # 27315 Zbl 367 # 02011 • ID 31353

BAILHACHE, P. [1980] *Several possible systems of deontic weak and strong norms* (J 0047) Notre Dame J Formal Log 21*89-100
◇ A05 B45 ◇ REV MR 81h:03034 Zbl 363 # 02004 • ID 53171

BAILHACHE, P. [1981] *Analytical deontic logic: Authorities and addressees* (J 0079) Logique & Anal, NS 24*65-80
◇ A05 B45 ◇ REV MR 82k:03013 Zbl 463 # 03002 • ID 54542

BAILHACHE, P. [1984] *Quelques foncteurs faussement primitifs en logique deontique (trivalence et action)* (J 0079) Logique & Anal, NS 27*393-405
◇ B45 ◇ REV Zbl 563 # 03009 • ID 45521

BAILHACHE, P. [1985] *Logique deontique et temporelle (Polish summary)* (J 3815) Acta Univ Lodz Folia Philos 3*15-28
◇ B45 ◇ ID 49650

BAIN, A. [1889] *Logic: deductive and inductive* (X 1228) Appleton-Century-Crofts: New York 731pp
◇ A05 B98 ◇ REM revised edition of "Logic, deduction" and "Logic, induction" (1870) • ID 22445

BAIRD, D. [1983] *The Fisher/Pearson chi-squared controversy: a turning point for inductive inference* (J 0013) Brit J Phil Sci 34*105-118
◇ B48 ◇ REV MR 85a:62002 • ID 38845

BAJRAMOV, R.A. [1967] *A series of precomplete classes in k-valued logic (Russian)* (J 0040) Kibernetika, Akad Nauk Ukr SSR 1967/1*7-9
• TRANSL [1967] (J 0021) Cybernetics 3/1*6-7
◇ B50 ◇ REV MR 43 # 27 Zbl 189.281 • ID 00671

BAJRAMOV, R.A. [1967] *On a question of functional completeness in many-valued logic (Russian)* (J 0071) Met Diskr Analiz (Novosibirsk) 11*3-20
◇ B50 ◇ REV MR 37 # 5083 Zbl 182.317 • ID 00672

BAJRAMOV, R.A. [1967] *Sheffer functions in many-valued logic (Russian)* (C 1963) Probl in Th of Electr Digital Comp, No 3 (Kiev) 62-71
◇ B50 ◇ REV MR 37 # 5084 • ID 41925

BAJRAMOV, R.A. [1968] *Letter to the editors (Russian)* (J 0071) Met Diskr Analiz (Novosibirsk) 12*115-116
◇ B50 ◇ REV MR 46 # 23 Zbl 182.317 • ID 70591

BAJRAMOV, R.A. [1968] *Predicate stabilizers and Sheffer functions in a finite-valued logic (Russian)* (J 0134) Dokl Akad Nauk Azerb SSR 24/2*3-6
◇ B50 ◇ REV MR 42 # 44 Zbl 182.317 • ID 00673

BAJRAMOV, R.A. [1969] *Criteria of fundamentality for permutation groups and for transformation semigroups (Russian)* (J 0023) Dokl Akad Nauk SSSR 189*455-457
• TRANSL [1969] (J 0062) Sov Math, Dokl 10*1414-1416
◇ B50 ◇ REV MR 41 # 5194 • ID 00675

BAJRAMOV, R.A. [1969] *The predicative characterizability of subalgebras of many-valued logic (Russian)* (J 0135) Izv Akad Nauk Azerb SSR, Ser Fiz-Tekh Mat 1969/1*100-104
◇ B50 ◇ REV MR 41 # 5195 Zbl 193.297 • ID 00674

BAJRAMOV, R.A. [1971] *The classes of multiplace functions on finite sets (Russian, Azerbaijani and English summaries)* (J 0134) Dokl Akad Nauk Azerb SSR 27/7*6-9
◇ B50 ◇ REV MR 46 # 22 Zbl 229 # 02046 • ID 41454

BAJRAMOV, R.A. [1972] *Deduction of criteria for basicness of permutation groups and transformation semigroups from the Rosenberg theorem (Russian) (English summary)* (J 0040) Kibernetika, Akad Nauk Ukr SSR 1972/3*71-76
• TRANSL [1972] (J 0021) Cybernetics 8*429-433
◇ B50 G15 ◇ REV MR 51 # 8221 Zbl 265 # 02012 • ID 17571

BAJRAMOV, R.A. [1978] *On sub-algebras, maximal in P_k, with a given σ_k-component and on subalgebras, maximal in P_k, with ideal $\Omega_{k,k-1}$-component (Russian)* (J 3166) Fkts Anal Primen, Akad Nauk Azerb SSR 3*92-100
◇ B50 G20 ◇ REV MR 80i:08004 Zbl 408 # 03047 • ID 56285

Author Index

BAJRAMOV, R.A. [1981] *Some new results in the theory of function algebras of finite-valued logics* (**P** 2552) Conf Finite Algeb & Multi-Val Log;1979 Szeged 41-67
◇ B50 G20 ◇ REV MR 84c:03105 Zbl 485 # 03035 • ID 34039

BAJRAMOV, R.A. [1982] *On independent sets of functions in finite-valued logics* (**P** 4002) Int Symp Multi-Val Log (12);1982 Paris 114-116
◇ B50 ◇ REV Zbl 544 # 03005 • ID 40988

BAJRAMOV, R.A. see Vol. III for further entries

BAKER, J.R. [1978] *Essentialism and the modal semantics of K.J.J. Hintikka* (**J** 0047) Notre Dame J Formal Log 19*81-91
◇ B45 ◇ REV MR 81e:03011 Zbl 258 # 02024 Zbl 373 # 02024 • ID 27080

BAKER, J.R. [1978] *Some remarks on Quine's arguments against modal logic* (**J** 0047) Notre Dame J Formal Log 19*663-673
◇ A05 B45 ◇ REV MR 80a:03021 Zbl 283 # 02023 • ID 53605

BAKER, J.R. see Vol. I for further entries

BAKER, K.A. [1976] *Equational axioms for classes of Heyting algebras* (**J** 0004) Algeb Universalis 6*105-120
◇ B55 C05 G05 G10 ◇ REV MR 54 # 12520 Zbl 355 # 02039 • ID 23893

BAKER, K.A. see Vol. III, V for further entries

BAKER, L.R. & WALD, J.D. [1979] *Indexical reference and de re belief* (**J** 0095) Philos Stud 36*317-327
◇ A05 B45 ◇ REV MR 81g:03013 • ID 70607

BAKHMUTOVA, I.V. [1981] *A variant of the theory of prediction for the case of four-valued logic (Russian)* (**S** 0507) Vychisl Sist (Akad Nauk SSSR Novosibirsk) 91*91-95
◇ B50 ◇ REV MR 84j:03053 Zbl 509 # 03011 • ID 34645

BALABAEV, V.M. & GERASIMOV, I.N. [1979] *A numerical method for the minimization of ternary logic functions (Russian)* (**J** 2743) Izv Sev-Kavk Nauch Tsentra, Tekh (Rostov nD) 1979/2*7-10
◇ B50 ◇ REV MR 81m:94030 Zbl 416 # 94025 • ID 53237

BALDWIN, J.F. & PILSWORTH, B.W. [1979] *A model of fuzzy reasoning through multi-valued logic and set theory* (**J** 1741) Int J Man-Mach Stud 11*351-380
◇ B52 E72 ◇ REV MR 81h:03051 Zbl 413 # 03015 • ID 53009

BALDWIN, J.F. [1979] *A new approach to approximate reasoning using a fuzzy logic* (**J** 2720) Fuzzy Sets Syst 2*309-325
◇ B52 ◇ REV Zbl 413 # 03017 • ID 53011

BALDWIN, J.F. & GUILD, N.C.F. [1979] *Fuzlog: A computer program for fuzzy reasoning* (**P** 3003) Int Symp Multi-Val Log (9);1979 Bath 38-45
◇ B52 ◇ ID 35938

BALDWIN, J.F. [1979] *Fuzzy logic and approximate reasoning for mixed input arguments* (**J** 1741) Int J Man-Mach Stud 11*381-396
◇ B52 ◇ REV MR 81h:03052 Zbl 413 # 03016 • ID 53010

BALDWIN, J.F. [1979] *Fuzzy logic and fuzzy reasoning* (**J** 1741) Int J Man-Mach Stud 11*465-480
◇ B52 ◇ REV MR 81c:03018 Zbl 418 # 03014 • ID 53299

BALDWIN, J.F. [1979] *Fuzzy logic and its application to fuzzy reasoning* (**C** 3514) Adv Fuzzy Sets & Appl 93-115
◇ B52 ◇ REV MR 81a:03023 • ID 70618

BALDWIN, J.F. & PILSWORTH, B.W. [1979] *Fuzzy truth definition of possibility measure for decision classification* (**J** 1741) Int J Man-Mach Stud 11*447-463
◇ B52 E72 ◇ REV MR 81b:94082 Zbl 407 # 94021 • ID 80615

BALDWIN, J.F. & GUILD, N.C.F. [1979] *On the satisfaction of a fuzzy relation by a set of inputs* (**J** 1741) Int J Man-Mach Stud 11*397-404
◇ B52 E72 ◇ REV MR 81h:03102 Zbl 405 # 03028 • ID 70620

BALDWIN, J.F. & PILSWORTH, B.W. [1980] *Axiomatic approach to implication for approximate reasoning with fuzzy logic* (**J** 2720) Fuzzy Sets Syst 3*193-219
◇ B52 ◇ REV MR 80m:03058 Zbl 434 # 03021 • ID 55725

BALDWIN, J.F. & GUILD, N.C.F. [1980] *Feasible algorithms for approximate reasoning using fuzzy logic* (**J** 2720) Fuzzy Sets Syst 3*225-251
◇ B52 ◇ REV MR 81h:03053 Zbl 435 # 03019 • ID 55780

BALDWIN, J.F. & GUILD, N.C.F. [1980] *The resolution of two paradoxes by approximate reasoning using a fuzzy logic* (**J** 0154) Synthese 44*397-420
◇ A05 B52 ◇ REV MR 81m:03031 Zbl 444 # 03011 • ID 56459

BALDWIN, J.F. [1981] *Fuzzy logic knowledge bases and automated fuzzy reasoning* (**P** 4050) Appl Syst & Cybern;1980 Acapulco 2859-2865
◇ B52 ◇ REV MR 84b:00009 • ID 46261

BALDWIN, J.F. & GUILD, N.C.F. [1981] *Fuzzy reasoning applied to control models* (**P** 3844) IMA Conf Control Th (3);1980 Sheffield 647-660
◇ B52 ◇ REV Zbl 509 # 93002 • ID 36803

BALDWIN, J.F. & PILSWORTH, B.W. [1981] *Fuzzy reasoning with probability* (**P** 3705) Int Symp Multi-Val Log (11);1981 Oklahoma City & Norman 100-108
◇ B52 ◇ REV MR 83m:94033 Zbl 543 # 94032 • ID 40990

BALDWIN, J.F. [1982] *An automated fuzzy reasoning algorithm* (**P** 4051) Fuzzy Set & Possibility Th;1980 Acapulco 169-195
◇ B52 ◇ REV MR 84b:03004 • ID 46325

BALDWIN, J.F. & ZHOU, S.Q. [1984] *A fuzzy relational inference language* (**J** 2720) Fuzzy Sets Syst 14*155-174
◇ B52 ◇ REV MR 85j:68098 • ID 44018

BALDWIN, THOMAS [1975] *The philosophical significance of intensional logic. II* (**P** 3584) Joint Sess Aristot Soc & Mind Assoc;1975 Canterbury 45-65
◇ A05 B45 ◇ REV MR 58 # 21390b • REM Part I 1975 by Kamp,H. • ID 70640

BALDWIN, THOMAS [1984] *E.J.Lowe on modalities de re: "On the alleged necessity of true identity statements"* (**J** 0094) Mind 93(370)*252-255
◇ A05 B45 ◇ REV MR 86e:03017 • REM See also Lowe,E.J. 1982 • ID 45048

BALLARD, D. [1985] *Independence in higher-order subclassical logic* (**J** 0047) Notre Dame J Formal Log 26*444-454
◇ B55 ◇ ID 47538

BALLARD, K.E. [1972] *Study guide for Copi: Introduction to logic. A self-instructional supplement* (**X** 0843) Macmillan : New York & London 274pp
◇ B98 ◇ ID 22457

BALLESTREM, K.G. [1965] *Dialectical logic* (**J** 2051) Stud Soviet Thought 5*139-172
◇ B53 ◇ ID 42804

BALLMER, T.T. [1976] *Prolegomena to a logic of changing the context by operations and commands (A case study of objects of intention.)* (**J** 0079) Logique & Anal, NS 19*427-456
◇ A05 B60 ◇ REV Zbl 355#02001 • ID 50199

BALLMER, T.T. & PINKAL, M. (EDS.) [1983] *Approaching vagueness* (**C** 3095) Approaching Vagueness x+429pp
◇ B52 ◇ REV MR 85j:03003 • ID 44583

BALUKA, E. [1965] *On verification of the expressions of many-valued sentential calculi I (Polish and Russian summaries)* (**J** 0063) Studia Logica 17*53-73
◇ B50 ◇ REV MR 34#4120 Zbl 309#02016 • ID 00747

BANACHOWSKI, L. & KRECZMAR, L. & MIRKOWSKA, G. & RASIOWA, H. & SALWICKI, A. [1977] *An introduction to algorithmic logic* (**C** 4723) Math Founds of Comput Sci 2*7-99
◇ B75 ◇ REV MR 58#16198 Zbl 358#68035 • ID 31445

BANACHOWSKI, L. [1977] *Investigations of properties of programs by means of the extended algorithmic logic. I,II* (**J** 2095) Fund Inform, Ann Soc Math Pol, Ser 4 1*93-119,167-193
◇ B75 ◇ REV MR 57#14570 Zbl 358#68037 Zbl 372#68004 • ID 50539

BANACHOWSKI, L. see Vol. I, IV for further entries

BANAI, M. [1981] *Propositional systems in field theories and lattice-valued quantum logic* (**P** 3820) Curr Iss in Quantum Log;1979 Erice 425-435
◇ B51 ◇ REV MR 84m:81032 Zbl 476#03060 • ID 39548

BANAI, M. [1981] *Propositional systems in local field theories* (**J** 2736) Int J Theor Phys 20*147-169
◇ B51 G12 ◇ REV MR 82g:81007 Zbl 476#03060 • ID 55572

BANDEMER, H. [1985] *Evaluating explicit functional relationships from fuzzy observations* (**J** 2720) Fuzzy Sets Syst 16*41-52
◇ B52 ◇ ID 46404

BANDLER, W. & KOHOUT, L.J. [1979] *Application of fuzzy logics to computer protection structures* (**P** 3003) Int Symp Multi-Val Log (9);1979 Bath 200-207
◇ B52 ◇ ID 35953

BANDLER, W. & KOHOUT, L.J. [1980] *Fuzzy power sets and fuzzy implication operators* (**J** 2720) Fuzzy Sets Syst 4*13-30
◇ B52 E72 ◇ REV MR 82d:03086 Zbl 433#03013 • ID 70651

BANDLER, W. & KOHOUT, L.J. [1982] *Axioms for conditional inference: probabilistic and possibilistic* (**P** 3845) Conf Math Service of Man (2,Proc)(Feriet);1982 Las Palmas 413-414
◇ B48 ◇ REV Zbl 515#03037 • ID 37848

BANDLER, W. & KOHOUT, L.J. [1983] *Probabilistic vs. fuzzy production rules in expert systems* (**J** 3919) BUSEFAL 14*105-114
◇ B52 ◇ REV Zbl 529#68069 • ID 38494

BANDLER, W. see Vol. V for further entries

BAR-HILLEL, Y. [1953] *A note on comparative inductive logic* (**J** 0013) Brit J Phil Sci 3*308-310
◇ A05 B48 ◇ REV MR 15.190 JSL 19.300 • ID 00804

BAR-HILLEL, Y. & DALEN VAN, D. & FRAENKEL, A.A. & LEVY, A. [1958] *Foundations of set theory* (**X** 0809) North Holland: Amsterdam x+415pp
◇ A05 B98 E30 E70 E98 F98 ◇ REV MR 21#648 MR 49#10546 Zbl 248#02071 Zbl 82.262 JSL 29.141 • REM 2nd ed. 1973; x+404pp; 1st ed. by Fraenkel,A.A. & Bar-Hillel,Y. • ID 00805

BAR-HILLEL, Y. [1964] *Language and information: selected essays on their theory and application* (**X** 0833) Acad Pr: Jerusalem x+388pp
• LAST ED [1964] (**X** 0832) Addison-Wesley: Reading x+388pp
◇ A05 B65 B98 ◇ REV MR 29#6959 Zbl 158.241 JSL 30.382 • ID 37308

BAR-HILLEL, Y. [1969] *Is modal logic different from the semantics of modal terms ?* (**J** 0147) An Univ Bucuresti, Acta Logica 12*223-230
◇ B45 ◇ REV MR 45#31 • ID 00810

BAR-HILLEL, Y. see Vol. I, III, IV, V for further entries

BARAK, A. [1975] see ARON, E.

BARCAN, R.C. [1946] *A functional calculus of first order based on strict implication* (**J** 0036) J Symb Logic 11*1-16
◇ B45 ◇ REV MR 8.96 JSL 11.96 • ID 00792

BARCAN, R.C. [1946] *The deduction theorem in a functional calculus of first order based on strict implication* (**J** 0036) J Symb Logic 11*115-118
◇ B45 ◇ REV MR 8.95 JSL 12.95 • ID 00791

BARCAN, R.C. also published under the name BARCAN MARCUS, R.

BARCAN, R.C. see Vol. I for further entries

BARCAN MARCUS, R. [1950] *The elimination of contextually defined predicates in a modal system* (**J** 0036) J Symb Logic 15*92
◇ B45 ◇ REV MR 12.2 Zbl 38.6 JSL 16.73 • ID 33288

BARCAN MARCUS, R. [1953] *Strict implication, deducibility, and the deduction theorem* (**J** 0036) J Symb Logic 18*234-236
◇ B46 ◇ REV MR 15.385 Zbl 53.341 JSL 19.294 • ID 08705

BARCAN MARCUS, R. [1963] *Classes and attributes in extended modal systems* (**J** 0096) Acta Philos Fenn 16*123-136
◇ B45 ◇ REV MR 31#2137 Zbl 135.6 JSL 37.180 • ID 08707

BARCAN MARCUS, R. [1966] *Iterated deontic modalities* (**J** 0094) Mind 75*580-582
◇ B45 ◇ REV JSL 34.500 • ID 43290

BARCAN MARCUS, R. [1967] *Essentialism and modal logic* (**J** 0097) Nous, Quart J Phil 1*91-96
◇ B45 ◇ ID 27086

BARCAN MARCUS, R. [1967] *Modalities and intensional languages* (**C** 0721) Contemp Readings in Log Th 278-293
◇ B45 ◇ ID 08708

BARCAN MARCUS, R. [1968] *Modal logic* (**C** 0552) Phil Contemp - Chroniques 87-101
◇ B45 ◇ ID 14647

BARCAN MARCUS, R. [1975] see ANDERSON, A.R.

Author Index

BARCAN MARCUS, R. also published under the name BARCAN, R.C.

BARENDREGT, H.P. [1981] *The lambda calculus, its syntax and semantics* (**S** 3303) Stud Logic Found Math 103*xiv+615pp
• TRANSL [1985] (**X** 0885) Mir: Moskva 606pp
◇ B40 B98 ◇ REV MR 83b:03016 MR 86a:03012 Zbl 467 # 03010 Zbl 551 # 03007 JSL 49.301 • REM Revised ed.1984, xv+621pp • ID 55008

BARENDREGT, H.P. see Vol. IV, V, VI for further entries

BARKER, C.C.H. [1957] *Some calculations in logic* (**J** 0148) Math Gaz 41*108-111
◇ B50 ◇ REV MR 19.3 Zbl 83.1 JSL 22.379 • ID 00812

BARKER, J.A. [1976] *Presupposition and entailment* (**J** 0047) Notre Dame J Formal Log 17*272-278
◇ B46 ◇ REV Zbl 262 # 02007 • ID 30407

BARKER, S.F. [1958] *Induction and hypothesis, A study of the logic of confirmation* (**X** 0992) Cornell Univ Pr: Ithaca xvi+203pp
◇ B48 ◇ REV JSL 27.122 • ID 43023

BARKER, S.F. [1965] *The elements of logic* (**X** 0822) McGraw-Hill: New York xii+336pp
◇ B98 ◇ REV Zbl 168.3 • ID 22458

BARLINGAY, S.S. [1965] *A modern introduction to Indian logic* (**X** 4322) Nat Publ House: New Delhi xv+238pp
◇ A10 B98 ◇ REV JSL 33.603 • ID 43203

BARNES, D.W. & MACK, J.M. [1975] *An algebraic introduction to mathematical logic* (**X** 0811) Springer: Heidelberg & New York viii+121pp
◇ B98 G98 ◇ REV MR 52 # 10362 Zbl 311 # 02001 • ID 21669

BARNES, D.W. see Vol. III, VI for further entries

BARNES, K.T. [1972] *Hintikka's argument for the need for quantifying into opaque contexts* (**J** 0095) Philos Stud 23*385-392
◇ A05 B45 ◇ REV MR 54 # 12472 • ID 70688

BARNES, R.F. & GUMB, R.D. [1979] *The completeness of presupposition-free tense logic* (**J** 0068) Z Math Logik Grundlagen Math 25*193-208
◇ B45 ◇ REV MR 80i:03027 Zbl 421 # 03014 • ID 53413

BARNES, R.F. [1981] *Interval temporal logic: a note* (**J** 0122) J Philos Logic 10*395-397
◇ B45 ◇ REV MR 84h:03032 • ID 33044

BARNETT, P.M. [1983] *Rational behavior in bargaining situations* (**J** 0097) Nous, Quart J Phil 17*621-635
◇ B45 ◇ REV MR 85e:03032 • ID 40603

BARONE, F. & GALDI, G.P. [1979] *On the question of atomicity and determinism in boolean systems* (**J** 2778) Lett Nuovo Cimento Ser 2 24*179-182
◇ B51 G12 ◇ REV MR 80e:03074 • ID 70690

BARONE, F. [1981] *On implemented state automorphisms with the logico-algebraic approach to deterministic mechanics* (**J** 2778) Lett Nuovo Cimento Ser 2 30*155-159
◇ B51 ◇ REV MR 83d:81007 • ID 40078

BARONE, F. see Vol. I for further entries

BARRINGER, H. & KUIPER, R. [1985] *Hierarchical development of concurrent systems in a temporal logic framework* (**P** 4580) Semin Concurrency;1984 Pittsburgh 35-61
◇ B45 ◇ REV Zbl 565 # 68017 • ID 49029

BARTH, E.M. [1972] *Implication et necessite* (**J** 0082) Bull Soc Math Belg 24*20-32
◇ A10 B45 ◇ REV MR 52 # 13294 Zbl 324 # 02013 • ID 21756

BARTH, E.M. [1982] *A normative-pragmatical foundation of the rules of some systems of formal$_3$ dialectics* (**P** 3754) Argumentation;1978 Groningen 159-170
◇ B53 ◇ ID 45907

BARTH, E.M. & MARTENS, J.L. (EDS.) [1982] *Argumentation* (**X** 2257) Benjamins: Amsterdam xvii+333pp
◇ B97 ◇ REV MR 83k:03003 Zbl 532 # 00004 • ID 45900

BARTH, E.M. see Vol. V, VI for further entries

BARTNICK, J. [1980] *Predicate logic without bound variables* (**X** 2865) Lang: Frankfurt vi+84pp
◇ B98 ◇ REV MR 82k:03010 • ID 70695

BARTON, S. [1979] *The functional completeness of Post's m-valued propositional calculus* (**J** 0068) Z Math Logik Grundlagen Math 25*445-446
◇ B50 G20 ◇ REV MR 80j:03036 Zbl 421 # 03015 • ID 53414

BARWISE, J. [1975] *Admissible sets and structures. An approach to definability theory* (**X** 0811) Springer: Heidelberg & New York xiii+394pp
◇ B98 C40 C70 C98 D60 D98 E30 E98 ◇ REV MR 54 # 12519 Zbl 316 # 02047 JSL 43.139 • ID 60316

BARWISE, J. (ED.) [1977] *Handbook of mathematical logic* (**X** 0809) North Holland: Amsterdam xi+1165pp
• TRANSL [1982] (**X** 2027) Nauka: Moskva
◇ B98 C98 D98 E98 F98 H98 ◇ REV MR 56 # 15351 MR 84g:03004 MR 84j:03006 Zbl 443 # 03001 JSL 49.968 JSL 49.971 JSL 49.975 JSL 49.980 • REM 3rd ed 1982. Transl. in 4 parts. Russian suppl. by Mints,G.E. & Orevkov,V.P • ID 70117

BARWISE, J. & KEISLER, H.J. & KUNEN, K. (EDS.) [1980] *The Kleene Symposium* (**S** 3303) Stud Logic Found Math 101*xx+425pp
◇ B97 ◇ REV MR 81j:03009 Zbl 436 # 00007 • ID 55837

BARWISE, J. & COOPER, R. [1981] *Generalized quantifiers and natural language* (**J** 2130) Linguist Philos 4*159-219
◇ A05 B60 C80 ◇ REV Zbl 473 # 03033 • ID 55362

BARWISE, J. see Vol. I, III, IV, V, VI for further entries

BARZDINS, J. (ED.) [1974] *Theory of algorithms and programs I (Russian)* (**X** 0895) Latv Valsts (Gos) Univ : Riga
◇ B97 D97 ◇ REM Vol.II 1975 • ID 46742

BARZDINS, J. (ED.) [1975] *Theory of algorithms and programs II (Russian)* (**X** 0895) Latv Valsts (Gos) Univ : Riga
◇ B97 D97 ◇ REM Vol.II 1974. Vol.III 1977 • ID 46743

BARZDINS, J. [1982] *Some rules of inductive inference and their application (Russian)* (**S** 2582) Semiotika & Inf, Akad Nauk SSSR 19*59-89
◇ B48 ◇ REV MR 84g:68010 Zbl 516 # 68015 • ID 39489

BARZDINS, J. see Vol. IV for further entries

BASSON, A.H. & O'CONNOR, D.J. [1953] *Introduction to symbolic logic* (**X** 0927) Univ Tutorial Pr: Slough 169pp
• LAST ED [1960] (**X** 0824) Free Press: New York viii + 175pp
◇ B98 ◇ REV MR 22 # 3672 Zbl 52.8 Zbl 89.6 JSL 20.84 JSL 28.169 • ID 22460

BATENS, D. [1971] *The paradoxes of confirmation* (**J** 2076) Rev Int Philos 25*101-118
◇ A05 B48 ◇ REV MR 58 # 27425 • ID 70739

BATENS, D. [1980] *Paraconsistent extensional propositional logics* (**J** 0079) Logique & Anal, NS 23*195-234
◇ A05 B53 ◇ REV MR 83g:03025 Zbl 459 # 03013 • ID 54453

BATENS, D. [1982] *A bridge between two-valued and many-valued semantic systems: n-tuple semantics* (**P** 4002) Int Symp Multi-Val Log (12);1982 Paris 318-322
◇ B50 ◇ REV Zbl 556 # 03020 • ID 46149

BATENS, D. & BENDEGEM VAN, J.P. [1985] *Relevant derivability and classical derivability in Fitch-style and axiomatic formulations of relevant logics* (**J** 0079) Logique & Anal, NS 28*21-31
◇ B46 ◇ ID 49167

BATENS, D. see Vol. I, V for further entries

BATTEN, L.M. & WALTON, D.N. [1984] *Games, graphs and circular arguments* (**J** 0079) Logique & Anal, NS 27*133-164
◇ B60 ◇ REV MR 86b:03011 Zbl 558 # 03001 • ID 44004

BATTICELLI, N.A. [1984] *Integral logic* (**J** 3099) Theoretic Papers 2*v + 161pp
◇ B60 ◇ ID 44928

BAUDISCH, A. & SEESE, D.G. & TUSCHIK, H.-P. & WEESE, M. [1980] *Decidability and generalized quantifiers* (**X** 0911) Akademie Verlag: Berlin xii + 235pp
◇ B25 B98 C10 C60 C80 C98 ◇ REV MR 82i:03048 Zbl 442 # 03011 JSL 47.907 • ID 56368

BAUDISCH, A. see Vol. III, IV, V for further entries

BAUMER, W.H. [1963] *Discussion: Von Wright's paradoxes* (**J** 0153) Phil of Sci (East Lansing) 30*165-172
◇ A05 B45 ◇ REV MR 27 # 33 • ID 00870

BAXANDALL, P.R. & BROWN, W.S. & ROSE, G.S.C. & WATSON, F.R. [1978] *Proof in mathematics ("If", "then" and "perhaps"). A collection of material illustrating the nature and variety of the idea of proof in mathematics* (**X** 3357) Inst Educ Univ Keele: Keele v + 130pp
◇ B98 ◇ REV MR 80g:00001 Zbl 426 # 03001 • ID 53597

BAXTER, RODNEY J. [1973] *On some models for modal logic* (**J** 0047) Notre Dame J Formal Log 14*121-122
◇ B45 ◇ REV MR 48 # 1883 Zbl 247 # 02026 • ID 29502

BAXTER, RODNEY J. see Vol. IV for further entries

BAYART, A. [1958] *Correction de la logique modale du premier et du second ordre S5* (**J** 0079) Logique & Anal, NS 1*28-45
• ERR/ADD ibid 2*54
◇ B45 ◇ REV JSL 30.373 • ID 00890

BAYART, A. [1959] *Quasi-adequation de la logique modale du second ordre S5 et adequation de la logique modale du premier ordre S5* (**J** 0079) Logique & Anal, NS 2*99-121
◇ B45 ◇ REV JSL 30.373 • ID 00891

BAYART, A. [1969] *Pour une logique de l'"entailment"* (**J** 0079) Logique & Anal, NS 12*353-360
◇ B46 ◇ REV MR 43 # 6052 Zbl 273 # 02018 • ID 30489

BAYART, A. [1970] *On truth tables for M, B, S4 and S5* (**J** 0079) Logique & Anal, NS 13*335-375
◇ B45 ◇ REV MR 45 # 1734 Zbl 218 # 02018 • ID 00892

BAYART, A. [1980] *Causalite,essai de definition* (**J** 0079) Logique & Anal, NS 23*331-343
◇ B45 ◇ REV MR 83c:03002 • ID 35117

BAYLIS, C.A. & BENNETT, A.A. [1939] *Formal logic: A modern introduction* (**X** 0819) Prentice Hall: Englewood Cliffs xviii + 407pp
◇ B98 ◇ REV Zbl 25.4 JSL 4.94 FdM 65.20 • ID 00893

BAYLIS, C.A. see Vol. I for further entries

BAZHANOV, V.A. [1984] *Logic of quantum mechanics and the problem of its decidability (Russian)* (**C** 4403) Logika, Pozn, Otrazh 58-65
◇ B51 D35 ◇ ID 46738

BEACH, J. [1970] *Introduction to logic* (**X** 0802) Allyn & Bacon: London x + 170pp
◇ B98 ◇ REV Zbl 215.318 • ID 22462

BEALER, G. [1982] *Quality and concept* (**X** 0815) Clarendon Pr: Oxford xii + 311pp
◇ A05 B45 ◇ REV JSL 50.554 • ID 47347

BEALER, G. see Vol. I for further entries

BEARD, J.B. [1976] *The modalities of* $KT4_nMG$ (**J** 0047) Notre Dame J Formal Log 17*462-464
◇ B45 ◇ REV MR 56 # 88 Zbl 316 # 02026 • ID 18101

BEARD, R.W. [1973] *A natural deduction variant of Systems T, S4, S5, and the brouwerian system* (**J** 0079) Logique & Anal, NS 16*311-314
◇ B45 ◇ REV MR 50 # 4246 Zbl 296 # 02011 • ID 03875

BEATTY, H. [1972] *On evaluating deontic logics* (**J** 0122) J Philos Logic 1*439-444
◇ B45 ◇ REV MR 58 # 21468b • ID 70803

BEATTY, H. [1973] *On evaluating deontic logics. Comments on van Fraassen's paper* (**P** 1556) Exact Philos: Probl, Tools & Goals;1971 Montreal 173-178
◇ B45 ◇ REV MR 58 # 21468b Zbl 316 # 02032 • ID 60380

BEAVER, O.R. [1983] *Regularity and decomposability of finitely additive functions on a quantum logic* (**P** 2410) Measure Th & Appl;1982 Sherbrooke 59-67
◇ B51 ◇ REV MR 85i:81008 Zbl 529 # 28006 • ID 44395

BEAZER, R. [1976] *Products of Post algebras with applications* (**S** 0019) Colloq Math (Warsaw) 35*179-187
◇ B50 G20 ◇ REV MR 53 # 13057 Zbl 341 # 06005 • ID 23205

BEAZER, R. & MACNAB, D.S. [1979] *Modal extensions of Heyting algebras* (**S** 0019) Colloq Math (Warsaw) 41*1-12
◇ B45 G10 G20 G25 ◇ REV MR 81d:06013 Zbl 436 # 06010 • ID 80656

BECCHIO, D. [1972] *Nouvelle demonstration de la completude du systeme de Wajsberg axiomatisant la logique trivalente Lukasiewicz* (**J** 2313) C R Acad Sci, Paris, Ser A-B 275*A679-A682
◇ B50 ◇ REV MR 46 # 6993 Zbl 248 # 02024 • ID 00926

Author Index

BECCHIO, D. [1976] *Calcul des sequents et deduction naturelle pour la logique trivalente de Lukasiewicz* (**J** 1934) Ann Sci Univ Clermont Math 13*55-73
◇ B50 ◇ REV MR 57 # 62 Zbl 355 # 02012 • ID 50210

BECCHIO, D. & PABION, J.F. [1977] *Gentzen's techniques in the three-valued logic of Lukasiewicz* (**J** 0036) J Symb Logic 42*123-124
◇ B50 ◇ ID 31354

BECCHIO, D. [1978] *Logique trivalente de Lukasiewicz* (**J** 1934) Ann Sci Univ Clermont Math 16*33-83
◇ B50 G10 G20 ◇ REV MR 82m:03034 Zbl 414 # 03013 • ID 31640

BECCHIO, D. see Vol. I for further entries

BECKER, O. [1930] *Zur Logik der Modalitaeten* (**J** 1118) Jbuch Phil Phaenomenol Forsch 11*496-548
◇ A05 B45 ◇ REV FdM 56.37 • ID 22127

BECKER, O. [1943] *Das formale System der ontologischen Modalitaeten* (**J** 0168) Blaetter Deutsch Philos 16*387-422
◇ B45 ◇ REV Zbl 60.19 • ID 47931

BECKER, O. [1951] *Einfuehrung in die Logistik, vorzueglich in den Modalkalkuel* (**X** 0825) Westkulturverlag : Meisenheim 92pp
◇ B98 ◇ REV MR 13.309 Zbl 45.1 JSL 17.59 • ID 00927

BECKER, O. [1952] *Untersuchungen ueber den Modalkalkuel* (**X** 0825) Westkulturverlag : Meisenheim 87pp
◇ B45 ◇ REV MR 14.3 Zbl 46.5 JSL 18.327 • ID 00928

BECKMAN, F.S. [1981] *Mathematical foundations of programming* (**X** 0832) Addison-Wesley: Reading xviii + 443pp
◇ B98 ◇ REV MR 83g:68004 Zbl 443.68021 • ID 39144

BECVAR, J. [1984] *Notes on vagueness and mathematics* (**P** 3096) Conf Math Service of Man (2,Pap);1982 Las Palmas 1-11
◇ B52 ◇ REV MR 86g:03086 • ID 44009

BECVAR, J. see Vol. I, IV, VI for further entries

BEDAU, H. & OPPENHEIM, P. [1961] *Complementarity in quantum mechanics: A logical analysis* (**J** 0154) Synthese 13*201-232
◇ A05 B51 G12 ◇ REV Zbl 129.416 JSL 38.340 • ID 00932

BEHMANN, H. [1927] *Mathematik und Logik* (**X** 0823) Teubner: Stuttgart 59pp
◇ A05 B98 ◇ REV FdM 53.41 • ID 16737

BEHMANN, H. [1953] *Die typenfreie Logik und die Modalitaet* (**P** 0645) Int Congr Philos (11);1953 Bruxelles 14*88-96
◇ A05 B45 ◇ REV MR 15.386 Zbl 53.343 JSL 22.326 • ID 01554

BEHMANN, H. see Vol. I, III, IV, V, VI for further entries

BEK, R. [1978] *Discourse on one way in which a quantum-mechanics language on the classical logical base can be built up* (**J** 0156) Kybernetika (Prague) 14*85-101
◇ B51 ◇ REV MR 58 # 4001 Zbl 383 # 03015 • ID 51997

BELDING, W.R. [1971] *Intuitionistic negation* (**J** 0047) Notre Dame J Formal Log 12*183-187
◇ B55 F55 ◇ REV MR 45 # 3160 Zbl 177.8 • ID 00957

BELDING, W.R. see Vol. V, VI for further entries

BELL, J.L. & MACHOVER, M. [1977] *A course in mathematical logic* (**X** 0809) North Holland: Amsterdam xviii + 599pp
◇ B98 ◇ REV MR 57 # 12155 Zbl 359 # 02001 JSL 45.378 • ID 28152

BELL, J.L. & HALLETT, M. [1982] *Logic, quantum logic and empiricism* (**J** 0153) Phil of Sci (East Lansing) 49*355-379
◇ B51 ◇ REV MR 84b:03082 • ID 35663

BELL, J.L. [1983] *Orthologic, forcing, and the manifestation of attributes* (**P** 3669) SE Asian Conf on Log;1981 Singapore 13-36
◇ B51 C90 ◇ REV MR 85m:03043 Zbl 553 # 03040 • ID 43055

BELL, J.L. see Vol. I, III, V for further entries

BELL, J.M. & HUMBERSTONE, I.L. [1977] *Two systems of presupposition logic* (**J** 0047) Notre Dame J Formal Log 18*321-339
◇ B45 ◇ REV MR 58 # 21438 Zbl 283 # 02020 • ID 23628

BELL, N. & PAGE, E.W. & THOMASON, M.G. [1972] *Extension of the Boolean difference concept to multi-valued logic systems* (**P** 2008) Symp Th & Appl of Multi-Val Log Design;1972 Buffalo 17-24
◇ B50 ◇ ID 42007

BELLISSIMA, F. [1978] *On the modal logic corresponding to diagonalizable algebra theory* (**J** 3495) Boll Unione Mat Ital, V Ser, B 15*915-930
◇ B45 G25 ◇ REV MR 81b:03016 Zbl 405 # 03012 • ID 54887

BELLISSIMA, F. [1980] *Considerazioni semantiche sulle logiche modali predicative* (**J** 3495) Boll Unione Mat Ital, V Ser, B 17*776-785
◇ B45 ◇ REV MR 81i:03018 Zbl 433 # 03009 • ID 54015

BELLISSIMA, F. & MIROLLI, M. [1983] *On the axiomatization of finite K-frames* (**J** 0063) Studia Logica 42*383-388
◇ B45 C13 C90 ◇ REV MR 86a:03014 Zbl 556 # 03017 • ID 42314

BELLISSIMA, F. [1984] *On the relationship between one-point frames and degrees of unsatisfiability of modal formulas* (**J** 0047) Notre Dame J Formal Log 25*117-126
◇ B45 ◇ REV Zbl 563 # 03007 • ID 42564

BELLISSIMA, F. [1985] *A test to determine distinct modalities in the extensions of* S4 (**J** 0068) Z Math Logik Grundlagen Math 31*57-62
◇ B45 ◇ REV Zbl 567 # 03006 • ID 42286

BELLISSIMA, F. see Vol. VI for further entries

BELLMAN, R.E. & ZADEH, L.A. [1970] *Decision-making in a fuzzy environment* (**J** 4697) Managmt Sci (Providence) 17/4*B141-B164
◇ B52 ◇ REV MR 46 # 769 Zbl 224 # 90032 • ID 41964

BELLMAN, R.E. & GIERTZ, M. [1973] *On the analytic formalism of the theory of fuzzy sets* (**J** 0191) Inform Sci 5*149-156
◇ B52 E72 ◇ REV MR 54 # 2417 Zbl 251 # 02059 • ID 24614

BELLMAN, R.E. [1975] *Local logics* (**P** 1805) Int Symp Multi-Val Log (5,Proc);1975 Bloomington 175
◇ B50 ◇ ID 35822

BELLMAN, R.E. & ZADEH, L.A. [1977] *Local and fuzzy logics* (**P** 1894) Int Symp Multi-Val Log (5,Inv Pap);1975 Bloomington 103-165
◇ B52 ◇ REV MR 58 # 16179 Zbl 382 # 03017 • ID 51941

BELLMAN, R.E. & ESOGBUE, A.O. [1982] *Constributions to fuzzy dynamic programming* (**P** 3845) Conf Math Service of Man (2,Proc)(Feriet);1982 Las Palmas 275-280
◇ B52 B75 ◇ REV Zbl 505 # 90079 • ID 38218

BELLUCE, L.P. & CHANG, C.C. [1963] *A weak completeness theorem for infinite valued first-order logic* (**J** 0036) J Symb Logic 28*43-50
◇ B50 ◇ REV MR 34 # 43 Zbl 121.12 JSL 36.332 • ID 01000

BELLUCE, L.P. [1964] *Further results on infinite valued predicate logic* (**J** 0036) J Symb Logic 29*69-78
◇ B50 ◇ REV MR 31 # 1183 Zbl 127.8 JSL 36.332 • ID 01002

BELNAP JR., N.D. [1959] see ANDERSON, A.R.

BELNAP JR., N.D. [1960] *Entailment and relevance* (**J** 0036) J Symb Logic 25*144-146
◇ B46 ◇ REV MR 25 # 4988 Zbl 107.8 JSL 34.120 • ID 01004

BELNAP JR., N.D. [1960] *EQ and the first order functional calculus* (**J** 0068) Z Math Logik Grundlagen Math 6*217-218
◇ B46 ◇ REV Zbl 95.8 JSL 36.520 • ID 01003

BELNAP JR., N.D. [1960] see ANDERSON, A.R.

BELNAP JR., N.D. [1961] see ANDERSON, A.R.

BELNAP JR., N.D. [1962] see ANDERSON, A.R.

BELNAP JR., N.D. [1963] *An analysis of questions: Preliminary report* (**X** 4477) Syst Devel: Santa Monica 160pp
◇ A05 B46 B60 ◇ REV JSL 37.420 • ID 43930

BELNAP JR., N.D. [1963] see ANDERSON, A.R.

BELNAP JR., N.D. & WALLACE, J.R. [1965] *A decision procedure for the system $E_{\bar{I}}$ of entailment with negation* (**J** 0068) Z Math Logik Grundlagen Math 11*277-289
◇ B46 ◇ REV MR 31 # 4714 Zbl 143.249 • ID 01012

BELNAP JR., N.D. [1967] *Intensional models for first degree formulas* (**J** 0036) J Symb Logic 32*1-22
◇ B46 ◇ REV MR 39 # 1296 Zbl 166.4 • ID 01016

BELNAP JR., N.D. [1970] *Conditional assertion and restricted quantification* (**J** 0097) Nous, Quart J Phil 4*1-13
◇ B45 ◇ REV MR 54 # 2430 • ID 24626

BELNAP JR., N.D. & MCCALL, S. [1970] *Every functionally complete m-valued logic has a post-complete axiomatization* (**J** 0047) Notre Dame J Formal Log 11*106
◇ B50 ◇ REV MR 43 # 3096 Zbl 187.266 • ID 01560

BELNAP JR., N.D. [1972] *Foreword to: A general interpreted modal calculus, by Aldo Bressan* (**X** 0875) Yale Univ Pr: New Haven xiii-xxv
◇ B45 ◇ REV JSL 39.352 • ID 04149

BELNAP JR., N.D. [1973] *Restricted quantification and conditional assertion* (**P** 0783) Truth, Syntax & Modal;1970 Philadelphia 48-75
◇ A05 B45 ◇ REV MR 56 # 5194 Zbl 278 # 02010 JSL 42.314 • ID 29054

BELNAP JR., N.D. [1973] *S-P interrogatives* (**P** 1556) Exact Philos: Probl, Tools & Goals;1971 Montreal 65-80
◇ A05 B45 ◇ REV MR 57 # 15988 Zbl 327 # 02026 • ID 60430

BELNAP JR., N.D. [1975] see ANDERSON, A.R.

BELNAP JR., N.D. & STEEL JR., T.B. [1976] *The logic of questions and answers* (**X** 0875) Yale Univ Pr: New Haven vi + 209pp
◇ B60 B65 ◇ REV MR 58 # 27358 Zbl 345 # 02015 JSL 43.379 • ID 29775

BELNAP JR., N.D. [1977] *A useful four-valued logic* (**P** 1894) Int Symp Multi-Val Log (5,Inv Pap);1975 Bloomington 5-37
◇ B50 ◇ REV MR 58 # 5021 Zbl 417 # 03009 • ID 53246

BELNAP JR., N.D. & MCROBBIE, M.A. [1977] *Relevant analytic tableaux* (**J** 1893) Relevance Logic Newslett 2*46-49
◇ B46 ◇ REV Zbl 349 # 02019 • ID 63849

BELNAP JR., N.D. & MCROBBIE, M.A. [1979] *Relevant analytic tableaux* (**J** 0063) Studia Logica 38*187-200
◇ B46 ◇ REV MR 81e:03015 Zbl 406 # 03032 • ID 56115

BELNAP JR., N.D. & DUNN, J.M. & GUPTA, A. [1980] *A consecutive calculus for positive relevant implication with necessity* (**J** 0122) J Philos Logic 9*343-362 • ERR/ADD ibid 10*291
◇ A05 B46 ◇ REV MR 82g:03022a MR 82g:03022b Zbl 448 # 03010 • ID 56610

BELNAP JR., N.D. & DUNN, J.M. [1981] *Entailment and the disjunctive syllogism* (**C** 2078) Contemp Philos 337-366
◇ B46 ◇ ID 42808

BELNAP JR., N.D. [1981] *Modal and relevance logics: 1977* (**C** 2617) Modern Log Survey 131-151
◇ B46 ◇ REV MR 82f:03002 Zbl 464 # 03001 • ID 42767

BELNAP JR., N.D. [1982] *Display logic* (**J** 0122) J Philos Logic 11*375-417
◇ B45 ◇ REV MR 86f:03094 Zbl 509 # 03008 • ID 36521

BELNAP JR., N.D. & MCROBBIE, M.A. & NUEL JR., D. [1984] *Proof tableau formulations of some first-order relevant orthologics* (**J** 0387) Bull Sect Logic, Pol Acad Sci 13*233-240
◇ B51 ◇ REV Zbl 569 # 03007 • ID 44666

BELNAP JR., N.D. see Vol. I, VI for further entries

BELTRAMETTI, E.G. & CASSINELLI, G. [1975] *Ideal, first-kind measurements in a proposition-state structure* (**J** 1113) Commun Math Phys 40*7-13
◇ B51 ◇ REV MR 50 # 15619 Zbl 317 # 06008 • ID 60849

BELTRAMETTI, E.G. & CASSINELLI, G. [1977] *On state transformations induced by yes-no experiments, in the context of quantum logic* (**J** 0122) J Philos Logic 6*369-379
◇ B51 ◇ REV MR 82f:81007 Zbl 383 # 03043 • ID 52025

BELTRAMETTI, E.G. & CASSINELLI, G. [1979] *Properties of states in quantum logic* (**P** 3234) Probl Founds of Physics;1977 Varenna 29-70
◇ B51 ◇ REV MR 82m:81003 Zbl 445 # 03037 • ID 56501

BELTRAMETTI, E.G. & FRAASSEN VAN, B.C. (EDS.) [1981] *Current issues in quantum logic* (**X** 1332) Plenum Publ: New York ix + 492pp
◇ B51 B97 ◇ REV MR 84i:03005 Zbl 537 # 03044 • ID 34491

BELTRAMETTI, E.G. & CASSINELLI, G. [1981] *Problems of the proposition-state structure of quantum mechanics* (**C** 3515) Ital Studies in Phil of Sci 215-235
◇ B51 ◇ REV MR 83i:03097 Zbl 491 # 03023 • ID 35544

BELTRAMETTI, E.G. & CASSINELLI, G. [1981] *The logic of quantum mechanics* (**X** 0832) Addison-Wesley: Reading xxvi + 305pp
◇ B51 ◇ REV MR 83d:81008 Zbl 504 # 03026 • ID 36972

Author Index

BELZER, M. & LOEWER, B.M. [1983] *Dyadic deontic detachment* (J 0154) Synthese 54*295-318
◇ B45 ◇ REV MR 85d:03035 Zbl 515 # 03010 • ID 38139

BEN, D.M. & CLARK, C.M. & KANDEL, A. [1981] *On the enumeration of distinct fuzzy switching functions* (J 2720) Fuzzy Sets Syst 5*69-81
◇ B52 ◇ REV MR 82a:94133 Zbl 448 # 94010 • ID 80933

BEN-ARI, M. & HALPERN, J.Y. & PNUELI, A. [1981] *Finite models for deterministic propositional dynamic logic* (P 2903) Automata, Lang & Progr (8);1981 Akko 249-263
◇ B75 ◇ REV MR 83f:03013 Zbl 465 # 68011 • ID 54944

BEN-ARI, M. & HALPERN, J.Y. & PNUELI, A. [1982] *Deterministic propositional dynamic logic: finite models, complexity, and completeness* (J 0119) J Comp Syst Sci 25*402-417
◇ B75 ◇ REV MR 84c:03033 Zbl 512 # 03013 • ID 35690

BEN-ARI, M. & MANNA, Z. & PNUELI, A. [1983] *The temporal logic of branching time* (J 1431) Acta Inf 20*207-226
◇ B45 ◇ REV MR 85a:03021 Zbl 533 # 68036 • ID 34772

BEN-ARI, M. see Vol. I, IV, VI for further entries

BENCIVENGA, E. [1976] *Set theory and free logic* (J 0122) J Philos Logic 5*1-15
◇ A05 B60 E70 ◇ REV MR 54 # 12523 Zbl 345 # 02009 • ID 29770

BENCIVENGA, E. [1977] *Are arithmetical truths analytic? New results in free set theory* (J 0122) J Philos Logic 6*319-330
◇ A05 B60 E70 ◇ REV MR 58 # 27482 Zbl 368 # 02063 • ID 51280

BENCIVENGA, E. [1978] *A semantics for a weak free logic* (J 0047) Notre Dame J Formal Log 19*646-652
◇ B60 ◇ REV MR 80b:03029 Zbl 368 # 02029 • ID 52128

BENCIVENGA, E. [1980] *A weak free logic with the existence sign* (J 0047) Notre Dame J Formal Log 21*572-576
◇ B60 ◇ REV MR 81i:03013 Zbl 419 # 03014 • ID 53923

BENCIVENGA, E. [1980] *Free semantics for definite descriptions* (J 0079) Logique & Anal, NS 23*393-405
◇ B60 ◇ REV MR 83f:03024 Zbl 454 # 03009 • ID 54222

BENCIVENGA, E. & WOODRUFF, P.W. [1981] *A new modal language with the λ operator* (J 0063) Studia Logica 40*383-389
◇ B45 ◇ REV MR 84g:03023 Zbl 498 # 03010 • ID 34137

BENCIVENGA, E. [1981] *Free semantics* (C 3515) Ital Studies in Phil of Sci 31-48
◇ B60 ◇ REV MR 83i:03045 Zbl 446 # 03019 • ID 56548

BENCIVENGA, E. [1984] *A possibility-free logic of descriptions* (J 0009) Arch Math Logik Grundlagenforsch 24*159-166
◇ B60 ◇ REV MR 86c:03021 Zbl 553 # 03016 • ID 42403

BENCIVENGA, E. [1984] *Il primo libro di logica. Introduzione ai metodi della logica contemporanea* (X 0905) Boringhieri: Torino 228pp
◇ B98 ◇ REV Zbl 557 # 03001 • ID 46185

BENCIVENGA, E. [1985] *Strong completeness of a pure free logic* (J 0068) Z Math Logik Grundlagen Math 31*35-38
◇ B60 ◇ REV Zbl 567 # 03009 • ID 42284

BENCIVENGA, E. see Vol. I, VI for further entries

BENDALL, K. [1979] *Belief-theoretic formal semantics for first-order logic and probability* (J 0122) J Philos Logic 8*375-397 • ERR/ADD ibid 9*342
◇ A05 B48 ◇ REV MR 81k:03026a MR 81k:03026b Zbl 421 # 03025 • ID 53424

BENDALL, K. [1982] *A "definitive" probabilistic semantics for first-order logic* (J 0122) J Philos Logic 11*255-278 • ERR/ADD ibid 12*101
◇ B48 ◇ REV MR 84g:03029a MR 84g:03029b Zbl 505 # 03002 • ID 34143

BENDALL, K. see Vol. I for further entries

BENDEGEM VAN, J.P. [1985] see BATENS, D.

BENDOVA, K. & HAJEK, P. & RENC, Z. [1971] *The GUHA method and the three-valued logic (Czech summary)* (J 0156) Kybernetika (Prague) 7*425-433
◇ B50 ◇ REV MR 48 # 3289 Zbl 232 # 68034 • ID 31133

BENDOVA, K. [1981] *On the relation of three-valued logic to modal logic* (J 0140) Comm Math Univ Carolinae (Prague) 22*637-653
◇ B50 ◇ REV MR 84h:03033 Zbl 481 # 03012 • ID 34244

BENDOVA, K. & HAJEK, P. [1982] *A logical analysis of the truth-reaction paradox* (J 0140) Comm Math Univ Carolinae (Prague) 23*699-713
◇ B45 F30 ◇ REV MR 84e:03068 Zbl 514 # 03037 • ID 34402

BENDOVA, K. see Vol. IV for further entries

BENGY-PUYVALLEE DE, R. [1945] *Sur les regles de composabilite dans les logiques de la complementarite de Mme. Destouches-Fevrier* (J 0109) C R Acad Sci, Paris 220*589-591
◇ B51 ◇ REV MR 7.186 Zbl 61.471 JSL 12.134 • ID 01032

BENGY-PUYVALLEE DE, R. [1949] *Sur la relation de composabilite dans les logiques de complementarite* (J 0109) C R Acad Sci, Paris 228*624-626
◇ B51 G12 ◇ REV MR 10.422 Zbl 32.386 • ID 01034

BENGY-PUYVALLEE DE, R. see Vol. VI for further entries

BENHALCEN, D. & HIRSCH, G. & LAMOTTE, M. [1977] *Codage et minimisation des fonctions floues dans une algebre floue* (J 2831) RAIRO Autom 11*17-31,117
◇ B52 ◇ REV MR 58 # 4655 Zbl 363 # 94005 • ID 50927

BENIOFF, P.A. [1973] *On definitions of validity applied to quantum theories* (J 1678) Found Phys 3*359-379
◇ B51 G12 ◇ REV MR 55 # 7150 • ID 30649

BENIOFF, P.A. see Vol. I, III, IV, V for further entries

BENNETT, A.A. [1939] see BAYLIS, C.A.

BENNETT, A.A. see Vol. I, V for further entries

BENNETT, J.H. [1982] *Psychology and semantics: comments on Schiffer's "Intention-based semantics"* (J 0047) Notre Dame J Formal Log 23*258-262
◇ B60 ◇ REV MR 84i:03043b • REM The article was published ibid 23*119-156 • ID 34940

BENNETT, J.H. see Vol. I for further entries

BENTHEM VAN, J.F.A.K. [1974] *Some correspondance results in modal logic* (X 2791) Univ Amsterdam Math Inst: Amsterdam 74-05*62pp
◇ B45 ◇ REV Zbl 308 # 02023 • ID 20840

BENTHEM VAN, J.F.A.K. [1975] *A note on modal formulae and relational properties* (**J** 0036) J Symb Logic 40*55-58
◇ B45 ◇ REV MR 55 # 2495 Zbl 308 # 02024 JSL 47.440 • ID 04247

BENTHEM VAN, J.F.A.K. [1976] *Modal formulas are either elementary or not ΣΔ-elementary* (**J** 0036) J Symb Logic 41*436-438
◇ B45 ◇ REV MR 53 # 7726 Zbl 337 # 02015 JSL 47.441 • ID 14774

BENTHEM VAN, J.F.A.K. [1976] *Modal reduction principles* (**J** 0036) J Symb Logic 41*301-312
◇ B45 ◇ REV MR 53 # 12873 Zbl 337 # 02014 • ID 14775

BENTHEM VAN, J.F.A.K. [1977] *Modal logic as second-order logic* (**X** 2791) Univ Amsterdam Math Inst: Amsterdam 77-04*65pp
◇ B45 ◇ ID 43335

BENTHEM VAN, J.F.A.K. [1977] *Tense logic and standard logic* (**J** 0079) Logique & Anal, NS 20*395-437
◇ A05 B45 ◇ REV MR 58 # 27316 Zbl 392 # 03019 • ID 52386

BENTHEM VAN, J.F.A.K. & BLOK, W.J. [1978] *Transitivity follows from Dummett's axiom* (**J** 0105) Theoria (Lund) 44*117-118
◇ B45 ◇ REV MR 81j:03029 Zbl 422 # 03004 JSL 48.488 • ID 53465

BENTHEM VAN, J.F.A.K. [1978] *Two simple incomplete modal logics* (**J** 0105) Theoria (Lund) 44*25-37
◇ B45 ◇ REV MR 80h:03021 Zbl 405 # 03010 JSL 48.488 • ID 54885

BENTHEM VAN, J.F.A.K. [1979] *Canonical modal logics and ultrafilter extensions* (**J** 0036) J Symb Logic 44*1-8
◇ B45 C90 G25 ◇ REV MR 80i:03028 Zbl 405 # 03011 JSL 47.441 • ID 28157

BENTHEM VAN, J.F.A.K. [1979] *Critical notice to: D. M. Gabbay's "Investigations in modal and tense logics with applications to problems in philosophy and linguistics"* (**J** 0154) Synthese 40*353-373
◇ A05 B45 ◇ REV Zbl 433 # 03008 • ID 54014

BENTHEM VAN, J.F.A.K. [1979] *Minimal deontic logics* (**J** 0387) Bull Sect Logic, Pol Acad Sci 8*36-42
◇ A05 B45 ◇ REV MR 80h:03022 Zbl 408 # 03016 • ID 56255

BENTHEM VAN, J.F.A.K. [1979] *Syntactic aspects of modal incompleteness theorems* (**J** 0105) Theoria (Lund) 45*63-77
◇ B45 ◇ REV MR 82g:03023 Zbl 452 # 03013 JSL 48.488 • ID 54077

BENTHEM VAN, J.F.A.K. [1979] *What is dialectical logic?* (**J** 0158) Erkenntnis (Dordrecht) 14*333-347
◇ A05 B53 ◇ ID 42809

BENTHEM VAN, J.F.A.K. [1980] *Some kinds of modal completeness* (**J** 0063) Studia Logica 39*125-141
◇ B45 ◇ REV MR 82b:03044 Zbl 459 # 03008 • ID 54448

BENTHEM VAN, J.F.A.K. [1981] *Tense logic, second-order logic and natural language* (**P** 3113) Aspects Philos Logic;1977 Tuebingen 1-20
◇ A05 B45 ◇ REV MR 83f:03014 Zbl 476 # 03029 • ID 55541

BENTHEM VAN, J.F.A.K. [1983] *Determiners and logic* (**J** 2130) Linguist Philos 6*447-478
◇ B45 ◇ REV Zbl 535 # 03010 • ID 38314

BENTHEM VAN, J.F.A.K. [1983] *Five easy pieces* (**C** 4638) Stud in Modelth Seman 1*1-17
◇ B45 ◇ REV Zbl 573 # 03007 • ID 48955

BENTHEM VAN, J.F.A.K. & HUMBERSTONE, I.L. [1983] *Hallden-completeness by gluing of Kripke frames* (**J** 0047) Notre Dame J Formal Log 24*426-430
◇ B45 C30 C90 ◇ REV MR 85i:03040 Zbl 487 # 03008 • ID 37277

BENTHEM VAN, J.F.A.K. [1983] *Logical semantics as an empirical science* (**J** 0063) Studia Logica 42*299-313
◇ A05 B45 ◇ REV MR 85i:03006 Zbl 542 # 03002 • ID 42316

BENTHEM VAN, J.F.A.K. [1983] *The logic of time* (**S** 3307) Synth Libr 156*xvi+260pp
◇ B45 ◇ REV MR 84i:03035 Zbl 508 # 03008 • ID 34519

BENTHEM VAN, J.F.A.K. [1984] *Foundations of conditional logic* (**J** 0122) J Philos Logic 13*303-349
◇ B60 B65 ◇ REV MR 85j:03017 Zbl 544 # 03011 • ID 40997

BENTHEM VAN, J.F.A.K. [1984] *Possible worlds semantics: a research program that cannot fail?* (**J** 0063) Studia Logica 43*379-394
◇ B45 ◇ REV Zbl 573 # 03005 • ID 42387

BENTHEM VAN, J.F.A.K. [1984] *Tense logic and time* (**J** 0047) Notre Dame J Formal Log 25*1-16
◇ B45 ◇ REV MR 85e:03033 Zbl 536 # 03008 • ID 37095

BENTHEM VAN, J.F.A.K. see Vol. I, III, V, VI for further entries

BENTON, R.A. [1979] *A descending chain of classical logics for which necessitation implies regularity* (**J** 0068) Z Math Logik Grundlagen Math 25*289-291
◇ B45 ◇ REV MR 81e:03012 Zbl 426 # 03026 • ID 53622

BERAN, H. [1982] see ADAM, ADOLF

BERG, J. [1960] *A note on deontic logic* (**J** 0094) Mind 69*566-567
◇ B45 ◇ REV JSL 36.182 • ID 01051

BERG, J. see Vol. I, IV, VI for further entries

BERGHAMMER, R. & SCHMIDT, GUNTHER [1983] *A relational view on gotos and dynamic logic (German summary)* (**P** 3908) Graph-Gram & Appl to Comput Sci (2);1982 Neunkirchen 13-24
◇ B75 ◇ REV MR 84e:68008 Zbl 537 # 68022 • ID 41317

BERGHAMMER, R. see Vol. V for further entries

BERGMANN, E. & NOLL, H. [1977] *Mathematische Logik mit Informatik-Anwendungen* (**X** 0811) Springer: Heidelberg & New York xv+324pp
◇ A05 A10 B98 ◇ REV MR 58 # 21353 Zbl 384 # 03001 • ID 52043

BERGMANN, G. [1949] *A syntactical characterization of S5* (**J** 0036) J Symb Logic 14*173-174
◇ B45 ◇ REV MR 11.303 Zbl 36.7 JSL 14.260 • ID 01059

BERGMANN, G. [1949] *The finite representations of S5* (**J** 0175) Methodos 1*217-219
◇ B45 ◇ REV MR 11.303 Zbl 36.7 JSL 15.224 • ID 01257

BERGMANN, G. [1956] *The representations of S5* (**J** 0036) J Symb Logic 21*257-260
◇ B45 ◇ REV MR 18.551 Zbl 75.232 JSL 25.184 • ID 01061

Author Index

BERGMANN, G. see Vol. I, V for further entries

BERGMANN, MERRIE & MOOR, J. & NELSON, JACK [1980] *The logic book* (**X** 0981) Random House: New York ix+459pp
◇ B98 ◇ REV JSL 47.915 • ID 47407

BERGMANN, MERRIE [1981] *'Only', 'even', and clefts in two-dimensional logic* (**P** 3705) Int Symp Multi-Val Log (11);1981 Oklahoma City & Norman 117-123
◇ B50 ◇ REV MR 83m:94033 • ID 45933

BERGMANN, MERRIE [1981] *Presupposition and two-dimensional logic* (**J** 0122) J Philos Logic 10*27-53
◇ A05 B60 ◇ REV MR 82i:03006 • ID 70957

BERGMANN, MERRIE [1982] *Expressibility in two-dimensional languages for presupposition* (**J** 0047) Notre Dame J Formal Log 23*459-470
◇ B60 ◇ REV MR 83m:03031 Zbl 464 # 03022 • ID 54612

BERGSTRA, J.A. & MEYER, J.-J.C. [1982] *On the elimination of iteration quantifiers in a fragment of algorithmic logic* (**J** 1426) Theor Comput Sci 21*269-279
◇ B75 ◇ REV MR 84d:03024 Zbl 507 # 03008 • ID 34062

BERGSTRA, J.A. see Vol. I, III, IV, VI for further entries

BERGSTROEM, L. [1974] *Hintikka on "prima facie" obligations* (**J** 0105) Theoria (Lund) 40*163-165
◇ A05 B45 ◇ REV MR 58 # 27317a Zbl 313 # 02019
• ID 29617

BERKA, K. [1969] *Remarks on the relations between the peripatetic and modern theories of modality* (**J** 0147) An Univ Bucuresti, Acta Logica 12*261-267
◇ A10 B45 ◇ REV MR 44 # 3843 Zbl 247 # 02009
• ID 01065

BERKA, K. see Vol. I for further entries

BERKELEY, E.C. [1952] *Symbolic logic. Twenty problems and solutions* (**X** 0851) Berkeley: New York ii+28pp
◇ B98 ◇ REV Zbl 48.244 JSL 20.287 • ID 01544

BERKELEY, E.C. see Vol. I for further entries

BERKLING, K.J. & FEHR, E. [1982] *A consistent extension of the lambda-calculus as a base for functional programming languages* (**J** 0194) Inform & Control 55*89-101
◇ B40 B60 B75 ◇ REV Zbl 553 # 68025 • ID 43431

BERKLING, K.J. see Vol. VI for further entries

BERMAN, F. [1979] *A completeness technique for D-axiomatizable semantics* (**P** 3542) ACM Symp Th of Comput (11);1979 Atlanta 160-166
◇ B45 ◇ REV MR 82c:03018 • ID 70969

BERMAN, F. & PATERSON, M.S. [1981] *Propositional dynamic logic is weaker without tests* (**J** 1426) Theor Comput Sci 16*321-328
◇ B75 ◇ REV MR 83e:68009 Zbl 468 # 68039 • ID 55122

BERMAN, F. [1982] *Compactness in models of propositional dynamic logic* (**J** 2293) Comp Linguist & Comp Lang 15*7-19
◇ B45 ◇ REV MR 83k:03023 Zbl 494 # 03015 • ID 36163

BERMAN, F. [1982] *Semantics of looping programs in propositional dynamic logic* (**J** 0041) Math Syst Theory 15*285-294
◇ B75 ◇ REV MR 84a:68012 Zbl 497 # 03012 • ID 38109

BERMAN, F. [1983] *Nonstandard models in propositional dynamic logic* (**P** 3830) Logics of Progr & Appl;1980 Poznan 81-85
◇ B75 ◇ REV MR 84b:68001 Zbl 523 # 03024 • ID 37026

BERMAN, J. [1980] *Algebraic properties of k-valued logics* (**P** 3673) Int Symp Multi-Val Log (10);1980 Evanston 195-204
◇ B50 ◇ REV MR 83m:94032 Zbl 539 # 03009 • ID 43743

BERMAN, J. & MUKAIDONO, M. [1984] *Enumerating fuzzy switching functions and free Kleene algebras* (**J** 2687) Comp Math Appl 10*25-35
◇ B52 ◇ REV MR 84m:94045 Zbl 543 # 94029 • ID 40987

BERMAN, J. see Vol. III, V for further entries

BERMAN, P. & HALPERN, J.Y. & TIURYN, J. [1982] *On the power of nondeterminism in dynamic logic* (**P** 3836) Automata, Lang & Progr (9);1982 Aarhus 48-60
◇ B75 ◇ REV MR 83m:68149 Zbl 494 # 68032 • ID 36658

BERMAN, P. [1983] *Deterministic dynamic logic of recursive programs is weaker than dynamic logic* (**P** 3864) FCT'83 Found of Comput Th;1983 Borgholm 14-25
◇ B75 ◇ REV Zbl 545 # 68020 • ID 41206

BERMAN, P. see Vol. III, IV for further entries

BERNARDI, C. [1975] *The fixed-point theorem for diagonalizable algebras (the algebraization of the theories which express Theor. III)* (**J** 0063) Studia Logica 34*239-251
◇ B45 F30 G05 G25 ◇ REV MR 57 # 106
Zbl 318 # 02031 • REM Part II 1975 by Magari,R. Part IV 1975 by Magari,R. • ID 31011

BERNARDI, C. [1976] *The uniqueness of the fixed-point in every diagonalizable algebra (The algebraization of the theories which express Theor. VIII.)* (**J** 0063) Studia Logica 35*335-343
◇ B45 F30 G05 G25 ◇ REV MR 57 # 111
Zbl 345 # 02020 • REM Part VII 1976 by Magari,R. Part IX 1976 by Sambin,G. • ID 60478

BERNARDI, C. (ED.) [1983] *Atti degli incontri di logica matematica* (**X** 3812) Univ Siena, Dip Mat: Siena 398pp
◇ B97 C97 ◇ REV MR 84k:03006 Zbl 505 # 00007
• ID 34945

BERNARDI, C. [1984] *A shorter proof of a recent result by R. Di Paola* (**J** 0047) Notre Dame J Formal Log 25*390-393
◇ B45 F30 ◇ REV MR 85m:03038 Zbl 562 # 03029
• ID 42581

BERNARDI, C. see Vol. III, IV, VI for further entries

BERNARDO DI, G. [1972] *Introduzione alla logica dei sistemi normativi* (**X** 0881) Il Mulino: Bologna 183pp
◇ B98 ◇ REV MR 50 # 57 JSL 40.466 • ID 17148

BERNARDO DI, G. [1982] *Antinomies and predicaments in normative systems* (**P** 4041) Log, Inform, Law;1981 Firenze 2*127-159
◇ B45 ◇ REV MR 85h:03020 • ID 43399

BERNAYS, P. [1934] see HILBERT, D.

BERNAYS, P. [1939] see HILBERT, D.

BERNAYS, P. & FRAENKEL, A.A. [1958] *Axiomatic set theory* (**X** 0809) North Holland: Amsterdam viii+227pp
◇ B98 E98 ◇ REV MR 21 # 4912 Zbl 82.263 JSL 24.224
• REM 2nd ed. 1968. With a historical introduction by A.A.Fraenkel • ID 16969

BERNAYS, P. & DOCKX, S. (EDS.) [1965] *Information and prediction in science* (**X** 0801) Academic Pr: New York xi+272pp
◇ A05 B48 ◇ REV Zbl 127.241 • ID 48629

BERNAYS, P. see Vol. I, III, IV, V, VI for further entries

BERNERT, J. & BIELA, A. [1981] *On two different modal logics denoted by S9* (**J** 0302) Rep Math Logic, Krakow & Katowice 13*3-9
◇ B45 ◇ REV MR 84a:03018 Zbl 485 # 03004 • ID 35565

BERNERT, J. see Vol. I for further entries

BERNHARDT, K. [1977] *Abschliessbarkeit von Kripke-Semantiken* (**J** 0115) Wiss Z Humboldt-Univ Berlin, Math-Nat Reihe 26*615-622
◇ B55 C90 ◇ REV MR 80e:03039 Zbl 423 # 03040 • ID 53552

BERNHARDT, K. see Vol. I, III, V for further entries

BERNINI, S. [1981] *A temporalization of modal semantics* (**C** 3515) Ital Studies in Phil of Sci 49-58
◇ A05 B45 ◇ REV MR 82d:03026 Zbl 451 # 03008 • ID 54023

BERNINI, S. [1981] *Quantum logic as an extension of classical logic* (**P** 3820) Curr Iss in Quantum Log;1979 Erice 161-171 • REPR [1981] (**P** 3092) Congr Naz Logica;1979 Montecatini Terme 303-312
◇ B51 ◇ REV MR 84j:03119 Zbl 537 # 03044 • ID 34709

BERNINI, S. see Vol. VI for further entries

BERTOLINI, F. [1970] *Kripke models and many-valued logics II* (**J** 0144) Rend Sem Mat Univ Padova 44*355-381
◇ B50 ◇ REV MR 46 # 1552 Zbl 266 # 02014 • REM Part I 1971 • ID 01145

BERTOLINI, F. [1971] *Kripke models and manyvalued logics* (**P** 0669) Conv Teor Modelli & Geom;1969/70 Roma 113-131
◇ B50 F50 ◇ REV MR 45 # 29 Zbl 212.19 • REM Part I. Part II 1970 • ID 01565

BERTOLINI, F. see Vol. VI for further entries

BESSLICH, P.W. [1981] *Incompletely specified fuzzy switching function minimization: an in-place transform approach* (**P** 3705) Int Symp Multi-Val Log (11);1981 Oklahoma City & Norman 35-40
◇ B52 ◇ REV MR 83m:94033 • ID 45948

BESSONOV, A.V. [1977] *New operations in intuitionistic calculus (Russian)* (**J** 0087) Mat Zametki (Akad Nauk SSSR) 22*23-28
• TRANSL [1977] (**J** 1044) Math Notes, Acad Sci USSR 22*503-506
◇ B55 F50 ◇ REV MR 56 # 11762 Zbl 365 # 02014 • ID 53417

BESSONOV, A.V. [1982] *Logical semantics and ontology (Russian)* (**C** 3743) Probl Log & Metodol Nauk 115-139
◇ B60 ◇ REV MR 84e:03002 • ID 34342

BESSONOV, A.V. see Vol. I for further entries

BETH, E.W. [1940] *Einfuehrung in die Philosophie der Mathematik (Dutch)* (**X** 2110) Dekker & van de Vegt: Nijmegen 269pp
◇ A05 B98 ◇ REV Zbl 25.293 JSL 8.144 FdM 66.23 FdM 68.20 • REM 2nd ed. 1942 • ID 43798

BETH, E.W. [1942] *Summulae logicales. Supplement to formal logic (Dutch)* (**X** 0812) Wolters-Noordhoff : Groningen 55pp
◇ A05 B98 ◇ REV MR 7.185 Zbl 26.242 • ID 01151

BETH, E.W. [1950] *Les fondements logiques des mathematiques* (**X** 0834) Gauthier-Villars: Paris 222pp
• TRANSL [1963] (**X** 0844) Feltrinelli: Milano xiii+335pp (Italian)
◇ A05 B98 ◇ REV MR 12.71 MR 26 # 6035 Zbl 40.290 Zbl 65.1 JSL 16.153 JSL 25.269 JSL 36.325 • REM 2nd rev. ed. 1955, xv+241pp • ID 01155

BETH, E.W. [1955] *Die Stellung der Logik im Gebaeude der heutigen Wissenschaft* (**J** 0178) Stud Gen 8*425-431
◇ B98 ◇ REV Zbl 66.247 JSL 23.66 • ID 01164

BETH, E.W. [1956] *L'existence en mathematiques* (**X** 0834) Gauthier-Villars: Paris 60pp
◇ A05 B98 F55 ◇ REV MR 19.625 Zbl 75.231 • ID 01166

BETH, E.W. [1959] *The foundation of mathematics. A study in the philosophy of science* (**X** 0809) North Holland: Amsterdam xxvi+741pp
◇ A05 B98 ◇ REV MR 22 # 9445 Zbl 85.241 JSL 27.73 JSL 33.618 • REM 2nd ed. 1965 • ID 01170

BETH, E.W. [1962] *Formal methods. An introduction to symbolic logic and to the study of effective operations in arithmetic and logic* (**X** 0835) Reidel: Dordrecht xiv+170pp
◇ B28 B98 F30 F50 F98 ◇ REV MR 28 # 3920 Zbl 105.245 JSL 30.235 • ID 01176

BETH, E.W. & NIELAND, J.J.F. [1965] *Semantic construction of Lewis' systems* S4 *and* S5 (**P** 0614) Th Models;1963 Berkeley 17-24
◇ B45 ◇ REV MR 33 # 7244 • ID 01180

BETH, E.W. [1970] *Aspects of modern logic* (**X** 0835) Reidel: Dordrecht xi+176pp
◇ B98 ◇ REV Zbl 209.6 • ID 25631

BETH, E.W. see Vol. I, III, IV, V, VI for further entries

BHAT, K.V.S. [1980] *Comments on: "Simplification of fuzzy switching functions" by Thomas P. Neff and Abraham Kandel* (**J** 2338) IEEE Trans Syst Man & Cybern 10*637-640
◇ B52 B70 ◇ REV MR 81m:94041 Zbl 452 # 94037 • REM The paper was published in J0435 6*55-70 (1977), MR 55 # 2386 • ID 80716

BHAT, K.V.S. [1981] *Minimization of disjunctive normal forms of fuzzy logic functions* (**J** 0219) J Franklin Inst 311*171-185
◇ B52 ◇ REV MR 82b:94050 Zbl 455 # 94044 • ID 80715

BHAT, K.V.S. [1982] *Further remarks on Neff-Kandel algorithm for minimization of fuzzy conjunctive forms: "Simplification of fuzzy switching functions" (Internat. J. Comput. Information Sci. 6/1*55-70, 1977)* (**J** 2338) IEEE Trans Syst Man & Cybern 12/2*430-431
◇ B52 ◇ REV Zbl 477 # 94035 • ID 46459

BHAT, K.V.S. [1982] *On the notion of fuzzy consensus* (**J** 2720) Fuzzy Sets Syst 8*285-289
◇ B52 ◇ REV MR 83j:94034 Zbl 498 # 94028 • ID 38539

BIAGIOLI, C. & NATALI, F.S. & SPINOSA, P.L. & TRIVISONNO, G. [1982] *Experiments on the "model" of Sanchez-Mazas* (**P** 4041) Log, Inform, Law;1981 Firenze 2*215-226
◇ B45 ◇ REV MR 85j:03021d • ID 45212

Author Index

BIBEL, W. [1984] *First-order reasoning about knowledge and belief* (**P** 3538) Artif Intell & Inf-Control Syst Robot (3);1984 Smolenice 9-16
◇ A05 B48 ◇ REV Zbl 554 # 68066 • ID 40319

BIBEL, W. see Vol. I, IV, VI for further entries

BIELA, A. [1975] *Note on the structural incompleteness of some modal propositional calculi* (**J** 0302) Rep Math Logic, Krakow & Katowice 4*3-6
◇ B45 ◇ REV MR 52 # 69 Zbl 316 # 02023 • ID 17979

BIELA, A. [1981] see BERNERT, J.

BIELA, A. [1983] *Introduction to algorithmic logic. Part I (Polish)* (**X** 1425) Univ Slaski: Katowice 190pp
◇ B75 ◇ ID 46349

BIELA, A. [1984] *The program-substitution in algorithmic logic and algorithmic logic with non-deterministic programs* (**J** 0387) Bull Sect Logic, Pol Acad Sci 13*69-74
◇ B75 ◇ REV Zbl 553 # 68024 • ID 43427

BIELA, A. see Vol. I, VI for further entries

BIELAK, P. [1974] *A note on weak functional completeness* (**J** 0387) Bull Sect Logic, Pol Acad Sci 3/2*38-40
◇ B55 ◇ REV MR 53 # 2639 • ID 21526

BIELAK, P. [1983] *Criteria of definability for positive fragments of some intermediate logics (Polish) (English summary)* (**J** 0481) Acta Univ Wroclaw 517(29)(Logika 8)*3-14
◇ B55 ◇ REV Zbl 545 # 03008 • ID 41223

BIGELOW, J.C. [1976] *If-then meets the possible worlds* (**J** 1807) Phil Quart Israel 6*215-235
◇ B45 ◇ ID 33860

BIGELOW, J.C. [1976] *Possible worlds foundations for probability* (**J** 0122) J Philos Logic 5*299-320
◇ B48 C90 ◇ REV Zbl 347 # 02020 • ID 60510

BIGELOW, J.C. [1977] *Semantics of probability* (**J** 0154) Synthese 36*459-472
◇ A05 B48 ◇ REV MR 58 # 27318 Zbl 373 # 02013 • ID 51477

BIGELOW, J.C. [1979] *Quantum probability in logical space* (**J** 1724) Kagaku Tetsugaku 46*223-243
◇ B51 G12 ◇ REV MR 80f:81010 • ID 33184

BINFORD, F. [1965] *Solution to the exercises in: "First course in mathematical logic"* (**X** 0841) Blaisdell: New York ix + 173pp
◇ B98 ◇ REV Zbl 126.8 JSL 32.422 • REM The book was published by Hill,S. & Suppes,P. 1964 • ID 01271

BINKLEY, R. [1976] *The logic of action* (**P** 2326) Action Th;1975 Winnipeg 87-104
◇ A05 B60 ◇ REV Zbl 335 # 02017 • ID 60513

BINKLEY, R. see Vol. I for further entries

BIRKHOFF, GARRETT & NEUMANN VON, J. [1936] *The logic of quantum mechanics* (**J** 0120) Ann of Math, Ser 2 37*823-843
◇ B51 ◇ REV Zbl 15.146 FdM 62.1061 • ID 40918

BIRKHOFF, GARRETT see Vol. I, III, IV, V for further entries

BIRNBAUM, L. [1976] *Algebre et logique tripolaire* (**J** 0047) Notre Dame J Formal Log 17*551-564
◇ B50 ◇ REV MR 55 # 2491 Zbl 254 # 02016 • ID 21943

BIRNBAUM, L. [1980] *N-polar class logic. Class calculus, judgements, syllogisms* (**J** 0047) Notre Dame J Formal Log 21*365-379
◇ B60 ◇ REV MR 81i:03027 Zbl 363 # 02035 • ID 50862

BIRYUKOV, B.V. & SPIRKIN, A.G. (EDS.) [1978] *Cybernetics and logic. The growth of cybernetical ideas and the development of computing machinery in their mathematical-logical aspects (Russian) (English summary)* (**X** 2027) Nauka: Moskva 334pp
◇ B97 ◇ REV Zbl 477 # 00045 • ID 55596

BISWAS, N.N. [1975] *Introduction to logic and switching theory* (**X** 0836) Gordon & Breach: New York xiii + 354pp
◇ B98 ◇ REV Zbl 314 # 94021 • ID 60523

BISWAS, N.N. see Vol. I for further entries

BITTINGER, M.L. [1970] *Logic and proof* (**X** 0832) Addison-Wesley: Reading 129pp
◇ B98 ◇ REV MR 41 # 3229 Zbl 217.6 • ID 01245

BITTINGER, M.L. [1982] *Logic, proof, and sets* (**X** 0832) Addison-Wesley: Reading viii + 131pp
◇ B98 ◇ REV JSL 50.860 • REM 2nd ed. • ID 47360

BIZAM, G. & HERCZEG, J. [1971] *Play and logic (Hungarian)* (**X** 1466) Mueszaki: Budapest 339pp
• TRANSL [1975] (**X** 0885) Mir: Moskva 359pp (Russian) [1976] (**X** 0928) Akad Kiado: Budapest 391 pp (German)
◇ B98 ◇ REV MR 52 # 13295 MR 54 # 2410 Zbl 347 # 00001 • ID 21757

BLACK, F. [1964] *A deductive question-answering system* (**C** 4454) Seman Inf Process 354-402
◇ B60 ◇ ID 47290

BLACK, M. [1949] *Language and philosophy: studies in method* (**X** 0992) Cornell Univ Pr: Ithaca xiii + 264pp
◇ A05 B98 ◇ REV JSL 15.210 • ID 25304

BLACK, M. [1964] *Critical thinking: an introduction to logic and scientific method* (**X** 0819) Prentice Hall: Englewood Cliffs xv + 402pp
◇ A05 B98 ◇ ID 25087

BLACK, M. [1966] *Notes on the "paradoxes of confirmation"* (**C** 1107) Aspects Inductive Log 175-197
◇ B48 ◇ REV Zbl 202.299 • ID 27707

BLACK, M. see Vol. III for further entries

BLACKWELL, P. [1970] *The equivalence problem for Post logic functions* (**P** 0702) Combin Struct & Appl;1969 Calgary 13-14
◇ B50 ◇ REV Zbl 251 # 94034 • ID 60524

BLANCHARD, N. [1982] *Fuzzy-lip functions - the fuzzy-lip category* (**P** 4051) Fuzzy Set & Possibility Th;1980 Acapulco 301-315
◇ B52 ◇ REV MR 84b:03004 • ID 46434

BLANCHARD, N. see Vol. V for further entries

BLANCHE, R. [1952] *Quantity, modality, and other kindred systems of categories* (**J** 0094) Mind 61*369-375
◇ B45 ◇ REV JSL 22.325 • ID 42151

BLANCHE, R. [1955] *L'axiomatique* (**X** 0840) Pr Univ France: Paris 102pp
◇ B98 ◇ REV MR 26 # 1259 JSL 23.438 • ID 01273

BLANCHE, R. [1957] *Introduction a la logique contemporaine* (**X** 0850) Colin: Paris 208pp
◇ A05 B98 F50 ◇ REV Zbl 292 # 02001 JSL 24.71 • ID 01538

BLANCHE, R. [1972] *Sur la trivalence* (**J** 0079) Logique & Anal, NS 15*569-582
◇ B50 ◇ REV Zbl 264 # 02014 • ID 27502

BLANCHE, R. [1974] *Modalite et temporalite* (**J** 0286) Int Logic Rev 9*103-110
◇ A05 B45 ◇ REV Zbl 326 # 02007 • ID 60528

BLANCHE, R. see Vol. I, III for further entries

BLASI, A.A. [1971] *Difficulties of a quantum mechanics over finite and p-adic fields* (**P** 3114) Conv Geom Combin & Appl;1970 Perugia 63-67
◇ B51 ◇ REV MR 49 # 4452 Zbl 272 # 06009 • ID 60531

BLASZCZUK, J.J. & DZIOBIAK, W. [1975] *Modal systems related to* $S4_n$ *of Sobocinski* (**J** 0387) Bull Sect Logic, Pol Acad Sci 4*103-108
◇ B45 ◇ REV MR 52 # 13330 • ID 21792

BLASZCZUK, J.J. & DZIOBIAK, W. [1975] *Modal systems placed in the "triangle"* $S4 - T_1^* - T$ (**J** 0387) Bull Sect Logic, Pol Acad Sci 4*138-142
◇ B45 ◇ REV MR 54 # 9978 • ID 25744

BLASZCZUK, J.J. & DZIOBIAK, W. [1975] *Remarks on Perzanowski's modal system* (**J** 0387) Bull Sect Logic, Pol Acad Sci 4*57-64
◇ B45 ◇ REV MR 52 # 13329 • ID 21790

BLASZCZUK, J.J. & DZIOBIAK, W. [1976] *An axiomatization of* M^n*-counterparts for some modal calculi* (**J** 0302) Rep Math Logic, Krakow & Katowice 6*3-6
◇ B45 ◇ REV MR 58 # 140 Zbl 371 # 02012 • ID 21908

BLASZCZUK, J.J. & DZIOBIAK, W. [1977] *Modal logics connected with systems* $S4_n$ *of Sobocinski* (**J** 0063) Studia Logica 36*151-164
◇ B45 ◇ REV MR 57 # 9482 Zbl 363 # 02018 • ID 31619

BLASZCZUK, J.J. [1977] *Remarks on* M^n*-counterparts of some normal calculi* (**J** 0387) Bull Sect Logic, Pol Acad Sci 6(2)*82-86
◇ B45 ◇ REV MR 57 # 12173 Zbl 407 # 03025 • ID 31620

BLASZCZUK, J.J. [1978] *Normal modal logics and its extensions (Polish)* (1111) Preprints, Manuscr., Techn. Reports etc. 114pp
◇ B45 ◇ REM Preprint No. 5/78, Torun • ID 31621

BLASZCZUK, J.J. [1978] *Weakest normal calculi with respect to* M^n*-counterparts* (**J** 0387) Bull Sect Logic, Pol Acad Sci 7(3)*102-106
◇ B45 ◇ REV MR 80a:03022 Zbl 407 # 03026 • ID 31622

BLASZCZUK, J.J. [1984] *Some paraconsistent sentential calculi* (**J** 0063) Studia Logica 43*51-62
◇ B53 ◇ ID 42358

BLAU, U. [1983] *Three-valued analysis of precise, vague and presupposing quantifiers* (**C** 3095) Approaching Vagueness 79-129
◇ B50 ◇ REV MR 85j:03003 • ID 44929

BLISHUN, A.F. [1980] *Truth of fuzzy set similarity (Russian and English)* (**J** 2819) Probl Contr Inf Th, Akad Nauk SSSR & Acad Sci Hung 9*381-392
◇ B52 ◇ REV MR 81k:03028 Zbl 443 # 94044 • ID 56447

BLOK, W.J. [1978] *On the degree of incompleteness of modal logics* (**J** 0387) Bull Sect Logic, Pol Acad Sci 7*167-175
◇ B45 G10 G25 ◇ REV MR 81a:03017 Zbl 418 # 03012 • ID 53297

BLOK, W.J. [1978] see BENTHEM VAN, J.F.A.K.

BLOK, W.J. [1979] *An axiomatization of the modal theory of the veiled recession frame* (**J** 0063) Studia Logica 38*37-47
◇ B45 ◇ REV MR 80m:03106 Zbl 415 # 03014 • ID 53116

BLOK, W.J. [1980] *The lattice of modal logics: An algebraic investigation* (**J** 0036) J Symb Logic 45*221-236
◇ B45 G10 ◇ REV MR 81g:03014 Zbl 436 # 03010 JSL 49.1419 • ID 55850

BLOK, W.J. & KOEHLER, P. [1983] *Algebraic semantics for quasiclassical modal logics* (**J** 0036) J Symb Logic 48*941-964
◇ B45 C90 G25 ◇ REV MR 85f:03014 Zbl 562 # 03011 • ID 40651

BLOKHINA, G.N. [1970] *The predicative description of Post classes (Russian)* (**J** 0071) Met Diskr Analiz (Novosibirsk) 16*16-29
◇ B50 ◇ REV MR 44 # 1547 • ID 01311

BLOKHINA, G.N. & KUDRYAVTSEV, V.B. [1971] *On basic groups in k-valued logics (Russian)* (**J** 0023) Dokl Akad Nauk SSSR 201*9-11
• TRANSL [1971] (**J** 0062) Sov Math, Dokl 12*1583-1585
◇ B50 ◇ REV MR 52 # 5349 Zbl 243 # 02012 • ID 30033

BLOKHINA, G.N. & BUROSCH, G. & KUDRYAVTSEV, V.B. [1973] *Das Problem der Vollstaendigkeit fuer boolesche Funktionen ueber zwei Dualmengen mit nichtleerem Durchschnitt I* (**J** 0068) Z Math Logik Grundlagen Math 19*163-180
◇ B50 ◇ REV MR 49 # 8848 Zbl 301 # 02063 • REM Part II 1974 • ID 03914

BLOKHINA, G.N. & KUDRYAVTSEV, V.B. [1974] *A criterion for the basis property of groups in k-valued logic (Russian)* (**C** 3189) Issl Oper Mod, Sist, Reshen, Vyp 4 103-110
◇ B50 ◇ REV MR 58 # 27302 Zbl 346 # 02009 • ID 60554

BLOOM, S.L. [1971] *A completeness theorem for "theories of kind W" (Polish and Russian summaries)* (**J** 0063) Studia Logica 27*43-56
◇ B45 ◇ REV MR 45 # 6595 Zbl 249 # 02014 • ID 01328

BLOOM, S.L. & SUSZKO, R. [1971] *Semantics for the sentential calculus with identity (Polish and Russian summaries)* (**J** 0063) Studia Logica 28*77-82
◇ B60 G05 ◇ REV MR 45 # 6570 Zbl 243 # 02016 • ID 01329

BLOOM, S.L. & SUSZKO, R. [1972] *Investigations into the sentential calculus with identity* (**J** 0047) Notre Dame J Formal Log 13*289-308 • ERR/ADD ibid 17*640
◇ B60 G05 ◇ REV MR 51 # 12478 MR 53 # 12874 Zbl 188.12 • ID 01332

BLOOM, S.L. [1982] *A note on the logic of signed equations* (**J** 0063) Studia Logica 41*75-81
◇ B60 C05 C07 ◇ REV MR 85e:03069 Zbl 509 # 03015 • ID 36525

Author Index

BLOOM, S.L. & WAGNER, E.G. [1985] *Many-sorted theories and their algebras with some applications to data types* (**P** 4651) Algeb Methods in Semantics;1982 Fontainebleau 133-168
◇ B75 B97 ◇ REV Zbl 575 # 18004 • ID 48870

BLOOM, S.L. see Vol. I, III, IV, V, VI for further entries

BLUE, N. [1981] *A metalinguistic interpretation of counterfactual conditionals* (**J** 0122) J Philos Logic 10*179-200
◇ A05 B45 ◇ REV MR 82j:03016 Zbl 473 # 03013
• ID 55344

BLUM, A. [1972] *Isomorphisms between C1 and C2* (**J** 0068) Z Math Logik Grundlagen Math 18*237-240
◇ B45 ◇ REV MR 46 # 3268 Zbl 251 # 02023 • ID 01336

BLUM, A. [1976] *A logic of belief* (**J** 0047) Notre Dame J Formal Log 17*344-348
◇ B48 ◇ REV MR 56 # 5221 Zbl 331 # 02015 • ID 15247

BLUM, A. [1977] *Two observations about S5* (**J** 0068) Z Math Logik Grundlagen Math 23*485-486
◇ B45 ◇ REV MR 57 # 15957 Zbl 382 # 03015 • ID 51939

BLUM, A. see Vol. I for further entries

BLUM, L. & BLUM, M. [1973] *Inductive inference: a recursion theoretic approach* (**P** 3062) IEEE Symp Switch & Automata Th (14);1973 Iowa City 200-208
◇ B48 D15 D80 ◇ REV MR 55 # 5411 • ID 71122

BLUM, L. & BLUM, M. [1975] *Toward a mathematical theory of inductive inference* (**J** 0194) Inform & Control 28*125-155
◇ B48 D10 D15 ◇ REV MR 52 # 16109 Zbl 375 # 02028
• ID 51605

BLUM, L. see Vol. III for further entries

BLUM, M. [1973] see BLUM, L.

BLUM, M. [1975] see BLUM, L.

BLUM, M. see Vol. IV for further entries

BLUMBERG, A.E. [1967] *Modern logic* (**C** 0601) Encycl of Philos 5*12-34
◇ A05 B98 ◇ REV JSL 35.299 • ID 01345

BLUMBERG, A.E. [1976] *Logic: A first course* (**X** 4512) Knopf: New York xiv + 462pp
◇ B98 ◇ REV JSL 44.281 • ID 44584

BLUTNER, R. [1984] *Eine Erlaeuterung zu Freges Begriff der "Erlaeuterung"; defaults und ihre Logik* (**P** 3621) Frege Konferenz (2);1984 Schwerin 114-120
◇ A10 B60 ◇ REV MR 86g:03005 Zbl 556 # 03008
• ID 44930

BLYTH, J.W. [1957] *A modern introduction to logic* (**X** 0847) Houghton Mifflin: Boston xvi + 426pp
◇ B98 ◇ ID 01518

BLYTH, J.W. & JACOBSON JR., J.H. [1963] *Class logic. A programmed text* (**X** 0863) Harcourt: New York & London xxi + 392pp
◇ B98 ◇ REM Review in J0075 1968, 29*292-294 • ID 04257

BLYUMIN, S.L. [1974] *The geometric aspect of multivalued threshold logic (Russian) (English summary insert)* (**J** 0040) Kibernetika, Akad Nauk Ukr SSR 1974/2*140-141
• TRANSL [1974] (**J** 0021) Cybernetics 10*357-358
◇ B50 ◇ REV MR 53 # 12759 Zbl 301 # 02020 • ID 23114

BOBROV, A.E. & ZHURKIN, V.A. [1972] *Minimierung von Funktionen einer mehrwertigen Logik in einem System, das die zyklische Negation enthaelt (Russisch)* (**J** 0474) Avtom Vychis Tekh, Akad Nauk Latv SSR 1972/5*19
◇ B50 ◇ REV Zbl 243 # 02013 • ID 28828

BOBROV, A.E. & KOSTYANKO, N.F. & UGAROV, B.N. [1974] *The Y-complexity of systems of functions of many-valued logic (Russian)* (**J** 0040) Kibernetika, Akad Nauk Ukr SSR 1974/5*147-149
◇ B50 ◇ REV MR 53 # 5251 Zbl 293 # 02015 • ID 22882

BOBROV, A.E. see Vol. I for further entries

BOBROVA, L.A. [1970] *The completeness of systems of degenerate inference and quasiinference (Russian)* (**C** 0668) Neklass Log 107-112
◇ B60 ◇ REV MR 47 # 6451 • ID 01557

BOBROW, D.G. (ED.) [1980] *Special issue on nonmonotonic logic* (**J** 0503) Artif Intell 13*1-174
◇ B60 ◇ REV MR 81d:68119 • ID 80429

BOCHENSKI, I.M. [1938] *Nove lezioni di logica simbolica* (**X** 2109) Angelicum: Roma 183pp
◇ A05 B98 ◇ REV FdM 65.1100 • ID 43782

BOCHENSKI, I.M. [1949] *Precis de logique mathematique* (**X** 0849) Kroonder: Bussum 90pp
◇ B98 ◇ REV MR 13.811 JSL 15.199 • ID 01537

BOCHENSKI, I.M. & MENNE, A. [1954] *Grundriss der Logistik* (**X** 0846) Schoeningh: Paderborn 124pp
◇ B98 ◇ REV JSL 24.220 • REM 2nd ed. 1962, 3rd ed. 1965, 4th ed. 1973 • ID 33428

BOCHENSKI, I.M. [1956] *Formale Logik* (**X** 0826) Alber: Freiburg xv + 640pp
• TRANSL [1961] (**X** 0845) Univ Notre Dame Pr: Notre Dame xxii + 567pp [1970] (**X** 0848) Chelsea: New York xxii + 567pp
◇ A10 B98 ◇ REV MR 22 # 10899 Zbl 70.7 Zbl 98.241 JSL 25.57 • REM Transl. title: A history of formal logic
• ID 42191

BOCHENSKI, I.M. [1959] *A precis of mathematical logic* (**X** 0835) Reidel: Dordrecht x + 100pp
• TRANSL [1982] (**X** 3560) Paraninfo: Madrid 120pp
◇ B98 ◇ REV MR 31 # 2123 Zbl 90.9 JSL 25.78 • REM 2nd ed. 1963 • ID 22545

BOCHVAR, D.A. [1938] *On a three-valued logical calculus and its application to the analysis of contradictions (Russian)* (**J** 0142) Mat Sb, Akad Nauk SSSR, NS 4(46)*287-308
• TRANSL [1981] (**J** 2028) Hist & Phil Log 2*87-112
◇ A05 B50 ◇ REV MR 83f:03022 Zbl 20.194 Zbl 512 # 03004 JSL 4.98 FdM 64.928 • ID 01347

BOCHVAR, D.A. [1943] *On the consistency of a three-valued logical calculus (Russian) (English summary)* (**J** 0142) Mat Sb, Akad Nauk SSSR, NS 12(54)*353-369
◇ B50 ◇ REV MR 5.197 Zbl 61.8 JSL 11.129 • ID 01349

BOCHVAR, D.A. [1961] *On a three-valued calculus and its application to analysis of the paradoxes of the classical extended functional calculus (Romanian)* (**J** 0210) Stud Sti Tehn Acad Romina Timisoara 15*200-222
◇ B50 ◇ REV MR 37 # 6162 • ID 02007

BOCHVAR, D.A. & FINN, V.K. [1974] *Quasilogical functions (Russian)* (**C** 2577) Issl Formaliz Yazyk & Neklass Log 200-213
◇ B50 ◇ REV MR 56 # 5212 • ID 71135

BOCHVAR, D.A. (ED.) [1974] *Studies in formalized languages and nonclassical logics (Russian)* (**X** 2027) Nauka: Moskva 275pp
◇ B97 ◇ REV MR 54 # 7191 • ID 70160

BOCHVAR, D.A. [1976] *On a general theory of logical matrices with a continuum of valences (Russian)* (**C** 3271) Issl Teor Mnozh & Neklass Logik 198–220
◇ B50 ◇ REV MR 57 # 15952 Zbl 404 # 03017 • ID 54804

BOCHVAR, D.A. & FINN, V.K. [1976] *Some supplements to articles on multivalued logics (Russian)* (**C** 3271) Issl Teor Mnozh & Neklass Logik 265–325
◇ B50 ◇ REV MR 57 # 15951 Zbl 406 # 03047 • ID 56130

BOCHVAR, D.A. & GRISHIN, V.N. (EDS.) [1976] *Studies in set theory and nonclassical logics. Work collection (Russian)* (**X** 2235) VINITI: Moskva 328pp
◇ B97 E97 ◇ REV MR 55 # 7721 Zbl 392 # 00004 • ID 52367

BOCHVAR, D.A. & FINN, V.K. [1977] *A note to the article "Some supplements to articles on multivalued logics" (Russian) (English and German summaries)* (**J** 0338) Nauch-Tekh Inf, Ser 2, Akad Nauk SSSR 1977/9*7,31,32
◇ B50 ◇ REV MR 58 # 16135 Zbl 406 # 03048 • ID 56131

BOCHVAR, D.A. & SHOKUROV, V.V. [1979] *The behaviour of the plausibility functional for "tertium non datur" in some sequences of logical matrices (Russian)* (**C** 2581) Issl Neklass Log & Teor Mnozh 330–344
◇ B50 ◇ REV MR 81j:03043 Zbl 438 # 03027 • ID 55939

BOCHVAR, D.A. [1982] *Some aspects of the investigations of reification paradoxes* (**P** 3800) Intens Log: Th & Appl;1979 Moskva 229–238
• TRANSL [1984] (**C** 4366) Modal & Intens Log & Primen Probl Metodol Nauk 99–107
◇ B45 ◇ REV MR 84k:03071 Zbl 538 # 03009 • ID 35011

BOCHVAR, D.A. [1984] *On the consistency of a three-valued logical calculus* (**J** 3781) Topoi 3*3–12
◇ B50 ◇ REM Translated form the Russian • ID 44656

BOCHVAR, D.A. see Vol. I, III, V for further entries

BODDENBERG, E. [1975] *Logik. I* (**X** 1039) Diesterweg: Frankfurt a.M. viii + 103pp
◇ B98 ◇ REV MR 51 # 40 Zbl 323 # 02001 • ID 15248

BODE, J.R. [1977] *Knowledge and the evidential conditional* (**J** 0095) Philos Stud 31*337–344
◇ A05 B45 ◇ REV MR 58 # 27202 • ID 71142

BODE, J.R. [1979] *The possibility of a conditional logic* (**J** 0047) Notre Dame J Formal Log 20*147–154
◇ A05 B45 ◇ REV MR 82k:03014 Zbl 322 # 02026 • ID 52389

BODENDIEK, R. [1974] *Ueber das Repraesentantenproblem in der k-wertigen Logik* (**J** 0160) Math-Phys Sem-ber, NS 21*46–66
◇ B50 ◇ REV MR 50 # 6788 Zbl 285 # 02016 • ID 03925

BOECH, P. [1983] *"Vagueness" is context-dependence. A solution to the sorites paradox* (**C** 3095) Approaching Vagueness 189–210
◇ B52 ◇ REV MR 85j:03003 • ID 44914

BOEHME, G. [1981] *Einstieg in die mathematische Logik* (**X** 3223) Hanser: Muenchen 208pp
◇ B98 ◇ REV MR 83f:03002 Zbl 469 # 03001 • ID 55129

BOERNER, W. [1969] *Eine Klasse von Darstellungen der Funktionen der k-wertigen Logik* (**J** 0192) Wiss Z Univ Jena, Math-Nat Reihe 18*313–316
◇ B50 ◇ REV MR 43 # 28 Zbl 222 # 02015 • ID 01513

BOGDAN, R.J. & NIINILUOTO, I. (EDS.) [1975] *Logic, language, and probability. A selection of papers contributed to sections IV, VI and XI of logic, methodology and philosophy of science* (**X** 0835) Reidel: Dordrecht x + 323pp
◇ B97 ◇ REV MR 55 # 7658 Zbl 266 # 00003 • ID 80462

BOGDAN, R.J. (ED.) [1976] *Local induction* (**X** 0835) Reidel: Dordrecht xiv + 339pp
◇ A05 B97 ◇ REV MR 58 # 27192 Zbl 312 # 00003 • ID 70074

BOGDAN, R.J. see Vol. IV for further entries

BOICESCU, V. [1971] *Sur les systemes deductifs dans la logique Θ-valente* (**J** 0056) Publ Dep Math, Lyon 8/1*123–133
◇ B50 G20 ◇ REV MR 48 # 10760 Zbl 262 # 02016 • ID 01412

BOICESCU, V. [1973] *On θ-valent logics (Romanian)* (**S** 1613) Probl Logic (Bucharest) 5*241–255
◇ B50 ◇ REV MR 58 # 5022 • ID 71188

BOLL, M. [1942] *Elements de logique scientifique* (**X** 0856) Dunod: Paris vii + 243pp
◇ A05 B98 ◇ REV JSL 14.266 • ID 22393

BONDAREV, V.M. [1979] *A decomposition theorem in predicate algebra (Russian)* (**S** 2024) Probl Bioniki 22*87–93,145
◇ B50 ◇ REV MR 83e:03106 • ID 34831

BONEVAC, D. [1983] *Chellas on conditional obligation* (**J** 0095) Philos Stud 44*247–255
◇ B45 ◇ REV MR 85d:03028 • ID 44804

BONEVAC, D. see Vol. I for further entries

BONISSONE, P.P. [1982] *A fuzzy sets based linguistic approach: theory and applications* (**C** 3786) Approx Reason in Decis Anal 329–339
◇ B52 ◇ REV MR 84i:03055 Zbl 482 # 03018 • ID 34534

BONOMI, A. & MANGIONE, C. (EDS.) [1977] *Recent philosophical logic in Italy* (**J** 0122) J Philos Logic 6*239–365
◇ A05 B97 ◇ REV MR 58 # 4977 • ID 70092

BONOTTO, C. [1982] *Synonymy for Bressan's modal calculus MC^{ν}. I,II* (**J** 0330) Atti Ist Veneto, Fis Mat Nat 140*11–24,85–99
◇ B45 ◇ REV Zbl 574 # 03009 Zbl 574 # 03010 • ID 48830

BONOTTO, C. see Vol. I for further entries

BOOLE, G. [1947] *The mathematical analysis of logic, being an essay toward a calculus of deductive reasoning* (**X** 1096) Blackwell: Oxford 82pp
◇ A05 A10 B98 G05 ◇ REM 1st ed. 1847 • ID 16747

BOOLE, G. [1952] *Studies in logic and probability* (**X** 1368) Watts: London 500pp
◇ A05 B98 ◇ REV Zbl 49.8 JSL 24.203 • ID 25635

BOOLE, G. [1958] *An investigation of the laws of thought, on which are founded the mathematical theories of logic and probabilities* (**X** 0813) Dover: New York xi + 424pp
◇ A05 B98 G05 ◇ REV JSL 16.224 • REM 1st ed. 1854 • ID 16748

BOOLE, G. see Vol. I for further entries

Author Index

BOOLOS, G. & JEFFREY, R.C. [1974] *Computability and logic* (X 0805) Cambridge Univ Pr: Cambridge, GB x+262pp
◇ B98 D98 ◇ REV MR 49 # 7120 MR 82d:03001 Zbl 298 # 02003 JSL 42.585 • ID 03933

BOOLOS, G. [1976] *On deciding the truth of certain statements involving the notion of consistency* (J 0036) J Symb Logic 41*779-781
◇ B25 B45 F30 ◇ REV MR 56 # 8353 Zbl 359 # 02050 • ID 14578

BOOLOS, G. [1977] *On deciding the provability of certain fixed point statements* (J 0036) J Symb Logic 42*191-193
◇ B25 B45 F30 ◇ REV MR 58 # 192 Zbl 381 # 03013 • ID 26442

BOOLOS, G. [1979] *Reflection principles and iterated consistency assertions* (J 0036) J Symb Logic 44*33-35
◇ B45 F30 ◇ REV MR 80k:03064 Zbl 409 # 03010 • ID 56313

BOOLOS, G. [1979] *The unprovability of consistency. An essay in modal logic* (X 0805) Cambridge Univ Pr: Cambridge, GB viii+184pp
◇ B25 B28 B45 B98 F30 F98 ◇ REV MR 81c:03013 Zbl 409 # 03009 JSL 46.871 • ID 56312

BOOLOS, G. [1980] *On systems of modal logic with provability interpretations* (J 0105) Theoria (Lund) 46*7-18
◇ B45 F30 ◇ REV MR 83c:03013 Zbl 473 # 03010 • ID 55341

BOOLOS, G. [1980] *Provability, truth, and modal logic* (J 0122) J Philos Logic 9*1-7
◇ B45 ◇ REV MR 81h:03035 Zbl 426 # 03024 • ID 53620

BOOLOS, G. [1980] *Provability in arithmetic and a schema of Grzegorczyk* (J 0027) Fund Math 106*41-45
◇ B45 ◇ REV MR 81m:03021 Zbl 438 # 03021 • ID 55933

BOOLOS, G. [1982] *On the nonexistence of certain normal forms in the logic of provability* (J 0036) J Symb Logic 47*638-640
◇ B45 F30 ◇ REV MR 83j:03031 Zbl 491 # 03005 • ID 35340

BOOLOS, G. [1984] *The logic of provability* (J 0005) Amer Math Mon 91*470-480
◇ B45 F30 ◇ REV MR 85m:03010 Zbl 562 # 03007 • ID 43982

BOOLOS, G. & SAMBIN, G. [1985] *An incomplete system of modal logic* (J 0122) J Philos Logic 14*351-358
◇ B45 ◇ ID 47640

BOOLOS, G. [1985] *1-consistency and the diamond* (J 0047) Notre Dame J Formal Log 26*341-347
◇ B45 F30 ◇ ID 47529

BOOLOS, G. see Vol. I, III, IV, V, VI for further entries

BOON, L. [1979] *Repeated tests and repeated testing: how to corroborate low level hypotheses* (J 0989) Z Allg Wissth 10*1-10
◇ A05 B48 ◇ REV MR 80h:03006 • ID 71221

BORDAT, J. [1978] *Resolution des equations de la logique a p valeurs* (J 0060) Rev Roumaine Math Pures Appl 23*507-531
◇ B50 G10 G20 ◇ REV MR 58 # 10634 Zbl 383 # 03042 • ID 52024

BOREL, E. [1939] *Valeur pratique et philosophie des probabilites* (X 0834) Gauthier-Villars: Paris ix+169pp
◇ A05 B48 ◇ REV FdM 65.545 • ID 25398

BOREL, E. see Vol. IV, V, VI for further entries

BORGA, M. [1983] *On some proof theoretical properties of the modal logic GL* (J 0063) Studia Logica 42*453-459
◇ B45 F05 F30 ◇ REV MR 86a:03015 Zbl 547 # 03016 • ID 42318

BORGA, M. see Vol. I, IV, VI for further entries

BORICIC, B.R. [1981] *Equational reformulations of intuitionistic propositional calculus and classical first-order predicate calculus* (J 0400) Publ Inst Math, NS (Belgrade) 29(43)*23-28
◇ B55 F50 ◇ REV MR 83e:03022 Zbl 498 # 03050 • ID 35211

BORICIC, B.R. [1983] *A decision procedure for certain disjunction-free intermediate propositional calculi* (J 0400) Publ Inst Math, NS (Belgrade) 34(48)*19-26
◇ B25 B55 ◇ REV MR 86d:03027 Zbl 556 # 03025 • ID 44659

BORICIC, B.R. [1983] *One of the possible formal descriptions of deducibility* (J 0400) Publ Inst Math, NS (Belgrade) 34(48)*13-18
◇ B60 ◇ REV MR 86f:03019 Zbl 554 # 03017 • ID 44939

BORICIC, B.R. [1984] *A note on some intermediate propositional calculi* (J 0036) J Symb Logic 49*329-333
◇ B55 ◇ REV MR 86b:03026 Zbl 555 # 03013 • ID 42424

BORICIC, B.R. [1985] *On some subsystems of Dummett's LC* (J 0068) Z Math Logik Grundlagen Math 31*243-247
◇ B45 ◇ ID 47573

BORICIC, B.R. see Vol. VI for further entries

BORILLO, M. (ED.) [1980] *Representation of knowledge and reason in the human sciences* (X 2732) INRIA: Le Chesnay Cedex iii+607pp
◇ B97 ◇ REV MR 82d:03007 • ID 70024

BORISOV, A.N. & LEVCHENKOV, A.S. [1982] *Methods of interactive evaluation of decisions (Russian)* (X 2230) Zinatne: Riga 140pp
◇ B52 ◇ REV Zbl 494 # 90002 • ID 36660

BORISOV, YU.F. (ED.) [1979] *Methodological problems of mathematics (Russian)* (X 2642) Nauka: Novosibirsk 303pp
◇ B97 ◇ REV MR 81a:03002 • ID 70053

BORISOV, YU.F. see Vol. I for further entries

BORKOWSKI, L. [1958] *On modal terms (Polish, Russian and English summaries)* (J 0063) Studia Logica 7*7-41
◇ B45 ◇ REV MR 21 # 1269 • ID 01478

BORKOWSKI, L. & SLUPECKI, J. [1963] *Elements of mathematical logic and set theory (Polish)* (X 1034) PWN: Warsaw 285pp • TRANSL [1967] (X 0869) Pergamon Pr: Oxford xii+349pp
◇ A05 B98 E98 ◇ REV MR 37 # 3904 Zbl 171.248 • REM 2nd ed.1967, xii+349pp; 3rd corr. ed.1969, 306pp (errata insert) • ID 19509

BORKOWSKI, L. [1976] *Formale Logik. Logische Systeme. Einfuehrung in die Metalogik* (X 0911) Akademie Verlag: Berlin xiv+578pp
◇ A05 B98 ◇ REV Zbl 357 # 02001 • REM Translated from Polish • ID 50349

BORKOWSKI, L. see Vol. I, III, V, VI for further entries

BORN, R. [1982] *Kausalitaet und Quantenlogik* (J 0455) Phil Naturalis 19*583-600
◇ B51 G12 ◇ ID 45736

Borowik, P. [1981] *On Gentzen axiomatization of the reducts of multivalued logic (Polish) (English summary)* (**J** 2412) Prace Nauk, Ser Mat-Przyr, Wyz Szk Ped, Czestochowa 4*5-9
◇ B50 ◇ REV Zbl 552 # 03016 • ID 43361

Bosanquet, B. [1911] *Logic. Vol. II* (**X** 3333) Unknown Publisher: See Remarks
◇ B98 ◇ REM 2nd ed., Oxford • ID 16751

Bossu, G. & Siegel, P. [1985] *Saturation, nonmonotonic reasoning and the closed-world assumption* (**J** 0503) Artif Intell 25*13-63
◇ B60 ◇ REV Zbl 569 # 68078 • ID 45074

Bossuet, G. [1977] *phi -algebras: some concepts* (**P** 2013) Int Symp Multi-Val Log (7);1977 Charlotte 95-97
◇ B50 G25 ◇ REV MR 57 # 12170 • ID 71256

Botezatu, P. [1967] *La logique extensionnaliste et la double negation* (**J** 0147) An Univ Bucuresti, Acta Logica 10*179-183
◇ B60 ◇ ID 60660

Botezatu, P. [1969] *L'implication probable et l'implication certaine* (**J** 0147) An Univ Bucuresti, Acta Logica 12*121-123
◇ B48 ◇ REV MR 44 # 3835 • ID 48001

Botezatu, P. [1975] *Assertoric-problematic syllogistics (Romanian) (French summary)* (**S** 1613) Probl Logic (Bucharest) 6*73-91
◇ B45 ◇ REV MR 57 # 2881 • ID 71261

Botta, O. & Delorme, M. [1982] *An f-sets universe* $\tilde{V}$ (**C** 3778) Fuzzy Inform & Decis Processes 133-141
◇ B52 E72 ◇ REV MR 84m:03073 Zbl 524 # 03046 • ID 35770

Botta, O. see Vol. V for further entries

Botusharov, O.I. [1983] *A recursion theory characterization of inductive inference with additional information classes* (**J** 2774) Koezlem MTA Szam & Autom: Kutat Intez 147*101-109
◇ B48 D20 ◇ ID 45089

Botusharov, O.I. [1985] *On inductive inference with additional information (Bulgarian summary)* (**P** 4391) Mat & Mat Obrazov (14);1985 Sl"nchev Bryag 330-335
◇ B48 D20 ◇ ID 48979

Bouchon, B. [1981] *Information transmitted by a system of fuzzy events* (**P** 4050) Appl Syst & Cybern;1980 Acapulco 2767-2772
◇ B52 ◇ REV MR 84b:00009 • ID 46299

Bouchon, B. see Vol. V for further entries

Boudot, M. [1972] *Logique inductive et probabilite. Philosophie pour l'age de la science* (**X** 0850) Colin: Paris 333pp
◇ A05 B48 ◇ REV MR 57 # 9499 Zbl 253 # 02025 • ID 28931

Bouligand, G. [1937] *Structure des theories. Problemes infinis* (**X** 0859) Hermann: Paris 58pp
◇ A05 B98 ◇ REV FdM 63.819 • ID 25518

Bouligand, G. see Vol. V for further entries

Bourbaki, N. [1949] *Foundations of mathematics for the working mathematician* (**J** 0036) J Symb Logic 14*1-8
◇ A05 B98 ◇ REV MR 11.73 Zbl 34.1 JSL 14.258 • ID 01499

Bourbaki, N. see Vol. V for further entries

Bourrelly, L. & Chouraqui, E. & Ricard, M. [1984] *Formalisation of an approximate reasoning: the analogical reasoning* (**P** 3081) IFAC Symp Fuzzy Inf, Knowl Repr & Decis. Anal;1983 Marseille 135-141
◇ B52 ◇ REV Zbl 556 # 68050 • ID 48162

Boussuet, G. [1976] *Random-valued switching algebra* (**P** 2011) Int Symp Multi-Val Log (6);1976 Logan 263
◇ B50 B70 ◇ ID 35897

Bowen, K.A. [1971] *An extension of the intuitionistic propositional calculus* (**J** 0028) Indag Math 33*287-294
◇ B55 F50 ◇ REV MR 45 # 3161 Zbl 221 # 02009 • ID 01504

Bowen, K.A. [1975] *Normal modal model theory* (**J** 0122) J Philos Logic 4*97-131
◇ B45 C90 ◇ REV MR 57 # 5733 Zbl 317 # 02015 • ID 29636

Bowen, K.A. [1978] *Model theory for modal logic. Kripke models for modal predicate calculi* (**X** 0835) Reidel: Dordrecht x + 128pp
◇ B45 C90 ◇ REV MR 80j:03023 Zbl 395 # 03022 JSL 46.415 • ID 30659

Bowen, K.A. [1980] *Interpolation in loop-free logic* (**J** 0063) Studia Logica 39*297-310
◇ B75 C40 ◇ REV MR 82c:03020 Zbl 457 # 03012 • ID 54337

Bowen, K.A. see Vol. I, III, V, VI for further entries

Boyarskaya, Z.A. [1980] *A comparison of the expressive powers of propositional modal logics (Russian)* (**C** 2620) Teor Model & Primen 7-14
◇ B45 ◇ REV MR 82c:03019 Zbl 541 # 03008 • ID 41372

Boyer, R.S. & Moore, J.S. [1979] *A computational logic* (**X** 0801) Academic Pr: New York xiv + 397pp
◇ B98 D70 ◇ REV MR 81d:68127 Zbl 448 # 68020 • ID 56665

Boyer, R.S. see Vol. I, IV for further entries

Bozic, M. & Dosen, K. [1983] *Axiomatizations of intuitionistic double negation* (**J** 0387) Bull Sect Logic, Pol Acad Sci 12*99-104
◇ B55 F50 ◇ REV Zbl 542 # 03007 • ID 46894

Bozic, M. & Dosen, K. [1984] *Models for normal intuitionistic modal logics* (**J** 0063) Studia Logica 43*217-245
◇ B45 C90 ◇ ID 40033

Bozic, M. [1984] *Positive logic with double negation* (**J** 0482) Publ Math Univ Belgrade 35(49)*21-31
◇ B55 ◇ REV Zbl 569 # 03010 • ID 44537

Bozic, M. [1985] *Semantics for some intermediate logics* (**J** 0400) Publ Inst Math, NS (Belgrade) 37*7-15
◇ B55 C90 ◇ ID 48301

Bozic, M. see Vol. V for further entries

Bozzi, S. & Meloni, G.C. [1977] *Completezza proposizionale per l'operatore modale "accade localmente che" (English summary)* (**J** 3285) Boll Unione Mat Ital, V Ser, A 14*489-497
◇ B45 ◇ REV MR 57 # 9491 Zbl 376 # 02021 • ID 51663

BOZZI, S. & MELONI, G.C. [1980] *Representation of Heyting algebras with covering and propositional intuitionistic logic with local operator* (J 3285) Boll Unione Mat Ital, V Ser, A 17*436-442
◇ B25 B55 C90 F50 G10 ◇ REV MR 82k:03099 Zbl 454 # 03008 • ID 54221

BOZZI, S. see Vol. III, VI for further entries

BRABEC, J. & PTAK, P. [1982] *On compatibility in quantum logics* (J 1678) Found Phys 12*207-212
◇ B51 G12 ◇ REV MR 84h:81009 • ID 39746

BRADLEY, R. & SWARTZ, N. [1979] *Possible worlds* (X 2725) Hackett Publ : Indianapolis xxi + 391pp
◇ A05 B45 ◇ REV MR 80m:03002 • ID 71290

BRADY, R.T. [1971] *The consistency of the axioms of abstraction and extensionality in a three-valued logic* (J 0047) Notre Dame J Formal Log 12*447-453
◇ B50 E35 E70 ◇ REV MR 45 # 6589 Zbl 185.14 • ID 01576

BRADY, R.T. [1972] *The relative consistency of the class axioms of abstraction and extensionality and the axioms of NBG in a three-valued logic* (J 0047) Notre Dame J Formal Log 13*161-176
◇ B50 E35 E70 ◇ REV MR 47 # 3173 Zbl 234 # 02043 • ID 01577

BRADY, R.T. & ROUTLEY, R. [1973] *Don't care was made to care* (J 0273) Australasian J Phil 51*211-225
◇ B50 ◇ ID 30637

BRADY, R.T. [1976] *Significance logics* (J 0047) Notre Dame J Formal Log 17*161-183
◇ B50 ◇ REV MR 54 # 57 Zbl 226 # 02029 • ID 18102

BRADY, R.T. [1980] *A theory of classes and individuals based on a 3-valued significance logic* (J 0047) Notre Dame J Formal Log 21*385-414
◇ B50 E35 E70 ◇ REV MR 81i:03028 Zbl 315 # 02021 • ID 53774

BRADY, R.T. [1980] *Significance range theory* (J 0047) Notre Dame J Formal Log 21*319-345
◇ B50 E70 ◇ REV MR 81i:03029 Zbl 271 # 02015 Zbl 442 # 03037 • ID 56394

BRADY, R.T. [1980] *Two remarks on "The logic of significance and context"* (J 0047) Notre Dame J Formal Log 21*263-272
◇ A05 B50 ◇ REV MR 81e:03013 Zbl 349 # 02023 • ID 53775

BRADY, R.T. [1982] *Completeness proofs for the systems RM3 and BN4* (J 0079) Logique & Anal, NS 25*9-32
◇ B46 ◇ REV MR 83j:03041 Zbl 498 # 03012 • ID 35346

BRADY, R.T. & MEYER, R.K. & PLUMWOOD, V. & ROUTLEY, R. [1982] *Relevant logics and their rivals. Part I* (X 3905) Ridgeview: Atascadero xv + 460pp
◇ B46 ◇ REV MR 85k:03013 • ID 44960

BRADY, R.T. [1984] *Depth relevance of some paraconsistent logics* (J 0063) Studia Logica 43*63-74
◇ B53 ◇ ID 42359

BRADY, R.T. [1984] *Natural deduction systems for some quantified relevant logics* (J 0079) Logique & Anal, NS 27*355-377
◇ B46 C90 ◇ REV Zbl 559 # 03011 • ID 45522

BRADY, R.T. & GIAMBRONE, S. & MEYER, R.K. [1984] *Where gamma fails* (J 0063) Studia Logica 43*247-256
◇ B46 ◇ ID 42373

BRADY, R.T. see Vol. I, III, V for further entries

BRAITHWAITE, R.B. [1953] *Scientific explanation: A study of the function of theory, probability and law in science* (X 0805) Cambridge Univ Pr: Cambridge, GB xii + 176pp
◇ A05 B48 ◇ REV MR 14.1097 Zbl 52.4 JSL 22.404 • ID 25400

BRAITHWAITE, R.B. see Vol. I for further entries

BRANDOM, R.B. & RESCHER, N. [1980] *The logic of inconsistency* (X 1096) Blackwell: Oxford x + 184pp
◇ B53 ◇ REV JSL 47.233 • ID 42883

BRANDOM, R.B. see Vol. I for further entries

BRAUN, E.L. [1963] *Digital computer design: logic, circuitry, and synthesis* (X 0801) Academic Pr: New York xiii + 606pp
◇ B70 B98 ◇ REV MR 28 # 2655 • ID 23515

BREITKOPF, A. & KUTSCHERA VON, F. [1971] *Einfuehrung in die moderne Logik* (X 0826) Alber: Freiburg 175pp
◇ B98 ◇ REV MR 83e:03002 Zbl 478 # 03001 • REM 4th edition 1979; 194pp • ID 22414

BRENNAN, J.G. [1957] *A handbook of logic* (X 0837) Harper & Row: New York x + 222pp
◇ B98 ◇ REV JSL 24.186 JSL 25.334 • ID 22465

BRENNENSTUHL, W. [1983] *A door's closing-a contribution to the vagueness semantics of tense and aspect* (C 3095) Approaching Vagueness 293-316
◇ B52 ◇ REV MR 85j:03003 MR 86e:03028 • ID 44931

BRENY, H. [1983] *A propos du terme "probabilite"* (J 0079) Logique & Anal, NS 26*129-155
◇ B48 ◇ REV MR 85j:03025 • ID 45327

BRESSAN, A. [1972] *A general interpreted modal calculus* (X 0875) Yale Univ Pr: New Haven xxviii + 327pp
◇ B45 ◇ REV MR 53 # 5260 Zbl 255 # 02015 JSL 39.352 • ID 04161

BRESSAN, A. [1973] *Intensional descriptions and relative completeness in the general interpreted modal calculus MC^{ν}* (P 0580) Int Congr Log, Meth & Phil of Sci (4,Sel Pap);1971 Bucharest 29-40
◇ B45 ◇ REV MR 57 # 65 Zbl 279 # 02009 • ID 28143

BRESSAN, A. [1973] *The interpreted type-free modal calculus MC^{∞}. I: The type free extensional calculus MC^{∞} involving individuals, and the interpreted language ML^{∞} on which EC^{∞} is based* (J 0144) Rend Sem Mat Univ Padova 49*157-194
◇ B45 E70 ◇ REV MR 51 # 12479a Zbl 279 # 02010 • REM Part II 1973 • ID 60703

BRESSAN, A. [1973] *The interpreted type-free modal calculus MC^{∞}. II. Foundations of MC^{∞}* (J 0144) Rend Sem Mat Univ Padova 50*19-57
◇ B45 E70 ◇ REV MR 51 # 12479b Zbl 285 # 02020 • REM Part I 1973. Part III 1974 • ID 17268

BRESSAN, A. [1974] *Comment's on Suppes' paper: the essential but implicit role of modal concepts in science* (P 1791) Proc Bienn Meet Phil of Sci Ass;1972 East Lansing 315-321
◇ A05 B45 ◇ REV Zbl 317 # 02005 • ID 29632

BRESSAN, A. [1974] *On the semantics for the language ML^{ν} based on a type system, and those for the type-free language ML^{∞}* (J 0122) J Philos Logic 3*171-194 • ERR/ADD ibid 3*469
• REPR [1977] (P 2116) Probl in Log & Ontology;1973 Salzburg 5-28
◇ B45 E70 ◇ REV MR 55 # 2496 Zbl 285 # 02019 • ID 29947

BRESSAN, A. [1974] *On the usefulness of modal logic in axiomatizations of physics* (P 1791) Proc Bienn Meet Phil of Sci Ass;1972 East Lansing 285-303
◇ B45 B80 ◇ REV Zbl 322 # 02014 • ID 60704

BRESSAN, A. [1974] *The interpreted type-free modal calculus MC^{∞} III. Ordinals and cardinals in MC^{∞} (Italian summary)* (J 0144) Rend Sem Mat Univ Padova 51*1-25
◇ B45 E70 ◇ REV MR 51 # 12479c Zbl 304 # 02009 • REM Part II 1973 • ID 17269

BRESSAN, A. & MONTANARO, A. [1981] *Contributions to foundations of probability calculus on the basis of the modal logical calculus MC^{ν} or MC_{*}^{ν} I* (J 0144) Rend Sem Mat Univ Padova 64*109-126
◇ B48 ◇ REV MR 83h:03032c Zbl 485 # 03006 • REM Part II 1982 • ID 36051

BRESSAN, A. & MONTANARO, A. [1981] *Contributions to foundations of probability calculus on the basis of the modal logical calculus MC^{ν} or MC_{*}^{ν} II* (J 0144) Rend Sem Mat Univ Padova 65*263-270
◇ B45 ◇ REV MR 83h:03032d Zbl 501 # 03010 • REM Part I 1981. Part III 1983 • ID 36052

BRESSAN, A. [1981] *Extensions of the modal calculi MC^{ν} and MC^{∞}. Comparison of them with similar calculi endowed with different semantics. Application to probability theory* (P 3113) Aspects Philos Logic;1977 Tuebingen 21-66
◇ B45 ◇ REV MR 83h:03032a Zbl 476 # 03028 • ID 55540

BRESSAN, A. & ZANARDO, A. [1981] *General operators binding variables in the interpreted modal calculus MC* (J 0149) Atti Accad Naz Lincei Fis Mat Nat, Ser 8 70*191-197
◇ B45 ◇ REV Zbl 525 # 03005 • ID 38246

BRESSAN, A. & MONTANARO, A. [1983] *Contributions to foundations of probability calculus on the basis of the modal logic calculus MC^{ν} or MC_{*}^{ν} III: An analysis of the notions of random variables and probability spaces, based on modal logic* (J 0144) Rend Sem Mat Univ Padova 70*1-11
◇ B48 ◇ REV MR 85m:03014 Zbl 533 # 03006 • REM Part II 1982 • ID 36530

BRESSAN, A. see Vol. I for further entries

BRICKHILL, C.J. [1972] see ASH, C.J.

BRIDGES, D.S. [1981] *Towards a constructive foundation for quantum mechanics* (P 3146) Constr Math;1980 Las Cruces 260-273
◇ B51 G12 ◇ REV MR 83a:81003 Zbl 463 # 03038 • ID 54578

BRIDGES, D.S. see Vol. IV, V, VI for further entries

BRILL JR., E.D. & CHANG, SHOOUYUH & HOPKINS, L.D. [1983] *Modeling to generate alternatives: a fuzzy approach* (J 2720) Fuzzy Sets Syst 9*137-151
◇ B52 ◇ REV Zbl 503 # 90056 • ID 36677

BRIM, L. [1980] *Algebraic semantics for propositional dynamic logic* (J 0667) Scripta Fac Sci Math, Brno 10*389-397
◇ B75 ◇ REV MR 83d:03017 • ID 35171

BRINK, C. [1981] *An eight-valued logic* (P 3705) Int Symp Multi-Val Log (11);1981 Oklahoma City & Norman 109-116
◇ B50 ◇ REV MR 83m:94033 Zbl 549 # 03017 • ID 43106

BRINK, C. see Vol. V for further entries

BRISKMAN, L. [1975] *Classical semantics and entailment* (J 0103) Analysis (Oxford) 35*118-126
◇ B46 ◇ REV Zbl 345 # 02012 • ID 29773

BROAD, C.D. [1968] *Induction, probability and causation: Selected papers* (X 0835) Reidel: Dordrecht xi + 296pp
◇ A05 B48 ◇ REV MR 39 # 4882 Zbl 172.286 • ID 25402

BROADIE, A. [1982] *The logical syntax of deontic operators* (J 2811) Phil Quart (St Andrews) 32*116-126
◇ B45 ◇ REV MR 84h:03034 • ID 34245

BRODSKIJ, I.N. [1972] *Elementare Einfuehrung in die Logik. 2.ueberarb. Aufl (Russian)* (X 0938) Leningrad Univ: Leningrad 64pp
◇ B98 ◇ REV Zbl 259 # 02001 • ID 30337

BRODSKIJ, I.N. see Vol. I for further entries

BRODY, B.A. [1973] *Logic, theoretical and applied* (X 0819) Prentice Hall: Englewood Cliffs viii + 280pp
◇ B98 ◇ REV JSL 42.319 • ID 44487

BRODY, T.A. [1984] *On quantum logic* (J 1678) Found Phys 14*409-430
◇ B51 G12 ◇ ID 44772

BROIDO, J. [1975] *Von Wright's principle of predication - some clarifications* (J 0122) J Philos Logic 4*1-11 • ERR/ADD ibid 4*381
◇ B45 ◇ REV MR 58 # 5044a MR 58 # 5044b Zbl 316 # 02025 • ID 17987

BROIDO, J. [1976] *Of the eliminability of de re modalities in some systems* (J 0047) Notre Dame J Formal Log 17*79-88
◇ B45 ◇ REV MR 55 # 82 Zbl 257 # 02016 • ID 17988

BROOKES, S.D. & ROUNDS, W.C. [1983] *Behavioural equivalence relations induced by programming logics* (P 3851) Automata, Lang & Progr (10);1983 Barcelona 97-108
◇ B75 ◇ REV MR 85d:68024 Zbl 536 # 68042 • ID 36777

BROOKS, R.B.S. [1961] see ALLEN, L.E.

BROWN, CHARLES D. [1978] *The ontological theorem* (J 0047) Notre Dame J Formal Log 19*591-592
◇ A05 B45 ◇ REV MR 80a:03023 Zbl 332 # 02025 • ID 52123

BROWN, F.M. [1979] *A sequent calculus for modal quantificational logic* (P 3108) AISB/GI Conf Artif Intel;1978 Hamburg 56-65
◇ B45 ◇ REV Zbl 423 # 03012 • ID 53524

BROWN, F.M. see Vol. I for further entries

BROWN, J.G. [1971] *A note on fuzzy sets* (J 0194) Inform & Control 18*32-39
◇ B52 E72 ◇ REV Zbl 217.14 • ID 04263

Author Index

BROWN, M.A. [1982] *Generalized S2-like systems of propositional modal logic* (J 0047) Notre Dame J Formal Log 23*53-61
◇ B45 ◇ REV MR 83d:03018 Zbl 442 # 03018 • ID 55077

BROWN, M.A. see Vol. III for further entries

BROWN, P.L. & STUBERMAN, W.E. [1965] *Elementary modern logic* (X 0880) Ronald Press: New York vii+269pp
◇ B98 ◇ ID 22467

BROWN, W.S. [1978] see BAXANDALL, P.R.

BROWNE, S.S.S. [1964] *Fundamentals of deductive logic* (X 1243) Brown: Dubuque vii+197pp
◇ B98 ◇ ID 22469

BROY, M. & GNATZ, R. & WIRSING, M. [1979] *Semantics of nondeterministic and noncontinuous constructs* (P 3246) Progr Constr;1978 Marktoberdorf 553-592
◇ B45 B75 ◇ REV MR 83m:68014 Zbl 406 # 03059
• ID 40370

BROY, M. see Vol. I, IV for further entries

BRUCE, BERTRAM [1976] *A logic for unknown outcomes* (J 0047) Notre Dame J Formal Log 17*542-550
◇ B50 ◇ REV MR 56 # 15364 Zbl 271 # 02016 • ID 21942

BRUDZY, K. & KISZKA, J.B. [1983] *The reproducibility property of fuzzy control systems* (J 2720) Fuzzy Sets Syst 9*161-177
◇ B52 B75 ◇ REV MR 84d:93020 Zbl 513 # 93027
• ID 37816

BRUIJN DE, N.G. [1975] *Exact finite models for minimal propositional calculus over a finite alphabet* (1111) Preprints, Manuscr., Techn. Reports etc. 75-W5K-02*30pp
◇ B50 ◇ REV Zbl 317 # 02023 • REM Technological University, Department of Mathematics. Eindhoven
• ID 29639

BRUIJN DE, N.G. see Vol. I, III, V, VI for further entries

BRUSENTSOV, N.P. & SHAUMAN, A.M. (EDS.) [1970] *Computing techniques and questions of cybernetics No.7 (Russian)* (X 0898) Moskov Gos Univ: Moskva 162pp
◇ B50 B97 ◇ REV MR 47 # 1322 • ID 70241

BRUSENTSOV, N.P. see Vol. I for further entries

BRYANT, J. [1980] *The logic of relative modality and the paradoxes of deontic logic* (J 0047) Notre Dame J Formal Log 21*78-88
◇ B45 ◇ REV MR 81b:03017 Zbl 421 # 03004 • ID 71362

BRYLL, G. [1966] *Ein neuer Beweis fuer die Vollstaendigkeit eines vielwertigen Systems des Aussagenkalkuels (Polnisch) (Deutsche und russische Zusammenfassung)* (J 0063) Studia Logica 19*111-116
◇ B50 ◇ REV MR 33 # 5469 Zbl 292 # 02017 • ID 60735

BRYLL, G. [1966] *Ueber die halbvollen Systeme des Aussagenkalkuels (Polnisch) (Deutsche und russische Zusammenfassung)* (J 0063) Studia Logica 19*117-126
◇ B50 ◇ REV MR 33 # 5468 Zbl 292 # 02018 • ID 60733

BRYLL, G. & PRUCNAL, T. & SLUPECKI, J. [1967] *Some remarks on three-valued logic of J. Lukasiewicz (Polish and Russian summaries)* (J 0063) Studia Logica 21*45-70
◇ B50 ◇ REV MR 38 # 02012 Zbl 309 # 02017 • ID 65383

BRYLL, G. & SLUPECKI, J. [1972] *The proof of T-decidability of Lewi's system S5* (J 0387) Bull Sect Logic, Pol Acad Sci 1/1*32-34
◇ B45 ◇ REV MR 50 # 12658 • ID 12491

BRYLL, G. & SLUPECKI, J. [1973] *A certain proof of the completeness of Lewis's system S5 (Polish) (English summary)* (S 1454) Zesz Nauk Wyz Szk Ped Mat, Opole 13*57-74
◇ B45 ◇ REV MR 54 # 7206 Zbl 291 # 02013 • ID 24998

BRYLL, G. & ROSIEK, M. [1973] *A note on the L-decidability of a certain modal logic of Lukasiewicz (Polish) (English summary)* (S 1454) Zesz Nauk Wyz Szk Ped Mat, Opole 13*153-157
◇ B45 ◇ REV MR 54 # 2441 Zbl 291 # 02008 • ID 24637

BRYLL, G. & SLUPECKI, J. [1973] *Proof of L-decidability of Lewis system S5 (Polish and Russian summaries)* (J 0063) Studia Logica 32*99-107
◇ B45 ◇ REV MR 50 # 1845 Zbl 313 # 02012 • ID 12498

BRYLL, G. & SLUPECKI, J. [1973] *Syntactic proof of the L-decidability of classical logic and Lukasiewicz's three-valued logic (Polish) (English summary)* (S 1454) Zesz Nauk Wyz Szk Ped Mat, Opole 13*131-152
◇ B25 B50 ◇ REV MR 54 # 4954 Zbl 291 # 02007
• ID 24728

BRYLL, G. & HALKOWSKA, K. [1984] *Slupecki's fragmentary systems* (J 0387) Bull Sect Logic, Pol Acad Sci 13*252-254
◇ B60 ◇ REV Zbl 557 # 03007 • ID 44651

BRYLL, G. see Vol. I, III, IV for further entries

BUB, J. [1977] *Von Neumann's projection postulate as a probability conditionalization rule in quantum mechanics* (J 0122) J Philos Logic 6*381-390
◇ A05 B51 ◇ REV MR 58 # 19993 Zbl 366 # 02016
• ID 51099

BUB, J. [1979] *Some reflections on quantum logic and Schroedinger's cat* (J 0013) Brit J Phil Sci 30*27-39
◇ A05 B51 ◇ REV MR 80k:81004 Zbl 442 # 03040
• ID 56397

BUB, J. [1979] *The measurement problem of quantum mechanics* (P 3234) Probl Founds of Physics;1977 Varenna 71-124
◇ B51 ◇ REV MR 82m:81003 Zbl 446 # 00019 • ID 56528

BUB, J. [1981] *What does quantum logic explain?* (P 3820) Curr Iss in Quantum Log;1979 Erice 89-100
◇ B51 G12 ◇ REV MR 84j:81014 Zbl 537 # 03044
• ID 39224

BUB, J. [1982] *Quantum logic, conditional probability, and interference* (J 0153) Phil of Sci (East Lansing) 49*402-421
◇ B51 ◇ REV MR 84h:03138 • ID 34324

BUCKLEY, J.J. [1985] *Ranking alternatives using fuzzy numbers* (J 2720) Fuzzy Sets Syst 15*21-31
◇ B52 ◇ REV MR 86e:90001 Zbl 567 # 90057 • ID 45107

BUDBAEVA, S.P. & PYATNITSYN, B.N. [1974] *The investigation and construction of pragmatic logics (Russian)* (C 2578) Filos & Logika 220-278
◇ B45 ◇ REV MR 58 # 27320 • ID 71382

BUENO SANCHEZ, E. [1969] *Polyvalent modal logics (Spanish)* (X 0221) Univ Havanna: Havanna 33*123-133
◇ B45 ◇ REV MR 41 # 5196 • ID 02034

BUENO SANCHEZ, E. [1970] *A certain class of many-valued modal systems (Russian)* (C 0668) Neklass Log 276-285
◇ B50 ◇ REV MR 47 # 6440 • ID 01712

BUENO SANCHEZ, E. & RUIZ SHULCLOPER, J. [1972] *Construction of many-valued logics. Method of multiplication of matrices (Spanish)* (J 0388) Human, Ser 4 Logica Mat 1*1-29
◇ B50 ◇ REV MR 51 # 12474 • ID 17262

BUGAJSKI, S. [1978] *Probability implication in the logics of classical and quantum mechanics* (J 0122) J Philos Logic 7*95-106
◇ B51 ◇ REV MR 80m:03098 Zbl 378 # 02015 • ID 51797

BUGAJSKI, S. [1981] *The inner language of operational quantum mechanics* (P 3820) Curr Iss in Quantum Log;1979 Erice 283-299
◇ B51 ◇ REV MR 84j:03120 Zbl 537 # 03044 • ID 34710

BUGAJSKI, S. [1982] *What is quantum logic?* (J 0063) Studia Logica 41*311-316
◇ B51 G12 ◇ REV MR 85i:03187 Zbl 539 # 03043 • ID 41262

BUGAJSKI, S. [1983] *Languages of similarity* (J 0122) J Philos Logic 12*1-18
◇ B45 ◇ REV MR 85i:03041 Zbl 539 # 03044 • ID 41263

BUGAJSKI, S. [1983] *Semantics in Banach spaces* (J 0063) Studia Logica 42*81-88
◇ B51 C65 C90 ◇ REV MR 86c:03036 Zbl 559 # 03017 • ID 42320

BUGAJSKI, S. see Vol. I for further entries

BULL, R.A. [1962] *The implicational fragment of Dummett's LC* (J 0036) J Symb Logic 27*189-194
◇ B55 ◇ REV MR 31 # 3320 Zbl 113.3 JSL 33.305 • ID 01730

BULL, R.A. [1964] *A note on the modal calculi S4.2 and S4.3* (J 0068) Z Math Logik Grundlagen Math 10*53-55
◇ B45 ◇ REV MR 28 # 3922 Zbl 119.12 JSL 33.136 • ID 01731

BULL, R.A. [1964] *An axiomatization of Prior's modal calculus Q* (J 0047) Notre Dame J Formal Log 5*211-214
◇ B45 ◇ REV MR 34 # 1168 Zbl 154.5 • ID 01732

BULL, R.A. [1964] *Some results for implicational calculi* (J 0036) J Symb Logic 29*33-39
◇ B55 ◇ REV MR 33 # 43 JSL 33.306 • ID 01733

BULL, R.A. [1965] *A class of extensions of the modal system S4 with the finite model property* (J 0068) Z Math Logik Grundlagen Math 11*127-132
◇ B45 ◇ REV MR 33 # 35 JSL 33.136 • ID 01736

BULL, R.A. [1965] *A modal extension of intuitionist logic* (J 0047) Notre Dame J Formal Log 6*142-146
◇ B55 ◇ REV MR 36 # 2475 Zbl 137.249 • ID 01735

BULL, R.A. [1965] *An algebraic study of diodorean modal systems* (J 0036) J Symb Logic 30*58-64
◇ B45 ◇ REV MR 33 # 3912 Zbl 129.257 JSL 32.245 • ID 01734

BULL, R.A. [1965] *Some modal calculi based on IC* (P 0688) Logic Colloq;1963 Oxford 3-7
◇ B45 ◇ REV MR 36 # 1300 • ID 02691

BULL, R.A. [1966] *MIPC as the formalisation of an intuitionist concept of modality* (J 0036) J Symb Logic 31*609-616
◇ B45 ◇ REV MR 35 # 5297 Zbl 192.30 • ID 01738

BULL, R.A. [1966] *That all normal extensions of S4.3 have the finite model property* (J 0068) Z Math Logik Grundlagen Math 12*341-344
◇ B45 ◇ REV MR 36 # 4966 Zbl 154.4 JSL 33.136 • ID 01737

BULL, R.A. [1967] *On the extension of S4 with CLMpMLp* (J 0047) Notre Dame J Formal Log 8*325-329
◇ B45 ◇ REV MR 44 # 2579 Zbl 189.7 • ID 01740

BULL, R.A. [1967] *On three related extensions of S4* (J 0047) Notre Dame J Formal Log 8*330-334
◇ B45 ◇ REV MR 44 # 2580 Zbl 189.7 • ID 01739

BULL, R.A. [1968] *An algebraic study of tense logics with linear time* (J 0036) J Symb Logic 33*27-38
◇ B45 ◇ REV MR 41 # 35 Zbl 155.14 JSL 36.173 • ID 01741

BULL, R.A. [1968] *On possible worlds in propositional calculi* (J 0105) Theoria (Lund) 34*171-182
◇ B45 ◇ REV MR 49 # 4752 • ID 03954

BULL, R.A. [1969] *Note on a paper in tense logic* (J 0036) J Symb Logic 34*215-218
◇ B45 ◇ REV MR 41 # 36 Zbl 179.313 JSL 36.173 • ID 01742

BULL, R.A. [1969] *On a paper of Akira Nakamura* (J 0068) Z Math Logik Grundlagen Math 15*155-156
◇ B45 ◇ REV MR 44 # 6454 Zbl 179.313 JSL 39.351 • ID 01744

BULL, R.A. [1969] *On modal logic with propositional quantifiers* (J 0036) J Symb Logic 34*257-263
◇ B45 ◇ REV Zbl 184.281 • ID 01743

BULL, R.A. [1970] *An approach to tense logic* (J 0105) Theoria (Lund) 36*282-300
◇ B45 ◇ REV MR 47 # 22 Zbl 227 # 02012 JSL 39.173 • ID 01745

BULL, R.A. [1972] *Mathematical logic* (J 0329) Math Chron (Auckland) 2*17-27
◇ B98 ◇ REV MR 48 # 3689 Zbl 238 # 02001 • ID 27895

BULL, R.A. & SEGERBERG, K. [1984] *Basic modal logic* (C 4085) Handb Philos Log 2*1-88
◇ B45 ◇ ID 44975

BULYGIN, E. [1971] see ALCHOURRON, C.E.

BULYGIN, E. [1981] see ALCHOURRON, C.E.

BULYGIN, E. [1984] see ALCHOURRON, C.E.

BUNCH, B.L. [1979] *Presupposition: An alternative approach* (J 0047) Notre Dame J Formal Log 20*341-354
◇ B60 ◇ REV MR 80f:03010 Zbl 332 # 02037 • ID 52637

BUNDER, M.W. [1979] *Λ-elimination in illative combinatory logic* (J 0047) Notre Dame J Formal Log 20*628-630
◇ B40 B50 ◇ REV MR 80h:03020 Zbl 394 # 03024 • ID 56104

BUNDER, M.W. [1979] *A more relevant relevance logic* (J 0047) Notre Dame J Formal Log 20*701-704
◇ B46 ◇ REV MR 83c:03014 Zbl 394 # 03027 • ID 56107

BUNDER, M.W. [1979] *Deduction theorems in significance logics* (J 0047) Notre Dame J Formal Log 20*695-700
◇ B45 ◇ REV MR 80h:03023 Zbl 394 # 03026 • ID 56106

BUNDER, M.W. [1979] *Paraconsistent combinatory logic* (J 0387) Bull Sect Logic, Pol Acad Sci 8*177-181
◇ B53 ◇ REV MR 81a:03025 Zbl 422 # 03003 • ID 53464

Author Index

BUNDER, M.W. [1979] *Some definitions of negation leading to paraconsistent logics* (1111) Preprints, Manuscr., Techn. Reports etc. 14*
◇ B53 ◇ REM The University of Wollongong,Department of Mathematic Preprint Series • ID 33488

BUNDER, M.W. [1980] *A new hierarchy of paraconsistent logics* (**P** 3006) Brazil Conf Math Log (3);1979 Recife 13-22
◇ B53 ◇ REV MR 82b:03055 Zbl 446 # 03020 • ID 56549

BUNDER, M.W. [1980] *A note on quantified significance logics* (**J** 0387) Bull Sect Logic, Pol Acad Sci 9*159-162
◇ B40 B45 ◇ REV MR 82d:03041 Zbl 452 # 03009 • ID 54073

BUNDER, M.W. [1980] *Quantified relevance logic and generalised restricted generality* (**J** 0079) Logique & Anal, NS 23*319-321
◇ B46 ◇ REV MR 84i:03036 Zbl 452 # 03010 • ID 54074

BUNDER, M.W. [1980] *Significance and illative combinatory logics* (**J** 0047) Notre Dame J Formal Log 21*380-384
◇ B40 B45 ◇ REV MR 81g:03011 Zbl 394 # 03025 • ID 53770

BUNDER, M.W. [1983] *On Arruda and Da Costa's logics J_1 to J_5* (**J** 3783) J Non-Classical Log (Univ Campinas) 2*43-48
◇ B53 ◇ REV Zbl 558 # 03012 • ID 46624

BUNDER, M.W. [1984] *Some definitions of negation leading to paraconsistent logics* (**J** 0063) Studia Logica 43*75-78
◇ B53 ◇ REV MR 86g:03045 Zbl 574 # 03012 • ID 42361

BUNDER, M.W. & MEYER, R.K. [1985] *A result for combinators, BCK logics and BCK algebras* (**J** 0079) Logique & Anal, NS 28*33-40
◇ B50 ◇ REV Zbl 574 # 03004 • ID 48829

BUNDER, M.W. see Vol. I, V, VI for further entries

BUNGE, MARIO (ED.) [1972] *Proceedings of the First Symposium on Exact Philosophy* (**J** 0122) J Philos Logic 1*263-480
◇ B46 B97 ◇ REV MR 52 # 7838 • ID 70188

BURALI-FORTI, C. [1894] *Logica matematica* (**X** 1283) Hoepli: Milano vi+158pp
◇ B98 ◇ REV FdM 25.115 • ID 22446

BURALI-FORTI, C. see Vol. I, V for further entries

BURGESS, J.P. [1969] *Probability logic* (**J** 0036) J Symb Logic 34*264-274
◇ B48 ◇ REV Zbl 184.9 • ID 01773

BURGESS, J.P. [1978] *The unreal future* (**J** 0105) Theoria (Lund) 44*157-179
◇ A05 B45 ◇ REV MR 82f:03003 Zbl 435 # 03017 • ID 55778

BURGESS, J.P. [1979] *Logic and time* (**J** 0036) J Symb Logic 44*566-582
◇ B45 ◇ REV MR 80m:03032 Zbl 423 # 03018 • ID 53530

BURGESS, J.P. [1980] *Decidability for branching time* (**J** 0063) Studia Logica 39*203-218
◇ B25 B45 C85 ◇ REV MR 82e:03020 Zbl 467 # 03006 • ID 55004

BURGESS, J.P. [1981] *Quick completeness proofs for some logics of conditionals* (**J** 0047) Notre Dame J Formal Log 22*76-84
◇ B45 ◇ REV MR 83d:03019 Zbl 416 # 03020 • ID 56470

BURGESS, J.P. [1981] *Relevance: A fallacy?* (**J** 0047) Notre Dame J Formal Log 22*97-104
◇ A05 B46 ◇ REV MR 82j:03017 Zbl 438 # 03008 • ID 54066

BURGESS, J.P. [1981] *The completeness of intuitionistic propositional calculus for its intended interpretation* (**J** 0047) Notre Dame J Formal Log 22*17-28
◇ B55 F50 ◇ REV MR 82g:03015 Zbl 445 # 03030 • ID 56494

BURGESS, J.P. [1982] *Axioms for tense logic I: "Since" and "until". II: Time periods* (**J** 0047) Notre Dame J Formal Log 23*367-374,375-383
◇ B45 ◇ REV MR 84j:03031 Zbl 452 # 03021 Zbl 452 # 03022 • ID 54085

BURGESS, J.P. [1983] *Common sense and "relevance"* (**J** 0047) Notre Dame J Formal Log 24*41-53
◇ A05 B46 ◇ REV MR 84a:03007 Zbl 476 # 03006 • ID 55518

BURGESS, J.P. [1984] *Beyond tense logic* (**J** 0122) J Philos Logic 13*235-248
◇ B45 ◇ REV MR 86g:03036 • ID 42645

BURGESS, J.P. [1984] *Read on relevance: a rejoinder* (**J** 0047) Notre Dame J Formal Log 25*217-223
◇ B46 ◇ REV MR 85i:03042 Zbl 569 # 03006 • ID 42569

BURGESS, J.P. & GUREVICH, Y. [1985] *The decision problem for linear temporal logic* (**J** 0047) Notre Dame J Formal Log 26*115-128
◇ B25 B45 ◇ REV Zbl 573 # 03004 • ID 42591

BURGESS, J.P. see Vol. III, IV, V, VI for further entries

BURGHARDT, F.J. [1980] *Modal quantum logic and its dialogic foundation* (**J** 2736) Int J Theor Phys 19*843-866
◇ B51 G12 ◇ REV MR 82i:03074 Zbl 453 # 03065 • ID 54195

BURGHARDT, F.J. [1984] *Modalities and quantum mechanics* (**J** 2736) Int J Theor Phys 23*1171-1196
◇ B51 ◇ ID 44740

BURKAMP, W. [1932] *Logik* (**X** 1308) Mittler: Herford vi+175pp
◇ B98 ◇ REV FdM 58.54 • ID 25089

BURKS, A.W. [1977] *Chance, cause, reason. An inquiry into the nature of scientific evidence* (**X** 0862) Univ Chicago Pr: Chicago xvi+694pp
◇ A05 B98 ◇ REV MR 58 # 10268 Zbl 421 # 03002 JSL 45.373 • ID 53402

BURKS, A.W. see Vol. I, IV for further entries

BURLE, G.A. [1967] *Classes of k-valued logic which contain all functions of a single variable (Russian)* (**J** 0071) Met Diskr Analiz (Novosibirsk) 10*3-7
◇ B50 ◇ REV MR 36 # 1297 Zbl 147.253 • ID 01782

BUROSCH, G. [1973] see BLOKHINA, G.N.

BUROSCH, G. see Vol. I for further entries

BUSTO DEL, E.H. [1955] *Probability and Carnap's inductive logic (Spanish)* (**J** 0169) Theoria (Madrid) 3*119-128
◇ B48 ◇ REV MR 17.818 • ID 01798

BUTLER, J.T. [1977] *Fanout-free networks of multivalued gates* (**P** 2013) Int Symp Multi-Val Log (7);1977 Charlotte 39-46
◇ B50 ◇ REV MR 56 # 18132 • ID 35852

BUTLER, J.T. & GINZER, J.A. [1979] *Multiple-valued logic: 1974 - 1978 -- survey and analysis* (**P** 3003) Int Symp Multi-Val Log (9);1979 Bath 1-13
◇ B50 ◇ REV MR 81c:03017 • ID 73378

BUTLER, J.W. [1957] *On complete and independent sets of truth functions in multi-valued logics* (**P** 1675) Summer Inst Symb Log;1957 Ithaca 78-80
◇ B50 ◇ REV JSL 30.246 • ID 01799

BUTLER, J.W. see Vol. III for further entries

BUTNARIU, D. [1978] *Fuzzy games: A description of the concept* (**J** 2720) Fuzzy Sets Syst 1*181-192
◇ B52 ◇ REV MR 58 # 20540 Zbl 389 # 90100 • ID 52326

BUTNARIU, D. see Vol. IV, V for further entries

BUTTERFIELD, J. [1982] *Dummett on temporal operators* (**J** 2811) Phil Quart (St Andrews) 34*31-42
◇ B45 ◇ REV MR 85j:03018 • ID 45257

BUTTS, R.E. & HINTIKKA, K.J.J. (EDS.) [1977] *Proceedings of the fifth international congress of logic, methodology and philosophy of science. I,II,III,IV* (**S** 3308) Univ Western Ontario Ser in Philos of Sci 9*x+406pp,10*x+427pp,11*x+321pp,12*x+336pp
◇ A05 A10 B97 ◇ REV MR 56 # 15352 Zbl 363 # 00005 • ID 50826

BYERLY, H.C. [1973] *A primer of logic* (**X** 0837) Harper & Row: New York xiii+560pp
◇ B98 ◇ REV MR 53 # 5238 • ID 22870

BYRD, M. [1973] *Knowledge and true belief in Hintikka's epistemic logic* (**J** 0122) J Philos Logic 2*181-192
◇ B45 ◇ REV MR 54 # 12483 Zbl 266 # 02019 • ID 60809

BYRD, M. [1975] *On incoherent quantification in languages without constants* (**J** 0079) Logique & Anal, NS 18*155-169
◇ B45 ◇ REV MR 53 # 7727 Zbl 344 # 02012 • ID 22978

BYRD, M. [1976] *Single variable formulas in* $S4_{\rightarrow}$ (**J** 0122) J Philos Logic 5*439-456
◇ B45 ◇ REV MR 55 # 10240 Zbl 343 # 02014 • ID 60808

BYRD, M. & ULLRICH, D. [1977] *The extensions of* $BAlt_3$ (**J** 0122) J Philos Logic 6*109-117
◇ B45 ◇ REV MR 58 # 10324 Zbl 398 # 03008 • ID 52736

BYRD, M. [1978] *On the addition of weakened L-reduction axioms to the Brouwer system* (**J** 0068) Z Math Logik Grundlagen Math 24*405-408
◇ B45 ◇ REV MR 80e:03013 Zbl 395 # 03016 • ID 52567

BYRD, M. & HENRY, DENNIS [1978] *Sugihara's criterion and some structural parallels between* $E_{\rightarrow}$ *and* $S3_{\rightarrow}$ (**J** 0068) Z Math Logik Grundlagen Math 24*187-191
◇ B46 ◇ REV MR 58 # 21463 Zbl 422 # 03006 • ID 53467

BYRD, M. [1978] *The extensions of* $BAlt_3$ *- revisited* (**J** 0122) J Philos Logic 7*407-413
◇ B45 ◇ REV MR 80j:03024 Zbl 405 # 03006 • ID 54881

BYRD, M. [1979] *A formal interpretation of Lukasiewicz' logics* (**J** 0047) Notre Dame J Formal Log 20*366-368
◇ B50 ◇ REV MR 80m:03050 Zbl 351 # 02014 • ID 52633

BYRD, M. & HENRY, DENNIS [1979] *On defining necessity in terms of entailment* (**J** 0063) Studia Logica 38*95-104
◇ B46 ◇ REV MR 80m:03036 Zbl 406 # 03026 • ID 56109

BYRD, M. [1979] *On Lukasiewiczian logics with several designated values* (**J** 0079) Logique & Anal, NS 22*489-504
◇ B50 ◇ REV MR 82c:03028 Zbl 442 # 03021 • ID 56378

BYRD, M. [1980] *Eventual permanence* (**J** 0047) Notre Dame J Formal Log 21*591-601
◇ B45 ◇ REV MR 82c:03021 Zbl 426 # 03019 • ID 55775

BYSTROV, P.I. & SEREBRYANNIKOV, O.F. [1982] *Criteria for the quality of inferences in modal logic (Russian)* (**C** 3792) Log & Filos Kategorii 97-112
◇ B45 ◇ REV MR 84h:03046 • ID 37395

CABANES, A. & PICHAT, E. [1973] *Minimal representation of a function of several finite variables - applications to data bank* (**P** 2009) Int Symp Multi-Val Log (3);1973 Toronto 237-238
◇ B50 ◇ ID 42052

CAI, QINGSHENG & HUANG, KECHENG & TANG, CHESHAN [1982] *Four-valued logic and a formulation of the CKP (Chinese)* (**J** 3816) Zhongguo Kexue Jishu Daxue Xuebao 1982, Suppl. II, 89-95
◇ B50 ◇ REV MR 84k:03074 • ID 35012

CAIN, R.G. & HONG, SEJUNE & OSTAPKO, D.L. [1974] *A practical approach to two-level minimization of multivalued logic* (**P** 1385) Int Symp Multi-Val Log (4);1974 Morgantown 168-182
◇ B50 ◇ ID 42101

CAIN, R.G. see Vol. I for further entries

CALL, R.L. [1968] *A minimal canonically complete M-valued proper logic for each M* (**J** 0036) J Symb Logic 33*108-110
◇ B50 ◇ REV MR 41 # 38 Zbl 191.287 JSL 35.326 • ID 01819

CALL, R.L. see Vol. I, III, VI for further entries

CALUDE, C. [1976] *Une axiomatisation de la logique deontique a une infinite de valeurs* (**J** 0060) Rev Roumaine Math Pures Appl 21*267-273
◇ B45 ◇ REV MR 53 # 7728 Zbl 347 # 02019 • ID 22979

CALUDE, C. see Vol. I, IV, V, VI for further entries

CALUWE DE, R.M.M. & KERRE, E.E. & ZENNER, R.B.R.C. [1985] *A new approach to information retrieval systems using fuzzy expressions* (**J** 2720) Fuzzy Sets Syst 17*9-22
◇ B52 ◇ ID 49324

CAMPRI, P. [1971] *Un sistema di logica deontica al livello del calcolo die predicati* (**J** 3069) Atti Accad Sci Bologna Fis Ser 12 9*45-53
◇ B45 ◇ REV MR 49 # 8813 Zbl 279 # 02012 • ID 17189

CANAK, B. & GILEZAN, K. [1978] *Quelques formes generales des fonctions pseudo-booleennes* (**J** 0400) Publ Inst Math, NS (Belgrade) 24(38)*45-52
◇ B50 ◇ REV MR 80d:06013 Zbl 449 # 03071 • ID 56738

CANGELOSI, V.E. [1967] *Compound statements and mathematical logic* (**X** 1164) Merrill: Columbus xii+114pp
◇ B98 ◇ ID 22470

CANTINI, A. [1980] *A note on three-valued logic and Tarski theorem on truth definitions* (**J** 0063) Studia Logica 39*405-414
◇ B50 C40 F30 ◇ REV MR 83i:03019 Zbl 457 # 03024 • ID 54349

CANTINI, A. see Vol. I, III, IV, V, VI for further entries

CANTY, J.T. [1964] *A natural deduction system for modal logic* (**J** 0047) Notre Dame J Formal Log 5*199-210
◇ B45 ◇ REV MR 33 # 7245 Zbl 173.4 • ID 01833

CANTY, J.T. [1965] *A note on the axiomatization of Rubin's system (S)* (**J** 0047) Notre Dame J Formal Log 6*190-192
◇ B45 ◇ REV MR 33 # 1233 Zbl 133.244 • ID 01834

CANTY, J.T. [1965] *Systems classically axiomatized and properly contained in Lewis's S3* (**J** 0047) Notre Dame J Formal Log 6*309-318
◇ B45 ◇ REV MR 36 # 2476 Zbl 163.244 JSL 33.309 • ID 01835

CANTY, J.T. & SCHARLE, T.W. [1966] *Note on the singularies of S5* (**J** 0047) Notre Dame J Formal Log 7*108
◇ B45 ◇ REV MR 34 # 5659 Zbl 168.4 • ID 30954

CANTY, J.T. [1968] *On symbolizing singulary S5 functions* (**J** 0047) Notre Dame J Formal Log 9*340-342
◇ B45 ◇ REV MR 39 # 2620 Zbl 167.9 • ID 01837

CANTY, J.T. [1969] *Lesniewski's terminological explanations as recursive concepts* (**J** 0047) Notre Dame J Formal Log 10*337-369
◇ B60 ◇ REV MR 40 # 29 Zbl 167.10 • ID 01839

CANTY, J.T. & KUENG, G. [1970] *Substitutional quantification and Lesniewskian quantifiers* (**J** 0105) Theoria (Lund) 36*165-182
◇ A05 B60 C80 E70 ◇ REV Zbl 213.14 • ID 27980

CANTY, J.T. [1971] *Elementary logic without referential quantification* (**J** 0047) Notre Dame J Formal Log 12*441-446
◇ B60 ◇ REV MR 46 # 1549 Zbl 224 # 02010 • ID 01840

CANTY, J.T. see Vol. I, III for further entries

CAO, ZHIJIANG [1982] *Structure of solution sets of fuzzy linear equations* (**P** 3845) Conf Math Service of Man (2,Proc)(Feriet);1982 Las Palmas 683-685
◇ B52 ◇ REV Zbl 505 # 03024 • ID 38136

CAO, ZHIJIANG [1982] *The eigen fuzzy sets of a fuzzy matrix* (**C** 3786) Approx Reason in Decis Anal 61-63
◇ B52 ◇ REV MR 84i:15021 Zbl 505 # 15007 • ID 38209

CAO, ZHIJIANG see Vol. V for further entries

CAPALDI, N. [1966] *Introduction to deductive logic* (**X** 1310) Monarch Pr: New York 166pp
◇ B98 ◇ ID 22471

CAPINSKA, E. [1979] *On standard consequence operations in the implicationless language* (**J** 0387) Bull Sect Logic, Pol Acad Sci 8*202-205
◇ B55 ◇ REV MR 80m:03060b Zbl 449 # 03018 • ID 56685

CAPINSKA, E. [1981] *On intermediate logics which can be axiomatized by means of implicationless formulas* (**J** 0302) Rep Math Logic, Krakow & Katowice 13*11-16
◇ B55 ◇ REV MR 84g:03030 Zbl 521 # 03016 • ID 34144

CARACCIOLO, A. [1966] *M-valued logics and m-ary selection functions* (**P** 0746) Automata Th;1964 Ravello 107-114
◇ B50 ◇ REV Zbl 199.3 • ID 27598

CARACCIOLO, A. see Vol. IV, V for further entries

CARDWELL, C.E. [1979] *Arguments and inference. An introduction to symbolic logic* (**X** 1164) Merrill: Columbus xi + 414pp
◇ B98 ◇ REV MR 83f:03003 Zbl 532 # 03002 • ID 34832

CARLSSON, C. [1983] *An approach to handling fuzzy problem structures* (**J** 2694) Cybern & Syst 14*33-54
◇ B52 ◇ REV Zbl 541 # 93002 • ID 45023

CARLSSON, C. [1983] *Some features of a methodology for dealing with multiple- criteria problems* (**P** 3859) Th & Pract Multiple Criteria Decis Making;1981 Moskva 1-49
◇ B52 ◇ REV MR 85h:90122a Zbl 521 # 90095 • ID 37477

CARLSTROM, I.F. [1975] *Truth and entailment for a vague quantifier* (**J** 0154) Synthese 30*461-495
◇ B52 C80 ◇ REV Zbl 309 # 02023 • ID 60828

CARMO, J. [1985] *The INFOLOG branching logic of events* (**P** 4605) Inform Syst Th & Formal Aspects;1985 Sitges 159-174
◇ B45 ◇ ID 48974

CARNAP, R. [1929] *Abriss der Logistik, mit besonderer Berueecksichtigung der Relationstheorie und ihrer Anwendungen* (**X** 0902) Springer: Wien vi + 114pp
◇ A05 B98 ◇ REV FdM 55.30 • ID 16764

CARNAP, R. [1939] *Foundations of logic and mathematics* (**X** 0862) Univ Chicago Pr: Chicago viii + 71pp
◇ A05 B98 ◇ REV Zbl 23.97 JSL 4.117 FdM 65.1099 • ID 02724

CARNAP, R. [1943] *Formalization of logic* (**X** 0858) Harvard Univ Pr: Cambridge xviii + 159pp
◇ A05 B98 ◇ REV MR 4.209 Zbl 61.7 JSL 8.81 • ID 02228

CARNAP, R. [1945] *On inductive logic* (**J** 0153) Phil of Sci (East Lansing) 12*72-97
◇ A05 B48 ◇ REV MR 7.46 JSL 11.19 • ID 01846

CARNAP, R. [1946] *Modalities and quantification* (**J** 0036) J Symb Logic 11*33-64
◇ A05 B45 ◇ REV MR 8.429 JSL 13.218 • ID 01848

CARNAP, R. [1946] *Remarks on induction and truth* (**J** 0075) Phil Phenom Research 6*590-602
◇ A05 B48 ◇ REV MR 8.245 JSL 11.124 • ID 01847

CARNAP, R. [1947] *Meaning and necessity. A study in semantics and modal logic* (**X** 0862) Univ Chicago Pr: Chicago vii + 210pp
◇ A05 B45 ◇ REV MR 8.430 Zbl 34.1 JSL 14.237 • ID 02229

CARNAP, R. [1947] *On the application of inductive logic* (**J** 0075) Phil Phenom Research 8*133-148
◇ A05 B48 ◇ REV MR 9.323 • ID 01849

CARNAP, R. [1950] *Logical foundations of probability* (**X** 0862) Univ Chicago Pr: Chicago xvii + 607pp
◇ A05 B48 ◇ REV MR 12.664 Zbl 40.70 JSL 16.205 • ID 02230

CARNAP, R. [1950] *Testability and meaning* (**X** 0875) Yale Univ Pr: New Haven 100pp
◇ A05 B48 ◇ REV JSL 16.137 • ID 25266

CARNAP, R. [1951] *The problem of relations in inductive logic* (**J** 0095) Philos Stud 2*75-89
◇ B48 ◇ ID 33248

CARNAP, R. [1952] *The continuum of inductive methods* (**X** 0862) Univ Chicago Pr: Chicago v + 92pp
◇ A05 B48 ◇ REV MR 14.4 Zbl 47.372 JSL 18.168 • ID 02231

CARNAP, R. [1953] *On the comparative concept of confirmation* (**J** 0013) Brit J Phil Sci 3*311-318
◇ A05 B48 ◇ REV MR 15.190 • ID 01851

CARNAP, R. [1954] *Einfuehrung in die symbolische Logik mit besonderer Beruecksichtigung ihrer Anwendungen* (**X** 0902) Springer: Wien x+209pp
• TRANSL [1959] (**X** 0813) Dover: New York xiv+241pp
◇ A05 B30 B98 ◇ REV MR 16.208 MR 21 # 2578 Zbl 56.6 Zbl 83.1 JSL 20.274 JSL 31.287 • ID 01852

CARNAP, R. [1962] *The aim of inductive logic* (**P** 0612) Int Congr Log, Meth & Phil of Sci (1,Proc);1960 Stanford 303-318
◇ A05 B48 ◇ REV Zbl 143.12 JSL 32.104 • ID 01856

CARNAP, R. [1963] *Donald Davidson on modalities and semantics* (**C** 0673) Phil of Carnap 911-914
◇ A05 B45 ◇ ID 04276

CARNAP, R. [1963] *John Myhill on modal logic and semantics* (**C** 0673) Phil of Carnap 908-911
◇ A05 B45 ◇ ID 04275

CARNAP, R. [1963] *Robert Feys on modalities* (**C** 0673) Phil of Carnap 905-908
◇ A05 B45 ◇ ID 04277

CARNAP, R. [1968] *Inductive logic and inductive intuition* (**P** 2268) Int Colloq Philos of Sci;1965 London 2*258-267
◇ A05 B48 ◇ REV Zbl 181.6 JSL 40.449 • ID 21288

CARNAP, R. [1971] *A basic system of inductive logic part I* (**C** 0799) Stud Induct Logic & Probab, Vol 1 33-165
◇ A05 B48 ◇ REV MR 58 # 166 JSL 40.581 • REM Part II 1980 • ID 14636

CARNAP, R. [1971] *Inductive logic and rational decisions* (**C** 0799) Stud Induct Logic & Probab, Vol 1 5-31
◇ A05 B48 ◇ REV MR 58 # 165 • ID 71536

CARNAP, R. & JEFFREY, R.C. (EDS.) [1971] *Studies in inductive logic and probability I* (**X** 0926) Univ Calif Pr: Berkeley vi+264pp
◇ B30 B48 ◇ REV MR 48 # 49 Zbl 246 # 02024 • ID 25461

CARNAP, R. [1972] *Bedeutung und Notwendigkeit: Eine Studie zur Semantik und modalen Logik* (**X** 0902) Springer: Wien xv+325pp
◇ A05 B45 ◇ REV MR 50 # 4242 Zbl 247 # 02022 • ID 03964

CARNAP, R. [1972] *Notes on probability and induction* (**J** 0154) Synthese 25*269-298
◇ B48 ◇ REV MR 57 # 15989 Zbl 279 # 60003 • ID 71537

CARNAP, R. [1973] *Grundlagen der Logik und Mathematik* (**X** 0917) Nymphenburger: Muenchen 106pp
◇ A05 B98 ◇ REV MR 52 # 13308 Zbl 333 # 02002 • ID 21772

CARNAP, R. [1980] *A basic system of inductive logic. II* (**C** 3537) Stud Induct Logic & Probab, Vol 2 7-155
◇ B48 ◇ REV MR 81m:03035a Zbl 539 # 03004 JSL 49.1409 • REM Part I 1971 • ID 71533

CARNAP, R. see Vol. I, VI for further entries

CARNEY, J.D. & SCHEER, R.K. [1964] *Fundamentals of logic* (**X** 0843) Macmillan : New York & London xv+474pp
◇ B98 ◇ ID 22474

CARNEY, J.D. [1970] *Introduction to symbolic logic* (**X** 0819) Prentice Hall: Englewood Cliffs ix+252pp
◇ B98 ◇ ID 22473

CARNIELLI, W.A. [1984] see ALCANTARA DE, L.P.

CARNIELLI, W.A. see Vol. I, V for further entries

CARO DE, F. & FATTOROSI-BARNABA, M. [1985] *Grade modalities I* (**J** 0063) Studia Logica 44*197-221
◇ B45 ◇ ID 47511

CARRIER, L.S. [1977] *The irreducibility of knowledge* (**J** 0079) Logique & Anal, NS 20*167-176
◇ A05 B45 ◇ REV MR 58 # 4981 • ID 71546

CARROLL, M.J. [1976] *The logical modalities and the system S5* (**J** 0079) Logique & Anal, NS 19*457-468
◇ A05 B45 ◇ REV MR 57 # 68 Zbl 361 # 02026 • ID 50661

CARROLL, M.J. [1979] *Reduction to first degree in quantificational S5* (**J** 0036) J Symb Logic 44*207-214
◇ B45 ◇ REV MR 80h:03024 Zbl 409 # 03008 • ID 56311

CARVALHO DE, R.L. & VELOSO, P.A.S. [1980] *Towards a logic of limited perception* (**P** 3006) Brazil Conf Math Log (3);1979 Recife 147-159
◇ A05 B60 ◇ REV MR 82e:68013 Zbl 448 # 68010 • ID 56664

CARVALHO DE, R.L. see Vol. I for further entries

CARVALLO, M. [1968] *Logique a trois valeurs. Logique a seuils* (**X** 0834) Gauthier-Villars: Paris 152pp
◇ B50 ◇ REV MR 39 # 39 Zbl 175.263 • ID 01880

CARVALLO, M. [1968] *Sur la resolution des equations de Post a v valeurs* (**J** 2313) C R Acad Sci, Paris, Ser A-B 267*A628-A630
◇ B50 G20 ◇ REV MR 38 # 4285 Zbl 169.304 • ID 01879

CARVALLO, M. see Vol. I, V for further entries

CASANUEVA, S.P. [1979] *Investigaciones en logica modal. Desarollo y sistemas axiomaticos de Aristoteles a Clarence I.Lewis* (**X** 4484) Licenciat Uni Concepcion: Concepcion xvi+297pp
◇ A10 B45 ◇ ID 31662

CASARI, E. [1960] *Lineamenti di logica matematica* (**X** 0844) Feltrinelli: Milano 324pp
◇ B98 ◇ REV Zbl 115.5 JSL 27.76 • ID 22395

CASARI, E. [1983] *Intermediate logics* (**P** 3829) Atti Incontri Log Mat (1);1982 Siena 243-298
◇ B55 ◇ REV MR 84k:03006 Zbl 531 # 03010 • ID 37674

CASARI, E. & MINARI, P. [1983] *Negation-free intermediate predicate logics (Italian summary)* (**J** 2100) Boll Unione Mat Ital, VI Ser, B 2*499-536
◇ B55 ◇ REV MR 85b:03035 Zbl 525 # 03014 • ID 38252

CASARI, E. see Vol. I, III, V for further entries

CASE, J. & SMITH, C.H. [1983] *Comparison of identification criteria for machine inductive inference* (**J** 1426) Theor Comput Sci 25*193-220
◇ B35 B48 D20 ◇ REV MR 84j:68029 Zbl 524 # 03025 • ID 37600

CASE, J. see Vol. I, IV for further entries

CASEY, H. & CLARK, M. [1970] *Logic: a practical approach* (**X** 1340) Regnery: Chicago 314pp
◇ B98 ◇ ID 22476

Author Index

CASEY, M.R. [1977] *Productive logic and congruency tables* (J 0286) Int Logic Rev 15*3-6
◇ B60 ◇ ID 51796

CASSINELLI, G. [1975] see BELTRAMETTI, E.G.

CASSINELLI, G. [1977] see BELTRAMETTI, E.G.

CASSINELLI, G. [1979] see BELTRAMETTI, E.G.

CASSINELLI, G. & TRUINI, P. [1979] *Toward a generalized probability theory: Conditional probabilities* (P 3234) Probl Founds of Physics;1977 Varenna 125-133
◇ B51 G12 ◇ REV MR 82m:81003 Zbl 445 # 03036
• ID 56500

CASSINELLI, G. [1981] see BELTRAMETTI, E.G.

CASTANEDA, H.-N. [1960] *Obligation and modal logic* (J 0079) Logique & Anal, NS 3*40-48
◇ B45 ◇ REV JSL 33.612 • ID 43220

CASTANEDA, H.-N. [1964] *A note on S5* (J 0036) J Symb Logic 29*191-192
◇ B45 ◇ REV MR 33 # 5470 JSL 31.275 • ID 01890

CASTANEDA, H.-N. [1975] *Thinking and doing. The philosophical foundations of institutions* (X 0835) Reidel: Dordrecht xviii + 366pp
◇ A05 B45 ◇ REV JSL 50.248 • ID 47351

CASTANEDA, H.-N. [1976] *The twofold structure and the unity of practical thinking* (P 2326) Action Th;1975 Winnipeg 105-130
◇ A05 B45 ◇ REV Zbl 345 # 02017 • ID 29777

CASTANEDA, H.-N. [1978] *Philosophical method and the theory of predication and identity* (J 0097) Nous, Quart J Phil 12*189-210
◇ A05 B45 ◇ REV MR 80c:03010 • ID 71569

CASTANEDA, H.-N. [1981] *The paradoxes of deontic logic: the simplest solution to all of them in one fell swoop* (C 3731) New Studies Deontic Log 37-85
◇ B45 ◇ REV MR 83a:03016 • ID 35059

CASTANEDA, H.-N. see Vol. I for further entries

CATON, C.E. [1962] *A stipulation of logical truth in a modal propositional calculus* (J 0154) Synthese 14*196-199
◇ B45 ◇ REV MR 40 # 1266 Zbl 168.5 JSL 39.611
• ID 01829

CATON, C.E. [1963] *A stipulation of a modal propositional calculus in terms of modalized truth values* (J 0047) Notre Dame J Formal Log 4*224-226
◇ B45 ◇ REV MR 29 # 2183 Zbl 119.11 JSL 39.611
• ID 01894

CATTANEO, G. [1980] *Fuzzy events and fuzzy logics in classical information systems* (J 0034) J Math Anal & Appl 75*523-548
◇ B52 G12 ◇ REV MR 82c:03031 Zbl 447 # 94062
• ID 56601

CATTANEO, G. [1983] *Canonical embedding of an abstract quantum logic into the partial Baer*-ring of complex fuzzy events* (J 2720) Fuzzy Sets Syst 9*179-198
◇ B52 ◇ REV MR 84e:03076 Zbl 537 # 03046 • ID 34408

CATTANEO, G. & NISTICO, G. [1984] *Orthogonality and orthocomplementations in the axiomatic approach to quantum mechanics: Remarks about some critiques* (J 0209) J Math Phys 25*513-531
◇ B51 G12 ◇ REV MR 85e:81009 Zbl 547 # 03044
• ID 43246

CAVALLI, A.R. & FARINAS DEL CERRO, L. [1984] *A decision method for linear temporal logic* (P 2633) Autom Deduct (7);1984 Napa 113-127
◇ B25 B45 ◇ REV MR 86g:03050 Zbl 547 # 03008
• ID 43192

CAVALLI, A.R. & FARINAS DEL CERRO, L. [1984] *Specification and verification of networks protocols using temporal logic* (P 4302) Symp on Progr (6);1984 Toulouse 59-73
◇ B45 ◇ REV MR 85k:68008 Zbl 549 # 68013 • ID 44577

CAVALLO, R.E. & KLIR, G.J. [1982] *Decision making in reconstructability analysis* (J 1743) Int J Gen Syst 8*243-255
◇ B52 ◇ REV MR 84b:00024 Zbl 493 # 93005 • ID 38447

CAYROL, M. & FARRENY, H. & PRADE, H. [1980] *Fuzzy reasoning based on multivalent logics in the framework of production-rules systems* (P 3673) Int Symp Multi-Val Log (10);1980 Evanston 143-148
◇ B52 ◇ REV MR 83m:94032 • ID 45949

CAZACU, C. [1968] *Sur les logiques m-valentes* (J 0147) An Univ Bucuresti, Acta Logica 11*154-155
◇ B50 ◇ REV MR 44 # 3840 Zbl 229 # 02019 • ID 01897

CAZACU, C. see Vol. IV for further entries

CECCHERINI, P.V. & D'ANDREA, R. [1978] *Nuove operazioni primitive in logiche polivalenti* (J 2311) Rend Mat, Ser 6 11*533-541
◇ B50 ◇ REV MR 80i:03035 Zbl 406 # 03075 • ID 56157

CEGLA, W. & JADCZYK, A.Z. [1976] *Logics generated by causality structures. Covariant representations of the Galilean logic* (J 1546) Rep Math Phys (Warsaw) 9*377-385
◇ B60 ◇ REV MR 55 # 5021 Zbl 359 # 02016 • ID 50564

CEGLA, W. & JADCZYK, A.Z. [1977] *Causal logic of Minkowski space* (J 1113) Commun Math Phys 57*213-217
◇ B51 G12 ◇ REV MR 57 # 8528 Zbl 393 # 03046
• ID 52466

CEGLA, W. [1981] *Causal logic of Minkowski space* (P 3820) Curr Iss in Quantum Log;1979 Erice 419-424
◇ B51 ◇ REV MR 84j:83007 Zbl 537 # 03044 • ID 39253

CERLA, G. [1983] *Contralogical modalities (Italian)* (P 3829) Atti Incontri Log Mat (1);1982 Siena 365-367
◇ B45 ◇ REV MR 84k:03006 • ID 44642

CERNY, E. [1980] *Characteristic functions in multivalued logic systems (French and German summaries)* (J 2701) Digit Processes 6*167-174
◇ B50 ◇ REV MR 83g:94045 Zbl 443 # 94035 • ID 69303

CERNY, E. see Vol. I for further entries

CERNY, M. & NEKOLA, J. & NOVAK, VILEM [1984] *Fuzzy sets - perspectives, problems and applications (Czech)* (J 1527) Pokroky Mat Fyz Astron (Prague) 29*126-137
◇ B52 E72 ◇ REV MR 85i:03169 Zbl 557 # 94021
• ID 44291

CHAGROV, A.V. [1981] *Superintuitionist fragments of nonnormal modal logics (Russian)* (C 3747) Mat Log & Mat Lingvistika 144-162
◇ B55 ◇ REV MR 84b:03030 • ID 34909

CHAGROV, A.V. [1983] *Polynomial finite approximability of modal and superintuitionistic logics (Russian)* (**C** 3798) Mat Log, Mat Ling & Teor Algor 75-83
◇ B45 B55 ◇ REV MR 85i:03043 • ID 44083

CHAKRABORTY, B.C. [1982] *Algebraic representations of n-valued logics, n being prime* (**J** 3763) Ganit (Dacca) 2*31-33
◇ B50 ◇ REV MR 84j:03054 • ID 34646

CHAKRAVARTI, S.S. [1979] *A note on Kripke's distinction between rigid designators and non-rigid designators* (**J** 0047) Notre Dame J Formal Log 20*309-313
◇ A05 B45 ◇ REV MR 80j:03007a Zbl 321 # 02006 • ID 52617

CHANDLER, M. [1982] *The logic of "unless"* (**J** 0095) Philos Stud 41*383-405
◇ B45 ◇ REV MR 84h:03062 • ID 34259

CHANDRA, A.K. & HALPERN, J.Y. & MEYER, A.R. & PARIKH, R. [1985] *Equations between regular terms and application to process logic* (**J** 1428) SIAM J Comp 14*935-942
◇ B45 D05 ◇ ID 49198

CHANDRA, A.K. see Vol. I, IV for further entries

CHANG, C.C. [1957] *Algebraization of infinitely many-valued logics* (**P** 1675) Summer Inst Symb Log;1957 Ithaca 144-146
◇ B50 G20 G25 ◇ REV JSL 36.159 • ID 01933

CHANG, C.C. [1958] *Algebraic analysis of many valued logics* (**J** 0064) Trans Amer Math Soc 88*467-490
◇ B50 G20 ◇ REV MR 20 # 821 Zbl 84.7 JSL 36.159 • ID 01934

CHANG, C.C. [1958] *Proof of an axiom of Lukasiewicz* (**J** 0064) Trans Amer Math Soc 87*55-56
◇ B50 ◇ REV MR 20 # 820 Zbl 85.244 JSL 24.248 • ID 01935

CHANG, C.C. [1959] *A new proof of the completeness of the Lukasiewicz axioms* (**J** 0064) Trans Amer Math Soc 93*74-80
◇ B50 ◇ REV MR 23 # A58 Zbl 93.11 JSL 36.159 • ID 01938

CHANG, C.C. & KEISLER, H.J. [1962] *Model theories with truth values in a uniform space* (**J** 0015) Bull Amer Math Soc 68*107-109
◇ B50 C20 C50 C90 ◇ REV MR 25 # 12 Zbl 104.241 • ID 01946

CHANG, C.C. [1963] see BELLUCE, L.P.

CHANG, C.C. [1963] *Logic with positive and negative truth values* (**J** 0096) Acta Philos Fenn 16*19-39
◇ B50 ◇ REV MR 28 # 2966 Zbl 129.256 JSL 36.331 • ID 02675

CHANG, C.C. [1963] *The axiom of comprehension in infinite valued logic* (**J** 0132) Math Scand 13*9-30
◇ B50 E70 ◇ REV MR 29 # 2181 JSL 32.128 • ID 01961

CHANG, C.C. [1965] *Infinite valued logic as a basis for set theory* (**P** 0623) Int Congr Log, Meth & Phil of Sci (2,Proc);1964 Jerusalem 93-100
◇ B50 E70 ◇ REV MR 34 # 2445 Zbl 149.15 JSL 32.129 • ID 01968

CHANG, C.C. & KEISLER, H.J. [1966] *Continuous model theory* (**X** 0857) Princeton Univ Pr: Princeton xi + 165pp
• TRANSL [1971] (**X** 0885) Mir: Moskva 184pp
◇ B50 C05 C20 C50 C90 C98 ◇ REV MR 38 # 36 Zbl 149.4 • ID 02410

CHANG, C.C. [1973] *Modal model theory* (**P** 0713) Cambridge Summer School Math Log;1971 Cambridge GB 599-617
◇ B45 C90 ◇ REV MR 50 # 9531 Zbl 276 # 02012 • ID 03978

CHANG, C.C. [1974] see ADDISON, J.W.

CHANG, C.C. see Vol. I, III, IV, V for further entries

CHANG, C.L. & LEE, R.C.T. [1971] *Some properties of fuzzy logic* (**J** 0194) Inform & Control 19*417-431
◇ B52 ◇ REV MR 46 # 6994 Zbl 237 # 02002 • ID 07964

CHANG, C.L. & LEE, R.C.T. [1973] *Symbolic logic and mechanical theorem proving* (**X** 0801) Academic Pr: New York xiii + 331pp
• TRANSL [1983] (**X** 2027) Nauka: Moskva 358pp
◇ B35 B98 ◇ REV MR 55 # 13894 MR 84k:03003 Zbl 263 # 68046 Zbl 518 # 03004 • ID 03980

CHANG, C.L. see Vol. I for further entries

CHANG, S.S.L. & ZADEH, L.A. [1972] *On fuzzy mapping and control* (**J** 2338) IEEE Trans Syst Man & Cybern SMC-2*30-34
◇ B52 ◇ REV MR 50 # 15986 Zbl 305 # 94001 • ID 60908

CHANG, SHOOUYUH [1983] see BRILL JR., E.D.

CHANKVETADZE, O.E. [1981] *The addition rule for implications and its applications (Russian) (English and Georgian summaries)* (**J** 0233) Soobshch Akad Nauk Gruz SSR 104*301-304
◇ B60 ◇ REV MR 83k:03012 Zbl 475 # 03007 • ID 55461

CHANKVETADZE, O.E. see Vol. I, V for further entries

CHAPMAN, F.M. & HENLE, P. [1933] *The fundamentals of logic* (**X** 1347) Scribner: New York xiii + 384pp
◇ B98 ◇ REV FdM 59.65 • ID 22578

CHAPMAN, T. [1975] *Prior's criticism of the Barcan formula* (**J** 0047) Notre Dame J Formal Log 16*116-118
◇ A05 B45 ◇ REV Zbl 254 # 02021 • ID 60914

CHAPMAN, T. [1978] *A modal logic with temporal variables* (**J** 0047) Notre Dame J Formal Log 19*558-578
◇ A05 B45 ◇ REV MR 80a:03024 Zbl 271 # 02019 Zbl 403 # 03012 • ID 54726

CHAPMAN, T. [1981] *Quantum logic and modality* (**J** 0079) Logique & Anal, NS 24*99-111
◇ B45 B51 ◇ REV MR 83h:81017 Zbl 489 # 03025 • ID 37209

CHAPMAN, T. see Vol. VI for further entries

CHARLWOOD, G. [1981] *An axiomatic version of positive semilattice relevance logic* (**J** 0036) J Symb Logic 46*233-239
◇ B46 ◇ REV MR 82j:03018 Zbl 479 # 03012 • ID 55674

CHATTERJI, B.N. [1982] *Character recognition using fuzzy similarity relations* (**C** 3786) Approx Reason in Decis Anal 131-137
◇ B52 ◇ REV MR 84d:90003 • ID 46110

CHAUVIN, A. [1960] *Sur les modeles du calcul K_0 de Bochvar, avec ou sans egalite, et l'interpretation des paradoxes de la logique dans la theorie des ensembles elementaires arithmetiques* (**J** 0150) Acad Roy Belg Bull Cl Sci (5) 46*124-131
◇ B50 D55 ◇ REV MR 22#7935 • ID 01998

CHAUVIN, A. see Vol. I, III, IV, V, VI for further entries

CHAUVINEAU, J. [1957] *La logique moderne* (**X** 0840) Pr Univ France: Paris 128pp
◇ B98 ◇ REV MR 19.1 JSL 24.70 • ID 22396

CHAVCHANIDZE, V.V. [1970] *Logik von Interessen (Phasenlogik auf der Basis von Logik-Informationsfunktionen.) (Russian) (English summary)* (**J** 0233) Soobshch Akad Nauk Gruz SSR 59*557-560
◇ B60 ◇ REV Zbl 246#94008 • ID 60863

CHAVCHANIDZE, V.V. see Vol. I for further entries

CHEIFETZ, P.M. [1970] see AVENOSO, F.J.

CHELLAS, B.F. [1969] *The logical form of imperatives* (**X** 1355) Stanford Univ Pr: Stanford 121pp
• LAST ED (**X** 2045) Perry Lane Pr: Stanford v+115pp
◇ A05 B98 ◇ ID 25267

CHELLAS, B.F. [1971] *Imperatives* (**J** 0105) Theoria (Lund) 37*114-129
◇ B45 ◇ REV MR 45#6581 Zbl 299#02034 • ID 02064

CHELLAS, B.F. [1972] *Notions of relevance* (**J** 0122) J Philos Logic 1*287-293
◇ A05 B46 ◇ REV MR 54#12487b • ID 31680

CHELLAS, B.F. [1973] *Notions of relevance.Comments on Leblanc's paper* (**P** 1556) Exact Philos: Probl, Tools & Goals;1971 Montreal 21-27
◇ A05 B46 ◇ REV MR 54#12487b Zbl 291#02011 • ID 31681

CHELLAS, B.F. [1974] *Conditional obligation* (**C** 1936) Log Th & Semant Anal (Kanger) 23-33
◇ A05 B45 ◇ REV Zbl 288#02016 • ID 31679

CHELLAS, B.F. [1975] *Basis conditional logic* (**J** 0122) J Philos Logic 4*133-153
• TRANSL [1978] (**C** 2049) Leggi di Nat, Modalita, Ipotesi 282-302 (Italian)
◇ A05 B45 ◇ REV MR 57#69 Zbl 317#02029 • ID 31676

CHELLAS, B.F. & MCKINNEY, A. [1975] *The completeness of monotonic modal logics* (**J** 0068) Z Math Logik Grundlagen Math 21*379-383
◇ B45 ◇ REV MR 51#12480 Zbl 331#02007 • ID 03984

CHELLAS, B.F. [1979] *Modalities in normal systems containing the S5 axiom* (**C** 3670) Intention & Intentionalities 261-265
◇ B45 ◇ ID 33657

CHELLAS, B.F. [1980] *Modal logic. An introduction* (**X** 0805) Cambridge Univ Pr: Cambridge, GB xii+295pp
◇ B45 B98 ◇ REV MR 81i:03019 Zbl 431#03009 JSL 46.670 • ID 53915

CHELLAS, B.F. [1983] *$KG^{k,l,m,n}$ and the EFMP* (**J** 0079) Logique & Anal, NS 26*255-262
◇ B45 ◇ REV MR 85i:03044 Zbl 562#03010 • ID 44090

CHELLAS, B.F. see Vol. I for further entries

CHEN, C.S. & SIY, P. [1974] *Fuzzy logic for handwritten numeral character recognition* (**J** 2338) IEEE Trans Syst Man & Cybern SMC-4*570-575
◇ B52 ◇ REV Zbl 287#68066 • ID 65334

CHEN, GUOQUAN & LEE, S.C. & PAN, WEIJIE [1981] *D-fuzzy logic* (**P** 3705) Int Symp Multi-Val Log (11);1981 Oklahoma City & Norman 263-274
◇ B52 ◇ REV MR 83m:94033 Zbl 557#94025 • ID 46214

CHEN, GUOQUAN & LEE, S.C. & PAN, WEIJIE [1982] *D-fuzzy logic (Chinese summary)* (**J** 3732) Mohu Shuxue 2/3*33-45
◇ B52 ◇ REV MR 84a:03026 • ID 34895

CHEN, HUAITING & WU, XUEMOU [1980] *Transition logic, fuzzy logic and pansystem logic I (Chinese) (English summary)* (**J** 2754) Huazhong Gongxueyuan Xuebao 1980/2*II-III,15-23
◇ B52 ◇ REV MR 81i:68088 • REM Special Issue on Fuzzy Mathematics • ID 80891

CHEN, QIANGYE & HU, SHIHUA [1951] *A note on the four-valued propositional calculus and the four colour problem (Chinese summary)* (**J** 0418) Shuxue Xuebao 1*243-246
◇ B50 ◇ REV MR 17.570 • ID 48689

CHEN, XINGGOU & WANG, PEIZHUANG [1980] *An improvement and strengthening of the Negoita-Ralescu representation theorem (Chinese) (English summary)* (**J** 2521) Beijing Shifan Daxue Xuebao, Ziran Kexue 1980/3-4*41-46
◇ B52 ◇ REV MR 84k:03134 • ID 35026

CHEN, YONGYI [1982] *An approach to fuzzy operators I (Chinese) (English summary)* (**J** 3732) Mohu Shuxue 2/2*1-10
◇ B52 ◇ REV MR 83j:03088 • ID 34867

CHEN, YONGYI [1982] *An approach to fuzzy operators* (**J** 3919) BUSEFAL 9*59-65
◇ B52 ◇ REV Zbl 525#03011 • ID 38249

CHEN, YONGYI & WANG, PEIZHUANG [1983] *Optimum fuzzy implication and direct method of approximate reasoning* (**J** 3919) BUSEFAL 16*107-113
◇ B52 ◇ REV Zbl 567#03008 • ID 49210

CHEN, YONGYI & WANG, PEIZHUANG [1985] *Optimum fuzzy implication and a direct method of approximate reasoning (Chinese) (English summary)* (**J** 3732) Mohu Shuxue 5/1*29-40
◇ B52 ◇ ID 49007

CHEN, YONGYI see Vol. V for further entries

CHEN, ZHAOKUAN [1983] *Fixed points for a class of fuzzy mappings (Chinese)* (**J** 3732) Mohu Shuxue 3/1*121
◇ B52 ◇ REV MR 84m:03075 • ID 35772

CHENDOV, B. [1969] *A system of dyadic (relative) modalities and its relation to the aristotelian syllogistic and some modern logical theories* (**J** 0147) An Univ Bucuresti, Acta Logica 12*125-144
◇ A10 B45 ◇ REV MR 46#5105 Zbl 324#02012 • ID 02069

CHENG, WEIMIN & WU, WANGMING & XU, WENLI [1982] *An algorithm to solve the max-min composite fuzzy relational equations* (**C** 3786) Approx Reason in Decis Anal 47-49
◇ B52 ◇ REV MR 84m:03084 • ID 35781

CHENG, ZHAOZHONG & LIU, DSOSU [1982] *Some problem in fuzzy probability* (**C** 3778) Fuzzy Inform & Decis Processes 67-70
◇ B52 ◇ REV MR 84d:90004 Zbl 501#60004 • ID 46111

CHENIQUE, F. [1974] *Comprendre la logique moderne. Tome I: Classes, propositions et predicats* (**X** 0856) Dunod: Paris xx+294+6pp
◇ B98 E98 ◇ REV MR 53#10539 Zbl 276#02001 • REM Vol. II 1974 • ID 23053

CHENIQUE, F. [1974] *Comprendre la logique moderne. Tome II: Logiques non classiques, relations et structures* (**X** 0856) Dunod: Paris xx+(295-480)+6pp
◇ B98 ◇ REV MR 53#10540 Zbl 276#02002 • REM Vol. I 1974 • ID 23054

CHENTSOV, N.N. & MOROZOVA, E.A. [1981] *Noncommutative quantum logic (finite-dimensional theory) (Russian) (English summary)* (**S** 2651) Prepr Inst Prikl Mat, Akad Nauk SSSR 1981/57*28pp
◇ B51 ◇ REV MR 83a:81005 • ID 38901

CHEREPOV, A.N. [1983] *Description of the structure of closed classes in P_k containing the class of polynomials (Russian)* (**J** 0052) Probl Kibern 40*5-18
◇ B50 ◇ REV MR 84j:03055 Zbl 522#03011 • ID 34647

CHERNIAVSKY, J.C. [1973] *The complexity of some non-classical logics* (**P** 3062) IEEE Symp Switch & Automata Th (14);1973 Iowa City 209-213
◇ B45 D15 F20 F50 ◇ REV MR 56#2801 • ID 71692

CHERNIAVSKY, J.C. see Vol. IV, VI for further entries

CHEUNG, P.T. & SU, S.Y.H. [1971] *Algebraic properties of multi-valued symmetric switching functions* (**P** 2007) Symp Th & Appl of Multi-Val Log Design;1971 Buffalo 26-35
◇ B50 ◇ ID 41973

CHEUNG, P.T. & SU, S.Y.H. [1971] *Computer-orientated algorithm for minimizing multi-valued switching functions* (**P** 2007) Symp Th & Appl of Multi-Val Log Design;1971 Buffalo 140-152
◇ B50 B70 ◇ ID 41996

CHEUNG, P.T. & PURVIS, D.M. [1975] *A computer-oriented heuristic minimization algorithm for multiple-output multi-valued switching functions* (**P** 1805) Int Symp Multi-Val Log (5,Proc);1975 Bloomington 112-120
◇ B50 ◇ ID 35817

CHEUNG, P.T. & SU, S.Y.H. [1976] *Cubical representation for computer-aided processing of multiple-valued switching functions* (**P** 2011) Int Symp Multi-Val Log (6);1976 Logan 24-29
◇ B50 B70 ◇ REV MR 58#20879 • ID 35846

CHEUNG, P.T. [1976] *Identification of different functional properties of multiple-valued switching functions* (**P** 2011) Int Symp Multi-Val Log (6);1976 Logan 74-78
◇ B50 ◇ REV MR 58#20831 • ID 35873

CHIDGEY, J.R. & PARKS, R.Z. [1972] *Necessity and ticket entailment* (**J** 0047) Notre Dame J Formal Log 13*224-226
◇ B46 ◇ REV MR 45#6585 Zbl 234#02014 • ID 16364

CHIDGEY, J.R. [1973] *A note on transitivity* (**J** 0047) Notre Dame J Formal Log 14*273-275 • ERR/ADD ibid 17*640
◇ B45 ◇ REV MR 51#2853 MR 55#77 Zbl 226#02003 • ID 17390

CHIDGEY, J.R. [1975] *Independence* (**C** 1852) Entailment - Log of Relev & Nec 143-144,322-323,452-460
◇ B46 ◇ REV MR 53#10542 JSL 42.311 • ID 44467

CHIDGEY, J.R. [1979] *On the non-availability of Dawson-modeling into certain relevance alethic modal logics* (**J** 0063) Studia Logica 38*89-94
◇ B46 ◇ REV MR 80m:03034 Zbl 406#03025 • ID 56108

CHILIN, V.I. [1978] *Continuous valuations on logics (Russian)* (**J** 0024) Dokl Akad Nauk Uzb SSR 1978/6*6-8
◇ B50 E65 G05 ◇ REV MR 80j:03088 Zbl 449#03021 • ID 56688

CHIMEV, K.N. [1965] *Some systems of functions of P_k (Bulgarian) (Russian and French summaries)* (**J** 0224) God Vissh Tekh Ucheb Zaved Mat, Sofiya 2/3*1-6
◇ B50 ◇ REV MR 36#1298 Zbl 311#02027 • ID 02186

CHIMEV, K.N. [1966] *The essential dependence of the functions from P_k (Bulgarian) (Russian and English summaries)* (**J** 0224) God Vissh Tekh Ucheb Zaved Mat, Sofiya 3/2*5-12
◇ B50 ◇ REV MR 48#5731 Zbl 315#02018 • ID 02187

CHIMEV, K.N. [1967] *A certain kind of dependence of certain functions of P_k on their arguments (Bulgarian) (Russian and French summaries)* (**J** 0224) God Vissh Tekh Ucheb Zaved Mat, Sofiya 4/1*5-12
◇ B50 ◇ REV MR 45#3166 Zbl 318#02018 • ID 02192

CHIMEV, K.N. [1967] *The dependence of the functions of P_k on their arguments (Bulgarian) (Russian and French summaries)* (**J** 0224) God Vissh Tekh Ucheb Zaved Mat, Sofiya 4/3*5-13
◇ B50 ◇ REV MR 49#2255 Zbl 314#02029 • ID 04005

CHIMEV, K.N. [1968] *Separable pairs and the representability of the functions of P_k (Bulgarian) (Russian and French summaries)* (**J** 0224) God Vissh Tekh Ucheb Zaved Mat, Sofiya 5/2*163-182
◇ B50 ◇ REV MR 49#2257 Zbl 318#02015 • ID 04002

CHIMEV, K.N. & KUTIRKOV, G.A. [1968] *The first order variables of the functions of many-valued logic (Bulgarian) (Russian and French summaries)* (**J** 0224) God Vissh Tekh Ucheb Zaved Mat, Sofiya 5/3*111-118
◇ B50 ◇ REV MR 45#3168 Zbl 318#02013 • ID 02189

CHIMEV, K.N. [1968] *The strongly essential variables of the functions of P_k (Bulgarian) (Russian and French summaries)* (**J** 0224) God Vissh Tekh Ucheb Zaved Mat, Sofiya 5/2*155-162
◇ B50 ◇ REV MR 49#2256 Zbl 318#02014 • ID 04001

CHIMEV, K.N. [1968] *The various kinds of dependence of the functions of many-valued logic on their arguments (Bulgarian) (Russian and French summaries)* (**J** 0224) God Vissh Tekh Ucheb Zaved Mat, Sofiya 5/3*101-109
◇ B50 ◇ REV MR 45#3167 Zbl 318#02017 • ID 02188

CHIMEV, K.N. [1970] *A certain class of functions (Bulgarian) (Russian and French summaries)* (**J** 0224) God Vissh Tekh Ucheb Zaved Mat, Sofiya 6/3*15-29
◇ B50 ◇ REV MR 50#1838 Zbl 323#02026 • ID 04006

CHIMEV, K.N. & KUTIRKOV, G.A. [1970] *Certain discrete functions (Bulgarian) (Russian and French summaries)* (**J** 0224) God Vissh Tekh Ucheb Zaved Mat, Sofiya 6/2*63-72
◇ B50 ◇ REV MR 48#10763 Zbl 322#02012 • ID 07707

Author Index

CHIMEV, K.N. [1970] *Certain functions of many-valued logic (Bulgarian) (Russian and French summaries)* (J 0224) God Vissh Tekh Ucheb Zaved Mat, Sofiya 6/3*5-13
◇ B50 ◇ REV MR 50 # 54 Zbl 322 # 02013 • ID 04000

CHIMEV, K.N. [1970] *Separable pairs for a certain class of functions (Bulgarian) (Russian and French summaries)* (J 0224) God Vissh Tekh Ucheb Zaved Mat, Sofiya 6/1*25-38
◇ B50 ◇ REV MR 50 # 9528 Zbl 318 # 02016 • ID 03999

CHIMEV, K.N. [1970] *The dependence of the function of k-valued logic on their arguments (Bulgarian) (Russian and French summaries)* (J 0224) God Vissh Tekh Ucheb Zaved Mat, Sofiya 6/2*53-62
◇ B50 ◇ REV MR 48 # 10761 Zbl 319 # 02014 • ID 02191

CHIMEV, K.N. [1971] *Certain properties of functions (Bulgarian) (Russian and French summaries)* (J 0224) God Vissh Tekh Ucheb Zaved Mat, Sofiya 7/1*23-32
◇ B50 ◇ REV MR 50 # 52 Zbl 318 # 02012 • ID 04003

CHIMEV, K.N. [1971] *Separable pairs of functions (Bulgarian) (Russian and French summaries)* (J 0224) God Vissh Tekh Ucheb Zaved Mat, Sofiya 7/3*7-12
◇ B50 ◇ REV MR 50 # 53 Zbl 324 # 02011 • ID 04004

CHIMEV, K.N. [1972] *On second order variables (Bulgarian) (Russian and French summaries)* (J 0224) God Vissh Tekh Ucheb Zaved Mat, Sofiya 8/2*49-56
◇ B50 ◇ REV MR 53 # 10548 Zbl 314 # 02013 • ID 23062

CHIMEV, K.N. [1972] *The representability of certain functions (Bulgarian) (Russian and French summaries)* (J 0224) God Vissh Tekh Ucheb Zaved Mat, Sofiya 8/4*107-119
◇ B50 ◇ REV MR 53 # 10549 Zbl 314 # 02030 • ID 23063

CHIMEV, K.N. [1973] *On certain classes of functions (Bulgarian) (Russian and French summaries)* (J 0224) God Vissh Tekh Ucheb Zaved Mat, Sofiya 9/2*51-60
◇ B50 ◇ REV MR 58 # 21433 Zbl 312 # 02015 • ID 30301

CHIMEV, K.N. [1973] *Subfunctions,and the representability of certain functions of k-valued logic (Bulgarian) (Russian and French summaries)* (J 0224) God Vissh Tekh Ucheb Zaved Mat, Sofiya 9/2*61-74
◇ B50 ◇ REV MR 58 # 21434 Zbl 312 # 02016 • ID 30302

CHIMEV, K.N. [1973] *Subfunctions and strongly essential variables of functions (Bulgarian) (Russian and French summaries)* (J 0224) God Vissh Tekh Ucheb Zaved Mat, Sofiya 9/4*43-55
◇ B50 ◇ REV MR 58 # 21431 Zbl 312 # 02017 • ID 30303

CHIMEV, K.N. [1973] *The structural properties of functions in relation to their separable pairs (Bulgarian) (Russian and French summaries)* (J 0224) God Vissh Tekh Ucheb Zaved Mat, Sofiya 9/1*37-50
◇ B50 ◇ REV MR 58 # 21432 Zbl 312 # 02014 • ID 29094

CHIMEV, K.N. [1975] *Die Invarianz gewisser Eigenschaften von Funktionen bei Identifizierung ihrer Argumente (Bulgarian) (Russian and English summaries)* (J 3171) God Vissh Ucheb Zaved, Prilozhna Mat, Sofiya 10/1*51-68
◇ B50 ◇ REV Zbl 339 # 02006 • ID 60989

CHIMEV, K.N. [1975] *Ueber separierte Teilmengen und stark wesentliche Variablen von Funktionen (Bulgarian) (Russian and English summaries)* (J 3171) God Vissh Ucheb Zaved, Prilozhna Mat, Sofiya 10/4*7-14
◇ B50 ◇ REV Zbl 338 # 02005 • ID 60986

CHIMEV, K.N. [1976] *Representability of some classes of functions of many-valued logic (Bulgarian) (Russian and French summaries)* (J 3171) God Vissh Ucheb Zaved, Prilozhna Mat, Sofiya 12/2*129-142
◇ B50 G20 G25 ◇ REV MR 80b:03023 Zbl 414 # 94037 • ID 53100

CHIMEV, K.N. [1982] *On a class of functions (Russian)* (S 2829) Rostocker Math Kolloq 19*9-17
◇ B50 ◇ REV MR 83i:03040 Zbl 511 # 03021 • ID 35507

CHIMEV, K.N. see Vol. I, IV, V for further entries

CHIRKOV, M.K. [1971] *Ueber die Methode der wahrscheinlichkeitstheoretischen Logik (Russian)* (S 0716) Vychisl Tekh Vopr Kibern (Univ Leningrad) 8*52-66
◇ B48 ◇ REV MR 48 # 10679 Zbl 277 # 94028 • ID 61000

CHIRKOV, M.K. see Vol. I, IV for further entries

CHISHOLM, R.M. [1955] *Law statements and counterfactual inference* (J 0103) Analysis (Oxford) 15*97-105
◇ B48 B65 ◇ REV JSL 21.86 • ID 42609

CHISHOLM, R.M. [1956] *Sentences about believing* (J 0113) Proc Aristotelian Soc 56*125-148
◇ B48 ◇ REV JSL 32.404 • ID 43110

CHISHOLM, R.M. [1963] *Contrary-to-duty imperatives and deontic logic* (J 0103) Analysis (Oxford) 24*33-36
◇ B45 ◇ REV JSL 32.243 • ID 43094

CHISHOLM, R.M. & KEIM, R.G. [1972] *A system of epistemic logic* (J 1022) Ratio (Oxford) 14*99-115
◇ B45 ◇ REV MR 56 # 2787 • ID 71746

CHISHOLM, R.M. [1975] *The intrinsic value in disjunctive states of affairs* (J 0097) Nous, Quart J Phil 9*295-308
◇ A05 B45 ◇ REV MR 58 # 27321a • ID 71745

CHISHOLM, R.M. [1976] *Knowledge and belief: "de dicto" and "de re"* (J 0095) Philos Stud 29*1-20
◇ A05 B45 ◇ REV MR 58 # 4982 • ID 71741

CHISHOLM, R.M. see Vol. I for further entries

CHLEBUS, B.S. [1979] *On decidability of propositional algorithmic logic (Polish) (English summary)* (S 3270) Spraw Inst Inf, Uniw Warsaw 83*28pp
◇ B25 B75 D10 D15 ◇ REV Zbl 454 # 03006 • ID 54219

CHLEBUS, B.S. [1981] *On the computational complexity of satisfiability in propositional logic programs* (S 3270) Spraw Inst Inf, Uniw Warsaw 99*1-40
◇ B75 D15 ◇ REV MR 83k:68026 Zbl 508 # 68019 • ID 38156

CHLEBUS, B.S. [1982] *Completeness proofs for some logics of programs* (J 0068) Z Math Logik Grundlagen Math 28*49-62
◇ B75 ◇ REV MR 83e:68021 Zbl 491 # 03008 • ID 36850

CHLEBUS, B.S. [1982] *On the computational complexity of satisfiability in propositional logics of programs* (J 1426) Theor Comput Sci 21*179-212
◇ B75 D15 ◇ REV MR 83k:68026 Zbl 496 # 68020 • ID 37748

CHLEBUS, B.S. [1982] *On the decidability of propositional algorithmic logic* (J 0068) Z Math Logik Grundlagen Math 28*247-261
◇ B25 B75 ◇ REV MR 84e:03022 Zbl 502 # 03012 • ID 34361

CHLEBUS, B.S. see Vol. I, III, IV, V for further entries

CHOMSKY, N. [1965] *Aspects of the theory of syntax* (**X** 0865) MIT Pr: Cambridge, MA x+251pp
◇ B98 ◇ REV JSL 32.385 JSL 35.167 • ID 02429

CHOMSKY, N. [1968] *Language and mind* (**X** 0863) Harcourt: New York & London vii+88pp
◇ A05 B65 B98 ◇ ID 25312

CHOMSKY, N. [1972] *Studies on semantics in generative grammars* (**X** 0873) Mouton: Paris 207pp
◇ B65 B98 D05 ◇ ID 25314

CHOMSKY, N. see Vol. I, IV for further entries

CHONG, C.T. & WICKS, M.J. (EDS.) [1983] *Southeast Asian conference on logic* (**P** 3669) SE Asian Conf on Log;1981 Singapore xiv+210pp
◇ B97 ◇ REV MR 84i:03004 Zbl 532 # 00005 • ID 34490

CHONG, C.T. see Vol. I, IV, V for further entries

CHOURAQUI, E. [1984] see BOURRELLY, L.

CHRESTENSON, H.E. [1955] *A class of generalized Walsh functions* (**J** 0048) Pac J Math 5*17-31
◇ B50 ◇ REV MR 16.920 Zbl 65.53 • ID 41718

CHRISTENSEN, D. [1983] *Glymour on evidential relevance* (**J** 0153) Phil of Sci (East Lansing) 50*471-481
◇ B46 ◇ REV MR 84j:03049 • ID 34641

CHRISTENSEN, N.E. [1979] *Mathematical logic and mathematical reasoning* (**P** 2615) Scand Logic Symp (5);1979 Aalborg 175-187
◇ A05 B60 ◇ REV MR 82f:03004 Zbl 458 # 03002 • ID 54408

CHRISTIAN, C.C. [1955] *Die Elimination des Unendlichkeitstheorems in einem intensionalen Kalkuel* (**J** 0169) Theoria (Madrid) 3*129-133
◇ B45 ◇ REV MR 17.815 • ID 02112

CHRISTIAN, C.C. [1972] *Modalkalkuel als formale Theorie und das Problem einer Praedikatoidenlogik* (**J** 0455) Phil Naturalis 13*113-156
◇ B45 ◇ REV MR 56 # 5222 • ID 71772

CHRISTIAN, C.C. [1976] *Die Bedeutung der Mengentheorie als Grundlagenwissenschaft* (**J** 0455) Phil Naturalis 16*238-271
◇ A05 B98 E10 E30 ◇ REV MR 56 # 5292 • ID 71770

CHRISTIAN, C.C. see Vol. I, IV, V, VI for further entries

CHRISTIAN, R.R. [1958] *Introduction to logic and sets* (**X** 0943) Ginn: Boston 69pp
• LAST ED [1965] (**X** 0841) Blaisdell: New York xii+116pp
◇ B98 E98 ◇ REV Zbl 156.4 JSL 33.631 • ID 22478

CHRZ, T. [1984] *A system for knowledge representation based on intensional logic I,II* (**J** 3932) Comput Artif Intell (Bratislava) 3*193-209,305-317
◇ B45 ◇ REV Zbl 565 # 68087 Zbl 565 # 68088 • ID 48671

CHUANG, H.Y.H. & DAS, S. [1974] *On the design of reliable and fail-safe digital systems using multiple-valued logic* (**P** 1385) Int Symp Multi-Val Log (4);1974 Morgantown 183-206
◇ B50 ◇ ID 42080

CHUAQUI, R.B. [1976] *Introduccion a la metamatematica y sus aplicaciones* (**J** 3685) Publ Inst Mat Univ Catolica Chile 83pp
◇ B98 ◇ ID 33786

CHUAQUI, R.B. [1977] see ARRUDA, A.I.

CHUAQUI, R.B. [1978] see ARRUDA, A.I.

CHUAQUI, R.B. [1980] *Foundations of statistical methods using a semantical definition of probability* (**P** 2958) Latin Amer Symp Math Log (4);1978 Santiago 103-120
◇ B48 C90 ◇ REV MR 82i:62010 Zbl 469 # 03024 • ID 55151

CHUAQUI, R.B. [1980] see ARRUDA, A.I.

CHUAQUI, R.B. [1985] *How to decide between different methods of statistical inference* (**C** 4181) Math Log & Formal Syst (Costa da) 43-56
◇ B48 ◇ ID 48202

CHUAQUI, R.B. [1985] *Models for probability* (**P** 4317) Anal, Geom & Probab;1981 Valparaiso 89-120
◇ B48 ◇ ID 48164

CHUAQUI, R.B. see Vol. I, III, V for further entries

CHUDNOVSKY, D.V. [1968] *Probability on first order logics (Ukrainian) (Russian and English summaries)* (**J** 0270) Dokl Akad Nauk Ukr SSR, Ser A 1968*914-916
◇ B48 C90 ◇ REV MR 39 # 5338 • ID 02732

CHUDNOVSKY, D.V. [1970] *Logical probability and conditional probability on Boolean algebras (Russian) (English summary)* (**J** 0295) Teor Veroyat i Mat Stat (Kiev) 2*221-225
• TRANSL [1974] (**J** 3456) Th Probab & Math Stat 2*225-229
◇ B48 C90 G05 ◇ REV MR 44 # 1063 Zbl 222 # 02066 • ID 26322

CHUDNOVSKY, D.V. see Vol. I, III, V for further entries

CHUPAKHIN, I.YA. & PLOTNIKOV, A.M. (EDS.) [1982] *Logic and philosophical categories (Russian)* (**X** 0938) Leningrad Univ: Leningrad 152pp
◇ A05 B97 ◇ REV MR 84c:03004 • ID 35671

CHURCH, A. [1936] *A bibliography of symbolic logic* (**J** 0036) J Symb Logic 1*121-218 • ERR/ADD ibid 3*178-192
◇ A10 B98 ◇ REV Zbl 16.97 FdM 62.1046 • ID 02129

CHURCH, A. [1936] *Mathematical logic (mimeographed notes)* (**X** 0857) Princeton Univ Pr: Princeton iii+113pp
◇ B40 B98 D20 ◇ REV JSL 2.39 FdM 62.1048 • ID 23466

CHURCH, A. [1940] *Elementary topics in mathematical logic* (**X** 0915) Galois Inst Math Art: Brooklyn 102pp
◇ B98 ◇ REV Zbl 27.148 JSL 7.91 • ID 02460

CHURCH, A. [1944] *Introduction to mathematical logic. Part I* (**X** 0857) Princeton Univ Pr: Princeton vi+118pp
• TRANSL [1960] (**X** 1656) Izdat Inostr Lit: Moskva
◇ B98 ◇ REV MR 18.631 MR 6.29 Zbl 60.20 JSL 10.19 JSL 22.286 JSL 23.362 • REM 2nd ed. 1956; x+376pp • ID 02181

CHURCH, A. [1951] *A formulation of the logic of sense and denotation* (**C** 0585) Struct, Meth & Meaning (Sheffer) 3-24
◇ B45 ◇ REV Zbl 54.6 JSL 17.133 • ID 31038

CHURCH, A. [1952] *Brief bibliography of formal logic* (**J** 0251) Proc Amer Acad Arts Sci 80*155-172
◇ A10 B98 ◇ REV Zbl 48.1 JSL 18.178 • ID 02458

CHURCH, A. [1953] *Non-normal truth-tables for the propositional calculus* (**J** 0250) Bol Soc Mat Mexicana 10*41-52
◇ B50 ◇ REV MR 15.385 Zbl 53.341 JSL 19.233 • ID 02401

CHURCH, A. [1953] *Tablas de verdad no normales para el calculo proposicional* (**P** 1809) Congr Cient Mexicano;1951 Mexico City 1*81-89
◇ B50 ◇ REV Zbl 53.341 • ID 32291

Author Index

Church, A. [1954] *Intensional isomorphism and identity of belief* (J 0095) Philos Stud 5*65-73
◇ A05 B45 ◇ rev JSL 20.294 • id 31041

Church, A. [1954] *On Carnap's analysis of statements of assertion and belief* (C 1811) Phil & Analysis 125-128
• repr [1971] (C 1812) Refer & Modal 168-170
◇ A05 B48 ◇ id 32289

Church, A. [1973] *Isomorfismo intensional e identidad de creencia (Spanish)* (C 1810) Semant Filos, Probl & Disc 171-181
◇ A05 B45 ◇ id 32284

Church, A. [1973] *Logica, analisis y intensionalidad (Spanish)* (C 1810) Semant Filos, Probl & Disc 395-399
◇ A05 B45 ◇ id 32288

Church, A. [1973] *Outline of a revised formulation of the logic of sense and denotation I* (J 0097) Nous, Quart J Phil 7*24-33
◇ B45 ◇ rev MR 57 # 15959a • rem Part II 1974
• id 31046

Church, A. [1973] *Postscript 1968* (C 1810) Semant Filos, Probl & Disc 147-152
◇ A05 B45 ◇ id 32283

Church, A. [1973] *Proposiciones y oraciones (Spanish)* (C 1810) Semant Filos, Probl & Disc 343-352
◇ B60 ◇ id 32286

Church, A. [1973] *Sobre el analisis del discurso indirecto propuesto por Scheffler* (C 1810) Semant Filos, Probl & Disc 363-369
◇ B45 B65 ◇ id 32287

Church, A. [1974] *Outline of a revised formulation of the logic of sense and denotation II* (J 0097) Nous, Quart J Phil 8*135-156
◇ B45 ◇ rev MR 57 # 15959b JSL 23.22 • rem Part I 1973
• id 31045

Church, A. see Vol. I, III, IV, V, VI for further entries

Churchman, C.W. [1938] *On finite and infinite modal systems* (J 0036) J Symb Logic 3*77-82
◇ B45 ◇ rev Zbl 21.290 JSL 3 JSL 3.163 FdM 64.29
• id 02153

Churchman, C.W. [1940] *Elements of logic and formal science* (X 1297) Lippincott: Philadelphia iv + 337pp
◇ B98 ◇ rev JSL 6.169 • id 22479

Chwistek, L.B. [1938] *Remarques critiques concernant la notion de la variable dans le systeme de la semantique rationnelle (Polish) (French summary)* (S 0281) Arch Towarz Nauk Lwow, Sect 3 9*283-334
◇ A05 B60 E70 ◇ rev Zbl 19.145 JSL 5.166 FdM 64.929 • id 02672

Chwistek, L.B. see Vol. I, V, VI for further entries

Cignoli, R. [1966] *Boolean multiplicative closures I,II* (J 0081) Proc Japan Acad 42*1168-1171,1172-1174
◇ B50 G05 G10 G20 ◇ rev MR 36 # 5043 Zbl 149.257
• id 03995

Cignoli, R. [1980] *Some algebraic aspects of many-valued logics* (P 3006) Brazil Conf Math Log (3);1979 Recife 49-69
◇ B50 G20 ◇ rev MR 82b:03111 Zbl 455 # 06005
• id 54293

Cignoli, R. [1982] *Proper n-valued Lukasiewicz algebras as S-algebras of Lukasiewicz n-valued propositional calculi* (J 0063) Studia Logica 41*3-16
◇ B50 ◇ rev MR 85e:03158 Zbl 509 # 03012 • id 36523

Cignoli, R. see Vol. V for further entries

Cirulis, J. [1970] *Quasilinear functions and precomplete classes (Russian) (Latvian and English summaries)* (J 0337) Mat Ezheg, Akad Nauk Latv SSR 7*283-288
◇ B50 ◇ rev MR 42 # 7484 Zbl 286 # 02024 • id 17179

Cirulis, J. [1975] *Lectures on mathematical logic and set theory. Part I: Mathematical logic (Russian)* (X 0895) Latv Valsts (Gos) Univ : Riga 143pp
◇ B98 ◇ rev MR 58 # 27169a • rem Part II 1975
• id 71815

Cirulis, J. [1979] *Protothetics without typical ambiguity of expressions (Russian)* (J 0337) Mat Ezheg, Akad Nauk Latv SSR 23*166-178,275
◇ B60 ◇ rev MR 81b:03028a Zbl 438 # 03017 • id 55929

Cirulis, J. [1979] *The traditional means of inference in protothetics (Russian)* (J 0337) Mat Ezheg, Akad Nauk Latv SSR 23*179-193,275
◇ B60 ◇ rev MR 81b:03028b Zbl 438 # 03018 • id 55930

Cirulis, J. [1982] *Axioms of protothetics with a strengthened capacity rule (Russian)* (J 0337) Mat Ezheg, Akad Nauk Latv SSR 26*264-270,285
◇ B15 B60 ◇ rev MR 84a:03028 Zbl 495 # 03007
• id 35570

Cirulis, J. [1984] *Variations on the theme of quantum logic (Russian)* (C 4091) Algeb & Diskret Mat (Riga) 146-158
◇ B51 ◇ rev MR 86g:03110 • id 44319

Cirulis, J. [1985] *A system of logic permitting vacuous and plural terms (Russian)* (J 0068) Z Math Logik Grundlagen Math 31*263-274
◇ B60 ◇ rev Zbl 553 # 03002 • id 48165

Cirulis, J. see Vol. I, V for further entries

Clark, C.M. & Kandel, A. [1981] *New results in the enumeration of minimized fuzzy-valued switching functions* (P 3705) Int Symp Multi-Val Log (11);1981 Oklahoma City & Norman 191-195
◇ B52 ◇ rev MR 83m:94033 Zbl 543 # 94026 • id 45952

Clark, C.M. [1981] see Ben, D.M.

Clark, C.M. & Kandel, A. [1982] *The enumeration of distinct fuzzy-valued switching functions* (J 2720) Fuzzy Sets Syst 8*291-310
◇ B52 B70 ◇ rev MR 84h:94032 Zbl 493 # 94016
• id 39521

Clark, I.D. [1973] *An axiomatisation of quantum logic* (J 0036) J Symb Logic 38*389-392
◇ B51 ◇ rev MR 50 # 3785 Zbl 276 # 02015 • id 17325

Clark, M. [1970] see Casey, H.

Clark, M. [1980] *The equivalence of tautological and "strict" entailment: Proof of an amended conjecture of Lewy's* (J 0122) J Philos Logic 9*9-15
◇ B46 ◇ rev MR 81f:03022 Zbl 427 # 03014 • id 53701

Clark, R. & Welsh Jr., P.J. [1962] *Introduction to logic* (X 0864) Van Nostrand: New York xii + 268pp
◇ B98 ◇ rev JSL 33.479 • id 02437

CLARK, R. [1976] *Old foundations for a logic of perception* (J 0154) Synthese 33*75-99
◇ B60 ◇ ID 61008

CLARK, R. [1978] *Not every object of thought has being: a paradox in naive predication theory* (J 0097) Nous, Quart J Phil 12*181-188
◇ A05 B45 ◇ REV MR 80c:03009 • ID 71830

CLARK, R. see Vol. I for further entries

CLARKE, B.L. [1981] *A calculus of individuals based on "connection"* (J 0047) Notre Dame J Formal Log 22*204-218
◇ B60 G25 ◇ REV MR 82i:03036 Zbl 438 # 03032 • ID 55547

CLARKE, E.M. & EMERSON, E.A. [1982] *Design and synthesis of synchronization skeletons using branching time temporal logic* (P 3738) Log of Progr;1981 Yorktown Heights 52-71
◇ B45 B75 ◇ REV MR 83i:68048 Zbl 546 # 68014 • ID 39313

CLARKE, E.M. & MISHRA, B. [1985] *Hierarchical verification of asynchronous circuits using temporal logic* (J 1426) Theor Comput Sci 38*269-291
◇ B45 ◇ ID 49197

CLARKE, E.M. & SISTLA, A.P. [1985] *The complexity of propositional linear temporal logics* (J 0037) ACM J 32*733-749
◇ B45 ◇ ID 47648

CLARKE, E.M. see Vol. I for further entries

CLARKE JR., E.M. & FRANCEZ, N. & GUREVICH, Y. & STULA, P.S. [1982] *Can message buffers be characterized in linear temporal logic?* (P 3715) Symp Princ of Distrib Computing 148-156
◇ B45 ◇ ID 33762

CLAY, R.E. [1962] *A simple proof of functional completeness in many-valued logics based on Lukasiewicz's C and N* (J 0047) Notre Dame J Formal Log 3*114-117
◇ B50 ◇ REV MR 26 # 4898 Zbl 118.13 JSL 30.105 • ID 02248

CLAY, R.E. [1962] *Note on Slupecki T-functions* (J 0036) J Symb Logic 27*53-54
◇ B50 ◇ REV MR 27 # 34 Zbl 107.8 JSL 30.105 • ID 02250

CLAY, R.E. [1963] *A standard form for Lukasiewicz many-valued logics* (J 0047) Notre Dame J Formal Log 4*59-66
◇ B50 ◇ REV MR 26 # 6044 Zbl 118.248 JSL 30.105 • ID 02249

CLAY, R.E. [1972] *Note on induvtive finiteness in mereology* (J 0047) Notre Dame J Formal Log 13*88-90
◇ B60 ◇ REV MR 45 # 6582 Zbl 228 # 02039 • ID 29130

CLAY, R.E. [1975] *Single axioms for atomistic and atomless mereology* (J 0047) Notre Dame J Formal Log 16*345-351
◇ B60 E70 ◇ REV MR 51 # 10068 Zbl 276 # 02009 • ID 17538

CLAY, R.E. see Vol. I, V, VI for further entries

CLEAVE, J.P. [1970] *The notion of validity in logical systems with inexact predicates* (J 0013) Brit J Phil Sci 21*269-274
◇ B52 ◇ REV MR 44 # 5205 Zbl 242 # 02033 • ID 02268

CLEAVE, J.P. [1974] *An account of entailment based on classical semantics* (J 0103) Analysis (Oxford) 34*118-122
◇ B46 ◇ REV Zbl 276 # 02007 • ID 29038

CLEAVE, J.P. [1974] *The notion of logical consequences in the logic of inexact predicates* (J 0068) Z Math Logik Grundlagen Math 20*307-324
◇ B52 ◇ REV MR 51 # 10028 Zbl 299 # 02015 • ID 04009

CLEAVE, J.P. [1976] *Comment: does many valued logic have any use* (P 1502) Phil of Logic;1976 Bristol 88-91
◇ A05 B50 ◇ REV MR 58 # 10307 • ID 30667

CLEAVE, J.P. [1976] *Quasi-boolean algebras, empirical continuity and 3-valued logic* (J 0068) Z Math Logik Grundlagen Math 22*481-500
◇ B50 G10 ◇ REV MR 58 # 27469 Zbl 364 # 02036 • ID 30668

CLEAVE, J.P. [1979] *The axiomatisation of theories of material necessity* (J 0047) Notre Dame J Formal Log 20*180-190
◇ B45 ◇ REV MR 80c:03020 Zbl 332 # 02027 • ID 52383

CLEAVE, J.P. [1980] *Some remarks on the interpretation of 3-valued logics* (J 1022) Ratio (Oxford) 22*52-60
◇ B50 ◇ REV MR 82g:03035 • ID 71844

CLEAVE, J.P. see Vol. I, III, IV, V, VI for further entries

CLIFFORD, J.E. [1966] *Tense logic and the logic of change* (J 0079) Logique & Anal, NS 9*219-230
◇ A05 B45 ◇ REV MR 35 # 6529 Zbl 166.252 JSL 36.327 • ID 02275

CLIFFORD, J.E. [1975] *Tense and tense logic* (X 0873) Mouton: Paris 173pp
◇ B45 ◇ REV JSL 43.381 • ID 44544

COATES, C.L. & LEWIS II, P.M. [1967] *Threshold logic* (X 0827) Wiley & Sons: New York xii+483pp
◇ B70 B98 ◇ ID 19520

COCCHIARELLA, N.B. [1969] *A completeness theorem in second order modal logic* (J 0105) Theoria (Lund) 35*81-103
◇ A05 B45 ◇ REV MR 40 # 2523 Zbl 205.6 • ID 02289

COCCHIARELLA, N.B. [1974] *La semantica della logica del tempo* (C 1808) Logica del Tempo
◇ A05 B45 ◇ ID 32276

COCCHIARELLA, N.B. [1974] *Logical atomism and modal logic* (J 1807) Phil Quart Israel 4*41-66
◇ A05 B45 ◇ ID 32275

COCCHIARELLA, N.B. [1975] *Logical atomism, nominalism, and modal logic* (J 0154) Synthese 31*23-62
◇ A05 B45 ◇ REV Zbl 317 # 02006 • ID 32279

COCCHIARELLA, N.B. [1975] *On the primary and secondary semantics of logical necessity* (J 0122) J Philos Logic 4*13-27
◇ A05 B45 ◇ REV MR 58 # 21464 Zbl 316 # 02022 • ID 32277

COCCHIARELLA, N.B. [1977] *Sortals, natural kinds and re-identification* (J 0079) Logique & Anal, NS 20*439-474
◇ A05 B45 ◇ REV MR 84i:03037 Zbl 389 # 03006 • ID 52295

COCCHIARELLA, N.B. see Vol. I, III, V, VI for further entries

CODE, A. [1976] *Aristotle's response to Quine's objections to modal logic* (J 0122) J Philos Logic 5*159-186
◇ A05 B45 ◇ REV MR 57 # 15961 Zbl 328 # 02004 • ID 61035

COFFA, J.A. [1975] *Fallacies of modality* (C 1852) Entailment - Log of Relev & Nec 244-252
◇ B45 ◇ REV MR 53 # 10542 • ID 44472

Author Index

COHEN, L.J. [1970] *The implications of induction* (**X** 0816) Methuen: London & New York vii+248pp
◇ A05 B48 B98 ◇ REV Zbl 298 # 02020 • ID 25404

COHEN, L.J. [1971] *The inductive logic of progressive problem-shifts* (**J** 2076) Rev Int Philos 25*62-77
◇ A05 B48 ◇ REV MR 58 # 27212 • ID 71862

COHEN, L.J. [1976] *A conspectus of the neo-classical theory of induction* (**C** 1702) Local Induction 235-262
◇ A05 B48 ◇ REV Zbl 331 # 02016 • ID 61039

COHEN, L.J. [1981] *Inductive logic, 1945-1977* (**C** 2617) Modern Log Survey 353-375
◇ B48 ◇ REV MR 82f:03002 Zbl 464 # 03001 • ID 42776

COHEN, L.J. & LOS, J. & PFEIFFER, H. & PODEWSKI, K.-P. (EDS.) [1982] *Logic, methodology and philosophy of science VI* (**S** 3303) Stud Logic Found Math 104*856pp
◇ B97 D98 ◇ REV MR 83k:03004 Zbl 489 # 00005 • ID 36588

COHEN, M.R. & NAGEL, E. [1934] *An introduction to logic and scientific method* (**X** 0863) Harcourt: New York & London xii+467pp
◇ B98 ◇ REV JSL 11.100 FdM 60.845 • ID 25096

COHEN, R.S. & WARTOFSKY, M.W. (EDS.) [1983] *Language, logic and method* (**S** 3311) Boston St Philos Sci 31*464pp
◇ B97 ◇ REV Zbl 493 # 00003 • ID 38427

COHEN, R.S. see Vol. I, IV for further entries

COHN, M. [1961] *Canonical forms of functions in p-valued logics* (**P** 0624) Switch Circ Th & Log Design (1,2);1960 Chicago;1961 Detroit 169-177
◇ B50 ◇ ID 04282

COLE, R. [1979] *Possibility matrices* (**J** 0105) Theoria (Lund) 45*8-39
◇ A05 B45 G25 ◇ REV MR 82g:03008 Zbl 449 # 03004 • ID 56671

COLE, R. see Vol. I for further entries

COLLIER, K.W. [1973] *Physical modalities and the system E* (**J** 0047) Notre Dame J Formal Log 14*185-194
◇ B45 ◇ REV MR 48 # 8200 Zbl 251 # 02028 • ID 02321

COLLIER, K.W. [1974] *A result of extending Bochvar's 3-valued logic* (**J** 0047) Notre Dame J Formal Log 15*344-346
◇ B50 ◇ REV MR 50 # 1839 Zbl 236 # 02015 • ID 02322

COLLIER, K.W. [1976] *The story of NR and E* (**J** 1893) Relevance Logic Newslett 1*8-10
◇ B46 ◇ REV Zbl 355 # 02013 • ID 50211

COLMERAUER, A. [1979] *Un sous-ensemble interessant du francais* (**J** 3441) RAIRO Inform Theor 13*309-336
◇ B50 B65 ◇ REV Zbl 423 # 68044 • ID 69344

COMBES, M. [1971] *Fondements des mathematiques* (**X** 0840) Pr Univ France: Paris 104pp
◇ A05 A10 B98 E98 ◇ REV MR 44 # 1537 Zbl 259 # 02003 • ID 02334

CONTE, A.G. & HILPINEN, R. & WRIGHT VON, G.H. (EDS.) [1977] *Deontische Logik und Semantik* (**X** 1169) Akad Verlagsges: Wiesbaden 215pp
◇ B45 ◇ ID 48640

COOK, T.A. [1978] *The geometry of generalized quantum logics* (**J** 2736) Int J Theor Phys 17*941-955
◇ B51 ◇ REV MR 80k:81009 Zbl 434 # 03044 • ID 55748

COOK, T.A. & RUETTIMANN, G.T. [1985] *Symmetries on quantum logics* (**J** 1546) Rep Math Phys (Warsaw) 21*121-126
◇ B51 G12 ◇ ID 48279

COOKE, R.M. & HILGEVOORD, J. [1981] *A new approach to equivalence in quantum logic* (**P** 3820) Curr Iss in Quantum Log;1979 Erice 101-113
◇ B51 G12 ◇ REV MR 84j:81015 Zbl 537 # 03044 • ID 39252

COOKE, R.M. [1983] *A result in Renyi's conditional probability theory with application to subjective probability* (**J** 0122) J Philos Logic 12*19-32
◇ B48 ◇ REV MR 85e:03053 Zbl 512 # 60003 • ID 40621

COOKE, R.M. see Vol. V for further entries

COOLEY, J.C. [1942] *A primer of formal logic* (**X** 0843) Macmillan : New York & London xi+378pp
◇ B98 ◇ REV MR 4.125 Zbl 60.20 JSL 8.80 • ID 02359

COOLS, M. & DUBOIS, T. & KAUFMANN, A. [1975] *Exercices avec solutions sur la theorie des sous-ensembles flous* (**X** 1752) Masson: Paris viii+166pp
• TRANSL [1982] (**X** 2686) CECSA: Mexico City 182pp (Spanish)
◇ B52 E72 E98 ◇ REV MR 58 # 27500 MR 84e:03064b Zbl 306 # 02002 • ID 62910

COOPER, D.E. [1977] *Bivalence, determinism, and realism* (**J** 0079) Logique & Anal, NS 20*148-155
◇ A05 B45 ◇ REV MR 58 # 4983 • ID 71916

COOPER, D.E. [1980] *Referential occurrence* (**J** 0047) Notre Dame J Formal Log 21*182-188
◇ A05 B45 ◇ REV MR 81h:03008 Zbl 351 # 02014 • ID 53190

COOPER, G.A. & PATT, Y.N. [1972] *Toward a characterization of logically complete switching functions in k-valued logic* (**P** 2008) Symp Th & Appl of Multi-Val Log Design;1972 Buffalo 100-103
◇ B50 ◇ ID 42036

COOPER, R. [1981] see BARWISE, J.

COOPER, R. see Vol. V for further entries

COOPER, W.S. [1974] *Set theory and syntactic description* (**X** 0873) Mouton: Paris 52pp
◇ B98 E75 ◇ REV MR 53 # 2685 Zbl 279 # 68051 • ID 21633

COOPER, W.S. [1978] *Foundations of logico-linguistics: A unified theory of information, language, and logic* (**X** 0835) Reidel: Dordrecht 2*xvi+249pp
◇ A05 B98 ◇ REV Zbl 406 # 03001 • ID 28182

COPELAND, B.J. [1977] *The system RS. An amendation of E* (**J** 1893) Relevance Logic Newslett 2*3-25
◇ B46 ◇ REV Zbl 363 # 02022 • ID 50849

COPELAND, B.J. [1979] *On when a semantics is not a semantics: some reasons for disliking the Routley-Meyer semantics for relevance logic* (**J** 0122) J Philos Logic 8*399-413
◇ A05 B46 ◇ REV MR 81h:03036 Zbl 416 # 03022 JSL 49.994 • ID 53188

COPELAND, B.J. [1980] *The trouble Anderson and Belnap have with relevance* (**J** 0095) Philos Stud 37*325-334
◇ A05 B46 ◇ REV MR 81k:03017 • ID 71924

COPELAND, B.J. [1982] *A note on the Barcan formula and substitutional quantification* (**J** 0079) Logique & Anal, NS 25*83-86
◇ B45 ◇ REV MR 83i:03028 Zbl 498 # 03009 • ID 35500

COPELAND, B.J. [1983] *A rejoinder to: "On the philosophical bases of relevant logic semantics"* (**J** 3783) J Non-Classical Log (Univ Campinas) 2*61
◇ A05 B46 ◇ REM For the article see Martin,E.P. (1982) • ID 49001

COPELAND, B.J. [1983] *Pure semantics and applied semantics: a response to R.Routley, V.Routley, R.K.Meyer and E.P.Martin: "On the philosophical bases of relevant logic semantics" (J. Non-Classical Logic 1 (1982), no. 1, 71-105)* (**J** 3781) Topoi 2*197-204
◇ A05 B46 ◇ REV MR 85i:03056b • ID 44109

COPELAND, B.J. [1983] *Tense trees: a tree system for* K_t (**J** 0047) Notre Dame J Formal Log 24*318-322
◇ B45 ◇ REV MR 85f:03015 Zbl 488 # 03011 • ID 36520

COPI, I.M. [1953] *Introduction to logic* (**X** 0843) Macmillan : New York & London xvi+472pp
◇ A10 B98 ◇ REV Zbl 481 # 03001 Zbl 57.5 JSL 19.147 JSL 29.92 JSL 35.166 • REM 6th edition 1982; xiv+604pp • ID 02371

COPI, I.M. [1954] *Symbolic logic* (**X** 0843) Macmillan : New York & London xiii+355pp
◇ B98 ◇ REV MR 17.223 MR 50 # 1824 MR 80c:03002 Zbl 281 # 02012 JSL 19.282 JSL 39.177 • ID 02372

COPI, I.M. & GOULD, JAMES A. (EDS.) [1967] *Contemporary readings in logical theory* (**X** 0843) Macmillan : New York & London viii+344pp
◇ A05 A10 B97 ◇ REV MR 35 # 14 • ID 25224

COPI, I.M. & GOULD, JAMES A. [1967] *Deontic logic, introduction* (**C** 0721) Contemp Readings in Log Th 301pp
◇ B45 B98 ◇ ID 04030

COPI, I.M. & GOULD, JAMES A. [1967] *Many valued logic, introduction* (**C** 0721) Contemp Readings in Log Th 333pp
◇ B50 ◇ ID 04032

COPI, I.M. & GOULD, JAMES A. [1967] *Modal logic, introduction* (**C** 0721) Contemp Readings in Log Th 257-258
◇ B45 ◇ ID 04028

COPI, I.M. also published under the name COPILOWISH, I.M.

COPI, I.M. see Vol. I, V for further entries

CORBETT, J. [1978] *Probability in quantum mechanics* (**J** 3104) Adv Appl Probab 10*725-729
◇ B51 ◇ REV Zbl 395 # 60099 • ID 52598

CORCORAN, J. & WEAVER, G.E. [1969] *Logical consequences in modal logic: natural deduction in S5* (**J** 0047) Notre Dame J Formal Log 10*370-384
◇ B45 F07 ◇ REV MR 40 # 2524 Zbl 187.266 • REM Part I. Part II 1974 • ID 02388

CORCORAN, J. & WOOD, S. [1973] *Discussion: the switches "paradox"* (**J** 0075) Phil Phenom Research 34*102-108
◇ B45 ◇ ID 28172

CORCORAN, J. & WEAVER, G.E. [1974] *Logical consequence in modal logic II: some semantic systems for S4* (**J** 0047) Notre Dame J Formal Log 15*370-378
◇ B45 ◇ REV MR 50 # 4253 Zbl 281 # 02029 • REM Part I 1969 • ID 14098

CORCORAN, J. see Vol. I, III, VI for further entries

CORISH, D. [1978] *Aristotle on temporal order: "Now," "Before," and "After"* (**J** 0256) Isis (Philadelphia) 69*68-74
◇ A05 A10 B45 ◇ REV Zbl 385 # 01006 • ID 52108

CORNIDES, T. [1974] *Ordinale Deontik* (**X** 0902) Springer: Wien x+210pp
◇ B45 B98 ◇ REV MR 57 # 15962 JSL 44.121 • ID 71948

CORNMAN, J.W. [1966] *Metaphysics, reference and language* (**X** 0875) Yale Univ Pr: New Haven xxi+288pp
◇ A05 B98 ◇ ID 25316

CORNMAN, J.W. see Vol. III for further entries

CORSI, G. [1981] *Deduzione naturale e logica quantistica* (**P** 3092) Congr Naz Logica;1979 Montecatini Terme 313-324
◇ B51 ◇ ID 48724

COSGROVE, R.J. [1980] *A three-valued free logic for presuppositional languages* (**J** 0047) Notre Dame J Formal Log 21*549-571
◇ A05 B50 ◇ REV MR 81h:03049 Zbl 416 # 03026 • ID 53918

COSTA DA, N.C.A. [1958] *Note on the concept of contradiction (Portuguese) (English summary)* (**J** 1831) Anuar Soc Paranaense Mat, Ser 2 1*6-8
◇ A05 B53 ◇ REV MR 21 # 7146 Zbl 85.241 • ID 04202

COSTA DA, N.C.A. [1959] *Observations on the concept of existence in mathematics (Portuguese) (English summary)* (**J** 1831) Anuar Soc Paranaense Mat, Ser 2 2*16-19
◇ A05 B53 ◇ REV MR 24 # A654 Zbl 91.7 • ID 04204

COSTA DA, N.C.A. [1963] *Calculs propositionnels pour les systemes formels inconsistants* (**J** 0109) C R Acad Sci, Paris 257*3790
◇ B53 ◇ REV MR 28 # 1123 Zbl 118.12 • ID 02462

COSTA DA, N.C.A. [1963] see ARRUDA, A.I.

COSTA DA, N.C.A. [1963] *The present situation in set theory (Portuguese)* (**J** 1957) Bol Soc Paranaense Mat 6*40-43
◇ B53 E98 ◇ REV Zbl 112.10 • ID 42829

COSTA DA, N.C.A. [1964] *Calculs des predicats pour les systemes formels inconsistants* (**J** 0109) C R Acad Sci, Paris 258*27-29
◇ B53 ◇ REV MR 28 # 2960 Zbl 126.10 • ID 02463

COSTA DA, N.C.A. [1964] *Calculs de predicats avec egalite pour les systemes formels inconsistants* (**J** 0109) C R Acad Sci, Paris 258*1111-1113
◇ B53 ◇ REV MR 29 # 2164 Zbl 126.10 • ID 02465

COSTA DA, N.C.A. [1964] *Calculs des descriptions pour les systemes formels inconsistants* (**J** 0109) C R Acad Sci, Paris 258*1366-1368
◇ B53 ◇ REV MR 29 # 2165 Zbl 126.10 • ID 02464

COSTA DA, N.C.A. & GUILLAUME, M. [1964] *Sur les calculs* $\mathscr{C}_n$ (**J** 0110) Anais Acad Bras Cienc 36*379-382
◇ B53 ◇ REV MR 32 # 5498 Zbl 134.15 • ID 04048

COSTA DA, N.C.A. [1964] *Sur un systeme inconsistant de theorie des ensembles* (**J** 0109) C R Acad Sci, Paris 258*3144-3147
◇ B53 E70 ◇ REV MR 28 # 4987 Zbl 134.15 • ID 02466

COSTA DA, N.C.A. [1964] see ARRUDA, A.I.

COSTA DA, N.C.A. & GUILLAUME, M. [1965] *Negations composees et loi de Peirce dans les systemes $\mathscr{C}_n$* (J 0050) Port Math 24*201-210
◇ B53 ◇ REV MR 36 # 32 Zbl 192.29 • ID 02468

COSTA DA, N.C.A. [1965] *Sur les systemes formels $\mathscr{C}_i, \mathscr{C}_i^*, \mathscr{C}_i^=, \mathscr{D}_i$ et NF_i* (J 0109) C R Acad Sci, Paris 260*5427-5430
◇ B53 ◇ REV MR 32 # 7394 Zbl 134.15 • ID 02467

COSTA DA, N.C.A. [1967] *Two formal systems of set theory* (J 0028) Indag Math 29*45-51
◇ B53 E30 E55 E70 ◇ REV MR 35 # 6541 Zbl 209.15 • ID 02476

COSTA DA, N.C.A. [1967] *Une nouvelle hierarchie de theories inconsistantes* (J 0056) Publ Dep Math, Lyon 4/3*2-8
◇ B53 E70 ◇ REV MR 39 # 52 Zbl 189.286 • ID 02475

COSTA DA, N.C.A. & DUBIKAJTIS, L. [1968] *Sur la logique discoursive de Jaskowski* (J 0014) Bull Acad Pol Sci, Ser Math Astron Phys 16*551-557
◇ B45 ◇ REV MR 38 # 982 Zbl 165.308 • ID 02477

COSTA DA, N.C.A. & SETTE, A.-M. [1969] *Les algebres $\mathscr{C}_\omega$* (J 2313) C R Acad Sci, Paris, Ser A-B 268*A1011-A1014
◇ B53 G15 ◇ REV MR 39 # 5332 Zbl 184.11 • ID 02482

COSTA DA, N.C.A. [1970] *On the underlying logic of two systems of set theory* (J 0028) Indag Math 32*1-8
◇ B53 E70 ◇ REV MR 41 # 8206 Zbl 193.304 • ID 02481

COSTA DA, N.C.A. [1970] see ARRUDA, A.I.

COSTA DA, N.C.A. & D'OTTAVIANO, I.M.L. [1970] *Sur un probleme de Jaskowski* (J 2313) C R Acad Sci, Paris, Ser A-B 270*A1349-A1353
◇ B53 ◇ REV MR 41 # 6670 Zbl 198.17 • ID 03092

COSTA DA, N.C.A. [1971] *Remarques sur le systeme NF_1* (J 2313) C R Acad Sci, Paris, Ser A-B 272*A1149-A1151
◇ B53 E70 ◇ REV MR 44 # 2599 Zbl 215.326 • ID 02484

COSTA DA, N.C.A. [1972] *On the theory of inconsistent formal systems* (J 1573) Notas Commun Mat (Recife) 41*27pp
◇ B53 ◇ REV Zbl 239 # 02017 • ID 27920

COSTA DA, N.C.A. [1974] see ARRUDA, A.I.

COSTA DA, N.C.A. [1974] *On the theory of inconsistent formal systems* (J 0047) Notre Dame J Formal Log 15*497-510 • ERR/ADD ibid 16*608
◇ B53 E35 E70 ◇ REV MR 50 # 6841 MR 51 # 10085 Zbl 286 # 02032 • ID 02487

COSTA DA, N.C.A. [1974] *Remarques sur les calculs $\mathscr{C}_n$, $\mathscr{C}_n^*$, $\mathscr{C}_n^=$ et $\mathscr{D}_n$* (J 2313) C R Acad Sci, Paris, Ser A-B 278*A819-A821
◇ B25 B53 ◇ REV MR 53 # 5291 Zbl 278 # 02021 • ID 22926

COSTA DA, N.C.A. [1975] *Remarks on Jaskowski's discussive logic* (J 0302) Rep Math Logic, Krakow & Katowice 4*7-15
◇ B53 ◇ REV MR 52 # 2821 Zbl 313 # 02018 • ID 17620

COSTA DA, N.C.A. & PINTER, C. [1976] *α-logic and infinitary languages* (J 0068) Z Math Logik Grundlagen Math 22*105-112
◇ B60 C75 ◇ REV MR 58 # 21426 Zbl 336 # 02039 • ID 14829

COSTA DA, N.C.A. [1976] see ALVES, E.H.

COSTA DA, N.C.A. [1977] see ALVES, E.H.

COSTA DA, N.C.A. [1977] see ARRUDA, A.I.

COSTA DA, N.C.A. & KOTAS, J. [1977] *On some modal logical systems defined in connexion with Jaskowski's problem* (P 1076) Latin Amer Symp Math Log (3);1976 Campinas 57-73
◇ B45 ◇ REV MR 58 # 5055 Zbl 359 # 02011 • ID 16598

COSTA DA, N.C.A. & DUBIKAJTIS, L. [1977] *On Jaskowski's discursive logic* (P 1076) Latin Amer Symp Math Log (3);1976 Campinas 37-56
◇ B53 ◇ REV MR 58 # 141 Zbl 359 # 02012 • ID 16594

COSTA DA, N.C.A. [1978] see ARRUDA, A.I.

COSTA DA, N.C.A. & KOTAS, J. [1979] *A new formulation of discussive logic* (J 0063) Studia Logica 38*429-445
◇ B53 ◇ REV MR 81m:03023 Zbl 431 # 03016 • ID 53922

COSTA DA, N.C.A. [1980] see ARRUDA, A.I.

COSTA DA, N.C.A. & KOTAS, J. [1980] *Some problems on logical matrices and valorizations* (P 3006) Brazil Conf Math Log (3);1979 Recife 131-146
◇ B50 ◇ REV MR 82h:03021 Zbl 472 # 03020 • ID 55286

COSTA DA, N.C.A. & WOLF, R.G. [1980] *Studies in paraconsistent logic. I* (J 1807) Phil Quart Israel 9*189-218
◇ B53 ◇ ID 42831

COSTA DA, N.C.A. [1981] see ALVES, E.H.

COSTA DA, N.C.A. [1982] see ALVES, E.H.

COSTA DA, N.C.A. [1982] see ARRUDA, A.I.

COSTA DA, N.C.A. [1982] *The philosophical import of paraconsistent logic* (J 3783) J Non-Classical Log (Univ Campinas) 1*1-19
◇ B53 ◇ REV MR 85h:03007 Zbl 508 # 03003 • ID 36930

COSTA DA, N.C.A. [1984] see ARRUDA, A.I.

COSTA DA, N.C.A. & LOPARIC, A. [1984] *Paraconsistency, paracompleteness and valuations* (J 0079) Logique & Anal, NS 27*119-131
◇ B53 ◇ REV MR 86c:03024 Zbl 549 # 03023 • ID 42854

COSTA DA, N.C.A. see Vol. I, III, V, VI for further entries

COSTANTINI, D. & GEYMONAT, L. [1981] *For an objectivist notion of probability (Italian) (English and French summaries)* (J 2839) Statistica (Bologna) 41*519-533
◇ B48 ◇ REV MR 83m:60001 • ID 40353

COSTANTINI, D. [1981] *Inductive logic and inductive statistics* (C 3515) Ital Studies in Phil of Sci 169-183
◇ A05 B48 ◇ REV MR 82e:03022 Zbl 445 # 62005 • ID 56521

COSTANTINI, D. & MONARI, P. [1982] *Inferenza predittiva e "fatti"* (J 2839) Statistica (Bologna) 42*513-528
◇ B48 ◇ REV MR 84d:62014 Zbl 519 # 62002 • ID 36728

COSTANTINI, D. [1983] *Il contributo di R. Carnap alla metodologia statistica* (J 2839) Statistica (Bologna) 43*347-350
◇ A10 B48 ◇ REV Zbl 516 # 62008 • ID 38587

COSTANTINI, D. & MONARI, P. [1983] *Inferenze statistiche finitarie* (J 2839) Statistica (Bologna) 43*19-30
◇ B48 ◇ REV MR 85a:62010 Zbl 519 # 62003 • ID 36729

COSTANTINI, D. [1985] *Finitary inductive inferences (Italian)* (**P** 4646) Atti Incontri Log Mat (2);1983/84 Siena 329-332 ◇ B48 ◇ ID 49659

COURVOISIER, M. & LAGASSE, J. & RICHARD, J. [1976] *Logique combinatoire. 3e ed* (**X** 0856) Dunod: Paris vi+73pp ◇ B40 B98 ◇ REV MR 57#15759 Zbl 376#02045 • ID 51687

COUTURAT, L. [1904] *L'algebre de la logique* (**X** 1324) Open Court: LaSalle 100pp • TRANSL [1914] (**X** 1324) Open Court: LaSalle xiv+98pp (English) • REPR [1980] (**X** 0872) Blanchard: Paris 100pp ◇ B98 G05 ◇ REV MR 81j:01014 Zbl 426#03064 FdM 45.120 • REM 2nd ed. 1908. Reprint of the 2nd ed. • ID 22581

COUTURAT, L. see Vol. I, V for further entries

COX, A. [1978] *Hintikka and the interdefinability of obligation and forbiddance* (**J** 2837) Southwest J Phil, Sect 1, Yokohama Univ 9*7-10 ◇ B45 ◇ REV MR 80j:03025 • ID 71968

COX, R.T. [1979] *Of inference and inquiry, an essay in inductive logic* (**P** 2964) Max Entropy Formalism;1978 Cambridge MA 119-167 ◇ B48 ◇ REV MR 80c:03029 Zbl 467#94001 • ID 71970

COY, W. & MORAGA, C. [1979] *Description and detection of faults in multiple-valued logic circuits* (**P** 3003) Int Symp Multi-Val Log (9);1979 Bath 74-81 ◇ B50 B70 ◇ REV MR 80m:94114 • ID 80987

COY, W. [1981] *A common approach to the description, implementation and test generation of multivalued functions* (**P** 3705) Int Symp Multi-Val Log (11);1981 Oklahoma City & Norman 90-94 ◇ B50 ◇ REV MR 83m:94033 • ID 45950

COY, W. see Vol. I, IV for further entries

CRAIG, W. [1974] see ADDISON, J.W.

CRAIG, W. see Vol. I, III, IV, V, VI for further entries

CRESSWELL, M.J. [1965] *Another basis for S4* (**J** 0079) Logique & Anal, NS 8*191-195 ◇ B45 ◇ REV MR 36#1302 JSL 35.581 • ID 02533

CRESSWELL, M.J. [1965] *On the logic of incomplete answers* (**J** 0036) J Symb Logic 30*65-68 ◇ B60 ◇ REV MR 33#5471 Zbl 146.9 JSL 31.498 • ID 02532

CRESSWELL, M.J. [1965] *The logic of interrogatives* (**P** 0688) Logic Colloq;1963 Oxford 8-11 ◇ B60 ◇ REV MR 36#1301 Zbl 129.258 JSL 31.668 • ID 04210

CRESSWELL, M.J. [1966] *The completeness of SO.5* (**J** 0079) Logique & Anal, NS 9*263-266 ◇ B45 ◇ REV MR 44#5210 Zbl 154.256 JSL 38.328 • ID 02534

CRESSWELL, M.J. [1967] *A Henkin completeness theorem for T* (**J** 0047) Notre Dame J Formal Log 8*186-190 ◇ B45 ◇ REV MR 43#7303 Zbl 183.8 JSL 35.581 • ID 02537

CRESSWELL, M.J. [1967] *Alternative completeness theorems for modal systems* (**J** 0047) Notre Dame J Formal Log 8*339-345 ◇ B45 ◇ REV MR 44#6455 Zbl 182.317 JSL 35.581 • ID 02538

CRESSWELL, M.J. [1967] *Note on a system of Aaqvist* (**J** 0036) J Symb Logic 32*58-60 ◇ B45 ◇ REV MR 38#5592 Zbl 163.5 JSL 35.137 • ID 02539

CRESSWELL, M.J. [1967] *Some further semantics for deontic logic* (**J** 0079) Logique & Anal, NS 10*179-191 ◇ B45 ◇ REV MR 37#2584 Zbl 166.2 • ID 02536

CRESSWELL, M.J. [1967] *The interpretation of some Lewis systems of modal logic* (**J** 0273) Australasian J Phil 45*198-206 ◇ B45 ◇ REV JSL 37.417 • ID 02725

CRESSWELL, M.J. & HUGHES, G.E. [1968] *An introduction to modal logic* (**X** 0816) Methuen: London & New York xii+388pp • TRANSL [1978] (**X** 1174) Gruyter: Berlin x+340pp [1973] (**X** 1781) Tecnos: Madrid 316pp [1973] (**X** 3615) Il Saggiatore: Milano xxxvii+436pp [1981] (**X** 3711) Koseisha Koseikaku: Tokyo xii+440pp ◇ B45 B98 ◇ REV MR 55#12472 Zbl 205.5 JSL 36.328 • ID 19031

CRESSWELL, M.J. [1968] *Completeness without the Barcan formula* (**J** 0047) Notre Dame J Formal Log 9*75-80 ◇ B45 ◇ REV MR 39#1307 Zbl 175.265 • ID 02541

CRESSWELL, M.J. [1968] *Some proofs of relative completeness in modal logic* (**J** 0047) Notre Dame J Formal Log 9*62-66 ◇ B45 ◇ REV MR 44#6456 Zbl 182.318 JSL 35.581 • ID 02543

CRESSWELL, M.J. [1968] *The representation of intensional logic* (**J** 0068) Z Math Logik Grundlagen Math 14*289-298 ◇ B45 ◇ REV MR 38#3121 Zbl 169.302 • ID 02542

CRESSWELL, M.J. [1969] *The conjunctive normal form for S3.5* (**J** 0036) J Symb Logic 34*253-255 ◇ B45 ◇ REV Zbl 184.280 • ID 02545

CRESSWELL, M.J. [1969] *The elimination of de re modalities* (**J** 0036) J Symb Logic 34*329-330 ◇ B45 ◇ REV MR 40#4086 Zbl 184.9 • ID 02544

CRESSWELL, M.J. [1970] *Classical intensional logics* (**J** 0105) Theoria (Lund) 36*347-372 ◇ B45 ◇ REV MR 47#14 Zbl 217.301 • ID 02547

CRESSWELL, M.J. [1970] *Note on the interpretation of S0.5* (**J** 0079) Logique & Anal, NS 13*376-378 ◇ B45 ◇ REV MR 44#3844 Zbl 215.46 JSL 38.328 • ID 02546

CRESSWELL, M.J. [1971] *The semantics for a logic of "better"* (**J** 0079) Logique & Anal, NS 14*775-782 ◇ B45 ◇ REV MR 47#8257 Zbl 239#02013 • ID 02549

CRESSWELL, M.J. [1972] *Intensional logics and logical truth* (**J** 0122) J Philos Logic 1*2-15 ◇ B45 ◇ REV MR 47#6453 Zbl 239#02012 • ID 02726

CRESSWELL, M.J. [1972] *Second-order intensional logic* (**J** 0068) Z Math Logik Grundlagen Math 18*297-320 ◇ B15 B45 ◇ REV MR 46#5114 Zbl 246#02018 • ID 02551

Author Index

CRESSWELL, M.J. [1972] *S0.5 is alive and well* (**J** 0079) Logique & Anal, NS 15*503
◇ B45 ◇ REV MR 48 # 67 Zbl 267 # 02011 • ID 29881

CRESSWELL, M.J. [1972] *The completeness of S1 and some related systems* (**J** 0047) Notre Dame J Formal Log 13*485-496
◇ B45 ◇ REV MR 50 # 6794 Zbl 242 # 02019 • ID 02550

CRESSWELL, M.J. [1973] *Logics and languages* (**X** 0816) Methuen: London & New York xi + 273pp
• TRANSL [1979] (**X** 1174) Gruyter: Berlin xi + 431pp (German)
◇ A05 B98 ◇ REV MR 55 # 2497 Zbl 408 # 03001 JSL 42.425 • REM See also 1975 • ID 02552

CRESSWELL, M.J. [1973] *Physical theories and possible worlds* (**J** 0079) Logique & Anal, NS 16*495-511
◇ B45 ◇ REV MR 51 # 7825 Zbl 287 # 02009 • ID 17320

CRESSWELL, M.J. [1975] *Frames and models in modal logic* (**P** 0765) Algeb & Log;1974 Clayton 63-86
◇ B45 ◇ REV MR 53 # 2629 Zbl 315 # 02029 • ID 21518

CRESSWELL, M.J. [1975] *Hamblin on time* (**J** 0097) Nous, Quart J Phil 9*193-204
◇ A05 B45 ◇ REV MR 58 # 21465a • ID 71983

CRESSWELL, M.J. [1975] *Hyperintensional logic* (**J** 0063) Studia Logica 34*25-38
◇ B45 ◇ REV MR 52 # 43 Zbl 307 # 02017 • ID 17212

CRESSWELL, M.J. [1975] *Identity and intensional objects* (**J** 1807) Phil Quart Israel 5*47-68
◇ B45 ◇ ID 33879

CRESSWELL, M.J. [1975] *Omnitemporal logic and tense logic* (**J** 0302) Rep Math Logic, Krakow & Katowice 4*17-23
◇ B45 ◇ REV MR 54 # 2421 Zbl 312 # 02021 • ID 21935

CRESSWELL, M.J. & HUGHES, G.E. [1975] *Omnitemporal logic and converging time* (**J** 0105) Theoria (Lund) 41*11-34
◇ B45 ◇ REV MR 58 # 21474 Zbl 322 # 02020 • ID 31028

CRESSWELL, M.J. [1977] *Interval semantics and logical words* (**C** 3619) Log Anal of Tense & Aspect 7-29
◇ B45 B65 ◇ ID 33881

CRESSWELL, M.J. [1979] *B Seg has the finite model property* (**J** 0387) Bull Sect Logic, Pol Acad Sci 8*154-160
◇ B45 ◇ REV MR 81d:03016 Zbl 423 # 03017 • ID 53529

CRESSWELL, M.J. [1979] *Interval semantics for some event expressions* (**C** 3611) Semantics from Diff Points of View 90-116
◇ B45 ◇ ID 33867

CRESSWELL, M.J. [1980] *Possible worlds* (**C** 3613) Literary Semant & Possible Worlds 6-16
◇ B45 ◇ ID 33869

CRESSWELL, M.J. [1982] *A canonical model for S2* (**J** 0079) Logique & Anal, NS 25*3-7
◇ B45 ◇ REV MR 84a:03019 Zbl 494 # 03011 • ID 33870

CRESSWELL, M.J. & HUGHES, G.E. [1982] *K1.1 is not canonical* (**J** 0387) Bull Sect Logic, Pol Acad Sci 11*109-113
◇ B45 ◇ REV MR 85d:03033 Zbl 525 # 03003 • ID 33752

CRESSWELL, M.J. & GOLDBLATT, R.I. & MEYERHOFF CRESSWELL, M. & SEGERBERG, K. [1982] *Symbolic logic. Proceedings of the 1981 Annual Conference of the Australian Association of Symbolic Logic held in Wellington, New Zealand, from 2-5 July 1981* (**J** 0063) Studia Logica 41*93-307
◇ B97 ◇ REV Zbl 529 # 00015 • ID 38476

CRESSWELL, M.J. [1982] *Urn models: A classical exposition* (**J** 0063) Studia Logica 41*109-130
◇ B60 ◇ REV MR 85i:03079 Zbl 536 # 03012 • ID 33876

CRESSWELL, M.J. [1983] *KM and the finite model property* (**J** 0047) Notre Dame J Formal Log 24*323-327
◇ B45 ◇ REV MR 84h:03035 Zbl 488 # 03009 • ID 33873

CRESSWELL, M.J. [1983] *The completeness of KW and K1.1* (**J** 0079) Logique & Anal, NS 26*123-127
◇ B45 ◇ REV MR 85e:03034 Zbl 533 # 03003 • ID 33875

CRESSWELL, M.J. & HUGHES, G.E. [1984] *A companion to modal logic* (**X** 0816) Methuen: London & New York
◇ B45 ◇ ID 33748

CRESSWELL, M.J. [1984] *An incomplete decidable modal logic* (**J** 0036) J Symb Logic 49*520-527
◇ B45 ◇ REV MR 85i:03045 • ID 42436

CRESSWELL, M.J. [1985] *The decidable normal modal logics are not recursively enumerable* (**J** 0122) J Philos Logic 14*231-233
◇ B25 B45 D25 D35 D80 ◇ ID 42657

CRESSWELL, M.J. see Vol. I, IV, V for further entries

CRIST, S.C. [1978] *A tri-state logic family* (**P** 2014) Int Symp Multi-Val Log (8);1978 Rosemont 1-6
◇ B50 ◇ ID 35906

CROSSLEY, J.N. & DUMMETT, M. (EDS.) [1965] *Formal systems and recursive functions* (**X** 0809) North Holland: Amsterdam 320pp
◇ B97 D20 D97 ◇ REV Zbl 126.2 • ID 31683

CROSSLEY, J.N. [1972] see ASH, C.J.

CROSSLEY, J.N. [1975] *What is mathematical logic ?* (**J** 2053) Math Medley 1*3-6
◇ B98 F30 ◇ ID 31689

CROSSLEY, J.N. & HUMBERSTONE, L.L. [1977] *The logic of "actually"* (**J** 0302) Rep Math Logic, Krakow & Katowice 8*11-29
◇ B45 ◇ REV MR 58 # 5046 Zbl 389 # 03010 • ID 24186

CROSSLEY, J.N. see Vol. I, III, IV, V, VI for further entries

CSAKANY, B. & ROSENBERG, I.G. (EDS.) [1981] *Finite algebra and multiple-valued logic (Colloquium held in Szeged, August 27-31, 1979)* (**S** 3312) Coll Math Soc Janos Bolyai 28*880pp
◇ B50 ◇ REV MR 83a:08001 Zbl 471 # 00008 • ID 55193

CSAKANY, B. see Vol. III for further entries

CULBERTSON, J.T. [1958] *Mathematics and logic for digital devices* (**X** 0864) Van Nostrand: New York x + 224pp
◇ B98 G05 ◇ REV MR 19.1200 Zbl 80.335 JSL 23.366 • ID 23519

CURD, M.V. [1980] *The logic of discovery: an analysis of three approaches* (**P** 3764) Sci Discover, Log & Ration (Leonard);1978 Reno 201-219
◇ A05 B60 ◇ REV MR 84c:03009 • ID 34929

Current, K.W. & Mow, D.A. [1978] *Applications of multi-valued logic in LSI digital signal processing circuits* (**P** 2014) Int Symp Multi-Val Log (8);1978 Rosemont 187
◇ B50 B70 ◇ ID 35925

Current, K.W. [1979] *Simultaneous analog-to-quaternary conversion* (**P** 3003) Int Symp Multi-Val Log (9);1979 Bath 62-66
◇ B50 ◇ ID 35940

Currie, G. & Oddie, G. [1980] *Changing numbers* (**J** 0105) Theoria (Lund) 46*148-164
◇ B45 ◇ REV MR 83e:03029 • ID 35215

Curry, H.B. [1952] *Lecons de logique algebrique* (**X** 0834) Gauthier-Villars: Paris 163pp
◇ B98 G15 G98 ◇ REV MR 13.613 Zbl 48.2 JSL 19.146 • ID 02741

Curry, H.B. [1952] *The elimination theorem when modality is present* (**J** 0036) J Symb Logic 17*249-265
◇ B45 ◇ REV MR 14.527 Zbl 48.3 JSL 20.67 • ID 02621

Curry, H.B. [1954] *Remarks on the definition and nature of mathematics* (**J** 0076) Dialectica 8*228-233
• TRANSL [1967] (**C** 2141) Filos Matematica 153-159 (Italian) • REPR [1964] (**C** 1105) Phil of Math. Sel Readings 152-156
◇ A05 B98 ◇ REV JSL 22.85 JSL 34.108 JSL 5.26 • ID 37345

Curry, H.B. [1957] *Combinatory logic* (**P** 1675) Summer Inst Symb Log;1957 Ithaca 90-99
• REPR [1968] (**C** 0552) Phil Contemp - Chroniques 295-307
◇ B40 B98 ◇ REV Zbl 158.247 • ID 29324

Curry, H.B. & Feys, R. [1958] *Combinatory logic. Volume I* (**X** 0809) North Holland: Amsterdam xvi+417pp
• TRANSL [1967] (**X** 1781) Tecnos: Madrid
◇ B40 B98 ◇ REV MR 20 # 817 MR 39 # 5368 Zbl 81.241 JSL 32.267 • REM With two sections by Craig,W.. 2nd ed. 1968. For vol.II 1972 see Curry,H.B. & Hindley,J.R. & Seldin,J.P. • ID 02627

Curry, H.B. [1959] *The interpretation of formalized implication* (**J** 0105) Theoria (Lund) 25*1-26
◇ B45 ◇ ID 33830

Curry, H.B. [1963] *Foundations of mathematical logic* (**X** 0822) McGraw-Hill: New York xii+408pp
• TRANSL [1969] (**X** 0885) Mir: Moskva 567pp • LAST ED [1977] (**X** 0813) Dover: New York viii+408pp
◇ B98 ◇ REV MR 26 # 6036 Zbl 163.242 Zbl 172.8 JSL 38.149 • ID 02634

Curry, H.B. [1965] *Remarks on inferential deduction* (**C** 0749) Contrib Logic & Methodol (Bochenski) 45-72
◇ B48 ◇ REV MR 51 # 127 • ID 15246

Curry, H.B. [1967] *Combinatory logic* (**C** 0601) Encycl of Philos 4*504-509
◇ A05 B40 B98 ◇ REV JSL 35.299 • ID 02637

Curry, H.B. see Vol. I, IV, VI for further entries

Cusmariu, A. [1977] *"About belief de re."* (**J** 0079) Logique & Anal, NS 20*138-147
◇ B45 ◇ REV MR 58 # 5047 • ID 72027

Cutland, N.J. & Gibbins, P. [1982] *A regular sequent calculus for quantum logic in which ∧ and ∨ are dual* (**J** 0079) Logique & Anal, NS 25*221-248
◇ B51 G12 ◇ REV MR 85g:03091 Zbl 518 # 03029 • ID 37526

Cutland, N.J. [1985] *Partially observed stochastic controls based on a cumulative digital read out of the observations* (**P** 4460) Stoch Diff Syst;1984 Marseille 261-269
◇ B48 ◇ REV Zbl 561 # 93067 • ID 47496

Cutland, N.J. see Vol. I, III, IV, V for further entries

Czaja-Pospiech, D. & Czogala, E. & Pedrycz, W. [1978] *Fuzzy-control as a mathematical formalisation of a heuristic approach to the control of complex processes* (**J** 2814) Podstawy Sterowania 8*293-302
◇ B52 ◇ REV Zbl 408 # 93003 • ID 56300

Czaja-Pospiech, D. [1980] *On some rules of fuzzy logics (Polish) (English and Russian summaries)* (**S** 2890) Zesz Nauk, Mat Fiz, Politech Slask (Gliwice) 31*15-29
◇ B52 ◇ REV MR 82m:03036 Zbl 439 # 03006 • ID 55998

Czaja-Pospiech, D. & Czogala, E. & Pedrycz, W. [1980] *The multiple implications and compositional rules of inference in multivalued logics (fuzzylogics) (Polish) (English and Russian summaries)* (**S** 2890) Zesz Nauk, Mat Fiz, Politech Slask (Gliwice) 31*31-47
◇ B52 ◇ REV MR 82m:03037 Zbl 439 # 03007 • ID 55999

Czelakowski, J. [1981] *Partial boolean algebras in a broader sense as a semantics for quantum logic* (**J** 0302) Rep Math Logic, Krakow & Katowice 11*49-56
◇ B51 G12 ◇ REV MR 84i:03112 Zbl 472 # 03049 • ID 55312

Czelakowski, J. [1981] *Partial referential matrices for quantum logics* (**P** 3820) Curr Iss in Quantum Log;1979 Erice 131-146
◇ B51 ◇ REV MR 84j:03121 Zbl 537 # 03044 JSL 50.558 • ID 34711

Czelakowski, J. [1983] *Some theorems on structural entailment relations* (**J** 0063) Studia Logica 42*417-429
◇ B46 ◇ REV MR 85m:03021 Zbl 547 # 03014 • ID 42323

Czelakowski, J. see Vol. I, III for further entries

Czermak, J. [1974] *A logical calculus with descriptions* (**J** 0122) J Philos Logic 3*211-228
• REPR [1977] (**P** 2116) Probl in Log & Ontology;1973 Salzburg 45-62
◇ A05 B60 ◇ REV MR 55 # 2519 MR 81d:03030 Zbl 285 # 02024 • ID 29951

Czermak, J. [1974] *Matrix calculi SS1M and SS1I compared with axiomatic systems* (**J** 0047) Notre Dame J Formal Log 15*312-315
◇ B55 ◇ REV MR 50 # 4247 Zbl 254 # 02017 • ID 02667

Czermak, J. [1975] *Ein Vollstaendigkeitsbeweis fuer die aussagenlogischen Modalitaetensysteme M, S4, BR und S5* (**J** 0009) Arch Math Logik Grundlagenforsch 17*45-50
◇ B45 ◇ REV MR 52 # 13331 Zbl 313 # 02013 • ID 02668

Czermak, J. [1975] *Embeddings of classical logic in S4* (**J** 0063) Studia Logica 34*87-100
◇ B45 ◇ REV MR 52 # 2822 Zbl 345 # 02014 • REM Part I. Part II 1976 • ID 17621

CZERMAK, J. [1975] *Interpolation theorem for some modal logics* (**P** 0775) Logic Colloq;1973 Bristol 381-393
◇ B45 ◇ REV MR 52 # 7851 Zbl 311 # 02030 • ID 18113

CZERMAK, J. [1976] *Distinct modalities are not equivalent in T* (**J** 0068) Z Math Logik Grundlagen Math 22*123-125
◇ B45 ◇ REV MR 56 # 89 Zbl 328 # 02011 • ID 18496

CZERMAK, J. [1976] *Embeddings of classical logic in S4. Part II* (**J** 0063) Studia Logica 35*257-271
◇ B45 ◇ REV MR 55 # 83 Zbl 359 # 02013 • REM Part I 1975 • ID 50561

CZERMAK, J. & MORSCHER, E. & WEINGARTNER, P. (EDS.) [1977] *Problems in logic and ontology* (**X** 2596) Akad Druck-& Verlagsanstalt: Graz 310pp
◇ A05 B97 ◇ REV MR 57 # 12158 Zbl 359 # 00007 • ID 50548

CZERMAK, J. [1978] *The reducibility of a logical space with many-valued dimensions* (**P** 3693) Int Wittgenstein Symp (2);1977 Kirchberg 162-164
◇ B50 ◇ ID 39757

CZERMAK, J. & WEINGARTNER, P. [1983] *Criteria for distinguishing logically necessary propositions* (**P** 4381) Int Wittgenstein Symp (7);1982 Kirchberg 48-51
◇ B45 B60 ◇ ID 39760

CZERMAK, J. see Vol. I, V, VI for further entries

CZERWINSKI, Z. [1957] *The problem of probabilistic justification of enumerative induction (Polish) (Russian and English summaries)* (**J** 0063) Studia Logica 5*91-107
• TRANSL [1977] (**C** 3174) Twenty-five Years Log Meth Poland 65-80
◇ A05 B48 ◇ REV MR 19.237 Zbl 378 # 02003 JSL 22.408 • ID 02665

CZERWINSKI, Z. [1958] *On the relation of statistical inference to traditional induction and deduction (Polish and Russian summaries)* (**J** 0063) Studia Logica 7*243-264
◇ B48 ◇ REV MR 20 # 7323 • ID 33083

CZERWINSKI, Z. [1960] *Degree of confirmation and critical region* (**J** 0063) Studia Logica 10*119-122
◇ B48 ◇ REV MR 24 # A1161 • ID 33096

CZERWINSKI, Z. [1960] *Enumerative induction and the theory of games (Polish and Russian summaries)* (**J** 0063) Studia Logica 10*29-38
• TRANSL [1977] (**C** 3174) Twenty-five Years Log Meth Poland 81-91
◇ A05 B48 ◇ REV MR 26 # 34 Zbl 373 # 02007 • ID 02666

CZERWINSKI, Z. [1960] *On the concept of cause and Mill's method (Polish) (Russian and English summaries)* (**J** 0063) Studia Logica 9*37-62
◇ B48 ◇ REV MR 24 # A678 Zbl 121.252 • ID 33088

CZOGALA, E. [1978] see CZAJA-POSPIECH, D.

CZOGALA, E. [1980] see CZAJA-POSPIECH, D.

CZOGALA, E. & PEDRYCZ, W. [1981] *Identification and control in fuzzy systems* (**J** 3453) Syst Sci Politech Univ Wroclaw 7*285-300
◇ B52 ◇ REV Zbl 521 # 93004 • ID 37478

CZOGALA, E. & PEDRYCZ, W. [1981] *Some problems concerning the construction of algorithms of decision-making in fuzzy systems* (**J** 1741) Int J Man-Mach Stud 15*201-211
◇ B52 ◇ REV MR 83a:90006 Zbl 483 # 93006 • ID 46242

CZOGALA, E. & GOTTWALD, S. & PEDRYCZ, W. [1982] *Aspects for the evaluation of decision situations* (**C** 3778) Fuzzy Inform & Decis Processes 41-49
◇ B52 ◇ REV MR 84j:90012 Zbl 503 # 90001 • ID 36670

CZOGALA, E. & GOTTWALD, S. & PEDRYCZ, W. [1982] *Quality indices of fuzzy sets for decision making in fuzzy environments* (**P** 3845) Conf Math Service of Man (2,Proc)(Feriet);1982 Las Palmas 230-235
◇ B52 ◇ REV Zbl 511 # 90002 • ID 38549

CZOGALA, E. [1983] *An approach to probabilistic L-valued logic* (**J** 3919) BUSEFAL 15*79-87
◇ B48 ◇ REV Zbl 516 # 03030 • ID 37261

CZOGALA, E. & GOTTWALD, S. & PEDRYCZ, W. [1983] *Logical connectives of probabilistic sets* (**J** 2720) Fuzzy Sets Syst 10*299-308
◇ B52 E72 ◇ REV MR 84j:03103 Zbl 522 # 06004 • ID 34693

CZOGALA, E. [1984] *An introduction to probabilistic L-valued logic* (**J** 2720) Fuzzy Sets Syst 13*179-185
◇ B48 ◇ REV MR 85i:03074 Zbl 551 # 03014 • ID 44175

CZOGALA, E. see Vol. I, V for further entries

DAHN, B.I. [1973] *Generalized Kripke-models* (**J** 0014) Bull Acad Pol Sci, Ser Math Astron Phys 21*1073-1077
◇ B45 C90 ◇ REV MR 50 # 12663 Zbl 279 # 02034 • ID 04057

DAHN, B.I. [1974] *A note on generalized Kripke-models* (**J** 0387) Bull Sect Logic, Pol Acad Sci 3/1*8-12
◇ B45 C90 ◇ REV MR 53 # 5262 • ID 22893

DAHN, B.I. [1974] *Contributions to the model theory for non-classical logics* (**J** 0068) Z Math Logik Grundlagen Math 20*473-479
◇ B45 C90 ◇ REV MR 54 # 12491a Zbl 306 # 02052 • ID 04058

DAHN, B.I. [1974] *Kripke-style semantics for some many-valued propositional calculi* (**J** 0014) Bull Acad Pol Sci, Ser Math Astron Phys 22*99-102
◇ B50 ◇ REV MR 52 # 13320 Zbl 281 # 02026 • ID 02747

DAHN, B.I. [1974] *Meta-mathematics of some many-valued calculi* (**J** 0014) Bull Acad Pol Sci, Ser Math Astron Phys 22*747-750
◇ B50 C40 C90 ◇ REV MR 50 # 6789 Zbl 293 # 02019 • ID 02746

DAHN, B.I. [1975] *Eine Anwendung des Basistheorems in der nichtklassischen Logik* (**J** 0115) Wiss Z Humboldt-Univ Berlin, Math-Nat Reihe 24*794-796
◇ B55 C40 C90 ◇ REV MR 58 # 5048 Zbl 339 # 02015 • ID 61216

DAHN, B.I. [1976] *Neighbourhood semantics and generalized Kripke semantics* (**J** 0387) Bull Sect Logic, Pol Acad Sci 5*2-8
◇ B45 ◇ REV MR 54 # 4925 • ID 24100

DAHN, B.I. see Vol. III, IV, V, VI for further entries

DALE, A.I. [1976] *Probability logic and $\mathfrak{F}$-coherence* (**J** 0153) Phil of Sci (East Lansing) 43*254-265
◇ B48 ◇ REV MR 58 # 5023 • ID 72095

DALE, A.I. see Vol. V for further entries

Dale, A.J. [1980] *Smiley's matrices and Dunn's semantics for tautological entailment* (**J** 0079) Logique & Anal, NS 23*323-325
◇ B46 ◇ rev MR 84i:03038 • id 34520

Dale, A.J. [1982] *Material equivalence and tautological entailment* (**J** 0047) Notre Dame J Formal Log 23*435-442
◇ A05 B46 ◇ rev MR 84h:03036 Zbl 464 # 03010 • id 54600

Dale, A.J. [1985] *A completeness property of negationless intuitionist propositional logics* (**J** 0079) Logique & Anal, NS 28*79-81
◇ B55 ◇ id 49169

Dale, A.J. see Vol. I for further entries

Dalen van, D. [1958] see Bar-Hillel, Y.

Dalen van, D. [1972] *Logik und formale Theorien (Niederlaendisch)* (**J** 0290) Euclides 48*384-394
◇ B98 ◇ rev MR 54 # 12468 Zbl 264 # 02002 • id 27492

Dalen van, D. [1974] *Variants of Rescher's semantics for preference logic and some completeness theorems* (**J** 0063) Studia Logica 33*163-181
◇ B45 ◇ rev MR 50 # 6795 Zbl 291 # 02018 • id 04062

Dalen van, D. [1980] *Logic and structure* (**X** 0811) Springer: Heidelberg & New York viii + 172pp
◇ B98 C98 ◇ rev MR 81f:03001 MR 84k:03002 Zbl 434 # 03001 • rem 2nd ed. 1983; X + 207pp • id 55705

Dalen van, D. see Vol. I, III, IV, V, VI for further entries

Daley, R.P. & Manders, K.L. [1982] *The complexity of the validity problem for dynamic logic* (**J** 0194) Inform & Control 54*48-69
◇ B75 D15 ◇ rev MR 85d:03089 Zbl 521 # 03023 • id 33516

Daley, R.P. see Vol. I, IV for further entries

Dalla Chiara Scabia, M.L. [1973] *Instanti e individui nelle logiche temporali* (**J** 0945) Riv Filos 64*95-122
◇ B45 ◇ id 31063

Dalla Chiara Scabia, M.L. [1977] *Logical selfreference, set theoretical paradoxes and the measurement problem in quantum mechanics* (**J** 0122) J Philos Logic 6*331-347
◇ A05 B51 E30 G12 ◇ rev MR 58 # 32482 Zbl 374 # 02003 • id 31066

Dalla Chiara Scabia, M.L. [1977] *Quantum logic and physical modalities* (**J** 0122) J Philos Logic 6*391-404
◇ B51 ◇ rev MR 58 # 21466 Zbl 368 # 02033 • id 31065

Dalla Chiara Scabia, M.L. & Toraldo di Francia, G.G. [1979] *Formal analysis of physical theories* (**P** 3234) Probl Founds of Physics;1977 Varenna 134-201
◇ A05 B30 B51 ◇ rev MR 82m:81003 Zbl 446 # 00020 • id 56529

Dalla Chiara Scabia, M.L. [1981] *Is there a logic of empirical sciences?* (**C** 3515) Ital Studies in Phil of Sci 187-196
◇ A05 B51 G12 ◇ rev MR 82b:03010 • id 72108

Dalla Chiara Scabia, M.L. [1981] *Logical foundations of quantum mechanics* (**C** 2617) Modern Log Survey 331-351
◇ B51 ◇ rev MR 82f:03002 Zbl 464 # 03001 • id 42775

Dalla Chiara Scabia, M.L. [1981] *Some metalogical pathologies of quantum logic* (**P** 3820) Curr Iss in Quantum Log;1979 Erice 147-159
◇ B51 ◇ rev MR 84j:03122 Zbl 537.03044 • id 34712

Dalla Chiara Scabia, M.L. [1981] *Some metalogical pathologies of non distributive logics* (**P** 3092) Congr Naz Logica;1979 Montecatini Terme 295-302
◇ B60 ◇ id 48723

Dalla Chiara Scabia, M.L. [1983] *Physical implications in a Kripkian semantical approach to physical theories* (**C** 3811) Logic 20th Century 37-52
◇ B45 ◇ rev MR 85b:03022 • id 40579

Dalla Chiara Scabia, M.L. (ed.) [1985] *Present state of the problem of the foundations of mathematics* (**J** 0154) Synthese 62*123-315
◇ B98 ◇ rev MR 86d:03003 • id 45510

Dalla Chiara Scabia, M.L. see Vol. I, III, V, VI for further entries

D'Amore, B. [1971] *Analisi ed ampliamenti della logica deontica proposizionale di von Wright* (**J** 0321) Atti Accad Sci Napoli Fis Mat 12*102-118
◇ A05 B45 ◇ rev MR 46 # 5106 Zbl 286 # 02025 • id 04172

D'Amore, B. & Matteuzzi, M.L.M. [1972] *Sulla definizione di una implicazione nelle logiche polivalenti che soddisfi alla transitivita a che rispetti la struttura di reticoli di Boole (English summary)* (**J** 0012) Boll Unione Mat Ital, IV Ser 6*385-396
◇ B50 G05 G20 ◇ rev MR 48 # 193 Zbl 263 # 02013 • id 02765

D'Amore, B. see Vol. V for further entries

D'Andrea, R. [1978] see Ceccherini, P.V.

Danecki, R. [1984] *Propositional dynamic logic with strong loop predicate* (**P** 3658) Math Founds of Comput Sci (11);1984 Prague 573-581
◇ B75 ◇ rev Zbl 552 # 68036 • id 43465

Danecki, R. [1985] *Nondeterministic propositional dynamic logic with intersection is decidable* (**P** 4670) Comput Th (5);1984 Zaborow 34-53
◇ B75 ◇ id 49663

Danecki, R. see Vol. IV for further entries

Daniel, W. [1984] *The entropy of observables on quantum logic* (**J** 1546) Rep Math Phys (Warsaw) 19*325-334
◇ B51 G12 ◇ rev MR 86d:81010 • id 45189

Daniels, C.B. & Freeman, J.B. [1978] *Maximal propositions and the coherence theory of truth* (**J** 0488) Dialogue (Ottawa) 17*56-71
◇ A05 B45 ◇ rev MR 80a:03026 • id 73040

Daniels, C.B. & Freeman, J.B. [1979] *A second-order relevance logic with modality* (**J** 0063) Studia Logica 38*113-135
◇ B46 ◇ rev MR 80j:03026 Zbl 406 # 03028 • id 56111

Daniels, C.B. & Freeman, J.B. [1980] *An analysis of the subjunctive conditional* (**J** 0047) Notre Dame J Formal Log 21*639-655
◇ A05 B45 ◇ rev MR 82b:03045 Zbl 419 # 03004 • id 55922

Daniels, C.B. see Vol. I, III for further entries

Danielsson, S. [1967] *Modal logic based on probability theory* (**J** 0105) Theoria (Lund) 33*189-197
◇ B45 ◇ rev MR 38 # 30 Zbl 167.9 • id 02769

DANIL'CHENKO, A.F. [1977] *Parametric expressibility of functions of three-valued logic (Russian)* (J 0003) Algebra i Logika 16*397-416,493
• TRANSL [1977] (J 0069) Algeb and Log 16*266-280
◇ B50 ◇ REV MR 80c:03024 Zbl 423 # 03020 • ID 29224

DANIL'CHENKO, A.F. [1978] *Parametrically closed classes of functions of three-valued logic (Russian)* (J 0967) Izv Akad Nauk Mold SSR, Ser Fiz-Tekh Mat 1978/2*13-20,93
◇ B50 G25 ◇ REV MR 80c:03025 Zbl 387 # 03031
• ID 52247

DANIL'CHENKO, A.F. [1981] *On parametrical expressibility of the functions of k-valued logic* (P 2552) Conf Finite Algeb & Multi-Val Log;1979 Szeged 147-159
◇ B50 ◇ REV MR 83k:03031 Zbl 491 # 03007 • ID 36169

DANKO, W. [1978] *Algorithmic properties of programs with tables* (J 2095) Fund Inform, Ann Soc Math Pol, Ser 4 1*379-398
◇ B75 ◇ REV MR 58 # 16199 Zbl 386 # 68042 • ID 72129

DANKO, W. [1979] *Definability in algorithmic logic* (J 2095) Fund Inform, Ann Soc Math Pol, Ser 4 2*277-287
◇ B75 D05 D20 ◇ REV MR 82f:68031 Zbl 453 # 03026
• ID 54156

DANKO, W. [1983] *Interpretability of algorithmic theories* (J 2095) Fund Inform, Ann Soc Math Pol, Ser 4 6*217-233
◇ B25 B75 C35 F25 ◇ REV MR 84h:68022
Zbl 559 # 68048 • ID 39522

DANKO, W. [1984] *Algorithmic logic with stacks and its model-theoretic properties* (J 2095) Fund Inform, Ann Soc Math Pol, Ser 4 7*391-428
◇ B75 ◇ ID 47685

DANKO, W. see Vol. III, IV for further entries

DANTZIG VAN, D. [1953] *Carnap's foundation of probability theory* (J 0154) Synthese 8*459-470
◇ B48 ◇ REV MR 17.227 • ID 02777

DANTZIG VAN, D. see Vol. I, V, VI for further entries

DARDZHANIYA, G.K. [1977] *Intuitionistic system without contraction* (J 0387) Bull Sect Logic, Pol Acad Sci 6*2-8
◇ B55 F50 ◇ REV MR 55 # 12476 Zbl 416 # 03030
• ID 53196

DARDZHANIYA, G.K. [1979] *On the complexity of countermodels for intuitionistic propositional calculus (Russian) (English summary)* (J 0233) Soobshch Akad Nauk Gruz SSR 95*17-20
◇ B55 F50 ◇ REV MR 81d:03012 Zbl 417 # 03032
• ID 53268

DARDZHANIYA, G.K. & PKHAKADZE, N.M. [1982] *On some universal modal theory (Russian) (English and Georgian summaries)* (J 0233) Soobshch Akad Nauk Gruz SSR 108*269-272
◇ B45 ◇ REV MR 85e:03035 Zbl 507 # 03010 • ID 37235

DARDZHANIYA, G.K. [1984] *A sequential variant of Grzegorczyk's modal system (Russian) (Grus. and English summaries)* (J 0233) Soobshch Akad Nauk Gruz SSR 116*29-32
◇ B45 ◇ ID 46594

DARDZHANIYA, G.K. see Vol. I, VI for further entries

DARSOW, W. & KITTEL, P.M. [1980] *On intermediate consequence relations* (J 0068) Z Math Logik Grundlagen Math 26*33-34
◇ B55 ◇ REV MR 81m:03036 Zbl 467 # 03018 • ID 55016

DAS, S. [1974] see CHUANG, H.Y.H.

DASSOW, J. [1976] *Kleene-Mengen und trennende Mengen* (J 0114) Math Nachr 74*89-97
◇ B50 D05 ◇ REV MR 55 # 2400 Zbl 344 # 94027
• ID 61236

DASSOW, J. see Vol. I, IV for further entries

DAUBECHIES, I. [1978] see AERTS, D.

DAUBECHIES, I. [1979] see AERTS, D.

DAVAL, R. & GUILBAUD, G.T. [1945] *Le raisonnement mathematique* (X 0840) Pr Univ France: Paris 152pp
◇ A05 B98 ◇ REV MR 8.4 • ID 25653

DAVIDSON, B. & ELLIS, B. [1976] *Logic and strict coherence* (J 0302) Rep Math Logic, Krakow & Katowice 6*29-40
◇ B45 ◇ REV MR 57 # 2883 Zbl 371 # 02011 • ID 21912

DAVIDSON, B. & JACKSON, F.C. & PARGETTER, R. [1977] *Modal trees for T and S5* (J 0047) Notre Dame J Formal Log 18*602-606
◇ B25 B45 ◇ REV MR 58 # 16145 Zbl 299 # 02022
• ID 23697

DAVIDSON, B. [1982] *Modal trees for modal predicate logics* (J 0079) Logique & Anal, NS 25*47-56
◇ B45 ◇ REV MR 84j:03032 Zbl 485 # 03005 • ID 34624

DAVIDSON, D. [1963] *The method of extension and intension* (C 0673) Phil of Carnap 311-349
◇ A05 B45 ◇ ID 04286

DAVIDSON, D. & HARMAN, G. (EDS.) [1972] *Semantics of natural language* (X 0835) Reidel: Dordrecht x+769pp
◇ B97 ◇ ID 25378

DAVIDSON, D. see Vol. I for further entries

DAVIES, M.K. [1978] *Weak necessity and truth theories* (J 0122) J Philos Logic 7*415-439
◇ A05 B45 ◇ REV MR 80a:03025 Zbl 413 # 03003
• ID 52997

DAVIES, M.K. & HUMBERSTONE, L.L. [1980] *Two notions of necessity* (J 0095) Philos Stud 38*1-30
◇ A05 B45 ◇ REV MR 81k:03018 • ID 72149

DAVIES, M.K. see Vol. I for further entries

DAVIES, R.O. [1966] *Two theorems on essential variables* (J 0039) J London Math Soc 41*333-335
◇ B50 ◇ REV MR 33 # 7246 Zbl 151.6 • ID 02802

DAVIES, R.O. [1979] *On n-valued Sheffer functions* (J 0068) Z Math Logik Grundlagen Math 25*293-298
◇ B50 G20 ◇ REV MR 81a:03021 Zbl 427 # 03055
JSL 50.1073 • ID 53742

DAVIES, R.O. & ROSENBERG, I.G. [1985] *Precomplete classes of operations on an uncountable set* (S 0019) Colloq Math (Warsaw) 50*1-12
◇ B50 ◇ ID 49491

DAVIES, R.O. see Vol. IV, V for further entries

DAVIO, M. & DESCHAMPS, J.-P. [1978] *Addition in signed digit number systems* (P 2014) Int Symp Multi-Val Log (8);1978 Rosemont 104-113
◇ B50 B70 ◇ REV MR 81b:94057 • ID 35919

DAVIO, M. see Vol. I for further entries

DAVIS, CHANDLER [1954] *Modal operations, equivalence relations, and projective algebras* (J 0100) Amer J Math 76*747-762
◇ B45 ◇ REV MR 16.324 Zbl 57.23 JSL 24.253 • ID 02814

DAVIS, CHARLES C. [1976] *A note on the axiom of choice in Lesniewski's ontology* (J 0047) Notre Dame J Formal Log 17*35-43
◇ B60 E25 E70 ◇ REV MR 55 # 5397 Zbl 299 # 02081 • ID 18114

DAVIS, CHARLES C. & MCKIM, V.R. [1976] *Temporal modalities and the future* (J 0047) Notre Dame J Formal Log 17*233-238
◇ A05 B45 ◇ REV MR 53 # 12875 Zbl 299 # 02024 • ID 18290

DAVIS, CHARLES C. see Vol. I, IV, V for further entries

DAVIS, H.W. & WINSLOW, L.E. [1982] *Computational power in query languages* (J 1428) SIAM J Comp 11*547-554
◇ B52 ◇ REV MR 83h:68153 Zbl 486 # 68116 • ID 38396

DAVIS, MARTIN D. [1959] *Lecture notes on mathematical logic* (X 0888) New York Univ Inst Math Sci: New York 91pp
◇ B98 ◇ REV JSL 35.167 • ID 17204

DAVIS, MARTIN D. [1974] *Computability* (X 1214) New York Univ Cour Inst Math: New York 248pp
◇ B98 D98 ◇ REV MR 50 # 77 Zbl 281 # 02038 • ID 21268

DAVIS, MARTIN D. see Vol. I, IV, VI for further entries

DAWSON, E.E. [1959] *A model for deontic logic* (J 0103) Analysis (Oxford) 19*73-78
◇ B45 ◇ REV JSL 31.666 • ID 02848

DELIYANNIS, P.C. [1973] *Vector space models of abstract quantum logics* (J 0209) J Math Phys 14*249-253
◇ B51 ◇ REV MR 48 # 3401 Zbl 259 # 02025 • ID 30348

DELIYANNIS, P.C. [1976] *Conditioning of states* (J 0209) J Math Phys 17*653-659
◇ B51 ◇ REV MR 54 # 1909 • REM Part I. Part II 1978 • ID 45443

DELIYANNIS, P.C. [1976] *Superposition of states and the structure of quantum logics* (J 0209) J Math Phys 17*248-254
◇ B51 ◇ REV MR 52 # 16317 Zbl 339 # 02026 • ID 61284

DELIYANNIS, P.C. [1984] *Geometrical models for quantum logics with conditioning* (J 0209) J Math Phys 25*2939-2946
◇ B51 ◇ REV MR 86a:81005 • ID 46840

DELIYANNIS, P.C. see Vol. III, V for further entries

DELONG, H. [1970] *A profile of mathematical logic* (X 0832) Addison-Wesley: Reading xiii + 304pp
◇ A05 A10 B98 ◇ REV MR 41 # 3230 Zbl 248 # 02001 JSL 40.101 • ID 02907

DELORME, M. [1982] see BOTTA, O.

DELORME, M. see Vol. V for further entries

DEMETROVICS, J. [1972] *On cardinal numbers of sets of precomplete classes in limit logics (Russian) (English summary)* (J 0380) Acta Cybern (Szeged) 1*233-239
◇ B50 ◇ REV MR 51 # 2865 Zbl 251 # 02020 • ID 17375

DEMETROVICS, J. [1974] *On the properties of the minimal limit logic Q (Russian)* (J 0411) Studia Sci Math Hung 9*233-245
◇ B50 ◇ REV MR 58 # 21435 Zbl 304 # 02008 • ID 61289

DEMETROVICS, J. [1974] *The number of pairewise nonisomorphic limit logics (Russian)* (J 0071) Met Diskr Analiz (Novosibirsk) 24*21-29,95-96
◇ B50 ◇ REV MR 51 # 2864 Zbl 295 # 02012 • ID 17327

DEMETROVICS, J. [1975] *Certain homomorphisms and relations for limit logics (Russian)* (J 0052) Probl Kibern 30*5-42
◇ B50 ◇ REV MR 55 # 10233 Zbl 414 # 03020 • ID 53066

DEMETROVICS, J. [1975] *On homomorphism of limit logics (Hungarian) (Russian summary)* (J 1458) Alkalmaz Mat Lapok 1*125-138
◇ B50 ◇ REV MR 53 # 12869 Zbl 365 # 02007 • ID 23129

DEMETROVICS, J. [1975] *On the principal diagonal of Sheffer functions* (J 2774) Koezlem MTA Szam & Autom: Kutat Intez 14*39-46
◇ B50 ◇ REV MR 51 # 12471 • ID 72279

DEMETROVICS, J. [1976] *The comparison of limit logics when finitely valued logics are modelled in them (Russian) (English summary)* (J 0380) Acta Cybern (Szeged) 2*307-312
◇ B50 ◇ REV MR 54 # 58 Zbl 349 # 02010 • ID 23941

DEMETROVICS, J. [1976] see BAGYINSZKI, J.

DEMETROVICS, J. [1976] *The M maximal limit-logic (Hungarian) (English summary)* (J 1458) Alkalmaz Mat Lapok 2*57-66
◇ B50 ◇ REV MR 55 # 10234 Zbl 365 # 02008 • ID 51009

DEMETROVICS, J. [1978] *On the main diagonal of Sheffer functions* (J 0193) Discr Math 21*1-5
◇ B50 ◇ REV MR 80e:03018 Zbl 371 # 02010 • ID 51371

DEMETROVICS, J. & HANNAK, L. [1979] *On the cardinality of self-dual closed classes in k-valued logics* (J 2774) Koezlem MTA Szam & Autom: Kutat Intez 23*7-17
◇ B50 ◇ REV MR 80j:03037 • ID 72274

DEMETROVICS, J. & HANNAK, L. [1979] *The cardinality of closed sets in pre-complete classes in k-valued logics* (J 0380) Acta Cybern (Szeged) 4*273-277
◇ B50 G20 ◇ REV MR 80m:03051 Zbl 427 # 03015 • ID 53702

DEMETROVICS, J. & HANNAK, L. & MARCHENKOV, S.S. [1980] *On closed classes of self-dual functions in P_3 (Russian)* (J 0071) Met Diskr Analiz (Novosibirsk) 34*38-73,101
◇ B50 G10 G25 ◇ REV MR 82j:03029 Zbl 469 # 03046 • ID 55173

DEMETROVICS, J. & HANNAK, L. & MARCHENKOV, S.S. [1980] *Some remarks on the structure of P_3* (J 2128) C R Math Acad Sci, Soc Roy Canada 2*215-219
◇ B50 G10 G25 ◇ REV MR 81m:03028 Zbl 469 # 03047 • ID 55174

DEMETROVICS, J. [1980] see BAGYINSZKI, J.

DEMETROVICS, J. & HANNAK, L. & MARCHENKOV, S.S. [1981] *On closed classes of selfdual functions in P_3* (P 2552) Conf Finite Algeb & Multi-Val Log;1979 Szeged 183-189
◇ B50 ◇ REV MR 83f:03068 Zbl 481 # 03013 • ID 36828

DEMETROVICS, J. & HANNAK, L. [1982] *How to construct a large set of nonequivalent functionally complete algebras (Russian and Hungarian summaries)* (J 2774) Koezlem MTA Szam & Autom: Kutat Intez 25*49-52
◇ B50 C05 ◇ REV MR 84b:08005a • ID 38910

DEMETROVICS, J. & HANNAK, L. & RONYAI, L. [1982] *On functional completeness of prime-element algebras (Russian and Hungarian summaries)* (J 2774) Koezlem MTA Szam & Autom: Kutat Intez 25*53-59
◇ B50 C05 ◇ REV MR 84b:08005b • ID 38912

DEMETROVICS, J. [1982] see BAGYINSZKI, J.

DEMETROVICS, J. & HANNAK, L. [1983] *Cardinality of sets of closed classes contained in precomplete classes in P_k (Russian) (English summary)* (J 2774) Koezlem MTA Szam & Autom: Kutat Intez 147*23-32
◇ B50 ◇ ID 45057

DEMETROVICS, J. & MAL'TSEV, I.A. [1984] *Essentially minimal TC-clones on a three-element set (Russian) (English and Hungarian summaries)* (J 2774) Koezlem MTA Szam & Autom: Kutat Intez 31*115-151
◇ B50 C05 ◇ REV Zbl 566#03012 • ID 47619

DEMETROVICS, J. see Vol. I for further entries

DEMOPOULOS, W. [1979] *Boolean representations of physical magnitudes and locality* (J 0154) Synthese 42*101-119
◇ B51 G12 ◇ REV MR 80j:81005 Zbl 435#03039 • ID 55799

DENECKE, H.-M. [1977] *Quantum logic of quantifiers* (J 0122) J Philos Logic 6*405-413
◇ B51 ◇ REV Zbl 365#02018 • ID 51019

DENECKE, K. [1982] *Preprimal algebras* (X 0911) Akademie Verlag: Berlin 162pp
◇ B50 C05 C13 ◇ REV MR 85a:08003 Zbl 506#08003 • ID 37824

DENEV, I.D. & GYUDZHENOV, I.D. [1983] *Separable subsets of arguments of functions from P_k (Russian) (English summary)* (J 2774) Koezlem MTA Szam & Autom: Kutat Intez 147*47-50
◇ B50 ◇ ID 45058

DENG, JULONG & PENG, GUOZHONG [1982] *The digital simulation and computer-aided design of a fuzzy integrated system (Chinese) (English summary)* (J 3732) Mohu Shuxue 2/4*61-72
◇ B52 ◇ ID 46098

DENISOVA, R.A. [1984] *A class of functions of k-valued logic (Russian)* (J 0199) Zh Vychisl Mat i Mat Fiz 24*167-169 • TRANSL [1984] (J 1049) USSR Comput Math & Math Phys 24/1*108-110
◇ B50 ◇ REV MR 85k:03017 Zbl 539#03007 • ID 43739

DENISOVA, R.A. [1984] *A method of synthesis of irredundant representative collections for k-valued tables (Russian)* (X 2265) Akad Nauk Vychis Tsentr: Moskva 30pp
◇ B50 ◇ REV MR 85i:03066 • ID 44168

DENNING, P.J. & DENNIS, J.B. & QUALITZ, J.E. [1978] *Machines, languages, and computation* (X 0819) Prentice Hall: Englewood Cliffs xxii+601pp
◇ B98 D05 D10 D98 ◇ REV Zbl 492#68003 JSL 45.630 • ID 37736

DENNIS, J.B. [1978] see DENNING, P.J.

DERUS, K.H. & HANSEN, J.C. [1979] *Propositions with multiple dispositions and multiple truth values* (J 1456) SIGACT News 11/2*30-35
◇ B50 ◇ REV Zbl 412#03008 • ID 52939

DESCHAMPS, J.-P. & THAYSE, A. [1975] *Representation of discrete functions* (P 1805) Int Symp Multi-Val Log (5,Proc);1975 Bloomington 99-111
◇ B50 ◇ REV MR 58#9826 • ID 35816

DESCHAMPS, J.-P. [1978] see DAVIO, M.

DESCHAMPS, J.-P. see Vol. I, IV for further entries

DESCLES, J.-P. [1976] *Operations sur des operateurs types (English summary)* (J 2313) C R Acad Sci, Paris, Ser A-B 283*A987-A990
◇ B60 ◇ REV MR 55#1781 Zbl 358#02037 • ID 50491

DESCLES, J.-P. see Vol. V for further entries

DESTOUCHES, J.-L. [1949] *Intervention d'une logique de modalite dans une theorie physique* (J 0154) Synthese 7*411-417
◇ A05 B45 ◇ REV MR 12.73 JSL 15.232 • ID 02958

DESTOUCHES, J.-L. [1954] *Allgemeine Theorie der Voraussagen* (J 0009) Arch Math Logik Grundlagenforsch 2*10-14
◇ A05 B48 ◇ REV MR 17.702 Zbl 55.172 • ID 33307

DESTOUCHES, J.-L. [1956] *Ueber den Aussagenkalkuel der Experimentalaussagen* (J 0009) Arch Math Logik Grundlagenforsch 2*104-105
◇ B51 G12 ◇ REV MR 18.95 Zbl 70.245 • ID 02969

DESTOUCHES, J.-L. see Vol. I, IV, VI for further entries

DESTOUCHES-FEVRIER, P. [1945] *Logique adaptee aux theories quantiques* (J 0109) C R Acad Sci, Paris 221*287-288
◇ B51 G12 ◇ REV MR 7.356 Zbl 61.471 • ID 02961

DESTOUCHES-FEVRIER, P. [1948] *Logique de l'intuitionisme sans negation et logique de l'intuitionisme positif* (J 0109) C R Acad Sci, Paris 226*38-39
◇ B55 F50 ◇ REV MR 10.94 Zbl 35.149 • ID 02965

DESTOUCHES-FEVRIER, P. [1949] *Logique et theories physiques* (J 0154) Synthese 7*400-410
◇ A05 B51 ◇ REV MR 12.73 • ID 02967

DESTOUCHES-FEVRIER, P. [1954] *La logique des propositions experimentales* (P 0646) Appl Sci de Log Math;1952 Paris 115-118
◇ B51 G12 ◇ REV MR 16.437 Zbl 58.245 • ID 27970

DESTOUCHES-FEVRIER, P. also published under the name FEVRIER, P.

DESTOUCHES-FEVRIER, P. see Vol. I, VI for further entries

DETLOVS, V.K. [1970] *Elements of mathematical logic and set theory (Latvian)* (X 0895) Latv Valsts (Gos) Univ : Riga 292pp
◇ B98 E98 ◇ REV MR 47#8230 • ID 17027

DETLOVS, V.K. [1974] *Mathematical logic (Latvian)* (X 2230) Zinatne: Riga 279pp
◇ B98 ◇ REV MR 55#2465 • ID 72316

DETLOVS, V.K. see Vol. IV for further entries

DEUTSCH, H. [1977] *Relevant analytic entailment* (J 1893) Relevance Logic Newslett 2*26-45
◇ B46 ◇ REV Zbl 363#02023 • ID 50850

DEUTSCH, H. [1979] *The completeness of S* (J 0063) Studia Logica 38*137-147
◇ B45 ◇ REV MR 80m:03035 Zbl 406#03029 • ID 56112

DEUTSCH, H. [1984] *Paraconsistent analytic implication* (J 0122) J Philos Logic 13*1-12
◇ B53 ◇ REV MR 85i:03046 Zbl 553#03010 • ID 42641

DEUTSCH, H. [1985] *A note on the decidability of a strong relevant logic* (**J** 0063) Studia Logica 44*159-164
◇ B25 B46 ◇ ID 47508

DEUTSCH, H. [1985] *Relevance and conformity* (**J** 0047) Notre Dame J Formal Log 26*455-462
◇ B46 ◇ ID 47539

DEUTSCH, H. [1985] *Relevance and first-degree entailments* (**J** 0079) Logique & Anal, NS 28*3-20
◇ B46 ◇ ID 49170

DEVIDE, V. [1964] *Mathematical logic. Part I (The classical logic of propositions) (Croatian) (English summary)* (**X** 4380) Posebna Izd Mat Inst: Belgrade 228pp
◇ B05 B98 ◇ REV MR 29#5711 Zbl 115.4 JSL 35.326 • ID 43409

DEVIDE, V. see Vol. I, IV, V for further entries

DIAZ, M.R. [1980] *Deductive completeness and conditionalization in systems of weak implication* (**J** 0047) Notre Dame J Formal Log 21*119-130
◇ B45 ◇ REV MR 81d:03017 Zbl 368#02030 • ID 53197

DIAZ, M.R. [1981] *Topics in the logic of relevance* (**X** 3111) Philosophia: Muenchen 144pp
◇ A05 B46 ◇ REV Zbl 471#03004 • ID 55199

DICKOFF, J.W. [1961] see ALLEN, L.E.

DICKOFF, J.W. & JAMES, P.A. [1965] *Symbolic logic and language: a programmed text* (**X** 0822) McGraw-Hill: New York 390pp
◇ B98 ◇ ID 22481

DICKOFF, J.W. see Vol. I for further entries

DIEGO, A. [1966] *Sur les algebres de Hilbert* (**X** 0834) Gauthier-Villars: Paris viii+55pp
◇ B55 G25 ◇ REV MR 33#7236 Zbl 144.1 JSL 35.139 • ID 03005

DIEGO, A. see Vol. VI for further entries

DIENES, P. [1949] *On an implication function in many-valued systems of logic* (**J** 0036) J Symb Logic 14*95-97
◇ B50 ◇ REV MR 11.1 Zbl 37.3 JSL 15.69 • ID 03010

DIENES, P. [1949] *On ternary logic* (**J** 0036) J Symb Logic 14*85-94
◇ B50 ◇ REV MR 11.1 JSL 15.225 • ID 03011

DIENES, P. see Vol. I, V, VI for further entries

DIETTERICH, T.G. & MICHALSKI, R.S. [1981] *Inductive learning of structural descriptions: evaluation criteria and comparative review of selected methods* (**J** 0503) Artif Intell 16*257-294
◇ B48 ◇ REV MR 82i:68054 • ID 81096

DILLER, J. [1978] *Klassische Praedikatenlogik* (**X** 3176) Fernuniv Hagen: Hagen 573pp
◇ B98 ◇ REV MR 80c:03003 • ID 72386

DILLER, J. see Vol. III, IV, VI for further entries

DIMITROV, V. [1982] *Creative decision-making through fuzzy catastrophes* (**C** 3786) Approx Reason in Decis Anal 391-399
◇ B52 ◇ REV Zbl 503#90002 • ID 36671

DIMITROV, V. see Vol. IV, V for further entries

DINI, G. [1982] *A modified version of an arithmetic model for legal informatics of Sanchez-Mazas* (**P** 4041) Log, Inform, Law;1981 Firenze 2*205-214
◇ B45 ◇ REV MR 85j:03021c • ID 45215

DINKINES, F. [1964] *Elementary concepts of modern mathematics. Part 1. Elementary theory of sets. Part 2. Introduction to mathematical logic* (**X** 1228) Appleton-Century-Crofts: New York x+457pp
◇ B98 E98 ◇ REV Zbl 128.10 JSL 32.422 • ID 22482

DISHKANT, G.P. [1972] *Semantics of the minimal logic of quantum mechanics (Polish and Russian summaries)* (**J** 0063) Studia Logica 30*23-32
◇ B51 ◇ REV MR 48#10329 Zbl 268#02018 • ID 03052

DISHKANT, G.P. [1974] *The first order predicate calculus based on the logic of quantum mechanics* (**J** 0302) Rep Math Logic, Krakow & Katowice 3*9-17
◇ B51 ◇ REV MR 53#12242 Zbl 356#02027 JSL 48.206 • ID 23113

DISHKANT, G.P. [1974] *Weak decidability of some modal extension of trigonometry* (**J** 0027) Fund Math 81*255-260
◇ B45 ◇ REV MR 50#64 Zbl 285#02021 • ID 03053

DISHKANT, G.P. [1976] *Logic of quantum mechanics* (**P** 1804) Form Meth in Methodol of Emp Sci;1974 Warsaw 368-370
◇ A05 B51 ◇ REV MR 58#32484 Zbl 367#02010 • ID 51170

DISHKANT, G.P. [1977] *Imbedding of the quantum logic in the modal system of Brouwer* (**J** 0036) J Symb Logic 42*321-328
◇ B51 ◇ REV MR 58#16146 Zbl 381#03016 • ID 51884

DISHKANT, G.P. [1977] *The connective "becoming" and the paradox of electron diffraction* (**J** 0302) Rep Math Logic, Krakow & Katowice 9*15-21
◇ B51 ◇ REV MR 80c:03061 Zbl 393#03045 • ID 52465

DISHKANT, G.P. [1978] *An extension of the Lukasiewicz logic to the modal logic of quantum mechanics* (**J** 0063) Studia Logica 37*149-155
◇ B51 ◇ REV MR 58#5049 Zbl 377#02026 • ID 51743

DISHKANT, G.P. [1979] *Formalization of a soft conjunction in the theory of approximate decision-making (Russian)* (**S** 2959) Mat Met Optim & Struct Sist 1979*98-105,184
◇ B52 ◇ REV MR 82a:03022 • ID 72395

DISHKANT, G.P. [1979] *The Lotfi Zadeh logic of taking approximate decisions as a game (Russian)* (**C** 3484) Diff, Beskoal, Koop & Stat Igry 43-49
◇ B52 ◇ REV MR 83e:90191 Zbl 463#03010 • ID 54550

DISHKANT, G.P. [1980] *Set theory as modal logic* (**J** 0063) Studia Logica 39*335-345
◇ B45 ◇ REV MR 83a:03054 Zbl 463#03007 • ID 54547

DISHKANT, G.P. [1980] *Three propositional calculi of probability* (**J** 0063) Studia Logica 39*49-61
◇ B48 ◇ REV MR 82d:03031 Zbl 443#03015 • ID 56421

DISHKANT, G.P. [1981] *About membership functions estimation* (**J** 2720) Fuzzy Sets Syst 5*141-147
◇ B52 E72 ◇ REV MR 83a:03056 Zbl 469#03038 • ID 55165

DISHKANT, G.P. see Vol. IV for further entries

DOCKX, S. [1965] see BERNAYS, P.

Author Index

DODGE, C.W. [1969] *Sets,logic and numbers* (**X** 1337) Prindle Weber Schmidt: Boston xiii+346pp
◇ B98 E98 ◇ ID 22633

DODGSON, C.L. [1887] *The game of logic* (**X** 1253) Crowell Collier & Macmillan: New York 96pp
◇ B98 ◇ ID 22449

DODGSON, C.L. [1895] *Symbolic logic. Part I. Elementary* (**X** 1253) Crowell Collier & Macmillan: New York xxxi+188pp
• LAST ED [1977] (**X** 3555) Potter: New York xxv+496pp
• REPR [1955] (**X** 0851) Berkeley: New York xxxi+203pp
◇ B20 B98 ◇ REV MR 19.1 MR 80j:03002 JSL 22.309 JSL 25.264 JSL 29.135 • REM Last ed. contains part II: advanced, which was never published previously • ID 02237

DOEMOELKI, B. & GERGELY, T. (EDS.) [1981] *Mathematical logic in computer science (Colloquium held in Salgotarjan, Hungary, September 10-15, 1978)* (**S** 3312) Coll Math Soc Janos Bolyai 26*758pp
◇ B97 ◇ REV MR 83g:68007 Zbl 476#00024 • ID 55510

DOERBAND, W. [1972] *Ein Vorschlag zur systematischen Bezeichnung aller n-wertigen und k-stelligen booleschen Funktionen und ein Algorithmus, nach dem auf dieser Grundlage die Funktionswerte dieser Funktionen berechnet werden koennen* (**J** 0129) Elektr Informationsverarbeitung & Kybern 8*519-522
◇ B50 ◇ REV Zbl 252#94024 • ID 61378

DOERBAND, W. [1973] *Ein Algorithmus zur Berechnung aller Funktionswerte aller n-wertigen und k-stelligen booleschen Funktionen* (**J** 1670) Mitt Math Ges DDR 1973/2-3*24
◇ B50 ◇ REV Zbl 264#94025 • ID 61377

DOI, T. [1982] see AIHARA, R.

DOMBI, J. [1982] *Basic concepts for a theory of evaluation: the aggregative operator* (**J** 3154) Eur J Oper Res 10*282-293
◇ B50 ◇ REV MR 83i:90003 Zbl 488#90003 • ID 46681

DOMBI, J. & VAS, Z. [1983] *Basic theoretical treatment of fuzzy connectives* (**J** 0380) Acta Cybern (Szeged) 6*191-201
◇ B52 ◇ REV MR 85a:03034 Zbl 517#94026 • ID 34785

DOMBI, J. see Vol. V for further entries

DOMINGO, X. & ESTEVA, F. [1980] *On weak negation functions for the theory of fuzzy sets (Catalan)* (**P** 3512) Jorn Mat Luso-Espanol (7);1980 St Feliu de Guixois 1*73-77
◇ B52 E72 ◇ ID 44665

DOMINGO, X. & TRILLAS, E. & VALVERDE, L. [1981] *Pushing Lukasiewicz-Tarski implication a little further* (**P** 3705) Int Symp Multi-Val Log (11);1981 Oklahoma City & Norman 232-234
◇ B50 ◇ REV MR 83m:94033 Zbl 564#03022 • ID 45935

DOMINGO, X. & ESTEVA, F. & TRILLAS, E. [1981] *Weak and strong negation functions for fuzzy set theory* (**P** 3705) Int Symp Multi-Val Log (11);1981 Oklahoma City & Norman 23-26
◇ B52 E72 ◇ REV Zbl 548#03036 • ID 43204

DOMINGO, X. see Vol. V for further entries

DOMOTOR, Z. [1976] *The probability structure of quantum-mechanical systems* (**C** 2953) Log & Probab in Quant Mech 147-177
◇ B51 G12 ◇ REV Zbl 362#02056 • ID 50778

DOMOTOR, Z. [1980] *Probability kinematics and representation of belief change* (**J** 0153) Phil of Sci (East Lansing) 47*384-403
◇ B48 ◇ REV MR 82c:03034 • ID 72414

DOMOTOR, Z. [1981] *Higher order probabilities* (**J** 0095) Philos Stud 40*31-46
◇ B48 ◇ REV MR 83e:03034 • ID 35219

DOMOTOR, Z. [1981] *Probabilistic and causal dependence structures* (**J** 0472) Theory Decis 13*275-292
◇ B48 ◇ REV MR 82m:03025 Zbl 481#03005 • ID 72413

DOMOTOR, Z. [1985] *Probability kinematics, conditionals,and entropy principles* (**J** 0154) Synthese 63*75-114
◇ B48 ◇ ID 47242

DOMOTOR, Z. see Vol. I for further entries

DOMPERE, K.K. [1982] *The theory of fuzzy decisions* (**C** 3786) Approx Reason in Decis Anal 365-379
◇ B52 ◇ REV MR 84d:90003 Zbl 506#90002 • ID 37831

DONCHENKO, V.V. [1963] *Some questions connected with the decision problem for Ackermann's system of strict implication (Russian)* (**C** 1654) Probl Logiki 18-24
◇ B45 ◇ REV Zbl 125.6 JSL 33.484 • ID 43184

DONG, PING [1983] *The optimization of GMC over GF(p)* (**P** 4054) Int Symp Multi-Val Log (13);1983 Kyoto 342-347
◇ B50 ◇ REV Zbl 562#94012 • ID 47447

DONNADIEU, M.-R. & RAMBAUD, C. [1978] *Theories conditionnelles (English summary)* (**J** 2313) C R Acad Sci, Paris, Ser A-B 286*A849-A852
◇ B60 ◇ REV MR 58#16122 Zbl 379#02003 • ID 51836

DONNADIEU, M.-R. see Vol. III, V for further entries

DOPP, J. [1962] *Logiques construites par une methode de deduction naturelle* (**X** 0834) Gauthier-Villars: Paris 191pp
◇ B98 F05 F07 F50 F98 ◇ REV MR 26#1243 Zbl 143.8 JSL 34.502 • ID 03085

DOPP, J. [1967] *Notions de logique formelle* (**X** 1313) Nauwelaerts: Louvain 304pp
• LAST ED [1980] (**X** 3734) Cabay: Louvain-la-Neuve 304pp
◇ B98 ◇ REV MR 15.384 MR 83h:03001 Zbl 524#03001 • ID 22397

DOPP, J. see Vol. VI for further entries

DORDEVIC, L.N. [1975] *Structural syntheses of sequential machines in three-valued logic* (**S** 1003) Publ Elektroteh, Ser Mat Fiz, Beograde 498-541*137-140
◇ B50 ◇ REV MR 55#14421 Zbl 331#94015 • ID 61388

DORLING, J. [1981] *How to rewrite a stochastic dynamical theory so as to generate a measurement paradox* (**P** 3820) Curr Iss in Quantum Log;1979 Erice 115-118
◇ B51 G12 ◇ REV MR 84i:03005 Zbl 537#03044 • ID 44875

DORNINGER, D. & LAENGER, H. & MACZYNSKI, M.J. [1983] *Zur Darstellung von Observablen auf σ-stetigen Quantenlogiken* (**J** 0238) Sitzb Oesterr Akad Wiss, Math-Nat Kl, Abt 2 192*169-176
◇ B51 G12 ◇ REV Zbl 557#03041 • ID 45417

DOROSLOVACKI, R. & TOSIC, R. [1982] *A combinatorial identity and its applications* (**J** 2887) Zbor Radova, NS 12*319-326
◇ B50 ◇ REV MR 85j:05004 Zbl 532#05004 • ID 36817

DORROUGH, D. [1970] *A logical calculus of analogy involving functions of order 2* (J 0047) Notre Dame J Formal Log 11*321-336
◇ A05 B60 ◇ REV MR 43 # 3092 Zbl 184.280 • ID 03087

DOSEN, K. [1978] *A note on the law of identity and the converse parry property* (J 0047) Notre Dame J Formal Log 19*174-176
◇ B45 ◇ REV MR 58 # 167 Zbl 368 # 02027 • ID 27104

DOSEN, K. [1981] *A reduction of classical propositional logic to the conjunction-negation fragment of an intuitionistic relevant logic* (J 0122) J Philos Logic 10*399-408
◇ B46 B55 F50 ◇ REV MR 83b:03017 Zbl 473 # 03008 • ID 55339

DOSEN, K. [1981] *Minimal modal systems in which Heyting and classical logic can be embedded* (J 0400) Publ Inst Math, NS (Belgrade) 30(44)*41-52
◇ B45 F50 ◇ REV MR 84c:03034 Zbl 494 # 03008 • ID 35691

DOSEN, K. [1983] see BOZIC, M.

DOSEN, K. [1984] *Intuitionistic double negation as a necessity operator* (J 0482) Publ Math Univ Belgrade 35(49)*15-20
◇ B45 F50 ◇ REV Zbl 555 # 03012 • ID 44540

DOSEN, K. [1984] see BOZIC, M.

DOSEN, K. [1984] *Negative modal operators in intuitionistic logic* (J 0482) Publ Math Univ Belgrade 35(49)*3-14
◇ B55 F50 ◇ REV Zbl 555 # 03011 • ID 44539

DOSEN, K. [1985] *An intuitionistic Sheffer function* (J 0047) Notre Dame J Formal Log 26*479-482
◇ B55 F50 ◇ ID 47541

DOSEN, K. [1985] *Models for stronger normal intuitionistic modal logics* (J 0063) Studia Logica 44*39-70
◇ B45 ◇ ID 47522

DOSEN, K. [1985] *Sequent-systems for modal logic* (J 0036) J Symb Logic 50*149-168
◇ B45 ◇ REV Zbl 562 # 03009 • ID 42533

DOSEN, K. see Vol. III, IV for further entries

D'OTTAVIANO, I.M.L. [1970] see COSTA DA, N.C.A.

D'OTTAVIANO, I.M.L. & LOPEZ-ESCOBAR, E.G.K. [1985] *The conditional and paraconsistent logics* (C 4181) Math Log & Formal Syst (Costa da) 141-159
◇ B53 ◇ REV Zbl 571 # 03006 • ID 48166

D'OTTAVIANO, I.M.L. [1985] *The model extension theorems for Π_3-theories* (P 2160) Latin Amer Symp Math Log (6);1983 Caracas 157-173
◇ B50 C07 C52 ◇ ID 41799

DOYLE, J. & MCDERMOTT, D. [1980] *Non-monotonic logic. I* (J 0503) Artif Intell 13*41-72
◇ B53 ◇ REV MR 81j:03045 Zbl 435 # 68074 • REM Part II 1982 by McDermott,D. • ID 55827

DOZ, A. & DUBARLE, D. [1972] *Logique et Dialectique* (X 0891) Larousse: Paris
◇ A05 B53 ◇ ID 42833

DRABBE, J. [1967] *Une propriete des matrices characteristiques des systemes S1, S2 et S3* (J 2313) C R Acad Sci, Paris, Ser A-B 265*A1
◇ B45 ◇ REV MR 37 # 5086 Zbl 162.311 • ID 03101

DRABBE, J. see Vol. I, III, IV, V, VI for further entries

DRAKE, F.R. [1962] *On Mckinsey's syntactical characterizations of systems of modal logic* (J 0036) J Symb Logic 27*400-406
◇ B45 ◇ REV MR 31 # 39 Zbl 139.6 JSL 36.691 • ID 03118

DRAKE, F.R. [1985] *How recent work in mathematical logic relates to the foundations of mathematics* (J 2789) Math Intell 7*27-35
◇ A05 B98 E98 ◇ ID 47704

DRAKE, F.R. see Vol. IV, V for further entries

DREBEN, B. [1957] *Relation of m-valued quantificational logic to 2-valued quantificational logic* (P 1675) Summer Inst Symb Log;1957 Ithaca 303-304
◇ B10 B50 ◇ REV JSL 30.375 • ID 29366

DREBEN, B. see Vol. I, III, IV, VI for further entries

DRECHSEL, P. [1979] *Traditional logic and quantum logic* (X 1272) Gutenberg: Mainz 29pp
◇ B51 ◇ REV MR 84g:03032 • ID 34146

DRIESCHNER, M. [1977] *Is (quantum) logic empirical?* (J 0122) J Philos Logic 6*415-423
◇ A05 B51 ◇ REV MR 58 # 21378 Zbl 369 # 02009 • ID 51309

DRIESSCHE VAN DEN, R. [1949] *Sur le "de syllogismo hypothetico" de Boece* (J 0175) Methodos 1*293-307
◇ A05 A10 B45 ◇ REV MR 11.636 Zbl 40.145 • ID 03614

DRIVER, J. [1983] *Promises, obligations, and abilities* (J 0095) Philos Stud 44*221-223
◇ B45 ◇ REV MR 84k:03012 • ID 34952

DROZDOVA, G.D. & GRIGOR'EV, V.V. [1984] *Combinatorial enumeration in the set of many-valued logic functions (Russian)* (X 2235) VINITI: Moskva 6992-84
◇ B50 ◇ ID 46896

DRUGUSH, YA.M. [1978] *A class of logics without disjunction property (Russian)* (J 0288) Vest Ser Mat Mekh, Univ Moskva 1978/6*9-14
• TRANSL [1978] (J 0510) Moscow Univ Math Bull 33/6*6-9
◇ B55 F50 ◇ REV MR 80j:03039 Zbl 396 # 03025 • ID 52635

DRUGUSH, YA.M. [1982] *Forest superintuitionistic logics (Russian)* (J 0023) Dokl Akad Nauk SSSR 262*523-526
• TRANSL [1982] (J 0062) Sov Math, Dokl 25*64-67
◇ B55 ◇ REV MR 83b:03025 Zbl 521 # 03015 • ID 35091

DRUGUSH, YA.M. [1982] *Union of logics modeled by finite trees* (J 0003) Algebra i Logika 21*149-161
• TRANSL [1982] (J 0069) Algeb and Log 21*97-106
◇ B55 ◇ REV MR 84h:03054 Zbl 522 # 03015 • ID 37402

DRUGUSH, YA.M. [1984] *Finite approximability of forest superintuitionistic logics (Russian)* (J 0087) Mat Zametki (Akad Nauk SSSR) 36*755-764
◇ B25 B55 ◇ REV MR 86h:03041 • ID 45347

DRUMMOND, P.D. [1983] *Violations of Bell's inequality in cooperative states* (J 2730) Phys Rev Lett 50/19*1407-1410
◇ B51 ◇ REV MR 85a:81006 • ID 38889

DUBARLE, D. [1969] *Sur une formalisation de la logique hegelienne* (J 2086) Epistem Sociol 7*48-61
◇ B53 ◇ ID 42832

DUBARLE, D. [1972] see DOZ, A.

DUBARLE, H.D. [1957] *Initiation a la logique* (**X** 0834) Gauthier-Villars: Paris 91pp
◇ B98 ◇ REV MR 20 # 1621 Zbl 77.11 JSL 23.30 • ID 03139

DUBIKAJTIS, L. [1968] see COSTA DA, N.C.A.

DUBIKAJTIS, L. [1977] see COSTA DA, N.C.A.

DUBIKAJTIS, L. & DUDEK, E. & KONIOR, J. [1980] *On axiomatics of Jaskowski's discussive propositional calculus* (**P** 3006) Brazil Conf Math Log (3);1979 Recife 109-117
◇ B53 ◇ REV MR 82b:03031 Zbl 459 # 03014 • ID 54454

DUBIKAJTIS, L. [1981] see ACHTELIK, G.

DUBIKAJTIS, L. & MORAES DE, L. [1981] *On single operator for Lewis S5 modal logic* (**J** 0302) Rep Math Logic, Krakow & Katowice 11*57-61
◇ B45 ◇ REV MR 82j:03019 Zbl 465 # 03006 • ID 54909

DUBIKAJTIS, L. see Vol. I for further entries

DUBOIS, DIDIER & PRADE, H. [1979] *Operations in a fuzzy-valued logic* (**J** 0194) Inform & Control 43*224-240
◇ B52 E72 ◇ REV MR 80m:94120 Zbl 434 # 03020 • ID 55724

DUBOIS, DIDIER & PRADE, H. [1982] *The use of fuzzy numbers in decision analysis* (**C** 3778) Fuzzy Inform & Decis Processes 309-321
◇ B52 ◇ REV MR 84i:90004 Zbl 507 # 90006 • ID 37724

DUBOIS, DIDIER & PRADE, H. [1982] *Towards fuzzy differential calculus I: Integration of fuzzy mappings. II:Integration on fuzzy intervals. III: Differentiation* (**J** 2720) Fuzzy Sets Syst 8*1-17,105-116,225-233
◇ B52 ◇ REV MR 84b:26010 Zbl 493 # 28002 Zbl 493 # 28003 Zbl 499 # 28009 • ID 38433

DUBOIS, DIDIER & PRADE, H. [1983] *Unfair coins and necessity measures towards a possibilistic interpretation of histograms* (**J** 2720) Fuzzy Sets Syst 10*15-20
◇ B48 E72 ◇ REV MR 84e:60005 Zbl 515 # 60005 • ID 39264

DUBOIS, DIDIER & PRADE, H. [1984] *Fuzzy logics and the generalized modus ponens revisted* (**J** 2694) Cybern & Syst 15*293-331
◇ B52 ◇ ID 49397

DUBOIS, DIDIER & PRADE, H. [1985] *A review of fuzzy set aggregation connectives* (**J** 0191) Inform Sci 36*85-121
◇ B51 E72 ◇ ID 49478

DUBOIS, DIDIER & PRADE, H. [1985] *Fuzzy cardinality and the modeling of imprecise quantification* (**J** 2720) Fuzzy Sets Syst 16*199-230
◇ B52 E72 ◇ ID 48203

DUBOIS, DIDIER see Vol. V for further entries

DUBOIS, T. [1975] see COOLS, M.

DUCASSE, E.G. & METZE, G. [1973] *Hazard-free realizations of Boolean functions using Post functions* (**P** 2009) Int Symp Multi-Val Log (3);1973 Toronto 59-67
◇ B50 ◇ REV MR 53 # 12762 • ID 42055

DUCASSE, E.G. & SU, S.Y.H. [1974] *Application of a multi-valued gate to the design of fault-tolerant digital systems* (**P** 1385) Int Symp Multi-Val Log (4);1974 Morgantown 423-436
◇ B50 B70 ◇ ID 42108

DUCASSE, E.G. [1975] *Reducibility of Post functions* (**P** 1805) Int Symp Multi-Val Log (5,Proc);1975 Bloomington 8-17
◇ B50 ◇ REV MR 58 # 5193 • ID 72470

DUDEK, E. [1980] see DUBIKAJTIS, L.

DUDEK, E. [1981] see ACHTELIK, G.

DUERR, K. [1954] *Lehrbuch der Logistik* (**X** 0804) Birkhaeuser: Basel vii + 181pp
◇ B98 ◇ REV MR 16.986 Zbl 56.5 JSL 21.88 • ID 03162

DUESBERG, K.J. [1980] *Zur Kritik einer neuen Definition von "Wahrscheinlichkeit"* (**J** 0989) Z Allg Wissth 11*103-107
◇ B48 ◇ REV MR 82d:60006 • ID 81128

DUFFY, M.J. [1979] *Modal interpretations of three valued logic. I,II* (**J** 0047) Notre Dame J Formal Log 20*647-657,658-673
◇ B50 ◇ REV MR 81i:03030 Zbl 332 # 02019 Zbl 349 # 02011 Zbl 406 # 03046 • ID 56128

DUGGAN, R.W. & SMITH, W.R. [1976] *Generation of value-consistent multi-valued prime implicants* (**P** 2011) Int Symp Multi-Val Log (6);1976 Logan 258-262
◇ B50 ◇ ID 35896

DUGUNDJI, J. [1940] *Note on a property of matrices for Lewis and Langford's calculi of propositions* (**J** 0036) J Symb Logic 5*150-151
◇ B45 ◇ REV MR 2.209 Zbl 24.97 JSL 6.37 FdM 66.1194 • ID 03164

DUJMOVIC, J.J. [1975] *Extended continuous logic and the theory of complex criteria* (**S** 1003) Publ Elektroteh, Ser Mat Fiz, Beograde 498-541*197-216
◇ B50 ◇ REV MR 53 # 2437 Zbl 365 # 02009 • ID 51010

DULIN, S.K. [1982] *Introduction to dissonant logic (Russian) (English summary)* (**J** 3932) Comput Artif Intell (Bratislava) 1*291-299
◇ B60 ◇ REV Zbl 511 # 03010 • ID 37385

DUMITRIU, A. [1971] *Logica polivalenta* (**X** 0922) Stiintifica Encicl: Bucharest
◇ B50 ◇ REM 2nd ed • ID 41974

DUMITRIU, A. see Vol. I, VI for further entries

DUMMETT, M. [1959] *A propositional calculus with denumerable matrix* (**J** 0036) J Symb Logic 24*97-106
◇ B55 ◇ REV MR 23 # A801 Zbl 89.243 JSL 33.305 • ID 03169

DUMMETT, M. & LEMMON, E.J. [1959] *Modal logics between S4 and S5* (**J** 0068) Z Math Logik Grundlagen Math 5*250-264
◇ B45 ◇ REV MR 28 # 27 JSL 32.396 • ID 03167

DUMMETT, M. [1965] see CROSSLEY, J.N.

DUMMETT, M. see Vol. VI for further entries

DUN VAN, F. [1982] *On the philosophy of argument and the logic of common morality* (**P** 3754) Argumentation;1978 Groningen 281-294
◇ B60 ◇ ID 45903

DUNDERDALE, H. [1969] *Current-mode circuits for ternary-logic realisation* (**J** 2712) Electronics Lett (London) 5*575-577
◇ B50 B70 ◇ ID 41945

DUNN, J.M. & MEYER, R.K. [1969] *E, R and γ* (J 0036) J Symb Logic 34*460-474
◇ B46 ◇ REV MR 40 # 5428 Zbl 274 # 02008 JSL 36.521 • ID 03632

DUNN, J.M. [1970] *Algebraic completeness results for R-mingle and its extensions* (J 0036) J Symb Logic 35*1-13
◇ B46 ◇ REV MR 44 # 5206 Zbl 231 # 02024 • ID 03178

DUNN, J.M. [1970] *Comments on N.D.Belnap,jr.'s conditional assertion and restricted quantification* (J 0097) Nous, Quart J Phil 4*13
◇ B45 ◇ ID 31734

DUNN, J.M. [1970] *Extensions of RM and LC* (J 0036) J Symb Logic 35*360
◇ B46 ◇ ID 41531

DUNN, J.M. & MEYER, R.K. [1971] *Algebraic completeness results for Dummett's LC and its extensions* (J 0068) Z Math Logik Grundlagen Math 17*225-230
◇ B45 ◇ REV MR 45 # 1723 Zbl 252 # 02018 • ID 03179

DUNN, J.M. [1972] *A modification of Parry's analytic implication* (J 0047) Notre Dame J Formal Log 13*195-205
◇ B46 ◇ REV MR 45 # 6583 Zbl 234 # 02012 • ID 03181

DUNN, J.M. [1973] *A truth value semantics for modal logic* (P 0783) Truth, Syntax & Modal;1970 Philadelphia 87-100
◇ B45 ◇ REV MR 53 # 93 Zbl 261 # 02013 JSL 42.314 • ID 16557

DUNN, J.M. & LEBLANC, H. & MEYER, R.K. [1974] *Completeness of relevant quantification theories* (J 0047) Notre Dame J Formal Log 15*97-121
◇ B46 ◇ REV MR 49 # 2333 Zbl 226 # 02022 • ID 09204

DUNN, J.M. [1975] *Axiomatizing Belnap's conditional assertion* (J 0122) J Philos Logic 4*383-397
◇ B45 ◇ REV MR 57 # 12174 Zbl 329 # 02006 • ID 31739

DUNN, J.M. [1975] *Consecution formulation of positive R* (C 1852) Entailment - Log of Relev & Nec PAR.28.5*381391
◇ B46 ◇ REV MR 53 # 10542 • ID 31743

DUNN, J.M. [1975] *Extensions of RM* (C 1852) Entailment - Log of Relev & Nec PAR.29.3*420-429
◇ B46 ◇ REV MR 53 # 10542 JSL 42.311 • ID 31740

DUNN, J.M. & MEYER, R.K. [1975] *The admissibility of (γ) in E* (C 1852) Entailment - Log of Relev & Nec 300-314
◇ B46 ◇ REV JSL 42.311 • ID 33941

DUNN, J.M. [1975] *The algebra of R* (C 1852) Entailment - Log of Relev & Nec PAR.28.2*352-371
◇ B46 G25 ◇ REV MR 53 # 10542 JSL 42.311 • ID 31742

DUNN, J.M. [1976] *A variation on the binary semantics for RM* (J 1893) Relevance Logic Newslett 1*56-67
◇ B46 ◇ REV Zbl 355 # 02014 • ID 31736

DUNN, J.M. [1976] *A Kripke-style semantics for R-mingle using a binary accessibility relation* (J 0063) Studia Logica 35*163-172
◇ B46 ◇ REV MR 56 # 90 Zbl 328 # 02010 • ID 31737

DUNN, J.M. [1976] *Intuitive semantics for first-degree entailments and coupled trees* (J 0095) Philos Stud 29*149-168
◇ B46 ◇ REV MR 58 # 10311 • ID 31738

DUNN, J.M. [1976] *Quantification and RM* (J 0063) Studia Logica 35*315-322
◇ B46 ◇ REV MR 56 # 2788 Zbl 359 # 02014 • ID 31735

DUNN, J.M. & EPSTEIN, G. (EDS.) [1977] *Modern uses of multiple-valued logic. Invited papers from the fifth international symposium on multiple-valued logic held at Indiana University, Bloomington* (X 0835) Reidel: Dordrecht 338pp
◇ B50 B97 ◇ REV MR 57 # 9454 Zbl 355 # 00008 • ID 31744

DUNN, J.M. [1979] *A theorem in 3-valued model theory with connections to number theory, type theory, and relevant logic* (J 0063) Studia Logica 38*149-169
◇ B50 ◇ REV MR 80h:03056 Zbl 406 # 03030 • ID 56113

DUNN, J.M. [1979] *R-mingle and beneath* (J 0047) Notre Dame J Formal Log 20*369-376
◇ B46 ◇ REV MR 80f:03023 Zbl 292 # 02021 • ID 31745

DUNN, J.M. [1979] *Relevant Robinson's arithmetic* (J 0063) Studia Logica 38*407-418
◇ B46 ◇ REV MR 81i:03090 Zbl 434 # 03018 • ID 55722

DUNN, J.M. [1980] see BELNAP JR., N.D.

DUNN, J.M. [1980] *A sieve for entailments* (J 0122) J Philos Logic 9*41-57
◇ B46 ◇ REV MR 81i:03020 Zbl 428 # 03011 • ID 53771

DUNN, J.M. [1981] see BELNAP JR., N.D.

DUNN, J.M. [1982] *Anderson and Belnap, and Lewy on entailment* (P 3622) Int Congr Log, Meth & Phil of Sci (6,Proc);1979 Hannover 291-297
◇ B46 ◇ REV MR 84f:03014 Zbl 543 # 03009 • ID 34432

DUNN, J.M. see Vol. I, V for further entries

DURAND, D. [1975] *Nouvelle demonstration de la completude fonctionelle du systeme $\{C,N,T\}$ de Rosser-Turquette, en calcul propositionnel m-valent (English summary)* (J 2313) C R Acad Sci, Paris, Ser A-B 281*A681-A682
◇ B50 ◇ REV MR 52 # 2866 Zbl 339 # 02019 • ID 17652

DURNEV, V.G. [1978] *Elements of set theory and mathematical logic (Russian)* (X 2766) Yaroslav Gos Univ: Yaroslavl' 116pp
◇ B98 E98 ◇ REV MR 80g:03001 • ID 72511

DURNEV, V.G. see Vol. III, IV for further entries

DUTA, D.L. & FABIAN, C. [1982] *The cutting stock problem with several objective functions and in fuzzy conditions* (J 3369) Econ Comput & Econ Cybern (Bucharest) 1982/2*43-47
◇ B52 ◇ REV Zbl 503 # 90074 • ID 36678

DUTHIE, G.D. [1970] *Intensional propositional logic* (J 0112) Phil Quart (Calcutta) 20*41-52
◇ B45 ◇ ID 04104

DUTHIE, G.D. see Vol. I for further entries

DUTTON, J.D. (ED.) [1966] *Logics: an introduction with exercises* (X 1248) Chandler: San Francisco 251pp
◇ B98 ◇ ID 22483

DVURECHENSKIJ, A. [1979] *Laws of large numbers and the central limit theorems on a logic* (J 1522) Math Slovaca 29*397-410
◇ B51 ◇ REV MR 81a:81011 Zbl 419 # 60022 • ID 53391

DVURECHENSKIJ, A. [1980] *On a sum of observables in a logic* (J 1522) Math Slovaca 30*187-196
◇ B51 ◇ REV MR 82d:03099 Zbl 435 # 03040 • ID 55800

Author Index

DVURECHENSKIJ, A. & RIECHAN, B. [1980] *On the individual ergodic theorem on a logic* (J 0140) Comm Math Univ Carolinae (Prague) 21*385-391
◇ B51 ◇ REV MR 81g:28024 Zbl 443 # 28014 • ID 56439

DVURECHENSKIJ, A. & PULMANNOVA, S. [1980] *On the sum of observables in a logic* (J 1522) Math Slovaca 30*393-399
◇ B51 ◇ REV MR 82f:03058 Zbl 454 # 03030 • ID 54242

DVURECHENSKIJ, A. [1981] *On extension properties for observables (Russian summary)* (J 1522) Math Slovaca 31*149-153
◇ B51 ◇ REV MR 83j:03105 Zbl 474 # 03032 • ID 35387

DVURECHENSKIJ, A. & PULMANNOVA, S. [1981] *Random measures on a logic* (J 1008) Demonstr Math (Warsaw) 14*305-320
◇ B51 ◇ REV MR 82k:81005 Zbl 474 # 03034 • ID 55438

DVURECHENSKIJ, A. & PULMANNOVA, S. [1984] *Connection between joint distribution and compatibility* (J 1546) Rep Math Phys (Warsaw) 19*349-359
◇ B51 ◇ REV MR 85m:81017 Zbl 552 # 03041 • ID 43420

DVURECHENSKIJ, A. & PULMANNOVA, S. [1985] *Uncertainty principle and joint distributions of observables* (J 2658) Ann Inst Henri Poincare, Sect A 42*253-265
◇ B51 ◇ ID 48053

DVURECHENSKIJ, A. see Vol. V for further entries

DWYER, R. & MARTIN, E.P. & MEYER, R.K. [1983] *The fundamental S-theorem - a corollary* (J 0047) Notre Dame J Formal Log 24*509-516
◇ B45 ◇ REV MR 85d:03036 Zbl 538 # 03017 • ID 41004

DYCKHOFF, H. & PEDRYCZ, W. [1984] *Generalized means as a model of compensative connectives* (J 2720) Fuzzy Sets Syst 14*143-154
◇ B52 E72 ◇ REV MR 85i:03166 Zbl 551 # 03035 • ID 43917

DYWAN, Z. [1981] *The necessity of modal logic* S5 *is metalogical* (J 0387) Bull Sect Logic, Pol Acad Sci 10*162-169
◇ B45 ◇ REV MR 84j:03033 Zbl 515 # 03008 • ID 34625

DYWAN, Z. [1983] *The connective of necessity of modal logic* S5 *is metalogical* (J 0047) Notre Dame J Formal Log 24*410-414
◇ B45 ◇ REV MR 84j:03034 Zbl 561 # 03008 • ID 34626

DYWAN, Z. [1984] *On Lemmon's interpretations of the connective of necessity* (J 0387) Bull Sect Logic, Pol Acad Sci 13*92-98
◇ B45 ◇ REV MR 86g:03031 Zbl 565 # 03010 • ID 45748

DYWAN, Z. [1985] *On a certain method of producing logical matrices* (J 0387) Bull Sect Logic, Pol Acad Sci 14*2-7
◇ B50 ◇ REV Zbl 575 # 03009 • ID 48821

DYWAN, Z. [1985] *On Lemmon's interpretations of the connective of necessity* (J 0079) Logique & Anal, NS 28*369-373
◇ B45 ◇ ID 49668

DYWAN, Z. see Vol. I, III, V, VI for further entries

DZIK, W. & WRONSKI, A. [1973] *Structural completeness of Goedel's and Dummett's propositional calculi (Polish and Russian summaries)* (J 0063) Studia Logica 32*69-75
◇ B55 ◇ REV MR 50 # 82 Zbl 345 # 02036 • ID 04092

DZIK, W. see Vol. I, V, VI for further entries

DZIOBIAK, W. [1975] see BLASZCZUK, J.J.

DZIOBIAK, W. [1976] see BLASZCZUK, J.J.

DZIOBIAK, W. [1976] *Classically axiomatizable modal propositional calculi containing the system T of Feys-von Wright* (J 0387) Bull Sect Logic, Pol Acad Sci 5*20-24
◇ B45 ◇ REV MR 53 # 5263 • ID 22894

DZIOBIAK, W. [1976] *Semantics of Kripke's style for some modal systems* (J 0387) Bull Sect Logic, Pol Acad Sci 5*63-67
◇ B45 ◇ REV MR 55 # 2499 • ID 27111

DZIOBIAK, W. [1977] see BLASZCZUK, J.J.

DZIOBIAK, W. [1977] *On detachment-substitutional formalization in normal modal logics* (J 0063) Studia Logica 36*165-171
◇ B45 ◇ REV MR 57 # 15963 Zbl 363 # 02017 • ID 50844

DZIOBIAK, W. [1978] *A note on incompleteness of modal logics with respect to neighbourhood semantics* (J 0387) Bull Sect Logic, Pol Acad Sci 7*185-190
◇ B45 C90 ◇ REV MR 80b:03018 Zbl 415 # 03020 • ID 53122

DZIOBIAK, W. [1980] *The degrees of maximality of the intuitionistic propositional logic and of some of its fragments* (J 0387) Bull Sect Logic, Pol Acad Sci 9*136-140
◇ B55 ◇ REV MR 82b:03032 Zbl 448 # 03013 • ID 56613

DZIOBIAK, W. [1981] *Nonexistence of a countable strongly adequate matrix semantics for neighbours of E* (J 0387) Bull Sect Logic, Pol Acad Sci 10*170-176
◇ B45 ◇ REV MR 84g:03024 Zbl 507 # 03007 • ID 34138

DZIOBIAK, W. [1981] *On matrices characteristic of relevant logics* (J 0387) Bull Sect Logic, Pol Acad Sci 10*113-115
◇ B46 ◇ REV MR 83b:03018 Zbl 483 # 03008 • ID 35086

DZIOBIAK, W. [1981] *Strong completeness with respect to finite Kripke models* (J 0063) Studia Logica 40*249-252
◇ B55 ◇ REV MR 84i:03039 Zbl 491 # 03027 • ID 34521

DZIOBIAK, W. [1981] *The degrees of maximality of the intuitionistic propositional logic and of some of its fragments* (J 0063) Studia Logica 40*195-198
◇ B55 ◇ REV MR 83c:03027 Zbl 474 # 03009 • ID 55413

DZIOBIAK, W. [1982] *On finite approximability of ψ-intermediate logics* (J 0063) Studia Logica 41*67-73
◇ B55 ◇ REV MR 84k:03079 Zbl 519 # 03015 • ID 35016

DZIOBIAK, W. [1983] *Cardinalities of proper ideals in some lattices of strengthenings of the intuitionistic propositional logic* (J 0063) Studia Logica 42*173-177
◇ B55 ◇ REV MR 85i:03077 Zbl 477 # 94033 Zbl 549 # 03055 • ID 42326

DZIOBIAK, W. [1983] *Structural completeness of modal logics containing K4* (J 0387) Bull Sect Logic, Pol Acad Sci 12*32-36
◇ B45 ◇ REV Zbl 548 # 03006 • ID 43190

DZIOBIAK, W. [1983] *There are $2^{\aleph_0}$ logics with the relevance principle between R and RM* (J 0063) Studia Logica 42*49-61
◇ B46 ◇ REV MR 85d:03029 Zbl 538 # 03018 • ID 41466

DZIOBIAK, W. see Vol. I, III for further entries

EARLE, J.N.F. [1973] *Logic* (X 0843) Macmillan : New York & London ix+131pp
◇ B98 ◇ ID 22484

EATON, R.M. [1931] *General logic: an introductory survey* (X 1347) Scribner: New York xii+630pp
◇ B98 ◇ ID 22582

EBBINGHAUS, H.-D. [1969] *Ueber eine Praedikatenlogik mit partiell definierten Praedikaten und Funktionen* (**J** 0009) Arch Math Logik Grundlagenforsch 12*39-53
◇ B60 C90 ◇ REV MR 40 # 7092 JSL 37.617 • ID 03219

EBBINGHAUS, H.-D. & FLUM, J. & THOMAS, WOLFGANG [1978] *Einfuehrung in die mathematische Logik* (**X** 0890) Wiss Buchges: Darmstadt ix+288pp
• TRANSL [1984] (**X** 0811) Springer: Heidelberg & New York ix+216pp
◇ B98 C95 C98 ◇ REV MR 81h:03001 Zbl 399 # 03001 Zbl 556 # 03001 • ID 28201

EBBINGHAUS, H.-D. see Vol. I, III, IV, V, VI for further entries

EBERHARD, P.H. [1982] *Constraints of determinism and of Bell's inequalities are not equivalent* (**J** 2730) Phys Rev Lett 49*1474-1477
◇ B51 ◇ REV MR 84m:81028c • ID 39542

EBERLE, R.A. [1967] *Some complete calculi of individuals* (**J** 0047) Notre Dame J Formal Log 8*267-278
◇ A05 B60 ◇ ID 03221

EBERLE, R.A. [1970] *Nominalistic systems* (**X** 0835) Reidel: Dordrecht ix+217pp
◇ A05 B60 ◇ REV MR 56 # 5196 Zbl 209.300 • ID 25100

EBERLE, R.A. [1982] *The relativization of truth to functions: its expressive power and ontic import* (**J** 0097) Nous, Quart J Phil 16*443-451
◇ B46 ◇ REV MR 84m:03023 • ID 35726

EBERLE, R.A. [1984] *Logic with a relative truth predicate and "that"-terms* (**J** 0154) Synthese 59*151-185
◇ B60 ◇ REV MR 86b:03029 Zbl 537 # 03004 • ID 43717

EBERLE, R.A. see Vol. I, V for further entries

ECK VAN, J.A. [1982] *A system of temporally relative modal and deontic predicate logic and its philosophical applications I, II* (**J** 0079) Logique & Anal, NS 25*249-290,339-381
◇ B45 ◇ REV MR 85d:03030 Zbl 549 # 03004 • ID 40991

EDEL'MAN, S.L. [1975] *Mathematical logic (Russian)* (**X** 3407) "Vysshaya Shkola": Moskva 178pp
◇ B98 ◇ ID 33333

EDELSTEIN, R. & SCHUMM, G.F. [1979] *Negation-free modal logics* (**J** 0068) Z Math Logik Grundlagen Math 25*281-288
◇ B45 ◇ REV MR 80h:03033 Zbl 421 # 03012 • ID 53411

EDELSTEIN, R. see Vol. I for further entries

EDIDIN, A. [1984] *Inductive reasoning and the uniformity of nature* (**J** 0122) J Philos Logic 13*285-302
◇ A05 B48 ◇ REV MR 85f:03023 Zbl 556 # 03018 • ID 40669

EDMONDS, E.A. [1969] *Independence of Rose's axioms for m-valued implication* (**J** 0036) J Symb Logic 34*283-284
◇ B50 ◇ REV Zbl 182.316 • ID 03224

EDMONDS, E.A. [1974] *Model formation - an application of m-valued logic* (**P** 3103) Adv Cybern Syst;1972 Oxford 3*1201-1206
◇ B50 ◇ REV Zbl 327 # 02017 • ID 61470

EDMONDS, E.A. [1980] *Lattice fuzzy logics* (**J** 1741) Int J Man-Mach Stud 13*455-465
◇ B52 E72 ◇ REV MR 82b:03102 Zbl 447 # 94063 • ID 72572

EDMONDS, E.A. see Vol. V for further entries

EDWARDS, R.E. [1979] *A formal background to mathematics. Ia, Ib: Logic, sets and numbers* (**X** 0811) Springer: Heidelberg & New York xxxiv+933pp
◇ B98 E98 ◇ REV MR 80h:03001a MR 80h:03001b Zbl 413 # 03001 • REM Vol.II 1980 • ID 52995

EDWARDS, R.E. see Vol. I for further entries

EELLS, E. [1982] *Rational decision and causality* (**X** 0805) Cambridge Univ Pr: Cambridge, GB x+234pp
◇ B48 ◇ REV MR 84j:03050 • ID 34642

EELLS, E. [1984] *Metatickles and the dynamics of deliberation* (**J** 0472) Theory Decis 17*71-95
◇ B48 ◇ REV MR 85f:90005 • ID 39938

EFIMOV, E.I. [1976] *Purposeful system behaviour logic (Russian) (English summary)* (**J** 2819) Probl Contr Inf Th, Akad Nauk SSSR & Acad Sci Hung 5*247-261
◇ B60 ◇ REV MR 55 # 10231 Zbl 354 # 68114 • ID 72578

EFIMOV, E.I. see Vol. I for further entries

EFSTATHIOU, J. & TONG, R.M. [1980] *Rule-based decomposition of fuzzy relational models* (**P** 4254) Joint Autom Control Conf;1980 San Francisco 3pp
◇ B52 ◇ ID 46078

EFSTATHIOU, J. & TONG, R.M. [1982] *A critical assessment of truth functional modification and its use in approximate reasoning* (**J** 2720) Fuzzy Sets Syst 7*103-108
◇ B52 ◇ REV MR 83e:03035 Zbl 483 # 03017 • ID 35220

EFSTATHIOU, J. see Vol. I for further entries

EHLERS, F. [1968] *Logic by way of set theory* (**X** 0818) Holt Rinehart & Winston: New York xi+386pp
◇ B98 E98 ◇ ID 22634

EHRENFEUCHT, A. [1974] *Logic without iterations* (**P** 0610) Tarski Symp;1971 Berkeley 265-268
◇ B60 F65 ◇ REV MR 50 # 9546 Zbl 345 # 02042 • ID 03261

EHRENFEUCHT, A. see Vol. I, III, IV, V, VI for further entries

EHSAKIA, L.L. [1965] *A system and algebra with derivation (Russian) (Georgian summary)* (**J** 0233) Soobshch Akad Nauk Gruz SSR 40*537-543
◇ B45 ◇ REV MR 35 # 5300 Zbl 199.4 • ID 03555

EHSAKIA, L.L. [1972] *On the category of complete Kripke models (Russian)* (**P** 2585) All-Union Conf Math Log (2);1972 Moskva 54
◇ B45 ◇ ID 33252

EHSAKIA, L.L. [1973] *On some new results in the theory of modal and superintuitionistic systems (Russian)* (**C** 1662) Teor Log Vyvoda 2*174-184
◇ B45 ◇ ID 33251

EHSAKIA, L.L. [1974] *Topological Kripke models (Russian)* (**J** 0023) Dokl Akad Nauk SSSR 214*298-301
• TRANSL [1974] (**J** 0062) Sov Math, Dokl 15*147-151
◇ B45 B55 G05 G25 G30 ◇ REV MR 49 # 4751 Zbl 296 # 02030 • ID 04119

EHSAKIA, L.L. [1976] *On modal partners of superintuitionistic logics (Russian)* (**P** 2566) All-Union Symp Log & Method of Sci (7);1976 Kiev 135-136
◇ B45 B55 ◇ ID 33253

EHSAKIA, L.L. & MESKHI, V.YU. [1977] *Five critical modal systems* (**J** 0105) Theoria (Lund) 43*52-60
◇ B45 ◇ REV MR 58 # 10312 Zbl 372 # 02013 JSL 50.231
• ID 32646

EHSAKIA, L.L. [1978] *Semantic analysis of bimodal (temporal) systems (Russian)* (**C** 2568) Sb Log, Semant, Metodol 87-99
◇ B45 ◇ ID 33255

EHSAKIA, L.L. [1979] *On the variety of Grzegorczyk algebras (Russian)* (**C** 2581) Issl Neklass Log & Teor Mnozh 257-287
• TRANSL [1983] (**S** 3489) Sel Math Sov 3*343-366
◇ B45 E70 G25 ◇ REV MR 81j:03097 Zbl 435 # 03012
• ID 55773

EHSAKIA, L.L. [1979] *To the theory of modal and superintuitionistic systems (Russian)* (**P** 2554) All-Union Symp Th Log Infer;1974 Moskva 147-172
◇ B45 B55 ◇ REV MR 84d:03052 MR 84k:03052
• ID 33256

EICHHORN, H. [1983] *Conceptual and conventional definitions in the mathematical sciences* (**J** 0455) Phil Naturalis 20*147-159
◇ A05 B98 ◇ REV MR 84k:00018 • ID 39171

EIKMEYER, H.-J. & RIESER, H. [1983] *A formal theory of context dependence and context change* (**C** 3095) Approaching Vagueness 131-188
◇ B52 ◇ REV MR 85j:03003 MR 86f:03049 • ID 45348

ELLIS, B. [1973] *The logic of subjective probability* (**J** 0013) Brit J Phil Sci 24*125-152
◇ A05 B48 ◇ REV MR 57 # 2882 Zbl 356 # 02007
• ID 03332

ELLIS, B. [1976] see DAVIDSON, B.

ELLIS, B. & JACKSON, F.C. & PARGETTER, R. [1977] *An objection to possible-world semantics for counterfactual logics* (**J** 0122) J Philos Logic 6*355-357
◇ A05 B45 ◇ REV MR 58 # 27323 Zbl 376 # 02002
• ID 31381

ELLIS, B. [1978] *A unified theory of conditionals* (**J** 0122) J Philos Logic 7*107-124
◇ A05 B45 ◇ REV MR 81f:03023 Zbl 376 # 02004
• ID 31382

ELLOZY, H.A. [1973] *On the generation of all the one variable functions in m-valued logic* (**P** 2009) Int Symp Multi-Val Log (3);1973 Toronto 119-126
◇ B50 ◇ REV MR 51 # 7820 • ID 17304

ELLOZY, H.A. & PATT, Y.N. [1975] *The linearity property and functional completeness in M-valued logic* (**P** 1805) Int Symp Multi-Val Log (5,Proc);1975 Bloomington 44-52
◇ B50 ◇ REV MR 58 # 10295 • ID 72656

EMCH, A.F. [1936] *Implication and deducibility* (**J** 0036) J Symb Logic 1*26-35 • ERR/ADD ibid 1*58
◇ B45 ◇ REV Zbl 14.193 JSL 1.67 FdM 62.32 • ID 03340

EMCH, A.F. [1937] *Deducibility with respect to necessary and impossible propositions* (**J** 0036) J Symb Logic 2*78-81
◇ B45 ◇ REV Zbl 17.49 FdM 63.21 • ID 03341

EMDE, G. [1957] *Kriterien fuer die Herleitbarkeit in Modalitaetenstrukturen* (**J** 0009) Arch Math Logik Grundlagenforsch 3*79-111
◇ B45 ◇ REV Zbl 79.244 • ID 03342

EMDE, H. & REYERSBACH, W. & STROMBACH, W. [1972] *Mathematische Logik. Ihre Grundprobleme in Theorie und Anwendung* (**X** 0995) Beck'sche Verlagsbuchh: Muenchen 227pp
◇ B98 G05 G10 ◇ REV MR 48 # 1869 Zbl 238 # 02002
• ID 13231

EMDE BOAS VAN, P. [1978] *The connection between modal and algorithmic logic* (**P** 1707) Math Founds of Comput Sci (7);1978 Zakopane 64*1-15
◇ B75 ◇ REV MR 81j:68017 Zbl 379 # 68048 • ID 32521

EMDE BOAS VAN, P. & GROENENDIJK, J. & STOKHOF, M. [1981] *The Conway paradox: Its solution in an epistemic framework* (**P** 3373) Form Meth in Stud of Lang;1980 Amsterdam 1*87-111
◇ A05 B45 ◇ REV MR 84i:03017 Zbl 466 # 03005
• ID 54956

EMDE BOAS VAN, P. see Vol. III, IV, V, VI for further entries

EMEL'YANOV, N.R. [1985] see ALEKSEEV, V.B.

EMEL'YANOV, N.R. [1985] *Complexity of the problem of expressibility in many-valued logics (Russian)* (**J** 0023) Dokl Akad Nauk SSSR 282*525-529
• TRANSL [1985] (**J** 0062) Sov Math, Dokl 31*477-488
◇ B50 ◇ ID 47622

EMERSON, E.A. [1982] see CLARKE, E.M.

EMERSON, E.A. [1983] *Alternative semantics for temporal logics* (**J** 1426) Theor Comput Sci 26*121-130
◇ B45 ◇ REV MR 85d:68044 Zbl 559 # 68050 • ID 38983

EMERSON, E.A. & SISTLA, A.P. [1984] *Deciding full branching time logic* (**J** 0194) Inform & Control 61*175-201
◇ B25 B45 D15 ◇ REV MR 86h:03021 • ID 45349

EMERSON, E.A. [1985] *Automata, tableaux, and temporal logics* (**P** 4571) Log of Progr;1985 Brooklyn 79-88
◇ B45 D05 ◇ ID 49191

EMERSON, E.A. & HALPERN, J.Y. [1985] *Decision procedures and expressiveness in the temporal logic of branching time* (**J** 0119) J Comp Syst Sci 30*1-24
◇ B45 ◇ REV Zbl 559 # 68051 • ID 46930

EMERSON, E.A. see Vol. IV for further entries

ENDERTON, H.B. [1972] *A mathematical introduction to logic* (**X** 0801) Academic Pr: New York xiii + 295pp
◇ B98 C98 F30 ◇ REV MR 49 # 2239 Zbl 298 # 02002 JSL 38.340 • ID 03355

ENDERTON, H.B. see Vol. I, III, IV, V, VI for further entries

ENGEL, P. & NEF, F. [1982] *Quelques remarques sur la logique des phrases d'action* (**J** 0079) Logique & Anal, NS 25*291-319
◇ B60 B65 ◇ REV MR 85d:03059 • ID 41096

ENGELER, E. (ED.) [1971] *Symposium on semantics of algorithmic languages* (**X** 0811) Springer: Heidelberg & New York vi + 372pp
◇ B75 B97 D05 ◇ REV MR 43 # 1477 Zbl 215.560
• ID 23569

ENGELER, E. [1974] *The logic of "can do"* (**P** 1511) Int Symp Th Progr;1972 Novosibirsk 5*17-28
◇ B45 ◇ REV MR 54 # 9132 Zbl 289 # 68022 • ID 25828

ENGELER, E. [1975] *Algorithmic logic* (**P** 1430) Adv Course Founds Computer Sci;1974 Amsterdam 55-85
◇ B25 B75 D15 ◇ REV MR 52 # 12380 Zbl 314 # 68009 JSL 42.420 • ID 21738

ENGELER, E. (ED.) [1981] *Logic of programs* (**S** 3302) Lect Notes Comput Sci 125*ii+245pp
◇ B75 ◇ REV MR 84m:68002 Zbl 463 # 00025 • ID 39517

ENGELER, E. & LAEUCHLI, H. & STRASSEN, V. (EDS.) [1982] *Logic and algorithmic. An International Symposium held in Honnour of Ernst Specker (Zuerich, February 5-11, 1980)* (**X** 3718) Enseign Math, Univ Geneve: Geneve 392pp
◇ B97 ◇ REV Zbl 471 # 00009 • ID 55194

ENGELER, E. [1983] *Metamathematik der Elementarmathematik* (**X** 0811) Springer: Heidelberg & New York VII*132pp
◇ B98 ◇ REV MR 85d:03024 Zbl 515 # 03001 • ID 37835

ENGELER, E. see Vol. I, III, IV, V, VI for further entries

ENGLEBRETSEN, G. [1973] *Epistemic logic and mere belief* (**J** 0079) Logique & Anal, NS 16*375-378
◇ B45 ◇ REV MR 50 # 58 Zbl 337 # 02020 • ID 04112

ENGLEBRETSEN, G. [1973] *Suggested truth-tables for a three-valued sentential logic* (**J** 0286) Int Logic Rev 8*255-259
◇ B50 ◇ REV Zbl 343 # 02012 • ID 61546

ENGLEBRETSEN, G. [1974] *A note on contrariety* (**J** 0047) Notre Dame J Formal Log 15*613-614
◇ B45 ◇ REV MR 50 # 6796 Zbl 232 # 02006 • ID 04114

ENGLEBRETSEN, G. see Vol. I for further entries

ENGLISH, W.R. [1977] *Canonical representation of nonlinear finite memory functionals in a Galois field* (**P** 2013) Int Symp Multi-Val Log (7);1977 Charlotte 13
◇ B50 ◇ ID 35849

ENJALBERT, P. [1983] *Algebraic semantics and program logics: algorithmic logic for program trees* (**P** 3830) Logics of Progr & Appl;1980 Poznan 132-147
◇ B75 ◇ REV MR 84b:68001 Zbl 516 # 68036 • ID 38589

ENJALBERT, P. & MICHEL, M. [1984] *Many-sorted temporal logic for multiprocesses systems* (**P** 3658) Math Founds of Comput Sci (11);1984 Prague 273-281
◇ B45 ◇ REV Zbl 558 # 68026 • ID 44742

ENJALBERT, P. see Vol. III for further entries

ENNIS, R.H. [1969] *Ordinary logic* (**X** 0819) Prentice Hall: Englewood Cliffs vi+151pp
◇ B98 ◇ ID 22485

ENOMOTO, H. & YONEZAKI, N. [1980] *Database system based on intensional logic* (**P** 3367) Proc Comput Linguistics;1980 Tokyo 220-227
◇ B45 ◇ REV Zbl 469 # 68088 • ID 66312

ENTA, Y. [1982] *Fuzzy decision theory* (**P** 4051) Fuzzy Set & Possibility Th;1980 Acapulco 439-449
◇ B52 ◇ REV MR 84b:03004 • ID 46331

ENTRESS, G. [1968] *Bemerkungen ueber eine Logik fuer kontinuierliche Groessen* (**J** 0412) Wiss Z Tech Hochsch Ilmenau 14*273-275
◇ B50 G05 ◇ REV MR 43 # 7261 Zbl 182.25 • ID 17937

EPSTEIN, G. & FRIEDER, G. & RINE, D.C. [1974] *A summary of the development of multiple-valued logic as related to computer science* (**P** 1385) Int Symp Multi-Val Log (4);1974 Morgantown 303-314
◇ B50 ◇ ID 42083

EPSTEIN, G. & HORN, A. [1974] *Propositional calculi based on subresiduation* (**J** 0387) Bull Sect Logic, Pol Acad Sci 3/1*41-42
◇ B50 F50 G10 ◇ REV MR 53 # 5264 • ID 22895

EPSTEIN, G. & FRIEDER, G. & RINE, D.C. [1974] *The development of multiple-valued logic as related to computer science* (**J** 2015) Computer 7*20-33
◇ B50 ◇ REV Zbl 287 # 68033 • ID 61553

EPSTEIN, G. [1976] *Decisive Venn diagrams* (**P** 2011) Int Symp Multi-Val Log (6);1976 Logan 142-149
◇ B50 ◇ ID 35882

EPSTEIN, G. & HORN, A. [1976] *Logics which are characterized by subresiduated lattices* (**J** 0068) Z Math Logik Grundlagen Math 22*199-210
◇ B50 F50 G10 ◇ REV MR 56 # 8360 Zbl 347 # 02040 • ID 18456

EPSTEIN, G. [1977] see DUNN, J.M.

EPSTEIN, G. [1978] *A summary of investigations into three and four valued logics* (**P** 2014) Int Symp Multi-Val Log (8);1978 Rosemont 257
◇ B50 ◇ ID 35932

EPSTEIN, G. & MILLER, D.MICHAEL & MUZIO, J.C. [1980] *Selecting don't-care sets for symmetric many-valued functions: a pictorial approach using matrices* (**P** 3673) Int Symp Multi-Val Log (10);1980 Evanston 219-225
◇ B50 ◇ REV MR 83m:94032 • ID 45951

EPSTEIN, G. & LIU, YOWEI [1982] *Positive multiple-valued switching functions - an extension of Dedekind's problem* (**P** 4002) Int Symp Multi-Val Log (12);1982 Paris 248-252
◇ B50 B70 ◇ REV Zbl 545 # 94028 • ID 41214

EPSTEIN, G. [1983] *The underlying ground for hypothetical propositions* (**C** 3811) Logic 20th Century 101-124
◇ B50 ◇ REV MR 85g:03038 • ID 43872

EPSTEIN, G. see Vol. I for further entries

EPSTEIN, R.L. [1979] *Relatedness and implication* (**J** 0095) Philos Stud 36*137-173
◇ A05 B45 ◇ REV MR 80k:03028a • ID 72679

EPSTEIN, R.L. see Vol. IV, VI for further entries

ERIMBETOV, M.M. [1980] *Model-theoretic properties of languages of algorithmic logics (Russian)* (**C** 2620) Teor Model & Primen 3-6
◇ B75 ◇ REV MR 83g:68019 Zbl 536 # 03014 • ID 37099

ERIMBETOV, M.M. see Vol. III, IV for further entries

ERMOLAEVA, N.M. & MUCHNIK, A.A. [1974] *Modal extensions of logical calculi of Hao Wang type (Russian)* (**C** 2577) Issl Formaliz Yazyk & Neklass Log 172-193
◇ B45 ◇ REV MR 56 # 11747 • ID 72713

ERMOLAEVA, N.M. & MUCHNIK, A.A. [1976] *Modal logics, defined by endomorphisms of distributive lattices (Russian)* (**C** 3271) Issl Teor Mnozh & Neklass Logik 229-246
◇ B45 G10 ◇ REV MR 58 # 142 Zbl 403 # 03013 • ID 54727

ERMOLAEVA, N.M. & MUCHNIK, A.A. [1979] *Functionally closed 4-valued extensions of the Boolean algebra and corresponding logics (Russian)* (**C** 2581) Issl Neklass Log & Teor Mnozh 298–315
◇ B50 G25 ◇ REV MR 82k:03025 Zbl 452 # 03023
• ID 54087

ERMOLAEVA, N.M. & MUCHNIK, A.A. [1979] *Pretabular temporal logic (Russian)* (**C** 2581) Issl Neklass Log & Teor Mnozh 288–297
◇ B45 ◇ REV MR 81i:03021 Zbl 452 # 03019 • ID 54083

ERMOLAEVA, N.M. see Vol. I, IV for further entries

ERSHOV, A.P. & KNUTH, D.E. (EDS.) [1979] *Algorithms in modern mathematics and their applications I,II* (**X** 2652) Akad Nauk Sibirsk Otd Inst Mat: Novosibirsk 364pp,316pp
◇ B75 B97 D97 ◇ ID 45742

ERSHOV, A.P. see Vol. IV for further entries

ERSHOV, YU.L. & PALYUTIN, E.A. & TAJTSLIN, M.A. [1973] *Mathematical logic (Russian)* (**X** 0913) Novosibirsk Gos Univ: Novosibirsk 159pp
◇ B98 C98 ◇ REV MR 57 # 9448 • ID 32036

ERSHOV, YU.L. & PALYUTIN, E.A. [1979] *Mathematical logic (Russian)* (**X** 2027) Nauka: Moskva 320pp
• TRANSL [1984] (**X** 0885) Mir: Moskva 303pp (English)
◇ B98 ◇ REV MR 81f:03002 MR 86a:03001 • ID 72749

ERSHOV, YU.L. [1983] *Dynamic logic over admissible sets (Russian)* (**J** 0023) Dokl Akad Nauk SSSR 273*1045–1048
• TRANSL [1983] (**J** 0062) Sov Math, Dokl 28*739–742
◇ B75 C70 ◇ REV MR 85d:03075 • ID 40496

ERSHOV, YU.L. & LAVROV, I.A. & PAVILENIS, R.I. & PETROV, V.V. & SMIRNOV, V.A. [1984] *Logic, the foundations of mathematics and linguistics (Russian) (English summary)* (**J** 2871) Vopr Fil, Moskva 1984/1*45–58
◇ B97 ◇ REV MR 85k:03004 • ID 45202

ERSHOV, YU.L. see Vol. I, III, IV, V, VI for further entries

ERWIN, E. [1978] *Quantum logic and the status of classical logic* (**J** 0079) Logique & Anal, NS 21*279–292
◇ B51 G12 ◇ REV MR 80b:03033 • ID 72754

ESCHBACH, M. [1982] *The logic of fuzzy Bayesian inference.* (**P** 4620) Model & Simulation (13);1982 Pittsburgh 13/3*1217–1222
◇ B52 ◇ ID 49344

ESHRAGH, F. & MAMDANI, E.H. [1979] *A general approach to linguistic approximation* (**J** 1741) Int J Man-Mach Stud 11*501–519
◇ B52 ◇ REV MR 80g:94115 Zbl 403 # 68075 • ID 81188

ESOGBUE, A.O. [1982] see BELLMAN, R.E.

ESSLER, W.K. [1966] *Einfuehrung in die Logik* (**X** 1292) Kroener: Stuttgart 239pp
◇ B98 ◇ REV JSL 45.381 JSL 45.382 • ID 22398

ESSLER, W.K. [1970] *Induktive Logik, Grundlagen und Voraussetzungen* (**X** 0826) Alber: Freiburg 376pp
◇ B48 ◇ ID 25409

ESSLER, W.K. [1970] *Ueber Intensionen und Modalitaeten* (**J** 0047) Notre Dame J Formal Log 11*416–424
◇ B45 ◇ REV MR 44 # 6457 Zbl 177.8 • ID 03558

ESSLER, W.K. [1984] *Induktion und induktive Logik I,II* (**J** 2839) Statistica (Bologna) 44*3–19,183–196
◇ B48 ◇ REV MR 86a:00008 MR 86a:00009 • ID 44129

ESSLER, W.K. see Vol. I for further entries

ESTEVA, F. [1980] see DOMINGO, X.

ESTEVA, F. [1981] *On the form of negations in posets* (**P** 3705) Int Symp Multi-Val Log (11);1981 Oklahoma City & Norman 228–231
◇ B50 ◇ REV MR 83m:94033 Zbl 549 # 03054 • ID 43135

ESTEVA, F. [1981] see DOMINGO, X.

ESTEVA, F. see Vol. V for further entries

ETIEMBLE, D. & ISRAEL, M. [1974] *A new concept for ternary logic elements* (**P** 1385) Int Symp Multi-Val Log (4);1974 Morgantown 437–456
◇ B50 ◇ ID 42084

EVANS, H.P. & KLEENE, S.C. [1939] *A postulational basis for probability* (**J** 0005) Amer Math Mon 46*141–148
◇ B30 B48 ◇ REV Zbl 21.145 JSL 4.120 FdM 65.549
• ID 30933

EVANS, T. & HARDY, L. [1957] *Sheffer stroke functions in many-valued logics* (**J** 0050) Port Math 16*83–93
◇ B50 ◇ REV MR 21 # 1271 Zbl 86.245 JSL 24.67
• ID 03562

EVANS, T. & SCHWARTZ, P.B. [1958] *On Slupecki T-functions* (**J** 0036) J Symb Logic 23*267–270
◇ B50 ◇ REV MR 21 # 6326 Zbl 88.10 JSL 24.249
• ID 03564

EVANS, T. see Vol. III, IV for further entries

EVENDEN, J. [1974] *Generalized logic* (**J** 0047) Notre Dame J Formal Log 15*35–44
◇ B50 ◇ REV MR 49 # 2259 Zbl 202.7 • ID 42085

EVENDEN, J. see Vol. I for further entries

EVES, H.W. [1965] *An introduction to the foundations and fundamental concepts of mathematics* (**X** 0818) Holt Rinehart & Winston: New York xv+398pp
◇ B98 ◇ REV Zbl 125.277 • ID 23350

EXNER, R.M. & ROSSKOPF, M.F. [1959] *Logic in elementary mathematics* (**X** 0822) McGraw-Hill: New York 274pp
◇ B98 ◇ REV JSL 39.179 • ID 22192

EYTAN, M. [1977] *Logique modale propositionnelle: une vue cavaliere (English summary)* (**J** 0392) Math Sci Hum 57*27–42
◇ A10 B45 ◇ REV MR 56 # 8309 Zbl 374 # 02012
• ID 31068

EYTAN, M. [1977] *Semantique preordonnee des ensembles flous* (**P** 1825) AFCET Congr Econ Tech;1977 2*601–608
◇ B52 E72 ◇ ID 32319

EYTAN, M. see Vol. III, V for further entries

EZAWA, Y. [1983] *Homomorphic linguistic hedges and reasonable fuzzy inferences* (1111) Preprints, Manuscr., Techn. Reports etc. 24*255–265
◇ B52 ◇ REV MR 84m:03037 • REM Kansai University. Osaka • ID 35741

EZAWA, Y. & MIZUMOTO, M. [1984] *Linguistic hedges and reasonable fuzzy inferences* (**P** 3081) IFAC Symp Fuzzy Inf, Knowl Repr & Decis. Anal;1983 Marseille 243–248
◇ B52 ◇ ID 48167

EZAWA, Y. see Vol. IV for further entries

EZHKOVA, I.V. & POSPELOV, D.A. [1977] *Decision making with fuzzy premises. I: Universel scale (Russian)* (**J** 0977) Izv Akad Nauk SSSR, Tekh Kibern 1977/6*3-11
◇ B52 ◇ REV Zbl 379 # 02006 • REM Part II 1978 • ID 51839

EZHKOVA, I.V. & POSPELOV, D.A. [1978] *Decision making with fuzzy premises. II: The inference schemes (Russian)* (**J** 0977) Izv Akad Nauk SSSR, Tekh Kibern 1978/2*5-11
◇ B52 ◇ REV Zbl 406 # 03050 • REM Part I 1977 • ID 56133

FABIAN, C. [1982] see DUTA, D.L.

FACIONE, P.A. [1972] *A modal truth-tabular interpretation for necessary and sufficient conditions* (**J** 0047) Notre Dame J Formal Log 13*270-272
◇ B45 ◇ REV MR 45 # 3175 Zbl 234 # 02015 • ID 03639

FACIONE, P.A. [1977] *The entailment operator* (**J** 0047) Notre Dame J Formal Log 18*415-420
◇ B46 ◇ REV MR 58 # 5024 Zbl 271 # 02017 • ID 24242

FACIONE, P.A. & SCHERER, D. [1978] *Logic and logical thinking: a modular approach* (**X** 0822) McGraw-Hill: New York xii + 495pp
◇ A05 B98 ◇ REV JSL 46.672 • ID 44754

FADINI, A. [1962] *Il calcolo delle classi in una logica a tre valori di verita* (**J** 0336) Giorn Mat Battaglini 10(90)*72-92
◇ B50 E70 ◇ REV MR 30 # 1037 Zbl 122.9 • ID 17180

FADINI, A. [1978] *Indecisive sets and their algebra* (**J** 3523) Period Mat, Ser 5 54*3-40
◇ B52 E70 ◇ REV MR 80g:03055 • ID 72779

FADINI, A. see Vol. I, V for further entries

FAJARDO, S. [1985] *Completeness theorems for the general theory of stochastic processes* (**P** 2160) Latin Amer Symp Math Log (6);1983 Caracas 174-194
◇ B48 C65 C90 ◇ ID 41798

FAJARDO, S. [1985] *Probability logic with conditional expectation* (**J** 0073) Ann Pure Appl Logic 28*137-161
◇ B48 C40 C90 ◇ REV Zbl 564 # 03019 • ID 39901

FAJARDO, S. see Vol. III, V for further entries

FALMAGNE, J.-C. [1983] *A random utility model for a belief function* (**J** 0154) Synthese 57*35-48
◇ B48 ◇ REV MR 85b:60002 Zbl 528 # 62004 • ID 39406

FALMAGNE, J.-C. see Vol. I, V for further entries

FANG, ZHENGXIAN [1982] *Ternary algebra and fault diagnosis in ternary logic (Chinese) (English summary)* (**J** 3793) Jisuanjii Xuebao 5*411-418
◇ B50 ◇ REV MR 84i:94067 • ID 40237

FARINAS DEL CERRO, L. [1982] *A simple deduction method for modal logic* (**J** 0232) Inform Process Lett 14*49-51
◇ B35 B45 ◇ REV MR 83k:03019 Zbl 515 # 03009 • ID 36159

FARINAS DEL CERRO, L. [1982] *Prolegomena for programming in modal logic* (**P** 3846) Cybern & Systems Research (6);1982 Wien 917-920
◇ B45 ◇ REV Zbl 528 # 68015 • ID 36761

FARINAS DEL CERRO, L. [1982] *Valid reasoning: classical logic = analogical reasoning: nonclassical logic* (**P** 4041) Log, Inform, Law;1981 Firenze 2*161-165
◇ A05 B45 ◇ REV MR 85h:03021 • ID 43413

FARINAS DEL CERRO, L. & RAGGIO, A.R. [1983] *Some results in intuitionistic modal logic* (**J** 0079) Logique & Anal, NS 26*219-224
◇ B45 ◇ REV MR 85e:03037 Zbl 544 # 03001 • ID 40441

FARINAS DEL CERRO, L. [1984] see CAVALLI, A.R.

FARINAS DEL CERRO, L. [1984] *Un principe de resolution en logique modale* (**J** 3441) RAIRO Inform Theor 18*161-170
◇ B45 ◇ REV Zbl 566 # 03007 • ID 43988

FARINAS DEL CERRO, L. see Vol. I for further entries

FARRELL, R.J. [1979] *Material implication, confirmation, and counterfactuals* (**J** 0047) Notre Dame J Formal Log 20*383-394
◇ B48 ◇ REV MR 80d:03004 Zbl 321 # 02005 • ID 52618

FARRELL, R.J. see Vol. I for further entries

FARRENY, H. [1980] see CAYROL, M.

FATTOROSI-BARNABA, M. [1985] see CARO DE, F.

FAY, G. & TOEROES, R. [1978] *Quantum logic (Hungarian)* (**X** 1465) Gondolat: Budapest 386pp
◇ B51 ◇ REV MR 80b:81009 • ID 81204

FEFERMAN, S. [1980] see ACZEL, P.

FEFERMAN, S. see Vol. I, III, IV, V, VI for further entries

FEHR, E. [1982] see BERKLING, K.J.

FEHR, E. see Vol. IV, VI for further entries

FEJZULLAEV, R.B. [1983] *Algebras of polymorphisms of n-ary models (Russian) (English and Azerbaijani summaries)* (**J** 0135) Izv Akad Nauk Azerb SSR, Ser Fiz-Tekh Mat 4/1*3-6
◇ B50 ◇ REV MR 85a:03046 Zbl 538 # 03048 • ID 34797

FELDMAN, Y.A. & HAREL, D. [1984] *A probabilistic dynamic logic* (**J** 0119) J Comp Syst Sci 28*193-215
◇ B75 ◇ REV Zbl 537 # 68036 • ID 41325

FENG, YINGJUN [1983] *A method using fuzzy mathematics to solve the vectormaximum problem* (**J** 2720) Fuzzy Sets Syst 9*129-136
◇ B52 ◇ REV MR 84b:90092 Zbl 504 # 90076 • ID 38423

FENG, YULIN [1982] *A basis for programming logic and a strong termination theorem (Chinese) (English summary)* (**J** 3793) Jisuanjii Xuebao 5*11-21
◇ B75 ◇ REV MR 84i:03040 • ID 34522

FENSTAD, J.E. [1959] *Notes on normative logic* (**J** 0974) Norsk Vid-Akad Oslo Mat-Natur Kl Skr 1*25
◇ B45 ◇ ID 33312

FENSTAD, J.E. [1964] *On the consistency of the axiom of comprehension in the Lukasiewicz infinite valued logic* (**J** 0132) Math Scand 14*64-74
◇ B50 E70 ◇ REV MR 30 # 14 Zbl 134.16 JSL 32.128 • ID 03740

FENSTAD, J.E. [1968] *The structure of logical probabilities* (**J** 0154) Synthese 18*1-23
◇ B48 ◇ REV Zbl 203.296 • ID 33221

FENSTAD, J.E. [1969] *Logic and probability (Norwegian) (English summary)* (**J** 0311) Nordisk Mat Tidskr 17*71-81
• TRANSL [1981] (**C** 2617) Modern Log Survey 223-233
◇ B48 ◇ REV MR 42 # 32 MR 82f:03002 Zbl 185.7 Zbl 464 # 03001 • ID 33222

FENSTAD, J.E. [1980] *The structure of probabilities defined on first-order languages* (**C** 3537) Stud Induct Logic & Probab, Vol 2 251-262
◇ B48 ◇ REV MR 82d:03032 Zbl 539 # 03004 JSL 49.1409
• ID 72853

FENSTAD, J.E. see Vol. I, III, IV, V, VI for further entries

FERENCZI, M. [1977] *On valid assertions - in probability logic* (**J** 0411) Studia Sci Math Hung 12*101-116
◇ B25 B48 C90 ◇ REV MR 82c:03026 Zbl 435 # 03022
• ID 55783

FERENCZI, M. see Vol. I for further entries

FERREIRA BARBOSA, J.E. [1970] *On the operational logic of junctors (Portuguese and English)* (**S** 1558) Bol Anal Log Mat (Rio de Janeiro) 2*29-40,41-52
◇ B60 ◇ ID 61675

FERREIRA BARBOSA, J.E. see Vol. V for further entries

FETZER, J.H. [1981] *Probability and explanation* (**J** 0154) Synthese 48*371-408
◇ B48 ◇ REV MR 84e:03027 Zbl 483 # 03011 • ID 34366

FETZER, J.H. [1983] *Probability and objectivity in deterministic and indeterministic situations* (**J** 0154) Synthese 57*367-386
◇ B48 ◇ REV MR 85e:03054 • ID 40622

FEYERABEND, P. [1958] *Reichenbach's interpretation of quantum-mechanics* (**J** 0095) Philos Stud 9*49-59
◇ B51 ◇ REV JSL 25.289 • ID 03755

FEYS, R. [1937] *Les logiques nouvelles des modalites* (**J** 1720) Rev Neoscolast Philos, Ser 2 40*517-553
◇ B20 B48 ◇ REV FdM 63.826 • REM See also 1938
• ID 41048

FEYS, R. [1938] *Les logiques nouvelles des modalites* (**J** 1720) Rev Neoscolast Philos, Ser 2 41*217-242
◇ B45 ◇ REV JSL 3.120 FdM 64.928 • REM See also 1937
• ID 41413

FEYS, R. [1939] *Principes de logistique, premier volume* (**X** 0879) Inst Sup Philos: Louvain 129pp
◇ A05 B98 ◇ REV JSL 5.38 • ID 03757

FEYS, R. [1944] *Logistic. Formal logic I. General survey. Logic of propositions and classes (Dutch)* (**X** 2110) Dekker & van de Vegt: Nijmegen 340pp
◇ B98 ◇ REV MR 7.185 Zbl 60.20 JSL 10.100 • ID 21325

FEYS, R. [1953] *A simplified proof of the reduction of all modalities to 42 in* S3 (**J** 0250) Bol Soc Mat Mexicana 10*53-57
◇ B45 ◇ REV MR 15.386 Zbl 53.343 JSL 20.66 • ID 03763

FEYS, R. [1958] see CURRY, H.B.

FEYS, R. [1961] *Modeles a variables de differentes sortes pour les logiques modales M" ou S5* (**P** 0711) Concept & Role of Model in Math & Sci;1960 Utrecht 58-72
◇ B45 ◇ ID 17199

FEYS, R. [1963] *Carnap on modalities* (**C** 0673) Phil of Carnap 283-297
◇ A05 B45 ◇ ID 04292

FEYS, R. [1965] *Modal logics* (**X** 0834) Gauthier-Villars: Paris xiv+219pp
• TRANSL [1974] (**X** 2027) Nauka: Moskva 520pp
◇ B45 ◇ REV MR 30 # 3008 MR 50 # 9527 Zbl 128.12
Zbl 291 # 02009 JSL 34.501 • ID 03765

FEYS, R. & FITCH, F.B. [1969] *Dictionary of symbols of mathematical logic* (**X** 0809) North Holland: Amsterdam xiv+171pp
• TRANSL [1980] (**X** 3560) Paraninfo: Madrid 189pp
◇ B98 ◇ REV MR 40 # 7082 Zbl 179.9 Zbl 489 # 03001
• ID 03766

FEYS, R. see Vol. I, V, VI for further entries

FICHTE, J. [1978] *Ueber den Wahrheitswertbreich von fuzzy Aussagenlogiken sowie Bemerkungen zu deren Axiomatisierung* (**P** 3059) Algeb Method Automatenth;1978 Altenberg 3-25
◇ B52 ◇ REV MR 83i:03042 • ID 35511

FIDEL, M.M. [1977] *The decidability of the calculi* $\mathscr{C}_n$ (**J** 0302) Rep Math Logic, Krakow & Katowice 8*31-40
◇ B25 B53 G25 ◇ REV MR 58 # 158 Zbl 378 # 02011
• ID 24188

FIDEL, M.M. [1978] *An algebraic study of a propositional system of Nelson* (**P** 1800) Brazil Conf Math Log (1);1977 Campinas 99-117
◇ B55 F50 G10 G25 ◇ REV MR 80a:03073
Zbl 389 # 03025 • ID 52314

FIDEL, M.M. [1980] *An algebraic study of logic with constructible negation* (**P** 3006) Brazil Conf Math Log (3);1979 Recife 119-129
◇ B25 B55 G25 ◇ REV MR 82i:03075 Zbl 453 # 03024
• ID 54154

FIELD, H.H. [1978] *A note on Jeffrey conditionalization* (**J** 0153) Phil of Sci (East Lansing) 45*361-367
◇ B48 ◇ REV MR 80f:62006 • ID 81218

FILIPOIU, A. [1978] *Analytic tableaux for* θ*-valued propositional logic* (**J** 1508) Math Sem Notes, Kobe Univ 6*517-526
◇ B50 G20 ◇ REV MR 80i:03036 Zbl 403 # 03021
• ID 54735

FILIPOIU, A. see Vol. I for further entries

FILLMORE, C.J. [1971] *Entailment rules in a semantic theory* (**C** 1029) Readings Phil of Lang 533-548
◇ A05 B46 ◇ ID 15152

FINCH, P.D. [1969] *On the structure of quantum logic* (**J** 0036) J Symb Logic 34*275-282
• REPR [1975] (**C** 3045) Log-Algeb Appr to Quant Mech 1*415-425
◇ B51 G12 ◇ REV Zbl 205.8 JSL 50.558 • ID 03778

FINCH, P.D. [1976] *Quantum mechanical physical quantities as random variables* (**P** 2411) Found Probab Th, Stat Inf & Stat Th Sci;1973 London ON 3*81-103
◇ B51 ◇ REV MR 58 # 19904 Zbl 349 # 02025 • ID 61682

FINE, A.I. [1982] *Comments on the significance of Bell's theorem* (**J** 2730) Phys Rev Lett 49*1536
◇ B51 ◇ REV MR 84m:81028d • ID 39543

FINE, A.I. see Vol. V for further entries

FINE, K. [1970] *Propositional quantifiers in modal logic* (**J** 0105) Theoria (Lund) 36*336-346
◇ B45 ◇ REV MR 48 # 5824 Zbl 302 # 02005 JSL 38.329
• ID 03782

FINE, K. [1971] *The logics containing S4.3* (**J** 0068) Z Math Logik Grundlagen Math 17*371-376
◇ B45 ◇ REV MR 45 # 1735 Zbl 228 # 02011 • ID 03783

FINE, K. [1972] *In so many possible worlds* (**J** 0047) Notre Dame J Formal Log 13*516–520
◇ B45 ◇ REV MR 47 # 8248 Zbl 242 # 02025 • ID 03784

FINE, K. [1972] *Logics containing S4 without the finite model property* (**P** 2080) Conf Math Log;1970 London 98–102
◇ B45 ◇ REV MR 48 # 10764 Zbl 239 # 02008 • ID 17175

FINE, K. [1974] *An ascending chain of S4 logics* (**J** 0105) Theoria (Lund) 40*110–116
◇ B45 ◇ REV MR 58 # 27326 Zbl 307 # 02013 • ID 30691

FINE, K. [1974] *An incomplete logic containing S4* (**J** 0105) Theoria (Lund) 40*23–29
◇ B45 ◇ REV MR 58 # 27325 Zbl 287 # 02011 • ID 04293

FINE, K. [1974] *Logics containing* K4 *I* (**J** 0036) J Symb Logic 39*31–42
◇ B45 ◇ REV MR 49 # 8814 Zbl 287 # 02010 • REM Part II 1985 • ID 03785

FINE, K. [1974] *Models for entailment* (**J** 0122) J Philos Logic 3*347–372
◇ B46 ◇ REV MR 55 # 10241 Zbl 296 # 02013 • ID 30692

FINE, K. [1975] *Normal forms in modal logic* (**J** 0047) Notre Dame J Formal Log 16*229–237
◇ B45 ◇ REV MR 51 # 2869 Zbl 245 # 02025 • ID 03786

FINE, K. [1975] *Some connections between elementary and modal logic* (**P** 0757) Scand Logic Symp (3);1973 Uppsala 15–31
◇ B45 C07 C90 ◇ REV MR 53 # 5265 Zbl 316 # 02021 • ID 22897

FINE, K. [1975] *Vagueness, truth and logic* (**J** 0154) Synthese 30*265–300
◇ A05 B45 ◇ REV Zbl 311 # 02011 • ID 29596

FINE, K. [1977] *Properties, propositions and sets* (**J** 0122) J Philos Logic 6*135–191
◇ A05 B45 ◇ REV MR 58 # 5050 Zbl 385 # 03006 • ID 52114

FINE, K. [1978] *Model theory for modal logic. part I: the de re/de dicto distinction. part II: the elimination of de re modality* (**J** 0122) J Philos Logic 7*125–156,277–306
◇ A05 B45 C40 C90 ◇ REV MR 80c:03021 MR 81f:03024 Zbl 375 # 02008 Zbl 409 # 03007 JSL 50.1083 • REM Part III 1981 • ID 30694

FINE, K. [1979] *Failures of the interpolation lemma in quantified modal logic* (**J** 0036) J Symb Logic 44*201–206
◇ B45 ◇ REV MR 80i:03029 Zbl 415 # 03015 JSL 48.486 • ID 53117

FINE, K. [1980] *First-order modal theories II* (**J** 0063) Studia Logica 39*159–202
◇ B45 ◇ REV MR 82g:03024 Zbl 468 # 03011 • REM Part I 1981. Part III 1982 • ID 55076

FINE, K. [1981] *First-order modal theories. I. Sets* (**J** 0097) Nous, Quart J Phil 15*177–205
◇ B45 E70 ◇ REV MR 82j:03020 • REM Part II 1980 • ID 72893

FINE, K. [1981] *Model theory for modal logic. III. Existence and predication* (**J** 0122) J Philos Logic 10*293–307
◇ A05 B45 C90 ◇ REV MR 83c:03016 Zbl 464 # 03017 JSL 50.1083 • REM Parts I,II 1978 • ID 54607

FINE, K. [1982] *First-order modal theories III. Facts* (**J** 0154) Synthese 53*43–122
◇ B45 ◇ REV MR 84j:03035 Zbl 522 # 03009 • REM Part II 1980 • ID 34627

FINE, K. [1983] *The permutation principle in quantificational logic* (**J** 0122) J Philos Logic 12*33–37
◇ B45 ◇ REV MR 85g:03016 Zbl 527 # 03007 • ID 37507

FINE, K. [1985] *Logics containing K4. II* (**J** 0036) J Symb Logic 50*619–651
◇ B45 ◇ REV Zbl 574 # 03008 • REM Part I 1974 • ID 47369

FINE, K. see Vol. III, VI for further entries

FINE, T.L. & WALLEY, P. [1979] *Varieties of modal (classificatory) and comparative probability* (**J** 0154) Synthese 41*321–374
◇ A05 B45 ◇ REV MR 80j:03016 Zbl 442 # 60005 • ID 56403

FINE, T.L. see Vol. I, IV for further entries

FINETTI DE, B. [1972] *Probability, induction and statistics: The art of guessing* (**X** 0827) Wiley & Sons: New York xxiv + 266pp
◇ A05 B48 ◇ REV MR 55 # 13512 Zbl 275 # 60001 • ID 25405

FINETTI DE, B. [1974] *Theory of probability. Vol.I* (**X** 0827) Wiley & Sons: New York xix + 300pp
◇ B48 ◇ REV MR 55 # 13514a Zbl 328 # 60002 • REM Part II 1975 • ID 25406

FINETTI DE, B. [1975] *Theory of probability. Vol.II* (**X** 0827) Wiley & Sons: New York xviii + 375pp
◇ B48 ◇ REV MR 55 # 13514b Zbl 328 # 60003 • REM Part I 1974 • ID 48287

FINETTI DE, B. [1982] *Probability: the different views and terminologies in a critical analysis* (**P** 3622) Int Congr Log, Meth & Phil of Sci (6,Proc);1979 Hannover 391–394
◇ B48 ◇ REV MR 83m:03030 • ID 35435

FINKELSTEIN, D. [1976] *Classical and quantum probability and set theory* (**P** 2411) Found Probab Th, Stat Inf & Stat Th Sci;1973 London ON 3*111–119
◇ A05 B51 E70 G12 ◇ REV Zbl 322 # 02002 • ID 61694

FINKELSTEIN, D. [1977] *The Leibnitz project* (**J** 0122) J Philos Logic 6*425-439
◇ B51 G12 G25 ◇ REV MR 58 # 32486 Zbl 394 # 03058 • ID 52535

FINKELSTEIN, D. [1981] *Quantum sets, assemblies and plexi* (**P** 3820) Curr Iss in Quantum Log;1979 Erice 323–331
◇ B51 E70 ◇ REV MR 84j:03123 • ID 34713

FINKELSTEIN, D. & RODRIGUEZ, E. [1985] *Application of quantum set theory to quantum time-space* (**P** 4401) Rect Devel in Quant Log;1984 Koeln 315–318
◇ B52 E70 ◇ ID 49064

FINKELSTEIN, D. see Vol. I for further entries

FINN, V.K. [1969] *The precompleteness of a class of functions that corresponds to the three-valued logic of J.Lukasiewicz (Russian)* (**J** 0338) Nauch-Tekh Inf, Ser 2, Akad Nauk SSSR 1969/10*35–38
◇ B50 ◇ REV MR 44 # 3841 • ID 17176

FINN, V.K. [1970] *Certain characteristic truth-tables of classical logic and of the three-valued logic of J.Lukasiewicz (Russian)* (**C** 1530) Issl Log Sist (Yanovskaya) 215–261
◇ B50 ◇ REV MR 48 # 1881 Zbl 252 # 02007 • ID 17177

FINN, V.K. [1971] *Axiomatization of some 3-valued propositional calculi and of their algebras (Russian)* (**X** 2235) VINITI: Moskva 4471–72
◇ B50 ◇ ID 41976

FINN, V.K. [1971] *Ueber die Axiomatisierung einiger dreiwertiger Logiken (Russisch)* (**J** 0338) Nauch-Tekh Inf, Ser 2, Akad Nauk SSSR 1971/11*16-20
◇ B50 ◇ REV Zbl 242 # 02015 • ID 26339

FINN, V.K. [1974] *A criterion of functional completeness for* $\mathfrak{B}^3$ (**J** 0063) Studia Logica 33*121-125
◇ B50 ◇ REV MR 50 # 9529 Zbl 295 # 02011 • ID 04237

FINN, V.K. [1974] *A test for functional completeness for* $\mathfrak{B}^3$ *(Russian)* (**C** 2577) Issl Formaliz Yazyk & Neklass Log 194-199
◇ B50 ◇ REV MR 56 # 5214 • ID 72911

FINN, V.K. [1974] *Axiomatization of some three-valued propositional calculi and their algebras (Russian)* (**C** 2578) Filos & Logika 398-438
◇ B50 G20 ◇ ID 72910

FINN, V.K. [1974] see BOCHVAR, D.A.

FINN, V.K. [1976] see BOCHVAR, D.A.

FINN, V.K. [1977] see BOCHVAR, D.A.

FINN, V.K. & GRIGOLIYA, R.SH. [1979] *Bochvar algebras and corresponding propositional calculi (Russian)* (**C** 2581) Issl Neklass Log & Teor Mnozh 345-372
• TRANSL [1980] (**J** 0387) Bull Sect Logic, Pol Acad Sci 9*39-45
◇ B50 G25 ◇ REV MR 81f:03031 MR 82c:03029 Zbl 441 # 03023 Zbl 464 # 03023 • ID 54613

FINN, V.K. [1980] see ANSHAKOV, O.M.

FINN, V.K. [1981] see ANSHAKOV, O.M.

FINN, V.K. [1982] see ANSHAKOV, O.M.

FINN, V.K. see Vol. I for further entries

FIORENTINI, M. & MARRUCCELLI, A. [1970] *Complementi di matematiche moderne; Logica matematicea, teorie degli insiemi, strutture algebriche* (**X** 0909) Cedam: Padova 150pp
◇ B98 E98 ◇ REV MR 44 # 1531 • ID 17901

FISCH, M.H. & TURQUETTE, A.R. [1966] *Peirce's triadic logic* (**J** 0327) Trans Pierce Soc 2*71-85
◇ A10 B50 ◇ REV MR 35 # 29 • ID 17110

FISCHER, MICHAEL J. & LADNER, R.E. [1979] *Propositional dynamic logic of regular programs* (**J** 0119) J Comp Syst Sci 18*194-211
◇ B75 D15 ◇ REV MR 80f:68013 Zbl 408 # 03014 • ID 56253

FISCHER, MICHAEL J. see Vol. I, III, IV, VI for further entries

FISCHER SERVI, G. [1976] *Un'algebrizzazione del calcolo intuizionista monadico (English summary)* (**J** 0319) Matematiche (Sem Mat Catania) 31*262-276
◇ B55 F50 G10 ◇ REV MR 83b:03072 Zbl 437 # 03033 • ID 55901

FISCHER SERVI, G. [1977] *On modal logic with an intuitionistic base* (**J** 0063) Studia Logica 36*141-149
◇ B45 F50 ◇ REV Zbl 364 # 02015 • ID 50945

FISCHER SERVI, G. [1978] *Semantics for a class of intuitionistic modal calculi* (**J** 0387) Bull Sect Logic, Pol Acad Sci 7*26-30
◇ B55 G25 ◇ REV MR 80c:03022 Zbl 429 # 03009 • ID 53840

FISCHER SERVI, G. [1978] *The finite model property for MIPQ and some consequences* (**J** 0047) Notre Dame J Formal Log 19*687-692
◇ B55 F50 G10 ◇ REV MR 80g:03015 Zbl 368 # 02024 • ID 52127

FISCHER SERVI, G. [1981] *Completeness for nonnormal intuitionistic modal logics (Italian summary)* (**J** 3746) Note Math (Lecce) 1*203-212
◇ B45 ◇ REV MR 85d:03031 Zbl 542 # 03006 • ID 40274

FISCHER SERVI, G. [1981] *Remarks on Halmos' duality theory (Italian summary)* (**J** 3285) Boll Unione Mat Ital, V Ser, A 18*457-460
◇ B45 G05 ◇ REV MR 83e:03107 Zbl 467 # 03014 • ID 35253

FISCHER SERVI, G. [1981] *Semantics for a class of intuitionistic modal calculi* (**C** 3515) Ital Studies in Phil of Sci 59-72
◇ B45 B55 F50 ◇ REV MR 82f:03013 Zbl 452 # 03014 • ID 54078

FISCHER SERVI, G. [1981] *Teoremi di completezza per calcoli bimodali (English summary)* (**J** 0549) Riv Mat Univ Parma, Ser 4 7*347-350
◇ B45 ◇ REV MR 83j:03032 Zbl 499 # 03004 • ID 38117

FISHBURN, P.C. [1970] *Utility theory for decision making* (**X** 0827) Wiley & Sons: New York 234pp
◇ B48 ◇ REV MR 41 # 9401 Zbl 213.462 • ID 25412

FISHBURN, P.C. [1982] *The foundations of expected utility* (**X** 0835) Reidel: Dordrecht xii + 176pp
◇ B48 ◇ REV MR 85k:90021 Zbl 497 # 90001 • ID 44848

FISHBURN, P.C. see Vol. I, III, V for further entries

FISHER, A. [1982] *Formal number theory and computability* (**X** 0894) Oxford Univ Pr: Oxford xiii + 190pp
◇ B98 D20 F30 F98 ◇ REV MR 85g:03001 Zbl 504 # 03002 • ID 36967

FISHER, M. [1961] *A three-valued calculus for deontic logic* (**J** 0105) Theoria (Lund) 27*107-118
◇ B50 ◇ REV JSL 31.278 • ID 04316

FISHER, M. [1962] *A system of deontic-alethic modal logic* (**J** 0094) Mind 71*231-236
◇ B45 ◇ REV JSL 27.220 JSL 38.327 • ID 15011

FISHER, M. [1979] *Rules as nests of beliefs and intentions* (**J** 0079) Logique & Anal, NS 22*133-146
◇ A05 B45 ◇ REV MR 81g:03016 • ID 72927

FISK, M. [1964] *A modern formal logic* (**X** 0819) Prentice Hall: Englewood Cliffs xi + 116pp
◇ B98 ◇ REV JSL 30.87 • ID 22489

FISK, M. see Vol. I for further entries

FITCH, F.B. [1937] *Modal functions in two-valued logic* (**J** 0036) J Symb Logic 2*125-128 • ERR/ADD ibid 13*38-39
◇ B45 ◇ REV Zbl 17.337 JSL 3.50 FdM 63.825 • ID 04320

FITCH, F.B. [1939] *Note on modal functions* (**J** 0036) J Symb Logic 4*115-116 • ERR/ADD ibid 13*38-39
◇ B45 ◇ REV MR 1.131 Zbl 21.386 JSL 5.31 • ID 04322

FITCH, F.B. [1948] *Intuitionistic modal logic with quantifiers* (**J** 0050) Port Math 7*113-118
◇ B45 F50 ◇ REV MR 10.669 Zbl 34.153 JSL 14.261 • ID 04332

FITCH, F.B. [1952] *Symbolic logic. An introduction* (**X** 0879) Inst Sup Philos: Louvain x + 238pp
◇ B40 B98 ◇ REV MR 15.592 Zbl 49.5 JSL 17.266 • ID 04335

FITCH, F.B. [1963] *A logical analysis of some value concepts* (**J** 0036) J Symb Logic 28*135-142
◇ B45 ◇ REV MR 30 # 1027 • ID 04344

FITCH, F.B. [1966] *Natural deduction rules for obligation* (**J** 0325) Amer Phil Quart 3*27-38
◇ B45 F07 ◇ REV JSL 33.136 • ID 17115

FITCH, F.B. [1967] *A complete and consistent modal set theory* (**J** 0036) J Symb Logic 32*93-103 • ERR/ADD ibid 35*242
◇ B45 E70 ◇ REV MR 38 # 5593 Zbl 158.12 JSL 34.125 • ID 04347

FITCH, F.B. [1969] see FEYS, R.

FITCH, F.B. [1973] *A correlation between modal reduction principles and properties of relations* (**J** 0122) J Philos Logic 2*97-101
◇ B45 ◇ REV MR 54 # 2422 Zbl 259 # 02012 • ID 14513

FITCH, F.B. [1974] *Elements of combinatory logic* (**X** 0875) Yale Univ Pr: New Haven viii + 162pp
◇ B40 B98 ◇ REV MR 54 # 2429 JSL 41.789 • ID 24019

FITCH, F.B. see Vol. I, IV, V, VI for further entries

FITTING, M. [1969] *Intuitionistic logic, model theory, and forcing* (**X** 0809) North Holland: Amsterdam 191pp
◇ B98 C90 C98 E25 E35 E45 E50 F50 F98 ◇ REV MR 41 # 6666 Zbl 188.320 JSL 36.166 • ID 04349

FITTING, M. [1969] *Logics with several modal operators* (**J** 0105) Theoria (Lund) 35*259-266
◇ B45 ◇ REV MR 41 # 5199 Zbl 188.318 • ID 04350

FITTING, M. [1970] *An embedding of classical logic in S4* (**J** 0036) J Symb Logic 35*529-534
◇ B45 ◇ REV MR 43 # 3099 Zbl 219 # 02011 • ID 04351

FITTING, M. [1972] *ε-calculus based axiom systems for some propositional modal logics* (**J** 0047) Notre Dame J Formal Log 13*381-384
◇ B45 ◇ REV MR 52 # 44 Zbl 238 # 02023 • ID 17213

FITTING, M. [1972] *An ε-calculus system for first order S 4* (**P** 2080) Conf Math Log;1970 London 103-110
◇ B45 ◇ REV MR 50 # 59 Zbl 235 # 02020 • ID 17119

FITTING, M. [1972] *Non-classical logics and the independence results of set theory* (**J** 0105) Theoria (Lund) 38*133-142
◇ B45 C90 E35 ◇ REV MR 58 # 16285 Zbl 255 # 02070 • ID 31074

FITTING, M. [1972] *Tableau methods of proof for modal logics* (**J** 0047) Notre Dame J Formal Log 13*237-247
◇ B45 ◇ REV MR 45 # 3174 Zbl 184.281 • ID 04353

FITTING, M. [1973] *A modal logic analog of Smullyan's fundamental theorem* (**J** 0068) Z Math Logik Grundlagen Math 19*1-16
◇ B45 ◇ REV MR 49 # 2264 Zbl 257 # 02017 • ID 04356

FITTING, M. [1973] *Model existence theorems for modal and intuitionistic logics* (**J** 0036) J Symb Logic 38*613-627
◇ B45 F50 ◇ REV MR 52 # 7852 Zbl 286 # 02060 • ID 04355

FITTING, M. [1975] *A modal logic ε-calculus* (**J** 0047) Notre Dame J Formal Log 16*1-16
◇ B45 ◇ REV MR 50 # 12664 Zbl 226 # 02018 • ID 04357

FITTING, M. [1977] *A tableau system for propositional S5* (**J** 0047) Notre Dame J Formal Log 18*292-294
◇ B45 ◇ REV MR 58 # 5051 Zbl 314 # 02039 • ID 23619

FITTING, M. [1978] *Subformula results in some propositional modal logics* (**J** 0063) Studia Logica 37*387-391
◇ B45 ◇ REV MR 81b:03018 Zbl 397 # 03011 • ID 52677

FITTING, M. [1983] *Proof methods for modal and intuitionistic logics* (**S** 3307) Synth Libr viii + 555pp
◇ B45 B55 F07 F50 F98 ◇ REV MR 84j:03036 Zbl 523 # 03013 JSL 50.855 • ID 34628

FITTING, M. [1984] *A symmetric approach to axiomatizing quantifiers and modalities* (**J** 0154) Synthese 60*5-19
◇ B45 ◇ REV MR 85i:03047 • ID 44095

FITTING, M. [1984] *Linear reasoning in modal logic* (**J** 0036) J Symb Logic 49*1363-1378
◇ B45 C40 ◇ REV MR 86f:03027 • ID 42443

FITTING, M. [1985] *A Kripke-Kleene semantics for logic programs* (**J** 2551) J Log Progr 2/4*295-312
◇ B45 ◇ ID 49493

FITTING, M. see Vol. I, III, IV, V, VI for further entries

FLAGG, R.C. [1985] *Epistemic set theory is a conservative extension of intuitionistic set theory* (**J** 0036) J Symb Logic 50*895-902
◇ B45 E35 E70 F50 ◇ ID 49605

FLAGG, R.C. see Vol. I, VI for further entries

FLOWERS, P.L. & KANDEL, A. [1985] *Possibilistic search trees* (**J** 2720) Fuzzy Sets Syst 16*1-24
◇ B52 ◇ ID 46406

FLUM, J. [1978] see EBBINGHAUS, H.-D.

FLUM, J. see Vol. I, III, IV, V for further entries

FOLEY, R. [1979] *Justified inconsistent beliefs* (**J** 0325) Amer Phil Quart 16*247-257
◇ B53 ◇ ID 42834

FOLLESDAL, D. [1965] *Contributions to a discussion after Ruth Barcan Marcus: "modal logics"* (**C** 1826) Boston Colloq Philos Sci 1960-64 111-112,114
◇ B45 ◇ ID 32323

FOLLESDAL, D. [1965] *Quantification into causal contexts* (**C** 1826) Boston Colloq Philos Sci 1960-64 263-274 • REPR [1971] (**C** 1812) Refer & Modal 52-62
◇ A05 B45 ◇ REV Zbl 192.24 • ID 32326

FOLLESDAL, D. [1966] *A model theoretic approach to causal logic* (**J** 0239) Skr, K Nor Vidensk Selsk 2*13pp
◇ A05 B45 C90 ◇ REV Zbl 192.32 • ID 32328

FOLLESDAL, D. [1967] *Knowledge, identity and existence* (**J** 0105) Theoria (Lund) 33*1-27
◇ A05 B45 ◇ REV Zbl 174.9 • ID 32329

FOLLESDAL, D. [1969] *Quine on modality* (**C** 0556) Words & Objections (Quine) 175-185 • REPR [1974] (**C** 1828) Readings Semantics
◇ A05 B45 ◇ ID 14943

FOLLESDAL, D. & HILPINEN, R. [1971] *Deontic logic: an introduction* (**C** 1547) Deontic Log Readings 1-35
◇ B45 ◇ REV MR 56 # 2790 Zbl 228 # 02012 • ID 32330

FOLLESDAL, D. see Vol. I, III for further entries

Font i Llovet, J.M. & Verdu i Solans, V. [1979] *Abstract logics, interior operators and modal logics S4 (Spanish, Catalan)* (**P** 4097) Jorn Mat Luso-Espanol (6);1979 Santander 867-869,1003-1015
◇ B45 ◇ ID 45454

Font i Llovet, J.M. [1980] *Induction of interiors of order in abstract logics (Catalan) (English summary)* (**P** 3512) Jorn Mat Luso-Espanol (7);1980 St Feliu de Guixois 1*79-82
◇ B45 ◇ ID 44705

Font i Llovet, J.M. [1982] *An implication connective for classical and intuitionistic modal logics (Catalan)* (**P** 3870) Congr Catala de Log Mat (1);1982 Barcelona 83-84
◇ B45 ◇ REV MR 84i:03003 Zbl 523 # 03050 • ID 37037

Font i Llovet, J.M. [1982] *Modal logic without primitive modal operators (Spanish) (English summary)* (**P** 4096) Jorn Mat Luso-Espanol (9);1982 Salamanca 1*169-172
◇ B45 ◇ ID 45453

Font i Llovet, J.M. [1982] *On the definition of necessity and possibility in non-classical modal logic (Catalan)* (**P** 3870) Congr Catala de Log Mat (1);1982 Barcelona 85-88
◇ B45 ◇ REV MR 84i:03003 Zbl 508 # 03009 • ID 36933

Font i Llovet, J.M. [1982] *Some non-classical connectives for intuitionistic modal logic* (**P** 3845) Conf Math Service of Man (2,Proc)(Feriet);1982 Las Palmas 285-287
◇ B45 ◇ REV Zbl 515 # 03007 • ID 37837

Font i Llovet, J.M. [1983] *Intuitionistic implication in some non-classical modal logics of type S4* (**J** 0387) Bull Sect Logic, Pol Acad Sci 12*2-7
◇ B45 ◇ REV Zbl 544 # 03002 • ID 40985

Font i Llovet, J.M. [1984] *Implication and deduction in some intuitionistic modal logics* (**J** 0302) Rep Math Logic, Krakow & Katowice 17*27-38
◇ B45 ◇ REV MR 86e:03018 • ID 44022

Forbes, G. [1981] *On the philosophical basis of essentialist theories* (**J** 0122) J Philos Logic 10*73-99
◇ A05 B45 ◇ REV MR 82f:03014 • ID 72992

Forbes, G. [1983] *Physicalism, instrumentalism and the semantics of modal logic* (**J** 0122) J Philos Logic 12*271-298
◇ A05 B45 ◇ REV MR 84k:03013 Zbl 513 # 03004
• ID 34953

Forbes, G. [1983] *Thisness and vagueness* (**J** 0154) Synthese 54*235-259
◇ A05 B45 ◇ REV MR 84k:03053 Zbl 519 # 03003
• ID 34991

Forbes, G. see Vol. V for further entries

Forge, J. [1984] *Theoretical functions, theory and evidence* (**J** 0153) Phil of Sci (East Lansing) 51*443-463
◇ A05 B48 ◇ REV MR 85m:03012 • ID 43991

Fort, M. [1973] *Calcul des predicats trivalent* (**J** 2313) C R Acad Sci, Paris, Ser A-B 276*A1257-A1260
◇ B50 ◇ REV MR 47 # 4762 Zbl 266 # 02013 • ID 04446

Fort, M. [1973] *Deduction en logique trivalente* (**J** 2313) C R Acad Sci, Paris, Ser A-B 277*A823-A825
◇ B50 ◇ REV MR 48 # 3700 Zbl 281 # 02024 • ID 04445

Foster, A.L. [1950] *On n-ality theories in rings and their logical algebras, including triality principle and their three valued logics* (**J** 0100) Amer J Math 72*101-123
◇ B50 G25 ◇ REV MR 11.414 Zbl 37.18 JSL 15.230
• ID 04452

Foster, A.L. [1951] *p^k-rings and ring-logics* (**J** 0315) Ann Sc Norm Sup Pisa Fis Mat, Ser 3 5*279-300
◇ B50 ◇ REV MR 13.903 Zbl 44.262 JSL 17.279
• ID 04453

Foster, A.L. [1951] *Ring-logics and p-rings* (**S** 0183) Publ Math Univ California 1*385-395
◇ B50 ◇ REV MR 13.426 Zbl 45.319 JSL 17.205
• ID 04454

Foster, A.L. see Vol. III for further entries

Fotr, J. [1983] *Subjective probabilities and methods for constructing them (Czech) (English summary)* (**J** 2711) Ekonom Mat Obzor (Prague) 19*318-336
◇ B48 ◇ REV MR 85d:60002 • ID 38980

Foulis, D.J. [1969] *Fundamental concepts of mathematics* (**X** 1337) Prindle Weber Schmidt: Boston 212pp
◇ B98 ◇ REV Zbl 222 # 00001 • ID 26299

Foulis, D.J. & Randall, C.H. [1970] *An approach to empirical logic* (**J** 0005) Amer Math Mon 77*363-374
◇ B51 G12 ◇ REV MR 41 # 3334 Zbl 209.303 • ID 10981

Foulis, D.J. & Randall, C.H. [1974] *Empirical logic and quantum mechanics* (**J** 0154) Synthese 29*81-111
◇ B51 G12 ◇ REV Zbl 344 # 02018 • ID 61752

Foulis, D.J. & Randall, C.H. [1976] *A mathematical setting for inductive reasoning* (**P** 2411) Found Probab Th, Stat Inf & Stat Th Sci;1973 London ON 3*169-205
◇ B48 ◇ REV MR 58 # 16185 Zbl 344 # 02019 • ID 64742

Foulis, D.J. & Randall, C.H. [1976] *Empirical logic and quantum mechanics* (**C** 2953) Log & Probab in Quant Mech 73-103
◇ B51 G12 ◇ REV Zbl 359 # 02018 • ID 50566

Foulis, D.J. & Randall, C.H. [1979] *The operational approach to quantum mechanics* (**C** 3566) Phys Theor as Log-Operat Struct 167-201
◇ B51 G12 ◇ REV MR 80f:81013 Zbl 402 # 03053
• ID 82572

Foulis, D.J. & Randall, C.H. [1981] *Empirical logic and tensor products* (**P** 3185) Interpr & Found of Quantum Th;1979 Marburg 9-20
◇ B51 ◇ REV MR 85g:81005 Zbl 495 # 03041 • ID 36889

Foulis, D.J. & Randall, C.H. [1981] *What are quantum logics and what ought they to be?* (**P** 3820) Curr Iss in Quantum Log;1979 Erice 35-52
◇ B51 ◇ REV MR 84k:03142 Zbl 537 # 03044 • ID 35034

Foulis, D.J. & Randall, C.H. [1983] *Properties and operational propositions in quantum mechanics* (**J** 1678) Found Phys 13*843-857
◇ B51 G12 ◇ ID 45638

Foulkes, P. [1972] *The logic of "and" in the anatomy of proof: a new model for entailment* (**P** 2080) Conf Math Log;1970 London 255*339-341
◇ B46 ◇ REV Zbl 228 # 02014 • ID 29117

Fox, F. [1974] *A note on a consistency proof* (**J** 0047) Notre Dame J Formal Log 15*176
◇ B45 ◇ REV MR 50 # 47 Zbl 232 # 02005 • ID 04471

Fox, J. [1981] *Towards a reconciliation of fuzzy logic and standard logic* (**J** 1741) Int J Man-Mach Stud 15*213-220
◇ B52 ◇ REV MR 83b:03023 • ID 35089

FOXLEY, E. [1962] *The determination of all Sheffer functions in 3-valued logic, using a logical computer* (**J** 0047) Notre Dame J Formal Log 3*41-50
◇ B35 B50 ◇ REV MR 27 # 3508 Zbl 112.5 JSL 27.681
• ID 04474

FOXLEY, E. see Vol. I for further entries

FRAASSEN VAN, B.C. [1966] *The completeness of free logic* (**J** 0068) Z Math Logik Grundlagen Math 12*219-234
◇ B60 G15 ◇ REV MR 34 # 7338 Zbl 149.6 • ID 04477

FRAASSEN VAN, B.C. & LAMBERT, K. [1967] *On free description theory* (**J** 0068) Z Math Logik Grundlagen Math 13*225-240
◇ B60 G15 ◇ REV Zbl 164.309 • ID 33191

FRAASSEN VAN, B.C. [1968] *A topological proof of the Loewenheim-Skolem, compactness, and strong completeness theorems for free logic* (**J** 0068) Z Math Logik Grundlagen Math 14*245-254
◇ B60 C07 ◇ REV MR 38 # 35 Zbl 165.15 • ID 04480

FRAASSEN VAN, B.C. & LAMBERT, K. [1970] *Meaning relations, possible objects and possible worlds* (**P** 0559) Phil Probl in Logic;1968 Irvine 1-19
◇ B45 ◇ REV MR 43 # 7304 Zbl 188.320 • ID 22261

FRAASSEN VAN, B.C. [1971] *Formal semantics and logic* (**X** 0843) Macmillan : New York & London xi+225pp
◇ B98 ◇ REV Zbl 253 # 02002 JSL 45.376 • ID 61762

FRAASSEN VAN, B.C. [1972] *The logic of conditional obligation* (**J** 0122) J Philos Logic 1*417-438
◇ B45 ◇ REV MR 58 # 21468a Zbl 246 # 02015 • ID 61760

FRAASSEN VAN, B.C. [1973] *Semantic analysis of quantum logic* (**P** 2312) Contemp Res in Found & Philos of Quantum Th;1973 London ON 80-113
◇ A05 B51 G12 ◇ REV Zbl 279 # 02016 • ID 29542

FRAASSEN VAN, B.C. [1973] *The logic of conditional obligation* (**P** 1556) Exact Philos: Probl, Tools & Goals;1971 Montreal 151-172
◇ B45 ◇ REV MR 58 # 21468a Zbl 316 # 02031 • ID 61761

FRAASSEN VAN, B.C. [1974] *Hidden variables in conditional logic* (**J** 0105) Theoria (Lund) 40*176-190
◇ B51 G12 ◇ REV MR 58 # 27327 Zbl 327 # 02027
• ID 61759

FRAASSEN VAN, B.C. [1974] *The labyrinth of quantum logics* (**C** 4290) Log & Epistem Stud Contemp Phys 224-254
◇ B51 ◇ ID 42111

FRAASSEN VAN, B.C. [1975] *Comments: Lakoff's fuzzy propositional logic* (**P** 1647) Contemp Res in Phil Log & Ling Semant;1975 London ON 273-277
◇ A05 B52 ◇ REV Zbl 315 # 02024 • ID 29805

FRAASSEN VAN, B.C. [1975] *Facts and tautological entailments* (**C** 1852) Entailment - Log of Relev & Nec 221-230
◇ A05 B46 ◇ REV MR 53 # 10542 JSL 42.311 • ID 44476

FRAASSEN VAN, B.C. [1975] *Incomplete assertion and Belnap connectives* (**P** 1647) Contemp Res in Phil Log & Ling Semant;1975 London ON 43-70
◇ B51 G12 ◇ REV Zbl 319 # 02021 • ID 29674

FRAASSEN VAN, B.C. & HOOKER, C.A. [1976] *A semantic analysis of Niels Bohr's philosophy of quantum theory* (**P** 2411) Found Probab Th, Stat Inf & Stat Th Sci;1973 London ON 3*221-241
◇ A05 B51 G12 ◇ REV Zbl 323 # 02045 • ID 30420

FRAASSEN VAN, B.C. [1976] *Probabilities of conditionals* (**P** 2411) Found Probab Th, Stat Inf & Stat Th Sci;1973 London ON 1*261-308
◇ A05 B45 B48 ◇ REV MR 58 # 16148 Zbl 339 # 02025
• ID 61766

FRAASSEN VAN, B.C. [1976] *Report on conditionals* (**J** 0162) Teorema (Valencia) 6*5-25
◇ B45 ◇ REV MR 58 # 21469 • ID 79636

FRAASSEN VAN, B.C. [1976] *Representation of conditional probabilities* (**J** 0122) J Philos Logic 5*417-430
◇ B48 ◇ REV Zbl 341 # 60029 • ID 33046

FRAASSEN VAN, B.C. [1979] *Hidden variables and the modal interpretation of quantum theory* (**J** 0154) Synthese 42*155-165
◇ B45 G12 ◇ REV MR 81e:03059 Zbl 431 # 60003
• ID 79633

FRAASSEN VAN, B.C. & LEBLANC, H. [1979] *On Carnap and Popper probability functions* (**J** 0036) J Symb Logic 44*369-373
◇ A05 B48 ◇ REV MR 81g:03005 Zbl 419 # 03018
• ID 53361

FRAASSEN VAN, B.C. [1980] *Rational belief and probability kinematics* (**J** 0153) Phil of Sci (East Lansing) 47*165-187
◇ A05 B48 ◇ REV MR 82e:03011 • ID 79631

FRAASSEN VAN, B.C. [1981] *A modal interpretation of quantum mechanics* (**P** 3820) Curr Iss in Quantum Log;1979 Erice 229-258
◇ B45 ◇ REV MR 84j:03125 Zbl 537 # 03044 • ID 34715

FRAASSEN VAN, B.C. [1981] *A temporal framework for conditionals and chance* (**C** 4140) Ifs 323-340
◇ B48 ◇ REV MR 83a:03003 • ID 47844

FRAASSEN VAN, B.C. [1981] *Assumptions and interpretations of quantum logic* (**P** 3820) Curr Iss in Quantum Log;1979 Erice 17-31
◇ B51 ◇ REV MR 84j:03124 Zbl 537 # 03044 • ID 34714

FRAASSEN VAN, B.C. [1981] see BELTRAMETTI, E.G.

FRAASSEN VAN, B.C. [1981] *Essences and laws of nature* (**C** 4283) Reduct, Time & Reality 189-200
◇ A05 B46 ◇ REV MR 84g:03008 • ID 45730

FRAASSEN VAN, B.C. [1981] *Probabilistic semantics objectified: I. Postulated and logics. II. Implication in probabilistic model sets* (**J** 0122) J Philos Logic 10*371-394,495-510 • ERR/ADD ibid 11*465
◇ B48 ◇ REV MR 84k:03069 Zbl 481 # 03004
Zbl 493 # 03001 • ID 33048

FRAASSEN VAN, B.C. [1981] *Report on tense logic* (**C** 2617) Modern Log Survey 425-438
◇ B45 ◇ REV MR 82f:03002 Zbl 464 # 03001 • ID 42779

FRAASSEN VAN, B.C. [1982] *The Charybdis of realism: epistemological implications of Bell's inequality* (**J** 0154) Synthese 52*25-38
◇ A05 B51 ◇ REV MR 84c:81006 • ID 39569

FRAASSEN VAN, B.C. [1983] *Gentlemen's wagers: relevant logic and probability* (**J** 0095) Philos Stud 43*47-61
◇ B48 ◇ REV MR 84m:03030 • ID 35734

FRAASSEN VAN, B.C. [1983] *Shafer on conditional probability* (**J** 0122) J Philos Logic 12*467-470
◇ B48 ◇ REV MR 85g:60008b Zbl 539 # 60005 • ID 41249

FRAASSEN VAN, B.C. see Vol. I for further entries

FRAENKEL, A.A. [1958] see BERNAYS, P.

FRAENKEL, A.A. [1958] see BAR-HILLEL, Y.

FRAENKEL, A.A. [1959] *Mengenlehre und Logik* (**X** 1258) Duncker & Humblot: Berlin 110pp
• TRANSL [1966] (**X** 0832) Addison-Wesley: Reading 102pp (English)
◇ B98 E98 ◇ REV MR 22 # 1513 MR 34 # 24 Zbl 139.5 Zbl 86.244 JSL 34.112 • ID 28712

FRAENKEL, A.A. see Vol. I, V, VI for further entries

FRAISSE, R. [1967] *Cours de logique mathematique. Tome I. Relation, formule logique, compacite, completude* (**X** 0834) Gauthier-Villars: Paris xii + 187pp
• TRANSL [1973] (**X** 0835) Reidel: Dordrecht xvi + 186pp
◇ B98 C98 ◇ REV MR 37 # 3902 Zbl 247 # 02001 Zbl 247 # 02002 JSL 35.580 • REM Tome II 1972 • ID 04529

FRAISSE, R. [1972] *Cours de logique mathematique. Tome 2: Theorie des modeles* (**X** 0834) Gauthier-Villars: Paris xiv + 177pp
• TRANSL [1974] (**X** 0835) Reidel: Dordrecht xix + 191pp
◇ B98 C98 ◇ REV MR 49 # 10514 Zbl 247 # 02003 • REM Tome I 1967. Tome III 1975. • ID 04535

FRAISSE, R. [1982] *La zerologie, une recherche aux frontieres de la logique et de l'art; application a la logique des relations de base vide* (**J** 0286) Int Logic Rev 26*67-79
◇ B60 C07 ◇ REV MR 85g:03039 • ID 43874

FRAISSE, R. see Vol. I, III, IV, V for further entries

FRANCEZ, N. [1982] see CLARKE JR., E.M.

FRANCEZ, N. see Vol. IV for further entries

FRANCIONI, J.M. & KANDEL, A. [1983] *Decomposable fuzzy-valued switching functions* (**J** 2720) Fuzzy Sets Syst 9*41-68
◇ B52 ◇ REV MR 85e:94034 Zbl 518 # 94020 • ID 40066

FRANKSEN, O.I. [1979] *Group representations of finite polyvalent logic. A case study using APL notation* (**P** 3487) IFAC World Congr (7);1978 Helsinki 2*875-887
◇ B50 B75 ◇ REV Zbl 442 # 03041 • ID 56398

FRAZER, P.J. & HARDEGREE, G.M. [1981] *Charting the labyrinth of quantum logics: a progress report* (**P** 3820) Curr Iss in Quantum Log;1979 Erice 53-76
◇ B51 ◇ REV MR 84k:03144 Zbl 537 # 03044 JSL 50.558 • ID 35036

FREELING, A. [1984] *Possibilities versus fuzzy probabilities - two alternative decision aids* (**C** 3389) Fuzzy Sets & Decision Anal 67-81
◇ B48 ◇ REV MR 85i:90002 • ID 45757

FREEMAN, J.B. [1975] *Compactness for RQ* (**J** 0063) Studia Logica 34*269-274
◇ B45 ◇ REV MR 54 # 2423 Zbl 329 # 02005 • ID 24622

FREEMAN, J.B. [1976] *Algebraic semantics for modal predicate logic* (**J** 0068) Z Math Logik Grundlagen Math 22*523-552
◇ B45 G15 ◇ REV MR 56 # 11753 Zbl 354 # 02022 • ID 30695

FREEMAN, J.B. [1978] see DANIELS, C.B.

FREEMAN, J.B. [1979] see DANIELS, C.B.

FREEMAN, J.B. [1980] see DANIELS, C.B.

FREEMAN, J.B. see Vol. I, III for further entries

FREJVALD, R.V. [1966] *A completeness criterion for partial functions of logic and many valued logic algebras (Russian)* (**J** 0023) Dokl Akad Nauk SSSR 167*1249-1250
• TRANSL [1966] (**J** 0470) Sov Phys, Dokl 11*288-289
◇ B50 G15 ◇ REV MR 35 # 1448 Zbl 149.244 • ID 04598

FREJVALD, R.V. [1968] *Completeness up to coding of systems of functions of k-valued logic and the complexity of its determination (Russian)* (**J** 0023) Dokl Akad Nauk SSSR 180*803-805
• TRANSL [1968] (**J** 0062) Sov Math, Dokl 9*699-702
◇ B50 D15 ◇ REV MR 38 # 987 Zbl 199.302 • ID 04599

FREJVALD, R.V. [1970] *Coding of finite sets and a criterion of completeness up to coding in three-valued logic (Russian)* (**J** 0023) Dokl Akad Nauk SSSR 190*1034-1037
• TRANSL [1970] (**J** 0062) Sov Math, Dokl 11*249-253
◇ B50 ◇ REV MR 41 # 5197 Zbl 206.6 • ID 04600

FREJVALD, R.V. & KINBER, E.B. & WIEHAGEN, R. [1982] *Inductive inference and computable one-one numberings* (**J** 0068) Z Math Logik Grundlagen Math 28*463-479
◇ B48 D20 D45 ◇ REV MR 85g:03065 Zbl 541 # 03025 • ID 43888

FREJVALD, R.V. & KINBER, E.B. & WIEHAGEN, R. [1984] *On the power of probabilistic strategies in inductive inference* (**J** 1426) Theor Comput Sci 28*111-133
◇ B48 D20 ◇ REV MR 85j:03064 Zbl 555 # 68014 • ID 42396

FREJVALD, R.V. see Vol. I, IV for further entries

FRENCH, S. [1980] *Updating of belief in the light of someone else's opinion* (**J** 2439) J Roy Stat Soc Ser A 143*43-48
◇ B48 ◇ REV MR 83g:60007 Zbl 432 # 62004 • ID 39098

FRENCH, S. [1982] *On the axiomatisation of subjective probabilities* (**J** 0472) Theory Decis 14*19-33
◇ B48 ◇ REV MR 83m:60002 Zbl 475 # 60003 • ID 40359

FRESCURA, F.A.M. & HILEY, B.J. [1980] *The implicate order, algebras, and the spinor* (**J** 1678) Found Phys 10*7-31
◇ B51 G12 ◇ REV MR 81d:81007 • ID 81275

FREUDENTHAL, H. [1949] *Sur le besoin d'une logique materielle* (**J** 0154) Synthese 7*337-345
◇ A05 B45 ◇ REV MR 10.670 • ID 28212

FREUDENTHAL, H. [1953] *Context related interpretations in formal logic (Dutch)* (**J** 0358) Versl Gewone Vergad Afd Natuurkd 62*94-96
◇ A05 B45 ◇ REV MR 15.494 Zbl 52.7 • ID 17080

FREUDENTHAL, H. [1958] *Logique mathematique appliquee* (**X** 0834) Gauthier-Villars: Paris 58pp
◇ A05 B05 B45 B70 ◇ REV MR 20 # 5737 Zbl 84.5 JSL 24.256 • ID 04610

FREUDENTHAL, H. [1960] *Models in applied probability* (**J** 0154) Synthese 12*202-212
◇ A05 B48 ◇ REV MR 26 # 5598 • ID 28223

FREUDENTHAL, H. [1961] *Exact logic (Dutch)* (**X** 1408) Bohn: Amsterdam vii + 117pp
• TRANSL [1966] (**X** 0838) Amer Elsevier: New York vi + 105pp (English) [1969] (**X** 2027) Nauka: Moskva 135pp (Russian) [1975] (**X** 0814) Oldenbourg: Muenchen 106pp (German)
◇ A05 B98 ◇ REV MR 25 # 3807 MR 57 # 9449 Zbl 97.244 JSL 33.603 • ID 28713

FREUDENTHAL, H. see Vol. I, IV, VI for further entries

FREUNDLICH, R. [1984] *Zur Begruendung einer formalen Normenlogik* (**C** 4065) Theorie der Normen 373-392
◇ B45 ◇ REV MR 85k:03015 • ID 44543

FREY, G. [1953] *Bemerkungen zum Problem der mehrwertigen Logiken* (**P** 0645) Int Congr Philos (11);1953 Bruxelles 5*53-58
◇ B50 ◇ REV MR 15.278 Zbl 51.245 JSL 19.131 • ID 04612

FREY, G. see Vol. I, III for further entries

FREYTAG-LOERINGHOFF BARON VON, B. [1951] *Logik I. Das System der reinen Logik und ihr Verhaeltnis zur Logistik* (**X** 0808) Kohlhammer: Stuttgart 222pp
◇ A05 B98 ◇ REV MR 22 # 2537 MR 50 # 12554 Zbl 286 # 02005 • REM 5. ueberarbeitete Auflage 1972 • ID 61820

FRICKE, J. [1978] *Decomposition of multiple-valued logic functions* (**P** 2014) Int Symp Multi-Val Log (8);1978 Rosemont 208-212
◇ B50 ◇ ID 35927

FRIEDER, G. [1974] see EPSTEIN, G.

FRIEDMAN, H.M. [1975] *One hundred and two problems in mathematical logic* (**J** 0036) J Symb Logic 40*113-129
◇ B98 C98 D98 E98 F98 ◇ REV MR 51 # 5254 Zbl 318 # 02002 • ID 04296

FRIEDMAN, H.M. see Vol. I, III, IV, V, VI for further entries

FRIEDMAN, JOEL I. [1982] *Kripkean necessity and the ontological argument* (**C** 3799) Logic & Religion 173-183
◇ B45 ◇ REV MR 84j:03037 MR 85i:03008 • ID 34629

FRIEDMAN, JOEL I. see Vol. I, V for further entries

FRIEDMAN, M. & GLYMOUR, C. [1972] *If quanta had logic* (**J** 0122) J Philos Logic 1*16-28
◇ A05 B51 G12 ◇ REV MR 45 # 6584 Zbl 275 # 02030 • ID 16816

FRIEDMAN, M. & PUTNAM, H. [1978] *Quantum logic, conditional probability, and interference* (**J** 0076) Dialectica 32*305-315
◇ A05 B51 G12 ◇ REV MR 82g:03108 Zbl 402 # 03016 • ID 54663

FRINK, O. [1938] *New algebras of logic* (**J** 0005) Amer Math Mon 45*210-219
◇ B50 F50 G05 ◇ REV Zbl 18.337 JSL 3.117 FdM 64.30 • ID 04671

FRINK, O. see Vol. V for further entries

FU, K.S. & WEE, W.G. [1969] *A formulation of fuzzy automata and its application as a model of learning systems* (**J** 1561) IEEE Trans Syst & Sci Cybern 5*215-223
◇ B52 ◇ REV Zbl 188.332 • ID 41960

FU, K.S. & ISHIZUKA, M. & YAO, J.T.P. [1982] *Inference procedures under uncertainty for the problem-reduction method* (**J** 0191) Inform Sci 28*179-206
◇ B48 ◇ REV MR 84i:03056 • ID 34535

FU, K.S. see Vol. IV, V for further entries

FUJIMURA, T. [1964] *On T-principle (Japanese)* (**J** 1830) Stud Phil Sci (Tokyo) 4*22-27
◇ B45 ◇ ID 32334

FUJIMURA, T. see Vol. I for further entries

FUJITA, Y. & KITAHASHI, T. & TANAKA, K. [1971] *Homomorphisms of logical functions and their applications* (**P** 2007) Symp Th & Appl of Multi-Val Log Design;1971 Buffalo 36-41
◇ B50 ◇ ID 41977

FUJITA, Y. & KITAHASHI, T. & TANAKA, K. [1972] *A consideration on compositions of many-valued logical functions* (**J** 0116) Electr & Comm Japan 55*46-51
◇ B50 ◇ REV MR 56 # 11565 • ID 81291

FUJITA, Y. & KITAHASHI, T. & TANAKA, K. [1972] *An extension of the theorem of isomorphism concerning the compositions of many-valued switching functions* (**J** 0116) Electr & Comm Japan 55-D*399
◇ B50 ◇ ID 42014

FUKAMI, S. & MIZUMOTO, M. & TANAKA, K. [1978] *On fuzzy reasoning* (**J** 0464) Syst-Comp-Controls 9/4*44-53
◇ B52 ◇ REV MR 82g:03039a • ID 73144

FUKAMI, S. & MIZUMOTO, M. & TANAKA, K. [1979] *Some methods of fuzzy reasoning* (**C** 3514) Adv Fuzzy Sets & Appl 117-136
◇ B52 ◇ REV MR 81g:03027 • ID 76470

FUKAMI, S. & MIZUMOTO, M. & TANAKA, K. [1980] *Some considerations on fuzzy conditional inference* (**J** 2720) Fuzzy Sets Syst 4*243-273
◇ B52 ◇ REV MR 82g:03039b Zbl 453 # 03022 • ID 54152

FUKAYAMA, T. [1982] *On the state of research in many-valued logic (Japanese)* (**P** 4081) Many-Val Log & Appl;1982 Kyoto 260-280
◇ B50 ◇ ID 47637

FULTON, J.A. [1979] *An intensional logic of predicates and predicate modifiers without modal operators* (**J** 0047) Notre Dame J Formal Log 20*807-834
◇ B45 ◇ REV MR 80h:03045 Zbl 305 # 02038 • ID 61858

FULTON, J.A. see Vol. I for further entries

FURMANOWSKI, T. [1975] *Remarks on discussive propositional calculus* (**J** 0063) Studia Logica 34*39-43
◇ B53 ◇ REV MR 51 # 10032 Zbl 309 # 02020 • ID 17558

FURMANOWSKI, T. see Vol. I, III, VI for further entries

FURTEK, F.C. [1979] *Doing without values* (**P** 3003) Int Symp Multi-Val Log (9);1979 Bath 114-120
◇ B50 ◇ REV MR 81i:68024 • ID 35944

FURTEK, F.C. see Vol. V for further entries

GABBAY, D.M. & KASHER, A. [1967] *Improper definite descriptions* (**C** 0800) New Catholic Encycl 74-89
◇ A05 B45 ◇ ID 32160

GABBAY, D.M. [1969] *Montague type semantics for non-classical logic* (**X** 2039) Poly Inst New York: Brooklyn 1-25
◇ B45 ◇ ID 32150

GABBAY, D.M. [1970] *Selective filtration in modal logic, I: Semantic tableaux method* (**J** 0105) Theoria (Lund) 36*323-330
◇ B45 ◇ REV MR 47 # 1575 Zbl 222 # 02018 • ID 04720

GABBAY, D.M. [1970] *The decidability of the Kreisel-Putnam system* (**J** 0036) J Symb Logic 35*431-436
◇ B25 B55 F50 ◇ REV MR 58 # 16167 Zbl 228 # 02013 • ID 04721

GABBAY, D.M. [1971] *Decidability results in non-classical logic, III: Systems with statability operators* (**J** 0029) Israel J Math 10*135-146
◇ B25 B45 C90 ◇ REV MR 46 # 1553 Zbl 289 # 02033
• REM Part I 1975. Parts II,IV have not yet appeared
• ID 30000

GABBAY, D.M. [1971] *Montague type semantics for modal logics with propositional quantifiers* (**J** 0068) Z Math Logik Grundlagen Math 17*245-249
◇ B45 ◇ REV MR 47 # 15 Zbl 226 # 02015 • ID 04722

GABBAY, D.M. [1971] *On decidable, finitely axiomatizable, modal and tense logics without the finite model property I, II* (**J** 0029) Israel J Math 10*478-495,496-503
◇ B25 B45 C90 ◇ REV MR 45 # 4961 Zbl 231 # 02023
• ID 04723

GABBAY, D.M. [1971] *Semantic proof of the Craig interpolation theorem for intuitionistic logic and extensions I,II* (**P** 0638) Logic Colloq;1969 Manchester 391-401,403-410
◇ B55 C40 C90 F50 ◇ REV MR 43 # 3094a MR 43 # 3094b Zbl 234 # 02017 • REM Part III 1977
• ID 04726

GABBAY, D.M. [1972] *A general filtration method for modal logics* (**J** 0122) J Philos Logic 1*29-34
◇ B45 ◇ REV MR 50 # 12657 Zbl 248 # 02025 • ID 04734

GABBAY, D.M. [1972] *A general theory of the conditional in terms of a ternary operator* (**J** 0105) Theoria (Lund) 38*97-105
◇ B46 ◇ REV MR 58 # 16149 Zbl 254 # 02019 • ID 32153

GABBAY, D.M. [1972] *Application of trees to intermediate logics* (**J** 0036) J Symb Logic 37*135-138
◇ B55 C90 ◇ REV MR 47 # 8251 Zbl 243 # 02019
• ID 04732

GABBAY, D.M. [1972] *Craig's interpolation theorem for modal logics* (**P** 2080) Conf Math Log;1970 London 111-127
◇ B45 C40 C90 ◇ REV MR 49 # 8815 Zbl 233 # 02009
• ID 04733

GABBAY, D.M. [1972] *Model theory for intuitionistic logic* (**J** 0068) Z Math Logik Grundlagen Math 18*49-54
◇ B55 C20 C90 F50 ◇ REV MR 45 # 4959 Zbl 242 # 02059 • ID 04730

GABBAY, D.M. [1972] *Tense systems with discrete moments of time I* (**J** 0122) J Philos Logic 1*35-44
◇ B45 ◇ REV MR 46 # 5107 Zbl 245 # 02031 • ID 04728

GABBAY, D.M. [1973] *A survey of decidability results for modal, tense and intermediate logics* (**P** 0793) Int Congr Log, Meth & Phil of Sci (4,Proc);1971 Bucharest 29-43
◇ B25 B45 B55 C98 F50 ◇ REV MR 55 # 12470
• ID 17911

GABBAY, D.M. [1973] *Applications of Scott's notion of consequence to the study of general binary intensional connectives and entailment* (**J** 0122) J Philos Logic 2*340-351
◇ B22 B46 ◇ REV MR 54 # 7202 Zbl 272 # 02034
• ID 30381

GABBAY, D.M. & MORAVCSIK, J.M.E. [1973] *Sameness and individuation* (**J** 0301) J Phil 70*513-526
◇ A05 B45 ◇ ID 32155

GABBAY, D.M. [1974] *A generalization of the concept of intensional semantics with applications* (**J** 1807) Phil Quart Israel 4*251-270
◇ B45 ◇ ID 32156

GABBAY, D.M. & JONGH DE, D.H.J. [1974] *A sequence of decidable finitely axiomatizable intermediate logics with the disjunction property* (**J** 0036) J Symb Logic 39*67-78
◇ B22 B25 B55 ◇ REV MR 51 # 10038 Zbl 289 # 02032
• ID 17555

GABBAY, D.M. [1974] *Tense logics and the tenses of English* (**C** 2253) Log & Philos for Linguist Readings 177-186
◇ B45 ◇ ID 32158

GABBAY, D.M. [1975] *A normal logic that is complete for neighborhood frames but not for Kripke frames* (**J** 0105) Theoria (Lund) 41*148-153
◇ B45 ◇ REV MR 56 # 91 Zbl 347 # 02017 • ID 61874

GABBAY, D.M. [1975] *Decidability results in non-classical logics I* (**J** 0007) Ann Math Logic 8*237-295
◇ B25 B45 C90 F50 ◇ REV MR 52 # 68 Zbl 309 # 02053
• REM Parts II,IV have not yet appeared. Part III 1971
• ID 04738

GABBAY, D.M. [1975] *Model theory for tense logics* (**J** 0007) Ann Math Logic 8*185-236
◇ B45 C20 C90 ◇ REV MR 51 # 7826 Zbl 307 # 02014
• ID 04739

GABBAY, D.M. [1976] *Completeness properties of Heyting's predicate calculus with respect to RE models* (**J** 0036) J Symb Logic 41*81-94
◇ B55 F50 ◇ REV MR 53 # 12883 Zbl 328 # 02013
• ID 14748

GABBAY, D.M. [1976] *Investigations in modal and tense logics with applications to problems in philosophy and linguistics* (**X** 0835) Reidel: Dordrecht xi+306pp
◇ A05 B25 B46 C90 ◇ REV MR 58 # 27328 Zbl 374 # 02013 JSL 44.656 • ID 32148

GABBAY, D.M. [1976] *On Kreisel's notion of validity in Post systems* (**J** 0063) Studia Logica 35*285-295
◇ B50 D03 F50 ◇ REV MR 55 # 12477 Zbl 363 # 02027
• ID 50854

GABBAY, D.M. [1976] *Two-dimensional propositional tense logics* (**C** 1701) Lang in Focus (Bar-Hillel) 569-583
◇ B45 ◇ REV Zbl 378 # 02007 • ID 51789

GABBAY, D.M. [1977] *A tense system with split truth table* (**J** 0079) Logique & Anal, NS 20*359-393
◇ B45 ◇ REV MR 58 # 27329 Zbl 392 # 03018 • ID 52385

GABBAY, D.M. [1977] *Craig interpolation theorem for intuitionistic logic and extensions part III* (**J** 0036) J Symb Logic 42*269-271
◇ B55 C40 C90 F50 ◇ REV MR 58 # 16168 Zbl 372 # 02016 • REM Parts I,II 1971 • ID 26456

GABBAY, D.M. [1977] *On some new intuitionistic propositional connectives I* (**J** 0063) Studia Logica 36*127-139
◇ B45 B55 F50 ◇ REV MR 58 # 161 Zbl 363 # 02026
• ID 50853

GABBAY, D.M. [1978] *Tense logics with split truth tables* (**J** 0079) Logique & Anal, NS 21*1-37
◇ B45 ◇ ID 32162

GABBAY, D.M. [1981] *An irreflexivity lemma with applications to axiomatizations of conditions on tense frames* (**P** 3113) Aspects Philos Logic;1977 Tuebingen 67-89
◇ B45 ◇ REV MR 83f:03015 Zbl 519 # 03008 • ID 35296

GABBAY, D.M. [1981] *Expressive functional completeness in tense logic* (**P** 3113) Aspects Philos Logic;1977 Tuebingen 91-117
◇ B45 ◇ REV Zbl 523 # 03017 • ID 37024

GABBAY, D.M. [1981] *Semantical investigations in Heyting's intuitionistic logic* (**X** 0835) Reidel: Dordrecht x+287pp
◇ B55 F50 F98 ◇ REV MR 83b:03012 Zbl 453 # 03001 • ID 54131

GABBAY, D.M. [1982] *Intuitionistic basis for non-monotonic logic* (**P** 3840) Autom Deduct (6);1982 New York 260-273
◇ B55 C90 F50 ◇ REV MR 85e:03062 Zbl 481 # 68091 • ID 38464

GABBAY, D.M. & GUENTHNER, F. (EDS.) [1983] *Handbook of philosophical logic Vol. I: Elements of classical logic* (**S** 3307) Synth Libr 164*xi+493pp
◇ B98 C98 ◇ REV Zbl 538 # 03001 • ID 41457

GABBAY, D.M. & GUENTHNER, F. (EDS.) [1984] *Handbook of philosophical logic Vol II. Extensions of classical logic* (**S** 3307) Synth Libr 165*x+776pp
◇ B98 C90 C98 ◇ REV Zbl 572 # 03003 • REM Part I 1983 • ID 41831

GABBAY, D.M. [1985] *Theoretical foundations for nonmonotonic reasoning in expert systems* (**P** 4621) Log & Models of Concurrent Syst;1984 La Colle-sur-Loup 439-457
◇ B60 ◇ ID 49346

GABBAY, D.M. see Vol. I, III, IV, VI for further entries

GACS, P. [1980] *Exact expressions for some randomness tests* (**J** 0068) Z Math Logik Grundlagen Math 26*385-394
◇ B48 ◇ REV MR 81j:65019 Zbl 464 # 60004 • ID 33200

GACS, P. see Vol. I, III, IV for further entries

GAERDENFORS, P. & HANSSON, B. [1973] *A guide to intensional semantics* (**C** 1389) Modality, Morality, Probl of Sense & Nonsense (Hallden) 151-167
◇ A05 B45 ◇ REV MR 56 # 11757 • ID 31114

GAERDENFORS, P. [1973] *On the extensions of S5* (**J** 0047) Notre Dame J Formal Log 14*277-280
◇ B45 ◇ REV MR 47 # 8249 Zbl 236 # 02017 • ID 04741

GAERDENFORS, P. & HANSSON, B. [1975] *Filtrations and the finite frame property in boolean semantics* (**P** 0757) Scand Logic Symp (3);1973 Uppsala 32-39
◇ B25 B45 C90 G05 G25 ◇ REV MR 53 # 2633 Zbl 311 # 02031 JSL 43.373 • ID 21522

GAERDENFORS, P. [1975] *Qualitative probability as an intensional logic* (**J** 0122) J Philos Logic 4*171-185
◇ B48 ◇ REV MR 58 # 21471 Zbl 317 # 02030 • ID 29643

GAERDENFORS, P. [1975] *Some basic theorems of qualitative probability* (**J** 0063) Studia Logica 34*257-264
◇ B48 ◇ REV MR 52 # 13332 Zbl 316 # 02036 • ID 21794

GAERDENFORS, P. [1978] *Conditionals and changes of belief* (**J** 0096) Acta Philos Fenn 30*381-404
◇ B48 ◇ REV MR 82i:00024 Zbl 413 # 03011 • ID 53005

GAERDENFORS, P. [1978] *On the interpretation of deontic logic* (**J** 0079) Logique & Anal, NS 21*371-398
◇ B45 ◇ REV MR 81b:03019 Zbl 413 # 03012 • ID 53006

GAERDENFORS, P. [1978] *On the logic of relevance* (**J** 0154) Synthese 37*351-367
◇ A05 B46 ◇ REV MR 58 # 10391 Zbl 377 # 02025 Zbl 395 # 03005 • ID 31116

GAERDENFORS, P. & SAHLIN, N.-E. [1982] *Unreliable probabilities, risk taking, and decision making* (**J** 0154) Synthese 53*361-386
◇ B48 ◇ REV MR 84e:03028 Zbl 516 # 62011 • ID 34367

GAERDENFORS, P. [1984] *The dynamics of belief as a basis for logic* (**J** 0013) Brit J Phil Sci 35*1-10
◇ B48 ◇ REV Zbl 551 # 03005 • ID 43882

GAERDENFORS, P. [1985] *Epistemic importance and the logic of theory change* (**P** 4180) Int Congr Log, Meth & Phil of Sci (7,Pap);1983 Salzburg 345-367
◇ A05 B60 ◇ ID 47933

GAERDENFORS, P. [1985] see ALCHOURRON, C.E.

GAERDENFORS, P. [1985] *Propositional logic based on the dynamics of belief* (**J** 0036) J Symb Logic 50*390-394
◇ B48 ◇ REV Zbl 566 # 03016 • ID 42553

GAGNON, L.S. [1980] *Three theories of dialectic* (**J** 0047) Notre Dame J Formal Log 21*316-318
◇ A05 B53 ◇ REV MR 81d:03031 Zbl 321 # 02010 • ID 53780

GAGNON, L.S. see Vol. I for further entries

GAIFMAN, H. [1960] *Probability models and the completeness theorem* (**P** 1953) Int Congr Log, Meth & Phil of Sci (1;Abstr);1960 Stanford 77-78
◇ B48 C90 ◇ ID 16922

GAIFMAN, H. [1964] *Concerning measures on first order calculi* (**J** 0029) Israel J Math 2*1-18
◇ B48 C90 ◇ REV MR 31 # 31 Zbl 192.33 • ID 33352

GAIFMAN, H. [1971] *Applications of de Finetti's theorem to inductive logic* (**C** 0799) Stud Induct Logic & Probab, Vol 1 235-251
◇ B48 ◇ REV MR 58 # 5088b • ID 73192

GAIFMAN, H. see Vol. I, III, IV, V, VI for further entries

GAINES, B.R. & KOHOUT, L.J. [1975] *Possible automata* (**P** 1805) Int Symp Multi-Val Log (5,Proc);1975 Bloomington 183-196
◇ B50 D05 ◇ ID 35823

GAINES, B.R. & KOHOUT, L.J. [1975] *The logic of protection* (**P** 0784) GI Jahrestag (5);1975 Dortmund 736-751
◇ B60 ◇ REV MR 52 # 9725 Zbl 315 # 68047 • ID 18231

GAINES, B.R. & HAACK, S. & KOTAHASHI, T. & POPPELBAUM, W.J. & RINE, D.C. & SMITH, KENNETH C. [1976] *Applications of multiple-valued logic* (**P** 2011) Int Symp Multi-Val Log (6);1976 Logan
◇ B50 ◇ ID 35905

GAINES, B.R. [1976] *Foundations of fuzzy reasoning* (**J** 1741) Int J Man-Mach Stud 8*623-668
◇ B52 E72 ◇ REV MR 82k:03027 Zbl 342 # 68056 • ID 61897

GAINES, B.R. [1976] *Fuzzy reasoning and the logic of uncertainty* (**P** 2011) Int Symp Multi-Val Log (6);1976 Logan 179-188
◇ B52 ◇ ID 35888

GAINES, B.R. & KOHOUT, L.J. [1977] *The fuzzy decade: A bibliography of fuzzy systems and closely related topics* (**J** 1741) Int J Man-Mach Stud 9*1-68
◇ A10 B52 E72 ◇ REV Zbl 353 # 02011 • ID 50094

GAINES, B.R. [1978] *Fuzzy and probability uncertainty logics* (**J** 0194) Inform & Control 38*154-169
◇ B48 B52 ◇ REV MR 80m:03059 Zbl 391 # 03015
• ID 52342

GAINES, B.R. [1978] *General fuzzy logics* (**P** 3428) Prog in Cybern & Syst Res;1978 Wien III*270-275
◇ B52 ◇ REV Zbl 436 # 03017 • ID 55857

GAINES, B.R. & MAMDANI, E.H. (EDS.) [1981] *Fuzzy reasoning and its applications* (**X** 0801) Academic Pr: New York xviii+381pp
◇ B52 B98 ◇ REV MR 83d:03001 Zbl 488 # 03001
• ID 34823

GAINES, B.R. [1982] *Fuzzy and probability uncertainty logics (Chinese)* (**J** 2018) Yingyong Shuxue Yu Jisuan Shuxue 1982/1*36-44
◇ B52 ◇ REV MR 83j:03046 Zbl 391 # 03015 • ID 34864

GAINES, B.R. & SHAW, M.L.G. [1984] *Deriving the constructs underlying decision* (**C** 3389) Fuzzy Sets & Decision Anal 335-355
◇ B52 ◇ REV MR 85i:90002 Zbl 566 # 90051 • ID 48750

GAINES, B.R. [1984] *Fundamentals of decision: Probabilistic, possibilistic and other forms of uncertainty in decision analysis* (**C** 3389) Fuzzy Sets & Decision Anal 47-65
◇ B48 ◇ REV MR 86d:03023 Zbl 545 # 03007 • ID 41222

GAINES, B.R. see Vol. I, IV for further entries

GALDI, G.P. [1979] see BARONE, F.

GALE, S. [1975] *Conjectures on many-valued logic, regions, and criteria for conflict resolution* (**P** 1805) Int Symp Multi-Val Log (5,Proc);1975 Bloomington 212-225
◇ B50 ◇ ID 35824

GALLIN, D. [1975] *Intensional and higher-order modal logic. With applications to Montague semantics* (**X** 0809) North Holland: Amsterdam vii+148pp
◇ B45 C85 C90 E35 E50 E70 ◇ REV MR 58 # 21470 Zbl 341 # 02014 JSL 42.581 • ID 29739

GALLIN, D. see Vol. I for further entries

GAO, HENGSHAN [1964] *On the simple completeness of the predicate calculus $\mathscr{S}_{\varepsilon}{}^{*}$ (Chinese)* (**J** 1024) Zhongguo Kexue 13*213-215
◇ B45 G05 ◇ REV MR 34 # 4118 Zbl 154.255 • ID 15110

GAO, HENGSHAN [1964] *Simple completeness of the predicate calculus $\mathscr{S}_{\varepsilon}{}^{*}$ (Chinese)* (**J** 0418) Shuxue Xuebao 14*546-548
• TRANSL [1964] (**J** 0419) Chinese Math Acta 5*588-590
◇ B45 G05 ◇ REV MR 32 # 5503 Zbl 156.6 • ID 14542

GAO, HENGSHAN [1973] *The decision problem of modal predicate calculi. I: On Slomson's reductions (Chinese)* (**J** 3816) Zhongguo Kexue Jishu Daxue Xuebao 3/2*49-54
◇ B45 D35 ◇ REV MR 58 # 27330 Zbl 349 # 02016 • REM Part II 1976 • ID 46707

GAO, HENGSHAN [1976] *The decision problem of modal predicate calculi. II: On Slomson's reductions (Chinese)* (**J** 0418) Shuxue Xuebao 19*276-280
◇ B45 C13 D35 ◇ REV MR 58 # 27330 Zbl 349 # 02016
• REM Part I 1973 • ID 61914

GAO, HENGSHAN [1979] *Some remarks on Lewis modal system S5 (Chinese) (English summary)* (**J** 2754) Huazhong Gongxueyuan Xuebao 7/1*13-17
• TRANSL [1979] (**J** 2684) J Huazhong Inst Tech (Engl Ed) 1/1*31-37
◇ B45 ◇ REV MR 83k:03024b • ID 36164

GAO, HENGSHAN see Vol. I, III, IV, V, VI for further entries

GARBER, D. [1980] *Discussion: Field and Jeffrey conditionalization* (**J** 0153) Phil of Sci (East Lansing) 47*142-145
◇ B48 ◇ REV MR 82g:62012 • ID 81300

GARDEN, R.W. [1983] *Modern logic and quantum mechanics* (**X** 1073) Hilger: Bristol xi+177pp
◇ B51 G12 ◇ REV MR 86f:03104 • ID 45022

GARDIES, J.-L. [1971] *Les particularites du systeme propositionnel trivalent de Lukasiewicz s'expliquent-elles par le conflit de deux exigences ?* (**J** 0063) Studia Logica 29*149-154
◇ B50 ◇ REV MR 48 # 5821 Zbl 257 # 02005 • ID 04789

GARDIES, J.-L. [1973] *Logique deontique et theorie generale des fonctions completives* (**J** 0079) Logique & Anal, NS 16*143-220
◇ B45 ◇ REV MR 56 # 11754 Zbl 281 # 02031 • ID 29569

GARDIES, J.-L. [1973] *Vingt annees de recherches deontiques* (**J** 1834) Arch Rechts- und Sozialphil 18*385-424
◇ A10 B45 ◇ ID 32337

GARDIES, J.-L. [1975] *A propos de l'article de Georges Kalinowski "Sur un nouveau systeme de logique deontique"* (**J** 0079) Logique & Anal, NS 18*189-208
◇ A05 B45 ◇ REV MR 56 # 11755 Zbl 342 # 02010
• ID 61918

GARDIES, J.-L. [1975] *Esquisse d'une theorie des predicats completifs* (**J** 0079) Logique & Anal, NS 18*3-29
◇ B45 ◇ REV MR 53 # 2631 Zbl 331 # 02014 • ID 21520

GARDIES, J.-L. [1975] *La logique du temps* (**X** 0840) Pr Univ France: Paris 160pp
◇ B45 ◇ REV JSL 42.430 • ID 32339

GARDIES, J.-L. [1978] *Formalization of that-clauses* (**J** 0063) Studia Logica 37*89-101
◇ A05 B60 B65 ◇ REV MR 80m:03010 Zbl 381 # 03006
• ID 32341

GARDIES, J.-L. & MALINOWSKI, G. [1981] *D'un certain usage de la trivalence en logique deontique* (**J** 0079) Logique & Anal, NS 24*179-199
◇ B45 ◇ REV MR 83g:03018 Zbl 481 # 03006 • ID 35996

GARDIES, J.-L. [1983] *Logique de l'action (ou du changement) et logique deontique* (**J** 0079) Logique & Anal, NS 26*71-89
◇ B45 ◇ REV MR 85e:03038 • ID 40606

GARDIES, J.-L. [1985] *Comment articuler logique deontique et logique modale l'une sur l'autre? (Polish summary)* (**J** 3815) Acta Univ Lodz Folia Philos 3*3-14
◇ B45 ◇ ID 49673

GARDIES, J.-L. see Vol. I for further entries

GARDNER, MICHAEL R. [1972] *Two deviant logics for quantum theory: Bohr and Reichenbach* (**J** 0013) Brit J Phil Sci 23*89-109
◇ B51 G12 ◇ REV MR 55 # 13972 Zbl 262 # 02011
• ID 27476

GARDOS, E. [1978] *On the monotone classes of the M limit-logic* (**J** 2774) Koezlem MTA Szam & Autom: Kutat Intez 20*7–20
◇ B50 ◇ REV MR 58 # 10296 Zbl 485 # 03008 • ID 73246

GARDOS, E. [1981] *On the monotone classes of maximal limit-logic M* (**J** 0057) Publ Math (Univ Debrecen) 28*199–207
◇ B50 ◇ REV MR 83b:03022 Zbl 485 # 03008 • ID 34818

GARDOS, E. [1982] *Generalization of the self-dualism in the limit-logic M (German and Hungarian summaries)* (**J** 2774) Koezlem MTA Szam & Autom: Kutat Intez 25*61–71
◇ B50 ◇ REV MR 84k:03073 • ID 35008

GARGOV, G.K. [1975] *Intuitionistic analysis, and intermediate logics (Russian)* (**J** 0023) Dokl Akad Nauk SSSR 224*1245–1247
• TRANSL [1975] (**J** 0062) Sov Math, Dokl 16*1372–1374
◇ B55 F50 ◇ REV MR 52 # 10375 Zbl 342 # 02014 • ID 21677

GARGOV, G.K. & KIROV, K.A. [1982] *The logic of "strong box" in IGL is IS4Grz* (**P** 3876) Mat & Mat Obrazov (11);1982 Sl"nchev Bryag 154–160
◇ B45 ◇ REV Zbl 521 # 03043 • ID 37079

GARGOV, G.K. [1984] *A note on the provability logics of certain extensions of Heyting's arithmetic* (**P** 4392) Mat Logika (Markova);1980 Sofia 20–26
◇ B45 F30 ◇ ID 46969

GAROLA, C. [1980] *Propositions and orthocomplementation in quantum logic* (**J** 2736) Int J Theor Phys 19*369–378
◇ B51 G12 ◇ REV MR 81f:81009 Zbl 448 # 03049 • ID 56649

GAROLA, C. [1985] *Embedding of posets into lattices in quantum logic* (**J** 2736) Int J Theor Phys 24*869–873
◇ B51 ◇ ID 48052

GARRETT, B.J. [1984] *More on rigidity and scope* (**J** 0079) Logique & Anal, NS 27*97–101
◇ B45 ◇ REV MR 85j:03019 • ID 45745

GARSON, J.W. & RESCHER, N. [1967] *A note on chronical logic* (**J** 0105) Theoria (Lund) 33*51–52
◇ B45 ◇ REV MR 39 # 41 Zbl 204.3 • ID 16378

GARSON, J.W. & RESCHER, N. [1968] *Topological logic* (**J** 0036) J Symb Logic 33*537–548
◇ B45 ◇ REV MR 38 # 5595 JSL 40.252 • ID 04801

GARSON, J.W. [1972] *Two new interpretations of modality* (**J** 0079) Logique & Anal, NS 15*443–459
◇ B45 ◇ REV MR 51 # 5265 Zbl 267 # 02010 • ID 17451

GARSON, J.W. [1973] *Indefinite topological logic* (**J** 0122) J Philos Logic 2*102–118
◇ B45 ◇ REV MR 57 # 5683 Zbl 259 # 02022 • ID 30345

GARSON, J.W. [1973] *The completeness of an intensional logic: Definite topological logic* (**J** 0047) Notre Dame J Formal Log 14*175–184
◇ B45 ◇ REV MR 48 # 8201 Zbl 251 # 02032 • ID 04803

GARSON, J.W. [1974] *The substitution interpretation in topological logic* (**J** 0122) J Philos Logic 3*109–132
◇ B45 ◇ REV MR 55 # 12481 Zbl 278 # 02026 • ID 28257

GARSON, J.W. [1978] *Completeness of some quantified modal logics* (**J** 0079) Logique & Anal, NS 21*153–164
◇ B45 ◇ REV MR 80b:03019 Zbl 399 # 03014 • ID 28259

GARSON, J.W. [1979] *Free topological logic* (**J** 0079) Logique & Anal, NS 22*453–475
◇ A05 B45 ◇ REV MR 81k:03019 Zbl 432 # 03004 • ID 53962

GARSON, J.W. [1979] *The substitution interpretation and the expressive power of intensional logics* (**J** 0047) Notre Dame J Formal Log 20*858–864
◇ B45 ◇ REV MR 80j:03027 Zbl 363 # 02034 • ID 50861

GARSON, J.W. [1980] *The unaxiomatizability of a quantified intensional logic* (**J** 0122) J Philos Logic 9*59–72
◇ B45 ◇ REV MR 81d:03018 Zbl 426 # 03020 • ID 53616

GARSON, J.W. [1984] *Quantification in modal logic* (**C** 4085) Handb Philos Log 2*249–307
◇ B45 C90 C98 ◇ ID 41834

GAUTHIER, Y. [1967] *Logique hegelienne et formalisation* (**J** 0488) Dialogue (Ottawa) 6*151–165
◇ B53 ◇ ID 42835

GAUTHIER, Y. [1978] *Methodes et concepts de la logique formelle* (**X** 0893) Pr Univ Montreal: Montreal 238pp
◇ A05 B98 ◇ REV MR 80e:03001 Zbl 383 # 03004 • ID 51986

GAUTHIER, Y. see Vol. VI for further entries

GAVRILOV, G.P. [1959] *Certain conditions for completeness in countable-valued logic (Russian)* (**J** 0023) Dokl Akad Nauk SSSR 128*21–24
◇ B50 ◇ REV MR 21 # 6327 Zbl 104.5 • ID 04814

GAVRILOV, G.P. [1965] *On functional completeness in countable-valued logic (Russian)* (**J** 0052) Probl Kibern 15*5–64
◇ B50 ◇ REV MR 37 # 5085 Zbl 305 # 02028 • ID 04815

GAVRILOV, G.P. [1969] *On the power of a set of limit logics with a finite base (Russian)* (**J** 0052) Probl Kibern 21*115–126
• TRANSL [1969] (**J** 0471) Syst Th Res 21*112–123
◇ B50 ◇ REV MR 50 # 6790 Zbl 263 # 02012 • ID 04816

GAVRILOV, G.P. [1974] *The power of the set of classes of finite height in countable-valued logic (Russian)* (**J** 0052) Probl Kibern 29*5–26,245
◇ B50 ◇ REV MR 52 # 5350 Zbl 336 # 02017 • ID 18155

GAVRILOV, G.P. [1976] *Pre-complete classes of partial countably-valued logic that contain all functions of one variable (Russian)* (**J** 0071) Met Diskr Analiz (Novosibirsk) 28*12–24
◇ B50 ◇ REV MR 57 # 2879 Zbl 426 # 03029 • ID 53625

GAVRILOV, G.P. [1984] *Inductive representations of Boolean functions and the finite generation of the Post classes (Russian)* (**J** 0003) Algebra i Logika 23*1–26
• TRANSL [1984] (**J** 0069) Algeb and Log 23*1–19
◇ B50 ◇ REV Zbl 555 # 03028 • ID 46126

GAVRILOV, G.P. see Vol. I, V for further entries

GAZALE, M.J. [1957] *Multi-valued switching functions (and a number- theoretical approach to their synthesis)* (**P** 1675) Summer Inst Symb Log;1957 Ithaca 147
◇ B50 ◇ REV Zbl 145.407 • ID 29335

GAZALE, M.J. [1959] *Les structures de communication a m valeurs et les calculatrices numeriques* (**X** 0834) Gauthier-Villars: Paris vi + 80pp
◇ B50 ◇ REV Zbl 90.347 • ID 22400

GAZALE, M.J. see Vol. I for further entries

GEACH, P.T. [1973] *On certain modal systems connected with S4.3* (J 0387) Bull Sect Logic, Pol Acad Sci 2*8-11
◇ B45 ◇ REV MR 55 # 2500 • ID 73278

GEACH, P.T. [1975] *On entailment* (J 0103) Analysis (Oxford) 35*186-187
◇ B46 ◇ REV Zbl 345 # 02013 • ID 29774

GEACH, P.T. see Vol. I, III, V for further entries

GENRICH, H.J. & THIELER-MEVISSEN, G. [1976] *The calculus of facts* (P 1401) Math Founds of Comput Sci (5);1976 Gdansk 588-595
◇ B45 D15 ◇ REV Zbl 341 # 68039 • ID 61945

GENRICH, H.J. see Vol. I, IV for further entries

GENSLER, H.J. [1985] *Ethical consistency principles* (J 2811) Phil Quart (St Andrews) 35*156-170
◇ B45 ◇ ID 49172

GENSLER, H.J. see Vol. I, III for further entries

GEORGACARAKOS, G.N. [1976] *Semantics for S4.04, S4.4 and S4.3.2* (J 0047) Notre Dame J Formal Log 17*297-302
◇ B45 ◇ REV MR 54 # 62 Zbl 271 # 02020 • ID 18157

GEORGACARAKOS, G.N. [1977] *Abnormal worlds and the non-Lewis modal systems* (J 0047) Notre Dame J Formal Log 18*95-100
◇ B45 ◇ REV MR 56 # 5223 Zbl 338 # 02012 • ID 21951

GEORGACARAKOS, G.N. [1977] *Additional extensions of S4* (J 0047) Notre Dame J Formal Log 18*477-488
◇ B45 ◇ REV MR 58 # 143 Zbl 292 # 02024 • ID 23645

GEORGACARAKOS, G.N. [1977] *Semantics for S4.03* (J 0047) Notre Dame J Formal Log 18*504-506
◇ B45 ◇ REV MR 58 # 144 Zbl 305 # 02035 • ID 61949

GEORGACARAKOS, G.N. [1978] *A modal system properly independent of both the Brouwerian system and S4* (J 0047) Notre Dame J Formal Log 19*101-114
◇ B45 ◇ REV MR 57 # 15965 Zbl 305 # 02034 • ID 27091

GEORGACARAKOS, G.N. [1978] *A new family of modal systems* (J 0047) Notre Dame J Formal Log 19*271-281
◇ B45 ◇ REV MR 57 # 12175 Zbl 305 # 02033 • ID 51312

GEORGACARAKOS, G.N. [1978] *Dawson modelling, deontic logic and moral necessity* (J 0079) Logique & Anal, NS 21*165-184
◇ A05 B45 ◇ REV MR 80i:03030 Zbl 418 # 03013 • ID 53298

GEORGACARAKOS, G.N. [1979] *Orthomodularity and relevance* (J 0122) J Philos Logic 8*415-432
◇ B51 G12 ◇ REV MR 81c:03014 Zbl 426 # 03017 JSL 48.206 • ID 53613

GEORGACARAKOS, G.N. [1982] *The semantics of minimal intuitionism* (J 0079) Logique & Anal, NS 25*383-397
◇ B55 F50 ◇ REV MR 85g:03021 Zbl 545 # 03035 • ID 43512

GEORGE, F.H. [1977] *Precision, language and logic* (X 0869) Pergamon Pr: Oxford x+216pp
◇ A05 B98 ◇ REV Zbl 404 # 03005 • ID 54792

GEORGESCU, G. [1971] *Algebres de Lukasiewicz monadiques et polyadiques* (J 2313) C R Acad Sci, Paris, Ser A-B 272*A416-A419
◇ B50 G15 ◇ REV MR 44 # 44 Zbl 248 # 02067 • ID 27238

GEORGESCU, G. [1971] *Les θ-ultrafiltres validants d'algebre de Lindenbaum-Tarski d'un calcul des predicats a θ valeurs* (J 0070) Bull Soc Sci Math Roumanie, NS 15(63)*9-19
◇ B50 G25 ◇ REV MR 48 # 1920 Zbl 242 # 02067 • ID 26369

GEORGESCU, G. [1971] *Praedikatenkalkuel I* (J 0378) Gaz Mat (Bucharest) Ser A 76*361-366
◇ B98 ◇ REV Zbl 222 # 02006 • ID 26301

GEORGESCU, G. [1971] *The θ-valued Lukasiewicz algebras, I,II,III* (J 0060) Rev Roumaine Math Pures Appl 16*195-209,363-369,1365-1390
◇ B50 G20 ◇ REV MR 46 # 3299 MR 47 # 8292 MR 47 # 8293 Zbl 255 # 06004 Zbl 255 # 06006 • REM Part II in Spanish: Algebras de Lukasiewicz de orden θ; Part III: Duality theorems • ID 04876

GEORGESCU, G. [1971] *Theoreme de completude en logique θ-valente* (J 2313) C R Acad Sci, Paris, Ser A-B 272*A1076-A1078
◇ B50 ◇ REV MR 43 # 7301 Zbl 212.16 • ID 04875

GEORGESCU, G. [1973] *The theory of categories and mathematical logic (Romanian)* (J 0378) Gaz Mat (Bucharest) Ser A 78*121-125
◇ B98 ◇ REV MR 52 # 10415 Zbl 256 # 18001 • ID 21712

GEORGESCU, G. [1978] *Extension of probabilities defined on polyadic algebras* (J 0070) Bull Soc Sci Math Roumanie, NS 22(70)*15-26
◇ B48 G15 ◇ REV MR 58 # 10440 Zbl 382 # 60014 • ID 73299

GEORGESCU, G. [1979] *A representation theorem for tense polyadic algebras* (J 0517) Mathematica (Cluj) 21(44)*131-138
◇ B45 G15 ◇ REV MR 81m:03073 Zbl 448 # 03050 • ID 56650

GEORGESCU, G. [1979] *Modal polyadic algebras* (J 0070) Bull Soc Sci Math Roumanie, NS 23(71)*49-64
◇ B45 G15 ◇ REV MR 80m:03102 Zbl 397 # 03042 • ID 52708

GEORGESCU, G. [1982] *On Moisil's work in many-valued logics* (J 0447) An Univ Bucuresti, Mat 31*45-50
◇ A10 B50 ◇ REV MR 84i:01087 Zbl 501 # 03002 • ID 37774

GEORGESCU, G. [1983] *Chang's modal operators in algebraic logic* (J 0063) Studia Logica 42*43-48
◇ B45 ◇ REV MR 86b:03081 Zbl 562 # 03036 • ID 42329

GEORGESCU, G. see Vol. I, III, V for further entries

GEORGIEVA, N.V. [1974] *An extension of the Kreisel-Putnam System* (J 0009) Arch Math Logik Grundlagenforsch 16*187-189
◇ B55 ◇ REV MR 50 # 9542 Zbl 289 # 02014 • ID 04887

GEORGIEVA, N.V. [1975] *On the decidability of p', p'' and P''* (J 0009) Arch Math Logik Grundlagenforsch 17*51-53
◇ B55 ◇ REV MR 55 # 5426 Zbl 361 # 02058 • ID 04886

GEORGIEVA, N.V. [1983] *Some properties of conditional expressions (Russian)* (J 0137) C R Acad Bulgar Sci 36*725-728
◇ B60 ◇ REV MR 85i:03150 Zbl 525 # 03020 • ID 44223

GEORGIEVA, N.V. see Vol. I, III, IV, VI for further entries

GERASIMOV, I.N. [1979] see BALABAEV, V.M.

GERASIMOV, V.A. [1973] *A completeness theorem for Bochvar's predicate logic (Russian)* (**C** 3036) Vopr Teor Grupp & Kolets 23-36
◇ B50 ◇ REV MR 58 # 21564 • ID 73328

GERASIMOVA, I.A. [1984] *Neighborhood semantics for epistemic logic (Russian)* (**C** 4366) Modal & Intens Log & Primen Probl Metodol Nauk 207-219
◇ B45 ◇ ID 46718

GERCHIU, V.YA. & KUZNETSOV, A.V. [1970] *On finitely axiomatizable superintuitionistic logics (Russian)* (**J** 0023) Dokl Akad Nauk SSSR 195*1263-1266 • ERR/ADD ibid 199*1222
• TRANSL [1970] (**J** 0062) Sov Math, Dokl 11*1654-1658
◇ B55 ◇ REV MR 45 # 4954 Zbl 219 # 02015 • ID 27431

GERCHIU, V.YA. [1970] *Superindependent superintuitionistic logics (Russian)* (**S** 0166) Mat Issl, Mold SSR 5*24-37
◇ B55 ◇ REV MR 44 # 6446 Zbl 232 # 02021 • ID 04892

GERCHIU, V.YA. & KUZNETSOV, A.V. [1970] *Superintuitionistic logic and finite approximability (Russian)* (**J** 0023) Dokl Akad Nauk SSSR 195*1029-1032 • ERR/ADD ibid 199*1222
• TRANSL [1970] (**J** 0062) Sov Math, Dokl 11*1614-1619, Err: 12*IV
◇ B55 ◇ REV MR 45 # 4953 Zbl 219 # 02014 JSL 37.757
• REM Transl-err ibid 12/4*VI • ID 19140

GERCHIU, V.YA. [1972] *The finite approximability of superintuitionistic logics (Russian)* (**S** 0166) Mat Issl, Mold SSR 7/1*186-192 • ERR/ADD ibid 7*278
◇ B55 ◇ REV MR 47 # 20 MR 51 # 2874 Zbl 243 # 02018
• ID 04893

GERGELY, T. [1981] see DOEMOELKI, B.

GERGELY, T. see Vol. I, III, IV for further entries

GERLA, G. & NOLA DI, A. [1981] *A three valued doxastic logic* (**J** 0099) Ricerca, Riv Mat Pure & Appl 32/1*19-33
◇ B45 B50 ◇ REV MR 85i:03067 Zbl 508 # 03010
• ID 36934

GERLA, G. [1982] *A note on the principle of predication* (**J** 0047) Notre Dame J Formal Log 23*471-472
◇ A05 B45 ◇ REV MR 84a:03020 Zbl 464 # 03016
• ID 54606

GERLA, G. [1983] *Model theory and modal logic (Italian)* (**P** 3829) Atti Incontri Log Mat (1);1982 Siena 87-89
◇ B45 C25 C75 C90 ◇ REV MR 84k:03006 • ID 44834

GERLA, G. & VACCARO, V. [1984] *Modal logic and model theory* (**J** 0063) Studia Logica 43*203-216
◇ B45 C25 C75 C90 ◇ REV MR 86h:03023 • ID 39889

GERLA, G. [1985] *Pavelka's fuzzy logic and free L-subsemigroups* (**J** 0068) Z Math Logik Grundlagen Math 31*123-129
◇ B52 ◇ ID 42293

GERLA, G. see Vol. IV, V for further entries

GERSON, M.S. [1975] *An extension of S4 complete for the neighbourhood semantics but incomplete for the relational semantics* (**J** 0063) Studia Logica 34*333-342
◇ B45 ◇ REV MR 55 # 12471 Zbl 323 # 02033 • ID 30413

GERSON, M.S. [1975] *The inadequacy of the neighbourhood semantics for modal logic* (**J** 0036) J Symb Logic 40*141-148
◇ B45 ◇ REV MR 51 # 7827 Zbl 312 # 02019 • ID 04913

GERSON, M.S. [1976] *A neighbourhood frame for T with no equivalent relational frame* (**J** 0068) Z Math Logik Grundlagen Math 22*29-34
◇ B45 ◇ REV MR 53 # 2632 Zbl 325 # 02014 • ID 18491

GEYMONAT, L. [1981] see COSTANTINI, D.

GEYMONAT, L. see Vol. V, VI for further entries

GHILARDI, S. & MELONI, G.C. [1985] *Completeness for intermediate logics (Italian)* (**P** 4646) Atti Incontri Log Mat (2);1983/84 Siena 613-620
◇ B55 ◇ ID 49688

GHILARDI, S. & MELONI, G.C. & PIERSANTINI, C. [1985] *Some results on intermediate propositional logics (Italian)* (**P** 4646) Atti Incontri Log Mat (2);1983/84 Siena 487-498
◇ B55 ◇ ID 49682

GHOSE, A. [1974] *The logic of dynamic identity in "Meaning and structure of time"* (**J** 0286) Int Logic Rev 10*273-277
◇ A05 B48 ◇ REV Zbl 338 # 02002 • ID 62006

GHOSE, A. see Vol. VI for further entries

GIAMBRONE, S. & MEYER, R.K. [1980] *R_+ is contained in T_+* (**J** 0387) Bull Sect Logic, Pol Acad Sci 9*30-32
◇ B46 ◇ REV MR 81f:03027 Zbl 426 # 03015 • ID 53611

GIAMBRONE, S. & MEYER, R.K. [1981] *Strict implication in T* (**J** 0079) Logique & Anal, NS 24*267-269
◇ B46 ◇ REV MR 83f:03018 Zbl 482 # 03003 • ID 35298

GIAMBRONE, S. [1982] *Gentzen systems and decision procedures for relevant logics* (**J** 0387) Bull Sect Logic, Pol Acad Sci 11*169-175
◇ B46 ◇ REV MR 85e:03039 Zbl 523 # 03015 • ID 40610

GIAMBRONE, S. [1984] see BRADY, R.T.

GIAMBRONE, S. [1985] *TW_+ and RW_+ are decidable* (**J** 0122) J Philos Logic 14*235-254
◇ B25 B46 ◇ ID 42658

GIAMBRONE, S. [1985] *On purported Gentzen formulations of two positive relevant logics* (**J** 0063) Studia Logica 44*233-236
◇ B46 F07 ◇ ID 47499

GIBBARD, A. [1975] *Contingent identity* (**J** 0122) J Philos Logic 4*187-221
◇ B45 ◇ REV MR 58 # 21382 Zbl 317 # 02028 • ID 29641

GIBBARD, A. & HARPER, W.L. [1978] *Counterfactuals and two kinds of expected utility* (**P** 2931) Found & Appl of Decision Th;1975 London ON 1*125-162
• REPR [1981] (**C** 4140) Ifs 153-190
◇ B48 ◇ REV MR 81f:90017 Zbl 497 # 90005 • ID 81335

GIBBARD, A. [1981] *Two recent theories of conditionals* (**C** 4140) Ifs 211-247
◇ B48 ◇ REV MR 83a:03003 • ID 47743

GIBBINS, P. [1981] *A note on quantum logic and the uncertainty principle* (**J** 0153) Phil of Sci (East Lansing) 48*122-126
◇ B51 ◇ REV MR 83b:81018 • ID 38997

GIBBINS, P. [1982] see CUTLAND, N.J.

GIBBINS, P. [1982] *The strange modal logic of indeterminacy* (**J** 0079) Logique & Anal, NS 25*443-446
◇ B45 ◇ REV MR 85i:03048 • ID 44098

GIBBINS, P. [1983] *Quantum logic and ensembles* (**P** 4309) Space, Time & Causality;1981 Kiel 191-205
◇ B51 G12 ◇ REV Zbl 562 # 03033 • ID 47433

Gibbins, P. see Vol. I for further entries

Giedymin, J. [1960] *Confirmation, critical region and empirical content of hypotheses* (**J** 0063) Studia Logica 10*122-125
◇ B48 ◇ REV MR 24 # A1162 • ID 33097

Giertz, M. [1973] see Bellman, R.E.

Giles, R. [1974] *A non-classical logic for physics* (**J** 0063) Studia Logica 33*397-415
• REPR [1977] (**C** 3026) Lukasiewicz Sel Pap on Sent Calc 13-51
◇ B50 ◇ REV MR 50 # 12647 MR 58 # 21436 Zbl 324 # 02017 • REM The reprint is an expanded version • ID 04928

Giles, R. [1975] *Lukasiewicz logic and fuzzy set theory* (**P** 1805) Int Symp Multi-Val Log (5,Proc);1975 Bloomington 197-211
• REPR [1976] (**J** 1741) Int J Man-Mach Stud 8*313-327
◇ B50 E72 ◇ REV MR 58 # 10297 MR 82m:03038 Zbl 335 # 02037 • ID 73364

Giles, R. [1976] *A logic for subjective belief* (**P** 2411) Found Probab Th, Stat Inf & Stat Th Sci;1973 London ON 1*41-72
◇ B48 ◇ REV MR 58 # 16181 Zbl 323 # 02043 • ID 30418

Giles, R. [1979] *A formal system for fuzzy reasoning* (**J** 2720) Fuzzy Sets Syst 2*233-257
◇ A05 B52 ◇ REV MR 81g:03026 Zbl 411 # 03018 • ID 52872

Giles, R. [1979] see Adamson, A.

Giles, R. [1979] *The concept of a proposition in classical and quantum physics* (**J** 0063) Studia Logica 38*337-353
◇ A05 B51 G12 ◇ REV MR 81f:81005 Zbl 438 # 03058 • ID 55970

Giles, R. [1980] *A computer program for fuzzy reasoning* (**J** 2720) Fuzzy Sets Syst 4*221-234
◇ B52 ◇ REV MR 81j:68115 Zbl 445 # 03007 • ID 56471

Giles, R. [1982] *Foundations for a theory of possibility* (**C** 3778) Fuzzy Inform & Decis Processes 183-195
◇ B52 ◇ REV MR 84j:03051 Zbl 514 # 94028 • ID 34643

Giles, R. [1982] *Semantics for fuzzy reasoning* (**J** 1741) Int J Man-Mach Stud 17*401-415
◇ B52 ◇ REV Zbl 515 # 03013 • ID 37840

Giles, R. see Vol. I for further entries

Gilezan, K. [1978] see Canak, B.

Gilezan, K. & Rudeanu, S. [1979] *Interpolation formulas over finite sets* (**J** 0400) Publ Inst Math, NS (Belgrade) 25(39)*45-49
◇ B50 ◇ REV MR 81b:94064 Zbl 419 # 03011 • ID 53354

Gilezan, K. [1980] *Interpolation formulas over finite sets* (**P** 3355) Algeb Conf (1);1980 Skopje 71-72
◇ B50 ◇ REV MR 81b:94064 Zbl 473 # 03021 • ID 55351

Gilezan, K. see Vol. I for further entries

Gilmore, P.C. [1977] *Defining and computing many-valued functions* (**P** 4301) Parallel Comp, Parallel Math;1977 Muenchen 17-23
◇ B50 ◇ REV MR 82k:68003 • ID 47309

Gilmore, P.C. see Vol. I, III, V, VI for further entries

Gimon, V.V. [1979] *Quantifier-free formulas in A. A. Markov's ramified semantics (Russian)* (**S** 0554) Issl Teor Algor & Mat Logik (Moskva) 3*5-17
◇ B55 F50 ◇ REV MR 81j:03092 Zbl 422 # 03033 • ID 53494

Gimon, V.V. see Vol. VI for further entries

Gindikin, S.G. [1985] *Algebraic logic* (**X** 0811) Springer: Heidelberg & New York xviii+356pp
◇ B98 ◇ REM Transl. from the Russian • ID 49125

Gindikin, S.G. see Vol. I for further entries

Ginsberg, M. [1972] *The entailment-presupposition relationship* (**J** 0047) Notre Dame J Formal Log 13*511-515
◇ A05 B46 B65 ◇ REV Zbl 242 # 02027 • ID 26344

Ginzer, J.A. [1979] see Butler, J.T.

Girard, J.-Y. [1976] *Three-valued logic and cut-elimination: The actual meaning of Takeuti's conjecture* (**J** 0202) Diss Math (Warsaw) 136*45pp
◇ B50 C85 C90 F05 F35 F50 ◇ REV MR 56 # 5235 Zbl 357 # 02027 • ID 50375

Girard, J.-Y. see Vol. I, III, IV, V, VI for further entries

Girle, R.A. [1974] *Possibility pre-supposition free logics* (**J** 0047) Notre Dame J Formal Log 15*45-62
◇ B45 ◇ REV MR 49 # 8810 Zbl 272 # 02020 • ID 05001

Girle, R.A. [1975] $S1 \neq S0.9$ (**J** 0047) Notre Dame J Formal Log 16*339-344
◇ B45 ◇ REV MR 51 # 7828 Zbl 245 # 02026 • ID 05002

Girle, R.A. [1978] *Logics for knowledge, possibility and existence* (**J** 0047) Notre Dame J Formal Log 19*200-214
◇ A05 B45 ◇ REV MR 58 # 10314 Zbl 226 # 02019 • ID 31095

Girling, B. & Moring, H. [1973] *Logic and logic design* (**X** 3119) Intertext Books: Aylesbury vii+328pp
◇ B70 B98 ◇ REV Zbl 298 # 94033 • ID 62036

Givone, D.D. & Snelsire, R.W. [1968] *Final report on the design of multiple-valued logic systems* (**X** 2183) SUNY Elect Engrg Dept: Buffalo
◇ B50 ◇ ID 41934

Givone, D.D. see Vol. I for further entries

Gladkich, Yu. [1979] *Singular terms, existence and truth: Some remarks on a first order logic of existence* (**P** 1705) Scand Logic Symp (4);1976 Jyvaeskylae 405-411
◇ A05 B45 ◇ REV MR 81g:03007 Zbl 403 # 03009 • ID 54723

Gladkij, A.V. [1974] *Lecture notes on mathematical logic and set theory (Russian)* (**X** 1434) Kalinin Gos Univ: Kalinin 163pp
◇ B98 E98 ◇ REV MR 52 # 13298 • ID 21763

Gladkij, A.V. [1977] *The language of mathematical logic* (**X** 1434) Kalinin Gos Univ: Kalinin
◇ B98 ◇ ID 33958

Gladkij, A.V. see Vol. I, IV, V for further entries

Gladkij, M. [1981] *Mathematical logic and mathematical linguistics (Russian)* (**X** 1434) Kalinin Gos Univ: Kalinin 172pp
◇ B98 ◇ REV MR 83h:03004 • ID 45984

Glaewe, W. [1973] see Baginski, M.

GLAS DE, M. [1984] *Invariance and stability of fuzzy systems* (**J** 0034) J Math Anal & Appl 99*299-319
◇ B52 ◇ REV MR 85e:03131 Zbl 557 # 93060 • ID 40741

GLUBRECHT, J.-M. & OBERSCHELP, A. & TODT, G. [1983] *Klassenlogik* (**X** 0876) Bibl Inst: Mannheim 467pp
◇ B10 B98 E30 ◇ REV MR 85j:03039 Zbl 514 # 03001 • ID 36976

GLUBRECHT, J.-M. see Vol. I, III, VI for further entries

GLUSHKOV, V.M. [1971] *Einfuehrung in die technische Kybernetik Band 1, 2* (**X** 1553) Dokumentation Saur: Muenchen 126pp
◇ B98 ◇ REV Zbl 243 # 94001 • ID 62053

GLUSHKOV, V.M. see Vol. I, IV, VI for further entries

GLYMOUR, C. [1972] see FRIEDMAN, M.

GLYMOUR, C. [1983] *Discussion: revisions of bootstrap testing* (**J** 0153) Phil of Sci (East Lansing) 50*626-629
◇ B48 ◇ REV MR 85k:03016 • ID 45329

GNANI, G. [1974] *Insiemi dialettici generalizzati* (**J** 0319) Matematiche (Sem Mat Catania) 29*263-273
◇ B60 D25 D55 ◇ REV MR 55 # 5416 Zbl 343 # 02018 • ID 62056

GNANI, G. see Vol. IV for further entries

GNATZ, R. [1979] see BROY, M.

GNIAZDOWSKI, A. [1973] *A note on the strong adequacy of a certain matrix for the logic W (Polish) (English summary)* (**S** 1454) Zesz Nauk Wyz Szk Ped Mat, Opole 13*111-114
◇ B50 ◇ REV MR 54 # 4922 Zbl 314 # 02019 • ID 24098

GNIAZDOWSKI, A. [1973] *Nonexistence of finite full systems of rejected axioms for certain sentential logics (Polish) (English summary)* (**S** 1454) Zesz Nauk Wyz Szk Ped Mat, Opole 13*123-130
◇ B50 ◇ REV MR 54 # 2447 Zbl 314 # 02022 • ID 24035

GNIAZDOWSKI, A. [1983] *Two sentential logics determined by a four-element quasi-Boolean algebra (Polish) (English summary)* (**J** 0481) Acta Univ Wroclaw 605(34)(Logika 10)*3-15
◇ B50 ◇ REV Zbl 562 # 03014 • ID 47426

GNIDENKO, V.M. [1962] *Ermittlung der Ordnung der praevollstaendigen Klassen von Funktionen der dreiwertigen Logik (Russisch)* (**J** 0052) Probl Kibern 8*341-346
• TRANSL [1965] (**J** 0449) Probl Kybern 8*379-384
◇ B50 ◇ REV MR 29 # 4674 Zbl 192.31 • ID 16288

GNIDENKO, V.M. [1963] *Determinarea ordinelor claselor aproape complete in logica trivalenta (Romanian)* (**J** 4105) An Romano-Soviet, Ser Autom Cibern 2*9-14
◇ B50 ◇ ID 41892

GOAD, C.A. [1978] *Monadic infinitary propositional logic: a special operator* (**J** 0302) Rep Math Logic, Krakow & Katowice 10*43-50
◇ B25 B60 C75 F50 ◇ REV MR 81e:03024 Zbl 424 # 03029 • ID 73424

GOAD, C.A. see Vol. I, VI for further entries

GOBLE, L.F. [1966] *The iteration of deontic modalities* (**J** 0079) Logique & Anal, NS 9*197-209
◇ B45 ◇ REV MR 37 # 45 Zbl 158.247 • ID 05055

GOBLE, L.F. [1970] *Grades of modality* (**J** 0079) Logique & Anal, NS 13*323-334
◇ B45 ◇ REV MR 44 # 36 Zbl 221 # 02008 • ID 05056

GOBLE, L.F. [1971] *A system of modality* (**J** 0047) Notre Dame J Formal Log 12*225-237
◇ B45 ◇ REV MR 44 # 5211 Zbl 188.13 • ID 05057

GOBLE, L.F. [1973] *A new modal model* (**J** 0079) Logique & Anal, NS 16*301-309
◇ B45 C90 ◇ REV MR 50 # 9532 Zbl 291 # 02012 • ID 05058

GOBLE, L.F. [1973] *A simplified semantics for modal logic* (**J** 0047) Notre Dame J Formal Log 14*151-174
◇ B45 ◇ REV MR 54 # 63 Zbl 188.14 • ID 05059

GOBLE, L.F. [1974] *Gentzen systems for modal logic* (**J** 0047) Notre Dame J Formal Log 15*455-461
◇ B45 F50 ◇ REV MR 51 # 59 Zbl 226 # 02014 • ID 05060

GOCHET, P. [1969] *Logique modale quantifiee et essentialisme* (**J** 0147) An Univ Bucuresti, Acta Logica 12*185-192
◇ B45 ◇ REV MR 44 # 3845 • ID 22242

GOCHET, P. [1977] *Model theory and the pragmatics of indexicals* (**J** 0076) Dialectica 31*389-408
◇ A05 B45 ◇ REV Zbl 409 # 03002 • ID 56305

GODDARD, L. [1965] *An augmented modal logic* (**J** 0047) Notre Dame J Formal Log 6*81-98
◇ B45 D35 ◇ REV MR 36 # 1303 Zbl 245 # 02024 • ID 05062

GODDARD, L. [1966] *Predicates, relations and categories* (**J** 0273) Australasian J Phil 44*139-171
◇ A05 B10 B50 ◇ ID 15234

GODDARD, L. [1968] *Towards a logic of significance* (**J** 0047) Notre Dame J Formal Log 9*233-264
◇ A05 B45 ◇ REV Zbl 175.265 • ID 05063

GODDARD, L. [1970] *Nonsignificance* (**J** 0273) Australasian J Phil 48*10-16
◇ A05 B45 ◇ ID 15235

GODDARD, L. & ROUTLEY, R. [1973] *The logic of significance and context* (**X** 1080) Scottish Acad Pr: Edinburgh x+641pp
• LAST ED [1974] (**X** 0827) Wiley & Sons: New York xi+641pp
◇ A05 B45 ◇ REV MR 58 # 5052 Zbl 302 # 02004 JSL 49.1413 • ID 21236

GODDARD, L. & MEYER, R.K. & ROUTLEY, R. [1974] *Choice and descriptions in enriched intensional languages I* (**J** 0122) J Philos Logic 3*291-316
• REPR [1977] (**P** 2116) Probl in Log & Ontology;1973 Salzburg 147-172
◇ A05 B60 ◇ REV MR 58 # 16156a Zbl 285 # 02023 • REM Parts II,III 1977 by Routley,R • ID 29950

GODDARD, L. [1980] *Significance, necessity, and verification* (**J** 0047) Notre Dame J Formal Log 21*193-215
◇ A05 B45 ◇ REV MR 81d:03019 Zbl 416 # 03025 • ID 53191

GODDARD, L. see Vol. I, IV, VI for further entries

GODFREY-SMITH, W. [1980] *Change and actuality* (**J** 2811) Phil Quart (St Andrews) 30*350-355
◇ A05 B45 ◇ REV MR 81k:03005 • ID 73434

GODOWSKI, R. & GREECHIE, R.J. [1984] *Some equations related to states on orthomodular lattices* (**J** 1008) Demonstr Math (Warsaw) 17*241-250
◇ B51 ◇ REV MR 86a:06013 Zbl 553 # 06013 • ID 44162

GOE, G. [1968] *Modifications of Quine's ML and inclusive quantification systems* (**J** 0036) J Symb Logic 33*39-42
◇ B60 E70 ◇ REV MR 38 # 5582 Zbl 172.294 JSL 36.325 • ID 05067

GOE, G. [1983] *Lezioni di logica* (**X** 3777) Angeli: Milano xiii + 515pp
◇ B98 ◇ REV MR 84f:03002 Zbl 509 # 03002 JSL 50.860 • ID 34422

GOE, G. see Vol. I for further entries

GOEDEL, K. [1931] *Eine Interpretation des intuitionistischen Aussagenkalkuels* (**J** 1124) Ergebn Math Kolloquium 4*39-40
• TRANSL [1969] (**C** 0569) Phil of Math Oxford Readings 128-129
◇ A05 B45 F50 ◇ REV Zbl 7.193 FdM 59.866 • ID 22120

GOEDEL, K. [1944] *Russell's mathematical logic* (**C** 4106) Phil of Russell 123-153
• TRANSL [1967] (**C** 2141) Filos Matematica 81-112 [1969] (**J** 1030) Formalisation 10*84-107
◇ A10 B98 ◇ REV JSL 11.75 JSL 34.313 • ID 15163

GOEDEL, K. [1947] *What is Cantor's continuum problem?* (**J** 0005) Amer Math Mon 54*515-525 • ERR/ADD ibid 55*151
• TRANSL [1967] (**C** 2141) Filos Matematica 113-136
• REPR [1964] (**C** 1105) Phil of Math. Sel Readings 258-273
◇ A05 B98 E50 E98 ◇ REV MR 9.403 Zbl 38.30 JSL 13.116 JSL 34.313 • ID 05073

GOEDEL, K. see Vol. I, III, IV, V, VI for further entries

GOERNEMANN, S. [1971] *A logic stronger than intuitionism* (**J** 0036) J Symb Logic 36*249-261
◇ B55 C90 F50 ◇ REV MR 45 # 27 Zbl 276 # 02013 • ID 39863

GOERNEMANN, S. also published under the name KOPPELBERG, S.

GOETLIND, E. [1950] *A System of postulates for Lewis's calculus* (**J** 4510) Norsk Mat Tidsskr 32*89-92
◇ B45 ◇ REV MR 12.578 Zbl 40.146 JSL 14.231 • ID 24921

GOETLIND, E. [1951] *A Lesniewski-Mihailescu-theorem for m-valued propositional calculi* (**J** 0050) Port Math 10*97-102
◇ B50 ◇ REV MR 13.615 Zbl 43.249 JSL 22.329 • ID 05081

GOETLIND, E. [1952] *Some Sheffer functions in n-valued logic* (**J** 0050) Port Math 11*141-149
◇ B50 ◇ REV MR 14.834 Zbl 47.16 JSL 22.329 • ID 05082

GOETLIND, E. see Vol. I, VI for further entries

GOGUEN, J.A. [1969] *The logic of inexact concepts* (**J** 0154) Synthese 19*325-373
◇ A05 B52 ◇ REV Zbl 184.9 • ID 05094

GOGUEN, J.A. see Vol. I, III, IV, V for further entries

GOHEEN, J.D. & LEWIS, C.I. & MOTHERSHEAD, J.L. [1970] *Collected papers* (**X** 1355) Stanford Univ Pr: Stanford x + 444pp
◇ A05 B96 ◇ ID 25193

GOLDBERG, H. & LEBLANC, H. & WEAVER, G.E. [1974] *A strong completeness theorem for 3-valued logic* (**J** 0047) Notre Dame J Formal Log 15*325-330
◇ B50 ◇ REV MR 51 # 12475 Zbl 232 # 02014 • REM Part I. Part II 1977 by Leblanc,H. • ID 05099

GOLDBLATT, R.I. [1973] *A model-theoretic study of some systems containing S3* (**J** 0068) Z Math Logik Grundlagen Math 19*75-82
◇ B25 B45 C90 ◇ REV MR 47 # 6441 Zbl 255 # 02016 • ID 05104

GOLDBLATT, R.I. [1973] *A new extension of S4* (**J** 0047) Notre Dame J Formal Log 14*567-573
◇ B45 ◇ REV MR 48 # 8202 Zbl 265 # 02015 • ID 05105

GOLDBLATT, R.I. [1973] *Concerning the proper axiom for S4.04 and some related systems* (**J** 0047) Notre Dame J Formal Log 14*392-396 • ERR/ADD ibid 16*608
◇ B45 ◇ REV MR 51 # 7829 Zbl 225 # 02015 • ID 18165

GOLDBLATT, R.I. [1974] *A study of Z modal systems* (**J** 0047) Notre Dame J Formal Log 15*289-294
◇ B45 ◇ REV MR 50 # 9533 Zbl 275 # 02024 • ID 05107

GOLDBLATT, R.I. [1974] *Decidability of some extensions of J* (**J** 0068) Z Math Logik Grundlagen Math 20*203-205
◇ B25 B55 ◇ REV MR 51 # 60 Zbl 306 # 02045 • ID 05106

GOLDBLATT, R.I. [1974] *Metamathematics of modal logic* (**J** 0016) Bull Austral Math Soc 10*479-480
◇ B45 ◇ REV Zbl 273 # 02016 • ID 30487

GOLDBLATT, R.I. [1974] *Semantic analysis of orthologic* (**J** 0122) J Philos Logic 3*19-35
◇ B51 G10 G12 ◇ REV MR 55 # 5398 Zbl 278 # 02023 • ID 29059

GOLDBLATT, R.I. & THOMASON, S.K. [1975] *Axiomatic classes in propositional modal logic* (**P** 0765) Algeb & Log;1974 Clayton 163-173
◇ B45 ◇ REV MR 52 # 5360 Zbl 325 # 02012 JSL 47.440 • ID 18166

GOLDBLATT, R.I. [1975] *First order definability in modal logic* (**J** 0036) J Symb Logic 40*35-40
◇ B45 C20 C90 ◇ REV MR 55 # 5389 Zbl 311 # 02029 JSL 47.440 • ID 05108

GOLDBLATT, R.I. [1975] *Solution to a completeness problem of Lemmon and Scott* (**J** 0047) Notre Dame J Formal Log 16*405-408
◇ B45 ◇ REV MR 51 # 12481 Zbl 258 # 02025 • ID 05109

GOLDBLATT, R.I. [1976] *Metamathematics of modal logic part I* (**J** 0302) Rep Math Logic, Krakow & Katowice 6*41-78
◇ B45 ◇ REV MR 58 # 27331 Zbl 356 # 02016 JSL 47.440 • REM Part II 1976 • ID 21913

GOLDBLATT, R.I. [1976] *Metamathematics of modal logic part II* (**J** 0302) Rep Math Logic, Krakow & Katowice 7*21-52
◇ B45 ◇ REV MR 58 # 27331b Zbl 356 # 02017 JSL 47.441 • REM Part I 1976 • ID 21924

GOLDBLATT, R.I. [1978] *Arithmetical necessity, provability and intuitionistic logic* (**J** 0105) Theoria (Lund) 44*38-46
◇ B45 F50 ◇ REV MR 80h:03026 Zbl 409 # 03011 • ID 56314

GOLDBLATT, R.I. [1979] *Topoi. The categorial analysis of logic* (**S** 3303) Stud Logic Found Math 98*xv+486pp
• TRANSL [1983] (**X** 0885) Mir: Moskva 488pp
◇ B98 C98 E98 F35 F50 F98 G30 ◇ REV MR 81a:03063 Zbl 434#03050 JSL 47.445 • REM 2nd rev. ed. 1984; xvi+552pp • ID 55754

GOLDBLATT, R.I. [1980] *Diodorean modality in Minkowski spacetime* (**J** 0063) Studia Logica 39*219-236
◇ B45 ◇ REV MR 82a:03018 Zbl 457#03019 • ID 54344

GOLDBLATT, R.I. [1981] *Grothendieck topology as geometric modality* (**J** 0068) Z Math Logik Grundlagen Math 27*495-529
◇ B25 B45 C90 F50 G30 ◇ REV MR 83d:03069 Zbl 474#03018 • ID 55422

GOLDBLATT, R.I. [1982] see CRESSWELL, M.J.

GOLDBLATT, R.I. [1984] *Orthomodularity is not elementary* (**J** 0036) J Symb Logic 49*401-404
◇ B51 C65 G10 ◇ REV MR 85e:03154 • ID 40768

GOLDBLATT, R.I. see Vol. III, V for further entries

GOLDMAN, H.S. [1977] *David Lewis's semantics for deontic logic* (**J** 0094) Mind 86*242-248
◇ B45 ◇ REV MR 57#5684 • ID 73475

GOL'DMAN, R.S. [1980] *Theory of fuzzy sets (Russian) (English summary)* (**J** 0011) Avtom Telemekh 1980/10*146-153
• TRANSL [1981] (**J** 0010) Autom & Remote Control 41*1450-1455
◇ B52 E72 ◇ REV MR 83b:92091 Zbl 473#93003 • ID 46184

GOLL, P. [1973] see BAGINSKI, M.

GOLOTA, YA.YA. [1969] *Nets of marks and deducibility in the intuitionistic propositional calculus (Russian)* (**S** 0228) Zap Nauch Sem Leningrad Otd Mat Inst Steklov 16*28-43
• ERR/ADD ibid 20*292-294
• TRANSL [1969] (**J** 0521) Semin Math, Inst Steklov 16*11-19
◇ B55 B75 F50 ◇ REV MR 41#6667 Zbl 237#02003 • ID 05123

GOLOTA, YA.YA. see Vol. III, VI for further entries

GOLUNKOV, YU.V. [1964] *On the equivalence of formulae realizing transformations of functions of a k-valued logic (Russian)* (**J** 0968) Uch Zap Univ, Kazan 124/2*71-92
◇ B50 ◇ REV MR 34#2446 Zbl 154.258 • ID 24971

GOLUNKOV, YU.V. [1966] *Ueber die Klasse der durch Formeln realisierten Abbildungen des P_2^n auf sich selbst (Russian)* (**J** 0968) Uch Zap Univ, Kazan 126/6*34-44
◇ B50 ◇ REV MR 37#1190 Zbl 255#02004 • ID 28952

GOLUNKOV, YU.V. [1980] *A criterion for the completeness of systems of operations in operator algorithms realizing functions of k-valued logics (Russian)* (**J** 3937) Veroyat Met i Kibern (Kazan) 17*23-34
◇ B50 ◇ REV MR 84h:68032 Zbl 486#03014 • ID 38078

GOLUNKOV, YU.V. see Vol. I, IV for further entries

GOMEZ CALDERON, J. [1979] *Logica simbolica* (**X** 2686) CECSA: Mexico City 147pp
◇ B98 ◇ REV MR 82i:03001 Zbl 521#03001 • ID 73482

GONSETH, F. [1937] *Qu'est-ce que la logique* (**X** 0859) Hermann: Paris 91pp
◇ B98 ◇ REV JSL 3.165 FdM 63.836 • ID 25110

GONSETH, F. see Vol. I, V, VI for further entries

GOOD, I.J. [1962] *Subjective probability as the measure of a non-measurable set* (**P** 0612) Int Congr Log, Meth & Phil of Sci (1,Proc);1960 Stanford 319-329
◇ B48 E75 ◇ REV MR 34#5122 Zbl 192.21 • ID 16283

GOOD, I.J. [1981] *The weight of evidence provided by uncertain testimony or from an uncertain event* (**J** 2763) J Stat Comp & Simul 13*56-60
◇ B48 ◇ REV MR 82m:03026 • ID 73504

GOODMAN, NICOLAS D. [1981] *The logic of contradiction* (**J** 0068) Z Math Logik Grundlagen Math 27*119-126
◇ B53 E70 ◇ REV MR 82g:03043 Zbl 467#03019 • ID 55017

GOODMAN, NICOLAS D. [1984] *Epistemic arithmetic is a conservative extension of intuitionistic arithmetic* (**J** 0036) J Symb Logic 49*192-203
◇ B45 F05 F30 F50 ◇ REV MR 85g:03084 • ID 42452

GOODMAN, NICOLAS D. see Vol. IV, V, VI for further entries

GOODSTEIN, R.L. [1951] *Constructive formalism: Essays on the foundations of mathematics* (**X** 0886) Leicester Univ Pr: Leicester 91pp
◇ A05 B98 F30 F60 F98 ◇ REV MR 14.123 Zbl 45.150 JSL 18.258 • ID 25669

GOODSTEIN, R.L. [1957] *Mathematical logic* (**X** 0886) Leicester Univ Pr: Leicester viii+104pp
◇ B98 ◇ REV MR 19.1 Zbl 77.12 JSL 28.98 • ID 22132

GOODSTEIN, R.L. [1958] *Models of propositional calculi in recursive arithmetic* (**J** 0132) Math Scand 6*293-296
◇ B05 B50 F30 F50 ◇ REV MR 21#5570 Zbl 88.10 JSL 28.291 • ID 05182

GOODSTEIN, R.L. [1958] *On the formalisation of indirect discourse* (**J** 0036) J Symb Logic 23*417-419
◇ B60 B65 ◇ REV MR 21#7150 JSL 32.549 • ID 05181

GOODSTEIN, R.L. see Vol. I, III, IV, V, VI for further entries

GORAN, V.P. [1977] *Two aspects of the problem of interpreting necessity and randomness (Russian)* (**C** 3573) Log-Metodol Probl Estestv & Odshchestbennykh Nauk 92-104
◇ A05 B48 ◇ REV MR 81h:03010 • ID 73523

GORANKO, V.F. [1985] *Propositional logics with strong negation and the Craig interpolation theorem* (**J** 0137) C R Acad Bulgar Sci 38*825-827
◇ B55 C40 ◇ ID 49398

GORANKO, V.F. [1985] *The Craig interpolation theorem for propositional logics with strong negation* (**J** 0063) Studia Logica 44*291-317
◇ B55 C40 ◇ ID 47504

GORELIK, A.G. [1967] *Geometric interpretation of a three-valued algebra (Russian)* (**J** 0413) Izv Akad Nauk Belor SSR, Ser Fiz-Mat 1967/1*9-17
◇ B50 ◇ REV MR 34#7351 Zbl 204.4 • ID 41926

GORELIK, A.G. [1972] *An indexed three-valued calculus and its geometric interpretation (Russian)* (**J** 0199) Zh Vychisl Mat i Mat Fiz 12*822-827
• TRANSL [1972] (**J** 1049) USSR Comput Math & Math Phys 12/3*334-341
◇ B50 ◇ REV MR 50#15480 Zbl 242#68062 • ID 62127

GORELIK, A.G. [1978] *Geometric interpretations of indexed three-valued calculi (Russian)* (**J** 0199) Zh Vychisl Mat i Mat Fiz 18*799-803
• TRANSL [1978] (**J** 1049) USSR Comput Math & Math Phys 18/3*281-287
◇ B50 ◇ REV MR 58 # 16137 Zbl 429 # 03011 • ID 53842

GORGY, F.W. [1979] *The independence of the rule of syllogism in* S_2 (**J** 0068) Z Math Logik Grundlagen Math 25*481-484
◇ B45 F50 ◇ REV MR 80m:03094 Zbl 432 # 03034
• ID 53991

GORGY, F.W. see Vol. VI for further entries

GORLOV, V.V. [1973] *Congruences on closed Post classes (Russian)* (**J** 0087) Mat Zametki (Akad Nauk SSSR) 13*725-734
• TRANSL [1973] (**J** 1044) Math Notes, Acad Sci USSR 13*434-438
◇ B50 D03 ◇ REV MR 48 # 10781 Zbl 261 # 02037
• ID 05201

GORLOV, V.V. [1975] *Kongruenzen auf abgeschlossenen Postschen Klassen* (**J** 0115) Wiss Z Humboldt-Univ Berlin, Math-Nat Reihe 24*728-729
◇ B50 ◇ REV MR 48 # 10781 Zbl 322 # 02011 • ID 62131

GORLOV, V.V. [1977] *On closed classes of k-valued logic, all superclasses of which have only trivial congruences (Russian)* (**J** 0023) Dokl Akad Nauk SSSR 234*273-276
• TRANSL [1977] (**J** 0062) Sov Math, Dokl 18*625-628
◇ B50 ◇ REV MR 56 # 8323 Zbl 408 # 03017 • ID 56256

GORLOV, V.V. [1977] *The closed classes of k-valued logic, all of whose congruences are all trivial (Russian)* (**J** 0087) Mat Zametki (Akad Nauk SSSR) 22*499-509
• TRANSL [1977] (**J** 1044) Math Notes, Acad Sci USSR 22*769-775
◇ B50 C05 G20 ◇ REV MR 58 # 5025 Zbl 375 # 02006
• ID 51584

GORLOV, V.V. [1979] *A sufficient condition for closed classes of k-valued logic to have only trivial congruences (Russian)* (**J** 0142) Mat Sb, Akad Nauk SSSR, NS 110(152)*551-578
• TRANSL [1979] (**J** 0349) Math of USSR, Sbor 38*507-532
◇ B50 ◇ REV MR 81d:03024 Zbl 434 # 03019 • ID 54451

GORLOV, V.V. & LAU, D. [1982] *On a family of M-classes of k-valued logic and congruences on submaximal classes of* P_3 *(Russian) (English and German summaries)* (**J** 0129) Elektr Informationsverarbeitung & Kybern 18*669-686
◇ B50 ◇ REV MR 85k:03018 Zbl 529 # 03006 • ID 37648

GORLOV, V.V. & LAU, D. [1983] *Congruences on closed sets of self-dual functions in many-valued logics and on closed sets of linear functions in prime-valued logics* (**J** 2774) Koezlem MTA Szam & Autom: Kutat Intez 29*31-39
◇ B50 ◇ REV MR 85h:03025 • ID 43301

GORN, S. [1962] *An axiomatic approach to prefix languages* (**P** 1192) Symb Lang in Data Processing;1962 Roma 1-21
◇ B60 ◇ REV Zbl 139.8 • ID 21976

GORN, S. see Vol. IV for further entries

GORSKIJ, D.P. [1962] *Application of dialectical logic to the study of thought processes (Russian)* (**C** 4691) Dialektika & Logika, Formy Mysl 5-41
◇ A05 B53 ◇ REV JSL 30.372 • ID 43054

GOTESKY, R. [1968] *The uses of inconsistency* (**J** 0075) Phil Phenom Research 28*471-500
◇ B53 ◇ ID 42836

GOTO, M. & KAO, S. & NINOMIYA, T. [1982] *Improvement of the axioms of Wajsberg's three-valued logic, using a specific four-valued logic (Japanese)* (**P** 4081) Many-Val Log & Appl;1982 Kyoto 1-21
◇ B50 ◇ ID 47591

GOTO, M. & KAO, S. & NINOMIYA, T. [1983] *Synthesis of axiom systems for the three-valued predicate logic by means of the special four-valued logic* (**P** 4054) Int Symp Multi-Val Log (13);1983 Kyoto 228-234
◇ B50 ◇ REV Zbl 559 # 03014 • ID 46886

GOTO, M. & KAO, S. & OKADA, Y. [1984] *A formulation of a system of specific four-valued logic axioms for the analysis of multi-valued logic axioms (Japanese) (English summary)* (**J** 4278) Bull Electr Lab (Japan) 48*650-657
◇ B50 ◇ ID 46866

GOTO, M. see Vol. I for further entries

GOTTWALD, S. [1977] *Theoretische Betrachtungen ueber Fuzzy-Logik* (**C** 4327) Problemsemin.Algeb & Log Grundl Inform 3-12
◇ B52 ◇ ID 42782

GOTTWALD, S. [1977] *Untersuchungen zur mehrwertigen Mengenlehre. III* (**J** 0114) Math Nachr 79*207-217
◇ B50 E70 ◇ REV MR 57 # 9543 Zbl 362 # 02066 • REM Part II 1976 • ID 50788

GOTTWALD, S. [1979] *Eine Anwendungsvariante der mehrwertigen Logik* (**S** 2877) Wiss Z Univ Leipzig, Ges-Sprachwiss Reihe 28*303-312
◇ B50 E72 ◇ REV MR 81d:03025 Zbl 414 # 03015
• ID 53061

GOTTWALD, S. [1979] *Eine Anwendung des Formalismus der mehrwertigen Logik* (**P** 2539) Frege Konferenz (1);1979 Jena 130-141
◇ B52 E72 G20 ◇ REV MR 82h:03061 • ID 73554

GOTTWALD, S. [1980] *Fuzzy propositional logics* (**J** 2720) Fuzzy Sets Syst 3*181-192
◇ B52 ◇ REV MR 81a:03024 Zbl 426 # 03030 • ID 53626

GOTTWALD, S. [1982] see CZOGALA, E.

GOTTWALD, S. [1983] see CZOGALA, E.

GOTTWALD, S. [1984] *T-Normen und φ-Operatoren als Wahrheitswertfunktion mehrwertiger Junktoren* (**P** 3621) Frege Konferenz (2);1984 Schwerin 121-128
◇ B50 ◇ REV MR 86d:03024 Zbl 556 # 03023 • ID 42784

GOTTWALD, S. [1984] *Criteria for non-interactivity of fuzzy logic controller rules* (**P** 4053) IFAC Symp Large Scale Syst: Th & Appl;1983 Warsaw 229-233
◇ B52 ◇ REV Zbl 558 # 93014 • ID 46639

GOTTWALD, S. [1985] *A generalized Lukasiewicz-style identity logic* (**C** 4181) Math Log & Formal Syst (Costa da) 183-195
◇ B50 ◇ REV Zbl 571 # 03008 • ID 48168

GOTTWALD, S. see Vol. I, V for further entries

GOULD, JAMES A. [1967] see COPI, I.M.

GOZE, M. & LUTZ, R. [1981] *Nonstandard analysis. A pratical guide with applications* (**S** 3301) Lect Notes Math 881*xiv+261pp
◇ B98 H05 ◇ REV MR 83i:03103 Zbl 506 # 03021
• ID 34857

GOZE, M. see Vol. I for further entries

GRACZYNSKA, E. & WRONSKI, A. [1974] *Constructing denumerable matrices strongly adequate for prefinite logics* (**J** 0063) Studia Logica 33*417-423
◇ B55 G10 ◇ REV MR 52 # 7859 Zbl 312 # 02022
• ID 18174

GRADSHTEJN, I.S. [1972] *Direct and inverse theorems (Russian)* (**X** 2027) Nauka: Moskva 128pp
• TRANSL [1963] (**X** 0869) Pergamon Pr: Oxford xviii+173pp
◇ B98 ◇ REV MR 53 # 85 Zbl 102.247 • REM 5th edition
• ID 16549

GRAF, S. [1984] *On Lamport's comparison between linear and branching time temporal logic* (**J** 3441) RAIRO Inform Theor 18*345-353
◇ B45 ◇ REV MR 86g:03033 Zbl 551 # 68033 • ID 43956

GRAF, S. see Vol. V for further entries

GRAHAM, R.L. [1967] *On n-valued functionally complete truth functions* (**J** 0036) J Symb Logic 32*190-195
◇ B50 ◇ REV MR 36 # 6266 Zbl 148.243 JSL 36.691
• ID 05299

GRAHAM, R.L. see Vol. V for further entries

GRANA, N. [1983] *Logica paraconsistente: Una introduzione* (**X** 1010) Loffredo: Napoli
◇ B53 ◇ ID 42837

GRANDY, R.E. [1972] *A definition of truth for theories with intensional definite description operators* (**J** 0122) J Philos Logic 1*137-155
◇ A05 B45 ◇ REV MR 53 # 7730 Zbl 248 # 02028
• ID 15035

GRANDY, R.E. [1976] *Anadic logic and English* (**J** 0154) Synthese 32*395-402
◇ B60 B65 G15 ◇ REV MR 58 # 16282 Zbl 341 # 02003
• ID 29732

GRANDY, R.E. [1976] *On the relation between free description theories and standard quantification theory* (**J** 0047) Notre Dame J Formal Log 17*149-152
◇ B60 C20 ◇ REV MR 58 # 21428 Zbl 292 # 02041
• ID 18176

GRANDY, R.E. [1977] *Advanced logic for applications* (**S** 3307) Synth Libr 110*xi+167pp
◇ B98 ◇ REV MR 57 # 9450 Zbl 381 # 03003 JSL 47.714
• ID 51871

GRANDY, R.E. see Vol. IV, VI for further entries

GRANT, J. [1974] *A non-truth-functional 3-valued logic* (**J** 0497) Math Mag 47*221-223
◇ B50 ◇ REV MR 51 # 52 Zbl 293 # 02018 • ID 15237

GRANT, J. [1974] *Incomplete models* (**J** 0047) Notre Dame J Formal Log 15*601-607
◇ B53 C90 ◇ REV MR 50 # 4282 Zbl 254 # 02045
• ID 05305

GRANT, J. [1975] *Inconsistent and incomplete logics* (**J** 0497) Math Mag 48*154-159
◇ B22 B53 ◇ REV MR 52 # 13311 Zbl 309 # 02010
• ID 21774

GRANT, J. [1977] *Incompleteness and inconsistency in propositional logic* (**J** 1893) Relevance Logic Newslett 2*109-114
◇ B53 ◇ REV Zbl 411 # 03008 • ID 52863

GRANT, J. [1978] *Classifications for inconsistent theories* (**J** 0047) Notre Dame J Formal Log 19*435-444
◇ B53 C07 ◇ REV MR 58 # 10392 Zbl 305 # 02040
• ID 28232

GRANT, J. see Vol. III for further entries

GRASSMANN, G. [1966] *Die Formenlehre oder Mathematik. 1.die Groessenlehre. 2.die Begriffslehre der Logik. 3.die Bindelehre oder Kombinationslehre. 4.die Zahlenlehre oder Arithmetik. 5.die Aussenlehre oder Ausdehnungslehre (reprinted)* (**X** 0892) Olms: Hildesheim 229pp
◇ B98 ◇ ID 22622

GRATTAN-GUINNESS, I. [1976] *Fuzzy membership mapped onto intervals and many-valued quantities* (**J** 0068) Z Math Logik Grundlagen Math 22*149-160
◇ B52 E72 ◇ REV MR 57 # 113 Zbl 334 # 02011
• ID 18482

GRATTAN-GUINNESS, I. [1979] *Forays into the meta-theory of fuzzy set theory* (**J** 0079) Logique & Anal, NS 22*321-337
◇ B52 E72 ◇ REV MR 81j:03084 Zbl 426 # 03058
• ID 53654

GRATTAN-GUINNESS, I. see Vol. V, VI for further entries

GREECHIE, R.J. [1971] *Combinatorial quantum logic* (**P** 1548) Conf Convexity & Combin Geom (1);1971 Norman 6-14
◇ B51 G12 ◇ REV MR 49 # 1944 Zbl 248 # 02031
• ID 62169

GREECHIE, R.J. & GUDDER, S.P. [1971] *Is a quantum logic a logic?* (**J** 2146) Helv Phys Acta 44*238-240
◇ B51 ◇ REV MR 46 # 8551 JSL 48.206 • ID 47408

GREECHIE, R.J. & GUDDER, S.P. [1973] *Quantum logics* (**P** 2312) Contemp Res in Found & Philos of Quantum Th;1973 London ON 143-173
◇ B51 G12 ◇ REV Zbl 279 # 02015 • ID 29541

GREECHIE, R.J. [1974] *Some results from the combinatorial approach to quantum logic* (**J** 0154) Synthese 29*113-127
◇ B51 G12 ◇ REV Zbl 318 # 02029 • ID 62170

GREECHIE, R.J. [1975] *On three dimensional quantum proposition systems* (**P** 3016) Quant Th & Struct of Time & Space;1974 Feldafing 71-83
◇ B51 G12 ◇ REV MR 58 # 32624 • ID 81395

GREECHIE, R.J. [1976] *Some results from the combinatorial approach to quantum logic* (**C** 2953) Log & Probab in Quant Mech 105-119
◇ B51 G12 ◇ REV Zbl 361 # 02038 • ID 50673

GREECHIE, R.J. [1981] *A nonstandard quantum logic with a strong set of states* (**P** 3820) Curr Iss in Quantum Log;1979 Erice 375-380
◇ B51 G12 ◇ REV MR 84k:03143 Zbl 537 # 03044
• ID 35035

GREECHIE, R.J. & GUDDER, S.P. & RUETTIMANN, G.T. [1982] *Measurements, Hilbert space and quantum logics* (J 0209) J Math Phys 23*2381-2386
◇ B51 ◇ REV MR 84f:81007 Zbl 508 # 03028 • ID 36943

GREECHIE, R.J. [1984] see GODOWSKI, R.

GREEF DE, J. [1973] *Professor Halberstadt on counterfactual conditionals and modality* (J 0286) Int Logic Rev 7*126-134
◇ A05 B45 ◇ REV Zbl 357 # 02005 • ID 50353

GREEF DE, J. [1983] *La negation deontique dans "Directives and norms" d'Alf Ross* (J 0079) Logique & Anal, NS 26*203-217
◇ B45 ◇ REV MR 85e:03036 • ID 40604

GREEN, K. [1985] *Is a logic for belief sentences possible?* (J 0095) Philos Stud 47*29-55
◇ A05 B45 ◇ ID 44545

GRENIEWSKI, H. [1955] *Elements of the logic of induction (Russian)* (X 1034) PWN: Warsaw 90pp
◇ A05 B48 ◇ REV JSL 23.77 • ID 16855

GRENIEWSKI, H. [1955] *Elements of formal logic (Polish)* (X 1034) PWN: Warsaw 492pp
◇ B98 ◇ REV Zbl 68.10 JSL 21.188 • ID 42617

GRENIEWSKI, H. [1957] 2^{n+1} *logical values* (J 1182) Studia Filoz 3*3-28
◇ B50 ◇ REV JSL 29.109 • ID 22130

GRENIEWSKI, H. see Vol. I, III, VI for further entries

GRENIEWSKI, M. [1956] *Utilisation des logiques trivalentes dans la theorie des mecanismes automatiques. I. Realisation des functions fondamentales par des circuits (Roumanian) (Russian and French summaries)* (J 4188) Com Acad Romine 6*225-229
◇ B50 ◇ REV MR 18.712 Zbl 70.211 JSL 24.257 • REM Part II 1956 by Moisil,G.C. ibid. 231-234 • ID 48776

GRIFFIN, N. [1978] *Supervaluations and Tarski* (J 0047) Notre Dame J Formal Log 19*297-298
◇ A05 B50 ◇ REV MR 58 # 4987 Zbl 363 # 02013 • ID 51310

GRIFFIN, N. & ROUTLEY, R. [1979] *Towards a logic of relative identity* (J 0079) Logique & Anal, NS 22*65-83
◇ A05 B60 ◇ REV MR 80j:03018 Zbl 416 # 03006 • ID 53172

GRIGOLIYA, R.SH. [1973] *Algebraic analysis of n-valued Lukasiewicz-Tarski logical systems (Russian) (Gregorian summary)* (J 0954) Tr Inst Prikl Mat, Tbilisi A6-7*121-132
◇ B50 G20 ◇ REV MR 51 # 53 • ID 24789

GRIGOLIYA, R.SH. [1975] *On the algebras corresponding to the n-valued Lukasiewicz-Tarski logical systems* (P 1805) Int Symp Multi-Val Log (5,Proc);1975 Bloomington 234-239
◇ B50 G20 ◇ REV MR 58 # 10434 • ID 73623

GRIGOLIYA, R.SH. [1976] *The lattice of all finitely-approximable extensions of a countably-valued Lukasiewicz-logic (Russian)* (C 3271) Issl Teor Mnozh & Neklass Logik 221-228
◇ B50 G20 ◇ REV MR 57 # 5865 Zbl 403 # 03051 • ID 54764

GRIGOLIYA, R.SH. [1977] *Algebraic analysis of Lukasiewicz-Tarski's n-valued logical systems* (C 3026) Lukasiewicz Sel Pap on Sent Calc 81-92
◇ B50 G20 ◇ REV MR 58 # 21437 Zbl 381 # 03017 • ID 73624

GRIGOLIYA, R.SH. & MESKHI, V.YU. [1978] *Critical bimodal systems (Russian)* (C 2567) Modal & Intentsional Log 35-37
◇ B45 ◇ ID 32640

GRIGOLIYA, R.SH. [1979] see FINN, V.K.

GRIGOLIYA, R.SH. [1980] see ANSHAKOV, O.M.

GRIGOLIYA, R.SH. [1982] see ANSHAKOV, O.M.

GRIGOLIYA, R.SH. [1985] *Formulas of m variables in Lukasiewicz calculi (Russian) (Polish summary)* (J 3815) Acta Univ Lodz Folia Philos 3*29-36
◇ B50 ◇ ID 49694

GRIGOR'EV, V.V. [1984] see DROZDOVA, G.D.

GRISHIN, V.N. [1974] *A nonstandard logic, and its application to set theory (Russian)* (C 2577) Issl Formaliz Yazyk & Neklass Log 135-171
◇ B60 E70 F07 ◇ REV MR 56 # 8331 • ID 73648

GRISHIN, V.N. [1976] *On the algebraic semantics of a logic without reduction (Russian)* (C 3271) Issl Teor Mnozh & Neklass Logik 247-264
◇ B50 G20 ◇ REV MR 57 # 9533 Zbl 403 # 03023 • ID 54737

GRISHIN, V.N. [1976] see BOCHVAR, D.A.

GRISHIN, V.N. [1981] *Predicate and set-theoretic calculi based on a logic without contractions (Russian)* (J 0216) Izv Akad Nauk SSSR, Ser Mat 45*47-68,239
• TRANSL [1981] (J 0448) Math of USSR, Izv 18*41-59
◇ B53 B60 E30 E70 F07 ◇ REV MR 82g:03016 Zbl 464 # 03027 Zbl 478 # 03009 • ID 54617

GRISHIN, V.N. see Vol. I, V, VI for further entries

GRIZE, J.-B. [1969] *Logique moderne. fasc.I: Logique des propositions et des predicats, deduction naturelle* (X 0834) Gauthier-Villars: Paris ii + 90pp
◇ B98 ◇ REV MR 42 # 36 Zbl 194.306 • REM Part II 1971 • ID 05369

GRIZE, J.-B. [1971] *Logique moderne. fasc.II: Logique des propositions et des predicats, tables de verite et axiomatisation* (X 0834) Gauthier-Villars: Paris iii + 79pp
◇ B98 ◇ REV MR 48 # 3690 Zbl 264 # 02015 • REM Part I 1969. Part III 1973 • ID 05370

GRIZE, J.-B. see Vol. I, VI for further entries

GROENENDIJK, J. & STOKHOF, M. [1975] *Modality and conversational information* (J 1517) Theor Linguist 2*61-112
◇ B45 B65 ◇ REV MR 58 # 21472 • ID 73656

GROENENDIJK, J. [1981] see EMDE BOAS VAN, P.

GROFMAN, B. & HYMAN, G. [1973] *Probability and logic in belief systems* (J 0472) Theory Decis 4*179-195
◇ A05 B48 ◇ REV Zbl 288 # 02017 • ID 62201

GROSJEAN, P.V. [1973] *La logique sur le corps de rupture des paradoxes* (J 0079) Logique & Anal, NS 16*535-562
◇ B50 ◇ REV MR 50 # 55 Zbl 346 # 02005 • ID 05383

GROSJEAN, P.V. see Vol. I for further entries

GROVER, D.L. [1973] *Propositional quantification and quotation contexts* (P 0783) Truth, Syntax & Modal;1970 Philadelphia 101-110
◇ A05 B60 ◇ REV MR 52 # 7840 Zbl 278 # 02011 JSL 42.313 • ID 18179

GRUENWALD, N. [1983] *Bestimmung saemtlicher abgeschlossener Mengen aus $P_{3,2}$, deren Projektion F_8^n ist* (S 2829) Rostocker Math Kolloq 23*5-26
◇ B50 ◇ REV MR 85j:03026a Zbl 553 # 03011 • ID 43299

GRUENWALD, N. [1983] *Bestimmung aller abgeschlossenen Mengen aus $P_{3,2}$, deren Projektion F_8^2 ist, mit Hilfe von Relationen* (S 2829) Rostocker Math Kolloq 23*27-34
◇ B50 ◇ REV MR 85j:03026b Zbl 533 # 03012 • ID 43302

GRZEGORCZYK, A. & JASKOWSKI, S. & LOS, J. & MAZUR, S. & MOSTOWSKI, ANDRZEJ & RASIOWA, H. & SIKORSKI, R. [1955] *The present state of investigations on the foundations of mathematics (Polish)* (J 0051) Commentat Math, Ann Soc Math Pol, Ser 1 1*13-55
• TRANSL [1954] (J 0067) Usp Mat Nauk 9/3*3-38 (Russian) [1954] (P 1924) Polnisch Math Kongr;1953 Warsaw 11-44 (German) [1955] (J 0202) Diss Math (Warsaw) 9*48pp (English)
◇ A10 B98 E98 ◇ REV MR 16.552 Zbl 57.243 Zbl 59.12 JSL 21.372 • ID 31447

GRZEGORCZYK, A. [1961] *An outline of mathematical logic (Polish)* (X 1034) PWN: Warsaw 477pp
• TRANSL [1974] (S 3307) Synth Libr 70*x+596pp
◇ B98 ◇ REV MR 27 # 1347 MR 83f:03007 Zbl 132.245 JSL 48.220 • REM 5th ed. 1981, 510pp • ID 20967

GRZEGORCZYK, A. [1961] *Le traitement axiomatique de la notion de prolongement temporel (Polish) (French and Russian summaries)* (J 0063) Studia Logica 11*23-35
◇ B45 ◇ REV MR 23 # B2467 • ID 33098

GRZEGORCZYK, A. [1961] *Logika popularna* (X 1034) PWN: Warsaw 131pp
◇ B98 ◇ REV MR 24 # A675 Zbl 134.6 • ID 24867

GRZEGORCZYK, A. [1967] *Non-classical propositional calculi in relation to methodological patterns of scientific investigation (Polish) (Russian and English summaries)* (J 0063) Studia Logica 20*117-132
◇ A05 B45 F50 ◇ REV Zbl 295 # 02005 • ID 62216

GRZEGORCZYK, A. [1967] *Some relational systems and the associated topological spaces* (J 0027) Fund Math 60*223-231
◇ B45 C65 G25 ◇ REV MR 36 # 1304 Zbl 207.296 JSL 34.652 • ID 05412

GRZEGORCZYK, A. [1968] *Assertions depending on time and corresponding logical calculi* (J 0020) Compos Math 20*83-87
• REPR [1968] (C 0727) Logic Found of Math (Heyting) 83-87
◇ B45 ◇ REV MR 37 # 6159 Zbl 175.263 JSL 40.499 • ID 05415

GRZEGORCZYK, A. [1981] *Individualistic formal approach to deontic logic* (J 0063) Studia Logica 40*99-102
◇ B45 ◇ REV MR 83f:03016 Zbl 482 # 03004 • ID 35297

GRZEGORCZYK, A. see Vol. I, III, IV, V, VI for further entries

GU, WEINAN & ZHENG, QILUN [1983] *Lattice symmetric ternary logic system and its simplification (Chinese)* (J 3793) Jisuanjii Xuebao 6*317-321
◇ B50 ◇ REV MR 85e:03057 • ID 40626

GUASPARI, D. & SOLOVAY, R.M. [1979] *Rosser sentences* (J 0007) Ann Math Logic 16*81-99
◇ B45 F30 ◇ REV MR 80h:03083 Zbl 426 # 03062 • ID 53658

GUASPARI, D. see Vol. III, IV, V, VI for further entries

GUCCIONE, S. [1980] *A semantics for a dialectical logic* (J 0079) Logique & Anal, NS 23*461-470
◇ B53 ◇ REV MR 83g:03026 • ID 36000

GUCCIONE, S. & TORTORA, R. & VACCARO, V. [1981] *Deduction theorems in Lukasiewicz propositional calculi* (J 2038) Rend Sem Mat, Torino 39/1*53-65
◇ B50 ◇ REV MR 83h:03033 Zbl 499 # 03005 • ID 36053

GUCCIONE, S. [1981] *Quantum logic and the two-slit experiment* (C 3515) Ital Studies in Phil of Sci 237-247
◇ B51 ◇ REV MR 83j:03106 • ID 35388

GUCCIONE, S. & TORTORA, R. [1982] *A semantical hierarchy for modal formulas* (J 2840) Stochastica, Univ Politec Barcelona 6*71-77
◇ B45 ◇ REV MR 85j:03020 Zbl 513 # 03011 • ID 37219

GUCCIONE, S. & TORTORA, R. [1982] *Deducibility in many-valued logics* (P 4002) Int Symp Multi-Val Log (12);1982 Paris 117-121
◇ B50 ◇ REV Zbl 549 # 03016 • ID 43105

GUCCIONE, S. & TAMBURRINI, G. [1982] *The labyrinth of dialectical logics* (P 3845) Conf Math Service of Man (2,Proc)(Feriet);1982 Las Palmas 318-322
◇ B53 ◇ REV Zbl 505 # 03003 • ID 38130

GUCCIONE, S. & TERMINI, S. & TORTORA, R. [1983] *In the labyrinth of many valued logics* (P 4054) Int Symp Multi-Val Log (13);1983 Kyoto 47-53
◇ B50 ◇ REV Zbl 563 # 03010 • ID 47455

GUCCIONE, S. see Vol. I, IV for further entries

GUDDER, S.P. [1969] *On the quantum logic approach to quantum mechanics* (J 1113) Commun Math Phys 12*1-15
◇ A05 B51 G12 ◇ REV MR 40 # 3806 • ID 21182

GUDDER, S.P. [1970] *Axiomatic quantum mechanics and generalized probability theory* (C 4684) Probab Method in Appl Math, Vol 2 53-129
◇ B51 G12 ◇ REV MR 42 # 1455 • ID 33675

GUDDER, S.P. [1971] see GREECHIE, R.J.

GUDDER, S.P. [1973] see GREECHIE, R.J.

GUDDER, S.P. [1979] *A survey of axiomatic quantum mechanics* (C 3045) Log-Algeb Appr to Quant Mech 2*323-363
◇ B51 G12 ◇ REV MR 82k:81006 • ID 81421

GUDDER, S.P. [1981] *Comparison of the quantum logic, convexity, and algebraic approaches to quantum mechanics* (P 3185) Interpr & Found of Quantum Th;1979 Marburg 125-131
◇ B51 G12 ◇ REV MR 85g:81005 Zbl 463 # 46055 • ID 45875

GUDDER, S.P. [1981] *Measure and integration in quantum set theory* (P 3820) Curr Iss in Quantum Log;1979 Erice 341-352
◇ B51 G12 ◇ REV MR 84j:03126 Zbl 537.03045 • ID 34716

GUDDER, S.P. [1981] *Representations of Baer* *-*semigroups and quantum logics in Hilbert space* (**P** 3820) Curr Iss in Quantum Log;1979 Erice 365-373
◇ B51 G12 ◇ REV MR 84k:81017 • ID 44881

GUDDER, S.P. [1982] *Hilbertian interpretations of manuals* (**J** 0053) Proc Amer Math Soc 85*251-255
◇ B51 G12 ◇ REV MR 83f:81017 Zbl 498 # 03049 • ID 36908

GUDDER, S.P. [1982] see GREECHIE, R.J.

GUDDER, S.P. see Vol. I, V for further entries

GUDZHINSKAS, EH. [1982] *Syntactical proof of the elimination theorem for von Wright's temporal logic (Russian) (English and Lithuanian summaries)* (**J** 3939) Mat Logika Primen (Akad Nauk Litov SSR) 2*113-130
◇ B45 ◇ REV MR 85g:03028 Zbl 517 # 03006 • ID 37279

GUENTHER, A. [1982] *A set of concepts for the study of dialogic argumentation* (**P** 3754) Argumentation;1978 Groningen 175-190
◇ B60 ◇ REV MR 83k:03003 Zbl 549 # 03003 • ID 43099

GUENTHER, G. [1955] *Dreiwertige Logik und die Heisenbergsche Unbestimmtheitsrelation* (**P** 0664) Congr Int Union Phil of Sci (2);1954 Zuerich 2*53-59
◇ B50 ◇ REV MR 17.815 JSL 23.90 • ID 05429

GUENTHNER, F. [1975] see AAQVIST, L.

GUENTHNER, F. [1977] see AAQVIST, L.

GUENTHNER, F. & ROHRER, C. (EDS.) [1979] *Studies in formal semantics. Intensionality, temporality, negation. 2nd printing* (**X** 0809) North Holland: Amsterdam viii+265pp
◇ A05 B45 ◇ REV Zbl 435 # 03001 • ID 55762

GUENTHNER, F. [1983] see GABBAY, D.M.

GUENTHNER, F. [1984] see GABBAY, D.M.

GUENTHNER, F. see Vol. III for further entries

GUILBAUD, G.T. [1945] see DAVAL, R.

GUILD, N.C.F. [1979] see BALDWIN, J.F.

GUILD, N.C.F. [1980] see BALDWIN, J.F.

GUILD, N.C.F. [1981] see BALDWIN, J.F.

GUILLAUME, M. [1958] *Calculs de consequences et tableaux d'epreuve pour les classes algebriques generales d'anneaux booleiens a operateurs* (**J** 0109) C R Acad Sci, Paris 247*1542-1544
◇ B45 C05 G05 ◇ REV MR 21 # 654 Zbl 81.244 JSL 25.297 • ID 05434

GUILLAUME, M. [1958] *Rapports entre calculs propositionels modaux et topologie impliques par certaines extensions de la methode des tableaux semantiques. Systeme de Feys-von Wright* (**J** 0109) C R Acad Sci, Paris 246*1140-1142
◇ B45 ◇ REV MR 20 # 2280 Zbl 81.243 • ID 42817

GUILLAUME, M. [1958] *Rapports entre calculs propositionels modaux et topologie impliques par certaines extensions de la methode des tableaux semantiques. Systeme S4 de Lewis* (**J** 0109) C R Acad Sci, Paris 246*2207-2210
◇ B45 G05 ◇ REV MR 21 # 649 Zbl 81.243 JSL 25.296 • ID 05437

GUILLAUME, M. [1964] see COSTA DA, N.C.A.

GUILLAUME, M. [1965] see COSTA DA, N.C.A.

GUILLAUME, M. see Vol. I, V, VI for further entries

GUMANSKI, L. [1975] *Deontic logic without certain paradoxes* (**J** 0063) Studia Logica 34*343-365
◇ B45 ◇ REV MR 55 # 10242 Zbl 329 # 02007 • ID 62228

GUMANSKI, L. [1977] *An improvement of the deontic calculus DSC* (**J** 0063) Studia Logica 36*177-180
◇ B45 ◇ REV MR 58 # 145 Zbl 363 # 02031 • ID 50858

GUMANSKI, L. [1980] *On deontic logic* (**J** 0063) Studia Logica 39*63-75
◇ A05 B45 ◇ REV MR 81h:03037 Zbl 444 # 03013 • ID 56461

GUMANSKI, L. [1981] *The deontic calculus* DSC_3 (**J** 0387) Bull Sect Logic, Pol Acad Sci 10*63-67
◇ A05 B45 ◇ REV Zbl 464 # 03026 • ID 54616

GUMANSKI, L. [1983] *An extension of the deontic calculus* DSC_1 (**J** 0063) Studia Logica 42*129-137 • ERR/ADD ibid 43*202
◇ B45 ◇ REV MR 86f:03028ab Zbl 568 # 03007 • ID 42331

GUMB, R.D. [1979] *An extended joint consistency theorem for free logic with equality* (**J** 0047) Notre Dame J Formal Log 20*321-335
◇ B60 C40 ◇ REV MR 80i:03020 Zbl 321 # 02012 • ID 62230

GUMB, R.D. [1979] *Evolving theories* (**X** 4315) Haven Publ: New York xi+96pp
◇ B45 ◇ REV JSL 47.454 • ID 47361

GUMB, R.D. [1979] see BARNES, R.F.

GUMB, R.D. [1981] *An explication of the notion of a consistent evolving theory* (**J** 0079) Logique & Anal, NS 24*113-127
◇ A05 B45 ◇ REV MR 82k:03015 Zbl 482 # 03001 • ID 73701

GUMB, R.D. & WEAVER, G.E. [1982] *First order properties of relations with the monotonic closure property* (**J** 0068) Z Math Logik Grundlagen Math 28*1-5
◇ B45 E07 ◇ REV MR 83e:03053 Zbl 489 # 03007 • ID 35230

GUMB, R.D. [1984] *An extended joint consistency theorem for a family of free modal logics with equality* (**J** 0036) J Symb Logic 49*174-183
◇ B45 ◇ REV MR 86a:03016 • ID 42455

GUMB, R.D. see Vol. I, III for further entries

GUPTA, A. [1978] *Modal logic and truth* (**J** 0122) J Philos Logic 7*441-472
◇ A05 B45 ◇ REV MR 80b:03020 Zbl 405 # 03004 • ID 54879

GUPTA, A. [1980] see BELNAP JR., N.D.

GUPTA, A. [1980] *The logic of common nouns. An investigation in quantified modal logic* (**X** 0875) Yale Univ Pr: New Haven xi+142pp
◇ B45 ◇ REV Zbl 528 # 03008 JSL 48.500 • ID 37627

GUPTA, A. & THOMASON, R.H. [1981] *A theory of conditionals in the context of branching time* (**C** 4140) Ifs 299-322
◇ B45 ◇ REV MR 83a:03003 • ID 47843

GUPTA, A. [1982] *Truth and paradox* (**J** 0122) J Philos Logic 11*1-60
◇ A05 B53 ◇ REV MR 84i:03020a Zbl 512 # 03002 JSL 50.1068 • ID 42842

GUREVICH, Y. [1982] see CLARKE JR., E.M.

GUREVICH, Y. [1985] see BURGESS, J.P.

GUREVICH, Y. & SHELAH, S. [1985] *The decision problem for branching time logic* (**J** 0036) J Symb Logic 50*668-681
◇ B25 B45 C65 D05 D25 ◇ ID 47372

GUREVICH, Y. & SHELAH, S. [1985] *To the decision problem for branching time logic* (**P** 4180) Int Congr Log, Meth & Phil of Sci (7,Pap);1983 Salzburg 181-198
◇ B45 ◇ ID 47934

GUREVICH, Y. see Vol. I, III, IV, V, VI for further entries

GUSEV, L.A. [1963] see AJZERMAN, M.A.

GUSEV, L.A. see Vol. IV, V for further entries

GUTTENPLAN, S.D. & TAMNY, M. [1972] *Logic: a comprehensive introduction* (**X** 0837) Harper & Row: New York 384pp
• LAST ED [1978] (**X** 2671) Basic Books: New York xiv+401pp
◇ A05 B98 ◇ REV Zbl 457 # 03002 JSL 45.383 • ID 22490

GUZ, W. [1977] *Axioms for nonrelativistic quantum mechanics* (**J** 2736) Int J Theor Phys 16*299-306
◇ B51 ◇ REV MR 56 # 7520 Zbl 386 # 03028 • ID 52204

GUZ, W. [1978] *On the lattice structure of quantum logics* (**J** 2658) Ann Inst Henri Poincare, Sect A 28*1-7
◇ B51 G12 ◇ REV MR 80a:81005 • ID 81443

GUZ, W. [1978] *On the simultaneous verifiability of yes-no measurements* (**J** 2736) Int J Theor Phys 17*543-548
◇ B51 G12 ◇ REV MR 80d:81003 Zbl 398 # 03051
• ID 52779

GUZ, W. [1979] *An improved formulation of axioms for quantum mechanics* (**J** 2658) Ann Inst Henri Poincare, Sect A 30*223-230
◇ B51 ◇ REV MR 80f:81011 Zbl 427 # 03054 • ID 53741

GUZ, W. [1980] *Conditional probability in quantum axiomatics* (**J** 2658) Ann Inst Henri Poincare, Sect A 33*63-119
◇ B51 G12 ◇ REV MR 81k:81008 Zbl 454 # 03032
• ID 54244

GUZ, W. [1980] *Event-phase-space structure: An alternative to quantum logic* (**J** 2760) J Phys A Math & Gen 13*881-899
◇ B51 G12 ◇ REV MR 81d:81008 Zbl 431 # 03040
• ID 53946

GUZ, W. [1981] *Projection postulate and superposition principle in non-lattice quantum logics* (**J** 2658) Ann Inst Henri Poincare, Sect A 34*373-389
◇ B51 G12 ◇ REV MR 82j:81007 Zbl 473 # 03056
• ID 55384

GUZ, W. see Vol. I for further entries

GYUDZHENOV, I.D. [1983] see DENEV, I.D.

HAACK, S. [1974] *Deviant logic* (**X** 0805) Cambridge Univ Pr: Cambridge, GB xii+191pp
◇ A05 B60 ◇ REV Zbl 288 # 02007 JSL 43.377 • ID 32347

HAACK, S. [1976] see GAINES, B.R.

HAACK, S. [1976] *Recent publications in logic* (**J** 0489) Philosophy
◇ A05 B45 ◇ ID 32349

HAACK, S. [1979] *Do we need "fuzzy logic"?* (**J** 1741) Int J Man-Mach Stud 11*437-445
◇ A05 B52 ◇ REV MR 81c:03004 Zbl 415 # 03003
• ID 53105

HAAS, G. [1984] *Konstruktive Einfuehrung in die formale Logik* (**X** 0876) Bibl Inst: Mannheim 268pp
◇ B98 F98 ◇ REV MR 86g:03001 Zbl 562 # 03001
• ID 44527

HABASINSKI, Z. [1984] *Process logics: two decidability results* (**P** 3658) Math Founds of Comput Sci (11);1984 Prague 282-290
◇ B75 ◇ REV Zbl 569 # 68024 • ID 44745

HACKING, I. [1963] *What is strict implication ?* (**J** 0036) J Symb Logic 28*51-71
◇ A05 B45 ◇ REV MR 31 # 4717 Zbl 132.245 JSL 37.417
• ID 05470

HACKING, I. [1965] *Logic of statistical inference* (**X** 0805) Cambridge Univ Pr: Cambridge, GB xiv+338pp
◇ A05 B48 ◇ REV MR 34 # 3684 Zbl 359 # 02020
JSL 40.250 • ID 05471

HACKING, I. [1969] *Linguistically invariant inductive logic* (**J** 0154) Synthese 20*25-47
◇ A05 B48 ◇ REV MR 44 # 3836 Zbl 175.265 • ID 05472

HACKING, I. [1972] *A concise introduction to logic* (**X** 0981) Random House: New York viii+339pp
◇ B98 ◇ REV JSL 38.341 • ID 15009

HACKING, I. see Vol. I for further entries

HACKSTAFF, L.H. [1966] *Systems of formal logic* (**X** 0835) Reidel: Dordrecht xi+354pp
◇ B98 ◇ REV Zbl 205.3 • ID 22549

HAILPERIN, T. [1937] *Foundations of probability in mathematical logic* (**J** 0153) Phil of Sci (East Lansing) 4*125-150
◇ B50 ◇ REV FdM 63.825 • ID 41047

HAILPERIN, T. [1976] *Boole's logic and probability* (**S** 3303) Stud Logic Found Math 85*x+252pp
◇ A10 B48 G05 ◇ REV MR 56 # 2744 Zbl 352 # 02002
JSL 50.851 • ID 28265

HAILPERIN, T. [1984] *Probability logic* (**J** 0047) Notre Dame J Formal Log 25*198-212
◇ B25 B48 C90 ◇ REV MR 85m:03013 Zbl 547 # 03018
• ID 41842

HAILPERIN, T. see Vol. I, III, V, VI for further entries

HAILPERN, B.T. [1982] *Verifying concurrent processes using temporal logic* (**S** 3302) Lect Notes Comput Sci 129*208pp
◇ B45 ◇ REV MR 85i:68029 Zbl 476 # 68015 • ID 55586

HAJEK, P. [1971] see BENDOVA, K.

HAJEK, P. [1975] *Observationsfunktorenkalkuele und die Logik der automatisierten Forschung* (**J** 0129) Elektr Informationsverarbeitung & Kybern 11*579
◇ B60 ◇ REV MR 58 # 27413 Zbl 318 # 68056 • ID 62277

HAJEK, P. [1975] *On logics of discovery* (**P** 0454) Math Founds of Comput Sci (4);1975 Marianske Lazne 30-45
◇ B60 ◇ REV MR 54 # 71 Zbl 353 # 68090 • ID 23953

HAJEK, P. [1976] *Observationsfunktorenkalkuele und die Logik der automatisierten Forschung* (**J** 0129) Elektr Informationsverarbeitung & Kybern 12*181-186
◇ B60 C13 C80 C90 ◇ REV MR 58 # 27413
Zbl 356 # 68095 • ID 31127

HAJEK, P. [1977] *Experimental logics and Π^0_3 theories* (**J** 0036) J Symb Logic 42*515-522
◇ B60 C62 D55 F30 ◇ REV MR 58 # 16243
Zbl 428 # 03043 • ID 26853

HAJEK, P. [1977] *Generalized quantifiers and finite sets* (**P** 1639) Set Th & Hierarch Th (1);1974 Karpacz 14*91-104
◇ B25 B50 C13 C40 C80 ◇ REV MR 58 # 5086 Zbl 386 # 03017 • ID 31126

HAJEK, P. & KALASEK, P. & KURKA, P. [1980] *On dynamic logic (Czech)* (**J** 3524) Kybernetika Suppl (Prague) Suppl.16*42pp
◇ B75 ◇ REV MR 83e:68022 Zbl 473 # 03023 • ID 55353

HAJEK, P. & KURKA, P. [1981] *A second-order dynamic logic with array assignments* (**J** 2095) Fund Inform, Ann Soc Math Pol, Ser 4 4*919-933
◇ B75 C85 ◇ REV MR 84c:03035 Zbl 491 # 68033 • ID 35692

HAJEK, P. [1981] *Decision problems of some statistically motivated monadic modal calculi* (**J** 1741) Int J Man-Mach Stud 15*351-358
◇ B25 B45 ◇ REV MR 83j:03039 Zbl 464 # 03015 • ID 54605

HAJEK, P. [1981] *Making dynamic logic first-order* (**P** 3429) Math Founds of Comput Sci (10);1981 Strbske Pleso 287-295
◇ B75 ◇ REV MR 83e:03030 Zbl 465 # 68010 • ID 54943

HAJEK, P. [1981] *On interpretability in theories containing arithmetics II* (**J** 0140) Comm Math Univ Carolinae (Prague) 22*667-688
◇ B28 B45 F25 F30 ◇ REV MR 83j:03094 Zbl 262 # 02049 Zbl 487 # 03032 • REM Part I 1972 by Hajek,P. & Hajkova,M. • ID 35379

HAJEK, P. [1982] see BENDOVA, K.

HAJEK, P. [1983] *Arithmetical interpretations of dynamic logic* (**J** 0036) J Symb Logic 48*704-713
◇ B28 B45 ◇ REV MR 84j:03038 Zbl 546 # 03012 • ID 34630

HAJEK, P. & HAVRANEK, T. [1983] *Logic, statistics and computers* (**C** 3811) Logic 20th Century 55-76
◇ B48 ◇ REV MR 84m:03031 • ID 35735

HAJEK, P. see Vol. I, III, IV, V, VI for further entries

HAJNAL, M. & KOCZY, L.T. [1977] *A new attempt to axiomatize fuzzy algebra with an application example* (**J** 2819) Probl Contr Inf Th, Akad Nauk SSSR & Acad Sci Hung 6*47-66
◇ B52 E72 ◇ REV MR 56 # 11668 Zbl 358 # 02017 • ID 50471

HAJNAL, M. & KOCZY, L.T. [1977] *A new fuzzy calculus and its application as a pattern recognition technique* (**P** 3406) Congr Cybern & Syst (3);1975 Bucharest II*103-118
◇ B52 E72 G25 ◇ REV MR 56 # 8370 Zbl 428 # 03051 • ID 53810

HALBERSTADT, W.H. [1960] *Introduction to modern logic: an elementary textbook of symbolic logic* (**X** 0837) Harper & Row: New York 221pp
◇ B98 ◇ REV JSL 29.43 • ID 22492

HALKOWSKA, K. & PIROG-RZEPECKA, K. & SLUPECKI, J. [1976] *Mathematical logic (Polish)* (**X** 1034) PWN: Warsaw 289pp
◇ B98 C07 C98 E98 ◇ REV MR 57 # 15933 Zbl 393 # 03001 • ID 52421

HALKOWSKA, K. & PIROG-RZEPECKA, K. & SLUPECKI, J. [1978] *Logic and set theory (Polish)* (**X** 1034) PWN: Warsaw 309pp
◇ B98 E98 ◇ REV MR 80a:03001 Zbl 404 # 03002 • ID 54789

HALKOWSKA, K. [1982] *On some connections between conditional algebras and the systems of nonsense-logics* (**P** 3237) Colloq Universal Algeb;1977 Esztergom 361-371
◇ B60 ◇ REV MR 83h:03037 Zbl 526 # 03040 • ID 36056

HALKOWSKA, K. [1984] see BRYLL, G.

HALKOWSKA, K. see Vol. I, III for further entries

HALLDEN, S. [1948] *A note concerning the paradoxes of strict implication and Lewis's system S1* (**J** 0036) J Symb Logic 13*138-139
◇ B46 ◇ REV MR 10.229 Zbl 36.7 JSL 14.69 • ID 05584

HALLDEN, S. [1948] *A question concerning a logical calculus related to Lewis' system of strict implication, which is of special interest for the study of entailment* (**J** 0105) Theoria (Lund) 14*265-269
◇ B46 ◇ REV MR 11.304 Zbl 36.8 JSL 14.199 • ID 05583

HALLDEN, S. [1949] *A reduction of the primitive symbols of the Lewis calculi* (**J** 0050) Port Math 8*85-88
◇ B45 ◇ REV MR 12.385 Zbl 37.295 JSL 16.231 • ID 05589

HALLDEN, S. [1949] *An analogy in modal logic to the Lesniewski-Mihailescu theorem* (**J** 4510) Norsk Mat Tidsskr 31*4-9
◇ B45 ◇ REV MR 10.585 Zbl 40.147 JSL 15.70 • ID 05588

HALLDEN, S. [1949] *On the decision problem of Lewis' calculus S5* (**J** 4510) Norsk Mat Tidsskr 31*89-94
◇ B25 B45 ◇ REV MR 11.411 Zbl 40.146 JSL 15.224 • ID 05587

HALLDEN, S. [1949] *Results concerning the decision problem of Lewis's calculi S3 and S6* (**J** 0036) J Symb Logic 14*230-236
◇ B25 B45 ◇ REV MR 11.303 Zbl 36.7 JSL 15.70 • ID 05586

HALLDEN, S. [1949] *The logic of nonsense* (**S** 4519) Uppsala Univ Arsskrift 9*132pp
◇ B60 ◇ REV Zbl 40.292 JSL 15.225 • ID 41616

HALLDEN, S. [1950] *Some results in modal logic (Swedish)* (**X** 1163) Almqvist & Wiksell: Stockholm 34pp
◇ B45 ◇ REV Zbl 40.147 JSL 16.70 • ID 16878

HALLDEN, S. [1951] *On the semantic non-completeness of certain Lewis calculi* (**J** 0036) J Symb Logic 16*127-129
◇ B45 ◇ REV MR 13.97 Zbl 45.150 JSL 16.273 • ID 05590

HALLDEN, S. [1957] *On the logic of "better"* (**X** 1493) Gleerup: Lund 112pp
◇ A05 B45 ◇ REV JSL 31.278 • ID 28617

HALLDEN, S. [1963] *A pragmatic approach to modal theory* (**J** 0096) Acta Philos Fenn 16*53-64
◇ B45 ◇ REV JSL 40.601 • ID 14477

HALLDEN, S. [1980] *The foundations of decision logic* (**X** 1493) Gleerup: Lund 99pp
◇ B48 ◇ REV MR 82k:90005 Zbl 483 # 90001 JSL 48.502 • ID 81468

HALLDEN, S. see Vol. I for further entries

HALLERBERG, A.E. [1974] *Logic in mathematics: An elementary approach* (**X** 1274) Hafner: New York vi+90pp
◇ B98 ◇ REV Zbl 293 # 02001 • ID 62297

HALLERBERG, A.E. [1974] *Mathematical proof: An elementary approach* (**X** 1274) Hafner: New York viii+104pp
◇ B98 ◇ REV Zbl 293 # 02002 • ID 62298

HALLETT, M. [1982] see BELL, J.L.

HALLETT, M. see Vol. V for further entries

HALPERN, J.Y. [1981] see BEN-ARI, M.

HALPERN, J.Y. [1982] see BEN-ARI, M.

HALPERN, J.Y. [1982] see BERMAN, P.

HALPERN, J.Y. [1983] *Deterministic process logic is elementary* (**J** 0194) Inform & Control 57*56-89
◇ B45 ◇ REV MR 85k:68059 Zbl 537 # 68038 • ID 41328

HALPERN, J.Y. & REIF, J.H. [1983] *The propositioal dynamic logic of deterministic, well-structured programs* (**J** 1426) Theor Comput Sci 27*127-165
◇ B75 ◇ REV MR 85j:03037 Zbl 552 # 68035 • ID 43448

HALPERN, J.Y. [1985] see EMERSON, E.A.

HALPERN, J.Y. [1985] see CHANDRA, A.K.

HALPERN, M. [1983] *Inductive inference in finite algebraic structures* (**P** 3858) Adequate Modeling of Syst;1982 Bad Honnef 167-176
◇ B48 C13 ◇ REV Zbl 499 # 68035 • ID 45211

HAMACHER, H.C. [1978] *Ueber logische Verknuepfungen unscharfer Aussagen und deren zugehoerige Bewertungsfunktion* (**P** 3428) Prog in Cybern & Syst Res;1978 Wien III*276-288
◇ B52 ◇ REV Zbl 435 # 03018 • ID 55779

HAMACHER, H.C. [1979] *Ueber das Zadeh'sche Konzept der Fuzzy Sets und dessen Verhaeltnis zu den Wahrscheinlichkeitsmodellen von Koopman und Kolmogoroff* (**P** 3422) Oper Res DGOR Jahrestag (8);1978 Berlin 437-446
◇ A05 B52 E72 ◇ REV Zbl 434 # 03009 • ID 55713

HAMACHER, V.C. & VRANESIC, Z.G. [1972] *Ternary logic in parallel multipliers* (**J** 1193) Comput J (London) 15*254-258
◇ B50 B70 ◇ REV Zbl 241 # 94039 • ID 65946

HAMBLIN, C.L. [1959] *The modal "probably"* (**J** 0094) Mind 68*234-240
◇ A05 B45 ◇ REV JSL 35.582 • ID 05622

HAMBLIN, C.L. [1967] *Elementary formal logic* (**X** 0816) Methuen: London & New York 182pp
◇ B98 ◇ ID 22493

HAMBLIN, C.L. [1971] *Instants and intervals* (**J** 0178) Stud Gen 24*127-134
◇ B45 ◇ REV JSL 40.99 • ID 05623

HAMBLIN, C.L. [1971] *Starting and stopping* (**C** 0571) Basic Issues Philos of Time 86-101
◇ B45 ◇ REV Zbl 243 # 02034 JSL 40.99 • ID 22211

HAMBLIN, C.L. [1975] *Cresswell's colleague TLM* (**J** 0097) Nous, Quart J Phil 9*205-210
◇ A05 B45 ◇ REV MR 58 # 21465b • ID 73802

HAMBLIN, C.L. see Vol. I, III, IV for further entries

HAMILTON, A.G. [1978] *Logic for mathematicians* (**X** 0805) Cambridge Univ Pr: Cambridge, GB viii + 224pp
• TRANSL [1981] (**X** 3560) Paraninfo: Madrid 243pp
◇ B98 ◇ REV MR 80c:03005 MR 83f:03004
Zbl 383 # 03003 Zbl 491 # 03001 JSL 45.379 • ID 28266

HAMILTON, A.G. see Vol. I, III, IV for further entries

HANNA, J.F. [1981] *Single case propensities and the explanation of particular events* (**J** 0154) Synthese 48*409-436
◇ B48 ◇ REV MR 84c:03046 Zbl 483 # 03012 • ID 35698

HANNA, S.C. & SABER, J.C. [1971] *Sets and logic* (**X** 1290) Irwin: Homewood xi + 274pp
◇ B98 E98 ◇ ID 22641

HANNAK, L. [1979] see DEMETROVICS, J.

HANNAK, L. [1980] see DEMETROVICS, J.

HANNAK, L. [1981] see DEMETROVICS, J.

HANNAK, L. [1982] see DEMETROVICS, J.

HANNAK, L. [1983] see DEMETROVICS, J.

HANNAK, L. [1984] *On the construction of many-valued logics (Hungarian)* (**J** 2845) Tanulmanyok 161*134pp
◇ B50 ◇ ID 44428

HANNAK, L. [1984] *On the structure of many-valued logic (Hungarian)* (**J** 2774) Koezlem MTA Szam & Autom: Kutat Intez 161*136pp
◇ B50 ◇ ID 46612

HANSEL, G. & LEVY, G. [1969] *Construction d'un schema fonctionnel en logique a plusieurs valeurs* (**J** 2313) C R Acad Sci, Paris, Ser A-B 268*A681-A684
◇ B50 ◇ REV MR 44 # 1541 Zbl 193.291 • ID 08087

HANSEL, G. see Vol. IV for further entries

HANSEN, J.C. [1979] see DERUS, K.H.

HANSEN, J.C. see Vol. I for further entries

HANSON, W.H. [1966] *On formalizing the distinction between logical and factual truth* (**J** 0036) J Symb Logic 31*460-477
◇ A05 B45 ◇ REV MR 34 # 4102 Zbl 154.3 • ID 05647

HANSON, W.H. [1966] *On some alleged decision procedures for S4* (**J** 0036) J Symb Logic 31*641-643
◇ B25 B45 ◇ REV MR 35 # 5298 Zbl 166.3 JSL 35.326
• ID 05648

HANSON, W.H. [1980] *First degree entailments and information* (**J** 0047) Notre Dame J Formal Log 21*659-671
◇ A05 B46 ◇ REV MR 81k:03020 Zbl 423 # 03026
• ID 55946

HANSON, W.H. see Vol. VI for further entries

HANSSON, B. [1969] *An analysis of some deontic logics* (**J** 0097) Nous, Quart J Phil 3*373-398 • ERR/ADD ibid 4*390
◇ B45 ◇ REV MR 58 # 5053a MR 58 # 5053b
Zbl 229 # 02025 • ID 73828

HANSSON, B. [1970] *Deontic logic and different levels of generality* (**J** 0105) Theoria (Lund) 36*241-248
◇ B45 ◇ REV Zbl 231 # 02025 • ID 27430

HANSSON, B. [1973] see GAERDENFORS, P.

HANSSON, B. [1975] see GAERDENFORS, P.

HANSSON, B. see Vol. I for further entries

HARARY, F. [1958] *Note on Carnap's relational asymptotic relative frequencies* (**J** 0036) J Symb Logic 23*257-260
◇ A05 B48 ◇ REV MR 21 # 3339 • ID 05652

HARARY, F. see Vol. I, III, V for further entries

HARBECK, G. [1963] *Einfuehrung in die formale Logik* (**X** 0900) Vieweg: Wiesbaden vi+114pp
◇ B98 ◇ REV MR 29#1134 Zbl 249#02001 JSL 31.287 • ID 05654

HARDEGREE, G.M. [1974] *The conditional in quantum logic* (**J** 0154) Synthese 29*63-80
◇ B51 G12 ◇ REV Zbl 361#02039 • ID 50674

HARDEGREE, G.M. [1975] *Quasi-implicative lattices and the logic of quantum mechanics* (**J** 0948) Z Naturforsch 30a*1347-1360
◇ B51 G10 G12 ◇ REV MR 55#4968 • ID 81487

HARDEGREE, G.M. [1975] *Stalnaker conditionals and quantum logic* (**J** 0122) J Philos Logic 4*399-421
◇ B51 G12 ◇ REV MR 57#18526 Zbl 357#02024 • ID 50372

HARDEGREE, G.M. [1976] *The conditional in quantum logic* (**C** 2953) Log & Probab in Quant Mech 129-130
◇ B51 G12 ◇ REV Zbl 363#02037 • ID 50864

HARDEGREE, G.M. [1977] *Reichenbach and the logic of quantum mechanics* (**J** 0154) Synthese 35*3-40
◇ A05 B51 G12 ◇ REV MR 58#14455 Zbl 371#02006 • ID 51367

HARDEGREE, G.M. [1979] *The conditional in abstract and concrete quantum logic* (**C** 3045) Log-Algeb Appr to Quant Mech 2*49-108
◇ B51 G12 ◇ REV MR 82b:03110 Zbl 429#03046 JSL 48.206 • ID 73836

HARDEGREE, G.M. [1981] *An axiom system for orthomodular quantum logic* (**J** 0063) Studia Logica 40*1-12
◇ B51 G12 ◇ REV MR 82f:03061 Zbl 476#03059 • ID 55571

HARDEGREE, G.M. [1981] see FRAZER, P.J.

HARDEGREE, G.M. [1981] *Some problems and methods in formal quantum logic* (**P** 3820) Curr Iss in Quantum Log;1979 Erice 209-225
◇ B51 ◇ REV MR 84j:03127 Zbl 537#03044 • ID 34717

HARDEGREE, G.M. & LOCK, P.F. [1985] *Connections among quantum logics I: Quantum propositional logics. II: Quantum event logics* (**J** 2736) Int J Theor Phys 24*43-53,55-61
◇ B51 ◇ ID 46421

HARDY, L. [1957] see EVANS, T.

HAREL, D. [1978] *Arithmetical completeness in logics of programs* (**P** 1872) Automata, Lang & Progr (5);1978 Udine 268-288
◇ B75 ◇ REV MR 80g:68019 Zbl 382#68024 • ID 69630

HAREL, D. & PRATT, V.R. [1978] *Nondeterminism in logics of programs* (**P** 3539) ACM Symp Princ Progr Lang (5);1978 Tucson 203-213
◇ B75 ◇ ID 49867

HAREL, D. [1979] *First-order dynamic logic* (**S** 3302) Lect Notes Comput Sci 68*x+133pp
◇ B75 ◇ REV MR 81j:68019 Zbl 403#03024 JSL 47.453 • ID 54738

HAREL, D. [1979] *Recursion in logics of programs* (**P** 2918) ACM Symp Princ Progr Lang (6);1979 San Antonio 81-92
◇ B75 ◇ REV MR 81j:68018 Zbl 473#68004 • ID 81490

HAREL, D. [1979] *Two results on process logic* (**J** 0232) Inform Process Lett 8*195-198
◇ B45 ◇ REV MR 81b:68007a Zbl 417#68017 • ID 53281

HAREL, D. [1980] *Proving the correctness of regular deterministic programs: a unifying survey using dynamic logic* (**J** 1426) Theor Comput Sci 12*61-81
◇ B75 ◇ REV MR 81f:68022 Zbl 433#68019 JSL 50.552 • ID 47345

HAREL, D. & PNUELI, A. & STAVI, J. [1981] *Propositional dynamic logic of context-free programs* (**P** 4235) IEEE Symp Found of Comp Sci (22);1981 Nashville 310-321
◇ B75 ◇ REV MR 84a:68004 • ID 45812

HAREL, D. & PNUELI, A. & STAVI, J. [1982] *Further results on propositional dynamic logic of nonregular programs* (**P** 3738) Log of Progr;1981 Yorktown Heights 124-136
◇ B75 ◇ REV MR 83i:03030 Zbl 491#68032 • ID 34849

HAREL, D. & SHERMAN, R. [1982] *Looping versus repeating in dynamic logic* (**J** 0194) Inform & Control 55*175-192
◇ B75 ◇ REV MR 85i:03087 Zbl 541#68010 • ID 41397

HAREL, D. & SHERMAN, R. [1983] *Propositional dynamic logic of flowcharts* (**P** 3864) FCT'83 Found of Comput Th;1983 Borgholm 195-206
◇ B75 ◇ REV MR 85g:03044 Zbl 534#68021 • ID 38357

HAREL, D. & PNUELI, A. & STAVI, J. [1983] *Propositional dynamic logic of nonregular programs* (**J** 0119) J Comp Syst Sci 26*222-243
◇ B75 ◇ REV MR 85b:03041 Zbl 536#68041 • ID 36776

HAREL, D. [1984] see FELDMAN, Y.A.

HAREL, D. & PNUELI, A. & SHERMAN, R. [1984] *Is the interesting part of process logic uninteresting?: A translation from PL to PDL* (**J** 1428) SIAM J Comp 13*825-839
◇ B75 ◇ REV Zbl 551#68031 • ID 43950

HAREL, D. & PELEG, D. [1984] *On static logics, dynamic logics, and complexity classes* (**J** 0194) Inform & Control 60*86-102
◇ B75 D15 ◇ ID 44121

HAREL, D. & PELEG, D. [1985] *More on looping vs. repeating in dynamic logic* (**J** 0232) Inform Process Lett 20*87-90
◇ B75 ◇ REV Zbl 559#68049 • ID 45591

HAREL, D. see Vol. I, III, IV for further entries

HARKER, J.E. [1980] *A note on believing that one knows and Lehrer's proof that knowledge entails belief* (**J** 0095) Philos Stud 37*321-324
◇ A05 B45 ◇ REV MR 81f:03010 • ID 73838

HARMAN, G. [1972] see DAVIDSON, D.

HARNAU, W. [1974] *Die vertauschbaren Funktionen der k-wertigen Logik und ein Basisproblem* (**J** 0068) Z Math Logik Grundlagen Math 20*453-463
◇ B50 ◇ REV MR 52#5352 Zbl 301#02017 • ID 05660

HARNAU, W. [1974] *Die Definition von Vertauschbarkeitsmengen in der k-wertigen Logik und das Maximalitaetsproblem* (**J** 0068) Z Math Logik Grundlagen Math 20*339-352
◇ B50 ◇ REV MR 52#5351 Zbl 308#02020 • ID 05659

HARNAU, W. [1974] *Einige numerische Invarianten der teilweise geordneten Menge Φ_k von Vertauschbarkeitsmengen der k-wertigen Logik* (**J** 0129) Elektr Informationsverarbeitung & Kybern 10*543-551
◇ B50 ◇ REV MR 54#59 Zbl 301#02018 • ID 23942

HARNAU, W. [1975] *Ueber Kettenlaengen der teilweise geordneten Menge Φ_k der Vertauschbarkeitsmengen der k-wertigen Logik* (J 0114) Math Nachr 68*289-297
◇ B50 ◇ REV MR 54 # 12479 Zbl 323 # 02025 • ID 30410

HARNAU, W. [1975] *Vertauschbarkeitsmengen in mehrwertigen Logiken* (J 0129) Elektr Informationsverarbeitung & Kybern 11*622-623
◇ B50 ◇ REV Zbl 316 # 02019 • ID 62339

HARNAU, W. [1976] *Die teilweise geordnete Menge Φ_k der Vertauschbarkeitsmengen der k-wertigen Logik* (J 0068) Z Math Logik Grundlagen Math 22*19-28
◇ B50 ◇ REV MR 54 # 12480 Zbl 326 # 02015 • ID 18492

HARNAU, W. [1976] *Eine Verallgemeinerung der Vertauschbarkeit in der k-wertigen Logik* (J 0129) Elektr Informationsverarbeitung & Kybern 12*33-43
◇ B50 ◇ REV MR 55 # 12465 Zbl 334 # 02012 • ID 62337

HARNAU, W. see Vol. IV for further entries

HARNISH, R.M. [1980] *Searle and the logic of Moore's paradox* (J 0286) Int Logic Rev 21*72-76
◇ A10 B45 ◇ REV MR 83d:03007 • ID 35163

HARPER, W.L. [1974] *A note on universal instantiation in the Stalnaker Thomason conditional logic and M type modal systems* (J 0122) J Philos Logic 3*373-379
◇ B45 ◇ REV MR 55 # 5390 Zbl 296 # 02010 • ID 62347

HARPER, W.L. & HOOKER, C.A. (EDS.) [1975] *Foundations of probability theory, statistical inference, and statistical theories of science I,II,III* (S 3308) Univ Western Ontario Ser in Philos of Sci xi+309pp,x+455pp,xii+241pp
◇ B48 ◇ REV MR 56 # 15251 MR 56 # 15252 MR 56 # 15253 Zbl 311 # 00018 Zbl 311 # 00019 Zbl 311 # 00020 • ID 80456

HARPER, W.L. [1976] *Ramsey test conditionals and iterated belief change (A response to Stalnaker.)* (P 2411) Found Probab Th, Stat Inf & Stat Th Sci;1973 London ON 1*117-135
◇ A05 B45 B48 ◇ REV Zbl 352 # 02026 • ID 50025

HARPER, W.L. & MAY, S. [1976] *Toward an optimization procedure for applying minimum change principles in probability kinematics* (P 2411) Found Probab Th, Stat Inf & Stat Th Sci;1973 London ON 1*137-166
◇ B48 ◇ REV MR 58 # 31329 Zbl 336 # 90067 • ID 82207

HARPER, W.L. [1978] see GIBBARD, A.

HARPER, W.L. [1981] *A sketch of some recent developments in the theory of conditionals* (C 4140) Ifs 3-38
◇ B48 B98 ◇ REV MR 83a:03003 JSL 49.1411 • ID 47401

HARPER, W.L. & PEARCE, G. & STALNAKER, R.C. (EDS.) [1981] *Ifs* (X 0835) Reidel: Dordrecht 15*x+345pp
◇ B48 ◇ REV MR 83a:03003 • ID 47734

HARRAH, D. [1961] *A logic of questions and answers* (J 0153) Phil of Sci (East Lansing) 28*40-46
◇ B60 ◇ REV JSL 29.136 • ID 43030

HARRAH, D. [1975] *A system for erotetic sentences* (C 1856) Log Enterprise 235-245
◇ B60 ◇ REV Zbl 357 # 02025 • ID 50373

HARRAH, D. [1982] *What should we teach about questions?* (J 0154) Synthese 51*21-38
◇ B60 ◇ REV MR 84h:03056 • ID 34254

HARRAH, D. [1985] *A logic of message and reply* (J 0154) Synthese 63*275-294
◇ B60 ◇ ID 47613

HARRINGTON, L.A. & MORLEY, M.D. & SCEDROV, A. & SIMPSON, S.G. [1985] *Harvey Friedman's research on the foundations of mathematics* (X 0809) North Holland: Amsterdam xvi+408pp
◇ B97 C97 D97 E97 F97 ◇ ID 49810

HARRINGTON, L.A. see Vol. III, IV, V, VI for further entries

HARRISON, M.A. [1969] *Sur la classification des fonctions logiques a plusieurs valeurs* (J 0070) Bull Soc Sci Math Roumanie, NS 13*41-54
◇ B50 ◇ REV MR 42 # 1670 Zbl 193.335 • ID 05675

HARRISON, M.A. see Vol. I, IV for further entries

HARSANYI, J.C. [1983] *Bayesian decision theory, subjective and objective probabilities, and acceptance of empirical hypotheses* (J 0154) Synthese 57*341-365
◇ B48 ◇ REV MR 85c:03005 • ID 40471

HARSANYI, J.C. [1985] *Acceptance of empirical statements: a Bayesian theory without cognitive utilities* (J 0472) Theory Decis 18*1-30
◇ B48 ◇ ID 47245

HART, A.M. & HENDRY, H.E. [1978] *Some observations on a method of McKinsey* (J 0047) Notre Dame J Formal Log 19*395-396
◇ B50 F50 ◇ REV MR 58 # 5026 Zbl 368 # 02025 • ID 51664

HART, A.M. [1980] *Toward a logic of doubt* (J 0286) Int Logic Rev 21*31-45
◇ B45 ◇ REV MR 83d:03020 • ID 35172

HART, W.D. [1972] *Probability as degree of possibility* (J 0047) Notre Dame J Formal Log 13*286-288
◇ B48 ◇ REV Zbl 234 # 02016 • ID 27793

HART, W.D. & MCGINN, C. [1976] *Knowledge and necessity* (J 0122) J Philos Logic 5*205-208
◇ A05 A10 B45 ◇ REV MR 58 # 10315 Zbl 333 # 02016 • ID 62360

HART, W.D. & MCGINN, C. [1978] *On propositions* (J 0047) Notre Dame J Formal Log 19*299-306
◇ A05 B45 ◇ REV MR 80i:03007 Zbl 314 # 02035 • ID 51311

HARTHONG, J. [1984] *Etudes sur la mecanique quantique (French)* (X 2244) Soc Math France: Paris 209pp
◇ B51 ◇ REV MR 85i:81003 • ID 44399

HARTHONG, J. see Vol. I for further entries

HASEGAWA, T. [1972] *Many-valued logic and its applications (Japanese)* (J 0425) Managmt Sci (Tokyo) 16*125-138
◇ B50 ◇ REV MR 47 # 8242 • ID 05719

HASEGAWA, T. & OKURA, Y. & SHIMADA, R. [1982] *Symmetric ternary scaled cyclic AN-codes (Japanese)* (P 4081) Many-Val Log & Appl;1982 Kyoto 231-244
◇ B50 ◇ ID 47635

HASENJAEGER, G. [1961] see SCHOLZ, H.

HASENJAEGER, G. [1962] *Einfuehrung in die Grundbegriffe und Probleme der modernen Logik* (**X** 0826) Alber: Freiburg 202pp
• TRANSL [1968] (**X** 4056) Labor: Barcelona 184pp [1972] (**X** 0835) Reidel: Dordrecht 180pp
◇ A05 A10 B98 ◇ REV MR 29 # 4667 Zbl 122.243 JSL 40.627 • ID 05729

HASENJAEGER, G. see Vol. I, III, IV, V, VI for further entries

HASSE, M. [1966] *Grundbegriffe der Mengenlehre und Logik* (**X** 1079) Teubner: Leipzig 86pp
◇ B98 E98 ◇ REV MR 35 # 6560 Zbl 168.245 • ID 22643

HATANO, Y. & TSUKAMOTO, Y. [1984] *Fuzzy statement formation by means of linguistic measures* (**P** 3081) IFAC Symp Fuzzy Inf, Knowl Repr & Decis. Anal;1983 Marseille 129-134
◇ B52 ◇ REV Zbl 563 # 94031 • ID 48198

HATCHER, W.S. [1968] *Foundations of mathematics* (**X** 0810) Saunders: Philadelphia xiii + 327pp
◇ B98 E30 E70 E98 G30 ◇ REV MR 38 # 5610 Zbl 191.282 JSL 51.467 • ID 22644

HATCHER, W.S. [1982] *The logical foundations of mathematics* (**X** 0869) Pergamon Pr: Oxford x + 320pp
◇ B98 ◇ REV MR 84g:03003 Zbl 504 # 03001 JSL 51.467 • ID 34117

HATCHER, W.S. see Vol. I, III, IV, V, VI for further entries

HAUSCHILD, K. [1982] *Model-theoretic properties of cause-and-effect structures* (**J** 0140) Comm Math Univ Carolinae (Prague) 23*541-555
◇ B45 C52 D35 ◇ REV MR 84e:03023 Zbl 522 # 03005 • ID 34362

HAUSCHILD, K. see Vol. I, III, IV, V, VI for further entries

HAUSSER, R.R. [1983] *Vagueness and truth (German summary)* (**J** 2688) Conceptus (Wien) 17*29-52
◇ B48 ◇ REV MR 85a:03044 • ID 34795

HAUTAMAKI, A. [1983] *The logic of viewpoints* (**J** 0063) Studia Logica 42*187-196
◇ B45 ◇ REV MR 85i:03049 Zbl 568 # 03005 • ID 42332

HAVAS, K.G. [1981] *Some remarks on an attempt at formalizing dialectical logic* (**J** 2051) Stud Soviet Thought 22*257-265
◇ B53 ◇ ID 42845

HAVRANEK, T. [1974] *The computation of characteristic vectors of logical probabilistic expressions* (**J** 0156) Kybernetika (Prague) 10*80-94
◇ B48 ◇ REV MR 49 # 12213 Zbl 281 # 02018 • ID 31142

HAVRANEK, T. [1975] *Statistical quantifiers in observational calculi* (**J** 0472) Theory Decis 6*213-230
◇ B50 C80 C90 ◇ REV MR 57 # 12187 Zbl 313 # 68070 • ID 31144

HAVRANEK, T. [1978] *Enumeration calculi and rank methods* (**J** 1741) Int J Man-Mach Stud 10*59-65
◇ B50 D10 ◇ REV MR 80g:68118a Zbl 404 # 68094 • ID 31145

HAVRANEK, T. [1981] *Formal systems for mechanized statistical inference* (**J** 1741) Int J Man-Mach Stud 15*333-350
◇ B48 ◇ REV MR 84h:03050 Zbl 465 # 68051 • ID 37399

HAVRANEK, T. [1983] see HAJEK, P.

HAVRANEK, T. see Vol. I, III, IV for further entries

HAWRANEK, J. & TOKARZ, M. [1977] *Matrices for predicate logics (Polish)* (**J** 0302) Rep Math Logic, Krakow & Katowice 9*27-30
◇ B50 ◇ REV MR 58 # 27306 Zbl 387 # 03005 • ID 73945

HAWRANEK, J. [1980] *A matrix adequate for S5 with MP and RN* (**J** 0387) Bull Sect Logic, Pol Acad Sci 9*122-124
◇ B45 ◇ REV MR 82a:03019 Zbl 459 # 03009 • ID 54449

HAWRANEK, J. [1983] *A characterization of prime theories in Johannsson's minimal logic* (**J** 0387) Bull Sect Logic, Pol Acad Sci 12*122-125
◇ B55 ◇ REV Zbl 543 # 03015 • ID 40909

HAWRANEK, J. [1984] *On the degree of matrix complexity of Johansson's minimal logic* (**J** 0387) Bull Sect Logic, Pol Acad Sci 13*50-54
◇ B55 ◇ REV Zbl 568 # 03014 • ID 45750

HAWRANEK, J. see Vol. I for further entries

HAY, L. [1963] *Axiomatization of the infinitive-valued predicate calculus* (**J** 0036) J Symb Logic 28*77-86
◇ B50 ◇ REV MR 31 # 40 Zbl 127.7 JSL 29.110 • ID 05809

HAY, L. see Vol. I, III, IV, V for further entries

HAYASHI, T. [1978] *Disjunction property in McCarthy's propositional knowledge systems* (**J** 0970) Math Rep Coll Gen Educ, Kyushu Univ 11*145-148
◇ B45 ◇ REV MR 80a:03027 Zbl 391 # 03012 • ID 32209

HAYASHI, T. see Vol. IV for further entries

HAYES, S.E. [1972] *Extensions of T^o* (**J** 0047) Notre Dame J Formal Log 13*501-505
◇ B45 ◇ REV MR 48 # 64 Zbl 242 # 02022 • ID 05827

HAZEN, A. [1972] *Semantics for S4.2* (**J** 0047) Notre Dame J Formal Log 13*527-528
◇ B45 ◇ REV MR 48 # 1884 Zbl 242 # 02020 • ID 05828

HAZEN, A. [1976] *Expressive completeness in modal language* (**J** 0122) J Philos Logic 5*25-46
◇ B45 ◇ REV MR 54 # 7203 Zbl 347 # 02011 • ID 24994

HAZEN, A. [1978] *The eliminability of the actuality operator in propositional modal logic* (**J** 0047) Notre Dame J Formal Log 19*617-622
◇ B45 ◇ REV MR 80e:03014 Zbl 314 # 02037 • ID 52122

HAZEN, A. [1980] *Comments on the logic of constructible falsity (strong negation)* (**J** 0387) Bull Sect Logic, Pol Acad Sci 9*10-15
◇ B55 ◇ REV MR 81g:03029 Zbl 428 # 03017 • ID 73971

HE, ZHONGXIONG & XIAO, XIAN & YUAN, XIANGWAN & ZHU, WUJIA [1984] *Some opinions on fuzzy mathematics and its foundations (Chinese)* (**J** 3732) Mohu Shuxue 4/3*103-108
◇ A05 B52 E72 ◇ ID 49002

HE, ZHONGXIONG see Vol. V for further entries

HEELAN, P.A. [1970] *Quantum logic and classical logic: their respective roles* (**J** 0154) Synthese 21*2-33
◇ B51 ◇ REV MR 55 # 4966 Zbl 205.8 • ID 41967

HEELAN, P.A. [1974] *Quantum logic and classical logic: their respective roles* (**C** 4290) Log & Epistem Stud Contemp Phys 318-349
◇ B51 ◇ ID 42116

HEGENBERG, L. [1975] *Logica: Simbolizacao e deducao* (**X** 4328) Ed Pedag & Univ: Sao Paulo xiv+219pp
◇ B98 ◇ REV JSL 44.126 • ID 44580

HEGENBERG, L. see Vol. I for further entries

HEIJENOORT VAN, J. [1979] *Introduction a la semantique des logiques non-classiques* (**X** 3562) Ecole Norm Sup Jeunes Filles: Paris ix+108pp
◇ B45 C90 F50 ◇ REV MR 81f:03025 • ID 79641

HEIJENOORT VAN, J. see Vol. I, VI for further entries

HELLMAN, GEOFFREY [1981] *Quantum logic and the projection postulate* (**J** 0153) Phil of Sci (East Lansing) 48*469-486 • ERR/ADD ibid 49/2*301
◇ B51 ◇ REV MR 83e:03110a MR 83e:03110b • ID 35255

HELLMAN, GEOFFREY [1982] *Einstein and Bell: strengthening the case for microphysical randomness* (**J** 0154) Synthese 53*445-460
◇ B51 ◇ REV MR 84k:81012a • ID 39269

HELLMAN, GEOFFREY see Vol. VI for further entries

HELMER, O. & OPPENHEIM, P. [1945] *A syntactical definition of probability and of degree of confirmation* (**J** 0036) J Symb Logic 10*25-60
◇ B48 ◇ REV MR 7.45 Zbl 61.8 JSL 11.17 • ID 05885

HELMER, O. see Vol. I, III, VI for further entries

HEMPEL, C.G. [1937] *A purely topological form of non-aristotelian logic* (**J** 0036) J Symb Logic 2*97-112
◇ B50 ◇ REV Zbl 17.241 JSL 3.91 FdM 63.828 • ID 05888

HEMPEL, C.G. [1937] *Ein System verallgemeinerter Negationen* (**P** 0756) Congr Int Phil (9);1937 Paris 26-32
◇ B50 ◇ REV JSL 3.164 FdM 63.836 • ID 05889

HEMPEL, C.G. [1938] *On the logical form of probability-statements* (**J** 0748) Erkenntnis (Leipzig) 7*154-160
◇ A05 B48 ◇ REV JSL 4.26 FdM 64.1189 • ID 41418

HEMPEL, C.G. [1943] *A purely syntactical definition of confirmation* (**J** 0036) J Symb Logic 8*122-143
◇ B60 ◇ REV MR 5.85 Zbl 61.8 JSL 9.47 • ID 05890

HEMPEL, C.G. & OPPENHEIM, P. [1945] *A definition of "degree of confirmation"* (**J** 0153) Phil of Sci (East Lansing) 12*98-115
◇ B48 ◇ REV MR 7.46 Zbl 61.8 JSL 11.18 • ID 41513

HEMPEL, C.G. [1945] *Studies in the logic of confirmation* (**J** 0094) Mind 54*1-26,54*97-121
◇ A05 B48 ◇ REV MR 6.197 Zbl 60.19 JSL 10.104 • ID 41512

HENDRY, H.E. & MASSEY, G.J. [1969] *On the concepts of Sheffer functions* (**C** 1134) Log Way of Doing Things 279-293
◇ B50 ◇ REV Zbl 188.317 • ID 22103

HENDRY, H.E. [1978] see HART, A.M.

HENDRY, H.E. [1980] *Functional completeness and non-Lukasiewiczian truth functions* (**J** 0047) Notre Dame J Formal Log 21*536-538
◇ B50 ◇ REV MR 81h:03050 Zbl 419 # 03010 • ID 53919

HENDRY, H.E. [1981] *Does IPC have a binary indigenous Sheffer function?* (**J** 0047) Notre Dame J Formal Log 22*183-186
◇ B55 ◇ REV MR 82d:03038 Zbl 423 # 03021 • ID 54090

HENDRY, H.E. [1983] *Minimally incomplete sets of Lukasiewiczian truth functions* (**J** 0047) Notre Dame J Formal Log 24*146-150
◇ B50 ◇ REV MR 84b:03083 Zbl 544 # 03003 • ID 35664

HENDRY, H.E. & POKRIEFKA, M.L. [1985] *Carnapian extensions of S5* (**J** 0122) J Philos Logic 14*111-128
◇ B45 ◇ ID 42653

HENDRY, H.E. see Vol. I, III, VI for further entries

HENKIN, L. [1974] see ADDISON, J.W.

HENKIN, L. & JOJA, A. & MOISIL, G.C. & SUPPES, P. (EDS.) [1975] *Logic, methodology and philosophy of science, IV. Proceedings of the fourth international congress for logic, methodology and philosophy of science* (**X** 0809) North Holland: Amsterdam x+981pp
◇ B97 ◇ REV MR 55 # 7657 • ID 80463

HENKIN, L. see Vol. I, III, IV, V, VI for further entries

HENLE, P. [1933] see CHAPMAN, F.M.

HENLE, P. [1951] *n-valued boolean algebra* (**C** 0585) Struct, Meth & Meaning (Sheffer) 68-73
◇ B50 G25 ◇ REV Zbl 54.17 JSL 16.288 • ID 16877

HENLE, P. & KALLEN, H.M. & LANGER, S.K. (EDS.) [1951] *Structure, method, and meaning. Essays in honor of Henry M. Scheffer* (**X** 4563) Liberal Arts Pr: New York xvi+306pp
◇ B98 ◇ REV Zbl 44.1 • ID 48573

HENLE, P. see Vol. I for further entries

HENNO, J. [1976] *On the completeness of idempotent functions on infinite sets* (**J** 0498) Ann Univ Turku, Ser A I 170*9pp
◇ B50 ◇ REV MR 53 # 12870 Zbl 346 # 02010 • ID 23130

HENNO, J. [1979] *The depth of monotone functions in multivalued logic* (**J** 0232) Inform Process Lett 8*176-177
◇ B50 ◇ REV MR 81g:68067 Zbl 399 # 03024 • ID 52820

HENNO, J. [1979] *The depth of functions in m-valued logic* (**J** 0129) Elektr Informationsverarbeitung & Kybern 15*33-36
◇ B50 ◇ REV MR 80i:03037 Zbl 436 # 03014 • ID 55854

HENNO, J. [1981] *The depth of functions in many-sorted logic* (**P** 2552) Conf Finite Algeb & Multi-Val Log;1979 Szeged 333-343
◇ B50 G25 ◇ REV MR 83i:03100 Zbl 482 # 94032 • ID 35547

HENNO, J. [1982] *Equivalent sets of functions of multiple-valued logic* (**P** 4002) Int Symp Multi-Val Log (12);1982 Paris 27-29
◇ B50 ◇ REV Zbl 545 # 94029 • ID 41215

HENNO, J. [1982] *On equivalent sets of functions* (**J** 2702) Discr Appl Math 4*153-156
◇ B50 ◇ REV MR 84b:03038 Zbl 481 # 94016 • ID 35630

HENNO, J. see Vol. V for further entries

HENRY, DENNIS [1978] see BYRD, M.

HENRY, DENNIS [1979] see BYRD, M.

HERBUT, F. [1984] *On a possible empirical meaning of meets and joins for quantum propositions* (**J** 2730) Phys Rev Lett 8*397-402
◇ B51 ◇ REV MR 86c:81006 Zbl 557 # 03042 • ID 46229

HERCZEG, J. [1971] see BIZAM, G.

HERINGER, H.J. [1972] *Formale Logik und Grammatik* (**X** 0877) Niemeyer: Tuebingen vi + 104pp
◇ B65 B98 D05 ◇ REV MR 52 # 13299 Zbl 337 # 68003
• ID 21764

HERMAN, L. & PIZIAK, R. [1974] *Modal propositional logic on an orthomodular basis I* (**J** 0036) J Symb Logic 39*478-488
◇ B45 ◇ REV MR 51 # 61 Zbl 301 # 02023 • ID 17395

HERMES, H. & SCHOLZ, H. [1952] *Mathematische Logik* (**X** 1079) Teubner: Leipzig 82pp
◇ B98 ◇ REV MR 16.435 Zbl 47.248 JSL 19.278
• ID 06004

HERMES, H. [1957] *Einfuehrung in die mathematische Logik. Klassische Praedikatenlogik* (**X** 0910) Aschendorffsche Verlagsbuchh: Muenster v + 176pp
• TRANSL [1973] (**X** 0811) Springer: Heidelberg & New York xi + 242pp (English) • LAST ED [1969] (**X** 0823) Teubner: Stuttgart 204pp
◇ B10 B98 ◇ REV MR 28 # 2035 MR 40 # 1256 MR 49 # 10518 MR 55 # 68 Zbl 115.5 JSL 30.355 JSL 38.647 • ID 21083

HERMES, H. [1958] *Zum Einfachheitsprinzip in der Wahrscheinlichkeitsrechnung* (**J** 0076) Dialectica 12*317-331
◇ A05 B48 D20 D80 ◇ REV MR 21 # 350 Zbl 90.348
• ID 14545

HERMES, H. [1965] *Eine Termlogik mit Auswahloperator* (**X** 0811) Springer: Heidelberg & New York iv + 42pp
• TRANSL [1970] (**S** 3301) Lect Notes Math 6*iv + 55pp (English)
◇ B10 B60 ◇ REV MR 33 # 3906 MR 41 # 8207 Zbl 166.254 JSL 34.679 JSL 35.440 • ID 06014

HERMES, H. [1965] *Induktion und Wahrscheinlichkeit* (**P** 0797) Fonds des Math, Machines Math & Appl;1962 Tihany 37
◇ A05 B48 ◇ ID 32176

HERMES, H. [1968] *Methodik der Mathematik und Logik* (**C** 4678) Enzykl Geisteswiss Arbeitsmethoden 3-43
◇ A05 A10 B98 ◇ REV MR 38 # 3117 Zbl 185.6
• ID 28565

HERMES, H. [1976] *Dialog games* (**P** 1619) Coloq Log Simb;1975 Madrid 115-125
◇ A05 B60 F50 ◇ REV MR 56 # 5264 Zbl 366 # 02019
• ID 51102

HERMES, H. see Vol. I, III, IV, V, VI for further entries

HERZBERGER, H.G. [1970] *Truth and modality in semantically closed languages* (**P** 0731) Paradox of Liar;1969 Buffalo 25-46
◇ A05 B45 ◇ REV JSL 40.584 • ID 14621

HERZBERGER, H.G. [1975] *Canonical superlanguages* (**J** 0122) J Philos Logic 4*45-65
◇ A05 B45 ◇ REV MR 57 # 15950 Zbl 309 # 02061
• ID 62497

HERZBERGER, H.G. [1975] *Supervaluations in two dimensions* (**P** 1805) Int Symp Multi-Val Log (5,Proc);1975 Bloomington 429-435
◇ B50 ◇ ID 35839

HERZBERGER, H.G. [1977] *Tertium without plenum* (**P** 2013) Int Symp Multi-Val Log (7);1977 Charlotte 92-94
◇ B50 ◇ REV MR 57 # 5679 • ID 74086

HERZBERGER, H.G. [1982] *The algebra of supervaluations* (**J** 3781) Topoi 1*74-81
◇ B50 ◇ REV MR 84h:03025 • ID 34239

HEYTING, A. [1958] *Intuitionism in mathematics* (**C** 0742) Phil Mid-Century 101-115
• TRANSL [1967] (**C** 2141) Filos Matematica 249-267
◇ A05 B98 F50 F55 F98 ◇ REV JSL 34.313 JSL 39.609 JSL 40.472 • ID 06062

HEYTING, A. see Vol. I, III, IV, VI for further entries

HICKIN, K.K. & PLOTKIN, J.M. [1975] *Compactness and ρ-valued logics* (**P** 1805) Int Symp Multi-Val Log (5,Proc);1975 Bloomington 400-405
◇ B50 C75 C90 ◇ REV MR 58 # 10299 • ID 28362

HICKIN, K.K. see Vol. III, IV, V for further entries

HIKITA, T. [1978] *Completeness criterion for functions with delay defined over a domain of three elements* (**J** 3239) Proc Japan Acad, Ser A 54*335-339
◇ B50 G20 G25 ◇ REV MR 80f:94036 Zbl 418 # 94019
• ID 53339

HIKITA, T. [1981] *Completeness properties of k-valued functions with delays inclusions among closed spectra* (**J** 0114) Math Nachr 103*5-19
◇ B50 ◇ REV MR 83f:94054 Zbl 489 # 94038 • ID 36606

HIKITA, T. [1981] *On completeness for k-valued functions with delay* (**P** 2552) Conf Finite Algeb & Multi-Val Log;1979 Szeged 345-371
◇ B50 ◇ REV MR 83f:94053 Zbl 479 # 03013 • ID 55675

HIKITA, T. [1982] *Completeness of functions with delay (Japanese)* (**P** 4081) Many-Val Log & Appl;1982 Kyoto 167-175
◇ B50 ◇ ID 47630

HILBERT, D. & ACKERMANN, W. [1928] *Grundzuege der theoretischen Logik* (**X** 0811) Springer: Heidelberg & New York viii + 120pp
• TRANSL [1950] (**X** 0848) Chelsea: New York xii + 172pp (English) [1950] (**X** 1876) Kexue Chubanshe: Beijing
◇ B25 B98 D35 ◇ REV MR 50 # 4230 Zbl 239 # 02001 JSL 15.59 JSL 16.52 JSL 25.158 JSL 3.83 FdM 54.55
• REM 4th ed. 1959;viii + 188pp. • ID 00107

HILBERT, D. & BERNAYS, P. [1934] *Grundlagen der Mathematik I* (**X** 0811) Springer: Heidelberg & New York xii + 471pp
• TRANSL [1979] (**X** 2027) Nauka: Moskva 558pp
◇ A05 B98 F05 F30 F98 ◇ REV MR 81c:03002 Zbl 191.284 Zbl 478 # 03002 JSL 35.321 FdM 60.17 • REM 2nd edition 1968; xv + 473pp. Part II 1939 • ID 01098

HILBERT, D. & BERNAYS, P. [1939] *Grundlagen der Mathematik II* (**X** 0811) Springer: Heidelberg & New York xii + 498pp
• TRANSL [1982] (**X** 2027) Nauka: Moskva 556pp
◇ A05 B98 F05 F15 F30 F40 F98 ◇ REV MR 42 # 7477 Zbl 20.193 Zbl 211.9 Zbl 518 # 03001 JSL 5.16 FdM 65.21 • REM 2nd edition 1970; xiv + 561pp. Part I 1934 • ID 01082

HILBERT, D. see Vol. I, IV, V, VI for further entries

HILEY, B.J. [1980] see FRESCURA, F.A.M.

HILGEVOORD, J. [1981] see COOKE, R.M.

HILL, S. & SUPPES, P. [1964] *First course in mathematical logic* (**X** 0841) Blaisdell: New York ix+274pp ◇ B98 ◇ REV Zbl 126.8 JSL 32.421 • REM See 1965 by Binford,F. • ID 22542

HILPINEN, R. & HINTIKKA, K.J.J. [1966] *Knowledge, acceptance, and inductive logic* (**C** 1107) Aspects Inductive Log 1-20 ◇ A05 B48 ◇ ID 27703

HILPINEN, R. [1968] *Rules of acceptance and inductive logic* (**J** 0096) Acta Philos Fenn 22*134pp ◇ A05 B48 ◇ REV MR 41 # 5200 Zbl 226 # 02030 • ID 21064

HILPINEN, R. (ED.) [1971] *Deontic logic: introductory and systematic readings* (**X** 0835) Reidel: Dordrecht vii+182pp ◇ B45 ◇ REV MR 56 # 2790 MR 83f:03020 Zbl 214.11 Zbl 515 # 03005 • REM 2nd ed 1981; xvii+183pp • ID 25291

HILPINEN, R. [1971] see FOLLESDAL, D.

HILPINEN, R. [1971] *Relational hypotheses and inductive inference* (**J** 0154) Synthese 23*266-286 ◇ A05 B48 ◇ REV MR 47 # 5908 Zbl 229 # 02011 • ID 06091

HILPINEN, R. & HINTIKKA, K.J.J. [1971] *Rules of acceptance, indices of lawlikeness, and singular inductive inference: reply to a critical discussion* (**J** 0153) Phil of Sci (East Lansing) 38*303-307 ◇ A05 B48 ◇ REV MR 58 # 16102b • ID 74129

HILPINEN, R. [1973] *Carnap's new system of inductive logic* (**J** 0154) Synthese 25*307-333 ◇ B48 ◇ REV MR 57 # 15990 Zbl 265 # 02023 • ID 29821

HILPINEN, R. [1973] *On the semantics of personal directives* (**J** 0963) Ajatus (Helsinki) 35*140-157 ◇ B45 ◇ REV Zbl 301 # 02026 • ID 62527

HILPINEN, R. [1976] *Approximate truth and truthlikeness* (**P** 1804) Form Meth in Methodol of Emp Sci;1974 Warsaw 19-42 ◇ A05 B45 ◇ REV MR 58 # 27227 Zbl 374 # 02007 • ID 51531

HILPINEN, R. [1977] see CONTE, A.G.

HILPINEN, R. & HINTIKKA, M.P. & NIINILUOTO, I. & SAARINEN, E. (EDS.) [1979] *Essays in honour of Jaakko Hintikka. On the occasion of his fiftieth birthday on January 12, 1979* (**S** 3307) Synth Libr 124*x+386pp ◇ B97 ◇ REV Zbl 419 # 00001 • ID 53340

HILPINEN, R. (ED.) [1981] *New studies in deontic logic* (**X** 0835) Reidel: Dordrecht ix+256pp ◇ B45 ◇ REV MR 82m:03023 • ID 70001

HILPINEN, R. [1984] *Disjunctive permissions and conditionals with disjunctive antecedents (Russian)* (**C** 4366) Modal & Intens Log & Primen Probl Metodol Nauk 310-329 ◇ B45 ◇ ID 46720

HILPINEN, R. see Vol. I for further entries

HILTON, A.M. [1963] *Logic, computing machines, and automation* (**X** 1354) Spartan Books : Sutton xxi+427pp ◇ A05 B98 D05 ◇ REV MR 28 # 741 Zbl 109.101 JSL 38.341 • ID 23532

HINDE, C.J. [1979] *Generalisation of Bochvar's 3-valued logic* (**P** 3003) Int Symp Multi-Val Log (9);1979 Bath 257-261 ◇ B50 G25 ◇ REV MR 80m:03053 • ID 74132

HINDE, C.J. [1983] *Inference of fyzzy relational tableaux from fuzzy exemplifications* (**J** 2720) Fuzzy Sets Syst 11*91-101 ◇ B52 ◇ REV MR 85a:03035 Zbl 537 # 03015 • ID 34786

HINDLEY, J.R. & LERCHER, B. & SELDIN, J.P. [1972] *Introduction to combinatory logic* (**X** 0805) Cambridge Univ Pr: Cambridge, GB 179pp • TRANSL [1975] (**X** 0905) Boringhieri: Torino 153pp (Italian) ◇ B40 B98 F98 ◇ REV MR 49 # 25 Zbl 269 # 02005 JSL 38.518 • ID 23471

HINDLEY, J.R. see Vol. IV, VI for further entries

HINO, K. [1977] *On Yablonskii theorem concerning functional completeness of k-valued logic* (**J** 0047) Notre Dame J Formal Log 18*251-254 ◇ B50 ◇ REV MR 56 # 5215 Zbl 321 # 02009 • ID 21972

HINTIKKA, K.J.J. [1957] *Modality as referential multiplicity* (**J** 0963) Ajatus (Helsinki) 20*49-64 ◇ B45 ◇ ID 27084

HINTIKKA, K.J.J. [1957] *Quantifiers in deontic logic* (**J** 0990) Soc Sci Fennicae Comment Phys-Math 23/4*23pp ◇ B45 ◇ REV JSL 24.178 • ID 16971

HINTIKKA, K.J.J. [1960] *Aristotle's different possibilities* (**J** 0310) Inquiry (Oslo) 3*18-28 • REPR [1967] (**C** 0736) Aristotle: Coll Critical Essays 34-50 ◇ A05 A10 B45 ◇ ID 06119

HINTIKKA, K.J.J. [1961] *Modality and quantification* (**J** 0105) Theoria (Lund) 27*119-128 ◇ B45 ◇ REV MR 25 # 2953 JSL 31.122 • ID 06120

HINTIKKA, K.J.J. [1963] *The modes of modality* (**J** 0096) Acta Philos Fenn 16*65-81 ◇ B45 ◇ REV MR 29 # 3366 Zbl 135.5 JSL 31.122 • ID 06121

HINTIKKA, K.J.J. [1964] *The once and future sea fight: Aristotle's discussion of future contingents in De Interpretatione IX* (**J** 0101) Phil Rev 73*461-492 ◇ A05 A10 B45 ◇ REV JSL 32.280 • ID 43108

HINTIKKA, K.J.J. [1965] *On a combined system of inductive logic* (**J** 0096) Acta Philos Fenn 18*21-30 ◇ A05 B48 ◇ REV MR 32 # 7401 Zbl 144.246 JSL 35.454 • ID 06125

HINTIKKA, K.J.J. [1965] *Towards a theory of inductive generalization* (**P** 0623) Int Congr Log, Meth & Phil of Sci (2,Proc);1964 Jerusalem 274-288 ◇ A05 B48 ◇ REV MR 34 # 5123 Zbl 182.319 JSL 35.454 • ID 06124

HINTIKKA, K.J.J. [1966] *A two-dimensional continuum of inductive methods* (**C** 1107) Aspects Inductive Log 113-132 ◇ B48 ◇ REV Zbl 202.298 JSL 35.455 • ID 21191

HINTIKKA, K.J.J. & SUPPES, P. (EDS.) [1966] *Aspects of inductive logic* (**X** 0809) North Holland: Amsterdam 328pp ◇ A05 B48 ◇ REV Zbl 158.241 • ID 25463

HINTIKKA, K.J.J. [1966] see HILPINEN, R.

HINTIKKA, K.J.J. & PIETARINEN, J. [1966] *Semantic information and inductive logic* (**C** 1107) Aspects Inductive Log 96-112 ◇ B48 ◇ REV Zbl 202.297 • ID 27705

HINTIKKA, K.J.J. [1967] *Existence and identity in epistemic contexts: A comment on Follesdal's paper* (**J** 0105) Theoria (Lund) 33*138-148
◇ B45 ◇ REV Zbl 174.9 • ID 27082

HINTIKKA, K.J.J. [1967] *Individuals, possible worlds, and epistemic logic* (**J** 0097) Nous, Quart J Phil 1*33-62
◇ B45 ◇ ID 27083

HINTIKKA, K.J.J. [1968] *Induction by enumeration and induction by elimination* (**P** 2268) Int Colloq Philos of Sci;1965 London 2*191-216
◇ A05 B48 ◇ REV Zbl 182.320 JSL 40.448 • ID 21287

HINTIKKA, K.J.J. [1969] *Models for modalities: selected essays* (**X** 0835) Reidel: Dordrecht ix+220pp
◇ B45 ◇ REV Zbl 194.304 • ID 25117

HINTIKKA, K.J.J. [1970] *Existential presuppositions and uniqueness presuppositions* (**P** 0559) Phil Probl in Logic;1968 Irvine 20-55
◇ A05 B45 ◇ REV MR 45 # 1736 Zbl 265 # 02014 • ID 22262

HINTIKKA, K.J.J. & SUPPES, P. (EDS.) [1970] *Information and inference* (**X** 0835) Reidel: Dordrecht vii+336pp
◇ A05 B48 ◇ REV MR 42 # 7456 Zbl 209.218 • ID 25465

HINTIKKA, K.J.J. [1970] *The semantics of modal notions and the indeterminancy of ontology* (**J** 0154) Synthese 21*409-424
◇ B45 ◇ REV Zbl 211.305 • ID 27085

HINTIKKA, K.J.J. [1971] *Inductive generalization and its problems: A comment on Kronthaler's comments* (**J** 0472) Theory Decis 1*393-398
◇ B48 ◇ REV Zbl 215.318 • ID 28013

HINTIKKA, K.J.J. [1971] see HILPINEN, R.

HINTIKKA, K.J.J. [1971] *Some main problems of deontic logic* (**C** 1547) Deontic Log Readings 59-104
◇ B45 ◇ REV Zbl 219 # 02012 • ID 28079

HINTIKKA, K.J.J. [1971] *The "lottery paradox" and the concept of shared information* (**J** 0963) Ajatus (Helsinki) 33*266-270
◇ B48 ◇ REV Zbl 242 # 02003 • ID 26334

HINTIKKA, K.J.J. (ED.) [1972] *Collection of articles in honor of Rudolf Carnap* (**J** 0154) Synthese 25*255-436
◇ B96 B97 ◇ REV MR 48 # 48 • ID 70235

HINTIKKA, K.J.J. [1973] *Surface semantics: definition and its motivation* (**P** 0783) Truth, Syntax & Modal;1970 Philadelphia 128-147
◇ A05 B10 B45 ◇ REV MR 53 # 7717 Zbl 261 # 02007 JSL 42.315 • ID 44480

HINTIKKA, K.J.J. [1974] *Questions about questions (Russian)* (**C** 2578) Filos & Logika 303-362
◇ A05 B60 B65 ◇ REV MR 58 # 27359 • ID 74153

HINTIKKA, K.J.J. [1975] *Comment on Professor Bergstroem* (**J** 0105) Theoria (Lund) 41*35-38
◇ B45 ◇ REV MR 58 # 27317b Zbl 318 # 02028 • ID 62541

HINTIKKA, K.J.J. [1975] *Impossible possible worlds vindicated* (**J** 0122) J Philos Logic 4*475-484
• TRANSL [1980] (**C** 4675) Hintikka: Logiko-Epist Issled 228-242
◇ A05 B45 B53 ◇ REV MR 57 # 12176 Zbl 334 # 02003 JSL 51.240 • ID 62545

HINTIKKA, K.J.J. [1975] *The intentions of intentionality and other new models for modalities* (**X** 0835) Reidel: Dordrecht xviii+262pp
◇ A05 B45 ◇ REV MR 57 # 5685 Zbl 323 # 02029 • ID 30412

HINTIKKA, K.J.J. & NIINILUOTO, I. [1976] *An axiomatic foundation for the logic of inductive generalization* (**P** 1804) Form Meth in Methodol of Emp Sci;1974 Warsaw 57-81
◇ A05 B48 ◇ REV MR 58 # 27427 Zbl 365 # 02016 JSL 49.1409 • ID 32408

HINTIKKA, K.J.J. [1976] *Information, causality, and the logic of perception* (**J** 0963) Ajatus (Helsinki) 36*76-94
◇ A05 B48 ◇ REV Zbl 366 # 02009 • ID 51092

HINTIKKA, K.J.J. [1976] *The semantics of questions and the questions of semantics: Case studies in the interrelations of logic, semantics, and syntax* (**J** 0096) Acta Philos Fenn 28*200pp
◇ A05 B45 B65 ◇ REV MR 57 # 9483 Zbl 392 # 03004 • ID 74163

HINTIKKA, K.J.J. [1977] see BUTTS, R.E.

HINTIKKA, K.J.J. & NIINILUOTO, I. & SAARINEN, E. (EDS.) [1979] *Essays on mathematical and philosophical logic* (**X** 0835) Reidel: Dordrecht viii+462pp
◇ A05 B97 ◇ REV MR 80f:03006 Zbl 393 # 00002 • ID 52420

HINTIKKA, K.J.J. [1980] *Aristotelian induction* (**J** 2076) Rev Int Philos 34*422-439
◇ A05 B48 ◇ REV MR 82m:00016 • ID 81563

HINTIKKA, K.J.J. [1980] *Degrees and dimensions of intentionality* (**P** 2949) Int Wittgenstein Symp (4);1979 Kirchberg 69-82
◇ A05 B45 ◇ REV MR 82j:03001 • ID 74156

HINTIKKA, K.J.J. [1981] *Aristotle on the realization of possibilities in time* (**C** 4139) Reforging Great Chain of Being 57-72
◇ A10 B45 ◇ REV MR 83a:03010 • ID 47727

HINTIKKA, K.J.J. [1981] *Intuitions and philosophical method* (**J** 2076) Rev Int Philos 35*74-90
◇ A05 B48 ◇ REV MR 82m:03008 • ID 74154

HINTIKKA, K.J.J. [1981] *On the logic of an interrogative model of scientific inquiry* (**J** 0154) Synthese 47*69-83
◇ B60 ◇ REV MR 83i:00016 • ID 39290

HINTIKKA, K.J.J. [1981] *Phenomenology vs.possible-worlds semantics: apparent and real differences* (**J** 2076) Rev Int Philos 35*113-119
◇ A05 B45 ◇ REV MR 83a:03008b • ID 35054

HINTIKKA, K.J.J. [1981] *Standard vs. nonstandard logic: higher order, modal, and first-order logics* (**C** 2617) Modern Log Survey 283-296
◇ B10 B15 B45 ◇ REV MR 82f:03002 Zbl 464 # 03001 • ID 42772

HINTIKKA, K.J.J. [1982] *Game-theoretical semantics: Insights and prospects* (**J** 0047) Notre Dame J Formal Log 23*219-241
◇ A05 B60 C80 ◇ REV MR 84a:03029a Zbl 464 # 03008 • ID 55406

HINTIKKA, K.J.J. [1982] *Is alethic modal logic possible?* (**P** 3800) Intens Log: Th & Appl;1979 Moskva 89-105
◇ B45 ◇ REV MR 84m:03024 • ID 35728

HINTIKKA, K.J.J. & LEOPOLD-WILDBURGER, U. [1982] *Regeln und Nutzen von Dialogspielen* (**J** 2688) Conceptus (Wien) 16*61-72
◇ A05 B60 ◇ REV MR 84d:03033 • ID 34073

HINTIKKA, K.J.J. [1982] *Temporal discourse and semantical games* (**J** 2130) Linguist Philos 5*3-22
◇ B45 ◇ REV Zbl 514 # 68079 • ID 37453

HINTIKKA, K.J.J. [1983] *Situations, possible worlds, and attitudes* (**J** 0154) Synthese 54*153-162
◇ B45 ◇ REV MR 85d:03040 • ID 41020

HINTIKKA, K.J.J. [1984] *Are there nonexistent objects? Why not? But where are they?* (**J** 0154) Synthese 60*451-458
◇ A05 B45 ◇ REV MR 85j:03006 • ID 44612

HINTIKKA, K.J.J. [1984] *Is an aletic modal logic possible? (Russian)* (**C** 4366) Modal & Intens Log & Primen Probl Metodol Nauk 31-49
◇ B45 ◇ REV MR 84m:03024 • ID 46848

HINTIKKA, K.J.J. & KULAS, J. [1985] *Anaphora and definite descriptions. Two applications of game-theoretical semantics* (**X** 0835) Reidel: Dordrecht xiv+250pp
◇ B45 ◇ ID 49400

HINTIKKA, K.J.J. see Vol. I, III, V, VI for further entries

HINTIKKA, M.P. [1979] see HILPINEN, R.

HINTIKKA, M.P. see Vol. I for further entries

HIROSE, K. & NAGAOKA, K. & UMEZU, M. [1979] *Many-valued logic with truth values from finite lattices* (**S** 1459) Mem School Sci & Engin, Waseda Univ 43*147-162
◇ B50 G10 ◇ REV MR 82d:03033 Zbl 453 # 03020
• ID 54150

HIROSE, K. see Vol. III, IV, V for further entries

HIROTA, K. & IIJIMA, T. [1979] *Logical basis in probabilistic set theory -- probabilistic expression of ambiguity and subjectivity* (**J** 0464) Syst-Comp-Controls 10*45-54
◇ B52 E72 ◇ REV MR 82f:94030 • ID 81567

HIROTA, K. & PEDRYCZ, W. [1983] *Analysis and synthesis of fuzzy systems by the use of probabilistic interpretation of histograms* (**J** 2720) Fuzzy Sets Syst 10*1-13
◇ B52 ◇ REV MR 84g:93078 Zbl 523 # 93007 • ID 46100

HIROTA, K. see Vol. V for further entries

HIRSCH, G. [1977] see BENHALCEN, D.

HIRSCHFELD, J. & MACHOVER, M. [1969] *Lectures on non-standard analysis* (**X** 0811) Springer: Heidelberg & New York 79pp
◇ B98 H05 ◇ REV MR 40 # 2531 Zbl 182.559 • ID 08459

HIRSCHFELD, J. see Vol. I, III, IV, V, VI for further entries

HISDAL, E. [1979] *Possibilistically dependent variables and a general theory of fuzzy sets* (**C** 3514) Adv Fuzzy Sets & Appl 215-234
◇ B52 E72 ◇ REV MR 81b:03060 • ID 74198

HISDAL, E. [1981] *The IF THEN ELSE statement and interval-valued fuzzy sets of higher type* (**J** 1741) Int J Man-Mach Stud 15*385-455
◇ A05 B52 B75 E72 ◇ REV MR 82m:68059
Zbl 471 # 03013 • ID 55208

HISDAL, E. [1982] *A fuzzy IF THEN ELSE relation with guaranteed correct inference* (**P** 4051) Fuzzy Set & Possibility Th;1980 Acapulco 204-210
◇ B52 ◇ REV MR 84b:03004 • ID 46323

HISDAL, E. see Vol. V for further entries

HOARE, C.A.R. [1984] see ATIYAH, M.

HOARE, C.A.R. see Vol. I, IV for further entries

HOBBS, J.R. & ROSENSCHEIN, S.J. [1977] *Making computational sense of Montague's intensional logic* (**J** 0503) Artif Intell 9*287-306
◇ B45 ◇ REV MR 58 # 8523 Zbl 379 # 02005 • ID 51838

HOCKNEY, D.J. & WILSON, K. [1965] *In defence of a relevance condition* (**J** 0079) Logique & Anal, NS 8*211-220
◇ A05 B46 ◇ REV MR 33 # 5472 • ID 06164

HOCUTT, M.O. [1972] *Is epistemic logic possible?* (**J** 0047) Notre Dame J Formal Log 13*433-453
◇ A05 B45 ◇ REV Zbl 242 # 02030 • ID 26348

HODES, H.T. [1984] *Axioms for actuality* (**J** 0122) J Philos Logic 13*27-34
◇ A05 B45 ◇ REV MR 85i:03050b Zbl 538 # 03015
• ID 44101

HODES, H.T. [1984] *On modal logics which enrich first-orders S5* (**J** 0122) J Philos Logic 13*423-454
◇ B45 ◇ REV Zbl 533 # 03008 • ID 42650

HODES, H.T. [1984] *Some theorems on the expressive limitations of modal languages* (**J** 0122) J Philos Logic 13*13-26
◇ B45 ◇ REV MR 85i:03050a Zbl 538 # 03014 • ID 41465

HODES, H.T. [1984] *The modal theory of pure identity and some related decision problems* (**J** 0068) Z Math Logik Grundlagen Math 30*415-423
◇ B25 B45 D35 ◇ REV MR 86d:03018 Zbl 534 # 03007
• ID 41142

HODES, H.T. [1984] *Well-behaved modal logics* (**J** 0036) J Symb Logic 49*1393-1402
◇ B45 ◇ ID 39885

HODES, H.T. see Vol. IV, V for further entries

HODGES, W. (ED.) [1972] *Conference in Mathematical Logic -- London '70* (**S** 3301) Lect Notes Math 255*viii+351pp
◇ B97 C97 ◇ REV MR 48 # 5804 Zbl 227 # 02011
• ID 70226

HODGES, W. [1977] *Logic* (**X** 0868) Penguin Books: Harmondsworth & New York 331pp
◇ B98 ◇ REV JSL 45.382 • ID 30709

HODGES, W. see Vol. I, III, IV, V, VI for further entries

HOEHLE, U. [1981] *A mathematical theory of uncertainty (fuzzy experiments and their realizations)* (**P** 4050) Appl Syst & Cybern;1980 Acapulco 2728-2733
◇ B52 ◇ REV MR 84b:00009 • ID 46304

HOEHLE, U. [1982] *A mathematical theory of uncertainty* (**P** 4051) Fuzzy Set & Possibility Th;1980 Acapulco 344-355
◇ B52 ◇ REV MR 84b:03004 • ID 46322

HOEHLE, U. [1984] *Fuzzy plausability measures* (**C** 3389) Fuzzy Sets & Decision Anal 83-96
◇ B52 ◇ REV MR 86a:28023 Zbl 562 # 60004 • ID 47441

Hoehle, U. & Klement, E.P. [1984] *Plausibility measures - a general framework for possibility and fuzzy probability measures* (**P** 3096) Conf Math Service of Man (2,Pap);1982 Las Palmas 31-50
◇ B52 ◇ REV Zbl 562 # 60003 • ID 44255

Hoehle, U. see Vol. V for further entries

Hoepelman, J.P. [1974] *Tense-logic and the semantics of the Russian aspects* (**J** 1517) Theor Linguist 1*158-180
◇ B45 ◇ REV MR 58 # 21473 • ID 74226

Hoepelman, J.P. [1981] see Aaqvist, L.

Hoepelman, J.P. see Vol. III for further entries

Hofstadter, D.R. [1979] *Goedel, Escher, Bach: an eternal golden braid* (**X** 2671) Basic Books: New York xxi + 777pp
◇ A05 B98 D99 ◇ REV MR 80j:03009 Zbl 457 # 03001 JSL 48.864 • ID 74228

Holdsworth, D.G. & Hooker, C.A. [1983] *A critical survey of quantum logic* (**C** 3811) Logic 20th Century 127-246
◇ B51 ◇ REV MR 85f:03065 • ID 40816

Hong, Hengling [1983] *Method of fuzzy statistics and applications* (**J** 3919) BUSEFAL 13*90-107
◇ B52 ◇ REV Zbl 523 # 62098 • ID 36999

Hong, Sejune [1974] see Cain, R.G.

Hong, Sejune see Vol. I for further entries

Hoogewijs, A. [1979] *On a formalization of the non-definedness notion* (**J** 0068) Z Math Logik Grundlagen Math 25*213-217
◇ B60 C90 ◇ REV MR 80f:03017 Zbl 415 # 03019 • ID 53121

Hoogewijs, A. [1983] *A partial predicate calculus in a two-valued logic* (**J** 0068) Z Math Logik Grundlagen Math 29*239-243
◇ B60 C90 ◇ REV MR 85d:03017 Zbl 521 # 03006 • ID 37062

Hoogewijs, A. see Vol. III for further entries

Hooker, C.A. & Stove, D. [1968] *Relevance and the ravens* (**J** 0013) Brit J Phil Sci 18*305-315
◇ A05 B45 ◇ REV Zbl 228 # 02004 • ID 29111

Hooker, C.A. [1975] see Harper, W.L.

Hooker, C.A. (ed.) [1975] *The logico-algebraic approach to quantum mechanics. Vol. I: Historical evolution* (**S** 3308) Univ Western Ontario Ser in Philos of Sci xv + 607pp
◇ A10 B51 G12 ◇ REV MR 81i:81003a Zbl 429 # 03045 • REM Vol.II 1979 • ID 53876

Hooker, C.A. [1976] see Fraassen van, B.C.

Hooker, C.A. & Leach, J.J. [1978] *Theoretical foundations* (**X** 0835) Reidel: Dordrecht xxiii + 446pp
◇ B48 ◇ REV MR 81a:90002a Zbl 398 # 90002 • ID 52792

Hooker, C.A. (ed.) [1979] *Physical theory as logico-operational structure* (**X** 0835) Reidel: Dordrecht xvii + 334pp
◇ A05 B51 G12 ◇ REV MR 80b:81001 Zbl 402 # 03053 • ID 54700

Hooker, C.A. (ed.) [1979] *The logico-algebraic approach to quantum mechanics. Vol.II* (**X** 0835) Reidel: Dordrecht xx + 466pp
◇ B51 G12 ◇ REV MR 81i:81003b Zbl 429 # 03046 • REM Vol.I 1975 • ID 80423

Hooker, C.A. [1983] see Holdsworth, D.G.

Hooker, C.A. see Vol. I for further entries

Hoover, D.N. [1978] *Probability logic* (**J** 0007) Ann Math Logic 14*287-313
◇ B48 C80 C90 ◇ REV MR 80b:03044 Zbl 394 # 03033 • ID 29154

Hoover, D.N. [1980] *A note on regularity* (**C** 3537) Stud Induct Logic & Probab, Vol 2 295-297
◇ B48 ◇ REV MR 81m:03035b Zbl 539 # 03004 JSL 49.1409 • ID 74251

Hoover, D.N. see Vol. I, III, IV for further entries

Hopkins, L.D. [1983] see Brill Jr., E.D.

Horecky, J. [1982] *COLING 82. Proceedings of the Ninth International Conference on Computational Linguistics, Prague, July 5-10, 1982* (**X** 0809) North Holland: Amsterdam 432pp
◇ B97 ◇ REV Zbl 529 # 68041 • ID 38490

Horn, A. [1969] *Logic with truth values in a linearly ordered Heyting algebra* (**J** 0036) J Symb Logic 34*395-408
◇ B55 ◇ REV MR 40 # 7089 Zbl 181.299 • ID 06231

Horn, A. [1974] see Epstein, G.

Horn, A. [1976] see Epstein, G.

Horn, A. see Vol. I, III, V for further entries

Horwich, P. [1975] *A formalization of "nothing"* (**J** 0047) Notre Dame J Formal Log 16*363-368
◇ A05 B60 ◇ REV Zbl 236 # 02007 • ID 62599

Horwich, P. [1982] *Probability and evidence* (**X** 0805) Cambridge Univ Pr: Cambridge, GB vii + 146pp
◇ B48 ◇ REV MR 85j:60003 Zbl 537 # 60001 • ID 43855

Hosiasson-Lindenbaum, J. [1940] *On confirmation* (**J** 0036) J Symb Logic 5*133-148
◇ B48 ◇ REV MR 2.210 Zbl 24.97 JSL 6.63 FdM 66.1194 • ID 06238

Hosoi, T. [1966] *Algebraic proof of the seperation theorem on Dummett's LC* (**J** 0081) Proc Japan Acad 42*693-695
◇ B55 ◇ REV MR 34 # 7350 Zbl 168.3 JSL 33.128 • ID 06241

Hosoi, T. [1966] *On the separation theorem of intermediate propositional calculi* (**J** 0081) Proc Japan Acad 42*535-538
◇ B55 ◇ REV MR 34 # 7349 Zbl 168.3 JSL 34.505 • ID 06243

Hosoi, T. [1966] *The axiomatization of the intermediate propositional systems S_n of Goedel* (**J** 0434) J Fac Sci Univ Tokyo, Sect 1 13*183-187
◇ B55 ◇ REV MR 36 # 33 Zbl 156.8 JSL 34.505 • ID 31118

Hosoi, T. [1966] *The separable axiomatization of the intermediate propositional systems S_n of Goedel* (**J** 0081) Proc Japan Acad 42*1001-1006
◇ B55 ◇ REV MR 36 # 34 Zbl 156.8 JSL 34.505 • ID 06244

Hosoi, T. [1967] *A criterion for the separable axiomatization of Goedel's S_n* (**J** 0081) Proc Japan Acad 43*365-368
◇ B55 ◇ REV MR 37 # 33 Zbl 168.249 • ID 06245

Hosoi, T. [1967] *On intermediate logics* (**J** 0434) J Fac Sci Univ Tokyo, Sect 1 14*293-312
◇ B55 ◇ REV MR 36 # 4961 Zbl 188.316 JSL 36.329 • REM Part I. Part II 1969 • ID 06247

HOSOI, T. [1967] *On the axiomatic method and the algebraic method for dealing with propositional logics* (**J** 0434) J Fac Sci Univ Tokyo, Sect 1 14*131-169
◇ B55 ◇ REV MR 36 # 2470 Zbl 162.311 • REM Part I. Part II 1977 • ID 06246

HOSOI, T. [1969] *On intermediate logics II* (**J** 0434) J Fac Sci Univ Tokyo, Sect 1 16*1-12
◇ B55 ◇ REV MR 40 # 5420 Zbl 188.316 • REM Part I 1967. Part III 1974 • ID 06248

HOSOI, T. & ONO, H. [1970] *The intermediate logics on the second slice* (**J** 0434) J Fac Sci Univ Tokyo, Sect 1 17*457-461
◇ B55 ◇ REV MR 44 # 3837 Zbl 216.5 • ID 06249

HOSOI, T. & ONO, H. [1972] *Axiomatization of models for intermediate logics constructed with boolean models by piling up* (**J** 0390) Publ Res Inst Math Sci (Kyoto) 8*1-11
◇ B55 G05 G25 ◇ REV MR 51 # 12501 Zbl 246 # 02021 • ID 17288

HOSOI, T. & ONO, H. [1973] *Intermediate propositional logics (a survey)* (**J** 0381) J Tsuda College (Tokyo) 5*67-82
◇ B55 ◇ REV MR 49 # 4753 • ID 06251

HOSOI, T. [1974] *On intermediate logics III* (**J** 0381) J Tsuda College (Tokyo) 6*23-38
◇ B55 ◇ REV MR 58 # 27352 • REM Part II 1969 • ID 31120

HOSOI, T. [1975] *On the implicational fragment logics* (**J** 0381) J Tsuda College (Tokyo) 7*1-5
◇ B55 ◇ REV MR 51 # 2875 • ID 17376

HOSOI, T. [1976] *Axiomatizability of fragments of intermediate finite logics* (**J** 0381) J Tsuda College (Tokyo) 8*9-12
◇ B55 ◇ REV MR 57 # 5691 • ID 31121

HOSOI, T. [1976] *Non-separable intermediate propositional logics* (**J** 0381) J Tsuda College (Tokyo) 8*13-18
◇ B55 ◇ REV MR 57 # 9492 • ID 31122

HOSOI, T. & SHUNDO, S. [1977] *A finer classification of intermediate propositional logics* (**J** 0381) J Tsuda College (Tokyo) 9*5-15
◇ B55 ◇ REV MR 57 # 12184a • ID 31123

HOSOI, T. [1977] *On the axiomatic method and the algebraic method for dealing with propositional logics II* (**J** 2332) J Fac Sci, Univ Tokyo, Sect 1 A 24*373-380
◇ B50 ◇ REV MR 58 # 5027 Zbl 381 # 03010 • REM Part I 1967 • ID 32360

HOSOI, T. [1978] *Axiomatization of certain intermediate logics* (**J** 0381) J Tsuda College (Tokyo) 10*9-12
◇ B55 ◇ REV MR 58 # 5072 • ID 31124

HOSOI, T. [1978] *The delta operation on the intermediate logics and the modal logics* (**J** 0381) J Tsuda College (Tokyo) 10*13-20
◇ B55 ◇ REV MR 81f:03034 • ID 31125

HOSOI, T. [1979] *Evaluation of Peirce's axiom on intermediate Kripke models and its application* (**J** 3239) Proc Japan Acad, Ser A 55*364-366
◇ B55 ◇ REV MR 81e:03023 Zbl 453 # 03023 • ID 54153

HOSOI, T. see Vol. I, VI for further entries

HOTTOIS, G. [1981] *Logique deontique et logique de l'action chez G. H. von Wright* (**J** 2076) Rev Int Philos 35*143-152
◇ A05 B45 ◇ REV MR 82k:03016 • ID 74272

HOWSON, C. [1975] *The rule of succession, inductive logic and probability logic* (**J** 0013) Brit J Phil Sci 26*187-198
◇ B48 ◇ REV Zbl 364 # 02001 • ID 32208

HOWSON, C. [1976] *The development of logical probability* (**C** 2103) Develop of Log Probab (Lakatos) 277-298
◇ A05 B30 B48 ◇ REV Zbl 346 # 02004 • ID 62614

HROMKOVIC, J. [1985] *On the number of monotonic functions from two-valued logic to k-valued logic* (**J** 0156) Kybernetika (Prague) 21*228-234
◇ B50 ◇ REV Zbl 574 # 94023 • ID 48839

HROMKOVIC, J. see Vol. IV for further entries

HU, SHIHUA [1949] *M-valued sub-system of $(m+n)$-valued propositional calculus* (**J** 0036) J Symb Logic 14*177-181
◇ B50 ◇ REV MR 11.487 Zbl 36.6 JSL 14.261 • ID 06206

HU, SHIHUA [1950] *Construction of an $\aleph_0$-valued propositional calculus (Chinese)* (**J** 1024) Zhongguo Kexue 1/2-4*273-294
◇ B50 ◇ ID 48508

HU, SHIHUA [1951] see CHEN, QIANGYE

HU, SHIHUA [1955] *Die endlichwertigen und funktionell vollstaendigen Subsysteme eines $\aleph_0$-wertigen Aussagenkalkuels (Chinese) (German summary)* (**J** 0418) Shuxue Xuebao 5*173-191
◇ B50 ◇ REV MR 17.224 Zbl 66.256 • ID 06279

HU, SHIHUA [1957] *Mathematical logic, its fundamental properties and scientific significance (Chinese)* (**J** 4452) Zhexue Yanjiu 1*1-45
◇ B98 ◇ ID 48510

HU, SHIHUA & LU, ZHONGWAN [1981] *Foundation of mathematical logic I (Chinese)* (**X** 1876) Kexue Chubanshe: Beijing
◇ B98 ◇ REM Vol.II 1982 • ID 48514

HU, SHIHUA & LU, ZHONGWAN [1982] *Foundation of mathematical logic II (Chinese)* (**X** 1876) Kexue Chubanshe: Beijing
◇ B98 ◇ REM Vol.I 1981 • ID 48515

HU, SHIHUA see Vol. I, IV, VI for further entries

HU, SZETIEN [1965] *Threshold logic* (**X** 0926) Univ Calif Pr: Berkeley xiv+338pp
◇ B70 B98 ◇ REV MR 36 # 2435 JSL 40.250 • ID 22214

HUANG, H.K. & LEDLEY, R.S. [1975] *Multivalued logic design and Postian matrices* (**P** 1805) Int Symp Multi-Val Log (5,Proc);1975 Bloomington 67-75
◇ B50 B70 ◇ ID 35813

HUANG, KECHENG [1982] see CAI, QINGSHENG

HUANG, ZHENDE & TONG, ZHENGXIANG [1981] *Fuzzy probability and possibility spaces I (Chinese) (English summary)* (**J** 3732) Mohu Shuxue 1/1*43-52
◇ B52 ◇ REV MR 83g:60010 • ID 47228

HUBBELING, H.G. [1982] *A modern reconstruction of the cosmological argument* (**C** 3799) Logic & Religion 165-171
◇ A05 B60 ◇ REV MR 85a:03022 MR 85i:03012 • ID 34773

HUBER, O. [1974] *An axiomatic system for multidimensional preference* (**J** 0472) Theory Decis 5*161-184
◇ B50 ◇ REV Zbl 306 # 02025 • ID 62620

HUBER, O. see Vol. V for further entries

HUBIEN, H. [1971] *On two of Professor Rescher's modal theses* (**J** 0079) Logique & Anal, NS 14*669-674
◇ B45 ◇ REV MR 45 # 8504 Zbl 333 # 02015 • ID 62621

HUBIEN, H. see Vol. I for further entries

HUERTAS, J.A. & MARTINEZ RECIO, A. [1979] *Inductive logic and artificial intelligence boundaries (Spanish) (English summary)* (**P** 4097) Jorn Mat Luso-Espanol (6);1979 Santander 871-890
◇ B48 ◇ ID 45463

HUGHES, C.E. & SINGLETARY, W.E. [1975] *Triadic partial implicational propositional calculi* (**J** 0068) Z Math Logik Grundlagen Math 21*21-28
◇ B50 ◇ REV MR 52 # 13348 Zbl 307 # 02008 • ID 06303

HUGHES, C.E. see Vol. I, IV for further entries

HUGHES, G.E. & LONDEY, D.G. [1965] *The elements of formal logic* (**X** 0816) Methuen: London & New York xii + 398pp
◇ B98 ◇ ID 22495

HUGHES, G.E. [1968] see CRESSWELL, M.J.

HUGHES, G.E. [1975] *B(S4.3, S4) unveiled* (**J** 0105) Theoria (Lund) 41*85-88
◇ B45 ◇ REV Zbl 331 # 02009 • ID 62638

HUGHES, G.E. [1975] see CRESSWELL, M.J.

HUGHES, G.E. [1976] *Modal systems with no minimal proper extensions* (**J** 0302) Rep Math Logic, Krakow & Katowice 6*93-98
◇ B45 ◇ REV MR 57 # 9484 Zbl 355 # 02015 • ID 21917

HUGHES, G.E. [1977] *Omnitemporal logic and nodal time* (**J** 0302) Rep Math Logic, Krakow & Katowice 8*41-61
◇ B45 ◇ REV MR 58 # 27333 Zbl 396 # 03018 • ID 24189

HUGHES, G.E. [1980] *Equivalence relations and S5* (**J** 0047) Notre Dame J Formal Log 21*577-584
◇ B45 ◇ REV MR 81g:03017 Zbl 416 # 03019 • ID 53916

HUGHES, G.E. [1980] *Modal logics between S4.2 and S4.3* (**J** 0387) Bull Sect Logic, Pol Acad Sci 9*73-77
◇ B45 ◇ REV MR 82g:03025 Zbl 433 # 03010 • ID 74314

HUGHES, G.E. [1980] *Some extensions of the Brouwerian logic* (**J** 0387) Bull Sect Logic, Pol Acad Sci 9*78-84
◇ B45 ◇ REV MR 82d:03027 Zbl 433 # 03011 • ID 74315

HUGHES, G.E. [1982] see CRESSWELL, M.J.

HUGHES, G.E. [1982] *Some strong omnitemporal logics* (**J** 0154) Synthese 53*19-42
◇ B45 ◇ REV MR 84j:03039 Zbl 501 # 03008 • ID 34631

HUGHES, G.E. [1984] see CRESSWELL, M.J.

HUGHES, R.J.G. [1981] *Realism and quantum logic* (**P** 3820) Curr Iss in Quantum Log;1979 Erice 77-87
◇ A05 B51 G12 ◇ REV MR 84j:03128 Zbl 537 # 03044 • ID 34718

HUGHES, R.J.G. [1985] *Semantic alternatives in partial boolean quantum logic* (**J** 0122) J Philos Logic 14*411-446
◇ B51 ◇ ID 47644

HUGLY, P. & SAYWARD, C.W. [1981] *Completeness theorems for two propositional logics in which identity diverges from mutual entailment* (**J** 0047) Notre Dame J Formal Log 22*269-282
◇ A05 B45 ◇ REV MR 83m:03024 Zbl 417 # 03004 • ID 55520

HUGLY, P. see Vol. I, III for further entries

HUMBERSTONE, I.L. [1974] *Logic for saints and heroes* (**J** 1022) Ratio (Oxford) 16*103-114
◇ B45 ◇ REV MR 58 # 16150 • ID 74326

HUMBERSTONE, I.L. [1977] see BELL, J.M.

HUMBERSTONE, I.L. [1978] *Two merits of the circumstantial operator language for conditional logics* (**J** 0273) Australasian J Phil 56*21-24
◇ A05 B60 ◇ ID 33460

HUMBERSTONE, I.L. [1981] *From worlds to possibilities* (**J** 0122) J Philos Logic 10*313-339
◇ A05 B45 ◇ REV MR 83a:03019 Zbl 481 # 03011 • ID 33050

HUMBERSTONE, I.L. [1981] *Relative necessity revisited* (**J** 0302) Rep Math Logic, Krakow & Katowice 13*33-42
◇ A05 B45 ◇ REV MR 83k:03025 Zbl 519 # 03009 • ID 36165

HUMBERSTONE, I.L. [1983] see BENTHEM VAN, J.F.A.K.

HUMBERSTONE, I.L. [1983] *Inaccessible worlds* (**J** 0047) Notre Dame J Formal Log 24*346-352
◇ B45 ◇ REV MR 84j:03040 Zbl 487 # 03007 • ID 34632

HUMBERSTONE, I.L. see Vol. III, IV, V for further entries

HUMBERSTONE, L.L. [1977] see CROSSLEY, J.N.

HUMBERSTONE, L.L. [1979] *Interval semantics for tense logic: Some remarks* (**J** 0122) J Philos Logic 8*171-196
◇ B45 ◇ REV MR 80h:03027 Zbl 409 # 03012 • ID 56315

HUMBERSTONE, L.L. [1980] see DAVIES, M.K.

HUMBURG, J. [1971] *The principle of instantial relevance* (**C** 0799) Stud Induct Logic & Probab, Vol 1 225-233
◇ B48 ◇ REV MR 58 # 5088a JSL 40.581 • ID 74328

HUNG, CH'ENGWAN [1970] *Remarks on intermediate logics* (**J** 1598) Taita J Math (Univ Taipei) 2*31-43
◇ B55 ◇ REV Zbl 236 # 02023 • ID 27853

HUNG, CH'ENGWAN [1973] *On the axiomatic structures of intensional logic. III: $B^+, N^+, T^+, E^+, S4^+, R^+, J^+$* (**J** 0500) Tamkang J Math 4*69-81 • ERR/ADD ibid 6*103
◇ B45 ◇ REV MR 58 # 27334c MR 58 # 27334d Zbl 344 # 02014 Zbl 344 # 02015 • REM Part II 1974 • ID 62650

HUNG, CH'ENGWAN [1974] *On the axiomatic structures of intensional logics. II: Gentzen type system GX, X = E, EQ* (**J** 0409) Chinese J Math (Taipei) 2*1-30
◇ B45 ◇ REV MR 58 # 27334b Zbl 389 # 03013 • REM Part I 1975. Part III 1973 • ID 52302

HUNG, CH'ENGWAN [1975] *Independence of connectives in E, EM, R, RM* (**J** 0500) Tamkang J Math 6*197-204
◇ B45 ◇ REV MR 54 # 12485 Zbl 333 # 02017 • ID 62653

HUNG, CH'ENGWAN [1975] *On the axiomatic structures of intensional logics. I: E, EM, R, RM and Π′* (**J** 0409) Chinese J Math (Taipei) 3*27-68
◇ B45 ◇ REV MR 58 # 27334a Zbl 389 # 03012 • REM Part II 1974 • ID 52301

HUNG, CH'ENGWAN [1975] *The independence of the axiom schemata and the inference rules of E* (**J** 0500) Tamkang J Math 6*205-214
◇ B45 ◇ REV MR 54 # 12486 Zbl 333 # 02018 • ID 62648

HUNG, CH'ENGWAN see Vol. V for further entries

HUNG, H.C. [1979] *Entailment and proof* (**J** 0047) Notre Dame J Formal Log 20*921-933
◇ A05 B46 ◇ REV MR 80g:03016 Zbl 315#02007
• ID 56175

HUNTER, D. & RICHTER, R. [1978] *Counterfactuals and Newcomb's paradox* (**J** 0154) Synthese 39*249-261
◇ A05 B45 B48 ◇ REV MR 80m:03037 Zbl 386#03003
• ID 52179

HUNTER, G. [1971] *Metalogic: an introduction to the metatheory of standard firstorder logic* (**X** 0926) Univ Calif Pr: Berkeley xiii+288pp
◇ B10 B98 ◇ REV MR 56#5186 Zbl 284#02001
• ID 22356

HUNTINGTON, E.V. [1934] *Independent postulates related to C.I.Lewis' theory of strict implication* (**J** 0094) Mind 43*181-198
◇ B45 ◇ REV Zbl 10.49 FdM 60.26 FdM 60.849
• ID 06335

HUNTINGTON, E.V. [1934] *The relation between Lewis' strict implication and boolean algebra* (**J** 0015) Bull Amer Math Soc 40*729-735
◇ B45 G05 ◇ REV Zbl 10.49 FdM 60.27 FdM 60.849
• ID 06336

HUNTINGTON, E.V. [1935] *The mathematical structure of Lewis's theory of strict implication* (**J** 0027) Fund Math 25*147-156
◇ B45 ◇ REV Zbl 13.98 FdM 61.975 • ID 06338

HUNTINGTON, E.V. see Vol. I, III, V, VI for further entries

HUSZCZA, J. [1981] *Foundations of Jean-Louis Gardies' deontic and modal algebra (Polish) (French summary)* (**J** 3815) Acta Univ Lodz Folia Philos 1*93-112
◇ B45 ◇ REV MR 84k:03054 • ID 34992

HYMAN, G. [1973] see GROFMAN, B.

IBRAGIMOV, S.G. [1973] *Consequence logic of Paul Lorenzen (Russian)* (**J** 3543) Uch Zap Vopr Prikl Mat Kibern, Univ Baku 1973*3-91
◇ B60 F50 ◇ REV MR 57#5692 • ID 74359

IIJIMA, T. [1979] see HIROTA, K.

ILKKA, S. [1966] *A new arithmetization for finitely many-valued propositional calculi* (**J** 0990) Soc Sci Fennicae Comment Phys-Math 32/8*13pp
◇ B50 ◇ REV MR 35#1449 Zbl 143.8 JSL 34.304
• ID 21141

IMANISHI, S. & MURANAKA, N. [1982] *Applications of the clause expansion method to functions in three-valued logic (Japanese)* (**P** 4081) Many-Val Log & Appl;1982 Kyoto 94-107
◇ B50 ◇ ID 47623

IMANISHI, S. & MURANAKA, N. [1982] *Three-valued ordered circuits* (**P** 4081) Many-Val Log & Appl;1982 Kyoto 76-93
◇ B50 ◇ ID 47621

INHETVEEN, R. [1982] *Ein konstruktiver Weg zur Semantik der "moeglichen Welten"* (**P** 3754) Argumentation;1978 Groningen 133-141
◇ B45 ◇ REV MR 84a:03021 Zbl 554#03013 • ID 34894

INWAGEN VAN, P. & MCKAY, T.J. [1977] *Counterfactuals with disjunctive antecedents* (**J** 0095) Philos Stud 31*353-356
◇ A05 B45 ◇ REV MR 58#27245 • ID 76210

INWAGEN VAN, P. [1977] *Ontological arguments* (**J** 0097) Nous, Quart J Phil 11*375-395
◇ A05 B45 ◇ REV MR 58#4992 • ID 79646

INWAGEN VAN, P. [1979] *Laws and counterfactuals* (**J** 0097) Nous, Quart J Phil 13*439-453
◇ A05 B45 ◇ REV MR 80j:03028 • ID 79645

IOAN, P. [1971] *The paradoxes of implication* (**J** 0147) An Univ Bucuresti, Acta Logica 14*81-89
◇ A10 B45 ◇ REV MR 51#12482 Zbl 275#02011
• ID 17270

IRVING, T.A. & NAGLE JR., H.T. [1973] *An approach to multi-valued sequential logic* (**P** 2009) Int Symp Multi-Val Log (3);1973 Toronto 89-105
◇ B50 ◇ REV MR 52#5220 • ID 42060

ISARD, S.D. [1977] *A finitely axiomatizable undecidable extension of K* (**J** 0105) Theoria (Lund) 43*195-202
◇ B45 D05 D35 ◇ REV MR 58#27335 Zbl 391#03011
• ID 52338

ISARD, S.D. see Vol. IV for further entries

ISEKI, K. & TANAKA, S. [1966] *A pentavalued logic and its algebraic theory* (**J** 0081) Proc Japan Acad 42*763-764
◇ B50 ◇ REV MR 34#7352 Zbl 145.7 • ID 41916

ISEKI, K. [1966] *Some three valued logics and its algebraic representations* (**J** 0081) Proc Japan Acad 42*761-762
◇ B50 ◇ REV MR 35#1450 Zbl 145.7 • ID 06438

ISEKI, K. [1967] *On the classical propositional calculus of A.R.Anderson and N.D.Belnap* (**J** 0081) Proc Japan Acad 43*202-203
◇ B46 ◇ REV MR 35#5286 • ID 06443

ISEKI, K. see Vol. I, V, VI for further entries

ISEMINGER, G. [1968] *An introduction to deductive logic* (**X** 1228) Appleton-Century-Crofts: New York vii+184pp
◇ B98 ◇ ID 22497

ISEMINGER, G. [1980] *Is relevance necessary for validity?* (**J** 0094) Mind 89*196-213
◇ A05 B46 ◇ REV MR 81h:03011 • ID 74405

ISHIKAWA, K. & SASAO, T. & TERADA, H. [1982] *On simplified algorithms for equations in many-valued logic (Japanese)* (**P** 4081) Many-Val Log & Appl;1982 Kyoto 64-75
◇ B50 ◇ ID 47620

ISHIMOTO, A. [1954] *A set of axioms of the modal propositional calculus equivalent to S3* (**J** 0443) Sci Thought (Tokyo) 1*1-11
◇ B45 ◇ REV MR 16.435 JSL 20.169 • ID 06448

ISHIMOTO, A. [1956] *A formulation of the modal propositional calculus equivalent to S4* (**J** 0443) Sci Thought (Tokyo) 2*73-82
◇ B45 ◇ REV JSL 22.326 • ID 06449

ISHIMOTO, A. [1956] *A note on the paper "A set of axioms of the modal propositional calculus equivalent to S3"* (**J** 0443) Sci Thought (Tokyo) 2*69-72
◇ B45 ◇ REV JSL 22.326 • ID 06450

ISHIMOTO, A. [1978] *A bi-modal characterization of epistemic logic* (**J** 0260) Ann Jap Ass Phil Sci 5*135-155
◇ B45 ◇ REV MR 80b:03021 Zbl 398#03009 • ID 52737

ISHIMOTO, A. see Vol. I for further entries

ISHIZUKA, M. [1982] see FU, K.S.

ISHMURATOV, A.T. [1981] *Logical theories of temporal contexts. Temporal logic (Russian)* (**X** 2199) Naukova Dumka: Kiev 152pp
◇ B45 ◇ REV MR 84k:03055 • ID 34993

ISHMURATOV, A.T. [1982] *Logics with partial interpretations (Russian)* (**C** 3849) Modal & Relevant Log, Vyp 1 19–26
◇ B45 ◇ REV Zbl 528 # 03010 • ID 37628

ISRAEL, M. [1974] see ETIEMBLE, D.

ITO, MAKOTO [1955] *On the lattice of n-valued functions (theory of n-valued logics) I,II* (**J** 4442) Kyushu Daigaku Kagaku Syuho 28/2*96–99,99–101
◇ B50 ◇ REV JSL 22.100 • ID 32187

ITO, MAKOTO [1956] *General solution of 3-valued logical equations with m-variables* (**J** 4442) Kyushu Daigaku Kagaku Syuho 28/4*248–250
◇ B50 ◇ ID 32191

ITO, MAKOTO [1956] *On the general solution of the n-valued function-lattice (logical) equation in several variables (Japanese)* (**J** 4442) Kyushu Daigaku Kagaku Syuho 28*243–246
◇ B50 ◇ REV JSL 22.101 • ID 32189

ITO, MAKOTO [1959] *Modal logic and its application to sequential circuits* (**J** 4442) Kyushu Daigaku Kagaku Syuho
◇ B45 ◇ ID 32193

ITO, MAKOTO [1965] *Boolean recursive functions and closure algebra* (**P** 0614) Th Models;1963 Berkeley 431–432
◇ B25 B45 D20 ◇ REV Zbl 147.256 • ID 27530

ITO, MAKOTO see Vol. I, III, VI for further entries

ITO, MINORU [1935] *On the generation of all functions in the finitely many-valued algebra of logic (Japanese)* (**J** 4436) Osaka Shijo Danwakai 58*27–35,59*15–23,60*6–15
◇ B50 ◇ ID 48764

ITO, MINORU [1956] *On the general solution of the n-valued function-lattice (logical) equation in one variable (Japanese)* (**J** 4442) Kyushu Daigaku Kagaku Syuho 28/4*239–243
◇ B50 ◇ REV JSL 22.101 • ID 48777

ITO, MINORU see Vol. IV for further entries

ITURRIOZ, L. [1965] *Axiomas para il calculo proposicional trivalente de Lukasiewicz* (**J** 0188) Rev Union Mat Argentina 22*150
◇ B50 ◇ ID 41907

ITURRIOZ, L. [1977] *An axiom system for three-valued Lukasiewicz propositional calculus* (**J** 0047) Notre Dame J Formal Log 18*616–620
◇ B50 ◇ REV MR 58 # 5028 Zbl 336 # 02014 • ID 50470

ITURRIOZ, L. [1977] *Lukasiewicz and symmetrical Heyting algebras* (**J** 0068) Z Math Logik Grundlagen Math 23*131–136
◇ B50 G10 G20 ◇ REV MR 58 # 10437 Zbl 373 # 02042 • ID 26474

ITURRIOZ, L. [1977] *Two characteristic properties of monadic three-valued Lukasiewicz algebras* (**J** 0302) Rep Math Logic, Krakow & Katowice 8*63–69
◇ B50 G15 ◇ REV MR 58 # 10436 Zbl 382 # 03047 • ID 24190

IVAS'KIV, YU.L. & RABINOWICZ, Z.L. [1963] *A class of canonical forms for the representation of 3-valued functions* (**J** 0522) Engin Cybern 3*17
◇ B50 ◇ ID 41895

IVAS'KIV, YU.L. & POSPELOV, D.A. & TOSIC, Z. [1969] *Representations in many-valued logics (Russian) (English summary insert)* (**J** 0040) Kibernetika, Akad Nauk Ukr SSR 1969/2*35–47
◇ B50 ◇ REV MR 46 # 1461 Zbl 216.4 JSL 35.349 • ID 22012

IVAS'KIV, YU.L. [1971] *Principles of multivalued implementation schemes (Russian)* (**X** 2199) Naukova Dumka: Kiev
◇ B50 ◇ ID 41982

IVAS'KIV, YU.L. [1977] *Elementare Automaten mit einem Speicher mit k-wertigem strukturellem Alphabet (Russisch)* (**J** 0040) Kibernetika, Akad Nauk Ukr SSR 1977/1*36–46
◇ B50 ◇ REV Zbl 356 # 94047 • ID 50339

IVAS'KIV, YU.L. [1977] see AJZENBERG, N.N.

IVERT, P.A. & SJOEDIN, T. [1978] *On the impossibility of a finite propositional lattice for quantum mechanics* (**J** 2146) Helv Phys Acta 51*635–636
◇ B51 G12 ◇ REV MR 80g:06005 • ID 81677

IVIN, A.A. [1967] *Some problems of the theory of deontic modalities (Russian)* (**C** 0582) Log Semant & Modal Logika 162–232
◇ B45 ◇ REV MR 36 # 2477 Zbl 204.2 • ID 24950

IVIN, A.A. [1970] *Certain formulations of modal systems (Russian)* (**C** 1530) Issl Log Sist (Yanovskaya) 137–141
◇ B45 ◇ REV MR 47 # 6443 Zbl 217.8 • ID 06469

IVIN, A.A. [1970] *Definitions of aletic and deontic modal functors in terms of material implication and constants (Russian)* (**C** 0668) Neklass Log 113–123
◇ B45 ◇ REV MR 47 # 6442 • ID 06470

IVIN, A.A. [1970] *Temporal logic (Russian)* (**C** 0668) Neklass Log 124–190
◇ B45 ◇ REV MR 48 # 3709 • ID 06471

IVIN, A.A. [1973] *Modale Logik und die Theorie der Implikation (Russian)* (**C** 1662) Teor Log Vyvoda 151–196
◇ A05 B45 ◇ REV Zbl 272 # 02030 • ID 30379

IVLEV, YU.V. [1980] *Intuitive semantics of modal logic (Russian)* (**C** 3038) Log-Metodol Issl 356–374
◇ B45 ◇ REV MR 83m:03025 • ID 35430

IVLEV, YU.V. [1985] *Intensional semantics of modal logic (Russian)* (**X** 0898) Moskov Gos Univ: Moskva 171pp
◇ B45 ◇ ID 49010

IWAI, S. & NAKAMURA, K. [1982] *A representation of analogical inference by fuzzy sets and its application to information retrieval system* (**C** 3778) Fuzzy Inform & Decis Processes 373–386
◇ B52 E72 ◇ ID 46116

IWAI, S. & NAKAMURA, K. [1982] *Topological fuzzy sets as a quantitative description of analogical inference and its application to question-answering systems for information retrieval* (**J** 2338) IEEE Trans Syst Man & Cybern 12*193–204
◇ B52 E72 ◇ REV MR 83k:68105 • ID 46458

IWANUS, B. [1973] *On Lesniewski's elementary ontology (Polish and Russian summaries)* (**J** 0063) Studia Logica 31*73-125
◇ B25 B60 E70 ◇ REV MR 51 # 10041 Zbl 275 # 02019
• ID 17553

IWANUS, B. [1974] *Concerning the so-called system BD of Aqvist* (**J** 0063) Studia Logica 33*339-343
◇ B45 ◇ REV MR 51 # 62 Zbl 306 # 02020 • ID 15229

IWANUS, B. see Vol. I, III for further entries

IWIN, A.A. [1975] *Grundlagen der Logik von Wertungen* (**X** 0911) Akademie Verlag: Berlin 319pp
◇ A05 B45 ◇ REV Zbl 217.302 • ID 29675

IZUMI, K. [1983] see ASAI, K.

IZUMI, K. [1984] see ASAI, K.

JACKSON, F.C. [1977] see ELLIS, B.

JACKSON, F.C. [1977] see DAVIDSON, B.

JACKSON, F.C. & PARGETTER, R. [1982] *Physical probability as a propensity* (**J** 0097) Nous, Quart J Phil 16*567-583
◇ A05 B48 ◇ REV MR 84c:03015 • ID 35679

JACOBS, B.E. [1980] see ARONSON, A.R.

JACOBS, B.E. see Vol. IV for further entries

JACOBSON, A. & KLEMKE, E.D. & ZABECH, F. (EDS.) [1974] *Readings in semantics* (**X** 1285) Univ Ill Pr: Urbana v+853pp
◇ A05 B98 ◇ ID 48637

JACOBSON JR., J.H. [1963] see BLYTH, J.W.

JADCZYK, A.Z. [1971] *Quantum logic and indefinite metric spaces* (**J** 1546) Rep Math Phys (Warsaw) 1*285-297
◇ B51 G12 ◇ REV MR 44 # 2443 Zbl 219 # 02017
• ID 28081

JADCZYK, A.Z. [1976] see CEGLA, W.

JADCZYK, A.Z. [1977] see CEGLA, W.

JAGER, T. [1982] *An actualistic semantics for quantified modal logic* (**J** 0047) Notre Dame J Formal Log 23*335-349
◇ B45 ◇ REV MR 83m:03026 • ID 35431

JAHN, K.-U. [1975] *Eine auf den Intervall-Zahlen fussende 3-wertige lineare Algebra* (**J** 0114) Math Nachr 65*105-116
◇ B50 ◇ REV MR 58 # 24894 Zbl 301 # 15002 • ID 81684

JAHN, K.-U. [1975] *Intervall-wertige Mengen* (**J** 0114) Math Nachr 68*115-132
◇ B50 E70 ◇ REV MR 52 # 10371 Zbl 317 # 02075
• ID 21673

JAHN, K.-U. [1975] *Mehrwertige Mengenlehre und Klassifikationsprozesse* (**J** 0129) Elektr Informationsverarbeitung & Kybern 11*622
◇ B50 E70 ◇ REV Zbl 316 # 02067 • ID 62724

JAHN, K.-U. see Vol. V for further entries

JAMES, E.W. [1975] *On preserving entailment* (**J** 0094) Mind 84*443-449
◇ B46 ◇ REV MR 56 # 92 • ID 74460

JAMES, P.A. [1961] see ALLEN, L.E.

JAMES, P.A. [1965] see DICKOFF, J.W.

JAMES, P.A. see Vol. I for further entries

JAMMER, M. [1982] *A note on Peter Gibbins':"A note on quantum logic and the uncertainty principle"* (**J** 0153) Phil of Sci (East Lansing) 49*478-479
◇ B51 ◇ REV MR 84f:81008 • ID 39669

JANCZEWSKI, L.J. & SMITH, KENNETH C. & VRANESIC, Z.G. [1971] *Circuit implementations of multivalued logic* (**P** 2007) Symp Th & Appl of Multi-Val Log Design;1971 Buffalo 133-139
◇ B50 ◇ ID 41994

JANCZEWSKI, L.J. [1973] *Geometrical approach to multivalued logic function synthesis* (**P** 2009) Int Symp Multi-Val Log (3);1973 Toronto 106-118
◇ B50 ◇ REV MR 52 # 2761 • ID 42061

JANICKA-ZUK, I. [1980] *Finite axiomatization for some intermediate logics* (**J** 0063) Studia Logica 39*415-423
◇ B55 ◇ REV MR 82g:03041 Zbl 488 # 03015 • ID 74462

JANICKA-ZUK, I. see Vol. VI for further entries

JANIN, P. [1969] *Bases de l'algebrisation de la logique trivalente* (**J** 0447) An Univ Bucuresti, Mat 18*57-88
◇ B50 ◇ REV MR 42 # 5785 Zbl 206.276 • ID 06511

JANKOWSKI, A.W. [1985] *Disjunctions in closure spaces* (**J** 0063) Studia Logica 44*11-24
◇ B45 E75 ◇ ID 47520

JANKOWSKI, A.W. see Vol. I, III, V, VI for further entries

JANTKE, K.P. [1984] *The main proof theoretic problems in inductive inference* (**P** 3621) Frege Konferenz (2);1984 Schwerin 321-330
◇ B48 ◇ REV MR 85m:03006 • ID 45400

JANTKE, K.P. see Vol. IV for further entries

JASKOWSKI, S. [1936] *Recherches sur le systeme de la logique intuitionniste* (**P** 0632) Congr Int Phil des Sci;1935 Paris 6*58-61
• TRANSL [1967] (**C** 0615) Polish Logic 1920-39 259-263 (English) [1975] (**J** 0063) Studia Logica 34*117-120 (English)
◇ B55 F50 ◇ REV MR 51 # 7832 Zbl 168.247 Zbl 313 # 02014 JSL 2.55 JSL 35.442 FdM 62.1045
• ID 06544

JASKOWSKI, S. [1948] *Un calcul des propositions pour les systemes deductifs contradictoires (Polish) (French summary)* (**J** 0451) Studia Soc Sci Torunensis Sect A 1*57-74,75-77
◇ B53 ◇ REV MR 10.175 Zbl 41.351 JSL 14.66 • ID 33684

JASKOWSKI, S. [1949] *Sur la conjonction discursive dans le calcul des propositions pour les systemes contradictoires (Polish) (French summary)* (**J** 0451) Studia Soc Sci Torunensis Sect A 1*171-172
◇ B53 ◇ REV JSL 18.345 • ID 06550

JASKOWSKI, S. [1955] see GRZEGORCZYK, A.

JASKOWSKI, S. [1969] *Propositional calculus for contradictory deductive systems (French summary)* (**J** 0063) Studia Logica 24*143-160
◇ B53 ◇ REV MR 40 # 2521 Zbl 244 # 02004 • ID 06559

JASKOWSKI, S. see Vol. I, IV, V for further entries

JAUCH, J.M. [1968] *Foundation of quantum theory* (**X** 0832) Addison-Wesley: Reading xii+299pp
◇ B51 G12 ◇ REV MR 36 # 1151 Zbl 166.233 • ID 25415

JAUCH, J.M. & PIRON, C. [1970] *What is "quantum-logic"?* (**C** 4318) Quant, Essay in Th Phys (Wentzel) 166-181
◇ B51 ◇ REV JSL 48.206 • ID 47409

JAUCH, J.M. see Vol. I for further entries

JEFFCOTT, B. [1972] *The center of an orthologic* (**J** 0036) J Symb Logic 37*641-645
◇ B51 G12 G25 ◇ REV MR 48 # 189 Zbl 259 # 02026 • ID 06584

JEFFCOTT, B. [1974] *A note on commutivity in the composite product of orthologics* (**J** 0246) J Nat Sci Math 14*137-148
◇ B51 ◇ REV MR 55 # 7869 Zbl 354 # 02026 • ID 50166

JEFFREY, R.C. [1965] *The logic of decision* (**X** 0822) McGraw-Hill: New York xiv + 201pp
• LAST ED [1983] (**X** 0862) Univ Chicago Pr: Chicago xiv + 231pp
◇ B48 B98 ◇ REV MR 38 # 1770 MR 85g:62009 Zbl 167.471 • ID 25416

JEFFREY, R.C. [1967] *Formal logic: Its scope and limits* (**X** 0822) McGraw-Hill: New York xii + 238pp
◇ B98 ◇ REV JSL 38.646 JSL 49.1408 • REM 2nd ed. 1981; xvi + 198pp • ID 47400

JEFFREY, R.C. [1971] see CARNAP, R.

JEFFREY, R.C. [1972] *Carnap's inductive logic* (**J** 0154) Synthese 25*299-306
◇ B48 ◇ REV MR 57 # 15991 Zbl 265 # 02022 • ID 29820

JEFFREY, R.C. [1974] see BOOLOS, G.

JEFFREY, R.C. [1978] *Axiomatizing the logic of decision* (**P** 2931) Found & Appl of Decision Th;1975 London ON I*227-231
◇ B48 ◇ REV MR 81e:90006 Zbl 417 # 90006 • ID 53286

JEFFREY, R.C. (ED.) [1980] *Studies in inductive logic and probability II* (**X** 0926) Univ Calif Pr: Berkeley i + 305pp
◇ B48 ◇ REV MR 81j:03008 Zbl 539 # 03004 JSL 49.1409 • ID 70043

JEFFREYS, H. [1931] *Scientific inference* (**X** 0805) Cambridge Univ Pr: Cambridge, GB vi + 247pp
◇ A05 B98 ◇ REV JSL 29.194 • ID 25417

JEFFREYS, H. [1942] *Does a contradiction entail every proposition?* (**J** 0094) Mind 51*90
◇ B53 ◇ ID 42847

JEFFREYS, H. [1983] *Theory of probability* (**X** 0815) Clarendon Pr: Oxford xi + 459pp
◇ B48 ◇ REV MR 85f:60005 Zbl 531 # 60003 • ID 39910

JENNINGS, R.E. [1974] *A utilitarian semantics for deontic logic* (**J** 0122) J Philos Logic 3*445-456
◇ B45 ◇ REV MR 57 # 15966 Zbl 296 # 02015 • ID 62758

JENNINGS, R.E. & SCHOTCH, P.K. [1979] *Multiple-valued consistency* (**P** 3003) Int Symp Multi-Val Log (9);1979 Bath 110-113
◇ A05 B50 ◇ REV MR 81a:03027 • ID 78237

JENNINGS, R.E. & SCHOTCH, P.K. [1980] *Inference and necessity* (**J** 0122) J Philos Logic 9*327-340
◇ A05 B45 ◇ REV MR 83j:03036 Zbl 442 # 03020 • ID 56377

JENNINGS, R.E. & JOHNSTON, D.K. & SCHOTCH, P.K. [1980] *Universal first-order definability in modal logic* (**J** 0068) Z Math Logik Grundlagen Math 26*327-330
◇ B45 ◇ REV MR 81h:03038 Zbl 467 # 03013 • ID 55011

JENNINGS, R.E. [1981] *A note on the axiomatisation of Brouwersche modal logic* (**J** 0122) J Philos Logic 10*341-343
◇ B45 ◇ REV MR 83a:03020 Zbl 466 # 03007 • ID 54958

JENNINGS, R.E. & SCHOTCH, P.K. [1981] *Epistemic logic, skepticism, and nonnormal modal logic* (**J** 0095) Philos Stud 40*47-67
◇ A05 B45 ◇ REV MR 82j:03022 • ID 78236

JENNINGS, R.E. & SCHOTCH, P.K. [1981] *Non-Kripkean deontic logic* (**C** 3731) New Studies Deontic Log 149-162
◇ B45 ◇ REV MR 84h:03045 • ID 37394

JENNINGS, R.E. & SCHOTCH, P.K. [1981] *Probabilistic considerations on modal semantics* (**J** 0047) Notre Dame J Formal Log 22*227-238
◇ A05 B45 ◇ REV MR 83j:03037 Zbl 443 # 03005 • ID 55521

JENNINGS, R.E. & SCHOTCH, P.K. [1981] *Some remarks on (weakly) weak modal logics* (**J** 0047) Notre Dame J Formal Log 22*309-314
◇ A05 B45 ◇ REV MR 84c:03036 Zbl 472 # 03014 • ID 55280

JENNINGS, R.E. & JOHNSTON, D.K. & SCHOTCH, P.K. [1981] *The n-adic first-order undefinability of the Geach formula* (**J** 0047) Notre Dame J Formal Log 22*375-378
◇ A05 B45 ◇ REV MR 83f:03017 Zbl 472 # 03015 • ID 55281

JENNINGS, R.E. & SCHOTCH, P.K. [1984] *The preservation of coherence* (**J** 0063) Studia Logica 43*89-106
◇ B45 ◇ ID 42363

JENNINGS, R.E. [1985] *Can there be a natural deontic logic?* (**J** 0154) Synthese 65*257-273
◇ B45 ◇ ID 49404

JENSEN, F.V. & MAYOH, B.H. & MOELLER, K.K. (EDS.) [1979] *Proceedings from 5th Scandinavian Logic Symposium, Aalborg, 17-19, January 1979* (**X** 2646) Aalborg Univ Pr: Aalborg vii + 361pp
◇ B97 ◇ REV MR 81m:03006 Zbl 419 # 00002 • ID 53341

JENSEN, F.V. see Vol. III, IV, V for further entries

JENSEN, J.B. & LARSEN, P.F. & SCHOTCH, P.K. [1977] *A strong completeness theorem for Lukasiewicz finitely many-valued logics* (**P** 2013) Int Symp Multi-Val Log (7);1977 Charlotte 89-91
◇ B50 ◇ REV MR 56 # 15366 • ID 78239

JENSEN, J.B. & LARSEN, P.F. & MACLELLAN, E.J. & SCHOTCH, P.K. [1978] *A note on three-valued modal logic* (**J** 0047) Notre Dame J Formal Log 19*63-68
◇ B45 ◇ REV MR 80k:03026 Zbl 332 # 02018 • ID 27078

JEROSLOW, R. [1975] *Experimental logics and Δ_2^0-theories* (**J** 0122) J Philos Logic 4*253-267
◇ B60 D55 ◇ REV MR 58 # 21558 Zbl 319 # 02024 • ID 28273

JEROSLOW, R. see Vol. III, VI for further entries

JERVELL, H.R. [1972] *An Herbrand theorem for a modal logic* (**S** 1626) Oslo Preprint Ser 5
◇ B45 ◇ ID 33264

JERVELL, H.R. see Vol. I, III, IV, V, VI for further entries

JEVONS, W.S. [1880] *Studies in deductive logic: a manual for students* (**X** 1253) Crowell Collier & Macmillan: New York xxviii + 304pp
◇ B98 ◇ ID 22566

JHA, J.S. & LAL, R.N. [1983] *n-valued logic of contacts* (**J** 1868) J Bihar Math Soc 7*15-19
◇ B50 ◇ ID 44750

JOBE, E.K. [1985] *Explanation, causality, and counterfactuals* (**J** 0153) Phil of Sci (East Lansing) 52*357-389
◇ A05 B45 ◇ ID 49406

JOBE, W.H. [1962] *Functional completeness and canonical forms in many-valued logics* (**J** 0036) J Symb Logic 27*409-422
◇ B50 ◇ REV MR 31 # 3321 Zbl 117.253 JSL 29.143
• ID 06616

JOHNSON, F.A. [1976] *A three-valued interpretation for a relevance logic* (**J** 1893) Relevance Logic Newslett 1*123-128
◇ B46 ◇ REV Zbl 345 # 02011 • ID 29772

JOHNSON, F.A. [1977] *A natural deduction relevance logic* (**J** 0387) Bull Sect Logic, Pol Acad Sci 6*164-170
◇ B46 ◇ REV MR 58 # 146 Zbl 405 # 03008 • ID 54883

JOHNSON, F.A. see Vol. I for further entries

JOHNSON WU, K. [1973] *A new approach to formalization of a logic of knowledge and belief* (**J** 0079) Logique & Anal, NS 16*513-525
◇ B45 ◇ REV MR 50 # 12670 Zbl 291 # 02019 • ID 14340

JOHNSON WU, K. [1975] *On $(C.KK^*)$ and the KK-thesis* (**J** 0122) J Philos Logic 4*91-95
◇ A05 B45 ◇ REV Zbl 307 # 02016 • ID 66162

JOHNSON WU, K. [1979] *Natural deduction rules for free logic* (**J** 0079) Logique & Anal, NS 22*435-445
◇ B60 ◇ REV MR 82f:03009 Zbl 471 # 03020 • ID 55215

JOHNSON WU, K. see Vol. I for further entries

JOHNSTON, D.K. [1980] see JENNINGS, R.E.

JOHNSTON, D.K. [1981] see JENNINGS, R.E.

JOHNSTONE, H.W. [1960] *The law of contradiction* (**J** 0079) Logique & Anal, NS 3*3-10
◇ B53 ◇ ID 42848

JOHNSTONE, P.T. [1979] *Conditions related to de Morgan's law* (**P** 2901) Appl Sheaves;1977 Durham 479-491
◇ B55 C65 C90 F50 G30 ◇ REV MR 81e:03062 Zbl 445 # 03041 • ID 56505

JOHNSTONE, P.T. see Vol. III, V, VI for further entries

JOHNSTONE JR., H.W. [1954] *Elementary deductive logic* (**X** 1253) Crowell Collier & Macmillan: New York viii + 241pp
◇ B98 ◇ REV Zbl 59.14 JSL 20.165 • ID 22499

JOHNSTONE JR., H.W. [1962] see ANDERSON, J.M.

JOHNSTONE JR., H.W. see Vol. I, III for further entries

JOJA, A. [1968] *Essai sur la modalite du jugement* (**J** 0147) An Univ Bucuresti, Acta Logica 11*5-74
◇ B45 ◇ REV MR 43 # 4642 • ID 48013

JOJA, A. [1969] *La theorie de la modalite dans le de interpretatione* (**J** 0147) An Univ Bucuresti, Acta Logica 12*235-260
◇ A10 B45 ◇ REV MR 44 # 3846 Zbl 247 # 02006
• ID 06690

JOJA, A. [1970] *The modality of judgments (Romanian)* (**S** 1613) Probl Logic (Bucharest) 2*7-78
◇ B45 ◇ ID 47814

JOJA, A. [1975] see HENKIN, L.

JONES, A.J.I. & POERN, I. [1985] *Ideality, subideality and deontic logic* (**J** 0154) Synthese 65*275-290
◇ A05 B45 ◇ ID 49408

JONES, JOHN [1982] *A formalisation of an $\aleph_0$-valued propositional calculus with variable functors* (**J** 0068) Z Math Logik Grundlagen Math 28*505-510
◇ B50 ◇ REV MR 84b:03039 Zbl 542 # 03008 • ID 35631

JONES, JOHN [1983] *A formalization of an m-valued propositional calculus with variable functors* (**J** 0068) Z Math Logik Grundlagen Math 29*377-378
◇ B50 ◇ REV MR 85d:03050 • ID 41093

JONES, JOHN [1983] *Implication and iterated implication* (**J** 0068) Z Math Logik Grundlagen Math 29*543-556
◇ B50 ◇ REV MR 85a:03032 Zbl 508 # 03011 • ID 34783

JONES, JOHN [1984] *Some propositional calculi with constant and variable functors* (**J** 0068) Z Math Logik Grundlagen Math 30*477-479
◇ B50 ◇ REV MR 86e:03024 Zbl 526 # 03008 • ID 41147

JONES, JOHN [1985] *Simplified axiom schemes for implication and iterated implication* (**J** 0068) Z Math Logik Grundlagen Math 31*31-33
◇ B50 ◇ REV Zbl 556 # 03021 • ID 42283

JONES, JOHN [1985] *The rule of procedure Re in Lukasiewicz's many-valued propositional calculi* (**J** 0047) Notre Dame J Formal Log 26*423-428
◇ B50 ◇ ID 47535

JONES, JOHN see Vol. I for further entries

JONES, ROBERT M. [1965] *Discussion: the non-reducibility of Koopman's theorems of probability in Carnaps system for MC* (**J** 0153) Phil of Sci (East Lansing) 32*368-369
◇ A05 B48 ◇ REV MR 34 # 1154 • ID 06703

JONES, ROGER [1977] *Causal anomalies and the completeness of quantum theory* (**J** 0154) Synthese 35*41-78
◇ A05 B51 G12 ◇ REV Zbl 366 # 02007 • ID 51090

JONGH DE, D.H.J. [1974] see GABBAY, D.M.

JONGH DE, D.H.J. [1980] *A class of intuitionistic connectives* (**P** 2058) Kleene Symp;1978 Madison 103-111
◇ B55 C75 C90 F50 ◇ REV MR 81m:03013 Zbl 479 # 03015 • ID 55677

JONGH DE, D.H.J. see Vol. III, IV, V, VI for further entries

JORDAN, I.B. & MOUFTAH, H.T. [1974] *Integrated circuits for ternary logic* (**P** 1385) Int Symp Multi-Val Log (4);1974 Morgantown 285-302
◇ B50 ◇ ID 42099

JORDAN, I.B. see Vol. I for further entries

JORDAN, P. [1950] *Zur Quanten-Logik* (**J** 0008) Arch Math, (Basel) 2*166-171
◇ B51 G12 ◇ REV MR 12.5 Zbl 36.296 JSL 15.283 • ID 06744

JORDAN, P. [1959] *Quantenlogik und das kommutative Gesetz* (**P** 0651) Axiomatic Method;1957 Berkeley 365-375
◇ A05 B51 G12 ◇ REV MR 22#3698 Zbl 119.249 JSL 39.353 • ID 22195

JORDAN, P. [1962] *Bemerkungen zur Quantenlogik* (**P** 1606) Colloq Math (Pascal);1962 Clermont-Ferrand 159-166
◇ B51 G12 ◇ ID 27946

JORDAN, P. see Vol. III for further entries

JORGENSEN, C.K. [1967] *The extensional classification of categorical propositions and trivalent logic* (**J** 0079) Logique & Anal, NS 10*141-156
◇ B50 ◇ REV MR 35#5293 Zbl 155.338 JSL 33.483 • ID 31788

JORGENSEN, J. [1931] *A treatise of formal logic: its evolution and main branches, with its relations to mathematics and philosophy. I,II,III* (**X** 0894) Oxford Univ Pr: Oxford xvi+266pp,vi+273pp,vi+321pp
• LAST ED (**X** 1343) Russell: New York
◇ A05 A10 B98 ◇ REV FdM 57.50 • ID 22584

JORGOVIC, M.M. [1969] *A remark concerning a base for many-valued propositional algebra (Serbo-Croatian) (English summary)* (**J** 0042) Mat Vesn, Drust Mat Fiz Astron Serb 6(21)*451-452
◇ B50 ◇ REV MR 41#1512 Zbl 187.269 • ID 41948

JOSHI, A.K. & WEISCHEDEL, R.M. [1976] *Some frills for modal tic-tac-toe: Semantics of predicate complement constructions* (**J** 0187) IEEE Trans Comp C-25*374-389
◇ B45 ◇ REV Zbl 326#68047 • ID 62811

JOSHI, A.K. see Vol. IV for further entries

JOSHI, V.M. [1982] *A counter example of Hacking against the long run rule* (**J** 0013) Brit J Phil Sci 33*287-289
◇ B48 ◇ REV MR 84j:62006 • ID 39216

JOUANNAUD, J.-P. (ED.) [1985] *Rewriting techniques and applications* (**S** 3302) Lect Notes Comput Sci 202*vi+441pp
◇ B75 B97 ◇ REV Zbl 568#00022 • ID 49794

JOUANNAUD, J.-P. see Vol. I, IV for further entries

JUHOS VON, B. [1953] *Wahrscheinlichkeitsschluesse als syntaktische Schlussformen* (**P** 0645) Int Congr Philos (11);1953 Bruxelles 5*105-108
◇ B48 ◇ REV MR 15.386 Zbl 50.134 • ID 06765

JUHOS VON, B. [1954] *Elemente der neuen Logik* (**X** 1026) Humboldt: Frankfurt & Wien 256pp
◇ B98 ◇ ID 22410

KAAZ, M.A. [1977] *Elemente der mathematischen Logik fuer den Gebrauch in Physik und Technik* (**X** 0814) Oldenbourg: Muenchen 243pp
◇ B98 ◇ REV MR 58#16076 Zbl 361#02022 • ID 50657

KAAZ, M.A. see Vol. V for further entries

KABBARA, G. [1982] *New utilization of fuzzy optimization method* (**C** 3778) Fuzzy Inform & Decis Processes 239-246
◇ B52 ◇ REV Zbl 503#90075 • ID 36679

KABULOV, A.V. [1977] *Minimization of correcting functions in three-valued logic (Russian)* (**J** 0430) Vopr Kibern (Akad Nauk Uzb SSR) 99*3-20
◇ B50 ◇ REV MR 81m:94033 • ID 81750

KABULOV, A.V. [1978] *On functions correcting collections of incorrect algorithms (Russian)* (**J** 0430) Vopr Kibern (Akad Nauk Uzb SSR) 100*3-12
◇ B50 ◇ REV MR 80a:68048 Zbl 471#03016 • ID 55211

KABULOV, A.V. [1979] *The principle of the independence of coding for the continuation of functions of k-valued logic (Russian)* (**J** 0430) Vopr Kibern (Akad Nauk Uzb SSR) 104*3-10,133
◇ B50 G10 ◇ REV MR 81g:94075 Zbl 471#03015 • ID 55210

KABULOV, A.V. & SHESTEROVA, N.A. [1981] *A representation of functions of the k-valued logic by formulas (Russian)* (**J** 0430) Vopr Kibern (Akad Nauk Uzb SSR) 116*5-11
◇ B50 ◇ REV Zbl 478#03006 • ID 55621

KABULOV, A.V. [1982] *Synthesis of bases of complete systems of logical functions (Russian)* (**J** 0024) Dokl Akad Nauk Uzb SSR 1982/4*3-5
◇ B50 ◇ REV MR 83k:03032 • ID 36170

KABULOV, A.V. [1982] *The method of invariant extension of partially defined functions of k-valued logic (Russian)* (**S** 2829) Rostocker Math Kolloq 19*69-76
◇ B50 ◇ REV MR 83j:03042 Zbl 498#03014 • ID 35347

KABULOV, A.V. see Vol. I, V for further entries

KABZINSKI, J.K. [1972] *Some derivational formalizations of propositional modal logics (Polish) (English summary)* (**S** 0458) Zesz Nauk, Prace Log, Uniw Krakow 7*7-21
◇ B45 ◇ REV MR 47#6444 • ID 06786

KABZINSKI, J.K. [1979] *On equivalential fragment of the three-valued logic of Lukasiewicz* (**J** 0387) Bull Sect Logic, Pol Acad Sci 8*182-187
◇ B50 G20 ◇ REV MR 83m:03071 Zbl 432#03037 • ID 53994

KABZINSKI, J.K. [1980] *Complete many-valued systems of propositonal logic* (**P** 4400) Hist of Log (24); Cracow 31-33
◇ B50 ◇ ID 40110

KABZINSKI, J.K. [1980] *What is the equivalence connective?* (**J** 0387) Bull Sect Logic, Pol Acad Sci 9*184-188
◇ B55 C05 G25 ◇ REV MR 82b:03065b Zbl 463#03014 • ID 54554

KABZINSKI, J.K. [1981] *Many-valued logic* (**C** 2098) Dict Log Appl Stud of Language 201-209
◇ B50 ◇ ID 40117

KABZINSKI, J.K. & POREBSKA, M. & WRONSKI, A. [1981] *On the* $\{\leftrightarrow, \sim\}$*-reduct of the intuitionistic consequence operation* (**J** 0063) Studia Logica 40*55-66
◇ B55 ◇ REV MR 82e:03023 Zbl 484#03010 • ID 74585

KABZINSKI, J.K. [1982] *Basic properties of the equivalence* (**J** 0063) Studia Logica 41*17-40
◇ B55 G25 ◇ REV MR 84h:03141 Zbl 528#03038 • ID 34327

KABZINSKI, J.K. [1984] *An axiomatization of the equivalential fragment of the three-valued logic of Lukasievicz* (**J** 0047) Notre Dame J Formal Log 25*354-355
◇ B50 ◇ REV MR 86a:03023 Zbl 571#03007 • ID 40041

KABZINSKI, J.K. see Vol. I, VI for further entries

Kac, M. & Ulam, S.M. [1968] *Mathematics and logic: Retrospect and prospects* (**X** 1334) Praeger: New York ix + 170pp
• repr [1979] (**X** 0868) Penguin Books: Harmondsworth & New York 204pp
⋄ A05 A10 B98 ⋄ rev MR 38 # 964 MR 82a:00007 Zbl 205.289 Zbl 479 # 00011 JSL 36.677 • rem Rep. abridged • id 25688

Kacprzyk, J. [1982] *Multistage decision processes in a fuzzy environment: a survey* (**C** 3778) Fuzzy Inform & Decis Processes 251-263
⋄ B52 ⋄ rev MR 84d:90004 Zbl 503 # 90004 • id 36673

Kacprzyk, J. [1983] *Multistage decision-making under fuzziness. Theory and applications* (**X** 3890) TUEV Rheinland: Koeln x + 128pp
⋄ B52 ⋄ rev MR 84d:90059 Zbl 507 # 90023 • id 37725

Kaegi-Romano, U. [1977] *Quantum logic and generalized probability theory* (**J** 0122) J Philos Logic 6*455-462
⋄ B51 G12 ⋄ rev Zbl 383 # 03044 • id 52026

Kaemmerer, W. [1972] *Ein Algorithmus zur Berechnung der Werte einer durch ihre charakteristische Zahl gegebenen n-wertigen und k-stelligen booleschen Funktion* (**J** 0129) Elektr Informationsverarbeitung & Kybern 8*523-524
⋄ B50 ⋄ rev MR 47 # 6398 Zbl 252 # 94025 • id 62823

Kaesbauer, M. & Kutschera von, F. (eds.) [1962] *Logik und Logikkalkuel* (**X** 0826) Alber: Freiburg 248pp
⋄ B98 ⋄ rev Zbl 119.242 • id 22441

Kaesbauer, M. see Vol. I for further entries

Kaga, Y. & Muraki, M. [1982] *Properties of many-valued logical functions and their structure (Japanese)* (**P** 4081) Many-Val Log & Appl;1982 Kyoto 22-39
⋄ B50 ⋄ id 47614

Kahane, H. [1969] *Logic and philosophy. A modern introduction* (**X** 0821) Wadsworth Publ: Belmont xv + 450pp
⋄ B98 ⋄ rev JSL 39.613 • id 06792

Kaklugin, B.A. [1965] *Decomposition of three-valued logical functions (Russian)* (**C** 1413) Tekh Kibernetika 373-381
⋄ B50 ⋄ rev JSL 35.348 • id 28695

Kalasek, P. [1980] see Hajek, P.

Kalicki, C. [1980] *Infinitary propositional intuitionistic logic* (**J** 0047) Notre Dame J Formal Log 21*216-228
⋄ B55 C75 F50 ⋄ rev MR 81e:03025 Zbl 336 # 02021 • id 53776

Kalicki, J. [1948] *On Tarski's matrix method* (**J** 0459) C R Soc Sci Lett Varsovie Cl 3 41*130-142
⋄ B50 ⋄ rev MR 14.344 Zbl 39.7 JSL 17.67 • id 06810

Kalicki, J. [1950] *A test for the existence of tautologies according to many-valued truth-tables* (**J** 0036) J Symb Logic 15*182-184
⋄ B50 ⋄ rev MR 12.663 Zbl 39.6 JSL 16.65 • id 06811

Kalicki, J. [1954] *On equivalent truth-tables of many-valued logics* (**J** 3420) Proc Edinburgh Math Soc, Ser 2 10*56-61
⋄ B50 ⋄ rev MR 19.3 Zbl 57.9 • id 31566

Kalicki, J. see Vol. I, III, IV for further entries

Kalinin, V.V. [1975] *Dimension functions on Boolean logics I,II (Russian)* (**J** 0031) Izv Vyssh Ucheb Zaved, Mat (Kazan) 1975/9*92-94,10*89-90
• transl [1975] (**J** 3449) Sov Math 19/9*80-82,10*80-81
⋄ B60 ⋄ rev MR 54 # 181 MR 54 # 182 Zbl 347 # 06009 • id 62836

Kalinowski, G. [1964] *Obligation derivee et logique deontique relationnelle* (**J** 0047) Notre Dame J Formal Log 5*181-190
⋄ A05 B45 ⋄ rev MR 38 # 3135 Zbl 192.32 • id 06821

Kalinowski, G. [1965] *Les themes actuels de la logique deontique. Esquisse de l'etat actuel des recherches dans la logique des normes (Polish and Russian summaries)* (**J** 0063) Studia Logica 17*75-113
⋄ B45 ⋄ rev MR 33 # 3914 Zbl 292 # 02026 • id 06822

Kalinowski, G. [1971] *Sur la syllogistique deontique de Zdzislaw Ziemba (Polish and Russian summaries)* (**J** 0063) Studia Logica 29*125-147
⋄ B45 ⋄ rev MR 46 # 1554 Zbl 267 # 02015 • id 06823

Kalinowski, G. [1972] *Einfuehrung in die Normenlogik* (**X** 2665) Athenaeum: Frankfurt xix + 162pp
⋄ A10 B45 ⋄ rev MR 57 # 5686 Zbl 295 # 02002 • id 62841

Kalinowski, G. [1977] *Du prescriptif et du descriptif en logique deontique* (**J** 0079) Logique & Anal, NS 20*317-328
⋄ B45 ⋄ rev Zbl 362 # 02013 • id 50735

Kalinowski, G. [1981] *Sur les semantiques des mondes possibles pour les systemes de logique deontique* (**J** 0079) Logique & Anal, NS 24*81-98
⋄ A05 B45 ⋄ rev MR 82k:03017 Zbl 463 # 03003 • id 54543

Kalinowski, G. [1984] *Presuppositions, verite et normes* (**C** 4065) Theorie der Normen 393-406
⋄ B45 ⋄ rev MR 85k:03015 • id 44550

Kalinowski, G. [1985] *Sur le fondement des normes et des enonces normatifs: a propos des idees de von Wright et de Castaneda* (**J** 4604) Theoria, Ser 2 (San Sebastian) 1*59-85
⋄ A05 B45 ⋄ id 48943

Kalinowski, J. [1963] *La norme, l'action et la theorie des propositions normatives (Polish and Russian summaries)* (**J** 0063) Studia Logica 14*99-111
⋄ B45 ⋄ rev MR 37 # 6164 • id 48137

Kalish, D. & Montague, R. [1964] *Logic. Techniques of formal reasoning* (**X** 0863) Harcourt: New York & London x + 350pp
⋄ B98 ⋄ rev JSL 34.641 • id 06824

Kalish, D. see Vol. I, VI for further entries

Kallen, H.M. [1951] see Henle, P.

Kalman, J.A. [1976] *Axiomatizations of logics with values in groups* (**J** 3172) J London Math Soc, Ser 2 14*193-199
⋄ B50 ⋄ rev MR 55 # 2484 Zbl 353 # 02004 • id 50087

Kalman, J.A. [1976] *Computer studies of* $T_{\rightarrow} - W - I$ (**J** 1893) Relevance Logic Newslett 1*181-188
⋄ B35 B46 ⋄ rev Zbl 346 # 02003 • id 62846

Kalman, J.A. see Vol. I, V for further entries

KALMAR, L. (ED.) [1965] *Colloque sur les fondements des mathematiques, les machines mathematiques, et leurs applications* (**X** 0928) Akad Kiado: Budapest 320pp
◇ B97 D05 D10 D97 ◇ REV MR 32 # 5493 Zbl 148.1
• ID 48630

KALMAR, L. see Vol. I, III, IV, V, VI for further entries

KALMBACH, G. [1974] *Orthomodular logic* (**J** 0068) Z Math Logik Grundlagen Math 20*395-406
◇ B51 G10 G12 ◇ REV MR 53 # 10556 Zbl 373 # 02030
• ID 06872

KALMBACH, G. [1981] *Omologic as a Hilbert type calculus* (**P** 3820) Curr Iss in Quantum Log;1979 Erice 333-340
◇ B51 ◇ REV MR 84j:03129 Zbl 537 # 03044 • ID 34719

KALUZHNIN, L.A. & POESCHEL, R. [1979] *Funktionen- und Relationenalgebren. Ein Kapitel der diskreten Mathematik* (**X** 0806) Dt Verlag Wiss: Berlin 259pp
• LAST ED [1979] (**X** 0804) Birkhaeuser: Basel 259pp
◇ B50 E07 G15 G20 G98 ◇ REV MR 81f:03075 Zbl 418 # 03044 Zbl 421 # 03049 • ID 53329

KALUZHNIN, L.A. see Vol. I, IV for further entries

KAMP, H. [1971] *Formal properties of "now"* (**J** 0105) Theoria (Lund) 37*227-273
◇ B45 ◇ REV MR 49 # 2265 Zbl 269 # 02008 • ID 06882

KAMPE DE FERIET, J. [1980] *Une interpretation des mesures de plausibilite et de credibilite au sens de G. Shafer et de la fonction d'appartenance definissant un ensemble flou de L. Zadeh* (**J** 2822) Publ UER Math Pures Appl IRMA 2*exp
◇ B52 E72 ◇ REV MR 82m:60003 • ID 81766

KAMPE DE FERIET, J. see Vol. V for further entries

KANDA, A. [1985] *Numeration models of λ-calculus* (**J** 0068) Z Math Logik Grundlagen Math 31*209-220
◇ B40 B60 C57 D45 ◇ ID 47568

KANDA, A. see Vol. IV, V, VI for further entries

KANDEL, A. [1973] *Comment on an algorithm that generates fuzzy prime implicants by Lee and Chang* (**J** 0194) Inform & Control 22*279-282
◇ B52 ◇ REV MR 48 # 5414 Zbl 295 # 94061 • ID 62858

KANDEL, A. [1973] *On the analysis of fuzzy logic* (**P** 3244) Hawaii Int Conf Syst Sci (6);1973 Honolulu 242-245
◇ B52 ◇ REV Zbl 356 # 02014 • ID 50286

KANDEL, A. [1974] *Fuzzy represantation, CNF minimization and their application to fuzzy transmission structures* (**P** 1385) Int Symp Multi-Val Log (4);1974 Morgantown 361-380
◇ B52 ◇ ID 42087

KANDEL, A. [1974] *On the minimization of incompletely specified fuzzy functions* (**J** 0194) Inform & Control 26*141-153
◇ B52 ◇ REV MR 50 # 6657 Zbl 295 # 94062 • ID 62857

KANDEL, A. [1976] *Block decomposition of imprecise models* (**P** 2975) Asilomar Conf Circ, Syst & Comput (9);1975 Pacific Grove 522-530
◇ B52 ◇ REV MR 58 # 33610 Zbl 363 # 94002 • ID 50924

KANDEL, A. [1976] *Fuzzy maps and their applications in the simplification of fuzzy switching functions* (**P** 2011) Int Symp Multi-Val Log (6);1976 Logan 189-195
◇ B52 ◇ REV MR 58 # 20848 • ID 35889

KANDEL, A. [1976] *On the decomposition of fuzzy functions* (**J** 0187) IEEE Trans Comp C-25*1124-1130
◇ B52 ◇ REV MR 56 # 11580 Zbl 363 # 94003 • ID 50925

KANDEL, A. & RICKMAN, S.M. [1977] *Column table approach for the minimization of fuzzy functions* (**J** 0191) Inform Sci 12*111-128
◇ B52 ◇ REV MR 80g:94118 Zbl 363 # 94004 • ID 50926

KANDEL, A. [1977] *Comments on comments by Lee - Author's reply* (**J** 0194) Inform & Control 35*109-113
◇ B52 ◇ REV MR 58 # 9853b Zbl 362 # 94044 • ID 50823

KANDEL, A. [1978] *On the compactification and enumeration of distinct fuzzy switching functions* (**P** 2014) Int Symp Multi-Val Log (8);1978 Rosemont 87-90
◇ B52 ◇ REV MR 81b:94084 • ID 35916

KANDEL, A. [1979] *Evaluating procedures and fuzzy logic* (**P** 3003) Int Symp Multi-Val Log (9);1979 Bath 46-52
◇ B52 ◇ ID 35939

KANDEL, A. [1979] *On the theory of fuzzy switching mechanisms (FSM's)* (**C** 3514) Adv Fuzzy Sets & Appl 247-262
◇ B52 ◇ REV MR 81j:94053 • ID 81769

KANDEL, A. [1980] *On the control and evaluation of uncertain processes* (**P** 4254) Joint Autom Control Conf;1980 San Francisco 5pp
◇ B52 ◇ REV MR 82f:62014 Zbl 496 # 93080 • ID 46077

KANDEL, A. [1980] *On the theory of fuzzy logics* (**P** 4254) Joint Autom Control Conf;1980 San Francisco 9pp
◇ B52 ◇ ID 46072

KANDEL, A. & MARGOLIS, I.B. [1981] *Absolutely dispensable prime implicants of fuzzy switching functions* (**P** 3705) Int Symp Multi-Val Log (11);1981 Oklahoma City & Norman 19-22
◇ B52 ◇ REV MR 83m:94033 Zbl 543 # 94027 • ID 45938

KANDEL, A. [1981] see CLARK, C.M.

KANDEL, A. [1981] see BEN, D.M.

KANDEL, A. [1982] see CLARK, C.M.

KANDEL, A. & LUGINBUHL, D.R. [1983] *A comparison of fuzzy switching functions and multiple-valued switching functions* (**P** 4054) Int Symp Multi-Val Log (13);1983 Kyoto 264-272
◇ B52 ◇ REV Zbl 562 # 94014 • ID 47449

KANDEL, A. [1983] see FRANCIONI, J.M.

KANDEL, A. & THUM, M. [1984] *On the complexity of growth of the number of distinct fuzzy switching functions* (**J** 2720) Fuzzy Sets Syst 13*125-137
◇ B52 ◇ REV MR 85m:94052 Zbl 552 # 94025 • ID 45288

KANDEL, A. [1985] see FLOWERS, P.L.

KANDEL, A. see Vol. V for further entries

KANE, R. [1984] *The modal ontological argument* (**J** 0094) Mind 93*336-350
◇ A05 B45 ◇ REV MR 85h:03022 • ID 43293

KANEYORI, S. & KITAHASHI, T. & TANAKA, K. [1976] *On multi-valued perfect codes* (**P** 2011) Int Symp Multi-Val Log (6);1976 Logan 169-174
◇ B50 ◇ REV MR 58 # 20787 • ID 35886

KANG, A.N.C. & SHEN, S.N.T. [1974] *On the Boolean functions over 4-element Boolean algebra* (**P** 1385) Int Symp Multi-Val Log (4);1974 Morgantown 381-392
◇ B50 ◇ ID 42088

KANGER, S. [1957] *A note on quantification and modalities* (J 0105) Theoria (Lund) 23*133-134
◇ B45 ◇ REV MR 20#3069 Zbl 83.2 JSL 34.305
• ID 06886

KANGER, S. [1957] *On the characterisation of modalities* (J 0105) Theoria (Lund) 23*152-155
◇ B45 ◇ REV Zbl 87.252 JSL 23.38 • ID 06887

KANGER, S. [1973] *Entailment* (C 1389) Modality, Morality, Probl of Sense & Nonsense (Hallden) 168-179
◇ B46 ◇ REV MR 56#8325 • ID 74651

KANGER, S. (ED.) [1975] *Proceedings of the third scandinavian logic symposium* (X 0809) North Holland: Amsterdam vi+214pp
◇ B97 ◇ REV MR 51#5256 Zbl 299#00012 • ID 70196

KANGER, S. see Vol. I, III, VI for further entries

KANTOR, P.B. [1981] *The logic of weighted queries* (J 2338) IEEE Trans Syst Man & Cybern 11*816-821
◇ B60 ◇ REV MR 83k:68102 • ID 40338

KAO, S. [1982] see GOTO, M.

KAO, S. [1983] see GOTO, M.

KAO, S. [1984] see GOTO, M.

KAO, S. see Vol. I for further entries

KAPETANOVIC, M. [1977] *On a class of sentential functions* (J 0400) Publ Inst Math, NS (Belgrade) 22(36)*127-129
◇ B50 ◇ REV MR 57#9479 Zbl 378#02006 • ID 51788

KAPETANOVIC, M. [1980] *On a many-valued sentential calculus* (J 0400) Publ Inst Math, NS (Belgrade) 27(41)*103-106
◇ B50 ◇ REV MR 82j:03028 Zbl 467#03016 • ID 55014

KAPETANOVIC, M. [1981] *A tableaux system in modal logic* (J 0400) Publ Inst Math, NS (Belgrade) 30(44)*65-67
◇ B45 ◇ REV MR 84c:03037 Zbl 494#03009 • ID 35693

KAPETANOVIC, M. see Vol. I, III for further entries

KAPLAN, DAVID [1979] *On the logic of demonstratives* (J 0122) J Philos Logic 8*81-98
◇ A05 B45 ◇ REV MR 80g:03017 Zbl 415#03004
• ID 53106

KAPLAN, DAVID see Vol. I for further entries

KARAKHANYAN, L.M. & SAPOZHENKO, A.A. [1981] *On the negative effects of eliminating unessential variables (Russian)* (J 0474) Avtom Vychis Tekh, Akad Nauk Latv SSR 1981/3*28-35
• TRANSL [1981] (J 2666) Autom Control Comput Sci 15/3*24-30
◇ B50 ◇ REV MR 84h:94026 Zbl 504#03025 • ID 36971

KARAKHANYAN, L.M. [1982] *Metric comparison of minimal (in various senses) disjunctive normal forms for partial functions of logic algebra (Russian)* (J 0071) Met Diskr Analiz (Novosibirsk) 38*19-36
◇ B50 ◇ REV MR 85e:03153 Zbl 516#94028 • ID 38600

KARAKHANYAN, L.M. see Vol. I for further entries

KARAVAEV, E.F. [1973] *The tensed qualification of propositions and theory of logical inference (Russian)* (C 1662) Teor Log Vyvoda 1*35-37
◇ B45 ◇ ID 33330

KARAVAEV, E.F. [1976] *A note on iteration of tense-logical operators (Russian)* (P 2064) All-Union Conf Math Log (4);1976 Kishinev 56
◇ B45 ◇ ID 33012

KARAVAEV, E.F. [1976] *On entering of an operator of the present tense into tense-logic (Russian)* (P 2566) All-Union Symp Log & Method of Sci (7);1976 Kiev 109-111
◇ B45 ◇ ID 33331

KARAVAEV, E.F. [1978] *Combining calculi of modal logic as a need of methodology of knowledge* (C 2567) Modal & Intentsional Log 63-65
◇ A05 B45 ◇ ID 33013

KARAVAEV, E.F. [1982] *Temporal logic as a means of realization of the logical function of the category of time (Russian)* (C 3792) Log & Filos Kategorii 121-133
◇ B45 ◇ REV MR 85e:03042 • ID 40614

KARAVAEV, E.F. [1983] *Foundations of tense-logic (Russian)* (X 0938) Leningrad Univ: Leningrad 177pp
◇ B45 ◇ ID 45152

KARPENKO, A.S. [1982] *Aristotle, Lukasiewicz and factor-semantics* (P 3800) Intens Log: Th & Appl;1979 Moskva 7-21
• TRANSL [1984] (C 4366) Modal & Intens Log & Primen Probl Metodol Nauk 107-124
◇ A05 A10 B50 ◇ REV MR 84k:03075 Zbl 525#03008
• ID 35013

KARPENKO, A.S. [1982] *Matrix logic for prime numbers and its generalization (Russian)* (C 3849) Modal & Relevant Log, Vyp 1 81-90
◇ B50 ◇ REV Zbl 528#03011 • ID 37629

KARPENKO, A.S. & SMIRNOV, V.A. (EDS.) [1982] *Modal and relevant logics (Russian)* (X 0899) Akad Nauk SSSR : Moskva 107pp
◇ B45 B97 ◇ REV Zbl 512#00010 • ID 37567

KARPENKO, A.S. [1983] *Factor semantics for n-valued logics* (J 0063) Studia Logica 42*179-185
◇ B50 ◇ REV MR 85i:03069 Zbl 549#03015 • ID 42334

KARPENKO, A.S. [1984] *Factor-semantics for n-valued logic (Russian)* (C 4366) Modal & Intens Log & Primen Probl Metodol Nauk 124-140
◇ B50 ◇ REV MR 85i:03069 • ID 46680

KARPOV, V.G. & MOSHCHENSKIJ, V.A. [1977] *Mathematical logic and discrete mathematics (Russian)* (X 1574) Vyssheyshaya Shkola: Minsk 254pp
◇ B98 E98 ◇ REV MR 57#51 Zbl 371#02001 • ID 51362

KARPOVICH, V.N. [1974] *Hypotheses introduced by reduction sentences (Russian)* (C 2578) Filos & Logika 168-176
◇ A05 B48 B65 ◇ REV MR 58#27428 • ID 74693

KARPOVSKY, M.G. & MOSKALEV, EH.S. [1970] *Realization of partially-defined boolean functions by expansion in orthogonal series (Russian)* (J 0011) Avtom Telemekh 1970/8*89-99
• TRANSL [1970] (J 0010) Autom & Remote Control 31*1278-1287
◇ B50 ◇ REV MR 46#6948 Zbl 204.322 • ID 19056

KARPOVSKY, M.G. [1981] *Spectral methods for decomposition, design, and testing of multiple-valued logical networks* (P 3705) Int Symp Multi-Val Log (11);1981 Oklahoma City & Norman 1-9
◇ B50 ◇ REV MR 83m:94033 Zbl 542#94031 • ID 43722

KARTSAKLIS, A. [1973] *On the Carnap's C-T axiomatic system of space-time and the Basri's deductive theory of space and time* (**P** 0715) Topol & Primen (2);1972 Budva 142-146
◇ A05 B48 ◇ REV MR 51 # 9729 Zbl 279 # 02004
• ID 29536

KARTTUNEN, L. [1974] *Presuppositions of compound sentences* (**J** 3676) Ling Inquiry 4*169-193
◇ B45 ◇ ID 42089

KASAHARA, Y. & KITAHASHI, T. & NOMURA, H. & TEZUKA, Y. [1969] *The characterizing parameters of ternary logical functions and their applications to threshold logical functions* (**J** 0116) Electr & Comm Japan 52*195-201
◇ B50 ◇ REV MR 42 # 5781 • ID 07129

KASAHARA, Y. & KITAHASHI, T. & NOMURA, H. & TEZUKA, Y. [1970] *Monotonicity of ternary logical functions and its applications to the analysis of ternary threshold functions (Japanese)* (**J** 0979) Denshi Tsushin Gakkai Ronbunshi, Sect A-D 53-C/1*73-79
• TRANSL [1970] (**J** 0464) Syst-Comp-Controls 1/1*55-63
◇ B50 ◇ REV MR 48 # 56 • ID 07132

KASHER, A. [1967] see GABBAY, D.M.

KASHER, A. (ED.) [1976] *Language in focus: foundations, methods and systems. Essays in memory of Yehoshua Bar-Hillel* (**X** 0835) Reidel: Dordrecht xxviii + 679pp
◇ B97 ◇ REV Zbl 367 # 00005 • ID 51156

KASHER, A. see Vol. III for further entries

KATERINOCHKINA, N.N. [1975] *Location of the maximal upper zero of a monotonic logic-algebra function (Russian)* (**J** 0023) Dokl Akad Nauk SSSR 224*557-560
• TRANSL [1975] (**J** 0470) Sov Phys, Dokl 20*610-611
◇ B50 ◇ REV MR 52 # 10256 Zbl 335 # 02004 • ID 62906

KATERINOCHKINA, N.N. [1977] *Search for a maximal upper zero for a class of monotone functions in k-valued logic (Russian)* (**J** 0023) Dokl Akad Nauk SSSR 234*746-749
• TRANSL [1977] (**J** 0062) Sov Math, Dokl 18*751-754
◇ B50 ◇ REV MR 56 # 11748 Zbl 392 # 03027 • ID 74704

KATERINOCHKINA, N.N. [1981] *Location of the maximal upper zero for a certain class of monotone functions of k-valued logic (Russian)* (**J** 0199) Zh Vychisl Mat i Mat Fiz 21*470-481,527
• TRANSL [1981] (**J** 1049) USSR Comput Math & Math Phys 21/2*212-224
◇ B50 ◇ REV MR 82g:03036 Zbl 479 # 03014 • ID 55676

KATERINOCHKINA, N.N. [1984] *Some relations for subsets of layers of an n-dimensional k-valued lattice (Russian) (English summary)* (**J** 0199) Zh Vychisl Mat i Mat Fiz 24*782-786
• TRANSL [1984] (**J** 1049) USSR Comput Math & Math Phys 24/5*103-105
◇ B50 G10 ◇ REV MR 86h:05021 Zbl 566 # 03011
• ID 48734

KATERINOCHKINA, N.N. see Vol. I, IV for further entries

KATZ, JERROLD J. [1962] *The problem of induction and its solution* (**X** 0862) Univ Chicago Pr: Chicago xiii + 125pp
◇ B48 ◇ REV JSL 36.320 • ID 06972

KATZ, M. [1980] *Inexact geometry* (**J** 0047) Notre Dame J Formal Log 21*521-535
◇ B50 ◇ REV MR 82c:03016 Zbl 416 # 03027 • ID 53920

KATZ, M. [1981] *Lukasiewicz logic and the foundations of measurement* (**J** 0063) Studia Logica 40*209-225
◇ B50 ◇ REV MR 84b:03040 Zbl 484 # 03008 • ID 35632

KATZ, M. [1981] *Two systems of many-valued logic for science* (**P** 3705) Int Symp Multi-Val Log (11);1981 Oklahoma City & Norman 175-182
◇ B50 ◇ REV MR 83m:94033 Zbl 549 # 03019 • ID 43111

KATZ, M. [1982] *Measures of proximity and dominance* (**P** 3845) Conf Math Service of Man (2,Proc)(Feriet);1982 Las Palmas 370-377
◇ B50 ◇ REV Zbl 505 # 03011 • ID 38133

KATZ, M. [1982] *Metric similarities in the logic of approximation* (**J** 2840) Stochastica, Univ Politec Barcelona 6*265-281
◇ B48 ◇ REV MR 85d:03051 Zbl 516 # 03015 • ID 37251

KATZ, M. [1982] *The logic of approximation in quantum theory* (**J** 0122) J Philos Logic 11*215-228
◇ B51 ◇ REV MR 83m:03070 Zbl 497 # 03013 • ID 35469

KATZ, M. [1983] *Quotient algebras for logics of imprecision* (**P** 4054) Int Symp Multi-Val Log (13);1983 Kyoto 42-46
◇ B50 ◇ REV Zbl 562 # 03038 • ID 47434

KATZ, M. [1983] *Towards a formal multi-valued utility theory* (**P** 4054) Int Symp Multi-Val Log (13);1983 Kyoto 219-221
◇ B50 ◇ REV Zbl 558 # 90016 • ID 46638

KATZ, M. [1984] *Controlled-error theories of proximity and dominance* (**P** 3096) Conf Math Service of Man (2,Pap);1982 Las Palmas 51-74
◇ B50 ◇ REV MR 86g:03041 Zbl 542 # 03009 • ID 43672

KATZ, M. see Vol. III for further entries

KAUFMANN, A. [1973] *Introduction a la theorie des sous-ensembles flous a l'usage des ingenieurs. Tome I: Elements theoriques de base* (**X** 1752) Masson: Paris xxi + 410pp
• TRANSL [1975] (**X** 0801) Academic Pr: New York xvi + 416pp [1982] (**X** 2686) CECSA: Mexico City 491pp (Spanish) [1982] (**X** 3775) Radio i Svyaz: Moskva 432pp (Russian)
◇ B52 E70 E72 E98 ◇ REV MR 58 # 5245 MR 84e:03064a MR 84e:03065 Zbl 302 # 02023 • REM 2nd ed. 1977 • ID 62912

KAUFMANN, A. [1975] see COOLS, M.

KAUFMANN, A. see Vol. V for further entries

KAWAI, H. [1981] *On the completeness theorem for bimodal systems LT_Ω and LET_Ω* (**J** 0352) Math Jap 26*377-383
◇ B45 ◇ REV MR 83j:03033 Zbl 498 # 03013 • ID 35342

KAWAI, H. [1982] *Eine Logik erster Stufe mit einem infinitaeren Zeitoperator* (**J** 0068) Z Math Logik Grundlagen Math 28*173-180
◇ B45 C40 C75 C90 ◇ REV MR 83h:03023 Zbl 511 # 03006 • ID 36043

KAYSER, D. [1980] *Vers une modelisation du raisonnement "approximatif"* (**P** 3580) Repr des Conn & Raison dans Sci Homme;1979 St Maximin 440-456
◇ B48 ◇ REV MR 82m:00017 • ID 81813

KEARNS, J.T. [1968] *The logical concept of existence* (**J** 0047) Notre Dame J Formal Log 9*313-324
◇ A05 B60 ◇ REV MR 39 # 2596 Zbl 181.4 • ID 06981

KEARNS, J.T. [1969] *Deductive logic: a programmed introduction* (**X** 1228) Appleton-Century-Crofts: New York
◇ B98 ◇ ID 22500

KEARNS, J.T. [1979] *The strong completeness of a system for Kleene's three-valued logic* (**J** 0068) Z Math Logik Grundlagen Math 25*61-68
◇ B50 ◇ REV MR 80d:03020 Zbl 407 # 03030 • ID 56196

KEARNS, J.T. [1981] *Modal semantics without possible worlds* (**J** 0036) J Symb Logic 46*77-86
◇ B45 ◇ REV MR 82f:03015 Zbl 479 # 03011 • ID 55673

KEARNS, J.T. see Vol. I, III, VI for further entries

KEDROV, B.M. [1962] *Operating with scientific concepts in dialectical and formal logic (Russian)* (**C** 4691) Dialektika & Logika, Formy Mysl 42-141
◇ A05 B53 ◇ REV JSL 30.372 • ID 43064

KEDROV, B.M. see Vol. V for further entries

KEENE, G.B. [1964] *First-order functional calculus* (**X** 0866) Routledge & Kegan Paul: Henley on Thames vi + 82pp
◇ B98 ◇ REV JSL 36.167 • ID 06993

KEENE, G.B. [1976] *The logical analysis of universal conditionals* (**J** 0162) Teorema (Valencia) 6*427-433
◇ B45 ◇ REV MR 55 # 12473 • ID 74751

KEENE, G.B. see Vol. I, V for further entries

KEIM, R.G. [1972] see CHISHOLM, R.M.

KEISLER, H.J. [1962] see CHANG, C.C.

KEISLER, H.J. [1966] see CHANG, C.C.

KEISLER, H.J. [1977] *Hyperfinite model theory* (**P** 1075) Logic Colloq;1976 Oxford 5-110
◇ B50 C65 C75 C80 C90 H20 ◇ REV MR 58 # 10421 Zbl 423 # 03041 • ID 16614

KEISLER, H.J. [1980] see BARWISE, J.

KEISLER, H.J. see Vol. I, III, IV, V, VI for further entries

KELLER, P. [1984] *Zum Verhaeltnis von Zeitlogik und Modallogik* (**P** 3621) Frege Konferenz (2);1984 Schwerin 129-135
◇ B45 ◇ REV MR 85m:03006 Zbl 551 # 03010 • ID 43889

KEMENY, J.G. [1955] *Fair bets and inductive probabilities* (**J** 0036) J Symb Logic 20*263-273
◇ B48 ◇ REV MR 17.633 Zbl 66.110 JSL 33.481 • ID 07052

KEMENY, J.G. [1956] *Semantics as a branch of logic* (**C** 0647) Encycl Britannica 20*313,313A-313D
◇ B98 ◇ REV JSL 23.23 • ID 16857

KEMENY, J.G. see Vol. I, III, VI for further entries

KENELLY, J.W. [1967] *Informal logic* (**X** 0802) Allyn & Bacon: London viii + 134pp
◇ B98 ◇ REV Zbl 162.11 • ID 22502

KERGALL, E. [1974] *Monomes premiers d'une fonction de plusieurs variables sur une algebre a p valeurs. Trois methodes de determination* (**J** 2313) C R Acad Sci, Paris, Ser A-B 278*A1483-A1485
◇ B50 ◇ REV MR 49 # 10622 Zbl 361 # 08006 • ID 07063

KERNTOPF, P. [1979] *On finding roots on finite-valued functions* (**J** 2845) Tanulmanyok 99*19-35
◇ B50 ◇ REV MR 81h:94032 Zbl 486 # 68002 • ID 81828

KERNTOPF, P. see Vol. I for further entries

KERRE, E.E. [1985] see CALUWE DE, R.M.M.

KESSLER, G. [1975] *A note on future branching time* (**J** 0105) Theoria (Lund) 41*89-95
◇ B45 ◇ REV Zbl 331 # 02010 • ID 62943

KESSLER, G. [1978] *Mathematics and modality* (**J** 0097) Nous, Quart J Phil 12*421-441
◇ A05 B45 ◇ REV MR 80c:00014 • ID 81832

KEUTH, H. [1973] *Comments on Dr. Vetter's paper "Deontic logic without deontic operators"* (**J** 0472) Theory Decis 3*298-310
◇ B45 ◇ REV Zbl 337 # 02017 • ID 62948

KEYNES, J.N. [1884] *Studies and exercises in formal logic, including a generalisation of logical processes in their application to complex inferences* (**X** 1253) Crowell Collier & Macmillan: New York xii + 414pp
◇ B98 ◇ REV JSL 29.135 • ID 22551

KFOURY, A.J. [1985] *Definability by deterministic and nondeterministic programs (with applications to first-order dynamic logic)* (**J** 0194) Inform & Control 65*98-121
◇ B75 ◇ ID 49499

KFOURY, A.J. see Vol. I, IV for further entries

KHARIN, N.N. [1963] *Mathematical logic and the theory of sets (Russian)* (**X** 1649) Rusvozizdat: Leningrad 192pp
◇ A05 B98 ◇ REV MR 30 # 10 • ID 42867

KHARSHING, L. [1980] *Criteria for acceptance of scientific hypotheses (Russian)* (**C** 3038) Log-Metodol Issl 52-70
◇ B48 ◇ REV MR 83i:03039 • ID 35506

KHASIN, L.S. [1969] *The realization of monotone symmetric functions by formulae in the base* $\vee$, $\wedge$, $\neg$ *(Russian)* (**J** 0052) Probl Kibern 21*253-257
• TRANSL [1969] (**J** 0471) Syst Th Res 21*254-259
◇ B05 B50 ◇ REV MR 44 # 1575 Zbl 203.490 • ID 05733

KHASIN, L.S. see Vol. I, IV for further entries

KHOMICH, V.I. [1976] *A separability theorem for superintuitionistic propositional calculi (Russian)* (**J** 0023) Dokl Akad Nauk SSSR 229*1327-1329
• TRANSL [1976] (**J** 0062) Sov Math, Dokl 17*1214-1216
◇ B55 ◇ REV MR 54 # 2427 Zbl 376 # 02008 • ID 51168

KHOMICH, V.I. [1976] *A separability theorem for a certain pre-truth table superintuitionistic propositional logic (Russian)* (**S** 0554) Issl Teor Algor & Mat Logik (Moskva) 2*140-153,160
◇ B55 ◇ REV MR 58 # 16170 • ID 74242

KHOMICH, V.I. [1979] *Separability of superintuitionistic propositional calculi (Russian)* (**S** 0554) Issl Teor Algor & Mat Logik (Moskva) 3*98-114
◇ B20 B55 ◇ REV MR 82d:03039 Zbl 422 # 03007 Zbl 464 # 03025 • ID 53468

KHOMICH, V.I. & MARKOV, A.A. (EDS.) [1979] *Studies in the theory of algorithms and mathematical logic (Russian)* (**X** 2027) Nauka: Moskva 135pp
◇ B97 D97 ◇ REV MR 81b:03003 • ID 70052

KHOMICH, V.I. [1980] *On the problem of separability for superintuitionistic propositional logics (Russian)* (**J** 0023) Dokl Akad Nauk SSSR 254*820-823
• TRANSL [1980] (**J** 0062) Sov Math, Dokl 22*476-479
◇ B55 ◇ REV MR 82a:03023 Zbl 464 # 03025 • ID 74241

KHOMICH, V.I. see Vol. I, IV, VI for further entries

KHROMOJ, YA.V. [1978] *A collection of exercises and problems in mathematical logic (Ukrainian)* (**X** 2645) Vishcha Shkola: Kiev 159pp
◇ B98 ◇ REV MR 58 # 16075 • ID 74291

KICKERT, W.J.M. & MAMDANI, E.H. [1978] *Analysis of a fuzzy logic controller* (**J** 2720) Fuzzy Sets Syst 1*29-44
◇ B52 ◇ REV Zbl 364 # 93022 • ID 50994

KICKERT, W.J.M. see Vol. V for further entries

KIELAK, W. [1976] *ENE-logic* (**J** 0387) Bull Sect Logic, Pol Acad Sci 5*84-86
◇ B60 ◇ REV MR 55 # 78 • ID 27113

KIELKOPF, C.F. [1971] *Provability as a deontic notion* (**J** 0472) Theory Decis 2*1-15
◇ B45 ◇ REV MR 46 # 1555 Zbl 267 # 02016 • ID 07083

KIELKOPF, C.F. [1971] *Semantics for a utilitarian deontic logic* (**J** 0079) Logique & Anal, NS 14*783-802 • ERR/ADD ibid 15*667
◇ B45 ◇ REV MR 47 # 3144 Zbl 238 # 02025 • ID 19099

KIELKOPF, C.F. [1972] *Kripke's axiomatization of S2* (**J** 0047) Notre Dame J Formal Log 13*379-380
◇ B45 ◇ REV MR 46 # 5108 Zbl 238 # 02021 JSL 38.661 • ID 07084

KIELKOPF, C.F. [1974] *K1 as a Dawson modeling of A.R.Anderson's sense of "ought"* (**J** 0047) Notre Dame J Formal Log 15*402-410
◇ B45 ◇ REV MR 50 # 9535 Zbl 281 # 02030 • ID 07085

KIELKOPF, C.F. [1975] *Adjunction and paradoxical derivations* (**J** 0103) Analysis (Oxford) 35*127-129
◇ B45 ◇ REV Zbl 362 # 02012 • ID 50734

KIELKOPF, C.F. [1975] *Kant's deontic logic* (**J** 0302) Rep Math Logic, Krakow & Katowice 5*43-51
• REPR [1976] (**J** 0286) Int Logic Rev 13*66-75
◇ A05 B45 ◇ REV MR 58 # 27336a Zbl 335 # 02010 • ID 21900

KIELKOPF, C.F. [1977] *Formal sentential entailment* (**X** 1946) Univ Pr Amer: Lanham
◇ B46 ◇ ID 42850

KIELKOPF, C.F. [1978] *A favorable mark for strict implication* (**J** 0079) Logique & Anal, NS 21*249-252
◇ A05 B46 ◇ REV MR 80d:03016 Zbl 403 # 03016 • ID 54730

KIELKOPF, C.F. [1982] *A completeness proof for Porte's S_a^0 and S_a* (**J** 0079) Logique & Anal, NS 25*435-441
◇ B45 ◇ REV MR 85g:03029 Zbl 512 # 03012 • ID 36508

KIELKOPF, C.F. see Vol. I for further entries

KILGORE, W.J. [1968] *An introductory logic* (**X** 0818) Holt Rinehart & Winston: New York xiv + 352pp
◇ B98 ◇ ID 22503

KILMISTER, C.W. [1967] *Language, logic and mathematics* (**X** 1165) Hodder & Stroughton: London v + 124pp
◇ A05 B98 D35 ◇ REV MR 37 # 1223 Zbl 162.11 • ID 22338

KINBER, E.B. [1982] see FREJVALD, R.V.

KINBER, E.B. [1984] see FREJVALD, R.V.

KINBER, E.B. see Vol. IV for further entries

KIRIN, V.G. [1963] *On the polynomial representation of operators in the n-valued propositional calculus* (**J** 0371) Glas Mat-Fiz Astron, Ser 2 (Zagreb) 18*3-12
◇ B50 ◇ REV MR 29 # 2184 Zbl 134.8 JSL 37.756 • ID 41894

KIRIN, V.G. [1966] *Gentzen's method for the many-valued propositional calculi* (**J** 0068) Z Math Logik Grundlagen Math 12*317-332
◇ B50 ◇ REV MR 34 # 1170 Zbl 147.250 JSL 32.538 • ID 07113

KIRIN, V.G. [1968] *Post algebras as semantic bases of some many-valued logics* (**J** 0027) Fund Math 63*279-294
◇ B50 G20 ◇ REV MR 39 # 1302 Zbl 175.266 • ID 07114

KIRIN, V.G. see Vol. I, V for further entries

KIRK, R.E. [1979] *Some classes of Kripke frames characteristic for the intuitionistic logic* (**J** 0068) Z Math Logik Grundlagen Math 25*409-410
◇ B55 C90 F50 ◇ REV MR 81b:03011 Zbl 423 # 03042 • ID 53554

KIRK, R.E. [1980] *A characterization of the classes of finite tree frames which are adequate for the intuitionistic logic* (**J** 0068) Z Math Logik Grundlagen Math 26*497-501
◇ B55 C90 F50 ◇ REV MR 81m:03014 Zbl 446 # 03017 • ID 56546

KIRK, R.E. see Vol. I for further entries

KIROV, K.A. [1982] see GARGOV, G.K.

KIROV, K.A. [1984] *An intuitionistic analogue of the modal logic of provability in Peano's arithmetic* (**P** 4392) Mat Logika (Markova);1980 Sofia 65-72
◇ B45 F30 ◇ ID 46974

KISZKA, J.B. [1983] see BRUDZY, K.

KISZKA, J.B. & KOCHANSKA, M.E. & SLIWINSKA, D.S. [1985] *The influence of some fuzzy implication operators on the accuracy of a fuzzy model. I,II* (**J** 2720) Fuzzy Sets Syst 15*111-128,223-240
◇ B52 ◇ ID 45526

KISZKA, J.B. see Vol. V for further entries

KITAHASHI, T. [1969] see KASAHARA, Y.

KITAHASHI, T. [1970] see KASAHARA, Y.

KITAHASHI, T. & TANAKA, K. [1971] *An orthogonal expansion of multiple-valued logical functions* (**P** 2007) Symp Th & Appl of Multi-Val Log Design;1971 Buffalo 54-61
◇ B50 ◇ ID 41983

KITAHASHI, T. [1971] see FUJITA, Y.

KITAHASHI, T. [1972] see FUJITA, Y.

KITAHASHI, T. [1976] see KANEYORI, S.

KITAHASHI, T. see Vol. I for further entries

KITTEL, P.M. [1980] see DARSOW, W.

KLAUA, D. [1965] *Ueber einen Ansatz zur mehrwertigen Mengenlehre* (**J** 0342) Monatsber Dt Akad Wiss 7*859-867
◇ B50 E70 ◇ REV MR 33 # 1231 Zbl 154.260 • ID 07147

KLAUA, D. [1973] *Zur Arithmetik mehrwertiger Zahlen* (**J** 0114) Math Nachr 57*275-306
◇ B50 ◇ REV MR 50 # 4243 Zbl 251 # 02058 • ID 07159

KLAUA, D. [1978] *Partielle boolesche Algebren in der Intervallmathematik* (J 0114) Math Nachr 83*311-336
◇ B50 G05 ◇ REV MR 81f:06014a Zbl 403 # 03049 • REM Part I. Part II 1979 • ID 54762

KLAUA, D. [1979] *Partielle boolesche Algebren in der Intervallmathematik. II* (J 0114) Math Nachr 88*141-173
◇ B50 G05 ◇ REV MR 81f:06014b Zbl 434 # 03043 • REM Part I 1978 • ID 55747

KLAUA, D. see Vol. I, III, V, VI for further entries

KLAUS, G. [1964] *Moderne Logik. Abriss der formalen Logik* (X 0806) Dt Verlag Wiss: Berlin xii + 452pp
◇ B98 ◇ ID 14564

KLEENE, S.C. [1939] see EVANS, H.P.

KLEENE, S.C. [1939] *On the term "analytic" in logical syntax* (J 0748) Erkenntnis (Leipzig) 9*189-192
◇ A05 B60 ◇ REV JSL 5.157 • ID 30932

KLEENE, S.C. [1952] *Introduction to metamathematics* (X 0809) North Holland: Amsterdam x + 550pp
• TRANSL [1957] (X 1656) Izdat Inostr Lit: Moskva 526pp (Russian) [1984] (X 1876) Kexue Chubanshe: Beijing xii + 234pp,x + 235-688pp (Chinese) [1974] (X 1781) Tecnos: Madrid (Spanish)
◇ A05 B98 D98 F30 F50 F98 ◇ REV MR 14.525 Zbl 47.7 JSL 19.215 JSL 25.280 JSL 33.290 JSL 35.350 JSL 38.333 • REM Co-publisher: Wolters-Noordhoff; 8th revised ed. 1980. Chinese transl. in 2 parts • ID 07173

KLEENE, S.C. [1960] *Mathematical logic: constructive and non-constructive operations* (P 0660) Int Congr Math (II, 8);1958 Edinburgh 137-153 • ERR/ADD [1963] (J 0064) Trans Amer Math Soc 108*142
◇ B98 D98 ◇ REV MR 22 # 5569 Zbl 126.19 JSL 27.78 • ID 07186

KLEENE, S.C. [1967] *Mathematical logic* (X 0827) Wiley & Sons: New York xiii + 398pp
• TRANSL [1967] (X 3636) Tokyo Tosho: Tokyo 200pp + 266pp [1971] (X 0850) Colin: Paris [1973] (X 0885) Mir: Moskva 480pp
◇ B98 D20 D25 D55 D98 F30 F98 ◇ REV MR 36 # 25 Zbl 149.243 JSL 35.438 • ID 45895

KLEENE, S.C. [1969] *The new logic* (J 1843) Amer Sci 57*333-347
◇ B98 ◇ ID 32371

KLEENE, S.C. see Vol. I, IV, VI for further entries

KLEIN, E. [1979] *On formalizing the referential/attributive distinction* (J 0122) J Philos Logic 8*333-337
◇ A05 B45 ◇ REV MR 80i:03009 • ID 74854

KLEMENT, E.P. & LOWEN, R. & SCHWYHLA, W. [1981] *Fuzzy probability measures* (J 2720) Fuzzy Sets Syst 5*21-30
◇ B52 E72 ◇ REV MR 83g:28014 Zbl 447 # 28005 • ID 56587

KLEMENT, E.P. [1984] see HOEHLE, U.

KLEMENT, E.P. see Vol. V for further entries

KLEMKE, D. [1971] *Ein Henkin-Beweis fuer die Vollstaendigkeit eines Kalkuels relativ zur Grzegorczyk-Semantik* (J 0009) Arch Math Logik Grundlagenforsch 14*148-161
◇ B55 C90 F50 ◇ REV MR 47 # 8253 Zbl 233 # 02020 • ID 07229

KLEMKE, E.D. [1974] see JACOBSON, A.

KLESHCHEV, A. [1984] *Frames (Russian)* (P 4288) Predst Znan Chelov-Mash & Robot Sist 122-135
◇ B45 ◇ ID 46587

KLEVACHEV, V.I. [1970] *Certain systems that are complete in P_k (Russian) (English summary)* (J 0040) Kibernetika, Akad Nauk Ukr SSR 1970/5*139-140
◇ B50 ◇ REV MR 45 # 3169 Zbl 218 # 02015 JSL 40.465 • ID 07230

KLING, R. [1974] *Fuzzy-PLANNER: Reasoning with inexact concepts in a procedural problem-solving language* (J 0383) J Cybern 4*105-122
◇ B52 ◇ REV MR 56 # 13823 Zbl 304 # 68085 • ID 63000

KLIR, G.J. [1982] see CAVALLO, R.E.

KLIR, G.J. see Vol. V for further entries

KLOETZER, G. & RAUTENBERG, W. [1972] *Im Grenzbereich Algebra, Logik, Maschinen. 10 Jahre Forschungsarbeit der Nowosibirsker Schule A.I. Malcev's (eine Studie)* (J 1670) Mitt Math Ges DDR 1972/1-2*43-102
◇ A10 B98 C05 D05 D98 ◇ REV Zbl 243 # 02001 • ID 30031

KLOSS, B.M. & NECHIPORUK, EH.I. [1963] *Ueber eine Klassifikation der Funktionen der mehrwertigen Logik (Russisch)* (J 0052) Probl Kibern 9*27-36
• TRANSL [1963] (J 0449) Probl Kybern 6*36-46
◇ B50 ◇ REV Zbl 192.83 • ID 16247

KLOSS, B.M. see Vol. IV for further entries

KLUKOWSKI, J. [1977] *Operators in multi-valued logic* (J 1008) Demonstr Math (Warsaw) 10*329-338
◇ B50 G20 ◇ REV MR 58 # 21439 Zbl 401 # 03006 • ID 74889

KNAPP, H.G. [1974] *Bemerkungen zu "Normative Wissenschaft ohne Normen -- aber mit Werten" von M. Bunge* (J 2688) Conceptus (Wien) 8*57-62
◇ B45 ◇ REV MR 57 # 67 • REM The article was published ibid. 7(1973)/21-22*57-64 • ID 74891

KNEEBONE, G.T. [1963] *Mathematical logic and the foundations of mathematics: an introduction* (X 0864) Van Nostrand: New York xiv + 435pp
◇ A05 A10 B98 ◇ REV MR 27 # 26 Zbl 166.247 • ID 19088

KNEEBONE, G.T. [1971] *Mathematical logic in relation to ordinary mathematics* (J 0178) Stud Gen 24*946-959
◇ A05 B98 ◇ REV Zbl 245 # 02005 • ID 28859

KNEEBONE, G.T. see Vol, V for further entries

KNOEBEL, A. [1978] *Further conclusions on functional completeness* (J 0027) Fund Math 99*93-112
◇ B50 ◇ REV MR 58 # 5447 Zbl 382 # 08005 • ID 81882

KNOEBEL, A. [1982] *The finite interpolation property for small sets of classical polynomials* (J 0002) Acta Sci Math (Szeged) 44*287-297
◇ B50 C05 ◇ REV MR 85f:03024 Zbl 535 # 08001 • ID 40672

KNOX JR., J. [1971] *Material implication and "if...then"* (J 0286) Int Logic Rev 3*90-92
◇ A05 B46 ◇ REV JSL 37.185 • ID 07262

KNUTH, D.E. [1979] see ERSHOV, A.P.

KNUTH, D.E. see Vol. IV, V for further entries

KNUUTTILA, S. (ED.) [1981] *Reforging the great chain of being* (**X** 0835) Reidel: Dordrecht xiv + 320pp
◇ A05 B45 ◇ REV MR 83a:03010 • ID 47733

KOCHANSKA, M.E. [1985] see KISZKA, J.B.

KOCHEN, S. & SPECKER, E. [1965] *Logical structures arising in quantum theory* (**P** 0614) Th Models;1963 Berkeley 177-189 • REPR [1975] (**C** 3045) Log-Algeb Appr to Quant Mech 1*263-276
◇ B51 G12 ◇ REV MR 34 # 5409 Zbl 171.254 JSL 50.558 • ID 27525

KOCHEN, S. see Vol. I, III, V for further entries

KOCZY, L.T. [1977] see HAJNAL, M.

KOCZY, L.T. [1979] *Some questions of b-algebras of fuzzy objects of type N* (**P** 3432) Prog in Cybern & Syst Res;1972 Wien V*536-541
◇ B52 E72 G25 ◇ REV MR 80j:94055 Zbl 409 # 03037 • ID 56339

KOCZY, L.T. see Vol. V for further entries

KODANDAPANI, K.L. & SETLUR, R.V. [1974] *Multi-valued algebraic generalizations of Reed-Muller canonical forms* (**P** 1385) Int Symp Multi-Val Log (4);1974 Morgantown 505-527
◇ B50 ◇ ID 42090

KODANDAPANI, K.L. & SETLUR, R.V. [1975] *Reed-Muller canonical forms in multivalued logic* (**J** 0187) IEEE Trans Comp C-24*628-636
◇ B50 ◇ REV MR 51 # 12429 Zbl 325 # 02009 • ID 17504

KOEHLER, P. [1983] see BLOK, W.J.

KOFLER, E. & MENGES, G. [1976] *Cognitive decisions under partial information* (**C** 1702) Local Induction 183-189
◇ B48 ◇ REV Zbl 352 # 02021 • ID 50020

KOGA, Y. & MINE, H. [1966] *A property of three-valued functions of one variable* (**J** 0116) Electr & Comm Japan 49*128-130
◇ B50 ◇ REV MR 35 # 5295 • ID 09298

KOHN, R.V. [1975] *Generalization of a result of Hallden* (**J** 0047) Notre Dame J Formal Log 16*605-606
◇ B45 ◇ REV MR 52 # 5361 Zbl 292 # 02023 • ID 18229

KOHN, R.V. [1977] *Some Post-complete extensions of S2 and S3* (**J** 0047) Notre Dame J Formal Log 18*467-470
◇ B45 ◇ REV MR 58 # 16152 Zbl 292 # 02022 • ID 23642

KOHOUT, L.J. [1974] *The Pinkava many-valued complete logic systems and their applications in the design of many-valued switching circuits* (**P** 1385) Int Symp Multi-Val Log (4);1974 Morgantown 261-284
◇ B50 ◇ ID 42091

KOHOUT, L.J. [1975] see GAINES, B.R.

KOHOUT, L.J. [1977] see GAINES, B.R.

KOHOUT, L.J. [1978] *Analysis of computing protection structures by means of multi-valued logic systems* (**P** 2014) Int Symp Multi-Val Log (8);1978 Rosemont 260-268
◇ B50 ◇ REV MR 81d:68007 • ID 81900

KOHOUT, L.J. [1979] see BANDLER, W.

KOHOUT, L.J. [1980] see BANDLER, W.

KOHOUT, L.J. & PINKAVA, V. [1980] *The algebraic structure of the Spencer Brown and Varela calculi* (**J** 1743) Int J Gen Syst 6*155-171
◇ B60 G25 ◇ REV MR 83c:00012 Zbl 441 # 03026 • ID 56079

KOHOUT, L.J. [1982] see BANDLER, W.

KOHOUT, L.J. [1983] see BANDLER, W.

KOHOUT, L.J. see Vol. I, IV, V for further entries

KOKOSZYNSKA, M. [1957] *On "good" and "bad" induction (Polish) (Russian and English summaries)* (**J** 0063) Studia Logica 5*43-70 • ERR/ADD ibid 5*5
◇ B48 ◇ ID 33076

KOKOSZYNSKA, M. see Vol. I for further entries

KOLESNIKOV, M.A. & SHEJNBERGAS, I.M. [1971] *Sheffer functions in four valued logic (Russian)* (**C** 1499) Raboty Tekhn Kibern 3*50-100
◇ B50 ◇ REV MR 48 # 10762 • ID 28626

KOLETSOS, G. [1981] *Notational and logical completness in three-valued logic* (**J** 0465) Bull Greek Math Soc (NS) 22*121-141
◇ B50 ◇ ID 49353

KOLETSOS, G. see Vol. IV, VI for further entries

KOL'MAN, EH. [1959] *On material and formal implication (Russian)* (**P** 0607) All-Union Math Conf (3);1956 Moskva 4*86
◇ B46 ◇ ID 30600

KOLPAKOV, V.I. [1974] *Subalgebras of monotone functions in Post algebras (Russian)* (**J** 0071) Met Diskr Analiz (Novosibirsk) 24*30-45,96
◇ B50 G20 ◇ REV MR 51 # 2861 Zbl 301 # 02062 • ID 22043

KOLPAKOV, V.I. see Vol. I for further entries

KOMORI, Y. [1975] *The finite model property of the intermediate propositional logics on finite slices* (**J** 2332) J Fac Sci, Univ Tokyo, Sect 1 A 22*117-120
◇ B55 ◇ REV MR 57 # 5693 Zbl 317 # 02021 • ID 29638

KOMORI, Y. [1978] *Logics without Craig's interpolation property* (**J** 3239) Proc Japan Acad, Ser A 54*46-48
◇ B55 C40 ◇ REV MR 58 # 10328 Zbl 399 # 03048 • ID 52844

KOMORI, Y. [1978] *Super-Lukasiewicz implicational logics* (**J** 0111) Nagoya Math J 72*127-133
◇ B50 ◇ REV MR 80d:03021 Zbl 363 # 02015 • ID 52184

KOMORI, Y. [1978] *The separation theorem of the $\aleph_0$-valued Lukasiewicz propositional logic* (**J** 1005) Rep Fac Sci, Shizuoka Univ 12*1-5
◇ B50 ◇ REV MR 80k:03024 Zbl 377 # 02021 • ID 51738

KOMORI, Y. [1981] *Super-Lukasiewicz proposotional logics* (**J** 0111) Nagoya Math J 84*119-133
◇ B55 ◇ REV MR 83i:03041 Zbl 482 # 03007 • ID 35508

KOMORI, Y. & ONO, H. [1982] *Logics without the contradiction rule* (**P** 3833) Symp Semigroups (6);1982 Kyoto 78-79
◇ B20 B53 ◇ REV Zbl 501 # 03007 • ID 37775

KOMORI, Y. [1983] *Some results on the super-intuitionistic predicate logics* (**J** 0302) Rep Math Logic, Krakow & Katowice 15*13-31
◇ B55 ◇ REV MR 84j:03059 Zbl 568 # 03013 • ID 34651

KOMORI, Y. see Vol. I, III for further entries

KONDAKOV, N.I. [1978] *Woerterbuch der Logik* (**X** 2600) DEB - Verlag: Berlin 554pp
◇ A05 A10 B98 ◇ REV Zbl 382 # 03001 • ID 51925

KONDO, M. [1983] *Completeness theorem for some logic in terms of interval semantics* (**J** 0407) Comm Math Univ St Pauli (Tokyo) 32*195-202
◇ B45 ◇ REV MR 86f:03031 Zbl 524 # 03012 • ID 37593

KONDO, M. [1984] *A note on a Brown's conjecture* (**J** 0407) Comm Math Univ St Pauli (Tokyo) 33*219-221
◇ B45 ◇ REV MR 86a:03017 Zbl 551 # 03009 • ID 43886

KONDO, M. [1984] *The completeness theorems for some intuitionistic epistemic logics in terms of interval semantics* (**J** 0390) Publ Res Inst Math Sci (Kyoto) 20*671-681
◇ B45 F50 ◇ REV MR 85i:03052 • ID 44181

KONDO, M. see Vol. I, IV, V, VI for further entries

KONG, YOUKUN & ZHANG, WENGQIAN [1982] *A fuzzy control model for weather forecasting (Chinese) (English summary)* (**J** 3732) Mohu Shuxue 2/4*91-105
◇ B52 ◇ ID 46099

KONIOR, J. [1980] see DUBIKAJTIS, L.

KONIOR, J. [1981] see ACHTELIK, G.

KONONOV, B.P. [1977] *The distinction relation: duality semantics (Russian) (English and German summaries)* (**J** 0338) Nauch-Tekh Inf, Ser 2, Akad Nauk SSSR 1977/9*8-12,31,32
◇ B60 ◇ REV MR 57 # 12214 Zbl 381 # 03020 • ID 51888

KONONOV, B.P. [1978] *Eine stationaere Logik zur Entscheidungsfindung (Russian) (German and English summaries)* (**J** 0338) Nauch-Tekh Inf, Ser 2, Akad Nauk SSSR 1978/9*18-22
◇ B60 ◇ REV MR 80b:90170 Zbl 432 # 03016 • ID 53973

KOPPEL, M. [1985] *Towards a new theory of confirmation* (**J** 0029) Israel J Math 50*207-218
◇ B48 ◇ ID 47216

KOPPEL, M. see Vol. III for further entries

KOPPELBERG, S. [1985] *Booleschwertige Logik* (**J** 0157) Jbuchber Dtsch Math-Ver 87*19-38
◇ B50 C90 C98 E35 E45 E50 E65 G05 ◇ REV MR 86e:03039 • ID 39673

KOPPELBERG, S. also published under the name GOERNEMANN, S.

KOPPELBERG, S. see Vol. III, V for further entries

KORDIG, C.R. [1975] *Relativized deontic modalities* (**C** 1856) Log Enterprise 221-233
◇ B45 ◇ REV Zbl 357 # 02019 • ID 50367

KORDIG, C.R. [1981] *A deontic argument for God's existence* (**J** 0097) Nous, Quart J Phil 15*207-208
◇ A05 B45 ◇ REV MR 82i:03009 • ID 74978

KOREN, T. & PNUELI, A. [1984] *There exist decidable context free propositional dynamic logics* (**P** 2989) Log of Progr; 1983 Pittsburgh 290-312
◇ B75 ◇ REV Zbl 542 # 68020 • ID 43709

KORFHAGE, R. [1966] *Logic and algorithms with applications to the computer and information sciences* (**X** 0827) Wiley & Sons: New York xii + 194pp
◇ B98 D98 ◇ REV MR 35 # 5279 Zbl 148.7 JSL 36.344 • ID 07376

KORNEJCHUK, V.I. & TARASENKO, V.P. & TESLENKO, A.K. [1972] *Synthesis of combination schemes in a certain class of multivalued algebras (Russian)* (**J** 0474) Avtom Vychis Tekh, Akad Nauk Latv SSR 1972/5*9-13
• TRANSL [1972] (**J** 2666) Autom Control Comput Sci 6/5*8-12
◇ B50 ◇ REV MR 47 # 8202 Zbl 243 # 94024 • ID 63057

KOROVIN, S.YA. & MELIKHOV, A.N. & TSELYKH, A.N. [1981] *Some principles of program realization for fuzzy control algorithms (Russian)* (**J** 2743) Izv Sev-Kavk Nauch Tsentra, Tekh (Rostov nD) 1981/3*9-13,101
◇ B52 ◇ REV MR 82m:94074 • ID 82230

KOSOK, M. [1966] *The formalization of Hegel's dialectical logic* (**J** 2340) Int Phil Quart 6*596-631
◇ A05 B53 ◇ ID 42851

KOSOVSKIJ, N.K. [1981] *Elements of mathematical logic and its applications to the theory of subrecursive algorithms (Russian)* (**X** 0938) Leningrad Univ: Leningrad 192pp
◇ B98 D20 D98 F30 F60 F98 ◇ REV MR 83m:03013 Zbl 479 # 03001 • ID 55663

KOSOVSKIJ, N.K. see Vol. I, III, IV, VI for further entries

KOSTYANKO, N.F. [1974] see BOBROV, A.E.

KOSTYUK, V.N. [1974] *The theory of inductive reasoning and its formalization (Russian)* (**C** 2578) Filos & Logika 279-302
◇ A05 B48 ◇ REV MR 58 # 27360 • ID 75000

KOSTYUK, V.N. [1975] *The theory of inductive reasoning (Russian)* (**S** 2847) Uch Zap Univ Tomsk 90/8*11-23
◇ A05 B48 ◇ REV MR 57 # 12162 • ID 75001

KOSTYUK, V.N. [1980] *Explanation, prediction and comprehension (Russian)* (**C** 2638) Log-Gnoseolog Issl Kategorial Strukt Myshleniya 246-261
◇ A05 B48 ◇ REV MR 82g:03009 • ID 74999

KOTAHASHI, T. [1976] see GAINES, B.R.

KOTARBINSKI, T. [1929] *Elements of epistemology, formal logic, and methodology of the sciences (Polish)* (**X** 3333) Unknown Publisher: See Remarks 483pp
• LAST ED [1961] (**X** 2885) Zakl Narod Wyd Pol Ak: Wroclaw 648pp
◇ A05 B98 ◇ REM Lwow • ID 37375

KOTAS, J. [1963] *Axiom for Birkhoff-v.Neumann quantum logic* (**J** 0014) Bull Acad Pol Sci, Ser Math Astron Phys 11*629-632
◇ B51 G12 ◇ REV MR 28 # 2060 JSL 40.463 • ID 07396

KOTAS, J. [1967] *An axiom system for the modular logic (Polish and Russian summaries)* (**J** 0063) Studia Logica 21*17-38
◇ B51 G12 ◇ REV MR 37 # 115 Zbl 333 # 02023 JSL 48.206 • ID 07399

KOTAS, J. [1971] *On the algebra of classes of formulae of Jaskowski's discussive system (Polish and Russian summaries)* (**J** 0063) Studia Logica 27*81-91
◇ B53 G15 ◇ REV MR 46 # 52 Zbl 286 # 02066 • ID 07405

KOTAS, J. [1973] *The axiomatization of S.Jaskowski's discussive systems* (**J** 0387) Bull Sect Logic, Pol Acad Sci 2*12-17
◇ B53 ◇ REV MR 52 # 7853 Zbl 295 # 02015 • ID 24811

KOTAS, J. [1974] *On quantity of logical values in the discussive D_2 system and in modular logic* (**J** 0063) Studia Logica 33*273-275
◇ B53 G12 ◇ REV MR 50 # 9547 Zbl 317 # 02032 • ID 07410

KOTAS, J. [1974] *The axiomatization of S.Jaskowski's discussive system* (**J** 0063) Studia Logica 33*195-200
◇ B53 ◇ REV MR 51 # 64 Zbl 295 # 02015 • ID 15228

KOTAS, J. [1975] *Discussive sentential calculus of Jaskowski* (**J** 0063) Studia Logica 34*149-168
◇ B53 ◇ REV MR 52 # 5362 Zbl 315 # 02033 • ID 18233

KOTAS, J. (ED.) [1975] *Stanislaw Jaskowski's achievements in mathematical logic* (**J** 0063) Studia Logica 34*107-214
◇ A10 B97 ◇ REV MR 51 # 7800 • ID 70194

KOTAS, J. [1977] see COSTA DA, N.C.A.

KOTAS, J. [1979] see COSTA DA, N.C.A.

KOTAS, J. [1980] see COSTA DA, N.C.A.

KOTAS, J. see Vol. I, III, VI for further entries

KOTS, V.S. [1971] *Synthesis of schemes of k-valued logic (Russian)* (**J** 0474) Avtom Vychis Tekh, Akad Nauk Latv SSR 1971/3*9-13
• TRANSL [1971] (**J** 2666) Autom Control Comput Sci 5/3*8-12
◇ B50 ◇ REV MR 46 # 9214 • ID 07270

KOTS, V.S. see Vol. I for further entries

KOVALEV, M.M. & MILANOV, P. [1984] *Monotone functions of many-valued logic and supermatroids (Russian)* (**J** 0199) Zh Vychisl Mat i Mat Fiz 24*786-790
◇ B50 ◇ REV MR 85f:03025 • ID 40682

KOZEN, D. [1980] *A representation theorem for models of *-free PDL* (**P** 2904) Automata, Lang & Progr (7);1980 Noordwijkerhout 351-362
◇ B75 G25 ◇ REV MR 82c:03077 Zbl 451 # 03005 • ID 54020

KOZEN, D. & PARIKH, R. [1981] *An elementary proof of the completeness of PDL* (**J** 1426) Theor Comput Sci 14*113-118
◇ B75 ◇ REV MR 82d:03028 Zbl 451 # 03006 JSL 51.225 • ID 54021

KOZEN, D. [1981] *On the duality of dynamic algebras and Kripke models* (**P** 3497) Log of Progr;1979 Zuerich 1-11
◇ B75 C90 ◇ REV MR 85b:03042 Zbl 482 # 03008 • ID 36840

KOZEN, D. [1982] *On induction vs. *-continuity* (**P** 3738) Log of Progr;1981 Yorktown Heights 167-176
◇ B75 ◇ REV MR 83j:03108 Zbl 495 # 03013 • ID 34869

KOZEN, D. see Vol. I, III, IV, VI for further entries

KRABBE, E.C.W. [1978] *Note on a completeness theorem in the theory of counterfactuals* (**J** 0122) J Philos Logic 7*91-93
◇ A05 B45 ◇ REV MR 57 # 12177 Zbl 378 # 02008 • ID 51790

KRABBE, E.C.W. [1978] *The adequacy of material dialogue-games* (**J** 0047) Notre Dame J Formal Log 19*321-330
◇ A05 B60 C07 ◇ REV MR 58 # 10341 Zbl 316 # 02011 • ID 51481

KRABBE, E.C.W. [1982] *Formal dialectics as immanent criticism of philosophical systems* (**P** 3754) Argumentation;1978 Groningen 233-243
◇ A05 B53 ◇ ID 45908

KRABBE, E.C.W. [1985] *Noncumulative dialectical models and formal dialectics* (**J** 0122) J Philos Logic 14*129-168
◇ A05 B46 F99 ◇ REV Zbl 571 # 03003 • ID 42654

KRAJEWSKI, S. [1982] *On relatedness logic of Richard L. Epstein* (**J** 0387) Bull Sect Logic, Pol Acad Sci 11*24-30
◇ B60 ◇ REV Zbl 526 # 03009 • ID 38167

KRAJEWSKI, S. see Vol. III, IV, V, VI for further entries

KRAJICEK, J. [1984] *Modal set theory* (**S** 3382) Sem-ber, Humboldt-Univ Berlin, Sekt Math 60*87-99
◇ B45 E35 E70 ◇ REV Zbl 552 # 03014 • ID 43351

KRAJICEK, J. see Vol. VI for further entries

KRAMER, H.P. [1962] *A note on the self-consistency of definitions of generalization and inductive inference* (**J** 0037) ACM J 9*280-281
◇ B48 ◇ REV MR 24 # B2521 Zbl 109.103 • ID 48039

KRAMOSIL, I. [1977] *A classification of inconsistent theories* (**J** 0387) Bull Sect Logic, Pol Acad Sci 6*35-41
◇ B53 ◇ REV MR 55 # 7761 Zbl 415 # 68001 • ID 53162

KRAMOSIL, I. [1978] *A boolean-valued probability theory* (**J** 0156) Kybernetika (Prague) 14*5-17
◇ B50 G05 ◇ REV MR 58 # 220 Zbl 383 # 60002 • ID 75053

KRAMOSIL, I. & SINDELAR, J. [1978] *Statistical deducibility testing with stochastic parameters* (**J** 0156) Kybernetika (Prague) 14*385-396
◇ B25 B48 ◇ REV MR 81b:68114 Zbl 403 # 62014 • ID 54777

KRAMOSIL, I. see Vol. I, IV, VI for further entries

KRASZEWSKI, Z. [1956] *Logic of extensional relations (Polish, Russian) (English summary)* (**J** 0063) Studia Logica 4*63-87,88-116
◇ B60 ◇ REV Zbl 74.247 • ID 33071

KRASZEWSKI, Z. see Vol. I, V for further entries

KRATZER, A. [1981] *Partition and revision: the semantics of counterfactuals* (**J** 0122) J Philos Logic 10*201-216
◇ A05 B45 ◇ REV MR 82m:03016 Zbl 473 # 03014 • ID 55345

KRAUSKOPF, R. [1969] *Ein Entscheidungskalkuel fuer den Lewisschen Modalkalkuel S4* (**J** 0068) Z Math Logik Grundlagen Math 15*193-210
◇ B45 ◇ REV MR 40 # 2526 Zbl 186.7 JSL 86.7 • ID 07458

KRAUSS, P.H. & SCOTT, D.S. [1966] *Assigning probabilities to logical formulas* (**C** 1107) Aspects Inductive Log 219-264
◇ B48 C75 C90 ◇ REV Zbl 202.299 • ID 27709

KRAUSS, P.H. see Vol. III for further entries

KRAUT, R. [1983] *There are no de dicto attitudes* (**J** 0154) Synthese 54*275-294
◇ B45 ◇ REV MR 84h:03037 • ID 34246

KRAWIETZ, W. & SCHELSKY, H. & SCHRAMM, A. & WINKLER, G. (EDS.) [1984] *Theorie der Normen* (**X** 1258) Duncker & Humblot: Berlin xiii + 627pp
◇ A05 B45 ◇ REV MR 85k:03015 • ID 44529

KRECZMAR, A. [1971] *The set of all tautologies of algorithmic logic is hyperarithmetical* (**J** 0014) Bull Acad Pol Sci, Ser Math Astron Phys 19*781-783
◇ B75 D35 D55 ◇ REV MR 46 # 10248 Zbl 221 # 02030 • ID 28106

KRECZMAR, A. [1972] *Degree of recursive unsolvability of algorithmic logic (Russian summary)* (**J** 0014) Bull Acad Pol Sci, Ser Math Astron Phys 20*615-617
◇ B75 D30 D35 ◇ REV MR 47 # 4776 Zbl 245 # 02038 • ID 07466

KRECZMAR, A. [1974] *Effectivity problems of algorithmic logic* (**P** 1869) Automata, Lang & Progr (2);1974 Saarbruecken 584-600
◇ B75 ◇ REV MR 54 # 12509 Zbl 294 # 68017 • ID 75072

KRECZMAR, A. [1977] *Effectivity problems of algorithmic logic* (**J** 2095) Fund Inform, Ann Soc Math Pol, Ser 4 1*19-32
◇ B75 ◇ REV MR 56 # 7300 Zbl 361 # 02056 • ID 50691

KRECZMAR, A. see Vol. III, IV for further entries

KRECZMAR, L. [1977] see BANACHOWSKI, L.

KREISEL, G. [1954] *Applications of mathematical logic to various branches of mathematics* (**P** 0646) Appl Sci de Log Math;1952 Paris 37-49
◇ B98 C65 ◇ REV MR 16.782 Zbl 57.245 JSL 24.236 • ID 16973

KREISEL, G. & PUTNAM, H. [1957] *Eine Unableitbarkeitsbeweismethode fuer den intuitionistischen Aussagenkalkuel* (**J** 0009) Arch Math Logik Grundlagenforsch 3*74-78
◇ B55 F50 ◇ REV MR 19.934 Zbl 79.7 JSL 23.229 • ID 07492

KREISEL, G. [1965] *Mathematical logic* (**C** 1602) Lect on Modern Math 3*95-195
• TRANSL [1965] (**C** 4652) Lect on Modern Math (Japanese) 128-291
◇ B98 F10 F35 F50 F98 ◇ REV MR 31 # 2124 Zbl 147.247 JSL 32.419 • ID 27577

KREISEL, G. & KRIVINE, J.-L. [1967] *Elements de logique mathematique. Theorie des modeles* (**X** 0856) Dunod: Paris viii+213pp
• TRANSL [1972] (**X** 0811) Springer: Heidelberg & New York xv+274pp [1967] (**X** 0809) North Holland: Amsterdam xi+222pp (English)
◇ B98 C98 ◇ REV MR 34 # 7331 MR 36 # 2463 MR 50 # 4231 Zbl 146.7 Zbl 238 # 02003 JSL 34.112 • ID 22411

KREISEL, G. [1975] *Was hat die Logik in den letzten 25 Jahren fur die Mathematik geleistet?* (**J** 2688) Conceptus (Wien) 9*40-45
◇ A05 B98 C75 D40 ◇ REV MR 58 # 21356 • ID 75092

KREISEL, G. [1985] *Mathematical logic: tool and object lesson for science* (**J** 0154) Synthese 62*139-151
◇ A05 B98 ◇ ID 45516

KREISEL, G. see Vol. I, III, IV, V, VI for further entries

KREMPA, J. & MAZBIC-KULMA, B. [1977] *Elements of logic, set theory and algebra (Elementy logiki, teorii mnogosci i algebry) (Polish)* (**X** 2880) Wydawn Nauk Techn: Warsaw 170pp
◇ B98 E98 ◇ REV MR 58 # 4968 Zbl 383 # 03001 • ID 51983

KREMPA, J. see Vol. IV for further entries

KRETZMANN, N. [1965] *Elements of formal logic* (**X** 1238) Bobbs-Merril: Indianapolis xi+243pp
◇ B98 ◇ ID 22505

KREYCHE, R.J. [1954] *Logic for undergraduates* (**X** 1257) Dryden Pr: Hinsdale 308pp
• LAST ED [1970] (**X** 0818) Holt Rinehart & Winston: New York xi+224pp
◇ B98 ◇ ID 22506

KRICHEVSKIJ, R.E. [1959] *Realization of functions by superpositions (Russian)* (**J** 0052) Probl Kibern 2*123-138
• TRANSL [1959] (**J** 1195) Probl Cybernet 2*458-477 [1963] (**J** 0449) Probl Kybern 2*139-159 (German)
◇ B50 B70 ◇ REV MR 23 # B1616 Zbl 131.11 JSL 37.626 • ID 21291

KRICHEVSKIJ, R.E. see Vol. IV for further entries

KRIPKE, S.A. [1959] *A completeness theorem in modal logic* (**J** 0036) J Symb Logic 24*1-14
• TRANSL [1974] (**C** 4138) Feys: Modal Logika
◇ B45 ◇ REV MR 22 # 1514 Zbl 91.9 JSL 31.276 • ID 07554

KRIPKE, S.A. [1962] *The undecidability of monadic modal quantification theory* (**J** 0068) Z Math Logik Grundlagen Math 8*113-116
• TRANSL [1974] (**C** 4138) Feys: Modal Logika
◇ B45 D35 ◇ REV MR 28 # 2975 Zbl 111.11 JSL 31.277 • ID 07555

KRIPKE, S.A. [1963] *Semantical analysis of modal logic I. Normal modal propositional calculi* (**J** 0068) Z Math Logik Grundlagen Math 9*67-96
• TRANSL [1974] (**C** 4138) Feys: Modal Logika
◇ B45 G15 ◇ REV MR 26 # 3579 Zbl 118.13 JSL 31.120 • REM Part II 1965 • ID 07558

KRIPKE, S.A. [1963] *Semantical considerations on modal logic* (**J** 0096) Acta Philos Fenn 16*83-94
◇ B45 ◇ REV MR 30 # 1035 Zbl 131.6 JSL 34.501 • ID 24953

KRIPKE, S.A. [1965] *Semantical analysis of intuitionistic logic I* (**P** 0688) Logic Colloq;1963 Oxford 92-130
◇ B55 C90 F50 ◇ REV MR 34 # 1184 Zbl 137.7 JSL 35.330 • REM Part II never appeared • ID 07557

KRIPKE, S.A. [1965] *Semantical analysis of modal logic II. Non-normal modal propositional calculi* (**P** 0614) Th Models;1963 Berkeley 206-220
• TRANSL [1974] (**C** 4138) Feys: Modal Logika
◇ B45 ◇ REV MR 33 # 7247 Zbl 163.5 JSL 35.135 • REM Part I 1963 • ID 07559

KRIPKE, S.A. see Vol. I, III, IV, V, VI for further entries

KRIVINE, J.-L. [1967] see KREISEL, G.

KRIVINE, J.-L. [1974] *Langages a valeurs reelles et applications* (**J** 0027) Fund Math 81*213-253
◇ B50 C40 C60 C90 ◇ REV MR 50 # 1873 Zbl 292 # 02019 • ID 07566

KRIVINE, J.-L. see Vol. I, III, V for further entries

KRNIC, L. [1963] *A note on enumeration of bases of algebra of logic generated by functions of one and two variables (Russian)* (**J** 0371) Glas Mat-Fiz Astron, Ser 2 (Zagreb) 18*13-16
◇ B50 ◇ REV MR 29 # 5712 Zbl 154.259 • ID 31409

KRNIC, L. [1973] *Cardinals of bases in the 3-valued logic I (Serbo-Croatian summary)* (J 3519) Glas Mat, Ser 3 (Zagreb) 8(28)*169-174
◇ B50 ◇ REV MR 51 # 12476 Zbl 293 # 02016 • ID 17264

KRNIC, L. see Vol. I for further entries

KROEGER, F. [1977] *LAR: A logic of algorithmic reasoning* (J 1431) Acta Inf 8*243-266
◇ B60 B75 ◇ REV MR 58 # 19300 Zbl 347 # 68016 • ID 50063

KROEGER, F. [1984] *A generalized nexttime operator in temporal logic* (J 0119) J Comp Syst Sci 29*80-98
◇ B45 ◇ REV MR 86c:03013 Zbl 551 # 68032 • ID 43952

KROEGER, F. see Vol. I, VI for further entries

KROLIKOSKI, S.J. [1978] *A proposed interpretation of Lukasiewicz's four-valued system of modal logic* (P 2014) Int Symp Multi-Val Log (8);1978 Rosemont 258
◇ B45 ◇ REV MR 81h:03039 • ID 75108

KROLIKOSKI, S.J. [1979] *A deduction theorem for rejection theses in Lukasiewicz's system of modal logic* (J 0047) Notre Dame J Formal Log 20*461-464
◇ B45 ◇ REV MR 80e:03016 Zbl 332 # 02021 • ID 52632

KROLIKOSKI, S.J. [1979] *A second deduction theorem for rejection theses in Lukasiewicz's system of modal logic* (J 0047) Notre Dame J Formal Log 20*545-548
◇ B45 ◇ REV MR 80i:03038 Zbl 332 # 02022 • ID 56127

KROLIKOSKI, S.J. [1979] *Did many-valued logic begin with a mistake?* (P 3003) Int Symp Multi-Val Log (9);1979 Bath 104-109
◇ A05 B50 ◇ REV MR 81b:03026 • ID 75110

KROLIKOSKI, S.J. [1979] *Lukasiewicz's twin possibility functors* (J 0047) Notre Dame J Formal Log 20*458-460
◇ B50 ◇ REV MR 80e:03015 Zbl 332 # 02020 • ID 52631

KROLIKOSKI, S.J. [1980] *On substitution for variable one-place functors* (J 0047) Notre Dame J Formal Log 21*243-250
◇ B50 ◇ REV MR 81g:03018 Zbl 363 # 02021 • ID 53773

KRON, A. [1972] *A note on E* (J 0047) Notre Dame J Formal Log 13*424-426 • ERR/ADD ibid 15*648
◇ B46 ◇ REV MR 58 # 10342 Zbl 238 # 02022 • ID 27900

KRON, A. [1973] *Deduction theorems for relevant logics* (J 0068) Z Math Logik Grundlagen Math 19*85-92
◇ B46 ◇ REV MR 47 # 4769 Zbl 261 # 02014 • ID 07577

KRON, A. & MILOVANOVIC, V. [1975] *Preference and choice* (J 0472) Theory Decis 6*185-196
◇ A05 B48 ◇ REV MR 57 # 18756 Zbl 311 # 02041 • ID 29607

KRON, A. [1976] *Deduction theorems for T, E and R reconsidered* (J 0068) Z Math Logik Grundlagen Math 22*261-264
◇ B46 ◇ REV MR 56 # 15367 Zbl 344 # 02013 • ID 18449

KRON, A. [1978] *Decision procedures for two positive relevance logics* (J 0302) Rep Math Logic, Krakow & Katowice 10*61-78
◇ B25 B46 ◇ REV MR 81b:03020 Zbl 432 # 03013 • ID 53971

KRON, A. [1980] *Gentzen formulations of two positive relevance logics* (J 0063) Studia Logica 39*381-403 • ERR/ADD ibid 40*311
◇ B46 ◇ REV MR 82g:03026 MR 83i:03031 Zbl 472 # 03016 • ID 55282

KRON, A. & MARIC, Z. & VUJOSEVIC, S.T. [1981] *Entailment and quantum logic* (P 3820) Curr Iss in Quantum Log;1979 Erice 193-207
◇ B51 ◇ REV MR 84j:03130 Zbl 537 # 03044 • ID 34720

KRON, A. [1985] *A constructive proof of a theorem in relevance logic* (J 0068) Z Math Logik Grundlagen Math 31*423-430
◇ B46 ◇ ID 47548

KRONTHALER, E. [1971] *Bemerkungen zum zweidimensionalen Kontinuum induktiver Methoden von J.Hintikka* (J 0472) Theory Decis 1*387-392
◇ A05 B48 ◇ REV Zbl 215.318 • ID 28012

KRUGER, A.N. & MANICAS, P.T. [1968] *Essentials of logic* (X 0864) Van Nostrand: New York xi+483PP
◇ B98 ◇ ID 22520

KRZYSTEK, P.S. & ZACHOROWSKI, S. [1977] *Lukasiewicz logics have not the interpolation property* (J 0302) Rep Math Logic, Krakow & Katowice 9*39-40
◇ B50 C40 ◇ REV MR 58 # 21440 Zbl 384 # 03011 • ID 52053

KRZYSTEK, P.S. [1980] *Equivalential fragment of the infinite valued logic of Lukasiewicz and the intermediate logics* (J 0387) Bull Sect Logic, Pol Acad Sci 9*170-175
◇ B50 ◇ REV MR 82b:03054 Zbl 463 # 03009 • ID 54549

KRZYSZTOF, R. (ED.) [1985] *Logics and models of concurrent systems* (X 0811) Springer: Heidelberg & New York viii+498pp
◇ B53 ◇ ID 49284

KUBIN, W. [1979] *Eine Axiomatisierung der mehrwertigen Logiken von Goedel* (J 0068) Z Math Logik Grundlagen Math 25*549-558
◇ B50 ◇ REV MR 80k:03025 Zbl 424 # 03011 • ID 75129

KUBINSKI, T. [1956] *On a method of constructing modal logics (Polish)(Russian and English summaries)* (J 0063) Studia Logica 4*213-240
◇ B45 ◇ REV MR 19.239 Zbl 74.247 JSL 23.348 • ID 16859

KUBINSKI, T. [1966] *A review of some problems of the logic of questions (Polish) (Russian and English summaries)* (J 0063) Studia Logica 18*105-137
◇ A05 B60 ◇ REV MR 33 # 5473 Zbl 309 # 02025 • ID 63167

KUBINSKI, T. [1971] *Introduction to the logical theory of questions* (X 1034) PWN: Warsaw 116pp
◇ B60 ◇ REV JSL 42.426 • ID 44497

KUBINSKI, T. [1979] *Completeness theorems for a class of elementary action theories (Polish)* (J 1093) Ruch Filoz 37*177-180
◇ B60 ◇ ID 33449

KUBINSKI, T. [1979] *On semantics of change logics (Polish)* (J 1093) Ruch Filoz 37*47-53
◇ B60 ◇ ID 33448

KUBINSKI, T. [1980] *An outline of the logical theory of questions* (**X** 0911) Akademie Verlag: Berlin 143pp
◇ A05 B60 ◇ REV MR 83b:03028 Zbl 432 # 03002 JSL 48.874 • ID 53960

KUBINSKI, T. [1980] *On a realisation of the von Wright programme of construction of deontic theories* (**J** 1093) Ruch Filoz 38*233-238
◇ B45 ◇ ID 33443

KUBINSKI, T. [1981] *Theorems on some classes of change theories with quantifiers* (**J** 1093) Ruch Filoz 39*77-83
◇ B60 ◇ ID 46659

KUBINSKI, T. [1982] *On Diodorean theories* (**J** 0387) Bull Sect Logic, Pol Acad Sci 11*31-34
◇ B60 ◇ ID 46660

KUBINSKI, T. [1982] *Tense theories taking into account many change periods (Polish)* (**J** 1093) Ruch Filoz 40/1-2*55-60
◇ B45 ◇ ID 48500

KUBINSKI, T. [1983] *Some classes of elementary action theories (Polish) (English summary)* (**J** 0481) Acta Univ Wroclaw 517(29)(Logika 8)*15-38
◇ B60 ◇ REV Zbl 532 # 03009 • ID 38274

KUBINSKI, T. see Vol. I, IV, V, VI for further entries

KUDLEK, M. & NOVOTNY, M. [1977] *Reducing operators for normed general formal systems* (**P** 1635) Math Founds of Comput Sci (6);1977 Tatranska Lomnica 350-358
◇ B60 ◇ REV MR 57 # 4650 Zbl 361 # 94033 • ID 50721

KUDLEK, M. see Vol. IV for further entries

KUDRYAVTSEV, V.B. & YABLONSKIJ, S.V. & ZAKHAROVA, E.YU. [1969] *Precomplete classes in k-valued logics (Russian)* (**J** 0023) Dokl Akad Nauk SSSR 186*509-512 • ERR/ADD ibid 199(1971)*90)
• TRANSL [1969] (**J** 0062) Sov Math, Dokl 10*618-622
◇ B50 ◇ REV MR 39 # 6738 Zbl 191.288 • ID 18085

KUDRYAVTSEV, V.B. [1970] *The coverings of precomplete classes of k-valued logic (Russian)* (**J** 0071) Met Diskr Analiz (Novosibirsk) 17*32-44
◇ B50 ◇ REV MR 45 # 1730 Zbl 253 # 02013 JSL 40.98 • ID 07604

KUDRYAVTSEV, V.B. [1971] see BLOKHINA, G.N.

KUDRYAVTSEV, V.B. [1971] *On S-systems of functions of k-valued logic (Russian)* (**J** 0023) Dokl Akad Nauk SSSR 199*20-22
• TRANSL [1971] (**J** 0062) Sov Math, Dokl 12*1013-1016
◇ B50 ◇ REV MR 44 # 33 Zbl 233 # 02008 • ID 07605

KUDRYAVTSEV, V.B. [1971] *Ueber Eigenschaften der S-Systeme von Funktionen der k-wertigen Logik (Russisch)* (**J** 0071) Met Diskr Analiz (Novosibirsk) 19*15-47
◇ B50 ◇ REV MR 46 # 5102 Zbl 253 # 02014 • ID 63174

KUDRYAVTSEV, V.B. [1973] see BLOKHINA, G.N.

KUDRYAVTSEV, V.B. [1973] *The properties of S-sytems of functions of k-valued logic (Russian) (German and English summary)* (**J** 0129) Elektr Informationsverarbeitung & Kybern 9*81-105
◇ B50 ◇ REV MR 50 # 12648 Zbl 275 # 02054 • ID 19159

KUDRYAVTSEV, V.B. [1974] see BLOKHINA, G.N.

KUDRYAVTSEV, V.B. [1975] *Ueber einige allgemeine Eigenschaften des Funktionalsystems P_Σ* (**J** 0115) Wiss Z Humboldt-Univ Berlin, Math-Nat Reihe 24*719-720
◇ B50 ◇ REV MR 58 # 5195 Zbl 335 # 02008 • ID 63178

KUDRYAVTSEV, V.B. [1980] *Functional systems (Russian)* (**X** 0898) Moskov Gos Univ: Moskva 158pp
◇ B50 C05 D05 ◇ REV Zbl 491 # 03024 • ID 36855

KUDRYAVTSEV, V.B. [1981] *Function systems (Russian)* (**X** 2265) Akad Nauk Vychis Tsentr: Moskva 64pp
◇ B50 E20 ◇ REV MR 83i:03101 Zbl 533 # 03042 • ID 35548

KUDRYAVTSEV, V.B. [1981] *On completeness for function systems* (**J** 0023) Dokl Akad Nauk SSSR 257*274-278
• TRANSL [1981] (**J** 0062) Sov Math, Dokl 23*263-267
◇ B50 ◇ REV MR 82i:68041 Zbl 469 # 68066 • ID 81956

KUDRYAVTSEV, V.B. [1982] *Truth functional systems (Russian)* (**S** 2874) Vopr Kibern, Akad Nauk SSSR 86*49-77
◇ B50 B75 ◇ REV Zbl 528 # 68038 • ID 36762

KUDRYAVTSEV, V.B. [1984] *Concerning functional systems (Russian)* (**J** 0052) Probl Kibern 41*5-40
◇ B50 ◇ REV MR 85h:68061 Zbl 545 # 03040 • ID 43471

KUDRYAVTSEV, V.B. see Vol. I, IV for further entries

KUENG, G. [1970] see CANTY, J.T.

KUENG, G. [1983] *The difficulty with the well-formedness of ontoligical statements* (**J** 3781) Topoi 2*111-119
◇ A05 B60 ◇ REV MR 84j:03062 • ID 34654

KUGEL, P. [1977] *Induction, pure and simple* (**J** 0194) Inform & Control 35*276-336
◇ B48 ◇ REV MR 57 # 16030 Zbl 366 # 02023 • ID 31147

KUHN, STEVEN T. [1979] *The pragmatics of tense* (**J** 0154) Synthese 40*231-263
◇ B45 ◇ REV MR 83c:03017 • ID 35126

KUHN, STEVEN T. [1980] *Quantifiers as modal operators* (**J** 0063) Studia Logica 39*145-158
◇ B45 ◇ REV MR 81k:03021 Zbl 457 # 03015 • ID 54340

KUHN, STEVEN T. see Vol. I for further entries

KUIPER, R. [1985] see BARRINGER, H.

KUIPERS, T.A.F. [1973] *A generalization of Carnap's inductive logic* (**J** 0154) Synthese 25*334-336
◇ A05 B48 ◇ REV MR 57 # 15992 Zbl 265 # 02024 • ID 29822

KUIPERS, T.A.F. [1978] *On the generalization of the continuum of inductive methods to universal hypotheses* (**J** 0154) Synthese 37*255-284
◇ A05 B48 ◇ REV MR 80m:03061 Zbl 381 # 03007 • ID 51875

KUIPERS, T.A.F. [1978] *Studies in inductive probability and rational expectation* (**X** 0835) Reidel: Dordrecht xii + 145pp
◇ B48 ◇ REV MR 82f:03018 Zbl 404 # 60005 • ID 75165

KUIPERS, T.A.F. [1980] *A survey of inductive systems* (**C** 3537) Stud Induct Logic & Probab, Vol 2 183-192
◇ B48 ◇ REV MR 82f:03019 Zbl 539 # 03004 JSL 49.1409 • ID 75164

KUIPERS, T.A.F. see Vol. I for further entries

KULAS, J. [1985] see HINTIKKA, K.J.J.

KULIKOWSKI, J.L. [1974] *An application of a tense-logic system to formal planning* (**J** 2690) Control & Cybern, Pol Akad Nauk 2/3-4*43-52
◇ B45 ◇ REV MR 57 # 18897 Zbl 337 # 02021 • ID 63192

KULIKOWSKI, J.L. [1976] *Logical and algebraical models of the networks of activities* (**P** 4677) Math Syst Th;1975 Udine 375-392
◇ B45 ◇ REV MR 58 # 33225 Zbl 351 # 94015 • ID 81962

KUMAR, D. [1967] *Logic and inexact predicates* (**J** 0013) Brit J Phil Sci 18*211-222
◇ B52 ◇ ID 41927

KUMAR, D. [1969] *Neutrality, contingency and undecidability* (**J** 0013) Brit J Phil Sci 19*353-356
◇ B48 ◇ ID 41950

KUMAR, D. [1971] *Vagueness and subjunctivity* (**J** 0094) Mind 80*127-131
◇ B52 ◇ ID 41984

KUNEN, K. [1980] see BARWISE, J.

KUNEN, K. see Vol. I, III, IV, V for further entries

KUNSEMUELLER, H. [1964] *Zur Axiomatik der Quantenlogik* (**J** 0455) Phil Naturalis 8*363-376
◇ B51 G12 ◇ REV JSL 40.464 • ID 14552

KUPPERMAN, J. & MCGRADE, A.S. [1966] *Fundamentals of logic* (**X** 0878) Doubleday: London 272pp
◇ B98 ◇ ID 22508

KURBATOV, V.I. [1984] *Logical semantics of norms (Russian)* (**C** 4403) Logika, Pozn, Otrazh 23-33
◇ B45 ◇ ID 46731

KURIHARA, T. [1955] *On the representation of finitely many-valued logics by electric circuits (Japanese)* (**J** 4442) Kyushu Daigaku Kagaku Syuho 28/2*102-106
◇ B50 ◇ REV JSL 22.102 • ID 48774

KURKA, P. [1980] see HAJEK, P.

KURKA, P. [1981] see HAJEK, P.

KURKA, P. see Vol. VI for further entries

KUTIRKOV, G.A. [1968] see CHIMEV, K.N.

KUTIRKOV, G.A. [1970] see CHIMEV, K.N.

KUTIRKOV, G.A. see Vol. I for further entries

KUTSCHERA VON, F. [1962] see KAESBAUER, M.

KUTSCHERA VON, F. [1967] *Elementare Logik* (**X** 0902) Springer: Wien viii + 392pp
◇ B98 ◇ REV MR 37 # 3903 Zbl 105.3 • ID 07711

KUTSCHERA VON, F. [1971] see BREITKOPF, A.

KUTSCHERA VON, F. [1976] *Einfuehrung in die intensionale Semantik* (**X** 1174) Gruyter: Berlin xii + 187pp
◇ A05 B45 ◇ REV Zbl 396 # 03017 • ID 28280

KUTSCHERA VON, F. [1976] *Epistemic interpretation of conditionals* (**C** 1701) Lang in Focus (Bar-Hillel) 487-501
◇ A05 B45 ◇ REV Zbl 385 # 03005 • ID 52113

KUTSCHERA VON, F. [1980] *Grundbegriffe der Handlungslogik* (**C** 4528) Handlg-Th Interdiszipl, Vol 1 1*67-106
◇ B60 ◇ ID 40451

KUTSCHERA VON, F. [1984] *Eine Logik vager Saetze* (**J** 0009) Arch Math Logik Grundlagenforsch 24*101-118
◇ B52 ◇ REV MR 86g:03046 Zbl 553 # 03018 • ID 40454

KUTSCHERA VON, F. see Vol. I, VI for further entries

KUZ'MENKO, V.M. & KUZ'MIN, I.V. & SAMOJLENKO, N.I. [1976] *Determination of the minimal disjunctive normal form of logical functions of k-valued logic (Russian)* (**J** 2668) Avtom Sist Upravl & Prib Avtom, Khar'kov 37*3-6,1
◇ B50 ◇ REV MR 56 # 15365 • ID 75242

KUZ'MIN, I.V. [1976] see KUZ'MENKO, V.M.

KUZNETSOV, A.V. [1970] see GERCHIU, V.YA.

KUZNETSOV, A.V. [1971] *The functional expressibility in superintuitionistic logics (Russian)* (**S** 0166) Mat Issl, Mold SSR 6/4(22)*75-122
◇ B55 ◇ REV MR 45 # 28 Zbl 239 # 02014 • ID 27919

KUZNETSOV, A.V. [1975] *On superintuitionistic logics* (**P** 1521) Int Congr Math (II,12);1974 Vancouver 1*243-249
◇ B25 B55 ◇ REV MR 55 # 2509 Zbl 342 # 02015 • ID 63246

KUZNETSOV, A.V. [1975] *Superintuitionistic logics (Russian)* (**S** 0166) Mat Issl, Mold SSR 10/2(36)*150-158,284
◇ B55 ◇ REV MR 53 # 7734 Zbl 357 # 02022 JSL 37.757 • ID 22981

KUZNETSOV, A.V. (ED.) [1976] *Fourth All-Union Conference on Mathematical Logic (Russian)* (**X** 2741) Shtiintsa: Kishinev 170pp
◇ B97 ◇ REV MR 56 # 15246 • ID 80457

KUZNETSOV, A.V. & MURAVITSKIJ, A.YU. [1980] *Provability as modality (Russian)* (**C** 2583) Aktual Probl Log & Metodol Nauki 193-230
◇ B45 ◇ REV Zbl 535 # 03008 • ID 38313

KUZNETSOV, A.V. see Vol. I, III, IV, VI for further entries

KUZYURIN, N.N. [1979] *On the approximate search for the maximal upper zero for monotone functions of k-valued logics (Russian)* (**J** 0071) Met Diskr Analiz (Novosibirsk) 33*31-40,109
◇ B50 ◇ REV MR 81m:03069 Zbl 449 # 03069 • ID 56736

KYBURG JR., H.E. [1961] *Probability and the logic of rational belief* (**X** 1370) Wesleyan Univ Pr : Middletown 346pp
◇ A05 B48 ◇ REV JSL 35.127 • ID 25424

KYBURG JR., H.E. [1965] *Probability, rationality, and a rule of detachment* (**P** 0623) Int Congr Log, Meth & Phil of Sci (2,Proc);1964 Jerusalem 301-310
◇ A05 B48 ◇ REV MR 34 # 6818 Zbl 198.15 • ID 16201

KYBURG JR., H.E. [1970] *Probability and inductive logic* (**X** 0843) Macmillan : New York & London 272pp
◇ A05 B48 ◇ ID 25426

KYBURG JR., H.E. [1974] *Propensities and probabilities* (**J** 0013) Brit J Phil Sci 25*358-375
◇ A05 B48 ◇ REV MR 58 # 27239 • ID 75252

KYBURG JR., H.E. [1974] *The logical foundations of statistical inference* (**S** 3307) Synth Libr 65*ix + 427pp
◇ A05 B48 ◇ REV MR 58 # 27238 Zbl 335 # 02001 • ID 63250

KYBURG JR., H.E. [1976] *Local and global induction* (**C** 1702) Local Induction 191-215
◇ A05 B48 ◇ REV Zbl 339 # 02002 • ID 63248

KYBURG JR., H.E. see Vol. I for further entries

LAARHOVEN VAN, P.J.M. & PEDRYCZ, W. [1983] *A fuzzy extension of Saaty's priority theory* (**J** 2720) Fuzzy Sets Syst 11*229-241
◇ B52 ◇ REV MR 85f:03056 Zbl 528 # 90054 • ID 36764

LABUNETS, V.G. & SITNIKOV, O.P. [1975] *Harmonic analysis of boolean functions and k-ary logic functions on finite fields (Russian)* (**J** 0977) Izv Akad Nauk SSSR, Tekh Kibern 1975/1*141-148
• TRANSL [1975] (**J** 0522) Engin Cybern 13/1*112-119
◇ B50 ◇ REV MR 54 # 4841 Zbl 305 # 94043 • ID 63255

LACEY, H.M. [1980] see ANDERSON, E.

LACEY, H.M. see Vol. VI for further entries

LACHLAN, A.H. [1974] *A note on Thomason's refined structures for tense logics* (**J** 0105) Theoria (Lund) 40*117-120
◇ B45 ◇ REV MR 58 # 27337 Zbl 302 # 02006 JSL 47.440 • ID 63266

LACHLAN, A.H. see Vol. I, III, IV, V, VI for further entries

LADNER, R.E. [1977] *The computational complexity of provability in systems of modal propositional logic* (**J** 1428) SIAM J Comp 6*467-480
◇ B45 D15 ◇ REV MR 56 # 8326 Zbl 373 # 02025 • ID 51489

LADNER, R.E. [1979] see FISCHER, MICHAEL J.

LADNER, R.E. see Vol. I, III, IV, V for further entries

LADRIERE, J. [1957] *Les limitations internes des formalismes. Etude sur la signification du theoreme de Goedel et des theoremes apparentes dans la theorie des fondements des mathematiques* (**X** 0834) Gauthier-Villars: Paris xv + 715pp
◇ A05 B98 F30 F98 ◇ REV MR 20 # 2 Zbl 78.242 JSL 25.270 • ID 07777

LADRIERE, J. see Vol. III, IV, VI for further entries

LAENGER, H. [1983] see DORNINGER, D.

LAEUCHLI, H. [1982] see ENGELER, E.

LAEUCHLI, H. see Vol. I, III, V, VI for further entries

LAGASSE, J. [1976] see COURVOISIER, M.

LAHTI, P.J. [1977] *Complementary logic in the description of physical systems* (**J** 1108) Arkhimedes (Helsinki) 29*15-24
◇ B51 G12 ◇ REV MR 55 # 4967 Zbl 353 # 70002 • ID 50138

LAHTI, P.J. [1979] *On the expectation value of an observable in quantum logic* (**J** 3293) Bull Acad Pol Sci, Ser Math 27*631-636
◇ B51 G12 ◇ REV MR 81i:81007 Zbl 437 # 03036 • ID 55904

LAHTI, P.J. [1980] *Characterization of quantum logics* (**J** 2736) Int J Theor Phys 19*905-923
◇ B51 G12 ◇ REV MR 82d:81014 Zbl 473 # 03055 • ID 55383

LAKATOS, I. (ED.) [1968] *The problem of inductive logic* (**X** 0809) North Holland: Amsterdam viii + 420pp
◇ A05 B48 ◇ REV MR 37 # 3886 Zbl 167.1 • ID 25467

LAKATOS, I. see Vol. I, VI for further entries

LAKOFF, G. [1973] *Hedges: A study in meaning criteria and the logic of fuzzy concepts* (**J** 0122) J Philos Logic 2*458-508
◇ B52 ◇ REV MR 54 # 9977 Zbl 272 # 02047 • ID 25743

LAKOFF, G. [1975] *Hedges: a study in meaning criteria and the logic of fuzzy concepts* (**P** 1647) Contemp Res in Phil Log & Ling Semant;1975 London ON 221-271
◇ A05 B52 B65 ◇ REV MR 54 # 9977 Zbl 315 # 02023 • ID 29804

LAL, R.N. [1983] see JHA, J.S.

LAL, R.N. see Vol. V for further entries

LAM, C.I. & VRANESIC, Z.G. [1974] *Multiple-valued pseudorandom number generators* (**P** 1385) Int Symp Multi-Val Log (4);1974 Morgantown 45-58
◇ B50 ◇ ID 42092

LAMACCHIA, S.E. [1967] see ABIAN, A.

LAMACCHIA, S.E. see Vol. V for further entries

LAMBERT, K. & SCHARLE, T.W. [1967] *A translation theorem for two systems of free logic* (**J** 0079) Logique & Anal, NS 10*328-341
◇ B20 B60 F25 ◇ REV MR 37 # 1229 Zbl 164.309 • ID 07823

LAMBERT, K. [1967] see FRAASSEN VAN, B.C.

LAMBERT, K. & MEYER, R.K. [1968] *Universally free logic and standard quantification theory* (**J** 0036) J Symb Logic 33*8-26
◇ B60 G15 ◇ REV MR 39 # 37 Zbl 175.261 • ID 09184

LAMBERT, K. & LEBLANC, H. & MEYER, R.K. [1969] *A liberated version of S5* (**J** 0009) Arch Math Logik Grundlagenforsch 12*151-154
◇ B45 ◇ REV MR 41 # 8214 Zbl 193.292 • ID 07828

LAMBERT, K. [1970] see FRAASSEN VAN, B.C.

LAMBERT, K. see Vol. I for further entries

LAMBROS, C.H. [1975] *Notes toward an axiomatization of recklessness* (**J** 0079) Logique & Anal, NS 18*133-142
◇ B60 ◇ REV MR 56 # 8316 Zbl 342 # 02020 • ID 63303

LAMBROS, C.H. see Vol. I for further entries

LAMOTTE, M. [1977] see BENHALCEN, D.

LAMOUCHE, A. [1959] *Logique de la simplicite* (**X** 0856) Dunod: Paris xiv + 536pp + xxi
◇ A05 B60 ◇ ID 07832

LAMPORT, L. [1982] *Timesets. A new method for temporal reasoning about programs* (**P** 3738) Log of Progr;1981 Yorktown Heights 177-196
◇ B45 ◇ REV MR 83i:68029 Zbl 485 # 68031 • ID 39310

LANDMAN, F. & VELTMAN, F. (EDS.) [1984] *Varieties of formal semantics* (**X** 4217) Foris: Dordrecht xii + 425pp
◇ B97 ◇ REV Zbl 556 # 00006 • REM Proceedings of the fourth Amsterdam colloquium 1982 • ID 46159

LANE, D.A. [1980] *Fisher, Jeffreys, and the nature of probability* (**C** 3019) Appreciation (Fisher) 148-160
◇ A05 B48 ◇ REV MR 82m:01061 • ID 81233

LANGER, S.K. [1937] *An introduction to symbolic logic* (**X** 0847) Houghton Mifflin: Boston 363pp
• LAST ED [1953] (**X** 0813) Dover: New York 367pp
◇ B98 ◇ REV MR 14.1051 Zbl 51.5 JSL 3.83 FdM 63.819 • REM 2nd ed.1953 • ID 07848

LANGER, S.K. [1951] see HENLE, P.

LANGER, S.K. see Vol. I for further entries

LANGFORD, C.H. & LEWIS, C.I. [1932] *Symbolic logic* (**X** 1228) Appleton-Century-Crofts: New York xi+506pp
• LAST ED [1959] (**X** 0813) Dover: New York ix+518pp
◇ B25 B98 C10 C98 ◇ REV MR 21 # 4091 Zbl 87.8 FdM 58.56 • ID 22592

LANGFORD, C.H. see Vol. I, III, V for further entries

LANTSOV, A.L. & RAKOV, M.A. [1976] *Synthesis of many-valued functions in a modular algebra with a composite modulus (Russian)* (**J** 0474) Avtom Vychis Tekh, Akad Nauk Latv SSR 1976/3*10-16
• TRANSL [1976] (**J** 2666) Autom Control Comput Sci 10/3*7-13
◇ B50 ◇ REV MR 58 # 5464 Zbl 327 # 94050 • ID 63306

LAPARA, N. [1976] *Semantics for a natural notion of entailment* (**J** 0095) Philos Stud 29*91-113
◇ B46 ◇ REV MR 58 # 21495 • ID 75342

LARGEAULT, J. [1979] *Chance, probabilities and induction (French)* (**X** 2859) Univ Toulouse-Le Mirail: Toulouse 202pp
◇ A05 B48 ◇ REV MR 81i:03007 • ID 75344

LARGEAULT, J. see Vol. VI for further entries

LARNER, A. [1979] *A matrix decision procedure for three modal logics* (**J** 0047) Notre Dame J Formal Log 20*599-602
◇ B45 ◇ REV MR 80h:03028 Zbl 394 # 03030 • ID 56126

LARSEN, J. [1976] *A multi-step formation of variable valued logic hypotheses* (**P** 2011) Int Symp Multi-Val Log (6);1976 Logan 157-163
◇ B50 ◇ ID 35883

LARSEN, P.F. [1977] see JENSEN, J.B.

LARSEN, P.F. [1978] see JENSEN, J.B.

LAU, D. [1975] *Praevollstaendige Klassen von* $P_{(k,l)}$ (**J** 0129) Elektr Informationsverarbeitung & Kybern 11*624-626
◇ B50 C05 ◇ REV Zbl 325 # 02008 • ID 30434

LAU, D. [1977] *Kongruenzen auf gewissen Teilklassen von* $P_{k,l}$ (**S** 2829) Rostocker Math Kolloq 3*37-43
◇ B50 ◇ REV MR 57 # 12213 Zbl 412 # 03005 • ID 52936

LAU, D. [1978] *Bestimmung der Ordnung maximaler Klassen von Funktionen der k-wertigen Logik* (**J** 0068) Z Math Logik Grundlagen Math 24*79-96
◇ B50 G20 ◇ REV MR 80i:03039 Zbl 401 # 03008 • ID 75367

LAU, D. [1978] *Ueber die Anzahl von abgeschlossenen Mengen linearer Funktionen der n-wertigen Logik* (**J** 0129) Elektr Informationsverarbeitung & Kybern 14*561-563
◇ B50 G25 ◇ REV MR 80c:03026 Zbl 412 # 03007 • ID 52938

LAU, D. [1979] *Automorphismen auf den maximalen Klassen der k-wertigen Logik* (**S** 2829) Rostocker Math Kolloq 12*13-16
◇ B50 G25 ◇ REV MR 81j:08006 Zbl 412 # 03006 • ID 52937

LAU, D. [1981] *Congruences on closed sets of k-valued logic* (**P** 2552) Conf Finite Algeb & Multi-Val Log;1979 Szeged 417-440
◇ B50 ◇ REV MR 83h:03092 Zbl 485 # 03007 • ID 36090

LAU, D. [1982] *Die maximalen Klassen von* $Pol_k(0)$ (**S** 2829) Rostocker Math Kolloq 19*29-47
◇ B50 ◇ REV MR 83g:03071 Zbl 494 # 03046 • ID 36025

LAU, D. [1982] see GORLOV, V.V.

LAU, D. [1982] *Submaximale Klasse von* P_3 *(English and Russian summaries)* (**J** 0129) Elektr Informationsverarbeitung & Kybern 18*227-243
◇ B50 ◇ REV MR 85b:30113 Zbl 502 # 03033 • ID 36899

LAU, D. [1983] see GORLOV, V.V.

LAU, D. [1984] *Die maximalen Klassen von* $Pol_k\{(x,x+1 \bmod k) \mid x \in E_k\}$ (**S** 2829) Rostocker Math Kolloq 25*23-30
◇ B50 ◇ REV MR 86f:08005 Zbl 553 # 03013 • ID 43305

LAU, D. [1984] *Unterhalbgruppen von* $(P_3^1,*)$ (**S** 2829) Rostocker Math Kolloq 26*55-62
◇ B50 ◇ REV Zbl 565 # 03030 • ID 44514

LAVENDHOMME, R. & LUCAS, T. [1984] *Une interpretation modale de la logique intuitionniste (English summary)* (**J** 3364) C R Acad Sci, Paris, Ser 1 298*193-196
◇ B45 F50 ◇ REV MR 85b:03016 Zbl 563 # 03008 • ID 40558

LAVENDHOMME, R. see Vol. III, V, VI for further entries

LAVROV, I.A. [1970] *Logic and algorithms (Russian)* (**X** 0913) Novosibirsk Gos Univ: Novosibirsk 173pp
◇ B98 D98 ◇ REV MR 48 # 8188 • ID 07889

LAVROV, I.A. & MAKSIMOVA, L.L. [1970] *Problems in logic (Russian)* (**X** 0913) Novosibirsk Gos Univ: Novosibirsk 112pp
◇ B98 E98 ◇ REV MR 48 # 8189 • ID 07888

LAVROV, I.A. & MAKSIMOVA, L.L. [1975] *Problems in set theory, mathematical logic and the theory of algorithms (Russian)* (**X** 2027) Nauka: Moskva 240pp
◇ B98 D98 E98 ◇ REV MR 52 # 13300 Zbl 307 # 02001 • REM 2nd ed. 1984; 224pp • ID 21765

LAVROV, I.A. [1984] see ERSHOV, YU.L.

LAVROV, I.A. see Vol. III, IV, VI for further entries

LAZEROWITZ, M. [1948] see AMBROSE, A.

LAZEROWITZ, M. [1961] see AMBROSE, A.

LAZEROWITZ, M. [1974] *Necessity and probability* (**J** 0094) Mind 83*282-285
◇ A05 B48 ◇ REV MR 58 # 21399a • ID 75392

LE, HUILING & ZHANG, WENXIU [1982] *Extended fuzzy operators and fuzzy truth-valued possibility (Chinese) (English summary)* (**J** 3795) Xi'an Jiaotong Daxue Xuebao 16/5*1-12
◇ B52 ◇ REV MR 84i:03060 Zbl 498 # 04003 • ID 34539

LE, HUILING & ZHANG, WENXIU [1984] *Fuzzy truth possibility degrees (Chinese) (English summary)* (**J** 3732) Mohu Shuxue 4/1*7-14
◇ B52 ◇ REV MR 85j:03029 Zbl 567 # 94032 • ID 44681

LE, HUILING & ZHANG, WENXIU [1984] *The extended fuzzy operators and the fuzzy truth-valued possibility degree* (**P** 3081) IFAC Symp Fuzzy Inf, Knowl Repr & Decis. Anal;1983 Marseille 103-106
◇ B52 ◇ REV Zbl 557 # 94024 • ID 48200

LEACH, J.J. [1978] see HOOKER, C.A.

LEACH, J.J. see Vol. I for further entries

LEBEDEV, S.A. [1980] *The role of induction in the functioning of modern scientific knowledge (Russian) (English Summary)* (J 2871) Vopr Fil, Moskva 1980/6*79-88,186
◇ A05 B48 ◇ REV MR 81j:03016 • ID 75394

LEBIEDIEWA, S. [1969] *The systems of modal calculus of names I (Polish and Russian summaries)* (J 0063) Studia Logica 24*83-107
◇ B45 ◇ REV MR 41 # 42 Zbl 252 # 02010 • REM Part II 1969 • ID 08135

LEBLANC, A. [1985] *Axioms for mereology* (J 0047) Notre Dame J Formal Log 26*429-436
◇ B60 ◇ ID 47536

LEBLANC, A. [1985] *Investigations in protothetic* (J 0047) Notre Dame J Formal Log 26*483-490
◇ B60 ◇ ID 47542

LEBLANC, A. [1985] *New axioms for mereology* (J 0047) Notre Dame J Formal Log 26*437-443
◇ B60 ◇ ID 47537

LEBLANC, H. [1955] *An introduction to deductive logic* (X 0827) Wiley & Sons: New York xii+244pp
◇ B98 ◇ REV MR 16.661 Zbl 64.5 JSL 23.210 • ID 07909

LEBLANC, H. [1957] *On logically false evidence statements* (J 0036) J Symb Logic 22*345-349
◇ A05 B48 ◇ REV MR 21 # 3319 Zbl 124.248 JSL 25.86 • ID 07911

LEBLANC, H. [1959] *On chances and estimated chances of being true* (J 0252) Rev Philos Louvain 57*225-239
◇ B48 ◇ ID 07912

LEBLANC, H. [1960] *On a recent allotment of probabilities to open and closed sentences* (J 0047) Notre Dame J Formal Log 1*171-175
◇ B48 ◇ ID 07914

LEBLANC, H. [1960] *On requirements for conditional probability functions* (J 0036) J Symb Logic 25*238-242
◇ B05 B30 B48 ◇ REV MR 33 # 750 Zbl 243 # 02010 • ID 07913

LEBLANC, H. [1961] *A new interpretation of c(h,e)* (J 0075) Phil Phenom Research 21*373-376
◇ A05 B48 ◇ ID 07917

LEBLANC, H. [1961] *Probabilities as truth-value estimates* (J 0153) Phil of Sci (East Lansing) 28*414-417
◇ B48 ◇ ID 07916

LEBLANC, H. [1962] *Statistical and inductive probabilities* (X 0819) Prentice Hall: Englewood Cliffs xii+148pp
◇ A05 B48 ◇ REV MR 31 # 5225 Zbl 128.375 • ID 25427

LEBLANC, H. [1966] *Techniques of deductive inference* (X 0819) Prentice Hall: Englewood Cliffs vii+216pp
◇ B98 ◇ ID 22510

LEBLANC, H. [1968] *On Meyer and Lambert's quantificational calculus FQ* (J 0036) J Symb Logic 33*275-280
◇ B45 G15 ◇ REV MR 39 # 38 Zbl 165.304 • ID 07928

LEBLANC, H. [1969] see LAMBERT, K.

LEBLANC, H. & WISDOM, W.A. [1972] *Deductive logic* (X 0802) Allyn & Bacon: London 368pp
◇ B98 ◇ REV JSL 40.628 • ID 22358

LEBLANC, H. (ED.) [1972] *Truth, syntax and modality* (X 0809) North Holland: Amsterdam 260pp
◇ A05 B46 B97 ◇ REV MR 49 # 7113 Zbl 254 # 00003 • ID 25292

LEBLANC, H. [1973] *On dispensing with things and worlds* (C 1222) Existence & Possible Worlds 241-259
◇ A05 B45 ◇ ID 20988

LEBLANC, H. [1973] *Semantic deviations* (P 0783) Truth, Syntax & Modal;1970 Philadelphia 1-16
◇ A05 B45 ◇ REV MR 52 # 10370 Zbl 275 # 02016 JSL 42.313 • ID 14976

LEBLANC, H. [1974] see GOLDBERG, H.

LEBLANC, H. [1974] see DUNN, J.M.

LEBLANC, H. [1975] *A Henkin-type completeness proof for 3-valued logic with quantifiers* (P 1805) Int Symp Multi-Val Log (5,Proc);1975 Bloomington 388-398
◇ B50 ◇ REV MR 58 # 10300 • ID 75399

LEBLANC, H. & MCARTHUR, R.P. [1976] *A completeness result for quantificational tense logic* (J 0068) Z Math Logik Grundlagen Math 22*89-96
◇ B45 ◇ REV MR 53 # 2634 Zbl 329 # 02008 • ID 18486

LEBLANC, H. [1976] *Truth-value semantics* (X 0809) North Holland: Amsterdam xii+319pp
◇ B98 ◇ REV MR 56 # 15360 Zbl 348 # 02001 JSL 43.376 • ID 63340

LEBLANC, H. [1977] *A strong completeness theorem for 3-valued logic. part II* (J 0047) Notre Dame J Formal Log 18*107-116
◇ B50 ◇ REV MR 55 # 2492 Zbl 283 # 02018 • REM Part I 1974 by Goldberg,H. & Leblanc,H. & Weaver,G.E. • ID 21953

LEBLANC, H. [1979] see FRAASSEN VAN, B.C.

LEBLANC, H. [1979] *Probabilistic semantics for first-order logic* (J 0068) Z Math Logik Grundlagen Math 25*497-509
◇ B48 ◇ REV MR 80i:03031 Zbl 424 # 03006 • ID 75398

LEBLANC, H. [1981] *What price substitutivity? A note on probability theory* (J 0153) Phil of Sci (East Lansing) 48*317-322
◇ B48 ◇ REV MR 83d:03025 • ID 35174

LEBLANC, H. & MORGAN, C.G. [1983] *Probabilistic semantics for intuitionistic logic* (J 0047) Notre Dame J Formal Log 24*161-180
◇ B48 F50 ◇ REV MR 84c:03027 Zbl 466 # 60003 • ID 54992

LEBLANC, H. & MORGAN, C.G. [1983] *Probability theory, intuitionism, semantics, and the Dutch book argument* (J 0047) Notre Dame J Formal Log 24*289-304
◇ B48 C90 F50 ◇ REV MR 85d:03048 Zbl 531 # 03037 • ID 37685

LEBLANC, H. [1983] *Probability functions and their assumption sets - the singulary case* (J 0122) J Philos Logic 12*379-402
◇ B48 ◇ REV MR 85g:03035 Zbl 539 # 60007 • ID 43871

LEBLANC, H. & MORGAN, C.G. [1984] *Probability functions and their assumption sets - the binary case* (J 0154) Synthese 60*91-106
◇ B48 ◇ REV MR 86e:03023 • ID 44037

LEBLANC, H. [1985] *Unary probabilistic semantics* (**P** 4180) Int Congr Log, Meth & Phil of Sci (7,Pap);1983 Salzburg 319–344
◇ B48 ◇ ID 47938

LEBLANC, H. see Vol. I, III, VI for further entries

LEDLEY, R.S. [1975] see HUANG, H.K.

LEE, E.S. & SMITH, KENNETH C. & VRANESIC, Z.G. [1970] *A many-valued algebra for switching systems* (**J** 0187) IEEE Trans Comp C-19*964–971
◇ B50 B70 ◇ REV Zbl 223 # 94017 • ID 41970

LEE, E.T. & LEE, S.C. [1971] *Multivalued symmetric functions* (**P** 2007) Symp Th & Appl of Multi-Val Log Design;1971 Buffalo 62–83
◇ B50 ◇ ID 41985

LEE, E.T. [1977] *Comments on two theorems by Kandel* (**J** 0194) Inform & Control 35*106–108
◇ B52 ◇ REV MR 58 # 9853a Zbl 362 # 94043 • ID 50822

LEE, E.T. see Vol. IV for further entries

LEE, H.N. [1961] *Symbolic logic* (**X** 0981) Random House: New York ix + 356pp
◇ B98 ◇ REV JSL 25.262 • ID 22511

LEE, R.C.T. [1971] see CHANG, C.L.

LEE, R.C.T. [1972] *Fuzzy logic and the resolution principle* (**J** 0037) ACM J 19*109–119
◇ B35 B52 ◇ REV MR 52 # 4736 Zbl 245 # 02020 • ID 28860

LEE, R.C.T. [1973] see CHANG, C.L.

LEE, R.C.T. see Vol. I for further entries

LEE, S.C. [1971] see LEE, E.T.

LEE, S.C. [1981] see CHEN, GUOQUAN

LEE, S.C. [1982] see CHEN, GUOQUAN

LEE, S.C. see Vol. I for further entries

LEHMAN, R.S. [1955] *On confirmation and rational betting* (**J** 0036) J Symb Logic 20*251–262
◇ A05 B48 ◇ REV Zbl 66.110 JSL 33.481 • ID 07977

LEHMAN, R.S. see Vol. VI for further entries

LEHMANN, D.J. & SHELAH, S. [1982] *Reasoning with time and chance* (**J** 0194) Inform & Control 53*165–198
◇ B45 B75 ◇ REV MR 85c:68046 Zbl 523 # 03016 • ID 37023

LEHMANN, D.J. & SHELAH, S. [1983] *Reasoning with time and chance* (**P** 3851) Automata, Lang & Progr (10);1983 Barcelona 445–457
◇ B25 B45 ◇ REV MR 84k:68004 Zbl 546 # 03010 • ID 41143

LEHMANN, D.J. see Vol. IV for further entries

LEHMANN, S.K. [1976] *A first-order logic of knowledge and belief with identity. I,II* (**J** 0047) Notre Dame J Formal Log 17*59–77,207–221
◇ A05 B48 ◇ REV MR 58 # 147 Zbl 316 # 02034 Zbl 322 # 02019 • ID 18255

LEHMANN, S.K. [1976] *An interpretation of "finite" modal first-order languages in classical second-order languages* (**J** 0036) J Symb Logic 41*337–340
◇ B15 B45 ◇ REV MR 53 # 10551 Zbl 341 # 02015 • ID 14765

LEHMANN, S.K. [1979] *A general propositional logic of conditionals* (**J** 0047) Notre Dame J Formal Log 20*77–83
◇ B60 ◇ REV MR 80f:03026 Zbl 423 # 03022 • ID 53534

LEHMANN, S.K. [1980] *Slightly nonstandard logic* (**J** 0079) Logique & Anal, NS 23*379–392
◇ B60 ◇ REV MR 84h:03026 • ID 34240

LEHRER, K. [1963] *Descriptive completeness and inductive methods* (**J** 0036) J Symb Logic 28*157–160
◇ B48 ◇ REV MR 30 # 3836 Zbl 121.252 • ID 07979

LEHRER, K. [1970] *Induction, reason and consistency* (**J** 0013) Brit J Phil Sci 21*103–114
◇ B48 ◇ REV MR 44 # 4794 Zbl 246 # 02025 • ID 63362

LEHRER, K. [1983] *Sellars on induction reconsidered* (**J** 0097) Nous, Quart J Phil 17*469–473
◇ B48 ◇ REV MR 85d:03046 • ID 41090

LEHRER, K. & WAGNER, C.G. [1985] *Intransitive indifference: the semi-order problem* (**J** 0154) Synthese 65*249–256
◇ B60 ◇ ID 49411

LEIGH, V. [1972] *Self-dual sets of unary and binary connectives for the 3-valued propositional calculus* (**J** 0068) Z Math Logik Grundlagen Math 18*201–204
◇ B50 ◇ REV MR 54 # 12481 Zbl 249 # 02009 • ID 07980

LEININGER, C.W. [1969] *Concerning some proposals for quantum logic* (**J** 0047) Notre Dame J Formal Log 10*95–96
◇ B51 G12 ◇ REV MR 41 # 3347 Zbl 191.290 • ID 07981

LEIVANT, D. [1981] *On the proof theory of the modal logic for arithmetic provability* (**J** 0036) J Symb Logic 46*531–538
◇ B45 F05 F30 ◇ REV MR 83a:03022 Zbl 464 # 03019 • ID 54609

LEIVANT, D. [1981] *Proof theoretic methodology for propositional dynamic logic* (**P** 2930) Formal of Progr Concepts;1981 Peniscola 356–373
◇ B75 F99 ◇ REV MR 83a:68030 Zbl 467 # 03015 • ID 55013

LEIVANT, D. see Vol. I, IV, VI for further entries

LEJEWSKI, C. [1955] *A contribution to Lesniewski's mereology* (**J** 1586) Rocz Pol Towarz Nauk Obczyznie 8*43–50
◇ B60 E70 ◇ REV JSL 21.325 • ID 27074

LEJEWSKI, C. [1974] *A system of logic for bicategorial ontology* (**J** 0122) J Philos Logic 3*265–283
• REPR [1977] (**P** 2116) Probl in Log & Ontology;1973 Salzburg 99–117
◇ B60 E70 ◇ REV MR 55 # 74 MR 58 # 27299 Zbl 285 # 02025 • ID 29952

LEJEWSKI, C. [1977] *Systems of Lesniewski's ontology with the functor of weak inclusion as the only primitive term* (**J** 0063) Studia Logica 36*323–349
◇ B60 ◇ REV MR 58 # 16182 Zbl 407 # 03013 • ID 31835

LEJEWSKI, C. see Vol. I, V for further entries

LEMAN, M. [1984] *On the logic and criteriology of causality* (**J** 0079) Logique & Anal, NS 27*245–266
◇ B60 ◇ ID 44432

LEMMON, E.J. [1956] *Alternative postulate sets for Lewi's S5* (J 0036) J Symb Logic 21*347-349
◇ B45 ◇ REV MR 20 # 3068 Zbl 78.5 JSL 22.380
• ID 07998

LEMMON, E.J. & MEREDITH, C.A. & MEREDITH, D. & PRIOR, A.N. & THOMAS, I. [1957] *Calculi of pure strict implication* (1111) Preprints, Manuscr., Techn. Reports etc. i + 22pp
◇ B46 ◇ REV MR 19.626 • REM Philos. Dept., Canterburry Univ. College, Christchurch, NZ • ID 19178

LEMMON, E.J. [1957] *New foundations for Lewis modal systems* (J 0036) J Symb Logic 22*176-186
◇ B45 ◇ REV MR 24 # A1817 Zbl 80.242 JSL 23.346
• ID 07999

LEMMON, E.J. [1958] *Quantifiers and modal operators* (J 0113) Proc Aristotelian Soc 58*245-268
◇ B45 ◇ REV JSL 31.275 • ID 08000

LEMMON, E.J. [1959] see DUMMETT, M.

LEMMON, E.J. [1959] *Symposium: Is there only one correct system of modal logic? I* (J 2119) Suppl Aristotelian Soc 33*23-40
◇ A05 B45 ◇ REV JSL 34.306 • ID 43283

LEMMON, E.J. [1960] *An extension algebra and the modal system T* (J 0047) Notre Dame J Formal Log 1*3-12
◇ B45 ◇ REV MR 32 # 5504 Zbl 114.246 JSL 35.136
• ID 08001

LEMMON, E.J. [1963] *A theory of attributes based on modal logic* (J 0096) Acta Philos Fenn 16*95-122
◇ B45 ◇ REV MR 29 # 10 Zbl 135.6 JSL 37.180 • ID 08003

LEMMON, E.J. [1965] *Beginning logic* (X 1035) Nelson: Walton on Thames x + 225pp
◇ B98 ◇ REV Zbl 158.244 JSL 40.287 • ID 15171

LEMMON, E.J. [1965] *Deontic logic and the logic of imperatives* (J 0079) Logique & Anal, NS 8*39-71
◇ B45 ◇ REV MR 33 # 5474 Zbl 129.258 • ID 48102

LEMMON, E.J. [1965] *Some results on finite axiomatizability in modal logic* (J 0047) Notre Dame J Formal Log 6*301-308
◇ B45 ◇ REV MR 35 # 6530 Zbl 156.249 JSL 34.648
• ID 08004

LEMMON, E.J. [1966] *A note on Hallden-incompleatness* (J 0047) Notre Dame J Formal Log 7*296-300
◇ B45 ◇ REV MR 38 # 3136 Zbl 192.32 JSL 34.648
• ID 08006

LEMMON, E.J. [1966] *Algebraic semantics for modal logics I,II* (J 0036) J Symb Logic 31*46-65,191-218
◇ B45 ◇ REV MR 34 # 5660 MR 34 # 5661 Zbl 147.248 JSL 35.136 • ID 22020

LEMMON, E.J. & MEREDITH, C.A. & MEREDITH, D. & PRIOR, A.N. & THOMAS, I. [1969] *Calculi of pure strict implication* (C 1101) Phil Logic 215-250
◇ B46 ◇ ID 31935

LEMMON, E.J. [1977] *An introduction to modal logic* (X 1096) Blackwell: Oxford x + 94pp
◇ A10 B45 B98 ◇ REV MR 57 # 15931 Zbl 388 # 03006 JSL 44.653 • ID 52261

LEMMON, E.J. [1978] *Beginning logic* (X 2725) Hackett Publ : Indianapolis x + 225pp
◇ B98 ◇ REV MR 58 # 107 JSL 46.421 • ID 75449

LEMMON, E.J. see Vol. I, V for further entries

LEMPICKI, A. [1974] *A new look at optical logic* (P 1385) Int Symp Multi-Val Log (4);1974 Morgantown 207-208
◇ B60 ◇ ID 42093

LENK, H. [1978] *Varieties of commitment: approaches to the symbolization of conditional obligations* (J 0472) Theory Decis 9*17-37
◇ A05 B45 ◇ REV MR 58 # 10317 Zbl 379 # 02009
• ID 51842

LENZEN, W. [1978] *A rare accident* (J 0047) Notre Dame J Formal Log 19*249-250
◇ B45 ◇ REV MR 57 # 15967 Zbl 368 # 02020 • ID 53115

LENZEN, W. [1978] *On some substitution instances of R1 and L1* (J 0047) Notre Dame J Formal Log 19*159-164
◇ B45 ◇ REV MR 58 # 5056 Zbl 351 # 02015 • ID 27101

LENZEN, W. [1978] *Recent work in epistemic logic* (J 0096) Acta Philos Fenn 1*219pp
◇ A05 B45 ◇ REV MR 81b:03021 Zbl 397 # 03002
• ID 52668

LENZEN, W. [1978] *S4.1.4. = S4.1.2 and S4.021 = S4.04* (J 0047) Notre Dame J Formal Log 19*465-466
◇ B45 ◇ REV MR 81e:03014 Zbl 368 # 02021 • ID 51661

LENZEN, W. [1980] *Beschraenkte und unbeschraenkte Reduktion von Konjunktionen von Modalitaeten in S4* (J 0068) Z Math Logik Grundlagen Math 26*131-143
◇ B45 ◇ REV MR 81f:03026 Zbl 434 # 03015 • ID 55719

LENZEN, W. [1980] *Glauben, Wissen und Wahrscheinlichkeit. Systeme der epistemischen Logik* (X 0902) Springer: Wien xvi + 360pp
◇ A05 B45 B48 ◇ REV Zbl 466 # 03001 • ID 54952

LENZEN, W. [1981] *Doxastic logic and the Burge-Buridan-paradox* (J 0095) Philos Stud 39*43-49
◇ B45 ◇ REV MR 82i:03027 • ID 75454

LENZEN, W. [1983] *On the representation of classificatory value structures* (J 0472) Theory Decis 15*349-369
◇ B48 ◇ REV MR 84k:03070 Zbl 537 # 90009 • ID 35007

LEONARD, H.S. [1951] *Two-valued tables for modal functions* (C 0585) Struct, Meth & Meaning (Sheffer) 42-67
◇ B45 ◇ REV Zbl 54.6 JSL 16.288 • ID 16876

LEONARD, H.S. [1967] *Principles of reasoning: an introduction to logic, methodology, and the theory of signs* (X 0813) Dover: New York xviii + 620pp
◇ A05 B98 ◇ REV JSL 23.435 • ID 22513

LEONARD, H.S. see Vol. V for further entries

LEONE, M. [1985] *On temporal invariance of menu systems* (P 2999) Proc Conf Databasis (Calzone);1985 Heidelberg 1-12
◇ B05 B75 C90 ◇ ID 49912

LEONT'EV, V.K. [1974] *A theorem on monotone functions (Russian)* (J 0071) Met Diskr Analiz (Novosibirsk) 26*61-64,85
◇ B50 ◇ REV MR 57 # 16063 Zbl 305 # 94044 • ID 63377

LEOPOLD-WILDBURGER, U. [1982] see HINTIKKA, K.J.J.

LEOPOLD-WILDBURGER, U. & SELTEN, R. [1982] *Subjunctive conditionals in decision and game theory* (P 3759) Phil of Econ;1981 Muenchen 191-200
◇ B45 ◇ REV MR 84b:03034 • ID 34910

LERCHER, B. [1972] see HINDLEY, J.R.

LERCHER, B. see Vol. VI for further entries

LESCHINE, T.M. [1978] *Propositional logic for topology-like matrices: A calculus with restricted substitution* (J 0063) Studia Logica 37*161-165
◇ B50 G10 ◇ REV MR 58 # 16183 Zbl 414 # 03012
• ID 53058

LESSI, O. [1981] *Corrispondenza tra funzioni di conferma e distribuzioni di probilita condizionate e identificazione di una classe di funzioni rappresentative (English and French summaries)* (J 2839) Statistica (Bologna) 41*237-253
◇ B48 ◇ REV MR 83c:03020 Zbl 483 # 62002 • ID 35127

LEVCHENKOV, A.S. [1982] see BORISOV, A.N.

LEVI, I. [1959] *Putnam's three truth values* (J 0095) Philos Stud 10*65-69
◇ B50 ◇ REV JSL 25.289 • ID 08043

LEVI, I. [1970] *Probability and evidence* (C 1577) Induct, Accept & Rat Belief 134-156
◇ A05 B48 ◇ REV Zbl 217.6 JSL 39.166 • ID 28033

LEVI, I. [1976] *Acceptance revisited* (C 1702) Local Induction 1-71
◇ A05 B48 ◇ REV Zbl 319 # 02003 • ID 29663

LEVI, I. [1978] *Irrelevance* (P 2931) Found & Appl of Decision Th;1975 London ON 1*263-273
◇ B48 ◇ REV MR 81h:90005 Zbl 405 # 62006 • ID 82058

LEVI, I. [1981] *Direct inference and confirmational conditionalization* (J 0153) Phil of Sci (East Lansing) 48*532-552
◇ B48 ◇ REV MR 83a:03024 • ID 35063

LEVI, I. [1982] *Ignorance, probability and rational choice* (J 0154) Synthese 53*387-417,433-438
◇ B48 ◇ REV MR 84e:03029 Zbl 516 # 62012 • ID 34368

LEVI, I. [1985] *Imprecision and indeterminacy in probability judgement* (J 0153) Phil of Sci (East Lansing) 52*390-409
◇ B48 ◇ ID 49413

LEVI, I. see Vol. I for further entries

LEVIN, L.A. [1969] *Some syntactic theorems on the calculus of finite problems of Ju.T.Medvedev (Russian)* (J 0023) Dokl Akad Nauk SSSR 185*32-33
• TRANSL [1969] (J 0062) Sov Math, Dokl 10*288-290
◇ B55 F50 ◇ REV MR 39 # 1299 • ID 08046

LEVIN, L.A. see Vol. IV, VI for further entries

LEVIN, V.A. [1972] *Infinite-valued logic and transient processes in finite automata (Russian)* (J 0474) Avtom Vychis Tekh, Akad Nauk Latv SSR 1972/6*1-9
• TRANSL [1972] (J 2666) Autom Control Comput Sci 6/6*1-8
◇ B50 D05 ◇ REV MR 49 # 4681 Zbl 245 # 94023
• ID 63410

LEVIN, V.A. [1974] *Equations in infinite-valued logic, and transition processes in finite automata (Russian)* (J 0474) Avtom Vychis Tekh, Akad Nauk Latv SSR 1974/5*12-15
• TRANSL [1974] (J 2666) Autom Control Comput Sci 8/5*12-15
◇ B50 D05 ◇ REV MR 50 # 16166 Zbl 295 # 94060
• ID 63412

LEVIN, V.A. [1975] *Equations in infinite-valued logic with deviating arguments (Russian)* (J 0474) Avtom Vychis Tekh, Akad Nauk Latv SSR 1975/1*17-20
• TRANSL [1975] (J 2666) Autom Control Comput Sci 9/1*15-18
◇ B50 ◇ REV MR 57 # 12171 Zbl 347 # 02009 • ID 63408

LEVIN, V.A. [1975] *Equations of infinite-valued logic that contain all logical operations (Russian)* (S 0764) Teor Konech Avtom & Prilozh (Riga) 5*5-14,113
◇ B50 ◇ REV MR 58 # 10301a • ID 75499

LEVIN, V.A. [1975] *Equations of infinite-valued logic with deviating arguments and with all possible logical operations (Russian)* (S 0764) Teor Konech Avtom & Prilozh (Riga) 5*15-31,113
◇ B50 ◇ REV MR 58 # 10301b • ID 75498

LEVIN, V.A. [1981] *The infinite-valued logic and the problem of making group decisions (Russian)* (S 0764) Teor Konech Avtom & Prilozh (Riga) 12*20-35
◇ B50 ◇ REV Zbl 487 # 90008 • ID 36714

LEVIN, V.A. see Vol. I, IV for further entries

LEVY, A. [1958] see BAR-HILLEL, Y.

LEVY, A. see Vol. III, IV, V, VI for further entries

LEVY, E. [1982] *Causal-relevance explanation: Salmon's theory and its relation to Reichenbach* (J 0154) Synthese 50*423-445
◇ A05 B48 ◇ REV MR 84k:03020 • ID 34959

LEVY, G. [1969] see HANSEL, G.

LEVY, G. [1969] *Formes disjonctives minimales dans une logique a plusieurs valeurs* (J 2313) C R Acad Sci, Paris, Ser A-B 269*A613-A615
◇ B50 ◇ REV MR 40 # 4083 Zbl 207.5 • ID 08089

LEVY, G. [1970] *Sur la partition de certains treillis en spheres concentriques* (J 2313) C R Acad Sci, Paris, Ser A-B 271*A61-A64
◇ B50 ◇ REV MR 56 # 11592 Zbl 212.316 • ID 27235

LEVY, G. see Vol. I for further entries

LEVY, P. [1927] *Logique classique, logique Brouwerienne et logique mixte* (J 0150) Acad Roy Belg Bull Cl Sci (5) 13*256-266
◇ A05 B55 F50 ◇ REV FdM 53.40 • ID 08092

LEVY, P. see Vol. III, V, VI for further entries

LEWIS, C.I. [1918] *A survey of symbolic logic* (X 0926) Univ Calif Pr: Berkeley vi + 406pp
• REPR [1960] (X 0813) Dover: New York
◇ B45 B46 B98 ◇ REV JSL 16.225 FdM 47.870
• ID 22590

LEWIS, C.I. [1920] *Strict implication - an emendation* (J 0301) J Phil 17*300-302
◇ A05 B45 B46 ◇ ID 28496

LEWIS, C.I. [1932] *Alternative systems of logic* (J 0320) Monist 42*481-507
◇ B60 ◇ REV Zbl 5.337 • ID 48762

LEWIS, C.I. [1932] see LANGFORD, C.H.

LEWIS, C.I. [1936] *Emch's calculus and strict implication* (J 0036) J Symb Logic 1*77-86
◇ B45 B46 ◇ REV Zbl 15.145 JSL 2.46 FdM 62.1051
• ID 08100

LEWIS, C.I. [1960] *A survey of symbolic logic (Reprinted edition)* (**X** 0813) Dover: New York x + 327pp
◇ B45 B98 ◇ ID 25502

LEWIS, C.I. [1967] *The structure of system of strict implication* (**C** 0721) Contemp Readings in Log Th 259-267
◇ B45 B46 ◇ ID 08103

LEWIS, C.I. [1970] see GOHEEN, J.D.

LEWIS, D. [1971] *Completeness and decidability of three logics of counterfactual conditionals* (**J** 0105) Theoria (Lund) 37*74-85
◇ B25 B45 ◇ REV MR 49 # 2278 Zbl 229 # 02023 • ID 08107

LEWIS, D. [1973] *Counterfactuals* (**X** 0858) Harvard Univ Pr: Cambridge x + 150pp
◇ A05 B45 ◇ REV MR 54 # 9979 JSL 44.278 • ID 25746

LEWIS, D. [1973] *Counterfactuals and comparative possibility* (**J** 0122) J Philos Logic 2*418-446
◇ B45 ◇ REV MR 54 # 9980 Zbl 272 # 02048 • ID 30383

LEWIS, D. [1974] *Intensional logics without iterative axioms* (**J** 0122) J Philos Logic 3*457-466
◇ B45 ◇ REV MR 54 # 9981 Zbl 296 # 02014 • ID 25748

LEWIS, D. [1974] *Semantic analyses for dyadic deontic logic* (**C** 1936) Log Th & Semant Anal (Kanger) 1-14
◇ B45 ◇ REV Zbl 331 # 02013 • ID 63431

LEWIS, D. [1975] *Counterfactuals and comparative possibility* (**P** 1647) Contemp Res in Phil Log & Ling Semant;1975 London ON 1-29
◇ B45 ◇ REV MR 54 # 9980 Zbl 308 # 02029 • ID 63433

LEWIS, D. [1977] *Possible-world semantics for counterfactual logics: A rejoinder* (**J** 0122) J Philos Logic 6*359-363
◇ A05 B45 ◇ REV MR 58 # 27324 Zbl 376 # 02003 • ID 51645

LEWIS, D. [1981] *Counterfactuals and comparative possibility* (**C** 4140) Ifs 57-85
◇ B45 ◇ REV MR 83a:03003 • ID 47736

LEWIS, D. [1981] *Index, context, and content* (**P** 3739) Phil & Grammar;1978 Uppsala 79-100
◇ B45 B65 ◇ REV MR 84j:03065 • ID 34657

LEWIS, D. [1981] *Ordering semantics and premise semantics for counterfactuals* (**J** 0122) J Philos Logic 10*217-234
◇ A05 B45 ◇ REV MR 82m:03017 Zbl 473 # 03015 • ID 55346

LEWIS, D. [1981] *Probabilities of conditionals and conditional probabilities* (**C** 4140) Ifs 129-147
◇ B48 ◇ REV MR 83a:03003 • ID 47739

LEWIS, D. [1982] *Logic for equivocators* (**J** 0097) Nous, Quart J Phil 16*431-441
◇ B45 B53 ◇ REV MR 85d:03034 • ID 41001

LEWIS, D. see Vol. III, V for further entries

LEWIS, H.A. [1985] *Substitutional quantification and nonstandard quantifiers* (**J** 0097) Nous, Quart J Phil 19*447-476
◇ B60 ◇ ID 49011

LEWIS, H.A. see Vol. VI for further entries

LEWIS II, P.M. [1967] see COATES, C.L.

LEWIS II, P.M. see Vol. IV for further entries

LEWY, C. [1976] *Meaning and modality* (**X** 0805) Cambridge Univ Pr: Cambridge, GB xi + 157pp
◇ A05 B46 ◇ REV MR 56 # 78 Zbl 368 # 02002 JSL 47.909 • ID 51219

LI, BANGRONG & LIU, GUOHAN [1980] *Some properties of fuzzy space and fuzzy logic (Chinese) (English summary)* (**J** 2754) Huazhong Gongxueyuan Xuebao 2*24-38
◇ B52 E72 ◇ REV MR 81i:94049 • REM Special issue on fuzzy mathematics, II • ID 82094

LI, BIQIANG & WANG, PEIZHUANG [1982] *Fuzzy logistic reasoning of second order (Chinese) (English summary)* (**J** 3733) Shuxue Zazhi 2*37-44
◇ B52 ◇ REV MR 83g:03024 Zbl 529 # 03007 • ID 34838

LI, CHULIN & WU, XUEMOU & YU, HONGZU [1980] *A binary comparison average method for fuzzy pattern recognition (Chinese)* (**J** 2754) Huazhong Gongxueyuan Xuebao 2*67-70
• TRANSL [1981] (**J** 2684) J Huazhong Inst Tech (Engl Ed) 3/1*26-29
◇ B52 ◇ REV MR 83h:68149 • ID 46686

LI, TAIHANG [1984] *Degree of abstraction of concepts and longitudinal reasoning* (**P** 3081) IFAC Symp Fuzzy Inf, Knowl Repr & Decis. Anal;1983 Marseille 143-146
◇ B52 ◇ ID 48190

LI, TAIHANG [1984] *Fuzzy longitudinal reasoning and selfgeneration of membership degrees (Chinese) (English summary)* (**J** 3732) Mohu Shuxue 4/2*9-16
◇ B52 ◇ REV MR 85m:03017 Zbl 563 # 94030 • ID 44355

LI, WENGUANG & WU, XUNWEI [1984] *Three-valued unary functions and their CMOS realization (Chinese) (English summary)* (**J** 2161) Hangzhou Daxue Xuebao, Ziran Kexue 11*202-206
◇ B50 ◇ REV MR 85k:94064 • ID 44136

LI, XIANG [1980] *The first-order theory of Kripke's semantical analysis of nonclassical logics (Chinese) (English summary)* (**J** 2754) Huazhong Gongxueyuan Xuebao 8/1*1,1-8
◇ B45 ◇ REV MR 84h:03038 • ID 34247

LI, XIANG see Vol. IV for further entries

LIAO, QUN [1985] *Solutions of fuzzy equations and the inverse problem of fuzzy reasoning based on $\bar{t}$ (Chinese) (English summary)* (**J** 3732) Mohu Shuxue 5/1*9-18
◇ B52 E72 ◇ ID 49012

LICHTBLAU, D.E. [1976] *Prior and the Barcan formula* (**J** 0047) Notre Dame J Formal Log 17*622-624
◇ A05 B45 ◇ REV MR 56 # 93 Zbl 315 # 02025 • ID 63441

LIDDELL, G.F. [1982] *A logic for propositions with indefinite truth values* (**J** 0063) Studia Logica 41*197-226
◇ B48 ◇ REV MR 85i:03062 Zbl 564 # 03020 • ID 42304

LIEBER, L.R. [1947] *Mits, wits, and logic* (**X** 0942) Norton: New York 240pp
◇ B60 B98 ◇ REV Zbl 223 # 02031 JSL 13.55 • ID 22514

LIEBER, L.R. see Vol. V for further entries

LIEBLER, M. & ROESSER, R. [1971] *Multiple-real-valued Walsh functions* (**P** 2007) Symp Th & Appl of Multi-Val Log Design;1971 Buffalo 84-102
◇ B50 ◇ ID 41986

LIGHTSTONE, A.H. [1964] *The axiomatic method: an introduction to mathematical logic* (**X** 0819) Prentice Hall: Englewood Cliffs x+246pp
◇ B98 ◇ REV MR 29 # 1133 Zbl 129.256 JSL 31.106 • ID 22360

LIGHTSTONE, A.H. [1978] *Mathematical logic. An introduction to model theory. Edited by H. B. Enderton* (**X** 1332) Plenum Publ: New York xiii+338pp
◇ B98 C98 H05 ◇ REV MR 80i:03002 Zbl 382 # 03002 • ID 51926

LIGHTSTONE, A.H. see Vol. I, III for further entries

LIN, FENGBU [1965] *On the congruenziation of k-valued logics (Chinese)* (**P** 4564) Math Logic;1963 Xi-An 117-118
◇ B50 ◇ ID 49349

LINDEMANS, J.F. [1983] *The reduction of deontic logic to the first-order predicate calculus* (**J** 0079) Logique & Anal, NS 26*191-202
◇ B45 ◇ REV MR 85h:03023 Zbl 534 # 03008 • ID 36556

LINDLEY, D.V. [1982] *Scoring rules and the inevitability of probability (French summary)* (**J** 3946) Int Stat Rev 50*1-26
◇ B52 ◇ REV MR 83m:62006 Zbl 497 # 62004 • ID 36639

LINDNER, R. & WAGNER, K. [1974] *The axiomatisation of a certain sequential propositional calculus (Russian)* (**J** 0052) Probl Kibern 28*43-80,278 • ERR/ADD ibid 29*248
◇ B60 B70 D05 D10 ◇ REV MR 51 # 89 Zbl 313 # 94021 • ID 24790

LINDNER, R. [1975] *On the theory of inference operators* (**P** 1521) Int Congr Math (II,12);1974 Vancouver 2*471-475
◇ B48 F99 ◇ REV MR 55 # 5417 Zbl 375 # 02029 • ID 51606

LINDNER, R. see Vol. I, IV, V for further entries

LINDSTROEM, T.L. [1982] *A Loeb-measure approach to theorems by Prohorov, Sazonov and Gross* (**J** 0064) Trans Amer Math Soc 269*521-534
◇ B50 H05 ◇ REV MR 83m:60008 Zbl 484 # 60007 • ID 36617

LINDSTROEM, T.L. see Vol. I for further entries

LINKE, P.F. [1948] *Die mehrwertigen Logiken und das Wahrheitsproblem* (**J** 0357) Z Phil Forsch 3*378-398,530-546
◇ B50 ◇ REV JSL 17.276 • ID 41848

LINN, R. [1983] *Kompositionsvollstaendigkeit von Algebren mit (m,n)-stelligen Operationen* (**S** 0405) Mitt Math Sem Giessen 160*115pp
◇ B50 ◇ REV MR 85h:08007 Zbl 526 # 08002 • ID 44859

LINSKY, L. [1971] *Reference and modality* (**X** 0894) Oxford Univ Pr: Oxford v+177pp
◇ A05 B45 ◇ ID 25137

LINSKY, L. [1977] *Names and descriptions* (**X** 0862) Univ Chicago Pr: Chicago xxi+184pp
◇ A05 B45 ◇ REV MR 80d:03017 • ID 75564

LINSKY, L. [1979] *Believing and necessity* (**J** 0472) Theory Decis 11*81-94
◇ A05 B45 ◇ REV MR 80m:03017 • ID 75563

LIPSKI JR., W. [1977] *On the logic of incomplete information* (**P** 1635) Math Founds of Comput Sci (6);1977 Tatranska Lomnica 53*374-381
◇ B60 ◇ REV MR 57 # 102 Zbl 363 # 02025 • ID 50852

LIPSKI JR., W. see Vol. V for further entries

LIST, G. [1973] see BAGINSKI, M.

LIU, DSOSU [1982] see CHENG, ZHAOZHONG

LIU, GUOHAN [1980] see LI, BANGRONG

LIU, GUOHAN [1981] *Some transforming relations among the pansystem logic spaces with soft algebra shadow system (Chinese) (English summary)* (**J** 3594) Kexue Tansuo 1981/1*175-182
◇ B52 ◇ REV MR 82j:03030 • ID 75590

LIU, SHICHAO [1974] *A theory of first order logic with description* (**J** 0406) Bull Inst Math, Acad Sin (Taipei) 2*43-65
◇ B60 ◇ REV MR 50 # 50 Zbl 304 # 02006 • ID 08208

LIU, SHICHAO see Vol. I, IV, V, VI for further entries

LIU, XUHUA [1983] *Minimization of fuzzy logic formulas (Chinese) (English summary)* (**J** 3732) Mohu Shuxue 3/3*7-15
◇ B52 ◇ REV MR 85i:03075 Zbl 516 # 94036 • ID 44176

LIU, XUHUA see Vol. I for further entries

LIU, YOWEI [1982] see EPSTEIN, G.

LIU, YUNFENG [1982] *An axiomatic structure of fuzzy set theory and fuzzy logic (Chinese) (English summary)* (**J** 3732) Mohu Shuxue 2/3*73-82
◇ B52 E72 ◇ REV MR 84a:03066 • ID 34899

LOADER, J. [1972] *Universal decision elements in m-valued logic* (**J** 0068) Z Math Logik Grundlagen Math 18*205-216
◇ B50 ◇ REV MR 50 # 1840 Zbl 253 # 02011 • ID 08221

LOADER, J. [1974] *A method for finding formulae corresponding to first order universal decision elements in m-valued logic* (**J** 0068) Z Math Logik Grundlagen Math 20*1-18
◇ B50 B70 ◇ REV MR 50 # 1841 Zbl 336 # 02010 • ID 08222

LOADER, J. [1975] *An alternative concept of the universal decision element in m-valued logic* (**J** 0068) Z Math Logik Grundlagen Math 21*369-375
◇ B50 ◇ REV MR 51 # 7821 Zbl 344 # 02011 • ID 08223

LOADER, J. [1975] *Second order and higher order universal decision elements in m-valued logic* (**P** 1805) Int Symp Multi-Val Log (5,Proc);1975 Bloomington 53-57
◇ B50 ◇ REV MR 58 # 5029 • ID 35811

LOADER, J. [1976] *The binary representation of m-valued logic with applications to universal decision elements* (**P** 2011) Int Symp Multi-Val Log (6);1976 Logan 219-223
◇ B50 ◇ REV MR 58 # 16138 Zbl 336 # 02011 • ID 75608

LOADER, J. [1977] *Second order and higher order universal decision elements in m-valued logic* (**J** 0047) Notre Dame J Formal Log 18*313-317
◇ B50 ◇ REV MR 56 # 5216 Zbl 305 # 02027 • ID 23625

LOCK, P.F. [1985] see HARDEGREE, G.M.

LOEB, M.H. [1966] *Extensional interpretations of modal logics* (**J** 0036) J Symb Logic 31*23-45
◇ B45 ◇ REV MR 35 # 33 JSL 36.692 • ID 08228

LOEB, M.H. (ED.) [1968] *Proceedings of the summer school in logic, Leeds, 1967* (**S** 3301) Lect Notes Math 70*iv+331pp
◇ B97 C97 F97 ◇ ID 37293

LOEB, M.H. see Vol. I, III, IV, VI for further entries

LOESCHAU, G. [1973] see BAGINSKI, M.

LOEWER, B.M. [1979] *Cotenability and counterfactual logics* (**J** 0122) J Philos Logic 8*99-115
◇ A05 B45 ◇ REV MR 80k:03020 Zbl 413 # 03021
• ID 53015

LOEWER, B.M. [1983] see BELZER, M.

LOLLI, G. [1978] *Lezioni di logica matematica* (**X** 0905) Boringhieri: Torino 203pp
◇ B98 ◇ ID 31832

LOLLI, G. see Vol. I, III, IV, V, VI for further entries

LONDEY, D.G. [1965] see HUGHES, G.E.

LOPARIC, A. & ROUTLEY, R. [1978] *Semantical analysis of Arruda da Costa P systems and adjacent non-replacement relevant systems* (**J** 0063) Studia Logica 37*301-320
◇ B53 ◇ REV MR 80g:03020 Zbl 406 # 03034 • ID 56117

LOPARIC, A. [1978] *The method of valuations in modal logic* (**P** 1800) Brazil Conf Math Log (1);1977 Campinas 141-157
◇ B45 ◇ REV MR 80b:03022 Zbl 385 # 03013 • ID 52121

LOPARIC, A. & ROUTLEY, R. [1980] *Semantics for quantified relevant logics without replacement* (**P** 3006) Brazil Conf Math Log (3);1979 Recife 263-280
◇ B46 ◇ REV MR 82g:03044 Zbl 451 # 03009 • ID 54024

LOPARIC, A. [1980] see ALVES, E.H.

LOPARIC, A. & MORTARI, C.A. [1983] *Valuation in temporal logic* (**J** 3783) J Non-Classical Log (Univ Campinas) 2*49-60
◇ B45 ◇ ID 49013

LOPARIC, A. [1984] see COSTA DA, N.C.A.

LOPARIC, A. see Vol. I, III for further entries

LOPES, L.L. & ODEN, G.C. [1982] *On the internal structure of fuzzy subjective categories* (**P** 4051) Fuzzy Set & Possibility Th;1980 Acapulco 75-89
◇ B52 ◇ REV MR 84b:03004 • ID 46328

LOPES DOS SANTOS, L.H. [1977] *Some remarks on discussive logic* (**P** 1076) Latin Amer Symp Math Log (3);1976 Campinas 99-113
◇ B53 ◇ REV MR 57 # 12178 Zbl 366 # 02014 • ID 16596

LOPES DOS SANTOS, L. [1978] *Discussive versions of the modal calculi T, B, S4 and S5* (**P** 1800) Brazil Conf Math Log (1);1977 Campinas 159-167
◇ B53 ◇ REV MR 80k:03021 Zbl 388 # 03007 • ID 52262

LOPES DOS SANTOS, L. see Vol. I for further entries

LOPEZ-ESCOBAR, E.G.K. [1981] *On the interpolation theorem for the logic of constant domains* (**J** 0036) J Symb Logic 46*87-88
◇ B55 C40 C90 F50 ◇ REV MR 82i:03068
Zbl 469 # 03015 • REM See also 1983 • ID 55143

LOPEZ-ESCOBAR, E.G.K. [1981] *Variations on a system of Gentzen* (**J** 0068) Z Math Logik Grundlagen Math 27*385-389
◇ B55 F50 ◇ REV MR 83e:03037 Zbl 469 # 03014
• ID 55142

LOPEZ-ESCOBAR, E.G.K. [1982] *A natural deduction system for some intermediate logics* (**J** 3783) J Non-Classical Log (Univ Campinas) 1*21-41
◇ B55 ◇ REV MR 84g:03031 Zbl 537 # 03016 • ID 34145

LOPEZ-ESCOBAR, E.G.K. [1982] *Implicational logics in natural deduction systems* (**J** 0036) J Symb Logic 47*184-186
◇ B55 ◇ REV MR 83i:03044 Zbl 485 # 03010 • ID 35512

LOPEZ-ESCOBAR, E.G.K. [1983] *A second paper on the interpolation theorem for the logic of constant domains* (**J** 0036) J Symb Logic 48*595-599
◇ B55 C40 C90 F05 F50 ◇ REV MR 85b:03099
Zbl 547 # 03022 • REM See also 1981 • ID 39786

LOPEZ-ESCOBAR, E.G.K. [1985] *Proof functional connectives* (**P** 2160) Latin Amer Symp Math Log (6);1983 Caracas 208-221
◇ B55 F07 F30 ◇ ID 47264

LOPEZ-ESCOBAR, E.G.K. [1985] see D'OTTAVIANO, I.M.L.

LOPEZ-ESCOBAR, E.G.K. see Vol. I, III, V, VI for further entries

LOPTSON, P.J. [1980] *Q, entailment, and the Parry property* (**J** 0079) Logique & Anal, NS 23*305-317
◇ B46 ◇ REV MR 83e:03031 • ID 35216

LORENZ, K. [1968] *Dialogspiele als semantische Grundlage von Logikkalkuelen I,II* (**J** 0009) Arch Math Logik Grundlagenforsch 11*32-55,73-100
◇ B60 F50 ◇ REV MR 37 # 1233 MR 39 # 2613
Zbl 179.13 • ID 28288

LORENZ, K. [1973] *Rules versus theorems. A new approach for mediation between intuitionistic and two-valued logic* (**J** 0122) J Philos Logic 2*352-369
◇ A05 B55 F50 ◇ REV MR 55 # 7724 Zbl 262 # 02008
• ID 27473

LORENZ, K. & LORENZEN, P. [1978] *Dialogische Logik* (**X** 0890) Wiss Buchges: Darmstadt viii + 238pp
◇ A05 B60 F99 ◇ REV MR 83j:03010 Zbl 435 # 03011
• ID 28287

LORENZ, K. [1982] *On the criteria for the choice of the rules of dialogic logic* (**P** 3754) Argumentation;1978 Groningen 145-157
◇ A05 B60 ◇ REV MR 84d:03032 • ID 34072

LORENZ, K. see Vol. VI for further entries

LORENZEN, P. [1948] *Einfuehrung in die Logik* (**X** 0908) Univ Math Inst: Bonn i + 34pp
◇ B98 ◇ REV MR 10.1 Zbl 30.2 • ID 28576

LORENZEN, P. [1954] *Zur Begruendung der Modallogik* (**J** 1114) Arch Phil 5*95-108
• REPR [1954] (**J** 0009) Arch Math Logik Grundlagenforsch 2*15-28
◇ B45 ◇ REV MR 17.224 Zbl 58.246 JSL 24.174
• ID 20825

LORENZEN, P. [1955] *Einfuehrung in die operative Logik und Mathematik* (**X** 0811) Springer: Heidelberg & New York vii + 298pp
◇ A05 B98 F50 F65 F98 ◇ REV MR 17.223 Zbl 66.248
Zbl 68.8 JSL 22.289 JSL 35.330 • REM 2nd ed. 1969
• ID 08290

LORENZEN, P. [1956] *Protologik. Ein Beitrag zum Begruendungsproblem der Logik* (**J** 1103) Kant Studien 47
◇ B60 ◇ ID 33169

LORENZEN, P. [1962] *Das Problem einer Formalisierung der Hegelschen Logik* (**J** 4130) Hegel-Stud Beiheft 1
◇ A05 B53 ◇ ID 33640

LORENZEN, P. [1962] *Metamathematik* (**X** 0876) Bibl Inst: Mannheim 173pp
• TRANSL [1967] (**X** 0834) Gauthier-Villars: Paris 162pp (French) [1971] (**X** 1781) Tecnos: Madrid (Spanish)
◇ A05 B98 C60 D20 D35 D98 F98 ◇ REV MR 28 # 3932 Zbl 105.246 JSL 31.106 • ID 08303

LORENZEN, P. [1968] *On modal logic* (**J** 2173) J Japan Ass Phil Sci 15
◇ B45 ◇ ID 33171

LORENZEN, P. [1969] *Theophrastische Modallogik* (**J** 0009) Arch Math Logik Grundlagenforsch 12*72-75
◇ A10 B45 ◇ REV MR 42 # 7485 Zbl 182.5 • ID 08308

LORENZEN, P. [1970] *Logica formal* (**X** 1781) Tecnos: Madrid
◇ B98 ◇ ID 33353

LORENZEN, P. [1972] *Dialogkalkuele* (**J** 0009) Arch Math Logik Grundlagenforsch 15*99-102
◇ B60 F50 ◇ REV MR 50 # 12676 Zbl 288 # 02011 • ID 08311

LORENZEN, P. [1972] *Zur konstruktiven Deutung der semantischen Vollstaendigkeit klassischer Quantoren- und Modalkalkuele* (**J** 0009) Arch Math Logik Grundlagenforsch 15*103-117
◇ B45 F50 ◇ REV MR 49 # 4750 Zbl 305 # 02024 • ID 08310

LORENZEN, P. [1978] see LORENZ, K.

LORENZEN, P. [1978] *Praktische und theoretische Modalitaten* (**J** 0455) Phil Naturalis 17*261-279
◇ B45 ◇ REV MR 80m:03038 • ID 75669

LORENZEN, P. [1982] *Die dialogische Begruendung von Logikkalkuelen* (**P** 3754) Argumentation;1978 Groningen 23-54
◇ A05 B60 ◇ REV MR 84d:03031 Zbl 554 # 03007 • ID 34071

LORENZEN, P. see Vol. I, IV, V, VI for further entries

LOS, J. [1948] *Many-valued logics and the formalization of intensional functions (Polish)* (**J** 0479) Kwart Filoz 17*59-78
◇ B50 ◇ REV MR 10.1 JSL 14.64 • ID 08313

LOS, J. [1949] *On logical matrices (Polish)* (**S** 0502) Prace Wroclaw Tow Nauk, Ser B 19*42pp
◇ B22 B50 ◇ REV MR 19.724 Zbl 45.296 JSL 16.59 • ID 19296

LOS, J. [1955] see GRZEGORCZYK, A.

LOS, J. [1963] *Semantic representation of the probability of formulas in formalized theories (Polish summary)* (**J** 0063) Studia Logica 14*183-196
• REPR [1977] (**C** 3174) Twenty-five Years Log Meth Poland 327-340
◇ A05 B10 B48 B60 ◇ REV MR 32 # 3087 Zbl 292 # 02008 • ID 31415

LOS, J. [1982] see COHEN, L.J.

LOS, J. see Vol. I, III, V, VI for further entries

LOU, SHIBO & TONG, ZHENGXIANG [1982] *LSG powers of an L-fuzzy matrix* (**C** 3786) Approx Reason in Decis Anal 51-55
◇ B52 ◇ REV MR 84m:15020 Zbl 504 # 15008 • ID 38407

LOU, SHIBO see Vol. V for further entries

LOU, ZHUKAI [1982] *The problem of functional completeness in many-valued logics (Chinese) (English summary)* (**J** 3948) Xiangtan Daxue Ziran Kexue Xuebao 1982/1*1-13
◇ B50 ◇ REV Zbl 493 # 03005 • ID 38101

LOU, ZHUKAI [1983] *Galois theory of total and partial functions in many-valued logics (Chinese) (English summary)* (**J** 3948) Xiangtan Daxue Ziran Kexue Xuebao 1983/1*1-12
◇ B50 ◇ REV Zbl 533 # 03009 • ID 36534

LOUREIRO, I. [1982] *Axiomatisation et proprietes des algebres modales tetravalentes (English summary)* (**J** 3364) C R Acad Sci, Paris, Ser 1 295*555-557
◇ B45 G25 ◇ REV MR 83m:03072 Zbl 516 # 03010 • ID 34889

LOWE, E.J. [1982] *On the alleged necessity of true identity statements* (**J** 0094) Mind 91(364)*579-584
◇ A05 B45 ◇ REV MR 84k:03056 • REM See also Baldwin,T. 1984 • ID 34994

LOWE, E.J. [1983] *A simplification of the logic of conditionals* (**J** 0047) Notre Dame J Formal Log 24*357-366
◇ B45 ◇ REV MR 84h:03039 Zbl 487 # 03010 • ID 34248

LOWEN, R. [1981] see KLEMENT, E.P.

LOWEN, R. see Vol. V for further entries

LOWESMITH, B.J. [1981] *Three-valued commutative pseudo-Sheffer functions* (**J** 0068) Z Math Logik Grundlagen Math 27*161-180
◇ B50 ◇ REV MR 82h:03019 Zbl 501.03011 • ID 75697

LOWESMITH, B.J. & ROSE, A. [1984] *A generalisation of Slupecki's criterion for functional completeness* (**J** 0068) Z Math Logik Grundlagen Math 30*173-175
◇ B50 ◇ REV MR 85i:03070 • ID 42220

LU, YANGCI & LU, ZHONGWAN & TANG, ZHISONG & WAN, ZHEXIAN [1960] *Mathematical logic and its application (Chinese)* (**J** 2771) Kexue Tongbao 2*50-52
◇ B98 ◇ ID 48538

LU, ZHONGWAN [1960] see LU, YANGCI

LU, ZHONGWAN [1981] see HU, SHIHUA

LU, ZHONGWAN [1982] see HU, SHIHUA

LU, ZHONGWAN see Vol. I for further entries

LUCAS, T. [1984] see LAVENDHOMME, R.

LUCAS, T. see Vol. I, III, V, VI for further entries

LUCE, DAVID RANDALL [1966] *A calculus of "before"* (**J** 0105) Theoria (Lund) 32*25-44
◇ B45 ◇ REV JSL 34.646 • ID 08365

LUCHINS, A.S. & LUCHINS, E.H. [1965] *Logical foundation of mathematics for behavioral scientists* (**X** 0818) Holt Rinehart & Winston: New York xii + 436pp
◇ B98 ◇ ID 22516

LUCHINS, E.H. [1965] see LUCHINS, A.S.

LUCKHARDT, H. [1967] *Von Wright's relative Modalitaeten* (**J** 0009) Arch Math Logik Grundlagenforsch 10*97-112
◇ B45 ◇ REV MR 41 # 3249 Zbl 165.306 • ID 08377

LUCKHARDT, H. see Vol. I, III, IV, VI for further entries

LUE, QICI [1983] *Counterexample to Bowen's theorem and failures of interpolation theorem in modal logic (Chinese)* (J 2771) Kexue Tongbao 28*897-899
• TRANSL [1984] (J 3769) Sci Bull, Foreign Lang Ed 29*433-436
◇ B45 C40 ◇ REV MR 86b:03022 Zbl 543 # 03012
• ID 40907

LUE, QICI [1984] *Counterexample to Bowen's theorem and failures of interpolation theorem in modal logic (Chinese)* (J 2771) Kexue Tongbao 29*433-436
◇ B45 ◇ REV Zbl 543 # 03012 • ID 48695

LUE, QICI see Vol. I for further entries

LUGINBUHL, D.R. [1983] see KANDEL, A.

LUKASIEWICZ, J. [1920] *On three-valued logic (Polish)* (J 1093) Ruch Filoz 5*169-171
• TRANSL [1967] (C 0615) Polish Logic 1920-39 16-18
◇ A05 B50 ◇ REV JSL 35.442 • ID 21214

LUKASIEWICZ, J. [1921] *Two-valued logic (Polish)* (J 1125) Przeglad Filoz 23*189-205
◇ A05 B05 B50 ◇ REV FdM 48.1125 • ID 28632

LUKASIEWICZ, J. [1929] *Elements of mathematical logic (Polish)* (X 1034) PWN: Warsaw 99pp
• TRANSL [1963] (X 0869) Pergamon Pr: Oxford xi + 124pp
◇ B98 ◇ REV MR 22 # 6692 Zbl 126.7 JSL 30.237 JSL 31.284 • REM 2nd ed. 1958 • ID 20964

LUKASIEWICZ, J. [1930] *Philosophische Bemerkungen zu mehrwertigen Systemen des Aussagenkalkuels* (J 0459) C R Soc Sci Lett Varsovie Cl 3 23*51-77
• TRANSL [1967] (C 0615) Polish Logic 1920-39 40-65
◇ A05 B50 ◇ REV JSL 33.131 JSL 35.442 FdM 57.1319
• ID 08392

LUKASIEWICZ, J. [1935] *Zur vollen dreiwertigen Aussagenlogik* (J 0748) Erkenntnis (Leipzig) 5*176
◇ B50 ◇ ID 08395

LUKASIEWICZ, J. [1937] *In defense of logistic (Polish)* (S 4530) Stud Gnieseusia 15*22pp
• TRANSL [1970] (C 1786) Lukasiewicz: Select Works 236-249 [1961] (C 0609) Lukasiewicz: Log & Philos Papers 210-219
◇ A05 B98 ◇ REV JSL 3.43 JSL 33.132 • ID 41703

LUKASIEWICZ, J. [1941] *Die Logik und das Grundlagenproblem* (P 0652) Entretiens Zuerich Fond & Method Sci Math;1938 Zuerich 82-100
• TRANSL [1970] (C 1786) Lukasiewicz: Select Works 278-294
◇ A05 B98 ◇ REV MR 2.338 Zbl 61.8 JSL 7.35
• ID 16528

LUKASIEWICZ, J. [1951] *On variable functors of propositional arguments* (J 0215) Proc Irish Acad, Sect A 54*25-35
◇ B60 ◇ REV MR 13.3 Zbl 42.244 JSL 16.229 • ID 08398

LUKASIEWICZ, J. [1953] *A system of modal logic* (J 0093) J Comp Syst 1*111-149
• REPR [1953] (P 0645) Int Congr Philos (11);1953 Bruxelles 14*82-87
◇ B45 ◇ REV MR 15.189 MR 15.2 Zbl 52.10 Zbl 53.6 JSL 25.293 • REM Reprint is a shorted version • ID 00002

LUKASIEWICZ, J. [1954] *Arithmetic and modal logic* (J 0093) J Comp Syst 1*213-219
◇ B45 ◇ REV MR 16.554 Zbl 57.246 JSL 25.293
• ID 22324

LUKASIEWICZ, J. [1954] *On a controversial problem of Aristotle's modal syllogistic* (J 0505) Dominican Studies 7*114-128
◇ A05 A10 B45 ◇ REV JSL 25.293 • ID 08403

LUKASIEWICZ, J. [1961] *Logistic and philosophy (Polish)* (C 0609) Lukasiewicz: Log & Philos Papers 195-209
◇ A05 B98 ◇ REV JSL 33.132 • ID 45092

LUKASIEWICZ, J. [1970] *Selected works* (X 0809) North Holland: Amsterdam xii + 405pp
◇ A05 A10 B96 ◇ REV MR 45 # 3155 Zbl 212.9
• ID 08410

LUKASIEWICZ, J. see Vol. I, VI for further entries

LUNDBERG, B. [1982] *An axiomatization of events* (J 0130) BIT 22*291-299
◇ B45 ◇ REV MR 84f:94004 Zbl 484 # 68078 • ID 36624

LUNGARZO, C.A. [1978] *Superposition of states in quantum logic from a set theoretical point of view* (P 1800) Brazil Conf Math Log (1);1977 Campinas 169-177
◇ B51 E75 G12 ◇ REV MR 80h:03092 Zbl 385 # 03056
• ID 52163

LUO, ZHUKAI [1965] *On the maximal closed set in 4-valued logic (Chinese)* (P 4564) Math Logic;1963 Xi-An 100-110
◇ B50 ◇ ID 49333

LUO, ZHUKAI [1980] *The determination of all maximal closed sets of functions in multivalued logic (Chinese)* (J 0418) Shuxue Xuebao 23*152-156
◇ B50 ◇ REV MR 83d:03027 Zbl 433 # 03012 • ID 35176

LUO, ZHUKAI [1984] *Maximal closed sets in the set of partial many-valued logic functions (Chinese)* (J 0418) Shuxue Xuebao 27*795-800
◇ B50 ◇ ID 48961

LUO, ZHUKAI [1984] *The classification of normal relations in many-valued logics (Chinese) (English summary)* (J 3948) Xiangtan Daxue Ziran Kexue Xuebao 1984/2*7-21
◇ B50 ◇ ID 48960

LUO, ZHUKAI [1984] *The completeness theory of partial many-valued logic functions (Chinese)* (J 0418) Shuxue Xuebao 27*676-683
◇ B50 ◇ REV Zbl 552 # 03017 • ID 46832

LUSZCZEWSKA-ROMAHNOWA, S. [1957] *Induction and probability (Polish) (Russian and English summaries)* (J 0063) Studia Logica 5*71-90 • ERR/ADD ibid 5*5
◇ B48 ◇ REV MR 19.237 • ID 33077

LUSZCZEWSKA-ROMAHNOWA, S. see Vol. I, V for further entries

LUTZ, R. [1981] see GOZE, M.

LUTZ, R. see Vol. I for further entries

LYASHENKO, L.N. & VASHCHENKO, V.P. [1980] *Multiple functional decomposition in K-valued logic (Russian)* (S 2850) Tr Ehnerg Inst Moskva 496*195-198
◇ B50 ◇ REV MR 84i:03054 • ID 34533

LYNCH, E.P. [1980] *Applied symbolic logic* (X 0820) Intersci Publ: New York xi + 260pp
◇ B98 ◇ REV Zbl 528 # 94018 • ID 36772

MA, XIWEN [1981] *Formalization of the logical problem of "to know" (Chinese)* (J 4452) Zhexue Yanjiu 5*30-38
◇ B60 ◇ ID 48521

MA, XIWEN see Vol. I for further entries

MACBEATH, M. [1983] *Communication and time reversal* (J 0154) Synthese 56*27-46
◇ B45 ◇ REV MR 84j:03016 • ID 34610

MACCHI, P. [1980] *Funzioni verita per* S_5 (J 0319) Matematiche (Sem Mat Catania) 35*264-269
◇ B45 ◇ REV MR 85f:03016 Zbl 525 # 03004 • ID 45793

MACCOLL, H. [1906] *Symbolic logic and its applications* (X 1296) Longman: Harlow & New York xi + 141pp
◇ B05 B98 ◇ REV FdM 37.66 • ID 22595

MACCOLL, H. see Vol. I for further entries

MACE, C.A. [1933] *The principles of logic: an introductory survey* (X 1296) Longman: Harlow & New York xii + 388pp
◇ B98 ◇ REV FdM 59.860 • ID 22596

MACHIDA, H. [1981] *On closed sets of three-valued monotone logical functions* (P 2552) Conf Finite Algeb & Multi-Val Log;1979 Szeged 441-467
◇ B50 ◇ REV MR 83d:03028 Zbl 488 # 03013 • ID 35177

MACHIDA, H. [1982] *On the lattice of closed sets of functions of three-valued logic (Japanese)* (P 4081) Many-Val Log & Appl;1982 Kyoto 40-52
◇ B50 ◇ ID 47616

MACHIDA, H. [1982] *Toward a classification of minimal closed sets in 3-valued logic* (P 4002) Int Symp Multi-Val Log (12);1982 Paris 313-317
◇ B50 ◇ REV Zbl 556 # 03022 • ID 46150

MACHINA, K.F. [1976] *Truth, belief, and vagueness* (J 0122) J Philos Logic 5*47-78
◇ A05 B52 ◇ REV MR 55 # 7725 Zbl 336 # 02009 • ID 63543

MACHOVER, M. [1969] see HIRSCHFELD, J.

MACHOVER, M. [1977] see BELL, J.L.

MACHOVER, M. see Vol. I, III, IV, V for further entries

MACINTYRE, A. & PACHOLSKI, L. & PARIS, J.B. (EDS.) [1978] *Logic colloquium '77. Proceedings of the colloquium held in Wroclaw, august 1977* (S 3303) Stud Logic Found Math 96*x + 311pp
◇ B97 C97 ◇ REV MR 80a:03003 Zbl 426 # 00004 • ID 53596

MACINTYRE, A. see Vol. III, IV, V, VI for further entries

MACK, J.M. [1975] see BARNES, D.W.

MACKENZIE, J.D. [1979] *Question-begging in non-cumulative systems* (J 0122) J Philos Logic 8*117-133
◇ A05 B53 ◇ REV MR 80f:03030 Zbl 397 # 03004 • ID 52670

MACKIE, J.L. [1969] *The relevance criterion of confirmation* (J 0013) Brit J Phil Sci 20*27-40
◇ B48 ◇ REV Zbl 233 # 02002 • ID 63564

MACLELLAN, E.J. [1978] see JENSEN, J.B.

MACNAB, D.S. [1979] see BEAZER, R.

MACNAB, D.S. see Vol. III, V, VI for further entries

MACVICAR-WHELAN, P.J. [1977] *Fuzzy and multivalued logic* (P 2013) Int Symp Multi-Val Log (7);1977 Charlotte 98-102
◇ B52 ◇ ID 35858

MACVICAR-WHELAN, P.J. [1979] *Fuzzy-logic: An alternative approach* (P 3003) Int Symp Multi-Val Log (9);1979 Bath 152-158
◇ B52 ◇ REV MR 81j:03040 • ID 35946

MACZYNSKI, M.J. [1971] *Boolean properties of observables in axiomatic quantum mechanics* (J 1546) Rep Math Phys (Warsaw) 2*135-150
◇ B51 G05 G12 ◇ REV MR 44 # 5012 Zbl 222 # 02071 • ID 26325

MACZYNSKI, M.J. [1981] *Commutativity and generalized transition probability in quantum logic* (P 3820) Curr Iss in Quantum Log;1979 Erice 355-364
◇ B51 ◇ REV MR 84j:03131 Zbl 537 # 03045 • ID 34721

MACZYNSKI, M.J. [1983] see DORNINGER, D.

MACZYNSKI, M.J. see Vol. III for further entries

MADUCH, M. [1972] *A definability criterion for the operations in finite Lukasiewicz's implicational matrices* (J 0387) Bull Sect Logic, Pol Acad Sci 1/4*15-17
◇ B50 ◇ ID 42019

MADUCH, M. & PIROG-RZEPECKA, K. [1973] *An adequate matrix for a certain sentential calculus (Polish) (English summary)* (S 1454) Zesz Nauk Wyz Szk Ped Mat, Opole 13*75-78
◇ B50 ◇ REV MR 53 # 10547 Zbl 314 # 02017 • ID 23060

MADUCH, M. [1973] *Axiomatization of the logic W (Polish) (English summary)* (S 1454) Zesz Nauk Wyz Szk Ped Mat, Opole 13*79-83
◇ B60 ◇ REV MR 54 # 4961 Zbl 314 # 02020 • ID 24129

MADUCH, M. [1978] *On three-valued implicative systems* (J 0063) Studia Logica 37*351-385
◇ B50 ◇ REV MR 80m:03054 Zbl 397 # 03012 • ID 52678

MADUCH, M. see Vol. I for further entries

MAEHARA, S. [1966] *Introduction to mathematical logic (Japanese)* (X 1168) Kyoritsu Shuppan: Tokyo i + 3 + 214 + 5pp
◇ B98 ◇ REV MR 35 # 2716 • ID 22167

MAEHARA, S. see Vol. I, III, IV, VI for further entries

MAGARI, R. [1966] *Calcoli generali e spazi* V_α (J 0319) Matematiche (Sem Mat Catania) 21*83-108
◇ B22 B60 ◇ REV MR 33 # 7242 Zbl 145.245 • REM Part I. Part II 1966 • ID 49174

MAGARI, R. [1966] *Sull'introduzione dei connettivi in un calcolo generale. calcoli generali. III* (J 0319) Matematiche (Sem Mat Catania) 21*237-281
◇ B50 ◇ REV MR 36 # 2486 Zbl 166.3 • REM Part II 1966 • ID 08531

MAGARI, R. [1967] *Sun una questione riguardante le chiusure di Moore* (J 0144) Rend Sem Mat Univ Padova 38*189-198
◇ B60 ◇ REV MR 36 # 3639 Zbl 155.349 • ID 08533

MAGARI, R. [1974] *Sur certe teorie non enumerabili. Sulle limitazioni dei sistemi formali I* (J 3526) Ann Mat Pura Appl, Ser 4 98*119-152
◇ B53 B60 F30 ◇ REV MR 50 # 6802 Zbl 286 # 02051 • ID 08539

MAGARI, R. [1982] *Primi risultati sulla varieta di Boolos (English summary)* (**J** 2100) Boll Unione Mat Ital, VI Ser, B 1*359-367
◇ B45 F30 G25 ◇ REV MR 83m:03073 Zbl 487 # 03039
• ID 35470

MAGARI, R. see Vol. I, VI for further entries

MAJEWSKI, M. [1978] *On some matrix of the Birkhoff and v. Neumann quantum logic* (**J** 0387) Bull Sect Logic, Pol Acad Sci 7*133-136
◇ B51 G12 ◇ REV MR 80d:03061 Zbl 426 # 03067
• ID 53663

MAJEWSKI, M. [1983] *About the intermediate logics between modular logic and the classical propositional calculus* (**J** 0481) Acta Univ Wroclaw 517(29)(Logika 8)*39-46
◇ B51 ◇ REV Zbl 553 # 03015 • ID 43307

MAKINSON, D. [1965] *An alternative characterisation of first-degree entailment* (**J** 0079) Logique & Anal, NS 8*308-311 • ERR/ADD ibid 9*394
◇ B46 ◇ REV MR 33 # 5460 Zbl 166.2 JSL 36.521
• ID 19279

MAKINSON, D. [1966] *On some completeness theorems in modal logic* (**J** 0068) Z Math Logik Grundlagen Math 12*379-384
◇ B45 ◇ REV MR 36 # 6269 Zbl 295 # 02014 JSL 35.135
• ID 08567

MAKINSON, D. [1966] *There are infinitely many diodorean modal functions* (**J** 0036) J Symb Logic 31*406-408
◇ B45 ◇ REV MR 34 # 5662 Zbl 147.249 JSL 32.397
• ID 08566

MAKINSON, D. [1969] *A normal modal calculus between T and S4 without the finite model property* (**J** 0036) J Symb Logic 34*35-38
◇ B45 ◇ REV MR 43 # 4643 Zbl 184.8 JSL 36.692
• ID 08569

MAKINSON, D. [1970] *A generalisation of the concept of a relational model for modal logic* (**J** 0105) Theoria (Lund) 36*331-335
◇ B45 ◇ REV MR 52 # 2823 Zbl 222 # 02019 JSL 38.520
• ID 08570

MAKINSON, D. [1971] *Aspectos de la logica modal* (**S** 0889) Notas Logica Mat 28*iii + 131pp
◇ B45 ◇ REV MR 51 # 12483 Zbl 226 # 02013 JSL 38.330
• ID 17271

MAKINSON, D. [1971] *Some embedding theorems for modal logic* (**J** 0047) Notre Dame J Formal Log 12*252-254
◇ B45 ◇ REV MR 52 # 2824 Zbl 193.293 JSL 39.351
• ID 08571

MAKINSON, D. [1973] *A warning about the choice of primitive operators in modal logic* (**J** 0122) J Philos Logic 2*193-196
◇ B45 ◇ REV MR 54 # 2424 Zbl 266 # 02016 • ID 08572

MAKINSON, D. [1973] *Topics in modern logic* (**X** 0816) Methuen: London & New York viii + 107pp
◇ B98 ◇ REV MR 58 # 27172 • ID 22363

MAKINSON, D. & SEGERBERG, K. [1974] *Post completeness and ultrafilters* (**J** 0068) Z Math Logik Grundlagen Math 20*385-388
◇ B22 B45 G20 ◇ REV MR 52 # 2825 Zbl 298 # 02014
• ID 08573

MAKINSON, D. [1981] see ALCHOURRON, C.E.

MAKINSON, D. [1981] *Non-equivalent formulae in one variable in a strong omnitemporal modal logic* (**J** 0068) Z Math Logik Grundlagen Math 27*111-112
◇ B45 ◇ REV MR 82h:03014 Zbl 467 # 03012 • ID 55010

MAKINSON, D. [1981] *Quantificational reefs in deontic waters* (**C** 3731) New Studies Deontic Log 87-91
◇ B45 ◇ REV MR 84a:03023 • ID 35567

MAKINSON, D. [1982] see ALCHOURRON, C.E.

MAKINSON, D. [1983] *Individual actions are very seldom obligatory* (**J** 3783) J Non-Classical Log (Univ Campinas) 2*7-13
◇ B45 ◇ REV Zbl 561 # 03010 • ID 47489

MAKINSON, D. [1985] *How to give it up: a survey of some formal aspects of the logic of theory change* (**J** 0154) Synthese 62*347-363
◇ B60 ◇ ID 45528

MAKINSON, D. [1985] see ALCHOURRON, C.E.

MAKINSON, D. see Vol. I for further entries

MAKOWSKY, J.A. & MARCJA, A. [1977] *Completeness theorems for modal model theory with the Montague-Chang semantics I* (**J** 0068) Z Math Logik Grundlagen Math 23*97-104
◇ B45 C90 ◇ REV MR 58 # 5057 Zbl 362 # 02043
• ID 31849

MAKOWSKY, J.A. [1980] *Measuring the expressive power of dynamic logics: an application of abstract model theory* (**P** 2904) Automata, Lang & Progr (7);1980 Noordwijkerhout 409-421 • ERR/ADD [1981] (**P** 2903) Automata, Lang & Progr (8);1981 Akko 551
◇ B75 C95 ◇ REV MR 82c:03022 MR 82j:03021 Zbl 465 # 68012 • ID 54945

MAKOWSKY, J.A. & TIOMKIN, M.L. [1985] *Propositional dynamic logic with local assignments* (**J** 1426) Theor Comput Sci 36*71-87
◇ B75 ◇ REV Zbl 574 # 03011 • ID 47225

MAKOWSKY, J.A. see Vol. III, V for further entries

MAKSIMOVA, L.L. [1964] *On a set of axioms for the system of rigorous implication (Russian)* (**J** 0003) Algebra i Logika 3/3*59-68
◇ B46 ◇ REV MR 30 # 1032 Zbl 148.244 JSL 33.484
• ID 19277

MAKSIMOVA, L.L. [1966] *Formal deduction in the system of rigorous implication (Russian)* (**J** 0003) Algebra i Logika 5/6*33-39
◇ B46 ◇ REV MR 34 # 5663 Zbl 171.253 JSL 33.484
• ID 19276

MAKSIMOVA, L.L. [1967] *On models of the calculus E (Russian)* (**J** 0003) Algebra i Logika 6/6*5-20
◇ B46 ◇ REV MR 37 # 2585 Zbl 223 # 02016 JSL 36.521
• ID 08596

MAKSIMOVA, L.L. [1967] *Some problems of the Ackermann calculus (Russian)* (**J** 0023) Dokl Akad Nauk SSSR 175*1222-1224
• TRANSL [1967] (**J** 0062) Sov Math, Dokl 8*997-999
◇ B46 ◇ REV MR 36 # 4956 Zbl 162.11 JSL 33.608
• ID 19275

MAKSIMOVA, L.L. [1968] *The calculus of strict implication (Russian)* (J 0003) Algebra i Logika 7/2∗55–76
• TRANSL [1968] (J 0069) Algeb and Log 7∗102–116
◇ B46 ◇ REV MR 38 # 4286 Zbl 186.4 Zbl 201.5
• ID 08597

MAKSIMOVA, L.L. [1970] *E-theories (Russian)* (J 0003) Algebra i Logika 9∗530–538
• TRANSL [1970] (J 0069) Algeb and Log 9∗320–325
◇ B46 ◇ REV MR 44 # 2574 Zbl 217.8 • ID 19273

MAKSIMOVA, L.L. [1970] see LAVROV, I.A.

MAKSIMOVA, L.L. [1971] *On interpretation and separation theorems for the calculi E and R (Russian)* (J 0003) Algebra i Logika 10∗376–392
• TRANSL [1971] (J 0069) Algeb and Log 10∗232–241
◇ B46 ◇ REV MR 46 # 8833 Zbl 242 # 02028 • ID 63614

MAKSIMOVA, L.L. [1972] *Pretabular superintuitionistic logics (Russian)* (J 0003) Algebra i Logika 11∗558–570,615
• TRANSL [1972] (J 0069) Algeb and Log 11∗308–314
◇ B55 ◇ REV MR 48 # 3705 Zbl 265 # 02018 • ID 63611

MAKSIMOVA, L.L. [1973] *A semantics for the calculus E of entailment* (J 0387) Bull Sect Logic, Pol Acad Sci 2∗18–21
◇ B46 ◇ REV MR 52 # 13334 • ID 21796

MAKSIMOVA, L.L. & RYBAKOV, V.V. [1974] *A lattice of normal modal logics (Russian)* (J 0003) Algebra i Logika 13∗188–216,235
• TRANSL [1974] (J 0069) Algeb and Log 13∗105–122
◇ B45 G10 ◇ REV MR 51 # 65 Zbl 315 # 02027 • ID 21427

MAKSIMOVA, L.L. & VAKARELOV, D. [1974] *Semantics for ω^+-valued predicate calculi* (J 0014) Bull Acad Pol Sci, Ser Math Astron Phys 22∗765–771
◇ B50 ◇ REV MR 50 # 6838 Zbl 299 # 02020 • ID 08600

MAKSIMOVA, L.L. [1975] *Finite-level modal logics (Russian)* (J 0003) Algebra i Logika 14∗304–319,369
• TRANSL [1975] (J 0069) Algeb and Log 14∗188–197
◇ B45 ◇ REV MR 55 # 2501 Zbl 375 # 02007 • ID 26003

MAKSIMOVA, L.L. [1975] *Pretabular extensions of Lewis* S4 *(Russian)* (J 0003) Algebra i Logika 14∗28–55,117
• TRANSL [1975] (J 0069) Algeb and Log 14∗16–33
◇ B45 ◇ REV MR 53 # 94 Zbl 319 # 02019 • ID 16558

MAKSIMOVA, L.L. [1975] see LAVROV, I.A.

MAKSIMOVA, L.L. [1975] *The tautologies of ω^+-valued logic (Russian)* (J 0087) Mat Zametki (Akad Nauk SSSR) 17∗947–955
• TRANSL [1975] (J 1044) Math Notes, Acad Sci USSR 17∗568–573
◇ B50 ◇ REV MR 53 # 5253 Zbl 331 # 02006 • ID 22886

MAKSIMOVA, L.L. [1977] *Craig's interpolation theorem and amalgamable varieties (Russian)* (J 0023) Dokl Akad Nauk SSSR 237∗1281–1284
• TRANSL [1977] (J 0062) Sov Math, Dokl 18∗1550–1553
◇ B22 B55 C05 C40 C90 G10 ◇ REV MR 57 # 5698 Zbl 393 # 03013 • ID 52433

MAKSIMOVA, L.L. [1977] *Craig's theorem in superintuitionistic logics and amalgamable varieties of pseudo-boolean algebras (Russian)* (J 0003) Algebra i Logika 16∗643–681,741
• TRANSL [1977] (J 0069) Algeb and Log 16∗427–455
◇ B55 C05 C40 C90 G05 G10 ◇ REV MR 80c:03028 Zbl 403 # 03047 • ID 54760

MAKSIMOVA, L.L. [1979] *A classification of modal logics (Russian)* (J 0003) Algebra i Logika 18∗328–340,385–386
• TRANSL [1979] (J 0069) Algeb and Log 18∗202–210
◇ B45 ◇ REV MR 81i:03032 Zbl 436 # 03012 • ID 55852

MAKSIMOVA, L.L. [1979] *Interpolation properties of superintuitionistic logics* (J 0063) Studia Logica 38∗420–428
◇ B55 ◇ REV MR 81f:03035 Zbl 435 # 03021 • ID 55782

MAKSIMOVA, L.L. [1979] *Interpolation theorems in modal logics and amalgamable varieties of topological boolean algebras (Russian)* (J 0003) Algebra i Logika 18∗556–586,632
• TRANSL [1979] (J 0069) Algeb and Log 18∗348–370
◇ B45 C05 C40 C90 G05 ◇ REV MR 81j:03031 Zbl 436 # 03011 • ID 55851

MAKSIMOVA, L.L. & SHEKHTMAN, V.B. & SKVORTSOV, D.P. [1979] *The impossibility of a finite axiomatization of Medvedev's logic of finitary problems (Russian)* (J 0023) Dokl Akad Nauk SSSR 245∗1051–1054
• TRANSL [1979] (J 0062) Sov Math, Dokl 20∗394–398
◇ B55 ◇ REV MR 80e:03021 Zbl 439 # 03008 • ID 56000

MAKSIMOVA, L.L. [1980] *Interpolation theorems in modal logics. Sufficient conditions (Russian)* (J 0003) Algebra i Logika 19∗194–213,250–251
• TRANSL [1980] (J 0069) Algeb and Log 19∗120–132
◇ B45 G25 ◇ REV MR 82i:03028 Zbl 469 # 03011
• ID 55139

MAKSIMOVA, L.L. [1982] *Absence of the interpolation property in modal companions of a Dummett logic (Russian)* (J 0003) Algebra i Logika 21∗690–694
• TRANSL [1982] (J 0069) Algeb and Log 21∗460–462
◇ B45 C40 C90 ◇ REV MR 85e:03043 • ID 40502

MAKSIMOVA, L.L. [1982] *Interpolation properties of superintuitionistic, positive and modal logics* (P 3800) Intens Log: Th & Appl;1979 Moskva 70–78
• TRANSL [1984] (C 4366) Modal & Intens Log & Primen Probl Metodol Nauk 81–88
◇ B55 C40 C90 ◇ REV MR 85f:03017 Zbl 521 # 03014
• ID 37067

MAKSIMOVA, L.L. [1982] *Lyndon's interpolation theorem in modal logics (Russian)* (C 3953) Mat Log & Teor Algor 45–55
◇ B45 C40 C90 ◇ REV MR 85e:03031 Zbl 543 # 03011
• ID 40416

MAKSIMOVA, L.L. see Vol. I for further entries

MALACHOWSKI, A.R. [1982] *Carney's proposal* (J 0094) Mind 91∗281–284
◇ B60 ◇ REV MR 83i:03009 • ID 35487

MALAI, V.P. [1982] *Chain precomplete classes of functions of many-valued logic (Russian)* (J 0967) Izv Akad Nauk Mold SSR, Ser Fiz-Tekh Mat 1982/3∗19–25
◇ B50 ◇ REV MR 85c:03007 Zbl 544 # 03004 • ID 40485

MALATESTA, M. [1976] *Logistica I: Introduzione. La logica degli enunciati* (X 1229) Ateneo: Napoli 170pp
◇ B98 ◇ REM Part II 1978 • ID 35977

MALATESTA, M. see Vol. I for further entries

MALINOWSKI, G. & SPASOWSKI, M. [1974] *Dual counterparts of Lukasiewicz's sentential calculi* (**J** 0063) Studia Logica 33*153-162
• REPR [1977] (**C** 3026) Lukasiewicz Sel Pap on Sent Calc 161-170
◇ B50 ◇ REV MR 51 # 54 MR 58 # 21442 Zbl 299 # 02016 Zbl 381 # 03017 • ID 15212

MALINOWSKI, G. [1974] *S-algebras and the degrees of maximality of three and four valued logics of Lukasiewicz* (**J** 0063) Studia Logica 33*359-370
◇ B50 ◇ REV MR 52 # 2818 Zbl 336 # 02013 • ID 17618

MALINOWSKI, G. [1975] *Matrix representation for the dual counterparts of Lukasiewicz n-valued sentential calculi and the problem of their degrees of maximality* (**J** 0387) Bull Sect Logic, Pol Acad Sci 4*26-32
◇ B50 ◇ REV MR 52 # 13328 • ID 21789

MALINOWSKI, G. [1975] *Matrix representation for the dual counterparts of Lukasiewicz n-valued sentential calculi and the problem of their degrees of maximality* (**P** 1805) Int Symp Multi-Val Log (5,Proc);1975 Bloomington 252-261
◇ B50 ◇ REV MR 58 # 5030 • ID 75913

MALINOWSKI, G. [1977] *Classical characterization of n-valued Lukasiewicz calculi* (**J** 0302) Rep Math Logic, Krakow & Katowice 9*41-45
◇ B50 G20 ◇ REV MR 58 # 27308 Zbl 392 # 03020 • ID 52387

MALINOWSKI, G. [1977] *Degrees of maximality of Lukasiewicz-like sentential calculi* (**J** 0063) Studia Logica 36*213-228
◇ B50 ◇ REV MR 57 # 15955 Zbl 358 # 02018 • ID 50472

MALINOWSKI, G. [1977] *S-algebras for n-valued sentential calculi of Lukasiewicz. The degrees of maximality of some Lukasiewicz's logics* (**C** 3026) Lukasiewicz Sel Pap on Sent Calc 149-159
◇ B50 ◇ REV MR 58 # 21441 Zbl 381 # 03017 • ID 75910

MALINOWSKI, G. & WOJCICKI, R. (EDS.) [1977] *Selected papers on Lukasiewicz sentential caculi* (**X** 2882) Pol Akad Nauk: Wroclaw 199pp
◇ B50 G20 ◇ REV MR 55 # 12446 Zbl 381 # 03017 • ID 51885

MALINOWSKI, G. & ZYGMUNT, J. [1978] *Results in general theory of matrices for sentential calculi with applications to the Lukasiewicz logics* (**J** 0481) Acta Univ Wroclaw 352*43-57
◇ B50 ◇ ID 31570

MALINOWSKI, G. [1979] *S-algebras for n-valued Lukasiewicz propositional calculi (Russian)* (**P** 2554) All-Union Symp Th Log Infer;1974 Moskva 50-53
◇ B50 ◇ REV MR 84m:03034 • ID 35738

MALINOWSKI, G. [1981] see GARDIES, J.-L.

MALINOWSKI, G. & MICHALCZYK, M. [1981] *Interpolation properties for a class of many-valued propositional calculi* (**J** 0387) Bull Sect Logic, Pol Acad Sci 10*9-16
◇ B50 C40 ◇ REV MR 82g:03037 Zbl 457 # 03023 • ID 54348

MALINOWSKI, G. & MICHALCZYK, M. [1982] *That SCI has the interpolation property* (**J** 0063) Studia Logica 41*375-380
◇ B60 C40 ◇ REV MR 86e:03006 Zbl 549 # 03008 • ID 42305

MALINOWSKI, G. [1985] *Non-fregean logic and other formalizations of propositional identity* (**J** 0387) Bull Sect Logic, Pol Acad Sci 14*21-29
◇ B60 ◇ REV Zbl 575 # 03010 • ID 46502

MALINOWSKI, G. [1985] *Notes on sentential logic with identity* (**J** 0079) Logique & Anal, NS 28*341-352
◇ B10 B60 ◇ ID 49715

MALINOWSKI, G. see Vol. I for further entries

MALINOWSKI, J. [1985] *On the number of quasimodal algebras* (**J** 0387) Bull Sect Logic, Pol Acad Sci 14*99-102
◇ B45 ◇ ID 49036

MALISHEVSKIJ, A.V. [1982] see AJZERMAN, M.A.

MALISOFF, W.M. [1941] *Meanings in multi-valued logics* (**J** 0153) Phil of Sci (East Lansing) 8*271-274
◇ A05 B50 ◇ REV MR 2.340 Zbl 60.20 • ID 08625

MALITZ, J. [1979] *Introduction to mathematical logic. Set theory, computable functions, model theory* (**X** 0811) Springer: Heidelberg & New York xii + 198pp
◇ B98 C98 D98 ◇ REV MR 81h:03002 Zbl 407 # 03001 JSL 49.672 • ID 56167

MALITZ, J. see Vol. I, III, IV, V for further entries

MALLY, E. [1971] *Logische Schriften* (**X** 0835) Reidel: Dordrecht x + 347pp
◇ A05 A10 B96 ◇ REV MR 54 # 7187 Zbl 209.7 • ID 24980

MALMNAES, P.E. & PRAWITZ, D. [1968] *A survey of some connections between classical, intuitionistic and minimal logic* (**P** 0608) Logic Colloq;1966 Hannover 215-229
◇ B55 F05 F25 F50 ◇ REV MR 38 # 4289 Zbl 188.11 JSL 40.503 • ID 24815

MAL'TSEV, A.I. [1966] *Iterative algebras and Post varieties (Russian)* (**J** 0003) Algebra i Logika 5/2*5-24
• TRANSL [1971] (**C** 2621) Mal'tsev: Metamath of Algeb Syst 396-415
◇ B50 C05 G20 ◇ REV MR 34 # 7424 Zbl 275 # 08001 JSL 36.338 • ID 19236

MAL'TSEV, A.I. see Vol. I, III, IV, V, VI for further entries

MAL'TSEV, I.A. [1984] see DEMETROVICS, J.

MAL'TSEV, I.A. see Vol. I, IV for further entries

MAMDANI, E.H. [1976] *Application of fuzzy logic to approximate reasoning using linguistic synthesis* (**P** 2011) Int Symp Multi-Val Log (6);1976 Logan 196-202
◇ B52 ◇ REV Zbl 397 # 94025 • ID 35890

MAMDANI, E.H. [1978] see KICKERT, W.J.M.

MAMDANI, E.H. [1979] see ESHRAGH, F.

MAMDANI, E.H. & SEMBI, B.S. [1979] *On the nature of implication in fuzzy logic* (**P** 3003) Int Symp Multi-Val Log (9);1979 Bath 143-151
◇ B52 ◇ REV MR 81d:03027 • ID 78392

MAMDANI, E.H. (ED.) [1979] *Special issue on fuzzy reasoning. Papers presented at a Workshop held at Queen Mary College, London, September 1978* (**J** 1741) Int J Man-Mach Stud 11*405-545
◇ B52 ◇ REV MR 80e:00005 • ID 80438

MAMDANI, E.H. & SEMBI, B.S. [1980] *Process control using fuzzy logic* (**P** 2936) Fuzzy Sets;1980 Durham 249-265
◇ B52 ◇ REV MR 81k:93048 • ID 82160

MAMDANI, E.H. [1981] see GAINES, B.R.

MAMDANI, E.H. see Vol. I for further entries

MANASTER, A.B. [1975] *Completeness, compactness, and undecidability: an introduction to mathematical logic* (**X** 0819) Prentice Hall: Englewood Cliffs vi + 154pp
◇ B98 C07 D35 D98 ◇ REV MR 53 # 12857 Zbl 306 # 02001 JSL 42.320 • ID 23120

MANASTER, A.B. see Vol. III, IV, V for further entries

MANDERS, K.L. [1982] see DALEY, R.P.

MANDERS, K.L. see Vol. I, III, IV, V for further entries

MANES, E.G. [1975] see ARBIB, M.A.

MANES, E.G. [1982] *A class of fuzzy theories* (**J** 0034) J Math Anal & Appl 85*409-451
◇ B52 ◇ REV MR 83k:68010 Zbl 497 # 18011 • ID 36635

MANES, E.G. see Vol. IV, V for further entries

MANGANI, P. [1976] *Alcune questioni di teoria dei modelli per linguaggi predicativi generali (English summary)* (**J** 0149) Atti Accad Naz Lincei Fis Mat Nat, Ser 8 60*368-376
◇ B50 C20 C90 ◇ REV MR 57 # 2906 Zbl 367 # 02027 • ID 51187

MANGANI, P. [1984] *General calculi "with types" and generalized logics (Italian) (English summary)* (**J** 0149) Atti Accad Naz Lincei Fis Mat Nat, Ser 8 76*1-6
◇ B60 ◇ ID 49358

MANGANI, P. see Vol. I, III for further entries

MANGIONE, C. [1964] *Elementi di logica matematica* (**X** 0905) Boringhieri: Torino 127pp
◇ B98 ◇ REV Zbl 144.243 • ID 22418

MANGIONE, C. [1977] see BONOMI, A.

MANGIONE, C. see Vol. III, VI for further entries

MANIA, A. [1981] see ABBATI, M.C.

MANIA, A. see Vol. III for further entries

MANICAS, P.T. [1968] see KRUGER, A.N.

MANIN, YU.I. [1977] *A course in mathematical logic* (**X** 0811) Springer: Heidelberg & New York xiii + 286pp
◇ B98 C07 D98 E35 E50 F30 G12 ◇ REV MR 56 # 15345 Zbl 383 # 03002 • REM Translated from Russian • ID 51984

MANIN, YU.I. [1979] *Provable and unprovable (Russian)* (**X** 2643) Sovet Radio: Moskva 168pp
• TRANSL [1981] (**X** 0885) Mir: Moskva 265pp (Spanish)
◇ A05 B98 E35 E50 F30 F98 G12 ◇ REV Zbl 403 # 03002 • ID 54716

MANIN, YU.I. see Vol. IV, V, VI for further entries

MANNA, Z. & PNUELI, A. [1979] *The modal logic of programs* (**P** 1873) Automata, Lang & Progr (6);1979 Graz 385-409
◇ B45 B75 ◇ REV MR 81d:68022 Zbl 404 # 68011 • ID 82171

MANNA, Z. [1983] see BEN-ARI, M.

MANNA, Z. see Vol. I, IV for further entries

MANOR, R. & RESCHER, N. [1970] *On inference from inconsistent premisses* (**J** 0472) Theory Decis 1*179-217
◇ A05 B53 ◇ REV Zbl 212.311 • ID 30899

MANOR, R. & RESCHER, N. [1972] *Modal elaborations of propositional logics* (**J** 0047) Notre Dame J Formal Log 13*323-330
◇ B45 ◇ REV MR 46 # 3270 Zbl 188.12 • ID 16380

MANOR, R. [1974] *A semantic analysis of conditional assertion* (**J** 0122) J Philos Logic 3*37-52
◇ B45 ◇ REV MR 56 # 94 Zbl 278 # 02024 • ID 29060

MANSON, M. [1962] *Introduction a la semantica de los sistemas normativos* (**X** 1715) Edunsa Ed Univ: Santiago 58pp
◇ B45 ◇ ID 31918

MARC-WOGAU, K. [1950] *Modern logic. An elementary text-book (Swedish)* (**X** 4499) Ehlins: Stockholm 211pp
◇ B98 ◇ REV JSL 17.288 • ID 41850

MARCHENKOV, S.S. [1979] *On closed classes of self-dual functions in a many-valued logic (Russian)* (**J** 0052) Probl Kibern 36*5-22,279
◇ B50 G25 ◇ REV MR 81d:03064 Zbl 448 # 03011 • REM Part I. Part II 1983 • ID 56611

MARCHENKOV, S.S. [1980] see DEMETROVICS, J.

MARCHENKOV, S.S. [1981] *Cardinality of the set of precomplete classes in certain classes of functions of countably valued logic (Russian)* (**J** 0052) Probl Kibern 38*109-116,272
◇ B50 ◇ REV MR 83m:03032 Zbl 525 # 03010 • ID 35436

MARCHENKOV, S.S. [1981] see DEMETROVICS, J.

MARCHENKOV, S.S. [1983] *On closed classes of self-dual functions of many-valued logic II (Russian)* (**J** 0052) Probl Kibern 40*261-266
◇ B50 ◇ REV MR 85d:03052 Zbl 516 # 03012 • REM Part I 1979 • ID 37249

MARCHENKOV, S.S. [1984] *Closed classes in P_k containing homogeneous functions (Russian) (English summary)* (**S** 2651) Prepr Inst Prikl Mat, Akad Nauk SSSR 35*28pp
◇ B50 ◇ REV MR 86d:03025 • ID 44435

MARCHENKOV, S.S. see Vol. I, IV for further entries

MARCHINI, C. [1981] *Modelli di Kripke e modelli algebrici $S_{\pi\iota}$ (English summary)* (**J** 0549) Riv Mat Univ Parma, Ser 4 7*105-116
◇ B50 ◇ REV MR 83j:03022 Zbl 514 # 03012 • ID 35335

MARCHINI, C. see Vol. III, V, VI for further entries

MARCJA, A. [1977] see MAKOWSKY, J.A.

MARCJA, A. see Vol. I, III, IV, VI for further entries

MARCONI, D. [1979] *La formalizzazione della dialettica* (**C** 2041) Formal della Dialettica 9-85
◇ A05 B53 ◇ ID 42856

MARCONI, D. [1980] *A decision method for the calculus C_1* (**P** 3006) Brazil Conf Math Log (3);1979 Recife 211-223
◇ A05 B25 B53 ◇ REV MR 83b:03026 Zbl 448 # 03014 • ID 56614

MARCONI, D. [1981] *Types of non-Scotian logic* (**J** 0079) Logique & Anal, NS 24*407-414
◇ B60 ◇ REV MR 83h:03038 Zbl 494 # 03018 • ID 36057

MARDAEV, S.I. [1984] *Number of prelocally table superintuitionistic propositional logics (Russian)* (**J** 0003) Algebra i Logika 23*74-87,119
• TRANSL [1984] (**J** 0069) Algeb and Log 23*56-66
◇ B55 ◇ ID 44664

MAREK, W. & ONYSZKIEWICZ, J. [1972] *Elements of logic and foundations of mathematics in problems (Polish)* (**X** 1034) PWN: Warsaw 278pp
• TRANSL [1982] (**X** 0835) Reidel: Dordrecht viii + 276pp
◇ B98 C98 D98 E98 ◇ REV MR 84h:03001 Zbl 288 # 02001 Zbl 574 # 03001 • ID 34216

MAREK, W. & SREBRNY, M. & ZARACH, A. (EDS.) [1976] *Set theory and hierarchy theory. A memorial tribute to Andrzej Mostowski* (**X** 0811) Springer: Heidelberg & New York xiii + 345pp
◇ B97 D55 D97 E97 ◇ REV MR 54 # 2412 Zbl 324 # 00007 • ID 24424

MAREK, W. see Vol. I, III, IV, V, VI for further entries

MARGARIS, A. [1958] *A problem of Rosser and Turquette* (**J** 0036) J Symb Logic 23*271-279
◇ B50 ◇ REV MR 21 # 4914 Zbl 86.245 JSL 35.304 • ID 08761

MARGARIS, A. [1967] *First order mathematical logic* (**X** 0841) Blaisdell: New York 211pp
◇ B98 ◇ REV MR 36 # 26 Zbl 156.247 JSL 37.616 • ID 08762

MARGARIS, A. see Vol. I for further entries

MARGHIDANU, D. [1982] *Five logical modalities of exclusion (Romanian) (French summary)* (**J** 0197) Stud Cercet Mat Acad Romana 34*157-168
◇ B45 ◇ REV MR 84b:03032 Zbl 513 # 03013 • ID 35626

MARGHIDANU, D. [1983] *An exclusion logic for the complementary of a theory (Roumanian) (English summary)* (**J** 0197) Stud Cercet Mat Acad Romana 35*362-375
◇ B60 ◇ REV MR 86a:03026 Zbl 531 # 03012 • ID 37675

MARGOLIS, I.B. [1981] see KANDEL, A.

MARGOLIS, J. [1973] *Identity and the necessity operator* (**J** 0079) Logique & Anal, NS 16*527-533
◇ B45 ◇ REV Zbl 285 # 02006 • ID 29938

MARIC, Z. [1981] see KRON, A.

MARINOS, P. [1969] *Fuzzy logic and its application to switching systems* (**J** 0187) IEEE Trans Comp C-18*343-348
◇ B52 ◇ REV Zbl 172.301 • ID 41951

MARKIN, V.I. [1982] *The neighborhood semantics for modalities de re (Russian)* (**C** 3849) Modal & Relevant Log, Vyp 1 13-18
◇ B45 ◇ REV Zbl 538 # 03013 • ID 41464

MARKOV, A.A. [1979] see KHOMICH, V.I.

MARKOV, A.A. [1984] *Elements of mathematical logic (Russian)* (**X** 0898) Moskov Gos Univ: Moskva 80pp
◇ B98 ◇ REV MR 85h:03001 • ID 43264

MARKOV, A.A. see Vol. I, III, IV, V, VI for further entries

MARKOVA, V.P. [1979] *Harmonic analysis of functions of many-valued logic (Russian)* (**J** 0474) Avtom Vychis Tekh, Akad Nauk Latv SSR 1979/4*8-14
• TRANSL [1979] (**J** 2666) Autom Control Comput Sci 13/4*7-12
◇ B50 ◇ REV MR 83m:94028 Zbl 447 # 94026 • ID 69695

MARKOVA, V.P. see Vol. I for further entries

MARKWALD, W. [1971] *Praedikatenlogik mit partiell definierten Funktionen I* (**J** 0009) Arch Math Logik Grundlagenforsch 14*10-23
◇ B10 B60 ◇ REV MR 44 # 2567 Zbl 215.46 • REM Part II 1974 • ID 08782

MARKWALD, W. [1972] *Einfuehrung in die formale Logik und Metamathematik* (**X** 0918) Klett: Stuttgart 168pp
◇ B98 ◇ REV MR 49 # 7112 Zbl 304 # 02001 • ID 08783

MARKWALD, W. [1974] *Praedikatenlogik mit partiell definierten Funktionen. II* (**J** 0009) Arch Math Logik Grundlagenforsch 16*15-22
◇ B10 B60 ◇ REV MR 51 # 10024 Zbl 289 # 02012 • REM Part I 1971 • ID 08784

MARKWALD, W. see Vol. III, IV, VI for further entries

MARLOW, A.R. [1981] *Space-time structure from quantum logic* (**P** 3820) Curr Iss in Quantum Log;1979 Erice 413-418
◇ B51 G12 ◇ REV MR 84i:03005 Zbl 537 # 03044 • ID 44884

MARLOW, A.R. see Vol. I for further entries

MARQUETTY, A. [1974] *Calculs propositionnels a p negations (aspect functionnel)* (**J** 2313) C R Acad Sci, Paris, Ser A-B 278*A109-A112
◇ B50 ◇ REV MR 48 # 8196 Zbl 288 # 02014 • ID 08787

MARQUETTY, A. [1974] *Calculs propositionnels a p negations (aspect ensembliste)* (**J** 2313) C R Acad Sci, Paris, Ser A-B 278*A189-A192
◇ B50 ◇ REV MR 48 # 8197 Zbl 311 # 02026 • ID 08788

MARRUCCELLI, A. [1970] see FIORENTINI, M.

MARSDEN, E.L. [1974] *Reducible implicative models* (**J** 0246) J Nat Sci Math 14*23-34
◇ B20 B45 C05 G25 ◇ REV MR 56 # 2819 Zbl 334 # 02028 • ID 63732

MARTENS, J.L. [1982] see BARTH, E.M.

MARTIC, B. [1975] *A certain nonclassical system of logic (Serbo-Croatian) (English summary)* (**J** 0042) Mat Vesn, Drust Mat Fiz Astron Serb 12(27)*87-94
◇ B60 ◇ REV MR 52 # 2836 Zbl 318 # 02020 • ID 17630

MARTIC, B. [1976] *On the completeness of the non-classical system A'* (**J** 3519) Glas Mat, Ser 3 (Zagreb) 11(31)*187-198
◇ B60 ◇ REV MR 57 # 2880 Zbl 395 # 03018 • ID 52569

MARTIC, B. see Vol. IV for further entries

MARTIN, E.P. & MEYER, R.K. & ROUTLEY, R. & ROUTLEY, V. [1982] *On the philosophical bases of relevant logic semantics* (**J** 3783) J Non-Classical Log (Univ Campinas) 1*71-105
◇ A05 B46 ◇ REV MR 85i:03056a Zbl 513 # 03002 • ID 37212

MARTIN, E.P. & MEYER, R.K. [1982] *Solution to the P-W problem* (**J** 0036) J Symb Logic 47*869-887
◇ B46 ◇ REV MR 85e:03044 Zbl 498 # 03011 • ID 36902

MARTIN, E.P. [1983] see DWYER, R.

MARTIN, G. [1949] *Ueber ein zweiwertiges Modell einer vierwertigen Logik* (**J** 0175) Methodos 1*385-389
◇ A05 B50 ◇ REV MR 12.663 Zbl 37.294 JSL 16.150 • ID 08806

MARTIN, J.N. [1975] *A syntactic characterization of Kleene's strong connectives with two designated values* (J 0068) Z Math Logik Grundlagen Math 21*181-184
◇ B50 ◇ REV MR 51 # 5264 Zbl 307 # 02011 • ID 08808

MARTIN, J.N. [1976] *Classical indeterminacy, many-valued logic, and supervaluations* (P 2011) Int Symp Multi-Val Log (6);1976 Logan 115-122
◇ B50 ◇ REV MR 58 # 21444 • ID 76081

MARTIN, J.N. [1977] *An axiomatization of Herzberger's 2-dimensional presuppositional semantics* (J 0047) Notre Dame J Formal Log 18*378-382
◇ B50 ◇ REV MR 58 # 5032 Zbl 315 # 02022 • ID 24238

MARTIN, J.N. [1977] *Many-valued logic, classical logic and the horizontal* (P 2013) Int Symp Multi-Val Log (7);1977 Charlotte 84-88
◇ B50 ◇ REV MR 57 # 15954 • ID 76082

MARTIN, J.N. [1980] *The emergent nature of coherence and the possibility of truth-functional logic* (P 3673) Int Symp Multi-Val Log (10);1980 Evanston 233-237
◇ B50 ◇ REV MR 83m:94032 • ID 33659

MARTIN, J.N. [1984] *Epistemic semantics for classical and intuitionistic logic* (J 0047) Notre Dame J Formal Log 25*105-116
◇ B50 F50 ◇ REV MR 85e:03058 Zbl 562 # 03005 • ID 40627

MARTIN, J.N. see Vol. I for further entries

MARTIN, L. & REISCHER, C. & ROSENBERG, I.G. [1978] *Completeness problems for switching circuits constructed from delayed gates* (P 2014) Int Symp Multi-Val Log (8);1978 Rosemont 142-148
◇ B50 ◇ REV MR 81b:94069 • ID 35923

MARTIN, N.M. [1950] *Some analogues of the Sheffer Stroke function in n-valued logic* (J 0028) Indag Math 12*393-400
◇ B50 ◇ REV MR 12.385 Zbl 39.245 JSL 16.275 • ID 08809

MARTIN, N.M. [1951] *A note on Sheffer functions in n-valued logic* (J 0175) Methodos 3*240-242
◇ B50 ◇ REV JSL 17.204 • ID 16880

MARTIN, N.M. [1953] *On completeness of decision element sets* (J 0093) J Comp Syst 1*150-154
◇ B50 ◇ REV MR 15.90 Zbl 53.92 JSL 19.143 JSL 32.134 • ID 08810

MARTIN, N.M. [1954] *The Sheffer functions of 3-valued logic* (J 0036) J Symb Logic 19*45-51
◇ B50 ◇ REV MR 15.669 Zbl 55.5 JSL 21.199 • ID 08811

MARTIN, N.M. [1960] *Deduction and strict implication* (J 0154) Synthese 12*25-33
◇ B46 ◇ REV MR 24 # A677 Zbl 229 # 02022 • ID 08812

MARTIN, N.M. [1972] *Systems of modal logic* (1111) Preprints, Manuscr., Techn. Reports etc. 73pp
◇ B45 ◇ REM Mimeographed Notes • ID 32215

MARTIN, N.M. [1976] *Direct analogues of the Sheffer stroke in m-valued logic* (J 0047) Notre Dame J Formal Log 17*415-420
◇ B50 ◇ REV MR 56 # 5217 Zbl 314 # 02031 • ID 18278

MARTIN, N.M. see Vol. VI for further entries

MARTIN, R.M. & WOODGER, J.-H. [1951] *Toward an inscriptional semantics* (J 0036) J Symb Logic 16*191-203
◇ A05 B60 ◇ REV MR 13.310 Zbl 45.150 JSL 17.71 • ID 08818

MARTIN, R.M. [1958] *A formalization of inductive logic* (J 0036) J Symb Logic 23*251-256
◇ A05 B48 ◇ REV MR 21 # 2577 JSL 34.137 • ID 08824

MARTIN, R.M. [1964] *Toward a logic of intensions* (C 0567) Form & Strategy in Sci (Woodger) 146-167
◇ A05 B45 ◇ ID 14874

MARTIN, R.M. [1968] *Semantics, analytical truth and the new modalism* (C 0552) Phil Contemp - Chroniques 44-60
◇ A05 B45 ◇ ID 14649

MARTIN, R.M. [1969] *Belief, existence and meaning* (X 0924) New York Univ Pr: New York xvi+284pp
◇ A05 B45 ◇ ID 25140

MARTIN, R.M. [1975] see ANDERSON, A.R.

MARTIN, R.M. see Vol. I, IV, V for further entries

MARTINEZ RECIO, A. [1979] see HUERTAS, J.A.

MARTINI, G.B. [1983] see AMBROSIO, R.

MARTINI, G.B. see Vol. V for further entries

MARTINO, A.A. [1981] *Deontic logic, computational linguistics and legal information systems. Vol.II* (X 0809) North Holland: Amsterdam xli+518pp
◇ B45 ◇ REV MR 85g:03003b • ID 43474

MARTIROSYAN, A.A. [1983] *On the influence of the consistency of hypotheses with initial information on the complexity of inductive inference (Russian) (Armenian summary)* (J 0346) Dokl Akad Nauk Armyan SSR 77*154-157
◇ B48 ◇ REV MR 85i:03063 Zbl 535 # 03015 • ID 38319

MARTIROSYAN, A.A. see Vol. IV for further entries

MARTYNYUK, V.V. [1960] *Untersuchungen einiger Klassen von Funktionen in der mehrwertigen Logik (Russisch)* (J 0052) Probl Kibern 3*49-60
• TRANSL [1963] (J 0449) Probl Kybern 3*55-67
◇ B50 ◇ REV MR 23 # A3668 JSL 31.502 • ID 19346

MASHURYAN, A.S. [1966] *Distributivity axioms in the strict implication calculus of Ackermann (Russian) (Armenian and English summaries)* (J 0312) Izv Akad Nauk Armyan SSR, Ser Mat 1*226-230
◇ B45 ◇ REV MR 34 # 1171 Zbl 192.29 • ID 08863

MASHURYAN, A.S. see Vol. III, IV for further entries

MASLOV, S.YU. & RUSAKOV, E.D. [1972] *Probabilistic canonical calculi (Russian) (English summary)* (S 0228) Zap Nauch Sem Leningrad Otd Mat Inst Steklov 32*66-76,155 • ERR/ADD ibid 40*161
• TRANSL [1976] (J 1531) J Sov Math 6*401-409
◇ B48 D03 ◇ REV MR 49 # 8829 Zbl 344 # 02027 • ID 19328

MASLOV, S.YU. see Vol. I, III, IV, VI for further entries

MASON, R.O. & MITROFF, I.I. [1982] *On the structure of dialectical reasoning in the social and policy sciences* (J 0472) Theory Decis 14*331-350
◇ B53 ◇ REV MR 84b:03042 Zbl 489 # 03004 • ID 35634

MASON, R.O. & MITROFF, I.I. & QUINTON, H. [1983] *Beyound contradiction and consistency: a design for a dialectical policy system* (**J** 0472) Theory Decis 15*107-120
◇ B53 ◇ REV MR 84i:03062 Zbl 509 # 03014 • ID 34541

MASSEY, G.J. [1965] *Four simple systems of modal propositional logic* (**J** 0153) Phil of Sci (East Lansing) 32*342-355
◇ B45 ◇ REV MR 36 # 1305 JSL 37.754 • ID 08854

MASSEY, G.J. [1966] *The theory of truth tabular connectives, both truth functional and modal* (**J** 0036) J Symb Logic 31*593-608
◇ B45 ◇ REV MR 38 # 31 Zbl 168.4 JSL 37.183 • ID 08856

MASSEY, G.J. [1967] *Binary connectives functionally complete by themselves in S5 modal logic* (**J** 0036) J Symb Logic 32*91-92
◇ B45 ◇ REV MR 37 # 6165 Zbl 168.5 JSL 37.183 • ID 08857

MASSEY, G.J. [1968] *Normal form generation of S5 functions via truth functions* (**J** 0047) Notre Dame J Formal Log 9*81-85
◇ B45 ◇ REV MR 39 # 5322 Zbl 181.297 • ID 08858

MASSEY, G.J. [1969] see HENDRY, H.E.

MASSEY, G.J. [1969] *Sheffer functions for many-valued S5 modal logics* (**J** 0068) Z Math Logik Grundlagen Math 15*101-104
◇ B45 B50 ◇ REV MR 41 # 6673 Zbl 193.292 • ID 08859

MASSEY, G.J. [1970] *Binary closure-algebraic operations that are funtionally complete* (**J** 0047) Notre Dame J Formal Log 11*340-342
◇ B45 ◇ REV MR 43 # 6060 Zbl 188.19 JSL 36.691 • ID 08861

MASSEY, G.J. [1970] *Understanding symbolic logic* (**X** 0837) Harper & Row: New York 428pp
◇ A05 B98 ◇ REV MR 43 # 1808 Zbl 237 # 02001 JSL 36.678 • ID 08860

MASSEY, G.J. [1972] *The modal structure of the Prior-Rescher family of infinite product systems* (**J** 0047) Notre Dame J Formal Log 13*219-223
◇ B45 ◇ REV MR 45 # 4956 Zbl 234 # 02011 • ID 08862

MASSEY, G.J. [1975] *Concerning an alleged Sheffer function* (**J** 0047) Notre Dame J Formal Log 16*549-550
◇ B50 ◇ REV MR 52 # 7850c Zbl 315 # 02019 • ID 18282

MASSEY, G.J. see Vol. I, V for further entries

MATERNA, P. [1972] *Intensional semantics of vague constants. An application of Tichy's concept of semantics* (**J** 0472) Theory Decis 2*267-273
◇ A05 B60 ◇ REV Zbl 241 # 02010 • ID 28822

MATERNA, P. [1974] *Expressibility of propositions in L_μ-languages* (**J** 0063) Studia Logica 33*259-271
◇ B45 ◇ REV MR 50 # 9536 Zbl 291 # 02006 • ID 08871

MATERNA, P. [1976] *Pragmatic meaning and truth* (**P** 1804) Form Meth in Methodol of Emp Sci;1974 Warsaw 387-393
◇ A05 B45 ◇ REV MR 55 # 12474 MR 58 # 27338 Zbl 439 # 03001 • ID 55993

MATERNA, P. [1982] *Transparent intensional logic (Russian)* (**C** 3849) Modal & Relevant Log, Vyp 1 66-80
◇ B45 ◇ REV Zbl 532 # 03010 • ID 38275

MATERNA, P. see Vol. I, III for further entries

MATES, B. [1965] *Elementary logic* (**X** 0894) Oxford Univ Pr: Oxford x+227pp
• TRANSL [1969] (**X** 0903) Vandenhoeck & Ruprecht: Goettingen 269pp • LAST ED [1972] (**X** 0815) Clarendon Pr: Oxford viii+227pp
◇ A10 B98 ◇ REV MR 40 # 15 MR 48 # 45 Zbl 146.246 Zbl 375 # 02001 JSL 31.483 • REM 2nd ed. 1978, 305pp • ID 08872

MATSUMOTO, K. [1950] *Sur la structure concernant la logique moderne* (**J** 0442) J Osaka Inst Sci & Tech, Part 1 2*67-78
◇ A05 B45 F50 G05 ◇ REV MR 15.278 Zbl 41.4 JSL 23.443 • ID 08888

MATSUMOTO, K. [1955] *Reduction theorem in Lewis' sentential calculi* (**J** 0352) Math Jap 3*133-135
◇ B45 ◇ REV MR 17.701 Zbl 66.9 JSL 21.200 • ID 08889

MATSUMOTO, K. & OHNISHI, M. & RIDDER, J. [1957] *Die Gentzenschen Schlussverfahren in modalen Aussagenlogiken I* (**J** 0028) Indag Math 19*481-491
◇ B45 ◇ REV MR 22 # 9442a Zbl 81.247 JSL 40.97 • REM Parts II,III 1958 by Ridder,J. See also 1955: Ridder,J. • ID 16963

MATSUMOTO, K. & OHNISHI, M. [1957] *Gentzen method in modal calculi. I* (**J** 1770) Osaka Math J 9*113-130
◇ B25 B45 ◇ REV MR 22 # 9441a Zbl 80.7 JSL 40.466 • REM Part II 1959 • ID 16393

MATSUMOTO, K. & OHNISHI, M. [1959] *Gentzen method in modal calculi II* (**J** 1770) Osaka Math J 11*115-120
◇ B25 B45 ◇ REV MR 22 # 9441b Zbl 89.6 JSL 40.466 • REM Part I 1957 • ID 16395

MATSUMOTO, K. [1960] *Decision procedure for modal sentential calculus S3* (**J** 1770) Osaka Math J 12*167-175
◇ B25 B45 ◇ REV MR 25 # 2954 Zbl 99.8 JSL 40.468 • ID 08893

MATSUMOTO, K. & OHNISHI, M. [1963] *A system for strict implication* (**P** 1127) Symp Founds of Math;1962 Katada 99-108
◇ B45 ◇ REV MR 30 # 13 JSL 35.326 • ID 21297

MATSUMOTO, K. & OHNISHI, M. [1964] *A system for strict implication* (**J** 0260) Ann Jap Ass Phil Sci 2*183-188
◇ B45 ◇ REV MR 30 # 13 JSL 35.326 • ID 21299

MATSUMOTO, K. [1970] *Mathematical logic (Japanese)* (**X** 1168) Kyoritsu Shuppan: Tokyo 192PP
◇ B98 ◇ ID 90096

MATSUMOTO, K. [1980] *Foundation of modal logic (Japanese)* (**J** 2612) Math (Monthly J) 1*52-60
◇ B45 ◇ ID 90098

MATSUMOTO, K. see Vol. IV, VI for further entries

MATTEUZZI, M.L.M. [1972] see D'AMORE, B.

MATTEUZZI, M.L.M. see Vol. V for further entries

MATVEJCHUK, M.S. [1980] *A theorem on the states of quantum logics (Russian) (English summary)* (**J** 1552) Teor Mat Fiz, Akad Nauk SSSR 45/2*244-250
◇ B51 ◇ REV MR 82d:81015 Zbl 467 # 46067 • REM Part I. Part II 1981 • ID 49177

MATVEJCHUK, M.S. [1981] *A theorem on the states of quantum logics. II (Russian) (English summary)* (**J** 1552) Teor Mat Fiz, Akad Nauk SSSR 48*261-265
◇ B51 G12 ◇ REV MR 83a:81004 Zbl 503 # 46041 • REM Part I 1980 • ID 45958

MATVEJCHUK, M.S. [1983] *A theorem on states on quantum logics. States in Jordan algebras (Russian)* (**J** 1552) Teor Mat Fiz, Akad Nauk SSSR 57*465-468
◇ B51 G12 ◇ REV MR 85m:46063 Zbl 543 # 46045
• ID 45201

MAXWELL, G. [1975] see ANDERSON JR., R.M.

MAY, S. [1976] see HARPER, W.L.

MAYET, R. [1972] *Relations entre les anneaux Booleens, les anneaux monadiques et les algebres trivalentes de Lukasiewicz* (**J** 2313) C R Acad Sci, Paris, Ser A-B 275*A1-A3
◇ B50 G05 G15 G20 ◇ REV MR 46 # 3301
Zbl 291 # 02043 • ID 27247

MAYO, D.G. [1983] *An objective theory of statistical testing* (**J** 0154) Synthese 57*297-340
◇ B48 ◇ REV MR 85d:03047 • ID 41091

MAYOH, B.H. [1979] see JENSEN, F.V.

MAYOH, B.H. see Vol. I, III, IV, VI for further entries

MAZBIC-KULMA, B. [1977] see KREMPA, J.

MAZUR, S. [1955] see GRZEGORCZYK, A.

MAZUR, S. see Vol. V, VI for further entries

MCARTHUR, R.P. [1974] *Factuality and modality in the future tense* (**J** 0097) Nous, Quart J Phil 8*283-288
◇ B45 ◇ ID 15216

MCARTHUR, R.P. [1974] *The Makinson completeness of tense logic* (**J** 0079) Logique & Anal, NS 17*453-460
◇ B45 ◇ REV MR 55 # 5391 Zbl 311 # 02036 • ID 29606

MCARTHUR, R.P. [1975] S5 *with the* CBF (**J** 0047) Notre Dame J Formal Log 16*528-530
◇ B45 ◇ REV MR 55 # 2502 Zbl 254 # 02024 • ID 18287

MCARTHUR, R.P. [1976] see LEBLANC, H.

MCARTHUR, R.P. [1976] *On proving linearity* (**J** 0079) Logique & Anal, NS 19*405-412
◇ B45 ◇ REV MR 56 # 11728 Zbl 347 # 02013 • ID 63810

MCARTHUR, R.P. [1976] *Tense logic* (**X** 0835) Reidel: Dordrecht v+84pp
◇ B45 ◇ REV MR 58 # 27339 Zbl 371 # 02013 • ID 51374

MCARTHUR, R.P. [1977] *Three-valued free tense logic* (**J** 0047) Notre Dame J Formal Log 18*101-106
◇ B45 ◇ REV MR 56 # 15368 Zbl 245 # 02030 • ID 21952

MCARTHUR, R.P. [1981] *Anderson's deontic logic and relevant implication* (**J** 0047) Notre Dame J Formal Log 22*145-154
◇ B46 ◇ REV MR 82g:03027 Zbl 423 # 03016 • ID 53528

MCCALL, S. [1963] *Aristotle's modal syllogisms* (**X** 0809) North Holland: Amsterdam viii+100pp
◇ A10 B45 ◇ REV MR 27 # 5674 Zbl 105.247 JSL 27.418
• ID 16808

MCCALL, S. [1965] *A modal logic with eight modalities* (**C** 0749) Contrib Logic & Methodol (Bochenski) 84-90
◇ B45 ◇ REV MR 48 # 10765 Zbl 173.5 • ID 31173

MCCALL, S. [1966] *Connexive implication* (**J** 0036) J Symb Logic 31*415-433
◇ A05 B45 ◇ REV MR 36 # 2471 Zbl 161.4 JSL 42.311
• ID 08946

MCCALL, S. & MEYER, R.K. [1966] *Pure three-valued Lukasiewiczian implication* (**J** 0036) J Symb Logic 31*399-405
◇ B50 ◇ REV MR 34 # 4122 Zbl 154.255 JSL 33.133
• ID 08943

MCCALL, S. [1967] *Connexive implication and the syllogism* (**J** 0094) Mind 76*346-356
◇ B45 ◇ REV MR 41 # 1506 • ID 31175

MCCALL, S. & NAT VANDER, A. [1969] *The system S9* (**C** 1101) Phil Logic 194-214
◇ B45 ◇ ID 21178

MCCALL, S. [1970] see BELNAP JR., N.D.

MCCALL, S. [1975] *Connexive implication* (**C** 1852) Entailment - Log of Relev & Nec 434-452
◇ B46 ◇ REV MR 53 # 10542 • ID 32380

MCCALL, S. [1976] *Objective time flow* (**J** 0153) Phil of Sci (East Lansing) 43*337-362
◇ B45 ◇ ID 32381

MCCALL, S. [1979] *The strong future tense* (**J** 0047) Notre Dame J Formal Log 20*489-504
◇ B45 ◇ REV MR 80g:03018 Zbl 283 # 02024
Zbl 429 # 03010 • ID 53841

MCCALL, S. [1983] *If, since, and because: a study in conditional connection* (**J** 0079) Logique & Anal, NS 26*309-321
◇ B45 ◇ REV MR 86f:03032 • ID 45165

MCCALL, S. [1984] *Counterfactuals based on real possible worlds* (**J** 0097) Nous, Quart J Phil 18*463-477
◇ A05 B45 ◇ REV MR 86c:03014 • ID 44640

MCCALL, S. see Vol. III, V, VI for further entries

MCCARTHY, J. [1979] *First order theories of individual concepts and propositions* (**J** 0508) Machine Intelligence 9*129-147
◇ A05 B60 ◇ REV MR 81m:03024 • ID 76199

MCCARTHY, J. [1980] *Circumscription -- a form of nonmonotonic reasoning* (**J** 0503) Artif Intell 13*27-39 • ERR/ADD ibid 13*171-172
◇ B60 ◇ REV MR 81i:03036a MR 81i:03036b
Zbl 435 # 68073 • ID 76201

MCCARTHY, J. see Vol. I, IV for further entries

MCCAWLEY, J.D. [1975] *Truth functionality and natural deduction* (**P** 1805) Int Symp Multi-Val Log (5,Proc);1975 Bloomington 412-418
◇ B50 ◇ ID 35838

MCCAWLEY, J.D. [1981] *Everything that linguists have always wanted to know about logic, but were ashamed to ask* (**X** 1096) Blackwell: Oxford xv+508pp
◇ B98 ◇ REV JSL 49.1407 • ID 47399

MCCAWLEY, J.D. [1981] *Fuzzy logic and restricted quantifiers* (**P** 3739) Phil & Grammar;1978 Uppsala 101-118
◇ B52 ◇ REV MR 83i:03043 • ID 34850

MCCAWLEY, J.D. see Vol. IV for further entries

MCCULLOCH, W.S. & PITTS, W. [1943] *A logical calculus of the ideas immanent in nervous activity* (**J** 0515) Bull Math Biophys 5*115-133
◇ B60 D05 ◇ REV MR 6.12 JSL 9.49 • ID 08977

MCDERMOTT, D. [1980] see DOYLE, J.

McDermott, D. [1982] *Nonmonotonic logic II: nonmonotonic modal theories* (**J** 0037) ACM J 29*33-57
◇ B60 ◇ REV MR 83h:03027 Zbl 477 # 68099 • REM Part I 1980 by McDermott,D. & Doyle,J. • ID 55611

McGee, V. [1981] *Finite matrices and the logic of conditionals* (**J** 0122) J Philos Logic 10*349-351
◇ B48 ◇ REV MR 83k:03030 Zbl 479 # 03010 • ID 55672

McGee, V. see Vol. I for further entries

McGinn, C. [1976] see Hart, W.D.

McGinn, C. [1978] see Hart, W.D.

McGinn, C. [1981] *Modal reality* (**C** 4283) Reduct, Time & Reality 143-187
◇ B45 ◇ REV MR 84g:03008 • ID 45732

McGinn, C. [1981] *The mechanism of reference* (**J** 0154) Synthese 49*157-186
◇ B45 ◇ REV MR 83c:03005 • ID 35120

McGrade, A.S. [1966] see Kupperman, J.

Mchedlishvili, L.I. [1981] *Minimal temporal logic with a modalized temporal operator (Russian)* (**C** 3810) Log-Semat Issl (Tbilisi) 71-89
◇ B45 ◇ REV MR 84j:03042 • ID 34634

Mchedlishvili, L.I. [1981] *Modal interpretation of traditional relations between propositions (Russian)* (**C** 3810) Log-Semat Issl (Tbilisi) 5-15
◇ B45 ◇ REV MR 84m:03025 • ID 35729

McKay, C.G. [1967] *A note on the Jaskowski sequence* (**J** 0068) Z Math Logik Grundlagen Math 13*95-96
◇ B55 F50 ◇ REV MR 37 # 34 Zbl 148.7 JSL 38.520 • REM Result incorrect • ID 08983

McKay, C.G. [1967] *On finite logics* (**J** 0028) Indag Math 29*363-365
◇ B55 ◇ REV MR 35 # 6524 Zbl 153.7 JSL 36.330 • ID 08981

McKay, C.G. [1967] *Some completeness results for intermediate propositional logics* (**J** 0047) Notre Dame J Formal Log 8*191-194 • ERR/ADD ibid 9*388
◇ B55 ◇ REV MR 39 # 2611 MR 39 # 36 Zbl 183.9 • ID 08982

McKay, C.G. [1968] *The decidability of certain intermediate propositional logics* (**J** 0036) J Symb Logic 33*258-264
◇ B25 B55 ◇ REV MR 39 # 1300 Zbl 175.271 • ID 08986

McKay, C.G. [1968] *The non-separability of a certain finite extension of Heyting's propositional logic* (**J** 0028) Indag Math 30*312-315
◇ B55 ◇ REV MR 37 # 2578 Zbl 175.262 JSL 36.331 • ID 08987

McKay, C.G. [1971] *A class of decidable intermediate propositional logics* (**J** 0036) J Symb Logic 36*127-128
◇ B25 B55 ◇ REV MR 43 # 7297 Zbl 216.292 • ID 08988

McKay, C.G. [1985] *A consistent propositional logic without any finite models* (**J** 0036) J Symb Logic 50*38-41
◇ B55 ◇ REV Zbl 566 # 03008 • ID 42521

McKay, C.G. see Vol. VI for further entries

McKay, T.J. [1975] *Essentialism in quantified modal logic* (**J** 0122) J Philos Logic 4*423-438
◇ A05 B45 ◇ REV MR 57 # 15969 Zbl 391 # 03013 • ID 52340

McKay, T.J. [1977] see Inwagen van, P.

McKay, T.J. [1978] *The principle of predication and the elimination of de re modalities* (**J** 0122) J Philos Logic 7*19-26 • ERR/ADD ibid 7/1*387
◇ A05 B45 ◇ REV MR 58 # 5058a MR 58 # 5058b Zbl 377 # 02019 • ID 51736

McKee, T.A. [1985] *Generalized equivalence: A pattern of mathematical expression* (**J** 0063) Studia Logica 44*285-290
◇ B60 ◇ ID 47503

McKee, T.A. see Vol. I, III for further entries

McKeon, R. [1969] *Principles of modal logic, aristotelian and modern* (**J** 0147) An Univ Bucuresti, Acta Logica 12*231-234
◇ A10 B45 ◇ REV Zbl 247 # 02023 • ID 29501

McKim, V.R. [1976] see Davis, Charles C.

McKinney, A. [1975] see Chellas, B.F.

McKinsey, J.C.C. [1940] *A correction to Lewis and Langford's symbolic logic* (**J** 0036) J Symb Logic 5*149
◇ B50 ◇ REV MR 2.209 FdM 66.29 • ID 09017

McKinsey, J.C.C. [1940] *Proof that there are infinitely many modalities in Lewis's system* S_2 (**J** 0036) J Symb Logic 5*110-112
◇ B45 ◇ REV MR 2.66 Zbl 24.241 JSL 6.37 FdM 66.29 • ID 09019

McKinsey, J.C.C. [1941] *A solution of the decision problem for the Lewis systems S2 and S4, with an application to topology* (**J** 0036) J Symb Logic 6*117-134
◇ B25 B45 C65 ◇ REV MR 3.290 JSL 7.118 • ID 09020

McKinsey, J.C.C. [1944] *On the number of complete extensions of the Lewis systems of sentential calculus* (**J** 0036) J Symb Logic 9*42-45
◇ B45 ◇ REV MR 6.96 Zbl 61.8 JSL 9.29 • ID 09022

McKinsey, J.C.C. [1945] *On the syntactical construction of systems of modal logic* (**J** 0036) J Symb Logic 10*83-94
◇ B45 ◇ REV MR 7.186 JSL 11.98 • ID 09025

McKinsey, J.C.C. & Tarski, A. [1948] *Some theorems about the sentential calculi of Lewis and Heyting* (**J** 0036) J Symb Logic 13*1-15
◇ B45 B55 ◇ REV MR 9.486 JSL 13.171 • ID 09028

McKinsey, J.C.C. [1949] *Construction of systems of modal logic* (**P** 0682) Int Congr Philos (10);1948 Amsterdam 740
◇ B45 ◇ REV MR 10.423 Zbl 31.104 • ID 09032

McKinsey, J.C.C. [1953] *Systems of modal logic which are not unreasonable in the sense of Hallden* (**J** 0036) J Symb Logic 18*109-113
◇ B45 ◇ REV MR 15.2 Zbl 53.2 JSL 19.67 • ID 09033

McKinsey, J.C.C. see Vol. I, III, V, VI for further entries

McKinsey, M. [1979] *Levels of obligation* (**J** 0095) Philos Stud 35*385-395
◇ A05 B45 ◇ REV MR 80h:03013 • ID 76224

McLane Jr., H.E. [1979] *On the possibility of epistemic logic* (**J** 0047) Notre Dame J Formal Log 20*559-574
◇ A05 B45 ◇ REV MR 80j:03029 Zbl 336 # 02023 • ID 56134

McLaughlin, R.N. [1973] *On the logic of general conditionals* (**J** 2811) Phil Quart (St Andrews) 23*133-143
◇ B45 ◇ REV MR 58 # 10318 • ID 76227

McLelland, J. [1971] *Epistemic logic and the paradox of the surprise examination* (J 0286) Int Logic Rev 3*69-85
◇ B45 ◇ REV MR 46 # 3269 • ID 22281

McLelland, J. [1976] *Epistemic logic with identifiers* (J 0047) Notre Dame J Formal Log 17*321-343
◇ B45 ◇ REV MR 58 # 27340 Zbl 283 # 02026 • ID 18291

McMichael, A. & Zalta, E.N. [1980] *An alternative theory of nonexistent objects* (J 0122) J Philos Logic 9*297-313
◇ A05 B45 ◇ REV MR 82c:03008 • ID 76247

McMichael, A. [1983] *A new actualist modal semantics* (J 0122) J Philos Logic 12*73-99
◇ B45 ◇ REV MR 84i:03042 • ID 34524

McMichael, A. see Vol. V for further entries

McNaughton, R. [1951] *A theorem about infinite-valued sentential logic* (J 0036) J Symb Logic 16*1-13
◇ B50 ◇ REV MR 13.3 Zbl 43.9 JSL 16.227 • ID 09066

McNaughton, R. see Vol. I, III, IV, V, VI for further entries

McRobbie, M.A. [1977] see Belnap Jr., N.D.

McRobbie, M.A. & Meyer, R.K. [1979] *A note on the admissibility of cut in relevant tableau systems* (J 0068) Z Math Logik Grundlagen Math 25*511-512
◇ B46 ◇ REV MR 81e:03016 Zbl 423 # 03015 • ID 53527

McRobbie, M.A. & Meyer, R.K. [1979] *Firesets and relevant implication* (X 0904) ANU RSSS Philos: Canberra
◇ B46 ◇ ID 33816

McRobbie, M.A. [1979] see Belnap Jr., N.D.

McRobbie, M.A. [1984] see Belnap Jr., N.D.

McRobbie, M.A. see Vol. I, III for further entries

Medvedev, Yu.T. [1979] *Transformations of information and calculi that describe them: types of information and their possible transformations (Russian)* (S 2582) Semiotika & Inf, Akad Nauk SSSR 13*109-141
◇ B55 ◇ REV MR 81j:03041a Zbl 485 # 03002 • ID 76258

Medvedev, Yu.T. see Vol. IV, VI for further entries

Meixner, J. [1982] *Are statistical explanations really explanatory?* (J 0095) Philos Stud 42*201-207
◇ B48 ◇ REV MR 85c:03006 • ID 40478

Melikhov, A.N. [1981] see Korovin, S.Ya.

Melikhov, A.N. see Vol. IV, V for further entries

Mellema, G. [1979] *An alternative semantics for knowledge* (J 0047) Notre Dame J Formal Log 20*265-278
◇ B45 ◇ REV MR 80f:03024 Zbl 322 # 02018 • ID 52630

Melliar-Smith, P.M. & Schwartz, R.L. & Vogt, F.H. [1984] *An interval-based temporal logic* (P 2989) Log of Progr; 1983 Pittsburgh 443-457
◇ B45 ◇ REV Zbl 558 # 03011 • ID 44571

Melliar-Smith, P.M. see Vol. I for further entries

Meloni, G.C. [1977] see Bozzi, S.

Meloni, G.C. & Radaelli, G. [1979] *Calcolo naturale per categorie con residui* (J 0207) Ist Lombardo Accad Sci Rend, A (Milano) 113*45-53
◇ B60 F07 G30 ◇ REV MR 82h:18001 Zbl 449 # 03075 • ID 56742

Meloni, G.C. [1980] see Bozzi, S.

Meloni, G.C. [1985] see Ghilardi, S.

Meloni, G.C. see Vol. I, III, V, VI for further entries

Meltzer, B. [1970] *The semantics of induction and the possibility of complete systems of inductive inference* (J 0503) Artif Intell 1*189-192
◇ B48 ◇ REV Zbl 206.9 • ID 46565

Meltzer, B. see Vol. I for further entries

Mendelson, E. [1960] *A semantic proof of the eliminability of descriptions* (J 0068) Z Math Logik Grundlagen Math 6*199-200
◇ B60 ◇ REV MR 23 # A792 Zbl 116.5 JSL 38.660 • ID 09114

Mendelson, E. [1964] *Introduction to mathematical logic* (X 0864) Van Nostrand: New York x+300pp • TRANSL [1971] (X 2027) Nauka: Moskva 320pp
◇ B98 ◇ REV MR 29 # 2158 MR 50 # 1825 MR 80d:03001 Zbl 192.19 Zbl 498 # 03001 JSL 34.110 • REM 2nd ed. 1979; viii+328pp. 3rd ed. 1984 • ID 09105

Mendelson, E. see Vol. I, III, IV, V, VI for further entries

Mendez, J.M. [1985] *Systems with the converse Ackermann property* (J 4604) Theoria, Ser 2 (San Sebastian) 1*253-258
◇ B46 ◇ ID 48963

Menger, K. [1939] *A logic of the doubtful. On optative and imperative logic* (J 2132) Rep Math Colloq, Ser 2 1*53-64
◇ B45 ◇ REV Zbl 21.290 FdM 65.29 • ID 43756

Menger, K. [1951] *Probabilistic theories of relations* (J 0054) Proc Nat Acad Sci USA 37*178-180
◇ B48 E07 ◇ REV MR 13.51 Zbl 42.371 • ID 31954

Menger, K. [1964] *Superassociative systems and logical functors* (J 0043) Math Ann 157*278-295
◇ B50 C05 ◇ REV MR 31 # 2186 Zbl 126.36 • ID 31957

Menger, K. & Whitlock, H.I. [1966] *Two theorems on the generation of systems of functions* (J 0027) Fund Math 58*229-240
◇ B50 ◇ REV MR 34 # 2448 Zbl 168.267 • ID 09125

Menger, K. see Vol. I, IV, V, VI for further entries

Menges, G. [1974] *Elements of an objective theory of inductive behaviour* (C 4722) Inform, Infer & Decis 3-49
◇ A05 B48 ◇ REV Zbl 298 # 02021 • ID 63867

Menges, G. & Skala, H.J. [1974] *On the problem of vagueness in the social sciences* (C 4722) Inform, Infer & Decis 51-61
◇ A05 B52 ◇ REV Zbl 298 # 02022 • ID 63868

Menges, G. [1976] see Kofler, E.

Menges, G. [1983] *Unscharfe Konzepte im Operations Research* (J 3401) Meth Oper Res 46*353-379
◇ B52 ◇ REV Zbl 512 # 90064 • ID 37554

Menne, A. [1952] *Zu den triadischen bivalenten Aussagefunktoren* (J 0105) Theoria (Lund) 18*66-69
◇ B50 ◇ ID 33430

Menne, A. [1954] see Bochenski, I.M.

Menne, A. [1966] *Einfuehrung in die Logik* (X 1987) Francke: Muenchen
◇ B98 ◇ ID 33438

Menne, A. [1969] *Modalitaeten als Stufenfunktoren* (J 0147) An Univ Bucuresti, Acta Logica 12*109-119
◇ B45 ◇ REV MR 44 # 6458 Zbl 254 # 02020 • ID 09128

MENNE, A. see Vol. I for further entries

MEREDITH, C.A. [1951] *On a extended system of the propositional calculus* (J 0215) Proc Irish Acad, Sect A 54*37-47
◇ B60 ◇ REV MR 13.3 Zbl 42.244 JSL 16.229 • ID 09132

MEREDITH, C.A. [1957] see LEMMON, E.J.

MEREDITH, C.A. [1958] *The dependence of an axiom of Lukasiewicz* (J 0064) Trans Amer Math Soc 87*54
◇ B50 ◇ REV MR 20#819 Zbl 85.244 JSL 24.248 • ID 09135

MEREDITH, C.A. & PRIOR, A.N. [1964] *Investigations into implicational S5* (J 0068) Z Math Logik Grundlagen Math 10*203-220
◇ B45 ◇ REV MR 29#1142 Zbl 146.8 • ID 09138

MEREDITH, C.A. & PRIOR, A.N. [1965] *Modal logic with functorial variables and a contingent constant* (J 0047) Notre Dame J Formal Log 6*99-109
◇ B45 ◇ REV MR 36#3636 • ID 09140

MEREDITH, C.A. [1969] see LEMMON, E.J.

MEREDITH, C.A. see Vol. I, VI for further entries

MEREDITH, D. [1956] *A correction to von Wright's decision procedure for the deontic system P* (J 0094) Mind 65*548-550
◇ B25 B45 ◇ REV JSL 22.92 • ID 09147

MEREDITH, D. [1957] see LEMMON, E.J.

MEREDITH, D. [1969] see LEMMON, E.J.

MEREDITH, D. [1976] *A calculus of matrical descriptors* (J 0047) Notre Dame J Formal Log 17*517-525
◇ B50 ◇ REV MR 55#7740 Zbl 258#02012 • ID 21940

MEREDITH, D. [1978] *Are there many-valued Scotan logics?* (J 0387) Bull Sect Logic, Pol Acad Sci 7*2-3
◇ B50 ◇ REV MR 58#21445 • ID 76290

MEREDITH, D. [1983] *Separating minimal, intuitionist and classical logic* (J 0047) Notre Dame J Formal Log 24*485-490
◇ B55 ◇ REV MR 84k:03037 Zbl 476#03021 • ID 55533

MEREDITH, D. see Vol. I, III, VI for further entries

MERMIN, N.D. [1983] *Pair distributions and conditional independence: some hints about the structure of strange quantum correlations* (J 0153) Phil of Sci (East Lansing) 50*359-373
◇ B51 G12 ◇ REV MR 84m:81025 • ID 39533

MERRETT, T. [1978] *QT logic: Simpler and more expressive than predicate calculus* (J 0232) Inform Process Lett 7*251-255
◇ B60 B75 ◇ REV MR 80f:68111 Zbl 389#68056 • ID 52325

MERRILL, G.H. [1975] *A free logic with intensions as possible values of terms* (J 0122) J Philos Logic 4*293-326
◇ B45 ◇ REV MR 58#21476 Zbl 333#02019 • ID 63877

MERRILL, G.H. [1977] *On an enduring non sequitur of Quine's* (J 0047) Notre Dame J Formal Log 18*613-615
◇ A05 B45 ◇ REV Zbl 332#02026 • ID 23701

MERRILL, G.H. [1979] *Confirmation and prediction* (J 0153) Phil of Sci (East Lansing) 46*98-117
◇ B48 ◇ REV MR 83b:03008 • ID 35083

MERTENS, A. [1973] see BAGINSKI, M.

MESKHI, S. [1977] *Fuzzy propositional logic. Algebraic approach* (J 0063) Studia Logica 36*189-194
◇ B52 G10 G25 ◇ REV MR 58#5033 Zbl 363#02016 Zbl 409#03036 • ID 50843

MESKHI, S. see Vol. III for further entries

MESKHI, V.YU. [1971] *To algebraic semantics of temporal logic (Russian)* (X 2235) VINITI: Moskva 3022-71
◇ B45 ◇ ID 32636

MESKHI, V.YU. [1974] *Kripke semantics for modal systems including S4.3 (Russian)* (J 0087) Mat Zametki (Akad Nauk SSSR) 15*875-884
• TRANSL [1974] (J 1044) Math Notes, Acad Sci USSR 15*523-528
◇ B45 ◇ REV MR 50#4248 Zbl 311#02033 • ID 09163

MESKHI, V.YU. [1976] *Critical normal modal systems containing K4 (Russian)* (P 2064) All-Union Conf Math Log (4);1976 Kishinev 93
◇ B45 ◇ ID 32639

MESKHI, V.YU. [1976] *Propositional diffusion logic (Russian)* (P 2566) All-Union Symp Log & Method of Sci (7);1976 Kiev 146-148
◇ B45 ◇ ID 32638

MESKHI, V.YU. [1977] see EHSAKIA, L.L.

MESKHI, V.YU. [1978] *Algebraic analysis of modal fragments of tense systems (Russian)* (C 2568) Sb Log, Semant, Metodol 113-124
◇ B45 ◇ ID 32641

MESKHI, V.YU. [1978] see GRIGOLIYA, R.SH.

MESKHI, V.YU. [1979] *Critical modal logics containing Brouwer's axioms (Russian)* (P 2570) Intens Log & Log Analiz Estestv Yazyk;1979 Moskva 65
◇ B45 ◇ ID 32644

MESKHI, V.YU. [1983] *Critical modal logics containing the Brouwer axiom (Russian)* (J 0087) Mat Zametki (Akad Nauk SSSR) 33*131-139,159
• TRANSL [1983] (J 1044) Math Notes, Acad Sci USSR 33*65-69
◇ B45 ◇ REV MR 84h:03041 Zbl 516#03009 • ID 34249

METZE, G. [1955] *An application of multi-valued logic systems to circuits* (P 4136) Proc Symp Circuit Anal;1955 Urbana 11-1-11-14
◇ B50 B70 ◇ ID 41719

METZE, G. [1973] see DUCASSE, E.G.

MEYER, A.R. & PARIKH, R. [1981] *Definability in dynamic logic* (J 0119) J Comp Syst Sci 23*279-298
◇ B75 ◇ REV MR 84b:68011 Zbl 472#03013 JSL 49.1420 • ID 55279

MEYER, A.R. & MIRKOWSKA, G. & STREETT, R.S. [1981] *The deducibility problem in propositional dynamic logic* (P 3497) Log of Progr;1979 Zuerich 12-22
• REPR [1981] (P 2903) Automata, Lang & Progr (8);1981 Akko 238-248
◇ B75 D25 D35 ◇ REV MR 83i:03046 MR 85h:03029 Zbl 466#68024 • REM The reprint is a revised version • ID 54996

MEYER, A.R. & TIURYN, J. [1982] *A note on equivalences among logics of programms* (P 3738) Log of Progr;1981 Yorktown Heights 282-299
◇ B75 ◇ REV MR 83i:68030 Zbl 502#68005 • ID 37433

MEYER, A.R. & WINKLMANN, K. [1982] *Expressing program looping in regular dynamic logic* (**J** 1426) Theor Comput Sci 18∗301–323
◇ B75 ◇ REV MR 83g:68024 Zbl 478 # 68031 • ID 55655

MEYER, A.R. & TIURYN, J. [1983] *A note on equivalence among logic of programs* (**S** 3382) Sem-ber, Humboldt-Univ Berlin, Sekt Math 52∗47–64
◇ B75 ◇ REV MR 85f:68003 Zbl 533 # 68037 • ID 36577

MEYER, A.R. & TIURYN, J. [1984] *Equivalences among logics of programs* (**J** 1426) Theor Comput Sci 29∗160–170
◇ B75 ◇ REV Zbl 552 # 68037 • ID 45358

MEYER, A.R. [1985] see CHANDRA, A.K.

MEYER, A.R. see Vol. I, III, IV, VI for further entries

MEYER, J.-J.C. [1982] see BERGSTRA, J.A.

MEYER, J.-J.C. see Vol. IV for further entries

MEYER, R.K. [1966] *Pure denumerable Lukasiewiczian implication* (**J** 0036) J Symb Logic 31∗575–580
◇ B50 ◇ REV MR 35 # 5287 Zbl 148.6 JSL 33.308
• ID 09179

MEYER, R.K. [1966] see MCCALL, S.

MEYER, R.K. [1968] *An undecidability result in the theory of relevant implication* (**J** 0068) Z Math Logik Grundlagen Math 14∗255–262
◇ B46 D35 ◇ REV MR 38 # 4287 Zbl 172.297 • ID 09181

MEYER, R.K. [1968] *Entailment and relevant implication* (**J** 0079) Logique & Anal, NS 11∗472–479
◇ B46 ◇ REV MR 40 # 7094 Zbl 186.6 • ID 09182

MEYER, R.K. [1968] see LAMBERT, K.

MEYER, R.K. [1969] see LAMBERT, K.

MEYER, R.K. [1969] see DUNN, J.M.

MEYER, R.K. [1970] *R_i – the bounds of finitude* (**J** 0068) Z Math Logik Grundlagen Math 16∗385–387
◇ B46 ◇ REV MR 44 # 6444 Zbl 206.275 • ID 09190

MEYER, R.K. [1970] *E and S4* (**J** 0047) Notre Dame J Formal Log 11∗181–199
◇ B46 ◇ REV MR 45 # 6592 Zbl 182.5 • ID 09191

MEYER, R.K. [1971] see DUNN, J.M.

MEYER, R.K. [1971] *Entailment* (**J** 0301) J Phil 68∗808–818
◇ B46 ◇ ID 33919

MEYER, R.K. [1971] *On coherence in modal logics* (**J** 0079) Logique & Anal, NS 14∗658–668
• REPR [1975] (**C** 1852) Entailment – Log of Relev & Nec 263–271
◇ B46 ◇ REV MR 46 # 6997 Zbl 239 # 02011 JSL 42.311
• ID 09192

MEYER, R.K. & ROUTLEY, R. [1972] *Algebraic analysis of entailment I* (**J** 0079) Logique & Anal, NS 15∗407–428
◇ B46 ◇ REV MR 48 # 5857 Zbl 336 # 02020 • ID 09196

MEYER, R.K. & PARKS, R.Z. [1972] *Independent axioms for the implicational fragment of Sobocinski's three-valued logic* (**J** 0068) Z Math Logik Grundlagen Math 18∗291–295
◇ B50 ◇ REV MR 50 # 9530 Zbl 261 # 02011 • ID 09193

MEYER, R.K. [1972] *On relevantly derivable disjunctions* (**J** 0047) Notre Dame J Formal Log 13∗476–480
◇ B46 ◇ REV MR 47 # 6456 Zbl 226 # 02020 JSL 42.311
• ID 09195

MEYER, R.K. & ROUTLEY, R. [1972] *The semantics of entailment II,III* (**J** 0122) J Philos Logic 1∗53–73,192–208
◇ B46 ◇ REV MR 53 # 12877 MR 53 # 12878
Zbl 317 # 02018 Zbl 317 # 02019 • REM Part I 1973
• ID 11624

MEYER, R.K. & ROUTLEY, R. [1973] *An undecidable relevant logic* (**J** 0068) Z Math Logik Grundlagen Math 19∗389–397
◇ B46 D35 D40 ◇ REV MR 49 # 32 Zbl 301 # 02024
• ID 09201

MEYER, R.K. & ROUTLEY, R. [1973] *Classical relevant logics I (Polish and Russian summaries)* (**J** 0063) Studia Logica 32∗51–68
◇ B46 ◇ REV MR 51 # 44 Zbl 316 # 02029 • REM Part II 1974 • ID 19501

MEYER, R.K. [1973] *Conservative extension in relevant implication (Polish and Russian summaries)* (**J** 0063) Studia Logica 31∗39–48
◇ B46 ◇ REV MR 52 # 2826 Zbl 273 # 02019 JSL 42.311
• ID 17622

MEYER, R.K. [1973] *Intuitionism, entailment, negation* (**P** 0783) Truth, Syntax & Modal;1970 Philadelphia 168–198
◇ B46 F50 ◇ REV MR 53 # 95 Zbl 317 # 02016
JSL 42.315 • ID 09200

MEYER, R.K. & ROUTLEY, R. [1973] *The semantics of entailment I* (**P** 0783) Truth, Syntax & Modal;1970 Philadelphia 199–243
◇ B46 ◇ REV MR 53 # 12876 Zbl 317 # 02017 JSL 42.315
• REM Parts II,III 1972 • ID 11626

MEYER, R.K. [1974] see GODDARD, L.

MEYER, R.K. & ROUTLEY, R. [1974] *Classical relevant logics II* (**J** 0063) Studia Logica 33∗183–194
◇ B46 ◇ REV MR 51 # 44 Zbl 316 # 02030 • REM Part I 1973 • ID 49853

MEYER, R.K. [1974] see DUNN, J.M.

MEYER, R.K. & ROUTLEY, R. [1974] *E is a conservative extension of $E_{\overline{I}}$* (**J** 1807) Phil Quart Israel 4∗223–249
◇ B46 ◇ ID 33921

MEYER, R.K. [1974] *Entailment is not strict implication* (**J** 0273) Australasian J Phil 52∗212–231
◇ B46 ◇ ID 35606

MEYER, R.K. [1974] *New axiomatics for relevant logics. I* (**J** 0122) J Philos Logic 3∗53–86
◇ B46 ◇ REV MR 54 # 7204 Zbl 278 # 02025 • ID 24995

MEYER, R.K. [1975] *Conservative extensions in R* (**C** 1852) Entailment – Log of Relev & Nec 371–378
◇ B46 ◇ ID 33937

MEYER, R.K. [1975] *On conserving positive logics I* (**C** 1852) Entailment – Log of Relev & Nec 288–296
◇ B46 ◇ ID 33936

MEYER, R.K. [1975] *On relevantly derivable disjunctions* (**C** 1852) Entailment – Log of Relev & Nec 378–381
◇ B46 ◇ ID 33938

MEYER, R.K. [1975] *Relevance is not reducible to modality* (C 1852) Entailment - Log of Relev & Nec 462-471
◇ B46 ◇ ID 33940

MEYER, R.K. [1975] *Sugihara is a characteristic matrix for RM* (C 1852) Entailment - Log of Relev & Nec 393-420
◇ B46 ◇ ID 33939

MEYER, R.K. [1975] see DUNN, J.M.

MEYER, R.K. & ROUTLEY, R. [1975] *Towards a general semantical theory of implication and conditionals. I. Systems with normal conjunctions and disjunctions and aberrant and normal negations* (J 0302) Rep Math Logic, Krakow & Katowice 4*67-89
◇ B46 ◇ REV MR 52 # 2828 Zbl 314 # 02032 • REM Part II 1977 • ID 17623

MEYER, R.K. [1976] *A general Gentzen system for implicational calculi* (J 1893) Relevance Logic Newslett 1*189-201
◇ B46 ◇ REV Zbl 364 # 02010 • ID 50940

MEYER, R.K. [1976] *Ackermann, Takeuti, und Schnitt: γ for higher-order relevant logic* (J 0387) Bull Sect Logic, Pol Acad Sci 5*138-144
◇ B46 F05 F35 ◇ REV MR 58 # 27341 • ID 27122

MEYER, R.K. & ROUTLEY, R. [1976] *Dialectical logic, classical logic and the consistency of the world* (J 2051) Stud Soviet Thought 16*1-25
• TRANSL [1979] (C 2041) Formal della Dialettica 324-354
◇ A05 B53 ◇ ID 33922

MEYER, R.K. & ROUTLEY, R. [1976] *Every sentential logic has a two-valued worlds semantics* (J 0079) Logique & Anal, NS 19*345-365
◇ B45 ◇ REV MR 58 # 10321 Zbl 347 # 02016 • ID 64943

MEYER, R.K. [1976] *Metacompleteness* (J 0047) Notre Dame J Formal Log 17*501-516
◇ B45 F50 ◇ REV MR 55 # 12479 Zbl 232 # 02015 • ID 21939

MEYER, R.K. [1976] *Negation disarmed* (J 0047) Notre Dame J Formal Log 17*184-190
◇ B45 ◇ REV MR 53 # 10552 Zbl 322 # 02028 • ID 18294

MEYER, R.K. [1976] *On $\exists p(\exists qAq \rightarrow Ap)$* (J 1893) Relevance Logic Newslett 1*68-71
◇ B46 ◇ REV Zbl 361 # 02028 • ID 50663

MEYER, R.K. [1976] *Relevant arithmetic* (J 0387) Bull Sect Logic, Pol Acad Sci 5*133-137
◇ B46 F30 ◇ REV MR 58 # 21477 • ID 27121

MEYER, R.K. [1976] *Two questions from Anderson and Belnap* (J 0302) Rep Math Logic, Krakow & Katowice 7*71-86
◇ B46 ◇ REV MR 58 # 21478 Zbl 361 # 02027 • ID 21929

MEYER, R.K. & ROUTLEY, R. [1977] *Extensional reduction I* (J 0320) Monist 60*355-369
◇ B46 ◇ ID 33920

MEYER, R.K. [1977] *First degree formulas in Curry's LD* (J 0047) Notre Dame J Formal Log 18*181-191
◇ B46 F50 ◇ REV MR 56 # 8328 Zbl 339 # 02021 • ID 21964

MEYER, R.K. [1977] *On a purported implication* (J 1893) Relevance Logic Newslett 2*129-130
◇ B46 ◇ REV Zbl 377 # 02023 • ID 51740

MEYER, R.K. [1977] *S5--the poor man's connexive implication* (J 1893) Relevance Logic Newslett 2*117-124
◇ B46 ◇ REV Zbl 377 # 02024 • ID 51741

MEYER, R.K. & ROUTLEY, R. [1977] *Towards a general semantical theory of implication and conditionals. II. Improved negation theory and propositional identity* (J 0302) Rep Math Logic, Krakow & Katowice 9*47-62
◇ B46 ◇ REV MR 80f:03019 Zbl 394 # 03028 • REM Part I 1975 • ID 52505

MEYER, R.K. [1979] see MCROBBIE, M.A.

MEYER, R.K. [1979] *Career induction stops here (and here = 2)* (J 0122) J Philos Logic 8*361-371
◇ A05 B46 ◇ REV MR 80i:03032 Zbl 429 # 03035 • ID 53866

MEYER, R.K. [1980] see GIAMBRONE, S.

MEYER, R.K. [1980] *Career induction for quantifiers* (J 0047) Notre Dame J Formal Log 21*539-548
◇ B46 ◇ REV MR 81i:03022 Zbl 423 # 03027 • ID 53539

MEYER, R.K. [1980] *Sentential constants in relevant implication* (J 0387) Bull Sect Logic, Pol Acad Sci 9*33-38
◇ B46 ◇ REV MR 81f:03028 Zbl 426 # 03016 • ID 53612

MEYER, R.K. [1981] *Almost Skolem forms for relevant (and other) logics* (J 0079) Logique & Anal, NS 24*277-289
◇ B46 ◇ REV MR 83i:03032 Zbl 519 # 03010 • ID 35501

MEYER, R.K. [1981] see GIAMBRONE, S.

MEYER, R.K. [1982] see MARTIN, E.P.

MEYER, R.K. [1982] see BRADY, R.T.

MEYER, R.K. [1983] *A note on R-matrices* (J 0047) Notre Dame J Formal Log 24*450-472
◇ B46 ◇ REV MR 85c:03004 Zbl 536 # 03009 • ID 37096

MEYER, R.K. & ROUTLEY, R. [1983] *Relevant logics and their semantics remain viable and undamaged by Lewis's equivocation charge* (J 3781) Topoi 2*205-215
◇ B46 ◇ REV MR 85i:03056c • ID 44114

MEYER, R.K. [1983] see DWYER, R.

MEYER, R.K. & MORTENSEN, C. [1984] *Inconsistent models for relevant arithmetics* (J 0036) J Symb Logic 49*917-929
◇ B46 ◇ REV MR 86e:03021 • ID 42482

MEYER, R.K. [1984] see BRADY, R.T.

MEYER, R.K. [1985] *A farewell to entailment* (P 4180) Int Congr Log, Meth & Phil of Sci (7,Pap);1983 Salzburg 577-636
◇ B46 ◇ ID 47940

MEYER, R.K. [1985] see BUNDER, M.W.

MEYER, R.K. & MORTENSEN, C. [1985] *Relevant quantum arithmetic* (C 4181) Math Log & Formal Syst (Costa da) 221-226
◇ B46 F30 ◇ REV Zbl 567 # 03007 • ID 48205

MEYER, R.K. see Vol. I, III, V, VI for further entries

MEYERHOFF CRESSWELL, M. [1982] see CRESSWELL, M.J.

MICHAEL, E. [1979] *Some considerations in medieval tense logic* (J 0047) Notre Dame J Formal Log 20*794-800
◇ A05 A10 B45 ◇ REV Zbl 351 # 02002 • ID 56171

MICHAELS, A. [1974] *A uniform proof procedure for SCI tautologies* (J 0063) Studia Logica 33*299-310
◇ B60 ◇ REV MR 51 # 10025 • ID 17563

MICHAELS, A. & SUSZKO, R. [1974] *EN-logic* (**J** 0387) Bull Sect Logic, Pol Acad Sci 3/1*13
◇ B60 ◇ REV MR 52 # 13316 • ID 21779

MICHAELS, A. & SUSZKO, R. [1976] *Sentential calculus of identity and negation* (**J** 0302) Rep Math Logic, Krakow & Katowice 7*87-106
◇ B60 ◇ REV MR 58 # 21496 Zbl 361 # 02018 • ID 21930

MICHALCZYK, M. [1981] see MALINOWSKI, G.

MICHALCZYK, M. [1982] see MALINOWSKI, G.

MICHALOS, A.C. [1969] *Principles of logic* (**X** 0819) Prentice Hall: Englewood Cliffs xiii + 433pp
◇ B98 ◇ ID 22523

MICHALOS, A.C. [1971] *Hilpinen's rules of acceptance and inductive logic* (**J** 0153) Phil of Sci (East Lansing) 38*293-302
◇ A05 B48 ◇ REV MR 58 # 16102a • ID 76351

MICHALSKI, R.S. [1974] *Variable-valued logic* (**P** 1385) Int Symp Multi-Val Log (4);1974 Morgantown 323-340
◇ B50 ◇ ID 42095

MICHALSKI, R.S. [1976] *Learning by inductive inference* (**P** 3143) Comput Orient Learn Process;1974 Bonas 321-337
◇ B48 ◇ REV MR 56 # 17243 Zbl 376 # 68060 • ID 51711

MICHALSKI, R.S. [1980] *Detection of conceptual patterns through inductive inference* (**P** 3139) Classific Autom & Percept par Ordin;1978/79 Rocquencourt 297-339
◇ B48 ◇ REV MR 82g:92061 Zbl 449 # 68053 • ID 56754

MICHALSKI, R.S. [1981] see DIETTERICH, T.G.

MICHALSKI, R.S. [1983] *A theory and methodology of inductive learning* (**J** 0503) Artif Intell 20*111-161
◇ B48 ◇ REV MR 84e:68103 • ID 39293

MICHALSKI, R.S. see Vol. I for further entries

MICHEL, M. [1984] *Algebre de machines et logique temporelle* (**P** 3565) Symp of Th Aspects of Comput Sci (1);1984 Paris 287-298
◇ B45 D05 ◇ REV MR 86a:68026 Zbl 542 # 68021 • ID 43712

MICHEL, M. [1984] see ENJALBERT, P.

MIELNIK, B. [1981] *Motion and form* (**P** 3820) Curr Iss in Quantum Log;1979 Erice 465-477
◇ B51 G12 ◇ REV MR 84i:03005 Zbl 537 # 03044 • ID 44885

MIGLIOLI, P.A. & MOSCATO, U. & ORNAGHI, M. & USBERTI, G. [1983] *Alcuni calcoli intermedi costruttivi* (**P** 3829) Atti Incontri Log Mat (1);1982 Siena 377-386
◇ B55 F65 ◇ REV MR 84k:03006 Zbl 522 # 03014 • ID 37781

MIGLIOLI, P.A. & USBERTI, G. [1984] *A logic of knowing (Italian)* (**J** 4608) Teoria (Pisa) 4/2*111-132
◇ A05 B60 ◇ ID 49154

MIGLIOLI, P.A. see Vol. I, III, IV, VI for further entries

MIHAILESCU, E.G. [1961] *Sur la methode de J.Lukasiewicz* (**J** 0147) An Univ Bucuresti, Acta Logica 4*117-122
◇ B50 ◇ REV MR 26 # 3585 Zbl 101.11 • ID 09226

MIHAILESCU, E.G. [1969] *Mathematical logic. Elements of propositional and predicate calculus (Romanian)* (**X** 0871) Acad Rep Soc Romania: Bucharest 334pp
◇ B98 ◇ REV MR 41 # 5191 Zbl 199.297 • ID 21042

MIHAILESCU, E.G. see Vol. I, III, IV for further entries

MIKELADZE, Z.N. (ED.) [1981] *Studies in logic and semantics (Russian)* (**X** 3790) Metsniereba: Tbilisi 144pp
◇ B97 ◇ REV MR 84h:03009 • ID 34223

MIKELADZE, Z.N. see Vol. I for further entries

MIKHAJLOV, A.I. (ED.) [1979] *Studies in nonclassical logics and set theory (Russian)* (**X** 2027) Nauka: Moskva 374pp
◇ B97 ◇ REV MR 81f:03003 • ID 70049

MIKHAJLOV, A.I. (ED.) [1983] *Studies in nonclassical logics and formal systems (Russian)* (**X** 2027) Nauka: Moskva 360pp
◇ B97 ◇ REV MR 84h:03010 • ID 34224

MIKHAJLOVA, M. [1980] *Reduction of modalities in several intuitionistic modal logics* (**J** 0137) C R Acad Bulgar Sci 33*743-745
◇ B45 ◇ REV MR 82h:03015 Zbl 453 # 03019 • ID 54149

MILANOV, P. [1984] see KOVALEV, M.M.

MILBERGER, M. [1978] *The minimal modal logic: A cautionary tale about primitives and definitions* (**J** 0047) Notre Dame J Formal Log 19*486-488
◇ A05 B45 ◇ REV MR 58 # 148 Zbl 349 # 02014 • ID 51485

MILICIC, M.I. [1984] *Galois correspondences for closed classes of functions with delay I, II (Russian)* (**J** 0400) Publ Inst Math, NS (Belgrade) 36*119-124,125-136
◇ B50 ◇ REV MR 86i:94076 Zbl 567 # 94021 • ID 49220

MILLER, D.MICHAEL & MUZIO, J.C. [1976] *A ternary universal decision element* (**J** 0047) Notre Dame J Formal Log 17*632-637
◇ B50 ◇ REV MR 57 # 5471 Zbl 292 # 02013 • ID 21947

MILLER, D.MICHAEL [1977] *A canonical representation for many-valued symmetric functions* (**P** 3004) Manitoba Conf Num Math (6);1976 Winnipeg 303-313
◇ B50 ◇ REV MR 80g:94086 Zbl 469 # 94016 • ID 82252

MILLER, D.MICHAEL & MUZIO, J.C. [1979] *On the minimization of many-valued functions* (**P** 3003) Int Symp Multi-Val Log (9);1979 Bath 294-299
◇ B50 ◇ ID 35963

MILLER, D.MICHAEL & MUZIO, J.C. [1980] *A class of two-place three-valued unary generators* (**J** 0047) Notre Dame J Formal Log 21*148-154
◇ B50 ◇ REV MR 81a:03022 Zbl 363 # 02014 • ID 53194

MILLER, D.MICHAEL [1980] see EPSTEIN, G.

MILLER, D.MICHAEL [1980] *Spectral addition techniques for multiple-valued functions* (**S** 3726) Congressus Numerantium 31*141-151
◇ B55 ◇ REV MR 83d:94028 • ID 39657

MILLER, D.MICHAEL [1981] *Spectral symmetry tests* (**P** 3705) Int Symp Multi-Val Log (11);1981 Oklahoma City & Norman 130-134
◇ B50 ◇ REV MR 83m:94033 Zbl 543 # 94022 • ID 45939

MILLER, D.MICHAEL [1981] *The fanout-free realization of multiple-valued functions* (**P** 3705) Int Symp Multi-Val Log (11);1981 Oklahoma City & Norman 246-255
◇ B50 ◇ REV MR 83m:94033 Zbl 543 # 94023 • ID 45954

MILLER, D.MICHAEL see Vol. I for further entries

MILLER, DAVID [1976] *On probability measures for deductive systems I* (J 0387) Bull Sect Logic, Pol Acad Sci 5*87-96
◇ B48 ◇ REV MR 58 # 27431 Zbl 441 # 03001 • REM Parts II,III 1978 • ID 27114

MILLER, DAVID [1977] *On distance from the truth as a true distance (short version)* (J 0387) Bull Sect Logic, Pol Acad Sci 6*15-26
◇ A05 B50 G05 G10 ◇ REV MR 56 # 5199 Zbl 411 # 03003 • ID 32384

MILLER, DAVID [1978] *On probability measures for deductive systems II,III* (J 0387) Bull Sect Logic, Pol Acad Sci 7*12-19,51-57
◇ B48 ◇ REV MR 80b:03030 Zbl 441 # 03002 • REM Part I 1976 • ID 32386

MILLER, DAVID [1982] *Impartial truth* (J 2840) Stochastica, Univ Politec Barcelona 6*169-186
• REPR [1984] (P 3096) Conf Math Service of Man (2,Pap);1982 Las Palmas 75-90
◇ A05 B48 ◇ REV MR 84i:03026 MR 86a:03005 • ID 34510

MILLER, DAVID see Vol. I for further entries

MILLER, JAMES WILKIN [1951] *The logic of terms* (C 0585) Struct, Meth & Meaning (Sheffer) 35-41
◇ B60 ◇ REV Zbl 54.5 JSL 16.287 • ID 41735

MILLER, JAMES WILKIN [1955] *Exercises in introductory symbolic logic* (X 0828) Edwards Brothers: Ann Arbor ix + 59pp
◇ B98 ◇ REV JSL 22.310 • ID 09255

MILLER, JAMES WILKIN [1958] *Logic workbook* (X 0894) Oxford Univ Pr: Oxford vii + 88pp
◇ B98 ◇ REV Zbl 99.5 JSL 23.213 • ID 09256

MILLER, JAMES WILKIN [1975] *The logic of the synthetic a priori* (J 0047) Notre Dame J Formal Log 16*465-475
◇ B45 ◇ REV MR 53 # 7732 Zbl 236 # 02018 • ID 18298

MILLER, JAMES WILKIN see Vol. I for further entries

MILLER, JOHN R. [1978] *Use of an infinite-valued propositional calculus in a document retrieval system* (P 2014) Int Symp Multi-Val Log (8);1978 Rosemont 157-162
◇ B50 ◇ REV MR 81d:68132 • ID 82255

MILNER, R. [1981] *A modal characterisation of observable machine-behaviour* (P 2923) CAAP'81 Arbres en Algeb & Progr (6);1981 Genova 25-34
◇ B45 ◇ REV MR 82j:68037 Zbl 474 # 68074 • ID 82260

MILNER, R. see Vol. I, IV, VI for further entries

MILOVANOVIC, V. [1975] see KRON, A.

MINARI, P. [1983] *Completeness theorems for some intermediate predicate calculi* (J 0063) Studia Logica 42*431-441
◇ B55 C90 ◇ REV MR 86f:03045 Zbl 571 # 03011 • ID 42336

MINARI, P. [1983] see CASARI, E.

MINARI, P. see Vol. III, V for further entries

MINE, H. [1966] see KOGA, Y.

MINE, H. & SHIMADA, R. [1970] *Permutations and lattice operations in many-valued logic (Japanese)* (J 0979) Denshi Tsushin Gakkai Ronbunshi, Sect A-D 53-C/4*166-173
• TRANSL [1970] (J 0464) Syst-Comp-Controls 1/2*24-32
◇ B50 ◇ REV MR 46 # 8812 • ID 09291

MINE, H. & YAMAMOTO, Y. [1981] *Testing and realization of three-valued majority functions* (P 3705) Int Symp Multi-Val Log (11);1981 Oklahoma City & Norman 157-162
◇ B50 ◇ REV MR 83m:94033 Zbl 543 # 94025 • ID 45968

MINE, H. see Vol. V for further entries

MINICOZZI, E. [1976] *Some natural properties of strong-identification in inductive inference* (J 1426) Theor Comput Sci 2*345-360
◇ B48 ◇ REV MR 58 # 21528 Zbl 373 # 68051 • ID 51524

MINICOZZI, E. see Vol. I, IV for further entries

MINKER, J. [1980] see ARONSON, A.R.

MINKER, J. see Vol. I for further entries

MINTS, G.E. [1967] *Embedding operations related to S.Kripke's semantics (Russian)* (S 0228) Zap Nauch Sem Leningrad Otd Mat Inst Steklov 4*152-159
• TRANSL [1967] (J 0521) Semin Math, Inst Steklov 4*52-55
◇ B55 ◇ REV MR 40 # 7091 Zbl 166.250 • ID 37137

MINTS, G.E. [1968] *Cut-free calculi of the type S5 (Russian)* (S 0228) Zap Nauch Sem Leningrad Otd Mat Inst Steklov 8*166-174
• TRANSL [1968] (J 0521) Semin Math, Inst Steklov 8*79-82
◇ B45 ◇ REV MR 44 # 1543 Zbl 172.292 • ID 09281

MINTS, G.E. [1968] *Some calculi of modal logic (Russian)* (S 0066) Tr Mat Inst Steklov 98*88-111
• TRANSL [1968] (S 0055) Proc Steklov Inst Math 98*97-124
◇ B45 ◇ REV MR 39 # 40 Zbl 169.299 • ID 09283

MINTS, G.E. [1969] *On the semantics of modal logic (Russian)* (S 0228) Zap Nauch Sem Leningrad Otd Mat Inst Steklov 16*147-151
• TRANSL [1969] (J 0521) Semin Math, Inst Steklov 16*74-76
◇ B45 ◇ REV MR 41 # 5201 Zbl 231 # 02022 • ID 09284

MINTS, G.E. (ED.) [1970] *Mathematical logic (Russian)* (X 2673) Bibl Akad Nauk SSSR: Leningrad 174pp
◇ A10 B98 ◇ REV MR 53 # 12855 • ID 70167

MINTS, G.E. [1972] *Cut-elimination theorem for relevant logics (Russian) (English summary)* (S 0228) Zap Nauch Sem Leningrad Otd Mat Inst Steklov 32*90-97,156
• TRANSL [1976] (J 1531) J Sov Math 6*422-428
◇ B46 ◇ REV MR 49 # 8823 Zbl 374 # 02017 • ID 09289

MINTS, G.E. [1972] *Derivability of addmissible rules (Russian) (English summary)* (S 0228) Zap Nauch Sem Leningrad Otd Mat Inst Steklov 32*85-89,156
• TRANSL [1976] (J 1531) J Sov Math 6*417-421
◇ B55 F50 ◇ REV MR 49 # 8816 Zbl 358 # 02031 • ID 09288

MINTS, G.E. [1974] *Lewis' systems and T (a survey 1965-1973) (Russian)* (C 4138) Feys: Modal Logika 422-509
◇ B45 ◇ REV MR 50 # 9527 • ID 37141

MINTS, G.E. [1984] *Structural synthesis with independent subproblems and the modal logic S4 (Russian) (English and Estonian summaries)* (J 0080) Izv Akad Nauk Ehston SSR, Fiz, Mat 33*147-151
◇ B45 ◇ REV MR 86b:68009 • ID 45465

MINTS, G.E. see Vol. I, III, IV, V, VI for further entries

MIODUSZEWSKA, E. [1984] *Referential presupposition within Ulrich Blau's three-valued logic system* (**J** 1517) Theor Linguist 11/3*251-268
◇ B50 ◇ ID 49721

MIRCHEVA, M. & VAKARELOV, D. [1980] *Modal Post algebras and many-valued modal logics* (**J** 0137) C R Acad Bulgar Sci 33*591-594
◇ B45 G20 ◇ REV MR 83a:03026 Zbl 432 # 03036 • ID 53993

MIRKOWSKA, G. [1971] *On formalized systems of algorithmic logic (Russian summary)* (**J** 0014) Bull Acad Pol Sci, Ser Math Astron Phys 19*421-428
◇ B75 ◇ REV MR 46 # 18 Zbl 222 # 02010 • ID 28479

MIRKOWSKA, G. [1974] *Herbrands theorem in the algorithmic logic* (**J** 0014) Bull Acad Pol Sci, Ser Math Astron Phys 22*539-543
◇ B75 ◇ REV MR 50 # 1854 Zbl 343 # 68015 • ID 09310

MIRKOWSKA, G. & SALWICKI, A. [1976] *A complete characterization of algorithmic properties of block-structured programs with procedures* (**P** 1401) Math Founds of Comput Sci (5);1976 Gdansk 45*602-606
◇ B75 ◇ REV Zbl 347 # 68013 • ID 31915

MIRKOWSKA, G. [1977] *Algorithmic logic and its application in the theory of programs I,II* (**J** 2095) Fund Inform, Ann Soc Math Pol, Ser 4 1*1-17,147-165
◇ B40 B75 ◇ REV MR 58 # 32012 Zbl 358 # 68036 Zbl 384 # 68010 • ID 31914

MIRKOWSKA, G. [1977] see BANACHOWSKI, L.

MIRKOWSKA, G. & ORLOWSKA, E. [1978] *An elimination of iteration quantifiers in a certain class of algorithmic formulas* (**J** 2095) Fund Inform, Ann Soc Math Pol, Ser 4 1*347-355
◇ B75 ◇ REV MR 80a:03070 Zbl 386 # 68043 • ID 76438

MIRKOWSKA, G. [1979] *On the propositional algorithmic logic* (**P** 2059) Math Founds of Comput Sci (8);1979 Olomouc 381-389
◇ B75 ◇ REV Zbl 438 # 68002 • ID 55981

MIRKOWSKA, G. [1980] *Algorithmic logic with nondeterministic programs* (**J** 2095) Fund Inform, Ann Soc Math Pol, Ser 4 3*45-64
◇ B75 ◇ REV MR 83b:68009 Zbl 439 # 68024 • ID 69726

MIRKOWSKA, G. [1980] *Model existence theorem in algorithmic logic with nondeterministic programs* (**J** 2095) Fund Inform, Ann Soc Math Pol, Ser 4 3*157-170
◇ B75 ◇ REV MR 82e:03024 Zbl 439 # 68025 • ID 76437

MIRKOWSKA, G. [1981] *Algorithmic logic with nondeterministic programs* (**P** 3642) Colloq Math Log in Computer Sci;1978 Salgotarjan 547-560
◇ B75 ◇ REV MR 83b:68009 Zbl 494 # 68031 • ID 36657

MIRKOWSKA, G. [1981] *PAL-propositional algorithmic logic* (**P** 3497) Log of Progr;1979 Zuerich 23-101
• REPR [1981] (**J** 2095) Fund Inform, Ann Soc Math Pol, Ser 4 4*675-760
◇ B75 ◇ REV MR 84k:03083 MR 85i:03088 Zbl 487 # 03012 • ID 44186

MIRKOWSKA, G. [1981] see MEYER, A.R.

MIRKOWSKA, G. [1982] *The representation theorem for algorithmic algebras* (**P** 3738) Log of Progr;1981 Yorktown Heights 300-310
◇ B75 ◇ REV MR 83i:68054 Zbl 501 # 03043 • ID 36955

MIRKOWSKA, G. [1983] *Multimodal logic as a base of algorithmic logic* (**S** 3382) Sem-ber, Humboldt-Univ Berlin, Sekt Math 52*65-68
◇ B75 ◇ REV MR 85f:68003 Zbl 533 # 03004 • ID 36529

MIRKOWSKA, G. [1983] *On the propositional algorithmic theory of arithmetic* (**P** 3830) Logics of Progr & Appl;1980 Poznan 166-185
◇ B75 ◇ REV MR 84k:68023 Zbl 541 # 68012 • ID 39241

MIRKOWSKA, G. [1984] *On a certain property not expressible in PAL* (**J** 2095) Fund Inform, Ann Soc Math Pol, Ser 4 7*343-348
◇ B75 ◇ ID 44602

MIROLLI, M. [1980] *On the axiomatization of finite frames of the modal system GL (Italian summary)* (**J** 3495) Boll Unione Mat Ital, V Ser, B 17*1075-1085
◇ B45 ◇ REV MR 85k:03011 Zbl 493 # 03004 • ID 38100

MIROLLI, M. [1983] see BELLISSIMA, F.

MIROLLI, M. see Vol. I, VI for further entries

MISHRA, B. [1985] see CLARKE, E.M.

MITCHELL, D. [1962] *An introduction to logic* (**X** 0939) Hutchinson: London 192pp
◇ B98 ◇ ID 25144

MITCHELL, D. [1970] *An introduction to logic* (**X** 0878) Doubleday: London ix + 227pp
◇ B98 ◇ REV MR 50 # 12629 JSL 38.345 • ID 09320

MITROFF, I.I. [1982] see MASON, R.O.

MITROFF, I.I. [1983] see MASON, R.O.

MITTELSTAEDT, P. [1970] *Quantenlogische Interpretation orthokomplementaerer quasimodularer Verbaende* (**J** 0948) Z Naturforsch 25A*1773-1778
◇ B51 G10 G12 ◇ REV MR 43 # 4361 Zbl 221 # 02010 • ID 28104

MITTELSTAEDT, P. [1972] *On the interpretation of the lattice of subspaces of the Hlbert space as a propositional calculus* (**J** 0948) Z Naturforsch 27a*1358-1362
◇ B51 G12 ◇ REV MR 53 # 2158 JSL 48.206 • ID 47410

MITTELSTAEDT, P. [1976] *On the applicability of the probability concept to quantum theory* (**P** 2411) Found Probab Th, Stat Inf & Stat Th Sci;1973 London ON 3*155-167
◇ B51 G12 ◇ REV Zbl 323 # 02044 • ID 30419

MITTELSTAEDT, P. [1976] *Quantum logic* (**P** 3516) Proc Bienn Meet Phil of Sci Ass;1974 Notre Dame 501-514
◇ B51 G12 ◇ REV Zbl 359 # 02022 • ID 50570

MITTELSTAEDT, P. (ED.) [1977] *Special issue: Symposium on Quantum Logic* (**J** 0122) J Philos Logic 6*369-496
◇ B51 G12 ◇ REV MR 56 # 15357 • ID 70112

MITTELSTAEDT, P. [1977] *Time dependent propositions and quantum logic* (**J** 0122) J Philos Logic 6*463-472
◇ A05 B51 G12 ◇ REV MR 58 # 27361 Zbl 371 # 02027 • ID 51388

MITTELSTAEDT, P. [1978] *Quantum logic* (**X** 0835) Reidel: Dordrecht viii + 149pp
◇ B51 G10 G12 ◇ REV MR 82g:03072 Zbl 411 # 03059 • ID 52913

MITTELSTAEDT, P. & STACHOW, E.-W. [1978] *The principle of excluded middle in quantum logic* (**J** 0122) J Philos Logic 7*181-208
◇ B51 G12 ◇ REV MR 80a:03076 Zbl 374#02015 • ID 51539

MITTELSTAEDT, P. [1979] *Quantum logic* (**P** 3234) Probl Founds of Physics;1977 Varenna 264-299
◇ B51 G12 ◇ REV MR 82m:81003 Zbl 445#03038 • ID 56502

MITTELSTAEDT, P. [1979] *Quantum logic* (**C** 3566) Phys Theor as Log-Operat Struct 153-166
◇ B51 G12 ◇ REV MR 80f:03072 Zbl 402#03053 • ID 76457

MITTELSTAEDT, P. [1979] *The modal logic of quantum logic* (**J** 0122) J Philos Logic 8*479-504
◇ B51 G12 ◇ REV MR 81d:03060 Zbl 418#03042 • ID 53327

MITTELSTAEDT, P. [1981] *Classification of different areas of work afferent to quantum logic* (**P** 3820) Curr Iss in Quantum Log;1979 Erice 3-16
◇ B51 ◇ REV MR 84j:03133 Zbl 537#03044 • ID 34723

MITTELSTAEDT, P. [1981] *The dialogic approach to modalities in the language of quantum physics* (**P** 3820) Curr Iss in Quantum Log;1979 Erice 259-281
◇ B51 ◇ REV MR 84j:03134 Zbl 537#03044 • ID 34724

MITTELSTAEDT, P. & STACHOW, E.-W. [1983] *Analysis of the Einstein-Podolsky-Rosen experiment by relativistic quantum logic* (**J** 2736) Int J Theor Phys 22*517-540
◇ B51 G12 ◇ REV MR 85f:81008 MR 86e:03059 Zbl 518#03030 • ID 37527

MITTELSTAEDT, P. [1983] *Relativistic quantum logic* (**J** 2736) Int J Theor Phys 22*293-314
◇ B51 ◇ REV MR 84h:03139 Zbl 515#03040 • ID 34325

MITTELSTAEDT, P. [1983] *Wahrheit, Wirklichkeit und Logik in der Sprache der Physik* (**J** 0989) Z Allg Wissth 14*24-45
◇ B51 ◇ REV MR 84k:03044 • ID 34980

MITTELSTAEDT, P. & STACHOW, E.-W. [1985] *Recent developments in quantum logic* (**X** 0876) Bibl Inst: Mannheim 352pp
◇ B51 ◇ ID 48931

MIURA, S. & NAGATA, S. [1968] *Certain methods for generating series of logics* (**J** 0111) Nagoya Math J 31*125-129
◇ B50 ◇ REV MR 37#2583 Zbl 165.304 • ID 09334

MIURA, S. [1969] *Primitive logics with modalities and their characterization by evaluation* (**J** 0523) Bull Nagoya Inst Tech 21*97-99
◇ B45 ◇ REV MR 46#1556 • ID 76460

MIURA, S. [1972] *Probabilistic models of modal logics* (**J** 0523) Bull Nagoya Inst Tech 24*67-72
◇ B45 C90 ◇ REV MR 48#1885 • ID 09336

MIURA, S. & OHAMA, S. [1977] *A note on Thomason's representation of S5* (**J** 0047) Notre Dame J Formal Log 18*177-180
◇ B45 ◇ REV MR 58#5059 Zbl 323#02035 • ID 21962

MIURA, S. & OHAMA, S. [1977] *Modally complete minimal extension of S5* (**J** 0523) Bull Nagoya Inst Tech 29*161-164
◇ B45 ◇ REV MR 58#21479 • ID 76461

MIURA, S. [1983] *Embedding of modal predicate systems into lower predicate calculus* (**J** 0260) Ann Jap Ass Phil Sci 6*147-160
◇ B45 ◇ REV MR 85a:03023 Zbl 514#03013 • ID 34774

MIURA, S. [1983] *Embedding of modal predicate logics into lower predicate calculus I* (**J** 0286) Int Logic Rev 28*94-105
◇ B45 ◇ REV MR 85a:03023 • ID 45590

MIURA, S. see Vol. I for further entries

MIYAKAWA, M. [1981] *Enumerations of bases of three-valued logical functions* (**P** 2552) Conf Finite Algeb & Multi-Val Log;1979 Szeged 469-487
◇ B50 ◇ REV MR 83m:03033 Zbl 476#03031 • ID 35437

MIYAKAWA, M. [1982] *Enumeration of bases of a submaximal set of three-valued logical functions* (**S** 2829) Rostocker Math Kolloq 19*49-67
◇ B50 ◇ REV MR 83J:03043 Zbl 495#03014 • ID 33708

MIYAKAWA, M. [1985] *A note to the classification and base enumeration of three-valued logical functions* (**J** 4278) Bull Electr Lab (Japan) 49/3*197-210
◇ B50 ◇ REV Zbl 571#94026 • ID 49277

MIYAMA, T. [1973] *Another proof and an extension of Gill's interpolation theorem* (**J** 0350) Sci Rep Tokyo Kyoiku Daigaku Sect A 12*59-65
◇ B50 ◇ REV MR 49#2261 Zbl 279#02008 • ID 09337

MIYAMA, T. [1974] *The interpolation theorem and Beth's theorem in many-valued logics* (**J** 0352) Math Jap 19*341-355
◇ B50 C40 ◇ REV MR 54#60 Zbl 319#02015 • ID 29672

MIYAMA, T. [1976] *Eliminability of descriptive definitions in many-valued logics* (**J** 0537) Yokohama Math J 24(1-2)*103-115
◇ B50 ◇ REV MR 55#2493 Zbl 361#02023 • ID 50658

MIYAMA, T. [1978] *Note on functional completeness in many valued logics* (**J** 0352) Math Jap 23*427-432
◇ B50 ◇ REV MR 80c:03027 Zbl 423#03019 • ID 53531

MIYAMA, T. [1979] *On the study of many-valued logics. I (Japanese)* (**J** 2681) Bull Univ Osaka Prefecture, Ser A 13*47-57
◇ B50 G20 ◇ REV MR 82b:03052a • REM Part II 1980 • ID 76464

MIYAMA, T. [1980] *On the study of many-valued logics. II (Japanese)* (**J** 2681) Bull Univ Osaka Prefecture, Ser A 14*125-132
◇ B50 G20 ◇ REV MR 82b:03052b • REM Part I 1979. Part III 1982 • ID 76463

MIYAMA, T. [1982] *On the study of many-valued logics III (Japanese)* (**J** 2681) Bull Univ Osaka Prefecture, Ser A 16*127-137
◇ B50 ◇ REV MR 84k:03076 • REM Part II 1980 • ID 33396

MIZUMOTO, M. [1978] see FUKAMI, S.

MIZUMOTO, M. [1979] see FUKAMI, S.

MIZUMOTO, M. [1980] see FUKAMI, S.

MIZUMOTO, M. [1981] *Fuzzy reasonings with "if...then...else..."* (**P** 4050) Appl Syst & Cybern;1980 Acapulco 2927-2932
◇ B52 ◇ REV MR 84b:00009 • ID 46263

MIZUMOTO, M. [1981] *Note on the arithmetic rule by Zadeh for fuzzy conditional inference* (**J** 2694) Cybern & Syst 12*247-306
◇ B52 ◇ REV MR 84h:03053 Zbl 472#03018 • ID 55284

MIZUMOTO, M. [1981] *Various fuzzy reasoning methods: the case of "If...then..." (Japanese)* (**J** 0979) Denshi Tsushin Gakkai Ronbunshi, Sect A-D 64/5*379-386
• TRANSL [1981] (**J** 0464) Syst-Comp-Controls 12/3*9-19
◇ B52 ◇ REV MR 84c:03050 • ID 35702

MIZUMOTO, M. & SEKITA, Y. [1982] *A fuzzified relevance tree approach for solving the complex planning* (**C** 3786) Approx Reason in Decis Anal 401-408
◇ B52 ◇ REV Zbl 503 # 90003 • ID 36672

MIZUMOTO, M. & ZIMMERMANN, H.-J. [1982] *Comparison of fuzzy reasoning methods* (**J** 2720) Fuzzy Sets Syst 8*253-283
◇ B52 ◇ REV MR 83m:03035 Zbl 501 # 03013 • ID 35438

MIZUMOTO, M. [1982] *Fuzzy conditional inference under max-⊙ composition* (**J** 0191) Inform Sci 27*183-209
◇ B52 ◇ REV MR 84b:03041 Zbl 521 # 03013 • ID 35633

MIZUMOTO, M. [1982] *Fuzzy inference using max-∧ composition in the compositional rule of inference* (**C** 3786) Approx Reason in Decis Anal 67-76
◇ B52 ◇ REV MR 84i:03058 Zbl 503 # 94032 • ID 34537

MIZUMOTO, M. [1982] *Fuzzy reasoning with a fuzzy conditional proposition "if...then...else..."* (**P** 4051) Fuzzy Set & Possibility Th;1980 Acapulco 211-223
◇ B52 ◇ REV MR 84b:03004 • ID 46326

MIZUMOTO, M. [1982] *Fuzzy reasoning with various fuzzy premises (Japanese)* (**P** 4081) Many-Val Log & Appl;1982 Kyoto 176-190
◇ B52 ◇ ID 47631

MIZUMOTO, M. [1983] *Fuzzy inferences with various fuzzy inputs (Chinese summary)* (**J** 3732) Mohu Shuxue 3/1*45-54
◇ B52 ◇ REV MR 84i:03057 Zbl 529 # 03008 • ID 34536

MIZUMOTO, M. [1983] *Fuzzy reasoning with various fuzzy inputs (Japanese)* (**J** 0979) Denshi Tsushin Gakkai Ronbunshi, Sect A-D 65/9*1175-1182
• TRANSL [1983] (**J** 0464) Syst-Comp-Controls 13/5*47-55
◇ B52 ◇ REV MR 84j:03057 • ID 34649

MIZUMOTO, M. [1983] *Fuzzy reasoning under new compositonal rules of inference* (**P** 4054) Int Symp Multi-Val Log (13);1983 Kyoto 273-278
◇ B52 ◇ REV Zbl 566 # 03014 • ID 48735

MIZUMOTO, M. [1984] *Fuzzy reasoning with various fuzzy inputs* (**P** 3081) IFAC Symp Fuzzy Inf, Knowl Repr & Decis. Anal;1983 Marseille 153-158
◇ B52 ◇ REV Zbl 566 # 03015 • ID 48192

MIZUMOTO, M. [1984] see EZAWA, Y.

MIZUMOTO, M. [1985] *Fuzzy reasoning under new compositional rules of inference* (**J** 1429) Kybernetes 14*107-117
◇ B52 ◇ ID 46346

MIZUMOTO, M. see Vol. IV, V for further entries

MLEZIVA, M. [1961] *On the axiomatization of three-valued propositional logic (Czech) (Russian and English summaries)* (**J** 0086) Cas Pestovani Mat, Ceskoslov Akad Ved 86*392-404
◇ B50 ◇ REV MR 24 # A1818 Zbl 100.9 JSL 35.465
• ID 09339

MLEZIVA, M. see Vol. I, III for further entries

MO, SHAOKUI [1950] *The deduction theorems and two new logical systems* (**J** 0175) Methodos 2*56-75
◇ B20 B45 ◇ REV MR 12.662 JSL 17.153 • ID 12028

MO, SHAOKUI [1954] *Logical paradoxes for many-valued systems* (**J** 0036) J Symb Logic 19*37-40
◇ A05 B50 B53 ◇ REV MR 15.669 JSL 22.90 • ID 12030

MO, SHAOKUI [1957] *Axiomatization of many-valued logical systems (Chinese)* (**J** 2804) Nanjing Daxue Xuebao, Ziran Kexue 1*791-800
◇ B50 ◇ REV JSL 25.181 • ID 42547

MO, SHAOKUI [1957] *Modal systems with a finite number of modalities (Chinese) (English summary)* (**J** 0418) Shuxue Xuebao 7*1-27
• TRANSL [1958] (**J** 1024) Zhongguo Kexue 7*388-412
◇ B45 ◇ REV MR 21 # 3 JSL 25.183 • ID 09352

MO, SHAOKUI [1957] *The axiomatization of the finite many valued matrix systems (Chinese)* (**J** 2804) Nanjing Daxue Xuebao, Ziran Kexue 1957/3*67-76
◇ B50 ◇ ID 47117

MO, SHAOKUI [1958] *Fundamental finite modal systems (Chinese) (English summary)* (**J** 0418) Shuxue Xuebao 8*153-180
◇ B45 ◇ REV MR 21 # 4 • ID 09353

MO, SHAOKUI [1959] *The modal system and implication system (Chinese) (English summary)* (**J** 0418) Shuxue Xuebao 9*121-142
◇ B45 ◇ REV MR 21 # 5565 • ID 09354

MO, SHAOKUI [1963] *Notes on mathematical logic (Chinese)* (**J** 0418) Shuxue Xuebao 13*485-507
• TRANSL [1963] (**J** 0419) Chinese Math Acta 4*527-552
◇ B98 ◇ REV MR 29 # 1143 • ID 19399

MO, SHAOKUI [1963] *The axiomatization of m-valued matrix systems (Chinese)* (**J** 2804) Nanjing Daxue Xuebao, Ziran Kexue 1963/3*87-93
◇ B50 ◇ ID 47131

MO, SHAOKUI [1963] *Two complete n-valued matrix-systems (Chinese)* (**J** 2804) Nanjing Daxue Xuebao, Ziran Kexue 1963/2*19-28
◇ B50 ◇ ID 47144

MO, SHAOKUI [1965] *Introduction to mathematical logic (Chinese)* (**X** 0740) Kexue Jishu Chubanshe: Shanghai
◇ B98 ◇ ID 47171

MO, SHAOKUI & SHEN, BAIYING [1965] *Semi-complete quasi basic system (Chinese)* (**P** 4564) Math Logic;1963 Xi-An
◇ B60 ◇ ID 49365

MO, SHAOKUI [1980] *Introduction to mathematical logic (Chinese)* (**X** 4589) Shanghai Jiaoyu Chubanshe: Shanghai 172pp
◇ B98 ◇ ID 48906

MO, SHAOKUI & SHEN, BAIYING & XU, YONGSHEN [1984] *Mathematical logic (Chinese)* (**X** 4579) Gnodeng Jiaoyu Chubanshe: Beijing 250pp
◇ B98 ◇ ID 48527

MO, SHAOKUI see Vol. I, III, IV, V, VI for further entries

MOCH, F. [1956] *La logique des attitudes* (**J** 0109) C R Acad Sci, Paris 242*1943-1945
◇ A05 B60 ◇ REV MR 17.1038 Zbl 71.6 • ID 47960

MOCH, F. see Vol. V for further entries

MOELLER, K.K. [1979] see JENSEN, F.V.

Mohr, H.P. [1979] *Wahl einer induktiven Methode mit entscheidungslogischen Mitteln* (**S** 2877) Wiss Z Univ Leipzig, Ges-Sprachwiss Reihe 28*351-360
◇ A05 B48 ◇ REV MR 80i:03013 Zbl 418 # 03008
• ID 76489

Moisil, G.C. [1938] *Sur la theorie classique de la modalite des jugements* (**J** 0494) Bull Math Soc Sci Roumanie 40*235-240
◇ B45 ◇ REV Zbl 20.097 JSL 4.167 • ID 48765

Moisil, G.C. [1941] *Contribution a l'etude des logiques non-chrysippiennes I. Un nouveau systeme d'axiomes pour les algebres Lukasiewicziennes tetra-valentes* (**J** 0525) C R Acad Sci Roumanie 5*289-293
◇ A05 A10 B50 G20 ◇ REV Zbl 27.5 JSL 13.160 • REM Parts II,III 1942 • ID 09359

Moisil, G.C. [1941] *Recherches sur la theorie des chaines* (**J** 0269) Ann Sci Univ Jassy Sec 1 27*181-240
◇ B50 E07 G10 ◇ REV MR 8.309 Zbl 25.294 JSL 13.50
• ID 09362

Moisil, G.C. [1941] *Remarques sur la logique modale du concept* (**J** 0526) An Acad Romana Mem Sti Ser 3 16*975-1012
◇ B45 ◇ REV Zbl 26.245 JSL 13.161 • ID 09360

Moisil, G.C. [1941] *Sur la structure algebrique de la logique de M. Bochvar* (**J** 0524) Disq Math Phys 1*307-314
◇ B50 G15 G25 ◇ REV MR 8.557 Zbl 25.386 JSL 13.116
• ID 09357

Moisil, G.C. [1941] *Sur les anneaux de caracteristique 2 ou 3 et leurs applications* (**J** 0527) Bull Inst Politeh Bucuresti 12*66-90
◇ B50 ◇ REV MR 7.110 Zbl 26.246 JSL 13.160 • ID 09361

Moisil, G.C. [1941] *Sur les theories deductives a logique non-chrysippienne* (**J** 0525) C R Acad Sci Roumanie 5*21-24
• REPR [1972] (**C** 3094) Moisil: Essais Logiques Non Chrysippiennes 189-191
◇ B50 ◇ REV Zbl 27.5 FdM 67.43 • ID 41708

Moisil, G.C. [1942] *Contributions a l'etude des logiques non-chrysippiennes II,III: Anneaux engendres par les algebres Lukasiewicziennes tetravalentes centrees* (**J** 0525) C R Acad Sci Roumanie 6*9-13,14-17
• REPR [1972] (**C** 3094) Moisil: Essais Logiques Non Chrysippiennes 287-292,293-297
◇ B50 ◇ REV FdM 68.24 • REM Part I 1941. Part IV 1945
• ID 41712

Moisil, G.C. [1942] *Logique modale* (**J** 0524) Disq Math Phys 2*3-98
◇ B45 ◇ REV MR 8.557 JSL 13.162 • ID 09363

Moisil, G.C. [1945] *Contributions a l'etude des logiques non-chrysippiennes IV. Sur la logique de M. Becker* (**J** 0525) C R Acad Sci Roumanie 7*9-11
◇ A05 A10 B50 ◇ REV MR 9.1 Zbl 61.7 • REM Part III 1942 • ID 09364

Moisil, G.C. [1956] *Utilisation des logiques trivalentes dans la theorie des mecanismes automatiques. II. Equation caracteristique d'un relai-polarise (Roumanian) (Russian and French summaries)* (**J** 4188) Com Acad Romine 6*231-234
◇ B50 ◇ REV MR 18.712 Zbl 70.358 • REM Part I 1956 by Greniewski,M. ibid. 225-229 • ID 48778

Moisil, G.C. [1960] *On the logic of Bochvar (Roumanian)* (**J** 0518) Bul Sti Mat-Fiz, Acad Romina 14*19-25
• TRANSL [1972] (**C** 3094) Moisil: Essais Logiques Non Chrysippiennes 685-693
◇ B50 ◇ REV MR 23 # A796 • ID 41874

Moisil, G.C. [1961] *Les logiques a plusieurs valeurs et l'automatique* (**P** 0633) Infinitist Meth;1959 Warsaw 337-345
◇ B50 B70 ◇ REV MR 26 # 1233 Zbl 129.100 JSL 36.546
• ID 22022

Moisil, G.C. [1961] *The predicate calculus in three-valued logic* (**J** 0147) An Univ Bucuresti, Acta Logica 4*103-112
• TRANSL [1972] (**C** 3094) Moisil: Essais Logiques Non Chrysippiennes 442-454
◇ B50 ◇ REV Zbl 103.247 • ID 41883

Moisil, G.C. [1962] *Sur la logique a trois valeurs de Lukasiewicz* (**J** 0147) An Univ Bucuresti, Acta Logica 5*103-117
◇ B50 ◇ REV MR 26 # 6045 Zbl 121.11 JSL 27.368
• ID 09369

Moisil, G.C. [1963] *Les logiques non-chrysippiennes et leurs applications* (**J** 0096) Acta Philos Fenn 16*137-152
◇ B50 G20 ◇ REV MR 28 # 2969 Zbl 126.260 • ID 09370

Moisil, G.C. [1964] *Sur les logiques de Lukasiewicz a un nombre fini de valeurs* (**J** 0060) Rev Roumaine Math Pures Appl 9*905-920
◇ B50 ◇ REV MR 32 # 40 Zbl 149.4 • ID 09372

Moisil, G.C. [1965] *Essays old and new on nonclassical logic* (**X** 0922) Stiintifica Encicl: Bucharest 461pp
◇ B96 ◇ REV MR 35 # 30 • ID 09373

Moisil, G.C. [1966] *Application des algebres de Lukasiewicz a l'etude des schemas a contacts et relais. I (Polish)* (**X** 2882) Pol Akad Nauk: Wroclaw 122pp
◇ B50 B70 ◇ REV JSL 37.187 • REM Part II 1967
• ID 43904

Moisil, G.C. [1967] *Sur le calcul des propositions de type superieur dans la logique de Lukasiewicz a plusieurs valeurs* (**J** 0147) An Univ Bucuresti, Acta Logica 10*93-104
◇ B50 ◇ REV MR 38 # 5590 Zbl 281 # 02023 • ID 09375

Moisil, G.C. [1968] *Sur la logique des propositions a modeles partiellement ordonnes* (**J** 0060) Rev Roumaine Math Pures Appl 13*1413-1437
◇ B50 ◇ REV MR 39 # 3971 Zbl 191.285 • ID 09377

Moisil, G.C. (ED.) [1971] *Logique automatique, informatique* (**X** 0871) Acad Rep Soc Romania: Bucharest 456pp
◇ B75 B97 ◇ REV MR 48 # 5811 Zbl 224 # 00005
• ID 48621

Moisil, G.C. [1971] *Sur le mode ponendo ponens (Romanian summary)* (**J** 0230) An Univ Iasi, NS, Sect Ia 17*1-16
◇ B60 ◇ REV MR 48 # 8207 Zbl 238 # 02010 • ID 09380

Moisil, G.C. [1972] *Exemples de propositions a plusieurs valeurs* (**C** 3094) Moisil: Essais Logiques Non Chrysippiennes 149-156
◇ B50 ◇ ID 42023

Moisil, G.C. [1972] *La logique des concepts nuances* (**C** 3094) Moisil: Essais Logiques Non Chrysippiennes 157-163
◇ B50 ◇ ID 42024

MOISIL, G.C. [1972] *Les axiomes des algebres de Lukasiewicz n-valentes* (**C** 3094) Moisil: Essais Logiques Non Chrysippiennes 298–310
◇ B50 ◇ ID 42026

MOISIL, G.C. [1972] *Les ensembles flous et la logique a trois valeurs* (**C** 3094) Moisil: Essais Logiques Non Chrysippiennes 99–103
◇ B50 E70 E72 ◇ ID 42022

MOISIL, G.C. [1972] *Sur les algebres de Lukasiewicz θ-valentes* (**C** 3094) Moisil: Essais Logiques Non Chrysippiennes 311–324
◇ B50 ◇ ID 42027

MOISIL, G.C. [1975] see HENKIN, L.

MOISIL, G.C. see Vol. I, IV, V, VI for further entries

MOKLYAK, N.G. & POPOV, V.A. & SKIBENKO, I.T. [1973] *The number of types of systems of k-valued logical functions (Russian) (English summary)* (**J** 0040) Kibernetika, Akad Nauk Ukr SSR 1973/6*18–27
◇ B50 ◇ REV MR 51 # 55 Zbl 281 # 02025 • ID 15200

MOKLYAK, N.G. see Vol. I for further entries

MOLDAUER, P.A. [1983] *Comment on : "Bell's theorem as a nonlocality property of quantum theory" by H.P. Stapp* (**J** 2730) Phys Rev Lett 50*701–702
◇ B51 ◇ REV MR 84m:81028b • REM The article was published ibid 49*1470–1474 • ID 39540

MONARI, P. [1982] see COSTANTINI, D.

MONARI, P. [1983] see COSTANTINI, D.

MONDADORI, M. [1978] *The probability of "laws of nature": a case for Hintikka's inductive logic* (**J** 2839) Statistica (Bologna) 38*517–536
◇ B48 ◇ REV MR 80h:03040 Zbl 401 # 62004 • ID 76502

MONK, J.D. [1976] *Mathematical logic* (**X** 0811) Springer: Heidelberg & New York x+531pp
◇ B98 ◇ REV MR 57 # 5656 Zbl 354 # 02002 JSL 44.283 • ID 30765

MONK, J.D. see Vol. I, III, V for further entries

MONTAGNA, F. [1980] *Some modal logics with quantifiers (Italian summary)* (**J** 3495) Boll Unione Mat Ital, V Ser, B 17*1395–1410
◇ B45 C80 C90 ◇ REV MR 85m:03011 Zbl 452 # 03012 • ID 54076

MONTAGNA, F. & PASINI, L. [1980] *The equational class corresponding to the modal logic K43W (Italian summary)* (**J** 3128) Boll Unione Mat Ital, Suppl 2*373–385
◇ B45 C05 ◇ REV MR 84f:03015 Zbl 448 # 03016 • ID 56616

MONTAGNA, F. [1983] *Il teorema di Solovay e l'aritmetica di Heyting* (**P** 3829) Atti Incontri Log Mat (1);1982 Siena 99–103
◇ B45 F30 ◇ REV MR 84k:03006 Zbl 521 # 03042 • ID 37078

MONTAGNA, F. [1983] *ZFC-models as Kripke-models* (**J** 0068) Z Math Logik Grundlagen Math 29*163–168
◇ B45 C62 E30 F30 ◇ REV MR 84j:03043 Zbl 519 # 03013 • ID 34635

MONTAGNA, F. [1984] *The predicate modal logic of provability* (**J** 0047) Notre Dame J Formal Log 25*179–192
◇ B45 C90 F30 ◇ REV MR 86b:03023 Zbl 549 # 03013 • ID 42567

MONTAGNA, F. [1985] *First results on the modal predicate logic of provability (Italian)* (**P** 4646) Atti Incontri Log Mat (2);1983/84 Siena 353–355
◇ B45 ◇ ID 49722

MONTAGNA, F. see Vol. III, IV, VI for further entries

MONTAGUE, R. [1963] *Syntactical treatments of modality, with corollaries on reflexion principles and finite axiomatizability* (**J** 0096) Acta Philos Fenn 16*153–167
• REPR [1974] (**C** 4062) Montague: Formal Philos 286–302
◇ B45 B65 C62 F30 ◇ REV MR 29 # 1140 Zbl 117.13 JSL 40.600 JSL 47.210 • ID 14478

MONTAGUE, R. [1964] see KALISH, D.

MONTAGUE, R. [1970] *Pragmatics and intensional logic* (**J** 0154) Synthese 22*68–94
• REPR [1974] (**C** 4062) Montague: Formal Philos 119–147
◇ B45 B65 ◇ REV Zbl 228 # 02017 JSL 47.210 • ID 29119

MONTAGUE, R. [1970] *Pragmatics and intensional logic* (**J** 0076) Dialectica 24*277–302
◇ B45 ◇ REV Zbl 264 # 02020 • ID 64001

MONTAGUE, R. [1972] *Pragmatics and intensial logic* (**C** 0574) Semantics of Nat Lang 142–168
◇ B45 ◇ REV Zbl 228 # 02012 • ID 09434

MONTAGUE, R. see Vol. I, III, IV, V, VI for further entries

MONTANARO, A. [1981] see BRESSAN, A.

MONTANARO, A. [1981] *Sui fondamenti del calcolo delle probabilita basato sui calcoli modali MC^{ν} oppure MC^{ν}_{*}* (**P** 3092) Congr Naz Logica;1979 Montecatini Terme 487–499
◇ B45 ◇ ID 48728

MONTANARO, A. [1983] see BRESSAN, A.

MONTEIRO, A. [1962] *Linearisation de la logique positive de Hilbert-Bernays* (**J** 0188) Rev Union Mat Argentina 20*308–309
◇ B55 ◇ REV MR 28 # 4989 • ID 09437

MONTEIRO, A. see Vol. I, V, VI for further entries

MONTEIRO, L.F.T. [1970] *Les algebres de Heyting et de Lukasiewicz trivalentes* (**J** 0047) Notre Dame J Formal Log 11*453–466
◇ B50 G10 G20 ◇ REV MR 44 # 3842 Zbl 177.10 • ID 09450

MONTEIRO, L.F.T. [1972] *Algebre du calcul propositionel trivalent de Heyting* (**J** 0027) Fund Math 74*99–109
◇ B50 G10 ◇ REV MR 45 # 4957 Zbl 248 # 02070 • ID 09451

MONTEIRO, L.F.T. see Vol. V for further entries

MONTGOMERY, D. & ROUTLEY, R. [1968] *Non-contigency axioms for S4 and S5* (**J** 0079) Logique & Anal, NS 11*58–60
◇ B45 ◇ REV MR 38 # 4288 Zbl 169.300 • ID 09454

MONTGOMERY, D. see Vol. V for further entries

MONTGOMERY, H. & ROUTLEY, R. [1966] *Contingency and non-contingency bases for normal modal logics* (J 0079) Logique & Anal, NS 9*318-328
◇ B45 ◇ REV MR 36 # 6270 Zbl 294 # 02008 • ID 64017

MONTGOMERY, H. & ROUTLEY, R. [1968] *Modal reduction axioms in extension of SL* (J 0079) Logique & Anal, NS 11*492-501
◇ B45 ◇ REV MR 39 # 5323 Zbl 182.5 • ID 09456

MONTGOMERY, H. [1968] *The inadequacy of Kripke's semantical analysis of* D_2 *and* D_3 (J 0036) J Symb Logic 33*568
◇ B45 ◇ REV MR 39 # 5324 JSL 35.135 • ID 09458

MONTGOMERY, H. & ROUTLEY, R. [1969] *Modalities in a sequence of normal non-contigency modal systems* (J 0079) Logique & Anal, NS 12*225-227
◇ B45 ◇ REV MR 43 # 7305 Zbl 196.9 • ID 09459

MONTGOMERY, H. & ROUTLEY, R. [1976] *Algebraic semantics for* $S2^0$ *and necessitated extensions* (J 0047) Notre Dame J Formal Log 17*44-58
◇ B25 B45 ◇ REV MR 55 # 5392 Zbl 313 # 02010 • ID 18366

MONTGOMERY, H. see Vol. I for further entries

MOON, R.H. [1983] *Correction of the semantics for S4.03 and a note on literal disjunctive symmetry* (J 0047) Notre Dame J Formal Log 24*337-345
◇ B45 ◇ REV MR 85g:03030 Zbl 488 # 03010 • ID 37384

MOOR, J. [1980] see BERGMANN, MERRIE

MOORE, J.S. [1979] see BOYER, R.S.

MOORE, J.S. see Vol. I, IV for further entries

MOORE, O.K. [1967] see ANDERSON, A.R.

MOORE, R.C. [1985] *Semantical considerations on nonmonotonic logic* (J 0503) Artif Intell 25*75-94
◇ B60 ◇ REV Zbl 569 # 68079 • ID 45060

MORAES DE, L. [1981] see DUBIKAJTIS, L.

MORAGA, C. & NAZARALA, J. [1975] *Bilineal separability of ternary functions* (P 1805) Int Symp Multi-Val Log (5,Proc);1975 Bloomington 305-315
◇ B50 ◇ REV MR 58 # 9870 • ID 35830

MORAGA, C. [1975] *Polynomial separation of ternary functions* (P 0784) GI Jahrestag (5);1975 Dortmund 523-533
◇ B50 ◇ REV MR 52 # 7732 Zbl 322 # 94025 • ID 64020

MORAGA, C. [1977] *Ternary spectral logic* (P 2013) Int Symp Multi-Val Log (7);1977 Charlotte 7-12
◇ B50 ◇ REV MR 57 # 9332 • ID 82293

MORAGA, C. [1978] *Complex spectral logic* (P 2014) Int Symp Multi-Val Log (8);1978 Rosemont 149-156
◇ B50 ◇ REV MR 81f:94042 • ID 82292

MORAGA, C. [1979] see COY, W.

MORAGA, C. [1980] *Language inference using multiple-valued logic* (P 3673) Int Symp Multi-Val Log (10);1980 Evanston 108-114
◇ B50 ◇ REV MR 83m:94032 • ID 45941

MORAGA, C. see Vol. IV for further entries

MORAIS, R. [1976] *Projective logic* (J 0110) Anais Acad Bras Cienc 48*627-661
◇ B60 C75 G05 ◇ REV MR 58 # 135 Zbl 403 # 03028 • ID 54742

MORAIS, R. [1977] *Projective logics and projective boolean algebras* (P 1076) Latin Amer Symp Math Log (3);1976 Campinas 201-221
◇ B60 C75 E15 G05 G25 ◇ REV MR 57 # 16065 Zbl 366 # 02038 • ID 16605

MORAVCSIK, J.M.E. [1973] see GABBAY, D.M.

MORAVCSIK, J.M.E. see Vol. III for further entries

MORAWIEC, A. & PIROG-RZEPECKA, K. [1985] *New results in logic of formulas which lose sense* (J 0387) Bull Sect Logic, Pol Acad Sci 14*114-121
◇ B60 ◇ ID 49016

MORE, M.J. [1980] *Rigidity and scope* (J 0079) Logique & Anal, NS 23*327-330
◇ B45 ◇ REV MR 83e:03032 • ID 35217

MORGAN, C.G. [1973] *Systems of modal logic for impossible worlds* (J 0310) Inquiry (Oslo) 16*280-289
◇ B45 ◇ ID 31887

MORGAN, C.G. [1974] *A theory of equality for a class of many-valued predicate calculi* (J 0068) Z Math Logik Grundlagen Math 20*427-432
◇ B50 ◇ REV MR 52 # 5353 Zbl 309 # 02018 • ID 09484

MORGAN, C.G. [1974] *Liberated Brouwerian modal logic* (J 0488) Dialogue (Ottawa) 13*505-514
◇ B45 ◇ REV MR 58 # 5060 Zbl 322 # 02015 • ID 31883

MORGAN, C.G. [1974] *Notes on garden of eden configurations of higher degree* (J 1429) Kybernetes 3*129-134
◇ B45 ◇ REV MR 57 # 5513 Zbl 319 # 94032 • ID 31884

MORGAN, C.G. [1975] *Liberated versions of T, S4 and S5* (J 0009) Arch Math Logik Grundlagenforsch 17*85-90
◇ B45 ◇ REV MR 52 # 7854 Zbl 316 # 02028 • ID 09486

MORGAN, C.G. [1975] *Similarity as a theory of graded equality for a class of many-valued predicate calculi* (P 1805) Int Symp Multi-Val Log (5,Proc);1975 Bloomington 436-449
◇ B50 ◇ REV MR 58 # 10302 • ID 31890

MORGAN, C.G. [1975] *Weak liberated versions of T and S4* (J 0036) J Symb Logic 40*25-30
◇ B45 ◇ REV MR 55 # 2503 Zbl 305 # 02032 • ID 09485

MORGAN, C.G. [1976] *A resolution principle for a class of many-valued logics* (J 0079) Logique & Anal, NS 19*311-339
◇ B50 ◇ REV MR 56 # 11749 Zbl 352 # 02017 • ID 31880

MORGAN, C.G. [1976] *Many-valued propositional intuitionism* (P 2011) Int Symp Multi-Val Log (6);1976 Logan 150-156
◇ B50 F50 ◇ REV MR 58 # 16171 • ID 76542

MORGAN, C.G. & PELLETIER, F.J. [1977] *Some notes concerning fuzzy logics* (J 2130) Linguist Philos 1*79-97
◇ B52 ◇ REV Zbl 348 # 02023 • ID 64025

MORGAN, C.G. [1979] *Local and global operators and many-valued modal logics* (J 0047) Notre Dame J Formal Log 20*401-411
◇ B45 ◇ REV MR 83c:03018 Zbl 283 # 02019 Zbl 437 # 03009 • ID 55877

MORGAN, C.G. [1979] *Note on a strong liberated modal logic and its relevance to possible world skepticism* (J 0047) Notre Dame J Formal Log 20*718-722
◇ B45 ◇ REV MR 80m:03039 Zbl 314 # 02036 Zbl 437 # 03008 • ID 55876

MORGAN, C.G. [1982] *Simple probabilistic semantics for propositional K, T, B, S4, and S5* (J 0122) J Philos Logic 11*443-458
◇ B45 C90 ◇ REV MR 85d:03056 Zbl 512 # 03011 • ID 36507

MORGAN, C.G. [1982] *There is a probabilistic semantics for every extension of classical sentence logic* (J 0122) J Philos Logic 11*431-442
◇ B45 C90 ◇ REV MR 85d:03055 Zbl 515 # 60003 • ID 37860

MORGAN, C.G. [1983] *Probabilistic semantics for orthologic and quantum logic* (J 0079) Logique & Anal, NS 26*323-339
◇ B51 ◇ REV MR 86f:03105 • ID 45168

MORGAN, C.G. [1983] see LEBLANC, H.

MORGAN, C.G. [1984] see LEBLANC, H.

MORGAN, C.G. [1984] *Weak conditional comparative probability as a formal semantic theory* (J 0068) Z Math Logik Grundlagen Math 30*199-212
◇ B48 ◇ REV MR 86d:03022 • ID 42223

MORGAN, C.G. see Vol. I, IV, VI for further entries

MORGAN DE, A. [1924] *Formal logic; or, the calculus of inference, necessary and probable* (X 2637) Taylor & Walton: London xvi + 336pp
• LAST ED [1926] (X 1324) Open Court: LaSalle xvi + 336pp
◇ A10 B45 B48 ◇ REM Reprint. 1st ed. 1847 • ID 28580

MORGAN DE, A. see Vol. I for further entries

MORGENSTERN, S. [1962] *Un algorithme du type de Zegalkin pour le probleme de la decidabilite en S2 (Romanian) (Russian and French summaries)* (J 0525) C R Acad Sci Roumanie 12*503-508
◇ B45 ◇ REV MR 26 # 1260 Zbl 128.12 • ID 42872

MORI, S. [1978] *On S5 type modal logics based on intermediate logics with finitely many valued linear models* (J 0381) J Tsuda College (Tokyo) 10*35-43
◇ B45 B55 ◇ REV MR 58 # 149 • ID 76558

MORI, S. [1980] *Axiomatization of S4 type modal logics based on finite linear intermediate logics* (J 0407) Comm Math Univ St Pauli (Tokyo) 29*29-43
◇ B45 B55 ◇ REV MR 82b:03046 Zbl 435 # 03015 • ID 55776

MORING, H. [1973] see GIRLING, B.

MORLEY, M.D. [1985] see HARRINGTON, L.A.

MORLEY, M.D. see Vol. III, IV, V for further entries

MOROZ, B.Z. [1983] *Reflections on quantum logic* (J 2736) Int J Theor Phys 22*329-340 • ERR/ADD ibid 23*497-498
◇ B51 ◇ REV MR 84i:81009 MR 85g:81016 Zbl 515 # 03041 • ID 37849

MOROZ, B.Z. see Vol. I for further entries

MOROZOVA, E.A. [1981] see CHENTSOV, N.N.

MORSCHER, E. [1971] *A matrix method for deontic logic* (J 0472) Theory Decis 2*16-34
◇ B45 ◇ REV MR 46 # 5109 Zbl 268 # 02015 • ID 09503

MORSCHER, E. & ZECHA, G. [1972] *Wozu deontische Logik?* (J 1834) Arch Rechts- und Sozialphil 53*363-378
◇ B45 ◇ ID 32390

MORSCHER, E. [1977] see CZERMAK, J.

MORSCHER, E. & WEINGARTNER, P. (EDS.) [1979] *Ontology and logic* (X 1258) Duncker & Humblot: Berlin 286pp
◇ A05 B97 ◇ REV MR 80f:03007 • ID 70060

MORSCHER, E. [1982] *Antinomies and incompatibilities within normative languages. Some semantic considerations* (P 4041) Log, Inform, Law;1981 Firenze 2*83-102
◇ B45 ◇ REV MR 85k:03012 • ID 45217

MORSCHER, E. [1984] *Sein-Sollen-Schluesse und wie Schluesse sein sollen* (C 4065) Theorie der Normen 421-439
◇ B45 ◇ REV MR 85k:03015 • ID 44561

MORSCHER, E. see Vol. I, IV for further entries

MORTARI, C.A. [1982] *On a system of modal-temporal logic* (P 3860) Coll Papers to Farah on Retirement;1981 Sao Paulo 21-36
◇ B45 ◇ REV MR 85b:03025 Zbl 538 # 03016 • ID 40592

MORTARI, C.A. [1983] see LOPARIC, A.

MORTENSEN, C. [1976] *A sequence of normal modal systems with non-contingency bases* (J 0079) Logique & Anal, NS 19*341-344
◇ B45 ◇ REV MR 58 # 16154 Zbl 347 # 02014 • ID 64040

MORTENSEN, C. [1983] *Relevance and verisimilitude* (J 0154) Synthese 55*353-364
◇ B46 ◇ REV MR 85f:03019 Zbl 524 # 03013 • ID 37594

MORTENSEN, C. [1984] see MEYER, R.K.

MORTENSEN, C. [1985] see MEYER, R.K.

MORTENSEN, C. see Vol. I, V for further entries

MORTIMER, H. [1964] *Probabilistic definition as exemplified by a definition of genotype (Polish) (Russian and English summaries)* (J 0063) Studia Logica 15*103-162
◇ B48 ◇ REV MR 34 # 4123 • ID 33120

MORTIMER, H. [1967] *Consistency condition of two probabilistic postulates (Polish) (English and Russian summaries)* (J 0063) Studia Logica 21*91-101
◇ B48 ◇ REV MR 37 # 944 Zbl 294 # 02025 • ID 64043

MORTIMER, H. [1970] *Conditions for acceptance of probabilistic postulates (Polish and Russian summaries)* (J 0063) Studia Logica 26*87-97
◇ A05 B48 ◇ REV MR 44 # 3360 Zbl 265 # 02004 • ID 29815

MORTIMER, H. [1973] *A rule of acceptance based on logical probability* (J 0154) Synthese 26*259-263
◇ A05 B48 ◇ REV Zbl 277 # 02005 • ID 29526

MORTIMER, M. [1974] *Some results in modal model theory* (J 0036) J Symb Logic 39*496-508
◇ B45 C90 ◇ REV MR 50 # 9537 Zbl 306 # 02021 • ID 09506

MORTIMER, M. see Vol. I for further entries

MOSAHEB, G.-H. [1955] *Introduction to formal logic (Iranian)* (X 4502) Univ Tehran: Tehran 709pp
◇ B98 ◇ REV JSL 22.354 • ID 42153

MOSCATO, U. [1983] see MIGLIOLI, P.A.

MOSCATO, U. see Vol. III, VI for further entries

MOSHCHENSKIJ, V.A. [1973] *Lectures on mathematical logic (Russian)* (**X** 1212) Belor Gos Univ: Minsk 159pp
◇ B98 D10 D20 ◇ REV MR 50 # 6775 Zbl 275 # 02001
• ID 21271

MOSHCHENSKIJ, V.A. [1977] see KARPOV, V.G.

MOSHCHENSKIJ, V.A. [1981] *Special completeness of some languages of k-valued logics (Russian)* (**J** 0071) Met Diskr Analiz (Novosibirsk) 36*39-45,94
◇ B50 ◇ REV MR 82m:03035 Zbl 483 # 03015 • ID 76588

MOSHCHENSKIJ, V.A. see Vol. I, IV for further entries

MOSKALEV, EH.S. [1970] see KARPOVSKY, M.G.

MOSS, B.P. [1972] *A picture of a Kripke model for S4* (**P** 2080) Conf Math Log;1970 London 346-347
◇ B45 ◇ REV Zbl 227 # 02009 • ID 27354

MOSS, B.P. see Vol. IV, V for further entries

MOSS, J.M.B. & SCOTT, D.S. [1975] *Bibliography of logic books* (**X** 0894) Oxford Univ Pr: Oxford v+106pp
◇ A10 B98 ◇ ID 32398

MOSTERIN, J. [1976] *Logica de primer order* (**X** 1744) Ariel: Esplugas de Llobregat 140pp
◇ B98 ◇ ID 31171

MOSTERIN, J. see Vol. III, V for further entries

MOSTOWSKI, A.WLODZIMIERZ & PAWLAK, Z. [1970] *Logik fuer Ingenieure (Polnisch)* (**X** 1034) PWN: Warsaw 315pp
◇ B98 ◇ REV Zbl 213.11 • ID 27979

MOSTOWSKI, A.WLODZIMIERZ see Vol. I, III, IV, VI for further entries

MOSTOWSKI, ANDRZEJ [1948] *Logique mathematique. Cours donne a l'universite (Polish)* (**S** 0257) Monograf Mat 18*viii+388pp
◇ B98 ◇ REV MR 10.229 Zbl 34.147 JSL 14.189
• ID 28463

MOSTOWSKI, ANDRZEJ [1955] see GRZEGORCZYK, A.

MOSTOWSKI, ANDRZEJ [1961] *An example of a non-axiomatizable many valued logic* (**J** 0068) Z Math Logik Grundlagen Math 7*72-76
◇ B50 ◇ REV MR 30 # 1038 Zbl 124.248 JSL 35.143
• ID 09572

MOSTOWSKI, ANDRZEJ [1961] *Axiomatizability of some many valued predicate calculi* (**J** 0027) Fund Math 50*165-190
◇ B50 ◇ REV MR 24 # A3060 Zbl 99.7 JSL 35.143
• ID 09571

MOSTOWSKI, ANDRZEJ [1962] *L'espace des modeles d'une theorie formalisee et quelques-unes de ses applications* (**P** 1606) Colloq Math (Pascal);1962 Clermont-Ferrand 7*107-116
◇ B50 C07 C40 C57 C80 C85 C90 D45 ◇ REV MR 47 # 4785 JSL 40.501 • ID 09574

MOSTOWSKI, ANDRZEJ [1963] *The Hilbert ε function in many-valued logics* (**J** 0096) Acta Philos Fenn 16*169-188
◇ B50 ◇ REV MR 28 # 2963 Zbl 121.12 • ID 09576

MOSTOWSKI, ANDRZEJ [1965] *Thirty years of foundational studies. Lectures on the development of mathematical logic and the studies of the foundations of mathematics in 1930-1964* (**J** 0096) Acta Philos Fenn 17*1-180
• REPR [1966] (**X** 1096) Blackwell: Oxford 180pp
◇ A10 B98 C98 D98 E98 F98 ◇ REV MR 33 # 18 MR 33 # 5445 Zbl 146.245 JSL 33.111 • ID 09578

MOSTOWSKI, ANDRZEJ [1979] *An excerpt from the book Mathematical logic* (**J** 0519) Wiad Mat, Ann Soc Math Pol, Ser 2 22*77-78
◇ B98 ◇ REV MR 82c:01043 • ID 82306

MOSTOWSKI, ANDRZEJ see Vol. I, III, IV, V, VI for further entries

MOTHERSHEAD, J.L. [1970] see GOHEEN, J.D.

MOTIL, J.M. [1974] *A multiple-valued digital "saturation" system* (**P** 1385) Int Symp Multi-Val Log (4);1974 Morgantown 483-492
◇ B50 ◇ ID 42098

MOTOHASHI, N. [1972] *Some investigations on many valued logics* (**J** 0081) Proc Japan Acad 48*59-61
◇ B50 ◇ REV MR 52 # 5354 Zbl 252 # 02009 • ID 18304

MOTOHASHI, N. & SHIRAI, K. [1981] *On a generalisation of theorems of Skolem and Mints (Japanese)* (**P** 4153) B-Val Anal & Nonstand Anal;1981 Kyoto 43-50
◇ B50 ◇ ID 47755

MOTOHASHI, N. see Vol. I, III, V, VI for further entries

MOTT, P.L. [1973] *On Chisholm's paradox* (**J** 0122) J Philos Logic 2*197-211
◇ B45 ◇ REV MR 54 # 7205 Zbl 266 # 02018 • ID 24996

MOUFTAH, H.T. [1974] see JORDAN, I.B.

MOUFTAH, H.T. see Vol. I for further entries

MOURANT, J.A. [1963] *Formal logic: an introductory text book* (**X** 0843) Macmillan : New York & London 421pp
◇ B98 ◇ ID 22524

MOUTAFAKIS, N.J. [1975] *Imperatives and their logics* (**X** 4321) Sterling Publ: New Delhi xv+216pp
◇ B60 ◇ REV JSL 45.375 • ID 44711

MOUTAFAKIS, N.J. [1982] *Rescher's logic of preference and linguistic analysis* (**J** 0079) Logique & Anal, NS 25*135-165
◇ B45 B65 ◇ REV MR 84e:03024 • ID 34363

MOW, D.A. [1978] see CURRENT, K.W.

MUCHNIK, A.A. & YANOV, YU.I. [1959] *Existence of k-valued closed classes without a finite basis (Russian)* (**J** 0023) Dokl Akad Nauk SSSR 127*44-46
◇ B30 B50 ◇ REV MR 21 # 7174 Zbl 100.10 JSL 27.247
• ID 00382

MUCHNIK, A.A. [1974] see ERMOLAEVA, N.M.

MUCHNIK, A.A. [1976] see ERMOLAEVA, N.M.

MUCHNIK, A.A. [1979] see ERMOLAEVA, N.M.

MUCHNIK, A.A. [1983] *Supplement of the translator to the paper "On alternation. I, II" (Russian)* (**J** 3079) Kiber Sb Perevodov, NS 20*141-158
◇ B45 D05 D10 D15 F50 ◇ REV Zbl 545 # 68041 • REM For the papers see Chandra,A.K. et al. 1976 and 1981
• ID 43519

MUCHNIK, A.A. see Vol. I, III, IV for further entries

MUELLER, GERT H. [1957] *Zur operativen Begruendung von Logik und Mathematik* (**J** 0170) Ratio (Frankfurt) 1*77-86
• TRANSL [1957] (**J** 1022) Ratio (Oxford) 1*85-95 (English)
◇ B30 B60 ◇ ID 33928

MUELLER, GERT H. [1966] *Zur Einfuehrung der Logik* (**J** 0178) Stud Gen 19*493-508
◇ A05 B60 ◇ REV Zbl 192.19 • ID 16280

MUELLER, GERT H. & OBERSCHELP, A. & POTTHOFF, K. (EDS.) [1975] *ISILC Logic Conference* (**X** 0811) Springer: Heidelberg & New York iv+651pp
◇ B97 ◇ REV MR 52 # 10365 Zbl 307 # 00007 • ID 70183

MUELLER, GERT H. see Vol. I, III, IV, V, VI for further entries

MUKAIDONO, M. [1972] *On the B-ternary logical function - a ternary logic considering ambiguity (Japanese)* (**J** 0979) Denshi Tsushin Gakkai Ronbunshi, Sect A-D 55-D/6*355-362
• TRANSL [1972] (**J** 0464) Syst-Comp-Controls 3/3*27-36
◇ B50 ◇ REV MR 49 # 4891 • ID 09631

MUKAIDONO, M. [1975] *On some properties of fuzzy logic* (**J** 0464) Syst-Comp-Controls 6/2*36-43
◇ B52 ◇ REV MR 56 # 8324 • ID 76642

MUKAIDONO, M. [1977] *On some properties of a quantization in fuzzy logic* (**P** 2013) Int Symp Multi-Val Log (7);1977 Charlotte 103-106
◇ B52 ◇ REV MR 56 # 13828 • ID 35859

MUKAIDONO, M. [1979] *A necessary and sufficient condition for fuzzy logic functions* (**P** 3003) Int Symp Multi-Val Log (9);1979 Bath 159-166
◇ B52 ◇ REV MR 81d:03026 • ID 76641

MUKAIDONO, M. [1980] *Some kinds of functional completeness of ternary logic functions* (**P** 3673) Int Symp Multi-Val Log (10);1980 Evanston 81-87
◇ B50 ◇ REV MR 83m:94032 Zbl 542 # 94032 • ID 43725

MUKAIDONO, M. [1982] *Fuzzy inference of resolution style* (**P** 4051) Fuzzy Set & Possibility Th;1980 Acapulco 224-231
◇ B52 ◇ REV MR 84b:03004 • ID 46327

MUKAIDONO, M. [1982] *New canonical forms of fuzzy switching functions* (**P** 3845) Conf Math Service of Man (2,Proc)(Feriet);1982 Las Palmas 520-523
◇ B52 ◇ REV Zbl 523 # 94026 • ID 37002

MUKAIDONO, M. [1982] *New canonical forms of fuzzy switching functions* (**C** 3786) Approx Reason in Decis Anal 97-101
◇ B52 ◇ REV MR 84d:90003 Zbl 523 # 94027 • ID 37003

MUKAIDONO, M. [1982] *New canonical forms and their applications to enumerating fuzzy switching functions* (**P** 4002) Int Symp Multi-Val Log (12);1982 Paris 275-279
◇ B52 ◇ REV Zbl 555 # 94018 • ID 46145

MUKAIDONO, M. [1982] *New canonical forms for and their application to fuzzy switching functions* (**P** 4081) Many-Val Log & Appl;1982 Kyoto 207-230
◇ B52 ◇ ID 47634

MUKAIDONO, M. [1984] see BERMAN, J.

MUKHERJEE, M.K. [1984] *A generalized characterization theorem for quantum logics* (**J** 2778) Lett Nuovo Cimento Ser 2 40*453-456
◇ B51 ◇ REV MR 86c:06013 • ID 44283

MULLIGAN, K. & SMITH, BARRY [1983] *Framework for formal ontology* (**J** 3781) Topoi 2*73-85
◇ B60 ◇ REV MR 85a:03043 • ID 34794

MULLOCK, P. [1974] *The permissiveness of powers* (**J** 1022) Ratio (Oxford) 16*76-81
◇ B45 ◇ REV MR 58 # 16155 • ID 76659

MULLOCK, P. [1975] *The Stone-Tammelo deontic logic* (**J** 0079) Logique & Anal, NS 18*65-89
◇ B45 ◇ REV MR 53 # 2635 Zbl 335 # 02009 • ID 21523

MULLOCK, P. [1979] *Logic and liberty* (**J** 0095) Philos Stud 35*217-238
◇ A05 B45 ◇ REV MR 80d:03018 • ID 76658

MUNDICI, D. [1984] *Abstract model theory of many-valued logics and K-theory of certain C^*-algebras* (**P** 1545) Easter Conf on Model Th (2);1984 Wittenberg 157-204
◇ B50 C90 C95 ◇ REV Zbl 561 # 03011 • ID 44636

MUNDICI, D. see Vol. I, III, IV, V, VI for further entries

MUNITZ, M.K. (ED.) [1972] *Identity and individuation* (**X** 0924) New York Univ Pr: New York 288pp
◇ A05 B45 ◇ ID 25246

MUNOZ, V. [1961] *De la axiomatica a los sistemas formales* (**X** 1407) Inst Mat J.Juan: Madrid 80pp
◇ B30 B98 ◇ REV MR 27 # 3494 • ID 28705

MURAKI, M. [1982] see KAGA, Y.

MURANAKA, N. [1982] see IMANISHI, S.

MURAVITSKIJ, A.YU. [1980] see KUZNETSOV, A.V.

MURAVITSKIJ, A.YU. [1981] *Finite approximability of the calculus I^Δ and the nonmodelability of some of its extensions (Russian)* (**J** 0087) Mat Zametki (Akad Nauk SSSR) 29*907-916,957
• TRANSL [1981] (**J** 1044) Math Notes, Acad Sci USSR 29*463-468
◇ B55 ◇ REV MR 82m:03018 Zbl 484 # 03009 • ID 76665

MURAVITSKIJ, A.YU. [1981] *Strong equivalence on an intuitionistic Kripke model and assertorically equivolumetric logics (Russian)* (**J** 0003) Algebra i Logika 20*165-182,250
• TRANSL [1981] (**J** 0069) Algeb and Log 20*112-123
◇ B55 C90 F50 ◇ REV MR 83h:03087 Zbl 486 # 03015 • ID 36087

MURAVITSKIJ, A.YU. [1983] *Comparison of the topological and relational semantics of superintuitionistic logics (Russian)* (**J** 0003) Algebra i Logika 22*276-296
• TRANSL [1983] (**J** 0069) Algeb and Log 22*197-213
◇ B55 C90 ◇ REV MR 85j:03031 Zbl 543 # 03022 • ID 40399

MURAVITSKIJ, A.YU. [1983] *Extensions of the provability logic (Russian)* (**J** 0087) Mat Zametki (Akad Nauk SSSR) 33*915-927
• TRANSL [1983] (**J** 1044) Math Notes, Acad Sci USSR 33*469-475
◇ B45 F30 ◇ REV MR 84j:03044 Zbl 538 # 03023 • ID 34636

MURAVITSKIJ, A.YU. [1984] *A result on the completeness of superintuitionistic logics* (**J** 0087) Mat Zametki (Akad Nauk SSSR) 36*765-776,799
• TRANSL [1984] (**J** 1044) Math Notes, Acad Sci USSR 36*883-889
◇ B55 C90 ◇ REV MR 86f:03046 Zbl 566 # 03017 • ID 45360

MURAVITSKIJ, A.YU. [1984] *Superintuitionistic logics approximable by algebras with the descending chain condition (Russian)* (**J** 0087) Mat Zametki (Akad Nauk SSSR) 35*273-276
• TRANSL [1984] (**J** 1044) Math Notes, Acad Sci USSR 35*145-146
◇ B55 ◇ REV MR 86b:03027 Zbl 542 # 03012 • ID 43677

MURAVITSKIJ, A.YU. see Vol. VI for further entries

MUROGA, S. [1962] *Majority logic and problems of probabilistic behaviour* (**P** 0578) Self Organizing Systs;1962 Chicago 243–281
◇ B48 ◇ ID 09652

MUROGA, S. see Vol. I for further entries

MURRAY, W.J. [1969] *Les relations de raison et la modalite du jugement dans la logique d'Aristotle* (**J** 0147) An Univ Bucuresti, Acta Logica 12*199–205
◇ A10 B45 ◇ REV MR 45 # 6593 • ID 09653

MURSKIJ, V.L. [1965] *The existence in three-valued logic of a closed class with finite basis, not having a finite complete system of identities (Russian)* (**J** 0023) Dokl Akad Nauk SSSR 163*815–818
• TRANSL [1965] (**J** 0062) Sov Math, Dokl 6*1020–1024
◇ B22 B50 ◇ REV MR 32 # 3998 Zbl 154.255 JSL 37.762 • ID 09654

MURSKIJ, V.L. see Vol. III, IV for further entries

MURUNGI, R.W. [1974] *On a nonthesis of classical modal logic* (**J** 0047) Notre Dame J Formal Log 15*494–496
◇ B45 ◇ REV MR 50 # 4249 Zbl 246 # 02016 • ID 09657

MURUNGI, R.W. [1977] *Necessitas consequentis in a singleton possible world* (**J** 0047) Notre Dame J Formal Log 18*637–638
◇ B45 ◇ REV MR 58 # 5062 Zbl 283 # 02021 • ID 23706

MUSKARDIN, V. [1985] *Logic of guaranty* (**P** 4661) Algeb & Log;1984 Zagreb 111–121
◇ B60 ◇ ID 49017

MUZIO, J.C. [1970] *A decision process for 3-valued Sheffer functions I* (**J** 0068) Z Math Logik Grundlagen Math 16*271–280
◇ B50 ◇ REV MR 43 # 4636 Zbl 191.287 • REM Part II 1971 • ID 09662

MUZIO, J.C. [1971] *A decision process for 3-valued Sheffer functions II* (**J** 0068) Z Math Logik Grundlagen Math 17*97–114
◇ B50 ◇ REV MR 43 # 4637 Zbl 191.287 • REM Part I 1970 • ID 09663

MUZIO, J.C. [1971] *On the t-closure condition of Martin* (**J** 0043) Math Ann 195*143–148
◇ B50 ◇ REV MR 45 # 3170 Zbl 216.287 • ID 09664

MUZIO, J.C. [1973] *Some large classes of n-valued Sheffer functions* (**J** 0171) Proc Cambridge Phil Soc Math Phys 74*201–211
◇ B50 ◇ REV MR 48 # 5822 Zbl 266 # 02010 • ID 09665

MUZIO, J.C. [1973] *The cosubstitution condition* (**J** 0047) Notre Dame J Formal Log 14*87–94
◇ B50 ◇ REV MR 48 # 3701 Zbl 247 # 02020 • ID 09666

MUZIO, J.C. [1973] *The necessity of a condition of Shoenfield* (**J** 0068) Z Math Logik Grundlagen Math 19*199–204
◇ B50 ◇ REV MR 47 # 8243 Zbl 308 # 02019 • ID 09667

MUZIO, J.C. [1975] *Ternary two-place functions that are complete with constants* (**P** 1805) Int Symp Multi-Val Log (5,Proc);1975 Bloomington 27–33
◇ B50 ◇ REV MR 58 # 16139 • ID 76690

MUZIO, J.C. [1975] *The size of some large classes of n-valued Sheffer functions* (**J** 0302) Rep Math Logic, Krakow & Katowice 5*53–62
◇ B50 ◇ REV MR 57 # 9480 Zbl 357 # 02017 • ID 21901

MUZIO, J.C. [1976] see MILLER, D.MICHAEL

MUZIO, J.C. [1976] *Binary functions that are complete with constants over* $\{0,1,2\}$ (**P** 2995) Manitoba Conf Num Math (5);1975 Winnipeg 561–570
◇ B50 ◇ REV MR 55 # 7609 Zbl 357 # 02016 • ID 50364

MUZIO, J.C. [1976] *Concerning completeness and abelian semigroups* (**J** 0068) Z Math Logik Grundlagen Math 22*85–86
◇ B50 ◇ REV MR 53 # 5254 Zbl 325 # 02010 • ID 14978

MUZIO, J.C. [1977] *The class structure of complete binary operations on* $\{1,2,3\}$ (**J** 0302) Rep Math Logic, Krakow & Katowice 8*81–85
◇ B50 G25 ◇ REV MR 58 # 5034 Zbl 383 # 03045 • ID 24194

MUZIO, J.C. [1977] *Three-valued unary generators based on Mouftah's primitives* (**P** 2013) Int Symp Multi-Val Log (7);1977 Charlotte 35–38
◇ B50 ◇ REV MR 58 # 136 • ID 76703

MUZIO, J.C. [1979] see MILLER, D.MICHAEL

MUZIO, J.C. [1980] see MILLER, D.MICHAEL

MUZIO, J.C. & ROSENBERG, I.G. [1980] *Large classes of functionally complete operations. I* (**P** 3673) Int Symp Multi-Val Log (10);1980 Evanston 94–101
◇ B50 ◇ REV MR 83m:94032 Zbl 539 # 03050 • REM Part II 1981 by Rosenberg,I.G. • ID 41266

MUZIO, J.C. [1980] see EPSTEIN, G.

MUZIO, J.C. & ROSENBERG, I.G. [1982] *Large classes of functionally complete groupoids. I* (**J** 0529) Aequationes Math 25*274–288
◇ B50 ◇ REV MR 85j:08008 Zbl 525 # 08003 • ID 45228

MUZIO, J.C. see Vol. I for further entries

MYCROFT, A. [1984] *Logic programs and many-valued logic* (**P** 3565) Symp of Th Aspects of Comput Sci (1);1984 Paris 274–286
◇ B50 ◇ REV Zbl 544 # 68027 • ID 45930

MYHILL, J.R. [1958] *Problems arising in the formalization of intensional logic* (**J** 0079) Logique & Anal, NS 1*74–83
◇ B45 ◇ REV JSL 37.180 • ID 09717

MYHILL, J.R. [1963] *An alternative to the method of extension and intension* (**C** 0673) Phil of Carnap 299–310
◇ B45 ◇ ID 09726

MYHILL, J.R. see Vol. I, III, IV, V, VI for further entries

NAESS, A. [1969] *Introduction to logic and scientific method* (**X** 1233) Bedminster Pr: Totowa
◇ B98 ◇ ID 22525

NAGAI, SATORU [1973] *On a semantics for non-classical logics* (**J** 0081) Proc Japan Acad 49*337–340
◇ B55 ◇ REV MR 49 # 2266 Zbl 276 # 02037 • ID 09753

NAGAI, SATORU & ONO, H. [1974] *Applications of general Kripke models to intermediate logics* (**J** 0381) J Tsuda College (Tokyo) 6*9–21
◇ B55 ◇ REV MR 58 # 5073 • ID 30770

NAGAI, SHIGEO & OKUBO, T. [1980] *On the logic of a priori modalities* (**J** 0260) Ann Jap Ass Phil Sci 5*213–223
◇ A05 B45 ◇ REV MR 82b:03058 Zbl 438 # 03025 • ID 55937

NAGAOKA, K. [1979] see HIROSE, K.

NAGAOKA, K. see Vol. III, IV for further entries

NAGASHIMA, T. [1973] *An intermediate predicate logic* (**J** 0531) Hitotsubashi J Arts Sci (Tokyo) 14∗53-58
◇ B55 ◇ REV MR 49 # 4758 • ID 76743

NAGASHIMA, T. [1977] *A theorem on an intermediate predicate logic* (**J** 0531) Hitotsubashi J Arts Sci (Tokyo) 18∗73-76
◇ B55 ◇ REV MR 57 # 5694 • ID 76741

NAGASHIMA, T. see Vol. III, IV, V, VI for further entries

NAGATA, M. & NAKANISHI, M. & NISHIMURA, T. [1975] *Implementation of Lukasiewicz's, Kleene's and McCarthy's 3-valued logics* (**J** 0350) Sci Rep Tokyo Kyoiku Daigaku Sect A 13∗90-100
◇ B50 ◇ REV MR 53 # 2055 Zbl 339 # 02018 • ID 16698

NAGATA, M. see Vol. I, VI for further entries

NAGATA, S. [1966] *A series of successive modifications of Peirce's rule* (**J** 0081) Proc Japan Acad 42∗859-861
◇ B55 ◇ REV MR 35 # 1443 Zbl 192.28 • ID 09759

NAGATA, S. [1968] see MIURA, S.

NAGATA, S. see Vol. I, VI for further entries

NAGEL, E. [1934] see COHEN, M.R.

NAGEL, E. & SUPPES, P. & TARSKI, A. (EDS.) [1962] *Logic, methodology and philosophy of science* (**X** 1355) Stanford Univ Pr: Stanford ix+661pp
◇ B97 ◇ REV Zbl 128.241 • REM Reprinted 1965 • ID 32720

NAGEL, E. [1963] *Carnap's theory of induction* (**C** 0673) Phil of Carnap 785-825
◇ A05 B48 ◇ REV JSL 37.631 • ID 09770

NAGEL, E. see Vol. I, IV, VI for further entries

NAGLE, M.C. [1981] *The decidability of normal K5 logics* (**J** 0036) J Symb Logic 46∗319-328
◇ B45 ◇ REV MR 82k:03019 Zbl 473 # 03009 • ID 55340

NAGLE, M.C. & THOMASON, S.K. [1985] *The extensions of the modal logic* k5 (**J** 0036) J Symb Logic 50∗102-109
◇ B45 ◇ ID 42528

NAGLE JR., H.T. [1973] see IRVING, T.A.

NAGLE JR., H.T. & SHIVA, S.G. [1974] *On multiple-valued memory elements* (**P** 1385) Int Symp Multi-Val Log (4);1974 Morgantown 209-224
◇ B50 ◇ ID 42105

NAKAMURA, A. [1962] *On the infinitely many-valued threshold logics and von Wright's System M* (**J** 0068) Z Math Logik Grundlagen Math 8∗147-164
◇ B45 ◇ REV MR 26 # 3593 Zbl 105.5 JSL 30.374 • ID 09780

NAKAMURA, A. [1962] *On the many-valued logics and an axiomatic system for them (Japanese)* (**J** 2173) J Japan Ass Phil Sci 5∗165-171
◇ B50 ◇ REV JSL 31.274 • ID 41887

NAKAMURA, A. [1962] *On the relation between modal logics and many-valued logic (Japanese)* (**J** 2173) J Japan Ass Phil Sci 6/1∗34-40
◇ B45 ◇ REV JSL 35.582 • ID 41891

NAKAMURA, A. [1963] *A note on truth-value functions in the infinitely many-valued logics* (**J** 0068) Z Math Logik Grundlagen Math 9∗141-144
◇ B50 ◇ REV MR 28 # 1125 Zbl 113.4 JSL 30.374 • ID 09781

NAKAMURA, A. [1963] *On a simple axiomatic system of the infinitely many-valued logic based on* (∧ , →) (**J** 0068) Z Math Logik Grundlagen Math 9∗251-263
◇ B50 ◇ REV MR 27 # 3509 Zbl 121.11 JSL 30.374 • ID 19442

NAKAMURA, A. [1963] *On an axiomatic system of the infinitely many-valued threshold predicate calculi* (**J** 0068) Z Math Logik Grundlagen Math 9∗321-329
◇ B50 ◇ REV MR 27 # 4740 Zbl 126.259 JSL 30.374 • ID 09782

NAKAMURA, A. [1964] *On the relation between modal logic and many-valued logic (Japanese)* (**J** 2173) J Japan Ass Phil Sci 6/4∗15-20
◇ B45 B50 ◇ REV JSL 35.582 • ID 41888

NAKAMURA, A. [1964] *Truth-value stipulations for the von Wright system M′ and the Heyting system* (**J** 0068) Z Math Logik Grundlagen Math 10∗173-183
◇ B45 ◇ REV MR 28 # 3923 Zbl 126.9 JSL 30.174 • ID 09783

NAKAMURA, A. [1965] *On certain systems of modal logic* (**J** 0068) Z Math Logik Grundlagen Math 11∗203-207
◇ B45 ◇ REV MR 31 # 3322 Zbl 154.5 JSL 31.665 • ID 09785

NAKAMURA, A. [1965] *On the infinitely many-valued double-threshold logic* (**J** 0068) Z Math Logik Grundlagen Math 11∗93-101
◇ B50 ◇ REV MR 31 # 1184 Zbl 168.4 JSL 31.665 • ID 09784

NAKAMURA, A. [1967] *A remark on the truth-value stipulation for the modal system M′* (**J** 0068) Z Math Logik Grundlagen Math 13∗11-14
◇ B45 ◇ REV MR 37 # 2586 Zbl 239 # 02009 JSL 39.351 • ID 09786

NAKAMURA, A. [1968] *On an axiomatic system of modal logic* (**J** 0068) Z Math Logik Grundlagen Math 14∗61-66
◇ B45 ◇ REV MR 38 # 5594 Zbl 179.313 JSL 39.351 • ID 09787

NAKAMURA, A. [1970] *On the undecidability of monadic modal predicate logic* (**J** 0068) Z Math Logik Grundlagen Math 16∗257-260
◇ B45 D35 ◇ REV MR 43 # 6059 Zbl 207.5 • ID 09789

NAKAMURA, A. & ONO, H. [1980] *Decidability results on a query language for data bases with incomplete information* (**P** 3210) Math Founds of Comput Sci (9);1980 Rydzyna 452-459
◇ B25 B52 B75 ◇ REV MR 82e:68102 Zbl 452 # 68095 • ID 54126

NAKAMURA, A. & ONO, H. [1980] *On the size of refutation Kripke models for some linear modal and tense logics* (**J** 0063) Studia Logica 39∗325-333
◇ B45 D15 ◇ REV MR 82i:03030 Zbl 466 # 03008 • ID 54959

NAKAMURA, A. [1982] *On a three-valued logic connected with incomplete information data bases* (**P** 4081) Many-Val Log & Appl;1982 Kyoto 53-63
◇ B50 ◇ ID 47617

NAKAMURA, A. [1983] *Three-valued logic and its application to the query language of incomplete information* (**P** 4054) Int Symp Multi-Val Log (13);1983 Kyoto 214-218
◇ B50 ◇ REV Zbl 562 # 68074 • ID 47443

NAKAMURA, A. see Vol. I, III, IV, VI for further entries

NAKAMURA, K. [1982] see IWAI, S.

NAKAMURA, T. [1983] *Disjunction property for some intermediate predicate logics* (**J** 0302) Rep Math Logic, Krakow & Katowice 15*33-39
◇ B55 ◇ REV MR 84k:03080 Zbl 537 # 03017 • ID 35017

NAKANISHI, M. [1975] see NAGATA, M.

NAKANISHI, M. [1978] *Gentzen type formal system for many-valued propositional logic* (**J** 2770) Keio Math Sem Rep (Yokohama) 3*49-55
◇ B50 ◇ REV MR 80e:03019 Zbl 399 # 03016 • ID 52813

NAKANISHI, M. see Vol. I, VI for further entries

NALIMOV, V.V. [1979] *The probability distribution function as a method of defining fuzzy sets. Metatheoretic sketches (Russian)* (**J** 1884) Avtomatika, Akad Nauk Ukr SSR 1979/6*80-87
• TRANSL [1979] (**J** 2548) Sov Autom Control 12/6*67-73
◇ B52 E72 ◇ REV MR 81m:03032 • ID 76753

NARENS, L. [1981] *A general theory of ratio scalability with remarks about the measurement-theoretic concept of meaningfulness* (**J** 0472) Theory Decis 13*1-70
◇ B48 ◇ REV MR 82e:06014 Zbl 497 # 00021 • ID 36629

NARENS, L. see Vol. I, III for further entries

NARSKIJ, I.S. (ED.) [1968] *Problems in logic and epistemology (Russian)* (**X** 0898) Moskov Gos Univ: Moskva 319pp
◇ A05 B97 ◇ REV MR 51 # 7799 Zbl 185.7 • ID 70195

NAT VANDER, A. [1969] see MCCALL, S.

NAT VANDER, A. [1972] *Axiomatic, Sequenzen-Kalkul, and subordinate proof versions of S9* (**J** 0047) Notre Dame J Formal Log 13*309-322 • ERR/ADD ibid 17*640
◇ B45 ◇ REV MR 47 # 16 Zbl 231 # 02020 • ID 22288

NAT VANDER, A. & RESCHER, N. [1973] *On alternatives in epistemic logic* (**J** 0122) J Philos Logic 2*119-135
◇ B45 ◇ REV MR 54 # 64 Zbl 259 # 02019 • ID 24552

NAT VANDER, A. [1979] *Beyond non-normal possible worlds* (**J** 0047) Notre Dame J Formal Log 20*631-635
◇ B45 C90 ◇ REV MR 80k:03022 Zbl 368 # 02022
• ID 56121

NAT VANDER, A. [1979] *First-order indefinite and uniform neighbourhood semantics* (**J** 0063) Studia Logica 38*277-296
◇ B45 ◇ REV MR 80m:03040 Zbl 436 # 03009 • ID 55849

NATALI, F.S. [1982] see BIAGIOLI, C.

NATORP, P. [1904] *Logik. Grundlegung und logischer Aufbau der Mathematik und mathematischen Naturwissenschaft* (**X** 1263) Elwert: Marburg 71pp
◇ B98 ◇ REV FdM 35.79 • ID 25563

NATORP, P. see Vol. I for further entries

NATVIG, B. [1983] *Possibility versus probability* (**J** 2720) Fuzzy Sets Syst 10*31-36
◇ B52 ◇ REV MR 84e:60006 Zbl 511 # 60003 • ID 38544

NAVARA, M. [1984] *Two-valued states on a concrete logic and the additivity problem (Russian summary)* (**J** 1522) Math Slovaca 34*329-336
◇ B60 ◇ REV MR 85m:28005 • ID 45803

NAZARALA, J. [1975] see MORAGA, C.

NECHAEV, A.A. [1980] *Criterion for completeness of systems of functions of p^n-valued logic containing operations of addition and multiplication modulo p^n (Russian)* (**J** 0071) Met Diskr Analiz (Novosibirsk) 34*74-87,102
◇ B50 ◇ REV MR 83h:03034 Zbl 473 # 03020 • ID 36054

NECHIPORUK, EH.I. [1963] see KLOSS, B.M.

NECHIPORUK, EH.I. [1969] *The topological principles of autocorrection (Russian)* (**J** 0052) Probl Kibern 21*5-102
◇ B50 ◇ REV MR 45 # 3171 • REM Part I. Part II 1973
• ID 09832

NECHIPORUK, EH.I. [1973] *The topological principles of autocorrection. II (Russian)* (**J** 0052) Probl Kibern 26*19-26,327
◇ B50 ◇ REV MR 50 # 9425 Zbl 311 # 94033 • REM Part I 1969 • ID 82335

NECHIPORUK, EH.I. see Vol. I for further entries

NEEDHAM, P. [1981] *Temporal intervals and temporal order* (**J** 0079) Logique & Anal, NS 24*49-64
◇ B45 ◇ REV MR 82k:03020 Zbl 483 # 03004 • ID 76781

NEF, F. [1982] see ENGEL, P.

NEGOITA, C.V. & RALESCU, D.A. [1974] *Applications of fuzzy sets to systems analysis (Romanian)* (**X** 2709) Ed Tehn: Bucharest 218pp
• TRANSL [1975] (**X** 0827) Wiley & Sons: New York 191pp [1975] (**X** 0804) Birkhaeuser: Basel (English)
◇ B52 E70 E72 ◇ REV MR 55 # 7589 MR 58 # 9442a MR 58 # 9442b Zbl 326 # 94001 Zbl 326 # 94002 JSL 44.284 • ID 64172

NEGOITA, C.V. & RALESCU, D.A. [1975] *Representation theorems for fuzzy concepts* (**J** 1429) Kybernetes 4*169-174
◇ B52 E72 ◇ REV Zbl 352 # 02044 • ID 50043

NEGOITA, C.V. & RALESCU, D.A. [1976] *Comment on a "Comment on an algorithm that generates fuzzy prime implicants by Lee and Chang"* (**J** 0194) Inform & Control 30*199-201
◇ B52 ◇ REV MR 53 # 7116 Zbl 324 # 94033 • ID 64174

NEGOITA, C.V. see Vol. V for further entries

NEGRU, I.S. [1978] *On the identities on the structure of Post classes (Russian)* (**X** 2235) VINITI: Moskva 1556-78
◇ B50 ◇ ID 33015

NEGRU, I.S. see Vol. I for further entries

NEIDORF, R. [1967] *Deductive forms: An elementary logic* (**X** 0837) Harper & Row: New York xiii + 407pp
◇ B98 ◇ REV Zbl 202.5 • ID 22526

NEKOLA, J. & NOVAK, VILEM [1984] *Basic operations with fuzzy sets from the point of view of fuzzy logic* (**P** 3081) IFAC Symp Fuzzy Inf, Knowl Repr & Decis. Anal;1983 Marseille 249-253
◇ B52 E72 ◇ ID 46735

NEKOLA, J. [1984] see CERNY, M.

NELSON, D. [1984] see ALMUKDAD, A.

NELSON, D. see Vol. I, VI for further entries

NELSON, JACK [1980] see BERGMANN, MERRIE

NELSON, JOHN O. [1964] *A question of entailment* (J 0125) Rev of Metaphysics 18*364-377
◇ A05 B46 ◇ REV JSL 38.668 • ID 09854

NEMETI, I. [1980] *Some constructions of cylindric algebra theory applied to dynamic algebras of programs* (J 2293) Comp Linguist & Comp Lang 14*43-65
◇ B75 G15 ◇ REV MR 83i:03099 Zbl 523 # 03054
• ID 35546

NEMETI, I. [1981] *Dynamic algebras of programs* (P 3165) FCT'81 Fund of Comput Th;1981 Szeged 281-290
◇ B75 ◇ REV MR 83i:08009 Zbl 511 # 68005 • ID 38546

NEMETI, I. [1982] *Nonstandard dynamic logic* (P 3738) Log of Progr;1981 Yorktown Heights 311-348
◇ B75 ◇ REV MR 84d:68013 Zbl 493 # 68032 • ID 38438

NEMETI, I. see Vol. I, III, V for further entries

NEPEJVODA, N.N. [1974] *The language Δ with a weak three-valued logic (Russian)* (J 0023) Dokl Akad Nauk SSSR 219*1325-1327
• TRANSL [1974] (J 0062) Sov Math, Dokl 15*1818-1821
◇ B50 ◇ REV MR 52 # 7848 Zbl 312 # 02018 • ID 18310

NEPEJVODA, N.N. [1982] *Constructive logics (Russian)* (C 3849) Modal & Relevant Log, Vyp 1 91-106
◇ B60 C90 C95 F50 ◇ REV Zbl 557 # 03039 • ID 46228

NEPEJVODA, N.N. [1983] *Notes concerning constructive implicative logics (Russian)* (S 0507) Vychisl Sist (Akad Nauk SSSR Novosibirsk) 96*87-108
◇ B55 F50 ◇ REV MR 86e:03058 Zbl 547 # 03036
• ID 43233

NEPEJVODA, N.N. see Vol. I, III, IV, V, VI for further entries

NEUBRUNN, T. & PULMANNOVA, S. [1983] *On compatability in quantum logics (Russian and Slovak summaries)* (J 0128) Acta Math Univ Comenianae (Bratislava) 42-43*153-168
◇ B51 G12 ◇ REV MR 85k:81017 Zbl 539 # 03045
• ID 41264

NEUBRUNN, T. see Vol. I, V for further entries

NEUMANN, H. (ED.) [1981] *Interpretations and foundations of quantum theory* (X 0876) Bibl Inst: Mannheim 144pp
◇ B51 G12 ◇ REV MR 85g:810005 Zbl 455 # 00026
• ID 45876

NEUMANN, H. see Vol. I, III, IV for further entries

NEUMANN VON, J. [1936] see BIRKHOFF, GARRETT

NEUMANN VON, J. [1956] *Probabilistic logics and the synthesis of reliable organisms from unreliable components* (C 0717) Automata Studies 43-98
• TRANSL [1974] (C 1902) Stud Th Automaten 57-121
◇ B48 ◇ REV MR 17.1040 • ID 09911

NEUMANN VON, J. see Vol. I, IV, V, VI for further entries

NEVELN, R.C. [1981] see ALPS, R.A.

NGUYEN CAT HO [1972] *Generalized algebras of Post and their applications to many-valued logics with infinitely long formulas* (J 0387) Bull Sect Logic, Pol Acad Sci 1/1*4-12
◇ B50 C75 G20 ◇ REV MR 50 # 12650 • ID 09939

NGUYEN CAT HO [1973] *Generalized Post algebras and their application to some infinitary many-valued logics* (J 0202) Diss Math (Warsaw) 107*72pp
◇ B50 C75 G20 ◇ REV Zbl 281 # 02057 • ID 29577

NICOD, J. [1924] *The logical problem of induction (French)* (X 1071) Alcan:Paris
• LAST ED [1961] (X 0840) Pr Univ France: Paris
◇ B48 ◇ REV MR 42 # 33 Zbl 242 # 00023 • REM For transl. see Nicod,J. 1930 • ID 46495

NICOD, J. see Vol. I, VI for further entries

NICOLAU, E. [1965] *Synthese einer Klasse von Automaten, die Operatoren in der n-wertigen Logik modellieren mit gegebenen Zeitfunktionen (Rumaenisch)* (J 0440) Bul Inst Politeh Bucuresti, Ser Mec 26*101
◇ B50 ◇ ID 41908

NIDDITCH, P.H. [1957] *Introductory formal logic of mathematics* (X 0927) Univ Tutorial Pr: Slough vii + 188pp
◇ B98 ◇ REV MR 19.723 Zbl 82.13 JSL 25.77 • ID 09944

NIDDITCH, P.H. see Vol. I for further entries

NIELAND, J.J.F. [1965] see BETH, E.W.

NIELAND, J.J.F. [1967] *Beth's tableau-method* (P 0683) Log & Founds of Sci (Beth);1964 Paris 19-38
◇ B05 B25 B50 F07 ◇ REV MR 39 # 1287 Zbl 203.9
• ID 09951

NIINILUOTO, I. [1971] *Can we accept Lehrer's inductive rule?* (J 0963) Ajatus (Helsinki) 33*254-265
◇ A05 B48 ◇ REV Zbl 242 # 02031 • ID 26349

NIINILUOTO, I. [1972] *Inductive systematization: Definition and a critical survey* (J 0154) Synthese 25*25-81
◇ A05 B48 ◇ REV Zbl 255 # 02021 • ID 28959

NIINILUOTO, I. [1973] *Mathematical logic (Finnish)* (X 3333) Unknown Publisher: See Remarks 94 + 58pp
◇ B98 ◇ ID 32404

NIINILUOTO, I. & TUOMELA, R. [1973] *Theoretical concepts and hypothetico-inductive inference* (X 0835) Reidel: Dordrecht x + 264pp
◇ A05 B48 ◇ REV Zbl 281 # 02003 • ID 32403

NIINILUOTO, I. [1975] see BOGDAN, R.J.

NIINILUOTO, I. [1976] see HINTIKKA, K.J.J.

NIINILUOTO, I. [1977] *On a k-dimensional System of inductive logic* (P 1703) Proc Bienn Meet Phil of Sci Ass;1976 Chicago 425-447
◇ B48 ◇ ID 32411

NIINILUOTO, I. & TUOMELA, R. (EDS.) [1978] *The logic and epistemology of scientific change* (J 0096) Acta Philos Fenn 30*1-461
◇ A05 B97 ◇ REV MR 82i:00024 • ID 80405

NIINILUOTO, I. [1979] see HILPINEN, R.

NIINILUOTO, I. [1979] see HINTIKKA, K.J.J.

NIINILUOTO, I. & SAARINEN, E. (EDS.) [1979] *Intensional logic: theory and applications* (J 0096) Acta Philos Fenn 35*301pp
◇ B45 ◇ REV MR 84e:03001 • ID 34341

NIINILUOTO, I. [1979] *Truthlikeness in first-order languages* (P 1705) Scand Logic Symp (4);1976 Jyvaeskylae 437-458
◇ A05 B48 ◇ REV MR 80h:03014 Zbl 402 # 03021
• ID 32413

NIINILUOTO, I. [1982] *Remarks on the logic of perception* (**P** 3800) Intens Log: Th & Appl;1979 Moskva 117-130 • TRANSL [1984] (**C** 4366) Modal & Intens Log & Primen Probl Metodol Nauk 329-340 ◇ B45 ◇ REV MR 85e:03045 Zbl 519 # 03018 • ID 37536

NIINILUOTO, I. [1983] *Inductive logic as a methodological research programme* (**C** 3811) Logic 20th Century 77-100 ◇ B48 ◇ REV MR 86a:03022 • ID 44700

NIINILUOTO, I. & ZYGMUNT, J. (EDS.) [1983] *Proceedings of the Finnish-Polish-Soviet logic conference* (**J** 0063) Studia Logica 42*117-378 ◇ B97 ◇ REV MR 85g:03007 Zbl 529 # 00014 • REM Held at Polanica Zdroj, Sept. 7-12, 1981 • ID 43478

NIINILUOTO, I. see Vol. I, III, V, VI for further entries

NIKITINA, I.G. [1977] *Intersection of set of all precomplete classes in countable-valued logic (Russian)* (**J** 2869) Vest Ser Vychisl Mat Kibern, Univ Moskva 1977/1*84-87 • TRANSL [1977] (**J** 3221) Moscow Univ Comp Math Cybern 1977/1*67-70 ◇ B50 G25 ◇ REV Zbl 448 # 03012 • ID 56612

NILSON, P.S. [1976] *Productive logic and the principle of self-contradiction* (**J** 0286) Int Logic Rev 14*135-136 ◇ A05 B60 ◇ ID 64215

NINOMIYA, T. [1982] see GOTO, M.

NINOMIYA, T. [1983] see GOTO, M.

NINOMIYA, T. see Vol. I for further entries

NISHIMURA, H. [1979] *Is the semantics of branching structures adequate for chronological modal logics?* (**J** 0122) J Philos Logic 8*469-475 ◇ B45 ◇ REV MR 80m:03033 Zbl 415 # 03016 • ID 53118

NISHIMURA, H. [1979] *Is the semantics of branching structures adequate for non-metric Ockhamist tense logics?* (**J** 0122) J Philos Logic 8*477-478 ◇ B45 ◇ REV MR 81i:03023 Zbl 415 # 03017 • ID 53119

NISHIMURA, H. [1979] *On the completeness of chronological logics with modal operators* (**J** 0068) Z Math Logik Grundlagen Math 25*487-496 ◇ B45 ◇ REV MR 81b:03022 Zbl 421 # 03013 • ID 53412

NISHIMURA, H. [1979] *Sequential method in propositional dynamic logic* (**J** 1431) Acta Inf 12*377-400 ◇ B75 ◇ REV MR 81c:03019 Zbl 407 # 03023 • ID 56189

NISHIMURA, H. [1980] *A preservation theorem for tense logic* (**J** 0068) Z Math Logik Grundlagen Math 26*331-335 ◇ B45 C20 C40 C90 ◇ REV MR 81i:03024 Zbl 452 # 03020 • ID 54084

NISHIMURA, H. [1980] *A study of some tense logics by Gentzen's sequential method* (**J** 0390) Publ Res Inst Math Sci (Kyoto) 16*343-353 ◇ B45 ◇ REV MR 82g:03028 Zbl 446 # 03013 • ID 56542

NISHIMURA, H. [1980] *Descriptively complete process logic* (**J** 1431) Acta Inf 14*359-369 ◇ B45 B75 ◇ REV MR 82b:68012 Zbl 431 # 68041 • ID 82359

NISHIMURA, H. [1980] *Interval logics with applications to study of tense and aspect in English* (**J** 0390) Publ Res Inst Math Sci (Kyoto) 16*417-459 ◇ B45 ◇ REV MR 82k:03021 Zbl 446 # 03012 • ID 56541

NISHIMURA, H. [1980] *Saturated and special models in modal model theory with applications to the modal and de re hierarchies* (**J** 0068) Z Math Logik Grundlagen Math 26*481-490 ◇ B45 C40 C50 C90 ◇ REV MR 81k:03023 Zbl 443 # 03018 • ID 56424

NISHIMURA, H. [1980] *Sequential method in quantum logic* (**J** 0036) J Symb Logic 45*339-352 ◇ B51 G12 ◇ REV MR 81h:03122 Zbl 437 # 03034 • ID 55902

NISHIMURA, H. [1981] *Arithmetical completeness in first-order dynamic logic for concurrent programs* (**J** 0390) Publ Res Inst Math Sci (Kyoto) 17*297-309 ◇ B75 ◇ REV MR 83i:68032 Zbl 466 # 68023 • ID 54995

NISHIMURA, H. [1981] *Model theory for tense logic: Saturated and special models with applications to the tense hierarchy* (**J** 0063) Studia Logica 40*89-98 ◇ B45 C40 C50 C90 ◇ REV MR 83g:03035 Zbl 469 # 03019 • ID 55147

NISHIMURA, H. [1981] *The semantical characterization of de dicto in continuous modal model theory* (**J** 0068) Z Math Logik Grundlagen Math 27*233-240 ◇ B45 C40 C90 ◇ REV MR 84c:03038 Zbl 464 # 03021 • ID 54611

NISHIMURA, H. [1982] *Propositional dynamic logic for concurrent programs* (**J** 0390) Publ Res Inst Math Sci (Kyoto) 18*233-250 ◇ B75 ◇ REV MR 84j:03063 Zbl 507 # 03006 • ID 34655

NISHIMURA, H. [1982] *Semantical analysis of constructiove PDL* (**J** 0390) Publ Res Inst Math Sci (Kyoto) 18*847-858 ◇ B75 ◇ REV MR 84k:03084 Zbl 504 # 03010 • ID 35021

NISHIMURA, H. [1983] *A cut-free sequential system for the propositional modal logic of finite chains* (**J** 0390) Publ Res Inst Math Sci (Kyoto) 19*305-316 ◇ B45 B75 ◇ REV MR 85g:03031 Zbl 524 # 03011 • ID 37592

NISHIMURA, H. [1983] *Hauptsatz for higher-order modal logic* (**J** 0036) J Symb Logic 48*744-751 ◇ B45 ◇ REV MR 86a:03063 Zbl 527 # 03004 • ID 37505

NISHIMURA, H. see Vol. V for further entries

NISHIMURA, I. [1962] *On a classification of intermediate propositional calculi (Japanese) (English summary)* (**J** 1171) Sci Rep Fac Educ Gifu Univ, Nat Sci 3*5-9 ◇ B55 ◇ REV MR 29 # 4670 • ID 22072

NISHIMURA, I. see Vol. VI for further entries

NISHIMURA, T. [1975] see NAGATA, M.

NISHIMURA, T. & OHYA, T. [1975] *The formal system for various 3-valued logics. I* (**J** 0350) Sci Rep Tokyo Kyoiku Daigaku Sect A 13*7-22 • REPR [1975] (**P** 3299) Progr Kiso Riron, Algor Okeru Shomei Ron;1973/74 Kyoto 87-110 ◇ B50 ◇ REV MR 53 # 5255 MR 58 # 5035 Zbl 339 # 02017 Zbl 376 # 02018 • REM Part II 1977 • ID 22887

NISHIMURA, T. & OHYA, T. [1977] *The formal system for various 3-valued logics. II* (**J** 0090) J Math Soc Japan 29*513-527 ◇ B50 ◇ REV MR 56 # 2786 Zbl 353 # 02007 • REM Part I 1975 • ID 50090

NISHIMURA, T. see Vol. I, III, IV, V, VI for further entries

NISTICO, G. [1984] see CATTANEO, G.

NOLA DI, A. [1981] see GERLA, G.

NOLA DI, A. see Vol. V for further entries

NOLAN, P.C. [1977] *A semantics model for imperatives* (**J** 0047) Notre Dame J Formal Log 18*79-84
◇ B60 ◇ REV Zbl 271 # 02037 • ID 64221

NOLL, H. [1977] see BERGMANN, E.

NOLL, H. see Vol. I for further entries

NOMURA, H. [1969] see KASAHARA, Y.

NOMURA, H. [1970] see KASAHARA, Y.

NOVAK, VILEM [1983] *Basic operations with fuzzy sets from the point of fuzzy logic* (**J** 3919) BUSEFAL 14*75-86
◇ B52 E72 ◇ REV Zbl 525 # 03012 • ID 38250

NOVAK, VILEM [1984] see NEKOLA, J.

NOVAK, VILEM [1984] see CERNY, M.

NOVAK, VILEM see Vol. V for further entries

NOVIKOV, P.S. [1949] *On regularity classes (Russian)* (**J** 0023) Dokl Akad Nauk SSSR 64*293-295
◇ B10 B45 ◇ REV MR 11.1 Zbl 38.150 JSL 14.255 • ID 10018

NOVIKOV, P.S. [1959] *Elements of mathematical logic (Russian)* (**X** 3709) Izdat Fiz-Mat Lit: Moskva 400pp
• TRANSL [1973] (**X** 0900) Vieweg: Wiesbaden 286pp (German) [1964] (**X** 1323) Oliver & Boyd: Edinburgh xi + 296pp (English) • REPR [1973] (**X** 2027) Nauka: Moskva 400pp
◇ B98 ◇ REV MR 22 # 5565 MR 50 # 1827 Zbl 288 # 02002 Zbl 90.8 JSL 30.356 JSL 31.672 • ID 22364

NOVIKOV, P.S. see Vol. I, III, IV, V, VI for further entries

NOVIKOV, V.F. & VITYAEV, E.E. [1981] *Paradoxicality of methods of induction (Russian)* (**S** 0507) Vychisl Sist (Akad Nauk SSSR Novosibirsk) 91*76-90
◇ B48 ◇ REV MR 84j:03052 Zbl 529 # 03005 • ID 34644

NOVOTNY, M. [1977] see KUDLEK, M.

NOVOTNY, M. see Vol. III, IV, V for further entries

NOWAKOWSKA, M. [1981] *A new theory of time: generation of time from fuzzy temporal relations* (**P** 4050) Appl Syst & Cybern;1980 Acapulco 2742-2747
• REPR [1981] (**J** 0387) Bull Sect Logic, Pol Acad Sci 10*56-62
◇ B52 ◇ REV MR 84b:00009 Zbl 465 # 94052 • ID 46306

NOWAKOWSKA, M. [1982] *Knowledge and induced subjective temporal systems* (**P** 3845) Conf Math Service of Man (2,Proc)(Feriet);1982 Las Palmas 531-537
◇ B45 ◇ REV Zbl 508 # 03013 • ID 36936

NOWAKOWSKA, M. see Vol. V for further entries

NOZAKI, A. [1981] *Completeness criteria for a set of delayed functions with or without nonuniform compositions* (**P** 2552) Conf Finite Algeb & Multi-Val Log;1979 Szeged 489-519
◇ B50 ◇ REV MR 83f:94057 Zbl 474 # 94039 • ID 40222

NOZAKI, A. see Vol. I, IV for further entries

NUEL JR., D. [1984] see BELNAP JR., N.D.

NURLYBAEV, A.N. [1976] *A majorant local algorithm of index 1 for the construction of a sum of minimal normal forms of a function of k-valued logic (Russian) (Kazakh summary)* (**J** 0403) Izv Akad Nauk Kazak SSR, Ser Fiz-Mat 1976/3*60-64,91
◇ B50 ◇ REV MR 57 # 12172 Zbl 354 # 02020 • ID 50160

NURLYBAEV, A.N. [1976] *Normal forms of k-valued logic (Russian)* (**C** 3585) Sb Rabot Mat Kibernetike, Vyp 1 56-68
◇ B50 ◇ REV MR 58 # 27309 • ID 76899

NURLYBAEV, A.N. [1976] *Simplification of normal forms of k-valued logic (Russian)* (**C** 3585) Sb Rabot Mat Kibernetike, Vyp 1 69-80
◇ B50 ◇ REV MR 58 # 27310 • ID 76903

NURLYBAEV, A.N. [1977] *A class of functions of three-valued logic (Russian)* (**C** 2557) Teor & Priklad Zad Mat & Mekh 252-260
◇ B50 ◇ REV MR 80d:03022 Zbl 429 # 94035 • ID 76902

NURLYBAEV, A.N. [1977] *A local algorithm of index 1 for constructing the sum of dead-end disjunctive normal forms for functions of k-valued logic (Russian)* (**J** 0199) Zh Vychisl Mat i Mat Fiz 17*1556-1563
• TRANSL [1977] (**J** 1049) USSR Comput Math & Math Phys 17/6*203-210
◇ B50 ◇ REV MR 58 # 9875 Zbl 393 # 03044 • ID 52706

NURLYBAEV, A.N. see Vol. I for further entries

NUTE, D.E. [1975] *Counterfactuals and the similarity of worlds* (**J** 0122) J Philos Logic 72*773-778
◇ B45 ◇ ID 28302

NUTE, D.E. [1978] *Algebraic semantics for conditional logics* (**J** 0302) Rep Math Logic, Krakow & Katowice 10*79-101
◇ B45 ◇ REV MR 81k:03029 Zbl 422 # 03005 • ID 53466

NUTE, D.E. [1978] *An incompleteness theorem for conditional logic* (**J** 0047) Notre Dame J Formal Log 19*634-636
◇ B45 ◇ REV MR 80a:03028 Zbl 314 # 02034 • ID 28303

NUTE, D.E. [1980] *Topics in conditional logic* (**X** 0835) Reidel: Dordrecht x + 164pp
◇ A05 B45 ◇ REV MR 84c:03039 Zbl 453 # 03016 JSL 47.713 • ID 54146

NUTE, D.E. [1981] *Essential formal semantics* (**X** 0854) Littlefield Adams: Totowa xiii + 186pp
◇ B98 ◇ REV MR 83e:03003 Zbl 494 # 03001 JSL 51.252 • ID 35269

NUTE, D.E. [1981] *Introduction to the issue on the theory of conditionals* (**J** 0122) J Philos Logic 10*127-147
◇ A05 B45 ◇ REV MR 84b:03013 Zbl 473 # 03011 • ID 55342

NUTE, D.E. [1985] *Permission* (**J** 0122) J Philos Logic 14*169-190
◇ B45 ◇ ID 42655

NUTE, D.E. see Vol. III for further entries

NUTTER, R.S. & SWARTWOUT, R.E. [1971] *A ternary logic minimization technique* (**P** 2007) Symp Th & Appl of Multi-Val Log Design;1971 Buffalo 112-123
◇ B50 B70 ◇ ID 41989

NUTTER JR., R.S. & RINE, D.C. & SWARTWOUT, R.E. [1974] *Equivalence and transformations for Post multivalued algebras* (**J** 0187) IEEE Trans Comp C-23*294-300
◇ B50 ◇ REV MR 53 # 12871 Zbl 296 # 06008 • ID 23131

OBERSCHELP, A. [1970] *Mengenlehre und Logik* (**S** 1747) Christiana Albertina (Kiel) 10*47-55
◇ B98 E98 ◇ ID 31198

OBERSCHELP, A. [1974] *Elementare Logik und Mengenlehre I* (**X** 0876) Bibl Inst: Mannheim 259pp
◇ B98 E98 ◇ REV MR 55 # 2466 Zbl 295 # 02001 • REM Part II 1978 • ID 31195

OBERSCHELP, A. [1975] see MUELLER, GERT H.

OBERSCHELP, A. [1978] *Elementare Logik und Mengenlehre II* (**X** 0876) Bibl Inst: Mannheim 229pp
◇ B98 E98 ◇ REV MR 58 # 108 Zbl 367 # 04001 • REM Part I 1974 • ID 31196

OBERSCHELP, A. [1983] see GLUBRECHT, J.-M.

OBERSCHELP, A. see Vol. I, III, IV, V for further entries

O'CARROLL, M.J. [1967] *A three-valued, non-levelled logic consistent for all self-reference* (**J** 0079) Logique & Anal, NS 10*173-178
◇ B50 ◇ REV MR 35 # 5296 Zbl 174.8 JSL 37.422 • ID 10061

O'CONNOR, D.J. [1953] see BASSON, A.H.

ODDIE, G. [1980] see CURRIE, G.

ODDIE, G. & TICHY, P. [1982] *The logic of ability, freedom and responsibility* (**J** 0063) Studia Logica 41*227-248
◇ B60 ◇ REV MR 86c:03025 Zbl 537 # 03006 • ID 43723

ODEN, G.C. [1982] see LOPES, L.L.

OHAMA, S. [1977] see MIURA, S.

OHAMA, S. [1984] *Conjunctive normal forms and weak modal logics without the axiom of necessity* (**J** 0047) Notre Dame J Formal Log 25*141-151
◇ B45 ◇ REV MR 85i:03053 Zbl 556 # 03015 • ID 42566

OHAMA, S. see Vol. I, VI for further entries

OHLBACH, H.-J. & WRIGHTSON, G. [1984] *Solving a problem in relevance logic with an automated theorem prover* (**P** 2633) Autom Deduct (7);1984 Napa 496-508
◇ B35 B46 ◇ REV MR 86b:03024 Zbl 547 # 03013 • ID 43199

OHLBACH, H.-J. see Vol. I for further entries

OHNISHI, M. [1957] see MATSUMOTO, K.

OHNISHI, M. [1959] see MATSUMOTO, K.

OHNISHI, M. [1961] *Gentzen decision procedures for Lewis's systems S2 and S3* (**J** 1770) Osaka Math J 13*125-137
◇ B25 B45 ◇ REV MR 24 # A1210 Zbl 106.4 JSL 40.468 • ID 10085

OHNISHI, M. [1961] *Von Wright-Anderson's decision procedures for Lewis's systems* S2 *and* S3 (**J** 1770) Osaka Math J 13*139-142
◇ B45 ◇ REV MR 24 # A2535 Zbl 106.4 JSL 40.469 • ID 10086

OHNISHI, M. [1963] see MATSUMOTO, K.

OHNISHI, M. [1964] see MATSUMOTO, K.

OHNISHI, M. [1982] *A new version of the Gentzen decision procedure for modal sentential calculus S5* (**J** 1508) Math Sem Notes, Kobe Univ 10*161-170
◇ B45 ◇ REV MR 83m:03027 Zbl 495 # 03011 • ID 35432

OHNISHI, M. see Vol. VI for further entries

OHRSTROM, P. [1984] *Buridan on interval semantics for temporal logic* (**J** 0079) Logique & Anal, NS 27*211-215
◇ A10 B45 ◇ REV MR 86f:03033 • ID 43955

OHYA, T. [1968] *Many valued logics extended to simple type theory* (**J** 0350) Sci Rep Tokyo Kyoiku Daigaku Sect A 9*260-270
◇ B50 ◇ REV MR 37 # 38 Zbl 172.8 • ID 10087

OHYA, T. [1975] see NISHIMURA, T.

OHYA, T. [1977] see NISHIMURA, T.

OHYA, T. see Vol. VI for further entries

OKABE, M. [1981] *An interpretation of Aristotle's modal syllogism* (**J** 0260) Ann Jap Ass Phil Sci 6*19-41
◇ A05 B45 ◇ REV MR 82i:03029 Zbl 459 # 03002 • ID 54442

OKADA, Y. [1984] see GOTO, M.

OKOL'NISHNIKOVA, E.A.A. [1976] *The distribution of types of functions of q-valued logic according to cardinality (Russian)* (**J** 0071) Met Diskr Analiz (Novosibirsk) 28*65-77,79
◇ B50 ◇ REV MR 57 # 5682 Zbl 414 # 94036 • ID 53099

OKOL'NISHNIKOVA, E.A.A. see Vol. I for further entries

OKUBO, T. [1980] see NAGAI, SHIGEO

OKURA, Y. [1982] see HASEGAWA, T.

OMEL'YANCHIK, V.I. [1982] *Understanding and the semantics of classical and modal logic (Russian)* (**C** 3813) Ponim Kak Log-Gnoseolog 259-269
◇ B45 ◇ REV MR 84k:03023 • ID 34961

OMER, I.A. [1980] *Minimal law explanation* (**J** 0170) Ratio (Frankfurt) 22*155-166
◇ B45 ◇ REV MR 83b:03019 • ID 35087

OMODEO, E.G. [1976] *The elimination of descriptions from A. Bressan's modal language* ML^{ν} *on which the logical calculus* MC^{ν} *is based* (**J** 0144) Rend Sem Mat Univ Padova 56*269-292
◇ B45 ◇ REV MR 58 # 10319 Zbl 383 # 03013 • ID 51995

OMODEO, E.G. [1980] *Three existence principles in a modal calculus without descriptions contained in A. Bressan's* MC^{ν} (**J** 0047) Notre Dame J Formal Log 21*711-727
◇ B45 ◇ REV MR 82b:03047 Zbl 423 # 03013 • ID 53525

OMODEO, E.G. see Vol. I, III, V for further entries

OMYLA, M. & SUSZKO, R. [1972] *Definitions in theories of kind W* (**J** 0387) Bull Sect Logic, Pol Acad Sci 1/3*14-19
◇ B60 ◇ REV MR 51 # 76 • ID 15203

OMYLA, M. & SUSZKO, R. [1972] *Descriptions in theories of kind W* (**J** 0387) Bull Sect Logic, Pol Acad Sci 1/3*8-13
◇ B60 ◇ REV MR 51 # 75 • ID 15205

OMYLA, M. [1976] *Translatability in non-Fregean theories* (**J** 0063) Studia Logica 35*127-138
◇ B45 ◇ REV MR 56 # 5262 Zbl 328 # 02015 • ID 64270

OMYLA, M. [1977] *Barcan formulas in SCI with quantifiers* (**J** 0387) Bull Sect Logic, Pol Acad Sci 6*171-176
◇ B60 ◇ REV MR 57 # 15970 Zbl 405 # 03007 • ID 54882

OMYLA, M. [1979] *Propositional quantifiers in non-Fregean theories* (**P** 2539) Frege Konferenz (1);1979 Jena 299-306
◇ B60 ◇ REV MR 82c:03035 • ID 76975

ONICESCU, O. [1971] *Principes de logique et de philosophie mathematique* (X 0871) Acad Rep Soc Romania: Bucharest 229pp
◇ A05 B98 ◇ REV MR 54 # 12435 Zbl 235 # 02005 • ID 27815

ONICESCU, O. see Vol. V for further entries

ONO, H. [1970] *Kripke models and intermediate logics* (J 0390) Publ Res Inst Math Sci (Kyoto) 6*461-476
◇ B55 ◇ REV MR 45 # 3163 Zbl 226 # 02025 • ID 10110

ONO, H. [1970] see HOSOI, T.

ONO, H. [1971] *On the finite model property for Kripke models* (J 0390) Publ Res Inst Math Sci (Kyoto) 7*85-93
◇ B55 ◇ REV MR 47 # 21 Zbl 249 # 02022 • ID 10111

ONO, H. [1972] see HOSOI, T.

ONO, H. [1972] *Some results on the intermediate logics* (J 0390) Publ Res Inst Math Sci (Kyoto) 8*117-130
◇ B55 ◇ REV MR 54 # 2428 Zbl 253 # 02022 • ID 24018

ONO, H. [1973] *A study of intermediate predicate logics* (J 0390) Publ Res Inst Math Sci (Kyoto) 8*619-649
◇ B55 ◇ REV MR 47 # 8255 Zbl 281 # 02033 • ID 29570

ONO, H. [1973] *Incompleteness of semantics for intermediate predicate logics. I.Kripke's semantics* (J 0081) Proc Japan Acad 49*711-713
◇ B55 ◇ REV MR 48 # 8205 Zbl 299 # 02026 • ID 10113

ONO, H. [1973] see HOSOI, T.

ONO, H. [1974] see NAGAI, SATORU

ONO, H. [1977] *On some intuitionistic modal logics* (J 0390) Publ Res Inst Math Sci (Kyoto) 13*687-722
◇ B45 ◇ REV MR 57 # 15971 Zbl 373 # 02026 • ID 30772

ONO, H. [1980] see NAKAMURA, A.

ONO, H. [1982] see KOMORI, Y.

ONO, H. & RAUSZER, C. [1982] *On an algebraic and Kripke semantics for intermediate logics* (P 3831) Universal Algeb & Appl;1978 Warsaw 431-438
◇ B55 ◇ REV MR 85g:03037 Zbl 528 # 03012 • ID 37630

ONO, H. [1983] *Model extension theorem and Craig's interpolation theorem for intermediate predicate logics* (J 0302) Rep Math Logic, Krakow & Katowice 15*41-58
◇ B55 C20 C40 C90 ◇ REV MR 84j:03060 Zbl 519 # 03016 • ID 34652

ONO, H. see Vol. I, III, IV, V, VI for further entries

ONO, K. [1968] *A remark on Peirce's rule in many-valued logics* (J 0111) Nagoya Math J 31*69-71
◇ B50 ◇ REV MR 36 # 6267 Zbl 162.12 • ID 90074

ONO, K. [1968] *On tabooistic treatment of proposition logics* (J 0081) Proc Japan Acad 44*291-293
◇ B55 ◇ REV MR 37 # 3907 Zbl 169.296 • ID 10132

ONO, K. see Vol. I, V, VI for further entries

ONYSZKIEWICZ, J. [1972] see MAREK, W.

ONYSZKIEWICZ, J. see Vol. III, V for further entries

OPPENHEIM, P. [1945] see HEMPEL, C.G.

OPPENHEIM, P. [1945] see HELMER, O.

OPPENHEIM, P. [1961] see BEDAU, H.

ORENDUFF, J. [1975] *A partially truth-functional modal calculus* (J 0079) Logique & Anal, NS 18*91-102
◇ B45 ◇ REV MR 56 # 5224 Zbl 347 # 02015 • ID 64284

OREVKOV, V.P. [1967] *The undecidability of a class of formulas containing just one single place predicate variable in modal calculus (Russian)* (S 0228) Zap Nauch Sem Leningrad Otd Mat Inst Steklov 4*168-173
• TRANSL [1967] (J 0521) Semin Math, Inst Steklov 4*67-70
◇ B45 D35 ◇ REV MR 41 # 1514 Zbl 153.316 • ID 10149

OREVKOV, V.P. (ED.) [1968] *The calculi of symbolic logic I (Russian)* (S 0066) Tr Mat Inst Steklov 98*5-202 • ERR/ADD ibid 121*167
• TRANSL [1968] (S 0055) Proc Steklov Inst Math 98*iv+229pp
◇ B97 F97 ◇ REV MR 43 # 4620 MR 49 # 18 • ID 37197

OREVKOV, V.P. see Vol. I, III, IV, VI for further entries

ORLOV, YU.F. [1978] *Group theory approach to logic -- the wave logic* (J 0455) Phil Naturalis 17*120-129
◇ B50 ◇ REV MR 80b:03024 • ID 77007

ORLOV, YU.F. [1981] *A quantum model of doubt* (P 3814) Int Conf Collective Phenomena (4);1981 Moskva 84-92
• ERR/ADD [1981] (P 4001) Int Conf Collective Phenomena (5);1981 Moskva 361-363
◇ B51 ◇ REV MR 84k:03024 • REM Transl. from Russian • ID 34964

ORLOV, YU.F. [1982] *The wave logic of consciousness: a hypothesis* (J 2736) Int J Theor Phys 21*37-53
◇ B60 ◇ REV MR 83b:81015 Zbl 504 # 46056 • ID 38413

ORLOVSKY, S.A. [1982] *Effective alternatives for multiple fuzzy preference relations* (P 3846) Cybern & Systems Research (6);1982 Wien 185-189
◇ B52 ◇ REV Zbl 533 # 90004 • ID 36579

ORLOVSKY, S.A. see Vol. V for further entries

ORLOWSKA, E. [1976] *The Gentzen style axiomatization of* ω^+*-valued logic* (J 0063) Studia Logica 35*433-445
◇ B50 ◇ REV MR 56 # 86 Zbl 361 # 02024 • ID 50659

ORLOWSKA, E. [1977] *The Herbrand theorem for* ω^+*-valued logic* (J 0014) Bull Acad Pol Sci, Ser Math Astron Phys 25*1069-1071
◇ B50 ◇ REV MR 58 # 5037 Zbl 369 # 02037 • ID 51337

ORLOWSKA, E. [1978] see MIRKOWSKA, G.

ORLOWSKA, E. [1978] *Resolution system for* ω^+*-valued logic* (J 0387) Bull Sect Logic, Pol Acad Sci 7*68-74
◇ B50 ◇ REV MR 80a:03032 Zbl 419 # 03006 • ID 53349

ORLOWSKA, E. [1980] *Program logic with quantifiable propositional variables* (J 1927) Prace Inst Podstaw Inf, Pol Akad Nauk 422*24pp
◇ B75 ◇ REV Zbl 533 # 03005 • ID 36531

ORLOWSKA, E. [1982] *Logic of vague concepts* (J 0387) Bull Sect Logic, Pol Acad Sci 11*115-126
◇ B52 ◇ REV MR 85d:03057 Zbl 511 # 03003 • ID 37382

ORLOWSKA, E. [1982] *Logic of vague concepts. Applications of rough sets* (J 1927) Prace Inst Podstaw Inf, Pol Akad Nauk 474*16pp
◇ B52 ◇ REV Zbl 525 # 03016 • ID 38254

ORLOWSKA, E. [1982] *Representation of temporal information* (**J** 0435) Int J Comput & Inf Sci 11*397-408
◇ B45 ◇ REV MR 85e:03046 Zbl 513 # 03017 • ID 37223

ORLOWSKA, E. [1982] *Semantics of vague concepts. Applications of rough sets* (**J** 1927) Prace Inst Podstaw Inf, Pol Akad Nauk 469*20pp
◇ B52 ◇ REV Zbl 525 # 03015 • ID 38253

ORLOWSKA, E. [1982] *Tense logic for nondeterministic time* (**J** 0387) Bull Sect Logic, Pol Acad Sci 11*127-133
◇ B45 ◇ REV MR 84k:03057 Zbl 536 # 03007 • ID 34995

ORLOWSKA, E. [1983] *On some extensions of dynamic logic* (**P** 3830) Logics of Progr & Appl;1980 Poznan 205-212
◇ B75 ◇ REV MR 84b:68001 Zbl 525 # 03007 • ID 38247

ORLOWSKA, E. [1984] *Logic of nondeterministic information* (**J** 1927) Prace Inst Podstaw Inf, Pol Akad Nauk 545*18pp
◇ B60 B75 ◇ REV Zbl 566 # 68037 • ID 48747

ORLOWSKA, E. & PAWLAK, Z. [1984] *Logical foundations of knowledge representation. I* (**J** 1927) Prace Inst Podstaw Inf, Pol Akad Nauk 108pp
◇ B45 ◇ REV Zbl 549 # 68090 • ID 43165

ORLOWSKA, E. & WIERZCHON, S.T. [1984] *Mechanical reasoning in fuzzy logics* (**J** 1927) Prace Inst Podstaw Inf, Pol Akad Nauk 551*24pp
◇ B52 ◇ REV Zbl 553 # 03014 • ID 43306

ORLOWSKA, E. [1984] *Modal logics in the theory of information systems* (**J** 0068) Z Math Logik Grundlagen Math 30*213-222
◇ B45 ◇ REV MR 85i:03054 Zbl 556 # 68053 • ID 42224

ORLOWSKA, E. [1984] *Semantical analysis of inductive reasoning* (**J** 1927) Prace Inst Podstaw Inf, Pol Akad Nauk 547*20pp
◇ B48 ◇ REV Zbl 556 # 03019 • ID 46148

ORLOWSKA, E. [1985] *A logic of indiscernibility relations* (**P** 4670) Comput Th (5);1984 Zaborow 177-186
◇ B60 ◇ ID 49733

ORLOWSKA, E. [1985] *Logic of indiscernibility relations* (**J** 1927) Prace Inst Podstaw Inf, Pol Akad Nauk 546*20pp
◇ B60 ◇ REV Zbl 566 # 03009 • ID 48733

ORLOWSKA, E. [1985] *Logic of indiscernibility relations (Russian summary)* (**J** 3417) Bull Pol Acad Sci, Math 33*475-485
◇ B60 ◇ ID 49731

ORLOWSKA, E. [1985] *Logic of nondeterministic information* (**J** 0063) Studia Logica 44*91-100
◇ B60 ◇ REV Zbl 575 # 03018 • ID 47526

ORLOWSKA, E. [1985] *Relative induction (Russian summary)* (**J** 3417) Bull Pol Acad Sci, Math 33*469-473
◇ B48 ◇ ID 49730

ORLOWSKA, E. [1985] *Semantics of vague concepts* (**P** 4180) Int Congr Log, Meth & Phil of Sci (7,Pap);1983 Salzburg 465-482
◇ B52 ◇ ID 47944

ORLOWSKA, E. [1985] *Semantics of nondeterministic possible worlds* (**J** 3417) Bull Pol Acad Sci, Math 33*453-458
◇ B48 B60 ◇ ID 49726

ORLOWSKA, E. see Vol. I, VI for further entries

ORNAGHI, M. [1983] see MIGLIOLI, P.A.

ORNAGHI, M. see Vol. I, III, IV, VI for further entries

ORSIC, M. [1972] *Multiple-valued universal logic modules* (**P** 2008) Symp Th & Appl of Multi-Val Log Design;1972 Buffalo 95-99
◇ B50 ◇ ID 42035

ORSIC, M. [1973] *Incremental curve generation using mutiple-valued logic* (**P** 2009) Int Symp Multi-Val Log (3);1973 Toronto 166-172
◇ B50 ◇ ID 42070

OSIPOV, G.S. & ZHOLONDZ', V.Y. [1983] *Pseudophysical logics in semiotic models* (**J** 0977) Izv Akad Nauk SSSR, Tekh Kibern 1983/3*84-99
• TRANSL [1983] (**J** 0522) Engin Cybern 21/3*75-89
◇ B60 ◇ ID 44189

OSIS, YA.YA. & SHAJTSANE, V.A. [1979] *Determination of diagnostic parameters of the checking of good conditions for a complex fuzzy system with a continuous process of functioning (Russian)* (**C** 3902) Met & Modeli Upravleniya & Kontrolya 9-16
◇ B52 ◇ REV Zbl 497 # 90026 • ID 36644

OSTAPKO, D.L. [1974] see CAIN, R.G.

OSTAPKO, D.L. see Vol. I for further entries

OTEPANOV, V.I. & VERKHOZIN, O.M. [1972] *Problems in logic (Russian)* (**X** 1006) Irkutsk Gos Univ: Irkutsk 73pp
◇ B98 E98 ◇ REV MR 50 # 1826 • ID 13931

OTTE, R. [1981] *A critique of Suppes' theory of probabilistic causality* (**J** 0154) Synthese 48*167-189
◇ A05 B48 ◇ REV MR 83g:03022 Zbl 476 # 03010 • ID 55522

OUELLET, R. [1981] *Inclusive first-order logic* (**J** 0063) Studia Logica 40*13-28
◇ B60 ◇ REV MR 83k:03034 Zbl 468 # 03043 • ID 55108

OVCHINNIKOV, P.G. [1985] *Finitely additive functions on extensions of quantum logics (Russian)* (**J** 0031) Izv Vyssh Ucheb Zaved, Mat (Kazan) 1985/5*74-75
◇ B51 ◇ ID 46561

OVCHINNIKOV, S.V. [1982] *On fuzzy relational systems* (**P** 3845) Conf Math Service of Man (2,Proc)(Feriet);1982 Las Palmas 566-568
◇ B52 ◇ REV Zbl 515 # 03012 • ID 37839

OVCHINNIKOV, S.V. [1982] *Social choice and Lukasiewicz logic* (**P** 4002) Int Symp Multi-Val Log (12);1982 Paris 163-166
◇ B50 ◇ REV Zbl 544 # 90003 • ID 41041

OVCHINNIKOV, S.V. see Vol. V for further entries

OVER, D. [1981] *Game theoretical semantics and entailment* (**J** 0063) Studia Logica 40*67-74
◇ B46 ◇ REV MR 82k:03022 Zbl 472 # 03017 • ID 55283

OVSIEVICH, B.L. [1965] *Some properties of symmetric functions of the three-valued logic (Russian)* (**J** 2320) Probl Peredachi Inf, Akad Nauk SSSR 1/1*57-64
◇ B50 ◇ REV MR 33 # 5404 Zbl 261 # 02010 • ID 30457

PABION, J.F. [1976] *Logique mathematique* (**X** 0859) Hermann: Paris xxxii+263pp
◇ B98 ◇ REV MR 56 # 15346 Zbl 345 # 02001 JSL 44.282 • ID 64331

PABION, J.F. [1977] see BECCHIO, D.

PABION, J.F. [1979] *Beth's tableaux for relevant logic* (**J** 0047) Notre Dame J Formal Log 20*891-899
◇ B46 ◇ REV MR 80k:03023 Zbl 439 # 03005 • ID 55997

PABION, J.F. see Vol. I, III, V, VI for further entries

PACHOLSKI, L. [1978] see MACINTYRE, A.

PACHOLSKI, L. see Vol. I, III, V for further entries

PACKARD, D.J. [1975] *A preference logic minimally complete for expected utility maximization* (**J** 0122) J Philos Logic 4*223-235
◇ B60 ◇ REV MR 58 # 151 Zbl 317 # 02031 • ID 29644

PACKARD, D.J. [1982] *Cyclical preference logic* (**J** 0472) Theory Decis 14*415-426
◇ B60 ◇ REV MR 84d:03034 Zbl 493 # 90004 • ID 34074

PADOA, A. [1912] *La logique deductive dans sa derniere phase de developpement* (**X** 0834) Gauthier-Villars: Paris 106pp
◇ A10 B98 ◇ REV FdM 43.98 • ID 22597

PADOA, A. see Vol. I, III, V for further entries

PAEPPINGHAUS, P. & WIRSING, M. [1983] *Nondeterministic three-valued logic: isotonic and guarded truth functions* (**J** 0063) Studia Logica 42*1-22
◇ B50 B75 ◇ REV MR 86c:03019 Zbl 537 # 68021 • ID 42340

PAEPPINGHAUS, P. see Vol. I, IV, VI for further entries

PAGE, E.W. [1972] see BELL, N.

PAHI, B. [1971] see APPLEBEE, R.C.

PAHI, B. [1972] *Maximal full matrices* (**J** 0047) Notre Dame J Formal Log 13*142-144
◇ B50 ◇ REV MR 46 # 8806 Zbl 188.12 • ID 10247

PAHI, B. [1973] *Necessity and some non-modal propositional calculi* (**J** 0047) Notre Dame J Formal Log 14*401-404
◇ B45 ◇ REV MR 51 # 7830 Zbl 258 # 02022 • ID 10250

PAHI, B. [1975] *Jankow-theorems for some implicational calculi* (**J** 0068) Z Math Logik Grundlagen Math 21*193-198
◇ B55 ◇ REV MR 51 # 10034 Zbl 323 # 02039 • ID 10252

PAHI, B. see Vol. I, IV for further entries

PALYUTIN, E.A. [1973] see ERSHOV, YU.L.

PALYUTIN, E.A. [1979] see ERSHOV, YU.L.

PALYUTIN, E.A. see Vol. I, III, IV, V for further entries

PAN, WEIJIE [1981] see CHEN, GUOQUAN

PAN, WEIJIE [1982] see CHEN, GUOQUAN

PANCZAKIEWICZ, M. [1970] *Mathematical logic (Polish)* (**S** 0544) Prace Mat Uniw Katowice 9*269pp
◇ B98 ◇ REV MR 50 # 12631 • ID 19525

PANCZAKIEWICZ, M. see Vol. I for further entries

PAPKE, W. & POSPELOV, D.A. & VAGIN, V.N. [1977] *Application of fuzzy logic in control systems* (**J** 2716) Found Control Eng, Poznan 2*153-160
◇ B52 ◇ REV Zbl 364 # 93007 • ID 50993

PARGETTER, R. [1977] see ELLIS, B.

PARGETTER, R. [1977] see DAVIDSON, B.

PARGETTER, R. [1982] see JACKSON, F.C.

PARIKH, R. [1978] *The completeness of propositional dynamic logic* (**P** 1707) Math Founds of Comput Sci (7);1978 Zakopane 64*403-415
◇ B75 ◇ REV MR 81b:03029 Zbl 392 # 03017 • ID 32419

PARIKH, R. [1981] see KOZEN, D.

PARIKH, R. [1981] see MEYER, A.R.

PARIKH, R. [1981] *Propositional dynamic logics of programs: A survey* (**P** 3497) Log of Progr;1979 Zuerich 102-144
◇ B75 ◇ REV MR 85g:03045 Zbl 468 # 68038 • ID 55121

PARIKH, R. [1983] *Propositional logics of programs: New directions* (**P** 3864) FCT'83 Found of Comput Th;1983 Borgholm 347-359
◇ B45 ◇ REV MR 86b:03034 Zbl 539 # 68016 • ID 41252

PARIKH, R. [1983] *The problem of vague predicates* (**C** 3834) Lang, Logic and Method 241-261
◇ B52 ◇ REV Zbl 502 # 03006 • ID 36895

PARIKH, R. [1984] *Logics of knowledge, games and dynamic logic* (**P** 4010) Found of Softw Tech & Th Comput Sci (4);1984 Bangalore 202-222
◇ B75 ◇ REV Zbl 555 # 03008 • ID 45082

PARIKH, R. & RAMANUJAM, R. [1985] *Distributed processes and the logic of knowledge* (**P** 4571) Log of Progr;1985 Brooklyn 256-268
◇ B60 ◇ REV Zbl 565 # 68025 • ID 49199

PARIKH, R. [1985] see CHANDRA, A.K.

PARIKH, R. see Vol. I, III, IV, V, VI for further entries

PARIS, J.B. [1978] see MACINTYRE, A.

PARIS, J.B. see Vol. I, III, IV, V, VI for further entries

PARKER, J.H. [1984] *Social logics and moral logics* (**J** 2694) Cybern & Syst 15*209-218
◇ B60 ◇ ID 49418

PARKS, R.Z. [1972] *A note on R-Mingle and Sobocinski's three-valued logic* (**J** 0047) Notre Dame J Formal Log 13*227-228
◇ B50 ◇ REV MR 45 # 4949 Zbl 234 # 02010 • ID 10290

PARKS, R.Z. [1972] *Classes and change* (**J** 0122) J Philos Logic 1*162-169
◇ B45 ◇ REV MR 53 # 2636 Zbl 246 # 02019 • ID 21524

PARKS, R.Z. [1972] *Dependence of some axioms of Rose* (**J** 0068) Z Math Logik Grundlagen Math 18*189-192
◇ B50 ◇ REV MR 48 # 57 Zbl 253 # 02016 • ID 10292

PARKS, R.Z. [1972] see MEYER, R.K.

PARKS, R.Z. [1972] see CHIDGEY, J.R.

PARKS, R.Z. [1972] *On formalizing Aristotle's theory of modal syllogisms* (**J** 0047) Notre Dame J Formal Log 13*385-386
◇ A05 A10 B45 ◇ REV MR 46 # 1557 Zbl 238 # 02024 JSL 38.519 • ID 10291

PARKS, R.Z. & RESCHER, N. [1973] *Possible individuals, trans-world identity, and quantified modal logic* (**J** 0097) Nous, Quart J Phil 7*330-350
◇ A05 B45 ◇ REV MR 56 # 15355 • ID 77748

PARKS, R.Z. [1974] *Semantics for contingent identity systems* (**J** 0047) Notre Dame J Formal Log 15*333-334
◇ B45 ◇ REV MR 50 # 6797 Zbl 276 # 02011 • ID 10295

PARKS, R.Z. & SMITH, T. [1974] *The inadequacy of Hughes and Cresswells semantics for the CI systems* (**J** 0047) Notre Dame J Formal Log 15*331-332
◇ B45 ◇ REV MR 50 # 1844 Zbl 276 # 02010 • ID 10293

PARKS, R.Z. [1975] *Indefinability of necessity in* $R_{\rightarrow}$ (**C** 1852) Entailment - Log of Relev & Nec 99-100
◇ B46 ◇ REV MR 53 # 10542 JSL 42.311 • ID 44475

PARKS, R.Z. [1975] *Solution of $T_{\rightarrow}$* (**C** 1852) Entailment - Log of Relev & Nec 89-90
◇ B46 ◇ REV MR 53 # 10542 JSL 42.311 • ID 44473

PARKS, R.Z. [1976] *Investigations into quantified modal logic. I* (**J** 0063) Studia Logica 35*109-125
◇ B45 ◇ REV MR 55 # 10243 Zbl 332 # 02028 • ID 64386

PARKS, R.Z. [1977] *The theory of limp implication* (**J** 1893) Relevance Logic Newslett 2*125-128
◇ B46 ◇ REV Zbl 377 # 02022 • ID 51739

PARKS, R.Z. see Vol. I, VI for further entries

PARRY, W.T. [1939] *Modalities in the survey system of strict implication* (**J** 0036) J Symb Logic 4*137-154
◇ B46 ◇ REV MR 1.131 Zbl 23.99 JSL 5.37 FdM 65.1105 • ID 10299

PARRY, W.T. [1976] *Abstract: Entailment: Analytic implication vs. the entailment of Anderson & Belnap* (**J** 1893) Relevance Logic Newslett 1*11-15
◇ B46 ◇ REV Zbl 349 # 02017 • ID 64393

PARRY, W.T. [1976] *Comparison of entailment theories* (**J** 1893) Relevance Logic Newslett 1*16-26
◇ B46 ◇ REV Zbl 349 # 02018 • ID 64392

PARRY, W.T. see Vol. I, VI for further entries

PARSONS, C. [1966] *A propositional calculus intermediate between the minimal calculus and the classical* (**J** 0047) Notre Dame J Formal Log 7*353-358 • ERR/ADD ibid 10*336
◇ B55 ◇ REV MR 38 # 3133 MR 39 # 6732 Zbl 169.297 • ID 10305

PARSONS, C. [1970] *Axiomatization of Aaqvist's CS-logics* (**J** 0105) Theoria (Lund) 36*43-64
◇ B60 ◇ REV MR 45 # 3157 Zbl 223 # 02017 • ID 10309

PARSONS, C. [1982] *Intensional logic in extensional language* (**J** 0036) J Symb Logic 47*289-328
◇ A05 B45 ◇ REV MR 84e:03009 Zbl 561 # 03007 • ID 34348

PARSONS, C. see Vol. I, IV, V, VI for further entries

PARSONS, T. [1967] *Grades of essentialism in quantified modal logic* (**J** 0097) Nous, Quart J Phil 1*181-191
◇ B45 ◇ ID 27088

PARSONS, T. [1978] *Nuclear and extranuclear properties, Meinong, and Leibniz* (**J** 0097) Nous, Quart J Phil 12*137-151
◇ A05 B45 ◇ REV MR 58 # 16103 • ID 77158

PARSONS, T. [1979] *Referring to nonexistent objects* (**J** 0472) Theory Decis 11*95-110
◇ A05 B45 ◇ REV MR 81a:03007 • ID 77157

PASINI, L. [1980] see MONTAGNA, F.

PASINI, L. see Vol. III for further entries

PASQUINELLI, A. [1957] *Introduzione alla logica simbolica* (**X** 1262) Einaudi: Torino x + 120pp
◇ B98 ◇ REV Zbl 85.243 JSL 32.105 • ID 22424

PASSY, S. & TINCHEV, T. [1985] *Quantifiers in combinatory PDL: completeness, definability, incompleteness* (**P** 4647) FCT'85 Fund of Comput Th;1985 Cottbus 512-519
◇ B75 ◇ ID 49018

PASSY, S. see Vol. IV for further entries

PATERSON, M.S. [1981] see BERMAN, F.

PATERSON, M.S. see Vol. I, IV for further entries

PATRYAS, W. [1975] *Concretizationand entailment (Comments on L. Nowak's article on "Idealizational laws and explanation".)* (**J** 0079) Logique & Anal, NS 18*183-187
◇ A05 B46 ◇ REV MR 53 # 7719 Zbl 363 # 02010 • ID 50837

PATT, Y.N. [1972] see COOPER, G.A.

PATT, Y.N. [1973] *Nonlinearity: A new necessary condition for logical completeness in k-valued logic* (**P** 3244) Hawaii Int Conf Syst Sci (6);1973 Honolulu 246-248
◇ B50 ◇ REV Zbl 374 # 02011 • ID 51535

PATT, Y.N. [1975] see ELLOZY, H.A.

PATT, Y.N. [1977] *Independent necessary conditions for functional completeness in m-valued logic* (**J** 0047) Notre Dame J Formal Log 18*318-320
◇ B50 ◇ REV MR 56 # 5218 Zbl 323 # 02028 • ID 24232

PATT, Y.N. see Vol. IV for further entries

PAULI, T. (ED.) [1970] *Logic and value* (**X** 0882) Univ Filos Foeren: Uppsala vi + 247pp
◇ A05 B97 ◇ REV MR 48 # 3691 • ID 70231

PAULI, T. [1970] *Some different kinds of entailment relations (Swedish)* (**C** 0735) Logic & Value (Dahlquist) 241-247
◇ B46 ◇ REV MR 50 # 9548 • ID 23606

PAULOS, J.A. [1976] *A model-theoretic semantics for modal logic* (**J** 0047) Notre Dame J Formal Log 17*465-468
◇ B45 C07 C90 ◇ REV MR 56 # 2789 Zbl 314 # 02033 • ID 18327

PAULOS, J.A. [1979] *A model-theoretic account of confirmation* (**J** 0047) Notre Dame J Formal Log 20*451-457
◇ A05 B48 ◇ REV MR 80g:03006 Zbl 351 # 02039 • ID 52643

PAULOS, J.A. [1981] *Probabilistic, truth-value, and standard semantics and the primacy of predicate logic* (**J** 0047) Notre Dame J Formal Log 22*11-16
◇ A05 B10 B48 ◇ REV MR 82b:03051 Zbl 416 # 03012 • ID 56466

PAULOS, J.A. see Vol. I, III for further entries

PAUN, G. [1983] *An impossibility theorem for indicators aggregation* (**J** 2720) Fuzzy Sets Syst 9*205-210
◇ B52 ◇ REV MR 84b:90030 Zbl 503 # 90029 • ID 36676

PAUN, G. see Vol. I, IV, VI for further entries

PAVELKA, J. [1977] *On L-fuzzy semantic* (**P** 2896) Algeb Method & Anwendgn in Automatenth;1976 Weissig 12-25
◇ B52 ◇ REV MR 58 # 5190 • ID 77176

PAVELKA, J. [1979] *On fuzzy logic. I: Many-valued rules of inference. III: Semantical completeness of some many-valued propositional calculi* (**J** 0068) Z Math Logik Grundlagen Math 25*45-52,447-464
◇ B52 ◇ REV MR 80j:03038 Zbl 435 # 03020 Zbl 446 # 03016 • REM Part II 1979 • ID 55781

PAVELKA, J. [1979] *On fuzzy logic. II. Enriched residuated lattices and semantics of propositional calculi* (**J** 0068) Z Math Logik Grundlagen Math 25*119-134
◇ B52 G25 ◇ REV MR 80j:03038b Zbl 446 # 03015 • REM Parts I,III 1979 • ID 56544

PAVILENIS, R.I. [1984] see ERSHOV, YU.L.

PAWLAK, Z. [1970] see MOSTOWSKI, A.WLODZIMIERZ

PAWLAK, Z. [1984] see ORLOWSKA, E.

PAWLAK, Z. see Vol. I, IV, V for further entries

PAWLOWSKY, V. & TRILLAS, E. [1982] *On rejection connectives* (**P** 3870) Congr Catala de Log Mat (1);1982 Barcelona 99-100
◇ B60 ◇ REV MR 84i:03003 • ID 47583

PAWLOWSKY, V. see Vol. I for further entries

PEACOCKE, C. [1980] *Causal modalities and realism* (**C** 3680) Platts: Refer Truth & Reality 41-68
◇ B46 B65 ◇ REV JSL 48.208 • ID 47412

PEACOCKE, C. [1981] *Are vague predicates incoherent?* (**J** 0154) Synthese 46*121-141
◇ A05 B52 ◇ REV MR 82j:03005 • ID 77188

PEACOCKE, C. see Vol. VI for further entries

PEARCE, G. [1981] see HARPER, W.L.

PEARL, J. [1976] *On the complexity of deriving imprecise assertions* (**P** 2011) Int Symp Multi-Val Log (6);1976 Logan 266
◇ B52 ◇ ID 35900

PEDRYCZ, W. [1978] see CZAJA-POSPIECH, D.

PEDRYCZ, W. [1980] *On the use of a fuzzy Lukasiewicz logic for fuzzy control* (**J** 0141) Arch Autom & Telemech 25*301-314
◇ B52 ◇ REV Zbl 463 # 93007 • ID 54590

PEDRYCZ, W. [1980] see CZAJA-POSPIECH, D.

PEDRYCZ, W. [1981] see CZOGALA, E.

PEDRYCZ, W. [1982] see CZOGALA, E.

PEDRYCZ, W. [1982] *Some aspects of fuzzy decision-making* (**J** 1429) Kybernetes 11*297-301
◇ B52 ◇ REV Zbl 494 # 90008 • ID 36661

PEDRYCZ, W. [1983] see LAARHOVEN VAN, P.J.M.

PEDRYCZ, W. [1983] see HIROTA, K.

PEDRYCZ, W. [1983] *Fuzzy relational equations with generalized connectives and their applications* (**J** 2720) Fuzzy Sets Syst 10*185-201
◇ B52 ◇ REV MR 85c:03008 Zbl 525 # 04004 • ID 37501

PEDRYCZ, W. [1983] see CZOGALA, E.

PEDRYCZ, W. [1983] *Some applicational aspects of fuzzy relational equations in system analysis* (**J** 1743) Int J Gen Syst 9*125-132
◇ B52 ◇ REV Zbl 521 # 93005 • ID 37479

PEDRYCZ, W. [1984] see DYCKHOFF, H.

PEDRYCZ, W. [1985] *Applications for fuzzy relational equations for methods of reasoning in presence of fuzzy data* (**J** 2720) Fuzzy Sets Syst 16*163-175
◇ B52 ◇ ID 47597

PEDRYCZ, W. see Vol. I, V for further entries

PEIRCE, C.S. [1883] *A theory of probable inference. Note b. The logic of relatives* (**C** 1491) Stud in Logic (J. Hopkins Univ) 187-203
◇ A05 B48 ◇ ID 28619

PEIRCE, C.S. see Vol. I for further entries

PEKLO, B.T. [1967] *The logical inference from "it does not hold that x ought to be", to "it holds that x ought not to be"* (**J** 0147) An Univ Bucuresti, Acta Logica 10*175-177
◇ B45 ◇ REV MR 38 # 2016 Zbl 248 # 02026 • ID 64430

PEKLO, B.T. [1968] *Deontic and alethic modalities* (**J** 0147) An Univ Bucuresti, Acta Logica 11*127-132
◇ B45 ◇ REV MR 43 # 4628 Zbl 227 # 02011 • ID 27356

PEKLO, B.T. [1971] *Analytical sentences and normative or evaluating logics* (**J** 0147) An Univ Bucuresti, Acta Logica 14*41-49
◇ B45 ◇ REV MR 47 # 6445 Zbl 315 # 02034 • ID 10350

PEKLO, B.T. [1974] *Sind die deontischen Funktoren distributiv?* (**J** 0047) Notre Dame J Formal Log 15*301-311
◇ B45 ◇ REV MR 50 # 4250 Zbl 281 # 02011 • ID 10351

PEKLO, B.T. [1975] *Erweiterte deontische Logik (EDL). Ein Versuch um weitere formale Ausbildung der deontischen Modallogik* (**J** 0047) Notre Dame J Formal Log 16*71-80
◇ B45 ◇ REV MR 50 # 12665 Zbl 293 # 02023 • ID 10352

PEKLO, B.T. see Vol. I for further entries

PELEG, D. [1984] see HAREL, D.

PELEG, D. [1985] see HAREL, D.

PELEG, D. see Vol. V for further entries

PELLETIER, F.J. [1977] see MORGAN, C.G.

PELLETIER, F.J. [1984] *The not-so-strange modal logic of indeterminacy* (**J** 0079) Logique & Anal, NS 27*415-422
◇ B45 ◇ ID 45531

PENA, L. [1980] *The philosophical relevance of a contradictorial system of logic: Ap* (**P** 3673) Int Symp Multi-Val Log (10);1980 Evanston 238-252
◇ A05 B53 ◇ REV MR 83m:94032 Zbl 541 # 03012 • ID 41376

PENA, L. [1981] *Prenexation, comparatives, and non-Archimedean infinite-valued fuzzy logic* (**P** 3705) Int Symp Multi-Val Log (11);1981 Oklahoma City & Norman 168-174
◇ B52 ◇ REV MR 83m:94033 Zbl 541 # 03013 • ID 41377

PENA, L. [1983] *(Quasi)transitive algebras* (**P** 4054) Int Symp Multi-Val Log (13);1983 Kyoto 129-135
◇ B50 ◇ REV Zbl 559 # 03040 • ID 46908

PENA, L. [1984] *A neo-Fregean (onto)logical fuzzy framework* (**P** 3621) Frege Konferenz (2);1984 Schwerin 253-262
◇ B52 ◇ REV MR 86e:03026 • ID 45362

PENA, L. [1984] *Identity, fuzziness and noncontradiction* (**J** 0097) Nous, Quart J Phil 18*227-259
◇ B52 ◇ REV MR 85f:03026 • ID 40685

PENA, L. see Vol. V for further entries

PENG, GUOZHONG [1982] see DENG, JULONG

PENZOV, YU.E. [1968] *Elements of mathematical logic and set theory (Russian)* (**X** 0958) Saratov Univ: Saratov 143pp
◇ B98 E98 ◇ REV MR 53 # 86 • ID 16550

PENZOV, YU.E. see Vol. V for further entries

PEREZ LARAUDOGOITIA, Y. [1985] *Modal interpretation of quantum mechanics (Spanish) (English summary)* (**J** 4604) Theoria, Ser 2 (San Sebastian) 1*235-251
◇ B45 B51 ◇ ID 48965

PERFIL'EVA, I.G. [1979] *The construction of a hypercontinuum of precomplete classes of a countably many-valued logic (Russian)* (J 0052) Probl Kibern 35*29-44,208
◇ B50 ◇ REV MR 80m:03055 Zbl 458 # 03006 • ID 77219

PERFIL'EVA, I.G. [1983] *Representation of functions in $P_{\aleph_0}$ by superpositions of one-place functions and addition (Russian)* (J 0087) Mat Zametki (Akad Nauk SSSR) 34*727-733
• TRANSL [1983] (J 1044) Math Notes, Acad Sci USSR 34*855-859
◇ B50 ◇ REV MR 85i:03072 Zbl 538 # 03021 • ID 41467

PERKOWSKA, E. [1971] *Herbrand theorem for theories based on m-valued logics (Russian summary)* (J 0014) Bull Acad Pol Sci, Ser Math Astron Phys 19*893-899
◇ B50 ◇ REV MR 48 # 58 Zbl 243 # 02015 • ID 10375

PERKOWSKA, E. [1972] *On algorithmic m-valued logics (Russian summary)* (J 0014) Bull Acad Pol Sci, Ser Math Astron Phys 20*717-719
◇ B75 ◇ REV MR 48 # 59 Zbl 256 # 02025 • ID 10376

PERKOWSKA, E. [1972] *On algorithmic m-valued logic* (J 0387) Bull Sect Logic, Pol Acad Sci 1/3*28-37
◇ B75 ◇ REV MR 50 # 12651 Zbl 256 # 02025 • ID 22039

PERZANOWSKI, J. & WRONSKI, A. [1971] *The deduction theorems for the system T of Feys-Wright (Polish summary)* (S 0458) Zesz Nauk, Prace Log, Uniw Krakow 6*11-14
◇ B45 ◇ REV MR 47 # 17 • ID 10386

PERZANOWSKI, J. [1972] *The first list of the deduction theorems characteristic for several modal calculi formalized after the manner of Lemmon* (J 0387) Bull Sect Logic, Pol Acad Sci 1/4*21-31
◇ B45 ◇ REV MR 58 # 27342 • ID 77231

PERZANOWSKI, J. [1973] *The deduction theorems for the modal propositional calculi formalized after the manner of Lemmon I* (J 0302) Rep Math Logic, Krakow & Katowice 1*1-12
◇ B45 ◇ REV MR 49 # 4754 Zbl 278 # 02016 • ID 10389

PERZANOWSKI, J. [1975] *On M-fragments and L-fragments of normal modal propositional logics* (J 0302) Rep Math Logic, Krakow & Katowice 5*63-72
◇ B45 ◇ REV MR 57 # 9486 Zbl 328 # 02012 • ID 21902

PERZANOWSKI, J. [1975] *On homogenous fragments of normal modal propositional logics* (J 0387) Bull Sect Logic, Pol Acad Sci 4*44-51
◇ B45 ◇ REV MR 52 # 13335 • ID 21797

PERZANOWSKI, J. see Vol. I, V, VI for further entries

PESCHEL, K. [1984] *Argumentationsfelder als Grundlage einer Argumentationslogik* (C 4026) Unters Log & Methodol, Vol 1 1-23
◇ B60 ◇ REV Zbl 546 # 03001 • ID 41137

PESCHEL, K. see Vol. I for further entries

PETCU, A. [1968] *The definition of the trivalent Lukasiewicz algebras by three equations* (J 0060) Rev Roumaine Math Pures Appl 13*247-250
◇ B50 ◇ REV MR 38 # 5591 Zbl 155.15 • ID 10391

PETERMANN, U. [1983] *On algorithmic logic with partial operations* (P 3830) Logics of Progr & Appl;1980 Poznan 213-223
◇ B75 ◇ REV MR 84b:68001 Zbl 519 # 68047 • ID 36730

PETERS, F.E. [1974] *Einfuehrung in mathematische Methoden der Informatik* (X 0876) Bibl Inst: Mannheim 342pp
◇ B70 B98 D80 D98 ◇ REV MR 53 # 2499 Zbl 319 # 94001 • ID 64464

PETERS, S. [1979] *A truth-conditional formulation of Karttunen's account of presupposition* (J 0154) Synthese 40*301-316
◇ A05 B50 ◇ REV MR 80f:03012 • ID 77240

PETERSEN, U. [1980] *Die logischen Grundlagen der Dialektik* (X 2162) Fink: Muenchen
◇ A05 B53 ◇ ID 42876

PETERSEN, U. [1983] *What do the antinomies teach us? Elements of a dialectical approach towards the contradictions of comprehension* (J 3783) J Non-Classical Log (Univ Campinas) 2*67-97
◇ B53 E30 E70 ◇ REV Zbl 556 # 03005 • ID 46146

PETERSON, P.L. [1979] *On the logic of "few", "many", and "most"* (J 0047) Notre Dame J Formal Log 20*155-179
◇ B60 C80 ◇ REV MR 80g:03019 Zbl 299 # 02012 • ID 52390

PETERSON, P.L. [1985] *Higher quantity syllogisms* (J 0047) Notre Dame J Formal Log 26*348-360
◇ B60 ◇ ID 47530

PETERSON, T. [1980] *The law of time conversion: an algebraic property of time adverbs* (J 1517) Theor Linguist 7*111-119
◇ B45 ◇ REV MR 83d:03023 • ID 35173

PETKOV, P.P. [1971] *On the question of the possibility of introducing disjunction on the lower levels of A.A. Markov's hierarchy of mathematical logic (Russian)* (J 0137) C R Acad Bulgar Sci 24*851-853
◇ B60 F50 ◇ REV MR 45 # 1726 Zbl 242 # 02036 • ID 10430

PETKOV, P.P. [1984] *The uniqueness of syntactical analysis for some calculi that are similar to Post's calculus. A generalization of the cut elimination theorem for classical propositional calculus (Russian)* (P 4117) Conf Math Log (Markov);1980 Sofia 73-88
◇ B50 F05 ◇ ID 46576

PETKOV, P.P. see Vol. VI for further entries

PETROV, S. [1979] *Hegel's thesis of contradictory truths* (J 0286) Int Logic Rev 17-18*69-76
◇ A05 B53 ◇ ID 42880

PETROV, V.V. [1984] see ERSHOV, YU.L.

PFEIFFER, H. [1982] see COHEN, L.J.

PFEIFFER, H. [1985] *Some system of predicate tense logic* (J 0068) Z Math Logik Grundlagen Math 31*557-568
◇ B45 ◇ ID 47807

PFEIFFER, H. see Vol. V, VI for further entries

PHILIPP, P. [1981] *Ein System der deontischen Logik II. Bedingte Normen* (J 0424) Wiss Z Univ Halle-Wittenberg, Math-Nat Reihe 30*81-88
◇ B45 ◇ REV MR 84k:03059a • REM Part III 1983 • ID 34997

PHILIPP, P. [1983] *Ein System der deontischen Logik III. Handlungen im Bereich von Normen* (J 0424) Wiss Z Univ Halle-Wittenberg, Math-Nat Reihe 32*47-55
◇ B45 ◇ REV MR 84k:03059b • REM Part II 1981 • ID 34998

PICHAT, E. [1973] see CABANES, A.

PIECZKOWSKI, A. [1964] *A set of axioms for the system Q_f of factorial implication* (J 0014) Bull Acad Pol Sci, Ser Math Astron Phys 12*141-142
◇ B45 ◇ REV MR 29 # 2175 • ID 10472

PIECZKOWSKI, A. [1964] *On the systems Q and Q_f* (J 0014) Bull Acad Pol Sci, Ser Math Astron Phys 12*137-139
◇ B60 ◇ REV MR 29 # 2174 • ID 10471

PIECZKOWSKI, A. [1966] *The axiomatic system of the factorial implication (Polish and Russian summaries)* (J 0063) Studia Logica 18*41-64
◇ B45 ◇ REV MR 34 # 4124 Zbl 306 # 02024 • ID 28364

PIECZKOWSKI, A. [1968] *The efficient implication (Polish and Russian summaries)* (J 0063) Studia Logica 23*7-23
◇ B45 ◇ REV MR 39 # 3977 Zbl 342 # 02008 Zbl 356 # 02020 • ID 10473

PIECZKOWSKI, A. [1971] *On the definitive implication (Polish and Russian summaries)* (J 0063) Studia Logica 27*101-116
◇ B60 ◇ REV MR 46 # 1558 Zbl 266 # 02009 • ID 10475

PIECZKOWSKI, A. [1975] *Causal implications of Jaskowski* (J 0063) Studia Logica 34*169-185
◇ B45 ◇ REV MR 52 # 5340 Zbl 308 # 02030 • ID 18336

PIECZKOWSKI, A. see Vol. I, IV, VI for further entries

PIERSANTINI, C. [1985] see GHILARDI, S.

PIETARINEN, J. [1966] see HINTIKKA, K.J.J.

PIETARINEN, J. [1972] *Lawlikeness, analogy, and inductive logic* (J 0096) Acta Philos Fenn 26*143pp
◇ A05 B48 ◇ REV MR 57 # 15944 Zbl 248 # 02013 • ID 25432

PIETERS, J. [1972] *Et en meme temps* (J 0079) Logique & Anal, NS 15*583-591
◇ B45 ◇ REV MR 48 # 65 Zbl 268 # 02016 • ID 10490

PIETERS, J. see Vol. V for further entries

PILSWORTH, B.W. [1979] see BALDWIN, J.F.

PILSWORTH, B.W. [1980] see BALDWIN, J.F.

PILSWORTH, B.W. [1981] see BALDWIN, J.F.

PINKAL, M. [1983] see BALLMER, T.T.

PINKAL, M. [1983] *Towards a semantics of precisation* (C 3095) Approaching Vagueness 13-57
◇ B52 ◇ REV MR 85j:03003 • ID 45370

PINKAVA, V. [1975] *Some further properties of the π-logics* (P 1805) Int Symp Multi-Val Log (5,Proc);1975 Bloomington 20-26
◇ B50 ◇ REV MR 58 # 16140 • ID 77309

PINKAVA, V. [1976] *"Fuzzification" of binary and finite multivalued logical calculi* (J 1741) Int J Man-Mach Stud 8*717-730
◇ B52 ◇ REV Zbl 347 # 02010 • ID 64520

PINKAVA, V. [1978] *On a class of functionally complete multi-valued logical calculi* (J 0063) Studia Logica 37*205-212
◇ B50 ◇ REV MR 58 # 10304 Zbl 381 # 03018 • ID 51886

PINKAVA, V. [1979] *On some manipulative properties of the π-algebras* (P 3003) Int Symp Multi-Val Log (9);1979 Bath 139-142
◇ B50 G25 ◇ REV MR 80m:03056 • ID 77308

PINKAVA, V. [1980] *On potential tautologies in k-valued calculi and their assessment by means of normal forms* (P 2936) Fuzzy Sets;1980 Durham 77-86
◇ B50 ◇ REV MR 81m:03030 • ID 77307

PINKAVA, V. [1980] see KOHOUT, L.J.

PINKAVA, V. [1981] *On Sheffer functions in k-valued logical calculi* (P 2552) Conf Finite Algeb & Multi-Val Log;1979 Szeged 537-545
◇ B50 ◇ REV MR 83m:03034 Zbl 476 # 03030 • ID 55542

PINKAVA, V. see Vol. I for further entries

PINTER, C. [1976] see COSTA DA, N.C.A.

PINTER, C. [1980] *The logic of inherent ambiguity* (P 3006) Brazil Conf Math Log (3);1979 Recife 253-262
◇ B53 ◇ REV MR 82b:03059 Zbl 462 # 03005 • ID 54512

PINTER, C. see Vol. I, III, V, VI for further entries

PIOCHI, B. [1978] *Matrici adeguate per calcoli generali predicativi (English summary)* (J 3285) Boll Unione Mat Ital, V Ser, A 15*66-76
◇ B50 ◇ REV MR 58 # 21595 Zbl 385 # 03057 • ID 52164

PIOCHI, B. [1980] *Nota su matrici adeguate per calcoli generali predicativi* (J 3285) Boll Unione Mat Ital, V Ser, A 17*271-273
◇ B50 ◇ REV MR 81i:03034 Zbl 429 # 03047 • ID 53877

PIOCHI, B. see Vol. I for further entries

PIROG-RZEPECKA, K. [1966] *A propositional calculus in which expressions are loosing their sense (Polish) (Russian and English summaries)* (J 0063) Studia Logica 18*139-164
◇ B60 ◇ REV Zbl 294 # 02005 • ID 64540

PIROG-RZEPECKA, K. [1968] *The proof of the non-existence of a finite matrix adequate for the sentential calculus in which expressions become meaningless (Polish and Russian summaries)* (J 0063) Studia Logica 22*57-59
◇ B60 ◇ REV MR 38 # 2007 Zbl 309 # 02009 • ID 64541

PIROG-RZEPECKA, K. [1973] see MADUCH, M.

PIROG-RZEPECKA, K. & SLUPECKI, J. [1973] *An extension of the algebra of sets (Polish and Russian summaries)* (J 0063) Studia Logica 31*7-37
◇ B50 E70 ◇ REV MR 47 # 8301 Zbl 277 # 02017 • ID 12500

PIROG-RZEPECKA, K. [1973] *Post normal forms for the system W (Polish) (English summary)* (S 1454) Zesz Nauk Wyz Szk Ped Mat, Opole 13*85-109
◇ B60 ◇ REV MR 54 # 2414 Zbl 314 # 02018 • ID 24004

PIROG-RZEPECKA, K. [1976] *A theory of deduction systems strongly adequate in relation to the sentential calculus W (Polish) (English summary)* (S 1454) Zesz Nauk Wyz Szk Ped Mat, Opole 19(Logika Mat)*103-123
◇ B60 ◇ REV MR 56 # 2813 Zbl 357 # 02046 • ID 50394

PIROG-RZEPECKA, K. [1976] see HALKOWSKA, K.

PIROG-RZEPECKA, K. [1976] *The problem of completeness of a certain sentential calculus system in different meanings of the term "complete" (Polish) (English summary)* (S 1454) Zesz Nauk Wyz Szk Ped Mat, Opole 19*95-101
◇ B50 ◇ REV MR 56 # 87 Zbl 359 # 02010 • ID 50558

PIROG-RZEPECKA, K. [1978] see HALKOWSKA, K.

PIROG-RZEPECKA, K. [1985] see MORAWIEC, A.

PIROG-RZEPECKA, K. see Vol. I, III for further entries

PIRON, C. [1970] see JAUCH, J.M.

PIRON, C. [1977] *On the logic of quantum logic* (**J** 0122) J Philos Logic 6*481-484
◇ A05 B51 G12 ◇ REV Zbl 373 # 02029 • ID 51493

PITOWSKY, I. [1982] *Substitution and truth in quantum logic* (**J** 0153) Phil of Sci (East Lansing) 49*380-401
◇ B51 ◇ REV MR 84h:03140 • ID 34326

PITOWSKY, I. [1985] *On the status of statistical inferences* (**J** 0154) Synthese 63*233-247
◇ B48 ◇ ID 47222

PITTS, W. [1943] see MCCULLOCH, W.S.

PIZIAK, R. [1974] see HERMAN, L.

PIZZI, C. [1977] *Boethius' thesis and conditional logic* (**J** 0122) J Philos Logic 6*283-302
◇ A05 A10 B45 ◇ REV MR 58 # 27343 Zbl 368 # 02031 • ID 51248

PIZZI, C. [1983] *La definizione delle modalita fisiche nella logica dell'implicazione consequenziale* (**P** 3829) Atti Incontri Log Mat (1);1982 Siena 391-395
◇ B45 ◇ REV MR 84k:03006 Zbl 521 # 03009 • ID 37065

PIZZI, C. see Vol. IV for further entries

PIZZORNO, B. [1978] *Some logical aspects of the drama of Galileo* (**J** 1068) Physis 20*271-283
◇ A10 B45 ◇ REV MR 81j:03033 • ID 77351

PKHAKADZE, N.M. [1976] *On Gentzen's formulation of a bimodal system (Russian)* (**P** 2566) All-Union Symp Log & Method of Sci (7);1976 Kiev
◇ B45 ◇ ID 90058

PKHAKADZE, N.M. [1982] see DARDZHANIYA, G.K.

PLA I CARRERA, J. [1979] *A formalization of the Lewis system S1 without rules of substitution* (**J** 2840) Stochastica, Univ Politec Barcelona 3/1*39-45
◇ B45 G25 ◇ REV MR 81a:03019 Zbl 426 # 03027 • ID 53623

PLA I CARRERA, J. see Vol. I, III, IV, V, VI for further entries

PLAISTED, D.A. [1984] *A low level language for obtaining decision procedures for classes of temporal logics* (**P** 2989) Log of Progr; 1983 Pittsburgh 403-420
◇ B45 ◇ ID 44567

PLAISTED, D.A. see Vol. I, IV, VI for further entries

PLANAVERGNE, G. [1975] *Deduction et trivalence* (**J** 0079) Logique & Anal, NS 18*51-58
◇ B50 ◇ REV MR 58 # 137 Zbl 327 # 02016 • ID 64544

PLANTINGA, A. [1974] *The nature of necessity* (**X** 0815) Clarendon Pr: Oxford ix + 255pp
◇ B45 ◇ REV MR 57 # 70 • ID 77354

PLATO VON, J. [1982] *The significance of the ergodic decomposition of stationary measures for the interpretation of probability* (**J** 0154) Synthese 53*419-432
◇ B48 ◇ REV MR 84f:03021 Zbl 498 # 60001 • ID 34437

PLEDGER, K.E. [1972] *Modalities of systems containing S3* (**J** 0068) Z Math Logik Grundlagen Math 18*267-283
◇ B45 ◇ REV MR 53 # 2637 Zbl 251 # 02024 • ID 10535

PLEDGER, K.E. [1975] *Some extensions of S3* (**J** 0047) Notre Dame J Formal Log 16*271-272
◇ B45 ◇ REV MR 51 # 66 Zbl 301 # 02022 • ID 10536

PLEDGER, K.E. [1980] *Location of some modal systems* (**J** 0047) Notre Dame J Formal Log 21*683-684
◇ B45 ◇ REV MR 81k:03024 Zbl 419 # 03008 • ID 55936

PLIUSKEVICIENE, A. [1985] *Generalized disjunction and existence properties for the logic of provability (Russian)* (**J** 3939) Mat Logika Primen (Akad Nauk Litov SSR) 1985/4*16-24
◇ B35 B45 ◇ ID 49019

PLIUSKEVICIENE, A. see Vol. I, VI for further entries

PLIUSKEVICIUS, R. [1968] *Kanger's version of the predicate calculus with symbols for not everywhere defined functions (Russian)* (**S** 0228) Zap Nauch Sem Leningrad Otd Mat Inst Steklov 8*211-224
• TRANSL [1968] (**J** 0521) Semin Math, Inst Steklov 8*103-109
◇ B10 B60 ◇ REV MR 45 # 1727 Zbl 175.273 • ID 10543

PLIUSKEVICIUS, R. [1981] *Symmetric dynamic logics (Russian) (English summary)* (**J** 3939) Mat Logika Primen (Akad Nauk Litov SSR) 1*61-108
◇ B75 ◇ REV Zbl 521 # 03018 • ID 37068

PLIUSKEVICIUS, R. [1982] *Symmetrization of a regular and inverse propositional dynamic logic (Russian) (English and Lithuanian summaries)* (**J** 3939) Mat Logika Primen (Akad Nauk Litov SSR) 2*48-74
◇ B75 ◇ REV MR 85k:03019a Zbl 572 # 03008 • ID 45140

PLIUSKEVICIUS, R. [1982] *Two theorems for symmetric propositional dynamic logic (Russian) (English and Lithuanian summaries)* (**J** 3939) Mat Logika Primen (Akad Nauk Litov SSR) 2*75-80
◇ B75 ◇ REV MR 85k:03019b Zbl 516 # 03005 • ID 37246

PLIUSKEVICIUS, R. [1983] *Some disjunction properties for propositional classical and constructive dynamic logic (Russian) (English and Lithuanian summaries)* (**J** 3939) Mat Logika Primen (Akad Nauk Litov SSR) 3*61-100
◇ B75 F07 ◇ REV MR 85g:03046 Zbl 572 # 03009 • ID 43884

PLIUSKEVICIUS, R. [1984] *On sequential form of functional dynamic logics* (**J** 2095) Fund Inform, Ann Soc Math Pol, Ser 4 7*357-358
◇ B75 ◇ REV MR 86d:68050 Zbl 575 # 03012 • ID 44614

PLIUSKEVICIUS, R. [1985] *A conservative extension of the quantified modal logic S5 (Russian) (English and Lithuanian summaries)* (**J** 3939) Mat Logika Primen (Akad Nauk Litov SSR) 1985/4*25-38,137
◇ B45 ◇ ID 49020

PLIUSKEVICIUS, R. see Vol. VI for further entries

PLOTKIN, J.M. [1975] see HICKIN, K.K.

PLOTKIN, J.M. see Vol. I, III, IV, V, VI for further entries

PLOTNIKOV, A.M. [1982] see CHUPAKHIN, I.YA.

PLUMWOOD, V. [1982] see BRADY, R.T.

PNUELI, A. [1977] *The temporal logic of programs* (**P** 3572) IEEE Symp Found of Comput Sci (18);1977 Providence 46-57
◇ B45 B75 ◇ REV MR 58 # 19311 • ID 82500

PNUELI, A. [1979] see MANNA, Z.

PNUELI, A. [1981] see BEN-ARI, M.

PNUELI, A. [1981] see HAREL, D.

PNUELI, A. [1982] see BEN-ARI, M.

PNUELI, A. [1982] see HAREL, D.

PNUELI, A. [1983] see HAREL, D.

PNUELI, A. [1983] see BEN-ARI, M.

PNUELI, A. [1984] see HAREL, D.

PNUELI, A. [1984] see KOREN, T.

PNUELI, A. [1985] *In transition from global to modular temporal reasoning about programs* (**P** 4621) Log & Models of Concurrent Syst;1984 La Colle-sur-Loup 123-144
◇ B35 B45 B75 ◇ ID 49386

PNUELI, A. [1985] *Linear and branching structures in the semantics and logics of reactive systems* (**P** 4628) Automata, Lang & Progr (12);1985 Nafplion 15-32
◇ B35 B45 B75 ◇ ID 49524

PNUELI, A. see Vol. IV for further entries

PODEWSKI, K.-P. [1982] see COHEN, L.J.

PODEWSKI, K.-P. see Vol. III, V for further entries

POERN, I. [1974] *Some basic concepts of action* (**C** 1936) Log Th & Semant Anal (Kanger) 93-101
◇ A05 B45 ◇ REV Zbl 301 # 02010 • ID 64562

POERN, I. [1982] *Meaning and intension* (**P** 3800) Intens Log: Th & Appl;1979 Moskva 273-281
◇ A05 B45 ◇ REV MR 84k:03060 • ID 35009

POERN, I. [1984] *Deontic detachment* (**J** 0387) Bull Sect Logic, Pol Acad Sci 13*60-63
◇ B45 ◇ ID 45752

POERN, I. [1985] see JONES, A.J.I.

POESCHEL, R. [1973] *Postsche Algebren von Funktionen ueber einer Familie endlicher Mengen* (**J** 0068) Z Math Logik Grundlagen Math 19*37-74
◇ B50 G20 ◇ REV MR 49 # 10545 Zbl 274 # 02031 • ID 19595

POESCHEL, R. [1973] *Postsche Algebren von Funktionen ueber einer Familie endlicher Mengen* (**J** 1670) Mitt Math Ges DDR 1973/2-3*111-118
◇ B50 G20 ◇ REV Zbl 273 # 02042 • ID 64564

POESCHEL, R. [1974] *Die funktionale Vollstaendigkeit von Funktionenklassen ueber einer Familie endlicher Mengen* (**J** 0068) Z Math Logik Grundlagen Math 20*537-550
◇ B50 ◇ REV MR 52 # 7849 Zbl 304 # 02026 • ID 10580

POESCHEL, R. [1975] *The number of maximal classes of functions closed with respect to superposition over a finite family of finite sets (Russian) (English summary)* (**J** 0040) Kibernetika, Akad Nauk Ukr SSR 1975/5*43-48,1975/6*149
• TRANSL [1975] (**J** 0021) Cybernetics 11*713-719,1006-1007
◇ B50 ◇ REV MR 58 # 27472a Zbl 317 # 08001 • ID 64566

POESCHEL, R. [1975] *Zur Charakterisierung spezieller superpositionsabgeschlossener Funktionenklassen, die der Durchschnitt einer absteigenden Kette sind, durch invariante Relationen (English and Russian summaries)* (**J** 0129) Elektr Informationsverarbeitung & Kybern 11*703-708
◇ B50 E07 ◇ REV MR 53 # 2626 Zbl 323 # 02024 • ID 21515

POESCHEL, R. [1977] *Funktionale Vollstaendigkeit in mehrwertigen Logiken* (**C** 3233) Probl & Ergebn Math 1974/75 18/N*25-27
◇ B50 ◇ REV Zbl 446 # 03014 • ID 56543

POESCHEL, R. [1979] see KALUZHNIN, L.A.

POGORZELSKI, W.A. & SLUPECKI, J. [1962] *On mathematical proof (Polish)* (**X** 1166) PZW: Warsaw 128pp
◇ B98 ◇ REV JSL 31.284 • ID 31970

POGORZELSKI, W.A. [1964] *The deduction theorem for Lukasiewicz many-valued propositional calculi (Polish and Russian summmaries)* (**J** 0063) Studia Logica 15*7-23
◇ B50 ◇ REV MR 33 # 5475 Zbl 292 # 02014 JSL 40.605 • ID 10587

POGORZELSKI, W.A. [1966] *A note on models of two-valued implicational-negational logic in many-valued Lukasiewicz's logics (Polish)* (**J** 0481) Acta Univ Wroclaw 52*93-96
◇ B50 ◇ ID 31972

POGORZELSKI, W.A. & PRUCNAL, T. [1974] *Introduction to mathematical logic. Part I. Elements of the algebra of propositional logic (Polish)* (**X** 1425) Univ Slaski: Katowice 36pp
◇ B98 G05 ◇ REV MR 52 # 10361 • ID 21667

POGORZELSKI, W.A. [1981] *Classical calculus of quantifiers. Outline of the theory (Polish)* (**X** 1034) PWN: Warsaw 1981*228pp
◇ B98 ◇ REV MR 83e:03004 • ID 35270

POGORZELSKI, W.A. see Vol. I, III, VI for further entries

POGOSYAN, EH.M. [1975] *Comparative characteristics of matching inductors (Russian) (Armenian summary)* (**J** 0346) Dokl Akad Nauk Armyan SSR 60*129-132
◇ B48 ◇ REV MR 52 # 10379 Zbl 327 # 94007 • ID 77404

POGOSYAN, EH.M. [1975] *Inductive inference with feedback (Russian) (Armenian summary)* (**J** 0346) Dokl Akad Nauk Armyan SSR 60*193-197
◇ B48 D15 ◇ REV MR 56 # 8345 Zbl 327 # 94008 • ID 77403

POGOSYAN, EH.M. see Vol. I, IV for further entries

POINCARE, H. [1902] *Du role de l'intuition et de la logique en mathematiques* (**P** 1484) Int Congr Math (2);1900 Paris 115-130
◇ A05 B98 ◇ ID 38663

POINCARE, H. see Vol. I, V for further entries

POIRIER, R. [1952] *Logique et modalite du point de vue organique* (**X** 0859) Hermann: Paris 113pp
◇ A05 B45 ◇ REV MR 14.527 Zbl 49.243 • ID 10605

POIRIER, R. [1969] *Sur les logiques de la modalite* (**J** 0147) An Univ Bucuresti, Acta Logica 12*85-108
◇ B45 ◇ REV MR 44 # 3848 Zbl 248 # 02012 • ID 10606

POKRIEFKA, M.L. [1985] see HENDRY, H.E.

POLIFERNO, M.J. [1961] *Decision algorithms for some functional calculi with modality* (**J** 0079) Logique & Anal, NS 4*138-153
◇ B25 B45 ◇ REV JSL 32.244 • REM Cf. "Correction to a paper on modal logic", ibid 1964, 7*32-33 • ID 10611

POLLOCK, J.L. [1966] *Model theory and modal logic* (**J** 0079) Logique & Anal, NS 9*313-317
◇ B45 C90 ◇ REV MR 38 # 39 Zbl 192.31 • ID 10622

POLLOCK, J.L. [1966] *The paradoxes of strict implication* (**J** 0079) Logique & Anal, NS 9*180-196
◇ A05 B46 ◇ REV MR 34 # 4103 Zbl 154.3 • ID 10621

POLLOCK, J.L. [1967] *Basic modal logic* (**J** 0036) J Symb Logic 32*355-365
◇ B45 ◇ REV MR 43 # 7306 Zbl 149.244 • ID 10623

POLLOCK, J.L. [1967] *The logic of logical necessity* (**J** 0079) Logique & Anal, NS 10*307-323
◇ B45 ◇ REV MR 37 # 42 Zbl 166.251 • ID 10624

POLLOCK, J.L. [1969] *Introduction to symbolic logic* (**X** 0818) Holt Rinehart & Winston: New York xii + 241pp
◇ B98 ◇ REV JSL 40.101 • ID 10626

POLLOCK, J.L. [1976] *Subjunctive reasoning* (**X** 0835) Reidel: Dordrecht xi + 255pp
◇ A05 B45 ◇ REV Zbl 345 # 02016 JSL 46.170 • ID 29776

POLLOCK, J.L. [1981] *A refined theory of counterfactuals* (**J** 0122) J Philos Logic 10*239-266
◇ A05 B45 ◇ REV MR 82m:03019 Zbl 473 # 03017 • ID 55348

POLLOCK, J.L. [1981] *Indicative conditionals and conditional probability* (**C** 4140) Ifs 249-256
◇ B48 ◇ REV MR 83a:03003 • ID 47744

POLLOCK, J.L. [1983] *A theory of direct inference I. A classical theory of direct inference* (**J** 0472) Theory Decis 15*29-95
◇ B45 ◇ REV MR 84m:03033 Zbl 528 # 62002 • ID 35737

POLLOCK, J.L. [1984] *Foundations for direct inference* (**J** 0472) Theory Decis 17*221-255
◇ B48 ◇ REV MR 86f:03039 Zbl 559 # 62004 • ID 44671

POLLOCK, J.L. see Vol. I, V for further entries

POLYA, G. [1948] *On patterns of plausible inference* (**C** 4679) Stud & Essays to Courant 277-288
◇ A05 B48 ◇ REV MR 9.262 Zbl 34.155 • ID 24897

POLYA, G. see Vol. I for further entries

POMPER, G. [1979] see ARMSTRONG, J.R.

PONASSE, D. [1967] *Logique mathematique. Elements de base: calcul propositionnel, calcul des predicats* (**X** 1172) Office Central Lib (OCDL): Paris 164pp
◇ B98 ◇ REV MR 39 # 31 Zbl 153.5 JSL 35.579 • ID 22069

PONASSE, D. [1973] *Mathematical logic* (**X** 0836) Gordon & Breach: New York x + 126pp
◇ B98 ◇ REV MR 49 # 8805 Zbl 252 # 02001 JSL 41.790 • ID 10640

PONASSE, D. see Vol. I, III, V for further entries

PONCELET, G. [1976] *Sur la logique productive (P.L)* (**J** 0286) Int Logic Rev 14*137-141
◇ A05 B60 ◇ REV Zbl 349 # 02024 • ID 64619

POPA, C.P. [1977] *An axiomatic system for a logical theory of liberty and constraint (Romanian)* (**S** 1613) Probl Logic (Bucharest) 7*215-228
◇ B45 ◇ REV MR 80j:03031 • ID 77437

POPOV, A.I. [1959] *Introduction to mathematical logic (Russian)* (**X** 0938) Leningrad Univ: Leningrad 108pp
◇ B98 ◇ REV MR 22 # 3674 Zbl 97.244 • ID 10646

POPOV, P.S. [1951] *Dialectic and the subject matter of formal logic (Russian)* (**J** 2871) Vopr Fil, Moskva 1951/1*210-218 • TRANSL [1952] (**C** 1993) Form Logik & Dialektik 118-130 (German)
◇ A05 B53 ◇ REV JSL 17.124 • ID 41830

POPOV, S.V. [1977] *Undecidable interval arithmetic (Russian) (English Summary)* (**S** 2651) Prepr Inst Prikl Mat, Akad Nauk SSSR 95*55pp
◇ B55 D35 ◇ REV MR 58 # 10389 • ID 77439

POPOV, S.V. [1981] *Nondecidable intermediate calculus (Russian)* (**J** 0003) Algebra i Logika 20*654-706,728 • TRANSL [1981] (**J** 0069) Algeb and Log 20*424-461
◇ B55 D35 ◇ REV MR 84h:03103 Zbl 528 # 03028 JSL 50.1081 • ID 34293

POPOV, S.V. see Vol. I, III, VI for further entries

POPOV, V.A. [1973] see MOKLYAK, N.G.

POPOV, V.A. & SKIBENKO, I.T. [1974] *Ueber die Komplementaritaet der Aequivalenzklassen k-wertiger logischer Funktionen (Russian)* (**J** 0474) Avtom Vychis Tekh, Akad Nauk Latv SSR 1974/5*15-16
◇ B50 ◇ REV Zbl 292 # 02012 • ID 64623

POPOV, V.A. & SKIBENKO, I.T. [1977] *Self-complementary types of the algebra of multivalued functions for self-dual transformations (Russian) (English summary)* (**J** 0040) Kibernetika, Akad Nauk Ukr SSR 1977/5*30-32 • TRANSL [1977] (**J** 0021) Cybernetics 13*666-669
◇ B50 ◇ REV MR 58 # 5038 Zbl 375 # 02051 • ID 51628

POPOV, V.A. & SKIBENKO, I.T. [1979] *Combinatorial approach to the classification of k-valued logical functions (Russian)* (**J** 0040) Kibernetika, Akad Nauk Ukr SSR 1979/6*15-18 • TRANSL [1979] (**J** 0021) Cybernetics 15*789-793
◇ B50 ◇ REV MR 81j:05016 Zbl 466 # 94033 • ID 82518

POPOV, V.A. [1982] *Gentzen-type calculi for some intuitionistic modal logics* (**P** 3876) Mat & Mat Obrazov (11);1982 Sl"nchev Bryag 274-280
◇ B45 ◇ REV Zbl 521 # 03044 • ID 37080

POPOV, V.A. see Vol. I for further entries

POPOV, V.M. [1984] *Semantical and syntactical analysis of an implicative-negative fragment of the system RM (Russian)* (**C** 4366) Modal & Intens Log & Primen Probl Metodol Nauk 192-198
◇ B46 ◇ ID 46847

POPOV, V.M. see Vol. I for further entries

POPOVICI, N. [1968] *On the consensus theory for the trivalent argument functions* (**J** 0070) Bull Soc Sci Math Roumanie, NS 12*103-109
◇ B50 ◇ REV MR 40 # 1264 Zbl 176.282 • ID 10647

POPPELBAUM, W.J. [1976] see GAINES, B.R.

POPPER, K.R. [1940] *What is dialectic?* (**J** 0094) Mind 49*403-426
◇ A05 B53 ◇ ID 42881

POPPER, K.R. [1954] *Degree of confirmation* (**J** 0013) Brit J Phil Sci 5*143-149 • ERR/ADD ibid 5*359
◇ A05 B48 ◇ REV MR 16.376 JSL 20.304 • ID 42513

POPPER, K.R. [1985] *The nonexistence of probabilistic inductive support* (**P** 4180) Int Congr Log, Meth & Phil of Sci (7,Pap);1983 Salzburg 303-318
◇ B48 ◇ ID 48109

POPPER, K.R. see Vol. I, VI for further entries

PORAT, D.I. [1969] *Three-valued digital systems* (J 1077) IEEE Proc 116*947-954
◇ B50 ◇ ID 41955

POREBSKA, M. & WRONSKI, A. [1975] *A characterization of fragments of the intuitionistic propositional logic* (J 0302) Rep Math Logic, Krakow & Katowice 4*39-42
◇ B55 F50 ◇ REV MR 51 # 12488 Zbl 318 # 02027 • ID 17277

POREBSKA, M. [1981] see KABZINSKI, J.K.

POREBSKA, M. [1985] *Interpolation for fragments of intermediate logics* (J 0387) Bull Sect Logic, Pol Acad Sci 14*79-83
◇ B55 C40 ◇ ID 48967

POREBSKA, M. see Vol. VI for further entries

PORTE, J. [1958] *Recherches sur les logiques modales* (P 0576) Raisonn en Math & Sci Exper;1955 Paris 117-126
◇ B45 ◇ REV MR 21 # 4900 Zbl 88.247 JSL 25.288 • ID 10663

PORTE, J. [1972] *La logique mathematique et le calcul mecanique* (S 0889) Notas Logica Mat 8*105pp
◇ B98 D98 ◇ REV MR 51 # 5253 Zbl 279 # 02001 JSL 24.70 • ID 17471

PORTE, J. [1979] *The Ω-system and the L-system of modal logic* (J 0047) Notre Dame J Formal Log 20*915-920
◇ B45 ◇ REV MR 80h:03030 Zbl 394 # 03031 • ID 56194

PORTE, J. [1980] *A research in modal logics* (J 0079) Logique & Anal, NS 23*3-34
◇ B45 ◇ REV MR 82a:03021 Zbl 452 # 03015 • ID 54079

PORTE, J. [1980] *Congruences in Lemmon's S0.5* (J 0047) Notre Dame J Formal Log 21*672-678
◇ B45 ◇ REV MR 82b:03048 Zbl 426 # 03021 • ID 55934

PORTE, J. [1981] *Notes on modal logics* (J 0079) Logique & Anal, NS 24*399-406
◇ B45 ◇ REV MR 83g:03019 Zbl 494 # 03010 • ID 35997

PORTE, J. [1981] *The deducibilities of S5* (J 0122) J Philos Logic 10*409-422
◇ B45 ◇ REV MR 83d:03024 Zbl 475 # 03005 • ID 55459

PORTE, J. [1982] *Boll-Reinhart modal logic* (J 0079) Logique & Anal, NS 25*181-190
◇ B45 ◇ REV MR 85f:03020 Zbl 496 # 03007 • ID 36862

PORTE, J. [1982] *S5 and the predicate calculus* (J 0079) Logique & Anal, NS 25*321-326
◇ B45 ◇ REV MR 84k:03061 Zbl 504 # 03009 • ID 34999

PORTE, J. [1983] *Antitheses in systems of relevant implication* (J 0036) J Symb Logic 48*97-99
◇ B46 ◇ REV MR 85e:03047 • ID 40615

PORTE, J. [1983] *Axiomatization and independence in S4 and in S5* (J 0302) Rep Math Logic, Krakow & Katowice 16*23-35
◇ B45 ◇ REV Zbl 541 # 03009 • ID 41373

PORTE, J. [1984] *Lukasiewicz's L-modal system and classical refutability* (J 0079) Logique & Anal, NS 27*87-92
◇ B45 ◇ REV MR 85g:03032 Zbl 541 # 03010 • ID 41374

PORTE, J. see Vol. I, IV, VI for further entries

PORUS, V.N. [1973] *Ueber eine analytische Darstellungsweise mehrwertiger Matrizen (Russian)* (C 1662) Teor Log Vyvoda 253-258
◇ B50 ◇ REV Zbl 263 # 02011 • ID 29861

POSPELOV, D.A. & TOSIC, Z. [1968] *Polynomdarstellungen in mehrwertigen Logiken (Russisch)* (C 3263) Sint Diskr Avtom Upravl Ustr 132-139
◇ B50 ◇ REV Zbl 248 # 94050 • ID 64630

POSPELOV, D.A. [1969] see IVAS'KIV, YU.L.

POSPELOV, D.A. [1977] see PAPKE, W.

POSPELOV, D.A. [1977] see EZHKOVA, I.V.

POSPELOV, D.A. [1978] see EZHKOVA, I.V.

POSPELOV, D.A. & VAROSYAN, S.O. [1982] *Nonmetric spatial logic (Russian)* (J 0977) Izv Akad Nauk SSSR, Tekh Kibern 1982/5*86-99
• TRANSL [1982] (J 0522) Engin Cybern 20/5*64-76
◇ B60 ◇ REV Zbl 525 # 68061 • ID 37493

POSPELOV, D.A. [1984] *Pseudophysical logics (Russian)* (P 4288) Predst Znan Chelov-Mash & Robot Sist 48-57
◇ B60 ◇ ID 46967

POSPESEL, H. [1971] *Arguments: deductive logic exercises* (X 0819) Prentice Hall: Englewood Cliffs ix + 208pp
◇ B98 ◇ ID 22527

POSPESEL, H. [1972] *Scepticism and modal logic* (J 0079) Logique & Anal, NS 15*653-664
◇ A05 B45 ◇ REV MR 48 # 10766 Zbl 272 # 02009 • ID 10671

POSPESEL, H. & WERNER, C.G. [1974] *Deductive inferences from particular to general* (J 0047) Notre Dame J Formal Log 15*351-352
◇ A05 B48 ◇ REV Zbl 245 # 02015 • ID 10674

POSPESEL, H. [1974] *Introduction to logic. Propositional logic* (X 0819) Prentice Hall: Englewood Cliffs xii + 211pp
◇ B05 B98 ◇ REV JSL 43.383 • ID 44546

POSPESEL, H. [1976] *Introduction to logic. Predicate logic* (X 0819) Prentice Hall: Englewood Cliffs xiii + 205pp
◇ B10 B98 ◇ REV JSL 43.383 • ID 44547

POST, E.L. [1921] *Introduction to a general theory of elementary propositions* (J 0100) Amer J Math 43*163-185
• REPR [1967] (C 0675) From Frege to Goedel 265-283
◇ B05 B20 B50 ◇ REV FdM 48.1122 • ID 10681

POST, E.L. see Vol. I, IV for further entries

POSY, C.J. [1982] *A free IPC is a natural logic: strong completeness for some intuitionistic free logics* (J 3781) Topoi 1*30-43
◇ B60 ◇ REV MR 84h:03038 • ID 34256

POSY, C.J. see Vol. VI for further entries

POTTHOFF, K. [1975] see MUELLER, GERT H.

POTTHOFF, K. see Vol. I, III, V for further entries

POTTINGER, G. [1975] *The C_I systems: an irenic theory of implications* (C 1852) Entailment - Log of Relev & Nec 1*101-106
◇ B45 ◇ REV MR 53 # 10542 • ID 31426

POTTINGER, G. [1977] *A new classical relevant logic* (J 1893) Relevance Logic Newslett 2*131-137
◇ B46 ◇ REV Zbl 368 # 02032 • ID 31428

POTTINGER, G. [1979] *A new classical relevance logic* (**J** 0122) J Philos Logic 8*135–147
◇ B46 ◇ REV MR 80h:03031 Zbl 438 # 03026 • ID 55938

POTTINGER, G. [1979] *On analysing relevance constructively* (**J** 0063) Studia Logica 38*171–185
◇ B46 ◇ REV MR 81k:03025 Zbl 406 # 03031 • ID 56114

POTTINGER, G. [1985] *Intension, designation, and extension* (**J** 0047) Notre Dame J Formal Log 26*309–340
◇ B45 ◇ ID 47528

POTTINGER, G. see Vol. VI for further entries

POWERS, L. [1976] *Latest on a logic problem* (**J** 1893) Relevance Logic Newslett 1*143–172
◇ A05 B46 ◇ REV Zbl 391 # 03006 • ID 52333

POWERS, L. [1976] *On P-W* (**J** 1893) Relevance Logic Newslett 1*131–142
◇ B46 ◇ REV Zbl 391 # 03005 • ID 52332

PRADE, H. [1979] see DUBOIS, DIDIER

PRADE, H. [1980] see CAYROL, M.

PRADE, H. [1982] *Modal semantics and fuzzy set theory* (**P** 4051) Fuzzy Set & Possibility Th;1980 Acapulco 232–246
◇ B45 E72 ◇ REV MR 84b:03004 • ID 46444

PRADE, H. [1982] *Possibility sets, fuzzy sets and their relation to Lukasiewicz logic* (**P** 4002) Int Symp Multi-Val Log (12);1982 Paris 223–227
◇ B50 E70 ◇ REV Zbl 544 # 03007 • ID 40989

PRADE, H. [1982] see DUBOIS, DIDIER

PRADE, H. [1983] *Data bases with fuzzy information and approximate reasoning in expert systems* (**J** 3919) BUSEFAL 14*115–126
◇ B52 B75 ◇ REV Zbl 517 # 94029 • ID 36700

PRADE, H. [1983] see DUBOIS, DIDIER

PRADE, H. [1984] see DUBOIS, DIDIER

PRADE, H. [1985] see DUBOIS, DIDIER

PRADE, H. see Vol. I, V for further entries

PRADHAN, D.K. [1974] *A multivalued switch algebra based on finite fields* (**P** 1385) Int Symp Multi-Val Log (4);1974 Morgantown 95–112
◇ B50 ◇ ID 42102

PRAKEL, J.M. [1977] *Some preliminary suggestions for the mirroring of non-metaphysical modalities in Lesniewski's ontology* (**J** 0063) Studia Logica 36*363–376
◇ A05 B45 ◇ REV MR 80a:03033 Zbl 407 # 03014 • ID 56180

PRANK, R.K. [1979] *Expressibility in the elementary theory of recursive sets with realizability logic (Russian) (English summary)* (**S** 3468) Tr Mat & Mekh (Tartu) 25(500)*119–129
◇ B60 D20 F50 ◇ REV MR 81a:03043 Zbl 421 # 03038 • ID 53437

PRANK, R.K. [1980] *Semantics of realizability for a language with variables for recursively enumerable sets (Russian)* (**J** 2852) Tr Vychisl Tsentra, Univ Tartu 43*112–131
◇ B60 D25 F35 F50 ◇ REV MR 81h:03111 • ID 77510

PRANK, R.K. [1981] *Expressibility in the elementary theory of recursively enumerable sets with realizability logic (Russian)* (**J** 0003) Algebra i Logika 20*427–439,484–485
• TRANSL [1981] (**J** 0069) Algeb and Log 20*282–291
◇ B60 D25 F50 ◇ REV MR 83i:03068 • ID 35552

PRANK, R.K. see Vol. IV for further entries

PRASAD, J. & RAO, S.V.L.N. [1982] *Definition of kriging in terms of fuzzy logic* (**J** 2145) J Int Ass Math Geol 14*37–42
◇ B52 ◇ REV MR 83d:86015 • ID 39651

PRATT, V.R. [1978] *A practical decision method for propositional dynamic logic* (**P** 1740) ACM Symp Th of Comput (10);1978 San Diego 326–337
◇ B75 ◇ ID 49865

PRATT, V.R. [1978] see HAREL, D.

PRATT, V.R. [1979] *Dynamic logic* (**P** 3375) Found of Comput Sci (3);1978 Amsterdam 2*53–82
◇ B75 ◇ REV MR 81h:03056 Zbl 446 # 03018 • ID 56547

PRATT, V.R. [1979] *Models of program logics* (**P** 3535) IEEE Symp Founds of Comput Sci (20);1979 San Juan 115–122
◇ B75 ◇ ID 49866

PRATT, V.R. [1980] *Application of modal logic to programming* (**J** 0063) Studia Logica 39*257–274
◇ B75 ◇ REV MR 83b:03020 Zbl 457 # 03013 • ID 54338

PRATT, V.R. [1982] *Dynamic logic* (**P** 3622) Int Congr Log, Meth & Phil of Sci (6,Proc);1979 Hannover 251–261
◇ B75 ◇ REV MR 84e:68014 Zbl 497 # 03011 • ID 38108

PRATT, V.R. [1982] *Using graphs to understand PDL* (**P** 3738) Log of Progr;1981 Yorktown Heights 387–396
◇ B75 ◇ REV MR 83i:68035 Zbl 491 # 68031 • ID 37760

PRATT, V.R. see Vol. I, III, IV for further entries

PRAWITZ, D. [1965] *Natural deduction. A proof-theoretical study* (**X** 1163) Almqvist & Wiksell: Stockholm 113pp
◇ B15 B98 F05 F07 F50 F98 ◇ REV MR 33 # 1227 Zbl 173.2 JSL 32.255 • ID 22164

PRAWITZ, D. [1968] see MALMNAES, P.E.

PRAWITZ, D. see Vol. I, V, VI for further entries

PREPARATA, F.P. & YEH, R.T. [1971] *On a theory of continuosly valued logic* (**P** 2007) Symp Th & Appl of Multi-Val Log Design;1971 Buffalo 124–132
◇ B50 ◇ ID 41992

PREPARATA, F.P. & YEH, R.T. [1972] *Continuously valued logic* (**J** 0119) J Comp Syst Sci 6*397–418
◇ B50 G05 ◇ REV MR 47 # 8244 Zbl 262 # 02020 • ID 10747

PREPARATA, F.P. see Vol. I, IV for further entries

PRESIC, S.B. [1979] *A completeness theorem for one class of propositional calculi* (**J** 0400) Publ Inst Math, NS (Belgrade) 26(40)*249–254
◇ B50 ◇ REV MR 83j:03044 Zbl 455 # 03008 • ID 54269

PRESIC, S.B. see Vol. I, III, IV for further entries

PREVIALE, F. [1975] *Tavole semantiche per sistemi astratti di logica estensionale* (**J** 0144) Rend Sem Mat Univ Padova 54*31–57
◇ B50 F50 ◇ REV MR 55 # 2522 Zbl 357 # 02020 • ID 31421

PREVIALE, F. see Vol. I, III, VI for further entries

PRIEST, G. [1976] *Modality as a meta-concept* (**J** 0047) Notre Dame J Formal Log 17*401-414
◇ B45 ◇ REV MR 56#5225 Zbl 326#02017 • ID 18351

PRIEST, G. [1977] *A refoundation of modal logic* (**J** 0047) Notre Dame J Formal Log 18*340-354
◇ B45 ◇ REV MR 58#21480 Zbl 236#02016 • ID 23629

PRIEST, G. [1979] *The logic of paradox* (**J** 0122) J Philos Logic 8*219-241
◇ A05 B50 ◇ REV MR 80g:03007 Zbl 402#03012 • ID 54659

PRIEST, G. [1980] *Sense, entailment and modus ponens* (**J** 0122) J Philos Logic 9*415-435
◇ A05 B46 ◇ REV MR 83c:03007 Zbl 436#03008 • ID 55848

PRIEST, G. & ROUTLEY, R. [1982] *Lessons from Pseudo Scotus* (**J** 0095) Philos Stud 42*189-199
◇ A10 B45 ◇ REV MR 84f:03016 • ID 34433

PRIEST, G. [1982] *To be and not to be: dialectical tense logic* (**J** 0063) Studia Logica 41*249-268
◇ B53 ◇ REV MR 86a:03018 Zbl 536#03006 • ID 37094

PRIEST, G. & ROUTLEY, R. [1984] *Introduction: Paraconsistent logics* (**J** 0063) Studia Logica 43*3-16
◇ B53 ◇ REV Zbl 575#03017 • ID 42356

PRIEST, G. [1984] *Semantic closure* (**J** 0063) Studia Logica 43*117-129
◇ B53 ◇ REV Zbl 575#03002 • ID 42365

PRIEST, G. see Vol. I, III, V, VI for further entries

PRIJATELJ, N. [1960] *Introduction to mathematical logic (Slovenian)* (**X** 2799) Mladinska Knjiga: Ljubljana 150pp
◇ B98 ◇ REV MR 40#16 Zbl 123.245 JSL 33.480 • REM 2nd ed. 1969 • ID 43180

PRIJATELJ, N. [1982] *Foundations of mathematical logic I (Slovenian)* (**J** 2310) Obz Mat Fiz, Ljubljana 188pp
◇ B98 ◇ REV MR 83f:03005 Zbl 486#03002 • ID 35289

PRIJATELJ, N. see Vol. III, V for further entries

PRIMENKO, EH.A. [1977] *On the number of types of invertible transformations in multivalued logics (Russian) (English summary)* (**J** 0040) Kibernetika, Akad Nauk Ukr SSR 1977/5*27-29
• TRANSL [1977] (**J** 0021) Cybernetics 13*663-665
◇ B50 ◇ REV MR 58#10648 Zbl 375#02050 • ID 51627

PRIMENKO, EH.A. see Vol. I for further entries

PRIOR, A.N. [1952] *Modality de dicto and modality de re* (**J** 0105) Theoria (Lund) 18*174-180
◇ A05 B45 ◇ REV JSL 20.167 • ID 42448

PRIOR, A.N. [1953] *Negative quantifiers* (**J** 0273) Australasian J Phil 31*107-123
◇ B60 ◇ REV JSL 20.166 • ID 42446

PRIOR, A.N. [1953] *On propositions neither necessary nor impossible* (**J** 0036) J Symb Logic 18*105-108
◇ B45 ◇ REV Zbl 50.5 JSL 20.167 • ID 10769

PRIOR, A.N. [1953] *Three-valued logic and future contingents* (**J** 2811) Phil Quart (St Andrews) 3*317-326
◇ B50 ◇ ID 38671

PRIOR, A.N. [1954] *The interpretation of two systems of modal logic* (**J** 0093) J Comp Syst 1*201-208
◇ B45 ◇ REV MR 16.554 Zbl 57.8 JSL 25.293 • ID 10770

PRIOR, A.N. [1955] *Curry's paradox and 3-valued logic* (**J** 0273) Australasian J Phil 33*177-182
◇ B50 ◇ REV JSL 22.90 • ID 33699

PRIOR, A.N. [1955] *Formal logic* (**X** 0815) Clarendon Pr: Oxford ix+329pp
◇ A05 A10 B98 E30 ◇ REV MR 17.569 Zbl 124.2 Zbl 67.249 JSL 27.218 • REM 2nd ed.; 1962; 341pp • ID 22325

PRIOR, A.N. [1955] *Many-valued and modal systems: an intuitive approach* (**J** 0101) Phil Rev 64*626-630
◇ B45 B50 ◇ REV JSL 22.328 • ID 10771

PRIOR, A.N. [1956] *Modality and quantification in S5* (**J** 0036) J Symb Logic 21*60-62
◇ B45 ◇ REV MR 17.1038 Zbl 71.10 JSL 22.91 • ID 10774

PRIOR, A.N. [1957] see LEMMON, E.J.

PRIOR, A.N. [1957] *Many-valued logics. The last of three talks of "The logic game."* (**J** 4398) Listener 57*717-719
◇ B50 ◇ REV JSL 23.347 • ID 42170

PRIOR, A.N. [1957] *The necessary and the possible. The first of three talks on "The logic game."* (**J** 4398) Listener 57*627-628
◇ A05 B45 ◇ REV JSL 23.347 • ID 42168

PRIOR, A.N. [1957] *Time and modality. Being the John Locke lectures for 1955-6 delivered in the University of Oxford* (**X** 0894) Oxford Univ Pr: Oxford ix+148pp
◇ A10 B45 ◇ REV Zbl 79.6 JSL 25.342 • ID 10775

PRIOR, A.N. [1958] *The syntax of time-distinctions* (**J** 4475) Franciscan Stud 18*105-120
◇ B45 ◇ REV JSL 27.114 • ID 43017

PRIOR, A.N. [1959] *Notes on a group of new modal systems* (**J** 0079) Logique & Anal, NS 2*122-127
◇ B45 ◇ REV JSL 35.464 • ID 10779

PRIOR, A.N. [1961] *Some axiom-pairs for material and strict implication* (**J** 0068) Z Math Logik Grundlagen Math 7*61-65
◇ B05 B45 ◇ REV MR 25#11 Zbl 116.4 JSL 37.184 • ID 10782

PRIOR, A.N. [1962] *Quantification and L-modality* (**J** 0047) Notre Dame J Formal Log 3*142-147
◇ B45 ◇ REV MR 27#2402 Zbl 135.5 • ID 10783

PRIOR, A.N. [1962] *Tense-logic and the continuity of time (Polish and Russian summaries)* (**J** 0063) Studia Logica 13*133-151
◇ B45 ◇ REV MR 26#4876 JSL 32.245 • ID 10784

PRIOR, A.N. [1963] *The theory of implication* (**J** 0068) Z Math Logik Grundlagen Math 9*1-6 • ERR/ADD ibid 11*381-382
◇ A05 B46 ◇ REV MR 26#3596 MR 33#5462 Zbl 124.246 JSL 31.665 • ID 10785

PRIOR, A.N. [1964] *Axiomatisations of the modal calculus* Q (**J** 0047) Notre Dame J Formal Log 5*215-217
◇ B45 ◇ REV MR 34#1169 Zbl 166.251 JSL 35.464 • ID 10788

PRIOR, A.N. [1964] see MEREDITH, C.A.

PRIOR, A.N. [1964] *K1, K2 and related modal systems* (**J** 0047) Notre Dame J Formal Log 5*299-304
◇ B45 ◇ REV MR 31#3323 Zbl 137.6 JSL 37.182 • ID 10790

PRIOR, A.N. [1964] *Two additions to positive implications* (J 0036) J Symb Logic 29*31-32
◇ B20 B45 ◇ REV MR 33 # 42 • ID 10787

PRIOR, A.N. [1965] see MEREDITH, C.A.

PRIOR, A.N. [1966] *Postulates for tense-logic* (J 0325) Amer Phil Quart 3*153-161
◇ B45 ◇ REV JSL 32.245 • ID 10792

PRIOR, A.N. [1967] *Correspondence theory of truth* (C 0601) Encycl of Philos 1*223-232
◇ A05 B45 ◇ REV JSL 35.304 • ID 10794

PRIOR, A.N. [1967] *Modal logic* (C 0601) Encycl of Philos 5*5-12
◇ A05 B45 ◇ REV JSL 35.299 • ID 10797

PRIOR, A.N. [1967] *Stratified metric tense logic* (J 0105) Theoria (Lund) 33*28-38
◇ B45 ◇ REV MR 40 # 7095 Zbl 203.7 JSL 36.516 • ID 10793

PRIOR, A.N. [1968] *Modal logic and the logic of applicability* (J 0105) Theoria (Lund) 34*183-202
◇ B45 ◇ REV MR 49 # 10528 JSL 44.654 • ID 10800

PRIOR, A.N. [1968] *Papers on time and tense* (X 0815) Clarendon Pr: Oxford 176pp
◇ B45 ◇ ID 25278

PRIOR, A.N. [1969] see LEMMON, E.J.

PRIOR, A.N. [1969] *On the calculus MCC* (J 0047) Notre Dame J Formal Log 10*273-274
◇ B45 ◇ REV MR 40 # 25 Zbl 188.317 • ID 10803

PRIOR, A.N. [1971] *Recent advances in tense logic* (C 0571) Basic Issues Philos of Time 1-15
◇ B45 ◇ REV JSL 40.99 • ID 10804

PRIOR, A.N. see Vol. I, V, VI for further entries

PRISCO DI, C.A. (ED.) [1985] *Methods in mathematical logic. Proceedings of the 6th Latin American Symposium on Mathematical Logic held in Caracas, Venezuela, Aug. 1-6, 1983* (S 3301) Lect Notes Math 1130*vii+407pp
◇ B97 C97 ◇ REV MR 86d:03002 Zbl 556 # 00007 • ID 41792

PRISCO DI, C.A. see Vol. III, V for further entries

PRUCNAL, T. [1966] *A proof of completeness of the three-valued C-N sentential calculus of Lukasiewicz (Polish and Russian summaries)* (J 0063) Studia Logica 18*65-72
◇ B50 ◇ REV MR 33 # 7248 Zbl 292 # 02016 • ID 10805

PRUCNAL, T. [1967] *A proof of the axiomatizability of Lukasiewicz's three-valued implicational propositional calculus (Polish) (Russian and English summaries)* (J 0063) Studia Logica 20*133-144
◇ B50 ◇ REV MR 35 # 31 Zbl 294 # 02007 • ID 10806

PRUCNAL, T. [1967] see BRYLL, G.

PRUCNAL, T. [1968] *A definability criterion for the functions in Lukasiewicz's matrices (Polish) (Russian and English summaries)* (J 0063) Studia Logica 23*71-77
◇ B50 ◇ REV MR 40 # 75 Zbl 305 # 02029 • ID 24837

PRUCNAL, T. [1972] *On the structural completeness of some pure implicational propositional calculi (Polish and Russian summaries)* (J 0063) Studia Logica 30*45-52
◇ B55 ◇ REV MR 47 # 6427 Zbl 268 # 02013 • ID 10807

PRUCNAL, T. [1972] *Structural completeness of Lewi's system S5 (Russian summary)* (J 0014) Bull Acad Pol Sci, Ser Math Astron Phys 20*101-103 • ERR/ADD [1983] (J 0302) Rep Math Logic, Krakow & Katowice 15*67-70
◇ B45 ◇ REV MR 46 # 8807 MR 84g:03028 Zbl 229 # 02024 • REM The correction was published by Wojtylak,P. • ID 10808

PRUCNAL, T. [1973] *Proof of structural completeness of a certain class of implicative propositional calculi (Polish and Russian summaries)* (J 0063) Studia Logica 32*93-97
◇ B50 ◇ REV MR 50 # 84 Zbl 343 # 02037 • ID 10809

PRUCNAL, T. [1974] *Interpretations of classical implicational sentential calculus in nonclassical implicational calculi* (J 0063) Studia Logica 33*59-64
◇ B55 ◇ REV MR 50 # 4236 Zbl 309 # 02008 • ID 10810

PRUCNAL, T. [1974] see POGORZELSKI, W.A.

PRUCNAL, T. [1975] *Structural completeness and the disjunction property of intermediate logics* (J 0387) Bull Sect Logic, Pol Acad Sci 4*72-73
◇ B55 ◇ REV MR 53 # 12924 • ID 23172

PRUCNAL, T. [1976] *Structural completeness of Medvedev's propositional calculus* (J 0302) Rep Math Logic, Krakow & Katowice 6*103-105
◇ B55 ◇ REV MR 58 # 21486 Zbl 358 # 02024 • ID 21919

PRUCNAL, T. [1979] *On two problems of Harvey Friedman* (J 0063) Studia Logica 38*247-262
◇ B45 B55 ◇ REV MR 81h:03040 Zbl 436 # 03018 • ID 55858

PRUCNAL, T. [1983] *Structural completeness of some fragments of intermediate logics* (J 0387) Bull Sect Logic, Pol Acad Sci 12*41-44
◇ B55 ◇ REV Zbl 549 # 03021 • ID 43114

PRUCNAL, T. [1985] *Structural completeness of purely implicational intermediate logics* (P 4180) Int Congr Log, Meth & Phil of Sci (7,Pap);1983 Salzburg 31-41
◇ B55 ◇ ID 48111

PRUCNAL, T. see Vol. I, V for further entries

PRZYMUSINSKA, H. [1980] *Craig interpolation theorem and Hanf number for ν-valued infinitary predicate calculi (Russian summary)* (J 3293) Bull Acad Pol Sci, Ser Math 28*207-211
◇ B50 C40 C55 C75 G20 ◇ REV MR 82k:03053d Zbl 465 # 03012 • ID 54915

PRZYMUSINSKA, H. [1980] *Gentzen-type semantics for ν-valued infinitary predicate calculi (Russian summary)* (J 3293) Bull Acad Pol Sci, Ser Math 28*203-206
◇ B50 C75 G20 ◇ REV MR 82k:03053c Zbl 465 # 03011 • ID 54914

PRZYMUSINSKA, H. see Vol. III for further entries

PTAK, P. [1982] see BRABEC, J.

PTAK, P. & ROGALEWICZ, V. [1983] *Regularly full logics and the uniqueness problem for observables* (J 2658) Ann Inst Henri Poincare, Sect A 38*69-74
◇ B51 ◇ REV MR 84f:81009 Zbl 519 # 03051 • ID 37549

PTAK, P. [1984] *On centers and state spaces of logics* (P 4043) Winter School on Abstract Anal (11);1983 Zelezna Ruda 225-229
◇ B51 ◇ REV MR 85e:03155 Zbl 544 # 03034 • ID 40769

PTAK, P. & WRIGHT, J.D.M. [1985] *On the concreteness of quantum logics* (**J** 1666) Apl Mat, Cheskoslov Akad Ved 30*274-285
◇ B51 ◇ ID 47771

PULMANNOVA, S. [1977] *Symmetries in quantum logics* (**J** 2736) Int J Theor Phys 16*681-688
◇ B51 G12 ◇ REV MR 58 # 4006 Zbl 388 # 06007 • ID 69795

PULMANNOVA, S. [1978] *Joint distributions of observables on quantum logics* (**J** 2736) Int J Theor Phys 17*665-675
◇ B51 G12 ◇ REV MR 80e:81018 Zbl 417 # 06007 • ID 69796

PULMANNOVA, S. [1979] *Superposition principle and sectors in quantum logics* (**J** 2736) Int J Theor Phys 18*915-922
◇ B51 ◇ REV MR 81f:81011 Zbl 491 # 03022 • ID 36854

PULMANNOVA, S. [1980] see DVURECHENSKIJ, A.

PULMANNOVA, S. [1980] *Semiobservables on quantum logics* (**J** 1522) Math Slovaca 30*419-432
◇ B51 G12 ◇ REV MR 82f:03059 Zbl 454 # 03031 • ID 54243

PULMANNOVA, S. [1981] *On the observables on quantum logics* (**J** 1678) Found Phys 11*127-136
◇ B51 ◇ REV MR 83m:81011 • ID 40663

PULMANNOVA, S. [1981] see DVURECHENSKIJ, A.

PULMANNOVA, S. [1982] *Individual ergodic theorem on a logic (Russian summary)* (**J** 1522) Math Slovaca 32*413-416
◇ B51 ◇ REV MR 84f:81010 Zbl 503 # 28005 • ID 39672

PULMANNOVA, S. [1983] *Coupling of quantum logics* (**J** 2736) Int J Theor Phys 22*837-850
◇ B51 G12 ◇ REV MR 84k:81020 Zbl 538 # 06011 • ID 39278

PULMANNOVA, S. [1983] see NEUBRUNN, T.

PULMANNOVA, S. [1984] see DVURECHENSKIJ, A.

PULMANNOVA, S. [1985] see DVURECHENSKIJ, A.

PULTR, A. [1982] *Fuzzyness and fuzzy equality* (**J** 0140) Comm Math Univ Carolinae (Prague) 23*249-284
• REPR [1984] (**P** 3096) Conf Math Service of Man (2,Pap);1982 Las Palmas 119-135
◇ B52 E72 ◇ REV MR 83i:03089 MR 86b:03073 Zbl 498 # 04001 Zbl 541 # 04001 • ID 35536

PULTR, A. see Vol. V for further entries

PURICA, I.I. [1981] *The modal logic of events (Romanian)* (**S** 1613) Probl Logic (Bucharest) 8*257-273
◇ B45 ◇ ID 48195

PURTILL, R.L. [1970] *Four-valued tables and modal logic* (**J** 0047) Notre Dame J Formal Log 11*505-511
◇ B45 ◇ REV MR 46 # 1559 Zbl 169.300 • ID 10824

PURTILL, R.L. [1971] *Logic for philosophers* (**X** 0837) Harper & Row: New York xxii + 419pp
◇ B98 ◇ REV JSL 39.614 • ID 10825

PURTILL, R.L. [1972] *Logical thinking* (**X** 0837) Harper & Row: New York xv + 157pp
◇ B98 ◇ ID 22528

PURTILL, R.L. [1973] *Meinongian deontic logic* (**J** 2810) Phil Forum Quart 4*585-592
◇ B45 ◇ REV MR 58 # 27344 • ID 77598

PURTILL, R.L. [1975] *Paradox-free deontic logics* (**J** 0047) Notre Dame J Formal Log 16*483-490
◇ B45 ◇ REV MR 52 # 5363 Zbl 236 # 02020 • ID 18352

PURTILL, R.L. [1979] *Logic. Argument, refutation, and proof* (**X** 0837) Harper & Row: New York xix + 412pp
◇ B98 ◇ REV Zbl 473 # 03001 • ID 55332

PURTILL, R.L. see Vol. I for further entries

PURVIS, D.M. [1975] see CHEUNG, P.T.

PUTNAM, H. [1957] see KREISEL, G.

PUTNAM, H. [1957] *Three-valued logic* (**J** 0095) Philos Stud 8*73-80
◇ B50 ◇ REV JSL 25.289 • ID 10828

PUTNAM, H. [1963] *"Degree of confirmation" and inductive logic* (**C** 0673) Phil of Carnap 761-783
◇ A05 B48 ◇ REV JSL 37.631 • ID 10836

PUTNAM, H. [1974] *How to think quantum-logically* (**J** 0154) Synthese 29*55-61
◇ A05 B51 G12 ◇ REV Zbl 338 # 02003 • ID 64709

PUTNAM, H. [1976] *How to think quantum-logically* (**C** 2953) Log & Probab in Quant Mech 47-53
◇ A05 B51 G12 ◇ REV Zbl 328 # 02003 • ID 64708

PUTNAM, H. [1978] see FRIEDMAN, M.

PUTNAM, H. [1980] *Models and reality* (**J** 0036) J Symb Logic 45*464-482
• TRANSL [1982] (**J** 2688) Conceptus (Wien) 16*9-30
◇ A05 B98 C99 ◇ REV MR 81h:03016 Zbl 443 # 03003 • ID 36156

PUTNAM, H. see Vol. I, III, IV, V, VI for further entries

PYATNITSYN, B.N. [1967] *The problem of the semantics of probabilistic and inductive logic (Russian)* (**C** 0582) Log Semant & Modal Logika 101-118
◇ A05 B48 ◇ REV MR 36 # 1294 • ID 10524

PYATNITSYN, B.N. & SUBBOTIN, A.L. [1971] *De quelques approches de la classification des systemes logiques* (**J** 2076) Rev Int Philos 25*520-527
◇ B45 ◇ REV MR 58 # 27292 • ID 77276

PYATNITSYN, B.N. [1972] *Ueber die Semantik der Wahrscheinlichkeitslogik und der induktiven Logik* (**C** 1533) Quantoren, Modal, Paradox 318-337
◇ A05 B48 ◇ REV Zbl 244 # 02009 • ID 27388

PYATNITSYN, B.N. [1974] see BUDBAEVA, S.P.

PYATNITSYN, B.N. & TYAGNIBEDINA, O.S. [1980] *The role of the concepts of "probability" and "degree of certainty" in B. Bolzano's logic (Russian)* (**C** 3038) Log-Metodol Issl 167-181
◇ A05 B48 ◇ REV MR 82m:01048 • ID 82490

PYATNITSYN, B.V. [1984] *Modality and probability - means to express the notion of uncertainty (Russian)* (**C** 4366) Modal & Intens Log & Primen Probl Metodol Nauk 359-365
◇ B48 ◇ ID 46846

QUALITZ, J.E. [1978] see DENNING, P.J.

QUEILLE, J.P. & SIFAKIS, J. [1982] *A temporal logic to deal with fairness in transition systems* (**P** 4289) IEEE Symp Found of Comput Sci (23);1982 Chicago 217-225
◇ B45 ◇ REV MR 85k:68007 • ID 44616

QUEILLE, J.P. & SIFAKIS, J. [1983] *Fairness and related properties in transition systems - a temporal logic to deal with fairness* (J 1431) Acta Inf 19*195-220
◇ B45 B75 ◇ REV MR 85e:68076 Zbl 489 # 68024
• ID 38155

QUESADA, J.D. [1979] *Referential presuppositions and three-valued logic (Spanish)* (J 0162) Teorema (Valencia) 9*79-94
◇ A05 B50 ◇ REV MR 80m:03019 • ID 77606

QUINE, W.V.O. [1934] *A system of logistic* (X 0858) Harvard Univ Pr: Cambridge x+204pp
◇ A05 B15 B98 ◇ REV FdM 60.845 • ID 19580

QUINE, W.V.O. [1940] *Mathematical logic* (X 0942) Norton: New York xiii+348pp
• LAST ED [1951] (X 0858) Harvard Univ Pr: Cambridge xii+346pp
◇ A05 B98 E70 ◇ REV MR 13.613 MR 2.65 MR 25 # 6 Zbl 44.247 JSL 12.56 JSL 17.149 JSL 5.163 FdM 66.27
• REM 3rd ed. 1951, repr. 1981 • ID 10867

QUINE, W.V.O. [1941] *Elementary logic* (X 0943) Ginn: Boston vi+170pp
• LAST ED [1966] (X 0837) Harper & Row: New York x+129pp
◇ B98 ◇ REV MR 3.129 MR 34 # 4100 JSL 35.166 JSL 6.99 • REM 2nd ed. 1965 • ID 10869

QUINE, W.V.O. [1943] *Notes on existence and necessity* (J 0301) J Phil 40*113-127
◇ A05 B45 B65 ◇ REV JSL 8.45 • ID 41449

QUINE, W.V.O. [1944] *O sentido da nova logica* (X 0949) St Martin's Pr: New York 252pp
◇ B98 ◇ REV MR 7.45 Zbl 60.20 JSL 12.16 • ID 22427

QUINE, W.V.O. [1946] *A short course in logic* (X 0944) Harvard Coop Soc: Cambridge iv+130pp
◇ B98 ◇ REV JSL 12.60 • ID 10876

QUINE, W.V.O. [1947] *The problem of interpreting modal logic* (J 0036) J Symb Logic 12*43-48
• REPR [1967] (C 0721) Contemp Readings in Log Th 267-273
◇ B45 ◇ REV MR 8.557 JSL 12.139 • ID 10877

QUINE, W.V.O. [1948] *Theory of deduction. Parts I-IV* (X 0944) Harvard Coop Soc: Cambridge 156pp
◇ B98 ◇ REV Zbl 41.148 JSL 14.190 • ID 10880

QUINE, W.V.O. [1950] *Methods of logic* (X 0818) Holt Rinehart & Winston: New York xx+264pp
• LAST ED [1974] (X 0866) Routledge & Kegan Paul: Henley on Thames viii+280pp
◇ B98 ◇ REV MR 12.233 Zbl 38.148 JSL 15.203 JSL 24.219 • ID 10883

QUINE, W.V.O. [1953] *Three grades of modal involvement* (P 0645) Int Congr Philos (11);1953 Bruxelles 14*65-81
• REPR [1966] (C 1587) Quine: Ways of Paradox & Essays
◇ B45 ◇ REV MR 15.386 Zbl 53.342 JSL 20.168
• ID 10890

QUINE, W.V.O. [1953] *Two theorems about truth-functions* (J 0250) Bol Soc Mat Mexicana 10*64-70
◇ B05 B45 ◇ REV MR 15.277 Zbl 53.198 JSL 19.142
• ID 10892

QUINE, W.V.O. [1954] *Quantification and the empty domain* (J 0036) J Symb Logic 19*177-179
◇ B10 B60 ◇ REV MR 16.324 Zbl 56.10 JSL 20.284
• ID 10895

QUINE, W.V.O. [1956] *Quantifiers and propositional attitudes* (J 0301) J Phil 53*177-187
◇ B45 ◇ ID 30784

QUINE, W.V.O. [1963] *Set theory and its logic* (X 0858) Harvard Univ Pr: Cambridge xv+359pp
• TRANSL [1973] (X 0900) Vieweg: Wiesbaden xv+263pp [1978] (X 2866) Ullstein: Berlin (German)
◇ A10 B98 E30 E98 ◇ REV MR 43 # 37 Zbl 122.246 JSL 37.768 • REM 2nd rev. ed. 1969 • ID 23295

QUINE, W.V.O. [1967] *Reply to Prof. Marcus* (C 0721) Contemp Readings in Log Th 293-299
◇ B45 ◇ ID 10916

QUINE, W.V.O. [1969] *Reply to Foellesdal* (C 0556) Words & Objections (Quine) 336
◇ B45 ◇ ID 14944

QUINE, W.V.O. & ULLIAN, J.S. [1970] *The web of belief* (X 0981) Random House: New York v+95pp
◇ A05 B45 ◇ ID 25166

QUINE, W.V.O. [1974] *Truth and disquotation* (P 0610) Tarski Symp;1971 Berkeley 373-384
◇ B60 ◇ REV MR 50 # 9524 Zbl 322 # 02003 • ID 10921

QUINE, W.V.O. [1975] *On the individuation of attributes* (C 1856) Log Enterprise 3-13
◇ A05 B45 ◇ REV Zbl 352 # 02008 • ID 32231

QUINE, W.V.O. [1977] *Intensions revisited* (S 1687) Midwest Studies Philos 2*5-11
◇ A05 B45 ◇ ID 30791

QUINE, W.V.O. see Vol. I, III, IV, V, VI for further entries

QUINN, P.L. [1977] *Improved foundations for a logic of intrinsic value* (J 0095) Philos Stud 32*73-81
◇ A05 B45 ◇ REV MR 58 # 27321b • ID 77616

QUINN, P.L. [1979] *Existence throughout an interval of time and existence at an instant of time* (J 1022) Ratio (Oxford) 21*1-12
◇ A05 B45 ◇ REV MR 81d:03007 • ID 77615

QUINTON, H. [1983] see MASON, R.O.

RABINOWICZ, W. [1978] *Utilitarianism and conflicting obligations* (J 0105) Theoria (Lund) 44*19-23
◇ B45 ◇ REV MR 80j:03032 • ID 77624

RABINOWICZ, W. [1979] *Reasonable beliefs* (J 0472) Theory Decis 10*61-81
◇ B45 ◇ REV MR 81e:03026 Zbl 409 # 03014 • ID 56317

RABINOWICZ, W. [1980] *Some remarks about the family K of modal systems* (J 0047) Notre Dame J Formal Log 21*429-448
◇ B45 ◇ REV MR 81i:03025 Zbl 283 # 02025 Zbl 452 # 03011 • ID 54075

RABINOWICZ, W. see Vol. VI for further entries

RABINOWICZ, Z.L. [1963] see IVAS'KIV, YU.L.

RADAELLI, G. [1979] see MELONI, G.C.

RADECKI, T. [1981] *Outline of a fuzzy logic approach to information retrieval* (**J** 1741) Int J Man-Mach Stud 14∗169-178
◇ B52 ◇ REV MR 83k:68106 Zbl 457 # 68100 • ID 40314

RADECKI, T. see Vol. V for further entries

RADEV, S.R. [1984] *Interpolation in the infinitary propositional normal modal logic KL_{ω_1}* (**J** 0137) C R Acad Bulgar Sci 37∗715-716
◇ B45 C40 C75 ◇ REV Zbl 561 # 03009 • ID 44072

RADEV, S.R. [1985] *Extensions of PDL and consequence relations* (**P** 4670) Comput Th (5);1984 Zaborow 251-264
◇ B75 ◇ ID 49734

RADEV, S.R. see Vol. I for further entries

RADU, E. [1974] *On a certain theorem of J.Porte (Romanian) (French summary)* (**J** 0197) Stud Cercet Mat Acad Romana 26∗461-464
◇ B55 ◇ REV MR 50 # 1848 Zbl 284 # 02010 • ID 10968

RADU, E. [1978] *L'oeuvre de Grigore C. Moisil en logique mathematique I,II,III* (**J** 0060) Rev Roumaine Math Pures Appl 23∗463-477,605-610,1077-1092
◇ A05 A10 B96 ◇ REV MR 58 # 4925 MR 81h:01013 Zbl 375 # 02002 Zbl 375 # 02003 Zbl 389 # 03004 • REM Part III includes a bibliography • ID 51580

RADU, E. see Vol. VI for further entries

RADU, J.B. (ED.) [1982] *Henry E.Kyburg, Jr.& Isaac Levi* (**X** 0835) Reidel: Dordrecht xi + 322pp
◇ A05 B48 ◇ REV MR 83k:01049 • ID 40055

RAGAZ, M. [1981] *Arithmetische Klassifikation von Formelmengen der unendlichwertigen Logik* (**X** 2710) Eidgen Techn Hochsch: Zuerich 98pp
◇ B50 ◇ ID 37248

RAGAZ, M. [1983] *Die Unentscheidbarkeit der einstelligen unendlichwertigen Praedikatenlogik* (**J** 0009) Arch Math Logik Grundlagenforsch 23∗129-139
◇ B50 D35 D55 ◇ REV MR 85e:03059a Zbl 533 # 03007 • ID 36532

RAGAZ, M. see Vol. V for further entries

RAGGIO, A.R. [1968] *Propositional sequence-calculi for inconsistent systems* (**J** 0047) Notre Dame J Formal Log 9∗359-366
◇ B53 ◇ REV MR 39 # 2612 Zbl 184.7 • ID 10971

RAGGIO, A.R. [1983] see FARINAS DEL CERRO, L.

RAGGIO, A.R. see Vol. I, VI for further entries

RAKOV, M.A. [1976] see LANTSOV, A.L.

RALESCU, A.L. & RALESCU, D.A. [1984] *Probability and fuzziness* (**J** 0191) Inform Sci 34∗85-92
◇ B52 ◇ REV MR 86c:60007 Zbl 559 # 60002 • ID 44966

RALESCU, D.A. [1974] see NEGOITA, C.V.

RALESCU, D.A. [1975] see NEGOITA, C.V.

RALESCU, D.A. [1976] see NEGOITA, C.V.

RALESCU, D.A. [1982] *Fuzzy logic and statistical estimation* (**P** 3845) Conf Math Service of Man (2,Proc)(Feriet);1982 Las Palmas 605-606
◇ B52 ◇ REV Zbl 503 # 62009 • ID 36665

RALESCU, D.A. [1984] see RALESCU, A.L.

RALESCU, D.A. see Vol. V for further entries

RAMANUJAM, R. [1985] see PARIKH, R.

RAMBAUD, C. [1978] see DONNADIEU, M.-R.

RAMBAUD, C. see Vol. III for further entries

RAMIK, J. [1983] *Extension principle and fuzzy-mathematical programming* (**J** 0156) Kybernetika (Prague) 19∗516-525
◇ B52 ◇ REV MR 85a:90210 Zbl 528 # 90059 • ID 36765

RAMIK, J. see Vol. V for further entries

RAMSEY, F.P. [1926] *The foundations of mathematics* (**J** 1910) Proc London Math Soc, Ser 2 25∗338-384
◇ A05 B98 ◇ REV FdM 52.46 • ID 28767

RAMSEY, F.P. see Vol. I, III, V for further entries

RANDALL, C.H. [1970] see FOULIS, D.J.

RANDALL, C.H. [1974] see FOULIS, D.J.

RANDALL, C.H. [1976] see FOULIS, D.J.

RANDALL, C.H. [1979] see FOULIS, D.J.

RANDALL, C.H. [1981] see FOULIS, D.J.

RANDALL, C.H. [1983] see FOULIS, D.J.

RANTALA, V. [1979] *Possible worlds and formal semantics* (**C** 1706) Essays Honour J. Hintikka 177-188
◇ A05 B45 ◇ REV Zbl 429 # 03003 • ID 53834

RANTALA, V. [1982] *Impossible worlds semantics and logical omniscience* (**P** 3800) Intens Log: Th & Appl;1979 Moskva 106-115
• TRANSL [1984] (**C** 4366) Modal & Intens Log & Primen Probl Metodol Nauk 199-207
◇ B45 ◇ REV MR 84j:03045 Zbl 519 # 03002 • ID 34637

RANTALA, V. [1982] *Quantified modal logic: non-normal worlds and propositional attitudes* (**J** 0063) Studia Logica 41∗41-65
◇ B45 C90 ◇ REV MR 85d:03037 Zbl 529 # 03004 • ID 37647

RANTALA, V. [1984] *Facts and the choice of logical foundations* (**J** 0047) Notre Dame J Formal Log 25∗347-353
◇ B45 ◇ REV MR 86g:03010 Zbl 557 # 03004 • ID 42578

RANTALA, V. see Vol. I, III, V for further entries

RAO, A.P. [1969] *A note on universally free first order quantification theory* (**J** 0079) Logique & Anal, NS 12∗228-230
◇ B60 ◇ REV MR 44 # 6445 Zbl 274 # 02006 • ID 10988

RAO, A.P. [1975] *A note on universally free description theory* (**J** 0047) Notre Dame J Formal Log 16∗539-542
◇ B60 ◇ REV MR 52 # 5344 Zbl 245 # 02032 • ID 18353

RAO, A.P. [1977] *A more natural alternative to Mostowski's (MFL)* (**J** 0068) Z Math Logik Grundlagen Math 23∗387-392
◇ B60 C07 ◇ REV MR 58 # 5011 Zbl 443 # 03014 • ID 56420

RAO, S.V.L.N. [1982] see PRASAD, J.

RASIOWA, H. [1947] *Sur certaines matrices logiques* (**J** 0283) Ann Soc Pol Math 20∗402-403
◇ B50 ◇ ID 31446

RASIOWA, H. [1951] *Algebraic treatment of the functional calculi of Heyting and Lewis* (J 0027) Fund Math 38*99-126
◇ B45 F50 G10 G25 ◇ REV MR 15.385 Zbl 44.249
JSL 18.72 • ID 11001

RASIOWA, H. & SIKORSKI, R. [1953] *On satisfiability and decidability in non-classical functional calculi* (J 0014) Bull Acad Pol Sci, Ser Math Astron Phys 1*229-231
◇ B10 B25 B45 C90 F50 ◇ REV MR 15.668 Zbl 51.245
• ID 11004

RASIOWA, H. & SIKORSKI, R. [1955] *An application of lattices to logic* (J 0027) Fund Math 42*83-100
◇ B45 C90 G10 ◇ REV MR 19.240 Zbl 68.243
JSL 35.137 • ID 11014

RASIOWA, H. & SIKORSKI, R. [1955] *On existential theorems in non-classical functional calculi* (J 0027) Fund Math 41*21-28
◇ B45 C90 F50 G05 ◇ REV MR 16.987 Zbl 56.11
JSL 20.80 • ID 11011

RASIOWA, H. [1955] see GRZEGORCZYK, A.

RASIOWA, H. [1963] *On modal theories* (J 0096) Acta Philos Fenn 16*201-214
◇ B45 G05 G25 ◇ REV MR 29 # 5718 Zbl 117.10
• ID 11027

RASIOWA, H. & SIKORSKI, R. [1963] *The mathematics of metamathematics* (S 0257) Monograf Mat 41*522pp
• TRANSL [1972] (X 2027) Nauka: Moskva 591pp (Russian)
◇ B98 C98 F50 F98 G05 G10 G98 ◇ REV
MR 29 # 1149 MR 50 # 4232 Zbl 122.243 JSL 32.274
• REM 3rd ed 1979 • ID 11028

RASIOWA, H. [1965] *On non-classical calculi of classes* (P 0797) Fonds des Math, Machines Math & Appl;1962 Tihany 53-55
◇ B20 B50 E70 G10 G25 ◇ REV MR 33 # 3894
Zbl 265 # 02036 • ID 29826

RASIOWA, H. [1969] *A theorem on the existence of prime filters in Post algebras and the completeness theorem for some many-valued predicate calculi* (J 0014) Bull Acad Pol Sci, Ser Math Astron Phys 17*347-354
◇ B50 G20 ◇ REV MR 41 # 3246 Zbl 184.11 • ID 11030

RASIOWA, H. [1970] *Ultraproducts of m-valued models and a generalization of the Loewenheim-Skolem-Goedel-Mal'cev theorem for theories based on m-valued logics (Russian summary)* (J 0014) Bull Acad Pol Sci, Ser Math Astron Phys 18*415-420
◇ B50 C20 C90 ◇ REV MR 43 # 4638 Zbl 221 # 02044
• ID 11031

RASIOWA, H. [1971] *Introduction to modern mathematics (Polish)* (X 1034) PWN: Warsaw 302pp
• TRANSL [1973] (X 0809) North Holland: Amsterdam xi + 339pp (English)
◇ B30 B98 C05 E98 ◇ REV MR 48 # 8190b
Zbl 204.308 JSL 43.153 • ID 22289

RASIOWA, H. [1972] *Introduction to set theory and mathematical logic (Bulgarian)* (X 1925) Nauka i Izkustwo: Sofia 343pp
◇ B98 C05 E98 ◇ ID 31450

RASIOWA, H. [1972] *The Craig interpolation theorem for m-valued predicate calculi (Russian summary)* (J 0014) Bull Acad Pol Sci, Ser Math Astron Phys 20*341-346
◇ B50 C40 C90 ◇ REV MR 52 # 5355 Zbl 243 # 02014
• ID 11032

RASIOWA, H. [1973] *Formalized ω^+ valued algorithmic systems (Russian summary)* (J 0014) Bull Acad Pol Sci, Ser Math Astron Phys 21*559-565
◇ B50 B75 D15 ◇ REV MR 48 # 8198 Zbl 277 # 68025
• ID 19572

RASIOWA, H. [1973] *On a logical structure of mix-valued programs and the ω^+ valued algorithmic logic (Russian summary)* (J 0014) Bull Acad Pol Sci, Ser Math Astron Phys 21*451-458
◇ B75 ◇ REV MR 48 # 12919 Zbl 277 # 68024 • ID 19574

RASIOWA, H. [1973] *On generalized Post algebras of order ω^+ and ω^+ valued predicate calculi (Russian summary)* (J 0014) Bull Acad Pol Sci, Ser Math Astron Phys 21*209-219
◇ B50 G20 ◇ REV MR 47 # 4763 Zbl 261 # 02012
• ID 19573

RASIOWA, H. [1974] *A simplified formalization of ω^+ valued algorithmic logic. A formalized theory of programs* (J 0014) Bull Acad Pol Sci, Ser Math Astron Phys 22*595-603
◇ B75 ◇ REV MR 51 # 2867 Zbl 339 # 68008 • ID 19570

RASIOWA, H. [1974] *An algebraic approach to non-classical logics* (X 0809) North Holland: Amsterdam xv + 403pp
◇ B98 F50 G05 G10 G20 G25 G98 ◇ REV
MR 56 # 5285 Zbl 299 # 02069 JSL 42.432 • ID 31451

RASIOWA, H. [1974] *Extended ω^+ valued algorithmic logic. A formalized theory of programs with recursive procedures* (J 0014) Bull Acad Pol Sci, Ser Math Astron Phys 22*605-610
◇ B75 ◇ REV MR 51 # 2868 Zbl 339 # 68009 • ID 19571

RASIOWA, H. [1974] *On ω^+-valued algorithmic logic and related problems* (J 1929) Prace Centr Oblicz Pol Akad Nauk 150*23pp
◇ B75 D05 ◇ REV Zbl 313 # 02027 • ID 29624

RASIOWA, H. [1974] *Post algebras as a semantic foundation of m-valued logics* (C 0768) Stud Algeb Logic 92-142
◇ B50 G20 ◇ REV MR 52 # 13378 Zbl 345 # 02044
JSL 43.146 • ID 21827

RASIOWA, H. [1975] *Many-valued algorithmic logic* (P 1442) ⊧ ISILC Logic Conf;1974 Kiel 543-567
◇ B75 G20 ◇ REV MR 55 # 12467 Zbl 343 # 02013
• ID 27218

RASIOWA, H. [1975] *Mixed-value predicate calculi* (J 0063) Studia Logica 34*215-234
◇ B50 ◇ REV MR 54 # 61 Zbl 317 # 02008 • ID 23944

RASIOWA, H. [1977] *Algorithmic logic* (J 1927) Prace Inst Podstaw Inf, Pol Akad Nauk 281*205pp
◇ B75 G20 ◇ REV Zbl 382 # 03018 • ID 31444

RASIOWA, H. [1977] see BANACHOWSKI, L.

RASIOWA, H. [1977] *Completeness theorem for extended algorithmic logic* (P 1704) Int Congr Log, Meth & Phil of Sci (5);1975 London ON 3*13-15
◇ B75 ◇ ID 31443

RASIOWA, H. [1979] *Algorithmic logic and its extensions, a survey* (P 2615) Scand Logic Symp (5);1979 Aalborg 163-174
◇ B75 ◇ REV MR 82m:68029 Zbl 426 # 68003 • ID 53675

RASIOWA, H. [1979] *Algorithmic logic. Multiple-valued extensions* (J 0063) Studia Logica 38*317-335
◇ B75 ◇ REV MR 81j:68045 Zbl 466#03009 • ID 54960

RASIOWA, H. [1979] *Logic of complex algorithms* (P 2935) FCT'79 Fund of Comput Th;1979 Berlin/Wendisch-Rietz 370-381
◇ B75 ◇ REV MR 81d:68047 Zbl 433#68033 • ID 82580

RASIOWA, H. [1980] *Completeness in classical logic of complex algorithms* (P 3210) Math Founds of Comput Sci (9);1980 Rydzyna 488-503
◇ B75 ◇ REV MR 81k:68003 Zbl 459#68001 • ID 54476

RASIOWA, H. [1981] *On logic of complex algorithms* (J 0063) Studia Logica 40*289-310
◇ B75 ◇ REV MR 83j:68008 Zbl 479#68037 • ID 39895

RASIOWA, H. [1982] *Lectures on infinitary logic and logics of programs* (J 3434) Pubbl Ist Appl Calcolo, Ser 3 142*122pp
◇ B75 C75 C98 ◇ REV MR 85i:03089 • ID 45056

RASIOWA, H. [1983] *Logic LNCA of nondeterministic complex algorithms* (P 2943) Symp Math Found Comp Sci; 1982 Diedrichshagen 100-118
◇ B75 ◇ REV MR 85f:68003 Zbl 566#68038 • ID 44082

RASIOWA, H. [1985] *Topological representations of Post algebras of order ω^+ and open theories based on ω^+-valued Post logic* (J 0063) Studia Logica 44*353-368
◇ B50 C90 G20 ◇ ID 48433

RASIOWA, H. see Vol. I, III, IV, V, VI for further entries

RATSA, M.F. [1965] *Class of logic functions corresponding to the first matrix of Jaskowski (Russian)* (J 1037) Issl Obshch Algebr (Kishinev) 1965*99-110
◇ B50 ◇ REV MR 34#1172 • ID 10949

RATSA, M.F. [1966] *A criterion of functional completeness in logic corresponding to the first Jaskowski matrix (Russian)* (J 0023) Dokl Akad Nauk SSSR 168*524-527
• TRANSL [1966] (J 0062) Sov Math, Dokl 7*683-687
◇ B50 B55 ◇ REV MR 34#1173 Zbl 178.306 • ID 30830

RATSA, M.F. [1969] *On a class of ternary logic functions corresponding to the first Jaskowski matrix (Russian)* (J 0052) Probl Kibern 21*185-214
• TRANSL [1969] (J 0471) Syst Th Res 21*181-210
◇ B50 ◇ REV MR 46#6995 Zbl 299#02018 • ID 10951

RATSA, M.F. [1970] *Functional completeness in certain logics that are intermediate between the classical and intuitionistic ones (Russian)* (S 0166) Mat Issl, Mold SSR 4(18)*171-176
◇ B55 ◇ REV MR 46#6996 Zbl 232#02020 • ID 22283

RATSA, M.F. [1982] *Functional completeness in modal logic (Russian)* (S 2829) Rostocker Math Kolloq 19*19-28
◇ B45 ◇ REV MR 83i:03034 Zbl 509#03009 • ID 35503

RATSA, M.F. [1982] *Nontabularity of the logic S4 with respect to functional completeness (Russian)* (J 0003) Algebra i Logika 21*283-320
• TRANSL [1982] (J 0069) Algeb and Log 21*191-219
◇ B45 ◇ REV MR 85b:03028 Zbl 522#03008 • ID 37778

RATSA, M.F. [1983] *Functional completeness in the modal logic S5 (Russian)* (C 3807) Issl Neklass Log & Formal Sist 222-280
◇ B45 ◇ REV MR 85f:03021 • ID 40665

RATSA, M.F. [1983] *Functional completeness in modal logics (Russian)* (J 0967) Izv Akad Nauk Mold SSR, Ser Fiz-Tekh Mat 1983/2*37-39
◇ B45 ◇ REV MR 85i:03055 Zbl 509#03009 Zbl 574#03007 • ID 46003

RATSA, M.F. [1983] *Undecidability of the problem of functional expressibility in the modal logic S4 (Russian)* (J 0023) Dokl Akad Nauk SSSR 268*814-817
• TRANSL [1983] (J 0062) Sov Math, Dokl 27*182-186
◇ B45 ◇ REV MR 84g:03025 Zbl 542#03005 • ID 34139

RATSA, M.F. see Vol. I, VI for further entries

RAUCH, J. [1975] *Ein Beitrag zu der GUHA Methode in der dreiwertigen Logik* (J 0156) Kybernetika (Prague) 11*101-113
◇ B50 ◇ REV MR 58#27307 Zbl 309#02013 • ID 64765

RAUSZER, C. [1974] *A formalization of the propositional calculus of H-B logic* (J 0063) Studia Logica 33*23-34
◇ B55 F50 G10 ◇ REV MR 51#69 Zbl 289#02015 • ID 15188

RAUSZER, C. [1976] *On the strong semantical completeness of any extension of the intuitionistic predicate calculus* (J 0014) Bull Acad Pol Sci, Ser Math Astron Phys 24*81-87
◇ B55 F50 ◇ REV MR 53#2640 Zbl 343#02015 • ID 18356

RAUSZER, C. [1977] *Applications of Kripke models to Heyting-Brouwer logic* (J 0063) Studia Logica 36*61-71
◇ B55 C90 F50 ◇ REV MR 57#15977 Zbl 361#02033 • ID 50668

RAUSZER, C. [1977] *Model theory for an extension of intuitionistic logic* (J 0063) Studia Logica 36*73-87
◇ B55 C90 F50 ◇ REV MR 57#15978 Zbl 361#02034 • ID 50669

RAUSZER, C. [1980] *An algebraic and Kripke-style approach to a certain extension of intuitionistic logic* (J 0202) Diss Math (Warsaw) 167*62pp
◇ B55 C20 C90 F50 G25 ◇ REV MR 82k:03100 Zbl 442#03024 • ID 56381

RAUSZER, C. [1982] see ONO, H.

RAUSZER, C. [1985] *An equivalence between indiscernibility relations in information systems and a fragment of intuitionistic logic* (P 4670) Comput Th (5);1984 Zaborow 298-317
◇ B60 B75 ◇ ID 49737

RAUSZER, C. [1985] *An equivalence between theory of functional dependencies and a fragment of intuitionistic logic (Russian summary)* (J 3417) Bull Pol Acad Sci, Math 33*571-579
◇ B60 B75 ◇ ID 49777

RAUSZER, C. [1985] *Formalizations of certain intermediate logics. I* (P 2160) Latin Amer Symp Math Log (6);1983 Caracas 360-384
◇ B55 C40 C90 ◇ ID 41793

RAUSZER, C. see Vol. III, VI for further entries

RAUTENBERG, W. [1972] see KLOETZER, G.

RAUTENBERG, W. [1972] *Zur praktischen und theoretischen Wirksamkeit der mathematischen Logik* (C 1533) Quantoren, Modal, Paradox 95-106
◇ B98 ◇ ID 39979

RAUTENBERG, W. [1976] *Some properties of the hierarchy of modal logics (preliminary report)* (**J** 0387) Bull Sect Logic, Pol Acad Sci 5*103-105
◇ B45 ◇ REV MR 55 # 2504 • ID 27116

RAUTENBERG, W. [1977] *Der Verband der normalen verzweigten Modallogiken* (**J** 0044) Math Z 156*123-140
◇ B45 G10 ◇ REV MR 58 # 5064 Zbl 414 # 03011
• ID 53057

RAUTENBERG, W. [1979] *Klassische und nichtklassische Aussagenlogik* (**X** 0900) Vieweg: Wiesbaden xi+361pp
◇ B45 B50 B98 C90 ◇ REV MR 81i:03002
Zbl 424 # 03007 • ID 77698

RAUTENBERG, W. [1979] *More about the lattice of tense logics* (**J** 0387) Bull Sect Logic, Pol Acad Sci 8*21-26
◇ B45 ◇ REV MR 80h:03032 Zbl 411 # 03015 • ID 52869

RAUTENBERG, W. [1980] *Splitting lattices of logics* (**J** 0009) Arch Math Logik Grundlagenforsch 20*155-159
◇ B45 ◇ REV MR 82e:03021 Zbl 453 # 03017 • ID 54147

RAUTENBERG, W. [1983] *Modal tableau calculi and interpolation* (**J** 0122) J Philos Logic 12*403-423 • ERR/ADD ibid 14*229
◇ B45 C40 ◇ REV MR 85b:03029 Zbl 547 # 03015
• ID 39823

RAUTENBERG, W. [1985] *A note on implicational intermediate consequences* (**J** 0387) Bull Sect Logic, Pol Acad Sci 14*103-108
◇ B55 ◇ ID 49022

RAUTENBERG, W. see Vol. I, III, IV, VI for further entries

REACH, K. [1938] *The name relation and the logical antinomies* (**J** 0036) J Symb Logic 3*97-111
◇ A05 B45 ◇ REV Zbl 19.385 JSL 4.134 FdM 64.924
• ID 11053

READ, S. [1983] *A paradox of relevant implication* (**J** 3783) J Non-Classical Log (Univ Campinas) 2*25-41
◇ B46 ◇ REV Zbl 559 # 03010 • ID 46872

READ, S. [1983] *Burgess on relevance: a fallacy indeed* (**J** 0047) Notre Dame J Formal Log 24*473-481
◇ B46 ◇ REV MR 85b:03030 Zbl 569 # 03005 • ID 40611

REDDAM, J.P. [1982] *Van Fraassen on propositional attitudes* (**J** 0095) Philos Stud 42*101-110
◇ A05 B45 ◇ REV MR 84k:03028 • ID 34967

REHDER, W. [1980] *Quantum logic of sequential events and their objectivistic probabilities* (**J** 2736) Int J Theor Phys 19*221-237
◇ B51 G12 ◇ REV MR 81f:81012 Zbl 449 # 03072
• ID 56739

REHDER, W. [1980] *Thought-experiments and modal logics* (**J** 0079) Logique & Anal, NS 23*407-417
◇ B45 ◇ REV MR 83f:03019 • ID 35299

REHDER, W. [1980] *When do projections commute?* (**J** 0948) Z Naturforsch 35*437-441
◇ B51 G12 ◇ REV MR 81h:81013 • ID 82592

REHDER, W. [1982] *Conditions for probabilities of conditionals to be conditional probabilities* (**J** 0154) Synthese 53*439-443
◇ B48 ◇ REV MR 84i:03113 Zbl 503 # 60002 • ID 34590

REHDER, W. [1982] *Von Wright's "and next" versus a sequential tense-logic* (**J** 0079) Logique & Anal, NS 25*33-46
◇ B45 ◇ REV MR 83m:03028 Zbl 503 # 03030 • ID 35433

REICHBACH, J. [1962] *On the connection of the first-order functional calculus with many-valued propositional calculi* (**J** 0047) Notre Dame J Formal Log 3*102-107
◇ B10 B50 ◇ REV MR 26 # 4897 Zbl 242 # 02011
• ID 11065

REICHBACH, J. [1963] *About connection of the first-order functional calculus with many valued propositional calculi* (**J** 0068) Z Math Logik Grundlagen Math 9*117-124
◇ B10 B50 ◇ REV MR 28 # 1124 Zbl 163.243 • ID 11067

REICHBACH, J. [1964] *A note about connection of the first-order functional calculus with many valued propositional calculi* (**J** 0047) Notre Dame J Formal Log 5*158-160
◇ B10 B50 ◇ REV MR 31 # 2138 Zbl 163.4 • ID 11068

REICHBACH, J. [1965] *On the connection of the first-order functional calculus with* $\aleph_0$ *propositional calculus* (**J** 0047) Notre Dame J Formal Log 6*73-80
◇ B10 B50 ◇ REV MR 34 # 7353 Zbl 163.4 • ID 11070

REICHBACH, J. [1967] *Boolean many valued calculi with quantifiers and a homogeny proof of Goedel-Skolem-Loewenheim-Herbrand's theorems* (**X** 3333) Unknown Publisher: See Remarks 23pp
◇ B50 ◇ REV Zbl 189.7 • REM Tel-Aviv • ID 19565

REICHBACH, J. [1974] *About decidability, generalized models and main theorems of many valued predicate calculi* (**J** 1550) Creation Math 7*11-20
◇ B25 B50 C90 ◇ REV Zbl 372 # 02034 • ID 31999

REICHBACH, J. [1974] *Partitions in decidability and generalized models with many valued codes* (**P** 1385) Int Symp Multi-Val Log (4);1974 Morgantown 113-114
◇ B50 ◇ ID 42103

REICHBACH, J. see Vol. I, III, IV, VI for further entries

REICHENBACH, H. [1932] *Wahrscheinlichkeitslogik* (**J** 0277) Sitzb Preuss Akad Wiss Phys Math Kl 1932*476-488
◇ A05 B30 B48 ◇ REV Zbl 6.67 JSL 2.54 FdM 58.59
• ID 11080

REICHENBACH, H. [1935] *Wahrscheinlichkeitslogik* (**J** 0748) Erkenntnis (Leipzig) 5*37-43
◇ A05 B48 ◇ REV Zbl 11.124 FdM 61.57 • ID 11082

REICHENBACH, H. [1939] *Introduction a la logistique* (**X** 0859) Hermann: Paris 68pp
◇ B98 ◇ REV Zbl 21.290 JSL 12.86 FdM 65.1106
• ID 22598

REICHENBACH, H. [1939] *Ueber die semantische und die Objekt-Auffassung von Wahrscheinlichkeitsausdruecken* (**J** 0748) Erkenntnis (Leipzig) 8*50-68
◇ B30 B48 ◇ REV Zbl 21.385 FdM 65.1107 • ID 43787

REICHENBACH, H. [1947] *Elements of symbolic logic* (**X** 0843) Macmillan : New York & London xiii+444pp
• LAST ED [1980] (**X** 0813) Dover: New York iv+444pp
◇ A05 B98 ◇ REV MR 8.556 Zbl 34.3 Zbl 486 # 03003
JSL 14.50 • REM 2nd ed. 1980 • ID 11084

REICHENBACH, H. [1948] *Philosophic foundations of quantum mechanics* (**X** 0926) Univ Calif Pr: Berkeley x+182pp
◇ A05 B51 G12 ◇ REV JSL 10.97 • ID 19521

REICHENBACH, H. [1949] *The theory of probability* (**X** 0926) Univ Calif Pr: Berkeley xvi+492pp
◇ A05 B98 ◇ REV MR 11.152 Zbl 38.286 JSL 16.48
• ID 25436

REICHENBACH, H. [1951] *Ueber die erkenntnistheoretische Problemlage und den Gebrauch einer dreiwertigen Logik in der Quantenmechanik* (**J** 0948) Z Naturforsch 6A*569-575
◇ A05 B51 G12 ◇ REV MR 13.716 Zbl 54.4 • ID 11085

REICHENBACH, H. [1953] *Les fondements logiques de la mecanique des quanta* (**J** 0984) Ann Inst Henri Poincare, Sect B 13*109-158
◇ B51 G12 ◇ REV MR 15.844 Zbl 53.168 JSL 10.97 • ID 30884

REICHENBACH, H. [1954] *Les fondements logiques de la theorie des quanta. Utilisation d'une logique a trois valeurs* (**P** 0646) Appl Sci de Log Math;1952 Paris 103-114
◇ B50 B51 G12 ◇ REV MR 16.782 Zbl 58.3 • ID 11086

REICHENBACH, H. [1954] *Nomological statements and admissible operations* (**X** 0809) North Holland: Amsterdam 140pp
◇ B45 ◇ REV MR 15.845 Zbl 56.244 JSL 20.50 • ID 11087

REICHENBACH, H. see Vol. I for further entries

REIF, J.H. [1983] see HALPERN, J.Y.

REIF, J.H. & SISTLA, A.P. [1985] *A multiprocess network logic with temporal and spatial modalities* (**J** 0119) J Comp Syst Sci 30*41-53
◇ B45 ◇ REV Zbl 565 # 68031 • ID 48668

REIF, J.H. see Vol. IV, V for further entries

REINHARDT, W.N. [1980] *Necessity predicates and operators* (**J** 0122) J Philos Logic 9*437-450
◇ A05 B45 ◇ REV MR 83j:03034 Zbl 451 # 03011 • ID 54026

REINHARDT, W.N. [1980] *Satisfaction definitions and axioms of infinity in a theory of properties with necessity operator* (**P** 2958) Latin Amer Symp Math Log (4);1978 Santiago 267-303
◇ A05 B45 E47 E55 E70 ◇ REV MR 81j:03034 Zbl 451 # 03010 • ID 54025

REINHARDT, W.N. see Vol. III, V, VI for further entries

REISCHER, C. [1978] see MARTIN, L.

REISCHER, C. & SIMOVICI, D.A. [1985] *Iteration properties of transformations of finite sets with application to multivalued logic* (**P** 4629) Dynamical Syst & Cellular Autom;1983 Luminy 369-374
◇ B50 ◇ ID 49525

REISCHER, C. see Vol. I, IV for further entries

REITERMAN, J. & TRNKOVA, V. [1980] *Dynamic algebras which are not Kripke structures* (**P** 3210) Math Founds of Comput Sci (9);1980 Rydzyna 528-538
◇ B75 C90 G25 ◇ REV MR 83f:68027 Zbl 451 # 03004 • ID 54019

REITERMAN, J. & TRNKOVA, V. [1981] *On representation of dynamic algebras with reversion* (**P** 3429) Math Founds of Comput Sci (10);1981 Strbske Pleso 463-472
◇ B75 G25 ◇ REV MR 83e:08016 Zbl 467 # 03057 • ID 55055

REITERMAN, J. & TRNKOVA, V. [1984] *From dynamic algebras to test algebras* (**P** 3658) Math Founds of Comput Sci (11);1984 Prague 490-497
◇ B75 ◇ REV Zbl 551 # 68030 • ID 43947

REITERMAN, J. see Vol. III, V for further entries

RENARDEL DE LAVALETTE, G.R. [1981] *The interpolation theorem in fragments of logics* (**J** 0028) Indag Math 43*71-86
◇ B20 B55 C40 F50 ◇ REV MR 82d:03019 Zbl 471 # 03012 • ID 55207

RENARDEL DE LAVALETTE, G.R. see Vol. I, VI for further entries

RENC, Z. [1971] see BENDOVA, K.

RENC, Z. see Vol. V for further entries

RENNA E SOUZA DE, C. & SILVA DE, O. [1982] *Changing the fuzzy rule of detachment* (**P** 4002) Int Symp Multi-Val Log (12);1982 Paris 267-271
◇ B52 ◇ REV Zbl 544 # 03009 • ID 40994

RENNA E SOUZA DE, C. see Vol. V for further entries

RENNIE, M.K. [1968] S3(5)=S3.5 (**J** 0036) J Symb Logic 33*444-445
◇ B45 ◇ REV Zbl 167.8 JSL 35.137 • ID 11097

RENNIE, M.K. [1968] *A function which bounds truth-tabular calculations in S5* (**J** 0079) Logique & Anal, NS 11*425-439
◇ B25 B45 D15 ◇ REV MR 43 # 6062 Zbl 181.6 • ID 11098

RENNIE, M.K. [1970] *Models for multiply modal systems* (**J** 0068) Z Math Logik Grundlagen Math 16*175-186
◇ B45 ◇ REV MR 42 # 2930 Zbl 164.311 • ID 11101

RENNIE, M.K. [1971] *Remark on Cresswell on S0.5* (**J** 0079) Logique & Anal, NS 14*757-758
◇ B45 ◇ REV MR 48 # 66 Zbl 238 # 02020 JSL 38.328 • ID 11102

RENNIE, M.K. [1971] *Semantics for* RK_t (**J** 0036) J Symb Logic 36*97-107
◇ B45 ◇ REV MR 44 # 3838 Zbl 215.46 • ID 11104

RENNIE, M.K. see Vol. I for further entries

RENYI, A. [1970] *Foundations of probability* (**X** 1167) Holden-Day: San Francisco 366pp
◇ B98 ◇ REV MR 41 # 9314 • ID 25437

RESCHER, N. [1958] *An axiom system for deontic logic* (**J** 0095) Philos Stud 9*24-30,64
◇ A05 B45 ◇ REV JSL 24.180 • ID 11107

RESCHER, N. [1960] *Many-sorted quantification* (**P** 0560) Int Congr Philos (12);1958 Venezia 447-453
◇ B50 ◇ REV JSL 31.123 • ID 11108

RESCHER, N. [1961] *Non-deductive rules of inference and problems in the analysis of inductive reasoning* (**J** 0154) Synthese 13*242-251
◇ B60 ◇ REV Zbl 111.7 JSL 33.613 • ID 43229

RESCHER, N. [1961] *On the formalization of two modal theses* (**J** 0047) Notre Dame J Formal Log 2*154-157
◇ B45 ◇ REV JSL 37.181 • ID 11109

RESCHER, N. [1962] *Conditional permission in deontic logic* (**J** 0095) Philos Stud 13*1-6
◇ B45 ◇ REV JSL 27.113 • ID 43016

RESCHER, N. [1962] *Modality conceived as a status* (**J** 0079) Logique & Anal, NS 5*81-89
◇ B45 ◇ ID 30892

RESCHER, N. [1962] *Quasi-truth-functional systems of propositional logic* (**J** 0036) J Symb Logic 27*1-10
◇ B60 ◇ REV MR 27 # 35 Zbl 107.7 JSL 29.50 • ID 11110

RESCHER, N. [1963] *A probabilistic approach to modal logic* (J 0096) Acta Philos Fenn 16*215-226
◇ B45 ◇ REV MR 28 # 2043 Zbl 135.6 JSL 35.583
• ID 11111

RESCHER, N. [1964] *A quantificational treatment of modality* (J 0079) Logique & Anal, NS 7*34-42
◇ B45 ◇ REV MR 31 # 2139 Zbl 124.247 • ID 11114

RESCHER, N. [1964] *Introduction to logic* (X 0949) St Martin's Pr: New York xv + 360pp
◇ A05 B98 ◇ REV JSL 35.579 • ID 22009

RESCHER, N. [1964] *Quantifiers in many-valued logic* (J 0079) Logique & Anal, NS 7*181-184
◇ B50 ◇ REV MR 30 # 4676 Zbl 129.3 • ID 11115

RESCHER, N. [1965] *An intuitive interpretation of systems of four-valued logic* (J 0047) Notre Dame J Formal Log 6*154-156
◇ B50 ◇ ID 11116

RESCHER, N. [1966] *On modal renderings of intuitionistic propositional logic* (J 0047) Notre Dame J Formal Log 7*277-280
• REPR [1969] (C 0555) Topic Philos Logic 18-23
◇ B45 F50 ◇ REV MR 35 # 2729 Zbl 163.4 • ID 11118

RESCHER, N. [1966] *On the logic of chronological propositions* (J 0094) Mind 75*75-96
◇ B45 ◇ ID 11117

RESCHER, N. & ROBISON, J. [1966] *Temporally conditioned descriptions* (J 1022) Ratio (Oxford) 8*46-54
• TRANSL [1966] (J 0170) Ratio (Frankfurt) 8*40-47
◇ B45 ◇ REV JSL 40.252 • ID 44323

RESCHER, N. [1966] *The logic of commands* (X 0866) Routledge & Kegan Paul: Henley on Thames xii + 147pp
◇ B45 ◇ REV JSL 34.499 • ID 25280

RESCHER, N. [1967] see GARSON, J.W.

RESCHER, N. [1967] *Semantic foundations for the logic of preference. With comments by A. R. Anderson and R. Ackermann* (P 0619) Log of Decis & Action;1966 Pittsburgh 37-79
◇ A05 B48 ◇ REV Zbl 252 # 02022 JSL 38.135 • ID 64807

RESCHER, N. (ED.) [1967] *The logic of decision and action* (X 1331) Univ Pittsburgh Pr: Pittsburgh 219pp
◇ A05 B97 ◇ ID 25294

RESCHER, N. [1968] *Chronological logic* (C 0552) Phil Contemp - Chroniques 123-124
◇ B45 ◇ ID 14562

RESCHER, N. [1968] *Conspectus of recent work in many-valued logic* (C 0552) Phil Contemp - Chroniques 74-86
◇ B50 ◇ ID 14937

RESCHER, N. [1968] see GARSON, J.W.

RESCHER, N. [1969] *A contribution to modal logic* (C 0555) Topic Philos Logic 24-39
◇ B45 ◇ ID 14963

RESCHER, N. [1969] *Appendix on the logic of determination and determinism* (C 0555) Topic Philos Logic 224-228
◇ A05 B45 ◇ ID 14881

RESCHER, N. [1969] *Appendix I. A note on R-2* (C 0555) Topic Philos Logic 245
◇ B45 ◇ ID 14882

RESCHER, N. [1969] *Appendix III. A summary of modal systems* (C 0555) Topic Philos Logic 285-286
◇ A05 B45 ◇ ID 14887

RESCHER, N. [1969] *Assertion logic* (C 0555) Topic Philos Logic 250-281
◇ A05 B45 ◇ ID 14884

RESCHER, N. [1969] *Deontic logic* (C 0555) Topic Philos Logic 321-331
◇ A05 B45 ◇ ID 14889

RESCHER, N. [1969] *Epistemic modality: the problem of a logical theory of belief statements* (C 0555) Topic Philos Logic 40-53
◇ A05 B45 ◇ ID 14964

RESCHER, N. [1969] *Many-valued logic* (C 0555) Topic Philos Logic 54-125
◇ B50 ◇ ID 11119

RESCHER, N. [1969] *Many-valued logic* (X 0822) McGraw-Hill: New York xv + 359pp
◇ B50 ◇ REV Zbl 248 # 02023 JSL 42.432 • ID 15033

RESCHER, N. [1969] *Nonstandard quantificational logic* (C 0555) Topic Philos Logic 162-181
◇ A05 B45 C90 ◇ ID 14969

RESCHER, N. [1969] *Probability logic* (C 0555) Topic Philos Logic 182-195
◇ B48 ◇ ID 14880

RESCHER, N. [1969] *The logic of existence* (C 0555) Topic Philos Logic 138-161
◇ B60 ◇ ID 14968

RESCHER, N. [1969] *The logic of preference* (C 0555) Topic Philos Logic 287-320
◇ A05 B45 ◇ ID 14888

RESCHER, N. [1969] *Topics in philosophical logic* (S 3307) Synth Libr xiv + 347pp
◇ A05 B45 B98 C90 ◇ REV Zbl 175.264 • ID 25168

RESCHER, N. [1970] see MANOR, R.

RESCHER, N. & URQUHART, A.I.F. [1971] *Temporal logic* (X 0902) Springer: Wien xviii + 273pp
◇ A05 B45 B98 ◇ REV MR 49 # 2267 Zbl 229 # 02027 JSL 40.252 • ID 11120

RESCHER, N. [1972] see MANOR, R.

RESCHER, N. [1973] see NAT VANDER, A.

RESCHER, N. [1973] see PARKS, R.Z.

RESCHER, N. [1975] *A theory of possibility* (X 1331) Univ Pittsburgh Pr: Pittsburgh xvi + 255pp
◇ A05 B98 ◇ REV MR 55 # 12455 JSL 43.158 • ID 77745

RESCHER, N. [1979] *Mondi possibili non-standard* (C 2041) Formal della Dialettica 354-417
◇ B45 ◇ ID 42882

RESCHER, N. [1980] see BRANDOM, R.B.

RESCHER, N. see Vol. I, III, V, VI for further entries

RESNIK, M.D. [1962] *A decision procedure for positive implication* (J 0047) Notre Dame J Formal Log 3*179-186
◇ B55 ◇ REV MR 26 # 4889 • ID 11124

RESNIK, M.D. [1970] *Elementary logic* (X 0822) McGraw-Hill: New York 457pp
◇ B98 ◇ REV MR 42 # 7478 Zbl 235 # 02002 • ID 11128

RESNIK, M.D. see Vol. I, V, VI for further entries

REYERSBACH, W. [1972] see EMDE, H.

RICARD, M. [1984] see BOURRELLY, L.

RICE, L.C. [1977] *Von Wright, rationalism and modality* (**J** 0286) Int Logic Rev 15*53-56
◇ A05 B45 ◇ REV Zbl 373 # 02016 • ID 51480

RICHARD, J. [1976] see COURVOISIER, M.

RICHARD, J. see Vol. V for further entries

RICHARD, M. [1982] *Tense, propositions, and meanings* (**J** 0095) Philos Stud 41*337-351
◇ A05 B45 ◇ REV MR 84k:03029 • ID 34968

RICHARD, M. [1983] *Direct reference and ascriptions of belief* (**J** 0122) J Philos Logic 12*425-452
◇ B45 ◇ REV MR 85a:03024 • ID 34775

RICHTER, M.M. [1976] *Mathematische Logik I* (**S** 1642) Schr Inf Angew Math, Ber (Aachen) 31*iii + 101pp
◇ B98 ◇ ID 33426

RICHTER, M.M. [1978] *Logikkalkuele* (**X** 0823) Teubner: Stuttgart 232pp
◇ B35 B98 F05 G10 ◇ REV MR 80g:03002 Zbl 381 # 03002 • ID 51870

RICHTER, M.M. see Vol. I, III, IV, V, VI for further entries

RICHTER, R. [1978] see HUNTER, D.

RICKEY, V.F. [1977] *A survey of Lesniewski's logic* (**J** 0063) Studia Logica 36*407-426
◇ A05 A10 B60 E70 ◇ REV MR 80b:03031 Zbl 418 # 03004 • ID 53290

RICKEY, V.F. see Vol. I, III for further entries

RICKMAN, S.M. [1977] see KANDEL, A.

RIDDER, J. [1948] *Ueber mehrwertige Aussagenkalkuele und mehrwertige engere Praedikatenkalkuele* (**J** 0028) Indag Math 10*221-231,264-273,324-328
◇ B50 ◇ REV MR 10.499 Zbl 30.386 JSL 14.139 • ID 28462

RIDDER, J. [1949] *Sur quelques logiques multivalentes* (**P** 0682) Int Congr Philos (10);1948 Amsterdam 728-730
◇ B50 ◇ REV MR 10.585 Zbl 31.104 JSL 14.130 • ID 11182

RIDDER, J. [1952] *Ueber modale Aussagenlogiken und ihren Zusammenhang mit Strukturen I,II* (**J** 0028) Indag Math 14*213-223,459-467
◇ B45 ◇ REV MR 14.527 MR 16.2 Zbl 56.10 JSL 37.628 • REM Part III 1953 • ID 16967

RIDDER, J. [1953] *Ueber modale Aussagenlogiken und ihren Zusammenhang mit Strukturen III,IV* (**J** 0028) Indag Math 15*1-11,99-110,378-388
◇ B45 ◇ REV MR 15.90 MR 16.2 Zbl 56.10 JSL 37.628 • REM Parts I,II 1952. Part V 1954 • ID 28421

RIDDER, J. [1954] *Ueber modale Aussagenlogiken und ihren Zusammenhang mit Strukturen V,VI* (**J** 0028) Indag Math 16*2-8,117-128,389-396
◇ B45 ◇ REV MR 16.2 MR 17.224 Zbl 56.10 JSL 37.628 • REM Parts III,IV 1953 • ID 28424

RIDDER, J. [1955] *Die Gentzenschen Schlussverfahren in modalen Aussagenlogiken I,II,III* (**J** 0028) Indag Math 17*163-169,170-177,270-276
◇ B45 ◇ REV MR 17.224 Zbl 67.249 JSL 40.97 • REM See also 1957 by Matsumoto,K. • ID 28426

RIDDER, J. [1957] see MATSUMOTO, K.

RIDDER, J. [1958] *Die Gentzenschen Schlussverfahren in modalen Aussagenlogiken II,III* (**J** 0028) Indag Math 20*16-22,23-27
◇ B45 ◇ REV MR 22 # 9442b Zbl 228 # 02009 JSL 40.97 • REM Part I 1957 by Matsumoto,K. & Ohnishi,M. & Ridder,J. • ID 16965

RIDDER, J. see Vol. I, VI for further entries

RIEBER, C.H. [1918] *Footnotes to formal logic* (**X** 0826) Alber: Freiburg 177pp
◇ A05 B98 ◇ REV FdM 46.73 • ID 25580

RIECHAN, B. [1980] see DVURECHENSKIJ, A.

RIEGER, L. [1967] *Algebraic methods of mathematical logic* (**X** 0801) Academic Pr: New York 210pp
◇ A05 B98 G05 G15 ◇ REV Zbl 218 # 02001 JSL 35.440 • ID 11194

RIEGER, L. see Vol. I, III, IV, V, VI for further entries

RIERA, T. & TRILLAS, E. [1980] *On a special kind of variables in fuzzy environment* (**P** 3673) Int Symp Multi-Val Log (10);1980 Evanston 149-152
◇ B52 ◇ REV MR 83m:94032 • ID 45976

RIERA, T. & TRILLAS, E. [1982] *From measures of fuzzyness to Booleanity control* (**C** 3778) Fuzzy Inform & Decis Processes 3-16
◇ B52 ◇ REV MR 84m:28012 Zbl 516 # 94038 • ID 39490

RIERA, T. see Vol. V for further entries

RIESER, H. [1983] see EIKMEYER, H.-J.

RIJEN VAN, J. [1980] *On criticizing deviant logics* (**J** 0079) Logique & Anal, NS 23*235-262
◇ B60 ◇ REV MR 83g:03009 • ID 35989

RINE, D.C. [1973] *A proposed multi-valued extension to ALGOL 68* (**J** 1429) Kybernetes 2*107-111
◇ B50 ◇ REV Zbl 278 # 68014 • ID 64831

RINE, D.C. [1973] *Basic concepts of multiple-discrete valued logic and decision theory* (**J** 0194) Inform & Control 22*320-352
◇ B50 ◇ REV MR 57 # 19142 Zbl 255 # 02009 • ID 28957

RINE, D.C. [1974] see EPSTEIN, G.

RINE, D.C. [1974] *Conditional and Post algebra expressions* (**J** 0193) Discr Math 10*309-323
◇ B50 G20 ◇ REV Zbl 296 # 02008 • ID 64832

RINE, D.C. [1974] see NUTTER JR., R.S.

RINE, D.C. [1974] *General cost factors for multiple-valued logic design of arithmetic units* (**J** 1429) Kybernetes 3*241-246
◇ B50 B70 ◇ REV Zbl 301 # 94042 • ID 64835

RINE, D.C. [1974] *Multple-valued logic and computer sciences in the 20th century* (**J** 2015) Computer 7/9*18-19
◇ B50 ◇ ID 42104

RINE, D.C. [1974] *There is more to boolean algebra than you would have thought* (**J** 0302) Rep Math Logic, Krakow & Katowice 2*25-32
◇ B50 G05 ◇ REV MR 50 # 4424 • ID 11201

RINE, D.C. [1975] *Multiple-valued logic in programming and flowcharting* (J 3116) Austral Comp J 7*75-79
◇ B50 ◇ REV MR 52 # 7191 Zbl 328 # 68018 • ID 64834

RINE, D.C. [1976] *A note regarding multiple-valued algorithmic logics as a tool to investigate programs* (J 1929) Prace Centr Oblicz Pol Akad Nauk 244*22pp
◇ B75 ◇ REV Zbl 338 # 68019 • ID 64838

RINE, D.C. [1976] *A survey of multiple-valued algorithmic logics: from a practical point of view* (P 2011) Int Symp Multi-Val Log (6);1976 Logan 1-8
◇ B50 B75 ◇ ID 35844

RINE, D.C. [1976] see GAINES, B.R.

RINE, D.C. [1977] *A note on multi-valued interrogation logic of associative memories* (J 0302) Rep Math Logic, Krakow & Katowice 8*87-100
◇ B50 ◇ REV MR 58 # 8584 Zbl 383 # 94035 • ID 24195

RINE, D.C. (ED.) [1977] *Computer science and multiple-valued logic. Theory and applications* (X 0809) North Holland: Amsterdam xiv + 548pp
◇ B50 B97 ◇ REV MR 56 # 11751 Zbl 359 # 94002 Zbl 546 # 94020 • REM 2nd rev. ed. 1984; xxiv + 641pp • ID 50633

RINE, D.C. [1981] *Picture processing using multiple-valued logic* (P 3705) Int Symp Multi-Val Log (11);1981 Oklahoma City & Norman 73-78
◇ B50 ◇ REV MR 83m:94033 • ID 45971

RINE, D.C. see Vol. I, IV for further entries

ROBBEL, G. [1980] *Many-valued logics based on L-algebras or on P-algebras* (J 3293) Bull Acad Pol Sci, Ser Math 28*529-533
◇ B50 G10 G25 ◇ REV MR 83d:03029 Zbl 471 # 03014 • ID 55209

ROBBIN, J.W. [1969] *Mathematical logic: a first course* (X 0867) Benjamin: Reading xii + 212pp
◇ B98 ◇ REV MR 40 # 4078 Zbl 186.3 • ID 19552

ROBBIN, J.W. see Vol. III for further entries

ROBERT, A. [1985] *Analyse non standard* (X 4457) Pr Polytechn: Lausanne xvii + 119pp
◇ B98 H05 ◇ REV Zbl 565 # 03033 • ID 48149

ROBERT, A. see Vol. I for further entries

ROBINSON, A. [1958] *Outline of an introduction to mathematical logic I,II,III* (J 0018) Canad Math Bull 1*41-54,113-127,193-208
◇ B98 C07 G05 ◇ REV MR 20 # 5123 MR 20 # 5124 MR 21 # 6321 Zbl 80.5 Zbl 82.15 Zbl 84.4 • REM Part IV 1959 • ID 25442

ROBINSON, A. [1959] *Outline of an introduction to mathematical logic IV* (J 0018) Canad Math Bull 2*33-42
◇ B10 B98 C07 C98 G05 ◇ REV MR 21 # 6321 Zbl 84.4 • REM Parts I-III 1958 • ID 19551

ROBINSON, A. [1963] *Introduction to model theory and to the metamathematics of algebra* (X 0809) North Holland: Amsterdam ix + 284pp
• TRANSL [1967] (X 2027) Nauka: Moskva 376pp
◇ B98 C60 C98 H15 ◇ REV MR 27 # 3533 MR 36 # 3642 JSL 25.172 • REM 2nd rev. ed. 1974 • ID 28804

ROBINSON, A. [1966] *Non-standard analysis* (X 0809) North Holland: Amsterdam ix + 293pp
◇ A10 B98 H05 H10 H15 H98 ◇ REV MR 34 # 5680 Zbl 151.8 JSL 35.292 • REM 2d edition 1968; 291pp • ID 11263

ROBINSON, A. also published under the name ROBINSOHN, A.

ROBINSON, A. see Vol. I, III, IV, V for further entries

ROBINSON, JOHN ALAN [1979] *Logic: form and function. The mechanization of deductive reasoning* (X 1261) Edinburgh Univ Pr: Edinburgh vi + 312pp
◇ A10 B10 B35 B98 ◇ REV MR 84d:03022 Zbl 428 # 03009 JSL 51.227 • ID 53769

ROBINSON, JOHN ALAN see Vol. I, VI for further entries

ROBINSON, O. [1979] *A modal natural deduction system for S4* (J 0047) Notre Dame J Formal Log 20*625-627
◇ B45 ◇ REV MR 80i:03033 Zbl 351 # 02016 • ID 56124

ROBISON, G.B. [1969] *An introduction to mathematical logic* (X 0819) Prentice Hall: Englewood Cliffs xii + 212pp
◇ B98 ◇ REV JSL 36.679 • ID 11329

ROBISON, J. [1966] see RESCHER, N.

ROBITASHVILI, N.G. [1979] *Semantics of languages containing not everywhere defined expressions* (P 2539) Frege Konferenz (1);1979 Jena 358-360
◇ B50 ◇ REV MR 82i:03034 • ID 77830

ROBITASHVILI, N.G. see Vol. I, III, VI for further entries

ROCKINGHAM GILL, R.R. [1970] *The Craig-Lyndon interpolation theorem in 3-valued logic* (J 0036) J Symb Logic 35*230-238
◇ B50 C40 ◇ REV MR 43 # 29 Zbl 212.315 • ID 22231

ROCKINGHAM GILL, R.R. [1977] *On Wandschneider's way out* (J 1022) Ratio (Oxford) 19*85-87
◇ A05 B45 E30 ◇ REV MR 56 # 15356 • ID 77833

ROCKINGHAM GILL, R.R. see Vol. III, V for further entries

RODOS, V.B. [1976] *Semantics and modalities (Russian)* (P 2566) All-Union Symp Log & Method of Sci (7);1976 Kiev 87-90
◇ B45 ◇ ID 33040

RODRIGUEZ, E. [1985] see FINKELSTEIN, D.

RODRIGUEZ MARIN, J. [1973] *Deontic logic. Natural deduction and decision via semantic tables (spanish)* (J 0162) Teorema (Valencia) 3*511-522
◇ B45 ◇ REV MR 49 # 7119 • ID 11334

RODRIGUEZ MARIN, J. [1976] *The reduction of deontic logic to modal logic* (J 0162) Teorema (Valencia) 6*99-117
◇ B45 ◇ REV MR 55 # 10244 • ID 77841

ROEHLE, F. [1973] *Der Informationsgehalt der Geraden. Ansaetze zu einer kommunikationstheoretischen Grundlegung der Mathematik* (J 2688) Conceptus (Wien) 7/23*1-84
◇ A05 B60 ◇ REV MR 56 # 5241 • ID 77844

ROEPER, P. [1980] *Intervals and tenses* (J 0122) J Philos Logic 9*451-470
◇ A05 B30 B45 ◇ REV MR 83i:03035 Zbl 451 # 03007 • ID 54022

ROEPER, P. see Vol. I for further entries

ROESSER, R. [1971] see LIEBLER, M.

ROGALEWICZ, V. [1983] see PTAK, P.

ROGALEWICZ, V. [1984] *On the uniqueness problem for quite full logics (French summary)* (**J** 2658) Ann Inst Henri Poincare, Sect A 41*445-451
◇ B51 ◇ REV MR 86e:81025 • ID 44517

ROGAVA, M.G. [1975] *Cut elimination in SCI* (**J** 0387) Bull Sect Logic, Pol Acad Sci 4*119-124
◇ B60 F05 ◇ REV MR 52 # 7872 • ID 18361

ROGAVA, M.G. [1979] *A new decision procedure for SCI (Russian)* (**P** 2539) Frege Konferenz (1);1979 Jena 361-364
◇ B25 B60 F05 ◇ REV MR 82h:03012 • ID 77850

ROGAVA, M.G. [1982] *Nonlenghtening applications of the identity rules in the calculus SCI (Russian) (English and Georgian summaries)* (**J** 0954) Tr Inst Prikl Mat, Tbilisi 11*67-74
◇ B60 ◇ REV MR 85j:03094 • ID 45330

ROGAVA, M.G. [1982] *Proof of the elimination theorem for the calculus SCI (Russian) (English and Georgian summaries)* (**J** 0954) Tr Inst Prikl Mat, Tbilisi 11*51-66
◇ B60 ◇ REV MR 85j:03093 • ID 45333

ROGAVA, M.G. see Vol. I, VI for further entries

ROGERS, R. [1971] *Mathematical logic and formalized theories. A survey of basic concepts and results* (**X** 0809) North Holland: Amsterdam xi + 235pp
◇ B98 C98 ◇ REV MR 47 # 9 Zbl 245 # 02001 • ID 11365

ROHLEDER, H. [1954] *Der dreiwertige Aussagenkalkuel der theoretischen Logik und seine Anwendung zur Beschreibung von Schaltungen, die aus Elementen mit zwei stabilen Zustaenden bestehen* (**J** 1046) Z Angew Math Mech 34*308-311
◇ B50 ◇ REV Zbl 55.362 JSL 23.366 • ID 11369

ROHLEDER, H. [1975] *Ein vollstaendiger Ableitungsbegriff fuer die Aequivalenz in einem funktionell unvollstaendigen dreiwertigen Aussagenkalkuel* (**J** 0129) Elektr Informationsverarbeitung & Kybern 11*623
• REPR [1977] (**J** 0068) Z Math Logik Grundlagen Math 23*401-404
◇ B50 G10 ◇ REV MR 58 # 10305 Zbl 316 # 02020 Zbl 443 # 03012 • ID 56418

ROHLEDER, H. see Vol. I, IV for further entries

ROHRER, C. [1979] see GUENTHNER, F.

ROMANKEVICH, A.M. [1965] *Minimization methods for functions of many-valued logic (Russian)* (**J** 0040) Kibernetika, Akad Nauk Ukr SSR 1965/4*38-42
• TRANSL [1965] (**J** 0021) Cybernetics 1/4*45-50
◇ B50 ◇ REV MR 34 # 1111 • ID 26676

ROMANKEVICH, A.M. [1969] *Minimization of multi-valued functions in a system that includes all one-place functions* (**J** 0021) Cybernetics 5*296-299
◇ B50 ◇ ID 26674

ROMANKEVICH, A.M. see Vol. I for further entries

ROMOV, B.A. [1971] *The uniprimitive foundations of the maximal subalgebras of Post algebras (Russian) (English summary)* (**J** 0040) Kibernetika, Akad Nauk Ukr SSR 1971/6*21-30
◇ B50 D03 G20 ◇ REV MR 47 # 8666 Zbl 265 # 02011 • ID 11387

ROMOV, B.A. [1973] *The solving algorithm for the functional completeness problem in class of vector multiple-valued schemes (Russian)* (**P** 3654) Mat Model Slonykh Sist;1969 Brasov 151-155
◇ B50 ◇ REV MR 57 # 17740 • ID 33142

ROMOV, B.A. [1980] *Maximal subalgebras of algebras of partial multivalued logic functions (Russian)* (**J** 0040) Kibernetika, Akad Nauk Ukr SSR 1980/1*28-36
• TRANSL [1980] (**J** 0021) Cybernetics 16*31-41
◇ B50 G20 G25 ◇ REV MR 82c:03078 Zbl 453 # 03068 • ID 54198

ROMOV, B.A. [1985] *A local analoque of Slupecki's theorem for infinitely many valued logic (Russian) (English summary)* (**J** 0040) Kibernetika, Akad Nauk Ukr SSR 1985/6*116-118,135
◇ B50 ◇ ID 49738

ROMOV, B.A. see Vol. III, IV, V for further entries

RONYAI, L. [1982] see DEMETROVICS, J.

ROOTSELAAR VAN, B. & STAAL, J.F. (EDS.) [1968] *Logic, methodology and philosophy of science III* (**S** 3303) Stud Logic Found Math xii + 553pp
◇ B98 ◇ REV MR 40 # 14 Zbl 177.293 • ID 33328

ROOTSELAAR VAN, B. [1971] *A class of models for intermediate logics* (**J** 0047) Notre Dame J Formal Log 12*358-362
◇ B55 G10 ◇ REV MR 45 # 6586 Zbl 193.290 • ID 11399

ROOTSELAAR VAN, B. see Vol. IV, VI for further entries

ROPER, A. [1982] *Towards an eliminative reduction of possible worlds* (**J** 2811) Phil Quart (St Andrews) 32*45-59
◇ B45 ◇ REV MR 83g:03010 • ID 35990

ROSADO HADDOCK, G.E. [1982] *Identity statements in the semantics of sense and reference* (**J** 0079) Logique & Anal, NS 25*399-411
◇ B45 ◇ REV MR 84m:03026 • ID 35730

ROSE, A. [1951] *A lattice-theoretic characterization of the $\aleph_0$-valued propositional calculi* (**J** 0043) Math Ann 123*285-287
◇ B50 G10 ◇ REV MR 13.614 Zbl 42.245 JSL 17.147 • ID 11404

ROSE, A. [1951] *A new proof of a theorem of Dienes* (**J** 4510) Norsk Mat Tidsskr 33*27-29
◇ B50 ◇ REV MR 12.790 Zbl 42.245 JSL 16.276 • ID 11408

ROSE, A. [1951] *An axiom system for the three-valued logic* (**J** 0175) Methodos 3*233-239
◇ B50 ◇ REV JSL 18.344 • ID 11405

ROSE, A. [1951] *Axiom systems for three-valued logic* (**J** 0039) J London Math Soc 26*50-58
◇ B50 ◇ REV Zbl 43.7 JSL 16.277 • ID 11409

ROSE, A. [1951] *Conditioned disjunction as a primitive connective for the m-valued propositional calculus* (**J** 0043) Math Ann 123*76-78
◇ B50 ◇ REV MR 12.790 Zbl 42.7 JSL 16.275 • ID 11407

ROSE, A. [1951] *Remarques sur les notions d'independance et de noncontradiction* (**J** 0109) C R Acad Sci, Paris 233*512-513
◇ A05 B50 ◇ REV MR 13.199 Zbl 43.8 • ID 11410

ROSE, A. [1951] *Systems of logic whose truth-values form lattices* (**J** 0043) Math Ann 123*152-165
◇ B50 G10 ◇ REV Zbl 44.2 JSL 17.147 • ID 11406

ROSE, A. [1951] *The degree of completeness of some Lukasiewicz-Tarski propositional calculi* (J 0039) J London Math Soc 26*47-49 • ERR/ADD ibid 44*587-591
◇ B20 B50 ◇ REV MR 12.662 MR 39 # 5321 Zbl 44.251 JSL 17.147 JSL 39.350 • ID 11403

ROSE, A. [1952] *A formalisation of Post's m-valued propositional calculus* (J 0044) Math Z 56*94-104
◇ B50 ◇ REV MR 14.233 Zbl 48.244 JSL 21.400 • ID 11423

ROSE, A. [1952] *An extension on the calculus of non-contradiction* (J 3240) Proc London Math Soc, Ser 3 54*184-200
◇ B50 ◇ REV MR 14.123 Zbl 49.147 JSL 18.66 • ID 11424

ROSE, A. [1952] *Eight-valued geometry* (J 3240) Proc London Math Soc, Ser 3 2*30-44
◇ B50 ◇ REV MR 13.898 Zbl 48.5 • ID 11420

ROSE, A. [1952] *Extensions of some theorems of Schmidt and McKinsey. I* (J 4510) Norsk Mat Tidsskr 34*1-9
◇ B50 G05 ◇ REV MR 14.3 Zbl 49.148 JSL 36.690 • ID 11415

ROSE, A. [1952] *Le degre de saturation du calcul propositionnel implicatif a trois valeurs de Sobocinski* (J 0109) C R Acad Sci, Paris 235*1000-1002
◇ B50 ◇ REV MR 14.834 Zbl 49.147 JSL 19.56 • ID 11422

ROSE, A. [1952] *Self dual primitives for modal logic* (J 0043) Math Ann 125*284-286
◇ B45 ◇ REV MR 15.846 Zbl 53.199 JSL 18.282 JSL 28.282 • ID 11414

ROSE, A. [1952] *Some generalized Sheffer functions* (J 0171) Proc Cambridge Phil Soc Math Phys 48*369-373
◇ B50 ◇ REV MR 14.3 Zbl 49.147 JSL 18.344 • ID 11417

ROSE, A. [1952] *The degree of completeness of m-valued Lukasiewicz propositional calculus* (J 0039) J London Math Soc 27*92-102
◇ B50 ◇ REV MR 13.811 Zbl 48.244 JSL 21.327 • ID 11419

ROSE, A. [1953] *A formalization of an* $\aleph_0$*-valued propositional calculus* (J 0171) Proc Cambridge Phil Soc Math Phys 49*367-376
◇ B50 ◇ REV MR 14.1052 Zbl 53.200 JSL 29.213 • ID 11431

ROSE, A. [1953] *A formalization of Sobocinski's three-valued implicational propositional calculus* (J 0093) J Comp Syst 1*165-168
◇ B50 ◇ REV MR 15.91 Zbl 53.199 JSL 19.144 • ID 42258

ROSE, A. [1953] *Fragments of the m-valued propositional calculus* (J 0044) Math Z 59*206-210
◇ B50 ◇ REV MR 15.668 Zbl 52.12 JSL 23.64 • ID 11426

ROSE, A. [1953] *Some self-dual primitive functions for propositional calculi* (J 0043) Math Ann 126*144-148
◇ B50 ◇ REV MR 15.277 Zbl 53.199 JSL 19.294 • ID 11428

ROSE, A. [1953] *The m-valued calculus of non-contradiction* (J 0036) J Symb Logic 18*237-241
◇ B50 ◇ REV MR 15.189 Zbl 53.199 JSL 20.180 • ID 11430

ROSE, A. [1953] *The degree of completeness of the* $\aleph_0$*-valued Lukasiewicz propositional calculus* (J 0039) J London Math Soc 28*176-184
◇ B50 ◇ REV MR 14.834 Zbl 53.200 JSL 21.328 • ID 11427

ROSE, A. [1954] *A Goedel theorem for an infinite-valued erweiterter Aussagenkalkuel* (P 0575) Int Congr Math (II, 7);1954 Amsterdam 2*406-407
◇ B50 ◇ ID 29471

ROSE, A. [1954] *Sur les fonctions definissables dans une logique a un nombre infini de valeurs* (J 0109) C R Acad Sci, Paris 238*1462-1463
◇ B50 ◇ REV MR 15.925 Zbl 58.245 • ID 11432

ROSE, A. [1955] *A Goedel theorem for an infinite-valued erweiterter Aussagenkalkuel* (J 0068) Z Math Logik Grundlagen Math 1*89-90
◇ B50 F30 ◇ REV MR 18.866 Zbl 67.250 • ID 24903

ROSE, A. [1955] *Le degre de saturation du calcul propositionnel implicatif a m valeurs de Lukasiewicz* (J 0109) C R Acad Sci, Paris 240*2280-2281
◇ B50 ◇ REV MR 16.892 Zbl 66.255 JSL 22.379 • ID 11436

ROSE, A. [1956] *An alternative formalisation of Sobocinski's three-valued implicational propositional calculus* (J 0068) Z Math Logik Grundlagen Math 2*166-172
◇ B50 ◇ REV MR 20 # 3777 Zbl 74.10 JSL 22.380 • ID 11437

ROSE, A. [1956] *Formalisation du calcul propositionnel implicatif a* $\aleph_0$ *valeurs de Lukasiewicz* (J 0109) C R Acad Sci, Paris 243*1183-1185
◇ B50 ◇ REV MR 18.271 Zbl 73.249 JSL 35.142 • ID 41721

ROSE, A. [1956] *Formalisation du calcul propositionnel implicatif a m valeurs de Lukasiewicz* (J 0109) C R Acad Sci, Paris 243*1263-1264
◇ B50 ◇ REV MR 18.633 Zbl 73.249 JSL 36.546 • ID 11440

ROSE, A. [1956] *Some formalisations of* $\aleph_0$*-valued propositional calculi* (J 0068) Z Math Logik Grundlagen Math 2*204-209
◇ B50 ◇ REV MR 19.626 Zbl 74.10 JSL 29.213 • ID 11438

ROSE, A. & ROSSER, J.B. [1958] *Fragments of many-valued statement calculi* (J 0064) Trans Amer Math Soc 87*1-53
◇ B50 ◇ REV MR 20 # 818 Zbl 85.243 JSL 24.248 • ID 11567

ROSE, A. [1958] *Many-valued logical machines* (J 0171) Proc Cambridge Phil Soc Math Phys 54*307-321
◇ B50 ◇ REV MR 20 # 2285 Zbl 85.244 JSL 27.250 • ID 11443

ROSE, A. [1958] *Sur les definitions de l'implication et de la negation dans certains systemes de logique dont les valeurs forment des treillis* (J 0109) C R Acad Sci, Paris 246*2091-2094
◇ B50 G10 ◇ REV MR 20 # 1629 JSL 24.250 • ID 11447

ROSE, A. [1960] *An extension of a theorem of Margaris* (J 0036) J Symb Logic 25*209-211
◇ B50 ◇ REV MR 25 # 1987 Zbl 107.7 JSL 35.305 • ID 11461

ROSE, A. [1960] *Nouvelle formalisation du calcul propositionnel bivalent dont les foncteurs primitifs forment un ensemble qui constitue son propre dual* (J 0109) C R Acad Sci, Paris 250*4246-4248
◇ B50 ◇ REV MR 22 # 6694 JSL 29.213 • ID 11454

ROSE, A. [1960] *Sur les schemas d'axiomes pour les calculs propositionnels a m-valeurs ayant des valeurs sur designees* (J 0109) C R Acad Sci, Paris 250*790-792
◇ B50 ◇ REV MR 22 # 670 Zbl 87.251 JSL 36.546 • ID 11453

ROSE, A. [1960] *Sur un ensemble independent de foncteurs primitifs pour le calcul propositionnel, lequel constitue son propre dual* (J 0109) C R Acad Sci, Paris 250*4089-4091
◇ B50 ◇ REV MR 22 # 6693 Zbl 218 # 02008 JSL 29.213 • ID 11452

ROSE, A. [1961] *Self-dual binary and ternary connectives for m-valued propositional calculi* (J 0043) Math Ann 143*448-462
◇ B50 ◇ REV MR 23 # A797 Zbl 107.7 JSL 29.144 • ID 11457

ROSE, A. [1961] *Sur certains calculs propositionnels a m-valeurs ayant un seul foncteur primitif lequel constitue son propre dual* (J 0109) C R Acad Sci, Paris 252*3176-3178,3375-3376
◇ B50 ◇ REV MR 30 # 1920 Zbl 218 # 02016 JSL 29.144 • ID 11459

ROSE, A. [1962] *A simplified self m-al set of primitive functors for the m-valued propositional calculus* (J 0068) Z Math Logik Grundlagen Math 8*257-266
◇ B50 ◇ REV MR 26 # 2359 Zbl 108.2 JSL 29.144 • ID 11463

ROSE, A. [1962] *An alternative generalisation of the concept of duality* (J 0043) Math Ann 147*318-327
◇ B50 ◇ REV MR 27 # 3510 Zbl 105.247 JSL 36.690 • ID 11462

ROSE, A. [1962] *Extensions of some theorems of Anderson and Belnap* (J 0036) J Symb Logic 27*423-425
◇ B50 ◇ REV MR 30 # 1922 Zbl 118.248 JSL 40.466 • ID 11460

ROSE, A. [1962] *Sur les applications dè la logique polyvalente a la construction des machines Turing* (J 0109) C R Acad Sci, Paris 255*1836-1838
◇ B50 D10 ◇ REV MR 25 # 4994 Zbl 173.11 • ID 11465

ROSE, A. [1962] *Sur un ensemble de foncteurs primitifs pour le calcul propositionnel a m valeurs lequel constitue son propre m-al* (J 0109) C R Acad Sci, Paris 254*1897-1899
◇ B50 ◇ REV MR 24 # A3058 Zbl 114.244 JSL 29.144 • ID 11467

ROSE, A. [1962] *Sur un ensemble complet de functeurs primitifs independants pour le calcul propositionnel trivalent lequel constitue son propre trial* (J 0109) C R Acad Sci, Paris 254*2111
◇ B50 ◇ REV MR 24 # A3059 Zbl 116.5 JSL 29.144 • ID 44890

ROSE, A. [1963] *A formalisation of the propositional calculus corresponding to Wang's calculus of partial predicates* (J 0068) Z Math Logik Grundlagen Math 9*177-198
◇ B60 ◇ REV MR 27 # 2398 Zbl 253 # 02017 • ID 11468

ROSE, A. [1964] *Formalisations de calculus propositionnels polyvalents a foncteurs variables* (J 0109) C R Acad Sci, Paris 258*1951-1953
◇ B50 ◇ REV MR 28 # 2967 Zbl 149.3 • ID 11469

ROSE, A. [1964] *Note sur la formalisation de calculs propositionnels polyvalents a foncteurs variables* (J 0109) C R Acad Sci, Paris 259*967-968
◇ B50 ◇ REV MR 29 # 3354 Zbl 129.4 • ID 11471

ROSE, A. [1964] *Sur la saturation de l'axiome trouve par Lukasiewicz pour le calcul c-0 a foncteurs variables* (J 0109) C R Acad Sci, Paris 258*3799-3800
◇ B50 ◇ ID 26678

ROSE, A. [1965] *A formalisation of Post's m-valued propositional calculus with variable functors* (J 0068) Z Math Logik Grundlagen Math 11*221-226
◇ B50 ◇ REV MR 33 # 37 • ID 11472

ROSE, A. [1965] *Formalisations of certain propositional calculi with partially variable functors* (J 0068) Z Math Logik Grundlagen Math 11*177-180
◇ B50 ◇ REV MR 31 # 3312 • ID 11473

ROSE, A. [1966] *A formalisation of the m-valued Lukasiewicz implicational propositional calculus with variable functors* (J 0068) Z Math Logik Grundlagen Math 12*169-176
◇ B50 ◇ REV MR 34 # 44 Zbl 212.314 • ID 11475

ROSE, A. [1966] *A formalisation of an infinite-valued propositional calculus in which the truth values are lattice ordered* (J 0036) J Symb Logic 31*683-684
◇ B50 ◇ ID 41921

ROSE, A. [1966] *An algebraic model of the m-valued propositional calculus with variable functors* (J 0188) Rev Union Mat Argentina 23*195-197
◇ B50 G20 ◇ REV MR 41 # 3244 Zbl 185.10 • ID 11476

ROSE, A. [1966] *Formalisation d'un calcul propositionnel a un nombre infini de valeurs lesquelles forment un treillis* (J 2313) C R Acad Sci, Paris, Ser A-B 262*A1139-A1141
◇ B50 G10 ◇ REV MR 34 # 1157 Zbl 199.4 • ID 11478

ROSE, A. [1966] *Formalisation de certains calculs propositionnels polyvalents a foncteurs variables au cas ou certaines valeurs sont sur designees* (J 2313) C R Acad Sci, Paris, Ser A-B 262*A1233-A1235
◇ B50 ◇ REV MR 34 # 1174 Zbl 199.4 • ID 41922

ROSE, A. [1967] *A formalisation of the $\aleph_0$-valued Lukasiewicz implicational propositional calculus with variable functors* (J 0068) Z Math Logik Grundlagen Math 13*293-294
◇ B50 ◇ REV MR 36 # 2473 Zbl 161.4 JSL 35.143 • ID 11482

ROSE, A. [1967] *A formalisation of the m-valued Lukasiewicz propositional calculus with super-designated truth-values* (J 0068) Z Math Logik Grundlagen Math 13*295-298
◇ B50 ◇ REV MR 36 # 2474 Zbl 161.4 JSL 36.546 • ID 11480

ROSE, A. [1967] *Formalisation of the $\aleph_0$-valued Lukasiewicz propositional calculus with variable functors* (J 0068) Z Math Logik Grundlagen Math 13*289-292
◇ B50 ◇ REV MR 36 # 2472 Zbl 161.4 JSL 35.142 • ID 11481

ROSE, A. [1967] *Nouveau calcul de non-contradiction* (J 2313) C R Acad Sci, Paris, Ser A-B 265*A125-A128
◇ B50 ◇ REV MR 36 # 37 Zbl 203.7 • ID 11483

Rose, A. [1967] *Nouveau calcul de non-contradiction (methode alternative)* (**J** 2313) C R Acad Sci, Paris, Ser A-B 265*A157-A159
◇ B50 ◇ REV MR 37 # 3913 Zbl 166.251 • ID 24824

Rose, A. [1968] *Binary generators for the m-valued and $\aleph_0$-valued Lukasiewicz propositional calculi* (**J** 0020) Compos Math 20*153-169
• REPR [1968] (**C** 0727) Logic Found of Math (Heyting) 153-169
◇ B50 ◇ REV MR 38 # 29 Zbl 157.15 JSL 40.500
• ID 11485

Rose, A. [1968] *Formalisations of some $\aleph_0$-valued Lukasiewicz propositional calculi* (**P** 0692) Summer School in Logic;1967 Leeds 269-271
◇ B50 ◇ REV MR 39 # 1303 Zbl 182.316 • ID 26677

Rose, A. [1969] *Nouvelle formalisation du calcul propositionnel a $\aleph_0$ valeurs de Lukasiewicz dans le cas ou il y a des foncteurs variables* (**J** 2313) C R Acad Sci, Paris, Ser A-B 269*A265-A269
◇ B50 ◇ REV MR 40 # 1265 Zbl 184.8 • ID 11491

Rose, A. [1969] *Some many-valued propositional calculi without single generators* (**J** 0068) Z Math Logik Grundlagen Math 15*105-106
◇ B50 ◇ REV MR 41 # 39 Zbl 184.280 • ID 11490

Rose, A. [1971] *Note sur la formalisation des calculus propositionnels polyvalents* (**J** 2313) C R Acad Sci, Paris, Ser A-B 272*A1689-A1690
◇ B50 ◇ REV MR 44 # 34 Zbl 229 # 02018 • ID 11494

Rose, A. [1971] *Sur certains foncteurs a m arguments des calculus propositionnels a m valeurs* (**J** 2313) C R Acad Sci, Paris, Ser A-B 273*A273-A274
◇ B50 ◇ REV MR 44 # 35 Zbl 229 # 02020 • ID 11492

Rose, A. [1972] *Locally full $\aleph_0$-valued propositional calculi* (**J** 0068) Z Math Logik Grundlagen Math 18*217-226
◇ B50 ◇ REV MR 46 # 8813 Zbl 253 # 02018 • ID 11495

Rose, A. [1972] *Sur certains calculus propositionnels sans constantes ou tautologies ordinaires* (**J** 2313) C R Acad Sci, Paris, Ser A-B 275*A409-A411
◇ B60 ◇ REV MR 54 # 4920 Zbl 257 # 02011 • ID 24702

Rose, A. [1972] *Tautologies sans variables propositionnelles* (**J** 2313) C R Acad Sci, Paris, Ser A-B 275*A377-A379
◇ B60 ◇ REV MR 54 # 4919 Zbl 257 # 02010 • ID 24095

Rose, A. [1973] *Note sur "l'erweiterter Aussagenkalkuel" ayant un seul quantificateur primitif* (**J** 2313) C R Acad Sci, Paris, Ser A-B 277*A641-A643
◇ B50 ◇ REV MR 48 # 3703 Zbl 266 # 02012 • ID 11496

Rose, A. [1973] *Sur certains "erweiterter Aussagenkalkuel" dont le seul quantificateur primitif constitue son dual* (**J** 2313) C R Acad Sci, Paris, Ser A-B 277*A559-A561
◇ B50 ◇ REV MR 48 # 3702 Zbl 266 # 02011 • ID 11497

Rose, A. [1974] *Sur certains ensembles de foncteurs connectifs primtifs pour des extensions du calcul propositionnel a $\aleph_0$ valeurs de Lukasiewicz lesquels constituent leurs propres duals* (**J** 2313) C R Acad Sci, Paris, Ser A-B 278*A1535-A1538
◇ B50 ◇ REV MR 49 # 8812 Zbl 287 # 02008 • ID 11499

Rose, A. [1974] *Sur un calcul propositionnel a $\aleph_0$ valeurs ayant la disjonction conditionee comme foncteur primitif* (**J** 2313) C R Acad Sci, Paris, Ser A-B 278*A1379-A1382
◇ B50 ◇ REV MR 49 # 8811 Zbl 287 # 02006 • ID 11500

Rose, A. [1974] *Sur un foncteur a cinq arguments d'un calcul propositionnel a $\aleph_0$-valeurs lequel constitue son propre dual* (**J** 2313) C R Acad Sci, Paris, Ser A-B 278*A1465-A1467
◇ B50 ◇ REV MR 49 # 10526 Zbl 287 # 02007 • ID 11498

Rose, A. [1975] *A note on the existence of tautologies without constants* (**J** 0068) Z Math Logik Grundlagen Math 21*141-144
◇ B60 ◇ REV MR 52 # 5341 Zbl 309 # 02007 • ID 11501

Rose, A. [1975] *Sur un ensemble de calculs propositionnels localement sature a $\aleph_0$ valeurs ayant un seul foncteur primitif binaire (English summary)* (**J** 2313) C R Acad Sci, Paris, Ser A-B 280*A397-A399
◇ B50 ◇ REV MR 52 # 5356 Zbl 308 # 02018 • ID 18362

Rose, A. [1976] *A note on the existence of tautologies in certain propositional calculi without propositional variables* (**J** 0068) Z Math Logik Grundlagen Math 22*117-118
◇ B60 ◇ REV MR 58 # 131 Zbl 342 # 02006 • ID 18498

Rose, A. [1978] *A three-valued model for set theory* (**J** 0068) Z Math Logik Grundlagen Math 24*437-440
◇ B50 E35 E70 ◇ REV MR 80d:03051 Zbl 399 # 03039
• ID 52835

Rose, A. [1978] *Formalisations of further $\aleph_0$-valued Lukasiewicz propositional calculi* (**J** 0036) J Symb Logic 43*207-210
◇ B50 ◇ REV MR 58 # 138 Zbl 401 # 03007 • ID 29253

Rose, A. [1979] *Many-valued propositional calculi without constants* (**P** 3003) Int Symp Multi-Val Log (9);1979 Bath 128-134
◇ B50 ◇ REV MR 81f:03033 • ID 77884

Rose, A. [1981] *Extensions de quelques theoremes de McNaughton (English summary)* (**J** 3364) C R Acad Sci, Paris, Ser 1 292*979-981
◇ B50 ◇ REV MR 82k:03026 Zbl 467 # 03017 • ID 55015

Rose, A. [1981] *Many-valued logics* (**C** 2617) Modern Log Survey 113-129
◇ B50 ◇ REV MR 82f:03002 Zbl 464 # 03001 • ID 42766

Rose, A. [1983] *Completeness of sets of three-valued Sheffer functions* (**J** 0068) Z Math Logik Grundlagen Math 29*481-483
◇ B50 ◇ REV MR 85d:03053 Zbl 539 # 03005 • ID 41094

Rose, A. [1984] see Lowesmith, B.J.

Rose, A. [1984] *Generalised functional completeness of sets of m-valued Sheffer functions* (**J** 0068) Z Math Logik Grundlagen Math 30*177-182
◇ B50 ◇ REV MR 85i:03071 • ID 42221

Rose, A. see Vol. I, IV for further entries

Rose, G.S.C. [1978] see Baxandall, P.R.

Rose, H.E. & Shepherdson, J.C. (EDS.) [1975] *Proceedings of the Logic Colloquium (Bristol, July, 1973)* (**X** 0809) North Holland: Amsterdam viii+513pp
◇ B97 ◇ REV MR 51 # 10015 • ID 70192

Rose, H.E. see Vol. IV, VI for further entries

ROSENBAUM, I. [1950] *Introduction to mathematical logic and its applications* (**X** 1306) Univ Miami Pr: Austin iii+98pp
◇ B98 ◇ REV Zbl 41.347 JSL 16.156 • ID 22534

ROSENBAUM, I. see Vol. I for further entries

ROSENBERG, I.G. [1965] *La structure des fonctions de plusieurs variables sur un ensemble fini* (**J** 0109) C R Acad Sci, Paris 260*3817-3819
◇ B50 ◇ REV MR 31#1185 • ID 11514

ROSENBERG, I.G. [1966] *Zu einigen Fragen der Superposition von Funktionen mehrerer Veraenderlicher* (**J** 0198) Bul Inst Politeh Iasi NS 12/1-2*7-15
◇ B50 C05 ◇ REV MR 36#1299 Zbl 163.244 • ID 11515

ROSENBERG, I.G. [1969] *Ueber die Verschiedenheit maximaler Klassen in P_K* (**J** 0060) Rev Roumaine Math Pures Appl 14*431-438
◇ B50 ◇ REV MR 40#28 Zbl 193.291 • ID 11516

ROSENBERG, I.G. [1970] *Complete sets for finite algebras* (**J** 0114) Math Nachr 44*105-114
◇ B50 ◇ REV MR 43#4639 Zbl 212.318 • ID 11518

ROSENBERG, I.G. [1970] *Maximal clones on algebras A and A^r* (**J** 3522) Rend Circ Mat Palermo, Ser 2 18*329-333
◇ B50 ◇ REV Zbl 258#08006 • ID 64913

ROSENBERG, I.G. [1970] *Ueber die funktionale Vollstaendigkeit in den mehrwertigen Logiken. Struktur der Funktionen von mehreren Veraenderlichen auf endlichen Mengen* (**J** 2830) Rozpr, Mat Prirod, Cheskoslov Akad Ved 80*93pp
◇ B50 ◇ REV MR 45#1732 Zbl 199.302 • ID 77916

ROSENBERG, I.G. [1973] *Strongly rigid relations* (**J** 0308) Rocky Mountain J Math 3*631-639
◇ B50 C05 C50 ◇ REV MR 47#6588 Zbl 274#08001 • ID 29025

ROSENBERG, I.G. [1973] *The number of maximal closed classes in the set of functions over a finite domain* (**J** 0164) J Comb Th, Ser A 14*1-7
◇ B50 C05 ◇ REV MR 47#3189 • ID 31986

ROSENBERG, I.G. [1974] *Completeness, closed classes and relations in multiple-valued logics* (**P** 1385) Int Symp Multi-Val Log (4);1974 Morgantown
◇ B50 ◇ ID 31989

ROSENBERG, I.G. [1974] *Some maximal closed classes of operations on infinite sets* (**J** 0043) Math Ann 212*157-164
◇ B50 C05 ◇ REV MR 50#4452 Zbl 288#08003 • ID 31988

ROSENBERG, I.G. [1975] *Functional completeness in heterogeneous multiple-valued logics* (**P** 1805) Int Symp Multi-Val Log (5,Proc);1975 Bloomington 34-43
◇ B50 C05 ◇ REV MR 58#9882 • ID 31990

ROSENBERG, I.G. [1975] *Polynomial functions over finite rings (Serbo-Croatian summary)* (**J** 3519) Glas Mat, Ser 3 (Zagreb) 10(30)*25-33
◇ B50 ◇ REV MR 52#352 Zbl 308#12005 • ID 18363

ROSENBERG, I.G. [1975] *Special types of universal algebras preserving a relation* (**J** 0050) Port Math 34(3)*173-188
◇ B50 C05 ◇ REV MR 57#5876 Zbl 327#08008 • ID 31991

ROSENBERG, I.G. [1976] *Some algebraic and combinatorial aspects of multiple-valued circuits* (**P** 2011) Int Symp Multi-Val Log (6);1976 Logan 9-23
◇ B50 ◇ ID 35845

ROSENBERG, I.G. [1976] *The set of maximal closed classes of operations on an infinite set A has cardinality $2^{2^{|A|}}$* (**J** 0008) Arch Math (Basel) 27*561-568
◇ B50 C05 E20 ◇ REV MR 55#2711 Zbl 345#02010 • ID 26563

ROSENBERG, I.G. [1977] *On closed classes, basic sets and groups* (**P** 2013) Int Symp Multi-Val Log (7);1977 Charlotte 1-6
◇ B50 ◇ REV MR 58#28240 • ID 35848

ROSENBERG, I.G. [1978] see MARTIN, L.

ROSENBERG, I.G. [1978] *On generating large classes of Sheffer functions* (**J** 0529) Aequationes Math 17*164-181
◇ B50 C05 ◇ REV MR 58#10306 Zbl 386#08005 • ID 31985

ROSENBERG, I.G. [1978] *Subalgebra systems of direct powers* (**J** 0004) Algeb Universalis 8*221-227
◇ B50 C05 ◇ REV MR 58#5449 Zbl 381#08003 • ID 31993

ROSENBERG, I.G. [1980] see MUZIO, J.C.

ROSENBERG, I.G. [1981] see CSAKANY, B.

ROSENBERG, I.G. [1981] *Large classes of functionally complete groupoids II* (**P** 3705) Int Symp Multi-Val Log (11);1981 Oklahoma City & Norman 259-262
◇ B50 ◇ REV MR 83m:94033 Zbl 547#08004 • REM Part I 1980 by Muzio,J.C. & Rosenberg,I.G. • ID 45942

ROSENBERG, I.G. [1982] see MUZIO, J.C.

ROSENBERG, I.G. [1985] see DAVIES, R.O.

ROSENBERG, I.G. see Vol. I, V for further entries

ROSENBERG, J. [1968] *Functional completeness in one variable* (**J** 0036) J Symb Logic 33*105-106
◇ B50 ◇ REV MR 39#3976 Zbl 157.16 • ID 11520

ROSENBERG, J. [1969] *The application of ternary semi-groups to the study of n-valued Sheffer functions* (**J** 0047) Notre Dame J Formal Log 10*90-94
◇ B50 ◇ REV MR 41#1513 Zbl 191.288 JSL 38.520 • ID 11521

ROSENBERG, S. [1981] *On the modal version of the ontological argument* (**J** 0079) Logique & Anal, NS 24*129-133
◇ A05 B45 ◇ REV MR 82m:03020 Zbl 488#03005 • ID 77917

ROSENBERG, S. see Vol. I for further entries

ROSENBLOOM, P.C. [1950] *The elements of mathematical logic* (**X** 0813) Dover: New York 6+iv+214pp
◇ B98 ◇ REV MR 12.789 JSL 18.277 • ID 11524

ROSENBLOOM, P.C. see Vol. IV, VI for further entries

ROSENKRANTZ, R.D. [1981] *Foundations and applications of inductive probability* (**X** 3905) Ridgeview: Atascadero 296pp
◇ B48 ◇ REV MR 84k:60005 Zbl 533#62003 • ID 36575

ROSENSCHEIN, S.J. [1977] see HOBBS, J.R.

ROSIEK, M. [1973] see BRYLL, G.

ROSSER, J.B. [1939] *The introduction of quantification into a three-valued logic* (J 0158) Erkenntnis (Dordrecht) 9*6pp
◇ B50 ◇ REV JSL 4.170 • ID 28435

ROSSER, J.B. [1941] *On the many-valued logics* (J 0955) Amer J Phys 9*207-212
◇ B50 ◇ REV MR 3.130 Zbl 61.8 JSL 6.109 • ID 11549

ROSSER, J.B. & TURQUETTE, A.R. [1945] *Axiom schemes for m-valued propositional calculi* (J 0036) J Symb Logic 10*61-82
◇ B50 ◇ REV MR 7.185 JSL 11.86 • ID 11552

ROSSER, J.B. & TURQUETTE, A.R. [1948] *Axiom schemes for m-valued functional calculi of first order. Part I. Definition of axiom schemes and proof of plausibility* (J 0036) J Symb Logic 13*177-192
◇ B50 ◇ REV MR 10.420 Zbl 35.5 JSL 14.259 • REM Part II 1951 • ID 11554

ROSSER, J.B. & TURQUETTE, A.R. [1949] *A note on the deductive completeness of m-valued propositional calculi* (J 0036) J Symb Logic 14*219-225
◇ B50 ◇ REV MR 11.709 Zbl 36.7 JSL 15.137 • ID 11556

ROSSER, J.B. & TURQUETTE, A.R. [1951] *Axiom schemes for m-valued functional calculi of first order. Part II. Deductive completeness* (J 0036) J Symb Logic 16*22-34
◇ B50 ◇ REV MR 12.790 Zbl 42.244 JSL 16.269 • REM Part I 1948 • ID 11560

ROSSER, J.B. & TURQUETTE, A.R. [1952] *Many-valued logics* (X 0809) North Holland: Amsterdam vii+124pp
◇ A05 A10 B50 ◇ REV MR 14.526 Zbl 47.15 JSL 20.45 • REM 2nd edition 1958; 124pp • ID 11562

ROSSER, J.B. [1953] *Logic for mathematicians* (X 0822) McGraw-Hill: New York xiv+530pp
• LAST ED [1978] (X 0848) Chelsea: New York xv+574pp
◇ B15 B98 E70 E98 ◇ REV MR 14.935 MR 80b:03001 Zbl 404#03001 Zbl 68.7 JSL 18.326 JSL 45.631 • REM 2nd ed. 1978 • ID 11565

ROSSER, J.B. [1957] *Many-valued logics* (P 1675) Summer Inst Symb Log;1957 Ithaca 1-2
◇ B50 ◇ ID 29307

ROSSER, J.B. [1958] see ROSE, A.

ROSSER, J.B. [1960] *Axiomatization of infinite valued logics* (J 0079) Logique & Anal, NS 3*137-153
◇ B50 ◇ REV JSL 27.111 • ID 11568

ROSSER, J.B. & TURQUETTE, A.R. [1967] *Many valued logics* (C 0721) Contemp Readings in Log Th 335-342
◇ A05 B50 ◇ ID 11569

ROSSER, J.B. see Vol. I, III, IV, V, VI for further entries

ROSSKOPF, M.F. [1959] see EXNER, R.M.

ROUNDS, W.C. [1983] see BROOKES, S.D.

ROURE, M.-L. [1967] *Elements de logique contemporaine* (X 0840) Pr Univ France: Paris 127pp
◇ B98 ◇ ID 22428

ROUSSEAU, G. [1967] *Completeness in finite algebras with a single operation* (J 0053) Proc Amer Math Soc 18*1009-1013
◇ B50 C05 ◇ REV MR 36#97 Zbl 157.41 JSL 40.98 • ID 11594

ROUSSEAU, G. [1967] *Sequents in many valued logic I* (J 0027) Fund Math 60*23-33 • ERR/ADD ibid 61*313
◇ B50 ◇ REV MR 35#1451 Zbl 154.255 • REM Part II 1970 • ID 19621

ROUSSEAU, G. [1969] *Logical systems with finitely many truth-values (Russian summary)* (J 0014) Bull Acad Pol Sci, Ser Math Astron Phys 17*189-194
◇ B50 G20 ◇ REV MR 41#3247 Zbl 284#02009 • ID 11600

ROUSSEAU, G. [1970] *Sequents in many valued logic II* (J 0027) Fund Math 67*125-131
◇ B50 ◇ REV MR 41#6671 Zbl 194.307 • REM Part I 1967 • ID 11603

ROUSSEAU, G. [1970] *The separation theorem for fragments of the intuitionistic propositional calculus* (J 0068) Z Math Logik Grundlagen Math 16*469-474
◇ B55 F50 ◇ REV MR 43#1815 Zbl 211.10 • ID 11601

ROUSSEAU, G. see Vol. I, III, V, VI for further entries

ROUTLEY, R. [1966] see MONTGOMERY, H.

ROUTLEY, R. [1966] *On a signification theory* (J 0273) Australasian J Phil 44*172-209
◇ B50 ◇ ID 15191

ROUTLEY, R. [1968] *Decision procedures and semantics for* C1, E1, *and* $S0.5^0$ (J 0079) Logique & Anal, NS 11*468-471
◇ B25 B45 ◇ REV MR 40#1267 Zbl 195.297 JSL 38.329 • ID 11610

ROUTLEY, R. [1968] see MONTGOMERY, H.

ROUTLEY, R. [1968] see MONTGOMERY, D.

ROUTLEY, R. [1968] *The decidability and semantical incompleteness of Lemmon's system S0.5* (J 0079) Logique & Anal, NS 11*413-421
◇ B25 B45 ◇ REV MR 38#4290 Zbl 172.292 JSL 38.328 • ID 11609

ROUTLEY, R. & ROUTLEY, V. [1969] *A fallacy of modality* (J 0097) Nous, Quart J Phil 3*129-153
◇ B46 ◇ ID 30842

ROUTLEY, R. [1969] *Existence and identity in quantified modal logics* (J 0047) Notre Dame J Formal Log 10*113-149
◇ B45 ◇ REV MR 43#7307 Zbl 169.299 • ID 11612

ROUTLEY, R. [1969] see MONTGOMERY, H.

ROUTLEY, R. [1970] *Decision procedures and semantics for Feys' system* $S2^0$ *and surrounding systems* (J 0068) Z Math Logik Grundlagen Math 16*165-174
◇ B25 B45 ◇ REV MR 43#1819 Zbl 206.275 • ID 11613

ROUTLEY, R. [1970] *Extensions of Makinsons completeness theorems in modal logic* (J 0068) Z Math Logik Grundlagen Math 16*239-256
◇ B45 ◇ REV MR 44#2581 Zbl 165.16 • ID 11615

ROUTLEY, R. [1971] *Conventionalist and contingency-oriented modal logics* (J 0047) Notre Dame J Formal Log 12*131-152
◇ B45 ◇ REV MR 44#6459 Zbl 222#02017 • ID 11618

ROUTLEY, R. [1972] *A semantical analysis of implicational system I and of the first degree of entailment* (J 0043) Math Ann 196*58-84
◇ B46 ◇ REV MR 54#4926 Zbl 255#02017 • ID 24101

ROUTLEY, R. [1972] see MEYER, R.K.

ROUTLEY, R. & ROUTLEY, V. [1972] *The semantics of first degree entailment* (J 0097) Nous, Quart J Phil 6*335-359
◇ B46 ◇ REV MR 58 # 21481 • ID 11622

ROUTLEY, R. [1972] *Vrendenduin's system of strict implication* (J 0079) Logique & Anal, NS 15*435-437
◇ B46 ◇ REV MR 49 # 2268 Zbl 267 # 02012 • ID 11620

ROUTLEY, R. [1973] see MEYER, R.K.

ROUTLEY, R. [1973] see BRADY, R.T.

ROUTLEY, R. [1973] see GODDARD, L.

ROUTLEY, R. [1974] *A rival account of logical consequence* (J 0302) Rep Math Logic, Krakow & Katowice 3*41-51
◇ A05 B46 ◇ REV MR 51 # 12484 Zbl 308 # 02025 • ID 17272

ROUTLEY, R. [1974] see GODDARD, L.

ROUTLEY, R. [1974] see MEYER, R.K.

ROUTLEY, R. & WOLF, R.G. [1974] *No rational sentential logic has a finite characteristic matrix* (J 0079) Logique & Anal, NS 17*317-321
◇ B50 ◇ REV MR 55 # 2505 Zbl 314 # 02028 • ID 64934

ROUTLEY, R. [1974] *Semantical analyses of propositional systems of Fitch and Nelson* (J 0063) Studia Logica 33*283-298
◇ B25 B46 C90 F50 ◇ REV MR 51 # 70 Zbl 356 # 02022 • ID 15192

ROUTLEY, R. & ROUTLEY, V. [1975] *The role of inconsistent and incomplete theories in the logic of belief* (J 2705) Comm Cognition Monograph 8*185-235
◇ B45 ◇ ID 42888

ROUTLEY, R. [1975] see MEYER, R.K.

ROUTLEY, R. [1976] see MONTGOMERY, H.

ROUTLEY, R. [1976] see MEYER, R.K.

ROUTLEY, R. [1977] *Choice and descriptions in enriched intensional languages. II,III* (P 2116) Probl in Log & Ontology;1973 Salzburg 173-204,205-222
◇ A05 B60 ◇ REV MR 58 # 16156b,c Zbl 384 # 03013 Zbl 384 # 03014 • REM Part I 1974 by Goddard,L. & Meyer,R.K. & Routley,R. • ID 52055

ROUTLEY, R. [1977] see MEYER, R.K.

ROUTLEY, R. [1977] *Postscript: Some setbacks on the choice and descriptions adventure* (P 2116) Probl in Log & Ontology;1973 Salzburg 223-227
◇ A05 B45 ◇ REV MR 58 # 16156d Zbl 384 # 03015 • ID 52057

ROUTLEY, R. [1977] *Welding semantics for weak strict modal logics into the general framework of modal logic semantics* (J 0068) Z Math Logik Grundlagen Math 23*497-510
◇ B45 ◇ REV MR 58 # 152 Zbl 412 # 03004 • ID 52935

ROUTLEY, R. [1978] *An inadequacy in Kripke-semantics for intuitionistic quantificational logic* (J 0387) Bull Sect Logic, Pol Acad Sci 7*61-67
◇ B55 C90 F50 ◇ REV MR 58 # 16172 Zbl 414 # 03037 • ID 53083

ROUTLEY, R. [1978] *Constant domain semantics for quantified non-normal modal logics and for certain quantified quasi-entailment logics* (J 0302) Rep Math Logic, Krakow & Katowice 10*103-121
◇ B45 ◇ REV MR 80m:03041 Zbl 419 # 03009 • ID 53352

ROUTLEY, R. [1978] see LOPARIC, A.

ROUTLEY, R. [1978] *Semantics for connexive logics. I* (J 0063) Studia Logica 37*393-412
◇ A05 B45 ◇ REV MR 82i:03031 Zbl 405 # 03009 • ID 54884

ROUTLEY, R. [1979] *Alternative semantics for quantified first degree relevant logic* (J 0063) Studia Logica 38*211-231
◇ B46 ◇ REV MR 81g:03020 Zbl 406 # 03033 • ID 56116

ROUTLEY, R. [1979] *Dialectical logic, semantics, and metamathematics* (J 0158) Erkenntnis (Dordrecht) 14*301-331
◇ B53 ◇ ID 42885

ROUTLEY, R. [1979] *Repairing proofs of Arrow's general impossibility theorem and enlarging the scope of the theorem* (J 0047) Notre Dame J Formal Log 20*879-890
◇ B46 ◇ REV MR 81c:90013 Zbl 438 # 90001 • ID 33497

ROUTLEY, R. [1979] see GRIFFIN, N.

ROUTLEY, R. [1980] *Problems and solutions in the semantics of quantified relevant logics. I* (P 2958) Latin Amer Symp Math Log (4);1978 Santiago 305-340
◇ B46 ◇ REV MR 81h:03041 Zbl 426 # 03014 • ID 53610

ROUTLEY, R. [1980] see LOPARIC, A.

ROUTLEY, R. [1982] see PRIEST, G.

ROUTLEY, R. [1982] see MARTIN, E.P.

ROUTLEY, R. [1982] see BRADY, R.T.

ROUTLEY, R. [1983] see MEYER, R.K.

ROUTLEY, R. & ROUTLEY, V. [1983] *Semantical foundations for value theory* (J 0097) Nous, Quart J Phil 17*441-456
◇ B45 ◇ REV MR 85g:03033 • ID 43869

ROUTLEY, R. [1984] see PRIEST, G.

ROUTLEY, R. [1984] *The American plan completed: Alternative classical-style semantics, without stars, for relevant and paraconsistent logics* (J 0063) Studia Logica 43*131-158
◇ B46 B53 ◇ ID 42366

ROUTLEY, R. see Vol. I, IV, V for further entries

ROUTLEY, V. [1969] see ROUTLEY, R.

ROUTLEY, V. [1972] see ROUTLEY, R.

ROUTLEY, V. [1975] see ROUTLEY, R.

ROUTLEY, V. [1982] see MARTIN, E.P.

ROUTLEY, V. [1983] see ROUTLEY, R.

ROWE, I. [1972] *The generation of Gaussian-distributed pseudorandum noise sequences from multiple-valued logic* (P 2008) Symp Th & Appl of Multi-Val Log Design;1972 Buffalo 136-152
◇ B50 ◇ ID 42040

ROY, B.B. & SHUKLA, R. [1973] *Logical interpretation of probability calculus and derivation of many valued logic* (J 0957) Ranchi Univ Math J 4*11-14
◇ B48 ◇ REV MR 49 # 7117 Zbl 301 # 02021 • ID 11631

ROZEBOOM, W.W. [1980] *Nicod's criterion: subtler than you think* (**J** 0153) Phil of Sci (East Lansing) 47*638-643
◇ B48 ◇ REV MR 84c:03047 • ID 35699

ROZONOEHR, L.I. [1963] see AJZERMAN, M.A.

ROZONOEHR, L.I. [1983] *Proving contradictions in formal theories. I, II (Russian)* (**J** 0011) Avtom Telemekh 1983/6*113-124,7*97-104 • ERR/ADD ibid 1985/4*172 • TRANSL [1983] (**J** 0010) Autom & Remote Control 44*781-790,908-914
◇ B53 ◇ REV MR 85g:03041 Zbl 532#03007 Zbl 532#03008 • ID 38273

ROZONOEHR, L.I. see Vol. IV for further entries

RUBIN, H. & SUPPES, P. [1955] *A note on two-place predicates and fitting sequences of measure functions* (**J** 0036) J Symb Logic 20*121-122
◇ B48 ◇ REV MR 17.119 Zbl 65.1 JSL 39.341 • ID 11647

RUBIN, H. see Vol. I, V for further entries

RUBIN, J.E. [1962] *Bi-modal logic, double-closure algebras, and Hilbert space* (**J** 0068) Z Math Logik Grundlagen Math 8*305-322
◇ B45 G20 ◇ REV MR 27#1357 Zbl 111.7 JSL 37.184 • ID 11652

RUBIN, J.E. see Vol. III, V for further entries

RUBY, L. [1960] *Logic: An introduction* (**X** 1297) Lippincott: Philadelphia xviii+522pp
◇ B98 ◇ ID 22535

RUDEANU, S. [1979] see GILEZAN, K.

RUDEANU, S. see Vol. I, V for further entries

RUETTIMANN, G.T. [1981] *Detectable properties and spectral quantum logics* (**P** 3185) Interpr & Found of Quantum Th;1979 Marburg 35-47
◇ B51 G12 ◇ REV MR 85g:81005 Zbl 474#03036 • ID 55440

RUETTIMANN, G.T. [1982] see GREECHIE, R.J.

RUETTIMANN, G.T. [1985] see COOK, T.A.

RUIZ SHULCLOPER, J. [1972] see BUENO SANCHEZ, E.

RUSAKOV, E.D. [1972] see MASLOV, S.YU.

RUSPINI, E.H. [1982] *Possibilistic data structures for the representation of uncertainty* (**C** 3786) Approx Reason in Decis Anal 411-416
◇ B52 ◇ REV MR 84d:90003 • ID 46119

RUSPINI, E.H. see Vol. V for further entries

RUSSELL, B. [1932] see WHITEHEAD, A.N.

RUSSELL, B. see Vol. I, V for further entries

RUTKOWSKI, A. [1978] *Elements of mathematical logic (Polish)* (**X** 2881) Wydawn Szkol Ped: Warsaw 200pp
◇ B98 ◇ REV MR 58#4969 • ID 78008

RUTKOWSKI, A. see Vol. III, IV, V for further entries

RUTLEDGE, J.D. [1960] *On the definition of an infinitely-many valued predicate calculus* (**J** 0036) J Symb Logic 25*212-216
◇ B50 G15 ◇ REV MR 25#1993 Zbl 105.5 • ID 11698

RUTLEDGE, J.D. see Vol. IV, VI for further entries

RUTZ, P. [1973] *Zweiwertige und mehrwertige Logik - ein Beitrag zur Geschichte und Einheit der Logik* (**X** 2036) Ehrenwirth: Muenchen 108pp
◇ A10 B98 ◇ REV MR 55#10235 Zbl 454#03002 • ID 32236

RUZAVIN, G.I. [1967] *The semantic conception of inductive logic (Russian)* (**C** 0582) Log Semant & Modal Logika 84-100
◇ A05 B48 ◇ REV MR 36#1295 Zbl 204.2 • ID 11700

RUZAVIN, G.I. [1972] *Die semantische Konzeption der induktiven Logik* (**C** 1533) Quantoren, Modal, Paradox 299-317
◇ A05 B48 ◇ REV Zbl 265#02021 • ID 29819

RUZSA, I. [1965] *Axiomatischer Aufbau eines Systems der deontischen Logik* (**J** 0002) Acta Sci Math (Szeged) 26*253-267
◇ B45 ◇ REV MR 32#5505 Zbl 156.7 • ID 11702

RUZSA, I. [1968] *Ein neues formales System der deontischen Logik* (**J** 0001) Acta Math Acad Sci Hung 19*287-309
◇ B45 ◇ REV MR 45#1737 • ID 11691

RUZSA, I. [1969] *Random models of some quasi-intuitionistic logics* (**J** 0006) Ann Univ Budapest, Sect Math 12*77-93
◇ B55 C90 ◇ REV MR 47#1603 Zbl 193.302 • ID 11692

RUZSA, I. [1969] *Random models of logical systems I: Models of valuing logics* (**J** 0411) Studia Sci Math Hung 4*301-312
◇ B50 C90 ◇ REV MR 47#1600 Zbl 187.269 • REM Part II 1971 • ID 19613

RUZSA, I. [1971] *Random models of logical systems II: Models of some modal logics* (**J** 0411) Studia Sci Math Hung 5*255-265
◇ B45 C90 ◇ REV MR 47#1601 Zbl 237#02016 • REM Part I 1969. Part III 1971 • ID 19614

RUZSA, I. [1971] *Random models of logical systems III: Models of quantified logics* (**J** 0049) Period Math Hung 1*195-208
◇ B50 C90 ◇ REV MR 47#1602 Zbl 221#02046 • REM Part II 1971 • ID 11693

RUZSA, I. [1973] *Prior-type modal logic I,II* (**J** 0049) Period Math Hung 4*51-69,183-201
◇ B45 ◇ REV MR 48#10767 MR 51#10035 Zbl 264#02023 Zbl 285#02018 • ID 11694

RUZSA, I. [1975] *Two variants of the system of entailment* (**J** 0068) Z Math Logik Grundlagen Math 21*57-68
◇ B46 ◇ REV MR 51#2870 Zbl 315#02031 • ID 11695

RUZSA, I. [1976] *Semantics for von Wright's latest deontic logic* (**J** 0063) Studia Logica 35*297-314
◇ B45 ◇ REV MR 57#9487 Zbl 361#02029 • ID 31995

RUZSA, I. [1977] *Semantica per la logica deontica del primo ordine* (**P** 2149) Log Deont & Semant;1975 Bielefeld 49-68
◇ B45 ◇ ID 31996

RUZSA, I. [1978] *A new approach to modal logic* (**J** 2293) Comp Linguist & Comp Lang 12*9-29
◇ B45 ◇ REV MR 80b:03026 Zbl 426#03025 • ID 53621

RUZSA, I. [1981] *An approach to intensional logic* (**J** 0063) Studia Logica 40*269-287
◇ B45 ◇ REV MR 83j:03035 Zbl 491#03006 • ID 35343

RUZSA, I. [1981] *Modal logic with descriptions* (**X** 1316) Nijhoff: Leiden & s'Gravenhage 135pp
◇ A05 B45 ◇ REV MR 84f:03018 Zbl 479#03009 • ID 55671

RUZSA, I. [1984] *Intensional logic without intensional variables (Russian)* (C 4366) Modal & Intens Log & Primen Probl Metodol Nauk 220-245
◇ B45 ◇ ID 46678

RVACHEV, V.L. & SHKLYAROV, L.I. [1967] *Some functionally closed classes of functions in k-valued logic (Russian)* (J 0040) Kibernetika, Akad Nauk Ukr SSR 1967/2*42-44
• TRANSL [1967] (J 0021) Cybernetics 3/2*32-34
◇ B50 ◇ REV MR 43 # 1818 Zbl 216.287 • ID 19611

RVACHEV, V.L. & SHKLYAROV, L.I. [1968] *R_k-functions (Ukrainian) (Russian and English summaries)* (J 0270) Dokl Akad Nauk Ukr SSR, Ser A 1968*415-417
◇ B50 ◇ REV MR 39 # 1304 • ID 28473

RVACHEV, V.L. & SHKLYAROV, L.I. [1968] *Predicate description of domains of complex form (Russian)* (J 0040) Kibernetika, Akad Nauk Ukr SSR 1968/3*59-62
• TRANSL [1968] (J 0021) Cybernetics 4/3*51-54
◇ B50 ◇ REV MR 45 # 8482 • ID 30873

RVACHEV, V.L. [1971] *A certain extension of the concept of r-function (Russian) (English summary)* (J 0040) Kibernetika, Akad Nauk Ukr SSR 1971/4*86-92
◇ B50 ◇ REV MR 46 # 5103 • ID 11704

RVACHEV, V.L. & SHKLYAROV, L.I. & TONITSA, V.S. [1976] *The closing functions of three-valued logic (Russian) (English summary)* (J 0270) Dokl Akad Nauk Ukr SSR, Ser A 1976*499-502,574
◇ B50 ◇ REV MR 54 # 2418 Zbl 332 # 02016 • ID 24616

RVACHEV, V.L. & SHKLYAROV, L.I. & TONITSA, V.S. [1979] *A set of closing prime functions of three-valued logic (Ukrainian) (English and Russian summaries)* (J 0270) Dokl Akad Nauk Ukr SSR, Ser A 1979*89-94,155
◇ B50 ◇ REV MR 80j:03093 Zbl 395 # 03019 • ID 52570

RVACHEV, V.L. see Vol. I for further entries

RYBAKOV, V.V. [1974] see MAKSIMOVA, L.L.

RYBAKOV, V.V. [1976] *Hereditarily finitely axiomatizable extensions of logic S4 (Russian)* (J 0003) Algebra i Logika 15*185-204,245-246
• TRANSL [1976] (J 0069) Algeb and Log 15*115-128
◇ B45 ◇ REV MR 57 # 12181 Zbl 358 # 02022 • ID 26039

RYBAKOV, V.V. [1977] *Noncompact extensions of the logic S4 (Russian)* (J 0003) Algebra i Logika 16*472-490,494
• TRANSL [1977] (J 0069) Algeb and Log 16*321-334
◇ B45 C90 ◇ REV MR 58 # 27346 Zbl 406 # 03039
• ID 56122

RYBAKOV, V.V. [1978] *A decidable noncompact extension of the logic S4 (Russian)* (J 0003) Algebra i Logika 17*210-219
• TRANSL [1978] (J 0069) Algeb and Log 17*148-154
◇ B25 B45 ◇ REV MR 80m:03042 Zbl 415 # 03012
• ID 53114

RYBAKOV, V.V. [1978] *Modal logics with LM-axioms (Russian)* (J 0003) Algebra i Logika 17*455-467,491
• TRANSL [1978] (J 0069) Algeb and Log 17*302-310
◇ B25 B45 ◇ REV MR 80m:03043 Zbl 417 # 03007
• ID 53244

RYBAKOV, V.V. [1981] *Admissible rules for pretable modal logics (Russian)* (J 0003) Algebra i Logika 20*440-464,485
• TRANSL [1981] (J 0069) Algeb and Log 20*291-307
◇ B45 ◇ REV MR 83m:03029 Zbl 489 # 03005 • ID 35434

RYBAKOV, V.V. [1982] *Bases of quasiidentities of finite modal algebras (Russian)* (J 0003) Algebra i Logika 21*219-227
• TRANSL [1982] (J 0069) Algeb and Log 21*149-155
◇ B45 ◇ REV MR 84j:03137 Zbl 507 # 08005 • ID 34727

RYBAKOV, V.V. [1982] *Completeness of modal logics of prefinite width (Russian)* (J 0087) Mat Zametki (Akad Nauk SSSR) 32*223-228,270
• TRANSL [1982] (J 1044) Math Notes, Acad Sci USSR 32*591-593
◇ B45 ◇ REV MR 84h:03043 Zbl 494 # 03012 Zbl 504 # 03008 • ID 34251

RYBAKOV, V.V. [1984] *A criterion for admissibility of rules in the modal system S4 and the intuitionistic logic (Russian)* (J 0003) Algebra i Logika 23*546-572,600
• TRANSL [1984] (J 0069) Algeb and Log 23*369-384
◇ B45 ◇ ID 46551

RYBAKOV, V.V. [1984] *Admissible rules for logics containing S4.3 (Russian)* (J 0092) Sib Mat Zh 25/5*141-145
• TRANSL [1984] (J 0475) Sib Math J 25*795-798
◇ B45 ◇ REV MR 86g:03036 • ID 44005

RYBAKOV, V.V. [1984] *Decidability of the admissibility problem in layer-finite modal logics (Russian)* (J 0003) Algebra i Logika 23*100-116,120
• TRANSL [1984] (J 0069) Algeb and Log 23*75-87
◇ B25 B45 ◇ REV Zbl 576 # 03012 • ID 39654

RYBAKOV, V.V. [1985] *A criterion for admissibility of deduction rules in modal and intuitionistic logic (Russian)* (J 0023) Dokl Akad Nauk SSSR 284*538-541
• TRANSL [1985] (J 0062) Sov Math, Dokl 32*452-455
◇ B45 B55 ◇ ID 48969

RYBAKOV, V.V. [1985] *Bases of admissible rules of the modal system Grz and intuitionistic logic (Russian)* (J 0142) Mat Sb, Akad Nauk SSSR, NS 128*321-338
◇ B45 B55 ◇ ID 49363

RYBAKOV, V.V. [1985] *Bases of admisssible rules of the logics S4 and Int (Russian)* (J 0003) Algebra i Logika 24*87-107,123
• TRANSL [1985] (J 0069) Algeb and Log 24*55-68
◇ B45 ◇ ID 49368

RYCHKOV, S.V. [1982] see ANSHAKOV, O.M.

RYCHKOV, S.V. [1984] see ANSHAKOV, O.M.

RYCHKOV, S.V. see Vol. III, V for further entries

SAARINEN, E. [1979] *Backwards-looking operators in tense logic and in natural language* (P 1705) Scand Logic Symp (4);1976 Jyvaeskylae 341-367
◇ B45 B65 ◇ REV MR 80m:03044 Zbl 421 # 03005
• ID 31500

SAARINEN, E. [1979] see HILPINEN, R.

SAARINEN, E. [1979] see HINTIKKA, K.J.J.

SAARINEN, E. [1979] see NIINILUOTO, I.

SAARINEN, E. [1982] *Propositional attitudes are not attitudes towards propositions* (P 3800) Intens Log: Th & Appl;1979 Moskva 130-162
◇ A05 B45 ◇ REV MR 84h:03044 • ID 34252

SAARINEN, E. [1984] *Does inference with objects of propositional purpose exist? (Russian)* (C 4366) Modal & Intens Log & Primen Probl Metodol Nauk 249-273
◇ B45 ◇ ID 46717

SAARINEN, E. see Vol. I for further entries

SABBAGH, G. [1971] *Logique mathematique I. Generalites* (C 1495) Encycl Universalis 10*53-56
◇ B98 ◇ REV JSL 38.341 • REM Part II 1971 by Reznikoff,I. • ID 28607

SABBAGH, G. see Vol. I, III, IV, V for further entries

SABER, J.C. [1971] see HANNA, S.C.

SADE, A. [1969] *Fonctions propositionnelles monadiques dans la logique trivalente* (J 0962) Ann Soc Sci Bruxelles, Ser 1 83*203-214
◇ B50 ◇ REV MR 40 # 4085 Zbl 175.264 • ID 11758

SADE, A. [1969] *Sur le premier systeme de Lukasiewicz* (J 0322) Arch Math (Brno) 5*207-214
◇ B50 ◇ REV MR 43 # 7302 Zbl 256 # 02010 • ID 11759

SADE, A. [1970] *Sur les axiomes de Goetlind* (J 0047) Notre Dame J Formal Log 11*81-88
◇ B50 ◇ REV MR 45 # 25 Zbl 169.298 • ID 11761

SADE, A. see Vol. I, V for further entries

SADOWSKI, W. [1964] *A proof of axiomatizability of certain n-valued sentential calculi (Polish) (Russian and English summaries)* (J 0063) Studia Logica 15*25-36
◇ B50 ◇ REV MR 32 # 2318 Zbl 292 # 02015 JSL 31.501 • ID 24783

SADOWSKI, W. see Vol. I for further entries

SAELI, D. [1975] *Problemi di decisione per algebre connesse a logiche a piu valori (English summary)* (J 0149) Atti Accad Naz Lincei Fis Mat Nat, Ser 8 59*219-223
◇ B25 B50 C60 G20 ◇ REV MR 57 # 5722 Zbl 354 # 02037 • ID 78070

SAELI, D. see Vol. III for further entries

SAERNDAL, C.-E. [1968] *Some aspects of Carnap's theory of inductive inference* (J 0013) Brit J Phil Sci 19*225-246
◇ A05 B48 ◇ REV MR 38 # 6691 Zbl 251 # 02010 • ID 28903

SAHLIN, N.-E. [1982] *On counterfactual probabilities and causation: a note* (J 0079) Logique & Anal, NS 25*327-332
◇ B45 ◇ REV MR 84g:03026 • ID 34140

SAHLIN, N.-E. [1982] see GAERDENFORS, P.

SAHLQVIST, H. [1975] *Completeness and correspondence in the first and second order semantics for modal logic* (P 0757) Scand Logic Symp (3);1973 Uppsala 110-143
◇ B45 ◇ REV MR 52 # 7855 Zbl 319 # 02018 JSL 43.373 • ID 18372

SAIN, I. [1981] *First order dynamic logic with decidable proofs and workable model theory* (P 3165) FCT'81 Fund of Comput Th;1981 Szeged 334-340
◇ B75 ◇ REV MR 83g:03020 Zbl 494 # 03022 • ID 35998

SAIN, I. [1983] *Total correctness in nonstandard dynamic logic* (J 0387) Bull Sect Logic, Pol Acad Sci 12*64-70
◇ B75 H10 ◇ REV MR 85b:03043 Zbl 538 # 68017 • ID 40674

SAIN, I. [1984] *Structured nonstandard dynamic logic* (J 0068) Z Math Logik Grundlagen Math 30*481-497
◇ B75 ◇ REV MR 86a:03029 Zbl 552 # 68034 • ID 42276

SAIN, I. see Vol. I, III, V for further entries

SAINT-DIZIER, P. (ED.) [1985] *Natural language understanding and logic programming* (X 0809) North Holland: Amsterdam xi+234pp
◇ B65 B75 B97 ◇ REV Zbl 574 # 68075 • ID 48837

SAITO, S. [1966] *On the completeness of the Leibnizian modal system with a restriction* (J 0081) Proc Japan Acad 42*198-200
◇ B45 ◇ REV MR 38 # 3138 Zbl 166.3 • ID 11767

SAITO, S. [1968] *On the Leibnizian modal system* (J 0047) Notre Dame J Formal Log 9*92-96
◇ A05 B45 ◇ REV MR 39 # 5325 Zbl 177.8 • ID 11768

SAITO, S. [1973] *Modality and preference relation* (J 0047) Notre Dame J Formal Log 14*387-391
◇ B45 ◇ REV MR 49 # 4755 Zbl 258 # 02023 • ID 11770

SAITO, S. see Vol. I for further entries

SAKALAUSKAITE, J. [1985] *Completeness theorem for some propositional dynamic logics with infinite repetition (Russian) (English and Lithuanian summaries)* (J 3939) Mat Logika Primen (Akad Nauk Litov SSR) 1985/4*39-64,137
◇ B75 ◇ ID 49023

SALES, T. [1982] *A multivalued Boolean logic (Catalan)* (P 3870) Congr Catala de Log Mat (1);1982 Barcelona 113-116
◇ B50 ◇ REV MR 84i:03003 Zbl 526 # 03007 • ID 44702

SALES, T. see Vol. V for further entries

SALIJ, V.N. [1965] *Binary $\mathscr{L}$-relations (Russian)* (J 0031) Izv Vyssh Ucheb Zaved, Mat (Kazan) 1965/1(44)*133-145
◇ B50 E07 E70 G10 G15 ◇ REV MR 31 # 110 Zbl 199.22 • ID 16216

SALIJ, V.N. see Vol. III, V for further entries

SALMON, W.C. [1963] *Logic* (X 0819) Prentice Hall: Englewood Cliffs xiv+114pp
◇ B98 ◇ REV JSL 29.89 JSL 42.107 • ID 11778

SALMON, W.C. (ED.) [1977] *Hans Reichenbach, logical empiricist. I,II,III,IV,V* (J 0154) Synthese 34*1-132,34*133-248,34*249-360,35*1-126,35*127-254
◇ B48 ◇ REV MR 55 # 12442 • ID 70141

SALMON, W.C. [1977] *Laws, modalities and counterfactuals* (J 0154) Synthese 35*191-229
◇ B45 ◇ REV MR 57 # 15972 Zbl 362 # 02017 • ID 50739

SALOMAA, A. [1959] *On many-valued systems of logic* (J 0963) Ajatus (Helsinki) 22*115-159
◇ B50 ◇ REV Zbl 93.11 JSL 25.291 • ID 11779

SALOMAA, A. [1960] *A theorem concerning the composition of functions of several variables ranging over a finite set* (J 0036) J Symb Logic 25*203-208
◇ B50 ◇ REV MR 26 # 2356 Zbl 119.250 JSL 33.307 • ID 11781

SALOMAA, A. [1960] *On the composition of functions of several variables ranging over a finite set* (J 0498) Ann Univ Turku, Ser A I 1960*48pp
◇ B50 ◇ REV MR 22 # 6696 Zbl 91.9 JSL 25.291 • ID 28431

SALOMAA, A. [1962] *On the number of simple bases of the set of functions over a finite domain* (J 0498) Ann Univ Turku, Ser A I 52*4pp
◇ B50 ◇ REV MR 26 # 3612 Zbl 103.245 JSL 27.247 • ID 11780

SALOMAA, A. [1962] *Some completeness criteria for sets of functions over a finite domain* (J 0498) Ann Univ Turku, Ser A I 53*10pp
◇ B50 ◇ REV MR 26 # 3613 Zbl 103.245 JSL 27.247 • REM Part I. Part II 1963 • ID 11782

SALOMAA, A. [1963] *On basic groups for the set of functions over a finite domain* (J 3994) Ann Acad Sci Fennicae Ser A I 338*15pp
◇ B50 ◇ REV MR 30 # 23 Zbl 121.11 JSL 33.307 • ID 11787

SALOMAA, A. [1963] *Some analogues of Sheffer functions in infinite-valued logics* (J 0096) Acta Philos Fenn 16*227-235
◇ B50 ◇ REV MR 34 # 1175 Zbl 121.11 JSL 31.118 • ID 11783

SALOMAA, A. [1963] *Some completeness criteria for sets of functions over a finite domain II* (J 0498) Ann Univ Turku, Ser A I 63*19pp
◇ B50 ◇ REV MR 27 # 3539 Zbl 113.3 JSL 30.106 • REM Part I 1962 • ID 11785

SALOMAA, A. [1964] *On infinitely generated sets of operations in finite algebras* (J 0498) Ann Univ Turku, Ser A I 74*13pp
◇ B50 C05 ◇ REV MR 30 # 25 Zbl 123.5 JSL 31.119 • ID 11790

SALOMAA, A. [1965] *On some algebraic notions in the theory of truth-functions* (J 0096) Acta Philos Fenn 18*193-202
◇ B50 ◇ REV MR 32 # 7395 Zbl 133.243 • ID 11791

SALOMAA, A. [1974] *Some remarks concerning many-valued propositional logics* (C 1936) Log Th & Semant Anal (Kanger) 15-21
◇ B50 ◇ REV Zbl 299 # 02017 • ID 31493

SALOMAA, A. see Vol. I, IV, V for further entries

SALONI, Z. [1972] *Gentzen rules for the m-valued logic* (J 0014) Bull Acad Pol Sci, Ser Math Astron Phys 20*819-826
◇ B50 ◇ REV MR 47 # 13 Zbl 255 # 02013 • ID 11797

SALONI, Z. [1973] *The sequent Gentzen system for m-valued logic* (J 0387) Bull Sect Logic, Pol Acad Sci 2*30-37
◇ B50 ◇ REV MR 53 # 5256 • ID 22889

SALWICKI, A. [1976] see MIRKOWSKA, G.

SALWICKI, A. [1977] see BANACHOWSKI, L.

SALWICKI, A. [1979] *Algorithmic logics of programs* (P 2539) Frege Konferenz (1);1979 Jena 382-400
◇ B75 ◇ REV MR 83e:68007 • ID 40134

SALWICKI, A. see Vol. IV, V for further entries

SAMBIN, G. [1978] *Fixed points through the finite model property (The algebraization of the theories which express Theor. XI)* (J 0063) Studia Logica 37*287-289
◇ B45 F30 G25 ◇ REV MR 80b:03107 Zbl 398 # 03055 • REM Part X 1978 by Montagna,F. • ID 52783

SAMBIN, G. & VALENTINI, S. [1980] *A modal sequent calculus for a fragment of arithmetic* (J 0063) Studia Logica 39*245-256
◇ B45 F05 F30 ◇ REV MR 81m:03025 Zbl 457 # 03016 • ID 54341

SAMBIN, G. [1980] *A simpler proof of Sahlqvist's theorem on completeness of modal logics* (J 0387) Bull Sect Logic, Pol Acad Sci 9*49-56
◇ B45 ◇ REV MR 81h:03043 Zbl 442 # 03019 • ID 56376

SAMBIN, G. & VALENTINI, S. [1980] *Post completeness and free algebras* (J 0068) Z Math Logik Grundlagen Math 26*343-347
◇ B45 ◇ REV MR 81h:03042 Zbl 464 # 03020 • ID 54610

SAMBIN, G. & VALENTINI, S. [1982] *The modal logic of provability. The sequential approach* (J 0122) J Philos Logic 11*311-342
◇ B45 F05 F30 ◇ REV MR 84b:03033 Zbl 523 # 03014 • ID 35627

SAMBIN, G. [1985] see BOOLOS, G.

SAMBIN, G. see Vol. VI for further entries

SAMCHENKO, V.N. [1982] *On Quine's paradox (Russian)* (C 3743) Probl Log & Metodol Nauk 57-66
◇ A05 B45 ◇ REV MR 83m:03010 • ID 34885

SAMOFALOV, K.G. [1973] *Synthese mehrwertiger Schaltungen in einer Klasse redundanter vollstaendiger Systeme (Ukrainian) (English and Russian summaries)* (J 0270) Dokl Akad Nauk Ukr SSR, Ser A 1973*231-234
◇ B50 B70 ◇ REV MR 51 # 7744 Zbl 255 # 94016 • ID 65038

SAMOJLENKO, N.I. [1976] see KUZ'MENKO, V.M.

SAMOKHVALOV, K.F. [1976] *The impossibility theorem for universal theory of prediction* (P 1804) Form Meth in Methodol of Emp Sci;1974 Warsaw 417-430
◇ A05 B60 ◇ REV MR 54 # 7197 Zbl 363 # 02008 • ID 50835

SAMOKHVALOV, K.F. see Vol. I, VI for further entries

SANCHEZ, E. [1976] *Resolution of composite fuzzy relation equations* (J 0194) Inform & Control 30*38-48
◇ B52 E72 ◇ REV MR 55 # 10340 Zbl 326 # 02048 • ID 65045

SANCHEZ, E. see Vol. I, V for further entries

SANCHEZ-MAZAS, M. [1955] *Formalization of logic according to the perspective intensionality (Spanish) (French and English summaries)* (J 0169) Theoria (Madrid) 3*103-117
◇ B45 ◇ REV MR 17.1037 • ID 11816

SANCHEZ-MAZAS, M. [1965] *Sobre la estructura de la logica modal* (J 0966) Episteme (Caracas) 1961-1963*347-361
◇ B45 ◇ REV JSL 32.399 • ID 11818

SANCHEZ-MAZAS, M. [1973] *Calculo de las normas* (X 1744) Ariel: Esplugas de Llobregat 194pp
◇ B45 ◇ REV MR 58 # 5065 Zbl 277 # 02004 • REM Thesis • ID 29525

SANCHEZ-MAZAS, M. [1980] *Essai de representation des systemes normatifs par des systemes d'equations* (P 3580) Repr des Conn & Raison dans Sci Homme;1979 St Maximin 500-532
◇ B45 ◇ REV MR 84i:92082 • ID 40230

SANCHEZ-MAZAS, M. [1982] *Algebraic and arithmetical translations of normative systems and applications in legal informatics* (P 4041) Log, Inform, Law;1981 Firenze 2*169-201
◇ B45 ◇ REV MR 85j:03021a • ID 45219

SANCHEZ-MAZAS, M. see Vol. I for further entries

SANCHEZ P., J. [1980] *Logical entailment and semantic intension of boolean formulas in relevant logic (Russian)* (C 3038) Log-Metodol Issl 313-331
◇ B46 ◇ REV MR 82f:03016 • ID 78113

Sanford, D.H. [1980] *Notes on logics of vagueness and some applications* (**P** 2936) Fuzzy Sets;1980 Durham 123-126
◇ B52 ◇ rev MR 81m:03033 • id 78121

Sanmartin Esplugues, J. [1973] *Syllogistics, many-valued logic and model theory (Spanish)* (**J** 0162) Teorema (Valencia) 3*355-365
◇ A10 B20 B50 C90 ◇ rev MR 50 # 9522 • id 78125

Sanmartin Esplugues, J. see Vol. III, V for further entries

Santos, E.S. [1976] *Fuzzy and probabilistic programs* (**J** 0191) Inform Sci 10*331-345
◇ B52 B75 D10 ◇ rev MR 58 # 19335 Zbl 334 # 68013 • id 65059

Santos, E.S. see Vol. IV for further entries

Sapozhenko, A.A. [1981] see Karakhanyan, L.M.

Sapozhenko, A.A. see Vol. I for further entries

Sarabia Alvarez-Ude, E.J. [1983] *On metalogic and three-valued logic (Spanish)* (**J** 0162) Teorema (Valencia) 13*225-260
◇ B50 ◇ id 45457

Sarlet, H.J. [1977] *Hintikka's free logic is not free* (**J** 0047) Notre Dame J Formal Log 18*458
◇ B60 ◇ rev MR 58 # 16126 Zbl 336 # 02024 • id 24246

Sarnavskij, N.G. [1983] *Recognition of an ordering predicate as a function of three-valued logic (Russian)* (**S** 2024) Probl Bioniki 30*28-32
◇ B50 ◇ rev MR 85i:04001 • id 44308

Sarnavskij, N.G. see Vol. V for further entries

Sarris, A.A. & Su, S.Y.H. [1972] *The relationship between multivalued swiching algebra and boolean algebra under different definitions of complement* (**J** 0187) IEEE Trans Comp C-21*479-485
◇ B50 B70 G05 ◇ rev MR 48 # 13507 Zbl 241 # 02008 • id 28821

Sasao, T. [1978] *An application of multiple-valued logic to a design of programmable logic arrays* (**P** 2014) Int Symp Multi-Val Log (8);1978 Rosemont 65-72
◇ B50 ◇ rev MR 81b:94072 • id 35914

Sasao, T. & Terada, H. [1979] *Multiple-valued logic and the design of programmable logic arrays with decoders* (**P** 3003) Int Symp Multi-Val Log (9);1979 Bath 27-37
◇ B50 ◇ rev MR 80m:94115 • id 82714

Sasao, T. [1981] *Multiple-valued decomposition of generalized boolean functions and the complexity of programmable logic arrays* (**J** 0187) IEEE Trans Comp C-30*635-643
◇ B50 B70 ◇ rev Zbl 463 # 94011 • id 69892

Sasao, T. [1982] see Ishikawa, K.

Sasiada, E. [1976] *A notion of entropy which does not increase* (**J** 1546) Rep Math Phys (Warsaw) 10*129-130
◇ B51 ◇ rev Zbl 363 # 02036 • id 50863

Sato, M. [1973] *Characterization of pseudo-boolean models by boolean models and its applications to intermediate logics* (**J** 0390) Publ Res Inst Math Sci (Kyoto) 9*141-155
◇ B55 ◇ rev MR 49 # 4771 Zbl 274 # 02029 • id 11859

Sato, M. [1977] *A study of Kripke-type models for some modal logics by Gentzen's sequential method* (**J** 0390) Publ Res Inst Math Sci (Kyoto) 13*381-468
◇ B45 C90 ◇ rev MR 57 # 2884 Zbl 405 # 03013 • id 30800

Sato, M. [1980] *A cut-free Gentzen-type system for the modal logic S5* (**J** 0036) J Symb Logic 45*67-84
◇ B45 ◇ rev MR 81h:03044 Zbl 444 # 03010 • id 56458

Sato, M. see Vol. I, VI for further entries

Satre, T.W. [1972] *Natural deduction rules for modal logics* (**J** 0047) Notre Dame J Formal Log 13*461-475
◇ B45 ◇ rev MR 47 # 6446 Zbl 242 # 02023 • id 65079

Satre, T.W. [1972] *Natural deduction rules for* $S1^0 - S4^0$ (**J** 0047) Notre Dame J Formal Log 13*565-568
◇ B45 ◇ rev MR 50 # 4251 Zbl 242 # 02024 • id 11861

Sawamura, H. [1985] *Axiomatization of computer-oriented modal logic and decision procedure* (**J** 3957) Bull Inf & Cybern (Kyushu Univ) 21/3-4*57-66
◇ B25 B45 ◇ rev Zbl 575 # 03011 • id 47625

Sayward, C.W. [1981] see Hugly, P.

Sayward, C.W. see Vol. I, III for further entries

Scales, R. [1977] *A Russellian approach to truth* (**J** 0097) Nous, Quart J Phil 11*169-174
◇ A05 B45 ◇ rev MR 58 # 5066 • id 78155

Scarpellini, B. [1962] *Die Nichtaxiomatisierbarkeit des unendlichwertigen Praedikatenkalkuels von Lukasiewicz* (**J** 0036) J Symb Logic 27*159-170
◇ B50 D35 ◇ rev MR 27 # 3503 Zbl 112.245 JSL 29.145 • id 11877

Scarpellini, B. see Vol. I, III, IV, V, VI for further entries

Scedrov, A. [1984] *On some nonclassical extensions of second-order intuitionistic propositional calculus* (**J** 0073) Ann Pure Appl Logic 27*155-164
◇ B55 F50 ◇ rev MR 86f:03022 Zbl 569 # 03026 • id 44085

Scedrov, A. [1985] *Extending Goedel's modal interpretation to type theory and set theory* (**C** 3659) Intens Math 81-119
◇ B15 B45 E70 F35 F50 ◇ rev Zbl 569 # 03025 • id 44771

Scedrov, A. [1985] see Harrington, L.A.

Scedrov, A. see Vol. III, IV, V, VI for further entries

Schagrin, M.L. [1968] *The language of logic. A programmed text* (**X** 0981) Random House: New York vii + 247pp
◇ B98 ◇ rev JSL 39.612 • id 12857

Scharle, T.W. [1966] see Canty, J.T.

Scharle, T.W. [1967] see Lambert, K.

Scharle, T.W. [1975] *Axiomatization of fragments of S5* (**J** 0047) Notre Dame J Formal Log 16*45-70
◇ B45 ◇ rev MR 50 # 12666 Zbl 258 # 02031 • id 12862

Scharle, T.W. see Vol. I for further entries

Scheer, R.K. [1964] see Carney, J.D.

Schefe, P. [1980] *On foundations of reasoning with uncertain facts and vague concepts* (**J** 1741) Int J Man-Mach Stud 12*35-62
◇ B52 ◇ rev MR 82g:03040 Zbl 437 # 03010 • id 55878

SCHEIBE, E. [1974] *Popper and quantum logic* (J 0013) Brit J Phil Sci 25*319-328
◇ A05 B51 G12 ◇ REV MR 58 # 32491 Zbl 344 # 02004 • ID 65098

SCHEIBE, E. see Vol. I for further entries

SCHELSKY, H. [1984] see KRAWIETZ, W.

SCHENK, G. [1974] *Die Logik* (J 0424) Wiss Z Univ Halle-Wittenberg, Math-Nat Reihe 23*27-54
◇ A10 B98 ◇ REV MR 57 # 2859 • ID 78168

SCHERER, D. [1978] see FACIONE, P.A.

SCHERER, D. see Vol. I for further entries

SCHIFFER, S. [1982] *Intention-based semantics* (J 0047) Notre Dame J Formal Log 23*119-156
◇ B45 ◇ REV MR 84i:03043a • REM Comments by Bennett,J. ibid 23*258-262; Grandy,R.E. ibid 23*327-332 • ID 34939

SCHINDLER, P. [1970] *Tense logic for discrete future time* (J 0036) J Symb Logic 35*105-118
◇ A05 B45 ◇ REV MR 45 # 33 Zbl 193.293 • ID 12874

SCHIPPER, E.W. & SCHUH, E. [1959] *A first course in modern logic* (X 0818) Holt Rinehart & Winston: New York xvii + 398pp
• REPR [1960] (X 0866) Routledge & Kegan Paul: Henley on Thames xviii + 398pp
◇ B98 ◇ REV JSL 24.220 • ID 22536

SCHLESINGER, G.N. [1984] *A theorem of epistemic logic* (J 0095) Philos Stud 45*285-292
◇ B45 ◇ REV MR 85d:03041 • ID 41024

SCHLESSINGER, N. & ZIERLER, N. [1965] *Boolean embeddings of orthomodular sets and quantum logic* (J 0025) Duke Math J 32*251-262
◇ A05 B51 G05 G10 G12 ◇ REV MR 30 # 5704 Zbl 171.254 JSL 37.190 • ID 14417

SCHMIDT, GUNTHER [1983] see BERGHAMMER, R.

SCHMIDT, GUNTHER see Vol. I, IV, V for further entries

SCHMIDT, H.A. [1950] *Mathematische Grundlagenforschung* (X 1079) Teubner: Leipzig 48pp
◇ B98 ◇ REV MR 13.4 JSL 17.198 • ID 22429

SCHMIDT, H.A. [1950] *Systematische Basisreduktion der Modalitaeten bei Idempotenz der positiven Grundmodalitaeten* (J 0043) Math Ann 122*71-89
◇ B45 ◇ REV MR 12.579 Zbl 45.149 JSL 16.230 • ID 12884

SCHMIDT, H.A. [1954] *Ein rein aussagenlogischer Zugang zu den Modalitaeten der strikten Logik* (P 0575) Int Congr Math (II, 7);1954 Amsterdam 2*407-408
◇ B45 ◇ ID 29472

SCHMIDT, H.A. [1956] *Das fundamentale Implikationssystem einer implikativen Modalitaetenstruktur mit idempotenter Moeglichkeit* (J 1114) Arch Phil 5*353-374
• REPR [1956] (J 0009) Arch Math Logik Grundlagenforsch 2*33-54
◇ B45 ◇ REV MR 19.724 Zbl 71.6 • ID 12888

SCHMIDT, H.A. [1957] *Die Gesamtheit der idempotenten implikativen Modalitaetenstrukturen* (J 0009) Arch Math Logik Grundlagenforsch 3*29-49
◇ B45 ◇ REV MR 20 # 3778 Zbl 81.11 JSL 20.626 • ID 12889

SCHMIDT, H.A. [1958] *Ueber einige neuere Untersuchungen zur Modalitaetenlogik* (J 0076) Dialectica 12*408-421
◇ B45 ◇ REV Zbl 99.8 JSL 27.230 • ID 12890

SCHMIDT, H.A. [1960] *Mathematische Gesetze der Logik. I. Vorlesungen ueber Aussagenlogik* (X 0811) Springer: Heidelberg & New York xxiv + 555pp
◇ B05 B98 F50 F98 ◇ REV MR 29 # 1135 Zbl 253 # 08002 Zbl 88.246 • ID 12892

SCHMIDT, H.A. see Vol. I, III, VI for further entries

SCHMIDT, J. [1971] *A semigroup in function algebra* (J 0136) Semigroup Forum 2*119-129
◇ B50 C05 E20 ◇ REV MR 43 # 3199 Zbl 237 # 08003 • ID 12921

SCHMIDT, J. see Vol. I, III, V, VI for further entries

SCHMIERER, Z. [1936] *On characteristic functions in many-valued systems of logic (Polish)* (J 1125) Przeglad Filoz 39*437
◇ B50 ◇ REV JSL 2.92 • ID 41347

SCHMITT, P.H. [1984] *Diamond formulas: A fragment of dynamic logic with recursively enumerable validity problem* (J 0194) Inform & Control 61*147-158
◇ B75 ◇ ID 39523

SCHMITT, P.H. see Vol. III, IV for further entries

SCHMUCKER, K.J. [1984] *Fuzzy sets, natural language computations, and risk analysis, Foreword by Lotfi A. Zadeh* (X 3581) Comput Sci Press: Rockville xv + 192pp
◇ B52 B98 ◇ REV Zbl 552 # 90025 • ID 43453

SCHOCK, R. [1961] *Some definitions of subjunctive implication, of counterfactual implication, and of related concepts* (J 0047) Notre Dame J Formal Log 2*206-221
◇ A05 B45 ◇ REV Zbl 114.244 JSL 35.319 • ID 12936

SCHOCK, R. [1962] *A note on subjunctive and counterfactual implication* (J 0047) Notre Dame J Formal Log 3*289-290
◇ A05 B45 ◇ REV Zbl 114.244 JSL 35.319 • ID 12938

SCHOCK, R. [1964] *On denumerable many-valued logics* (J 0079) Logique & Anal, NS 7*190-195
◇ B50 ◇ REV MR 31 # 1187 Zbl 166.2 JSL 35.140 • ID 12941

SCHOCK, R. [1964] *On finitely many-valued logics* (J 0079) Logique & Anal, NS 7*43-58,195
◇ B50 ◇ REV MR 31 # 1186 Zbl 127.8 JSL 35.140 • ID 12940

SCHOCK, R. [1965] *On probability logics* (J 0047) Notre Dame J Formal Log 6*129-134
◇ B48 ◇ REV MR 33 # 4960 JSL 34.139 • ID 12949

SCHOCK, R. [1965] *Some theorems on the relative strenghts of many-valued logic* (J 0079) Logique & Anal, NS 8*101-104
◇ B50 ◇ REV MR 34 # 5665 Zbl 133.244 JSL 35.140 • ID 12944

SCHOCK, R. [1967] *Logik* (X 1163) Almqvist & Wiksell: Stockholm 105pp
◇ B98 ◇ REV Zbl 169.296 JSL 34.642 • ID 21980

SCHOCK, R. [1971] *Quasi-connectives definable in concept theory* (X 1493) Gleerup: Lund 75pp
◇ A05 B45 E70 ◇ REV MR 48 # 5823 Zbl 443 # 03010 • ID 28602

SCHOCK, R. [1980] *A complete system of indexical logic* (J 0047) Notre Dame J Formal Log 21*293-315
◇ B45 ◇ REV MR 82c:03023 Zbl 351 # 02022 • ID 53777

SCHOCK, R. [1980] *A natural deduction system of indexical logic* (J 0047) Notre Dame J Formal Log 21*351-364
◇ B45 ◇ REV MR 81f:03037 Zbl 351 # 02023 • ID 53778

SCHOCK, R. see Vol. I, III, V for further entries

SCHOFIELD, P. [1964] *On a correspondence between many-valued and two-valued logics* (J 0068) Z Math Logik Grundlagen Math 10*265-274
◇ B50 ◇ REV MR 30 # 15 Zbl 137.249 JSL 32.539
• ID 12960

SCHOFIELD, P. [1966] *Complete subsets of mappings over a finite domain* (J 0171) Proc Cambridge Phil Soc Math Phys 62*597-611
◇ B50 ◇ REV MR 34 # 2449 Zbl 178.306 JSL 32.539
• ID 12961

SCHOFIELD, P. [1969] *Independent conditions for completeness of finite algebra with a single generator* (J 3172) J London Math Soc, Ser 2 44*413-423
◇ B50 ◇ REV MR 39 # 1305 Zbl 192.337 JSL 40.98
• ID 12962

SCHOLZ, H. [1948] *Vorlesungen ueber Grundzuege der mathematischen Logik I* (X 0910) Aschendorffsche Verlagsbuchh: Muenster xiv+276pp
◇ B98 ◇ REV MR 12.661 Zbl 31.103 JSL 15.200 JSL 19.115 • REM Vol.II 1949, 2nd revised edition 1950
• ID 22431

SCHOLZ, H. [1949] *Vorlesungen ueber Grundzuege der mathematischen Logik II* (X 0910) Aschendorffsche Verlagsbuchh: Muenster x+311pp
◇ B98 ◇ REV Zbl 32.242 JSL 15.200 • REM Vol.I 1948, 2nd revised edition 1951 • ID 20807

SCHOLZ, H. [1952] see HERMES, H.

SCHOLZ, H. & HASENJAEGER, G. [1961] *Grundzuege der mathematischen Logik* (X 0811) Springer: Heidelberg & New York xvi+504pp
◇ B98 ◇ REV MR 23 # A2306 Zbl 137.4 JSL 28.245
• ID 22434

SCHOLZ, H. see Vol. I, III for further entries

SCHOTCH, P.K. [1975] *Fuzzy modal logic* (P 1805) Int Symp Multi-Val Log (5,Proc);1975 Bloomington 176-182
◇ B52 ◇ ID 27079

SCHOTCH, P.K. [1977] see JENSEN, J.B.

SCHOTCH, P.K. [1978] see JENSEN, J.B.

SCHOTCH, P.K. [1979] see JENNINGS, R.E.

SCHOTCH, P.K. [1980] see JENNINGS, R.E.

SCHOTCH, P.K. [1981] see JENNINGS, R.E.

SCHOTCH, P.K. [1984] *Remarks on the semantics of nonnormal modal logics* (J 3781) Topoi 3*85-90
◇ B45 ◇ ID 44679

SCHOTCH, P.K. [1984] see JENNINGS, R.E.

SCHOTT, H.F. [1970] *Generalizability of the propositional and predicate calculi to infinite valued calculi* (J 0047) Notre Dame J Formal Log 11*107-128
◇ B50 ◇ REV MR 44 # 6441 Zbl 212.315 • ID 12973

SCHRAMM, A. [1984] *Norm-Folgern ohne Normenlogik* (C 4065) Theorie der Normen 441-446
◇ B45 ◇ REV MR 85k:03015 • ID 44568

SCHRAMM, A. [1984] see KRAWIETZ, W.

SCHREIBER, P. [1977] *Grundlagen der Mathematik* (X 0806) Dt Verlag Wiss: Berlin 239pp
◇ B98 ◇ REV MR 57 # 15932 Zbl 362 # 02001 • REM 2nd ed. 1984 • ID 50723

SCHREIBER, P. see Vol. I, III, IV, VI for further entries

SCHROEDER, F.W.K.E. [1890] *Vorlesungen ueber die Algebra der Logik (exakte Logik) Vol.1* (X 1079) Teubner: Leipzig xii+717pp
• LAST ED [1966] (X 0848) Chelsea: New York 819pp
◇ A05 A10 B20 B98 G98 ◇ REV MR 33 # 1219 MR 33 # 1220 Zbl 188.307 JSL 40.609 FdM 22.73
• ID 19702

SCHROEDER, F.W.K.E. [1891] *Vorlesungen ueber die Algebra der Logik (exakte Logik) Vol.2* (X 1079) Teubner: Leipzig xiii+400pp
◇ A05 A10 B20 B98 G05 G15 G98 ◇ REV MR 33 # 1220 JSL 31.283 FdM 23.51 • ID 19703

SCHROEDER, F.W.K.E. [1895] *Vorlesungen ueber die Algebra der Logik (exakte Logik) Vol.3 Algebra und Logik der Relative* (X 1079) Teubner: Leipzig viii+649pp
• TRANSL [1966] (X 0848) Chelsea: New York vi+819pp
◇ A05 A10 B20 B98 G98 ◇ REV MR 33 # 1221 FdM 26.74 • ID 19704

SCHROEDER, F.W.K.E. see Vol. I, V for further entries

SCHUBERT, H. [1896] *Mathematical essays and recreations* (X 1324) Open Court: LaSalle 149pp
◇ B98 ◇ ID 25498

SCHUBERT, H. see Vol. I for further entries

SCHUETTE, K. [1960] *Beweistheorie* (X 0811) Springer: Heidelberg & New York x+355pp
• TRANSL [1977] (X 0811) Springer: Heidelberg & New York xii+299pp
◇ B98 F05 F15 F30 F35 F98 ◇ REV MR 22 # 9438 MR 58 # 21497 Zbl 102.247 Zbl 367 # 02012 JSL 25.243 JSL 47.218 • REM 2nd rev. ed. Comment 1960 by Frey,G.
• ID 13010

SCHUETTE, K. [1962] *Lecture notes in mathematical logic. Vol.I* (X 1328) Penn State Math: University Park 105pp
◇ B98 F15 ◇ REM Vol.II 1963 • ID 22382

SCHUETTE, K. [1963] *Lecture notes in mathematical logic. Vol.II* (X 1328) Penn State Math: University Park 109pp
◇ B98 F15 ◇ REM Vol.I 1962 • ID 22383

SCHUETTE, K. [1968] *Vollstaendige Systeme modaler und intuitionistischer Logik* (X 0811) Springer: Heidelberg & New York vii+87pp
• TRANSL [1974] (C 4138) Feys: Modal Logika
◇ B45 C90 F50 F98 ◇ REV MR 37 # 2587 Zbl 157.16 JSL 36.522 • ID 24822

SCHUETTE, K. see Vol. I, III, IV, V, VI for further entries

SCHUH, E. [1959] see SCHIPPER, E.W.

SCHUH, E. [1973] *Many-valued logics and the Lewis paradoxes* (J 0047) Notre Dame J Formal Log 14*250-252
◇ A05 B50 ◇ REV MR 48 # 61 Zbl 251 # 02021 • ID 13029

SCHULTZ, KONRAD [1970] *Keine kontingenten Identitaeten in Lemmons modaler Mengenlehre* (J 0068) Z Math Logik Grundlagen Math 16*261-262
◇ B45 E70 ◇ REV MR 43 # 6063 Zbl 205.7 • ID 13034

SCHULTZ, KONRAD [1970] *Modelle modaler Mengenlehren* (J 0068) Z Math Logik Grundlagen Math 16*327-339 ◇ B45 C62 C90 E35 E70 ◇ REV MR 44#3849 Zbl 213.15 • ID 13035

SCHULTZ, KONRAD [1984] *A generalization of Lindenbaum's theorem for predicate calculi* (J 0068) Z Math Logik Grundlagen Math 30*165-168 ◇ B50 ◇ REV Zbl 559#03006 • ID 42218

SCHULTZ, KONRAD see Vol. III, V, VI for further entries

SCHUMM, G.F. [1969] *On a modal system of D.C.Makinson and B.Sobocinski* (J 0047) Notre Dame J Formal Log 10*263-265 ◇ B45 ◇ REV MR 40#4089 Zbl 167.9 • ID 13040

SCHUMM, G.F. [1969] *On some open questions of B.Sobocinski* (J 0047) Notre Dame J Formal Log 10*261-262 ◇ B45 ◇ REV MR 40#4088 Zbl 167.9 JSL 36.329 • ID 13039

SCHUMM, G.F. [1971] *Solutions to four modal problems of Soboconski* (J 0047) Notre Dame J Formal Log 12*335-340 ◇ B45 ◇ REV MR 46#5110 Zbl 205.303 JSL 40.602 • ID 13041

SCHUMM, G.F. [1974] *K and Z* (J 0047) Notre Dame J Formal Log 15*295-297 ◇ B45 ◇ REV MR 50#9534 Zbl 258#02026 • ID 13043

SCHUMM, G.F. [1974] *S3.02 = S3.03* (J 0047) Notre Dame J Formal Log 15*147-148 ◇ B45 ◇ REV MR 48#10768 Zbl 272#02031 • ID 13042

SCHUMM, G.F. [1975] *Disjunctive extensions of S4 and a conjecture of Goldblatt's* (J 0068) Z Math Logik Grundlagen Math 21*81-86 ◇ B45 ◇ REV MR 51#2871 Zbl 357#02021 • ID 13044

SCHUMM, G.F. [1975] *Remark on a logic of preference* (J 0047) Notre Dame J Formal Log 16*509-510 ◇ B45 ◇ REV MR 52#5364 Zbl 311#02042 • ID 18406

SCHUMM, G.F. [1975] *Wajsberg normal forms for S5* (J 0122) J Philos Logic 4*357-360 ◇ B45 ◇ REV MR 58#21482 Zbl 323#02034 • ID 30414

SCHUMM, G.F. [1976] *Interpolation in S5 and some related systems* (J 0302) Rep Math Logic, Krakow & Katowice 6*107-110 ◇ B45 ◇ REV MR 58#16157 Zbl 361#02030 • ID 21920

SCHUMM, G.F. [1978] *An incomplete nonnormal extension of S3* (J 0036) J Symb Logic 43*211-212 ◇ B45 C90 ◇ REV MR 58#5067 Zbl 386#03007 • ID 29254

SCHUMM, G.F. [1978] *Modalities in the extensions of B* (J 0302) Rep Math Logic, Krakow & Katowice 10*123-128 ◇ B45 ◇ REV MR 81b:03023 Zbl 434#03016 • ID 55720

SCHUMM, G.F. [1978] *Putting 𝔎 in its place* (J 0047) Notre Dame J Formal Log 19*623-628 ◇ B45 C90 ◇ REV MR 80a:03029 Zbl 336#02019 • ID 52126

SCHUMM, G.F. [1979] *Modal logics with no minimal proper extensions* (J 0063) Studia Logica 38*233-235 ◇ B45 ◇ REV MR 81c:03015 Zbl 434#03017 • ID 55721

SCHUMM, G.F. [1979] see EDELSTEIN, R.

SCHUMM, G.F. [1981] *Bounded properties in modal logic* (J 0068) Z Math Logik Grundlagen Math 27*197-200 ◇ B45 ◇ REV MR 82g:03029 Zbl 465#03007 • ID 54910

SCHUMM, G.F. see Vol. I, V for further entries

SCHWANITZ, G. [1973] see BAGINSKI, M.

SCHWARTZ, DIETRICH [1976] *Das Homomorphietheorem fuer MV-Algebren endlicher Ordnung* (J 0068) Z Math Logik Grundlagen Math 22*141-148 ◇ B50 ◇ REV MR 58#21446 Zbl 362#02044 • ID 18483

SCHWARTZ, DIETRICH see Vol. I, III, IV, V for further entries

SCHWARTZ, J.T. (ED.) [1967] *Mathematical aspects of computer science* (X 0803) Amer Math Soc: Providence vi+226pp ◇ B35 B65 B75 B97 D10 D80 D97 ◇ REV MR 38#2975 Zbl 165.2 • ID 23572

SCHWARTZ, J.T. see Vol. I, III, IV, V for further entries

SCHWARTZ, P.B. [1958] see EVANS, T.

SCHWARTZ, R.J. [1981] *Approximate truth and confirmation* (J 0153) Phil of Sci (East Lansing) 48*606-610 ◇ B48 ◇ REV MR 82m:03030 • ID 78302

SCHWARTZ, R.L. [1984] see MELLIAR-SMITH, P.M.

SCHWARTZ, R.L. see Vol. I for further entries

SCHWARTZ, T. [1979] *Necessary truth as analyticity, and the eliminability of monadic de re formulas* (J 0047) Notre Dame J Formal Log 20*336-340 ◇ A05 B45 ◇ REV MR 80f:03013 Zbl 394#03032 • ID 54731

SCHWARTZ, T. see Vol. I, IV for further entries

SCHWEDE, G.W. [1976] *N-variable fuzzy maps with application to disjunctive decomposition of fuzzy switching functions* (P 2011) Int Symp Multi-Val Log (6);1976 Logan 203-216 ◇ B52 ◇ REV MR 58#20875 • ID 35891

SCHWEDE, G.W. see Vol. V for further entries

SCHWEIGERT, D. & SZYMANSKA, M. [1983] *On functionally complete modal algebras related to M and S4* (J 1008) Demonstr Math (Warsaw) 16*323-328 ◇ B45 G25 ◇ REV MR 85e:03160 Zbl 542#03042 • ID 40793

SCHWIND, C.B. [1979] *Representing actions by state logic* (P 3108) AISB/GI Conf Artif Intel;1978 Hamburg 304-308 ◇ B45 ◇ REV Zbl 404#68041 • ID 54856

SCHWIND, C.B. [1980] *Natural language analysis by theorem proving methods: disambiguating pronouns in natural language texts* (P 3580) Repr des Conn & Raison dans Sci Homme;1979 St Maximin 123-138 ◇ B35 B45 B65 ◇ REV MR 82g:03045 • ID 78311

SCHWIND, C.B. see Vol. I for further entries

SCHWYHLA, W. [1981] see KLEMENT, E.P.

SCHWYHLA, W. see Vol. V for further entries

SCOTT, D.S. [1966] see KRAUSS, P.H.

SCOTT, D.S. [1970] *Advice on modal logic* (P 0559) Phil Probl in Logic;1968 Irvine 143-173 ◇ B45 ◇ REV MR 43#1820 Zbl 295#02013 • ID 11922

SCOTT, D.S. [1973] *Background to formalization* (**P** 0783) Truth, Syntax & Modal;1970 Philadelphia 244-273
◇ A05 B45 ◇ REV MR 56 # 5202 Zbl 277 # 02003 JSL 42.316 • ID 29524

SCOTT, D.S. [1974] *Completeness and axiomatizability in many-valued logic* (**P** 0610) Tarski Symp;1971 Berkeley 411-435
◇ B50 ◇ REV MR 51 # 57 Zbl 318 # 02021 • ID 11924

SCOTT, D.S. [1974] see ADDISON, J.W.

SCOTT, D.S. [1975] see MOSS, J.M.B.

SCOTT, D.S. [1975] *Combinators and classes* (**P** 1603) λ-Calc & Comput Sci Th;1975 Roma 1-26
◇ B40 B50 E70 ◇ REV MR 58 # 21489 Zbl 342 # 02018 • ID 27594

SCOTT, D.S. [1976] *Does many-valued logic have any use?* (**P** 1502) Phil of Logic;1976 Bristol 64-95
◇ B50 ◇ REV MR 58 # 10307 • ID 78317

SCOTT, D.S. [1982] *Lectures on a mathematical theory of computation* (**P** 3906) Th Found of Progr Methodol;1981 Marktoberdorf 145-292
◇ B40 B98 D75 ◇ REV MR 85g:68043 Zbl 516 # 68064 • ID 38593

SCOTT, D.S. see Vol. I, III, IV, V, VI for further entries

SCOTT, L.L. [1980] *Necessary and sufficient conditions for the values of a function of fuzzy variables to lie in a specified subinterval of (0, 1)* (**P** 2936) Fuzzy Sets;1980 Durham 35-47
◇ B52 ◇ REV MR 81m:03065 • ID 78325

SCOZZAFAVA, R. [1980] *Bayesian inference and inductive "logic"* (**J** 2834) Scientia (Milano) 115*37-53
◇ B48 ◇ REV MR 82c:62007 • ID 82761

SCOZZAFAVA, R. see Vol. I, III for further entries

SCROGGS, S.J. [1951] *Extensions of the Lewis system S5* (**J** 0036) J Symb Logic 16*112-120
◇ B45 ◇ REV MR 13.97 Zbl 43.8 JSL 16.272 • ID 11926

SEESE, D.G. [1980] see BAUDISCH, A.

SEESE, D.G. see Vol. I, III, IV, V for further entries

SEESKIN, K.R. [1971] *Many-valued logic and future contingencies* (**J** 0079) Logique & Anal, NS 14*759-773 • ERR/ADD ibid 15*665
◇ A05 B50 ◇ REV MR 48 # 62 Zbl 333 # 02014 • ID 19638

SEESKIN, K.R. [1974] *Some remarks on truth and bivalence* (**J** 0079) Logique & Anal, NS 17*101-109
◇ B50 ◇ REV MR 51 # 77 Zbl 341 # 02010 • ID 15181

SEGERBERG, K. [1965] *A contribution to nonsense-logic* (**J** 0105) Theoria (Lund) 31*199-217
◇ B60 ◇ REV JSL 33.134 • ID 43155

SEGERBERG, K. [1967] *Some modal logics based on a three-valued logic* (**J** 0105) Theoria (Lund) 33*53-71
◇ B45 ◇ REV MR 44 # 6460 Zbl 189.7 JSL 33.309 • ID 11932

SEGERBERG, K. [1968] *Decidability of S4.1* (**J** 0105) Theoria (Lund) 34*7-20
◇ B25 B45 ◇ REV MR 39 # 1308 JSL 39.611 • ID 11935

SEGERBERG, K. [1968] *Decidability of four modal logics* (**J** 0105) Theoria (Lund) 34*21-25
◇ B25 B45 ◇ REV MR 39 # 1309 • ID 41673

SEGERBERG, K. [1968] *Propositional logics related to Heyting and Johansson* (**J** 0105) Theoria (Lund) 34*26-61
◇ B55 F50 ◇ REV MR 39 # 2617 • ID 11934

SEGERBERG, K. [1970] *A remark concerning the Brouwer system (Swedish)* (**P** 0785) Scand Logic Symp (1);1968 Aabo 166-169
◇ B45 ◇ REV MR 50 # 9539 Zbl 318 # 02023 • ID 11936

SEGERBERG, K. [1970] *Kripke-type semantics for preference logic* (**C** 0735) Logic & Value (Dahlquist) 128-134
◇ B45 ◇ REV MR 50 # 4252 • ID 11937

SEGERBERG, K. [1970] *Modal logics with linear alternative relations* (**J** 0105) Theoria (Lund) 36*301-322
◇ B45 C90 ◇ REV MR 47 # 18 Zbl 235 # 02019 • ID 11938

SEGERBERG, K. [1971] *An essay in classical modal logic I,II,III* (**X** 0882) Univ Filos Foeren: Uppsala v+250pp
◇ B45 ◇ REV MR 49 # 4756 Zbl 311 # 02028 • ID 32576

SEGERBERG, K. [1971] *Qualitative probability in a modal setting* (**P** 0604) Scand Logic Symp (2);1970 Oslo 341-352
◇ B48 ◇ REV MR 48 # 68 Zbl 223 # 02013 JSL 40.503 • ID 78354

SEGERBERG, K. [1971] *Some logics of commitment and obligation* (**C** 1547) Deontic Log Readings 148-158
◇ B45 ◇ REV Zbl 251 # 02033 • ID 28908

SEGERBERG, K. [1972] *Post completeness in modal logic* (**J** 0036) J Symb Logic 37*711-715
◇ B45 C90 ◇ REV MR 47 # 4765 Zbl 268 # 02012 • ID 11939

SEGERBERG, K. [1973] *Franzen's proof of Bull's theorem* (**J** 0963) Ajatus (Helsinki) 35*216-221
◇ B45 ◇ REV Zbl 299 # 02021 • ID 65184

SEGERBERG, K. [1973] *Two dimensional modal logic* (**J** 0122) J Philos Logic 2*77-96
◇ B45 ◇ REV MR 54 # 12488 Zbl 259 # 02013 • ID 11940

SEGERBERG, K. [1974] see MAKINSON, D.

SEGERBERG, K. [1974] *Proof of a conjecture of McKay* (**J** 0027) Fund Math 81*267-270
◇ B55 F50 ◇ REV MR 50 # 1849 Zbl 287 # 02013 • ID 11941

SEGERBERG, K. [1975] *That every extension of S4.3 is normal* (**P** 0757) Scand Logic Symp (3);1973 Uppsala 194-196
◇ B45 ◇ REV MR 52 # 5365 Zbl 319 # 02020 JSL 43.383 • ID 18376

SEGERBERG, K. [1976] *Discrete linear future time without axioms* (**J** 0063) Studia Logica 35*273-278,327
◇ B25 B45 C90 ◇ REV MR 55 # 10245 Zbl 343 # 02017 • ID 65185

SEGERBERG, K. [1976] *The truth about some Post numbers* (**J** 0036) J Symb Logic 41*239-244
◇ B45 ◇ REV MR 55 # 5393 Zbl 331 # 02008 • ID 14804

SEGERBERG, K. [1979] *Von Wright's temporal logic (Russian)* (**P** 2554) All-Union Symp Th Log Infer;1974 Moskva 173-205
◇ B45 ◇ REV MR 84k:03062 • ID 35000

SEGERBERG, K. [1980] *A note on the logic of elsewhere* (**J** 0105) Theoria (Lund) 46*183-187
◇ B45 ◇ REV MR 83a:03023 • ID 35062

SEGERBERG, K. [1980] *Applying modal logic* (J 0063) Studia Logica 39*275-295
◇ A05 B45 ◇ REV MR 82b:03050 Zbl 457 # 03014 • ID 54339

SEGERBERG, K. [1981] *Action-games* (J 0096) Acta Philos Fenn 32*220-231
◇ B45 ◇ REV MR 84j:03046 Zbl 484 # 03007 • ID 34638

SEGERBERG, K. [1982] *"After" and "during" in dynamic logic* (P 3800) Intens Log: Th & Appl;1979 Moskva 203-228 • TRANSL [1984] (C 4366) Modal & Intens Log & Primen Probl Metodol Nauk 58-81
◇ B75 ◇ REV MR 84k:03063 Zbl 564 # # 03018 • ID 35010

SEGERBERG, K. [1982] *A completeness theorem in the modal logic of programs* (P 3831) Universal Algeb & Appl;1978 Warsaw 31-46
◇ B75 C90 ◇ REV MR 85k:03014 Zbl 546 # 03011 • ID 41144

SEGERBERG, K. [1982] *A deontic logic of action* (J 0063) Studia Logica 41*269-282
◇ B45 ◇ REV MR 85i:03057 Zbl 537 # 03018 • ID 43735

SEGERBERG, K. [1982] see CRESSWELL, M.J.

SEGERBERG, K. [1982] *The logic of deliberate action* (J 0122) J Philos Logic 11*233-254
◇ B45 ◇ REV MR 83i:03036 Zbl 488 # 03012 JSL 51.476 • ID 35505

SEGERBERG, K. [1984] *A topological logic of action* (J 0063) Studia Logica 43*415-420
◇ B60 ◇ ID 42390

SEGERBERG, K. [1984] see BULL, R.A.

SEGERBERG, K. [1985] *Routines* (J 0154) Synthese 65*185-210
◇ B45 ◇ ID 49421

SEGERBERG, K. see Vol. I for further entries

SEIDENFELD, T. [1982] *Paradoxes of conglomerability and fiducial inference* (P 3622) Int Congr Log, Meth & Phil of Sci (6,Proc);1979 Hannover 395-412
◇ B48 ◇ REV MR 85f:62006 Zbl 501 # 60002 • ID 45943

SEIFFERT, H. [1973] *Einfuehrung in die Logik. Logische Propaedeutik und formale Logik* (X 0995) Beck'sche Verlagsbuchh: Muenchen 231pp
◇ B98 ◇ REV MR 52 # 13302 Zbl 346 # 02001 • ID 22435

SEKITA, Y. [1982] see MIZUMOTO, M.

SELDIN, J.P. [1972] see HINDLEY, J.R.

SELDIN, J.P. see Vol. I, VI for further entries

SELMAN, A.L. [1972] *Completeness of calculi for axiomatically defined classes of algebras* (J 0004) Algeb Universalis 2*20-32
◇ B55 C05 C07 ◇ REV MR 47 # 1725 Zbl 251 # 08005 • ID 11964

SELMAN, A.L. see Vol. III, IV, V for further entries

SELTEN, R. [1982] see LEOPOLD-WILDBURGER, U.

SEMBI, B.S. [1979] see MAMDANI, E.H.

SEMBI, B.S. [1980] see MAMDANI, E.H.

SEMENENKO, M.I. [1970] *Properties of some subsystems of classical and intuitionistic propositional calculi (Russian)* (J 0068) Z Math Logik Grundlagen Math 16*201-238 • TRANSL [1979] (J 0225) Amer Math Soc, Transl, Ser 2 113*1-35
◇ B55 F50 ◇ REV MR 44 # 29 Zbl 212.12 JSL 39.351 • ID 11972

SEMION, I.V. [1971] see AJZENBERG, N.N.

SEN, A. [1985] *Rationality and uncertainty* (J 0472) Theory Decis 18*109-127
◇ A05 B48 ◇ ID 47926

SEPER, K. [1971] *A decision problem concerning autoduality in k-valued logic* (J 0068) Z Math Logik Grundlagen Math 17*251-255
◇ B50 ◇ REV MR 44 # 6450 Zbl 193.291 • ID 11979

SEPER, K. [1985] *Contra-intuitionist logic and symmetric Skolem algebras* (P 4661) Algeb & Log;1984 Zagreb 155-163
◇ B60 ◇ ID 49040

SEPER, K. see Vol. IV, VI for further entries

SEREBRYANNIKOV, O.F. [1972] *Heuristic principles and logical calculi* (X 2737) Israel Progr Sci Transl: Jerusalem III*182pp
◇ B98 ◇ REV MR 58 # 16186 Zbl 302 # 02008 • REM Translated from Russian • ID 65215

SEREBRYANNIKOV, O.F. [1982] see BYSTROV, P.I.

SEREBRYANNIKOV, O.F. [1982] *Gentzen's Hauptsatz for modal logic with quantifiers* (P 3800) Intens Log: Th & Appl;1979 Moskva 79-88
◇ B45 ◇ REV MR 85e:03049 Zbl 547 # 03017 • ID 40618

SEREBRYANNIKOV, O.F. [1984] *Some generalizations of theorems about normal form in quantor modal logic (Russian)* (C 4366) Modal & Intens Log & Primen Probl Metodol Nauk 88-99
◇ B45 ◇ ID 46849

SEREBRYANNIKOV, O.F. see Vol. VI for further entries

SERFATI, M. [1977] *Une methode de resolution des equations postiennes a partir d'une solution particuliere* (J 0193) Discr Math 17*187-189
◇ B50 G20 ◇ REV MR 56 # 5286 Zbl 384 # 03046 • ID 52088

SERGANT, M. [1978] *Reflexions sur un systeme de matrices a huit valeurs propose par N. D. Belnap* (J 0079) Logique & Anal, NS 21*421-436
◇ A05 B45 G10 ◇ REV MR 80m:03045 Zbl 409 # 03004 • ID 56307

SERRUS, C. [1945] *Traite de logique* (X 1312) Aubier-Montaigne: Paris 383pp
◇ A05 B98 ◇ REV JSL 12.57 • ID 22436

SESELJA, B. & VOJVODIC, G. [1974] *Implication in three-valued logic as an exponential function (Serbo-Croatian) (English summary)* (J 0042) Mat Vesn, Drust Mat Fiz Astron Serb 11(26)*137-142
◇ B50 ◇ REV MR 52 # 5357 Zbl 294 # 02006 • ID 18379

SESELJA, B. see Vol. V for further entries

SESIC, B.V. [1971] *Foundations of the logic of change and development* (J 0286) Int Logic Rev 4*150-167
◇ B45 ◇ REV MR 46 # 1563 • ID 78523

SESIC, B.V. [1976] *Productive logic and foundations of a logic of variable predicates (LVP)* (J 0286) Int Logic Rev 14*142-159
◇ B60 ◇ REV Zbl 362 # 02019 • ID 50741

SESMAT, A. [1951] *Logique. vol.II: les raisonnements, la logistique* (X 0859) Hermann: Paris 361-776pp
◇ A05 B98 ◇ REM Vol.I 1950 • ID 22438

SETLUR, R.V. [1971] *Duality in finite many-valued logic* (J 0047) Notre Dame J Formal Log 12*188-194
◇ B50 ◇ REV MR 44 # 5209 Zbl 222 # 02016 • ID 11991

SETLUR, R.V. [1974] see KODANDAPANI, K.L.

SETLUR, R.V. [1975] see KODANDAPANI, K.L.

SETLUR, R.V. see Vol. I for further entries

SETTE, A.-M. [1969] see COSTA DA, N.C.A.

SETTE, A.-M. [1980] see ARRUDA, A.I.

SETTE, A.-M. see Vol. I, III, VI for further entries

SHAFER, G. [1981] *Constructive probability* (J 0154) Synthese 48*1-60
◇ B48 ◇ REV MR 82j:60006 Zbl 522 # 60001 • ID 82779

SHAFER, G. [1983] *A subjective interpretation of conditional probability* (J 0122) J Philos Logic 12*453-466
◇ B48 ◇ REV MR 85g:60008a Zbl 539 # 60004 • ID 41248

SHAH, H. & SMITH, W.R. [1976] *Multiple-output multi-valued prime implicants* (P 2011) Int Symp Multi-Val Log (6);1976 Logan 254-257
◇ B50 ◇ ID 35895

SHAJDA, J. [1979] *On probability of first order formulas in a given model* (J 0156) Kybernetika (Prague) 15*261-271
◇ B48 ◇ REV MR 81d:60002 Zbl 441 # 03008 • ID 56061

SHAJDA, J. [1983] *An application of fuzzy logic and approximate reasoning* (J 3919) BUSEFAL 15*10-12
◇ B52 ◇ REV Zbl 542 # 03010 • ID 43674

SHAJDA, J. [1984] *On some basic problems of propositional fuzzy logic* (J 3919) BUSEFAL 20*12-16
◇ B52 ◇ REV Zbl 559 # 03015 • ID 46898

SHAJTSANE, V.A. [1979] see OSIS, YA.YA.

SHANIN, N.A. [1977] *On the quantifier of limiting realizability* (P 3269) Set Th Found Math (Kurepa);1977 Beograd 127
◇ B55 C57 C80 F50 ◇ REV MR 58 # 21516 Zbl 361 # 02046 • ID 53812

SHANIN, N.A. see Vol. I, III, IV, V, VI for further entries

SHAPIRO, D.G. & TONG, R.M. [1984] *An experiment with multiple-valued logics in an expert system* (P 3081) IFAC Symp Fuzzy Inf, Knowl Repr & Decis. Anal;1983 Marseille 209-214
◇ B50 ◇ ID 46833

SHAPIRO, S. (ED.) [1985] *Intensional mathematics* (S 3303) Stud Logic Found Math 113*v+230pp
◇ B45 ◇ REV MR 85m:03004 Zbl 547 # 00010 • ID 43261

SHAPIRO, S. see Vol. I, III, IV, VI for further entries

SHATALOV, V.V. [1977] *Topological models for intuitionistic logic and the spectrum of refutions of a formula in the Fitch logic (Russian)* (S 2579) Teor Mnozhestv & Topol (Izhevsk) 1*93-95
◇ B55 ◇ REV Zbl 485 # 03016 • ID 36847

SHAUMAN, A.M. [1970] see BRUSENTSOV, N.P.

SHAW, M.L.G. [1984] see GAINES, B.R.

SHEJNBERGAS, I.M. [1971] see KOLESNIKOV, M.A.

SHEJNBERGAS, I.M. see Vol. I for further entries

SHEKHTMAN, V.B. [1977] *A remark on: "Models for multiply modal systems" by M.K.Rennie* (J 0068) Z Math Logik Grundlagen Math 23*555-558
◇ B45 ◇ REV MR 57 # 15973 Zbl 382 # 03016 • REM Published ibid. 16*175-186 (1970) • ID 51940

SHEKHTMAN, V.B. [1977] *On incomplete propositional logics (Russian)* (J 0023) Dokl Akad Nauk SSSR 235*542-545
• TRANSL [1977] (J 0062) Sov Math, Dokl 18*985-989
◇ B55 ◇ REV MR 57 # 71 Zbl 412 # 03011 • ID 52942

SHEKHTMAN, V.B. [1978] *An undecidable superintuitionistic propositional calculus (Russian)* (J 0023) Dokl Akad Nauk SSSR 240*549-552
• TRANSL [1978] (J 0062) Sov Math, Dokl 19*656-660
◇ B55 D35 ◇ REV MR 58 # 10330 Zbl 417 # 03010 JSL 50.1081 • ID 78453

SHEKHTMAN, V.B. [1978] *Rieger-Nishimura lattices (Russian)* (J 0023) Dokl Akad Nauk SSSR 241*1288-1291
• TRANSL [1978] (J 0062) Sov Math, Dokl 19*1014-1018
◇ B55 C90 G10 G25 ◇ REV MR 80a:03075 Zbl 412 # 03010 • ID 52941

SHEKHTMAN, V.B. [1978] *Two-dimensional modal logic (Russian)* (J 0087) Mat Zametki (Akad Nauk SSSR) 23*759-772
• TRANSL [1978] (J 1044) Math Notes, Acad Sci USSR 23*417-424
◇ B45 C90 ◇ REV MR 58 # 5068 Zbl 384 # 03010 • ID 52052

SHEKHTMAN, V.B. [1979] see MAKSIMOVA, L.L.

SHEKHTMAN, V.B. [1979] *Topological semantics of superintuitionistic logics (Russian)* (S 2582) Semiotika & Inf, Akad Nauk SSSR 12*62-67
◇ B55 C90 ◇ REV MR 82d:03040 • ID 78450

SHEKHTMAN, V.B. [1980] *Topological models of propositional logics (Russian)* (S 2582) Semiotika & Inf, Akad Nauk SSSR 15*74-98
◇ B45 C90 ◇ REV MR 82d:03029 Zbl 455 # 03013 • ID 54274

SHEKHTMAN, V.B. [1981] *Semantics of modal propositions (Russian)* (S 2582) Semiotika & Inf, Akad Nauk SSSR 17*148-169
◇ B45 ◇ REV MR 84m:03027 Zbl 502 # 03014 • ID 35731

SHEKHTMAN, V.B. [1982] *Undecidable propositional calculi (Russian)* (S 2874) Vopr Kibern, Akad Nauk SSSR 75*74-116
◇ B22 B55 D35 ◇ REV Zbl 499 # 03003 • ID 38116

SHEKHTMAN, V.B. [1983] *Denumerable approximability of superintuitionistic and modal logics (Russian)* (C 3807) Issl Neklass Log & Formal Sist 287-299
◇ B45 ◇ REV MR 85g:03034 • ID 43870

SHEKHTMAN, V.B. [1983] *Modal logics of domains on the real plane* (J 0063) Studia Logica 42*63-80
◇ B45 ◇ REV MR 86d:03020 Zbl 541 # 03011 • ID 41375

SHELAH, S. [1982] see LEHMANN, D.J.

SHELAH, S. [1983] see LEHMANN, D.J.

SHELAH, S. [1985] see GUREVICH, Y.

SHELAH, S. see Vol. I, III, IV, V, VI for further entries

SHELTON, L.V. [1983] *A diachronic semantics for inexact reference* (**J** 0047) Notre Dame J Formal Log 24*67-88
◇ B45 ◇ REV MR 84g:03027 • ID 34141

SHEN, BAIYING [1965] see MO, SHAOKUI

SHEN, BAIYING [1984] see MO, SHAOKUI

SHEN, BAIYING see Vol. IV, V, VI for further entries

SHEN, S.N.T. [1974] see KANG, A.N.C.

SHEN, YOUDING [1957] *The basic calculus (Chinese) (English summary)* (**J** 0418) Shuxue Xuebao 7*132-143
◇ B45 ◇ REV MR 20#3776 • ID 12062

SHEN, YOUDING see Vol. I, V, VI for further entries

SHENG, C.L. [1969] *Threshold logic* (**X** 0801) Academic Pr: New York 206pp
◇ B75 B98 ◇ ID 48529

SHENG, C.L. see Vol. I, IV for further entries

SHEPHERDSON, J.C. [1975] see ROSE, H.E.

SHEPHERDSON, J.C. [1984] see ATIYAH, M.

SHEPHERDSON, J.C. see Vol. I, III, IV, V, VI for further entries

SHERMAN, R. [1982] see HAREL, D.

SHERMAN, R. [1983] see HAREL, D.

SHERMAN, R. [1984] see HAREL, D.

SHESTAKOV, V.I. [1964] *On the relationship between certain three-valued logical calculi (Russian)* (**J** 0067) Usp Mat Nauk 19/2*177-181
◇ B50 ◇ REV MR 31#2140 Zbl 166.250 • ID 11986

SHESTAKOV, V.I. [1983] *Application of three-valued logic for analysis of relations between physical values (Russian)* (**J** 3801) Vest Ser Fiz Astron, Univ Moskva 24/4*40-45
◇ B50 ◇ REV MR 85a:03033 • ID 34784

SHESTAKOV, V.I. see Vol. I for further entries

SHESTEROVA, N.A. [1981] see KABULOV, A.V.

SHESTEROVA, N.A. [1981] *Some metric properties of functions of the k-valued logic (Russian)* (**J** 0430) Vopr Kibern (Akad Nauk Uzb SSR) 113*3-11
◇ B50 ◇ REV Zbl 471#03017 • ID 55212

SHEVCHENKO, V.N. [1984] *On some functions on many-valued logics related to integer programming (Russian)* (**X** 2235) VINITI: Moskva 6062-84
◇ B50 ◇ ID 46585

SHI, NIANDONG & YUE, YUJUN [1980] *Notes on the W_1L calculus (Chinese) (English summary)* (**J** 2754) Huazhong Gongxueyuan Xuebao 1*27-32
◇ B52 ◇ REV MR 84j:03058 • ID 34650

SHI, NIANDONG see Vol. III, IV for further entries

SHIMADA, R. [1970] see MINE, H.

SHIMADA, R. [1982] see HASEGAWA, T.

SHIMONY, A. [1955] *Coherence and the axioms of confirmation* (**J** 0036) J Symb Logic 20*1-28
◇ A05 B48 ◇ REV MR 16.1080 Zbl 64.244 JSL 33.481 • ID 12081

SHIMURA, M. [1975] *An approach to pattern recognition and associative memories using fuzzy logic* (**P** 0774) Fuzzy Sets & Appl;1974 Berkeley 449-476
◇ B52 ◇ REV MR 55#11749 Zbl 328#68085 • ID 65258

SHIMURA, M. see Vol. V for further entries

SHINOHARA, T. [1985] *Inductive inference from negtive data* (**J** 3957) Bull Inf & Cybern (Kyushu Univ) 21/3-4*67-70
◇ B48 ◇ ID 47709

SHIRAI, K. [1981] see MOTOHASHI, N.

SHIRAI, K. see Vol. III, VI for further entries

SHIRLEY, E.S. [1981] *An unnoticed flaw in Barker and Achinstein's solution to Goodman's new riddle of induction* (**J** 0153) Phil of Sci (East Lansing) 48*611-617
◇ B48 ◇ REV MR 82m:03031 • ID 78537

SHIVA, S.G. [1974] see NAGLE JR., H.T.

SHKLYAROV, L.I. [1967] see RVACHEV, V.L.

SHKLYAROV, L.I. [1968] see RVACHEV, V.L.

SHKLYAROV, L.I. [1976] see RVACHEV, V.L.

SHKLYAROV, L.I. [1979] see RVACHEV, V.L.

SHOENFIELD, J.R. [1967] *Mathematical logic* (**X** 0832) Addison-Wesley: Reading viii+344pp
• TRANSL [1975] (**X** 2027) Nauka: Moskva 527pp
◇ B98 C98 D98 E98 F98 ◇ REV MR 37#1224 MR 53#87 Zbl 155.11 JSL 40.234 • ID 22384

SHOENFIELD, J.R. see Vol. I, III, IV, V, VI for further entries

SHOESMITH, D.J. & SMILEY, T.J. [1971] *Deducibility and many-valuedness* (**J** 0036) J Symb Logic 36*610-622
◇ B50 ◇ REV MR 46#29 Zbl 253#02015 • ID 12106

SHOESMITH, D.J. see Vol. I, VI for further entries

SHOJLEV, KH. & TENEV, V. [1979] *Polynomial forms of many-valued logical functions. Minimization of polynomial form* (**J** 3360) Avtom Izchisl Tekh, Sofiya 13/4*31-37
◇ B50 ◇ REV Zbl 413#94017 • ID 69914

SHOKUROV, V.V. [1979] see BOCHVAR, D.A.

SHREJDER, YU.A. & VILENKIN, N.YA. [1977] *Majorant spaces and "majority" quantifier (Russian) (English summary)* (**S** 2582) Semiotika & Inf, Akad Nauk SSSR 8*45-82
• ERR/ADD ibid 212-213
◇ B45 C80 E20 ◇ REV MR 57#5677 • ID 79715

SHREJDER, YU.A. [1979] *Majorant models for modal (deontic) logics (Russian)* (**S** 2582) Semiotika & Inf, Akad Nauk SSSR 11*81-87
◇ B45 ◇ REV MR 81a:03020 • ID 78573

SHREJDER, YU.A. see Vol. IV, V for further entries

SHTRAKOV, S.V. [1985] *On some transformations on k-valued logic (Bulgarian summary)* (**P** 4391) Mat & Mat Obrazov (14);1985 Sl"nchev Bryag 305-309
◇ B50 ◇ ID 46558

SHU, YONGCHANG & WANG, YIZHI [1981] *Fuzzy languages and fuzzy grammars (Chinese)* (**J** 3732) Mohu Shuxue 1/2*113-123
◇ B52 D05 ◇ REV MR 84g:68071 • ID 47233

SHUKLA, A. [1967] *A note on the axiomatizations of certain modal systems* (**J** 0047) Notre Dame J Formal Log 8*118-120
◇ B45 ◇ REV MR 38 # 2017 • ID 12119

SHUKLA, A. [1970] *Decision procedures for Lewis system S1 and related modal systems* (**J** 0047) Notre Dame J Formal Log 11*141-180 • ERR/ADD ibid 14*584
◇ B25 B45 ◇ REV MR 44 # 1544 MR 48 # 8203 Zbl 182.6 JSL 37.754 • ID 19678

SHUKLA, A. [1971] *Finite model property for five modal calculi in the neighbourhood of S3* (**J** 0047) Notre Dame J Formal Log 12*69-74
◇ B45 ◇ REV MR 44 # 3850 Zbl 188.14 • ID 12121

SHUKLA, A. [1972] *Consistent, independent, and distinct propositions* (**J** 0047) Notre Dame J Formal Log 13*399-406
◇ B45 ◇ REV MR 46 # 5112 Zbl 231 # 02021 • REM Part I. Part II 1976 • ID 12122

SHUKLA, A. [1972] *The existence postulate and non-regular systems of modal logic* (**J** 0047) Notre Dame J Formal Log 13*369-378
◇ B45 ◇ REV MR 46 # 5111 Zbl 238 # 02019 • ID 12123

SHUKLA, A. [1976] *Consistent, independent, and distinct propositions. II* (**J** 0047) Notre Dame J Formal Log 17*135-136
◇ B45 ◇ REV MR 53 # 96 Zbl 313 # 02009 • REM Part I 1972. Part III 1983 • ID 18386

SHUKLA, A. [1983] *Consistent, independent, and distinct propositions III: Modalities in S6* (**J** 0047) Notre Dame J Formal Log 24*141-142
◇ B45 ◇ REV MR 84b:03035 Zbl 464 # 03018 • REM Part II 1976 • ID 54608

SHUKLA, A. see Vol. I for further entries

SHUKLA, R. [1973] see ROY, B.B.

SHUKLA, R. see Vol. I, V for further entries

SHUM, A.A. [1979] *Propositional semantic systems (Russian)* (**S** 2582) Semiotika & Inf, Akad Nauk SSSR 11*88-116
◇ B50 G25 ◇ REV MR 81h:03057 • ID 78587

SHUM, A.A. [1981] *Finite axiomatization of the logic of infinite problems in an extended language (Russian)* (**C** 3747) Mat Log & Mat Lingvistika 165-170
◇ B55 ◇ REV MR 84a:03027 • ID 34896

SHUM, A.A. [1983] *A pseudointuitionistic propositional calculus (Russian)* (**C** 3798) Mat Log, Mat Ling & Teor Algor 95-108
◇ B55 ◇ REV MR 85a:03017 • ID 34768

SHUM, A.A. [1984] *Varieties of algebraic systems and propositional calculi (Russian)* (**J** 0003) Algebra i Logika 23*341-359
• TRANSL [1984] (**J** 0069) Algeb and Log 23*237-251
◇ B45 C05 ◇ REV MR 86g:08008 • ID 44781

SHUNDO, S. [1977] see HOSOI, T.

SHUNDO, S. [1977] *Some results concerning the new classification of intermediate propositional logics* (**J** 0381) J Tsuda College (Tokyo) 9*17-27
◇ B55 ◇ REV MR 57 # 12184b • ID 78590

SHUNDO, S. [1978] *A procedure for classifying the intermediate propositional logics and its application to the finite slices* (**J** 0381) J Tsuda College (Tokyo) 10*1-8
◇ B55 ◇ REV MR 81f:03036 • ID 78589

SIDORENKO, E.A. [1970] *Certain variants of systems of logical inference (Russian)* (**C** 0668) Neklass Log 52-59
◇ B45 ◇ REV MR 47 # 8259 • ID 12143

SIDORENKO, E.A. [1970] *Independence in systems of logical inference (Russian)* (**C** 0668) Neklass Log 60-64
◇ B45 ◇ REV MR 47 # 8260 • ID 12142

SIDORENKO, E.A. [1979] *Some valid extensions of relevant systems (Russian)* (**P** 2554) All-Union Symp Th Log Infer;1974 Moskva 118-121
◇ B46 ◇ REV MR 84k:03064 • ID 35001

SIDORENKO, E.A. [1982] *On extensions of E* (**P** 3800) Intens Log: Th & Appl;1979 Moskva 195-202
◇ B46 ◇ REV MR 85a:03025 Zbl 519 # 03012 • ID 34776

SIDORENKO, E.A. [1982] *On various notions of the conclusion from hypotheses (Russian)* (**C** 3849) Modal & Relevant Log, Vyp 1 27-35
◇ B45 ◇ REV Zbl 526 # 03006 • ID 38166

SIDORENKO, E.A. [1983] *Logical entailment and conditional propositions (Russian)* (**X** 2027) Nauka: Moskva 174pp
◇ B46 ◇ REV MR 85j:03022 • ID 45142

SIDORENKO, E.A. [1983] *The strong proof from hypotheses and conditionals: some theorems of deduction for relevant systems* (**J** 0063) Studia Logica 42*165-171
◇ B46 ◇ REV MR 85i:03058 Zbl 569 # 03008 • ID 42347

SIDORENKO, E.A. [1984] *Relevance and amplification (Russian)* (**C** 4366) Modal & Intens Log & Primen Probl Metodol Nauk 158-182
◇ B46 ◇ ID 46671

SIEGEL, P. [1985] see BOSSU, G.

SIEMENS JR., D.F. [1977] *Fitch-style rules for many modal logics* (**J** 0047) Notre Dame J Formal Log 18*631-636
◇ B45 ◇ REV MR 58 # 16158 Zbl 315 # 02028 • ID 24311

SIEMENS JR., D.F. see Vol. I for further entries

SIERPINSKI, W. [1961] *Sur un probleme de la logique a n valeurs* (**J** 0027) Fund Math 49*167-170
◇ B50 ◇ REV MR 23 # A3671 Zbl 100.9 • ID 12270

SIERPINSKI, W. see Vol. IV, V, VI for further entries

SIFAKIS, J. [1982] see QUEILLE, J.P.

SIFAKIS, J. [1983] see QUEILLE, J.P.

SIGWART, C. [1895] *Logic. Vol.II: logical methods* (**X** 0959) Allen & Unwin: London viii+584pp
◇ B98 ◇ REM Vol.I 1904 • ID 22561

SIGWART, C. [1904] *Logik Vol.1* (**X** 1309) Mohr: Tuebingen
• TRANSL [1908] (**X** 4465) Obshch Pol'za & Provints: Leningrad
◇ B98 ◇ REM 3rd ed.; Vol.II 1895 • ID 43001

SIKIC, Z. [1984] *Multiple forms of Gentzen's rules and some intermediate forms* (**J** 0068) Z Math Logik Grundlagen Math 30*335-338
◇ B55 F07 ◇ REV MR 86a:03024 • ID 42266

SIKIC, Z. see Vol. VI for further entries

SIKORSKI, R. [1953] see RASIOWA, H.

SIKORSKI, R. [1955] see RASIOWA, H.

SIKORSKI, R. [1955] see GRZEGORCZYK, A.

SIKORSKI, R. [1956] *A theorem on non-classical functional calculi* (**J** 0014) Bull Acad Pol Sci, Ser Math Astron Phys 4*649-650
◇ B45 F50 G10 ◇ REV MR 19.830 Zbl 75.4 JSL 32.537 • ID 12298

SIKORSKI, R. [1958] *Some applications of interior mappings* (**J** 0027) Fund Math 45*200-212
◇ B45 E75 F50 G05 G10 G25 ◇ REV MR 24 # A2371 Zbl 83.386 JSL 32.537 • ID 12300

SIKORSKI, R. [1963] see RASIOWA, H.

SIKORSKI, R. see Vol. I, III, IV, V, VI for further entries

SILVA DE, O. [1982] see RENNA E SOUZA DE, C.

SILVERSTEIN, H.S. [1974] *Von Wright's deontic logics* (**J** 0095) Philos Stud 25*365-371
◇ B45 ◇ REV MR 58 # 153 • ID 78621

SILVESTRINI, D. [1981] *Alcune considerazioni sul metodo delle supervalutazioni e una semantica a supervalutazioni per il calcolo predicativo classico* (**P** 3092) Congr Naz Logica;1979 Montecatini Terme 335-350
◇ B35 B60 ◇ ID 48725

SIMI, G. [1980] *Algebre pre-modali. I: Su una classe di algebre atte all'introduzione di un operatore di possibilita. II: Sulla varieta delle algebre premodali* (**J** 3128) Boll Unione Mat Ital, Suppl 2*1-14,365-371
◇ B45 B60 G25 ◇ REV MR 84a:03080 Zbl 446 # 03048 Zbl 446 # 03049 • ID 56577

SIMMONS, H. [1975] *Topological aspects of suitable theories* (**J** 3420) Proc Edinburgh Math Soc, Ser 2 19*383-391
◇ B45 F30 G05 ◇ REV MR 52 # 7894 Zbl 327 # 02028 • ID 18388

SIMMONS, H. see Vol. III, IV, VI for further entries

SIMONS, L. [1953] *New axiomatizations of S3 and S4* (**J** 0036) J Symb Logic 18*309-316
◇ B45 ◇ REV MR 15.493 Zbl 52.11 JSL 19.293 • ID 12336

SIMONS, L. [1962] *A reduction in the number of independent axiom schemata for S4* (**J** 0047) Notre Dame J Formal Log 3*256-258
◇ B45 ◇ REV MR 33 # 5476 Zbl 131.246 JSL 32.245 • ID 12337

SIMONS, L. see Vol. I, V for further entries

SIMONS, P.M. [1981] *A note on Lesniewski and free logic* (**J** 0079) Logique & Anal, NS 24*415-420
◇ B60 ◇ REV MR 83g:03027 Zbl 494 # 03004 • ID 36001

SIMONS, P.M. [1983] *Class, mass and mereology* (**J** 2028) Hist & Phil Log 4*157-180
◇ A05 B60 ◇ REV MR 84k:03085 Zbl 524 # 03014 • ID 35022

SIMONS, P.M. [1984] *A Brentanian basis for Lesniewskian logic* (**J** 0079) Logique & Anal, NS 27*297-307
◇ B60 ◇ REV MR 86d:03028 Zbl 557 # 03016 • ID 44382

SIMONS, P.M. [1985] *Lesniewski's logic and its relation to classical and free logics* (**P** 4180) Int Congr Log, Meth & Phil of Sci (7,Pap);1983 Salzburg 369-400
◇ B60 ◇ ID 48114

SIMONS, P.M. see Vol. V for further entries

SIMOVICI, D.A. [1985] see REISCHER, C.

SIMOVICI, D.A. see Vol. I, IV for further entries

SIMPSON, S.G. [1985] see HARRINGTON, L.A.

SIMPSON, S.G. see Vol. I, III, IV, V, VI for further entries

SINDELAR, J. [1978] see KRAMOSIL, I.

SINDELAR, J. [1980] *Statistical theory of logical derivability* (**J** 0156) Kybernetika (Prague) 16*225-239
◇ B48 ◇ REV MR 83b:68098 Zbl 444 # 03008 • ID 56456

SINDELAR, J. see Vol. I for further entries

SINGER, F.R. [1967] *Some Sheffer functions for m-valued logics* (**J** 0287) Scripta Math 28*21-27
◇ B50 ◇ REV MR 35 # 6527 Zbl 154.256 JSL 34.520 • ID 12345

SINGLETARY, W.E. [1975] see HUGHES, C.E.

SINGLETARY, W.E. see Vol. I, IV for further entries

SINTONEN, L. [1972] *On the realization of functions in n-valued logic* (**J** 0187) IEEE Trans Comp C-21*610-612
◇ B50 B70 ◇ REV Zbl 236 # 94031 • ID 65330

SIOSON, F.M. [1964] *Further axiomatizations of the Lukasiewicz three-valued calculus* (**J** 0047) Notre Dame J Formal Log 5*62-70
◇ B50 ◇ REV MR 31 # 2141 Zbl 137.249 JSL 31.500 • ID 12370

SIOSON, F.M. see Vol. I for further entries

SISTLA, A.P. [1984] see EMERSON, E.A.

SISTLA, A.P. [1985] see REIF, J.H.

SISTLA, A.P. [1985] see CLARKE, E.M.

SISTLA, A.P. & VARDI, M.Y. & WOLPER, P. [1985] *The complementation problem for Buechi automata with applications to temporal logic (extended abstract)* (**P** 4628) Automata, Lang & Progr (12);1985 Nafplion 465-474
◇ B45 D05 ◇ ID 49529

SISTLA, A.P. see Vol. IV for further entries

SITNIKOV, O.P. [1975] see LABUNETS, V.G.

SIY, P. [1974] see CHEN, C.S.

SJOEDIN, T. [1978] see IVERT, P.A.

SKALA, H.J. [1974] see MENGES, G.

SKALA, H.J. [1976] *On the problem of imprecision* (**J** 0472) Theory Decis 7*159-170
◇ B52 ◇ REV MR 55 # 7741 Zbl 348 # 02022 • ID 65337

SKALA, H.J. [1978] *On many-valued logics, fuzzy sets, fuzzy logics and their applications* (**J** 2720) Fuzzy Sets Syst 1*129-149
◇ B52 E72 ◇ REV MR 58 # 5040 Zbl 396 # 03024 • ID 52634

SKALA, H.J. [1982] *Modelling vagueness* (**C** 3778) Fuzzy Inform & Decis Processes 101-109
◇ B52 ◇ REV MR 84m:03072 Zbl 537 # 03037 • ID 35769

SKALA, H.J. see Vol. V for further entries

SKIBENKO, I.T. [1973] see MOKLYAK, N.G.

SKIBENKO, I.T. [1974] see POPOV, V.A.

SKIBENKO, I.T. [1977] see POPOV, V.A.

SKIBENKO, I.T. [1979] see POPOV, V.A.

SKIBENKO, I.T. see Vol. I for further entries

SKOLEM, T.A. [1928] *Ueber die mathematische Logik* (**J** 4510) Norsk Mat Tidsskr 10*125-142
• TRANSL [1967] (**C** 0675) From Frege to Goedel 508-524
• REPR [1970] (**C** 1098) Skolem: Select Works in Logic 189-206
◇ A05 B25 B98 C07 C10 C35 ◇ REV FdM 54.58
• ID 12385

SKOLEM, T.A. [1952] *Consideraciones sobre los fundamentos de la matematica I* (**J** 0236) Rev Mat Hisp-Amer, Ser 4 12*169-200
◇ B98 F30 F50 ◇ REV MR 15.92 Zbl 52.8 JSL 21.373
• REM Part II 1953 • ID 21071

SKOLEM, T.A. [1953] *Consideraciones sobre los fundamentos de la matematica II* (**J** 0236) Rev Mat Hisp-Amer, Ser 4 13*149-174
◇ B98 F30 F50 ◇ REV MR 15.92 Zbl 52.8 JSL 21.373
• REM Part I 1952 • ID 21072

SKOLEM, T.A. [1954] *Results in investigations in the foundations (Norwegian)* (**P** 0788) Skand Mat Kongr (12);1953 Lund 273-289
◇ A05 B98 D03 F55 ◇ REV MR 16.553 Zbl 56.245
• ID 21133

SKOLEM, T.A. [1957] *Bemerkungen zum Komprehensionsaxiom* (**J** 0068) Z Math Logik Grundlagen Math 3*1-17
• REPR [1970] (**C** 1098) Skolem: Select Works in Logic 615-631
◇ B50 E70 ◇ REV MR 20#3067 Zbl 95.9 JSL 32.128
• ID 12420

SKOLEM, T.A. [1957] *Mengenlehre gegruendet auf einer Logik mit unendlich vielen Wahrheitswerten* (**J** 0366) Sitzber Berlin Math Ges 58*41-56
◇ B50 E70 ◇ REV MR 27#3511 Zbl 116.9 • ID 12418

SKOLEM, T.A. [1960] *A set theory based on a certain three-valued logic* (**J** 0132) Math Scand 8*127-136
• REPR [1970] (**C** 1098) Skolem: Select Works in Logic 653-662
◇ B50 E70 ◇ REV MR 23#A3669 Zbl 96.243 • ID 12425

SKOLEM, T.A. [1963] *Studies on the axiom of comprehension* (**J** 0047) Notre Dame J Formal Log 4*162-170
• REPR [1970] (**C** 1098) Skolem: Select Works in Logic 703-711
◇ A05 B50 E30 E70 ◇ REV MR 29#5716 Zbl 133.258
• ID 12431

SKOLEM, T.A. see Vol. I, III, IV, V, VI for further entries

SKORDEV, D.G. [1984] *On a modal-type language for the predicate calculus* (**J** 0387) Bull Sect Logic, Pol Acad Sci 13*111-119
◇ B45 ◇ REV Zbl 556#03014 • ID 44675

SKORDEV, D.G. see Vol. III, IV, V, VI for further entries

SKORUBSKIJ, V.I. [1980] *Arithmetical and logical foundations of digital machines. Textbook (Russian)* (**X** 3219) Tochnoj Mekhaniki Optiki: Leningrad 60pp
◇ B70 B98 ◇ REV Zbl 479#94029 • ID 55704

SKVORTSOV, D.P. [1979] *Logic of infinite problems and Kripke models on atomic semilattices of sets (Russian)* (**J** 0023) Dokl Akad Nauk SSSR 245*798-801
• TRANSL [1979] (**J** 0062) Sov Math, Dokl 20*360-363
◇ B55 C90 F50 G20 ◇ REV MR 80i:03040
Zbl 438#03028 • ID 55940

SKVORTSOV, D.P. [1979] *On some propositional logics connected with the concept of Yu. T. Medvedev's types of information (Russian)* (**S** 2582) Semiotika & Inf, Akad Nauk SSSR 13*142-149
◇ B55 ◇ REV MR 81j:03041b Zbl 462#03004 • ID 54511

SKVORTSOV, D.P. [1979] *Realizability and finite validity of propositional formulas with restructions on the occurrences of implication (Russian)* (**J** 0087) Mat Zametki (Akad Nauk SSSR) 25*919-931
• TRANSL [1979] (**J** 1044) Math Notes, Acad Sci USSR 25*474-481
◇ B55 ◇ REV MR 81j:03027 Zbl 417#03011 • ID 53248

SKVORTSOV, D.P. [1979] see MAKSIMOVA, L.L.

SKVORTSOV, D.P. [1980] *On the connection of finitary general validity of certain propositional formulas with derivability in the Kreisel-Putnam logic (Russian)* (**J** 0288) Vest Ser Mat Mekh, Univ Moskva 1980/3*29-32
• TRANSL [1980] (**J** 0510) Moscow Univ Math Bull 35/3*30-33
◇ B55 F50 ◇ REV MR 81i:03088 Zbl 438#03029
• ID 55941

SKVORTSOV, D.P. [1983] *The structure of extensions of the propositional fragment of the Ackermann logic (Russian)* (**C** 3807) Issl Neklass Log & Formal Sist 209-221
◇ B46 ◇ REV MR 85b:03017 • ID 40565

SKVORTSOV, D.P. [1985] *A superintuitionistic propositional calculus (Russian)* (**J** 0003) Algebra i Logika 24*195-204,250
• TRANSL [1985] (**J** 0069) Algeb and Log 24*119-125
◇ B55 ◇ ID 49372

SKVORTSOV, D.P. see Vol. I, III, V, VI for further entries

SKYRMS, B. [1966] *Choice and chance* (**X** 1201) Dickenson: Belmont viii+165pp
◇ A05 B48 B98 ◇ REV JSL 41.547 • ID 21285

SKYRMS, B. [1970] *Notes on quantification and self reference* (**P** 0731) Paradox of Liar;1969 Buffalo 67-74
◇ A05 B50 ◇ REV JSL 40.584 • ID 14624

SKYRMS, B. [1970] *Return of the liar: three-valued logic and the concept of truth* (**J** 0325) Amer Phil Quart 7*153-161
◇ A05 B50 ◇ REV JSL 40.584 • ID 14631

SKYRMS, B. [1976] *Definitions of semantical reference and self-reference* (**J** 0047) Notre Dame J Formal Log 17*147-148
◇ A05 B50 ◇ REV MR 55#2479 Zbl 272#02011
• ID 18394

SKYRMS, B. [1978] *An immaculate conception of modality* (**J** 0301) J Phil 75*368-387
◇ B45 ◇ ID 28381

SKYRMS, B. see Vol. I for further entries

SLAGHT, R.L. [1977] *Modal tree constructions* (**J** 0047) Notre Dame J Formal Log 18*517-526
◇ B45 ◇ REV MR 58#21483 Zbl 258#02030
Zbl 372#02014 • ID 23687

SLAGHT, R.L. see Vol. I for further entries

SLANEY, J.K. [1984] *A metacompleteness theorem for contraction-free relevant logics* (**J** 0063) Studia Logica 43*159-168
◇ B46 ◇ ID 42367

SLANEY, J.K. [1985] *3088 varieties: a solution to the Ackermann constant problem* (J 0036) J Symb Logic 50*487-501
◇ B46 C05 ◇ ID 42560

SLATER, B.H. [1972] *The foundations of logic* (J 0094) Mind 81*42-56
◇ A05 B45 ◇ REV MR 57 # 2867 • ID 78706

SLATER, B.H. [1980] *Logic without tears* (J 0286) Int Logic Rev 22*120-128
◇ B60 ◇ REV MR 83i:03022 Zbl 502 # 03008 • ID 35497

SLATER, B.H. see Vol. I, V for further entries

SLININ, YA.A. [1967] *Theory of modalities in contemporary logic (Russian)* (C 0582) Log Semant & Modal Logika 119-147
◇ B45 ◇ REV MR 36 # 2478 Zbl 204.2 • ID 12460

SLININ, YA.A. [1970] *Iterated modalities in contemporary logic (Russian)* (C 0668) Neklass Log 286-296
◇ B45 ◇ REV MR 47 # 6447 Zbl 204.2 • ID 12461

SLININ, YA.A. [1972] *Die Modalitaetentheorie in der modernen Logik* (C 1533) Quantoren, Modal, Paradox 362-401
◇ B45 ◇ REV Zbl 249 # 02010 • ID 28881

SLIWINSKA, D.S. [1985] see KISZKA, J.B.

SLOGEDAL, S. [1976] *Unzugaengliche Welten* (J 0239) Skr, K Nor Vidensk Selsk 1976/1*5pp
◇ A05 B45 ◇ REV Zbl 318 # 02009 • ID 65375

SLUPECKI, J. [1936] *Der volle dreiwertige Aussagenkalkuel* (J 0459) C R Soc Sci Lett Varsovie Cl 3 29*9-11
◇ B50 ◇ REV Zbl 15.51 JSL 2.46 FdM 62.1052 • ID 16519

SLUPECKI, J. [1939] *A criterion of fullness of many-valued systems of propositional logic (Polish) (German summary)* (J 0459) C R Soc Sci Lett Varsovie Cl 3 32*102-110
◇ B50 ◇ REV JSL 11.128 • ID 19655

SLUPECKI, J. [1939] *Proof of the axiomatizability of full many-valued systems of propositional calculus (Polish) (German summary)* (J 0459) C R Soc Sci Lett Varsovie Cl 3 32*110-128
◇ B50 ◇ REV JSL 11.92 • ID 19656

SLUPECKI, J. [1946] *The complete three-valued propositional calculus (Polish)* (J 4551) Ann Univ Lublin, Sect F 1*193-209
◇ B50 ◇ REV MR 10.1 JSL 13.165 • ID 41599

SLUPECKI, J. [1953] *St. Lesniewski's protothetics (Polish and Russian summaries)* (J 0063) Studia Logica 1*44-112,299
◇ B60 ◇ REV MR 16.892 Zbl 59.12 JSL 21.188 • ID 12478

SLUPECKI, J. [1955] *St. Lesniewski's calculus of names (Polish and Russian summaries)* (J 0063) Studia Logica 3*7-76
• ERR/ADD ibid 3*6
◇ B60 ◇ REV MR 17.1171 • ID 12481

SLUPECKI, J. [1962] see POGORZELSKI, W.A.

SLUPECKI, J. [1963] see BORKOWSKI, L.

SLUPECKI, J. [1967] see BRYLL, G.

SLUPECKI, J. [1967] *The full three-valued propositional calculus* (C 0615) Polish Logic 1920-39 335-337
◇ B50 ◇ REV JSL 35.442 • ID 12483

SLUPECKI, J. [1971] *A generalization of modal logic (Polish and Russian summaries)* (J 0063) Studia Logica 28*7-17
◇ B45 ◇ REV MR 45 # 8505 Zbl 253 # 02019 • ID 12486

SLUPECKI, J. [1971] *Proof of the axiomatizability of full many-valued systems of calculus of propositions* (J 0063) Studia Logica 29*155-168
◇ B50 ◇ REV MR 51 # 10030 Zbl 247 # 02021 • ID 17560

SLUPECKI, J. [1972] *A criterion of fullness of many-valued systems of propositional logic* (J 0063) Studia Logica 30*153-157
◇ B50 ◇ REV MR 47 # 8245 Zbl 273 # 02015 • ID 12495

SLUPECKI, J. [1972] see BRYLL, G.

SLUPECKI, J. [1973] see BRYLL, G.

SLUPECKI, J. [1973] see PIROG-RZEPECKA, K.

SLUPECKI, J. [1974] *Some remarks on the many-valued logics of Jan Lukasiewicz (Russian)* (C 2578) Filos & Logika 177-187
◇ A05 B50 ◇ REV MR 58 # 27311 • ID 78732

SLUPECKI, J. [1976] see HALKOWSKA, K.

SLUPECKI, J. [1978] see HALKOWSKA, K.

SLUPECKI, J. see Vol. I, III, V, VI for further entries

SMETS, P. [1981] *The degree of belief in a fuzzy event* (J 0191) Inform Sci 25*1-19
◇ B52 ◇ REV MR 83b:03024 Zbl 472 # 62005 • ID 35090

SMETS, P. [1982] *Probability of a fuzzy event: an axiomatic approach* (J 2720) Fuzzy Sets Syst 7*153-164
◇ B52 ◇ REV MR 83e:03086 Zbl 479 # 60009 • ID 35240

SMETS, P. [1982] *Subjective probability and fuzzy measures* (C 3778) Fuzzy Inform & Decis Processes 87-91
◇ B52 ◇ REV MR 84d:90004 • ID 46105

SMETS, P. see Vol. V for further entries

SMILEY, T.J. [1959] *Entailment and deducibility* (J 0113) Proc Aristotelian Soc 59*233-254
◇ A05 B46 ◇ REV JSL 30.240 • ID 12508

SMILEY, T.J. [1961] *On Lukasiewicz's L-modal System* (J 0047) Notre Dame J Formal Log 2*149-153
◇ B45 ◇ REV MR 26 # 3594 Zbl 113.5 JSL 27.113
• ID 12509

SMILEY, T.J. [1963] *Relative necessity* (J 0036) J Symb Logic 28*113-134
◇ A05 B45 ◇ REV MR 31 # 41 Zbl 125.5 JSL 32.401
• ID 12513

SMILEY, T.J. [1971] see SHOESMITH, D.J.

SMILEY, T.J. [1976] *Does many-valued logic have any use?* (P 1502) Phil of Logic;1976 Bristol 74-88
◇ A05 B50 ◇ REV MR 58 # 10307 • ID 33245

SMILEY, T.J. see Vol. I, IV, VI for further entries

SMIRNOV, G.A. [1967] *The proof of basic theorems in the theory of strong logical inference (Russian)* (C 0582) Log Semant & Modal Logika 148-161
◇ B45 ◇ REV MR 37 # 32 Zbl 203.6 • ID 12515

SMIRNOV, G.A. [1970] *The forms of logical inference (Russian)* (C 1530) Issl Log Sist (Yanovskaya) 204-214
◇ B45 ◇ REV MR 47 # 8239 Zbl 217.8 • ID 12516

SMIRNOV, V.A. [1970] *The so-called paradoxes of material implication, and systems of logic with a concept of strict implication (Russian)* (C 1530) Issl Log Sist (Yanovskaya) 122-136
◇ B46 ◇ REV MR 48 # 5814 Zbl 218 # 02007 • ID 12520

SMIRNOV, V.A. [1972] *Formal derivation and logical calculi (Russian)* (X 2027) Nauka: Moskva 272pp
◇ B98 C07 C40 C98 F05 F07 F98 ◇ REV Zbl 269 # 02004 • ID 30359

SMIRNOV, V.A. & TAVANETS, P.V. (EDS.) [1974] *Philosphy and logic (Russian)* (X 2027) Nauka: Moskva 479pp
◇ A05 B97 ◇ REV MR 58 # 21368 • ID 70080

SMIRNOV, V.A. [1979] *Formal inference, deduction theorems and theories of implication (Russian)* (P 2554) All-Union Symp Th Log Infer;1974 Moskva 54-68
◇ B46 ◇ REV MR 84k:03038 • ID 34974

SMIRNOV, V.A. (ED.) [1979] *Logical inference (Russian)* (X 2027) Nauka: Moskva 312pp
◇ B97 ◇ REV MR 84j:03007 • ID 34601

SMIRNOV, V.A. [1982] *An immersion of St. Lesniewski's elementary ontology into second-order one-place predicate calculus (Russian)* (C 3849) Modal & Relevant Log, Vyp 1 55-65
◇ B60 ◇ REV Zbl 528 # 03013 • ID 37631

SMIRNOV, V.A. [1982] see KARPENKO, A.S.

SMIRNOV, V.A. [1982] *The definition of modal operators by means of tense operators* (P 3800) Intens Log: Th & Appl;1979 Moskva 50-69
• TRANSL [1984] (C 4366) Modal & Intens Log & Primen Probl Metodol Nauk 14-31
◇ B45 ◇ REV MR 85f:03022 Zbl 517 # 03005 • ID 37278

SMIRNOV, V.A. [1984] see ERSHOV, YU.L.

SMIRNOV, V.A. [1984] *Logical systems with modal tense operators (Russian)* (C 4366) Modal & Intens Log & Primen Probl Metodol Nauk 49-58
◇ B45 ◇ ID 46677

SMIRNOV, V.A. see Vol. I, III, V, VI for further entries

SMIRNOVA, E.D. & TAVANETS, P.V. [1967] *Semantics in logic (Russian)* (C 0582) Log Semant & Modal Logika 3-53
• TRANSL [1972] (C 1533) Quantoren, Modal, Paradox 123-178
◇ A05 B98 ◇ REV MR 37 # 48 Zbl 204.1 • ID 12523

SMIRNOVA, E.D. see Vol. I, VI for further entries

SMIRNOVA, I.M. [1963] see AJZERMAN, M.A.

SMIRNOVA, I.M. see Vol. IV, V for further entries

SMITH, BARRY [1983] see MULLIGAN, K.

SMITH, C.H. [1983] see CASE, J.

SMITH, C.H. [1985] see ANGLUIN, D.

SMITH, C.H. see Vol. I, IV for further entries

SMITH, HENRY BRADFORD [1936] *The algebra of propositions* (J 0153) Phil of Sci (East Lansing) 3*551-578
◇ B05 B45 ◇ REV JSL 2.43 FdM 62.1052 • ID 41315

SMITH, HENRY BRADFORD [1936] *The law of transitivity* (J 0153) Phil of Sci (East Lansing) 3*123
◇ B45 ◇ REV JSL 1.43 FdM 62.1065 • ID 41320

SMITH, HENRY BRADFORD [1937] *Modal Logic - a revision* (J 0153) Phil of Sci (East Lansing) 4*383-384
◇ B45 ◇ REV JSL 2.172 • ID 41365

SMITH, HENRY BRADFORD see Vol. I for further entries

SMITH, KENNETH C. [1970] see LEE, E.S.

SMITH, KENNETH C. [1971] see JANCZEWSKI, L.J.

SMITH, KENNETH C. & VRANESIC, Z.G. [1974] *Engineering aspects of multi-valued logic systems* (J 2015) Computer 7/9*34-41
◇ B50 ◇ ID 42113

SMITH, KENNETH C. [1976] see GAINES, B.R.

SMITH, KENNETH C. [1981] *The prospects for multivalued logic: a technology and applications view* (J 0187) IEEE Trans Comp C-30*619-634
◇ B50 B70 ◇ REV MR 82m:94051 Zbl 463 # 94010 • ID 82836

SMITH, KENNETH C. see Vol. V for further entries

SMITH, T. [1974] see PARKS, R.Z.

SMITH, W.R. [1976] see DUGGAN, R.W.

SMITH, W.R. [1976] see SHAH, H.

SMITH III, W.R. [1974] *Minimization of multivalued functions* (P 1385) Int Symp Multi-Val Log (4);1974 Morgantown 27-44
◇ B50 ◇ ID 42106

SMITHSON, M. [1984] *Multivariate analysis using "and" and "or"* (J 3914) Math Soc Sci 7*231-251
◇ B52 ◇ REV MR 86a:62092 Zbl 562 # 62052 • ID 44128

SMOKLER, H. [1977] *Semantical questions in Carnap's inductive logic* (J 0013) Brit J Phil Sci 28*129-135
◇ A05 B48 ◇ REV MR 58 # 10322 Zbl 395 # 03006 • ID 52557

SMOKLER, H. [1977] *Three grades of probabilistic involvement* (J 0095) Philos Stud 32*129-142
◇ A05 B45 B48 ◇ REV MR 57 # 5662 • ID 78755

SMOKLER, H. [1979] *Single-case propensities, modality, and confirmation* (J 0154) Synthese 40*497-506
◇ A05 B45 ◇ REV MR 80e:03017 Zbl 406 # 03009 • ID 56092

SMOKLER, H. [1979] *The collapse of modal distinctions in probabilistic contexts* (J 0105) Theoria (Lund) 45*1-7
◇ A05 B45 ◇ REV MR 82g:03033 Zbl 453 # 03003 • ID 54133

SMOLENOV, KH. [1983] *Paraconsistency, paracompleteness and intentional contradictions* (J 0387) Bull Sect Logic, Pol Acad Sci 12*8-16
◇ B53 ◇ REV MR 84k:03086 Zbl 531 # 03004 • ID 36100

SMOLENOV, KH. [1984] *Truthfulness and nontrivial contraction-free relevant logics* (J 0387) Bull Sect Logic, Pol Acad Sci 13*144-153
◇ B46 ◇ REV Zbl 564 # 03023 • ID 44676

SMOLENOV, KH. [1984] *Zeno's paradoxes and temporal becoming in dialectical atomism* (J 0063) Studia Logica 43*169-180
◇ A05 B45 ◇ ID 42369

SMORYNSKI, C.A. [1973] *Investigations of intuitionistic formal systems by means of Kripke models* (0000) Diss., Habil. etc
◇ B55 C90 F50 ◇ REM Diss., University of Chicago • ID 27058

SMORYNSKI, C.A. [1978] *Beth's theorem and self-referential sentences* (P 1897) Logic Colloq;1977 Wroclaw 253-261
◇ B45 C40 F30 ◇ REV MR 80c:03023 Zbl 453 # 03018 • ID 54148

SMORYNSKI, C.A. [1978] *The axiomatization problem for fragments* (**J** 0007) Ann Math Logic 14*193-221
◇ B55 F50 ◇ REV MR 80e:03040 Zbl 395 # 03039
• ID 29151

SMORYNSKI, C.A. [1979] *Calculating self-referential statements. I: Explicit calculations* (**J** 0063) Studia Logica 38*17-36
◇ B45 F30 G25 ◇ REV MR 80m:03030 Zbl 412 # 03037
• ID 52968

SMORYNSKI, C.A. [1980] *Calculating self-referential statements* (**J** 0027) Fund Math 109*189-210
◇ B45 F30 G25 ◇ REV MR 82g:03099 Zbl 449 # 03067
• ID 56734

SMORYNSKI, C.A. [1982] *Commutativity and self-reference* (**J** 0047) Notre Dame J Formal Log 23*443-452
◇ B45 F30 ◇ REV MR 83k:03027 Zbl 528 # 03009
• ID 36167

SMORYNSKI, C.A. [1982] *The finite inseparability of the first-order theory of diagonalisable algebras* (**J** 0063) Studia Logica 41*347-349
◇ B45 C13 D35 F30 ◇ REV Zbl 542 # 03024 • ID 43687

SMORYNSKI, C.A. [1984] *Modal logic and self-reference* (**C** 4085) Handb Philos Log 2*441-496
◇ B45 F30 ◇ ID 44891

SMORYNSKI, C.A. [1985] *Self-reference and modal logic* (**X** 0811) Springer: Heidelberg & New York vii + 333pp
◇ B45 F30 F98 ◇ ID 44663

SMORYNSKI, C.A. see Vol. III, IV, V, VI for further entries

SMULLYAN, A. [1948] *Modality and description* (**J** 0036) J Symb Logic 13*31-37
◇ B45 ◇ REV MR 10.176 Zbl 36.6 JSL 13.149 • ID 12536

SMULLYAN, A. [1962] *Fundamentals of logic* (**X** 0819) Prentice Hall: Englewood Cliffs 131pp
◇ B98 ◇ REV JSL 29.41 • ID 12537

SMULLYAN, R.M. [1961] *Extended canonical systems* (**J** 0053) Proc Amer Math Soc 12*440-442
◇ B45 D03 D25 ◇ REV MR 23 # A3092 Zbl 101.11 JSL 32.524 • ID 12542

SMULLYAN, R.M. [1961] *Theory of formal systems* (**S** 3513) Ann Math Stud xi + 142pp
• TRANSL [1981] (**X** 2027) Nauka: Moskva 208pp
◇ B98 D03 D05 D20 D25 D98 ◇ REV MR 22 # 12042 MR 27 # 2409 MR 83h:03002 Zbl 529 # 03014 Zbl 97.245 JSL 30.88 • ID 21025

SMULLYAN, R.M. [1963] *First order logic* (**X** 0811) Springer: Heidelberg & New York xii + 158pp
◇ B98 F07 F98 ◇ REV MR 39 # 5311 Zbl 172.289 JSL 40.237 • REM 3rd edition 1971 • ID 28562

SMULLYAN, R.M. [1973] *A generalization of intuitionistic and modal logics* (**P** 0783) Truth, Syntax & Modal;1970 Philadelphia 274-293
◇ B45 F50 ◇ REV MR 53 # 7733 Zbl 268 # 02017 JSL 42.316 • ID 14740

SMULLYAN, R.M. [1978] *What is the name of this book? The riddle of Dracula and other logical puzzles* (**X** 0819) Prentice Hall: Englewood Cliffs xi + 241pp
◇ B98 ◇ REV Zbl 432 # 00028 • ID 53958

SMULLYAN, R.M. [1983] *Dame oder Tiger? Logische Denkspiele und eine mathematische Novelle ueber Goedels grosse Entdeckung* (**X** 1265) Fischer: Stuttgart 239pp
◇ B98 F99 ◇ REV Zbl 545 # 00007 • REM Transl. from English • ID 41200

SMULLYAN, R.M. [1985] *Modality and self-reference* (**C** 3659) Intens Math 191-211
◇ B45 F30 ◇ ID 44678

SMULLYAN, R.M. [1985] *Some principles related to Loeb's theorem* (**C** 3659) Intens Math 213-230
◇ B45 F30 ◇ ID 44783

SMULLYAN, R.M. see Vol. I, III, IV, V, VI for further entries

SNELSIRE, R.W. [1968] see GIVONE, D.D.

SNYDER, D.P. [1966] *Models for logical entailment* (**J** 0079) Logique & Anal, NS 9*344-359
◇ B46 ◇ REV MR 38 # 32 Zbl 161.5 • ID 12563

SNYDER, D.P. [1971] *Modal logic and its applications* (**X** 0864) Van Nostrand: New York iv + 335pp
◇ B45 ◇ REV JSL 40.251 • ID 12561

SNYDER, D.P. see Vol. I, VI for further entries

SOBOCINSKI, B. [1936] *Axiomatization of certain many-valued systems of the theory of deduction* (**J** 4468) Rocz Prac Nauk Zrzesz Asystentow Uniw Warszaw 1*399-419
◇ B50 ◇ REV JSL 2.93 • ID 41349

SOBOCINSKI, B. [1939] *Aus den Untersuchungen zur Protothetik (Polish)* (**J** 0490) Collecteana Logica 1*171-177
◇ B60 ◇ REV Zbl 23.289 FdM 65.1105 • ID 43786

SOBOCINSKI, B. [1949] *An investigation of protothetic* (**X** 4506) Inst Etudes Polon Belgique: Bruxelles v + 44pp
◇ B60 ◇ REV JSL 15.64 • ID 41611

SOBOCINSKI, B. [1952] *Axiomatization of a partial system of three-value calculus of propositions* (**J** 0093) J Comp Syst 1*23-55
◇ B50 ◇ REV MR 14.834 Zbl 49.292 JSL 18.283
• ID 12572

SOBOCINSKI, B. [1953] *Note on a modal system of Feys-von Wright* (**J** 0093) J Comp Syst 1*171-178
◇ B45 ◇ REV MR 15.925 Zbl 53.6 JSL 19.293 • ID 12575

SOBOCINSKI, B. [1961] *A note concerning the many-valued propositional calculi* (**J** 0047) Notre Dame J Formal Log 2*127-128
◇ B50 ◇ REV MR 28 # 2044 Zbl 121.11 JSL 31.117
• ID 12583

SOBOCINSKI, B. [1962] *A contribution to the axiomatization of Lewis' system S5* (**J** 0047) Notre Dame J Formal Log 3*51-60
◇ B45 ◇ REV MR 27 # 3535 Zbl 113.5 JSL 31.498
• ID 12587

SOBOCINSKI, B. [1962] *A note on the regular and irregular modal systems of Lewis* (**J** 0047) Notre Dame J Formal Log 3*109-113
◇ B45 ◇ REV MR 27 # 5677 Zbl 113.5 JSL 37.181
• ID 12590

SOBOCINSKI, B. [1962] *An axiom-system for* $\{K, N\}$*-propositional calculus related to Simon's axiomatization of* S3 (**J** 0047) Notre Dame J Formal Log 3*206-208
◇ B45 ◇ REV MR 27 # 3536 Zbl 128.12 JSL 32.245 • ID 12588

SOBOCINSKI, B. [1962] *On the generalized Brouwerian axioms* (**J** 0047) Notre Dame J Formal Log 3*123-128
◇ B45 ◇ REV MR 27 # 5678 Zbl 113.5 JSL 31.498 • ID 12589

SOBOCINSKI, B. [1963] *A note on modal systems* (**J** 0047) Notre Dame J Formal Log 4*155-157
◇ B45 ◇ REV MR 29 # 4684 Zbl 131.245 JSL 31.498 • ID 12591

SOBOCINSKI, B. [1964] *A note on Prior's system in "the theory of deduction"* (**J** 0047) Notre Dame J Formal Log 5*139-140
◇ B45 ◇ REV MR 31 # 2142 Zbl 134.7 JSL 31.665 • ID 12592

SOBOCINSKI, B. [1964] *Family K of non-Lewis modal systems* (**J** 0047) Notre Dame J Formal Log 5*313-318
◇ B45 ◇ REV MR 31 # 3325 Zbl 129.257 JSL 37.182 • ID 12594

SOBOCINSKI, B. [1964] *Modal system S4.4* (**J** 0047) Notre Dame J Formal Log 5*305-312
◇ B45 ◇ REV MR 31 # 3324 Zbl 129.257 JSL 37.182 • ID 12596

SOBOCINSKI, B. [1964] *On the propositional system A of Vuchkovich and its extension I,II* (**J** 0047) Notre Dame J Formal Log 5*141-153,223-237
◇ B20 B60 ◇ REV MR 36 # 4965 Zbl 133.254 JSL 31.118 • ID 21264

SOBOCINSKI, B. [1964] *Remarks about axiomatization of certain modal systems* (**J** 0047) Notre Dame J Formal Log 5*71-80
◇ B45 ◇ REV MR 31 # 43 Zbl 137.5 • ID 12593

SOBOCINSKI, B. [1970] *Certain extensions of modal system S4* (**J** 0047) Notre Dame J Formal Log 11*347-368
◇ B45 ◇ REV MR 43 # 3100 Zbl 188.13 JSL 40.602 • ID 12601

SOBOCINSKI, B. [1970] *Note on G.J.Massey's closure-algebraic operation* (**J** 0047) Notre Dame J Formal Log 11*343-346 • ERR/ADD ibid 14*584
◇ B45 G05 G25 ◇ REV MR 43 # 6061 MR 48 # 8204 Zbl 188.19 JSL 36.691 • ID 19722

SOBOCINSKI, B. [1970] *Note on Zeman's modal system S4.o4* (**J** 0047) Notre Dame J Formal Log 11*383-384
◇ B45 ◇ REV MR 43 # 3097 Zbl 188.13 • ID 12600

SOBOCINSKI, B. [1971] *A new class of modal systems* (**J** 0047) Notre Dame J Formal Log 12*371-377
◇ B45 ◇ REV MR 45 # 8507 Zbl 216.4 JSL 40.602 • ID 12605

SOBOCINSKI, B. [1971] *A proper subsystem of S4.04* (**J** 0047) Notre Dame J Formal Log 12*381-384
◇ B45 ◇ REV MR 46 # 26 Zbl 215.46 JSL 40.602 • ID 12603

SOBOCINSKI, B. [1971] *Concerning some extensions of S4* (**J** 0047) Notre Dame J Formal Log 12*363-370
◇ B45 ◇ REV MR 45 # 8506 Zbl 216.4 JSL 40.602 • ID 12606

SOBOCINSKI, B. [1973] *A new axiomatization of modal system K1.2* (**J** 0047) Notre Dame J Formal Log 14*413-414
◇ B45 ◇ REV MR 48 # 69 Zbl 258 # 02027 • ID 12616

SOBOCINSKI, B. [1973] *Modal system S3 and the proper axioms of S4.02 and S4.04* (**J** 0047) Notre Dame J Formal Log 14*415-418 • ERR/ADD ibid 15*648
◇ B45 ◇ REV MR 48 # 70 MR 50 # 1846 Zbl 258 # 02028 • ID 12615

SOBOCINSKI, B. [1974] *Concerning the proper axioms of S4.02* (**J** 0047) Notre Dame J Formal Log 15*169-172
◇ B45 ◇ REV MR 48 # 10769 Zbl 272 # 02032 • ID 12620

SOBOCINSKI, B. [1976] *Pledger lemma and the modal system* $S3^0$ (**J** 0047) Notre Dame J Formal Log 17*253-256
◇ B45 ◇ REV MR 53 # 10553 Zbl 313 # 02011 • ID 18396

SOBOCINSKI, B. see Vol. I, III, V for further entries

SOBOLEV, S.K. [1977] *On finite approximability of superintuitionistic logics (Russian)* (**J** 0142) Mat Sb, Akad Nauk SSSR, NS 102(144)*289-301,327
• TRANSL [1977] (**J** 0349) Math of USSR, Sbor 31*257-268
◇ B55 ◇ REV MR 56 # 11763 Zbl 352 # 02020 • ID 50019

SOBOLEV, S.K. [1977] *On finite-dimensional superintuitionistic logics (Russian)* (**J** 0216) Izv Akad Nauk SSSR, Ser Mat 41*963-986,1199
• TRANSL [1977] (**J** 0448) Math of USSR, Izv 11*909-935
◇ B55 D05 F50 G10 G25 ◇ REV MR 58 # 10333 Zbl 368 # 02062 • ID 51279

SOBOLEV, S.K. [1977] *The intuitionistic propositional calculus with quantifiers (Russian)* (**J** 0087) Mat Zametki (Akad Nauk SSSR) 22*69-76
• TRANSL [1977] (**J** 1044) Math Notes, Acad Sci USSR 22*528-532
◇ B55 C80 C90 D35 F50 ◇ REV MR 56 # 15371 Zbl 365 # 02013 • ID 51014

SOBOLEV, S.L. (ED.) [1982] *Mathematical logic and the theory of algorithms (Russian)* (**X** 2642) Nauka: Novosibirsk 176pp
◇ B97 C97 D97 ◇ REV MR 84i:03007 Zbl 539 # 00002 • ID 34492

SOETEMAN, A. [1973] *Some remarks about two famous paradoxes of deontic logic* (**J** 0079) Logique & Anal, NS 16*272-283
◇ B45 ◇ REV Zbl 269 # 02010 • ID 30362

SOKHATSKIJ, F.N. [1985] *A generalization of two theorems of Belousov for strongly dependent functions of k-valued logic (Russian)* (**S** 0166) Mat Issl, Mold SSR 83*105-115,137-138
◇ B50 ◇ ID 48970

SOKOLOV, N.A. [1981] *The search for the maximum upper zero for a class of monotonic functions of finite-valued logic (Russian)* (**J** 0199) Zh Vychisl Mat i Mat Fiz 21*1552-1565
• TRANSL [1981] (**J** 1049) USSR Comput Math & Math Phys 21/6*191-205
◇ B50 ◇ REV MR 83f:94058 Zbl 483 # 03016 • ID 38084

SOKOLOV, V.A. [1971] *Remarks on the class of partial functions of countably-valued logic (Russian)* (**J** 0071) Met Diskr Analiz (Novosibirsk) 19*56-62
◇ B50 ◇ REV MR 46 # 5104 Zbl 262 # 02019 • ID 41995

SOKOLOV, V.A. see Vol. IV for further entries

SOKULER, Z.A. [1980] *Different interpretations of quantification (Russian)* (**C** 3038) Log-Metodol Issl 340-355
◇ B45 ◇ REV MR 83j:03038 • ID 35344

SOLITRO, U. & VALENTINI, S. [1983] *The modal logic of consistency assertions of Peano arithmetic* (J 0068) Z Math Logik Grundlagen Math 29*25-32
◇ B45 F30 ◇ REV MR 84c:03043 Zbl 521 # 03010
• ID 35696

SOLODUKHIN, YU.N. [1970] *A concept of function-analogue and a criterion for functional incompleteness of systems of many-valued logic (Russian)* (C 1530) Issl Log Sist (Yanovskaya) 271-281
◇ B50 ◇ REV MR 47 # 6438 Zbl 265 # 02013 • ID 12636

SOLOVAY, R.M. [1976] *Provability interpretations of modal logic* (J 0029) Israel J Math 25*287-304
◇ B45 E30 F30 ◇ REV MR 56 # 15369 Zbl 352 # 02019 JSL 46.661 • ID 26088

SOLOVAY, R.M. [1979] see GUASPARI, D.

SOLOVAY, R.M. see Vol. III, IV, V, VI for further entries

SOLOV'EV, A.E. [1975] *A certain approach to the description of control systems (Russian)* (S 2809) Sbor Nauch Trud (Politekhn Inst, Perm) 176*86-89
◇ B45 ◇ REV MR 56 # 14851 • ID 82848

SOLT, K. [1984] *Deontic alternative worlds and the truth-value of "OA"* (J 0079) Logique & Anal, NS 27*349-351
◇ B45 ◇ REV MR 86c:03016 Zbl 558 # 03009 • ID 44438

SOMMERS, F. [1970] *The calculus of terms* (J 0094) Mind 79*1-39
◇ A05 B60 ◇ REV MR 41 # 1502 • ID 12647

SONG, DAHE [1983] *A mathematical model and analysis of the algorithm of a fuzzy controller (Chinese) (English summary)* (J 3732) Mohu Shuxue 3/2*35-42
◇ B52 ◇ REV MR 84m:93068 Zbl 525 # 93026 • ID 37497

SONOBE, O. [1975] *A Gentzen-type formulation of some intermediate propositional logics* (J 0381) J Tsuda College (Tokyo) 7*7-13
◇ B55 ◇ REV MR 51 # 2877 • ID 17377

SONOBE, O. [1975] *Bi-relational frameworks for minimal and intuitionistic logics* (P 3299) Progr Kiso Riron, Algor Okeru Shomei Ron;1973/74 Kyoto 174-189
◇ B45 C90 F50 ◇ REV MR 58 # 5075 Zbl 375 # 02009
• ID 51586

SONOBE, O. [1976] *Tableau methods of proof for LPn and LQn* (J 0381) J Tsuda College (Tokyo) 8*19-26
◇ B55 ◇ REV MR 57 # 2889 • ID 78860

SONOBE, O. [1977] *Beth-type tableau proof systems for some intermediate and modal logics* (J 0381) J Tsuda College (Tokyo) 9*49-60
◇ B45 B55 ◇ REV MR 58 # 21419 • ID 78858

SORENSEN, R.A. [1984] *Unique alternative guessing* (J 0079) Logique & Anal, NS 27*77-85
◇ B45 D20 ◇ REV MR 85h:03024 • ID 43297

SOSINSKIJ, L.M. [1964] *On the representation of functions by irreversible superpositions in three-valued logic (Russian)* (J 0052) Probl Kibern 12*57-68
◇ B50 ◇ REV MR 34 # 4125 • ID 12652

SOTIROV, V.KH. [1980] *Nonfinitely approximable intuitionistic modal logics (Russian)* (J 0087) Mat Zametki (Akad Nauk SSSR) 27*89-94,158
• TRANSL [1980] (J 1044) Math Notes, Acad Sci USSR 27*47-49
◇ B45 ◇ REV MR 81b:03024 Zbl 443 # 03011 • ID 56417

SOTIROV, V.KH. [1984] *Modal theories with intuitionistic logic* (P 4392) Mat Logika (Markova);1980 Sofia 139-171
◇ B45 ◇ ID 46975

SOTIROV, V.KH. see Vol. I for further entries

SPACEK, A. [1960] *Statistical estimation of provability in boolean logic* (P 1538) Inform Th, Stat Decis Fcts & Random Proc (2);1959 Liblice 609-626
◇ B48 G05 ◇ REV MR 23 # A802 Zbl 242 # 02034 JSL 27.101 • ID 26350

SPACEK, A. [1961] *Statistical estimation of semantic provability* (P 1539) Berkeley Symp Math Stat & Probab (4);1960 Berkeley 1*655-668
◇ B48 G15 ◇ REV MR 25 # 8 Zbl 242 # 02035 • ID 26351

SPADE, P.V. [1976] *An alternative to Brian Skyrms' approach to the liar* (J 0047) Notre Dame J Formal Log 17*137-146
◇ A05 B50 ◇ REV MR 55 # 2478 Zbl 258 # 02004
• ID 18399

SPADE, P.V. [1977] *General semantic closure* (J 0122) J Philos Logic 6*209-221
◇ A05 B45 ◇ REV MR 58 # 10308 Zbl 355 # 02007
• ID 50205

SPADE, P.V. [1982] *Three theories of obligations: Burley, Kilvington and Swyneshed on counterfactual reasoning* (J 2028) Hist & Phil Log 3*1-32
◇ A10 B45 ◇ REV MR 84h:01023 • ID 39502

SPASOWSKI, M. [1974] see MALINOWSKI, G.

SPASOWSKI, M. [1975] *The degrees of completeness of dual counterparts of Lukasiewicz sentential calculi* (J 0387) Bull Sect Logic, Pol Acad Sci 4*163-170
◇ B50 ◇ REV MR 55 # 12468 • ID 78875

SPASOWSKI, M. see Vol. I for further entries

SPECKER, E. [1965] see KOCHEN, S.

SPECKER, E. see Vol. I, III, IV, V, VI for further entries

SPELLER, J. [1984] *Filling a gap in Professor von Kutschera's decision procedure for deontic logic* (J 0079) Logique & Anal, NS 27*435-438
◇ B45 ◇ ID 45536

SPERANTIA, E. [1936] *Remarques sur les propositions interrogatives. Projet d'une "logique du probleme"* (P 0632) Congr Int Phil des Sci;1935 Paris 7*18-28
◇ B60 ◇ REV JSL 2.55 • ID 41343

SPIGLER, R. [1980] *The structure of the relations "much better" and "much worse" in fuzzy calculus* (J 0144) Rend Sem Mat Univ Padova 63*27-39
◇ B52 E72 ◇ REV MR 82h:03062 Zbl 443 # 06001
• ID 78882

SPINOSA, P.L. [1982] see BIAGIOLI, C.

SPIRKIN, A.G. [1978] see BIRYUKOV, B.V.

SPISANI, F. [1970] *Fondamenti di logica produttiva. Logica produttiva e logica formale* (J 0286) Int Logic Rev 1*5-12
• TRANSL [1970] (J 0286) Int Logic Rev 1*13-20 (English) [1970] (J 0286) Int Logic Rev 1*21-28 (French) [1970] (J 0286) Int Logic Rev 1*29-37 (German)
◇ A05 B53 ◇ REV MR 46 # 1567 • ID 30084

SPISANI, F. [1970] *Fondamenti di logica produttiva. La differenza dell'identico e le logiche dogmatiche* (J 0286) Int Logic Rev 2*118-126
• TRANSL [1970] (J 0286) Int Logic Rev 2*127-136 (English)
◇ B53 ◇ ID 65474

SPISANI, F. [1971] *Fondamenti di logica produttiva. Annibale alle porte* (J 0286) Int Logic Rev 4*120-134 • ERR/ADD ibid 5*128
• TRANSL [1971] (J 0286) Int Logic Rev 4*135-149 (English)
◇ A05 B53 ◇ ID 19717

SPISANI, F. [1972] *Foundations of productive logic (Italian, English)* (J 0286) Int Logic Rev 5*4-20,21-37 • ERR/ADD ibid 6*267
◇ A05 B53 ◇ REV MR 50#6778a MR 50#6778b
• ID 19718

SPISANI, F. [1972] *Outlines of productive logic (Italian and English)* (J 0286) Int Logic Rev
◇ B53 ◇ REV Zbl 378#02013 • REM 1973 4-15,142-161;1974 5-22 (Italian) and 1973 16-28,162-175;1974 23-40 (English) • ID 51795

SPISANI, F. [1975] *Outlines of productive logic (Italian and English)* (J 0286) Int Logic Rev 12*133-138,139-144
◇ B53 ◇ REV Zbl 341#02008 • ID 29737

SPISANI, F. [1976] *Outlines of productive logic (Italian)* (J 0286) Int Logic Rev 13*3-12,14*132-133
• TRANSL [1976] (J 0286) Int Logic Rev 13*13-22,14*133-134 (English)
◇ A05 B53 ◇ REV Zbl 387#03009 Zbl 398#03012
• ID 52225

SPISANI, F. [1977] *Implicazione, endometria, universo del discorso. Implication, endometry, universe of discourse* (X 2472) Cent Sup Log Sci Comp: Bologna 174pp
◇ A05 B60 ◇ REV Zbl 396#03002 • ID 52612

SPISANI, F. [1983] *Teoria generale dei numeri relativi. Vol.I (Italian and English)* (J 0286) Int Logic Rev 247pp
◇ B60 ◇ REV MR 85i:03035 • ID 44071

SPOHN, W. [1975] *An analysis of Hansson's dyadic deontic logic* (J 0122) J Philos Logic 4*237-252
◇ B45 ◇ REV MR 57#2885 Zbl 317#02020 • ID 29637

SPOHN, W. [1980] *Stochastic independence, causal independence, and shieldability* (J 0122) J Philos Logic 9*73-99
◇ B48 ◇ REV MR 82a:60007 Zbl 436#60004 • ID 82855

SREBRNY, M. [1976] see MAREK, W.

SREBRNY, M. see Vol. III, IV, V, VI for further entries

SRINI, V.P. [1975] *Realization of fuzzy forms* (J 0187) IEEE Trans Comp C-24*941-943
◇ B52 ◇ REV Zbl 309#68071 • ID 65479

STAAL, J.F. [1968] see ROOTSELAAR VAN, B.

STAAL, J.F. see Vol. I for further entries

STABLER, E.R. [1953] *An introduction to mathematical thought* (X 0832) Addison-Wesley: Reading xviii+268pp
◇ A05 B98 ◇ REV MR 14.441 Zbl 50.4 JSL 20.288
• ID 23429

STABLER, E.R. see Vol. I for further entries

STACHOW, E.-W. [1976] *Completeness of quantum logic* (J 0122) J Philos Logic 5*237-280
◇ B51 G12 ◇ REV MR 57#15994 Zbl 334#02018
• ID 65480

STACHOW, E.-W. [1977] *How does quantum logic correspond to physical reality?* (J 0122) J Philos Logic 6*485-496
◇ A05 B51 G12 ◇ REV Zbl 387#03027 • ID 52243

STACHOW, E.-W. [1978] *Quantum logical calculi and lattice structures* (J 0122) J Philos Logic 7*347-386
◇ B51 G12 ◇ REV MR 80i:81004 Zbl 387#03028
• ID 52244

STACHOW, E.-W. [1978] see MITTELSTAEDT, P.

STACHOW, E.-W. [1979] *An operational approach to quantum probability* (C 3566) Phys Theor as Log-Operat Struct 285-321
◇ B51 G12 ◇ REV MR 81a:81012 Zbl 402#03053
• ID 82858

STACHOW, E.-W. [1979] *Completeness of quantum logic* (C 3566) Phys Theor as Log-Operat Struct 203-243
◇ B51 G12 ◇ REV MR 80f:03073 Zbl 402#03053
• ID 78911

STACHOW, E.-W. [1979] *Quantum logical calculi and lattice structures* (C 3566) Phys Theor as Log-Operat Struct 245-284
◇ B51 G12 ◇ REV MR 80h:03093 • ID 78910

STACHOW, E.-W. [1980] *Logical foundation of quantum mechanics* (J 2736) Int J Theor Phys 19*251-304
◇ B51 G12 ◇ REV MR 81g:03076 Zbl 462#03014
• ID 54521

STACHOW, E.-W. [1981] *Der quantenlogische Wahrscheinlichkeitskalkuel* (C 3737) Grundl-probl Modern Physik (Mittelstaedt) 271-305
◇ B51 ◇ REV MR 83k:03076 • ID 34882

STACHOW, E.-W. [1981] *Sequential quantum logic* (P 3820) Curr Iss in Quantum Log;1979 Erice 173-191
◇ B51 ◇ REV MR 84j:03135 Zbl 537#03044 • ID 34725

STACHOW, E.-W. [1983] see MITTELSTAEDT, P.

STACHOW, E.-W. [1985] see MITTELSTAEDT, P.

STAHL, G. [1956] *Introduccion a la logica simbolica* (X 1715) Edunsa Ed Univ: Santiago 206pp
◇ B98 ◇ REV JSL 32.519 • REM 2nd ed. 1962 • ID 32460

STAHL, G. [1956] *La logica de las preguntas* (J 1717) An Univ Chile 102*71-75
◇ B60 ◇ REV JSL 22.93 • ID 32466

STAHL, G. [1959] *General considerations about modal sentences* (J 0068) Z Math Logik Grundlagen Math 5*280-290
◇ B45 ◇ REV Zbl 91.9 JSL 36.182 • ID 13088

STAHL, G. [1961] *Preguntas y premisas* (J 0988) Rev Filos (Santiago) 8*3-9
◇ B60 ◇ REV JSL 28.257 • ID 32473

STAHL, G. [1963] *Un developpement de la logique des questions* (J 0180) Rev Phil France & Etranger 153*293-301
◇ B60 ◇ REV JSL 32.548 JSL 40.599 • ID 14593

STAHL, G. [1964] *Elementos de la metalogica y metamatematica* (X 1715) Edunsa Ed Univ: Santiago 160pp
◇ B98 ◇ REV JSL 32.519 • ID 32462

STAHL, G. [1969] *The effectivity of questions* (J 0097) Nous, Quart J Phil 3*211-218
◇ B60 ◇ REV JSL 40.600 • ID 15090

STAHL, G. [1972] *Temporal terms in functional systems* (**P** 2080) Conf Math Log;1970 London 255*347-348
◇ B45 ◇ REV Zbl 228 # 02018 • ID 29120

STAHL, G. [1973] *Elementos de metamatematica* (**X** 1715) Edunsa Ed Univ: Santiago
◇ B98 ◇ ID 32465

STAHL, G. [1974] *Modal models corresponding to models* (**J** 0068) Z Math Logik Grundlagen Math 20*407-410
◇ B45 ◇ REV MR 51 # 7831 Zbl 299 # 02064 • ID 13093

STAHL, G. [1974] *Termes temporels dans des systemes fonctionnels* (**J** 0180) Rev Phil France & Etranger 3*293-303
◇ B45 ◇ ID 32474

STAHL, G. [1979] *Le dominateur et ses problemes* (**J** 0079) Logique & Anal, NS 22*121-131
◇ A05 B45 ◇ REV MR 80j:03033 Zbl 426 # 03008 • ID 53604

STAHL, G. [1981] *Divisions of intensional universes* (**J** 0079) Logique & Anal, NS 24*389-397
◇ B45 ◇ REV MR 83k:03028 • ID 36168

STAHL, G. [1982] *Analyse et applications des suites de questions* (**J** 0079) Logique & Anal, NS 25*167-179
◇ B60 ◇ REV MR 84e:03033 • ID 34372

STAHL, G. [1984] *Mythe et realite de la formule de Barcan* (**J** 0079) Logique & Anal, NS 27*343-348
◇ B45 ◇ REV MR 86g:03037 Zbl 557 # 03010 • ID 44439

STAHL, G. see Vol. I, IV, V, VI for further entries

STAIRS, A. [1982] *Discussion: quantum logic and the Lueders rule* (**J** 0153) Phil of Sci (East Lansing) 49*422-436
◇ B51 ◇ REV MR 84i:03114 • ID 34591

STAIRS, A. [1983] *On the logic of pairs of quantum systems* (**J** 0154) Synthese 56*47-60
◇ B51 ◇ REV MR 85e:81011 Zbl 518 # 03028 • ID 37525

STAIRS, A. [1983] *Quantum logic, realism, and value definiteness* (**J** 0153) Phil of Sci (East Lansing) 50*578-602
◇ B51 G12 ◇ REV MR 85i:03188 • ID 44300

STALNAKER, R.C. & THOMASON, R.H. [1968] *Abstraction in first order modal logic* (**J** 0105) Theoria (Lund) 34*203-207
◇ B45 ◇ REV MR 49 # 2269 • ID 13094

STALNAKER, R.C. [1970] *Probability and conditionals* (**J** 0153) Phil of Sci (East Lansing) 37*64-80
◇ A05 B48 ◇ REV MR 44 # 2565 • ID 13098

STALNAKER, R.C. [1981] *A defense of conditional excluded middle* (**C** 4140) Ifs 87-104
◇ B48 ◇ REV MR 83a:03003 • ID 47737

STALNAKER, R.C. [1981] *A theory of conditionals* (**C** 4140) Ifs 41-55
◇ B48 ◇ REV MR 83a:03003 • ID 47735

STALNAKER, R.C. [1981] see HARPER, W.L.

STALNAKER, R.C. [1981] *Indexical belief* (**J** 0154) Synthese 49*129-151
◇ A05 B45 ◇ REV MR 82i:03012 • ID 78922

STALNAKER, R.C. [1981] *Indicative conditionals* (**C** 4140) Ifs 193-210
◇ B48 ◇ REV MR 83a:03003 • ID 47742

STALNAKER, R.C. [1981] *Logical semiotic* (**C** 2617) Modern Log Survey 439-456
◇ B60 ◇ REV MR 82f:03002 Zbl 464 # 03001 • ID 42780

STALNAKER, R.C. [1981] *Probability and conditionals* (**C** 4140) Ifs 107-128
◇ B48 ◇ REV MR 83a:03003 • ID 47738

STANCIU, L. & STANCOVICI, V. [1975] *The structure of space and time in Franco Spisani's productive logic (Italian and English)* (**J** 0286) Int Logic Rev 11*23-50,53-80
◇ B53 ◇ REV Zbl 341 # 02009 • ID 29738

STANCOVICI, V. [1975] see STANCIU, L.

STANDLEY, G.B. [1966] *Testing singly quantified tautologies* (**J** 0036) J Symb Logic 31*478-480
◇ B55 ◇ REV Zbl 148.243 • ID 13103

STANDLEY, G.B. [1971] *New methods in symbolic logic* (**X** 0847) Houghton Mifflin: Boston 217pp
◇ B98 ◇ REV JSL 39.178 • ID 13104

STANDLEY, G.B. see Vol. I for further entries

STAPP, H.P. [1982] *Bell's theorem as a nonlocality property of quantum theory* (**J** 2730) Phys Rev Lett 49*1470-1474
◇ B51 ◇ REV MR 84m:81028a • ID 39537

STARCHENKO, A.A. (ED.) [1980] *Studies in logical methodology (Russian)* (**X** 0898) Moskov Gos Univ: Moskva 375pp
◇ A05 B97 ◇ REV MR 82c:03003 • ID 70025

STARK, W.R. [1981] *A logic of knowledge* (**J** 0068) Z Math Logik Grundlagen Math 27*371-374
◇ B45 ◇ REV MR 82j:03023 Zbl 498 # 03017 • ID 78940

STARK, W.R. see Vol. I, III, IV, V for further entries

STAVI, J. [1981] see HAREL, D.

STAVI, J. [1982] see HAREL, D.

STAVI, J. [1983] see HAREL, D.

STAVI, J. see Vol. I, III, IV, V, VI for further entries

STEBBING, L.S. [1930] *A modern introduction to logic* (**X** 0816) Methuen: London & New York xviii + 505pp • LAST ED [1961] (**X** 0837) Harper & Row: New York xvii + 525pp
◇ B98 ◇ REV MR 22 # 10900 FdM 56.829 • ID 22602

STEBBING, L.S. [1943] *A modern elementary logic* (**X** 0816) Methuen: London & New York viii + 214pp
◇ B98 ◇ REV JSL 9.63 • ID 13139

STEBBINS, S. [1980] *Necessity and natural language* (**J** 0095) Philos Stud 37*1-12
◇ A05 B45 B65 ◇ REV MR 80m:03021 • ID 78961

STEEL JR., T.B. [1976] see BELNAP JR., N.D.

STEEL JR., T.B. see Vol. IV for further entries

STEEN, S.W.P. [1972] *Mathematical logic, with special reference to the natural numbers* (**X** 0805) Cambridge Univ Pr: Cambridge, GB xvi + 638pp
◇ B98 F30 F98 ◇ REV MR 57 # 9451 Zbl 275 # 02002 • ID 22386

STEGMUELLER, W. [1973] *Probleme und Resultate der Wissenschaftstheorie und analytischen Philosophie. Band IV: Personelle und statistische Wahrscheinlichkeit. Erster Halbband: Personelle Wahrscheinlichkeit und rationale Entscheidung* (**X** 0811) Springer: Heidelberg & New York xxiv + 560pp
◇ A05 B48 ◇ REV MR 58 # 27266a Zbl 264 # 02006 • REM Part III 1984 by Stegmueller,W. & Varga von Kibed,M. Part IV,2 1973 • ID 27495

STEGMUELLER, W. [1973] *Probleme und Resultate der Wissenschaftstheorie und analytischen Philosophie. Band IV: Personelle und statistische Wahrscheinlichkeit. Zweiter Halbband: Statistisches Schliessen. Statistische Begruendung. Statistische Analyse* (**X** 0811) Springer: Heidelberg & New York xvi + 420pp
◇ A05 B48 ◇ REV Zbl 264 # 02006 • REM Part IV,I 1973 • ID 27496

STEGMUELLER, W. see Vol. I, III, IV, V, VI for further entries

STEINACKER, P. [1979] *Zur Semantik superschwacher modaler Kalkuele* (**S** 2877) Wiss Z Univ Leipzig, Ges-Sprachwiss Reihe 28*319-323
◇ B45 ◇ REV MR 80m:03046 Zbl 416 # 03018 • ID 53184

STEINACKER, P. [1984] *Ueber eine intensionale Alternative bei C.I. Lewis* (**P** 3621) Frege Konferenz (2);1984 Schwerin 189-193
◇ B45 ◇ REV MR 85m:03006 • ID 45364

STELZNER, W. [1979] *Funktorenvariable, Funktionenvariable und nichtklassische Logik* (**S** 2877) Wiss Z Univ Leipzig, Ges-Sprachwiss Reihe 28*313-318
◇ A05 B60 ◇ REV MR 81i:03012 Zbl 421 # 03007 • ID 53406

STEMMER, N. [1982] *A solution to the lottery paradox* (**J** 0154) Synthese 51*339-353
◇ B48 ◇ REV MR 84i:03052 • ID 34531

STENIUS, E. [1947] *Natural implication and material implication* (**J** 0105) Theoria (Lund) 13*136-156
◇ A05 B45 ◇ REV Zbl 36.8 JSL 14.198 • ID 13163

STENIUS, E. [1951] *Modern logic (Swedish)* (**J** 3665) Finsk Tidskrift 150*102-112
◇ B98 ◇ REV JSL 24.190 • ID 33917

STENIUS, E. [1963] *The principles of a logic of normative systems* (**J** 0096) Acta Philos Fenn 16*247-260
◇ A05 B45 ◇ REV MR 31 # 44 Zbl 122.7 JSL 36.519 • ID 33177

STENIUS, E. [1982] *Ross' paradox and well-formed codices* (**J** 0105) Theoria (Lund) 48*49-77
◇ B45 ◇ REV MR 84h:03021 • ID 34236

STENIUS, E. see Vol. III, VI for further entries

STENLUND, S. [1973] *The logic of description and existence* (**X** 0882) Univ Filos Foeren: Uppsala ii + 75pp
◇ A05 B60 ◇ REV MR 50 # 12668 Zbl 309 # 02012 • ID 13167

STENLUND, S. see Vol. IV, VI for further entries

STEPHAN, B.J. [1975] *Compactness and recursive enumerability in intensional logic* (**J** 0068) Z Math Logik Grundlagen Math 21*343-346
◇ B45 D25 ◇ REV MR 52 # 46 Zbl 312 # 02025 • ID 13174

STEPIEN, T. [1981] *System S* (**J** 0387) Bull Sect Logic, Pol Acad Sci 10*75-79
◇ B45 ◇ REV Zbl 471 # 03010 • ID 55205

STEPIEN, T. [1985] *Logic based on atomic entailment* (**J** 0387) Bull Sect Logic, Pol Acad Sci 14*65-71
◇ B22 B60 ◇ ID 48971

STEPIEN, T. see Vol. I, III for further entries

STERN, C.D. [1981] *Lewi's counterfactual analysis of causation* (**J** 0154) Synthese 48*335-345
◇ B45 ◇ REV MR 84e:03025 Zbl 483 # 03009 • ID 34364

STEWARD, M.H. [1978] *The arithmetic properties of certain number systems* (**P** 2014) Int Symp Multi-Val Log (8);1978 Rosemont 122-127
◇ B50 ◇ ID 35921

STICKEL, M.E. [1978] *Fuzzy four-valued logic for inconsistency and uncertainty* (**P** 2014) Int Symp Multi-Val Log (8);1978 Rosemont 91-94
◇ B52 ◇ REV MR 81d:03028 • ID 79023

STICKEL, M.E. see Vol. I for further entries

STILLWELL, J.C. [1972] see ASH, C.J.

STILLWELL, J.C. [1977] *Concise survey of mathematical logic* (**J** 3194) J Austral Math Soc, Ser A 24*139-161
◇ B98 C98 D10 D35 ◇ REV MR 57 # 5655 Zbl 393 # 03002 • ID 52422

STILLWELL, J.C. see Vol. III, IV for further entries

STINE, G.C. [1969] *Hintikka on quantification and belief* (**J** 0097) Nous, Quart J Phil 3*399-408
◇ A05 B45 ◇ REV MR 58 # 5070 • ID 79033

STIRLING, C. [1985] *A complete modal proof system for a subset of SCCS* (**P** 4627) CAAP'85 Arbres en Algeb & Progr (10);1985 Berlin 1*253-266
◇ B45 B75 ◇ REV Zbl 563 # 68025 • ID 49530

STIRLING, C. [1985] *A complete compositional modal proof system for a subset of CCS* (**P** 4628) Automata, Lang & Progr (12);1985 Nafplion 475-486
◇ B45 B75 ◇ ID 49531

STOCK, W.G. [1978] *Zur Bestimmung der Negation in der Handlungslogik* (**J** 0455) Phil Naturalis 17*99-104
◇ B50 ◇ REV MR 58 # 5041 • ID 79035

STOEVA, S. [1984] see ATANASSOV, K.

STOICHITA, R. [1968] *Der Wahrheitsbegriff. Beziehung der zweiwertigen zur mehrwertigen Logik* (**J** 0147) An Univ Bucuresti, Acta Logica 11*75-83
◇ B50 ◇ REV MR 43 # 4641 Zbl 223 # 02006 • ID 13194

STOICHITA, R. see Vol. I for further entries

STOIDE, M. [1976] *On the completeness theorem for some many-valued predicate calculi with countable conjunctions and disjunctions* (**P** 3001) Proc Inst Math;1974 Iasi 85-92
◇ B50 ◇ REV MR 56 # 5219 Zbl 368 # 02019 • ID 51236

STOJAKOVIC, M. [1966] *Matrizeninversion in der algebraischen Theorie der Wahrheitsfunktionen der mehrwertigen Logik* (**J** 0993) Bull Acad Serbe, Cl Sci Math Nat 35*39-43
◇ B50 ◇ REV MR 35 # 6528 Zbl 207.293 • ID 13195

STOJAKOVIC, M. see Vol. I, III, V for further entries

STOJMENOVIC, I. & TOSIC, R. [1983] *Enumeration of monotone symmetric functions of three-valued logic (Russian) (Serbo-Croatian summary)* (**J** 2855) Zbor Rad, Prir-Mat Fak, Ser Mat (Novi Sad) 13*367-373
◇ B50 ◇ ID 45067

STOJMENOVIC, I. [1983] *Enumeration of symmetric functions of precomplete classes of three-valued logic (Russian) (Serbo-Croatian summary)* (**J** 2855) Zbor Rad, Prir-Mat Fak, Ser Mat (Novi Sad) 13*375-387
◇ B50 ◇ ID 45069

STOJMENOVIC, I. [1984] *Classification of P_3 and the enumeration of bases of P_3* (J 2855) Zbor Rad, Prir-Mat Fak, Ser Mat (Novi Sad) 14*73-80
◇ B50 ◇ REV Zbl 564 # 94011 • ID 45537

STOJMENOVIC, I. [1984] *Enumeration of the bases of three-valued monotone logical functions* (J 2855) Zbor Rad, Prir-Mat Fak, Ser Mat (Novi Sad) 14*81-98
◇ B50 ◇ REV Zbl 564 # 94012 • ID 45538

STOKHOF, M. [1975] see GROENENDIJK, J.

STOKHOF, M. [1981] see EMDE BOAS VAN, P.

STOLBOUSHKIN, A.P. & TAJTSLIN, M.A. [1983] *Deterministic dynamic logic is strictly weaker than dynamic logic* (J 0194) Inform & Control 57*48-55
◇ B75 ◇ REV MR 85i:03090 Zbl 537 # 68037 • ID 41327

STOLBOUSHKIN, A.P. [1983] *Regular dynamic logic is not interpretable in deterministic context-free dynamic logic* (J 0194) Inform & Control 59*94-107
◇ B75 ◇ REV Zbl 567 # 03005 • ID 44203

STOLBOUSHKIN, A.P. & TAJTSLIN, M.A. [1983] *The comparison of the expressive power of first-order dynamic logics* (J 1426) Theor Comput Sci 27*197-209
◇ B75 ◇ REV MR 85j:03038 Zbl 536 # 03015 • ID 37100

STOLBOUSHKIN, A.P. [1984] *The expressive power of certain dynamic logics (Russian)* (J 0142) Mat Sb, Akad Nauk SSSR, NS 125(167)*410-419
• TRANSL [1986] (J 0349) Math of USSR, Sbor 53*411-419
◇ B75 ◇ REV MR 86g:68115 • ID 44087

STOLL, R.R. [1961] *Sets, logic and axiomatic theories* (X 0994) Freeman: San Francisco x + 206pp
◇ B98 E98 ◇ REV MR 23 # A791 Zbl 106.235 Zbl 292 # 02002 JSL 25.278 • ID 19683

STOLL, R.R. [1963] *Set theory and logic* (X 0994) Freeman: San Francisco xiv + 474pp
◇ B98 E98 ◇ REV MR 30 # 11 Zbl 112.244 JSL 29.40 • ID 13197

STOLYAR, A.A. [1970] *Introduction to elementary mathematical logic* (X 0865) MIT Pr: Cambridge, MA vii + 209pp
• TRANSL [1971] (X 1574) Vyssheyshaya Shkola: Minsk 224pp
◇ B98 ◇ REV Zbl 215.46 Zbl 234 # 02001 • ID 22539

STOUT, L.N. [1979] *Laminations, or how to build a quantum-logic-valued model of set theory* (J 0504) Manuscr Math 28*379-403
◇ B51 E70 G12 G30 ◇ REV MR 81i:03099 Zbl 409 # 03039 • ID 56341

STOUT, L.N. see Vol. V, VI for further entries

STOVE, D. [1968] see HOOKER, C.A.

STRALBERG, A.H. [1965] *On the reduction of classical logic. An extension of some theorems of Glivenko and Goedel* (J 0121) Kon Norske Vidensk Selsk Forh 38*94-99
◇ B55 F30 F50 ◇ REV MR 33 # 2536 Zbl 148.244 • ID 13218

STRASSEN, V. [1982] see ENGELER, E.

STRASSEN, V. see Vol. IV for further entries

STRAWSON, P.F. [1952] *Introduction to logical theory* (X 0816) Methuen: London & New York x + 266pp
◇ A05 B98 ◇ REV Zbl 49.146 JSL 18.273 • ID 25196

STRAZDINS, I.E. [1974] *The number of self-dual types in finite-valued logics (Russian)* (J 0087) Mat Zametki (Akad Nauk SSSR) 15*631-639
• TRANSL [1974] (J 1044) Math Notes, Acad Sci USSR 15*374-379
◇ B50 ◇ REV MR 50 # 12652 Zbl 336 # 02016 • ID 65555

STRAZDINS, I.E. [1975] *Selfdual transformations in many-valued logics (Russian) (English summary insert)* (J 0040) Kibernetika, Akad Nauk Ukr SSR 1975/1*9-21
• TRANSL [1975] (J 0021) Cybernetics 11*10-23
◇ B50 ◇ REV MR 53 # 15582 Zbl 309 # 02014 • ID 23234

STRAZDINS, I.E. [1981] *On fundamental transformation groups in the algebra of logic* (P 2552) Conf Finite Algeb & Multi-Val Log;1979 Szeged 669-691
◇ B05 B50 ◇ REV MR 84g:03105 Zbl 478 # 03005 • ID 55620

STRAZDINS, I.E. see Vol. I for further entries

STREETT, R.S. [1981] *Propositional dynamic logic of looping and converse* (P 3474) ACM Symp Th of Comput (13);1981 375-381
◇ B75 ◇ ID 49869

STREETT, R.S. [1981] see MEYER, A.R.

STREETT, R.S. [1982] *Propositional dynamic logic of looping and converse is elementarily decidable* (J 0194) Inform & Control 54*121-141
◇ B75 ◇ REV MR 85d:68051 Zbl 515 # 68062 • ID 37869

STREETT, R.S. [1985] *Fixpoints and program looping reductions from the propositional mu-calculus into propositional dynamic logics of looping* (P 4571) Log of Progr;1985 Brooklyn 359-372
◇ B75 ◇ ID 49250

STREETT, R.S. see Vol. IV for further entries

STREHLE, P. [1984] *Zur Notwendigkeit und zum Aufbau mehrdimensionaler mehrwertiger Logiken* (C 4026) Unters Log & Methodol, Vol 1 65-92
◇ B50 ◇ REV MR 85f:03027 Zbl 544 # 03006 • ID 40688

STROMBACH, W. [1972] see EMDE, H.

STUBERMAN, W.E. [1965] see BROWN, P.L.

STULA, P.S. [1982] see CLARKE JR., E.M.

STYAZHKIN, N.I. [1974] *Logic with the elements of mathematical logic (Russian)* (X 1453) Moskov Gos Istor-Arkh Inst: Moskva 163pp
◇ B98 ◇ REV MR 53 # 10541 • ID 23055

STYAZHKIN, N.I. see Vol. I for further entries

SU, S.Y.H. [1971] see CHEUNG, P.T.

SU, S.Y.H. [1972] see SARRIS, A.A.

SU, S.Y.H. [1974] see DUCASSE, E.G.

SU, S.Y.H. [1976] see CHEUNG, P.T.

SUBBOTIN, A.L. [1971] see PYATNITSYN, B.N.

SUBBOTIN, A.L. see Vol. I for further entries

SUBLET, J. [1954] *Essai de formalisation complete du raisonnement mathematique sur la base de trois operations* (P 0646) Appl Sci de Log Math;1952 Paris 91-94
◇ B50 E30 ◇ REV MR 16.437 Zbl 57.7 JSL 36.675 • ID 28005

SUCHON, W. [1972] *On non equivalence of two definitions of the algebras of Lukasiewicz* (J 0387) Bull Sect Logic, Pol Acad Sci 1/1*35-37
◇ B50 G20 ◇ REV MR 50 # 12653 • ID 12701

SUCHON, W. [1973] *Application of the theory of logical matrices in the independence proofs* (J 0387) Bull Sect Logic, Pol Acad Sci 2*121-126
◇ B50 ◇ REV MR 55 # 7732 • ID 79075

SUCHON, W. [1973] *On defining Moisil's functors in n-valued propositional logic* (J 0387) Bull Sect Logic, Pol Acad Sci 2*195-196
◇ B50 ◇ REV MR 52 # 13323 • ID 21784

SUCHON, W. [1974] *Definition des foncteurs modaux de Moisil dans le calcul n-valent des propositions de Lukasiewicz avec implication et negation* (J 0302) Rep Math Logic, Krakow & Katowice 2*43-47
◇ B50 ◇ REV MR 50 # 6793 Zbl 284 # 02008 • ID 12705

SUCHON, W. [1974] *La methode de Smullyan de construire le calcul n-valent de Lukasiewicz avec implication et negation* (J 0302) Rep Math Logic, Krakow & Katowice 2*37-42
◇ B50 ◇ REV MR 54 # 2419 Zbl 284 # 02007 • ID 21932

SUCHON, W. [1975] *Definition of Moisil functors in n-valued implicative-negative Lukasiewicz propositional calculus (Polish with English translation)* (S 1454) Zesz Nauk Wyz Szk Ped Mat, Opole 15*17-18,19-20
◇ B50 ◇ REV MR 54 # 2420 Zbl 334 # 02008 • ID 24014

SUCHON, W. [1975] *Matrix Lukasiewicz algebras* (J 0302) Rep Math Logic, Krakow & Katowice 4*91-104
◇ B50 ◇ REV MR 54 # 4973 Zbl 348 # 02021 • ID 21937

SUCHON, W. [1977] *Smullyan's method of contructing Llukasiewicz's n-valued implicational- negational sentential calculus* (C 3026) Lukasiewicz Sel Pap on Sent Calc 119-124
◇ B50 ◇ REV MR 58 # 21447 Zbl 381 # 03017 • ID 79074

SUCHON, W. [1984] *An elementary method of determining the degree of completeness of n-valued Lukasiewicz propositional calculus* (J 0387) Bull Sect Logic, Pol Acad Sci 13*226-229
◇ B50 ◇ ID 44721

SUCHON, W. [1984] *The deontic calculus* D_{KRZ} (J 0302) Rep Math Logic, Krakow & Katowice 18*61-66
◇ B45 ◇ ID 45539

SUEKI, T. [1952] *The formulization of two-valued and n-valued systems I (Japanese) (English summary)* (J 0979) Denshi Tsushin Gakkai Ronbunshi, Sect A-D 4*1-24
◇ B50 ◇ REV MR 16 # 782 • REM Part II 1953, Part III 1954 • ID 48770

SUEKI, T. [1953] *The formulization of two-valued and n-valued systems II (Japanese) (English summary)* (J 0979) Denshi Tsushin Gakkai Ronbunshi, Sect A-D 5*1-18
◇ B50 ◇ REV MR 16 # 782 • REM Part I 1952, Part III 1954 • ID 48771

SUEKI, T. [1954] *The formulization of two-valued and n-valued systems III (Japanese) (English summary)* (J 0979) Denshi Tsushin Gakkai Ronbunshi, Sect A-D 6*1-19
◇ B50 ◇ REV MR 16 # 782 • REM Part I 1952, Part II 1953 • ID 48772

SUEKI, T. [1955] *Formalization of modal systems (Japanese) (English summary)* (J 0979) Denshi Tsushin Gakkai Ronbunshi, Sect A-D 7*1-16
◇ B45 ◇ REV MR 17.933 • ID 12710

SUGENO, M. & TANG, DUOYUAN & TERANO, T. [1983] *Linguistic models of processes and "optimal" control (Chinese) (English summary)* (J 3732) Mohu Shuxue 3/2*19-24
◇ B52 ◇ REV Zbl 525 # 93004 • ID 37496

SUGENO, M. & TAKAGI, T. [1983] *Multi-dimensional fuzzy reasoning* (J 2720) Fuzzy Sets Syst 9*313-325
◇ B52 ◇ REV MR 85b:03033 Zbl 513 # 94035 • ID 37818

SUGENO, M. see Vol. V for further entries

SUGIHARA, T. [1950] *Many-valued logic (Japanese)* (J 4438) Tetsugaku Kenkyu 33*684-703
◇ B50 ◇ REV JSL 16.151 • ID 48768

SUGIHARA, T. [1951] *Many-valued logical characteristics of Brouwerian logic (Japanese)* (J 4439) Kagaku 21*294-295
◇ B50 G10 ◇ REV JSL 20.172 • ID 48769

SUGIHARA, T. [1952] *Negation in many-valued logic* (J 0980) Mem Fukui Univ Lib Arts College 1*1-5
◇ B50 ◇ REV MR 15.91 JSL 17.278 • ID 12713

SUGIHARA, T. [1953] *The axiomatization of the aristotelian modal logic* (J 0980) Mem Fukui Univ Lib Arts College 2*53-60
◇ B45 ◇ REV JSL 19.292 • ID 32457

SUGIHARA, T. [1955] *Strict implication free from implicational paradoxes* (J 0980) Mem Fukui Univ Lib Arts College 4*55-59
◇ A05 B46 ◇ REV JSL 20.303 • ID 12714

SUGIHARA, T. [1956] *Four-valued propositional calculus with one designated truth value* (J 0980) Mem Fukui Univ Lib Arts College 5*41-48
◇ B50 ◇ REV JSL 23.64 • ID 12715

SUGIHARA, T. [1958] *A three-valued logic with meaning-operator* (J 0980) Mem Fukui Univ Lib Arts College 8*59-60
◇ B50 ◇ REV JSL 25.293 • ID 12716

SUGIHARA, T. [1960] *Modern logic (Japanese)* (X 1187) Sankibo: Tokyo 134pp
◇ B98 ◇ REV JSL 32.543 • ID 28570

SUGIHARA, T. [1962] *The numbers of modalities in T supplemented by the axiom* CL^2pL^3p (J 0036) J Symb Logic 27*407-408
◇ B45 ◇ REV MR 31 # 45 Zbl 122.244 JSL 34.305 • ID 12717

SUGIHARA, T. [1964] *A study on modal logic (Japanese)* (X 1187) Sankibo: Tokyo ii + 188pp
◇ B45 ◇ REV JSL 32.544 • ID 28571

SUGIHARA, T. [1964] *Sequential truth-function* (J 0980) Mem Fukui Univ Lib Arts College 13*63-68
◇ B60 ◇ ID 32456

SUGIHARA, T. [1974] *Modalities in the entailment logic* (J 0980) Mem Fukui Univ Lib Arts College 24*1-14
◇ B46 ◇ ID 32455

SUGIHARA, T. [1974] *Temporal logic (Japanese)* (S 1459) Mem School Sci & Engin, Waseda Univ vi + 232pp
◇ B45 ◇ REV JSL 44.657 • ID 44609

SUHARA, K. [1984] *Another mode of metalinguistic speech: multi-modal logic on a new basis* (**J** 0286) Int Logic Rev 29*38-66
◇ B45 ◇ ID 45893

SULLIVAN, D.J. [1963] *Fundamentals of logic* (**X** 0822) McGraw-Hill: New York 288pp
◇ B98 ◇ ID 22540

SULTAN, L.H. [1983] *Fuzzy logic implementation in the realization of fuzzy algorithms* (**J** 3919) BUSEFAL 16*80-90
◇ B52 ◇ REV Zbl 552 # 68039 • ID 43450

SUMMERS, G.J. [1968] *New puzzles in logical deduction* (**X** 0813) Dover: New York vi + 121pp
◇ B98 ◇ REV Zbl 235 # 02001 • ID 27814

SUMNER, L.W. & WOODS, J. (EDS.) [1969] *Necessary truth: a book of readings* (**X** 0981) Random House: New York vii + 226pp
◇ A05 B45 ◇ ID 25255

SUNDHOLM, G. [1977] *A completeness proof for an infinitary tense-logic* (**J** 0105) Theoria (Lund) 43*47-51
◇ B45 C75 ◇ REV MR 56 # 15370 Zbl 364 # 02011
• ID 50941

SUNDHOLM, G. see Vol. VI for further entries

SUPPES, P. [1955] see RUBIN, H.

SUPPES, P. [1957] *Introduction to logic* (**X** 0864) Van Nostrand: New York xviii + 312pp
◇ B98 ◇ REV MR 26 # 4875 Zbl 77.11 JSL 22.353
• ID 12728

SUPPES, P. [1959] *Measurement,empirical meaningfulness and threevalued logic* (**C** 1190) Measurement 129-143
◇ A05 B50 ◇ REV JSL 35.129 • ID 21978

SUPPES, P. [1962] see NAGEL, E.

SUPPES, P. [1964] see HILL, S.

SUPPES, P. [1966] see HINTIKKA, K.J.J.

SUPPES, P. [1969] *Studies in the methodology and foundations of science. Selected papers from 1951-1969* (**X** 0835) Reidel: Dordrecht 483pp
◇ A05 B96 ◇ REV MR 40 # 4049 Zbl 188.310 • ID 12737

SUPPES, P. [1970] see HINTIKKA, K.J.J.

SUPPES, P. [1974] *The essential but implicit role of modal concepts in science* (**P** 1791) Proc Bienn Meet Phil of Sci Ass;1972 East Lansing 305-314
◇ A05 B45 ◇ REV Zbl 301 # 00004 • ID 29594

SUPPES, P. [1975] see HENKIN, L.

SUPPES, P. [1981] *Logique du probable. Demarche bayesienne et rationalite (French)* (**X** 2075) Flammarion: Paris ii + 136pp
◇ B48 ◇ REV MR 84e:03030 Zbl 507 # 00025 • ID 34369

SUPPES, P. see Vol. I, III, IV, V for further entries

SURMA, S.J. [1969] *On the question of natural deduction in modal logic* (**P** 3653) Modal du Jugement Chez Aristote & Log Moderne;1969 Brasov 181-184
◇ B45 ◇ REV MR 46 # 7000 Zbl 247 # 02024 • ID 33420

SURMA, S.J. [1970] *Concerning natural deduction in modal logic (Polish) (English summary)* (**S** 1454) Zesz Nauk Wyz Szk Ped Mat, Opole 10*21-29
◇ B45 ◇ REV MR 53 # 12879 • ID 23137

SURMA, S.J. [1970] *The doctoral dissertation of Emil L. Post (Polish) (English summary)* (**S** 0458) Zesz Nauk, Prace Log, Uniw Krakow 5*65-70
◇ A10 B50 ◇ REV MR 44 # 6439 • ID 12755

SURMA, S.J. [1971] *Deduction theorems in modal systems constructed by Goedel's method (Polish summary)* (**S** 0458) Zesz Nauk, Prace Log, Uniw Krakow 6*69-83
◇ B45 ◇ REV MR 45 # 1738 • ID 12758

SURMA, S.J. [1972] *The deduction theorems valid in certain fragments of the Lewis' system S2 and the system of Feys-von Wright* (**J** 0387) Bull Sect Logic, Pol Acad Sci 1/1*45-52
◇ B45 ◇ REV MR 50 # 12659 • ID 79117

SURMA, S.J. [1973] *A note on indirect deduction theorems valid in Lukasiewicz's finitely-valued propositional calculi (Polish) (English and Russian summaries)* (**J** 0063) Studia Logica 31*139-143
◇ B50 ◇ REV MR 53 # 12872 Zbl 267 # 02009 • ID 29880

SURMA, S.J. [1973] *Jaskowski's matrix criterion for the intuitionistic propositional calculus* (**C** 2084) Stud Hist of Math Log 87-121
◇ B55 F50 ◇ ID 42072

SURMA, S.J. [1973] *The deduction theorems valid in certain fragments of the Lewi's system S2 and the system T of Feys-von Wright (Polish and Russian summaries)* (**J** 0063) Studia Logica 31*127-138
◇ B45 ◇ REV MR 47 # 8250 Zbl 267 # 02013 • ID 12760

SURMA, S.J. [1974] *A method of the construction of finite Lukasiewiczian algebras and its application to a Gentzen-style characterization of finite logics* (**J** 0302) Rep Math Logic, Krakow & Katowice 2*49-54
• REPR [1977] (**C** 3026) Lukasiewicz Sel Pap on Sent Calc 93-99
◇ B50 G20 ◇ REV MR 52 # 91 MR 58 # 21448 Zbl 282 # 02024 Zbl 381 # 03017 • ID 18402

SURMA, S.J. & WRONSKI, A. & ZACHOROWSKI, S. [1974] *On Jaskowski-type semantics for the intuitionistic propositional logic* (**J** 0387) Bull Sect Logic, Pol Acad Sci 3/2*6-10
◇ B55 F50 ◇ REV MR 53 # 98 Zbl 341 # 02016 • ID 16561

SURMA, S.J. [1975] *A certain method for constructing Lukasiewicz algebras, and its application to the Gentzenisation of Lukasiewicz logics (Polish with English translation)* (**S** 1454) Zesz Nauk Wyz Szk Ped Mat, Opole 15*7-11,11-15
◇ B50 ◇ REV MR 54 # 4923 Zbl 311 # 02066 • ID 24099

SURMA, S.J. & WRONSKI, A. & ZACHOROWSKI, S. [1975] *On Jaskowski-type semantics for the intuitionistic propositional logic* (**J** 0063) Studia Logica 34*145-148
◇ B55 F50 ◇ REV MR 53 # 12885 Zbl 341 # 02016
• ID 23141

SURMA, S.J. (ED.) [1977] *On Lesniewski's systems. Proceedings of XXII conference on the history of logic* (**J** 0063) Studia Logica 36*245-426
◇ A10 B15 B97 ◇ REV MR 57 # 15936 Zbl 372 # 00005
• ID 51405

SURMA, S.J. [1981] *On closure operators with the deduction property* (**P** 3705) Int Symp Multi-Val Log (11);1981 Oklahoma City & Norman 235-243
◇ B50 ◇ REV MR 83m:94033 Zbl 566 # 03018 • ID 45973

SURMA, S.J. see Vol. I, III, V, VI for further entries

SUSSMAN, A.N. [1981] *Counterfactuals and nontrivial de re modalities* (**J** 0170) Ratio (Frankfurt) 23*139-140
◇ B45 ◇ REV MR 83b:03009 • ID 35084

SUSTAL, J. [1982] *On the uncertainty of fuzzy classifications* (**C** 3786) Approx Reason in Decis Anal 125-129
◇ B52 ◇ REV MR 84d:90003 Zbl 509 # 90050 • ID 36802

SUSZKO, R. [1957] *A formal theory of the logical values I (Polish) (Russian and English summaries)* (**J** 0063) Studia Logica 6*145-237
◇ B50 ◇ REV MR 21 # 1270 • ID 24892

SUSZKO, R. [1967] *A proposal concerning the formulation of the infinitistic axiom in the theory of logical probability* (**S** 0019) Colloq Math (Warsaw) 17*347-349
◇ B48 C90 ◇ REV MR 36 # 4595 Zbl 218 # 02021 • ID 26275

SUSZKO, R. [1967] *An essay in the formal theory of extension and of intension (Polish and Russian summaries)* (**J** 0063) Studia Logica 20*7-36
◇ A05 B45 ◇ REV MR 35 # 6531 Zbl 308 # 02053 • ID 12772

SUSZKO, R. [1968] *Non-Fregean logic and theories* (**J** 0147) An Univ Bucuresti, Acta Logica 11*105-125
◇ A05 B45 ◇ REV MR 44 # 2575 Zbl 233 # 02010 • ID 12773

SUSZKO, R. [1970] *Non-Frege logic and theories that are based on it (Russian)* (**C** 0668) Neklass Log 349-383
◇ B45 ◇ REV MR 47 # 6448 • ID 12775

SUSZKO, R. [1971] *Identity connective and modality (Polish and Russian summaries)* (**J** 0063) Studia Logica 27*7-41
◇ B45 ◇ REV MR 45 # 6594 Zbl 263 # 02015 • ID 12780

SUSZKO, R. & ZANDAROWSKA, W. [1971] *Lewi's system S4 and S5 and the identity connective (Polish) (Russian and English summaries)* (**J** 0063) Studia Logica 29*169-181
◇ B45 ◇ REV MR 46 # 3271 Zbl 244 # 02007 • ID 12778

SUSZKO, R. [1971] *Quasi-completeness in non-Fregean logic (Polish and Russian summaries)* (**J** 0063) Studia Logica 29*7-16
◇ B60 C07 ◇ REV MR 46 # 5115 Zbl 272 # 02029 • ID 12777

SUSZKO, R. [1971] see BLOOM, S.L.

SUSZKO, R. [1972] *A note on modal systems and SCI* (**J** 0387) Bull Sect Logic, Pol Acad Sci 1/4*38-41
◇ B45 B60 ◇ REV MR 58 # 27347 • ID 79129

SUSZKO, R. [1972] *A note on adequate models for non-Fregean sentential calculi* (**J** 0387) Bull Sect Logic, Pol Acad Sci 1/4*42-45
◇ B60 ◇ REV MR 58 # 27348 • ID 79128

SUSZKO, R. [1972] see OMYLA, M.

SUSZKO, R. [1972] see BLOOM, S.L.

SUSZKO, R. [1973] *Adequate models for the non-fregean sentential calculus (SCI)* (**P** 0580) Int Congr Log, Meth & Phil of Sci (4,Sel Pap);1971 Bucharest 49-54
◇ B60 ◇ REV MR 57 # 64 Zbl 293 # 02009 • ID 65611

SUSZKO, R. [1974] see MICHAELS, A.

SUSZKO, R. [1974] *Some notions and theorems of McKinsey and Tarski and SCI* (**J** 0477) Spis Bulgar Akad Nauk 3*3-5
◇ B45 C05 G05 ◇ REV MR 53 # 129 • ID 16651

SUSZKO, R. [1975] *Abolition of the Fregean axiom* (**C** 0758) Logic Colloq Boston 1972-73 169-239
◇ A05 B45 ◇ REV MR 56 # 11760 Zbl 308 # 02026 • ID 27699

SUSZKO, R. [1975] *Remarks on Lukasiewicz's three-valued logic* (**J** 0387) Bull Sect Logic, Pol Acad Sci 4*87-90
◇ B50 ◇ REV MR 53 # 5257 • ID 22890

SUSZKO, R. [1976] see MICHAELS, A.

SUSZKO, R. see Vol. I, III, V, VI for further entries

SUTHERLAND, R.I. [1982] *On Kochen and Specker's impossibility proof (Italian and Russian summaries)* (**J** 2775) Nuovo Cimento B, Ser 2 69/2*239-244
◇ B51 G12 ◇ REV MR 84c:81007 • ID 39653

SVEJDAR, V. [1983] *Modal analysis of generalized Rosser sentences* (**J** 0036) J Symb Logic 48*986-999
◇ B45 F25 F30 ◇ REV MR 85k:03041 Zbl 543 # 03010 • ID 40906

SVEJDAR, V. see Vol. V, VI for further entries

SWAIN, M. (ED.) [1970] *Induction, acceptance, and rational belief* (**X** 0835) Reidel: Dordrecht vii + 232pp
◇ A05 B48 ◇ REV Zbl 188.15 JSL 39.166 • ID 25470

SWAIN, M. [1970] *The consistency of rational belief* (**C** 1577) Induct, Accept & Rat Belief 27-54
◇ A05 B48 ◇ REV Zbl 217.6 JSL 39.166 • ID 28032

SWART DE, H.C.M. [1980] *Gentzen-type systems for C, K and several extensions of C and K; constructive completeness proofs and effective decision procedures for these systems* (**J** 0079) Logique & Anal, NS 23*263-284
◇ B25 B45 ◇ REV MR 83i:03037 Zbl 454 # 03007 • ID 54220

SWART DE, H.C.M. [1983] *A Gentzen- or Beth-type system, a practical decision procedure and a constructive completeness proof for the counterfactual logics VC and VCS* (**J** 0036) J Symb Logic 48*1-20
◇ B45 ◇ REV MR 84h:03048 Zbl 527 # 03005 • ID 37397

SWART DE, H.C.M. [1985] *Gentzen-type or Beth-type systems,constructive completeness proofs and practical decision procedures (with special attention to relevance logic)* (**P** 4180) Int Congr Log, Meth & Phil of Sci (7,Pap);1983 Salzburg 89-122
◇ B46 ◇ ID 48115

SWART DE, H.C.M. see Vol. III, V, VI for further entries

SWARTWOUT, R.E. [1971] see NUTTER, R.S.

SWARTWOUT, R.E. [1974] see NUTTER JR., R.S.

SWARTZ, N. [1979] see BRADLEY, R.

SWEET, A.M. [1963] *Toward a pragmatical explication of epistemic modalities* (**J** 0047) Notre Dame J Formal Log 4*145-150
◇ A05 B45 ◇ REV MR 30 # 1036 Zbl 122.9 • ID 12804

SWEET, A.M. [1969] *The pragmatics of monadic quantification* (**J** 0047) Notre Dame J Formal Log 10*31-46
◇ A05 B60 ◇ REV MR 39 # 3974 Zbl 165.14 • ID 12805

SWEET, A.M. see Vol. III for further entries

SWIFT, J.D. [1952] *Algebraic properties of n-valued propositional calculi* (**J** 0005) Amer Math Mon 59*612-621
◇ B50 ◇ REV MR 14.440 Zbl 49.6 JSL 18.282 • ID 12815

SWINBURNE, R. [1973] *An introduction to confirmation theory* (X 0816) Methuen: London & New York vi+218pp
◇ A05 B48 ◇ ID 25448

SWOYER, C. [1983] *Belief and predication* (J 0097) Nous, Quart J Phil 17*197-220
◇ B48 ◇ REV MR 85a:03026 • ID 34777

SZABO, M.E. [1983] *The continuous realizability of entailment* (J 0068) Z Math Logik Grundlagen Math 29*219-233
◇ B46 ◇ REV MR 85e:03050 Zbl 538 # 03019 • ID 39586

SZABO, M.E. see Vol. I, IV, VI for further entries

SZANIAWSKI, K. [1960] *A note on confirmation of statistical hypotheses* (J 0063) Studia Logica 10*111-118
◇ B48 ◇ REV MR 24 # A1160 • ID 33095

SZANIAWSKI, K. [1960] *Some remarks concerning the criterion of rational decision making (Polish and Russian summaries)* (J 0063) Studia Logica 9*221-239
◇ A05 B48 ◇ REV MR 25 # 5799 • ID 33092

SZANIAWSKI, K. [1961] *A method of deciding between N statistical hypotheses (Polish and Russian summaries)* (J 0063) Studia Logica 12*135-143
◇ B48 ◇ REV MR 26 # 7076 • ID 33104

SZANIAWSKI, K. [1961] *On some basic patterns of statistical inference (Polish and Russian summaries)* (J 0063) Studia Logica 11*77-89
◇ B48 ◇ ID 33100

SZANIAWSKI, K. [1962] *On the justification of inductive rules of inference (Polish) (Russian and English summaries)* (J 0063) Studia Logica 13*219-225
◇ B48 ◇ ID 33110

SZANIAWSKI, K. see Vol. I for further entries

SZATKOWSKI, M. [1981] *On fragments of Medvedev's logic* (J 0063) Studia Logica 40*39-54
◇ B55 ◇ REV MR 84c:03051 Zbl 488 # 03016 • ID 35703

SZATKOWSKI, M. [1984] *Concerning Kripke semantics for intermediate predicate logics* (S 1070) Probl Mat, Wyz Szk Ped, Bydgosszcz 4*43-62
◇ B55 C90 ◇ ID 44572

SZCZECH, W. [1973] *Axiomatizability of finite matrices* (J 0387) Bull Sect Logic, Pol Acad Sci 2*118-120
◇ B50 ◇ REV MR 55 # 7758 • ID 79169

SZCZECH, W. see Vol. I, V for further entries

SZENDREI, A. [1978] *On closed sets of linear operations over a finite set of square-free cardinality* (J 0129) Elektr Informationsverarbeitung & Kybern 14*547-559
◇ B50 ◇ REV MR 81e:08006 Zbl 399 # 03017 • ID 52814

SZENDREI, A. [1980] *On closed classes of quasilinear functions* (J 0022) Cheskoslov Mat Zh 30(105)*498-509
◇ B50 G10 ◇ REV MR 82b:03053 Zbl 451 # 03023
• ID 54038

SZENDREI, A. [1981] *Clones of linear operations on finite sets* (P 2552) Conf Finite Algeb & Multi-Val Log;1979 Szeged 693-738
◇ B50 C05 ◇ REV MR 83j:08007 Zbl 487 # 08002
• ID 39877

SZENDREI, A. [1982] *Algebras of prime cardinality with a cyclic automorphism* (J 0008) Arch Math (Basel) 39*417-424
◇ B50 C05 ◇ REV MR 84g:08015 Zbl 487 # 08003
• ID 39416

SZENDREI, A. see Vol. III for further entries

SZEWCZAK, E. [1974] *Iterated modalities and the parallel between deontic and modal logics* (J 0079) Logique & Anal, NS 17*323-333
◇ A05 B45 ◇ REV MR 55 # 85 Zbl 304 # 02010 • ID 65633

SZOSTEK, B. [1983] *A certain classification of the Heyting-Brouwer intermediate propositional logics* (J 1008) Demonstr Math (Warsaw) 16*945-960
◇ B55 ◇ REV MR 85m:03018 Zbl 571 # 03010 • ID 45144

SZYMANSKA, M. [1983] see SCHWEIGERT, D.

TAHARA, M. & TANAKA, S. [1970] *Functional completeness and polypheks in 3-valued logic (Japanese)* (J 0979) Denshi Tsushin Gakkai Ronbunshi, Sect A-D 53-C/1*111-118
• TRANSL [1970] (J 0464) Syst-Comp-Controls 1/1*64-73
◇ B50 ◇ REV MR 48 # 63 • ID 22340

TAJMANOV, V.A. [1983] *On cartesian powers of P_2 (Russian)* (J 0023) Dokl Akad Nauk SSSR 270*1327-1330
• TRANSL [1983] (J 0062) Sov Math, Dokl 27*773-776
◇ B50 ◇ REV MR 85i:03186 Zbl 549 # 03053 • ID 44299

TAJMANOV, V.A. [1983] *On function systems in k-valued logic with closure operations of program type (Russian)* (J 0023) Dokl Akad Nauk SSSR 268*1307-1310
• TRANSL [1983] (J 0062) Sov Math, Dokl 27*255-258
◇ B50 ◇ REV MR 84f:03022 Zbl 549 # 03018 • ID 34439

TAJTSLIN, M.A. [1970] *Model theory (Russian)* (X 0913) Novosibirsk Gos Univ: Novosibirsk 214pp
◇ B98 C98 ◇ REV MR 49 # 2350 • ID 28675

TAJTSLIN, M.A. [1973] see ERSHOV, YU.L.

TAJTSLIN, M.A. (ED.) [1981] *Investigations in theoretical programming (Russian)* (X 2769) Kazakh Gos Univ: Alma-Ata 104pp
◇ B97 ◇ REV MR 84j:03008 • ID 34602

TAJTSLIN, M.A. [1983] see STOLBOUSHKIN, A.P.

TAJTSLIN, M.A. see Vol. I, III, IV for further entries

TAKAGI, M. [1967] *On an extension of two-valued logic. Extended boolean function and its application (Japanese summary)* (J 0996) RAAG Res Notes 119*ii+15pp
◇ B50 ◇ REV MR 35 # 6593 Zbl 149.251 • ID 13288

TAKAGI, T. [1983] see SUGENO, M.

TAKAGI, T. see Vol. I for further entries

TAKAHASHI, MOTO-O [1968] *Many-valued logics of extended Gentzen style. I* (J 0350) Sci Rep Tokyo Kyoiku Daigaku Sect A 9*271-292
◇ B50 ◇ REV MR 37 # 39 Zbl 172.8 • REM Part II 1970
• ID 22253

TAKAHASHI, MOTO-O [1970] *Continuous λ-ε logics* (J 0260) Ann Jap Ass Phil Sci 3*205-215
◇ B50 ◇ REV MR 44 # 6451 Zbl 213.14 • ID 13299

TAKAHASHI, MOTO-O [1970] *Many-valued logics of extended Gentzen style II* (J 0036) J Symb Logic 35*493-528
◇ B50 ◇ REV MR 44 # 6452 Zbl 229 # 02021 • REM Part I 1968 • ID 22254

TAKAHASHI, MOTO-O [1977] *Many-valued logics and their algebras (Japanese)* (J 0091) Sugaku 29*135-147
◇ B50 G20 ◇ REV MR 58 # 21449 • ID 79197

TAKAHASHI, MOTO-O see Vol. I, III, IV, V, VI for further entries

TAKAMATSU, T. [1968] *On foundation of three-valued logic (Japanese)* (**J** 1724) Kagaku Tetsugaku 1*31-44
◇ B50 ◇ ID 32489

TAKAMATSU, T. see Vol. I, VI for further entries

TAKANO, M. [1980] *Valid sequents in many-valued logics* (**J** 0260) Ann Jap Ass Phil Sci 5*245-260
◇ B50 ◇ REV MR 82c:03030 Zbl 442 # 03022 • ID 56379

TAKANO, M. [1985] *A semantical investigation into Lesniewski's axiom and his ontology* (**J** 0063) Studia Logica 44*71-78
◇ B60 E70 ◇ ID 47523

TAKANO, M. see Vol. VI for further entries

TAKAOKA, T. [1971] *A construction method of fail-safe systems for many-valued logic* (**J** 4555) Trans IECE 54-C*41-49
◇ B50 ◇ ID 41997

TAKAOKA, T. see Vol. IV for further entries

TAKEDA, E. [1982] *Interactive identification of fuzzy outranking relations in a multicriteria decision problem* (**C** 3778) Fuzzy Inform & Decis Processes 301-307
◇ B52 ◇ REV MR 84d:90004 Zbl 503 # 90087 • ID 36680

TAKEKUMA, R. [1954] *On a nine-valued propositional calculus* (**J** 0093) J Comp Syst 1*225-228
◇ B50 ◇ REV MR 16.555 Zbl 57.9 JSL 22.330 • ID 13303

TAKEUTI, G. [1978] *Two applications of logic to mathematics* (**X** 3552) Iwanami Shoten: Tokyo viii + 137pp
• LAST ED [1978] (**X** 0857) Princeton Univ Pr: Princeton viii + 137pp
◇ B98 C90 E40 E75 F05 F30 F35 G12 ◇ REV MR 58 # 21591 Zbl 393 # 03027 • ID 52447

TAKEUTI, G. [1981] *Logic and set theory* (**C** 2617) Modern Log Survey 167-171
◇ B51 B98 E40 E70 F50 ◇ REV MR 82f:03002 Zbl 464 # 03001 • ID 39565

TAKEUTI, G. [1984] *Quantum logic and quantization* (**P** 4383) Found Quant Mech;1983 Kokubunji-shi 256-260
◇ B51 C90 E40 E75 ◇ REV MR 86g:03108 • ID 39581

TAKEUTI, G. see Vol. I, III, IV, V, VI for further entries

TAL', A.A. [1963] see AJZERMAN, M.A.

TAL', A.A. see Vol. IV for further entries

TALJA, J. [1985] *On the logic of omissions* (**J** 0154) Synthese 65*235-248
◇ B60 ◇ ID 49424

TALJA, J. see Vol. III, IV for further entries

TALLET, J. [1970] *On the symmetry of many-valued logical systems* (**J** 0079) Logique & Anal, NS 13*302-322
◇ B50 ◇ REV MR 44 # 1542 Zbl 216.287 • ID 13344

TALLET, J. [1976] *The class of possibilities* (**J** 0079) Logique & Anal, NS 19*367-389
◇ B45 ◇ REV MR 58 # 16159 • ID 79232

TAMAS, G. (ED.) [1983] *Studien zur Logik* (**X** 0928) Akad Kiado: Budapest 345pp
◇ B98 ◇ REV MR 85j:01005a • ID 46783

TAMBURINO, J. [1975] see ASENJO, F.G.

TAMBURRINI, G. [1982] see GUCCIONE, S.

TAMBURRINI, G. see Vol. V for further entries

TAMNY, M. [1972] see GUTTENPLAN, S.D.

TAMURA, S. [1971] *The implicational fragment of R-mingle* (**J** 0081) Proc Japan Acad 47*71-75
◇ B46 ◇ REV MR 44 # 30 Zbl 227 # 02007 • ID 13352

TAMURA, S. [1985] *Decision procedure for pseudocomplemented lattices* (**P** 4423) Semigroups (8);1984 Matsue 36-39
◇ B55 ◇ ID 48582

TAMURA, S. see Vol. I, III, IV for further entries

TANAKA, H. [1982] see ASAI, K.

TANAKA, H. [1983] see ASAI, K.

TANAKA, H. [1984] see ASAI, K.

TANAKA, H. see Vol. I, III, IV, V, VI for further entries

TANAKA, K. [1971] see KITAHASHI, T.

TANAKA, K. [1971] see FUJITA, Y.

TANAKA, K. [1972] see FUJITA, Y.

TANAKA, K. [1976] see KANEYORI, S.

TANAKA, K. [1978] see FUKAMI, S.

TANAKA, K. [1979] see FUKAMI, S.

TANAKA, K. [1980] see FUKAMI, S.

TANAKA, K. see Vol. I, IV, V for further entries

TANAKA, S. [1966] see ISEKI, K.

TANAKA, S. [1968] *On axioms of ontology* (**J** 0081) Proc Japan Acad 44*54-55
◇ B60 ◇ REV MR 37 # 2574 Zbl 185.11 JSL 37.192 • ID 13381

TANAKA, S. [1968] *On theorems of ontology* (**J** 0081) Proc Japan Acad 44*231-233
◇ B60 ◇ REV MR 37 # 2575 Zbl 185.11 JSL 37.192 • ID 13382

TANAKA, S. [1970] see TAHARA, M.

TANAKA, S. [1970] *On axiom systems of ontology. I* (**J** 0081) Proc Japan Acad 46*255-257
◇ B60 ◇ REV MR 43 # 24 Zbl 209.9 • REM Part II 1971 • ID 13384

TANAKA, S. [1971] *On axiom systems of ontology. II* (**J** 0081) Proc Japan Acad 47*177-179
◇ B60 ◇ REV MR 45 # 1728 Zbl 231 # 02031 • REM Part I 1970 • ID 13385

TANAKA, S. see Vol. I, V for further entries

TANG, CAOZHEN [1936] *A paradox of Lewis's strict implication* (**J** 0015) Bull Amer Math Soc 42*707-709
◇ B45 B46 ◇ REV Zbl 15.146 JSL 2.58 FdM 62.1051 • ID 41326

TANG, CAOZHEN [1936] *The theorem "p⥽q. = .pq = p" and Huntingtons's relation between Lewis's strict implication and boolean algebra* (**J** 0015) Bull Amer Math Soc 42*743-746
◇ B46 B75 G05 ◇ REV Zbl 15.146 JSL 2.58 FdM 62.1051 • ID 13701

TANG, CAOZHEN [1938] *Algebraic postulates and a geometric interpretation for Lewis calculus of strict implication* (**J** 0015) Bull Amer Math Soc 44*737-744
◇ B45 B46 G25 ◇ REV Zbl 19.385 JSL 4.27 FdM 64.30 • ID 13702

TANG, CHESHAN [1982] see CAI, QINGSHENG

TANG, DUOYUAN [1983] see SUGENO, M.

TANG, TONGGAO [1985] *A completeness theorem of first order temporal logic with equality (Chinese)* (**J** 3766) Zhongguo Kexue, Xi A 28*532-540
◇ B45 ◇ ID 48530

TANG, TONGGAO [1985] *Systems of first-order temporal logic with equality (Chinese) (English summary)* (**J** 3735) Fudan Xuebao, Ziran Kexue 24*177-181
◇ B45 ◇ ID 49426

TANG, TONGGAO see Vol. I, IV, VI for further entries

TANG, XUZHANG [1982] *The existence of fuzziness and its meaning (Chinese) (English summary)* (**J** 3732) Mohu Shuxue 2/1*121-124
◇ A05 B52 ◇ REV MR 83g:03061 • ID 34840

TANG, ZHISONG [1960] see LU, YANGCI

TANG, ZHISONG see Vol. IV for further entries

TAPSCOTT, B.L. [1976] *Elementary applied symbolic logic* (**X** 0819) Prentice Hall: Englewood Cliffs xii + 496pp
◇ B98 ◇ REV JSL 44.281 • ID 44589

TAPSCOTT, B.L. [1984] *Correcting the tableau procedure for S4* (**J** 0047) Notre Dame J Formal Log 25*241-249
◇ B45 ◇ REV MR 85i:03059 Zbl 556 # 03016 • ID 42571

TARASENKO, V.P. [1972] see KORNEJCHUK, V.I.

TARASOV, V.V. [1972] *Certain properties of the essential function of k-valued logic (Russian)* (**J** 0071) Met Diskr Analiz (Novosibirsk) 20*66-78
◇ B50 ◇ REV MR 47 # 8246 Zbl 262 # 02018 • ID 13390

TARASOV, V.V. [1972] *Letter to the editors (Diskret. Analiz 20 (1972), 66 - 78) (Russian)* (**J** 0071) Met Diskr Analiz (Novosibirsk) 21*98
◇ B50 ◇ REV MR 50 # 12655 • ID 79251

TARASOV, V.V. [1975] *A test for the completeness of not everywhere defined functions of the algebra of logic (Russian)* (**J** 0052) Probl Kibern 30*319-325
◇ B50 G20 ◇ REV MR 52 # 2819 Zbl 414 # 94040
• ID 17619

TARASOV, V.V. [1982] *Realization of not everywhere defined functions of the algebra of logic (Russian)* (**J** 0087) Mat Zametki (Akad Nauk SSSR) 32*89-96,126
• TRANSL [1982] (**J** 1044) Math Notes, Acad Sci USSR 32*524-528
◇ B50 ◇ REV MR 83j:03045 Zbl 505 # 03010
Zbl 548 # 03007 • ID 35348

TARASOV, V.V. see Vol. I for further entries

TARSKI, A. [1935] *Wahrscheinlichkeitslehre und mehrwertige Logik* (**J** 0290) Euclides 5*174-175
◇ B50 ◇ ID 13411

TARSKI, A. [1936] *On mathematical logic and the deductive method (Polish)* (**X** 1764) Ksiaznica Atlas: Warsaw 167pp
• TRANSL [1960] (**X** 0834) Gauthier-Villars: Paris xv + 224pp [1969] (**X** 1056) Valention Bompiani: Milano
◇ B98 ◇ REV MR 40 # 5410 Zbl 171.252 JSL 30.236
FdM 62.39 • REM For futher transl. see also 1937 • ID 32714

TARSKI, A. [1937] *Einfuehrung in die mathematische Logik und in die Methodologie der Mathematik* (**X** 0902) Springer: Wien x + 166pp
• TRANSL [1941] (**X** 0894) Oxford Univ Pr: Oxford xviii + 239pp (English) 3rd ed 1965 [1948] (**X** 2027) Nauka: Moskva 322pp (Russian) [1951] (**X** 1042) Espasa-Calpe: Madrid (Spanish) 2nd ed 1968 [1953] (**X** 0809) North Holland: Amsterdam (Dutch) 2nd ed 1964 [1956] (**X** 0929) Weizmann Science Pr Israel: Jerusalem (Hebrew) 2nd ed 1966 [1966] (**X** 1226) Academia: Prague (Czech) 2nd ed 1969 [1971] (**X** 1052) Tbilisi Univ: Tbilisi (Georgian) [1980] (**X** 4569) Commercial Pr: Beijing (Chinese) 2nd ed. • LAST ED [1970] (**X** 0903) Vandenhoeck & Ruprecht: Goettingen 252pp
◇ A05 B30 B98 ◇ REV MR 2.209 MR 40 # 5409
MR 55 # 7716 Zbl 18.1 Zbl 25.4 Zbl 352 # 02001
JSL 12.61 JSL 3.51 JSL 31.647 FdM 63.22 • REM For the Polish original and further transl. see 1936 • ID 19757

TARSKI, A. [1948] see MCKINSEY, J.C.C.

TARSKI, A. [1962] see NAGEL, E.

TARSKI, A. also published under the name TAJTELBAUM, A.

TARSKI, A. see Vol. I, III, IV, V, VI for further entries

TARSKI, J. [1974] *Quantum field theory; an unusual discipline* (**P** 0610) Tarski Symp;1971 Berkeley 447-464
◇ B51 G12 ◇ REV MR 55 # 7206 • ID 13472

TARSKI, J. see Vol. I for further entries

TAUTS, A. [1967] *The definition of truth values as formulae (Russian) (Estonian and German summaries)* (**S** 3468) Tr Mat & Mekh (Tartu) 7(206)*3-9
◇ B50 ◇ REV MR 41 # 3255 Zbl 207.291 • ID 13476

TAUTS, A. [1968] *Logic as a classification of formulae (Russian) (Estonian and German summaries)* (**S** 3468) Tr Mat & Mekh (Tartu) 8(220)*3-11
◇ B50 ◇ REV MR 42 # 4380 Zbl 318 # 02011 • ID 13477

TAUTS, A. [1969] *Nonregular first order predicate calculus (Russian)* (**J** 0003) Algebra i Logika 8*76-92
• TRANSL [1969] (**J** 0069) Algeb and Log 8*41-49
◇ B50 ◇ REV MR 43 # 6053 Zbl 225 # 02041 • ID 13478

TAUTS, A. [1970] *Nonfixed formulae (Russian) (Estonian and German summaries)* (**S** 3468) Tr Mat & Mekh (Tartu) 9(253)*45-54
◇ B50 ◇ REV MR 44 # 3839 Zbl 235 # 02016 • ID 13479

TAUTS, A. [1974] *A formal deduction of tautological formulas in pseudo-boolean algebras (Russian) (Estonian and German summaries)* (**S** 3468) Tr Mat & Mekh (Tartu) 13(336)*3-30
◇ B60 C90 F50 G05 G10 ◇ REV MR 51 # 7864
Zbl 356 # 02010 • ID 17694

TAUTS, A. [1975] *Search for deduction by means of a semantic model (Russian) (Estonian and German summaries)* (**S** 3468) Tr Mat & Mekh (Tartu) 17(374)*3-28
◇ B55 C75 C90 F50 ◇ REV MR 58 # 5076
Zbl 393 # 03039 • ID 52459

TAUTS, A. [1983] *The possibility of constructing a countermodel for an indeterminate formula (Russian) (German and Estonian summaries)* (**J** 0080) Izv Akad Nauk Ehston SSR, Fiz, Mat 32*257-262
◇ B60 ◇ REV MR 85d:03058 Zbl 559 # 03007 • ID 41095

TAUTS, A. see Vol. I, III, IV, V, VI for further entries

TAUTU, P. [1983] *Belief functions and analogous fuzzy measures (with applications to Bayes' theorem)* (**C** 4270) Stud in Probab & Rel Topics (Onicescu) 425-456
◇ B52 ◇ REV MR 86e:60005 Zbl 545 # 62003 • ID 44460

TAVANETS, P.V. (ED.) [1967] *Logical semantics and modal logic (Russian)* (**X** 2027) Nauka: Moskva 277pp
◇ B45 ◇ ID 48572

TAVANETS, P.V. [1967] see SMIRNOVA, E.D.

TAVANETS, P.V. (ED.) [1970] *Nonclassical logic (Russian)* (**X** 2027) Nauka: Moskva 384pp
◇ B97 ◇ REV MR 47 # 6450 • ID 70237

TAVANETS, P.V. (ED.) [1970] *Studies in systems of logic (Russian)* (**X** 2027) Nauka: Moskva 334pp
◇ B97 ◇ REV MR 47 # 6426 • ID 70238

TAVANETS, P.V. [1974] see SMIRNOV, V.A.

TAVANETS, P.V. see Vol. I for further entries

TAX, R.E. [1973] *On the intuitionistic equivalent calculus* (**J** 0047) Notre Dame J Formal Log 14*448-456
◇ B55 ◇ REV MR 48 # 8206 Zbl 226 # 02023 • ID 13483

TAYLOR, B. [1977] *Tense and continuity* (**J** 2130) Linguist Philos 1*199-220
◇ A05 B45 ◇ REV Zbl 394 # 03008 • ID 52485

TEH, H.H. [1963] *On 3-valued sentential calculus. An axiomatic approach* (**J** 0241) Bull Math Soc Nanyang Univ 1963*1-37
◇ B50 ◇ REV MR 29 # 11 • ID 13511

TEH, H.H. see Vol. V for further entries

TENEV, V. [1979] see SHOJLEV, KH.

TENNANT, N. [1978] *Entailment and proofs* (**J** 0113) Proc Aristotelian Soc 79*167-189
◇ B48 ◇ ID 42890

TENNANT, N. [1978] *Natural logic* (**X** 1261) Edinburgh Univ Pr: Edinburgh ix + 196pp
◇ B98 ◇ REV MR 81i:03004 Zbl 483 # 03001 JSL 48.215 • ID 79279

TENNANT, N. [1980] *A proof-theoretic approach to entailment* (**J** 0122) J Philos Logic 9*185-209
◇ B46 F05 ◇ REV MR 81j:03035 Zbl 432 # 03032 • ID 53989

TENNANT, N. [1984] *Perfect validity, entailment ond paraconsistency* (**J** 0063) Studia Logica 43*181-200
◇ B53 ◇ ID 42371

TENNANT, N. see Vol. I, III for further entries

TERADA, H. [1979] see SASAO, T.

TERADA, H. [1982] see ISHIKAWA, K.

TERANO, T. [1983] see SUGENO, M.

TERMINI, S. [1983] see GUCCIONE, S.

TERMINI, S. [1984] *Aspects of vagueness and some epistemological problems related to their formalization* (**P** 3096) Conf Math Service of Man (2,Pap);1982 Las Palmas 205-230
◇ B52 ◇ REV MR 86e:03005 Zbl 533 # 00005 • ID 44003

TERMINI, S. see Vol. V for further entries

TERRELL, D.B. [1967] *Logic: A modern introduction to deductive reasoning* (**X** 0818) Holt Rinehart & Winston: New York xii + 355pp
◇ B98 ◇ REV MR 35 # 2717 • ID 22306

TESLENKO, A.K. [1972] see KORNEJCHUK, V.I.

TETRAULT, G.E. [1978] *Deviant, many-valued modellings of paradox: Outlines of indefiniteness as the formal source of exact logic* (**J** 0286) Int Logic Rev 17-18*47-57
◇ A05 B50 ◇ REV MR 82d:03035 Zbl 435 # 03003 • ID 55764

TEVZADZE, D.D. [1981] *Analysis of a difficulty in quantified modal logic (Russian)* (**C** 3810) Log-Semat Issl (Tbilisi) 16-24
◇ B45 ◇ REV MR 84m:03028 • ID 35732

TEZUKA, Y. [1969] see KASAHARA, Y.

TEZUKA, Y. [1970] see KASAHARA, Y.

THAYSE, A. [1975] see DESCHAMPS, J.-P.

THAYSE, A. [1976] *Difference operators and extended truth vectors for discrete functions* (**J** 0193) Discr Math 14*171-202
◇ B50 ◇ REV MR 56 # 18168 Zbl 334 # 94012 • ID 65691

THAYSE, A. [1977] *Static-hazard detection in switching circuits by prime-implicant examination in fuzzy functions* (**J** 2712) Electronics Lett (London) 13*94-96
◇ B52 B70 ◇ REV MR 58 # 4705 • ID 82967

THAYSE, A. [1979] *Integer expansions of discrete functions and their use in optimization problems* (**P** 3003) Int Symp Multi-Val Log (9);1979 Bath 82-87
◇ B50 ◇ REV MR 80m:94104 • ID 35942

THAYSE, A. see Vol. I for further entries

THELLIEZ, S. [1973] *Introduction a l'etude des structures ternaires de commutation* (**X** 0836) Gordon & Breach: New York (French)
• TRANSL [1975] (**X** 0836) Gordon & Breach: New York xii + 186pp (English)
◇ B50 B70 ◇ REV Zbl 323 # 94024 • ID 42073

THELLIEZ, S. [1976] *M-valued sequential cellular networks* (**P** 2011) Int Symp Multi-Val Log (6);1976 Logan 264
◇ B50 ◇ ID 35898

THIEFFINE, F. [1983] *Compatible complement in Piron's system and ordinary modal logic* (**J** 2778) Lett Nuovo Cimento Ser 2 36*377-381
◇ B45 ◇ REV MR 85h:03066 • ID 43342

THIELE, H. [1958] *Theorie der endlichwertigen Lukasiewiczschen Praedikatenkalkuele der ersten Stufe* (**J** 0068) Z Math Logik Grundlagen Math 4*108-142
◇ B50 ◇ REV MR 21 # 7158 Zbl 88.247 JSL 32.260 • ID 13539

THIELE, H. [1968] *Ueber einen sequentiellen Praedikatenkalkuel als Grundlage fuer den Aufbau problemorientierter algorithmischer Sprachen* (**J** 0947) Wiss Z Tech Univ Dresden 17*1122-1126
◇ B45 B75 ◇ REV MR 40 # 1278 • ID 13540

THIELE, H. see Vol. I, IV for further entries

THIELE, R. [1979] *Mathematische Beweise* (**X** 1079) Teubner: Leipzig 172pp
◇ B98 ◇ REV MR 82k:03002 Zbl 405 # 00018 • REM 2nd ed. 1981; 176pp • ID 54872

THIELER-MEVISSEN, G. [1976] see GENRICH, H.J.

THIELER-MEVISSEN, G. see Vol. IV for further entries

THOMAS, G.G. [1982] *On permutographs* (**P** 4724) Winter School on Abstract Anal (10);1982 Srni 275-286
◇ B50 ◇ REV MR 84f:05049 Zbl 504 # 03011 • ID 36969

THOMAS, I. [1953] *Note on a modal system of Lukasiewicz* (**J** 0505) Dominican Studies 6*167-170
◇ B45 ◇ REV JSL 25.293 • ID 13542

THOMAS, I. [1957] see LEMMON, E.J.

THOMAS, I. [1962] *Finite limitations on Dummett's LC* (**J** 0047) Notre Dame J Formal Log 3*170-174
◇ B45 ◇ REV MR 27 # 2395 Zbl 131.245 JSL 33.305 • ID 13552

THOMAS, I. [1962] *Solutions of five modal problems of Sobocinski* (**J** 0047) Notre Dame J Formal Log 3*199-200
◇ B45 ◇ REV MR 33 # 3916 Zbl 126.10 JSL 31.498 • ID 13553

THOMAS, I. [1963] $S1^0$ *and Brouwerian axioms* (**J** 0047) Notre Dame J Formal Log 4*151-152
◇ B45 ◇ REV MR 29 # 4681 Zbl 125.5 JSL 31.498 • ID 13555

THOMAS, I. [1963] $S1^0$ *and generalized S5-axioms* (**J** 0047) Notre Dame J Formal Log 4*153-154
◇ B45 ◇ REV MR 29 # 4682 Zbl 125.5 Zbl 131.245 JSL 31.498 • ID 33508

THOMAS, I. [1963] *A final note on* $S1^0$ *and the Brouwerian axioms* (**J** 0047) Notre Dame J Formal Log 4*231-232
◇ B45 ◇ REV MR 29 # 4683 Zbl 131.245 JSL 31.498 • ID 13556

THOMAS, I. [1964] *Decision procedures for* $S2^2$ *and* T^0 (**J** 0047) Notre Dame J Formal Log 5*319-320
◇ B45 ◇ REV MR 31 # 3326 Zbl 129.258 • ID 33509

THOMAS, I. [1964] *Modal systems in the neighbourhood of T* (**J** 0047) Notre Dame J Formal Log 5*59-61
◇ B45 ◇ REV MR 31 # 1188 Zbl 137.5 JSL 31.499 • ID 13557

THOMAS, I. [1964] *Ten modal models* (**J** 0036) J Symb Logic 29*125-128
◇ B45 ◇ REV MR 31 # 46 Zbl 137.5 JSL 31.499 • ID 13558

THOMAS, I. [1967] *A theorem on S4.2 and S4.4* (**J** 0047) Notre Dame J Formal Log 8*335-336
◇ B45 ◇ REV MR 44 # 6461 Zbl 189.282 JSL 37.182 • ID 13562

THOMAS, I. [1967] *Three-valued propositional fragments with classical implication* (**J** 0047) Notre Dame J Formal Log 8*145-147
◇ B50 ◇ REV MR 38 # 2013 Zbl 189.281 • ID 13564

THOMAS, I. [1968] *Replacement in some modal systems* (**J** 0036) J Symb Logic 33*569-570
◇ B45 ◇ REV MR 39 # 5326 Zbl 181.297 • ID 13565

THOMAS, I. [1968] *The rule of Peirce* (**J** 0047) Notre Dame J Formal Log 9*34
◇ B50 ◇ REV MR 39 # 2601 Zbl 169.296 • ID 13566

THOMAS, I. [1969] see LEMMON, E.J.

THOMAS, I. [1971] *A proof of a theorem of Lukasiewicz* (**J** 0047) Notre Dame J Formal Log 12*507-508
◇ B50 ◇ REV MR 45 # 4942 Zbl 224 # 02008 • ID 13570

THOMAS, I. [1973] *Further extensions of* $S3^*$ (**J** 0047) Notre Dame J Formal Log 14*423-424
◇ B45 ◇ REV MR 51 # 10036 Zbl 258 # 02029 • ID 13572

THOMAS, I. [1973] *Unusual feature of* $S3^*$ (**J** 0047) Notre Dame J Formal Log 14*276
◇ B45 ◇ REV MR 48 # 1886 Zbl 245 # 02027 • ID 13571

THOMAS, I. [1976] *Axiom sets equivalent to syllogism and Peirce* (**J** 0047) Notre Dame J Formal Log 17*248
◇ B50 ◇ REV MR 57 # 9488 Zbl 315 # 02003 • ID 18417

THOMAS, I. see Vol. I, III, V for further entries

THOMAS, N.L. [1966] *Modern logic. An introduction* (**X** 1000) Barnes & Noble: Totowa xvii+236pp
◇ B98 ◇ REV JSL 36.544 • ID 13577

THOMAS, WOLFGANG [1978] see EBBINGHAUS, H.-D.

THOMAS, WOLFGANG see Vol. I, III, IV, V, VI for further entries

THOMASON, M.G. [1972] see BELL, N.

THOMASON, M.G. see Vol. IV for further entries

THOMASON, R.H. [1967] *A decision procedure for Fitch's propositional calculus* (**J** 0047) Notre Dame J Formal Log 8*101-117
◇ B25 B45 ◇ REV MR 38 # 3139 Zbl 183.15 • ID 13579

THOMASON, R.H. [1968] see STALNAKER, R.C.

THOMASON, R.H. [1969] *A semantical study of constructible falsity* (**J** 0068) Z Math Logik Grundlagen Math 15*247-257
◇ B55 ◇ REV MR 41 # 1510 Zbl 181.9 • ID 33195

THOMASON, R.H. [1969] *Modal logic and metaphysics. The logical way of doing things* (**C** 1134) Log Way of Doing Things 119-146
◇ A05 B45 ◇ REV Zbl 188.317 • ID 22089

THOMASON, R.H. [1970] *A Fitch-style formulation of conditional logic* (**J** 0079) Logique & Anal, NS 13*397-412
◇ B45 ◇ REV MR 45 # 3158 Zbl 228 # 02016 • ID 13583

THOMASON, R.H. [1970] *Indeterminist time and truth-value gaps* (**J** 0105) Theoria (Lund) 36*264-281
◇ A05 B45 ◇ REV MR 47 # 19 Zbl 269 # 02007 • ID 13582

THOMASON, R.H. [1970] *Some completeness results for modal predicate calculi* (**P** 0559) Phil Probl in Logic;1968 Irvine 56-76
◇ B45 ◇ REV MR 43 # 4644 Zbl 188.320 • ID 33681

THOMASON, R.H. [1970] *Symbolic logic. An introduction* (**X** 0843) Macmillan : New York & London 367pp
◇ B98 ◇ REV MR 43 # 4627 Zbl 214.6 JSL 36.678 • ID 13581

THOMASON, R.H. [1972] *A semantic theory of sortal incorrectness* (**J** 0122) J Philos Logic 1*209-258
◇ A05 B60 ◇ REV MR 58 # 16127 Zbl 248 # 02029 • ID 65712

THOMASON, R.H. [1973] *Philosophy and formal semantics* (**P** 0783) Truth, Syntax & Modal;1970 Philadelphia 294-307
◇ A05 B45 ◇ REV MR 53 # 5243 Zbl 264 # 02008 JSL 42.317 • ID 22876

THOMASON, R.H. [1975] *Decidability in the logic of conditionals* (**C** 1856) Log Enterprise 167-178
◇ B25 B45 ◇ REV Zbl 374 # 02014 • ID 51538

THOMASON, R.H. [1976] *Necessity, quotation, and truth: An indexical theory* (**C** 1701) Lang in Focus (Bar-Hillel) 119-138
◇ A05 B45 ◇ REV Zbl 397 # 03006 • ID 52672

THOMASON, R.H. [1980] *A note on syntactical treatments of modality* (**J** 0154) Synthese 44*391-395
◇ A05 B45 ◇ REV MR 81g:03021 Zbl 449 # 03005 • ID 56672

THOMASON, R.H. [1981] see GUPTA, A.

THOMASON, R.H. [1981] *Deontic logic as founded on tense logic* (**C** 3731) New Studies Deontic Log 165-176
◇ B45 ◇ REV MR 84i:03044 • ID 34525

THOMASON, R.H. [1981] *Deontic logic and the role of freedom in moral deliberation* (**S** 3307) Synth Libr 152*177-186
◇ B45 ◇ REV MR 83b:03021 • ID 35088

THOMASON, R.H. [1985] *Note on tense and subjunctive conditionals* (**J** 0153) Phil of Sci (East Lansing) 52*151-153
◇ B45 ◇ ID 45540

THOMASON, R.H. see Vol. I, III, VI for further entries

THOMASON, S.K. [1972] *Noncompactness in propositional modal logic* (**J** 0036) J Symb Logic 37*716-720
◇ B45 ◇ REV MR 47 # 4767 Zbl 262 # 02023 • ID 13592

THOMASON, S.K. [1972] *Semantic analysis of tense logics* (**J** 0036) J Symb Logic 37*150-158
◇ B45 ◇ REV MR 47 # 4766 Zbl 238 # 02027 • ID 13593

THOMASON, S.K. [1973] *A new representation of S5* (**J** 0047) Notre Dame J Formal Log 14*281-284
◇ B45 ◇ REV MR 48 # 5825 Zbl 225 # 02016 • ID 13594

THOMASON, S.K. [1974] *An incompleteness theorem in modal logic* (**J** 0105) Theoria (Lund) 40*30-34
◇ B45 ◇ REV MR 58 # 27349 Zbl 287 # 02012 • ID 13595

THOMASON, S.K. [1974] *Reduction of tense logic to modal logic, I* (**J** 0036) J Symb Logic 39*549-551
◇ B45 ◇ REV MR 50 # 12669 Zbl 317 # 02010 • REM Part II 1975 • ID 13596

THOMASON, S.K. [1975] see GOLDBLATT, R.I.

THOMASON, S.K. [1975] *Categories of frames for modal logic* (**J** 0036) J Symb Logic 40*439-442
◇ B45 C90 G30 ◇ REV MR 52 # 2829 Zbl 317 # 02012 JSL 47.440 • ID 29635

THOMASON, S.K. [1975] *Reduction of second-order logic to modal logic* (**J** 0068) Z Math Logik Grundlagen Math 21*107-114
◇ B45 ◇ REV MR 51 # 10037 Zbl 317 # 02011 • ID 29634

THOMASON, S.K. [1975] *Reduction of tense logic to modal logic II* (**J** 0105) Theoria (Lund) 41*154-169
◇ B45 ◇ REV MR 58 # 21484 Zbl 365 # 02010 • REM Part I 1974 • ID 30572

THOMASON, S.K. [1975] *The logical consequence relation of propositional tense logic* (**J** 0068) Z Math Logik Grundlagen Math 21*29-40
◇ B45 D20 D55 ◇ REV MR 51 # 67 Zbl 324 # 02014 • ID 13597

THOMASON, S.K. [1977] *Modal operators and functional completeness II* (**J** 0036) J Symb Logic 42*391-399
◇ B45 ◇ REV MR 81g:03022 Zbl 408 # 03015 • REM Part I unpublished • ID 30573

THOMASON, S.K. [1978] *Possible worlds and many truth values* (**J** 0063) Studia Logica 37*195-204
◇ B45 ◇ REV MR 80a:03030 Zbl 391 # 03014 • ID 52341

THOMASON, S.K. [1980] *Independent propositional modal logics* (**J** 0063) Studia Logica 39*143-144
◇ B45 ◇ REV MR 81m:03026 Zbl 457 # 03017 • ID 54342

THOMASON, S.K. [1982] *Undecidability of the completeness problem of modal logic* (**P** 3831) Universal Algeb & Appl;1978 Warsaw 341-345
◇ B45 ◇ REV MR 85e:03051 Zbl 522 # 03007 • ID 37440

THOMASON, S.K. [1983] *Finite matrices for quasi-classical modal logics* (**J** 0079) Logique & Anal, NS 26*341-344
◇ B45 ◇ REV MR 85j:03023 Zbl 552 # 03015 • ID 43357

THOMASON, S.K. [1984] *On constructing instants from events* (**J** 0122) J Philos Logic 13*85-96
◇ B45 ◇ REV MR 86f:03035 Zbl 556 # 03006 • ID 42642

THOMASON, S.K. [1985] see NAGLE, M.C.

THOMASON, S.K. see Vol. IV, V for further entries

THOMPSON, B. [1982] *Syllogisms using "few", "many", and "most"* (**J** 0047) Notre Dame J Formal Log 23*75-84
◇ B60 C80 ◇ REV MR 84j:03067 Zbl 452 # 03025 • ID 55079

THUM, M. [1984] see KANDEL, A.

TICHY, P. [1973] *On "de dicto" modalities in quantified S5* (**J** 0122) J Philos Logic 2*387-392
◇ B45 ◇ REV MR 55 # 2506 Zbl 268 # 02014 • ID 32496

TICHY, P. [1975] *What do we talk about?* (**J** 0153) Phil of Sci (East Lansing) 42*80-93
◇ A05 B45 ◇ REV MR 58 # 128 • ID 79353

TICHY, P. [1978] *A new theory of subjunctive conditionals* (**J** 0154) Synthese 37*433-457
◇ A05 B45 ◇ REV MR 58 # 16161 • ID 79352

TICHY, P. [1980] *The logic of temporal discourse* (**J** 2130) Linguist Philos 3*343-369
◇ A05 B45 ◇ REV Zbl 433 # 03005 • ID 54011

TICHY, P. [1980] *The transiency of truth* (**J** 0105) Theoria (Lund) 46*165-182
◇ B45 ◇ REV MR 83e:03033 • ID 35218

TICHY, P. [1982] see ODDIE, G.

TICHY, P. [1984] *Subjective conditionals: two parameters vs. three* (**J** 0095) Philos Stud 45*147-179
◇ B45 ◇ REV MR 85i:03060 • ID 44116

TICHY, P. see Vol. I, IV for further entries

TIDEN, E. [1985] see ARNBORG, S.

TIEN, P.S. [1971] *The analysis and synthesis of fuzzy switching circuits* (**P** 2007) Symp Th & Appl of Multi-Val Log Design;1971 Buffalo 165-172
◇ B52 ◇ ID 41999

TINCHEV, T. & VAKARELOV, D. [1985] *Propositional dynamic logics with counters and stacks* (**P** 4670) Comput Th (5);1984 Zaborow 364-374
◇ B75 ◇ ID 49788

TINCHEV, T. [1985] see PASSY, S.

TIOMKIN, M.L. [1985] see MAKOWSKY, J.A.

TIRNOVEANU, M. [1961] *Sur un domaine logique transfini (Romanian) (Russian and French summaries)* (**J** 4188) Com Acad Romine 11*1017-1023
◇ B50 ◇ REV MR 24 # A1824 Zbl 118.12 • ID 42846

TIRNOVEANU, M. [1962] *Sur les extensions des types P et Q de la logique $\mathfrak{L}_{\Omega^s}$ (Romanian) (Russian and French summaries)* (**J** 4188) Com Acad Romine 12*269-273
◇ B50 ◇ REV MR 26 # 1247 Zbl 117.253 • ID 42871

TIRNOVEANU, M. [1963] *Probabilistic methods in nonconstructive logical systems (Romanian) (Russian and French summaries)* (**J** 0197) Stud Cercet Mat Acad Romana 14*485-491
◇ B60 ◇ REV MR 32 # 2319 Zbl 118.12 • ID 42945

TIRNOVEANU, M. [1964] *Ueber die Modalitaet der Werte eines logischen nichtkonstruktiven Systems der Stufe I (rumaenisch)* (**J** 0440) Bul Inst Politeh Bucuresti, Ser Mec 26/6*29-33
◇ B45 ◇ REV MR 35 # 32 • ID 16210

TIRNOVEANU, M. [1966] *Sur quelques proprietes d'un systeme logistique semi-constructif modal I* (**J** 0147) An Univ Bucuresti, Acta Logica 9*187-208
◇ B45 ◇ REV MR 35 # 1452 Zbl 252 # 02014 • ID 13611

TIRNOVEANU, M. [1969] *Sur un systeme logistique deontique modal semi-constructif* (**J** 0147) An Univ Bucuresti, Acta Logica 12*275-293
◇ B45 ◇ REV MR 46 # 1560 Zbl 253 # 02021 • ID 13612

TIRNOVEANU, M. [1970] *Sur quelques notions de proto-logistique semi-constructive* (**J** 0147) An Univ Bucuresti, Acta Logica 13*133-145
◇ B60 F99 ◇ REV Zbl 272 # 02049 • ID 30384

TIRNOVEANU, M. [1971] *On certain properties of the deonic logical system D1 (Romanian)* (**S** 1613) Probl Logic (Bucharest) 4*33-61
◇ B45 ◇ ID 47828

TIRNOVEANU, M. [1971] *Sur quelques proprietes du systeme logistique modal deontique modal transfini L'_D (Romanian) (French summary)* (**S** 1613) Probl Logic (Bucharest) 3*205-235
◇ B45 ◇ REV Zbl 215.321 • ID 28015

TIRNOVEANU, M. [1973] *On some properties of the transfinite assertoric-deontic logistic systems $S^{\alpha 210}_{ad}$, $S^{\alpha 2n0}_{ad}$ (Romanian)* (**S** 1613) Probl Logic (Bucharest) 5*31-90
◇ B45 ◇ REV MR 58 # 27350 • ID 79362

TIRNOVEANU, M. [1975] *On some syntactic properties of transfinite assertoric-deontic-modal logical systems of the form $S^{\alpha 210}_{adm}$, $S^{\alpha 2n0}_{adm}$ (Romanian) (French summary)* (**S** 1613) Probl Logic (Bucharest) 6*35-66
◇ B45 ◇ REV MR 58 # 154 • ID 79363

TIRNOVEANU, M. [1977] *Some metalogical properties of transfinite assertorical, modal, deontic logistic systems of order zero (Romanian) (French summary)* (**S** 1613) Probl Logic (Bucharest) 7*75-94
◇ B45 ◇ REV MR 58 # 27351 • ID 79361

TIRNOVEANU, M. see Vol. I, VI for further entries

TITIEV, R.J. [1980] *On self-sustenance in systems of epistemic logic* (**J** 0047) Notre Dame J Formal Log 21*585-590
◇ A05 B45 ◇ REV MR 81h:03046 Zbl 416 # 03021 • ID 53917

TITIEV, R.J. see Vol. III for further entries

TIURYN, J. [1981] *A survey of the logic of effective definitions* (**P** 3497) Log of Progr;1979 Zuerich 198-245
◇ B60 B75 ◇ REV MR 85b:03044 Zbl 468 # 68037 • ID 55120

TIURYN, J. [1981] *Logic of effective definitions* (**J** 2095) Fund Inform, Ann Soc Math Pol, Ser 4 4*629-659
◇ B60 C55 C75 ◇ REV MR 84c:03068 Zbl 486 # 68017 • ID 34010

TIURYN, J. [1981] *Unbounded program memory adds to the expressive power of first-order dynamic logic* (**P** 4235) IEEE Symp Found of Comp Sci (22);1981 Nashville 335-339
◇ B75 ◇ ID 49868

TIURYN, J. [1982] see MEYER, A.R.

TIURYN, J. [1982] see BERMAN, P.

TIURYN, J. [1983] see MEYER, A.R.

TIURYN, J. [1984] see MEYER, A.R.

TIURYN, J. see Vol. III, IV for further entries

TKACHEV, G.A. [1977] *Complexity of realization of one sequence of functions of k-valued logic (Russian)* (**J** 2869) Vest Ser Vychisl Mat Kibern, Univ Moskva 1977/1*45-57
• TRANSL [1977] (**J** 3221) Moscow Univ Comp Math Cybern 1977/1*36-45
◇ B50 B70 D15 ◇ REV Zbl 447 # 94031 • ID 69960

TKACHEV, G.A. [1980] *On the influence of a basis on the behaviour of the Shanon function (Russian)* (**J** 0071) Met Diskr Analiz (Novosibirsk) 34*88-99
◇ B50 ◇ REV MR 83c:94035 Zbl 483 # 94036 • ID 36816

TODT, G. [1983] *Fuzzy logic and modal logic* (**C** 3095) Approaching Vagueness 213-260
◇ B45 ◇ REV MR 85j:03003 • ID 45371

TODT, G. [1983] see GLUBRECHT, J.-M.

TODT, G. see Vol. V for further entries

TOERNEBOHM, H. [1958] *Notes on modal operators* (**J** 0105) Theoria (Lund) 24*130-135
◇ B45 ◇ REV JSL 25.368 • ID 13625

TOERNEBOHM, H. [1964] *Information and confirmation* (**X** 1163) Almqvist & Wiksell: Stockholm 80pp
◇ A05 B48 B60 ◇ ID 25449

TOERNEBOHM, H. see Vol. I for further entries

TOEROES, R. [1978] see FAY, G.

TOGAI, M. & WANG, P.P. [1983] *A study of fuzzy relations and their inverse problem* (**P** 4054) Int Symp Multi-Val Log (13);1983 Kyoto 279-285
◇ B52 E72 ◇ REV Zbl 559 # 03016 • ID 46900

TOGAI, M. [1985] *A fuzzy inverse relation based on Goedelian logic and its applications* (**J** 2720) Fuzzy Sets Syst 17*211-219
◇ B52 ◇ ID 49427

TOKARZ, M. [1972] *On invariant systems and structural completeness of Lukasiewicz's logics* (J 0387) Bull Sect Logic, Pol Acad Sci 1/1*63-68
◇ B50 ◇ REV MR 50 # 12660 • ID 13630

TOKARZ, M. [1972] *On structural completeness of Lukasiewicz's logics (Polish and Russian summaries)* (J 0063) Studia Logica 30*53-61
◇ B50 ◇ REV MR 47 # 4764 Zbl 288 # 02015 • ID 13631

TOKARZ, M. [1973] *On mutual non-reconstructability of the Lukasiewicz calculi and their dual counterparts* (J 0387) Bull Sect Logic, Pol Acad Sci 2*51-53
◇ B50 ◇ REV MR 52 # 13324 • ID 21785

TOKARZ, M. [1974] *A method of axiomatization of Lukasiewicz logics* (J 0063) Studia Logica 33*333-338
• REPR [1977] (C 3026) Lukasiewicz Sel Pap on Sent Calc 113-117
◇ B50 ◇ REV MR 51 # 7822 MR 58 # 21450 Zbl 298 # 02013 Zbl 381 # 03017 • ID 17306

TOKARZ, M. [1974] *A new proof of "topographic" theorem on Lukasiewicz logics* (J 0302) Rep Math Logic, Krakow & Katowice 3*63-66
◇ B50 ◇ REV MR 51 # 58 Zbl 326 # 02016 • ID 15179

TOKARZ, M. [1974] *Binary functions definable in implicational Goedel algebras* (J 0387) Bull Sect Logic, Pol Acad Sci 3/1*22-24
◇ B55 ◇ REV MR 53 # 2628 • ID 21517

TOKARZ, M. [1974] *Invariant systems of Lukasiewicz calculi* (J 0068) Z Math Logik Grundlagen Math 20*221-228
◇ B50 ◇ REV MR 50 # 12661 MR 51 # 7823 Zbl 298 # 02012 • ID 22041

TOKARZ, M. [1975] *Functions definable in Sugihara algebras and their fragments I* (J 0063) Studia Logica 34*295-304
◇ B50 G25 ◇ REV MR 56 # 97 Zbl 347 # 02041 • REM Part II 1976 • ID 65739

TOKARZ, M. [1976] *Functions definable in Sugihara algebras and their fragments II* (J 0063) Studia Logica 35*279-283
◇ B50 G25 ◇ REV MR 56 # 97 Zbl 361 # 02025 • REM Part I 1975 • ID 49854

TOKARZ, M. [1977] *A remark on maximal matrix consequences* (J 0387) Bull Sect Logic, Pol Acad Sci 6*190-192
◇ B50 ◇ REV MR 58 # 16280 Zbl 404 # 03012 • ID 54799

TOKARZ, M. [1977] *Deduction theorems for RM and its extensions* (J 0387) Bull Sect Logic, Pol Acad Sci 6*67-69
◇ B46 G25 ◇ REV MR 58 # 21421 Zbl 403 # 03022 • ID 54736

TOKARZ, M. [1977] *Degrees of maximality of three- and four-valued RM-extensions* (J 0302) Rep Math Logic, Krakow & Katowice 9*63-69
◇ B46 ◇ REV MR 80m:03105 Zbl 398 # 03010 • ID 52738

TOKARZ, M. [1977] *Degrees of completeness of Lukasiewicz logics* (C 3026) Lukasiewicz Sel Pap on Sent Calc 127-134
◇ B50 ◇ REV MR 58 # 21451 Zbl 381 # 03017 • ID 79386

TOKARZ, M. [1977] see HAWRANEK, J.

TOKARZ, M. [1977] *On structural completeness of Lukasiewicz logics* (C 3026) Lukasiewicz Sel Pap on Sent Calc 171-173
◇ B50 ◇ REV MR 58 # 21452 Zbl 381 # 03017 • ID 79385

TOKARZ, M. [1979] *Deduction theorems for RM and its extensions* (J 0063) Studia Logica 38*105-111
◇ B46 G25 ◇ REV MR 82i:03032 Zbl 406 # 03027 • ID 56110

TOKARZ, M. [1979] *The existence of matrices strongly adequate for E, R and their fragments* (J 0063) Studia Logica 38*75-85
◇ B46 ◇ REV MR 80h:03034 Zbl 411 # 03016 • ID 52870

TOKARZ, M. [1979] *Theorems derivable from the McNaughton criterion (Russian)* (P 2554) All-Union Symp Th Log Infer;1974 Moskva 43-49
◇ B50 ◇ REV MR 84k:03077 • ID 35014

TOKARZ, M. [1980] *Essays in matrix semantics of relevant logics* (X 2733) Acad Sci Inst Phi Soc: Wroclaw iii + 108pp
◇ B46 G25 ◇ REV MR 82h:03016 Zbl 452 # 03048 • ID 54112

TOKARZ, M. see Vol. I, VI for further entries

TOKMEN, V.H. [1979] *Some properties of the spectra of ternary logic functions* (P 3003) Int Symp Multi-Val Log (9);1979 Bath 88-93
◇ B50 ◇ REV MR 80m:94116 • ID 82986

TOKMEN, V.H. [1980] *Disjoint decomposability of multivalued functions by spectral means* (P 3673) Int Symp Multi-Val Log (10);1980 Evanston 88-93
◇ B50 ◇ REV MR 83m:94032 • ID 45945

TOLSTOVA, YU.N. [1967] *Modelling l-valued logic in k-valued logic k>1 (Russian)* (J 0052) Probl Kibern 18*67-82
• ERR/ADD ibid 22*293-295
• TRANSL [1967] (J 0471) Syst Th Res 18*62-76
◇ B50 ◇ REV MR 41 # 41 Zbl 307 # 02010 • ID 16941

TOLSTOVA, YU.N. [1970] *Letter to the editors (Russian)* (J 0052) Probl Kibern 22*293-295
◇ B50 ◇ REV MR 45 # 1733 • ID 79412

TOLSTOVA, YU.N. [1971] *A weakening of intuitionistic logic (Russian) (English summary)* (S 0228) Zap Nauch Sem Leningrad Otd Mat Inst Steklov 20*208-219,288
• TRANSL [1973] (J 1531) J Sov Math 1*132-138
◇ B60 F50 ◇ REV MR 45 # 4951 MR 48 # 5808 Zbl 231 # 02028 • ID 13633

TOMATIS, P. [1972] *Il teorema di eliminazione per una logica modale S4* (J 0220) Atti Accad Sci Torino, Fis Mat Nat 106*787-797
◇ B45 ◇ REV MR 48 # 1887 Zbl 251 # 02026 • ID 13636

TOMBERLIN, J.E. [1971] *The sea battle tomorrow and fatalism* (J 0075) Phil Phenom Research 31*352-357
◇ A05 B45 ◇ REV JSL 40.254 • ID 13637

TOMBERLIN, J.E. [1981] *Contrary-to-duty imperatives and conditional obligation* (J 0097) Nous, Quart J Phil 15*357-375
◇ B45 ◇ REV MR 84a:03024 • ID 35568

TONDL, L. [1974] *Semantic evaluation of prognostic statements and the base of probabilistic parameters* (J 0156) Kybernetika (Prague) 10*199-220
◇ A05 B48 ◇ REV MR 51 # 11909 Zbl 279 # 02005 • ID 29537

TONDL, L. [1974] *Types of preference* (J 0156) Kybernetika (Prague) 10*389-408
◇ B45 ◇ REV MR 52 # 2809 Zbl 286 # 02033 • ID 17611

TONG, R.M. [1978] *Synthesis of fuzzy models for industrial processes - some recent results* (J 1743) Int J Gen Syst 4*143-162
◇ B52 ◇ REV Zbl 372 # 93025 • ID 51460

TONG, R.M. [1980] see EFSTATHIOU, J.

TONG, R.M. [1982] see EFSTATHIOU, J.

TONG, R.M. [1984] see SHAPIRO, D.G.

TONG, ZHENGXIANG [1981] see HUANG, ZHENDE

TONG, ZHENGXIANG [1982] see LOU, SHIBO

TONITSA, V.S. [1976] see RVACHEV, V.L.

TONITSA, V.S. [1979] see RVACHEV, V.L.

TORALDO DI FRANCIA, G.G. [1979] see DALLA CHIARA SCABIA, M.L.

TORALDO DI FRANCIA, G.G. see Vol. I for further entries

TORRANCE, S.B. [1981] *Prescriptivism and incompleteness* (J 0094) Mind 90*580-585
◇ A05 B45 ◇ REV MR 83e:03017 • ID 35207

TORTORA, R. [1981] see GUCCIONE, S.

TORTORA, R. [1982] see GUCCIONE, S.

TORTORA, R. [1983] see GUCCIONE, S.

TORTORA, R. see Vol. I, VI for further entries

TOSIC, R. [1981] *Types of bases for a modification of the algebra of logic (Russian) (Serbo-Croatian summary)* (J 2855) Zbor Rad, Prir-Mat Fak, Ser Mat (Novi Sad) 11*287-295
◇ B60 ◇ REV MR 84e:03077 Zbl 523 # 03007 • ID 34409

TOSIC, R. [1982] see DOROSLOVACKI, R.

TOSIC, R. [1983] see STOJMENOVIC, I.

TOSIC, R. see Vol. I for further entries

TOSIC, Z. [1968] see POSPELOV, D.A.

TOSIC, Z. [1969] see IVAS'KIV, YU.L.

TOSIC, Z. [1970] *Polynomial representations of m-valued logical functions (Russian)* (S 1003) Publ Elektroteh, Ser Mat Fiz, Beograde 302/319*43-48
◇ B50 ◇ REV MR 50 # 1775 Zbl 214.9 • ID 13649

TOSIC, Z. [1972] *Analytical representation of m-valued logical functions over the ring of integers modulo m* (S 1003) Publ Elektroteh, Ser Mat Fiz, Beograde 410/411*1-36
◇ B50 ◇ REV MR 48 # 10699 Zbl 262 # 02017 • ID 13650

TOTH, K. [1979] *Modal logics with function symbols* (J 0380) Acta Cybern (Szeged) 4*291-302
◇ B45 ◇ REV MR 80j:03034 Zbl 426 # 03022 • ID 53618

TOTH, K. [1980] *Completeness in non-simple and stable modal logics* (J 0380) Acta Cybern (Szeged) 4*377-382
◇ B45 ◇ REV MR 82c:03024 Zbl 448 # 03009 • ID 56609

TRAGESSER, R.S. & ZUCKER, J.I. [1978] *The adequacy problem for inferential logic* (J 0122) J Philos Logic 7*501-516
◇ A05 B55 F50 ◇ REV MR 80c:03031 Zbl 408 # 03020 JSL 47.689 • ID 30597

TRAKHTENBROT, B.A. [1979] *Completeness of algorithmic logic (Russian)* (J 0040) Kibernetika, Akad Nauk Ukr SSR 1979/2*6-11
• TRANSL [1979] (J 0021) Cybernetics 15*160-166
◇ B75 ◇ REV MR 81g:68014 Zbl 459 # 68002 • ID 54477

TRAKHTENBROT, B.A. see Vol. I, III, IV, V, VI for further entries

TRAN VAN TOAN [1971] *Le Strichkalkuel de B. von Freytag-Loeringhoff et la recherche des premisses sous-entendues* (J 0286) Int Logic Rev 4*219-241
◇ A05 B60 ◇ REV Zbl 339 # 02004 • ID 65773

TREW, A. [1968] *Incompleteness of a logic of Routley's* (J 0047) Notre Dame J Formal Log 9*385-387
◇ A05 B45 ◇ REV MR 41 # 8218 Zbl 191.290 • ID 13672

TREW, A. [1970] *Nonstandard theories of quantifications and identity* (J 0036) J Symb Logic 35*267-293
◇ B20 B60 ◇ REV MR 46 # 3264 Zbl 208.9 • ID 13673

TRILLAS, E. [1979] *Note on negation functions for fuzzy sets (Spanish)* (P 4097) Jorn Mat Luso-Espanol (6);1979 Santander 935-937
◇ B52 E72 ◇ ID 45484

TRILLAS, E. [1979] *On negation functions in the theory of fuzzy sets (Spanish)* (J 2840) Stochastica, Univ Politec Barcelona 3/1*47-59
◇ B52 E72 ◇ REV MR 81j:03085 Zbl 419 # 03035 • ID 53378

TRILLAS, E. [1980] see RIERA, T.

TRILLAS, E. [1980] see ALSINA, C.

TRILLAS, E. & VALVERDE, L. [1981] *On some functionally expressible implications for fuzzy set theory* (P 3168) Fuzzy Set Theory (3);1981 Linz 173-190
◇ B52 ◇ REV MR 83f:03023 Zbl 498 # 03015 • ID 35300

TRILLAS, E. [1981] see ALSINA, C.

TRILLAS, E. [1981] see DOMINGO, X.

TRILLAS, E. & VALVERDE, L. [1982] *A few remarks on some lattice-type properties of fuzzy connectives* (P 4002) Int Symp Multi-Val Log (12);1982 Paris 228-231
◇ B52 ◇ REV Zbl 546 # 03015 • ID 41152

TRILLAS, E. [1982] *An essay on indistinguishability relations (Catalan)* (P 3870) Congr Catala de Log Mat (1);1982 Barcelona 51-59
◇ B52 ◇ REV MR 84i:03003 Zbl 527 # 03006 • ID 37506

TRILLAS, E. [1982] see RIERA, T.

TRILLAS, E. [1982] see PAWLOWSKY, V.

TRILLAS, E. [1983] see ALSINA, C.

TRILLAS, E. see Vol. I, V for further entries

TRIVISONNO, G. [1982] see BIAGIOLI, C.

TRNKOVA, V. [1980] see REITERMAN, J.

TRNKOVA, V. [1981] see REITERMAN, J.

TRNKOVA, V. [1984] see REITERMAN, J.

TRNKOVA, V. see Vol. III, IV, V for further entries

TROELSTRA, A.S. [1965] *On intermediate propositional logics* (J 0028) Indag Math 27*141-152
◇ B55 ◇ REV MR 30 # 4674 Zbl 143.11 JSL 33.607 • ID 13676

TROELSTRA, A.S. [1981] *On a second order propositional operator in intuitionistic logic* (J 0063) Studia Logica 40*113-139
◇ B55 F35 F50 ◇ REV MR 84a:03015 Zbl 473 # 03022 • ID 55352

TROELSTRA, A.S. see Vol. I, III, IV, V, VI for further entries

TRUINI, P. [1979] see CASSINELLI, G.

TRZESICKI, K. [1984] *Gentzen-style axiomatization of tense logic* (**J** 0387) Bull Sect Logic, Pol Acad Sci 13*75-84
◇ B45 ◇ REV Zbl 551 # 03011 • ID 43890

TSELISHCHEV, V.V. [1977] *Philosophical problems of the semantics of possible worlds (Russian)* (**X** 2642) Nauka: Novosibirsk 191pp
◇ A05 B45 ◇ REV MR 58 # 16087 • ID 71598

TSELISHCHEV, V.V. [1978] *The concept of object in modal logic (Russian)* (**X** 3353) Akad Nauk SSSR Inst Istor Filol & Filos: Novosibirsk 175pp
◇ A05 B45 ◇ REV Zbl 474 # 03001 • ID 55405

TSELISHCHEV, V.V. [1979] *Hintikka's possible worlds and rigid designators* (**P** 1705) Scand Logic Symp (4);1976 Jyvaeskylae 387-399
◇ A05 B45 ◇ REV MR 80h:03007 Zbl 418 # 03005
• ID 53291

TSELISHCHEV, V.V. (ED.) [1982] *Problems of logic and methodology of science (Russian)* (**X** 2642) Nauka: Novosibirsk 336pp
◇ B98 ◇ REV MR 83h:03005 Zbl 505 # 00009 • ID 38203

TSELKOV, V. [1983] *Intuitionistic modal logics contradicting the Rieger-Nishimura logics (Bulgarian summary)* (**P** 3821) Mat & Mat Obrazov (12);1983 Sl"nchev Bryag 133-139
◇ B45 ◇ REV MR 84k:03065 Zbl 537 # 03012 • ID 35002

TSELYKH, A.N. [1981] see KOROVIN, S.YA.

TSERETELI, S.B. [1966] *The concept of dialectical logic* (**J** 2361) Sov Stud Philos 5*15-22
◇ A05 B53 ◇ ID 42891

TSICHRITZIS, D.C. [1969] *Fuzzy computability* (**P** 1129) Princeton Conf Inform Sci & Syst (3);1969 Princeton 157-161
◇ B52 D75 ◇ REV Zbl 286 # 68028 • ID 65798

TSICHRITZIS, D.C. [1973] *Approximate logic* (**P** 2009) Int Symp Multi-Val Log (3);1973 Toronto 205-216
◇ B50 ◇ REV MR 51 # 10031 • ID 17559

TSICHRITZIS, D.C. see Vol. IV for further entries

TSITKIN, A.I. [1971] see AJZENBERG, N.N.

TSITKIN, A.I. [1977] *On admissible rules of intuitionistic propositional logic (Russian)* (**J** 0142) Mat Sb, Akad Nauk SSSR, NS 102(144)*314-323,328
• TRANSL [1977] (**J** 0349) Math of USSR, Sbor 31*279-288
◇ B55 F50 ◇ REV MR 57 # 5689 Zbl 355 # 02016
• ID 50214

TSITKIN, A.I. [1978] *On structurally complete superintuitionistic logics (Russian)* (**J** 0023) Dokl Akad Nauk SSSR 241*40-43
• TRANSL [1978] (**J** 0062) Sov Math, Dokl 19*816-819
◇ B55 G10 ◇ REV MR 80d:03024 Zbl 412 # 03009
• ID 52940

TSITKIN, A.I. see Vol. I, VI for further entries

TSKHADADZE, O.S. [1982] *Equivalence of sequential variants of the calculus SCI (Russian) (English and Georgian summaries)* (**J** 0954) Tr Inst Prikl Mat, Tbilisi 11*114-119
◇ B60 ◇ REV MR 85j:03095 • ID 45331

TSUKAMOTO, Y. [1979] *An approach to fuzzy reasoning method* (**C** 3514) Adv Fuzzy Sets & Appl 137-149
◇ B52 ◇ REV MR 81e:03021 • ID 79492

TSUKAMOTO, Y. [1984] see HATANO, Y.

TSUKIYAMA, T. [1982] see ASAI, K.

TUERKSEN, I.B. & YAO, D.D.W. [1982] *Bounds for fuzzy inference* (**P** 3846) Cybern & Systems Research (6);1982 Wien 729-734
◇ B52 ◇ REV Zbl 547 # 03020 • ID 43207

TUERKSEN, I.B. [1983] *Inference regions for fuzzy propositions* (**C** 3582) Adv Fuzzy Sets, Possibility Th & Appl 137-148
◇ B52 ◇ REV MR 85j:03028 • ID 45834

TUERKSEN, I.B. [1984] *Measurement of fuzziness: an interpretation of the axioms of measurement* (**P** 3081) IFAC Symp Fuzzy Inf, Knowl Repr & Decis. Anal;1983 Marseille 97-102
◇ B52 ◇ ID 48206

TUERKSEN, I.B. & YAO, D.D.W. [1984] *Representations of connectives in fuzzy reasoning: the view through normal forms* (**J** 2338) IEEE Trans Syst Man & Cybern 14*146-151
◇ B52 ◇ REV MR 85b:03034 Zbl 547 # 03021 • ID 40652

TUERKSEN, I.B. see Vol. I, V for further entries

TUOMELA, R. [1966] *Inductive generalization in an ordered universe* (**C** 1107) Aspects Inductive Log 155-174
◇ A05 B48 ◇ REV Zbl 202.299 • ID 27706

TUOMELA, R. [1973] see NIINILUOTO, I.

TUOMELA, R. [1976] *Confirmation, explanation, and the paradoxes of transitivity* (**C** 1702) Local Induction 319-328
◇ A05 B48 ◇ REV Zbl 335 # 02018 • ID 65806

TUOMELA, R. (ED.) [1977] *Dispositions* (**J** 0154) Synthese 34*361-512
◇ A05 B97 ◇ REV MR 55 # 12459 • ID 70131

TUOMELA, R. [1978] see NIINILUOTO, I.

TUOMELA, R. [1981] *Inductive explanation* (**J** 0154) Synthese 48*257-294
◇ A05 B48 ◇ REV MR 82m:03032 Zbl 476 # 03015
• ID 55527

TUOMELA, R. [1982] *Action generation* (**P** 3800) Intens Log: Th & Appl;1979 Moskva 282-301
• TRANSL [1984] (**C** 4366) Modal & Intens Log & Primen Probl Metodol Nauk 341-359
◇ B45 ◇ REV MR 84i:03045 Zbl 538 # 03004 • ID 34526

TUOMELA, R. see Vol. III for further entries

TURNER, R. [1981] *Counterfactuals without possible worlds* (**J** 0122) J Philos Logic 10*453-493
◇ A05 B45 ◇ REV MR 84i:03046 Zbl 479 # 03008 JSL 50.556 • ID 55670

TURQUETTE, A.R. [1945] see ROSSER, J.B.

TURQUETTE, A.R. [1948] see ROSSER, J.B.

TURQUETTE, A.R. [1949] see ROSSER, J.B.

TURQUETTE, A.R. [1951] see ROSSER, J.B.

TURQUETTE, A.R. [1952] see ROSSER, J.B.

TURQUETTE, A.R. [1954] *Many-valued logics and systems of strict implication* (**J** 0101) Phil Rev 63*365-379
◇ A05 B50 ◇ REV JSL 22.328 • ID 13740

TURQUETTE, A.R. [1958] *Simplified axioms for many-valued quantification theory* (**J** 0036) J Symb Logic 23*139-148
◇ B50 ◇ REV MR 21 # 4913 Zbl 85.245 • ID 13741

TURQUETTE, A.R. [1961] *Solution to a problem of Rose and Rosser* (**J** 0053) Proc Amer Math Soc 12*253-255
◇ B50 ◇ REV MR 23 # A3672 Zbl 106.4 JSL 31.664
• ID 13742

TURQUETTE, A.R. [1963] *Independent axioms for infinite-valued logic* (**J** 0036) J Symb Logic 28*217-221
◇ B50 ◇ REV MR 31 # 3327 Zbl 128.12 JSL 31.665
• ID 13746

TURQUETTE, A.R. [1963] *Modality, minimality, and many valuedness* (**J** 0096) Acta Philos Fenn 16*261-276
◇ A05 B45 B50 ◇ REV MR 28 # 2965 Zbl 121.12
JSL 37.753 • ID 13744

TURQUETTE, A.R. [1966] see FISCH, M.H.

TURQUETTE, A.R. [1967] see ROSSER, J.B.

TURQUETTE, A.R. [1967] *Peirce's phi and psi operators for triadic logic* (**J** 0327) Trans Pierce Soc 3*66-73
◇ A05 A10 B50 ◇ REV MR 39 # 2619 • ID 13748

TURQUETTE, A.R. [1968] *Dualizable quasi-strokes for m-state automata* (**J** 0191) Inform Sci 1*131-142
◇ B50 D05 ◇ REV MR 42 # 46 • ID 13749

TURQUETTE, A.R. [1968] *The Pascal triangle of Post sets for Lukasiewicz's many-valued logics* (**P** 1571) Int Congr Philos (14);1968 Wien 3*102-108
◇ B50 ◇ ID 33695

TURQUETTE, A.R. [1969] *Generalizable Kleene logics* (**J** 0053) Proc Amer Math Soc 20*361-367
◇ B50 ◇ REV MR 39 # 1306 Zbl 182.315 • ID 13750

TURQUETTE, A.R. [1969] *Peirce's complete systems of triadic logic* (**J** 0327) Trans Pierce Soc 5*199-210
◇ A05 A10 B50 ◇ ID 14980

TURQUETTE, A.R. [1970] *Pascal triangles of Post sets* (**P** 1158) Int Congr Math (II,11,Comm Ind);1970 Nice 6
◇ A10 B50 ◇ ID 30569

TURQUETTE, A.R. [1972] *Generalized modal sets* (**J** 0068) Z Math Logik Grundlagen Math 18*261-266
◇ A10 B45 ◇ REV MR 46 # 8814 Zbl 245 # 02023
• ID 13751

TURQUETTE, A.R. [1976] *Minimal axioms for Peirce's triadic logic* (**J** 0068) Z Math Logik Grundlagen Math 22*169-176
◇ A10 B50 ◇ REV MR 56 # 5220 Zbl 331 # 02005
• ID 18479

TURQUETTE, A.R. [1978] *Alternative axioms for Peirce's triadic logic* (**J** 0068) Z Math Logik Grundlagen Math 24*443-444
◇ B50 ◇ REV MR 80b:03027 Zbl 387 # 03008 • ID 52224

TURQUETTE, A.R. [1981] *Quantification for Peirce's preferred system of triadic logic* (**J** 0063) Studia Logica 40*373-382
◇ B50 ◇ REV MR 84f:03023 Zbl 492 # 03007 • ID 34440

TURQUETTE, A.R. see Vol. I, IV for further entries

TUSCHIK, H.-P. [1980] see BAUDISCH, A.

TUSCHIK, H.-P. see Vol. III, IV, V for further entries

TUTUGAN, F. [1956] *On the existence of certain valid syllogistic modi other than those of classical logic (Romanian)* (**S** 1613) Probl Logic (Bucharest) 1956*315-362
◇ B60 ◇ ID 47813

TVERDOKHLEBOVA, N.N. [1974] *A number-theoretic method for the recognition of functions of k-valued logic (Russian)* (**J** 3118) Avtom Mezhvuzov Nauch Sb 1*64-77
◇ B50 ◇ REV MR 58 # 16141 Zbl 309 # 02015 • ID 65820

TVERDOKHLEBOVA, N.N. see Vol. I for further entries

TYAGNIBEDINA, O.S. [1980] see PYATNITSYN, B.N.

TYMIENIECKA, A.-T. (ED.) [1965] *Contributions to logic and methodology, in honor of I.M.Bochenski* (**X** 0809) North Holland: Amsterdam xviii + 326pp
◇ A05 B97 ◇ REV MR 47 # 6424 Zbl 135.242 • ID 25257

UCHII, S. [1973] *Inductive logic with causal modalities: a deterministic approach* (**J** 0154) Synthese 26*264-303
◇ A05 B48 ◇ REV Zbl 275 # 02029 • ID 29698

UCHII, S. [1974] *"Ought" and conditionals* (**J** 0079) Logique & Anal, NS 17*143-164
◇ B45 ◇ REV MR 50 # 9541 Zbl 307 # 02015 • ID 13758

UGAROV, B.N. [1974] see BOBROV, A.E.

UGOL'NIKOV, A.B. [1983] *Realization of functions from closed classes by schemes of functional elements in a complete basis* (**J** 0023) Dokl Akad Nauk SSSR 271*49-51
• TRANSL [1983] (**J** 0062) Sov Math, Dokl 28*45-46
◇ B50 ◇ REV MR 85b:03112 Zbl 562 # 94017 • ID 40757

UGOL'NIKOV, A.B. see Vol. I for further entries

ULAM, S.M. [1968] see KAC, M.

ULAM, S.M. see Vol. V for further entries

ULLIAN, J.S. [1970] see QUINE, W.V.O.

ULLIAN, J.S. see Vol. IV for further entries

ULLRICH, D. [1977] see BYRD, M.

ULRICH, D. [1975] *Finitely-many-valued logics with infinitely-many-valued extensions: Two examples* (**P** 1805) Int Symp Multi-Val Log (5,Proc);1975 Bloomington 406-411
◇ B50 ◇ REV MR 58 # 10309 • ID 22107

ULRICH, D. [1976] *Generalization of a result of Pahi's* (**J** 0068) Z Math Logik Grundlagen Math 22*437-438
◇ B45 ◇ REV MR 56 # 95 Zbl 361 # 02031 • ID 23683

ULRICH, D. [1976] *On a modal system of R.A.Bull's* (**J** 0047) Notre Dame J Formal Log 17*479-480
◇ B45 ◇ REV Zbl 315 # 02030 • ID 18420

ULRICH, D. [1978] *Semantics for S4.1.2* (**J** 0047) Notre Dame J Formal Log 19*461-464
◇ B45 ◇ REV MR 81e:03017 Zbl 332 # 02029 • ID 51486

ULRICH, D. [1981] *RMLC: Solution to a problem left open by Lemmon* (**J** 0047) Notre Dame J Formal Log 22*187-189
◇ B45 ◇ REV MR 82j:03024 Zbl 438 # 03030 • ID 54088

ULRICH, D. [1981] *Strict implication in a sequence of extensions of S4* (**J** 0068) Z Math Logik Grundlagen Math 27*201-212
◇ B45 ◇ REV MR 82h:03017 Zbl 465 # 03008 • ID 79533

ULRICH, D. [1983] *Models of three-valued calculi in implicational S5* (**J** 0387) Bull Sect Logic, Pol Acad Sci 12*73-75
◇ B50 ◇ REV MR 84k:03078 Zbl 568 # 03012 • ID 35015

ULRICH, D. [1984] *Answer to a question suggested by Schumm* (**J** 0068) Z Math Logik Grundlagen Math 30*385-387
◇ B45 ◇ REV Zbl 553 # 03009 • ID 42270

ULRICH, D. [1985] *A descending chain of incomplete extensions of implicational S5* (**J** 0068) Z Math Logik Grundlagen Math 31*201-208
◇ B45 ◇ ID 47567

ULRICH, D. see Vol. I, III, VI for further entries

UMEZAWA, T. [1955] *Ueber die Zwischensysteme der Aussagenlogik* (**J** 0111) Nagoya Math J 9*181-189
◇ B55 ◇ REV MR 17.446 Zbl 66.11 JSL 21.324 • ID 13779

UMEZAWA, T. [1959] *On intermediate many-valued logics* (**J** 0090) J Math Soc Japan 11*116-128
◇ B55 ◇ REV MR 22 # 6702 Zbl 86.245 JSL 24.250 • ID 13781

UMEZAWA, T. [1959] *On intermediate propositional logics* (**J** 0036) J Symb Logic 24*20-36
◇ B55 ◇ REV MR 22 # 4634 Zbl 113.242 JSL 25.180 • ID 13782

UMEZAWA, T. [1959] *On logics intermediate between intuitionistic and classical predicate logic* (**J** 0036) J Symb Logic 24*141-153
◇ B55 ◇ REV MR 22 # 6704 JSL 33.607 • ID 13784

UMEZAWA, T. [1959] *On some properties of intermediate logics* (**J** 0081) Proc Japan Acad 35*575-577
◇ B55 ◇ REV MR 22 # 6703 Zbl 113.243 JSL 25.297 • ID 13780

UMEZAWA, T. [1960] *On an application of intermediate logics* (**J** 0111) Nagoya Math J 16*119-133
◇ B55 ◇ REV MR 22 # 6705 Zbl 143.11 JSL 25.298 • ID 13783

UMEZAWA, T. [1980] *Cut elimination in intuitionistic and some intermediate predicate logics* (**J** 1005) Rep Fac Sci, Shizuoka Univ 14*1-12
◇ B55 F05 F50 ◇ REV MR 81h:03108 Zbl 442 # 03038 • ID 56395

UMEZAWA, T. see Vol. I, III, IV, V, VI for further entries

UMEZU, M. [1979] see HIROSE, K.

UNTERHOLZNER, P. [1981] *Algebraic and relational semantics for tense logics* (**J** 0144) Rend Sem Mat Univ Padova 65*119-128
◇ B45 ◇ REV MR 84m:03029 Zbl 482 # 03006 • ID 35733

URCHS, M.P. [1981] *Kripke-style semantics for Jaskowski's system Q_f* (**J** 0387) Bull Sect Logic, Pol Acad Sci 10*24-29
◇ B46 B53 ◇ REV MR 82g:03030 Zbl 457 # 03025 • ID 54350

URCHS, M.P. [1981] *Semantische Typen fuer klassische Modallogiken* (**J** 0424) Wiss Z Univ Halle-Wittenberg, Math-Nat Reihe 30/1*71-80
◇ B45 ◇ REV MR 82m:03021 Zbl 476 # 03027 • ID 55539

URCHS, M.P. [1984] *Eine Familie von Kausalsystemen* (**P** 3621) Frege Konferenz (2);1984 Schwerin 205-210
◇ B45 ◇ REV MR 85m:03006 Zbl 564 # 03024 • ID 42920

URCHS, M.P. [1984] *Kausallogik und Kripkesemantik* (**C** 4026) Unters Log & Methodol, Vol 1 42-64
◇ B45 ◇ REV MR 85e:03063 Zbl 564 # 03025 • ID 40678

URQUHART, A.I.F. [1971] *Completeness of weak implication* (**J** 0105) Theoria (Lund) 37*274-282
◇ B46 ◇ REV MR 57 # 72 Zbl 238 # 02018 • ID 27899

URQUHART, A.I.F. [1971] see RESCHER, N.

URQUHART, A.I.F. [1972] *Semantics for relevant logics* (**J** 0036) J Symb Logic 37*159-169
◇ B46 ◇ REV MR 48 # 1888 Zbl 245 # 02028 • ID 13793

URQUHART, A.I.F. [1973] *A semantical theory of analytic implication* (**J** 0122) J Philos Logic 2*212-219
◇ B46 ◇ REV MR 56 # 5227 Zbl 266 # 02015 • ID 65844

URQUHART, A.I.F. [1973] *An interpretation of many-valued logic* (**J** 0068) Z Math Logik Grundlagen Math 19*111-114
◇ B50 ◇ REV MR 47 # 1576 Zbl 301 # 02019 • ID 13796

URQUHART, A.I.F. [1983] *Relevant implication and projective geometry* (**J** 0079) Logique & Anal, NS 26*345-357
◇ B46 ◇ REV MR 86c:03017 Zbl 557 # 03012 • ID 40050

URQUHART, A.I.F. [1984] *The undecidability of entailment and relevant implication* (**J** 0036) J Symb Logic 49*1059-1073
◇ B46 D35 G10 ◇ ID 40056

URQUHART, A.I.F. see Vol. I, III, IV, VI for further entries

URSINI, A. [1979] *A modal calculus analogous to K4W, based on intuitionistic propositional logic, I^0* (**J** 0063) Studia Logica 38*297-311
◇ B45 G25 ◇ REV MR 81d:03022 Zbl 423 # 03014 • ID 53526

URSINI, A. [1979] *Two remarks on models in modal logics (Italian summary)* (**J** 3285) Boll Unione Mat Ital, V Ser, A 16*124-127
◇ B45 C90 ◇ REV MR 81b:03025 Zbl 395 # 03017 • ID 52568

URSINI, A. [1985] *Decision problems for classes of diagonalizable algebras* (**J** 0063) Studia Logica 44*87-90
◇ B25 B45 C05 D35 ◇ ID 47525

URSINI, A. see Vol. I, III, VI for further entries

URZYCZYN, P. [1983] *Deterministic context-free dynamic logic is more expressive than deterministic dynamic logic of regular programs* (**P** 3864) FCT'83 Found of Comput Th;1983 Borgholm 496-504
◇ B75 ◇ REV MR 85g:03047 Zbl 539 # 68017 • ID 41253

URZYCZYN, P. see Vol. III, IV for further entries

USBERTI, G. [1983] see MIGLIOLI, P.A.

USBERTI, G. [1984] see MIGLIOLI, P.A.

USBERTI, G. see Vol. VI for further entries

USHENKO, A.P. [1936] *The theory of logic* (**X** 0837) Harper & Row: New York xii + 197pp
◇ B98 ◇ REV JSL 1.113 FdM 62.1046 • ID 22603

USHENKO, A.P. [1941] *The problems of logic* (**X** 0857) Princeton Univ Pr: Princeton 225pp
◇ A05 B98 ◇ REV JSL 6.166 • ID 13802

USHENKO, A.P. see Vol. I for further entries

UZIEMBLO, A.O. & ZURAWSKA, L. [1978] *Definability of functions on a finite set by means of compositions of addition and multiplication modulo some prime numbers* (**J** 1008) Demonstr Math (Warsaw) 11*1029-1032
◇ B50 ◇ REV MR 80i:10010 Zbl 415 # 10007 • ID 53159

VACCARINO, G. [1953] *La logiche polivalenti e non aristoteliche* (**J** 1515) Archimede 5*226-231
◇ B50 ◇ REV MR 15.386 Zbl 51.245 • ID 47889

VACCARINO, G. see Vol. I, V for further entries

VACCARO, V. [1981] see GUCCIONE, S.

VACCARO, V. [1984] see GERLA, G.

VAGIN, V.N. [1977] see PAPKE, W.

VAIDYANATHASWAMY, R. [1938] *Quasi-boolean algebras and many-valued logics* (J 0545) Proc Indian Acad Sci, Sect A 8*165-170
◇ B50 ◇ REV Zbl 19.386 JSL 4.27 FdM 64.30 • ID 13806

VAIDYANATHASWAMY, R. see Vol. I, VI for further entries

VAKARELOV, D. [1965] *A propositional calculus with functors for "likelyhood" and "doubt". I* (J 0255) God Fak Mat & Mekh, Univ Sofiya 60*83-103
◇ B45 ◇ REV MR 36 # 3629 Zbl 155.338 • REM Part II 1966 • ID 13810

VAKARELOV, D. [1966] *A propositional calculus with functors for "likelyhood" and "doubt". II* (J 0255) God Fak Mat & Mekh, Univ Sofiya 61*47-70
◇ B45 ◇ REV MR 39 # 42 Zbl 207.5 • REM Part I 1965 • ID 13811

VAKARELOV, D. [1970] *The concept of dogma and some systems of many-valued logic (Russian) (English summery)* (J 1667) God Filos-Istor Fak, Univ Sofiya 64*249-264
◇ A05 B50 ◇ REV Zbl 247 # 02019 • ID 29500

VAKARELOV, D. [1972] *Notes on the semantics of three-valued Lukasiewicz logic (Russian)* (J 0137) C R Acad Bulgar Sci 25*1467-1469
◇ B50 G20 ◇ REV MR 48 # 54 Zbl 247 # 02018 Zbl 347 # 02008 • ID 00679

VAKARELOV, D. [1974] see MAKSIMOVA, L.L.

VAKARELOV, D. [1977] *Lattices related to Post algebras and their applications to some logical systems* (J 0063) Studia Logica 36*89-107
◇ B50 G10 G15 G20 ◇ REV MR 57 # 16066 Zbl 387 # 03026 • ID 31289

VAKARELOV, D. [1977] *Notes on $\mathcal{N}$-lattices and constructive logic with strong negation* (J 0063) Studia Logica 36*109-125
◇ B55 F50 G10 ◇ REV MR 57 # 12216 Zbl 385 # 03055 • ID 31280

VAKARELOV, D. [1980] see MIRCHEVA, M.

VAKARELOV, D. [1980] *Simple examples of incomplete logics* (J 0137) C R Acad Bulgar Sci 33*587-589
◇ B45 ◇ REV MR 81j:03036 Zbl 435 # 03013 • ID 55774

VAKARELOV, D. [1981] *Intuitionistic modal logics incompatible with the law of the excluded middle* (J 0063) Studia Logica 40*103-111
◇ B45 ◇ REV MR 83k:03029 Zbl 469 # 03009 • ID 55137

VAKARELOV, D. [1983] *Filtration theorem for dynamic algebras with tests and inverse operator* (P 3830) Logics of Progr & Appl;1980 Poznan 314-324
◇ B75 G25 ◇ REV MR 84e:68016 Zbl 534 # 68020 • ID 38356

VAKARELOV, D. [1984] *An application of the Rieger-Nishimura formulas to the intuitionistic modal logics* (J 0387) Bull Sect Logic, Pol Acad Sci 13*120-124
◇ B45 ◇ REV Zbl 568 # 03011 • ID 44730

VAKARELOV, D. [1985] *An application of Rieger-Nishimura formulas to the intuitionistic modal logics* (J 0063) Studia Logica 44*79-86
◇ B55 ◇ ID 47524

VAKARELOV, D. [1985] see TINCHEV, T.

VAKARELOV, D. see Vol. I, III, VI for further entries

VALENTINI, S. [1980] see SAMBIN, G.

VALENTINI, S. [1982] *Cut elimination in a modal sequent calculus for K (Italian summary)* (J 2100) Boll Unione Mat Ital, VI Ser, B 1*119-130
◇ B45 ◇ REV MR 84f:03019 Zbl 495 # 03012 • ID 34435

VALENTINI, S. [1982] see SAMBIN, G.

VALENTINI, S. [1983] *A "canonical" model for GL (Italian summary)* (J 2099) Boll Unione Mat Ital, VI Ser, A 2*361-368
◇ B45 F30 ◇ REV MR 85a:03027 Zbl 525 # 03006 • ID 34778

VALENTINI, S. [1983] see SOLITRO, U.

VALENTINI, S. [1983] *The modal logic of provability: cut-elimination* (J 0122) J Philos Logic 12*471-476
◇ B45 F05 F30 ◇ REV MR 86a:03019 Zbl 535 # 03031 • ID 38327

VALIEV, M.K. [1980] *Decision complexity of variants of propositional dynamic logic* (P 3210) Math Founds of Comput Sci (9);1980 Rydzyna 656-664
◇ B25 B75 D15 ◇ REV MR 81k:68003 Zbl 451 # 03003 • ID 54018

VALIEV, M.K. [1981] *The symmetrized von Wright temporal logic (Russian)* (C 3747) Mat Log & Mat Lingvistika 23-31
◇ B45 ◇ REV MR 84d:03026 • ID 34064

VALIEV, M.K. see Vol. III, IV for further entries

VALVERDE, L. [1980] see ALSINA, C.

VALVERDE, L. [1981] see TRILLAS, E.

VALVERDE, L. [1981] see DOMINGO, X.

VALVERDE, L. [1982] see TRILLAS, E.

VALVERDE, L. [1982] *Notes on a study of the "complements" in fuzzy logic (Catalan) (English summary)* (P 3870) Congr Catala de Log Mat (1);1982 Barcelona 125-128
◇ B52 ◇ REV MR 84i:03003 Zbl 533 # 03010 • ID 36535

VALVERDE, L. [1983] see ALSINA, C.

VALVERDE, L. see Vol. V for further entries

VANDAMME, F. [1972] *On negation* (J 0079) Logique & Anal, NS 15*39-101
◇ B45 ◇ REV MR 48 # 3692 • ID 32048

VANDAMME, R.W. [1977] *Productive logic and propositional paradoxes* (J 0286) Int Logic Rev 15*7-8
◇ A05 B60 ◇ REV Zbl 392 # 03024 • ID 52391

VANDERVEKEN, D.R. [1973] *La completude et la compacite de la pragmatique* (J 0079) Logique & Anal, NS 16*451-494
◇ B45 ◇ REV MR 50 # 1847 Zbl 292 # 02042 • ID 13822

VANDERVEKEN, D.R. [1975] *An extension of Lesniewski-Curry's formal theory of syntactical categories adequate for the categorially open functors* (J 0387) Bull Sect Logic, Pol Acad Sci 4*78-79
◇ B40 B60 ◇ REV MR 53 # 7722 • ID 22976

VANDERVEKEN, D.R. [1976] *A formal definition of the set of the logical connectors of pragmatics* (J 0068) Z Math Logik Grundlagen Math 22*513-516
◇ A05 B45 ◇ REV MR 58 # 5071 Zbl 352 # 02009
• ID 50008

VANDERVEKEN, D.R. [1981] *A strong completeness theorem for pragmatics* (J 0068) Z Math Logik Grundlagen Math 27*151-160
◇ A05 B45 ◇ REV MR 82h:03018 Zbl 469 # 03018
• ID 55146

VANDERVEKEN, D.R. see Vol. I, IV for further entries

VARDANYAN, V.A. [1979] *A formalization of the logic of decision making (Russian)* (S 2582) Semiotika & Inf, Akad Nauk SSSR 11*66-80
◇ B45 ◇ REV MR 80m:03047 • ID 79663

VARDANYAN, V.A. [1981] *The wise men:logic of knowledge and actions (Russian)* (C 3747) Mat Log & Mat Lingvistika 31-37
◇ B45 ◇ REV MR 84e:03026 • ID 34365

VARDI, M.Y. [1985] see SISTLA, A.P.

VARDI, M.Y. see Vol. IV for further entries

VARELA GARCIA, F.J. [1979] *The extended calculus of indications interpreted as a three-valued logic* (J 0047) Notre Dame J Formal Log 20*141-146
◇ A05 B50 ◇ REV MR 80g:03021 Zbl 309 # 02019
Zbl 466 # 03004 • ID 54955

VARELA GARCIA, F.J. see Vol. V, VI for further entries

VARLET, J.C. [1968] *Algebres de Lukasiewicz trivalentes* (J 0408) Bull Soc R Sci Liege 36*399-408
◇ B50 ◇ REV MR 38 # 5676 Zbl 175.266 • ID 15036

VAROSYAN, S.O. [1982] see POSPELOV, D.A.

VAROSYAN, S.O. [1983] *Space logic for robots (Russian) (English summary)* (J 3932) Comput Artif Intell (Bratislava) 2*119-126
◇ B60 ◇ REV Zbl 508 # 03014 • ID 36937

VARPAKHOVSKIJ, F.L. [1965] *The nonrealizability of a disjunction of nonrealizable formulas of propositional logic (Russian)* (J 0023) Dokl Akad Nauk SSSR 161*1257-1258
• TRANSL [1965] (J 0062) Sov Math, Dokl 6*568-570
◇ B55 F50 ◇ REV MR 31 # 2132 Zbl 236 # 02026
JSL 32.396 • ID 27855

VARPAKHOVSKIJ, F.L. see Vol. VI for further entries

VARSHAVSKIJ, V.I. [1969] *A majority operation in a multi-valued logic (Russian)* (J 0040) Kibernetika, Akad Nauk Ukr SSR 1969/2*64-67
• TRANSL [1969] (J 0021) Cybernetics 5*198-202
◇ B50 ◇ REV MR 46 # 1480 Zbl 211.12 JSL 35.349
• ID 13830

VAS, Z. [1983] see DOMBI, J.

VASHCHENKO, V.P. [1980] see LYASHENKO, L.N.

VASHCHENKO, V.P. see Vol. I, V for further entries

VASILACHE, S. [1977] *Ensembles, structures, categories, faisceaux* (X 2776) Pr Univ Laval: Ste-Foy xii + 315pp
• LAST ED [1977] (X 1752) Masson: Paris xii + 315pp
◇ B98 E98 G30 ◇ REV MR 58 # 5445 Zbl 375 # 18001
• ID 51636

VASILACHE, S. see Vol. V for further entries

VASILENKO, YU.A. [1972] *Multivalued structures (Russian)* (X 1218) Uzhgorod Gos Univ: Uzhgorod 148pp
◇ B50 ◇ REV MR 50 # 9431 • ID 20997

VASILENKO, YU.A. [1976] see AVRAMENKO, M.B.

VASIL'EV, N.A. [1912] *Imaginary (non-Aristotelian) logic (Russian)* (J 1395) Minist Narod Prosv Zh, NS (Petrograd) 40*207-246
◇ B60 ◇ ID 42893

VASIL'EV, N.A. see Vol. I for further entries

VASSAILS, G. [1966] *Elements de formalisation d'une logique dialectique* (J 1471) Ann Univ Madagascar, Ser Sci Nat Math 4*3-16
◇ B53 ◇ REV MR 54 # 50 • ID 23933

VASSAILS, G. [1967] *Les loqiques metriques divalentes* (J 1471) Ann Univ Madagascar, Ser Sci Nat Math 5*21-34
◇ B53 ◇ REV MR 54 # 52 • ID 23935

VASSAILS, G. [1967] *Note sur l'antinomie* (J 1471) Ann Univ Madagascar, Ser Sci Nat Math 5*39-40
◇ A05 B53 ◇ REV MR 54 # 53 • ID 23936

VASSAILS, G. see Vol. V for further entries

VAUGHT, R.L. [1974] see ADDISON, J.W.

VAUGHT, R.L. see Vol. I, III, IV, V for further entries

VEATCH, H.B. [1952] *Intentional logic: a logic based on philosophical realism* (X 0875) Yale Univ Pr: New Haven xxi + 440pp
◇ A05 B45 ◇ REV JSL 18.336 • ID 22369

VEATCH, H.B. [1968] *Intentional logic* (C 0552) Phil Contemp - Chroniques 41-43
◇ B45 ◇ ID 14650

VELOSO, P.A.S. [1980] see CARVALHO DE, R.L.

VELOSO, P.A.S. see Vol. IV for further entries

VELTMAN, F. [1984] see LANDMAN, F.

VENDLER, Z. [1968] *Adjectives and nominalizations* (X 0873) Mouton: Paris 134pp
◇ A05 B48 ◇ ID 25365

VENKATESH, G. [1985] *A decision method for temporal logic based on resolution* (P 4672) Found of Softw Tech & Th Comput Sci (5);1985 New Delhi 272-289
◇ B35 B45 ◇ ID 49739

VENN, J. [1881] *Symbolic logic* (X 1253) Crowell Collier & Macmillan: New York xxxix + 446pp
• LAST ED [1971] (X 0848) Chelsea: New York xxxviii + 540pp
◇ A10 B20 B98 ◇ REV MR 52 # 13290 Zbl 263 # 01030
JSL 37.614 • ID 21755

VERDU I SOLANS, V. [1979] see FONT I LLOVET, J.M.

VERDU I SOLANS, V. see Vol. I, III, V, VI for further entries

VERKHOZIN, O.M. [1972] see OTEPANOV, V.I.

VERLOREN VAN THEMAAT, W.A. [1975] *The confirmation of sentences by instances with different truth-values of its atoms* (J 0047) Notre Dame J Formal Log 16*421-424
◇ A05 B48 ◇ REV Zbl 272 # 02018 • ID 65694

VESSEL', KH.A. [1970] *Topological logic (Russian)* (C 0668) Neklass Log 238-261
◇ B50 ◇ REV MR 49 # 7118 • ID 13938

VESSEL', KH.A. see Vol. VI for further entries

VETTER, H. [1967] *Wahrscheinlichkeit und logischer Spielraum* (**X** 1309) Mohr: Tuebingen viii+112pp
◇ A05 B48 ◇ ID 25450

VETTER, H. [1971] *Deontic logic without deontic operators* (**J** 0472) Theory Decis 2*67-78
◇ B45 ◇ REV MR 46 # 5113 Zbl 337 # 02016 • ID 13940

VETTER, H. [1971] *Inductivism and falsificationism reconcilable* (**J** 0154) Synthese 23*226-233
◇ A05 B48 ◇ REV Zbl 227 # 02003 • ID 27352

VICKERS, J.M. [1970] *Probability and non-standard logics* (**P** 0559) Phil Probl in Logic;1968 Irvine 102-120
◇ B48 ◇ REV MR 43 # 6958 Zbl 224 # 02013 • ID 13941

VICKERS, J.M. [1976] *Belief and probability* (**S** 3307) Synth Libr 104*viii+202pp
◇ A05 B48 ◇ REV Zbl 366 # 02006 • ID 51089

VIERU, S. [1969] *Sur les systemes de syllogistique apodictique* (**J** 0147) An Univ Bucuresti, Acta Logica 12*215-222
◇ A10 B45 ◇ REV MR 45 # 6556 Zbl 247 # 02008 • ID 13942

VIERU, S. [1975] *Deontic logic, and the theory of modalities (Romanian) (French summary)* (**S** 1613) Probl Logic (Bucharest) 6*109-133
◇ B45 ◇ REV MR 57 # 5688 • ID 79713

VILENKIN, N.YA. [1977] see SHREJDER, YU.A.

VILENKIN, N.YA. see Vol. V for further entries

VILFAN, B. [1976] *Lower bounds for the size of expressions for certain functions in d-ary logic* (**J** 1426) Theor Comput Sci 2*249-269
◇ B50 ◇ REV MR 54 # 4852 Zbl 337 # 68030 • ID 24080

VILLA, G. [1970] *Calcolo proposizionale trivalente dialettico* (**J** 0012) Boll Unione Mat Ital, IV Ser 3*1026-1041
◇ A05 B50 ◇ REV MR 43 # 3093 Zbl 206.6 • ID 13943

VILLA, G. [1981] *Sul rapporto degli operatori deontici fra loro e sull'estensione alla logica deontica del teorema dello pseudo scoto* (**J** 3068) Atti Accad Sci Bologna Fis Ser 13 8/1-2*197-205
◇ B45 ◇ REV MR 84h:03049 Zbl 538 # 03005 • ID 37398

VILLA, G. see Vol. I for further entries

VINCENT, R.H. [1975] *Selective confirmation and the ravens* (**J** 0488) Dialogue (Ottawa) 14*3-49
◇ A05 B48 ◇ REV MR 58 # 5003 • ID 79718

VINCENZI, A. [1985] *Alcune proprieta della logica di Chang* (**P** 4646) Atti Incontri Log Mat (2);1983/84 Siena 197-200
◇ B45 C40 ◇ ID 49797

VINCENZI, A. [1985] *Effective logic II. Expressive power (Italian)* (**P** 4646) Atti Incontri Log Mat (2);1983/84 Siena 519-523
◇ B60 ◇ ID 49743

VINCENZI, A. [1985] *Logica effettiva L_e* (**P** 4646) Atti Incontri Log Mat (2);1983/84 Siena 387-390
◇ B60 ◇ ID 49741

VINCENZI, A. see Vol. IV for further entries

VISHIN, V.V. [1963] *Identical transformations in four-valued logic (Russian)* (**J** 0023) Dokl Akad Nauk SSSR 150*719-721
• TRANSL [1963] (**J** 0062) Sov Math, Dokl 4*724-726
◇ B50 ◇ REV MR 34 # 1176 Zbl 147.250 JSL 37.762 • ID 13951

VISHNEVSKIJ, A.P. [1973] *Decomposition and minimization of multilogical functions in a single submultilogic (Russian)* (**J** 0474) Avtom Vychis Tekh, Akad Nauk Latv SSR 1973/2*8-13
• TRANSL [1973] (**J** 2666) Autom Control Comput Sci 7/2*8-14
◇ B50 ◇ REV MR 48 # 3635 Zbl 259 # 94036 • ID 13952

VISSER, A. [1981] *A propositional logic with explicit fixed points* (**J** 0063) Studia Logica 40*155-175
◇ A05 B45 ◇ REV MR 83h:03039 Zbl 469 # 03012 • ID 55140

VISSER, A. [1984] *Four valued semantics and the Liar* (**J** 0122) J Philos Logic 13*181-212
◇ B50 ◇ REV MR 86f:03042 Zbl 546 # 03007 • ID 41141

VISSER, A. [1984] *The provability logic of recursively enumerable theories extending Peano arithmetic and arbitrary theories extending Peano arithmetic* (**J** 0122) J Philos Logic 13*97-113
◇ B45 C62 F30 ◇ REV MR 85j:03102 • ID 42643

VISSER, A. see Vol. IV, VI for further entries

VITEK, M. [1984] *Fuzzy information and fuzzy time* (**P** 3081) IFAC Symp Fuzzy Inf, Knowl Repr & Decis. Anal;1983 Marseille 159-162
◇ B52 ◇ REV Zbl 557 # 94022 • ID 48199

VITYAEV, E.E. [1979] *Detection of regularities expressed by universal formulas (Russian)* (**S** 0507) Vychisl Sist (Akad Nauk SSSR Novosibirsk) 79*57-59,124
◇ A05 B48 ◇ REV MR 82g:03034 Zbl 453 # 03031 • ID 54161

VITYAEV, E.E. [1981] see NOVIKOV, V.F.

VITYAEV, E.E. see Vol. I for further entries

VLACH, F. [1983] *On situation semantics for perception* (**J** 0154) Synthese 54*129-152
◇ B45 ◇ REV MR 85d:03039 • ID 41016

VOGT, F.H. [1984] see MELLIAR-SMITH, P.M.

VOJSHVILLO, E.K. [1979] *Natural variants of certain systems of relevant logic (on the problem of explication of the notions of logical consequence and conditional relations) (Russian)* (**P** 2554) All-Union Symp Th Log Infer;1974 Moskva 69-117
◇ B46 ◇ REV MR 84k:03066 • ID 35003

VOJSHVILLO, E.K. [1980] *Logical entailment and implication (semantic analysis) (Russian)* (**C** 2583) Aktual Probl Log & Metodol Nauki 173-193
◇ B46 ◇ REV Zbl 534 # 03004 • ID 36554

VOJSHVILLO, E.K. [1980] *The concepts of intensional information and the intensional relation of logical entailment (intensional analysis) (Russian)* (**C** 3038) Log-Metodol Issl 206-245
◇ B46 ◇ REV MR 84a:03025 • ID 35569

VOJSHVILLO, E.K. [1982] *Semantics of generalized state descriptions* (**P** 3622) Int Congr Log, Meth & Phil of Sci (6,Proc);1979 Hannover 315-323
◇ B46 ◇ REV MR 84f:03020 Zbl 501 # 03009 • ID 34436

VOJSHVILLO, E.K. [1983] *A decision procedure for the system E (of entailment) I* (J 0063) Studia Logica 42*139-164
◇ B25 B46 ◇ REV MR 86f:03036 Zbl 559 # 03009
• ID 48317

VOJSHVILLO, E.K. [1984] *Logical entailment and semantics of general state description (Russian)* (C 4366) Modal & Intens Log & Primen Probl Metodol Nauk 183-192
◇ B45 ◇ ID 46675

VOJSHVILLO, E.K. see Vol. I for further entries

VOJVODIC, G. [1974] see SESELJA, B.

VOJVODIC, G. [1978] *Some theorems for model theory of mixed-valued predicate calculi* (J 0400) Publ Inst Math, NS (Belgrade) 23(37)*229-234
◇ B50 C20 C35 C90 ◇ REV MR 80a:03050
Zbl 399 # 03018 • ID 52815

VOJVODIC, G. [1979] *On the II ε-theorem for mixed-valued predicate calculi* (J 2855) Zbor Rad, Prir-Mat Fak, Ser Mat (Novi Sad) 9*127-131
◇ B50 ◇ REV MR 82d:03036 Zbl 449 # 03016 • ID 56683

VOJVODIC, G. [1980] *The Craig interpolation theorem for mixed-valued predicate calculi (Serbo-Croatian summary)* (J 2855) Zbor Rad, Prir-Mat Fak, Ser Mat (Novi Sad) 10*173-175
◇ B50 C40 C90 ◇ REV MR 83h:03035 Zbl 531 # 03009
• ID 36055

VOJVODIC, G. see Vol. V for further entries

VOL'NEVICH, B. [1974] *The concept of a fact as a modal operator (Russian)* (C 2578) Filos & Logika 156-164
◇ B45 ◇ REV MR 58 # 27301 • ID 79752

VOROBEJ, M. [1982] *Deontic accessibility* (J 0095) Philos Stud 41*317-319
◇ B45 ◇ REV MR 84j:03047 • ID 34639

VOROB'EV, N.N. [1964] *A constructive calculus of statements with strong negation (Russian)* (S 0066) Tr Mat Inst Steklov 72*195-227
• TRANSL [1972] (J 0225) Amer Math Soc, Transl, Ser 2 99*40-82
◇ B55 F50 ◇ REV MR 33 # 44 Zbl 202.295 • ID 16331

VOROB'EV, N.N. see Vol. VI for further entries

VRANESIC, Z.G. [1970] see LEE, E.S.

VRANESIC, Z.G. [1971] see JANCZEWSKI, L.J.

VRANESIC, Z.G. [1972] see HAMACHER, V.C.

VRANESIC, Z.G. [1974] see SMITH, KENNETH C.

VRANESIC, Z.G. [1974] see LAM, C.I.

VRANESIC, Z.G. [1979] *Multivalued signaling in daisy chain bus control* (P 3003) Int Symp Multi-Val Log (9);1979 Bath 14-18
◇ B50 ◇ ID 35936

VRANESIC, Z.G. see Vol. I for further entries

VREDENDUIN, P.G.J. [1939] *A system of strict implication* (J 0036) J Symb Logic 4*73-76
◇ B45 ◇ REV Zbl 21.98 JSL 4.124 FdM 65.28 • ID 13889

VREDENDUIN, P.G.J. see Vol. VI for further entries

VUCKOVIC, V. [1967] *A recursive model for the extended system A of B. Sobocinski* (J 0047) Notre Dame J Formal Log 8*154-158
◇ B45 F30 ◇ REV MR 38 # 2015 Zbl 262 # 02022
• ID 27479

VUCKOVIC, V. [1967] *Recursive models for three-valued propositional calculi with classical implication* (J 0047) Notre Dame J Formal Log 8*148-153
◇ B50 C57 C90 F30 ◇ REV Zbl 262 # 02021 • ID 27478

VUCKOVIC, V. see Vol. I, III, IV, V, VI for further entries

VUILLEMIN, J. [1983] *Le carre chrysippeen des modalites (English and German summaries)* (J 0076) Dialectica 37*235-247
◇ B45 ◇ REV MR 85d:03042 • ID 41028

VUILLEMIN, J. see Vol. IV for further entries

VUJOSEVIC, S.T. [1981] see KRON, A.

VUJOSEVIC, S.T. see Vol. I for further entries

VUKOVIC, A. [1984] *On the bases of the three-valued logic (Serbo-Croatian summary)* (J 3519) Glas Mat, Ser 3 (Zagreb) 19(39)*3-11
◇ B50 ◇ REV MR 85i:03073 Zbl 536 # 03011 • ID 37097

VUKOVIC, A. [1984] *On three-values logic function types* (J 3944) Rad Jugosl Akad Znan Umjet, Mat Znan 3*61-70
◇ B50 ◇ REV Zbl 557 # 03013 • ID 46189

VUKOVIC, A. [1984] *Three-valued logic function inside the basic type ‹012›* (J 3944) Rad Jugosl Akad Znan Umjet, Mat Znan 3*71-78
◇ B50 ◇ REV Zbl 557 # 03014 • ID 46190

WAGNER, C.G. [1984] *Aggregating subjective probabilities: some limitative theorems* (J 0047) Notre Dame J Formal Log 25*233-240
◇ B48 ◇ REV MR 85e:03056 Zbl 544 # 62010 • ID 40624

WAGNER, C.G. [1985] see LEHRER, K.

WAGNER, E.G. [1985] see BLOOM, S.L.

WAGNER, E.G. see Vol. IV, V, VI for further entries

WAGNER, K. [1950] *Zum Repraesentantenproblem der Logik fuer Aussagenfunktionen mit beliebig endlich vielen Wahrheitswerten* (J 0044) Math Z 53*364-374
◇ B50 ◇ REV MR 12.469 Zbl 38.149 JSL 16.227
• ID 13967

WAGNER, K. [1974] see LINDNER, R.

WAGNER, K. see Vol. I, IV, V for further entries

WAGNER DECEW, J. [1981] *Conditional obligation and counterfactuals* (J 0122) J Philos Logic 10*55-72
◇ B45 ◇ REV MR 82i:03026 • ID 72213

WAISMANN, F. [1936] *Einfuehrung in das mathematische Denken. Die Begriffsbildung der modernen Mathematik* (X 1007) Gerold: Wien viii + 188pp
• TRANSL [1939] (X 1262) Einaudi: Torino 326pp
◇ A05 B98 ◇ REV Zbl 29.242 JSL 13.117 FdM 62.1046 FdM 65.20 • REM Foreword by K.Menger. 2nd ed. 1947, vii + 168pp • ID 40876

WAISMANN, F. [1951] *Introduction to mathematical thinking. The formation of concepts in modern mathematics* (X 1221) Ungar: New York & London xi + 260pp
◇ B98 ◇ REV MR 13.899 Zbl 45.148 JSL 17.208
• ID 25728

WAJSBERG, M. [1931] *Axiomatization of the three-valued propositional calculus (Polish) (German summary)* (J 0459) C R Soc Sci Lett Varsovie Cl 3 24*126-148
• TRANSL [1967] (C 0615) Polish Logic 1920-39 264-284
• REPR [1977] (C 4055) Wajsberg: Logical Works 12-29
◇ B50 ◇ REV JSL 35.442 JSL 48.873 FdM 58.998
• ID 13984

WAJSBERG, M. [1936] *Untersuchung ueber Unabhaengigkeitsbeweise nach der Matrizenmethode* (J 4710) Pol Tow Mat, Wiad Mat 41*33-70 • ERR/ADD ibid 43*166
• TRANSL [1977] (C 4055) Wajsberg: Logical Works 107-131
◇ B50 ◇ REV Zbl 13.289 JSL 1.75 JSL 48.873 FdM 62.40
• ID 13981

WAJSBERG, M. see Vol. I, III, V, VI for further entries

WALD, H. [1975] *Introduction to dialectical logic* (X 2178) Gruner: Amsterdam
◇ A05 B53 ◇ ID 42895

WALD, J.D. [1979] see BAKER, L.R.

WALIGORSKI, S. [1968] *Implications in boolean algebras with a two-valued closure operator (Polish and Russian summaries)* (J 0063) Studia Logica 23*25-34
◇ B45 G05 G25 ◇ REV MR 39 # 2621 Zbl 315 # 02026
• ID 13992

WALIGORSKI, S. see Vol. I, IV for further entries

WALK, K. [1966] *Simplicity, entropy and inductive logic* (C 1107) Aspects Inductive Log 66-80
◇ B48 ◇ REV Zbl 202.297 • ID 27704

WALLACE, J.R. [1960] see ANDERSON, A.R.

WALLACE, J.R. [1965] see BELNAP JR., N.D.

WALLEY, P. [1979] see FINE, T.L.

WALTER, M. [1973] see BAGINSKI, M.

WALTON, D.N. [1975] *Modal logic and agency* (J 0079) Logique & Anal, NS 18*103-111
◇ B45 ◇ REV MR 53 # 91 Zbl 323 # 02031 • ID 16555

WALTON, D.N. [1976] *Time and modality in the "can" of opportunity. With a comment by Keith Lehrer* (P 2326) Action Th;1975 Winnipeg 271-290
◇ B45 ◇ REV Zbl 343 # 02016 • ID 65978

WALTON, D.N. & WOODS, J. [1979] *Circular demonstration and von Wright-Geach entailment* (J 0047) Notre Dame J Formal Log 20*768-772
◇ B46 ◇ REV MR 81c:03009 Zbl 351 # 02018 • ID 56193

WALTON, D.N. [1979] *Philosophical basis of relatedness logic* (J 0095) Philos Stud 36*115-136
◇ A05 B45 ◇ REV MR 81g:03023 • ID 79812

WALTON, D.N. [1979] *Relatedness in intensional action chains* (J 0095) Philos Stud 36*175-223
◇ A05 B45 ◇ REV MR 81g:03024 • ID 79811

WALTON, D.N. & WOODS, J. [1982] *Question-begging and cumulativeness in dialectical games* (J 0097) Nous, Quart J Phil 16*585-605
◇ A05 B60 ◇ REV MR 84g:03015 • ID 34129

WALTON, D.N. [1984] see BATTEN, L.M.

WALTON, D.N. [1985] *New directions in the logic of dialogue* (J 0154) Synthese 63*259-274
◇ B60 ◇ ID 47236

WAN, ZHEXIAN [1960] see LU, YANGCI

WAN, ZHEXIAN see Vol. I for further entries

WAND, M. [1977] *A characterization of weakest preconditions* (J 0119) J Comp Syst Sci 15*209-212
◇ B45 ◇ REV MR 57 # 8165 Zbl 369 # 68004 • ID 32530

WAND, M. see Vol. IV, VI for further entries

WANG, HAO [1955] *On formalization* (J 0094) Mind 64*226-238
• LAST ED [1967] (C 0721) Contemp Readings in Log Th 29-40 • REPR [1963] (C 1009) Wang: Survey Math Logic 57-67
◇ A05 B98 ◇ REV JSL 22.292 • ID 38706

WANG, HAO [1958] *Eighty years of foundational studies (German and French summaries)* (J 0076) Dialectica 12*466-497
◇ A05 A10 B98 ◇ REV MR 21 # 9 Zbl 90.8 JSL 28.173
• ID 14029

WANG, HAO [1961] *The calculus of partial predicates and its extension to set theory I* (J 0068) Z Math Logik Grundlagen Math 7*283-288
◇ B50 E70 ◇ REV MR 26 # 11 Zbl 124.246 JSL 37.617
• ID 14036

WANG, HAO [1963] *The predicate calculus* (C 1009) Wang: Survey Math Logic 307-321
◇ B98 ◇ REV JSL 28.250 • ID 14053

WANG, HAO [1981] *Popular lectures on mathematical logic (Chinese)* (X 1876) Kexue Chubanshe: Beijing vii + 257pp
• TRANSL [1981] (X 0864) Van Nostrand: New York ix + 273pp
◇ A05 B98 C98 E98 F30 F98 ◇ REV MR 82e:03001 MR 84g:03002 JSL 47.908 • ID 34116

WANG, HAO see Vol. I, III, IV, V, VI for further entries

WANG, JINCHENG [1982] *An informal discussion of fuzzy mathematics (Chinese)* (J 4632) Qufu Shiyuan Xuebao 1982/4*80-81,89
◇ B52 E72 ◇ ID 49625

WANG, P.P. [1983] see TOGAI, M.

WANG, P.P. see Vol. V for further entries

WANG, PEIZHUANG [1980] see CHEN, XINGGOU

WANG, PEIZHUANG [1982] *Fuzzy contactability and fuzzy variables* (J 2720) Fuzzy Sets Syst 8*81-92
◇ B52 ◇ REV MR 83i:28010 Zbl 493 # 28001 • ID 38432

WANG, PEIZHUANG [1982] see LI, BIQIANG

WANG, PEIZHUANG [1983] see CHEN, YONGYI

WANG, PEIZHUANG [1985] see CHEN, YONGYI

WANG, PEIZHUANG see Vol. V for further entries

WANG, SHIQIANG & WU, WANGMING [1964] *On the classification of complemented lattices according to their tautologies. I (Chinese)(English summary)* (J 2521) Beijing Shifan Daxue Xuebao, Ziran Kexue 1964/2*125-133
◇ B50 ◇ ID 48760

WANG, SHIQIANG & WENG, JIAFENG [1980] *Normal forms in lattice-valued predicate calculi (Chinese) (English summary)* (J 2521) Beijing Shifan Daxue Xuebao, Ziran Kexue 1980/2*19-23
◇ B50 ◇ REV MR 84j:03056 • ID 34648

WANG, SHIQIANG & WU, WANGMING [1980] *On a classification problem for complemented lattices (Chinese)* (**J** 2771) Kexue Tongbao 25*725-726
• TRANSL [1981] (**J** 3769) Sci Bull, Foreign Lang Ed 26*289-292
◇ B50 G10 ◇ REV MR 84d:03029 Zbl 504 # 06007
• ID 34066

WANG, SHIQIANG see Vol. I, III, V for further entries

WANG, XIANJUN [1982] *Introduction to mathematical logic (Chinese)* (**X** 3123) Beijing Shifan Daxue: Beijing 372pp
◇ B98 ◇ ID 48542

WANG, XIANJUN see Vol. I for further entries

WANG, XUANYUAN [1984] *Mr.S and Mr.P Puzzle - a brief introduction to modal logic (Chinese)* (**J** 4430) Ziran Zazhi 7*446
◇ B45 ◇ ID 49395

WANG, YIZHI [1981] see SHU, YONGCHANG

WARAGAI, T. [1981] *The ontological law of contradiction and its logical structure* (**J** 0260) Ann Jap Ass Phil Sci 6*43-58
◇ A10 B53 ◇ REV MR 82m:03043 Zbl 486 # 03006
• ID 79822

WARAGAI, T. see Vol. V for further entries

WARD, M. [1935] *A determination of all possible systems of strict implication* (**J** 0100) Amer J Math 57*261-266
◇ B45 ◇ REV Zbl 11.242 FdM 61.53 • ID 40779

WARD, M. see Vol. V for further entries

WARMBROD, K. [1981] *An indexical theory of conditionals* (**J** 0488) Dialogue (Ottawa) 20*644-664
◇ B45 ◇ REV MR 83h:03029 • ID 36048

WARMBROD, K. [1981] *Counterfactuals and substitution of equivalent antecedents* (**J** 0122) J Philos Logic 10*267-289
◇ A05 B45 ◇ REV MR 82k:03023 Zbl 473 # 03018
• ID 55349

WARMBROD, K. [1982] *A defense of the limit assumption* (**J** 0095) Philos Stud 42*53-66
◇ B45 ◇ REV MR 84i:03048 • ID 34527

WARTOFSKY, M.W. [1983] see COHEN, R.S.

WASILEWSKA, A. [1971] *A formalization of the modal propositional S4 calculus (Polish and Russian summaries)* (**J** 0063) Studia Logica 27*133-149
◇ B45 ◇ REV MR 45 # 8508 Zbl 286 # 02027 • ID 14073

WASILEWSKA, A. [1972] *The diagrams of formulas of the modal propositional* $S4^*$ *calculus (Polish and Russian summaries)* (**J** 0063) Studia Logica 30*69-78
◇ B45 ◇ REV MR 48 # 3704 Zbl 279 # 02011 • ID 14074

WASILEWSKA, A. [1979] *A constructive proof of Craig's interpolation lemma for m-valued logic* (**J** 0063) Studia Logica 38*267-275
◇ B50 C40 ◇ REV MR 81e:03019 Zbl 442 # 03023
• ID 56380

WASILEWSKA, A. [1984] *DFC-algorithms for Suszko logic SCI and one-to-one Gentzen type formalizations* (**J** 0063) Studia Logica 43*395-404
◇ B60 ◇ ID 42388

WASILEWSKA, A. see Vol. I, III, IV, VI for further entries

WASSERMAN, H.C. [1976] *A note of evaluation mappings* (**J** 0047) Notre Dame J Formal Log 17*613-614
◇ B05 B45 ◇ REV MR 55 # 5440 Zbl 258 # 02009
• ID 65990

WASSERMAN, H.C. [1976] *An analysis of the counterfactual conditional* (**J** 0047) Notre Dame J Formal Log 17*395-400
◇ B25 B45 ◇ REV MR 56 # 96 Zbl 245 # 02017 • ID 18423

WASSERMAN, H.C. see Vol. I for further entries

WASZKIEWICZ, J. & WEGLORZ, B. [1969] *On products of structures for generalized logics (Polish and Russian summaries)* (**J** 0063) Studia Logica 25*7-15
◇ B50 C20 C30 C90 ◇ REV MR 41 # 8220
Zbl 264 # 02019 • ID 14081

WASZKIEWICZ, J. [1971] *The notions of isomorphism and identity for many-valued relational structures (Polish and Russian summaries)* (**J** 0063) Studia Logica 27*93-99
◇ B50 C90 ◇ REV MR 45 # 8503 Zbl 252 # 02053
• ID 14084

WASZKIEWICZ, J. see Vol. III for further entries

WATANABE, S. [1969] *Knowing and guessing: A quantitative study of inference and information* (**X** 0827) Wiley & Sons: New York xiii + 592pp
◇ A05 B48 ◇ REV MR 52 # 12977 Zbl 206.209 • ID 25459

WATANABE, S. [1974] *Logic of the empirical world. With reference to identity theory and reductionism* (**J** 0260) Ann Jap Ass Phil Sci 4*253-270
◇ A05 B60 ◇ REV Zbl 298 # 02008 • ID 65999

WATANABE, S. [1983] *Theory of propensity. A new foundation of logic* (**C** 3834) Lang, Logic and Method 283-308
◇ B60 ◇ REV Zbl 502 # 03005 • ID 36894

WATANABE, S. see Vol. III, IV, V for further entries

WATSON, F.R. [1978] see BAXANDALL, P.R.

WEAVER, G.E. [1969] see CORCORAN, J.

WEAVER, G.E. [1973] *Logical consequence in modal logic: alternative semantic systems for normal modal logics* (**P** 0783) Truth, Syntax & Modal;1970 Philadelphia 308-317
◇ B45 ◇ REV MR 55 # 2507 Zbl 273 # 02017 JSL 42.317
• ID 30488

WEAVER, G.E. [1974] see GOLDBERG, H.

WEAVER, G.E. [1974] see CORCORAN, J.

WEAVER, G.E. [1978] *Compactness theorems for finitely-many-valued sentential logics* (**J** 0063) Studia Logica 37*413-416
◇ B50 ◇ REV MR 80h:03037 Zbl 415 # 03018 • ID 53120

WEAVER, G.E. [1982] see GUMB, R.D.

WEAVER, G.E. see Vol. I, III, V for further entries

WEBB, D.L. [1935] *Generation of any n-valued logic by one binary operation* (**J** 0054) Proc Nat Acad Sci USA 21*252-254
◇ B50 ◇ REV Zbl 12.1 FdM 61.52 • ID 14099

WEBB, D.L. [1936] *Definition of Post's generalized negative and maximum in terms of one binary operation* (**J** 0100) Amer J Math 58*193-194
◇ B50 ◇ REV Zbl 13.243 JSL 1.42 FdM 62.33 • ID 14100

WEBB, D.L. [1937] *The algebra of n-valued logic* (**J** 0459) C R Soc Sci Lett Varsovie Cl 3 29*153-168
◇ B50 ◇ REV Zbl 17.145 JSL 3.52 FdM 63.826 • ID 14101

WEBER, S. [1983] *A general concept of fuzzy connectives, negations and implications based on t-norms and t-conorms* (**J** 2720) Fuzzy Sets Syst 11*115-134
◇ B52 ◇ REV MR 84m:03083 Zbl 543 # 03013 • ID 35780

WEBER, S. see Vol. V for further entries

WECHLER, W. [1978] *The concept of fuzziness in automata and language theory* (**X** 0911) Akademie Verlag: Berlin vii + 141pp
◇ B52 D05 E72 ◇ REV MR 80g:68105 Zbl 401 # 94048 • ID 54647

WECHLER, W. see Vol. IV, V for further entries

WEDIN, M.V. [1982] *Atistotle on the range of the principle of noncontradiction* (**J** 0079) Logique & Anal, NS 25*87-92
◇ A10 B53 ◇ REV MR 83i:030002 • ID 35482

WEE, W.G. [1969] see FU, K.S.

WEESE, M. [1980] see BAUDISCH, A.

WEESE, M. see Vol. III, IV, V, VI for further entries

WEGENER, I. [1982] *Best possible asymptotic bounds on the depth of monotone functions in multivalued logic* (**J** 0232) Inform Process Lett 15*81-83
◇ B50 ◇ REV MR 84c:03049 Zbl 488 # 94034 • ID 35701

WEGENER, I. see Vol. I, IV for further entries

WEGLORZ, B. [1969] see WASZKIEWICZ, J.

WEGLORZ, B. see Vol. I, III, V for further entries

WEINBERGER, O. [1970] *Normenlogik anwendbar im Recht* (**J** 0079) Logique & Anal, NS 13*93-106
◇ B45 ◇ REV Zbl 198.15 • ID 16204

WEINBERGER, O. [1972] *The concept of non-satisfaction and deontic logic* (**J** 1022) Ratio (Oxford) 14*16-35
◇ B45 ◇ REV MR 46 # 7001 • ID 79885

WEINBERGER, O. [1973] *Der Erlaubnisbegriff und der Aufbau der Normenlogik* (**J** 0079) Logique & Anal, NS 16*113-142
◇ B45 ◇ REV Zbl 269 # 02009 • ID 30361

WEINBERGER, O. [1973] *Erwiderung auf A.Soetemans Kritik und seine Reformversuche der deontischen Logik* (**J** 0079) Logique & Anal, NS 16*285-292
◇ B45 ◇ REV MR 48 # 5826 • ID 47906

WEINBERGER, O. [1974] *Contrary-to-fact and fact-transcendent conditionals* (**J** 1022) Ratio (Oxford) 16*15-32
◇ B45 ◇ REV MR 58 # 16162 • ID 79884

WEINGARTNER, P. [1968] *Modal logics with two kinds of necessity and possibility. In collaboration with Hans Knapp* (**J** 0047) Notre Dame J Formal Log 9*97-159
◇ B45 ◇ REV MR 39 # 2622 Zbl 247 # 02025 • ID 14126

WEINGARTNER, P. [1969] *Ein Kalkuel fuer die Begriffe "beweisbar" und "entscheidbar"* (**P** 1571) Int Congr Philos (14);1968 Wien 3*233-240
◇ A05 B45 ◇ ID 31537

WEINGARTNER, P. & WOLF, K. (EDS.) [1971] *Ernst Mally: Logische Schriften* (**X** 0835) Reidel: Dordrecht 347pp
◇ A10 B45 ◇ REV Zbl 209.7 • ID 32546

WEINGARTNER, P. [1973] *A predicate calculus for intensional logic* (**J** 0122) J Philos Logic 2*220-303
◇ A05 B45 ◇ REV MR 57 # 15974 Zbl 266 # 02017 • ID 32541

WEINGARTNER, P. (ED.) [1974] *Proceedings of a Colloquium on Logic and Ontology* (**J** 0122) J Philos Logic 3*171-343
◇ A05 B97 ◇ REV MR 49 # 10523 • ID 70209

WEINGARTNER, P. [1977] see CZERMAK, J.

WEINGARTNER, P. [1979] see MORSCHER, E.

WEINGARTNER, P. [1981] *A new theory of intensions* (**C** 4451) Sci Phil Today (Bunge) 439-464
◇ A05 B60 ◇ ID 40248

WEINGARTNER, P. [1982] *Conditions of rationality for the concepts belief, knowledge, and assumption (French and German summaries)* (**J** 0076) Dialectica 36*243-263
◇ A05 B45 ◇ REV MR 84k:03067 Zbl 511 # 03001 • ID 35004

WEINGARTNER, P. [1983] see CZERMAK, J.

WEINGARTNER, P. [1985] *A simple relevance criterion for natural language and its semantics* (**P** 4180) Int Congr Log, Meth & Phil of Sci (7,Pap);1983 Salzburg
◇ B46 ◇ ID 40258

WEINGARTNER, P. see Vol. I, III, V for further entries

WEINKE, K. (ED.) [1981] *Logik, Ethik und Sprache. Festschrift fuer Rudolf Freundlich* (**X** 0814) Oldenbourg: Muenchen 309pp
◇ B97 ◇ REV Zbl 458 # 00001 • ID 54406

WEIRICH, P. [1983] *Conditional probabilities and probabilities given knowledge of a condition* (**J** 0153) Phil of Sci (East Lansing) 50*82-95
◇ B48 ◇ REV MR 84i:60006 • ID 40135

WEISCHEDEL, R.M. [1976] see JOSHI, A.K.

WEISPFENNING, V. [1985] *Quantifier elimination for distributive lattices and measure algebras* (**J** 0068) Z Math Logik Grundlagen Math 31*249-261
◇ B25 B50 C10 C65 G05 G10 ◇ REV Zbl 547 # 03026 • ID 47574

WEISPFENNING, V. see Vol. III, IV for further entries

WEIZSAECKER VON, C.F. [1981] *In welchem Sinne ist die Quantenlogik eine zeitliche Logik?* (**C** 3737) Grundl-probl Modern Physik (Mittelstaedt) 311-317
◇ B51 ◇ REV MR 83h:03090 • ID 34847

WELSH JR., P.J. [1962] see CLARK, R.

WELSH JR., P.J. [1978] *Primitivity in mereology I,II* (**J** 0047) Notre Dame J Formal Log 19*25-62,355-385
◇ A05 B60 E70 ◇ REV MR 81e:03027a MR 81e:03027b Zbl 254 # 02027 Zbl 393 # 03015 Zbl 393 # 03016 • ID 27072

WELSH JR., P.J. see Vol. V for further entries

WENDEL, N. [1978] *The inconsistency of Bernini's very strong intuitionistic theory* (**J** 0063) Studia Logica 37*341-347
◇ B55 F35 F50 ◇ REV MR 80m:03096 Zbl 416 # 03051 • ID 53217

WENDEL, N. see Vol. VI for further entries

WENG, JIAFENG [1980] see WANG, SHIQIANG

WENSTOP, F. [1979] *Exploring linguistic consequences of assertions in social sciences* (C 3514) Adv Fuzzy Sets & Appl 501-518
◇ B52 ◇ REV MR 81h:92034 • ID 83134

WERNER, C.G. [1974] see POSPESEL, H.

WERNER, C.G. [1977] *Frequencies and beliefs* (J 0047) Notre Dame J Formal Log 18*496-498
◇ A05 B48 ◇ REV Zbl 306 # 02011 • ID 66053

WESSEL, H. [1972] *Probleme topologischer Logiken* (C 1533) Quantoren, Modal, Paradox 279-298
◇ A05 B60 ◇ REV Zbl 263 # 02016 • ID 29863

WESSEL, H. (ED.) [1972] *Quantoren-Modalitaeten-Paradoxien* (X 0806) Dt Verlag Wiss: Berlin 524pp
◇ A05 B45 ◇ REV MR 56 # 2775 Zbl 237 # 00011 • ID 70126

WESSEL, H. & ZINOV'EV, A.A. [1975] *Logische Sprachregeln. Eine Einfuehrung in die Logik* (X 0806) Dt Verlag Wiss: Berlin 592pp
◇ A05 B98 ◇ REV Zbl 307 # 02002 • REM Translation of the book of A. Sinowjew from Russian • ID 65327

WESSEL, H. [1979] *Ein System der strikten logischen Folgebeziehung* (P 2539) Frege Konferenz (1);1979 Jena 505-518
◇ B45 ◇ REV MR 82d:03030 • ID 79919

WESSEL, H. see Vol. I, VI for further entries

WESSELKAMPER, T.C. [1974] *A note on UDE's in an n-valued logic* (J 0047) Notre Dame J Formal Log 15*485-486
◇ B50 ◇ REV MR 52 # 2820 Zbl 272 # 02025 • ID 14148

WESSELKAMPER, T.C. [1974] *Some completeness results for abelian semigroups and groups* (P 1385) Int Symp Multi-Val Log (4);1974 Morgantown 393-400
◇ B50 ◇ ID 28634

WESSELKAMPER, T.C. [1975] *A sole sufficient operator* (J 0047) Notre Dame J Formal Log 16*86-88,551 • ERR/ADD ibid 16*551
◇ B50 ◇ REV MR 52 # 7850a MR 52 # 7850b Zbl 245 # 02021 Zbl 315 # 02020 • ID 14149

WESSELKAMPER, T.C. [1975] *Weak completeness and abelian semigroups* (J 0068) Z Math Logik Grundlagen Math 21*303-305
◇ B50 ◇ REV MR 51 # 9981 Zbl 308 # 02022 • ID 14150

WESSELKAMPER, T.C. [1976] *No abelian semigroup operation is complete* (J 0068) Z Math Logik Grundlagen Math 22*87-88
◇ B50 ◇ REV MR 53 # 5258 Zbl 325 # 02011 • ID 18487

WESSELKAMPER, T.C. [1979] *The algebraic representation of partial functions* (J 2702) Discr Appl Math 1*137-142
◇ B50 G25 ◇ REV MR 80m:94106 Zbl 425 # 94023 • ID 83141

WESSELS, L. [1981] *The "EPR" argument: a post-mortem* (J 0095) Philos Stud 40*3-30
◇ B51 ◇ REV MR 83c:03056 • ID 35149

WESSELS, L. see Vol. VI for further entries

WETTE, E. [1976] *On the formalization of productive logic* (J 0286) Int Logic Rev 13*23-33
◇ B60 D20 ◇ REV Zbl 362 # 02018 • ID 31554

WETTE, E. see Vol. I, III, IV, V, VI for further entries

WEYL, H. [1940] *The ghost of modality* (C 1013) Phil Essays (Husserl) 278-303
◇ A05 B45 ◇ REV JSL 8.31 • ID 14170

WEYL, H. [1981] *Le fantome de la modalite (English summary)* (J 0392) Math Sci Hum 74*37-60
◇ A05 B45 ◇ REV MR 83d:03012 Zbl 472 # 03001 • ID 55267

WEYL, H. see Vol. I, V, VI for further entries

WHEELER, R.F. [1966] *Complete connectives for the 3-valued propositional calculus* (J 3240) Proc London Math Soc, Ser 3 16*167-191
◇ B50 ◇ REV MR 32 # 7402 Zbl 143.10 JSL 33.127 • ID 14178

WHEELER, R.F. see Vol. I for further entries

WHITE, A.R. [1975] *Modal thinking* (X 0992) Cornell Univ Pr: Ithaca vii + 190pp
◇ B45 B65 ◇ REV JSL 42.428 • ID 44498

WHITE, B. [1974] *A note on natural deduction in many-valued logic* (J 0047) Notre Dame J Formal Log 15*167-168
◇ B50 ◇ REV MR 49 # 10527 Zbl 246 # 02014 Zbl 272 # 02027 • ID 14179

WHITE, M.J. [1976] *A suggestion regarding the semantical analysis of performatives* (J 0076) Dialectica 30*117-134
◇ A05 B45 ◇ REV Zbl 399 # 03015 • ID 52812

WHITE, M.J. [1979] *An S5 Diodorean modal system* (J 0079) Logique & Anal, NS 22*477-487
◇ B45 ◇ REV MR 82c:03025 Zbl 444 # 03009 • ID 56457

WHITE, M.J. [1980] *Necessity and unactualized possibilities in Aristotle* (J 0095) Philos Stud 38*287-298
◇ A05 B45 ◇ REV MR 82g:03004 • ID 79938

WHITE, M.J. [1981] *On some ascending chains of Brouwerian modal logics* (J 0063) Studia Logica 40*75-87
◇ B45 ◇ REV MR 82g:03031 Zbl 469 # 03010 • ID 55138

WHITE, M.J. [1984] *The necessity of the past and modal-tense logic incompleteness* (J 0047) Notre Dame J Formal Log 25*59-71
◇ B45 ◇ REV MR 86b:03025 Zbl 558 # 03010 • ID 42563

WHITE, R.B. [1979] *The consistency of the axiom of comprehension in the infinite-valued predicate logic of Lukasiewicz* (J 0122) J Philos Logic 8*509-534
◇ B50 E35 E70 F05 ◇ REV MR 81b:03059 Zbl 418 # 03037 • ID 53322

WHITE, R.B. [1980] *Natural deduction in the Lukasiewicz logics* (P 3673) Int Symp Multi-Val Log (10);1980 Evanston 226-232
◇ B50 ◇ REV MR 83m:94032 Zbl 566 # 03013 • ID 45947

WHITEHEAD, A.N. [1911] *Introduction to mathematics* (X 1236) Benn: Tunbridge 256pp
◇ B98 ◇ REV FdM 42.76 • ID 22606

WHITEHEAD, A.N. & RUSSELL, B. [1932] *Einfuehrung in die mathematische Logik* (X 1225) Drei Masken: Muenchen viii + 168pp
◇ A05 B98 ◇ REV Zbl 4.1 FdM 58.997 • REM German translation of the introduction of the "Principia Mathematica" • ID 28556

WHITEHEAD, A.N. see Vol. I, V for further entries

WHITLOCK, H.I. [1966] see MENGER, K.

WICKS, M.J. [1983] see CHONG, C.T.

WIDERKER, D. [1983] *The extensionality argument* (J 0097) Nous, Quart J Phil 17*457-468
◇ A05 B45 ◇ REV MR 84k:03068 • ID 35005

WIEHAGEN, R. [1982] see FREJVALD, R.V.

WIEHAGEN, R. [1984] see FREJVALD, R.V.

WIEHAGEN, R. see Vol. IV for further entries

WIELADEK, R. [1969] *A propositional calculus without the law of extensionality (Polish) (English and Russian summaries)* (J 0063) Studia Logica 24*187-208
◇ B60 ◇ REV Zbl 244 # 02017 • ID 14191

WIERZCHON, S.T. [1982] *Applications of fuzzy decision-making theory to coping with ill-defined problems* (J 2720) Fuzzy Sets Syst 7*1-18
◇ B52 ◇ REV MR 84f:90014 Zbl 468 # 90003 • ID 46471

WIERZCHON, S.T. [1984] see ORLOWSKA, E.

WIGGINS, D. [1976] *Identity, necessity and physicalism* (P 1502) Phil of Logic;1976 Bristol 96-182
◇ A05 B45 ◇ REV MR 58 # 16113 • ID 79964

WILCOX, W.C. [1970] *On infinite matrices and the paradoxes of material implication* (J 0047) Notre Dame J Formal Log 11*254-256
◇ A05 A10 B46 ◇ REV MR 43 # 3098 Zbl 167.8
• ID 14196

WILCOX, W.C. see Vol. I for further entries

WILDER, R.L. [1952] *Introduction to the foundations of mathematics* (X 0827) Wiley & Sons: New York xiv + 305pp
• LAST ED [1980] (X 2828) Krieger Publ Co: Melbourne xvi + 327pp
◇ A05 B98 ◇ REV MR 14.441 MR 32 # 35 MR 81j:03002 Zbl 442 # 03001 Zbl 49.9 JSL 19.225
• ID 56358

WILDGEN, W. [1983] *Modelling vagueness in catastrophe-theoretic semantics* (C 3095) Approaching Vagueness 317-360
◇ B52 ◇ REV MR 85j:03003 • ID 45379

WILHELMY, A. [1962] *Bemerkungen zur Semantik quantifizierter mehrwertiger logistischer Systeme* (C 0712) Logik & Logikkalkuel 179-188
◇ B50 ◇ REV Zbl 119.11 • ID 14203

WILHELMY, A. see Vol. I for further entries

WILL, U. [1982] *Eine pragmatische Rechtfertigung der Induktion* (J 0989) Z Allg Wissth 13*84-98
◇ A05 B48 ◇ REV MR 84i:03030 • ID 34514

WILLIAMS, A.G.P. [1982] *Applicable inductive logic* (X 3794) Edsall: London xi + 178pp
◇ B48 ◇ REV MR 84i:03051 Zbl 494 # 03002 • ID 36873

WILLIAMS, D.C. [1947] *The ground of induction* (X 0858) Harvard Univ Pr: Cambridge ix + 213pp
◇ A05 B48 ◇ REV JSL 12.141 • ID 25460

WILLIAMS, J.N. [1981] *Inconsistency and contradiction* (J 0094) Mind 90*600-602
◇ B53 ◇ ID 42896

WILLIAMS, J.N. [1982] *Believing the self-contradictory* (J 0325) Amer Phil Quart 19*279-286
◇ A05 B53 ◇ ID 42897

WILLIAMS, N.H. [1972] see ASH, C.J.

WILLIAMS, N.H. see Vol. V for further entries

WILLIAMSON, J. [1976] *The complete axiomatization of any many-valued propositional logic* (J 0068) Z Math Logik Grundlagen Math 22*299-306
◇ B50 ◇ REV MR 58 # 21453 Zbl 357 # 02018 • ID 18445

WILLIAMSON, J. [1978] *An ambiguity in modal logic* (J 0047) Notre Dame J Formal Log 19*475-485
◇ B45 ◇ REV MR 58 # 10325 Zbl 316 # 02027 • ID 31308

WILLIAMSON, J. [1979] *S5 without modal axioms* (J 0047) Notre Dame J Formal Log 20*593-594
◇ B45 ◇ REV MR 80i:03034 Zbl 394 # 03029 • ID 56125

WILLMOTT, R. [1980] *On the transitivity of implication and equivalence in some many-valued logics* (P 3673) Int Symp Multi-Val Log (10);1980 Evanston 253-262
◇ B50 ◇ REV MR 83m:94032 Zbl 539 # 03008 • ID 43742

WILLMOTT, R. [1980] *Two fuzzier implication operators in the theory of fuzzy power sets* (J 2720) Fuzzy Sets Syst 4*31-36
◇ B52 E72 ◇ REV MR 82d:03087 Zbl 433 # 03014
• ID 79990

WILLMOTT, R. [1985] *A probabilistic interpreation of a case of inference involving fuzzy quantification* (J 2720) Fuzzy Sets Syst 16*149-162
◇ B48 ◇ ID 47237

WILLMOTT, R. see Vol. IV, V for further entries

WILSON, K. [1965] see HOCKNEY, D.J.

WILSON, M. [1983] *Why contigent identity is necessary* (J 0095) Philos Stud 43*301-327
◇ B45 ◇ REV MR 84j:03022 • ID 34616

WILSON, N.L. [1983] *The transitivity of implication in tree logic* (J 0047) Notre Dame J Formal Log 24*106-114
◇ B60 ◇ REV MR 83m:03014 Zbl 476 # 03036 • ID 55548

WILSON, R.L. [1974] *Prenex normal form in the modal predicate logic PS*S and the Grosseteste algebra of sets GS*S* (J 0068) Z Math Logik Grundlagen Math 20*271-280
◇ B45 ◇ REV MR 52 # 5366 Zbl 317 # 02013 • ID 14233

WILSON, R.L. [1974] *The nine-valued modal logic ME*E* (J 0068) Z Math Logik Grundlagen Math 20*281-288
◇ B45 ◇ REV MR 52 # 5367 Zbl 317 # 02014 • ID 14232

WILSON, R.L. [1976] *On some modal logics related to the L-modal system* (J 0047) Notre Dame J Formal Log 17*191-206
◇ B45 ◇ REV MR 58 # 16163 Zbl 322 # 02016 • ID 18428

WILSON, R.L. [1976] *Some remarks on metaphysics and the modal logics F^*F* (J 0047) Notre Dame J Formal Log 17*349-360
◇ A05 B45 ◇ REV MR 58 # 16164 Zbl 326 # 02006
• ID 18429

WILSON, R.L. [1977] *A note on metaphysics and the foundations of mathematics* (J 0047) Notre Dame J Formal Log 18*355-362
◇ A05 B45 ◇ REV MR 58 # 27287 Zbl 225 # 02005
• ID 24236

WILSON, R.L. [1977] *The modal predicate logics PF^*F* (J 0047) Notre Dame J Formal Log 18*208-220
◇ B45 ◇ REV MR 58 # 16165 Zbl 225 # 02017 • ID 21967

WINKER, S.K. & WOS, L. [1978] *Automated generation of models and counter-examples and its application to open questions in ternary Boolean algebra* (**P** 2014) Int Symp Multi-Val Log (8);1978 Rosemont 251-256
◇ B50 ◇ ID 35931

WINKER, S.K. see Vol. I for further entries

WINKLER, G. [1984] see KRAWIETZ, W.

WINKLMANN, K. [1982] see MEYER, A.R.

WINKLMANN, K. see Vol. IV, VI for further entries

WINNIE, J.A. [1970] *The completeness of Copi's system of natural deduction* (**J** 0047) Notre Dame J Formal Log 11*379-382
◇ B10 B45 ◇ REV MR 44 # 3833 Zbl 197.277 • ID 14236

WINSKEL, G. [1983] *A note on power domains and modality* (**P** 3864) FCT'83 Found of Comput Th;1983 Borgholm 505-514
◇ B45 ◇ REV MR 85f:68063 Zbl 531 # 68001 • ID 38571

WINSKEL, G. see Vol. IV for further entries

WINSLOW, L.E. [1982] see DAVIS, H.W.

WIREDU, J.E. [1973] *On the real logical structure of Lewis' "independent proof"* (**J** 0047) Notre Dame J Formal Log 14*543-546
◇ B45 ◇ REV MR 49 # 19 Zbl 232 # 02004 • ID 29816

WIREDU, J.E. [1974] *A remark on a certain consequence of connexive logic for Zermelo's set theory* (**J** 0063) Studia Logica 33*127-130
◇ B45 E30 E70 ◇ REV MR 52 # 5421 Zbl 307 # 02019 • ID 18430

WIREDU, J.E. [1979] *On the necessity of S4* (**J** 0047) Notre Dame J Formal Log 20*689-694
◇ A05 B45 ◇ REV MR 80h:03035 Zbl 363 # 02019 • ID 56123

WIRSING, M. [1979] see BROY, M.

WIRSING, M. [1983] see PAEPPINGHAUS, P.

WIRSING, M. see Vol. I, III, IV for further entries

WISDOM, W.A. [1964] *Possibility-elimination in natural deduction* (**J** 0047) Notre Dame J Formal Log 5*295-298
◇ B45 ◇ REV MR 36 # 1306 • ID 14239

WISDOM, W.A. [1972] see LEBLANC, H.

WOJCICKI, R. [1970] *On reconstructability of the classical propositional logic in intuitionistic logic (Russian summary)* (**J** 0014) Bull Acad Pol Sci, Ser Math Astron Phys 18*421-422
◇ B55 F50 ◇ REV MR 43 # 1816 Zbl 218 # 02009 • ID 14254

WOJCICKI, R. [1972] *Some results concerning many valued Lukasiewicz's calculi* (**J** 0387) Bull Sect Logic, Pol Acad Sci 1/3*49-51
◇ B50 ◇ REV MR 50 # 12662 • ID 80039

WOJCICKI, R. [1972] *The degrees of completeness of the finitely-valued propositional calculi (Polish summary)* (**S** 0458) Zesz Nauk, Prace Log, Uniw Krakow 7*77-85
◇ B50 ◇ REV MR 47 # 8247 • ID 14257

WOJCICKI, R. [1973] *On matrix representations of consequence operations of Lukasiewicz's calculi* (**J** 0068) Z Math Logik Grundlagen Math 19*239-247
◇ B50 ◇ REV MR 48 # 1882 Zbl 313 # 02008 • ID 14258

WOJCICKI, R. [1973] *Set theoretic representations of empirical phenomena* (**P** 1457) Rep Semin Formal Methodol of Empir Sci;1973 Wroclaw 29-34
◇ A05 B60 ◇ REV MR 54 # 54 Zbl 281 # 02009 • ID 23937

WOJCICKI, R. [1974] *Note on deducibility and many-valuedness* (**J** 0036) J Symb Logic 39*563-566
◇ B50 ◇ REV MR 52 # 5358 Zbl 334 # 02007 • ID 14260

WOJCICKI, R. [1974] *The logics stronger than Lukasiewicz's three valued sentential calculus - The notion of degree of maximality versus the notion of degree of completeness* (**J** 0063) Studia Logica 33*201-214
• REPR [1977] (**C** 3026) Lukasiewicz Sel Pap on Sent Calc 135-148
◇ B50 ◇ REV MR 52 # 5359 MR 58 # 21456 Zbl 299 # 02019 Zbl 381 # 03017 • ID 18431

WOJCICKI, R. [1975] *A theorem on the finiteness of the degree of maximality of the n-valued Lukasiewicz logic* (**P** 1805) Int Symp Multi-Val Log (5,Proc);1975 Bloomington 240-251
◇ B50 ◇ REV MR 58 # 5042 • ID 80027

WOJCICKI, R. [1977] *On matrix representations of consequence operations of Lukasiewicz's sentencial calculi* (**C** 3026) Lukasiewicz Sel Pap on Sent Calc 101-111
◇ B50 ◇ REV MR 58 # 21455 Zbl 381 # 03017 • ID 80024

WOJCICKI, R. [1977] see MALINOWSKI, G.

WOJCICKI, R. [1980] *Entailment semantics for T* (**P** 3006) Brazil Conf Math Log (3);1979 Recife 309-336
◇ B46 ◇ REV MR 82g:03032 Zbl 459 # 03010 • ID 54450

WOJCICKI, R. [1980] *More about referential matrices* (**J** 0387) Bull Sect Logic, Pol Acad Sci 9*93-95
◇ B50 ◇ REV Zbl 445 # 03010 • ID 56474

WOJCICKI, R. [1982] *Referential matrix semantics for propositional calculi* (**P** 3622) Int Congr Log, Meth & Phil of Sci (6,Proc);1979 Hannover 325-334
◇ B22 B50 ◇ REV MR 84b:03050 Zbl 505 # 03007 • ID 35669

WOJCICKI, R. [1984] *Lectures on propositional calculi* (**X** 2212) Ossolineum: Wroclaw 292pp
◇ B98 ◇ ID 45741

WOJCICKI, R. [1984] *R. Suszko's situational semantics* (**J** 0063) Studia Logica 43*323-340
◇ B60 ◇ ID 42381

WOJCICKI, R. see Vol. I, III, V, VI for further entries

WOJCIECHOWSKA, A. [1979] *Elements of logic and set theory* (**X** 1034) PWN: Warsaw 200pp
◇ B98 E98 ◇ REV MR 81b:04002 • ID 80045

WOJCIECHOWSKA, A. see Vol. III, V for further entries

WOJCIK, A.S. [1977] *Digital system design using 4-valued logic devices* (**P** 4047) Allerton Conf Commun, Control & Comput (15);1977 Monticello 128-135
◇ B50 ◇ REV MR 84b:94001 • ID 46266

WOJCIK, A.S. see Vol. I for further entries

WOJTYLAK, P. [1976] *A new proof of structural completeness of Lukasiewicz's logics* (**J** 0387) Bull Sect Logic, Pol Acad Sci 5*145-152
◇ B50 ◇ REV MR 58 # 21457 • ID 27123

WOJTYLAK, P. [1976] *On structural completeness of the infinite-valued Lukasiewicz's propositional calculus* (**J** 0387) Bull Sect Logic, Pol Acad Sci 5*153-157
◇ B50 ◇ REV MR 58 # 21458 • ID 27124

WOJTYLAK, P. [1976] *Some generalisations of Makinson's theorem on structural completeness* (**J** 0302) Rep Math Logic, Krakow & Katowice 7*107-110
◇ B45 ◇ REV MR 57 # 9518 Zbl 358 # 02064 • ID 21931

WOJTYLAK, P. [1978] *On structural completeness of many-valued logics* (**J** 0063) Studia Logica 37*139-147
◇ B50 ◇ REV MR 80k:03027 Zbl 393 # 03011 • ID 52431

WOJTYLAK, P. see Vol. I, III, VI for further entries

WOLF, K. [1971] see WEINGARTNER, P.

WOLF, R.G. [1974] see ROUTLEY, R.

WOLF, R.G. [1975] *A critical survey of many-valued logics, 1966 - 1974* (**P** 1805) Int Symp Multi-Val Log (5,Proc);1975 Bloomington 468-474
◇ B50 ◇ REV MR 58 # 16142 • ID 80058

WOLF, R.G. [1977] *A survey of many-valued logic (1966-1974)* (**P** 1894) Int Symp Multi-Val Log (5,Inv Pap);1975 Bloomington 167-323
◇ B50 ◇ REV MR 58 # 16143 Zbl 372 # 02012 • ID 51420

WOLF, R.G. [1980] see COSTA DA, N.C.A.

WOLNIEWICZ, B. [1972] *The notion of fact as a modal operator* (**J** 0162) Teorema (Valencia) 1972/2*59-66
◇ B45 ◇ ID 28814

WOLNIEWICZ, B. [1982] *A closure system for elementary situations* (**J** 0387) Bull Sect Logic, Pol Acad Sci 11*134-139
◇ B48 ◇ REV MR 85d:03043 Zbl 522 # 03052 • ID 37800

WOLNIEWICZ, B. [1982] *A formal ontology of situations* (**J** 0063) Studia Logica 41*381-413
◇ B45 ◇ REV MR 85i:03082 Zbl 558 # 03003 • ID 42310

WOLNIEWICZ, B. [1984] *A topology for logical space* (**J** 0387) Bull Sect Logic, Pol Acad Sci 13*255-259
◇ B60 ◇ ID 44902

WOLNIEWICZ, B. see Vol. I for further entries

WOLPER, P. [1981] *Temporal logic can be more expressive* (**P** 4235) IEEE Symp Found of Comp Sci (22);1981 Nashville 340-348
◇ B45 ◇ REV MR 84a:68004 • ID 45813

WOLPER, P. [1983] *Temporal logic can be more expressive* (**J** 0194) Inform & Control 56*72-99
◇ B45 B75 ◇ REV MR 84a:68004 Zbl 534 # 03009 • ID 36557

WOLPER, P. [1985] see SISTLA, A.P.

WOOD, S. [1973] see CORCORAN, J.

WOODGER, J.-H. [1951] see MARTIN, R.M.

WOODGER, J.-H. see Vol. I for further entries

WOODRUFF, P.W. [1970] *A note on JP'* (**J** 0105) Theoria (Lund) 36*183-184
◇ B45 ◇ REV Zbl 212.13 • ID 14293

WOODRUFF, P.W. [1973] *On compactness in many-valued logic. I* (**J** 0047) Notre Dame J Formal Log 14*405-407
◇ B50 ◇ REV MR 48 # 8199 Zbl 259 # 02011 • ID 14294

WOODRUFF, P.W. [1973] *On constructive nonsense logic* (**C** 1389) Modality, Morality, Probl of Sense & Nonsense (Hallden) 192-205
◇ B60 ◇ REV MR 56 # 11764 • ID 80082

WOODRUFF, P.W. [1974] *A modal interpretation of three-valued logic* (**J** 0122) J Philos Logic 3*433-439
◇ B50 ◇ REV MR 54 # 7207 Zbl 296 # 02009 • ID 24999

WOODRUFF, P.W. [1981] see BENCIVENGA, E.

WOODRUFF, P.W. [1984] *Paradox, truth and logic. I. Paradox and truth* (**J** 0122) J Philos Logic 13*213-232
◇ A05 B60 ◇ REV MR 86d:03005d Zbl 546 # 03006 • ID 41140

WOODS, J. [1969] *Intensional relations (Polish and Russian summaries)* (**J** 0063) Studia Logica 25*61-77
◇ A05 B45 ◇ REV MR 45 # 3159 Zbl 253 # 02020 • ID 14296

WOODS, J. [1969] see SUMNER, L.W.

WOODS, J. [1973] *Descriptions, essences and quantified modal logic* (**J** 0122) J Philos Logic 2*304-321
◇ A05 B45 ◇ REV MR 55 # 12475 Zbl 266 # 02007 • ID 66150

WOODS, J. [1979] see WALTON, D.N.

WOODS, J. [1982] see WALTON, D.N.

WOODWARD, D. [1985] *Modal logic without essences* (**J** 0079) Logique & Anal, NS 28*301-315
◇ B45 ◇ ID 49750

WOOLHOUSE, R.S. [1973] *Tensed modalities* (**J** 0122) J Philos Logic 2*393-415
◇ B45 ◇ REV MR 56 # 11761 Zbl 272 # 02033 • ID 30380

WOS, L. [1978] see WINKER, S.K.

WOS, L. see Vol. I for further entries

WOZNIAKOWSKA, B. [1978] *Algebraic proof of the separation theorem for the infinite-valued logic of Lukasiewicz* (**J** 0302) Rep Math Logic, Krakow & Katowice 10*129-137
◇ B50 ◇ REV MR 80m:03057 Zbl 426 # 03069 • ID 31563

WOZNIAKOWSKA, B. [1978] *The representation theorem for the algebras determined by the fragments of infinite-valued logic of Lukasiewicz* (**J** 0387) Bull Sect Logic, Pol Acad Sci 7*176-178
◇ B20 B50 G20 G25 ◇ REV MR 80b:03108 Zbl 407 # 03029 • ID 31565

WRIGHT, J.D.M. [1985] see PTAK, P.

WRIGHT, J.D.M. see Vol. V for further entries

WRIGHT VON, G.H. [1951] *A treatise on induction and probability* (**X** 0866) Routledge & Kegan Paul: Henley on Thames 310pp
◇ A05 B48 ◇ REV MR 15.386 Zbl 54.4 • ID 25457

WRIGHT VON, G.H. [1951] *An essay in modal logic* (**X** 0809) North Holland: Amsterdam vii+90pp
◇ B25 B45 ◇ REV MR 13.614 Zbl 43.7 JSL 18.174 • ID 14305

WRIGHT VON, G.H. [1951] *Deontic logic* (**J** 0094) Mind 60*1-15
• REPR [1957] (**C** 1017) Logical Studies 58-74
◇ A05 B25 B45 ◇ REV Zbl 81.10 JSL 17.140 JSL 35.461 • ID 14313

WRIGHT VON, G.H. [1952] *Interpretations of modal logic* (**J** 0094) Mind 61*165-177
• REPR [1957] (**C** 1017) Logical Studies 75-88
◇ B45 ◇ REV Zbl 48.244 Zbl 81.10 JSL 18.176 JSL 35.461
• REM Revised and extended reprint • ID 14308

WRIGHT VON, G.H. [1953] *A new system of modal logic* (**P** 0645) Int Congr Philos (11);1953 Bruxelles 5*59-63
◇ B45 ◇ REV MR 15.189 Zbl 52.11 JSL 19.66 • ID 14310

WRIGHT VON, G.H. [1956] *A note on deontic logic and derived obligation* (**J** 0094) Mind 65*507-509
• REPR [1967] (**C** 0721) Contemp Readings in Log Th 316-318
◇ B45 ◇ REV JSL 22.91 • ID 14311

WRIGHT VON, G.H. [1957] *A new system of modal logic* (**C** 1017) Logical Studies 89-126
◇ B45 ◇ REV Zbl 81.10 JSL 35.461 • ID 14317

WRIGHT VON, G.H. [1957] *On conditionals* (**C** 1017) Logical Studies 127-165
◇ A05 B45 ◇ REV Zbl 81.10 JSL 35.461 • ID 14319

WRIGHT VON, G.H. [1957] *The concept of entailment* (**C** 1017) Logical Studies 166-191
◇ A05 B46 ◇ REV Zbl 81.10 JSL 35.461 • ID 14318

WRIGHT VON, G.H. [1957] *The logical problem of induction (2nd edition)* (**X** 1096) Blackwell: Oxford xii+249pp
◇ A05 B48 ◇ REV Zbl 78.5 • ID 25458

WRIGHT VON, G.H. [1959] *A note on entailment* (**J** 0112) Phil Quart (Calcutta) 9*363-365
◇ A05 B46 ◇ REV JSL 35.462 • ID 14321

WRIGHT VON, G.H. [1963] *The logic of preference* (**X** 1261) Edinburgh Univ Pr: Edinburgh 67pp
◇ B45 ◇ ID 25285

WRIGHT VON, G.H. [1964] *A new system of deontic logic* (**J** 0509) Danish Yearbook Philos 1*173-182
◇ B45 ◇ REV JSL 32.243 • ID 43096

WRIGHT VON, G.H. [1965] *A correction to a new system of deontic logic* (**J** 0509) Danish Yearbook Philos 2*103-107
◇ B45 ◇ REV JSL 32.243 • ID 43097

WRIGHT VON, G.H. [1966] *"and then"* (**J** 0990) Soc Sci Fennicae Comment Phys-Math 32/7*11pp
◇ B45 B65 ◇ REV MR 33 # 7249 Zbl 166.252 • ID 14323

WRIGHT VON, G.H. [1966] *The paradoxes of confirmation* (**C** 1107) Aspects Inductive Log 208-218
◇ A05 B48 ◇ REV Zbl 202.299 • ID 27708

WRIGHT VON, G.H. [1967] *Deontic logic* (**J** 0325) Amer Phil Quart 4*136-143
• REPR [1967] (**C** 0721) Contemp Readings in Log Th 303-315
◇ B45 ◇ REV JSL 35.462 • ID 14325

WRIGHT VON, G.H. [1968] *"Always"* (**J** 0105) Theoria (Lund) 34*208-221
◇ A05 B45 ◇ REV MR 48 # 10757 • ID 14327

WRIGHT VON, G.H. [1968] *An essay in deontic logic and the general theory of action. With a bibliography of deontic and imperative logic* (**J** 0096) Acta Philos Fenn 21*110pp
◇ A05 B45 ◇ REV MR 45 # 34 Zbl 172.292 • ID 14328

WRIGHT VON, G.H. [1969] *Time, change and contradiction* (**X** 0805) Cambridge Univ Pr: Cambridge, GB 32pp
◇ A05 B53 ◇ ID 16838

WRIGHT VON, G.H. [1970] *A note on confirmation theory and on the concept of evidence* (**C** 0735) Logic & Value (Dahlquist) 36-51
◇ A05 B48 ◇ REV MR 50 # 6780 • ID 14330

WRIGHT VON, G.H. [1972] *Some observations on modal logic and philosophical systems* (**C** 0558) Contemp Phil Scand 17-26
◇ A05 B45 ◇ ID 14898

WRIGHT VON, G.H. [1972] *The logic of preference reconsidered* (**J** 0472) Theory Decis 3*140-169
◇ B48 ◇ REV Zbl 252 # 02023 • ID 27740

WRIGHT VON, G.H. [1973] *On the logic and epistemology of the causal relation* (**P** 0793) Int Congr Log, Meth & Phil of Sci (4,Proc);1971 Bucharest 293-312
◇ A05 B45 ◇ REV MR 56 # 8320 • ID 79761

WRIGHT VON, G.H. [1973] *Remarks on the logic of predication* (**J** 0963) Ajatus (Helsinki) 35*158-167
◇ B45 ◇ REV Zbl 296 # 02007 • ID 66153

WRIGHT VON, G.H. [1977] see CONTE, A.G.

WRIGHT VON, G.H. [1979] *Diachronic and synchronic modalities* (**J** 0162) Teorema (Valencia) 9*231-245
◇ B45 ◇ REV MR 81f:03030 • ID 79765

WRIGHT VON, G.H. [1981] *Explanation and understanding of action* (**J** 2076) Rev Int Philos 35*127-142
◇ A05 B45 ◇ REV MR 82k:03028 • ID 79764

WRIGHT VON, G.H. [1981] *On the logic of norms and actions* (**C** 3731) New Studies Deontic Log 3-35
◇ B45 ◇ REV MR 83h:03030 • ID 36097

WRIGHT VON, G.H. [1981] *Problems and prospects of deontic logic: a survey* (**C** 2617) Modern Log Survey 399-423
◇ B45 ◇ REV MR 82f:03002 Zbl 464 # 03001 • ID 42778

WRIGHT VON, G.H. [1982] *Diachronic and synchronic modalities* (**P** 3800) Intens Log: Th & Appl;1979 Moskva 42-49
• TRANSL [1984] (**C** 4366) Modal & Intens Log & Primen Probl Metodol Nauk 8-14
◇ B45 ◇ REV MR 84j:03048 Zbl 511 # 03002 • ID 34640

WRIGHT VON, G.H. [1982] *Norms, truth and logic* (**P** 4041) Log, Inform, Law;1981 Firenze 2*3-20
◇ A05 B45 ◇ REV MR 85i:03061 • ID 44118

WRIGHT VON, G.H. [1983] *Norms of higher order* (**J** 0063) Studia Logica 42*119-127
◇ B45 ◇ REV MR 86a:03020 Zbl 568 # 03006 • ID 42353

WRIGHT VON, G.H. [1984] *Bedingungsnormen - ein Pruefstein fuer die Normenlogik* (**C** 4065) Theorie der Normen 447-456
◇ B45 ◇ REV MR 85k:03015 • ID 44574

WRIGHT VON, G.H. [1985] *Probleme des Erklaerens und Verstehens von Handlungen (English summary)* (**J** 2688) Conceptus (Wien) 19/47*3-19
◇ A05 B45 ◇ ID 49752

WRIGHT VON, G.H. see Vol. I, III for further entries

WRIGHTSON, G. [1984] *On some semantic tableau proof procedures for modal logic* (**J** 0658) Festschr-Ber VDI-Zeitschr, Reihe 10 iv+128pp
◇ B45 ◇ REV MR 85j:03024 Zbl 551 # 03008 • ID 43885

WRIGHTSON, G. [1984] see OHLBACH, H.-J.

WRIGHTSON, G. see Vol. I, VI for further entries

WROBLEWSKI, J. [1984] *Negation in law* (C 4065) Theorie der Normen 457-471
◇ B45 ◇ REV MR 85k:03015 • ID 44575

WRONSKI, A. [1971] *Axiomatization of the implicational Goedel's matrices by Kalmar's method (Polish summary)* (S 0458) Zesz Nauk, Prace Log, Uniw Krakow 6*89-98
◇ B55 F50 ◇ REV MR 45 # 6590 • ID 14337

WRONSKI, A. [1971] *Axiomatization of the implicational matrix* $\Gamma((\mathscr{L}_2)^2)$ *of Jaskowski's family* $(\mathscr{L}_1)_{\Pi\Gamma}$ *(Polish summary)* (S 0458) Zesz Nauk, Prace Log, Uniw Krakow 6*99-109
◇ B55 ◇ REV MR 46 # 3263 • ID 14338

WRONSKI, A. [1971] see PERZANOWSKI, J.

WRONSKI, A. [1972] *An algorithm for finding finite axiomatization of finite intermediate logics (Polish summary)* (S 0458) Zesz Nauk, Prace Log, Uniw Krakow 7*87-92
◇ B55 ◇ REV MR 47 # 8282 • ID 16531

WRONSKI, A. [1972] *Intermediate logics and the disjunction property. I,II* (J 0387) Bull Sect Logic, Pol Acad Sci 1/4*46-53,54-60
◇ B55 ◇ REV MR 58 # 27355 • ID 80102

WRONSKI, A. [1972] *Remarks on algebraic semantics for modal systems S0.9, S1 and S2* (J 0387) Bull Sect Logic, Pol Acad Sci 1/2*20-24
◇ B45 ◇ REV MR 51 # 2872 • ID 15175

WRONSKI, A. [1973] *Axiomatization of the implicational Goedel's matrices by Kalmar's method* (C 2084) Stud Hist of Math Log 123-132
◇ B55 ◇ ID 42076

WRONSKI, A. [1973] *Intermediate logics and the disjunction property* (J 0302) Rep Math Logic, Krakow & Katowice 1*39-51 • ERR/ADD ibid 2*83
◇ B55 ◇ REV MR 54 # 4927a MR 54 # 4927b Zbl 308 # 02027 • ID 24102

WRONSKI, A. [1973] *On the degree of completeness of positive logic* (J 0387) Bull Sect Logic, Pol Acad Sci 2*65-70
◇ B55 ◇ REV MR 53 # 5267 • ID 22898

WRONSKI, A. [1973] *Remarks on intermediate logics with axioms containing only one variable* (J 0387) Bull Sect Logic, Pol Acad Sci 2*58-64
◇ B55 ◇ REV MR 53 # 5270 Zbl 312 # 02024 • ID 22901

WRONSKI, A. [1973] see DZIK, W.

WRONSKI, A. [1974] see GRACZYNSKA, E.

WRONSKI, A. [1974] *On cardinalities of matrices strongly adequate for the intuitionistic propositional logic* (J 0302) Rep Math Logic, Krakow & Katowice 3*67-72
◇ B55 F50 ◇ REV MR 52 # 7858 Zbl 342 # 02011 • ID 18433

WRONSKI, A. [1974] *On equivalential fragments of some intermediate logics* (J 0387) Bull Sect Logic, Pol Acad Sci 3/2*11-14
◇ B55 ◇ REV MR 53 # 5271 • ID 22902

WRONSKI, A. [1974] see SURMA, S.J.

WRONSKI, A. [1974] *Remarks on intermediate logics with axioms containing only one variable* (J 0302) Rep Math Logic, Krakow & Katowice 2*63-75
◇ B55 ◇ REV MR 51 # 12519 Zbl 312 # 02024 • ID 17261

WRONSKI, A. [1975] see POREBSKA, M.

WRONSKI, A. [1975] *On equivalential fragments of some intermediate logics* (J 0302) Rep Math Logic, Krakow & Katowice 4*105-112
◇ B55 ◇ REV MR 52 # 2832 Zbl 334 # 02010 • ID 17626

WRONSKI, A. [1975] see SURMA, S.J.

WRONSKI, A. [1976] *Remarks on Hallden completeness of modal and intermediate logics* (J 0387) Bull Sect Logic, Pol Acad Sci 5*126-129
◇ B45 ◇ REV MR 58 # 21485 • ID 27119

WRONSKI, A. [1980] *On reducts of intermediate logics* (J 0387) Bull Sect Logic, Pol Acad Sci 9*176-179
◇ B55 ◇ REV MR 82c:03033 Zbl 471 # 03019 • ID 55214

WRONSKI, A. [1981] see KABZINSKI, J.K.

WRONSKI, A. see Vol. I, III, VI for further entries

WU, WANGMING [1964] see WANG, SHIQIANG

WU, WANGMING [1980] see WANG, SHIQIANG

WU, WANGMING [1982] see CHENG, WEIMIN

WU, WANGMING see Vol. V for further entries

WU, XUEMOU [1978] *Studies and applications of pansystems analysis Ia: A brief survey of pansystems analysis (Chinese)* (J 2879) Wuhan Daxue Xuebao, Ziran Kexue 1978/3*87-105
◇ B52 E72 ◇ REV MR 82j:93002a • REM Part Ib 1979 • ID 83180

WU, XUEMOU [1979] *Studies and applications of pansystems analysis Ib: A brief survey of pansystems analysis (Chinese)* (J 2879) Wuhan Daxue Xuebao, Ziran Kexue 1979/1*104-117
◇ B52 E72 ◇ REV MR 82j:93002b • REM Part Ia 1978 • ID 83179

WU, XUEMOU [1980] see LI, CHULIN

WU, XUEMOU & ZHU, YUANZHAO [1980] *Pansystem analysis of dynamic yinyang logic and fuzziness I (Chinese) (English summary)* (J 2754) Huazhong Gongxueyuan Xuebao 1980/2*II,39-41
◇ B52 ◇ REV MR 82d:03037 • REM Special issue on fuzzy mathematics • ID 80262

WU, XUEMOU [1980] *Studies and applications of pansystems analysis IV: Pansystem observocontrollability, pansystem logic and fuzziness (Chinese) (English summary)* (J 2754) Huazhong Gongxueyuan Xuebao 1980/2*I,1-14
◇ B52 E72 ◇ REV MR 82j:93002d • REM Special issue on fuzzy mathematics. Part III 1981 • ID 83177

WU, XUEMOU [1980] see CHEN, HUAITING

WU, XUEMOU [1981] *Fuzziness, reliability and pansystem V: Analytic pansystem logic and its applications to the analysis of fuzzy sets (Chinese) (English summary)* (J 3732) Mohu Shuxue 1/1*15-20
◇ B52 ◇ REV MR 83h:03084 • ID 34845

WU, XUEMOU [1981] *Pansystems analysis: some new investigations of logic, observocontrollability and fuzziness* (J 2684) J Huazhong Inst Tech (Engl Ed) 3/1*15-25
◇ B52 E72 ◇ REV MR 83g:03062 • REM Transl. from the Chinese • ID 36019

WU, XUEMOU [1981] *Studies and applications of pansystems analysis III: Pansystems analysis - a new exploration of interdisciplinary investigation (Chinese) (English summary)* (J 3594) Kexue Tansuo 1981/1*125-164
◇ B52 E72 ◇ REV MR 82j:93002c • REM Part IV 1980 • ID 83178

WU, XUEMOU see Vol. V for further entries

WU, XUNWEI [1984] see LI, WENGUANG

WU, YUNZENG [1979] *Two problems of modern mathematical logic and its philosophic significance (Chinese)* (J 4452) Zhexue Yanjiu 2*51-55
◇ A05 B98 ◇ ID 48545

WU, YUNZENG see Vol. IV, V for further entries

WUTTICH, K. [1979] *Fragen des Psychologismus in der epistemischen Logik* (P 2539) Frege Konferenz (1);1979 Jena 519-533
◇ A05 B60 ◇ REV MR 82b:03023 • ID 80131

XIAO, XIAN [1983] *Fuzzy expressions and the property of order preserving (Chinese) (English summary)* (J 3732) Mohu Shuxue 3/3*37-44
◇ B52 ◇ REV MR 85e:03136 Zbl 521 # 04003 • ID 37088

XIAO, XIAN & ZHU, WUJIA [1984] *Foundations of classical mathematics and fuzzy mathematics (Chinese) (English summary)* (J 4430) Ziran Zazhi 7*723-726
◇ B52 ◇ ID 46827

XIAO, XIAN [1984] see HE, ZHONGXIONG

XIAO, XIAN [1984] *The improvement in bounds on the number of fuzzy expressions (Chinese)* (J 3732) Mohu Shuxue 4/2*87-88
◇ B52 ◇ REV MR 86g:03042 Zbl 561 # 94021 • ID 44363

XIAO, XIAN & ZHENG, YUXIN & ZHU, WUJIA [1985] *Finite-valued of infinite-valued logical paradoxes (Chinese)* (J 3742) Shuxue Yanjiu yu Pinglun 5/3*14
◇ B50 ◇ ID 49378

XU, LIZHI & YUAN, XIANGWAN & ZHENG, YUXIN & ZHU, WUJIA [1982] *Antinomies and the foundational problem of mathematics II (Chinese) (English summary)* (J 3742) Shuxue Yanjiu yu Pinglun 2/4*121-134
◇ A05 B98 ◇ REV MR 84g:03016b Zbl 505 # 03005 • REM Part I 1982. Part III 1983. • ID 34131

XU, LIZHI see Vol. I, VI for further entries

XU, WENLI [1982] see CHENG, WEIMIN

XU, XIAOSHU [1982] *The difference of the maximal closed sets in P_k (Chinese) (English summary)* (J 3948) Xiangtan Daxue Ziran Kexue Xuebao 1982/2*14-23
◇ B50 ◇ REV Zbl 525 # 03009 • ID 38248

XU, YONGSHEN [1984] see MO, SHAOKUI

XU, ZHIWEI [1984] *Multivalued logic and fuzzy logic - their relationship, minimization, and application to fault diagnosis* (J 0187) IEEE Trans Comp C-33*679-681
◇ B50 B52 ◇ REV Zbl 543 # 94028 • ID 40984

YABLON, P. [1975] *A generalized propositional calculus* (J 0047) Notre Dame J Formal Log 16*295-297
◇ B60 D35 ◇ REV MR 51 # 5259 Zbl 236 # 02021 • ID 17472

YABLONSKIJ, S.V. [1954] *On functional completeness in a three-valued calculus (Russian)* (J 0023) Dokl Akad Nauk SSSR 95*1153-1155
◇ B50 ◇ REV MR 15.925 Zbl 59.15 JSL 20.175 • ID 18002

YABLONSKIJ, S.V. [1958] *Functional constructions in a k-valued logic (Russian)* (S 0066) Tr Mat Inst Steklov 51*5-142
◇ B50 ◇ REV MR 21 # 3331 Zbl 92.251 JSL 29.214 • ID 18003

YABLONSKIJ, S.V. [1958] *On limit logics (Russian)* (J 0023) Dokl Akad Nauk SSSR 118*657-660
◇ B50 ◇ REV MR 22 # 9449 Zbl 85.245 • ID 18004

YABLONSKIJ, S.V. [1959] *Functional constructions in many-valued logics (Russian)* (P 0607) All-Union Math Conf (3);1956 Moskva 2*71-73
◇ B50 ◇ REV JSL 23.65 • ID 00090

YABLONSKIJ, S.V. [1959] *Functional constructions in many-valued logic (Russian)* (P 0607) All-Union Math Conf (3);1956 Moskva 3*425-431
◇ B50 ◇ REV Zbl 86.245 • ID 28399

YABLONSKIJ, S.V. [1959] *Some properties of enumerable closed classes from $P_{\aleph_0}$ (Russian)* (J 0023) Dokl Akad Nauk SSSR 124*990-993
◇ B50 ◇ REV MR 21 # 1936 Zbl 91.14 • ID 18006

YABLONSKIJ, S.V. [1963] *On superpositions of functions in P_k (Russian)* (J 0052) Probl Kibern 9*337-340
• TRANSL [1963] (J 0449) Probl Kybern 6*345-349
◇ B50 ◇ REV Zbl 166.271 JSL 31.502 • ID 00077

YABLONSKIJ, S.V. & ZAKHAROVA, E.YU. [1964] *On some properties of essential functions from P_k (Russian)* (J 0052) Probl Kibern 12*247-252
◇ B50 E20 ◇ REV MR 32 # 7417 Zbl 262 # 94026 JSL 31.502 • ID 00078

YABLONSKIJ, S.V. [1969] see KUDRYAVTSEV, V.B.

YABLONSKIJ, S.V. & ZAKHAROVA, E.YU. [1972] *Certain properties of nondegenerate superpositions in P_k (Russian)* (J 0087) Mat Zametki (Akad Nauk SSSR) 12*3-12
• TRANSL [1972] (J 1044) Math Notes, Acad Sci USSR 12*435-440
◇ B50 ◇ REV MR 48 # 13509 Zbl 242 # 02017 • ID 18089

YABLONSKIJ, S.V. [1974] *The structure of the upper neighborhood for predicate-describable classes in P_k (Russian)* (J 0023) Dokl Akad Nauk SSSR 218*304-307
• TRANSL [1974] (J 0062) Sov Math, Dokl 15*1353-1356
◇ B50 ◇ REV MR 50 # 6792 Zbl 308 # 02021 • ID 06482

YABLONSKIJ, S.V. [1979] *Introduction to discrete mathematics (Russian)* (X 2027) Nauka: Moskva 272pp
• TRANSL [1983] (X 0885) Mir: Moskva 285pp
◇ B50 ◇ REV MR 82b:94001 Zbl 498 # 94001 • ID 38535

YABLONSKIJ, S.V. [1980] *On some results in the theory of function systems (Russian)* (P 1959) Int Congr Math (II,13);1978 Helsinki 2*963-971
• TRANSL [1981] (J 0225) Amer Math Soc, Transl, Ser 2 117*39-46
◇ B50 D20 G25 ◇ REV MR 81j:03038 Zbl 473 # 03019 • ID 55350

YABLONSKIJ, S.V. see Vol. I for further entries

YAGER, R.R. [1979] *A note on fuzziness in a standard uncertainty logic* (J 2338) IEEE Trans Syst Man & Cybern SMC-9*387-388
◇ B52 ◇ REV MR 81g:03028 Zbl 409 # 03013 • ID 56316

YAGER, R.R. [1980] *A foundation for a theory of possibility* (J 0383) J Cybern 10*177-204
◇ B52 ◇ REV MR 81e:03022 Zbl 438 # 94042 • ID 80144

YAGER, R.R. [1980] *An approach to inference in approximate reasoning* (J 1741) Int J Man-Mach Stud 13*323-338
◇ B52 ◇ REV MR 82a:94135 • ID 83183

YAGER, R.R. [1980] *Fuzzy thinking as quick and efficient* (J 2695) Cybernetica 23*265-298
◇ B52 E72 ◇ REV MR 82d:03089 Zbl 453 # 94029 • ID 54211

YAGER, R.R. [1980] *Generalized "and/or" operators for multivalued and fuzzy logic* (P 3673) Int Symp Multi-Val Log (10);1980 Evanston 214-218
◇ B50 B97 ◇ REV MR 83m:94032 Zbl 539 # 03010 • ID 43745

YAGER, R.R. [1980] *On a general class of fuzzy connectives* (J 2720) Fuzzy Sets Syst 4*235-242
◇ B52 ◇ REV MR 81m:03066 Zbl 443 # 04008 • ID 80140

YAGER, R.R. [1980] *On choosing between fuzzy subsets* (J 1429) Kybernetes 9*151-154
◇ B52 E72 ◇ REV Zbl 428 # 03050 • ID 53809

YAGER, R.R. [1980] *Some observations on probabilistic qualification in approximate reasoning* (J 0191) Inform Sci 22*217-234
◇ B52 ◇ REV MR 82c:03032 Zbl 452 # 94044 • ID 80139

YAGER, R.R. [1981] *Quantified propositions in a linguistic logic* (P 3167) Fuzzy Set Theory (2);1980 Linz 69-124
◇ B52 ◇ REV MR 83c:03024 Zbl 472 # 03019 • ID 55285

YAGER, R.R. (ED.) [1982] *Fuzzy set and possibility theory. Recent developments* (X 0869) Pergamon Pr: Oxford xiv + 633pp
◇ B52 ◇ REV MR 84b:03004 • ID 34904

YAGER, R.R. [1982] *Generalized probabilities of fuzzy events from fuzzy belief structures* (J 0191) Inform Sci 28*45-62
◇ B52 ◇ REV MR 84f:60006 Zbl 525 # 60006 • ID 37488

YAGER, R.R. [1982] *Some procedures for selecting fuzzy set-theoretic operators* (J 1743) Int J Gen Syst 8*115-124
◇ B52 ◇ REV MR 83i:03091 Zbl 488 # 04005 • ID 35537

YAGER, R.R. [1983] *Entropy and specificity in a mathematical theory of evidence* (J 1743) Int J Gen Syst 9*249-260
◇ B52 ◇ REV MR 85d:03049 Zbl 521 # 94008 • ID 41092

YAGER, R.R. [1983] *On the implication operator in fuzzy logic* (J 0191) Inform Sci 31*141-164
◇ B52 ◇ REV MR 85a:03036 Zbl 557 # 03015 • ID 34787

YAGER, R.R. [1983] *Presupposition in binary and fuzzy logics* (J 1429) Kybernetes 12*135-139
◇ B52 ◇ REV MR 85a:03037 Zbl 514 # 03014 • ID 34788

YAGER, R.R. [1983] *Some relationships between possibility, truth and certainty* (J 2720) Fuzzy Sets Syst 11*151-156
◇ B52 E72 ◇ REV MR 84k:03135 Zbl 524 # 94038 • ID 35027

YAGER, R.R. [1985] *Aggregation evidence using quantified statements* (J 0191) Inform Sci 36*179-206
◇ B52 B75 ◇ ID 49504

YAGER, R.R. [1985] *Knowledge trees in complex knowledge bases* (J 2720) Fuzzy Sets Syst 15*45-64
◇ B52 ◇ ID 45103

YAGER, R.R. [1985] *Reasoning with fuzzy quantified statements I* (J 1429) Kybernetes 14/4*233-240
◇ B52 ◇ ID 49432

YAGER, R.R. [1985] *Strong truth and rules of inference in fuzzy logic and approximate reasoning* (J 2694) Cybern & Syst 16*23-63
◇ B52 ◇ ID 49754

YAGER, R.R. see Vol. V for further entries

YAMAMOTO, A. [1977] *Algebraic proof of separability of certain modal logics* (J 0381) J Tsuda College (Tokyo) 9*29-41
◇ B45 ◇ REV MR 57 # 9489 • ID 80150

YAMAMOTO, A. [1978] *Cut-elimination theorem for multi-modal logics* (J 0381) J Tsuda College (Tokyo) 10*21-34
◇ B45 ◇ REV MR 58 # 156 • ID 80149

YAMAMOTO, Y. [1981] see MINE, H.

YANASE, M.M. [1980] *Hidden realism. I* (J 0260) Ann Jap Ass Phil Sci 244*8
◇ A05 B52 ◇ REV MR 81i:03010 • ID 80152

YANASE, M.M. [1985] *Fuzziness and probability* (J 0260) Ann Jap Ass Phil Sci 6*219-226
◇ A05 B48 B52 E72 ◇ REV Zbl 575 # 03014 • ID 49435

YANKOV, V.A. [1963] *Realizable formulas of propositional logic (Russian)* (J 0023) Dokl Akad Nauk SSSR 151*1035-1037
• TRANSL [1963] (J 0062) Sov Math, Dokl 4*1146-1148
◇ B55 F50 ◇ REV MR 27 # 4746 Zbl 143.251 JSL 35.138 • ID 19006

YANKOV, V.A. [1963] *Some superconstructive propositional calculi (Russian)* (J 0023) Dokl Akad Nauk SSSR 151*796-798
• TRANSL [1963] (J 0062) Sov Math, Dokl 4*1103-1105
◇ B55 ◇ REV MR 27 # 3499 Zbl 143.251 JSL 35.138 • ID 06512

YANKOV, V.A. [1963] *The relationship between deducibility in the intuitionistic propositional calculus and finite implicational structures (Russian)* (J 0023) Dokl Akad Nauk SSSR 151*1293-1294
• TRANSL [1963] (J 0062) Sov Math, Dokl 4*1203-1204
◇ B55 F50 G25 ◇ REV MR 27 # 5685 Zbl 143.252 JSL 35.138 • ID 16539

YANKOV, V.A. [1967] *Finite validity of formulas of a special form (Russian)* (J 0023) Dokl Akad Nauk SSSR 174*302-304
• TRANSL [1967] (J 0062) Sov Math, Dokl 8*648-650
◇ B55 C13 F50 ◇ REV MR 36 # 2468 Zbl 209.304 JSL 38.331 • ID 06513

YANKOV, V.A. [1968] *Constructing a sequence of strongly independent superintuitionistic propositional calculi (Russian)* (J 0023) Dokl Akad Nauk SSSR 181*33-34
• TRANSL [1968] (J 0062) Sov Math, Dokl 9*806-807
◇ B55 ◇ REV MR 38 # 984 Zbl 198.318 JSL 37.186 • ID 19009

YANKOV, V.A. [1968] *On the extension of the intuitionist propositional calculus to the classical calculus, and the minimal calculus to the intuitionist calculus* (J 0216) Izv Akad Nauk SSSR, Ser Mat 32*208-211
• TRANSL [1968] (J 0448) Math of USSR, Izv 2*205-208
◇ B55 F50 ◇ REV MR 37 # 31 Zbl 191.286 JSL 38.331
• ID 06514

YANKOV, V.A. [1968] *The calculus of the weak "law of excluded middle" (Russian)* (J 0216) Izv Akad Nauk SSSR, Ser Mat 32*1044-1051
• TRANSL [1968] (J 0448) Math of USSR, Izv 2*997-1004
◇ B55 ◇ REV MR 38 # 5588 Zbl 165.12 JSL 37.186
• ID 06515

YANKOV, V.A. [1968] *Three sequences of formulas with two variables in positive propositional logic (Russian)* (J 0216) Izv Akad Nauk SSSR, Ser Mat 32*880-893
• TRANSL [1968] (J 0448) Math of USSR, Izv 2*845-848
◇ B55 ◇ REV MR 38 # 5584 Zbl 209.302 • ID 06516

YANKOV, V.A. [1969] *Conjunctively indecomposable formulas in propositional calculi (Russian)* (J 0216) Izv Akad Nauk SSSR, Ser Mat 33*18-38
• TRANSL [1969] (J 0448) Math of USSR, Izv 3*17-35
◇ B55 F50 G10 ◇ REV MR 39 # 5318 Zbl 181.4
JSL 37.186 • ID 06518

YANKOV, V.A. see Vol. IV, V for further entries

YANOV, YU.I. [1959] see MUCHNIK, A.A.

YANOV, YU.I. see Vol. III, IV, VI for further entries

YAO, D.D.W. [1982] see TUERKSEN, I.B.

YAO, D.D.W. [1984] see TUERKSEN, I.B.

YAO, J.T.P. [1982] see FU, K.S.

YASHIN, A.D. [1982] *Intuitionistic predicate logic with the connective "tomorrow" (Russian) (English summary)* (J 0288) Vest Ser Mat Mekh, Univ Moskva 1982/4*19-22,84
• TRANSL [1982] (J 0510) Moscow Univ Math Bull 37/4*22-26
◇ B45 ◇ REV MR 84d:03027 Zbl 494 # 03017 • ID 34065

YASHIN, A.D. [1985] *Semantic characterization of intuitionistic logical connectives (Russian)* (J 0087) Mat Zametki (Akad Nauk SSSR) 38*157-167,171
◇ B55 ◇ ID 48584

YASHIN, A.D. see Vol. III, VI for further entries

YASHIN, A.L. [1977] *On an extension of intuitionistic logic and its interpretations (Russian)* (S 2579) Teor Mnozhestv & Topol (Izhevsk) 1*110-112
◇ B55 C90 ◇ REV Zbl 481 # 03036 • ID 36830

YASUGI, M. [1981] *Definitive valuation of set theory* (J 0407) Comm Math Univ St Pauli (Tokyo) 30*175-191
◇ B50 E70 ◇ REV MR 84b:03077 Zbl 473 # 03047
• ID 55376

YASUGI, M. [1982] *Continuous valuation and logic* (J 0111) Nagoya Math J 85*175-188
◇ B50 ◇ REV MR 85g:03036 Zbl 449 # 03022 • ID 55625

YASUGI, M. see Vol. III, V, VI for further entries

YASUHARA, A. [1971] *Recursive function theory and logic* (X 0801) Academic Pr: New York xv + 338pp
◇ B98 D10 D20 D98 ◇ REV MR 47 # 1582
Zbl 254 # 02002 JSL 40.619 • ID 18028

YASUHARA, A. see Vol. IV for further entries

YASUURA, K. [1955] *On the representation of many-valued propositional logics by relay circuits (Japanese)* (J 4442) Kyushu Daigaku Kagaku Syuho 28/2*94-96
◇ B50 B70 ◇ REV JSL 22.102 • ID 48775

YEH, R.T. [1971] see PREPARATA, F.P.

YEH, R.T. [1972] see PREPARATA, F.P.

YEH, R.T. see Vol. IV, V for further entries

YOKATA, S. [1985] *General charakterization results on intuitionistic modal propositional logics* (J 0407) Comm Math Univ St Pauli (Tokyo) 34/2*177-199
◇ B45 ◇ ID 49375

YONEMITSU, N. [1951] *On systems of strict implication* (J 0261) Tohoku Math J 3*48-58
◇ B45 ◇ REV MR 12.789 Zbl 44.2 JSL 16.278 • ID 18052

YONEMITSU, N. [1954] *A decision method and a topological interpretation for systems of logical implication* (S 1181) Mem Osaka Univ Lib Arts Educ, Ser B 3*6-20
◇ B45 ◇ REV MR 17.119 JSL 21.326 • ID 42130

YONEMITSU, N. [1954] *A note on systems of logical implication* (S 1181) Mem Osaka Univ Lib Arts Educ, Ser B 3*21-24
◇ B45 ◇ REV MR 17.119 JSL 21.327 • ID 42131

YONEMITSU, N. [1954] *Note on the completeness of m-valued propositional calculus* (J 0352) Math Jap 3*57-61
◇ B50 ◇ REV MR 17.224 Zbl 57.247 JSL 21.328
• ID 18053

YONEMITSU, N. [1955] *A note on modal systems, von Wright's M and Lewis's S1* (S 1181) Mem Osaka Univ Lib Arts Educ, Ser B 4*45
◇ B45 ◇ REV MR 22 # 6698 JSL 21.379 • REM Part I. Part II 1957 • ID 42135

YONEMITSU, N. [1957] *A note on modal systems II* (S 1181) Mem Osaka Univ Lib Arts Educ, Ser B 6*9-10
◇ B45 ◇ REV MR 22 # 6699 JSL 23.64 • REM Part I 1955
• ID 20821

YONEMITSU, N. [1960] *Systems of weak implications* (S 1181) Mem Osaka Univ Lib Arts Educ, Ser B 9*137-158
◇ B45 ◇ REV JSL 28.256 • ID 22131

YONEZAKI, N. [1980] see ENOMOTO, H.

YOURGRAU, P. [1983] *Knowledge and relevant alternatives* (J 0154) Synthese 55*175-190
◇ B48 ◇ REV MR 84i:03049 • ID 34528

YOURGRAU, P. [1985] *On the logic of indeterminist time. With a comment by Paul Fitzgerald* (J 0407) Comm Math Univ St Pauli (Tokyo) 82/10*548-562
◇ B45 ◇ ID 49377

YU, HONGZU [1980] see LI, CHULIN

YU, LIANSHENG [1982] *Research into the theory of fuzzy clustering: principle of the maximum element of a matrix (Chinese) (English summary)* (J 3750) Jilin Daxue, Ziran Kexue Xuebao 1982/4*106-114
◇ B52 E72 ◇ REV MR 84m:03085 • ID 35782

YUAN, XIANGWAN [1982] see XU, LIZHI

YUAN, XIANGWAN [1984] see HE, ZHONGXIONG

YUAN, XIANGWAN see Vol. I, VI for further entries

YUE, YUJUN [1980] see SHI, NIANDONG

ZABECH, F. [1974] see JACOBSON, A.

ZABEZHAJLO, M.I. [1980] see ANSHAKOV, O.M.

ZABEZHAJLO, M.I. [1982] see ANSHAKOV, O.M.

ZACHOROWSKI, S. [1973] *Wajsberg's semantics for the system S5 of Lewis* (J 0387) Bull Sect Logic, Pol Acad Sci 2*112-117
◇ B45 ◇ REV MR 55 # 7742 • ID 80188

ZACHOROWSKI, S. [1974] see SURMA, S.J.

ZACHOROWSKI, S. [1975] *A proof of a conjecture of Roman Suszko* (J 0063) Studia Logica 34*253-256
◇ B45 B50 F50 ◇ REV MR 54 # 65 Zbl 325 # 02013
• ID 23947

ZACHOROWSKI, S. [1975] *Consistency of the scheme of comprehension with the Lukasiewicz's logic (Polish and English)* (S 1454) Zesz Nauk Wyz Szk Ped Mat, Opole 15*21-28
◇ B50 E35 E70 ◇ REV MR 55 # 81 Zbl 334 # 02009
• ID 66185

ZACHOROWSKI, S. [1975] see SURMA, S.J.

ZACHOROWSKI, S. [1977] *Intermediate logics without the interpolation property* (J 0387) Bull Sect Logic, Pol Acad Sci 6*161-163
◇ B55 C40 ◇ REV MR 57 # 15980 Zbl 404 # 03019
• ID 54806

ZACHOROWSKI, S. [1977] see KRZYSTEK, P.S.

ZACHOROWSKI, S. [1978] *Dummett's LC has the interpolation property* (J 0387) Bull Sect Logic, Pol Acad Sci 7*58-60
◇ B55 C40 C90 ◇ REV MR 58 # 16174 Zbl 407 # 03046
• ID 56212

ZACHOROWSKI, S. [1978] *Remarks on interpolation property for intermediate logics* (J 0302) Rep Math Logic, Krakow & Katowice 10*139-146
◇ B55 ◇ REV MR 81b:03027 Zbl 419 # 03013 • ID 53356

ZADEH, L.A. [1965] *Fuzzy sets* (J 0194) Inform & Control 8*338-353
◇ B52 E72 ◇ REV MR 36 # 2509 Zbl 139.246 JSL 38.656
• ID 24934

ZADEH, L.A. [1968] *Probability measures of fuzzy events* (J 0034) J Math Anal & Appl 23*421-427
◇ B52 ◇ REV MR 37 # 6131 Zbl 174.490 • ID 49799

ZADEH, L.A. [1970] see BELLMAN, R.E.

ZADEH, L.A. [1971] *Quantitative fuzzy semantics* (J 0191) Inform Sci 3*159-176
◇ B52 E72 ◇ REV MR 44 # 3816 Zbl 218 # 02057
• ID 26291

ZADEH, L.A. [1972] see CHANG, S.S.L.

ZADEH, L.A. [1974] *Fuzzy logic and its application to approximate reasoning* (P 1691) Inform Processing (6);1974 Stockholm 591-594
◇ B52 ◇ REV MR 53 # 12123 Zbl 361 # 68126 • ID 50719

ZADEH, L.A. [1975] *Calculus of fuzzy restrictions* (P 0774) Fuzzy Sets & Appl;1974 Berkeley 1-39
◇ B52 ◇ REV MR 55 # 10239 Zbl 327 # 02018 • ID 66188

ZADEH, L.A. [1975] *Fuzzy logic and approximate reasoning* (J 0154) Synthese 30*407-428
◇ B52 ◇ REV Zbl 319 # 02016 • ID 29673

ZADEH, L.A. [1975] *The concept of a linguistic variable and its application to approximate reasoning. I,II,III* (J 0191) Inform Sci 8*199-249,8*301-357,9*43-80
◇ B52 B65 ◇ REV MR 52 # 7225 Zbl 397 # 68071
Zbl 404 # 680074 Zbl 404 # 68075 • ID 18434

ZADEH, L.A. [1976] *A fuzzy-algorithmic approach to the definition of complex or imprecise concepts* (J 1741) Int J Man-Mach Stud 8*249-291
◇ B52 E72 ◇ REV MR 82k:68021 Zbl 332 # 68068
• ID 83203

ZADEH, L.A. [1976] *Semantic influence from fuzzy premises* (P 2011) Int Symp Multi-Val Log (6);1976 Logan 217-218
◇ B52 ◇ ID 35892

ZADEH, L.A. [1977] see BELLMAN, R.E.

ZADEH, L.A. [1978] *PRUF - a meaning representation language for natural languages* (J 1741) Int J Man-Mach Stud 10*395-460
◇ B52 ◇ REV MR 58 # 25172 Zbl 406 # 68063 • ID 56162

ZADEH, L.A. [1979] *A theory of approximate reasoning* (J 0508) Machine Intelligence 9*149-194
◇ B52 ◇ REV MR 81m:03034 • ID 80192

ZADEH, L.A. [1980] *Inference in fuzzy logic* (P 3673) Int Symp Multi-Val Log (10);1980 Evanston 124-131
◇ B52 ◇ REV MR 83m:94032 Zbl 546 # 03014 • ID 41149

ZADEH, L.A. [1983] *A computational approach to fuzzy quantifiers in natural languages* (J 2687) Comp Math Appl 9*149-184
◇ B52 ◇ REV MR 85a:03039 Zbl 517 # 94028 • ID 34790

ZADEH, L.A. [1983] *The role of fuzzy logic in the management of uncertainty in expert systems* (J 2720) Fuzzy Sets Syst 11*199-227
◇ B52 ◇ REV MR 85a:03038 Zbl 553 # 68050 • ID 34789

ZADEH, L.A. [1984] *A theory of commonsense knowledge* (P 3096) Conf Math Service of Man (2,Pap);1982 Las Palmas 257-295
◇ B52 ◇ REV MR 86e:03027 • ID 45927

ZADEH, L.A. [1984] *Coping with the imprecision of the real world* (J 0212) ACM Commun 27*304-311
◇ B52 ◇ ID 45040

ZADEH, L.A. [1984] *Fuzzy probabilities* (J 1185) Inform Process & Managmt 20*363-372
◇ B52 ◇ REV Zbl 543 # 60007 • ID 40948

ZADEH, L.A. [1985] *Syllogistic reasoning in fuzzy logic and its applications to usuality and reasoning with dispositions* (J 2338) IEEE Trans Syst Man & Cybern 15*754-763
◇ B52 ◇ ID 49367

ZADEH, L.A. see Vol. V for further entries

ZAHN, P. [1982] *Ein argumentativer Weg zur Logik* (X 0890) Wiss Buchges: Darmstadt 212pp
◇ B98 ◇ REV MR 84f:03003 Zbl 502 # 03001 • ID 34423

ZAHN, P. see Vol. I, VI for further entries

ZAKHAROVA, E.YU. [1964] see YABLONSKIJ, S.V.

ZAKHAROVA, E.YU. [1966] *On a sufficient condition for completeness in* P_k *(Russian)* (J 0052) Probl Kibern 16*239-244
◇ B50 ◇ REV JSL 33.606 • ID 18082

ZAKHAROVA, E.YU. [1967] *Criteria of completeness for systems of functions from P_k (Russian)* (J 0052) Probl Kibern 18*5-10
◇ B50 ◇ REV Zbl 268 # 02010 JSL 33.606 • ID 18083

ZAKHAROVA, E.YU. [1969] see KUDRYAVTSEV, V.B.

ZAKHAROVA, E.YU. [1971] *A remark on the maximal order of a simple basis in P_k (Russian)* (J 0071) Met Diskr Analiz (Novosibirsk) 18*31-34
◇ B50 ◇ REV MR 45 # 3172 Zbl 242 # 02016 • ID 18086

ZAKHAROVA, E.YU. [1972] see YABLONSKIJ, S.V.

ZAKHAROVA, E.YU. [1972] *The realization of functions of the P_k by formulae ($k \geqslant 3$) (Russian)* (J 0087) Mat Zametki (Akad Nauk SSSR) 11*99-108
• TRANSL [1972] (J 1044) Math Notes, Acad Sci USSR 11*64-69
◇ B50 ◇ REV MR 45 # 3173 Zbl 268 # 02011 • ID 18087

ZAKHAR'YASHCHEV, M.V. [1981] *Some classes of intermediate logics (Russian) (English summary)* (S 2651) Prepr Inst Prikl Mat, Akad Nauk SSSR 1981*28pp
◇ B55 ◇ REV MR 83j:03047 • ID 35349

ZAKHAR'YASHCHEV, M.V. [1983] *On intermediate logics (Russian)* (J 0023) Dokl Akad Nauk SSSR 269*18-22
• TRANSL [1983] (J 0062) Sov Math, Dokl 27*274-277
◇ B55 C90 ◇ REV MR 85e:03060 Zbl 548 # 03009
• ID 40499

ZAKHAR'YASHCHEV, M.V. [1984] *Normal modal logics containing S4 (Russian)* (J 0023) Dokl Akad Nauk SSSR 275*537-540
• TRANSL [1984] (J 0062) Sov Math, Dokl 29*252-255
◇ B45 ◇ REV MR 85e:03052 Zbl 548 # 03009 • ID 40620

ZAKHAR'YASHCHEV, M.V. see Vol. I, III, VI for further entries

ZAKRZEWSKA, T. [1976] *Some remarks on three-valued implicative sentential calculi* (J 0387) Bull Sect Logic, Pol Acad Sci 5*9-12
◇ B50 ◇ REV MR 53 # 5259 • ID 22891

ZALTA, E.N. [1980] see MCMICHAEL, A.

ZALTA, E.N. see Vol. I for further entries

ZANARDO, A. [1981] *A completeness theorem for the general interpreted modal calculus MC^ν of A. Bressan* (J 0144) Rend Sem Mat Univ Padova 64*39-57
◇ B45 ◇ REV MR 83h:03032b Zbl 484 # 03006 • ID 36050

ZANARDO, A. [1981] see BRESSAN, A.

ZANARDO, A. [1981] *Teorema di completezza per il calcolo modale interpretato MC^ν* (P 3092) Congr Naz Logica;1979 Montecatini Terme 473-486
◇ B45 ◇ ID 48727

ZANARDO, A. [1983] *On the equivalence between the calculi MC^ν and $EC^{\nu+1}$ of A. Bressan* (J 0047) Notre Dame J Formal Log 24*367-388
◇ B45 ◇ REV MR 85d:03044 Zbl 524 # 03010 • ID 37591

ZANARDO, A. [1984] *Individual concepts as propositional variables in $ML^{\nu+1}$* (J 0047) Notre Dame J Formal Log 25*332-346
◇ B45 ◇ REV Zbl 555 # 03009 • ID 42577

ZANDAROWSKA, W. [1971] see SUSZKO, R.

ZANDAROWSKA, W. [1976] *The modal system* S3 *and* SCI (J 0387) Bull Sect Logic, Pol Acad Sci 5*13-15
◇ B45 ◇ REV MR 53 # 12881 • ID 23139

ZANDAROWSKA, W. see Vol. I for further entries

ZARACH, A. [1976] see MAREK, W.

ZARACH, A. see Vol. III, IV, V, VI for further entries

ZARNECKA-BIALY, E. [1966] *The intensional functions (Polish) (English summary)* (S 0458) Zesz Nauk, Prace Log, Uniw Krakow 2*47-54
◇ B45 ◇ ID 16509

ZARNECKA-BIALY, E. [1968] *A note on deduction theorem for Goedel's propositional calculus G4 (Polish and Russian summaries)* (J 0063) Studia Logica 23*35-41
◇ B45 ◇ REV MR 43 # 6049 Zbl 307 # 02012 • ID 14360

ZARNECKA-BIALY, E. [1969] *Undefinability of possibility and necessity functors in some poor purely implicational logical calculi (Polish) (English summary)* (S 0458) Zesz Nauk, Prace Log, Uniw Krakow 4*47-52
◇ B45 ◇ REV MR 41 # 43 • ID 14361

ZARNECKA-BIALY, E. [1970] *The deduction theorem for Goedel's propositional calculus G5 (Polish summary)* (S 0458) Zesz Nauk, Prace Log, Uniw Krakow 5*77-78
◇ B45 ◇ REV MR 44 # 1545 • ID 14362

ZARNECKA-BIALY, E. [1973] *Modal functors and their definability in propositional calculus systems (Polish) (English summary)* (J 0302) Rep Math Logic, Krakow & Katowice 1*68-98
◇ B45 F50 ◇ REV MR 49 # 4757 Zbl 292 # 02020
• ID 14364

ZARNECKA-BIALY, E. see Vol. I, VI for further entries

ZASLAVSKIJ, I.D. [1978] *Symmetric constructive logic (Russian)* (X 2225) Akad Nauk Armyan SSR : Erevan 281pp
◇ B55 F05 F30 F50 ◇ REV MR 81a:03059 • ID 80231

ZASLAVSKIJ, I.D. [1979] *The realization of three-valued logical functions by means of recursive and Turing operators (Russian)* (S 0554) Issl Teor Algor & Mat Logik (Moskva) 3*52-61
◇ B50 D10 D20 ◇ REV MR 81i:03031 Zbl 478 # 03007
• ID 55622

ZASLAVSKIJ, I.D. see Vol. IV, VI for further entries

ZAVISCA, E. [1971] see ALLEN, C.M.

ZAVISCA, E. [1972] *Multiple-valued combinational function synthesis using a single gate type* (P 2008) Symp Th & Appl of Multi-Val Log Design;1972 Buffalo 193-201
◇ B50 ◇ ID 42048

ZAWIRSKI, Z. [1934] *Bedeutung der mehrwertigen Logik fuer die Erkenntnis und ihr Zusammenhang mit der Wahrscheinlichkeitsrechnung (Polnisch)* (J 1125) Przeglad Filoz 37*393-398
• TRANSL [1936] (P 4508) Congr Int Phil (8); 175-180
◇ B48 ◇ REV JSL 2.139 • ID 14378

ZAWIRSKI, Z. [1935] *Ueber das Verhaeltnis der mehrwertigen Logik zur Wahrscheinlichkeitsrechnung* (J 4716) Stud Philos, Leopolis (Poznan) 1*407-442
◇ B50 ◇ REV Zbl 15.339 JSL 1.198 FdM 62.1053
• ID 45790

ZAWIRSKI, Z. see Vol. VI for further entries

ZECCA, A. [1981] *Products of logics* (**P** 3820) Curr Iss in Quantum Log;1979 Erice 405–412
◇ B51 ◇ REV MR 84k:03145 Zbl 537 # 03044 • ID 35037

ZECHA, G. [1972] see MORSCHER, E.

ZEMACH, E.M. [1982] *Schematic objects and relative identity* (**J** 0097) Nous, Quart J Phil 16*295–305
◇ B45 ◇ REV MR 83h:03031 • ID 36049

ZEMAN, J.J. [1963] *Bases for S4 and S4.2 without added axioms* (**J** 0047) Notre Dame J Formal Log 4*227–230
◇ B45 ◇ REV MR 29 # 4685 Zbl 147.249 JSL 38.328 • ID 14390

ZEMAN, J.J. [1967] *A system of implicit quantification* (**J** 0036) J Symb Logic 32*480–504
◇ B60 ◇ REV MR 37 # 1225 Zbl 175.261 • ID 14392

ZEMAN, J.J. [1967] *The deduction theorem in S4, S4.2 and S5* (**J** 0047) Notre Dame J Formal Log 8*56–59
◇ B45 ◇ REV MR 38 # 4291 Zbl 189.282 • ID 14391

ZEMAN, J.J. [1968] *Lemmon-style bases for the systems* $S1^o - S4^o$ (**J** 0036) J Symb Logic 33*458–461
◇ B45 ◇ REV MR 38 # 989 Zbl 169.6 • ID 14393

ZEMAN, J.J. [1968] *Some calculi with strong negation primitive* (**J** 0036) J Symb Logic 33*97–100
◇ B45 F50 ◇ REV MR 38 # 3140 Zbl 167.8 • ID 14395

ZEMAN, J.J. [1968] *The propositional calculus MC and its modal analog* (**J** 0047) Notre Dame J Formal Log 9*294–298
◇ B45 ◇ REV MR 41 # 1515 Zbl 182.5 • ID 14396

ZEMAN, J.J. [1968] *The semisubstitutivity of strict implication* (**J** 0036) J Symb Logic 33*462–464
◇ B45 ◇ REV MR 38 # 990 • ID 14394

ZEMAN, J.J. [1969] *Complete modalization in S4.4 and S4.04* (**J** 0047) Notre Dame J Formal Log 10*257–260
◇ B45 ◇ REV MR 40 # 1269 Zbl 184.8 • ID 14397

ZEMAN, J.J. [1969] *Decision procedures for* $S3^o$ *and* $S4^o$ (**J** 0009) Arch Math Logik Grundlagenforsch 12*155–158
◇ B25 B45 ◇ REV MR 43 # 4645 Zbl 195.12 • ID 14399

ZEMAN, J.J. [1969] *Modal systems in which necessity is "factorable"* (**J** 0047) Notre Dame J Formal Log 10*247–256
◇ B45 ◇ REV MR 40 # 1268 Zbl 184.8 • ID 14398

ZEMAN, J.J. [1971] *A study of some systems in the neighbourhood of* S4.4 (**J** 0047) Notre Dame J Formal Log 12*341–357
◇ B45 ◇ REV MR 45 # 6596 Zbl 206.7 JSL 40.602 • ID 14400

ZEMAN, J.J. [1972] *Semantics for* S4.3.2 (**J** 0047) Notre Dame J Formal Log 13*454–460
◇ B45 ◇ REV MR 47 # 6449 Zbl 226 # 02017 JSL 38.328 • ID 14401

ZEMAN, J.J. [1972] *S4.6 is S4.9* (**J** 0047) Notre Dame J Formal Log 13*118
◇ B45 ◇ REV MR 45 # 6597 Zbl 227 # 02010 • ID 27355

ZEMAN, J.J. [1973] *Modal logic. The Lewis-modal system* (**X** 0815) Clarendon Pr: Oxford x+302pp
◇ B45 ◇ REV MR 53 # 2638 Zbl 255 # 02014 JSL 42.581 • ID 21525

ZEMAN, J.J. [1978] *Generalized normal logic* (**J** 0122) J Philos Logic 7*225-243
◇ B45 G10 ◇ REV MR 80j:03035 Zbl 413 # 03014 JSL 48.206 • ID 53008

ZEMAN, J.J. [1979] *Normal implications, bounded posets, and the existence of meets* (**J** 0047) Notre Dame J Formal Log 20*685–688
◇ B45 G10 ◇ REV MR 80m:03048 Zbl 406 # 03036 • ID 56119

ZEMAN, J.J. [1979] *Normal, Sasaki, and classical implications* (**J** 0122) J Philos Logic 8*243–245
◇ B60 ◇ REV MR 80f:03028 Zbl 429 # 03044 • ID 53875

ZEMAN, J.J. [1979] *Quantum logic with implication* (**J** 0047) Notre Dame J Formal Log 20*723–728
◇ B51 G12 ◇ REV MR 80m:03099 Zbl 426 # 03066 • ID 29954

ZEMAN, J.J. [1979] *Two basic pure-implicational systems* (**J** 0047) Notre Dame J Formal Log 20*674–684
◇ B45 ◇ REV MR 80h:03036 Zbl 299 # 02023 • ID 56120

ZENNER, R.B.R.C. [1985] see CALUWE DE, R.M.M.

ZEUGMANN, T. [1983] *On the synthesis of fastest programs in inductive inference* (**J** 0129) Elektr Informationsverarbeitung & Kybern 19*625–642
◇ B48 ◇ REV Zbl 545 # 68034 • ID 41209

ZEUGMANN, T. see Vol. IV for further entries

ZHANG, HONGYU [1979] *A simplification for ternary functions (Chinese) (English summary)* (**J** 2754) Huazhong Gongxueyuan Xuebao 7/1*2,18–23
• TRANSL [1979] (**J** 2684) J Huazhong Inst Tech (Engl Ed) 1/2*12–17
◇ B50 ◇ REV MR 84h:03051 • ID 37400

ZHANG, JINWEN [1979] *Fuzzy logic and its formal system* (**P** 4588) Logic in China;1978 Beijing 408–416
◇ B52 ◇ ID 48550

ZHANG, JINWEN [1980] *A unified treatment of fuzzy set theory and boolean-valued set theory - fuzzy set structures and normal fuzzy set structures* (**J** 0034) J Math Anal & Appl 76*297–301
◇ B50 E40 E72 ◇ REV MR 83i:03092c Zbl 452 # 03043 • ID 54107

ZHANG, JINWEN [1981] *System logic and fuzzy set theory (Chinese) (English summary)* (**J** 3732) Mohu Shuxue 1/1*73–77
◇ B52 E72 ◇ REV MR 83k:03033 • ID 34874

ZHANG, JINWEN [1983] *Fuzzy set structure with strong implication* (**C** 3582) Adv Fuzzy Sets, Possibility Th & Appl 107–136
◇ B52 ◇ REV MR 85i:03076 • ID 44178

ZHANG, JINWEN see Vol. III, IV, V, VI for further entries

ZHANG, MINGYI [1981] *Some projection theorems in pansystem logic conservations (Chinese) (English summary)* (**J** 3594) Kexue Tansuo 1981/1*173–174
◇ B52 ◇ REV MR 83f:06024 • ID 40201

ZHANG, WENGQIAN [1982] see KONG, YOUKUN

ZHANG, WENXIU [1982] see LE, HUILING

ZHANG, WENXIU & ZHAO, QINPING [1982] *Possibility degree theories (Chinese)* (**J** 0420) Shuxue Jinzhan 11*220–227
◇ B52 ◇ REV MR 84e:03032 • ID 34371

ZHANG, WENXIU [1984] see LE, HUILING

ZHANG, WENXIU see Vol. V for further entries

ZHAO, QINPING [1982] see ZHANG, WENXIU

ZHAO, ZHEN [1982] *A new map approach for the minimization of fuzzy logic functions (Chinese) (English summary)* (**J** 3796) Lanzhou Daxue Xuebao, Ziran Kexue 18/3*48-56
◇ B52 ◇ REV MR 84i:03061 Zbl 514 # 03015 • ID 34540

ZHAO, ZHEN [1983] *A graphical algorithm for decomposing fuzzy logic functions (Chinese) (English summary)* (**J** 3796) Lanzhou Daxue Xuebao, Ziran Kexue 19/4*62-71
◇ B52 ◇ REV MR 85m:94053 Zbl 526 # 94018 • ID 45276

ZHAO, ZHEN [1984] *A discussion on graphical minimization and decomposition of multivariate fuzzy logic functions (Chinese) (English summary)* (**J** 3796) Lanzhou Daxue Xuebao, Ziran Kexue 20/2*39-48
◇ B52 ◇ REV MR 86g:03044 Zbl 537 # 94031 • ID 44123

ZHAO, ZONGGUAN [1980] *Discussion of the problem of logical contradiction and dialectical contradiction* (**J** 1779) Chinese Stud Phil 11*36-57
◇ B53 ◇ ID 42900

ZHENG, QILUN [1983] see GU, WEINAN

ZHENG, YUXIN [1982] see XU, LIZHI

ZHENG, YUXIN [1985] see XIAO, XIAN

ZHENG, YUXIN see Vol. I, VI for further entries

ZHOLONDZ', V.Y. [1983] see OSIPOV, G.S.

ZHOU, S.Q. [1984] see BALDWIN, J.F.

ZHU, WUJIA [1982] see XU, LIZHI

ZHU, WUJIA [1984] see XIAO, XIAN

ZHU, WUJIA [1984] see HE, ZHONGXIONG

ZHU, WUJIA [1985] see XIAO, XIAN

ZHU, WUJIA see Vol. I, VI for further entries

ZHU, YANJUN [1934] *A summary on mathematical logic (Chinese)* (**J** 2879) Wuhan Daxue Xuebao, Ziran Kexue 5/2*
◇ B98 ◇ ID 48556

ZHU, YANJUN [1936] *Introduction to mathematical logic* (**J** 0409) Chinese J Math (Taipei) 1/1*
◇ B98 ◇ ID 48558

ZHU, YUANZHAO [1980] see WU, XUEMOU

ZHUKOVIN, V.E. [1981] *Models and procedures in decision-making (Russian)* (**X** 3790) Metsniereba: Tbilisi 120pp
◇ B52 ◇ REV MR 83f:90014 Zbl 505 # 90041 • ID 38215

ZHUKOVIN, V.E. [1981] *On the relationship between multicriteria and fuzzy representations of decision making problems (Russian)* (**S** 3909) Mat Met Opt & Upravleniya Slozh Sist (Kalinin) 1981*52-60
◇ B52 ◇ REV MR 85e:90080 Zbl 514 # 90050 • ID 37455

ZHURKIN, V.A. [1972] see BOBROV, A.E.

ZHURKIN, V.A. see Vol. I for further entries

ZIAI, H. [1979] *Modal propositions in Islamic philosophy* (**J** 2680) Bull Iranian Math Soc 7*58-77
◇ A05 A10 B45 ◇ REV MR 82h:03002 • ID 80266

ZICH, O.V. [1938] *Sentential calculus with complex values (Czech)* (**J** 4437) Ceska Mysl 34*189-196
◇ B50 ◇ REV JSL 4.165 • ID 48766

ZICH, O.V. see Vol. I, V for further entries

ZIEMBA, Z. [1961] *Rational belief, probability and the justification of induction (Polish) (Russian and English summaries)* (**J** 0063) Studia Logica 12*99-124
• TRANSL [1977] (**C** 3174) Twenty-five Years Log Meth Poland 709-735
◇ A05 B48 ◇ REV Zbl 379 # 02001 • ID 33103

ZIEMBA, Z. [1971] *Deontic syllogistics (Polish and Russian summaries)* (**J** 0063) Studia Logica 28*139-159
◇ B45 ◇ REV MR 46 # 27 Zbl 267 # 02014 • ID 29882

ZIEMBINSKI, Z. [1961] *The propositional character of enacted rules (Polish) (Russian and English summaries)* (**J** 0063) Studia Logica 11*37-48
◇ B60 ◇ REV MR 23 # A3078 • ID 33099

ZIEMBINSKI, Z. [1976] *Practical logic (Polish)* (**X** 1034) PWN: Warsaw xv+437pp
• TRANSL [1976] (**X** 0835) Reidel: Dordrecht xv+437pp
◇ A05 B45 B98 ◇ REV Zbl 372 # 02001 JSL 23.73 JSL 47.231 • REM With an appendix on "Deontic logic" by Ziemba,Z. • ID 51409

ZIEMBINSKI, Z. [1984] *Kinds of discordance of norms* (**C** 4065) Theorie der Normen 473-484
◇ A05 B45 ◇ REV MR 85k:03015 • ID 44576

ZIEMBINSKI, Z. see Vol. I for further entries

ZIERLER, N. [1965] see SCHLESSINGER, N.

ZIMMERMANN, H.-J. [1982] see MIZUMOTO, M.

ZIMMERMANN, H.-J. [1985] *Fuzzy set theory - and its applications* (**X** 4624) Kluwer-Nijhoff: Boston xii+363pp
◇ B52 E72 ◇ ID 49392

ZIMMERMANN, H.-J. see Vol. V for further entries

ZINOV'EV, A.A. [1960] *Philosophical problems of many-valued logic (Russian)* (**X** 0899) Akad Nauk SSSR : Moskva 139pp
• TRANSL [1963] (**X** 0835) Reidel: Dordrecht xiv+155pp (English) [1968] (**X** 0900) Vieweg: Wiesbaden 127pp (German) [1968] (**X** 0806) Dt Verlag Wiss: Berlin 127pp (German) [1968] (**X** 0901) Winter: Braunschweig 127pp (German)
◇ A05 B50 ◇ REV MR 28 # 16 MR 51 # 7804 Zbl 119.249 JSL 28.255 JSL 29.213 • REM The English and German translations are revised versions. 2nd German ed. 1970 • ID 17298

ZINOV'EV, A.A. [1961] *A method of describing the truth-functions of the n-valued propositional calculus (Russian) (Polish and English summaries)* (**J** 0063) Studia Logica 11*217-222
◇ B50 ◇ REV MR 24 # A38 Zbl 121.253 JSL 36.691 • ID 14420

ZINOV'EV, A.A. [1962] *Propositional logic and theory of deduction (Russian)* (**X** 0899) Akad Nauk SSSR : Moskva 152pp
◇ B98 ◇ REV Zbl 105.4 JSL 30.373 • ID 19782

ZINOV'EV, A.A. [1968] *An outline of many-valued logic (Russian)* (**C** 0766) Probl Log & Teor Poznan 113-204
◇ A05 B50 ◇ REV MR 51 # 7805 • ID 17299

ZINOV'EV, A.A. [1970] *Complex logic (Russian)* (**C** 1530) Issl Log Sist (Yanovskaya) 166–203
◇ B45 ◇ REV MR 47 # 8261 Zbl 221 # 02004 • ID 14423

ZINOV'EV, A.A. [1970] *Complex logic (Russian)* (**X** 2027) Nauka: Moskva 203pp
◇ B45 ◇ REV MR 51 # 7811 Zbl 221 # 02004 • ID 80283

ZINOV'EV, A.A. [1970] *Complex logic (a formal construction) (Russian)* (**C** 0668) Neklass Log 13–47
◇ A05 B45 ◇ REV MR 47 # 8261 • ID 80289

ZINOV'EV, A.A. [1970] *Minimal tautologies and contradictions in finitely many-valued logic (Russian)* (**C** 1530) Issl Log Sist (Yanovskaya) 262–270
◇ B50 ◇ REV MR 47 # 6439 Zbl 243 # 02011 • ID 14425

ZINOV'EV, A.A. [1974] *Certain systems of formal arithmetic (Russian)* (**J** 0302) Rep Math Logic, Krakow & Katowice 3*73–90
◇ B45 F30 ◇ REV MR 51 # 7812 Zbl 324 # 02033
• ID 17312

ZINOV'EV, A.A. [1975] *Logik und Sprache der Physik* (**X** 0911) Akademie Verlag: Berlin 250pp
◇ A05 B50 ◇ REV Zbl 326 # 02005 • ID 32054

ZINOV'EV, A.A. [1975] see WESSEL, H.

ZINOV'EV, A.A. [1983] *Nonstandard logic and its applications* (**X** 4450) Meeuws: Oxford 36pp
◇ B60 ◇ ID 47238

ZINOV'EV, A.A. see Vol. I, III, V, VI for further entries

ZLATOS, P. [1981] *Two-levelled logic and model theory* (**P** 2552) Conf Finite Algeb & Multi-Val Log;1979 Szeged 825–872
◇ B50 C90 G25 ◇ REV MR 83e:03057 Zbl 478 # 03019
• ID 55634

ZLATOS, P. see Vol. III, V for further entries

ZOLL, E.J. [1968] *Logic: A programmed text for 2-valued and 3-valued logics* (**X** 1330) Pitman Publ: Belmont & London 137pp
◇ B50 B98 ◇ ID 22544

ZOLOTOVA, T.M. [1973] *Some representations of continuous-logic functions, conserving values of arguments and their complements (Russian)* (**J** 0011) Avtom Telemekh 1973/9*79–84
• TRANSL [1973] (**J** 0010) Autom & Remote Control 34*1435–1439
◇ B50 ◇ REV MR 56 # 11750 Zbl 275 # 94026 • ID 66250

ZUCKER, J.I. [1978] see TRAGESSER, R.S.

ZUCKER, J.I. see Vol. I, IV, VI for further entries

ZURAWSKA, L. [1978] see UZIEMBLO, A.O.

ZURAWSKA, L. see Vol. V for further entries

ZYGMUNT, J. [1974] *A note on direct products and ultraproducts of logical matrices* (**J** 0063) Studia Logica 33*349–357
◇ B50 C05 C20 ◇ REV MR 50 # 12656 Zbl 312 # 02039
• ID 14462

ZYGMUNT, J. [1977] *Research project on strongly finite sentential calculi* (**J** 0387) Bull Sect Logic, Pol Acad Sci 6*87–90
◇ B50 ◇ ID 31569

ZYGMUNT, J. [1978] see MALINOWSKI, G.

ZYGMUNT, J. [1979] *Entailment relations and matrices. I* (**J** 0387) Bull Sect Logic, Pol Acad Sci 8*112–115
◇ B50 ◇ REV MR 80h:03048 Zbl 409 # 03015 • ID 56318

ZYGMUNT, J. [1983] see NIINILUOTO, I.

ZYGMUNT, J. see Vol. I, III, VI for further entries

Source Index

Journals

J 0001 Acta Math Acad Sci Hung • H
Acta Mathematica Academiae Scientiarum Hungaricae [1950-1982] ISSN 0001-5954
• CONT AS (**J** 4729) Acta Math Hung

J 0002 Acta Sci Math (Szeged) • H
Acta Scientiarum Mathematicarum [1947ff] ISSN 0001-6969
• CONT OF (**J** 0460) Acta Univ Szeged, Sect Mat

J 0003 Algebra i Logika • SU
Algebra i Logika (Algebra and Logic) [1962ff] ISSN 0373-9252
• TRANSL IN (**J** 0069) Algeb and Log

J 0004 Algeb Universalis • CDN
Algebra Universalis [1970ff] ISSN 0002-5240

J 0005 Amer Math Mon • USA
American Mathematical Monthly [1894ff] ISSN 0002-9890

J 0006 Ann Univ Budapest, Sect Math • H
Annales Universitatis Scientiarum Budapestinensis. Sectio Mathematica [1958ff] ISSN 0524-9007

J 0007 Ann Math Logic • NL
Annals of Mathematical Logic [1970-1982] ISSN 0003-4843
• CONT AS (**J** 0073) Ann Pure Appl Logic

J 0008 Arch Math (Basel) • CH
*Archiv der Mathematik * Archives of Mathematics * Archives Mathematiques* [1948ff] ISSN 0003-889X

J 0009 Arch Math Logik Grundlagenforsch • D
Archiv fuer Mathematische Logik und Grundlagenforschung [1950ff] ISSN 0003-9268

J 0010 Autom & Remote Control • USA
Automation and Remote Control [1958ff] ISSN 0005-1179
• TRANSL OF (**J** 0011) Avtom Telemekh

J 0011 Avtom Telemekh • SU
Avtomatika i Telemekhanika (Automation and Telemechanics) [1934ff] ISSN 0005-2310
• TRANSL IN (**J** 0010) Autom & Remote Control

J 0012 Boll Unione Mat Ital, IV Ser • I
Bolletino della Unione Matematica Italiana. Serie IV [1968-1975] ISSN 0041-7084
• CONT OF (**J** 4408) Boll Unione Mat Ital, III Ser • CONT AS (**J** 3285) Boll Unione Mat Ital, V Ser, A & (**J** 3495) Boll Unione Mat Ital, V Ser, B

J 0013 Brit J Phil Sci • GB
British Journal for the Philosophy of Science [1950ff] ISSN 0007-0882

J 0014 Bull Acad Pol Sci, Ser Math Astron Phys • PL
Bulletin de l'Academie Polonaise des Sciences. Serie des Sciences Mathematiques, Astronomiques et Physiques [1953-1978] ISSN 0001-4117
• CONT AS (**J** 3293) Bull Acad Pol Sci, Ser Math

J 0015 Bull Amer Math Soc • USA
Bulletin of the American Mathematical Society [1894-1978] ISSN 0002-9904
• CONT AS (**J** 0589) Bull Amer Math Soc (NS)

J 0016 Bull Austral Math Soc • AUS
Bulletin of the Australian Mathematical Society [1969ff] ISSN 0004-9727

J 0018 Canad Math Bull • CDN
*Canadian Mathematical Bulletin * Bulletin Canadien de Mathematiques* [1958ff] ISSN 0008-4395

J 0020 Compos Math • NL
Compositio Mathematica [1933ff] ISSN 0010-437X

J 0021 Cybernetics • USA
Cybernetics [1965ff] ISSN 0011-4235
• TRANSL OF (**J** 0040) Kibernetika, Akad Nauk Ukr SSR

J 0022 Cheskoslov Mat Zh • CS
*Cheskoslovatskij Matematicheskij Zhurnal * Czechoslovak Mathematical Journal* [1951ff] ISSN 0011-4642
• REM From Vol. 19 (1969) on the title is only: Czechoslovak Mathematical Journal

J 0023 Dokl Akad Nauk SSSR • SU
Doklady Akademii Nauk SSSR (Reports of the Academy of Sciences of the USSR) [1933ff] ISSN 0002-3264
• TRANSL IN (**J** 0062) Sov Math, Dokl & (**J** 0470) Sov Phys, Dokl

J 0024 Dokl Akad Nauk Uzb SSR • SU
*Doklady Akademii Nauk Uzb SSR (DAN Uzb SSR) * UzSSR Fanlar Akademijasining Dokladlari (Reports of the Academy of Sciences of the Uzb SSR)* [1944ff] ISSN 0134-4307

J 0025 Duke Math J • USA
Duke Mathematical Journal [1935ff] ISSN 0012-7094

J 0027 Fund Math • PL
Fundamenta Mathematicae [1920ff] ISSN 0016-2736

J 0028 Indag Math • NL
*Indagationes Mathematicae * Nederlandse Akademie van Wetenschappen. Proceedings* [1939ff] ISSN 0019-3577, ISSN 0023-3358
• REM Until 1950 part of Koninklijke Nederlandsche Akademie van Wetenschappen, Proceedings of the Section of Sciences; vol n+41 with separate pagination. Since 1951 same as Koninklijke Nederlandse Akademie van Wetenschappen, Proceedings of the Section of Sciences, Series A; vol n+41. Before 1951 page numbers in Proceedings and Indagationes different. Since 1951 the same page numbers as Proceedings Series A.

J 0029 Israel J Math • IL
Israel Journal of Mathematics [1963ff] ISSN 0021-2172
• CONT OF (**J** 0493) Bull Res Counc Israel Sect F

Source Index

J 0031 Izv Vyssh Ucheb Zaved, Mat (Kazan) • SU
Izvestiya Vysshikh Uchebnykh Zavedenij. Matematika (Proceedings of the University. Mathematics) [1957ff] ISSN 0021-3446
• TRANSL IN (**J** 3449) Sov Math

J 0033 J Comb Th, Ser B • USA
Journal of Combinatorial Theory. Series B [1971ff] ISSN 0095-8956
• CONT OF (**J** 1669) J Comb Th

J 0034 J Math Anal & Appl • USA
Journal of Mathematical Analysis and Applications [1960ff] ISSN 0022-247X

J 0036 J Symb Logic • USA
The Journal of Symbolic Logic [1936ff] ISSN 0022-4812

J 0037 ACM J • USA
Journal of the ACM (= Association for Computing Machinery) [1954ff] ISSN 0004-5411

J 0038 J Austral Math Soc • AUS
Journal of the Australian Mathematical Society [1959-1975] ISSN 0004-9735
• CONT AS (**J** 3194) J Austral Math Soc, Ser A

J 0039 J London Math Soc • GB
The Journal of the London Mathematical Society [1926-1968]
• CONT AS (**J** 3172) J London Math Soc, Ser 2

J 0040 Kibernetika, Akad Nauk Ukr SSR • SU
Kibernetika. Akademiya Nauk Ukrainskoj SSR (Cybernetics. Academy of Sciences of the Ukrainian SSR) [1965ff] ISSN 0023-1274
• TRANSL IN (**J** 0021) Cybernetics

J 0041 Math Syst Theory • D
Mathematical Systems Theory. An International Journal [1967ff] ISSN 0025-5661

J 0042 Mat Vesn, Drust Mat Fiz Astron Serb • YU
Matematichki Vesnik. Drushtvo Matematichara, Fizichara i Astromoma SR Serbije, SFR Jugoslavija (Mathematical Publications. Society Serbe of Mathematicians, Physicists and Astronomers) [1964ff] ISSN 0025-5165
• CONT OF (**J** 4277) Vesn Drusht Mat Fiz Serbije

J 0043 Math Ann • D
Mathematische Annalen [1868ff] ISSN 0025-5831

J 0044 Math Z • D
Mathematische Zeitschrift [1918ff] ISSN 0025-5874

J 0047 Notre Dame J Formal Log • USA
Notre Dame Journal of Formal Logic [1960ff] ISSN 0029-4527

J 0048 Pac J Math • USA
Pacific Journal of Mathematics [1951ff] ISSN 0030-8730

J 0049 Period Math Hung • H
Periodica Mathematica Hungarica [1971ff] ISSN 0031-5303

J 0050 Port Math • P
Portugaliae Mathematica [1937ff] ISSN 0032-5155

J 0051 Commentat Math, Ann Soc Math Pol, Ser 1 • PL
*Annales Societatis Mathematicae Polonae. Series I. Commentationes Mathematicae * Roczniki Polskiego Towarzystwa Matematycznego. Seria I. Prace Matematyczne.* [1955ff] ISSN 0079-368X
• CONT OF (**J** 0611) Pol Tow Mat, Prace Mat-Fiz

J 0052 Probl Kibern • SU
Problemy Kibernetiki. Glavnaya Redaktsiya Fiziko-Matematicheskoj Literatury [1958ff] ISSN 0555-277X
• TRANSL IN (**J** 0471) Syst Th Res & (**J** 0449) Probl Kybern & (**J** 1195) Probl Cybernet

J 0053 Proc Amer Math Soc • USA
Proceedings of the American Mathematical Society [1950ff] ISSN 0002-9939

J 0054 Proc Nat Acad Sci USA • USA
Proceedings of the National Academy of Sciences of the United States of America [1915ff] ISSN 0027-8424

J 0056 Publ Dep Math, Lyon • F
Publications du Departement de Mathematiques. Faculte des Sciences de Lyon. [1964-1981] ISSN 0076-1656
• CONT AS (**J** 2107) Publ Dep Math, Lyon, NS

J 0057 Publ Math (Univ Debrecen) • H
Publicationes Mathematicae [1949ff] ISSN 0033-3883

J 0058 Rend Circ Mat Palermo • I
Rendiconti del Circolo Matematico di Palermo [1887-1940]
• CONT AS (**J** 3522) Rend Circ Mat Palermo, Ser 2

J 0060 Rev Roumaine Math Pures Appl • RO
Revue Roumaine de Mathematiques Pures et Appliquees. Academia Republicii Socialiste Romania [1956ff] ISSN 0035-3965

J 0062 Sov Math, Dokl • USA
Soviet Mathematics. Doklady. [1960ff] ISSN 0038-5573
• TRANSL OF (**J** 0023) Dokl Akad Nauk SSSR

J 0063 Studia Logica • PL
Studia Logica [1953ff] ISSN 0039-3215

J 0064 Trans Amer Math Soc • USA
Transactions of the American Mathematical Society [1900ff] ISSN 0002-9947

J 0067 Usp Mat Nauk • SU
Uspekhi Matematicheskikh Nauk (Advances in Mathematical Sciences) [1936ff] ISSN 0042-1316
• TRANSL IN (**J** 1399) Russ Math Surv

J 0068 Z Math Logik Grundlagen Math • DDR
Zeitschrift fuer Mathematische Logik und Grundlagen der Mathematik [1955ff] ISSN 0044-3050

J 0069 Algeb and Log • USA
Algebra and Logic [1968ff] ISSN 0002-5232
• TRANSL OF (**J** 0003) Algebra i Logika

J 0070 Bull Soc Sci Math Roumanie, NS • RO
Bulletin Mathematique de la Societe des Sciences Mathematiques de la Republique Socialiste de Roumanie. Nouvelle Serie. [1957ff] ISSN 0007-4691
• CONT OF (**J** 0494) Bull Math Soc Sci Roumanie

J 0071 Met Diskr Analiz (Novosibirsk) • SU
Metody Diskretnogo Analiza. Sbornik Trudov (Methods of Discrete Analysis. Collected Papers) [1963ff] ISSN 0419-4160, ISSN 0136-1228

J 0072 IRE Trans Electr Comp • USA
Transactions on Electronic Computers. IRE (= Institute of Radio Engineers) [1952-1962]
• CONT AS (**J** 4305) IEEE Trans Electr Comp

J 0073 Ann Pure Appl Logic • NL
Annals of Pure and Applied Logic [1983ff] ISSN 0168-0072
• CONT OF (**J** 0007) Ann Math Logic

J 0075 Phil Phenom Research • USA
Philosophy and Phenomenological Research [1940ff] ISSN 0031-8205

J 0076 Dialectica • CH
Dialectica. International Review of Philosophy of Knowledge [1947ff] ISSN 0012-2017

J 0077 Proc London Math Soc • GB
Proceedings of the London Mathematical Society [1865-1903]
• CONT AS (**J** 1910) Proc London Math Soc, Ser 2

J 0079 Logique & Anal, NS • B
Logique et Analyse. Nouvelle Serie. Publication Trimestrielle du Centre National Belge de Recherche de Logique [1958ff] ISSN 0024-5836

J 0080 Izv Akad Nauk Ehston SSR, Fiz, Mat • SU
*Izvestiya Akademiya Nauk Ehstonskoj SSR. Fizika. Matematika * Eesti NSV Teaduste Akadeemia Toimetised. Fueuesika-Matemaatika (Proceedings of the Academy of Sciences of the Estonian SSR. Physics. Mathematics)* [1956ff] ISSN 0002-3140

J 0081 Proc Japan Acad • J
Proceedings of the Japan Academy [1925-1977] ISSN 0021-4280
• CONT AS (**J** 3239) Proc Japan Acad, Ser A

J 0082 Bull Soc Math Belg • B
Bulletin de la Societe Mathematique de Belgique [1948-1976] ISSN 0037-9476
• CONT AS (**J** 3133) Bull Soc Math Belg, Ser B & (**J** 3824) Bull Soc Math Belg, Ser A

J 0086 Cas Pestovani Mat, Ceskoslov Akad Ved • CS
Casopis pro Pestovani Matematiky. Ceskoslovenska Akademie Ved (Journal for the Cultivation of Mathematics. Czechoslovak Academy of Sciences) [1872ff]

J 0087 Mat Zametki (Akad Nauk SSSR) • SU
Matematicheskie Zametki (Mathematical Notes) [1967ff] ISSN 0025-567X
• TRANSL IN (**J** 1044) Math Notes, Acad Sci USSR

J 0088 Ann Univ Ferrara, NS, Sez 7 • I
Annali dell'Universita di Ferrara. Nuova Serie. Sezione 7. Scienze Matematiche [1966ff]

J 0090 J Math Soc Japan • J
Journal of the Mathematical Society of Japan [1885ff] ISSN 0025-5645

J 0091 Sugaku • J
Sugaku (Mathematics) [1947ff] ISSN 0039-470X

J 0092 Sib Mat Zh • SU
Sibirskij Matematicheskij Zhurnal. Akademiya Nauk SSSR. Sibirskoe Otdelenie (Siberian Mathematical Journal. Academy of Sciences of the USSR. Siberian Section) [1960ff] ISSN 0037-4474
• TRANSL IN (**J** 0475) Sib Math J

J 0093 J Comp Syst • USA
The Journal of Computing Systems [1952-1954]

J 0094 Mind • GB
Mind. A Quarterly Review of Philosophy [1876ff] ISSN 0026-4423

J 0095 Philos Stud • NL
Philosophical Studies. An International Journal for Philosophy in the Analytic Tradition [1950ff] ISSN 0031-8116

J 0096 Acta Philos Fenn • SF
Acta Philosophica Fennica [1948ff] ISSN 0355-1792

J 0097 Nous, Quart J Phil • USA
Nous. A Quarterly Journal of Philosophy [1967ff] ISSN 0029-4624

J 0099 Ricerca, Riv Mat Pure & Appl • I
La Ricerca. Rivista di Matematiche Pure ed Applicate. [1950ff] ISSN 0048-8283

J 0100 Amer J Math • USA
American Journal of Mathematics [1878ff] ISSN 0002-9327

J 0101 Phil Rev • USA
The Philosophical Review [1896ff] ISSN 0031-8108

J 0103 Analysis (Oxford) • GB
Analysis [1933ff] ISSN 0003-2638

J 0104 Atti Accad Sci Lett Arti Palermo, Ser 4/I • I
Atti della Accademia di Scienze Lettere e Arti di Palermo. Serie Quarta. Parte I: Scienze [1940-1983]
• CONT AS (**J** 4186) Atti Accad Sci Lett Arti Palermo, Ser 5/I

J 0105 Theoria (Lund) • S
Theoria. A Swedish Journal of Philosophy [1934ff] ISSN 0040-5825

J 0108 Bol Soc Mat Sao Paulo • BR
Sociedade de Matematica de Sao Paulo. Boletim. [1946ff]

J 0109 C R Acad Sci, Paris • F
Academie des Sciences de Paris. Comptes Rendus Hebdomadaires des Seances [1835-1965]
• CONT AS (**J** 2313) C R Acad Sci, Paris, Ser A-B

J 0110 Anais Acad Bras Cienc • BR
Anais da Academia Brasileira de Ciencias. [1929ff] ISSN 0001-3765

J 0111 Nagoya Math J • J
Nagoya Sugaku Zashi (Nagoya Mathematical Journal) [1950ff] ISSN 0027-7630

J 0112 Phil Quart (Calcutta) • IND
Philosophical Quarterly [1954-1973] ISSN 0376-415X

J 0113 Proc Aristotelian Soc • GB
Aristotelian Society. Proceedings [1900ff] ISSN 0066-7374
• REL PUBL (**J** 2119) Suppl Aristotelian Soc

J 0114 Math Nachr • DDR
Mathematische Nachrichten [1948ff] ISSN 0025-584X

J 0115 Wiss Z Humboldt-Univ Berlin, Math-Nat Reihe • DDR
Wissenschaftliche Zeitschrift der Humboldt-Universitaet Berlin. Mathematisch-Naturwissenschaftliche Reihe [1951ff] ISSN 0043-6852

J 0116 Electr & Comm Japan • USA
*Electronics and Communications in Japan * Scripta Electronica Japonica* [1963ff] ISSN 0036-9683
• TRANSL OF (**J** 0979) Denshi Tsushin Gakkai Ronbunshi, Sect A-D

J 0119 J Comp Syst Sci • USA
Journal of Computer and System Sciences [1967ff] ISSN 0022-0000

J 0120 Ann of Math, Ser 2 • USA
Annals of Mathematics. 2nd Series [1899ff] ISSN 0003-486X

J 0121 Kon Norske Vidensk Selsk Forh • N
Kongelige Norske Videnskabers Selskabs. Forhandlinger. (Proceedings of the Royal Scandinavian Society of Sciences) [1926ff] ISSN 0368-6302

J 0122 J Philos Logic • NL
Journal of Philosophical Logic [1972ff] ISSN 0022-3611

J 0125 Rev of Metaphysics • USA
The Review of Metaphysics. A Philososphical Quarterly. [1947ff] ISSN 0034-6632

J 0128 Acta Math Univ Comenianae (Bratislava) • CS
Universitas Comeniana. Acta Facultatis Rerum Naturalium. Mathematica [1956ff]

J 0129 Elektr Informationsverarbeitung & Kybern • DDR
Elektronische Informationsverarbeitung und Kybernetik [1965ff] ISSN 0013-5712

J 0130 BIT • DK
BIT. Nordisk Tidskrift for Informationsbehandling (BIT. Scandinavian Journal for Informatics) [1961ff] ISSN 0006-3835

J 0132 Math Scand • DK
Mathematica Scandinavica [1953ff] ISSN 0025-5521

J 0134 Dokl Akad Nauk Azerb SSR • SU
Akademiya Nauk Azerbajdzhanskoj SSR. Doklady (Reports of The Academy of Sciences of the Azerbaijan SSR) [1945ff] ISSN 0002-3078

J 0135 Izv Akad Nauk Azerb SSR, Ser Fiz-Tekh Mat • SU
Izvestiya Akademii Nauk Azerbajdzhanskoj SSR. Seriya Fiziko-Tekhnicheskikh i Matematicheskikh Nauk (Proceedings of the Academy of Sciences of the Azerbaijan SSR. Series: Physical-Technical and Mathematical Sciences) [1958ff] ISSN 0002-3108

J 0136 Semigroup Forum • D
Semigroup Forum [1970ff] ISSN 0037-1912

J 0137 C R Acad Bulgar Sci • BG
Doklady Bolgarskoi Akademii Nauk (Comptes Rendus de l'Academie Bulgare des Sciences) [1948ff] ISSN 0001-3978

J 0140 Comm Math Univ Carolinae (Prague) • CS
Commentationes Mathematicae Universitatis Carolinae [1960ff] ISSN 0010-2628

J 0141 Arch Autom & Telemech • PL
Archiwum Automatyki i Telemechaniki (Archives of Automation and Telemechanics) [1956ff] ISSN 0004-072X

J 0142 Mat Sb, Akad Nauk SSSR, NS • SU
Matematicheskij Sbornik. Novaya Seriya. Akademiya Nauk SSSR i Moskovskoe Matematicheskoe Obshchestvo (Mathematical Collected Articles. New Series. Academy of Sciences of the USSR and Moskovian Mathematical Society) [1936ff] ISSN 0025-5157
• CONT OF (**J** 1404) Mat Sb, Akad Nauk SSSR • TRANSL IN (**J** 0349) Math of USSR, Sbor

J 0143 Mat Chasopis (Slov Akad Ved) • CS
Matematicky Chasopis (Journal of Mathematics) [1967-1975] ISSN 0025-5173
• CONT OF (**J** 4713) Mat Fyz Chasopis (Slov Akad Ved)
• CONT AS (**J** 1522) Math Slovaca

J 0144 Rend Sem Mat Univ Padova • I
Rendiconti del Seminario Matematico dell'Universita di Padova [1930ff] ISSN 0041-8994

J 0147 An Univ Bucuresti, Acta Logica • RO
Universitatea Bucuresti. Analele. Acta Logica (Universitaet Bukarest. Annalen. Acta Logica) [1958ff] ISSN 0524-823X
• REM From 1962-1967 Analele Universitatii Bucuresti. Seria Acta Logica. From 1958-1961 Analele Universitatii C.I. Parhon. Seria Acta Logica

J 0148 Math Gaz • GB
The Mathematical Gazette [1894ff] ISSN 0025-5572

J 0149 Atti Accad Naz Lincei Fis Mat Nat, Ser 8 • I
Atti della Accademia Nazionale dei Lincei. Rendiconti. Classe di Scienze Fisiche, Matematiche e Naturali. Serie VIII [1946ff] ISSN 0001-4435

J 0150 Acad Roy Belg Bull Cl Sci (5) • B
*Academie Royale des Sciences, des Lettres et des Beaux Arts de Belgique. Bulletin de la Classe des Sciences. Cinquieme Serie * Koninklijke Academie voor Wetenschappen. Mededeelingen van de Afdeeling Wetenschappen. 5. Serie* [1915ff] ISSN 0001-4141

J 0151 Arch Soc Belg Philos • B
Archives de la Societe Belge de Philosophie. [1928-1937]
• CONT AS (**J** 2076) Rev Int Philos

J 0153 Phil of Sci (East Lansing) • USA
Philosophy of Science [1934ff] ISSN 0031-8248

J 0154 Synthese • NL
Synthese. An International Journal for Epistemology, Methodology and Philosophy of Science [1936ff] ISSN 0039-7857

J 0156 Kybernetika (Prague) • CS
Kybernetika (Cybernetics) [1965ff] ISSN 0023-5954
• REL PUBL (**J** 3524) Kybernetika Suppl (Prague)

J 0157 Jbuchber Dtsch Math-Ver • D
Jahresbericht der Deutschen Mathematiker-Vereinigung [1890ff] ISSN 0012-0456

J 0158 Erkenntnis (Dordrecht) • NL
Erkenntnis. The Journal of Unified Science: An International Journal of Analytic Philosophy [1975ff] ISSN 0165-0106
• CONT OF (**J** 3597) J Unif Sci

J 0160 Math-Phys Sem-ber, NS • D
Mathematisch-Physikalische Semesterberichte: Zur Pflege des Zusammenhangs von Schule und Universitaet. Neue Folge [1950-1979] ISSN 0340-4897
• CONT AS (**J** 2790) Math Sem-ber

J 0162 Teorema (Valencia) • E
Teorema [1971ff]

J 0164 J Comb Th, Ser A • USA
Journal of Combinatorial Theory. Series A [1971ff] ISSN 0097-3165
• CONT OF (**J** 1669) J Comb Th

J 0168 Blaetter Deutsch Philos • D
Blaetter fuer Deutsche Philosophie [1927-1944]
• CONT OF (**J** 1095) Beitr Phil Deutsch Idealismus

J 0169 Theoria (Madrid) • E
Theoria [1953ff] ISSN 0040-5817

J 0170 Ratio (Frankfurt) • D
Ratio [1957ff] ISSN 0342-1848
• REL PUBL (**J** 1022) Ratio (Oxford) • REM German Edition

J 0171 Proc Cambridge Phil Soc Math Phys • GB
Proceedings of the Cambridge Philosophical Society. Mathematical and Physical Sciences [1843-1974] ISSN 0008-1981
• CONT AS (J 0332) Math Proc Cambridge Phil Soc

J 0174 Bull Sect Sci Acad Roumaine • RO
Academie Roumaine. Bulletin de la Section Scientifique [1918-1948]
• CONT AS (J 0518) Bul Sti Mat-Fiz, Acad Romina

J 0175 Methodos • I
Methodos [1949ff]
• REM Does not seem to appear anymore

J 0178 Stud Gen • D
Studium Generale: Zeitschrift fuer die Einheit der Wissenschaften. Zusammenhang ihrer Begriffsbildungen und Forschungsmethoden. [1947-1971] ISSN 0039-4149

J 0179 Ann Fac Sci Clermont • F
Universite de Clermont. Faculte des Sciences. Annales [1952-1972]
• CONT AS (J 1934) Ann Sci Univ Clermont Math

J 0180 Rev Phil France & Etranger • F
Revue Philosophique de la France et de l'Etranger [1876ff] ISSN 0035-3833

J 0187 IEEE Trans Comp • USA
Transactions on Computers. IEEE (= Institute of Electrical and Electronics Engineers) [1968ff] ISSN 0018-9340
• CONT OF (J 4305) IEEE Trans Electr Comp

J 0188 Rev Union Mat Argentina • RA
Revista de la Union Matematica Argentina [1936ff] ISSN 0041-6932

J 0191 Inform Sci • USA
Information Sciences. An International Journal [1957ff] ISSN 0020-0255

J 0192 Wiss Z Univ Jena, Math-Nat Reihe • DDR
Wissenschaftliche Zeitschrift der Friedrich-Schiller-Universitaet Jena. Mathematisch-Naturwissenschaftliche Reihe. [1951ff] ISSN 0043-6836

J 0193 Discr Math • NL
Discrete Mathematics [1971ff] ISSN 0012-365X

J 0194 Inform & Control • USA
Information and Control [1958ff] ISSN 0019-9958

J 0197 Stud Cercet Mat Acad Romana • RO
Studii si Cercetari Matematice. Academia Republicii Socialiste Romania. (Mathematische Studien und Untersuchungen. Akademie der Sozialistischen Republik Rumaenien) [1950ff] ISSN 0567-6401
• CONT OF (J 0524) Disq Math Phys

J 0198 Bul Inst Politeh Iasi NS • RO
*Buletinul Institutului Politehnic din Iasi. Serie Noua * Bulletin de l'Ecole Polytechnique de Jassy. Nouveau Serie* [1946-1976]
• CONT AS (J 3070) Bul Inst Politeh Iasi, Sect 1

J 0199 Zh Vychisl Mat i Mat Fiz • SU
Zhurnal Vychislitel'noj Matematiki i Matematicheskoj Fiziki (Journal of Computational Mathematical and Mathematical Physics) [1961ff] ISSN 0044-4669
• TRANSL IN (J 1049) USSR Comput Math & Math Phys

J 0202 Diss Math (Warsaw) • PL
*Dissertationes Mathematicae. Polska Akademia Nauk, Instytut Matematyczny * Rozprawy Matematyczne* [1952ff] ISSN 0012-3862

J 0205 Rev Franc Autom, Inf & Rech Operat • F
Revue Francaise d'Automatique, Informatique et Recherche Operationelle (RAIRO). Series: Bleue, Jaune, Rouge, Verte [1972-1976] ISSN 0399-0559
• CONT OF (J 3954) Rev Franc Inf & Rech Operat • CONT AS (J 4698) Rev Franc Autom, Inf & Rech Operat, Ser Rouge Inf Th & (J 2831) RAIRO Autom & (J 2832) RAIRO Inform
• REM In 1975 the Serie Rouge split into: Serie Rouge Analyse Numerique & J4698

J 0207 Ist Lombardo Accad Sci Rend, A (Milano) • I
Istituto Lombardo. Accademia di Science e Lettere. Rendiconti. A. Scienze Matematiche, Fisiche, Chimichze e Geologiche [1937ff] ISSN 0021-2504
• CONT OF (J 3986) Ist Lombardo Rend, Ser 2 (Milano)

J 0209 J Math Phys • USA
Journal of Mathematical Physics [1960ff] ISSN 0022-2488

J 0210 Stud Sti Tehn Acad Romina Timisoara • RO
Studii si Cercetari Stiinte Tehnice. Academia Republicii Populare Romine. Baza de Cercetari Stiintifice Timisoara (Technische Wissenschaften. Studien und Untersuchungen. Akademie der Volksrepublik Rumaenien. Basis fuer Wissenschaftliche Untersuchungen Timisoara) [1954ff] ISSN 0515-1406

J 0212 ACM Commun • USA
Communications of ACM (= Association for Computing Machinery) [1958ff] ISSN 0001-0782

J 0215 Proc Irish Acad, Sect A • IRL
Proceedings of the Royal Irish Academy. Section A. Mathematical and Physical Sciences [1899ff] ISSN 0557-4056

J 0216 Izv Akad Nauk SSSR, Ser Mat • SU
Izvestiya Akademii Nauk SSSR. Seriya Matematicheskaya (Proceedings of the Academy of Sciences of the USSR. Mathematical Series) [1937ff] ISSN 0373-2436
• CONT OF (J 4717) Izv Akad Nauk SSSR • TRANSL IN (J 0448) Math of USSR, Izv

J 0219 J Franklin Inst • GB
Journal of the Franklin Institute ISSN 0016-0032

J 0220 Atti Accad Sci Torino, Fis Mat Nat • I
*Atti della Accademia delle Scienze di Torino. Classi di Scienze Fisiche, Matematiche e Naturali. Parte 1. * Acta Academiae Scientiarum Taurinensis* [1940ff] ISSN 0373-3033
• CONT OF (J 1742) Atti Accad Sci Torino, Fis Mat Nat

J 0224 God Vissh Tekh Ucheb Zaved Mat, Sofiya • BG
Godishnik na Visshite Tekhnicheski Uchebni Zavedeniya Matematika (Annuaire des Ecoles Techniques Superieures:Mathematiques) [1964-1973] ISSN 0436-1083
• CONT AS (J 3171) God Vissh Ucheb Zaved, Prilozhna Mat, Sofiya

J 0225 Amer Math Soc, Transl, Ser 2 • USA
American Mathematical Society. Translations. Series 2 [1955ff] ISSN 0065-9290

J 0229 Ann Mat Pura Appl, Ser 3 • I
Annali di Matematica Pura ed Applicata [1898-1923]
• CONT AS (J 3526) Ann Mat Pura Appl, Ser 4

Source Index

J 0230 An Univ Iasi, NS, Sect Ia • RO
Analele Stiintifice ale Universitatii Al.I. Cuza din Iasi. (Serie Noua) Sectiunea 1a: Matematica (Wissenschaftliche Annalen der Al.I. Cuza Universitaet Iasi. (Neue Serie) Sektion 1a: Mathematik) [1955ff] ISSN 0041-9109

J 0232 Inform Process Lett • NL
Information Processing Letters. Devoted to the Rapid Publication of Short Contributions to Information Processing [1971ff] ISSN 0020-0190

J 0233 Soobshch Akad Nauk Gruz SSR • SU
*Soobshcheniya Akademii Nauk Gruzinskoj SSR * Sakaharth SSR Mecnierebatha Akademia Moambe (Communications of the Academy of Sciences of the Georgian SSR)* [1940ff] ISSN 0002-3167

J 0236 Rev Mat Hisp-Amer, Ser 4 • E
Revista Matematica Hispano-Americana. 4a Serie. Real Sociedad Matematica Espanola. [1941ff] ISSN 0373-0999
• CONT OF (**J** 3993) Rev Mat Hisp-Amer, Ser 2

J 0238 Sitzb Oesterr Akad Wiss, Math-Nat Kl, Abt 2 • A
Oesterreichische Akademie der Wissenschaften. Mathematisch-Naturwissenschaftliche Klasse. Sitzungsberichte. Abteilung II. Mathematische, Physikalische und Technische Wissenschaft [1846ff] ISSN 0029-8816

J 0239 Skr, K Nor Vidensk Selsk • N
Det Kongelige Norske Videnskabers Selskab. Skrifter. (Monographs of the Scandinavian Society of Sciences) [1791ff] ISSN 0368-6310

J 0241 Bull Math Soc Nanyang Univ • SGP
Shu Hsuen Nien K'An (Nan-Yang Ta Hsuen Shu Shueh Hui) (Nanyang University. Mathematical Society. Bulletin) [1959-1965]
• CONT AS (**J** 0245) Nanta Math

J 0245 Nanta Math • SGP
Nanta Mathematica [1966ff] ISSN 0077-2739
• CONT OF (**J** 0241) Bull Math Soc Nanyang Univ

J 0246 J Nat Sci Math • PAK
The Journal of Natural Sciences and Mathematics [1961ff] ISSN 0022-2941

J 0250 Bol Soc Mat Mexicana • MEX
Boletin de la Sociedad Matematica Mexicana [1944-1955]
• CONT AS (**J** 3127) Bol Soc Mat Mexicana, Ser 2

J 0251 Proc Amer Acad Arts Sci • USA
American Academy of Arts and Sciences. Proceedings [1846-1958]

J 0252 Rev Philos Louvain • B
Revue Philosophique de Louvain [1946ff] ISSN 0035-3841
• CONT OF (**J** 1720) Rev Neoscolast Philos, Ser 2

J 0255 God Fak Mat & Mekh, Univ Sofiya • BG
Godishnik na Sofijskiya Universitet. Fakultet po Matematika i Mekhanika (Annuaire de l'Universite de Sofia. Faculte de Mathematiques.)

J 0256 Isis (Philadelphia) • USA
Isis. [1912ff] ISSN 0021-1753

J 0259 Nyt Tidsskr Mat • DK
Nyt Tidsskrift for Matematik (New Journal for Mathematics) [1890-1918]
• CONT AS (**J** 4510) Norsk Mat Tidsskr

J 0260 Ann Jap Ass Phil Sci • J
Annals of the Japan Association for Philosophy of Science [1956ff]

J 0261 Tohoku Math J • J
Tohoku Mathematical Journal (Tohoku Sugaku Zashi) [1911ff] ISSN 0040-8735

J 0269 Ann Sci Univ Jassy Sec 1 • RO
Universite de Jassy. Annales Scientifiques. Premiere Section. Mathematiques, Physique, Chimie [1937-1941]
• CONT OF (**J** 4693) Ann Sci Univ Jassy

J 0270 Dokl Akad Nauk Ukr SSR, Ser A • SU
*Doklady Akademii Nauk Ukrainskoj SSR. Seriya A. Fiziko-Matematicheskie i Tekhnicheskie Nauki * Dopovidi Akademii Nauk Uk'rainskoj RSR. Seriya A. Fiziko-Matematichni Ta Tekhnichi Nauki (Reports of the Academy of Sciences of the Ukrainian SSR. Series A. Physical-Mathematical and Engineering Sciences)* [1939ff] ISSN 0002-3531, ISBN 0201-8446

J 0273 Australasian J Phil • AUS
Australasian Journal of Philosophy [1947ff] ISSN 0004-8402
• CONT OF (**J** 4731) Australasian J Psych & Phil

J 0277 Sitzb Preuss Akad Wiss Phys Math Kl • DDR
Die Preussische Akademie der Wissenschaften. Sitzungsberichte. Physikalisch-Mathematische Klasse [1922-1949]

J 0283 Ann Soc Pol Math • PL
*Societe Polonaise de Mathematique. Annales. * Rocznik i Polskiego Towarzystwa Matematycznego* [1922-1952]
• CONT AS (**J** 1405) Ann Pol Math

J 0286 Int Logic Rev • I
*International Logic Review. * Rassegna Internazionale di Logica* [1970ff] ISSN 0048-6779

J 0287 Scripta Math • USA
Scripta Mathematica. [1932ff] ISSN 0036-9713

J 0288 Vest Ser Mat Mekh, Univ Moskva • SU
Vestnik Moskovskogo Universiteta. Seriya I. Matematika, Mekhanika (Publications of the Moscow University. Series I. Mathematics. Mechanics) [1946ff] ISSN 0201-7385, ISSN 0579-9368
• TRANSL IN (**J** 0510) Moscow Univ Math Bull & Moscow University. Mechanics Bulletin

J 0290 Euclides • NL
Euclides: Maandblad voor de Didactiek van de Wiskunde [1925ff]

J 0294 Anuar Soc Paranaense Mat • BR
Sociedade Paranaense de Matematica. Anuario. [1954-1957]
• CONT AS (**J** 1831) Anuar Soc Paranaense Mat, Ser 2

J 0295 Teor Veroyat i Mat Stat (Kiev) • SU
Teoriya Veroyatnostej i Matematicheskaya Statistika. Mezhvedomstvennyj Nauchnyj Sbornik (Wahrscheinlichkeitsrechnung und Mathematische Statistik) [1970ff]
• TRANSL IN (**J** 3456) Th Probab & Math Stat

J 0299 Gac Mat, Ser 1a (Madrid) • E
Gaceta Matematica. Consejo Superior de Investigaciones Cientificas. Instituto "Jorge Juan". 1a Serie [1970ff] ISSN 0016-3805
• CONT OF (**J** 4443) Gac Mat (Madrid)

J 0301 J Phil • USA
The Journal of Philosophy [1904ff] ISSN 0022-362X

J 0302 Rep Math Logic, Krakow & Katowice • PL
Reports on Mathematical Logic. The Jagiellonian University of Cracow. The Silesian University of Katowice [1973ff] ISSN 0083-4432
• CONT OF (S 0458) Zesz Nauk, Prace Log, Uniw Krakow

J 0308 Rocky Mountain J Math • USA
The Rocky Mountain Journal of Mathematics [1971ff] ISSN 0035-7596

J 0310 Inquiry (Oslo) • N
Inquiry: An Interdisciplinary Journal of Philosophy and the Social Sciences [1958ff] ISSN 0020-174X

J 0311 Nordisk Mat Tidskr • N
Nordisk Matematisk Tidskrift (Scandinavian Mathematical Journal) [1953-1978] ISSN 0029-1412
• CONT OF (J 4510) Norsk Mat Tidsskr • CONT AS (J 3075) Normat

J 0312 Izv Akad Nauk Armyan SSR, Ser Mat • SU
Izvestiya Akademii Nauk Armyanskoj SSR. Seriya Matematika (Proceedings of the Academy of Sciences of the Armenian SSR. Series: Mathematics) [1965ff] ISSN 0002-3043
• TRANSL IN (J 3265) Sov J Contemp Math Anal, Armen Acad Sci

J 0315 Ann Sc Norm Sup Pisa Fis Mat, Ser 3 • I
Annali della Scuola Normale Superiore di Pisa. Classe di Science. Fisiche e Matematiche. Seria III [1947-1973] ISSN 0036-9918
• CONT OF (J 1568) Ann Sc Norm Sup Pisa, Fis Mat, Ser 2
• CONT AS (J 4702) Ann Sc Norm Sup Pisa Fis Mat, Ser 4

J 0319 Matematiche (Sem Mat Catania) • I
Le Matematiche [1946ff]

J 0320 Monist • USA
The Monist: International Quarterly of General Philosophical Inquiry [1890ff] ISSN 0026-9662

J 0321 Atti Accad Sci Napoli Fis Mat • I
Atti della Accademia delle Scienze Fisiche e Matematiche di Napoli

J 0322 Arch Math (Brno) • CS
Archivum Mathematicum [1965ff] ISSN 0044-8753

J 0325 Amer Phil Quart • GB
American Philosophical Quarterly [1964ff] ISSN 0003-0481

J 0327 Trans Pierce Soc • USA
Charles S. Pierce Society. Transactions [1965ff] ISSN 0009-1774

J 0329 Math Chron (Auckland) • NZ
Mathematical Chronicle [1969ff] ISSN 0581-1155

J 0330 Atti Ist Veneto, Fis Mat Nat • I
Istituto Veneto di Scienze, Lettere ed Arti. Venezia. Atti. Classe de Scienze Matematiche e Naturali

J 0332 Math Proc Cambridge Phil Soc • GB
Mathematical Proceedings of the Cambridge Philosophical Society [1975ff] ISSN 0305-0041
• CONT OF (J 0171) Proc Cambridge Phil Soc Math Phys

J 0336 Giorn Mat Battaglini • I
Giornale di Matematiche di Battaglini [1863ff] ISSN 0017-033X

J 0337 Mat Ezheg, Akad Nauk Latv SSR • SU
Latvijskij Matematicheskij Ezhegodnik. Latvijskij Ordena Trudovogo Krasnogo Znameni Gosudarstvennyj Universitet imeni P.Stuchki. Akademiya Nauk Latvijskoj SSR (Latvian Mathematical Yearbook) [1965ff] ISSN 0458-8223
• TRANSL IN Latvian Mathematical Yearbook

J 0338 Nauch-Tekh Inf, Ser 2, Akad Nauk SSSR • SU
Nauchno-Tekhnicheskaya Informatsiya. Seriya 2. Gosudarstvennyj Komitet SSSR po Nauke i Tekhnike. Akademiya Nauk SSSR. Vsesoyuznyj Institut Nauchnoj i Tekhnicheskoj Informatsii. Informatsionnye Protsesy i Sistemy (Scientific Technical Information. Series 2) [1967ff]
• TRANSL IN (J 2667) Autom Doc Math Linguist

J 0342 Monatsber Dt Akad Wiss • DDR
Die Deutsche Akademie der Wissenschaften zu Berlin. Monatsberichte: Mitteilungen aus Mathematik, Naturwissenschaft, Medizin und Technik [1959-1971] ISSN 0011-9814

J 0346 Dokl Akad Nauk Armyan SSR • SU
Doklady Akademii Nauk Armyanskoj SSR (Reports of the Academy of Sciences of the Armenian SSR) [1944ff] ISSN 0321-1339

J 0347 Atti Accad Sci Bologna Fis, Ser 11 • I
Atti della Accademia delle Scienze dell'Istituto di Bologna. Classe di Scienze Fisiche. Rendiconti. Serie XI [?-1962]
• CONT AS (J 3069) Atti Accad Sci Bologna Fis Ser 12

J 0349 Math of USSR, Sbor • USA
Mathematics of the USSR, Sbornik [1967ff] ISSN 0025-5734
• TRANSL OF (J 0142) Mat Sb, Akad Nauk SSSR, NS

J 0350 Sci Rep Tokyo Kyoiku Daigaku Sect A • J
Tokyo Kyoiku Daigaku (Tokyo University of Education. Science Reports. Section A.)

J 0351 Osaka J Math • J
Osaka Journal of Mathematics [1964ff] ISSN 0030-6126
• CONT OF (J 1770) Osaka Math J

J 0352 Math Jap • J
Mathematica Japonica [1948ff] ISSN 0025-5513

J 0354 Phil Trans Roy Soc London, Ser A • GB
Philosophical Transactions of the Royal Society of London. Series A. Mathematical and Physical Sciences. ISSN 0080-4614

J 0357 Z Phil Forsch • D
Zeitschrift fuer Philosophische Forschung [1947ff] ISSN 0044-3301

J 0358 Versl Gewone Vergad Afd Natuurkd • NL
Koninklijke Nederlandse Akademie van Wetenschappen. Verslagee van de Gewone Vergaderingen der Afdling Natuurkunde (Royal Dutch Academy of Sciences. Reports of the Meetings of the Physics Section) [1892ff] ISSN 0023-3382

J 0366 Sitzber Berlin Math Ges • DDR
Die Berliner Mathematische Gesellschaft. Sitzungsberichte

J 0371 Glas Mat-Fiz Astron, Ser 2 (Zagreb) • YU
Glasnik Matematichko-Fizichki i Astronomichki. Ser II (Publications of Mathematics, Physics, Astronomy. Ser II) [1946-1965]
• CONT AS (J 3519) Glas Mat, Ser 3 (Zagreb)

J 0372 Rad Jugosl Akad Znan Umjet, Mat Fiz Teh Znan • YU
Radovi Jugoslavenske Akademije Znanosti i Umjetnosti. Razred Za Matematichke, Fizichke i Tehnichke Znanosti. (Papers. Yugoslavian Academy of Sciences and Arts. Department of Mathematics, Physics and Engineering) [1940-1981]
• CONT OF (**J** 4735) Rad Jugosl Akad Znan Umjet • CONT AS (**J** 3944) Rad Jugosl Akad Znan Umjet, Mat Znan

J 0378 Gaz Mat (Bucharest) Ser A • RO
Gazeta Matematica. Seria A (Mathematische Zeitschrift. Serie A) [1964-1974] ISSN 0016-5433
• CONT AS (**J** 3377) Gaz Mat (Bucharest)
• REM Combined with: Gazeta Matematica.Seria B to form Gazeta Mathematica

J 0380 Acta Cybern (Szeged) • H
Acta Cybernetica. Forum Centrale Publicationum Cyberneticarum Hungaricum [1969ff] ISSN 0324-721X

J 0381 J Tsuda College (Tokyo) • J
Journal of Tsuda College [1969ff]

J 0383 J Cybern • USA
Journal of Cybernetics. Transactions of the American Society for Cybernetics [1971ff] ISSN 0022-0280
• REM Includes Translations from Appropriate Russian and Japanese Journals

J 0384 Rend Mat, Ser 5 • I
Rendiconti di Matematica. Serie 5 [1940-1967]
• CONT AS (**J** 2311) Rend Mat, Ser 6

J 0386 Proc Edinburgh Math Soc • GB
Proceedings of the Edinburgh Mathematical Society [1883-1926]
• CONT AS (**J** 3420) Proc Edinburgh Math Soc, Ser 2

J 0387 Bull Sect Logic, Pol Acad Sci • PL
Bulletin of the Section of Logic. Polish Academy of Sciences. Institute of Philosophy and Sociology. [1972ff]
• REM Papers Published in the Bulletin are Generally: 1. Abstracts or preprints of papers submitted to other journals e.g. Studia Logica. 2. Abstracts of papers read at seminars or local conferences

J 0388 Human, Ser 4 Logica Mat • C
Humanidades. Ser 4 Logica Matematica [1972ff]

J 0390 Publ Res Inst Math Sci (Kyoto) • J
Publications of the Research Institute for Mathematical Sciences [1965ff] ISSN 0034-5318
• REM Vols 1-4 Issued as: Kyoto Univ. Research Institute for Mathematical Sciences. Publications. Series A.

J 0391 Riv Mat Univ Parma • I
Rivista di Matematica della Universita di Parma [1891-1915]
• CONT AS (**J** 1526) Riv Mat Univ Parma, Ser 2

J 0392 Math Sci Hum • F
Mathematiques et Sciences Humaines [1962ff] ISSN 0025-5815

J 0400 Publ Inst Math, NS (Belgrade) • YU
Institut Mathematique. Publications de l'Institut Mathematique. Nouvelle Serie [1961ff] ISSN 0522-828X
• CONT OF (**J** 4706) Publ Inst Math (Belgrade)

J 0403 Izv Akad Nauk Kazak SSR, Ser Fiz-Mat • SU
Izvestiya Akademii Nauk Kazakhskoj SSR. Seriya Fiziko-Matematicheskaya (Proceedings of the Academy of Sciences of the Kazakh SSR. Series: Physics & Mathematics) [1963ff] ISSN 0002-3191

J 0406 Bull Inst Math, Acad Sin (Taipei) • RC
Chung Yang Yen Chui y Uan Shu Hs Ueh Yen Chiu So (Bulletin of the Institute of Mathematics. Academia Sinica.) [1973ff]

J 0407 Comm Math Univ St Pauli (Tokyo) • J
Commentarii Mathematici Universitatis Sancti Pauli [1952ff] ISSN 0010-258X

J 0408 Bull Soc R Sci Liege • B
Bulletin de la Societe Royale des Sciences de Liege [1932ff] ISSN 0037-9565

J 0409 Chinese J Math (Taipei) • RC
Chinese Journal of Mathematics [1936ff]

J 0411 Studia Sci Math Hung • H
Studia Scientiarum Mathematicarum Hungaria. Auxilio Consilii Instituti Mathematici. Academiae Scientiarum Hungaricae [1966ff] ISSN 0081-6906

J 0412 Wiss Z Tech Hochsch Ilmenau • DDR
Wissenschaftliche Zeitschrift der Technischen Hochschule Ilmenau [1954ff] ISSN 0043-6917

J 0413 Izv Akad Nauk Belor SSR, Ser Fiz-Mat • SU
*Vestsi Akademii Navuk BeSSR. Seriya Fizika-Matematychnykh Navuk * Izvestiya Akademii Nauk BSSR. Seriya Fiziko-Matematicheskikh Nauk (Proceedings of the Academy of Sciences of the Byelorussian SSR. Series: Physics, Mathematics)* [1964ff] ISSN 0002-3574

J 0418 Shuxue Xuebao • TJ
Shuxue Xuebao (Acta Mathematica Sinica) [1951ff]
• TRANSL IN (**J** 0419) Chinese Math Acta
• REM In 1951 published as: Journal of the Chinese Mathematical Society (N.S.)

J 0419 Chinese Math Acta • USA
Chinese Mathematics. Acta [1962-1967]
• TRANSL OF (**J** 0418) Shuxue Xuebao

J 0420 Shuxue Jinzhan • TJ
Shuxue Jinzhan (Advances in Mathematics) [1955ff] ISSN 0559-9326

J 0424 Wiss Z Univ Halle-Wittenberg, Math-Nat Reihe • DDR
Wissenschaftliche Zeitschrift der Martin-Luther-Universitaet Halle-Wittenberg. Mathematisch-Naturwissenschaftliche Reihe [1951ff]

J 0425 Managmt Sci (Tokyo) • J
Keiei-Kagaku (Management Science) [1956ff] ISSN 0451-5978

J 0430 Vopr Kibern (Akad Nauk Uzb SSR) • SU
Voprosy Kibernetiki i Vychislitel'noj Matematiki (Problems of Cybernetics and Numerical Mathematics) [1966ff] ISSN 0507-3502

J 0434 J Fac Sci Univ Tokyo, Sect 1 • J
Journal of the Faculty of Science. University of Tokyo. Section 1 Mathematics, Astronomy, Physics, Chemistry [1925-1970] ISSN 0040-8980
• CONT AS (**J** 2332) J Fac Sci, Univ Tokyo, Sect 1 A

J 0435 Int J Comput & Inf Sci • USA
International Journal of Computer and Information Sciences [1972ff] ISSN 0091-7036

J 0440 Bul Inst Politeh Bucuresti, Ser Mec • RO
Buletinul Institutului Politehnic "Gheorghe Gheorghiu-Dej" Bucuresti. Seria Mecanica (Bulletin des Polytechnischen Instituts "Gheorghe Gheorghiu-Dej" Bukarest. Serie Mechanik)

J 0442 J Osaka Inst Sci & Tech, Part 1 • J
Osaka Institute of Science and Technology (The Kinki University). Journal. Part 1. Mathematics and Physics [1949ff]

J 0443 Sci Thought (Tokyo) • J
Science of Thought [1954ff] ISSN 0559-1228

J 0446 Ann Acad Sci Fennicae, Ser A I, Diss • SF
Annales Academiae Scientiarum Fennicae. Series AI. Mathematica. Dissertationes [1975ff] ISSN 0066-1953
• CONT OF (**J** 3994) Ann Acad Sci Fennicae Ser A I

J 0447 An Univ Bucuresti, Mat • RO
Analele Universitatii Bucuresti. Matematica (Annalen der Universitaet Bukarest. Mathematik) [1951ff]

J 0448 Math of USSR, Izv • USA
Mathematics of the USSR, Izvestiya [1967ff] ISSN 0025-5726
• TRANSL OF (**J** 0216) Izv Akad Nauk SSSR, Ser Mat

J 0449 Probl Kybern • DDR
Probleme der Kybernetik [1958-1965]
• CONT AS (**J** 0471) Syst Th Res • TRANSL OF (**J** 0052) Probl Kibern

J 0451 Studia Soc Sci Torunensis Sect A • PL
Studia Societatis Scientiarum Torunensis, Sectio A

J 0455 Phil Naturalis • D
Philosophia Naturalis: Archiv fuer Naturphilosphie und die Philosophischen Grenzgebiete der Exakten Wissenschaften und Wissenschaftsgeschichte [1950ff] ISSN 0031-8027

J 0459 C R Soc Sci Lett Varsovie Cl 3 • PL
*Societe des Sciences et des Lettres de Varsovie. Comptes Rendus des Seances. Classe III: Sciences Mathematiques et Physiques * Towarzystwo Naukowe Warszawskie. Sprawozdania z Posiedze. Wydzialu III: Nauk Matematyczno-Fizycznych (Warschauer Sitzungsberichte)* [1908-1950]

J 0460 Acta Univ Szeged, Sect Mat • H
Acta Litterarum ac Scientiarum Regiae Universitatis Hungaricae Francisco-Josephinae, Sectio Scientiarum Mathematicarum [1922-1946]
• CONT AS (**J** 0002) Acta Sci Math (Szeged)

J 0462 Mat Fiz Oszt Koezlem, Acad Sci Hung • H
Magyar Tudomanyos Akademia. Matematikai es Fizikai Tudomanyok Osztalyanak Koezlemenyek. (Hungarian Academy of Sciences. Bulletin of the Mathematical and Physical Sciences) [1952-1974]
• CONT AS (**J** 1458) Alkalmaz Mat Lapok

J 0464 Syst-Comp-Controls • USA
Systems - Computers - Controls [1970ff] ISSN 0096-8765
• TRANSL OF (**J** 0979) Denshi Tsushin Gakkai Ronbunshi, Sect A-D

J 0465 Bull Greek Math Soc (NS) • GR
Bulletin of the Greek Mathematical Society.New Series (Hellenike Mathematike Hetaireia. Deltion. Nea Seira.) [1960ff] ISSN 0072-7466
• CONT OF (**J** 1699) Bull Soc Math Grece

J 0470 Sov Phys, Dokl • USA
Soviet Physics. Doklady. [1956ff] ISSN 0038-5689
• TRANSL OF (**J** 0023) Dokl Akad Nauk SSSR

J 0471 Syst Th Res • USA
Systems Theory Research [1966ff] ISSN 0082-1255
• CONT OF (**J** 1195) Probl Cybernet & (**J** 0449) Probl Kybern
• TRANSL OF (**J** 0052) Probl Kibern

J 0472 Theory Decis • NL
Theory and Decision. An International Journal for Philosophy and Methodology of the Social Sciences [1970ff] ISSN 0040-5833

J 0474 Avtom Vychis Tekh, Akad Nauk Latv SSR • SU
Avtomatika i Vychislitel'naya Tekhnika. Akademiya Nauk Latvijskoj SSR (Automation and Computer Science. Academy of Sciences of the Latvian SSR) [1967ff] ISSN 0572-4538
• TRANSL IN (**J** 2666) Autom Control Comput Sci

J 0475 Sib Math J • USA
Siberian Mathematical Journal [1966ff] ISSN 0037-4466
• TRANSL OF (**J** 0092) Sib Mat Zh

J 0477 Spis Bulgar Akad Nauk • BG
B"lgarski Akademija na Naukite. Spisanie (Bulgarian Academy of Sciences. Journal) ISSN 0015-3265

J 0479 Kwart Filoz • PL
Kwartalnik Filozoficzny (Philosophical Quaterly)

J 0481 Acta Univ Wroclaw • PL
Acta Universitatis Wratislaviensis

J 0482 Publ Math Univ Belgrade • YU
Universite de Belgrade. Publications Mathematiques [1932ff]

J 0488 Dialogue (Ottawa) • CDN
*Dialogue: Canadian Philosophical Review * Dialogue: Revue Canadienne de Philosophie.* [1962ff] ISSN 0012-2173

J 0489 Philosophy • GB
Philosophy [1925ff] ISSN 0031-8191

J 0490 Collecteana Logica • PL
Collecteana Logica [1939ff]

J 0493 Bull Res Counc Israel Sect F • IL
Research Council of Israel. Bulletin. Section F. Mathematics and Physics [1952-1962]
• CONT AS (**J** 0029) Israel J Math

J 0494 Bull Math Soc Sci Roumanie • RO
Societatea Romana de Stiinte, Sectia Mathematica. Bulletin Mathematiques de la Societe Roumaine des Sciences [1908-1956]
• CONT AS (**J** 0070) Bull Soc Sci Math Roumanie, NS

J 0497 Math Mag • USA
Mathematics Magazine [1947ff] ISSN 0025-570X
• CONT OF (**J** 1737) Nat Math Magazine (Louisiana)

J 0498 Ann Univ Turku, Ser A I • SF
Annales Universitatis Turkuensis. Series A.I: Astronomica, Chemica, Physica, Mathematica (Turun Yliopiston Julkaisuja. Sarja A.1.: Astronomica, Chemica, Physica, Mathematica) [1957ff] ISSN 0082-7002

J 0500 Tamkang J Math • RC
Tamkang Journal of Mathematics [1970ff] ISSN 0049-2930

J 0503 Artif Intell • NL
Artificial Intelligence [1970ff] ISSN 0004-3702

J 0504 Manuscr Math • D
Manuscripta Mathematica [1969ff] ISSN 0025-2611

J 0505 Dominican Studies
Dominican Studies

Source Index

J 0508 Machine Intelligence • GB
Machine Intelligence [1967ff] ISSN 0541-6418

J 0509 Danish Yearbook Philos • DK
Danish Yearbook of Philosophy [1964ff] ISSN 0070-2749

J 0510 Moscow Univ Math Bull • USA
Moscow University Mathematics Bulletin [1969ff] ISSN 0027-1322
• TRANSL OF (**J** 0288) Vest Ser Mat Mekh, Univ Moskva

J 0515 Bull Math Biophys • GB
Bulletin of Mathematical Biophysics [1939-1971] ISSN 0007-4985
• CONT AS (**J** 3073) Bull Math Biol

J 0517 Mathematica (Cluj) • RO
Mathematica. Revue d'Analyse Numerique et de Theorie de l'Approximation [1929ff] ISSN 0025-5505

J 0518 Bul Sti Mat-Fiz, Acad Romina • RO
*Academia Republicii Populare Romine. Buletinul Stiintific. Sectia de Stiinte Matematice si Fizice * Academie de la Republique Populaire Roumainie. Bulletin Scientifique. Section des Sciences Mathematiques et Physiques * Akademija Rumynskoi Respubliki. Nachnyi Vestnik. Otdelenie Matematicheskih i Fizicheskih Nauk* [1949-1965] ISSN 0515-1333
• CONT OF (**J** 0174) Bull Sect Sci Acad Roumaine

J 0519 Wiad Mat, Ann Soc Math Pol, Ser 2 • PL
*Annales Societatis Mathematicae Polonae. Seria 2. Wiadomosci Matematyczne * Roczniki Polskiego Towarzystwa Matematycznego. Seria 2. Wiadomosci Matematyczne* [1955ff] ISSN 0079-3698
• CONT OF (**J** 4710) Pol Tow Mat, Wiad Mat

J 0521 Semin Math, Inst Steklov • USA
Seminars in Mathematics. V.A.Steklov Mathematical Institute Leningrad [1967ff]
• TRANSL OF (**S** 0228) Zap Nauch Sem Leningrad Otd Mat Inst Steklov

J 0522 Engin Cybern • USA
Engineering Cybernetics. Soviet Journal of Computer and System Sciences. Essential Serials in Electronics and Systems Science [1963ff] ISSN 0013-788X
• TRANSL OF (**J** 0977) Izv Akad Nauk SSSR, Tekh Kibern

J 0523 Bull Nagoya Inst Tech • J
Bulletin of Nagoya Institute of Technology [1949ff]

J 0524 Disq Math Phys • RO
Disquisitiones Mathematicae et Physicae [1940-1949]
• CONT AS (**J** 0197) Stud Cercet Mat Acad Romana

J 0525 C R Acad Sci Roumanie • RO
Academie des Sciences de Roumanie. Comptes-Rendus des Seances [1937ff]
• REM Since 1939 called: Comptes-Rendus des Seances de l'Institut des Sciences de Roumanie

J 0526 An Acad Romana Mem Sti Ser 3 • RO
*Academia Romana.Analele. Sectiunea Stiintifica. Memorii * Academie Roumaine. Annales. Memoirs de la Section Scientifique. Serie 3* [1922-?]

J 0527 Bull Inst Politeh Bucuresti • RO
*Institutul Politehnic Bucuresti. Buletinul * Ecole Polytechnique de Bucarest. Bulletin* [1929-?]

J 0529 Aequationes Math • CH
Aequationes Mathematicae [1968ff] ISSN 0001-9054

J 0531 Hitotsubashi J Arts Sci (Tokyo) • J
Hitotsubashi Journal of Arts & Sciences [1960ff] ISSN 0073-2788

J 0537 Yokohama Math J • J
The Yokohama Mathematical Journal [1953ff] ISSN 0044-0523

J 0545 Proc Indian Acad Sci, Sect A • IND
Indian Academy of Sciences. Proceedings. Section A: Physical Sciences [1931ff] ISSN 0019-428X

J 0549 Riv Mat Univ Parma, Ser 4 • I
Rivista di Matematica della Universita di Parma. Serie 4 [1975ff] ISSN 0035-6298
• CONT OF (**J** 3254) Riv Mat Univ Parma, Ser 3

J 0589 Bull Amer Math Soc (NS) • USA
Bulletin of the American Mathematical Society. New Series [1979ff] ISSN 0273-0979
• CONT OF (**J** 0015) Bull Amer Math Soc

J 0611 Pol Tow Mat, Prace Mat-Fiz • PL
Polskie Towarzystwo Matematyczene. Prace Matematyczno-Fizyczne (Mathematische und Physikalische Abhandlungen) [1887-1954]
• CONT AS (**J** 0051) Commentat Math, Ann Soc Math Pol, Ser 1 & (**J** 2095) Fund Inform, Ann Soc Math Pol, Ser 4

J 0658 Festschr-Ber VDI-Zeitschr, Reihe 10 • D
Festschrift-Berichte der VDI-Zeitschriften. Reihe 10. Angewandte Informatik (Verein Deutscher Ingenieure)

J 0667 Scripta Fac Sci Math, Brno • CS
Scripta Facultatis Scientarium Naturalium Universitatis J.E. Purkyne Brunensis. Mathematica [1971ff]

J 0748 Erkenntnis (Leipzig) • DDR
Erkenntnis [1930-1939]
• CONT OF (**J** 1380) Ann Philos & Philos Kritik • CONT AS (**J** 3597) J Unif Sci

J 0945 Riv Filos • I
Rivista di Filosofia [1909ff] ISSN 0035-6239

J 0947 Wiss Z Tech Univ Dresden • DDR
Wissenschaftliche Zeitschrift der Technischen Universitaet Dresden [1951ff] ISSN 0043-6925

J 0948 Z Naturforsch • D
Zeitschrift fuer Naturforschung. [1946ff]

J 0954 Tr Inst Prikl Mat, Tbilisi • SU
*Tbilisskij Gosudarstvennyj Universitet. Institut Prikladnoj Matematiki. Trudy * Thbilisis Sahelmcipho Universiteti Gamoqenebithi Mathematikis Instituti Shromebi (State University of Tbilisi. Institute of Applied Mathematics. Publications)* [1969ff] ISSN 0082-2191

J 0955 Amer J Phys • USA
American Journal of Physics [1933ff] ISSN 0002-9505

J 0957 Ranchi Univ Math J • IND
Ranchi University Mathematical Journal [1970ff] ISSN 0079-9602

J 0962 Ann Soc Sci Bruxelles, Ser 1 • B
Annales de la Societe Scientifique de Bruxelles. Serie I. Sciences Mathematiques, Astronomiques et Physiques [1875ff] ISSN 0037-959X

J 0963 Ajatus (Helsinki) • SF
Ajatus. Suomen Filosofisen Yhdislyksen Vuosikirja (Yearbook of the Philosophical Society of Finland) [1926ff]

J 0966 Episteme (Caracas) • YV
Episteme: Anuario de Filosofia [1957ff] ISSN 0425-1547

J 0967 Izv Akad Nauk Mold SSR, Ser Fiz-Tekh Mat • SU
*Izvestiya Akademii Nauk Moldavskoj SSR. Seriya Fiziko-Tekhnicheskikh i Matematicheskikh Nauk * Buletinul Akademiei de Shtiince a RSS Moldovensht (Proceedings of the Academy of Sciences of the Moldavian SSR. Series: Physical-Technical and Mathematical Sciences)* [1951ff] ISSN 0321-169X

J 0968 Uch Zap Univ, Kazan • SU
Uchenye Zapiski Kazanskogo (Ordena Trudovogo Krasnogo Znameni) Gosudarstvennogo Universiteta Imeni V.I. Ul'yanova-Lenina (Scientific Notes. State Universirty Kazan) [1843ff]

J 0970 Math Rep Coll Gen Educ, Kyushu Univ • J
Mathematical Reports of College of General Education. Kyushu University

J 0973 Bull Acad Serbe Sci Math Nat, NS • YU
Academie Serbe des Sciences et des Arts. Classe des Sciences Mathematiques et Naturelles. Bulletin. Nouvelle Serie (Srpska Akademija Nauki Umetnosti, Otdeljenje Prirodno-Matematichkikh Nauk.) [1967ff] ISSN 0001-4184
• CONT OF (**J** 0993) Bull Acad Serbe, Cl Sci Math Nat

J 0974 Norsk Vid-Akad Oslo Mat-Natur Kl Skr • N
Norske Videnskaps - Akademi i Oslo. Matematisk-Naturvidenskapelig Klasse. Skrifter. (Monographs of the Scandinavian Academy of Sciences. Mathematical and Natural Sciences Class) [1929ff] ISSN 0029-2338
• CONT OF (**J** 1145) Vidensk Selsk Kristiana Skrifter Ser 1

J 0977 Izv Akad Nauk SSSR, Tekh Kibern • SU
Izvestiya Akademii Nauk SSSR. Tekhnicheskaya Kibernetika. Otdelenie Mekhaniki i Protsessov Upravlenija (Proceedings of the Academy of Sciences of the USSR. Engineering Cybernetics. Department of Mechanics and Control Processes) [1963ff] ISSN 0002-3388
• TRANSL IN (**J** 0522) Engin Cybern

J 0979 Denshi Tsushin Gakkai Ronbunshi, Sect A-D • J
Denshi Tsushin Gakkai Ronbunshi. Sect. A-D (Reports of the University of Electro-Communications) [1949ff] ISSN 0493-4253
• TRANSL IN (**J** 0116) Electr & Comm Japan & (**J** 0464) Syst-Comp-Controls

J 0980 Mem Fukui Univ Lib Arts College • J
Fukui University. Liberal Arts College. Memoirs [1952ff]

J 0984 Ann Inst Henri Poincare, Sect B • F
Annales de l'Institut Henri Poincare. Section B : Calcul des Probabilites et Statistique [1930ff] ISSN 0020-2347

J 0988 Rev Filos (Santiago) • RCH
Revista de Filosofia [1954ff] ISSN 0034-8236

J 0989 Z Allg Wissth • D
Zeitschrift fuer Allgemeine Wissenschaftstheorie (Journal for General Philosophy of Science) [1969ff] ISSN 0044-2216

J 0990 Soc Sci Fennicae Comment Phys-Math • SF
Societas Scientiarum Fennicae. Commentationes Physico-Mathematicae

J 0993 Bull Acad Serbe, Cl Sci Math Nat • YU
Academie Serbe des Sciences et des Arts. Bulletin. Classe des Sciences Mathematiques et Naturelles. Sciences Mathematiques [1952-1966] ISSN 0561-7332
• CONT OF (**J** 4720) Bull Acad Royal Serbe, Math Phys
• CONT AS (**J** 0973) Bull Acad Serbe Sci Math Nat, NS

J 0996 RAAG Res Notes • J
Research Notes and Memoranda of Applied Geometry for Prevenient Natural Philosophy. Post RAAG-Reports.

J 1005 Rep Fac Sci, Shizuoka Univ • J
Reports of the Faculty of Science. Shizuoka University. [1965ff] ISSN 0583-0923

J 1008 Demonstr Math (Warsaw) • PL
Demonstratio Mathematica [1969ff] ISSN 0420-1213

J 1021 Itogi Nauki Ser Mat • SU
Itogi Nauki. Seriya Matematiki (Progress in Science. Mathematical Series) [1962-1971] ISSN 0579-1731
• CONT AS (**J** 1488) Itogi Nauki Tekh, Ser Probl Geom & (**J** 1501) Itogi Nauki Tekh, Ser Algeb, Topol, Geom & (**J** 1452) Itogi Nauki Tekh, Ser Sovrem Probl Mat & (**J** 3188) Itogi Nauki Tekh, Ser Teor Veroyat Mat Stat Teor Kibern & (**J** 4387) Itogi Nauki Tekh, Ser Tekh Kibern • TRANSL IN (**J** 1531) J Sov Math
• REM J1531 contains only selected translations

J 1022 Ratio (Oxford) • GB
Ratio [1957ff] ISSN 0034-0006
• REL PUBL (**J** 0170) Ratio (Frankfurt) • REM English Edition

J 1024 Zhongguo Kexue • TJ
Zhongguo Kexue (Scientia Sinica) [1950-1981]
• CONT AS (**J** 3766) Zhongguo Kexue, Xi A

J 1030 Formalisation • F
La Formalisation: Cahier pour l'Analyse

J 1037 Issl Obshch Algebr (Kishinev) • SU
Issledovaniya po Obshchej Algebre (Studies in General Algebra) [1965ff]

J 1044 Math Notes, Acad Sci USSR • USA
Mathematical Notes of the Academy of Sciences of the USSR [1967ff] ISSN 0001-4346
• TRANSL OF (**J** 0087) Mat Zametki (Akad Nauk SSSR)

J 1046 Z Angew Math Mech • DDR
Zeitschrift fuer Angewandte Mathematik und Mechanik: Ingeneurwissenschaftliche Forschungsarbeiten. Applied Mathematics and Mechanics [1921ff] ISSN 0044-2267

J 1048 Kiber Sb Perevodov • SU
Kibernetícheskij Sbornik: Sbornik Perevodov. (Collected Articles on Cybernetics: Collected Translations) [1960-1964]
• CONT AS (**J** 3079) Kiber Sb Perevodov, NS

J 1049 USSR Comput Math & Math Phys • GB
USSR Computational Mathematics and Mathematical Physics [1962ff] ISSN 0041-5553
• TRANSL OF (**J** 0199) Zh Vychisl Mat i Mat Fiz

J 1068 Physis • I
Physis. Rivista Internazionale di Storia della Scienza [1959ff] ISSN 0031-9414

J 1077 IEEE Proc • USA
Proceedings. IEEE (= Institute of Electrical and Electronics Engineers) [1963ff] ISSN 0018-9219
• CONT OF (**J** 4711) IRE Proc

Source Index

J 1093 Ruch Filoz • PL
Ruch Filozoficzny. Polskie Towarzystwo Filozoficzne (Philosophical Movement. Polish Philosophical Society) ISSN 0035-9599

J 1095 Beitr Phil Deutsch Idealismus • D
Beitraege zur Philosophie des Deutschen Idealismus [1918-1926]
• CONT AS (**J** 0168) Blaetter Deutsch Philos

J 1103 Kant Studien • D
Kant Studien [1896ff] ISSN 0022-8877
• REM X1174 is the most recent publisher

J 1108 Arkhimedes (Helsinki) • SF
Arkhimedes [1949ff] ISSN 0004-1920

J 1113 Commun Math Phys • D
Communications in Mathematical Physics [1965ff] ISSN 0010-3616

J 1114 Arch Phil • D
Archiv fuer Philosophie [1911-1930, 1947-1964]

J 1118 Jbuch Phil Phaenomenol Forsch • D
Jahrbuch fuer Philosophie und Phaenomenologische Forschung [1913-1930]

J 1124 Ergebn Math Kolloquium • A
Ergebnisse eines Mathematischen Kolloquiums [1929-1936]

J 1125 Przeglad Filoz • PL
Przeglad Filozoficzny (Revue Philosophique) [1897-1949]

J 1145 Vidensk Selsk Kristiana Skrifter Ser 1 • N
Videnskaps Selskapet i Kristiana. Skrifter Utgit. 1 Matematisk-Naturvidenskapelig Klasse [?-1928]
• CONT AS (**J** 0974) Norsk Vid-Akad Oslo Mat-Natur Kl Skr

J 1171 Sci Rep Fac Educ Gifu Univ, Nat Sci • J
Science Reports of the Faculty of Education. Gifu University. Natural Science

J 1182 Studia Filoz • PL
Studia Filozoficzne (Philosophical Studies) [1957ff] ISSN 0039-3142

J 1185 Inform Process & Managmt • GB
Information Processing & Management. Libraries and Information Retrieval, Systems and Communication Networks [1963ff] ISSN 0306-4573

J 1193 Comput J (London) • GB
The Computer Journal [1958ff] ISSN 0010-4620

J 1195 Probl Cybernet • GB
Problems of Cybernetics [1958-1965]
• CONT AS (**J** 0471) Syst Th Res • TRANSL OF (**J** 0052) Probl Kibern

J 1377 Ann New York Acad Sci • USA
New York Academy of Sciences. Annals [1877ff] ISSN 0077-8923

J 1380 Ann Philos & Philos Kritik • DDR
Annalen der Philosophie und Philosophischen Kritik [1925-1929]
• CONT AS (**J** 0748) Erkenntnis (Leipzig)

J 1395 Minist Narod Prosv Zh, NS (Petrograd) • SU
Ministerstvo Narodnogo Prosveshcheniya. Zhurnal. NS (Ministry of Education. Journal. New Series) [1906-1917]

J 1399 Russ Math Surv • GB
Russian Mathematical Surveys [1946ff] ISSN 0036-0279
• TRANSL OF (**J** 0067) Usp Mat Nauk

J 1404 Mat Sb, Akad Nauk SSSR • SU
Matematicheskij Sbornik. Akademiya Nauk SSSR i Moskovskoe Matematicheskoe Obshchestvo (Mathematical Collected Articles. New Series. Academy of Sciences of the USSR and the Moscovian Mathematical Society) [1866-1935]
• CONT AS (**J** 0142) Mat Sb, Akad Nauk SSSR, NS

J 1405 Ann Pol Math • PL
Annales Polonici Mathematici. Polska Akademia Nauk, Instytut Matematyczny [1953ff] ISSN 0066-2216
• CONT OF (**J** 0283) Ann Soc Pol Math

J 1426 Theor Comput Sci • NL
Theoretical Computer Science [1975ff] ISSN 0304-3975

J 1428 SIAM J Comp • USA
Journal on Computing. SIAM (= Society for Industrial and Applied Mathematics) [1972ff] ISSN 0097-5397

J 1429 Kybernetes • GB
Kybernetes: An International Journal of Cybernetics and General Systems [1972ff] ISSN 0368-492X

J 1431 Acta Inf • D
Acta Informatica [1971ff] ISSN 0001-5903

J 1452 Itogi Nauki Tekh, Ser Sovrem Probl Mat • SU
Itogi Nauki i Tekhniki: Seriya Sovremennye Problemy Matematiki (Progress in Science and Technology: Series on Current Problems in Mathematics) [1972ff]
• CONT OF (**J** 1021) Itogi Nauki Ser Mat • TRANSL IN (**J** 1531) J Sov Math

J 1456 SIGACT News • USA
SIGACT (= ACM Special Interest Group on Automata and Computability Theory). News [1969ff]

J 1458 Alkalmaz Mat Lapok • H
Alkalmazott Matematikai Lapok (Papers in Applied Mathematics) [1951ff] ISSN 0133-3399
• CONT OF (**J** 0462) Mat Fiz Oszt Koezlem, Acad Sci Hung

J 1471 Ann Univ Madagascar, Ser Sci Nat Math • RM
Annales de l'Universite de Madagascar. Serie Sciences de la Nature et Mathematiques [1963ff]

J 1488 Itogi Nauki Tekh, Ser Probl Geom • SU
Itogi Nauki i Tekhnike. Seriya Problemy Geometrii (Progress in Science and Technology. Series Problems in Geometry) [1972ff] ISSN 0202-7461
• CONT OF (**J** 1021) Itogi Nauki Ser Mat • TRANSL IN (**J** 1531) J Sov Math

J 1501 Itogi Nauki Tekh, Ser Algeb, Topol, Geom • SU
Itogi Nauki i Tekhniki. Seriya Algebra, Topologiya, Geometriya. (Progress in Science and Technology. Series Algebra, Topology, Geometry) [1972ff] ISSN 0202-7445
• CONT OF (**J** 1021) Itogi Nauki Ser Mat • TRANSL IN (**J** 1531) J Sov Math & (**C** 4688) Prog in Math, Vol 12
• REM C4688 is Volume 1968

J 1508 Math Sem Notes, Kobe Univ • J
Mathematics Seminar Notes. Kobe University [1973-1983] ISSN 0385-633X
• CONT AS (**J** 4390) Kobe J Math

J 1515 Archimede • I
Archimede. Rivista per gli Insegnanti e i Cultori di Matematiche Pure e Applicate [1949ff] ISSN 0003-8369

J 1517 Theor Linguist • D
Theoretical Linguistics [1974ff] ISSN 0301-4428

J 1522 Math Slovaca • CS
Mathematica Slovaca [1976ff] ISSN 0025-5173
• CONT OF (**J** 0143) Mat Chasopis (Slov Akad Ved)

J 1526 Riv Mat Univ Parma, Ser 2 • I
Rivista di Matematica della Universita di Parma. Serie 2 [1960-1971] ISSN 0035-6298
• CONT OF (**J** 0391) Riv Mat Univ Parma • CONT AS (**J** 3254) Riv Mat Univ Parma, Ser 3

J 1527 Pokroky Mat Fyz Astron (Prague) • CS
Pokroky Matematiky, Fyziky a Astronomie (Progress in Mathematics, Physics and Astronomy) [1956ff] ISSN 0032-2423

J 1531 J Sov Math • USA
Journal of Soviet Mathematics [1973ff] ISSN 0090-4104
• TRANSL OF (**J** 1021) Itogi Nauki Ser Mat & (**S** 0228) Zap Nauch Sem Leningrad Otd Mat Inst Steklov & Problemy Matimaticheskogo Analiza & (**J** 1452) Itogi Nauki Tekh, Ser Sovrem Probl Mat & (**J** 1501) Itogi Nauki Tekh, Ser Algeb, Topol, Geom & (**J** 3188) Itogi Nauki Tekh, Ser Teor Veroyat Mat Stat Teor Kibern & (**J** 1488) Itogi Nauki Tekh, Ser Probl Geom
• REM This contains selected translations from each of the Russian Journals listed

J 1546 Rep Math Phys (Warsaw) • PL
Reports on Mathematical Physics [1970ff]

J 1550 Creation Math • IL
Creation in Mathematics [1970ff]

J 1552 Teor Mat Fiz, Akad Nauk SSSR • SU
Akademiya Nauk SSSR. Teoreticheskaya i Matematicheskaya Fizika (Academy of Sciences of the USSR. Theoretical and Mathematical Physics) [1969ff] ISSN 0564-6162
• TRANSL IN Theor Math Phys

J 1561 IEEE Trans Syst & Sci Cybern • USA
Transactions on Systems Science and Cybernetics. IEEE (= Institute of Electrical and Electronics Engineers) [1945-1970] ISSN 0018-9472
• CONT AS (**J** 2338) IEEE Trans Syst Man & Cybern

J 1568 Ann Sc Norm Sup Pisa, Fis Mat, Ser 2 • I
Annali della R. Scuola Normale Superiore di Pisa. Serie 2 Scienze Fisiche e Matematiche [1932-1946]
• CONT OF (**J** 1908) Ann Sc Norm Sup Pisa, Fis Mat • CONT AS (**J** 0315) Ann Sc Norm Sup Pisa Fis Mat, Ser 3

J 1573 Notas Commun Mat (Recife) • BR
Notas e Comunicacoes de Matematica [1965ff] ISSN 0085-5413

J 1586 Rocz Pol Towarz Nauk Obczyznie • PL
Rocznik Polskiego Towarzystwa Naukowego na Obczyznie (Yearbook of the Polish Society of Arts and Sciences Abroad) [1951ff]

J 1598 Taita J Math (Univ Taipei) • RC
Taita Journal of Mathematics [1969ff]

J 1666 Apl Mat, Cheskoslov Akad Ved • CS
Aplikace Matematiky. Cheskoslovenska Akademie Ved (Applied Mathematics. Czechoslovak Academy of Sciences) [1956ff] ISSN 0003-6501

J 1667 God Filos-Istor Fak, Univ Sofiya • BG
Sofijskija Universitet. Filosofosko-Istoricheski Fakultet. Godishnik. (Annuaire. Universite de Sofia. Faculte de Philosophie et Histoire) [1950ff] ISSN 0081-184X

J 1669 J Comb Th • USA
Journal of Combinatorial Theory [1966-1970] ISSN 0021-9800
• CONT AS (**J** 0164) J Comb Th, Ser A & (**J** 0033) J Comb Th, Ser B

J 1670 Mitt Math Ges DDR • DDR
Mitteilungen der Mathematischen Gesellschaft der DDR

J 1678 Found Phys • USA
Foundations of Physics [1970ff] ISSN 0015-9018

J 1699 Bull Soc Math Grece • GR
(Bulletin de la Societe Mathematique Grece) [1921-1959]
• CONT AS (**J** 0465) Bull Greek Math Soc (NS)

J 1717 An Univ Chile • RCH
Universidad de Chile. Anales ISSN 0041-8358

J 1720 Rev Neoscolast Philos, Ser 2 • B
Revue Neo-Scolastique de Philosophie. Serie 2 [1910-1945]
• CONT AS (**J** 0252) Rev Philos Louvain

J 1724 Kagaku Tetsugaku • J
Kagaku Tetsugaku (Philosophy of Science) [1968ff]

J 1737 Nat Math Magazine (Louisiana) • USA
National Mathematics Magazine [1926-1946]
• CONT AS (**J** 0497) Math Mag

J 1741 Int J Man-Mach Stud • USA
International Journal of Man-Machine Studies [1969ff] ISSN 0020-7373

J 1742 Atti Accad Sci Torino, Fis Mat Nat • I
Atti della Reale Accademia delle Scienze di Torino. Classe I: Scienze Fisiche, Matematiche e Naturali [1865-1939]
• CONT AS (**J** 0220) Atti Accad Sci Torino, Fis Mat Nat

J 1743 Int J Gen Syst • USA
International Journal of General Systems: Methodology, Applications, Education [1974ff] ISSN 0308-1079

J 1770 Osaka Math J • J
Osaka Mathematical Journal [1949-1963]
• CONT AS (**J** 0351) Osaka J Math

J 1779 Chinese Stud Phil • USA
Chinese Studies in Philosophy [1969ff] ISSN 0023-8627

J 1807 Phil Quart Israel • IL
Philosophia. Philosophical Quarterly of Israel [1971ff] ISSN 0048-3893

J 1830 Stud Phil Sci (Tokyo) • J
Studies in the Philosophy of Science

J 1831 Anuar Soc Paranaense Mat, Ser 2 • BR
Sociedade Paranaense de Matematica. Anuario. Serie 2 [1958ff]
• CONT OF (**J** 0294) Anuar Soc Paranaense Mat

J 1834 Arch Rechts- und Sozialphil • D
*Archives de Philosophie du Droit et de Philosophie Sociale * Archiv fuer Rechts- und Sozialphilosophie * Archives for Philosophy of Law and Social Philosophy* [1907ff] ISSN 0001-2343

J 1843 Amer Sci • USA
American Scientist. [1913ff] ISSN 0003-0996

Source Index

J 1868 J Bihar Math Soc • IND
Journal of the Bihar Mathematical Society [1977ff]

J 1884 Avtomatika, Akad Nauk Ukr SSR • SU
Avtomatika. Akademiya Nauk Ukrainskoj SSR. Nauchno-tekhnicheskij Kompleks "Institut Kibernetiki imeni V.M.Glushkova". Nauchno-tekhnicheskij Zhurnal (Automation. Academy of Sciences of the Ukrainian SSR. Scientific-Technical Complex: Glushkov Institute of Cybernetics) [1956ff] ISSN 0572-2691
• TRANSL IN (J 2548) Sov Autom Control

J 1893 Relevance Logic Newslett • USA
The Relevance Logic Newsletter. [1976ff]

J 1908 Ann Sc Norm Sup Pisa, Fis Mat • I
Annali della Reale Scuola Normale Superiore Universitaria di Pisa. Scienze Fisiche e Matematiche [1913-1931]
• CONT AS (J 1568) Ann Sc Norm Sup Pisa, Fis Mat, Ser 2

J 1910 Proc London Math Soc, Ser 2 • GB
Proceedings of the London Mathematical Society. Serie 2 [1904-1951]
• CONT OF (J 0077) Proc London Math Soc • CONT AS (J 3240) Proc London Math Soc, Ser 3

J 1927 Prace Inst Podstaw Inf, Pol Akad Nauk • PL
Prace Instytut Podstaw Informatyki Polskiej Akademii Nauk (Reports. Institute of Computer Science Polish Academy of Science)

J 1929 Prace Centr Oblicz Pol Akad Nauk • PL
Polska Akademija Nauk. Centrum Obliczeniowe. Prace. (Polish Academy of Sciences. Computation Centre. Reports) ISSN 0079-3175

J 1934 Ann Sci Univ Clermont Math • F
Annales Scientifiques de l'Universite de Clermont-Ferrand II, Section Mathematiques (Clermont Ferrand) [1973ff] ISSN 0069-472X
• CONT OF (J 0179) Ann Fac Sci Clermont

J 1957 Bol Soc Paranaense Mat • BR
Sociedade Paranaense de Matematica. Boletim. [1958-1979] ISSN 0037-8712
• CONT AS (J 4709) Bol Soc Paranaense Mat, Ser 2

J 2015 Computer • USA
Computer: A Monthly Magazine [1970ff] ISSN 0018-9162

J 2018 Yingyong Shuxue Yu Jisuan Shuxue • TJ
Yingyong Shuxue Yu Jisuan Shuxue (Journal on Numerical Methods and Computer Applications) [1980ff]

J 2026 J West Virginia Phil Soc • USA
Journal of the West Virginia Philosophical Society

J 2028 Hist & Phil Log • GB
History and Philosophy of Logic [1980ff] ISSN 0144-5340

J 2038 Rend Sem Mat, Torino • I
Rendiconti del Seminario Matematico (gia "Conferenze di Fisica e di Matematica"). Universita e Politecnico di Torino

J 2051 Stud Soviet Thought • CH
(Studies in Soviet Thought) [1961ff] ISSN 0039-3797

J 2053 Math Medley • SGP
Mathematical Medley [1975ff]

J 2076 Rev Int Philos • B
Revue Internationale de Philosophie [1938ff] ISSN 0048-8143
• CONT OF (J 0151) Arch Soc Belg Philos

J 2086 Epistem Sociol • F
Epistemologie Sociologique [1964ff] ISSN 0013-9645

J 2095 Fund Inform, Ann Soc Math Pol, Ser 4 • PL
*Fundamenta Informaticae. Annales Societatis Mathematicae Polonae. Series 4 * Roczniki Polskiego Towarzystwa Matematycznego. Seria 4* [1977ff] ISSN 0324-8429
• CONT OF (J 0611) Pol Tow Mat, Prace Mat-Fiz

J 2099 Boll Unione Mat Ital, VI Ser, A • I
Bolletino della Unione Matematica Italiana. Serie VI. A [1982ff] ISSN 0041-7084
• CONT OF (J 3285) Boll Unione Mat Ital, V Ser, A

J 2100 Boll Unione Mat Ital, VI Ser, B • I
Bolletino della Unione Matematica Italiana. Serie VI. B [1982ff] ISSN 0041-7084
• CONT OF (J 3495) Boll Unione Mat Ital, V Ser, B

J 2107 Publ Dep Math, Lyon, NS • F
Publications du Departement de Mathematiques. Nouvelle Serie. Faculte des Sciences de Lyon. [1982ff] ISSN 0076-1656
• CONT OF (J 0056) Publ Dep Math, Lyon

J 2119 Suppl Aristotelian Soc • GB
The Aristotelian Society Supplementary [1918ff]
• REL PUBL (J 0113) Proc Aristotelian Soc

J 2128 C R Math Acad Sci, Soc Roy Canada • CDN
*Comptes Rendus Mathematiques de l'Academie des Sciences. La Societe Royale du Canada * Mathematical Reports of the Academy of Sciences* [1979ff] ISSN 0706-1994

J 2130 Linguist Philos • NL
Linguistics and Philosophy [1977ff] ISSN 0165-0157

J 2132 Rep Math Colloq, Ser 2 • USA
Reports of a Mathematical Colloqium. 2nd Series. [1935-?]

J 2145 J Int Ass Math Geol • USA
Journal of the International Association for Mathematical Geology [1969ff] ISSN 0020-5958

J 2146 Helv Phys Acta • CH
Helvetica Physica Acta [1928ff] ISSN 0018-0238

J 2161 Hangzhou Daxue Xuebao, Ziran Kexue • TJ
Hangzhou Daxue Xuebao. Ziran Kexue Ban (Journal of Hangzhou University.Natural Science Edition)

J 2173 J Japan Ass Phil Sci • J
Kagaku Kisoron Kenkyo (Japanese Association for the Philosophy of Science. Journal) [1954ff] ISSN 0022-7668

J 2293 Comp Linguist & Comp Lang • H
Computational Linguistics and Computer Languages

J 2310 Obz Mat Fiz, Ljubljana • YU
Obzornik za Matematiko in Fiziko (Mathematical and Physical Reviews) [1951ff] ISSN 0473-7466

J 2311 Rend Mat, Ser 6 • I
Rendiconti di Matematica. Serie 6 [1968-1980] ISSN 0034-4427
• CONT OF (J 0384) Rend Mat, Ser 5

J 2313 C R Acad Sci, Paris, Ser A-B • F
Academie des Sciences de Paris. Comptes Rendus Hebdomadaires des Seances. Serie A: Sciences Mathematiques, Serie B: Sciences Physiques [1966-1980] ISSN 0001-4036
• CONT OF (J 0109) C R Acad Sci, Paris • CONT AS (J 3364) C R Acad Sci, Paris, Ser 1 & (J 2314) C R Acad Sci, Paris, Ser 2

J 2314 C R Acad Sci, Paris, Ser 2 • F
Comptes Rendus des Seances de l'Academie des Sciences. Serie II. Mecanique-Physique, Chimie, Sciences de la Terre, Sciences de l'Univers [1981ff]
• CONT OF (J 2313) C R Acad Sci, Paris, Ser A-B

J 2320 Probl Peredachi Inf, Akad Nauk SSSR • SU
Problemy Peredachi Informatsii. Zhurnal Akademii Nauk SSSR (Probleme der Informationsuebertragung) [1965ff] ISSN 0555-2923
• TRANSL IN (J 3419) Probl Inf Transmiss

J 2331 ACM Comp Surveys • USA
Computing Surveys: The Survey and Tutorial Journal of the ACM (= Association for Computing Machinery) [1969ff] ISSN 0360-0300

J 2332 J Fac Sci, Univ Tokyo, Sect 1 A • J
Journal of the Faculty of Science. University of Tokyo. Section 1 A. Mathematics [1971ff] ISSN 0040-8980
• CONT OF (J 0434) J Fac Sci Univ Tokyo, Sect 1

J 2338 IEEE Trans Syst Man & Cybern • USA
Transactions on Systems, Man and Cybernetics. IEEE (= Institute of Electrical and Electronics Engineers) [1971ff] ISSN 0018-9472
• CONT OF (J 1561) IEEE Trans Syst & Sci Cybern

J 2340 Int Phil Quart • USA
International Philosophical Quarterly [1961ff] ISSN 0019-0365

J 2361 Sov Stud Philos • USA
(Soviet Studies in Philosophy) [1962ff] ISSN 0038-5883

J 2412 Prace Nauk, Ser Mat-Przyr, Wyz Szk Ped, Czestochowa • PL
Wyzsza Szkola Pedagogiczna v Czestochowie. Prace Naukowe. Seria Matematyczno-Przyrodnicza (University of Education in Czestochowa. Scientific Publications. Series: Mathematics and Natural Sciences) [1978ff]

J 2439 J Roy Stat Soc Ser A • GB
Journal of the Royal Statistical Society. Series A. General [1838ff] ISSN 0035-9238

J 2521 Beijing Shifan Daxue Xuebao, Ziran Kexue • TJ
Beijing Shifan Daxue Xuebao. Ziran Kexue Ban (Journal of Natural Sciences of Beijing Normal University. Natural Science Edition)

J 2548 Sov Autom Control • USA
Soviet Automatic Control. Essential Serials in Electronics and Cybernetics [1968ff] ISSN 0038-5328
• TRANSL OF (J 1884) Avtomatika, Akad Nauk Ukr SSR

J 2551 J Log Progr • USA
Journal of Logic Programming [1984ff] ISSN 0743-1066

J 2562 Publ Sec Mat Univ Autonoma Barcelona • E
Publications de la Seccio de Matematiques. Universitat Autonoma de Barcelona (Veroeffentlichunge der Abteilung Mathematik)

J 2612 Math (Monthly J) • USA
Mathematics (Monthly Journal) [1980ff]

J 2658 Ann Inst Henri Poincare, Sect A • F
Annales de l'Institut Henri Poincare. Section A. Physique Theorique ISSN 0020-2339

J 2666 Autom Control Comput Sci • USA
Automatic Control and Computer Sciences [1969ff] ISSN 0146-4116
• TRANSL OF (J 0474) Avtom Vychis Tekh, Akad Nauk Latv SSR

J 2667 Autom Doc Math Linguist • USA
Automatic Documentation and Mathematical Linguistics [1967ff] ISSN 0005-1055
• TRANSL OF (J 0338) Nauch-Tekh Inf, Ser 2, Akad Nauk SSSR

J 2668 Avtom Sist Upravl & Prib Avtom, Khar'kov • SU
Avtomatizirovannye Sistemy Upravleniya i Pribory Avtomatiki (Automatisierte Steuersysteme und Automatisierungsvorrichtungen) [1965ff] ISSN 0135-1710

J 2680 Bull Iranian Math Soc • IR
Bulletin. Iranian Mathematical Society [1973ff]

J 2681 Bull Univ Osaka Prefecture, Ser A • J
Bulletin of University of Osaka Prefecture. Series A. Engineering and Natural Sciences [1952ff] ISSN 0474-7844

J 2684 J Huazhong Inst Tech (Engl Ed) • TJ
Journal of Huazhong Institute of Technology. English Edition [1979-1981]
• CONT AS (J 3218) J Huazhong Univ Sci Tech (Engl Ed)
• TRANSL OF (J 2754) Huazhong Gongxueyuan Xuebao

J 2687 Comp Math Appl • USA
Computers and Mathematics with Applications. [1975ff] ISSN 0097-4943

J 2688 Conceptus (Wien) • A
Conceptus. Zeitschrift fuer Philosophie [1967ff] ISSN 0010-5155

J 2690 Control & Cybern, Pol Akad Nauk • PL
Control and Cybernetics. Polish Academy of Sciences. Systems Research Institute [1972ff] ISSN 0324-8569

J 2694 Cybern & Syst • USA
Cybernetics and Systems. An International Journal. [1980ff] ISSN 0196-9722

J 2695 Cybernetica • B
Cybernetica. Revue Trimestrielle de l'Association Internationale de Cybernetique [1958ff] ISSN 0011-4227

J 2701 Digit Processes • CH
Digital Processes. An International Journal on the Theory and Design of Digital Systems [1975ff] ISSN 0301-4185

J 2702 Discr Appl Math • NL
Discrete Applied Mathematics [1979ff] ISSN 0166-218X

J 2705 Comm Cognition Monograph • B
Communication and Cognition [1970ff]
• REM Vol 1 - 3 with Dutch title 'Communicatie en Cognitie'

J 2711 Ekonom Mat Obzor (Prague) • CS
Ekonomicko-Matematicky Obzor (Economical & Mathematical Review) [1965ff] ISSN 0013-3027

J 2712 Electronics Lett (London) • GB
Electronics Letters [1965ff] ISSN 0013-5194

J 2716 Found Control Eng, Poznan • PL
Foundations of Control Engineering. Institute of Control Engineering. Technical University of Poznan [1975ff] ISSN 0324-8747

J 2720 Fuzzy Sets Syst • NL
Fuzzy Sets and Systems [1978ff] ISSN 0165-0114

J 2730 Phys Rev Lett • USA
Physical Review Letters [1958ff] ISSN 0031-9007

J 2736 Int J Theor Phys • USA
International Journal of Theoretical Physics [1968ff] ISSN 0020-7748

J 2743 Izv Sev-Kavk Nauch Tsentra, Tekh (Rostov nD) • SU
Izvestiya Severo-Kavkazskogo Nauchnogo Tsentra Vysshej Shkoly. Seriya Tekhnicheskie Nauki (Proceedings of the Scientific University Centre of the North-Caucasus. Series: Technical Sciences)

J 2754 Huazhong Gongxueyuan Xuebao • TJ
*Huazhong Gongxueyuan Xuebao * Zhongguo Kexue Shuxue Zhuanji (Journal Huazhong (Central China) University of Science and Technology)*
• TRANSL IN (J 2684) J Huazhong Inst Tech (Engl Ed) & (J 3218) J Huazhong Univ Sci Tech (Engl Ed)

J 2760 J Phys A Math & Gen • GB
Journal of Physics. A. Mathematical and General [1968ff] ISSN 0305-4470

J 2763 J Stat Comp & Simul • USA
Journal of Statistical Computation and Simulation [1972ff] ISSN 0094-9655

J 2770 Keio Math Sem Rep (Yokohama) • J
Keio Mathematical Seminar Reports [1976ff]

J 2771 Kexue Tongbao • TJ
Kexue Tongbao (Science Bulletin) [1950ff] ISSN 0023-074X
• TRANSL IN (J 3769) Sci Bull, Foreign Lang Ed

J 2774 Koezlem MTA Szam & Autom: Kutat Intez • H
Koezlemenyek. Magyar Tudomanyos Akademia Szamitastechnikai es Automatizalasi Kutato Intezet Budapest (Bulletin of the Hungarian Academy of Sciences Budapest, Research Institut of Computer Science and Automatization) [1968ff]

J 2775 Nuovo Cimento B, Ser 2 • I
Il Nuovo Cimento. B. Serie II

J 2777 Lett Math Phys • NL
Letters in Mathematical Physics. A Journal for the Rapid Dissemination of Short Contributions in the Field of Mathematical Physics [1975ff] ISSN 0377-9017

J 2778 Lett Nuovo Cimento Ser 2 • I
Lettere al Nuovo Cimento. Rivista Internazionale della Societa Italiana di Fisica. Serie 2 [1969ff] ISSN 0375-930X

J 2789 Math Intell • D
The Mathematical Intelligencer [1978ff] ISSN 0343-6993

J 2790 Math Sem-ber • D
Mathematische Semesterberichte [1980ff] ISSN 0720-728X
• CONT OF (J 0160) Math-Phys Sem-ber, NS

J 2804 Nanjing Daxue Xuebao, Ziran Kexue • TJ
Nanjing Daxue Xuebao. Ziran Kexue Ban (Nanjing University Journal. Natural Sciences Edition) [1957ff]

J 2810 Phil Forum Quart • USA
The Philosophical Forum. A Quarterly [1942ff] ISSN 0031-806X

J 2811 Phil Quart (St Andrews) • GB
The Philosophical Quarterly [1950ff] ISSN 0031-8094

J 2814 Podstawy Sterowania • PL
Podstawy Sterowania. Zaklad Systemow Automatyki Kompleksowej (Foundations of Cybernetics. Institute of Complex Automation Systems)

J 2819 Probl Contr Inf Th, Akad Nauk SSSR & Acad Sci Hung • H
*Problems of Control and Information Theory. Academy of Sciences of the USSR & Hungarian Academy of Sciences * Problemy Upravlenija i Teorii Informacii* [1971ff] ISSN 0370-2529

J 2822 Publ UER Math Pures Appl IRMA • F
Publications de UER (= Unites d'Enseignement et de Recherche) de Mathematiques Pures et Appliquees. IRMA (= Institut de Recherche de Mathematiques Avancees)

J 2830 Rozpr, Mat Prirod, Cheskoslov Akad Ved • CS
Rozpravy Cheskoslovenske Akademie Ved Rada Matematickykh a Prirodnikh Ved. (Studies of the Czechoslovak Academy of Sciences. Series: Mathematics and Natural Sciences) [1953ff] ISSN 0069-228X
• CONT OF (J 3989) Rozpr II, Tr Cesk Akad Ved

J 2831 RAIRO Autom • F
RAIRO Automatique. RAIRO (= Revue Francaise d'Automatique, d'Informatique et de Recherche Operationnelle). Series Automatique [1977ff] ISSN 0399-0524
• CONT OF (J 0205) Rev Franc Autom, Inf & Rech Operat Ser Jaune

J 2832 RAIRO Inform • F
RAIRO Informatique. Revue Francaise d'Automatique, d'Informatique et de Recherche Operationnelle. Series Informatique [1977ff] ISSN 0399-0532
• CONT OF (J 0205) Rev Franc Autom, Inf & Rech Operat Ser Bleue

J 2834 Scientia (Milano) • I
Scientia. Rivista Internazionale di Sintesi Scientifica. [1907ff] ISSN 0036-8687

J 2837 Southwest J Phil, Sect 1, Yokohama Univ • J
Southwestern Journal of Philosophy. Yokohama National University. Section I. Mathematics, Physics.

J 2839 Statistica (Bologna) • I
Statistica. Costituita Sotto gli Auspici delle Universita di Bologna, Padova e Palermo [1941ff] ISSN 0039-0380

J 2840 Stochastica, Univ Politec Barcelona • E
Stochastica. Revista de Matematica Pura y Aplicada del Departamento de Matematicas y Estadistica de la Escuela Tecnica Superior de Arquitectura [1975ff]

J 2844 Systems and Control • J
Systems and Control (Shisutemu to Seigyo) [1957ff]

J 2845 Tanulmanyok • H
Tanulmanyok. Szamitastechnikai es Automatizalasi Kutato Intezete (Studies. Research Institut for Computer Science and Automatization)

J 2852 Tr Vychisl Tsentra, Univ Tartu • SU
Trudy Vychislitel'nogo Tsentra. Tartuskij Gosudarstvennyj Universitet (Publications of the Computational Centre of the State University of Tartu) [1893ff]

J 2855 Zbor Rad, Prir-Mat Fak, Ser Mat (Novi Sad) • YU
Zbornik Radova Prirodno-Matematichkog Fakulteta. Serija za Matematiku (Collected Papers of the Faculty of Natural Sciences and Mathematics. Mathematical Series) [1971ff]

J 2869 Vest Ser Vychisl Mat Kibern, Univ Moskva • SU
Vestnik Moskovskogo Universiteta. Nauchnyj Zhurnal. Seriya XV. Vychislitel'naya Matematika i Kibernetika (Publications of the Moscow University. Scientific Journal. Series XV. Computational Mathematics and Cybernetics) [1946ff] ISSN 0201-7385, ISSN 0137-0782
• TRANSL IN (J 3221) Moscow Univ Comp Math Cybern

J 2871 Vopr Fil, Moskva • SU
Voprosy Filosofii (Problems of Philosophy) [1947ff] ISSN 0042-8744

J 2879 Wuhan Daxue Xuebao, Ziran Kexue • TJ
Wuhan Daxue Xuebao. Ziran Kexue Ban (Wuhan University Journal. Natural Sciences)

J 2887 Zbor Radova, NS • YU
Zbornik Radova. Nova Serija. (Collected Papers. New Series)

J 3068 Atti Accad Sci Bologna Fis Ser 13 • I
Atti della Accademia delle Scienze dell'Istituto di Bologna. Classe di Scienze Fisiche. Rendiconti. Serie XIII [1974ff]
• CONT OF (J 3069) Atti Accad Sci Bologna Fis Ser 12

J 3069 Atti Accad Sci Bologna Fis Ser 12 • I
Atti della Accademia delle Scienze dell'Istituto di Bologna. Classe di Scienze Fisiche. Rendiconti. Serie XII [1963-1973]
• CONT OF (J 0347) Atti Accad Sci Bologna Fis, Ser 11 • CONT AS (J 3068) Atti Accad Sci Bologna Fis Ser 13

J 3070 Bul Inst Politeh Iasi, Sect 1 • RO
Buletinul Institutului Politehnic din Iasi. Sectia 1. Matematica, Mecanica Teoretica, Fizica (Bulletin des Polytechnischen Instituts Jassy. Sektion 1. Mathematik, Theoretische Mechanik, Physik) [1977ff] ISSN 0304-5188
• CONT OF (J 0198) Bul Inst Politeh Iasi NS

J 3073 Bull Math Biol • GB
Bulletin of Mathematical Biology [1973ff] ISSN 0092-8240
• CONT OF (J 0515) Bull Math Biophys

J 3075 Normat • N
Normat. Nordisk Matematisk Tidskrift (Normat. Scandinavian Mathematical Journal) [1979ff]
• CONT OF (J 0311) Nordisk Mat Tidskr

J 3079 Kiber Sb Perevodov, NS • SU
Kiberneticheskij Sbornik. Novaya Seriya. Sbornik Perevodov (Collected Articles on Cybernetics: New Series. Collected Translations) [1965ff] ISSN 0453-8382
• CONT OF (J 1048) Kiber Sb Perevodov

J 3099 Theoretic Papers • N
Theoretic Papers [1984ff]

J 3104 Adv Appl Probab • GB
Advances in Applied Probability [1969ff] ISSN 0001-8678

J 3116 Austral Comp J • AUS
The Australian Computer Journal [1967ff] ISSN 0004-8917

J 3118 Avtom Mezhvuzov Nauch Sb • SU
Avtomaty, Mezhvuzovskij Nauchnyj Sbornik (Automatic Machines. Inter-University Scientific Collected Articles)

J 3127 Bol Soc Mat Mexicana, Ser 2 • MEX
Boletin de la Sociedad Matematica Mexicana. Segunda Serie [1956ff] ISSN 0037-8615
• CONT OF (J 0250) Bol Soc Mat Mexicana

J 3128 Boll Unione Mat Ital, Suppl • I
Bollettino della Unione Matematica Italiana. Supplemento

J 3133 Bull Soc Math Belg, Ser B • B
Bulletin de la Societe Mathematique de Belgique. Serie B [1977ff] ISSN 0037-9476
• CONT OF (J 0082) Bull Soc Math Belg

J 3154 Eur J Oper Res • NL
European Journal of Operational Research [1977ff] ISSN 0377-2217

J 3166 Fkts Anal Primen, Akad Nauk Azerb SSR • SU
Funktsional'nyj Analiz i Ego Primenenie. Akademiya Nauk Azerbajdzhanskoj SSR. Institut Matematiki i Mekhaniki (Functional Analysis and its Application. Academy of Sciences of the AzSSR. Institute of Mathematics and Mechanics)

J 3171 God Vissh Ucheb Zaved, Prilozhna Mat, Sofiya • BG
Godishnik na Visshite Uchebni Zavedeniya. Prilozhna Matematika (Annuaire de l'Universite, Mathematiques Appliquees) [1974ff]
• CONT OF (J 0224) God Vissh Tekh Ucheb Zaved Mat, Sofiya

J 3172 J London Math Soc, Ser 2 • GB
Journal of the London Mathematical Society. 2nd Series [1969ff] ISSN 0024-6107
• CONT OF (J 0039) J London Math Soc

J 3188 Itogi Nauki Tekh, Ser Teor Veroyat Mat Stat Teor Kibern • SU
Itogi Nauki i Tekhniki. Seriya Teoriya Veroyatnostej, Matematicheskaya Statistika, Teoreticheskaya Kibernetika (Progress in Science and Technology. Series Probability Theory, Mathematical Statistics, Theoretical Cybernetics) [1972ff] ISSN 0202-7488
• CONT OF (J 1021) Itogi Nauki Ser Mat • TRANSL IN (J 1531) J Sov Math

J 3194 J Austral Math Soc, Ser A • AUS
Journal of the Australian Mathematical Society. Series A [1976ff] ISSN 0263-6115
• CONT OF (J 0038) J Austral Math Soc

J 3218 J Huazhong Univ Sci Tech (Engl Ed) • TJ
Journal of Huazhong University of Science and Technology [1982ff]
• CONT OF (J 2684) J Huazhong Inst Tech (Engl Ed) • TRANSL OF (J 2754) Huazhong Gongxueyuan Xuebao

J 3221 Moscow Univ Comp Math Cybern • USA
Moscow University Computational Mathematics and Cybernetics ISSN 0278-6419
• TRANSL OF (J 2869) Vest Ser Vychisl Mat Kibern, Univ Moskva

J 3239 Proc Japan Acad, Ser A • J
Proceedings of the Japan Academy. Series A. Mathematical Sciences [1978ff] ISSN 0386-2194

J 3240 Proc London Math Soc, Ser 3 • GB
Proceedings of the London Mathematical Society. 3rd Series [1951ff] ISSN 0024-6115
• CONT OF (J 1910) Proc London Math Soc, Ser 2

J 3254 Riv Mat Univ Parma, Ser 3 • I
Rivista di Matematica della Universita di Parma. Serie 3 [1972-1974] ISSN 0035-6298
• CONT OF (J 1526) Riv Mat Univ Parma, Ser 2 • CONT AS (J 0549) Riv Mat Univ Parma, Ser 4

J 3265 Sov J Contemp Math Anal, Armen Acad Sci • USA
Soviet Journal of Contemporary Mathematical Analysis. Armenian Academy of Sciences [1979ff] ISSN 0735-2719
• TRANSL OF (**J** 0312) Izv Akad Nauk Armyan SSR, Ser Mat

J 3273 Stud Hist Philos Sci • GB
Studies in History and Philosophy of Science [1970ff] ISSN 0039-3681

J 3285 Boll Unione Mat Ital, V Ser, A • I
Bollettino della Unione Matematica Italiana. Serie V. A [1976-1981] ISSN 0041-7084
• CONT OF (**J** 0012) Boll Unione Mat Ital, IV Ser • CONT AS (**J** 2099) Boll Unione Mat Ital, VI Ser, A

J 3293 Bull Acad Pol Sci, Ser Math • PL
Bulletin de l'Academie Polonaise des Sciences. Serie des Sciences Mathematiques [1979-1982] ISSN 0001-4117
• CONT OF (**J** 0014) Bull Acad Pol Sci, Ser Math Astron Phys
• CONT AS (**J** 3417) Bull Pol Acad Sci, Math

J 3360 Avtom Izchisl Tekh, Sofiya • BG
Avtomatika i Izchislitelna Tekhnika (Automation and Computational Technology)

J 3364 C R Acad Sci, Paris, Ser 1 • F
Comptes Rendus des Seances de l'Academie des Sciences. Serie I: Science Mathematique [1981ff] ISSN 0151-0509
• CONT OF (**J** 2313) C R Acad Sci, Paris, Ser A-B

J 3369 Econ Comput & Econ Cybern (Bucharest) • RO
Economic Computation and Economic Cybernetics Studies and Research [1966ff] ISSN 0424-267X

J 3377 Gaz Mat (Bucharest) • RO
Gazeta Matematica (Mathematische Zeitschrift) [1975ff]
• CONT OF (**J** 0378) Gaz Mat (Bucharest) Ser A

J 3401 Meth Oper Res • D
Methods of Operations Research [1963ff] ISSN 0078-5318

J 3417 Bull Pol Acad Sci, Math • PL
Bulletin of the Polish Academy of Sciences. Mathematics [1983ff] ISSN 0001-4117
• CONT OF (**J** 3293) Bull Acad Pol Sci, Ser Math

J 3419 Probl Inf Transmiss • USA
Problems of Information Transmission [1965ff] ISSN 0032-9460
• TRANSL OF (**J** 2320) Probl Peredachi Inf, Akad Nauk SSSR

J 3420 Proc Edinburgh Math Soc, Ser 2 • GB
Proceedings of the Edinburgh Mathematical Society. Series 2. [1927ff] ISSN 0013-0915
• CONT OF (**J** 0386) Proc Edinburgh Math Soc

J 3434 Pubbl Ist Appl Calcolo, Ser 3 • I
Pubblicazioni. Serie III. Istituto per le Applicazioni del Calcolo "Mauro Picone" (IAC) Consiglio Nazionale delle Ricerche

J 3441 RAIRO Inform Theor • F
Revue Francaise d'Automatique, d'Informatique et de Recherche Operationnelle (RAIRO), Informatique Theorique [1977ff] ISSN 0399-0540
• CONT OF (**J** 4698) Rev Franc Autom, Inf & Rech Operat, Ser Rouge Inf Th

J 3449 Sov Math • USA
Soviet Mathematics [1974ff] ISSN 0197-7156
• TRANSL OF (**J** 0031) Izv Vyssh Ucheb Zaved, Mat (Kazan)

J 3453 Syst Sci Politech Univ Wroclaw • PL
Systems Science. Wroclaw Technical University [1975ff] ISSN 0137-1223

J 3456 Th Probab & Math Stat • USA
Theory of Probability and Mathematical Statistics [1956ff] ISSN 0040-585X
• TRANSL OF (**J** 0295) Teor Veroyat i Mat Stat (Kiev)

J 3495 Boll Unione Mat Ital, V Ser, B • I
Bolletino della Unione Matematica Italiana. Serie V. B [1976-1981] ISSN 0041-7084
• CONT OF (**J** 0012) Boll Unione Mat Ital, IV Ser • CONT AS (**J** 2100) Boll Unione Mat Ital, VI Ser, B

J 3519 Glas Mat, Ser 3 (Zagreb) • YU
Glasnik Matematichki. Serija 3 (Publications of Mathematics. Series 3) [1966ff] ISSN 0017-095X
• CONT OF (**J** 0371) Glas Mat-Fiz Astron, Ser 2 (Zagreb)

J 3522 Rend Circ Mat Palermo, Ser 2 • I
Rendiconti del Circolo Matematico di Palermo. Serie II [1952ff] ISSN 0009-725X
• CONT OF (**J** 0058) Rend Circ Mat Palermo

J 3523 Period Mat, Ser 5 • I
Periodico di Matematiche. Serie V. (Roma)

J 3524 Kybernetika Suppl (Prague) • CS
Kybernetika. Supplement. (Prague) (Cybernetics. Supplement. (Prague))
• REL PUBL (**J** 0156) Kybernetika (Prague)

J 3526 Ann Mat Pura Appl, Ser 4 • I
Annali di Matematica Pura ed Applicata. Serie Quarta. Sotto gli Auspici del Consiglio Nazionale delle Ricerche ISSN 0003-4622
• CONT OF (**J** 0229) Ann Mat Pura Appl, Ser 3

J 3543 Uch Zap Vopr Prikl Mat Kibern, Univ Baku • SU
Azerbajdzhanskij Gosudarstvennyj Universitet. Uchenye Zapiski. Voprosy Prikladnoj Matematiki i Kibernetiki (State University of Azerbaidjan. Scientific Notes. Problems of Applied Mathematics and Cybernetics)

J 3594 Kexue Tansuo • TJ
Kexue Tansuo (Science Exploration)

J 3597 J Unif Sci • USA
The Journal of Unified Science [1940-1974]
• CONT OF (**J** 0748) Erkenntnis (Leipzig) • CONT AS (**J** 0158) Erkenntnis (Dordrecht)

J 3665 Finsk Tidskrift • SF
Finsk Tidskrift. Kultur - Ekonomi - Politik (Finnische Zeitschrift. Kultur, Ekonomie, Politik) [1876ff] ISSN 0015-248X

J 3676 Ling Inquiry • USA
Linguistic Inquiry [1969ff] ISSN 0024-3892

J 3685 Publ Inst Mat Univ Catolica Chile • RCH
Publicaciones del Instituto de Matematica de la Universidad Catolica de Chile

J 3732 Mohu Shuxue • TJ
Mohu Shuxue (Fuzzy Mathematics) [1981ff]

J 3733 Shuxue Zazhi • TJ
Shuxue Zazhi (Journal of Mathematics)

J 3735 Fudan Xuebao, Ziran Kexue • TJ
Fudan Xuebao. Ziran Kexue Ban (Fudan Journal. Natural Science)

J 3742 Shuxue Yanjiu yu Pinglun • TJ
Shuxue Yanjiu yu Pinglun (Journal of Mathematical Research & Exposition) [1981ff]

J 3746 Note Math (Lecce) • I
Note di Matematica. Pubblicazione Semestrale [1981ff]

J 3750 Jilin Daxue, Ziran Kexue Xuebao • TJ
*Jilin Daxue. Ziran Kexue Xuebao (Acta Scientiarum Naturalium Universitatis Jilinensis * Jilin University. Natural Sciences Journal)* [1975ff]
• CONT OF (**J** 4364) Dongbei Renmin Daxue Ziran Kexue Xuebao

J 3763 Ganit (Dacca) • BD
Ganit. Journal of the Bangladesh Mathematical Society [1981ff]

J 3766 Zhongguo Kexue, Xi A • TJ
*Zhongguo Kexue. Xi A. Mathematical, Physical, Astronomical & Technical Sciences * Scientia Sinica. Series A* [1982ff]
• CONT OF (**J** 1024) Zhongguo Kexue

J 3769 Sci Bull, Foreign Lang Ed • TJ
Kexue Tongbao. Foreign Language Edition (Science Bulletin) [1980ff]
• TRANSL OF (**J** 2771) Kexue Tongbao

J 3781 Topoi • NL
Topoi. An International Review of Philosophy [1982ff] ISSN 0167-7411

J 3783 J Non-Classical Log (Univ Campinas) • BR
The Journal of Non-Classical Logic [1982ff]

J 3793 Jisuanjii Xuebao • TJ
Jisuanji Xuebao (Chinese Journal of Computers) [1978ff]

J 3795 Xi'an Jiaotong Daxue Xuebao • TJ
Xi'an Jiaotong Daxue Xuebao (Journal of Xi'an Jiaotong University)

J 3796 Lanzhou Daxue Xuebao, Ziran Kexue • TJ
Lanzhou Daxue Xuebao. Ziran Kexue Ban (Journal of Lanzhou University. Natural Sciences)

J 3801 Vest Ser Fiz Astron, Univ Moskva • SU
Vestnik Moskovskogo Universiteta. Seriya III. Fizika, Astronomiya (Publications of the Moscow University. Series III: Physics, Astronomy) [1960ff] ISSN 0579-9392
• TRANSL IN Mosc. Univ. Phys. Bull.

J 3815 Acta Univ Lodz Folia Philos • PL
Acta Universitatis Lodziensis Folia Philosophica [1981ff] ISSN 0208-6107

J 3816 Zhongguo Kexue Jishu Daxue Xuebao • TJ
Zhongguo Kexue Jishu Daxue Xuebao (Journal of China University of Science and Technology)

J 3824 Bull Soc Math Belg, Ser A • B
Bulletin de la Societe Mathematique de Belgique. Serie A [1977ff]
• CONT OF (**J** 0082) Bull Soc Math Belg

J 3914 Math Soc Sci • NL
Mathematical Social Sciences [1980ff] ISSN 0165-4896

J 3919 BUSEFAL • F
BUSEFAL (= Bulletin pour les Sous-Ensembles Flous et leurs Applications)

J 3932 Comput Artif Intell (Bratislava) • CS
Pocitace a Umela Inteligencia (Computers and Artificial Intelligence) [1982ff]

J 3937 Veroyat Met i Kibern (Kazan) • SU
Veroyatnostnye Metody i Kibernetika (Probabilistic Methods and Cybernetics)

J 3939 Mat Logika Primen (Akad Nauk Litov SSR) • SU
Matematicheskaya Logika i Ee Primeneniya (Mathematical Logic and its Applications) [1981ff]

J 3944 Rad Jugosl Akad Znan Umjet, Mat Znan • YU
Radovi Jugoslavenske Akademije Znanosti i Umjetnosti Matematichke Znanosti (Papers. Academy of Sciences and Arts of Yugoslavia. Mathematics) [1982ff]
• CONT OF (**J** 0372) Rad Jugosl Akad Znan Umjet, Mat Fiz Teh Znan

J 3946 Int Stat Rev • USA
*International Statistical Review * Revue Internationale de Statistique* [1933ff] ISSN 0306-7734

J 3948 Xiangtan Daxue Ziran Kexue Xuebao • TJ
Xiangtan Daxue Ziran Kexue Xuebao (Natural Science Journal of Xiangtan University)

J 3954 Rev Franc Inf & Rech Operat • F
Association Francaise pour la Cybernetique Economique et Technique. Revue Francaise d'Informatique et de Recherche Operationnelle [1967-1971]
• CONT AS (**J** 0205) Rev Franc Autom, Inf & Rech Operat

J 3957 Bull Inf & Cybern (Kyushu Univ) • J
Bulletin of Informatics and Cybernetics. Research Association of Statistical Sciences [1982ff] ISSN 0286-522X

J 3986 Ist Lombardo Rend, Ser 2 (Milano) • I
Reale Istituto Lombardo di Scienze e Lettre. Rendiconti. Series 2 [1868-1936]
• CONT AS (**J** 0207) Ist Lombardo Accad Sci Rend, A (Milano)

J 3989 Rozpr II, Tr Cesk Akad Ved • CS
Tridy Ceske Akademie Ved. Rozprawy II (Scientific Papers of the Royal Czech Academy of Sciences. Series: Mathematics & Natural Sciences.) [1891-1952]
• CONT AS (**J** 2830) Rozpr, Mat Prirod, Cheskoslov Akad Ved

J 3993 Rev Mat Hisp-Amer, Ser 2 • E
Revista Matematica Hispano-Americana. Serie 2 ISSN 0373-0999
• CONT AS (**J** 0236) Rev Mat Hisp-Amer, Ser 4

J 3994 Ann Acad Sci Fennicae Ser A I • SF
Annales Academiae Scientiarum Fennicae. Serie A I [1941-1974] ISSN 0066-1953
• CONT AS (**J** 0446) Ann Acad Sci Fennicae, Ser A I, Diss

J 3996 Boll Unione Mat Ital • I
Bollettino della Unione Matematica Italiana [1922-1945]
• CONT AS (**J** 4408) Boll Unione Mat Ital, III Ser

J 4105 An Romano-Soviet, Ser Autom Cibern • RO
Analele Romano-Sovietice, Seria Automatica Cibernetica (Rumaenisch-Sowjetische Annalen, Serie Automatik Kibernetik)

J 4130 Hegel-Stud • D
Hegel-Studien [1961ff] ISSN 0073-1587

J 4186 Atti Accad Sci Lett Arti Palermo, Ser 5/I • I
Atti della Accademia di Scienze. Lettere e Arti di Palermo. Serie Quinta. Parte I: Scienze [1984ff]
• CONT OF (**J** 0104) Atti Accad Sci Lett Arti Palermo, Ser 4/I

J 4188 Com Acad Romine • RO
*Comunicarile Academiei Republicii Populare Romine * Comptes Rendus de l'Academie de la Republique Populaire Roumaine * Doklady Akademii Rumynskoi Narodnoi Respubliki* [1951-1963]

J 4277 Vesn Drusht Mat Fiz Serbije • YU
*Vesnik Drushtva Matematichara i Fizichara Narodne Republike Serbije * Vestnik Obshchestva Matematikov i Fizikov N.R. Serbie * Bulletin de la Societe des Mathematiciens et Physiciens de la R.P. Serbie (Publications of the Society of Mathematicians and Physicists of the P.R. Serbia)* [1949-1963]
• CONT AS (**J** 0042) Mat Vesn, Drust Mat Fiz Astron Serb

J 4278 Bull Electr Lab (Japan) • J
Densi Qidzyutsu Sogo Kenkyuso Ikho (Bulletin of the Electrotechnical Laboratories)

J 4305 IEEE Trans Electr Comp • USA
Transactions of Electronic Computers. IEEE (= Institute of Electrical and Electronics Engineers) [1963-1967]
• CONT OF (**J** 0072) IRE Trans Electr Comp • CONT AS (**J** 0187) IEEE Trans Comp

J 4364 Dongbei Renmin Daxue Ziran Kexue Xuebao • TJ
*Dongbei Renmin Daxue Ziran Kexue Xuebao * Acta Scientiarum Naturalium (Natural Science Journal of Northeast Peoples's University)* [?-1974]
• CONT AS (**J** 3750) Jilin Daxue, Ziran Kexue Xuebao

J 4387 Itogi Nauki Tekh, Ser Tekh Kibern • SU
Itogi Nauki i Tekhniki. Seriya Tekhnicheskaya Kibernetika. Gosudarstvennyj Komitet SSSR po Nauke i Tekhnike. Akademiya Nauk SSSR. Vsesoyuznyj Institut Nauchnoj i Tekhnicheskoj Informatsii. (Progress in Science and Technology. Series: Engineering Cybernetics) [1972ff]
• CONT OF (**J** 1021) Itogi Nauki Ser Mat

J 4390 Kobe J Math • J
Kobe Journal of Mathematics [1984ff] ISSN 0289-9051
• CONT OF (**J** 1508) Math Sem Notes, Kobe Univ

J 4398 Listener • GB
The Listener

J 4408 Boll Unione Mat Ital, III Ser • I
Bolletino della Unione Matematica Italiana, Ser III [1946-1967]
• CONT OF (**J** 3996) Boll Unione Mat Ital • CONT AS (**J** 0012) Boll Unione Mat Ital, IV Ser

J 4430 Ziran Zazhi • TJ
Ziran Zazhi (Nature Journal) [1978ff]

J 4436 Osaka Shijo Danwakai • J
Osaka Shijo Danwakai

J 4437 Ceska Mysl • CS
Ceska Mysl (Czech Thinking)

J 4438 Tetsugaku Kenkyu • J
Tetsugaku Kenkyu

J 4439 Kagaku • J
Kagaku (Chemical Sciences)

J 4441 Rev Calc Autom Cibern
Revista de Calculo Automatico y Cibernetica

J 4442 Kyushu Daigaku Kagaku Syuho • J
Kyushu Daigaku Kagaku Syuho (Technology Reports of the Kyushu University)

J 4443 Gac Mat (Madrid) • E
Gaceta Matematica [1949-1969]
• CONT AS (**J** 0299) Gac Mat, Ser 1a (Madrid)

J 4452 Zhexue Yanjiu • TJ
Zhexue Yanjiu (Philosophical Research. Studies on Philosophy)

J 4468 Rocz Prac Nauk Zrzesz Asystentow Uniw Warszaw • PL
Roczniki Prac Naukowych Zrzeszenia Asystentow Uniwersytetu Jozefa Pilsudskiego w Warszawie (Annals of the Scientific Papers of the Union of the Assistents of the Jozef Pilsudski University in Warsaw) [1936ff]

J 4475 Franciscan Stud
Franciscan Studies

J 4510 Norsk Mat Tidsskr • N
Norsk Matematisk Tidsskrift (Scandinavian Mathematical Journal) [1919-1952]
• CONT OF (**J** 0259) Nyt Tidsskr Mat • CONT AS (**J** 0311) Nordisk Mat Tidskr

J 4551 Ann Univ Lublin, Sect F • PL
Annales Universitatis Mariae Curie-Sklodowska. Sectio F.

J 4555 Trans IECE
Transactions of the IECE

J 4604 Theoria, Ser 2 (San Sebastian) • E
Theoria (San Sebastian), Seria II [1985ff]

J 4608 Teoria (Pisa) • I
Teoria. Logica e Filosofia del Linguaggio [1981ff]

J 4632 Qufu Shiyuan Xuebao • TJ
Qufu Shiyuan Xuebao

J 4693 Ann Sci Univ Jassy • RO
Annales Scientifiques de l'Universite de Jassy [?-1936]
• CONT AS (**J** 0269) Ann Sci Univ Jassy Sec 1

J 4697 Managmt Sci (Providence) • USA
Management Science. Journal of the Institute of Management Sciences

J 4698 Rev Franc Autom, Inf & Rech Operat, Ser Rouge Inf Th • F
Revue Francaise d'Automatique, d'Informatique et de Recherche Operationnelle (RAIRO). Serie Rouge Informatique Theorique [1975-1976] ISSN 0399-0540
• CONT OF (**J** 0205) Rev Franc Autom, Inf & Rech Operat Ser Rouge • CONT AS (**J** 3441) RAIRO Inform Theor

J 4702 Ann Sc Norm Sup Pisa Fis Mat, Ser 4 • I
Annali della Scuola Normale Superiore di Pisa. Classe di Science. Fisiche e Matematiche. Seria IV [1974ff]
• CONT OF (**J** 0315) Ann Sc Norm Sup Pisa Fis Mat, Ser 3

J 4706 Publ Inst Math (Belgrade) • YU
Academie Serbe des Sciences. Publications de l'Institut Mathematique [1948-1960]
• CONT AS (**J** 0400) Publ Inst Math, NS (Belgrade)

J 4709 Bol Soc Paranaense Mat, Ser 2 • BR
Sociedade Paranaense de Matematica. Boletim. Seria 2 ISSN 0037-8712
• CONT OF (**J** 1957) Bol Soc Paranaense Mat

J 4710 Pol Tow Mat, Wiad Mat • PL
Polskie Towarzystwo Matematyczne. Wiadomosci Matematiczne [1899-1954]
• CONT AS (**J** 0519) Wiad Mat, Ann Soc Math Pol, Ser 2

J 4711 IRE Proc • USA
Proceedings. IRE (= Institute of Radio Engineers) [1911-1962]
• CONT AS (**J** 1077) IEEE Proc

J 4713 Mat Fyz Chasopis (Slov Akad Ved) • CS
Matematicky-Fyzikalny Chasopis (Journal of Mathematical Physics) [1951-1966]
• CONT AS (**J** 0143) Mat Chasopis (Slov Akad Ved)

J 4716 Stud Philos, Leopolis (Poznan) • PL
Studia Philosophica. Commentarii Societatis Polonorum. Leopolis [1935ff]

J 4717 Izv Akad Nauk SSSR • SU
Izvestiya Akademii Nauk SSSR (Bulletin de l'Academie des Sciences Mathematiques et Naturelles. Leningrad) [?-1936]
• CONT AS (**J** 0216) Izv Akad Nauk SSSR, Ser Mat

J 4720 Bull Acad Royal Serbe, Math Phys • YU
Academie Royale Serbe. Bulletin de l'Academie des Sciences Mathematiques et Naturelles. A: Sciences Mathematiques et Physique [1933-1951]
• CONT AS (**J** 0993) Bull Acad Serbe, Cl Sci Math Nat

J 4729 Acta Math Hung • H
Acta Mathematica Hungarica [1983ff] ISSN 0236-5294
• CONT OF (**J** 0001) Acta Math Acad Sci Hung

J 4731 Australasian J Psych & Phil • AUS
The Australasian Journal of Psychology and Philosophy [1923-1946]
• CONT AS (**J** 0273) Australasian J Phil

J 4735 Rad Jugosl Akad Znan Umjet • YU
Radovi Jugoslavenske Akademije Znanosti i Umjetnosti. (Papers. Yugoslavian Academy of Sciences and Arts) [/??/-1939]
• CONT AS (**J** 0372) Rad Jugosl Akad Znan Umjet, Mat Fiz Teh Znan

Source Index

Series

S 0019 Colloq Math (Warsaw) • PL
Colloquium Mathematicum [1947ff] • PUBL Academie Polonaise des Sciences, Institut Mathematique: Warsaw • ISSN 0010-1354

S 0055 Proc Steklov Inst Math • USA
Proceedings of the Steklov Institute of Mathematics [1967ff] • PUBL (**X** 0803) Amer Math Soc: Providence • TRANSL OF (**S** 0066) Tr Mat Inst Steklov • ISSN 0081-5438

S 0066 Tr Mat Inst Steklov • SU
Trudy Ordena Lenina Matematicheskogo Instituta imeni V.A.Steklova. Akademiya Nauk SSSR (Proceedings of the Mathematical Steklov-Institute of the Academy of Sciences SSSR) [1938ff] • PUBL (**X** 0899) Akad Nauk SSSR : Moskva • CONT OF (**S** 1644) Tr Fiz-Mat Inst Steklov • TRANSL IN (**S** 0055) Proc Steklov Inst Math

S 0166 Mat Issl, Mold SSR • SU
Matematicheskie Issledovaniya. Akademiya Nauk Moldavskoj SSR. Ordena Trudovogo Krasnogo Znameni Institut Matematiki s Vychislitel'nym Tsentrom (Mathematical Studies) [1966ff] • PUBL (**X** 2741) Shtiintsa: Kishinev • ISSN 0542-9994

S 0183 Publ Math Univ California • USA
University of California Publications in Mathematics PUBL (**X** 0926) Univ Calif Pr: Berkeley

S 0228 Zap Nauch Sem Leningrad Otd Mat Inst Steklov • SU
Zapiski Nauchnykh Seminarov Leningradskogo Otdeleniya Ordena Lenina Matematicheskogo Instituta imeni V.A.Steklova Akademii Nauk SSSR (LOMI) (Reports of the Scientific Seminars of the Leningrad Steklov Institute of Mathematics) PUBL (**X** 2641) Nauka: Leningrad • TRANSL IN (**J** 1531) J Sov Math & (**J** 0521) Semin Math, Inst Steklov • REM Transl. in J0521 up to vol. 19

S 0257 Monograf Mat • PL
Monografie Matematyczne (Mathematical Monography) [1932ff] • PUBL (**X** 1034) PWN: Warsaw • ALT PUBL (**X** 1758) Ars Polona: Warsaw • ISSN 0077-0507

S 0281 Arch Towarz Nauk Lwow, Sect 3 • PL
Archivum Towarzystwa Naukowego we Lwowie. Dzial 3. Matematyczno-Przyrodniczy (Archive of the Scientific Society of Lwow. Section 3. Mathematics and Natural Sciences) [1920-1939] • PUBL Tow Nauk: Lwow

S 0393 Uch Zap Univ, Tartu • SU
*Uchenye Zapiski Tartuskogo Gosudarstvennogo Universiteta. * Tartu Riikliku Uelikooli Toimetised. * Acta et Commentationes Universitatis Tartuensis (Scientific Notes of the Tartu State University.)* [1961ff] • PUBL (**X** 2463) Tartusk Gos Univ: Tartu

S 0405 Mitt Math Sem Giessen • D
Mitteilungen aus dem Mathematischen Seminar Giessen PUBL (**X** 2464) Math Sem Selbstverl: Giessen

S 0458 Zesz Nauk, Prace Log, Uniw Krakow • PL
Zeszyty Naukowe Uniwersytetu Jagiellonskiego Prace z Logiki (Scientific Papers. Jagielleonian University. Reports on Logic) [1965-1972] • PUBL (**X** 1034) PWN: Warsaw • CONT AS (**J** 0302) Rep Math Logic, Krakow & Katowice

S 0502 Prace Wroclaw Tow Nauk, Ser B • PL
Prace Wroclawskiego Towarzystwa Naukowe. Series B. Nauki Scisle (Papers of the Scientific Society of Wroclaw. Series B. Exact Sciences) [1947ff] • PUBL (**X** 1034) PWN: Warsaw • ISSN 0084-3024

S 0507 Vychisl Sist (Akad Nauk SSSR Novosibirsk) • SU
Vychislitel'nye Sistemy. Sbornik Trudov. Akademiya Nauk SSSR. Sibirskoe Otdelenie (Computer Systems. Collected Articles) [1962ff] • PUBL (**X** 2652) Akad Nauk Sibirsk Otd Inst Mat: Novosibirsk • ISSN 0568-661X

S 0544 Prace Mat Uniw Katowice • PL
Prace Naukowe Uniwersytetu Slaskiego w Katowicach. Prace Matematyczne (Scientific Publications of the Silesian University in Katowice. Mathematical Papers) PUBL (**X** 1425) Univ Slaski: Katowice

S 0554 Issl Teor Algor & Mat Logik (Moskva) • SU
Issledovaniya po Teorii Algorifmov i Matematicheskoj Logike (Studies in the Theory of Algorithms and Mathematical Logic) [1973,1976,1979] • ED: MARKOV, A.A. & PETRI, N.V. (VOL 1). MARKOV, A.A. & KUSHNER, B.A. (VOL 2). MARKOV, A.A. & KHOMICH, V.I. (VOL 3) • PUBL (**X** 2265) Akad Nauk Vychis Tsentr: Moskva 287pp,160pp,133pp • ISSN 0302-9085

S 0716 Vychisl Tekh Vopr Kibern (Univ Leningrad) • SU
Vychislitel'naya Tekhnika i Voprosy Kibernetiki (Computer Technology and Questions of Cybernetics) [1962ff] • ED: CHIRKOV, M.K. & MASLOV, S.P. & TSAR'KOVA, Z.I. (V 8). BRUSENTSOVA, N.P. & SHAUMAN, A.M. (V 7, V 15 - V 19) • PUBL (**X** 0938) Leningrad Univ: Leningrad • ISSN 0507-536X

S 0764 Teor Konech Avtom & Prilozh (Riga) • SU
Teoriya Konechnykh Avtomatov i ee Prilozheniya (Theory of Finite Automata and its Applications) [1972ff] • PUBL (**X** 2230) Zinatne: Riga • LC-No 74-304069

S 0889 Notas Logica Mat • RA
Notas de Logica Matematica. [1963ff] • PUBL Inst Mat, Univ Nacional del Sur: Bahia Blanca • ISSN 0078-2017

S 1003 Publ Elektroteh, Ser Mat Fiz, Beograde • YU
Univerzitet u Beogradu. Publikacije Elektrotehnichkog Fakulteta. Serija Matematika i Fizika (University of Belgrade. Publications of the Faculty of Electrical Engineering. Series: Mathematics and Physics) PUBL Univ Beograd: Belgrade

S 1070 Probl Mat, Wyz Szk Ped, Bydgosszcz • PL
Problemy Matematyczne. Bydgosszcz Wyzsza Szkola Pedagogiczna. Zeszyty Naukowe (Mathematical Problems. Institute of Education of Bydgosszcz. Scientific Papers) PUBL Uczel WSP: Bydgosszcz

S 1181 Mem Osaka Univ Lib Arts Educ, Ser B • J
Osaka University of Liberal Arts and Education. Memoirs. B. Natural Science [1952ff] • PUBL (**X** 2242) Osaka Univ Dept Math: Osaka

S 1454 Zesz Nauk Wyz Szk Ped Mat, Opole • PL
Zeszyty Naukowe Wyzszej Szkoly Pedagogicznej W Opolu Matematyka (Scientific Papers. University of Education in Opole. Mathematics) [1961ff] • PUBL Wyzsza Szkola Pedagogiczna: Opole
• ISSN 0078-5431

S 1459 Mem School Sci & Engin, Waseda Univ • J
Memoirs of the School of Sciences and Engineering. Waseda University [1922ff] • PUBL Waseda Univ.: Tokyo
• ISSN 0369-1950

S 1558 Bol Anal Log Mat (Rio de Janeiro) • BR
Boletim de Analise e Logica Matematica. Universidade Federal Fluminese. Instituto de Matematica PUBL Niteroi: Rio de Janeiro

S 1605 Math Centr Tracts • NL
Mathematical Centre Tracts [1963-1983] • PUBL (**X** 1121) Math Centr: Amsterdam

S 1613 Probl Logic (Bucharest) • RO
Probleme de Logica [1956ff] • PUBL (**X** 0871) Acad Rep Soc Romania: Bucharest
• ISSN 0556-1655

S 1626 Oslo Preprint Ser • N
Oslo Preprint Series [1970ff] • PUBL (**X** 2786) Univ Oslo Mat Inst: Oslo

S 1642 Schr Inf Angew Math, Ber (Aachen) • D
Schriften zur Informatik und Angewandten Mathematik. Bericht PUBL (**X** 3215) TH Aachen Math Nat Fak: Aachen

S 1644 Tr Fiz-Mat Inst Steklov • SU
Trudy Fiziko-Matematicheskogo Instituta imeni V.A.Steklova Akademiya Nauk SSSR (Travaux de l'Institut Physico-Mathematique V.A.Stekloff de l'Academie des Sciences SSSR) [1930-1937] • PUBL (**X** 0899) Akad Nauk SSSR : Moskva
• CONT AS (**S** 0066) Tr Mat Inst Steklov

S 1687 Midwest Studies Philos • USA
Midwest Studies in Philosophy [1976ff] • PUBL (**X** 1307) Univ Minnesota Pr: Minneapolis
• ISSN 0363-6550

S 1747 Christiana Albertina (Kiel) • D
Christiana Albertina. Kieler Universitaetszeitschrift [1965ff] • PUBL Christian Albrechts Universitaet, Presse und Informationsstelle: Kiel
• ISSN 0578-0160

S 1802 Colloq Int CNRS • F
Colloques Internationaux du Centre National de la Recherche Scientifique (CNRS) PUBL (**X** 0999) CNRS Inst B Pascal: Paris

S 1926 Banach Cent Publ • PL
Banach Center Publications. Polish Academy of Science. Institut of Mathematics [1976ff] • PUBL (**X** 1034) PWN: Warsaw
• ISSN 0137-6934

S 2024 Probl Bioniki • SU
Problemy Bioniki (Problems of Bionics) ED: SHABANOV-KUSHNARENKO, YU.P. • PUBL (**X** 2644) Vishcha Shkola: Khar'kov
• LC-No 70-473175

S 2073 Actualites Sci Indust • F
Actualites Scientifiques et Industrielles PUBL (**X** 0859) Hermann: Paris

S 2308 Symp Kyoto Univ Res Inst Math Sci (RIMS) • J
Surikaisekikenkyusho Kokyuroku (Kyoto University. Research Institute for Mathematical Sciences (RIMS). Proceedings of Symposia) PUBL (**X** 2441) Kyoto Univ Res Inst Math Sci: Kyoto

S 2579 Teor Mnozhestv & Topol (Izhevsk) • SU
Teoriya Mnozhestv i Topologiya (Set Theory and Topology) [1977,1979,1982] • ED: GRYZLOV, A.A. • PUBL (**X** 4562) Udmurtskij Gos Univ: Izhevsk 114pp,116pp,116pp
• REM Title of Vol.2: Sovremennaya Topologiya i Teoriya Mnozhestv, Vyp.2

S 2582 Semiotika & Inf, Akad Nauk SSSR • SU
Semiotika i Informatika. Gosudarstvennyj Komitet SSSR po Nauke i Tekhnike. Akademiya Nauk SSSR. Vsesoyuznyj Institut Nauchnoj i Tekhnicheskoj Informatsii (Semiotics and Information Science) ED: MIKHAJLOV, A.I. • PUBL (**X** 2235) VINITI: Moskva

S 2651 Prepr Inst Prikl Mat, Akad Nauk SSSR • SU
Preprint. Akademiya Nauk SSSR. Institut Prikladnoj Matematiki. (Preprint. Academy of Sciences of the USSR. Institute of Applied Mathematics) PUBL Akad Nauk SSSR, Inst Prikl Mat: Moskva

S 2809 Sbor Nauch Trud (Politekhn Inst, Perm) • SU
Sbornik Nauchnykh Trudov (Collected Scientific Papers) PUBL Permskij Politekhnicheskij Institut: Perm

S 2829 Rostocker Math Kolloq • DDR
Rostocker Mathematisches Kolloquium. PUBL Wilhelm-Pieck-Univ Rostock, Sekt Math: Rostock

S 2847 Uch Zap Univ Tomsk • SU
Uchenye Zapiski Tomskogo (Ordena Trudovogo Krasnogo Znameni) Gosudarstvennogo Universiteta imeni V.V.Kujbysheva (Scientific Notes of the Tomsk State University) PUBL (**X** 3606) Tomsk Univ: Tomsk

S 2850 Tr Ehnerg Inst Moskva • SU
Trudy Moskovskogo Ordena Lenina Ehnergeticheskogo Instituta Tematicheskij Sbornik (Proceedings of the Moscow Institute of Energetics) PUBL Moskovsk Ehnergetichesk Instituta: Moskva

S 2874 Vopr Kibern, Akad Nauk SSSR • SU
Akademiya Nauk SSSR Nauchnyj Sovet po Kompleksnoj Probleme. "Kibernetika". Voprosy Kibernetiki (Problems of Cybernetics. Academy of Sciences. Scientific Council for Complexity Problems. Cybernetics) PUBL Akad Nauk SSSR Nauch Sovet Komplek Probl Kibern: Moskva

S 2877 Wiss Z Univ Leipzig, Ges-Sprachwiss Reihe • DDR
Wissenschaftliche Zeitschrift der Karl-Marx-Universitaet Leipzig. Gesellschafts- und Sprachwissenschaftliche Reihe [1951ff] • PUBL (**X** 2373) Karl-Marx-Univ: Leipzig
• ISSN 0043-6879

S 2890 Zesz Nauk, Mat Fiz, Politech Slask (Gliwice) • PL
Zeszyty Naukowe Politechniki Slaskiej. Matematyka, Fizyka (Scientific Papers. Silesian Technical University. Mathematics. Physics) [1961ff] • PUBL Politech Slask: Gliwice
• ISSN 0072-470X

S 2959 Mat Met Optim & Struct Sist • SU
Matematicheskie Metody Optimizatsii i Strukturovaniya Sistem (Mathematical Methods of Optimization and Structuring of Systems) [1979,1980] • ED: ABRAMOV, YU.A. ET AL. • PUBL (**X** 1434) Kalinin Gos Univ: Kalinin 188pp,172pp

S 3270 Spraw Inst Inf, Uniw Warsaw • PL
Sprawozdania Instytutu Informatyki Uniwersytetu Warszawskiego (Reports. Institute of Informatics. University of Warsaw.) PUBL Uniwersytet Warszawska: Warsaw

S 3301 Lect Notes Math • D
Lecture Notes in Mathematics [1964ff] • PUBL (**X** 0811) Springer: Heidelberg & New York
• ISSN 0075-8434

S 3302 Lect Notes Comput Sci • D
Lecture Notes in Computer Science [1973ff] • PUBL (**X** 0811) Springer: Heidelberg & New York
• ISSN 0302-9743

S 3303 Stud Logic Found Math • NL
Studies in Logic and the Foundations of Mathematics [1954ff] • PUBL (**X** 0809) North Holland: Amsterdam
• ISSN 0049-237X

S 3304 Proc Symp Pure Math • USA
Proceedings of Symposia in Pure Mathematics PUBL (**X** 0803) Amer Math Soc: Providence

S 3305 Symposia Matematica • I
Symposia Matematica [1969ff] • PUBL (**X** 3604) INDAM: Roma
• ISSN 0082-0725

S 3307 Synth Libr • NL
Synthese Library. Studies in Epistemology, Logic, Methodology, and Philosophy of Science [1959ff] • PUBL (**X** 0835) Reidel: Dordrecht
• ISSN 0082-1128

S 3308 Univ Western Ontario Ser in Philos of Sci • NL
University of Western Ontario Series in Philosophy of Science [1972ff] • PUBL (**X** 0835) Reidel: Dordrecht

S 3310 Lect Notes Pure Appl Math • USA
Lecture Notes in Pure and Applied Mathematics [1971ff] • PUBL (**X** 1684) Dekker: New York
• ISSN 0075-8469

S 3311 Boston St Philos Sci • NL
Boston Studies in the Philosophy of Science [1963ff] • PUBL (**X** 0835) Reidel: Dordrecht
• ISSN 0068-0346

S 3312 Coll Math Soc Janos Bolyai • H
Colloquia Mathematica Societatis Janos Bolyai PUBL (**X** 0809) North Holland: Amsterdam

S 3314 Lect Notes Econ & Math Syst • D
Lecture Notes in Economics and Mathematical Systems [1968ff] • PUBL (**X** 0811) Springer: Heidelberg & New York
• ISSN 0075-8442

S 3382 Sem-ber, Humboldt-Univ Berlin, Sekt Math • DDR
Seminarberichte. Humboldt-Universitaet zu Berlin, Sektion Mathematik PUBL (**X** 2219) Humboldt-Univ Berlin: Berlin

S 3412 Prace Inst Mat, Politech Wroclaw, Ser Konf • PL
Prace Naukowe Instytutu Matematyki Politechniki Wroclaw. Serija: Konferencje (Scientific Papers of the Institute of Mathematics of Wroclaw Technical University. Series: Conferences) PUBL Politechnika Wroclawska: Wroclaw

S 3468 Tr Mat & Mekh (Tartu) • SU
*Matemaatika - ja Mekhaanika-Alaseid Toeid. * Trudy po Matematike i Mekhanike. (Works about Mathematics and Mechanics.)* SER (**S** 0393) Uch Zap Univ, Tartu • PUBL (**X** 2463) Tartusk Gos Univ: Tartu

S 3489 Sel Math Sov • CH
Selecta Mathematica Sovietica [1981ff] • PUBL (**X** 0804) Birkhaeuser: Basel
• ISSN 0272-9903

S 3513 Ann Math Stud • USA
Annals of Mathematics Studies [1940ff] • PUBL (**X** 0857) Princeton Univ Pr: Princeton

S 3726 Congressus Numerantium • CDN
Congressus Numerantium [1970ff] • PUBL (**X** 2420) Utilitas Mathematica Publ: Winnipeg

S 3909 Mat Met Opt & Upravleniya Slozh Sist (Kalinin) • SU
Matematicheskie Metody Optimizatsii i Upravleniya v Slozhnykh Sistemakh (Mathematical Methods of Optimization and Control in Complex Systems) [1981,1982] • ED: ABRAMOV, YU.A. • PUBL (**X** 1434) Kalinin Gos Univ: Kalinin 180pp,196pp

S 4519 Uppsala Univ Arsskrift • S
Uppsala Universitets Arsskrift PUBL (**X** 2113) Univ Uppsala: Uppsala

S 4530 Stud Gnieseusia
Studia Gnieseusia

S 4560 Groningen-Amsterdam Stud Semant • NL
Groningen - Amsterdam Studies in Semantic (GRASS) [1983ff] • PUBL (**X** 4217) Foris: Dordrecht

Proceedings

P 0454 Math Founds of Comput Sci (4);1975 Marianske Lazne • CS
[1975] *Mathematical Foundations of Computer Science. Proceedings of the 4th Symposium* ED: BECVAR, J. • SER (S 3302) Lect Notes Comput Sci 32 • PUBL (X 0811) Springer: Heidelberg & New York x+476pp
• DAT&PL 1975 Sep; Marianske Lazne, CS • ISBN 3-540-07389-2, LC-No 75-22406

P 0559 Phil Probl in Logic;1968 Irvine • USA
[1970] *Philosophical Problems in Logic: Some Recent Developments. Colloquium on Free Logic, Modal Logic and Related Areas.* ED: LAMBERT, K. • SER (S 3307) Synth Libr • PUBL (X 0835) Reidel: Dordrecht viii+176pp • ALT PUBL (X 2696) Humanities Pr: Atlantic Highlands
• DAT&PL 1968 May; Irvine, CA, USA • LC-No 76-490025

P 0560 Int Congr Philos (12);1958 Venezia • I
[1958-1961] *Atti del 12 Congresso Internazionale di Filosofia (Actes du 12eme Congres International de Philosophie * Proceedings of the 12th International Congress of Philosophy)* PUBL (X 1346) Sansoni: Firenze 12 Vols
• DAT&PL 1958 Sep; Venezia, I

P 0575 Int Congr Math (II, 7);1954 Amsterdam • NL
[1954-1957] *Proceedings of the International Congress of Mathematicians 1954* ED: GERRETSEN, J.C.H. & GROOT DE, J.
• PUBL (X 0809) North Holland: Amsterdam 3 Vols: 582pp,440pp,560pp • ALT PUBL (X 1317) Noordhoff: Groningen 1954-1957 & (X 3602) Kraus: Vaduz 1967
• DAT&PL 1954 Sep; Amsterdam, NL • LC-No 52-1808

P 0576 Raisonn en Math & Sci Exper;1955 Paris • F
[1958] *La Raisonnement en Mathematiques et en Sciences Experimentales.* SER (S 1802) Colloq Int CNRS 70 • PUBL (X 0999) CNRS Inst B Pascal: Paris 140pp
• DAT&PL 1955 Sep; Paris, F • LC-No 63-24106

P 0578 Self Organizing Systs;1962 Chicago • USA
[1962] *Self Organizing Systems* ED: YOVITIS, M.C. & JACOBI, G.T. & GOLDSTEIN, G.D. • PUBL (X 1354) Spartan Books : Sutton ix+563pp
• DAT&PL 1962 -?-; Chicago, IL, USA • LC-No 62-20444

P 0580 Int Congr Log, Meth & Phil of Sci (4,Sel Pap);1971 Bucharest • RO
[1973] *Logic, Language and Probability.* ED: BOGDAN, R.J. & NIINILUOTO, I. • SER (S 3307) Synth Libr • PUBL (X 0835) Reidel: Dordrecht x+323pp
• DAT&PL 1971 Aug; Bucharest, RO • ISBN 90-277-0312-4, LC-No 72-95892 • REL PUBL (P 0793) Int Congr Log, Meth & Phil of Sci (4,Proc);1971 Bucharest
• REM A Selection of Papers Contributed to Sections IV, VI and XI of P0793

P 0604 Scand Logic Symp (2);1970 Oslo • N
[1971] *Proceedings of the 2nd Scandinavian Logic Symposium* ED: FENSTAD, J.E. • SER (S 3303) Stud Logic Found Math 63 • PUBL (X 0809) North Holland: Amsterdam ii+405pp
• DAT&PL 1970 Jun; Oslo, N • ISBN 0-7204-2259-0, LC-No 71-153401

P 0607 All-Union Math Conf (3);1956 Moskva • SU
[1959] *Trudy 3'go Vsesoyuznogo Matematicheskogo S'ezda (Proceedings of the 3rd All Union Mathematical Conference)* ED: NIKOL'SKIJ, S.M. & ABRAMOV, A.A. & BOLTYANSKIJ, V.G.
• PUBL (X 0899) Akad Nauk SSSR : Moskva 4 Vols
• DAT&PL 1956 Jun; Moskva, SU

P 0608 Logic Colloq;1966 Hannover • D
[1968] *Contributions to Mathematical Logic. Proceedings of the Logic Colloquium* ED: SCHMIDT, H.A. & SCHUETTE, K. & THIELE, H.-J. • SER (S 3303) Stud Logic Found Math • PUBL (X 0809) North Holland: Amsterdam ix+298pp
• DAT&PL 1966 Aug; Hannover, D • LC-No 68-24434

P 0610 Tarski Symp;1971 Berkeley • USA
[1974] *Proceedings of the Tarski Symposium. An International Symposium held to Honor Alfred Tarski on the Occasion of His 70th Birthday* ED: HENKIN, L. & ADDISON, J. & CHANG, C.C. & CRAIG, W. & SCOTT, D.S. & VAUGHT, R. • SER (S 3304) Proc Symp Pure Math 25 • PUBL (X 0803) Amer Math Soc: Providence ix+498pp
• DAT&PL 1971 Jun; Berkeley, CA, USA • ISBN 0-8218-1425-7, LC-No 74-8666
• REM Corrected Reprint 1979; xx+498pp

P 0612 Int Congr Log, Meth & Phil of Sci (1,Proc);1960 Stanford • USA
[1962] *Proceedings of the 1st International Congress for Logic, Methodology and Philosophy of Science* ED: NAGEL, E. & SUPPES, P. & TARSKI, A. • PUBL (X 1355) Stanford Univ Pr: Stanford ix+661pp
• DAT&PL 1960 Aug; Stanford, CA, USA • LC-No 62-9620
• TRANSL IN [1965] (P 2251) Mat Log & Primen;1960 Stanford

P 0614 Th Models;1963 Berkeley • USA
[1965] *The Theory of Models.* ED: ADDISON, J.W. & HENKIN, L. & TARSKI, A. • SER (S 3303) Stud Logic Found Math
• PUBL (X 0809) North Holland: Amsterdam xv+494pp
• DAT&PL 1963 Jun; Berkeley, CA, USA • LC-No 66-7051

P 0619 Log of Decis & Action;1966 Pittsburgh • USA
[1967] *The Logic of Decision and Action* ED: RESCHER, N.
• PUBL (X 1331) Univ Pittsburgh Pr: Pittsburgh 226pp
• DAT&PL 1966 Mar; Pittsburgh, PA, USA • LC-No 67-18272

P 0623 Int Congr Log, Meth & Phil of Sci (2,Proc);1964 Jerusalem • IL
[1965] *Proceedings of the 2nd International Congress for Logic, Methodology and Philosophy of Science* ED: BAR-HILLEL, Y. • SER (S 3303) Stud Logic Found Math • PUBL (X 0809) North Holland: Amsterdam viii + 440pp
• DAT&PL 1964 Aug; Jerusalem, IL • ISBN 0-7204-2235-3, LC-No 66-7008
• REM 2nd ed. 1972

P 0624 Switch Circ Th & Log Design (1,2);1960 Chicago;1961 Detroit • USA
[1961] *Switching Circuit Theory and Logical Design. Proceedings of the 2nd Annual Symposium and Papers from the 1st Annual Symposium* ED: LEDLEY, R.S. • PUBL American Institute of Electrical Engineers: New York xi + 341pp
• DAT&PL 1960 Oct; Chicago, IL, USA, 1961 Oct; Detroit, MI, USA

P 0626 Ber Math-Tagung Berlin;1953 Berlin • DDR
[1953] *Bericht ueber die Mathematiker-Tagung* ED: GRELL, H. & SCHMID, H.L. • PUBL (X 0806) Dt Verlag Wiss: Berlin viii + 302pp
• DAT&PL 1953 Jan; Berlin, D • LC-No 78-235228

P 0632 Congr Int Phil des Sci;1935 Paris • F
[1936] *Actes du Congres International de Philosophie Scientifique* SER (S 2073) Actualites Sci Indust 388-395
• PUBL (X 0859) Hermann: Paris 8 Vol
• DAT&PL 1935 Sep; Paris, F
• REM Vol. II: Unite de la Science. Vol. III: Language et Pseudo-Problemes. Vol. IV: Induction et Probabilite. Vol. VI: Philosophie des Mathematiques. Vol. VII: Logique. Vol. VIII: Histoire de la Logique et de la Philosophie Scientifique.

P 0633 Infinitist Meth;1959 Warsaw • PL
[1961] *Infinitistic Methods. Proceedings of the Symposium on Foundations of Mathematics* PUBL (X 1034) PWN: Warsaw 362pp • ALT PUBL (X 0869) Pergamon Pr: Oxford 1961
• DAT&PL 1959 Sep; Warsaw, PL • LC-No 61-11351

P 0638 Logic Colloq;1969 Manchester • GB
[1971] *Logic Colloquium '69* ED: GANDY, R.O. & YATES, C.M.E. • SER (S 3303) Stud Logic Found Math 61 • PUBL (X 0809) North Holland: Amsterdam xiv + 457pp
• DAT&PL 1969 Aug; Manchester, GB • ISBN 0-7204-2261-2, LC-No 71-146188

P 0645 Int Congr Philos (11);1953 Bruxelles • B
[1953] *Actes du 11eme Congres International de Philosophie. * Proceedings of the 11th International Congress of Philosophy* PUBL (X 1313) Nauwelaerts: Louvain • ALT PUBL (X 0809) North Holland: Amsterdam 1953;14 Vols
• DAT&PL 1953 Aug; Bruxelles, B

P 0646 Appl Sci de Log Math;1952 Paris • F
[1954] *Applications Scientifiques de la Logique Mathematique. Actes du 2eme Colloque International de Logique Mathematique* ED: DESTOUCHES, J.-L. & DESTOUCHES-FEVRIER, P. • SER Collection de Logique Mathematique, Serie A 5 • PUBL (X 0834) Gauthier-Villars: Paris 176pp • ALT PUBL (X 1313) Nauwelaerts: Louvain 1954
• DAT&PL 1952 Aug; Paris, F

P 0651 Axiomatic Method;1957 Berkeley • USA
[1959] *The Axiomatic Method. With Special Reference to Geometry and Physics* ED: HENKIN, L. & SUPPES, P. & TARSKI, A. • SER (S 3303) Stud Logic Found Math • PUBL (X 0809) North Holland: Amsterdam xi + 488pp
• DAT&PL 1957 Dec; Berkeley, CA, USA • LC-No 58-63025

P 0652 Entretiens Zuerich Fond & Method Sci Math;1938 Zuerich • CH
[1941] *Les Entretiens de Zuerich sur les Fondements et la Methode des Sciences Mathematiques: Exposes et Discussions* ED: GONSETH, F. • PUBL (X 2220) Leemen: Zuerich 209pp
• DAT&PL 1938 Dec; Zuerich, CH • LC-No 42-650

P 0660 Int Congr Math (II, 8);1958 Edinburgh • GB
[1960] *Proceedings of the International Congress of Mathematicians* ED: TODD, J.A. • PUBL (X 0805) Cambridge Univ Pr: Cambridge, GB lxiv + 573pp
• DAT&PL 1958 Aug; Edinburgh, GB • LC-No 52-1808

P 0664 Congr Int Union Phil of Sci (2);1954 Zuerich • CH
[1955] *Actes du 2eme Congres International de L'Union Internationale de la Philosophie des Sciences * Proceedings of the 2nd International Congress of the International Union for the Philosophy of Science* PUBL (X 0272) Griffon: Neuchatel 5 Vols
• DAT&PL 1954 -?-; Zuerich, CH • LC-No 64-52910

P 0669 Conv Teor Modelli & Geom;1969/70 Roma • I
[1971] *Convegni Teoria dei Modelli & Geometria* SER (S 3305) Symposia Matematica 5 • PUBL (X 3604) INDAM: Roma 475pp • ALT PUBL (X 0801) Academic Pr: New York
• DAT&PL 1969 Nov; Rome, I, 1970 Apr; Rome, I

P 0682 Int Congr Philos (10);1948 Amsterdam • NL
[1949] *Library of the 10th International Congress of Philosophy* ED: BETH, E.W. & POS, H.J. & HOLLAK, H.J.A. • PUBL (X 0809) North Holland: Amsterdam Vol 1, L.J.Veen: Amsterdam Vol 2
• DAT&PL 1948 Aug; Amsterdam, NL • LC-No 50-35721

P 0683 Log & Founds of Sci (Beth);1964 Paris • F
[1967] *Logic and Foundations of Science. E.W.Beth Memorial Colloquium.* ED: DESTOUCHES, J.-L. • PUBL (X 0835) Reidel: Dordrecht viii + 140pp
• DAT&PL 1964 May; Paris, F • LC-No 68-107520

P 0688 Logic Colloq;1963 Oxford • GB
[1965] *Formal Systems and Recursive Functions. Proceedings of the 8th Logic Colloquium* ED: CROSSLEY, J.N. & DUMMETT, M.A.E. • SER (S 3303) Stud Logic Found Math • PUBL (X 0809) North Holland: Amsterdam 320pp
• DAT&PL 1963 Jul; Oxford, GB • LC-No 66-2289

P 0692 Summer School in Logic;1967 Leeds • GB
[1968] *Proceedings of the Summer School in Logic* ED: LOEB, M.H. • SER (S 3301) Lect Notes Math 70 • PUBL (X 0811) Springer: Heidelberg & New York iv + 331pp
• DAT&PL 1967 Aug; Leeds, GB • ISBN 3-540-04240-7, LC-No 68-56951

P 0702 Combin Struct & Appl;1969 Calgary • CDN
[1970] *Combinatorial Structures and Their Applications.* ED: GUY, R. & SAUER, N. & HANANI, H. & SCHONHEIM, J. • PUBL (X 0836) Gordon & Breach: New York xvi + 508pp
• DAT&PL 1969 Jun; Calgary, AL, CDN • LC-No 78-111302

P 0711 Concept & Role of Model in Math & Sci;1960 Utrecht • NL
[1961] *The Concept and the Role of the Model in Mathematics and Natural and Social Sciences* ED: FREUDENTHAL, H. • SER (S 3307) Synth Libr • PUBL (X 0835) Reidel: Dordrecht 194pp
• DAT&PL 1960 Jan; Utrecht, NL • LC-No 63-1436

P 0713 Cambridge Summer School Math Log;1971 Cambridge GB • GB
[1973] *Cambridge Summer School in Mathematical Logic* ED: ROGERS, H. & MATHIAS, A.R.D. • SER (S 3301) Lect Notes Math 337 • PUBL (X 0811) Springer: Heidelberg & New York ix + 660pp
• DAT&PL 1971 Aug; Cambridge, GB • ISBN 3-540-05569-X, LC-No 73-12410

P 0715 Topol & Primen (2);1972 Budva • YU
[1973] *Topologija i Ee Primenenija II (Topology and Its Applications II)* ED: KUREPA, D.R. • PUBL (X 2505) Sav Drush Mat Fiz Ju: Belgrade 272pp
• DAT&PL 1972 Aug; Budva, YU

P 0731 Paradox of Liar;1969 Buffalo • USA
[1970] *The Paradox of the Liar.* ED: MARTIN, R.L. • PUBL (X 0875) Yale Univ Pr: New Haven xv + 149pp
• DAT&PL 1969 Mar; Buffalo, NY, USA • ISBN 0-300-01355-8, LC-No 79-118732

P 0743 Int Congr Math (II,11,Proc);1970 Nice • F
[1971] *Actes du Congres International de Mathematiciens 1970* ED: BERGER, M. & DIEUDONNE, J. & LERAY, J. & LIONS, J.-L. & MALLIAVIN, M.P. & SERRE, J.-P. • PUBL (X 0834) Gauthier-Villars: Paris 3 Vols: xxxiii + 532pp,959pp,iii + 371pp
• DAT&PL 1970 Sep; Nice, F • REL PUBL (P 1158) Int Congr Math (II,11,Comm Ind);1970 Nice
• REM Vol 1: Documents.Medailles Fields.Conferences Generales.Logique.Algebre. Vol 2: Geometrie et Topologie. Analyse. Vol 3: Mathematiques Appliquees.Historie et Enseignement.

P 0746 Automata Th;1964 Ravello • I
[1966] *Automata Theory. International School of Physics* ED: CAIANIELLO, E.R. • PUBL (X 0801) Academic Pr: New York xiv + 342pp
• DAT&PL 1964 Jun; Ravello, I • LC-No 65-22775
• REM 2nd ed. 1968

P 0756 Congr Int Phil (9);1937 Paris • F
[1937] *Travaux du IXe Congres International de Philosophie: Congres Descartes* ED: BAYER, R. • SER (S 2073) Actualites Sci Indust 530-541 • PUBL (X 0859) Hermann: Paris 12 Vols
• DAT&PL 1937 -?-; Paris, F

P 0757 Scand Logic Symp (3);1973 Uppsala • S
[1975] *Proceedings of the 3rd Scandinavian Logic Symposium* ED: KANGER, S. • SER (S 3303) Stud Logic Found Math 82
• PUBL (X 0809) North Holland: Amsterdam vii + 214pp
• ALT PUBL (X 0838) Amer Elsevier: New York
• DAT&PL 1973 Apr; Uppsala, S • ISBN 0-444-10679-0 , LC-No 74-80113

P 0765 Algeb & Log;1974 Clayton • AUS
[1975] *Algebra and Logic. Papers from the 1974 Summer Research Institute of the Australian Mathematical Society* ED: CROSSLEY, J.N. • SER (S 3301) Lect Notes Math 450 • PUBL (X 0811) Springer: Heidelberg & New York viii + 307pp
• DAT&PL 1974 Jan; Clayton, Vic, AUS • ISBN 3-540-07152-0, LC-No 75-9903

P 0774 Fuzzy Sets & Appl;1974 Berkeley • USA
[1975] *Fuzzy Sets and Their Applications to Cognitive and Decision Processes. Proceedings of the US-Japan Seminar* ED: ZADEH, L.A. & FU, KING SUN & TANAKA, K. & SHIMARU, M.
• PUBL (X 0801) Academic Pr: New York x + 496pp
• DAT&PL 1974 Jul; Berkeley, CA, USA • ISBN 0-12-775260-9, LC-No 75-15772

P 0775 Logic Colloq;1973 Bristol • GB
[1975] *Logic Colloquium '73* ED: ROSE, H.E. & SHEPHERDSON, J.C. • SER (S 3303) Stud Logic Found Math 80 • PUBL (X 0809) North Holland: Amsterdam viii + 513pp • ALT PUBL (X 0838) Amer Elsevier: New York
• DAT&PL 1973 Jul; Bristol, GB • ISBN 0-444-10642-1, LC-No 74-79302

P 0783 Truth, Syntax & Modal;1970 Philadelphia • USA
[1973] *Truth, Syntax and Modality: Proceedings of the Temple University Conference on Alternative Semantics* ED: LEBLANC, H. • SER (S 3303) Stud Logic Found Math 68 • PUBL (X 0809) North Holland: Amsterdam vii + 317pp
• DAT&PL 1970 Dec; Philadelphia, PA, USA • ISBN 0-7204-2269-8, LC-No 72-79730

P 0784 GI Jahrestag (5);1975 Dortmund • D
[1975] *GI - 5. Jahrestagung (Gesellschaft fuer Informatik)* ED: MUEHLBACHER, J. • SER (S 3302) Lect Notes Comput Sci 34
• PUBL (X 0811) Springer: Heidelberg & New York x + 755pp
• DAT&PL 1975 Oct; Dortmund, D • ISBN 3-540-07410-4

P 0785 Scand Logic Symp (1);1968 Aabo • S
[1970] *Proceedings of the 1st Scandinavian Logic Symposium* SER Filosofiska Studier 8 • PUBL (X 0882) Univ Filos Foeren: Uppsala 171pp
• DAT&PL 1968 Sep; Aabo, S • LC-No 72-186670

P 0788 Skand Mat Kongr (12);1953 Lund • S
[1954] *12te Skandinaviska Matematikerkongressen (Comptes-Rendus du 12eme Congres des Mathematiciens Scandinaves)* PUBL Hakan Oh Issons Boktryckeri: Lund xvi + 337pp
• DAT&PL 1953 Aug; Lund, S • LC-No 55-58514

P 0793 Int Congr Log, Meth & Phil of Sci (4,Proc);1971 Bucharest • RO
[1973] *Proceedings of the 4th International Congress for Logic, Methodology and Philosophy of Science* ED: SUPPES, P. & HENKIN, L. & MOISIL, G.C. & JOJA, A. • SER (S 3303) Stud Logic Found Math 74 • PUBL (X 0809) North Holland: Amsterdam x + 981pp • ALT PUBL (X 0838) Amer Elsevier: New York & (X 1034) PWN: Warsaw
• DAT&PL 1971 Aug; Bucharest, RO • ISBN 0-444-10491-7, LC-No 72-88505 • REL PUBL (P 0580) Int Congr Log, Meth & Phil of Sci (4,Sel Pap);1971 Bucharest

P 0797 Fonds des Math, Machines Math & Appl;1962 Tihany • H
[1965] *Colloque sur les Fondements des Mathematiques, les Machines Mathematiques, et leurs Applications* ED: KALMAR, L. • SER Collection de Logique Mathematique, Serie A 19 • PUBL (**X** 0928) Akad Kiado: Budapest 320pp • ALT PUBL (**X** 0834) Gauthier-Villars: Paris & (**X** 1313) Nauwelaerts: Louvain
• DAT&PL 1962 Sep; Tihany, H

P 1075 Logic Colloq;1976 Oxford • GB
[1977] *Logic Colloquium 76* ED: GANDY, R.O. & HYLAND, J.M.E. • SER (**S** 3303) Stud Logic Found Math 87 • PUBL (**X** 0809) North Holland: Amsterdam x+612pp • ALT PUBL (**X** 0838) Amer Elsevier: New York
• DAT&PL 1976 Jul; Oxford, GB • ISBN 0-7204-0691-9, LC-No 77-8943

P 1076 Latin Amer Symp Math Log (3);1976 Campinas • BR
[1977] *Non-Classical Logic, Model Theory and Computability. 3rd Latin American Symposium on Mathematical Logic* ED: ARRUDA, A.I. & COSTA DA, N.C.A & CHUAQUI, R. • SER (**S** 3303) Stud Logic Found Math 89 • PUBL (**X** 0809) North Holland: Amsterdam xviii+307pp • ALT PUBL (**X** 0838) Amer Elsevier: New York
• DAT&PL 1976 Jul; Campinas, BR • ISBN 0-7204-0752-4, LC-No 77-7366

P 1127 Symp Founds of Math;1962 Katada • J
[1963] *Proceedings of the Symposium on the Foundations of Mathematics* ED: TAKEUTI, G. • PUBL Tokyo University of Education: Tokyo vi+144pp
• DAT&PL 1962 Oct; Katada, J

P 1129 Princeton Conf Inform Sci & Syst (3);1969 Princeton • USA
[1969] *Proceedings of the 3rd Annual Conference on Information Science and Systems* ED: THOMAS, J.B. & VALKENBURG VAN, M.E. & WEINER, P. • PUBL (**X** 2188) Princeton Univ Dept Elect Eng & Comp Sci: Princeton xiii+550pp
• DAT&PL 1969 Mar; Princeton, NJ, USA

P 1158 Int Congr Math (II,11,Comm Ind);1970 Nice • F
[1970] *Congres International des Mathematiciens 1970. Les 265 Communications Individuels* PUBL (**X** 0834) Gauthier-Villars: Paris vii+290pp
• DAT&PL 1970 Sep; Nice, F • LC-No 72-374601 • REL PUBL (**P** 0743) Int Congr Math (II,11,Proc);1970 Nice

P 1192 Symb Lang in Data Processing;1962 Roma • I
[1962] *Symbolic Languages in Data Processing* PUBL (**X** 0836) Gordon & Breach: New York xii+849pp
• DAT&PL 1962 Mar; Rome, I • LC-No 62-22085

P 1385 Int Symp Multi-Val Log (4);1974 Morgantown • USA
[1974] *Proceedings of the 1974 International Symposium on Multiple-Valued Logic.* PUBL (**X** 2179) IEEE: New York iv+551pp
• DAT&PL 1974 May; Morgantown, WV, USA • LC-No 79-641110

P 1401 Math Founds of Comput Sci (5);1976 Gdansk • PL
[1976] *Mathematical Foundations of Computer Science. Proceedings of the 5th Symposium* ED: MAZURKIEWICZ, A. • SER (**S** 3302) Lect Notes Comput Sci 45 • PUBL (**X** 0811) Springer: Heidelberg & New York xi+606pp
• DAT&PL 1976 Sep; Gdansk, PL • ISBN 3-540-07854-1, LC-No 76-25494

P 1430 Adv Course Founds Computer Sci;1974 Amsterdam • NL
[1975] *Foundations of Computer Science. Advanced Course* ED: BAKKER DE, J.W. • SER (**S** 1605) Math Centr Tracts 63 • PUBL (**X** 1121) Math Centr: Amsterdam 215pp
• DAT&PL 1974 May; Amsterdam, NL • ISBN 90-6196-111-4, LC-No 76-363070

P 1440 ⊧ ISILC Proof Th Symp (Schuette);1974 Kiel • D
[1975] ⊧ *ISILC Proof Theory Symposium. Dedicated to Kurt Schuette on the Occasion of His 65th Birthday. Proceedings of the International Summer Institute and Logic Colloquium* ED: DILLER, J. & MUELLER, GERT H. • SER (**S** 3301) Lect Notes Math 500 • PUBL (**X** 0811) Springer: Heidelberg & New York viii+383pp
• DAT&PL 1974 Jul; Kiel, D • ISBN 3-540-07533-X, LC-No 75-40482 • REL PUBL (**P** 1442) ⊧ ISILC Logic Conf;1974 Kiel
• REM This Volume Contains Only the Proof Theory Part of the Conference.

P 1442 ⊧ ISILC Logic Conf;1974 Kiel • D
[1975] ⊧ *ISILC Logic Conference. Proceedings of the International Summer Institute and Logic Colloquium* ED: MUELLER, GERT H. & OBERSCHELP, A. & POTTHOFF, K. • SER (**S** 3301) Lect Notes Math 499 • PUBL (**X** 0811) Springer: Heidelberg & New York iv+651pp
• DAT&PL 1974 Jul; Kiel, D • ISBN 3-540-07534-8, LC-No 75-40431 • REL PUBL (**P** 1440) ⊧ ISILC Proof Th Symp (Schuette);1974 Kiel

P 1457 Rep Semin Formal Methodol of Empir Sci;1973 Wroclaw • PL
[1973] *Reports of the Seminar on Formal Methodology of Empirical Sciences* ED: WOJCICKI, R. & PRZELECKI, M. • PUBL (**X** 2733) Acad Sci Inst Phi Soc: Wroclaw ii+39pp
• DAT&PL 1973 Mar-Oct; Wroclaw, PL

P 1484 Int Congr Math (2);1900 Paris • F
[1902] *Comptes Rendus du 2eme Congres International des Mathematiciens. Proces Verbaux et Communications* ED: DUPORCQ, E. • PUBL (**X** 0834) Gauthier-Villars: Paris 455pp
• DAT&PL 1900 Aug; Paris, F

P 1502 Phil of Logic;1976 Bristol • GB
[1976] *Philosophy of Logic. Proceedings of the 3rd Bristol Conference on Critical Philosophy.* ED: KOERNER, S. • PUBL (**X** 1096) Blackwell: Oxford 273pp • ALT PUBL (**X** 0926) Univ Calif Pr: Berkeley
• DAT&PL 1974 -?-; Bristol, GB • ISBN 0-631-16960-1 (1096), ISBN 0-520-03235-7 (0926), LC-No 77-359489 (1096), LC-No 76-020-2 (0926)

P 1511 Int Symp Th Progr;1972 Novosibirsk • SU
[1974] *International Symposium on Theoretical Programming* ED: ERSHOV, A.P. & NEPOMNYASHCHIJ, V.A. • SER (**S** 3302) Lect Notes Comput Sci 5 • PUBL (**X** 0811) Springer: Heidelberg & New York vi+407pp
• DAT&PL 1972 Aug; Novosibirsk, SU • ISBN 3-540-06720-5, LC-No 74-176124

P 1521 Int Congr Math (II,12);1974 Vancouver • CDN
[1975] *Proceedings of the International Congress of Mathematicians* ED: JAMES, R.D. • PUBL Canadian Mathematical Congress: Montreal 2 Vols: xlix+552pp,viii+600pp
• DAT&PL 1974 Aug; Vancouver, BC, CDN • ISBN 0-919558-04-6, LC-No 74-34533

P 1538 Inform Th, Stat Decis Fcts & Random Proc (2);1959 Liblice • CS
[1960] *Transactions of the 2nd Prague Conference on Information Theory, Statistical Decision Functions, Random Processes* ED: SHPACHEK, A. • PUBL (**X** 1226) Academia: Prague 843pp • ALT PUBL (**X** 0801) Academic Pr: New York 1961
• DAT&PL 1959 Jun; Liblice, CS

P 1539 Berkeley Symp Math Stat & Probab (4);1960 Berkeley • USA
[1961] *Proceedings of the 4th Berkeley Symposium on Mathematical Statistics and Probability.* ED: NEYMAN, J. • PUBL (**X** 0926) Univ Calif Pr: Berkeley 4 Vols: xii+767pp,xii+633pp, xi+335pp, xii+413pp • ALT PUBL (**X** 0805) Cambridge Univ Pr: Cambridge, GB
• DAT&PL 1960 Jun; Berkeley, CA, USA • LC-No 49-8189
• REM Vol 1: Contributions to the Theory of Statistics. Vol 2: Contributions to Probability Theory. Vol 3: Contributions to Astronomy, Meteorology, and Physics. Vol 4: Contributions to Biology and Problems of Medicine.

P 1545 Easter Conf on Model Th (2);1984 Wittenberg • DDR
[1984] *Proceedings of the 2nd Easter Conference on Model Theory* SER (**S** 3382) Sem-ber, Humboldt-Univ Berlin, Sekt Math 60 • PUBL (**X** 2219) Humboldt-Univ Berlin: Berlin ii+243pp
• DAT&PL 1984 Apr; Wittenberg, DDR

P 1548 Conf Convexity & Combin Geom (1);1971 Norman • USA
[1971] *Proceedings of the Conference on Convexity and Combinatorial Geometry* ED: KAY, D.C. • PUBL (**X** 2374) Univ Oklahoma Pr: Norman vi+138pp
• DAT&PL 1971 Jun; Norman, OK, USA • LC-No 73-622304

P 1556 Exact Philos: Probl, Tools & Goals;1971 Montreal • CDN
[1973] *Exact Philosophy: Problems, Tools, and Goals* ED: BUNGE, M. • SER (**S** 3307) Synth Libr, (**J** 0122) J Philos Logic 1/3-4 • PUBL (**X** 0835) Reidel: Dordrecht x+214pp
• DAT&PL 1971 Nov; Montreal, Que, CDN • ISBN 90-277-0251-9, LC-No 72-77872

P 1571 Int Congr Philos (14);1968 Wien • A
[1968-1971] *Akten des 14. Internationalen Kongresses fuer Philosophie (Proceedings of the 14th International Congress of Philosophy)* PUBL (**X** 1279) Herder: Freiburg 6 Vols
• DAT&PL 1968 Sep; Wien, A • LC-No 78-352552

P 1603 λ-Calc & Comput Sci Th;1975 Roma • I
[1975] *λ Calculus and Computer Science Theory* ED: BOEHM, C. • SER (**S** 3302) Lect Notes Comput Sci 37 • PUBL (**X** 0811) Springer: Heidelberg & New York xii+370pp
• DAT&PL 1975 Mar; Roma, I • ISBN 3-540-07416-3, LC-No 75-33375

P 1606 Colloq Math (Pascal);1962 Clermont-Ferrand • F
[1962] *Actes du Colloque de Mathematiques Reuni a Clermont a l'Occasion du Tricentenaire de la Mort de Blaise Pascal* SER (**J** 0179) Ann Fac Sci Clermont 7,8 • PUBL Univ Clermont, Fac. Sci.: Clermont 2 Vols: 123pp,189pp
• DAT&PL 1962 Jun; Clermont-Ferrand, F
• REM Vol 1: Introduction et Logique Mathematique. Vol 2: Calcul des Probabilites, Analyse Numerique et Calcul Automatique, Geometrie et Physique Mathematique

P 1619 Coloq Log Simb;1975 Madrid • E
[1976] *Coloquio Sobre Logica Simbolica* PUBL Centro Calculo Univ. Complutense: Madrid 176pp
• DAT&PL 1975 Feb; Madrid, E • LC-No 77-555677

P 1635 Math Founds of Comput Sci (6);1977 Tatranska Lomnica • CS
[1977] *Mathematical Foundations of Computer Science. Proceedings of the 6th Symposium* ED: GRUSKA, J. • SER (**S** 3302) Lect Notes Comput Sci 53 • PUBL (**X** 0811) Springer: Heidelberg & New York xi+595pp
• DAT&PL 1977 Sep; Tatranska Lomnica, CS • ISBN 3-540-08353-7, LC-No 77-10135

P 1639 Set Th & Hierarch Th (1);1974 Karpacz • PL
[1977] *Set Theory and Hierarchy Theory. Proceedings of the 1st Colloquium in Set Theory and Hierarchy Theory* SER (**S** 3412) Prace Inst Mat, Politech Wroclaw, Ser Konf 14/1 • PUBL Polytechnical Edition: Wroclaw 123pp
• DAT&PL 1974 Sep; Karpacz, PL

P 1647 Contemp Res in Phil Log & Ling Semant;1975 London ON • CDN
[1975] *Contemporary Research in Philosophical Logic and Linguistic Semantics* ED: HOCKNEY, D. & HARPER, W. & FREED, B. • SER (**S** 3308) Univ Western Ontario Ser in Philos of Sci 4 • PUBL (**X** 0835) Reidel: Dordrecht 332pp
• DAT&PL 1973 -?-; London, ON, CDN • ISBN 90-277-0511-9, LC-No 74-34079

P 1675 Summer Inst Symb Log;1957 Ithaca • USA
[1957] *Summaries of Talks Presented at the Summer Institute for Symbolic Logic* PUBL Institute for Defense Analyses, Communications Research Division: Princeton; xvi+427pp
• DAT&PL 1957 Jul; Ithaca, NY, USA • LC-No 65-4418
• REM 2nd ed. 1960

P 1691 Inform Processing (6);1974 Stockholm • S
[1974] *Information Processing '74. Proceedings of IFIP Congress* ED: ROSENFELD, J.L. • PUBL (**X** 0809) North Holland: Amsterdam xxi+1107pp • ALT PUBL (**X** 0838) Amer Elsevier: New York
• DAT&PL 1974 Aug; Stockholm, S • ISBN 0-444-10689-8, LC-No 74-76063

P 1703 Proc Bienn Meet Phil of Sci Ass;1976 Chicago • USA
[1977] *Proceedings of Biennial Meeting of Philosophy of Science Association (PSA)* ED: SUPPE, F. & ASQUITH, P. • PUBL (**X** 2203) Philos Sci Ass: East Lansing
• DAT&PL 1976 -?-; Chicago, IL, USA • LC-No 72-624169

P 1704 Int Congr Log, Meth & Phil of Sci (5);1975 London ON • CDN
[1977] *Proceedings of 5th International Congress of Logic, Methodology and Philosophy of Science* ED: BUTTS, R.E. & HINTIKKA, J. • SER (**S** 3308) Univ Western Ontario Ser in Philos of Sci 9-12 • PUBL (**X** 0835) Reidel: Dordrecht 4 Vols: x+406pp, x+427pp, x+321pp, x+336pp
• DAT&PL 1975 Aug; London, ON, CDN • ISBN 90-277-0708-1 (V1), ISBN 90-277-0710-3 (V2), ISBN 90-277-0829-0 (V3), ISBN 90-277-0831-2 (V4), ISBN 90-277-0706-5 (Set of the 4 Vols), LC-No 77-22429 (V1), LC-No 77-22431 (V2), LC-No 77-22432 (V3), LC-No 77-22433 (V4)
• REM Vol 1: Logic, Foundations of Mathematics, and Computability Theory. Vol 2: Foundational problems in the Special Sciences. Vol 3: Basic Problems in Methodology and Linguistics. Vol 4: Historical and Philosophical Dimensions of Logic, Methodology and Philosophy of Science.

P 1705 Scand Logic Symp (4);1976 Jyvaeskylae • SF
[1979] *Essays on Mathematical and Philosophical Logic. Proceedings of the 4th Scandinavian Logic Symposium and of the 1st Soviet-Finnish Logic Conference* ED: HINTIKKA, J. & NIINILUOTO, I. & SAARINEN, E. • SER (S 3307) Synth Libr 122 • PUBL (X 0835) Reidel: Dordrecht viii+462pp
• DAT&PL 1976 Jun; Jyvaeskylae, SF • ISBN 90-277-0879-7, LC-No 78-14736

P 1707 Math Founds of Comput Sci (7);1978 Zakopane • PL
[1978] *Mathematical Foundations of Computer Science. Proceedings of the 7th Symposium* ED: WINKOWOSKI, J • SER (S 3302) Lect Notes Comput Sci 64 • PUBL (X 0811) Springer: Heidelberg & New York ix+551pp
• DAT&PL 1978 Sep; Zakopane, PL • ISBN 3-540-08921-7, LC-No 78-14457

P 1740 ACM Symp Th of Comput (10);1978 San Diego • USA
[1978] *Conference Record of the 10th Annual ACM Symposium on Theory of Computing (Association for Computing Machinery)* PUBL (X 2205) ACM: New York 346pp
• DAT&PL 1978 May; San Diego, CA, USA • LC-No 79-101797

P 1791 Proc Bienn Meet Phil of Sci Ass;1972 East Lansing • USA
[1974] *Proceedings of the 1972 Biennial Meeting: Philosophy of Science Association (PSA)* ED: SCHAFFNER, K.F.& COHEN, R.S. • SER (S 3307) Synth Libr 64, (S 3311) Boston St Philos Sci 20 • PUBL (X 0835) Reidel: Dordrecht ix+445pp
• DAT&PL 1972 Oct; East Lansing, MI, USA • ISBN 90-277-0408-2 (Cloth), ISBN 90-277-0409-0 (Pb.), LC-No 72-624169

P 1800 Brazil Conf Math Log (1);1977 Campinas • BR
[1978] *Proceedings of 1st Brazilian Conference on Mathematical Logic* ED: ARRUDA, A.I. & CHAQUI, R. & COSTA DA, N.C.A. • SER (S 3310) Lect Notes Pure Appl Math 39 • PUBL (X 1684) Dekker: New York xii+303pp
• DAT&PL 1977 Jul; Campinas, BR • LC-No 78-14488

P 1804 Form Meth in Methodol of Emp Sci;1974 Warsaw • PL
[1976] *Formal Methods in the Methodology of Empirical Sciences* ED: PRZELECKI, M. & SZANIAWSKI, K. & WOJCICKI, R. & MALINOWSKI, G. • SER (S 3307) Synth Libr 103 • PUBL (X 2212) Ossolineum: Wroclaw 457pp • ALT PUBL (X 0835) Reidel: Dordrecht
• DAT&PL 1974 Jun; Warsaw, PL • ISBN 90-277-0698-0, LC-No 76-4586

P 1805 Int Symp Multi-Val Log (5,Proc);1975 Bloomington • USA
[1975] *Proceedings of the 1975 International Symposium on Multiple-Valued Logic* PUBL (X 2179) IEEE: New York iv+475pp
• DAT&PL 1975 May; Bloomington, IN, USA • LC-No 76-370321 • REL PUBL (P 1894) Int Symp Multi-Val Log (5,Inv Pap);1975 Bloomington

P 1809 Congr Cient Mexicano;1951 Mexico City • MEX
[1953] *Memoria del Congreso Cientifico Mexicano. IV Centenario de la Universidad de Mexico (1551-1951)* PUBL (X 1014) Univ Nac Auton: Mexico City 7 Vols
• DAT&PL 1951 Sep; Mexico City, MEX • LC-No 59-17745

P 1825 AFCET Congr Econ Tech;1977 • F
[1977] *Actes du Congres de l'Association Francaise pour la Cybernetique, Economique et Technique (AFCET)*
• DAT&PL 1977 Nov; -?-

P 1869 Automata, Lang & Progr (2);1974 Saarbruecken • D
[1974] *Automata, Languages and Programming: 2nd Colloquium* ED: LOECKX, J. • SER (S 3302) Lect Notes Comput Sci 14 • PUBL (X 0811) Springer: Heidelberg & New York viii+611pp
• DAT&PL 1974 Jul; Saarbruecken, D • ISBN 3-540-06841-4, LC-No 74-180345
• REM Also Abbreviated as ICALP 74

P 1872 Automata, Lang & Progr (5);1978 Udine • I
[1978] *Automata, Languages and Programming. 5th Colloquium* ED: AUSIELLO, G. & BOEHM, C. • SER (S 3302) Lect Notes Comput Sci 62 • PUBL (X 0811) Springer: Heidelberg & New York viii+508pp
• DAT&PL 1978 Jul; Udine, I • ISBN 3-540-08860-1, LC-No 79-303999
• REM Also Abbreviated as ICALP 78

P 1873 Automata, Lang & Progr (6);1979 Graz • A
[1979] *Automata, Languages and Programming. 6th Colloquium* ED: MAURER, H.A. • SER (S 3302) Lect Notes Comput Sci 71 • PUBL (X 0811) Springer: Heidelberg & New York ix+682pp
• DAT&PL 1979 Jul; Graz, A • ISBN 3-540-09510-1, LC-No 79-15859
• REM Also Abbreviated as ICALP 79

P 1894 Int Symp Multi-Val Log (5,Inv Pap);1975 Bloomington • USA
[1977] *Modern Uses of Multiple-Valued Logic. Invited Papers from the 5th International Symposium on Multiple-Valued Logic* ED: DUNN, J.M. & EPSTEIN, G. • SER Epistime 2
• PUBL (X 0835) Reidel: Dordrecht x+338pp
• DAT&PL 1975 May; Bloomington, IN, USA • ISBN 90-277-0747-2, LC-No 77-23098 • REL PUBL (P 1805) Int Symp Multi-Val Log (5,Proc);1975 Bloomington
• REM With a Bibliography of Many-Valued Logic

P 1897 Logic Colloq;1977 Wroclaw • PL
[1978] *Logic Colloquium 77* ED: MACINTYRE, A. & PACHOLSKI, L. & PARIS, J. • SER (S 3303) Stud Logic Found Math 96 • PUBL (X 0809) North Holland: Amsterdam x+311pp
• DAT&PL 1977 Aug; Wroclaw, PL • ISBN 0-444-85178-X, LC-No 78-13396

P 1924 Polnisch Math Kongr;1953 Warsaw • PL
[1954] *Die Hauptreferate des 8. Polnischen Mathematikerkongresses* ED: GRELL, H. • PUBL (X 0806) Dt Verlag Wiss: Berlin 125pp
• DAT&PL 1953 Sep; Warsaw, PL
• REM Autorisierte Uebersetzung

P 1953 Int Congr Log, Meth & Phil of Sci (1;Abstr);1960 Stanford • USA
[1960] *1st International Congress for Logic, Methodology and Philosophy of Science. Abstracts*
• DAT&PL 1960 Aug; Stanford, CA, USA

P 1959 Int Congr Math (II,13);1978 Helsinki • SF
[1980] *Proceedings of the International Congress of Mathematicians* ED: LEHTO, O. • PUBL Academia Scientiarum Fennica: Helsinki 2 Vols: 1022pp
• DAT&PL 1978 Aug; Helsinki, SF • ISBN 951-41-0352-1

P 2007 Symp Th & Appl of Multi-Val Log Design;1971 Buffalo • USA
[1971] *Conference Records of the 1971 Symposium on the Theory and Applications of Multiple-Valued Logic Design*
• DAT&PL 1971 -?-; Buffalo, NY, USA

P 2008 Symp Th & Appl of Multi-Val Log Design;1972 Buffalo • USA
[1972] *Conference Records of the 1972 Symposium on the Theory and Applications of Multiple-Valued Logic Design* PUBL (**X** 2179) IEEE: New York 209pp
• DAT&PL 1972 May; Buffalo, NY, USA

P 2009 Int Symp Multi-Val Log (3);1973 Toronto • CDN
[1973] *Conference Records of the 1973 International Symposium on Multiple-Valued Logic* ED: ALLEN, C.M. & GIVONE, D.D. • PUBL (**X** 2179) IEEE: New York i+245pp
• DAT&PL 1973 May; Toronto, ON, CDN

P 2011 Int Symp Multi-Val Log (6);1976 Logan • USA
[1976] *Proceedings of the 6th International Symposium on Multiple-Valued Logic* PUBL (**X** 2205) ACM: New York vii+273pp • ALT PUBL (**X** 2179) IEEE: New York
• DAT&PL 1976 May; Logan, UT, USA • LC-No 79-641110
• REM IEEE Publication No. 76CH1111-4C

P 2013 Int Symp Multi-Val Log (7);1977 Charlotte • USA
[1977] *Proceedings of the 7th International Symposium on Multiple-Valued Logic* PUBL (**X** 2179) IEEE: New York iii+155pp • ALT PUBL (**X** 2205) ACM: New York
• DAT&PL 1977 May; Charlotte, NC, USA • LC-No 79-641110
• REM IEEE Publication No 77CH1222-9C

P 2014 Int Symp Multi-Val Log (8);1978 Rosemont • USA
[1978] *Proceedings of the 8th International Symposium on Multiple- Valued Logic* PUBL (**X** 2179) IEEE: New York 298pp
• DAT&PL 1978 May; Rosemont, IL, USA • LC-No 78-107956

P 2047 Latin Amer Symp Math Log (5);
[1982] *Proceedings of the 5th Latin American Symposium on Mathematical Logic* ED: DEKKER, M.

P 2058 Kleene Symp;1978 Madison • USA
[1980] *The Kleene Symposium* ED: BARWISE, K.J. & KEISLER, H.J. & KUNEN, K. • SER (**S** 3303) Stud Logic Found Math 101 • PUBL (**X** 0809) North Holland: Amsterdam xx+425pp
• DAT&PL 1978 Jun; Madison, WI, USA • ISBN 0-444-85345-6, LC-No 79-20792

P 2059 Math Founds of Comput Sci (8);1979 Olomouc • CS
[1979] *Mathematical Foundations of Computer Science. Proceedings of the 8th Symposium* ED: BECVAR, J. • SER (**S** 3302) Lect Notes Comput Sci 74 • PUBL (**X** 0811) Springer: Heidelberg & New York ix+580pp
• DAT&PL 1979 Sep; Olomouc, CS • ISBN 3-540-09526-8, LC-No 79-17801

P 2064 All-Union Conf Math Log (4);1976 Kishinev • SU
[1976] *4 Vsesoyuznaya Konferentsiya po Matematicheskoj Logike. Tezitsy Doklady i Soobshcheniya (4th All-Union Conference on Mathematical Logic)* ED: KUZNETSOV, A.V.
• PUBL (**X** 2741) Shtiintsa: Kishinev 170pp
• DAT&PL 1976 -?-; Kishinev, SU • LC-No 78-410667

P 2080 Conf Math Log;1970 London • GB
[1972] *Conference on Mathematical Logic - London '70* ED: HODGES, W. • SER (**S** 3301) Lect Notes Math 255 • PUBL (**X** 0811) Springer: Heidelberg & New York viii+351pp
• DAT&PL 1970 Aug; London, GB • ISBN 3-540-05744-7, LC-No 70-189457

P 2116 Probl in Log & Ontology;1973 Salzburg • A
[1977] *Problems in Logic and Ontology. Internationales Forschungszentrum Salzburg. Forschungsgespraeche.* ED: MORSCHER, E. & CZERMAK, J. & WEINGARTNER, P. • PUBL (**X** 2596) Akad Druck-& Verlagsanstalt: Graz 310pp
• DAT&PL 1973 Sep; Salzburg, A • ISBN 3-201-01021-9, LC-No 80-487240
• REM Reprint of Vol. 3/3*171-343 of J0122

P 2149 Log Deont & Semant;1975 Bielefeld • D
[1977] *Logica Deontica e Semantica* ED: BERNARDO DI, G.
• SER Publicazioni dell'Universita Degli Studi di Trento 4
• PUBL (**X** 0881) Il Mulino: Bologna 447pp
• DAT&PL 1975 Mar; Bielefeld, D • LC-No 77-570684
• TRANSL OF [1977] (**P** 2150) Deont Log & Semant;1975 Bielefeld

P 2150 Deont Log & Semant;1975 Bielefeld • D
[1977] *Deontische Logik und Semantik.* ED: CONTE, A.G. & HILPINEN, R. & WRIGHT VON, G.H. • SER Linguistische Forschungen 15 • PUBL (**X** 1169) Akad Verlagsges: Wiesbaden 215pp
• DAT&PL 1975 Mar; Bielefeld, D • ISBN 3-7997-0645-3, LC-No 78-347955
• TRANSL IN [1977] (**P** 2149) Log Deont & Semant;1975 Bielefeld

P 2160 Latin Amer Symp Math Log (6);1983 Caracas • YV
[1985] *Methods in Mathematical Logic. Proceedings of the 6th Latin American Symposium on Mathematical Logic* ED: PRISCO DI, C.A. • SER (**S** 3301) Lect Notes Math 1130
• PUBL (**X** 0811) Springer: Heidelberg & New York vii+407pp
• DAT&PL 1983 Aug; Caracas, YV • ISBN 3-540-15236-9, LC-No 85-14779

P 2251 Mat Log & Primen;1960 Stanford • USA
[1965] *Matematicheskaya Logika i Ee Primeneniya: Sbornik Statei (Mathematical Logic and Its Applications. Logic, Methodology and Philosophy of Science)* ED: MAL'TSEV, A.I. & NAGEL, E. & SUPPES, P. & TARSKI, A. • PUBL (**X** 0885) Mir: Moskva 341pp
• DAT&PL 1960 Aug; Stanford, CA, USA
• TRANSL OF [1962] (**P** 0612) Int Congr Log, Meth & Phil of Sci (1,Proc);1960 Stanford
• REM Only Parts of P0612 are translated.

P 2268 Int Colloq Philos of Sci;1965 London • GB
[1967-1970] *Proceedings of the International Colloquium in the Philosophy of Science. 4 Volumes* ED: LAKATOS, I. (V 1 - V 4) & MUSGRAVE, A. (V 3, V 4) • SER (**S** 3303) Stud Logic Found Math V 1 - V 3 • PUBL (**X** 0809) North Holland: Amsterdam V 1: xv+241pp, V 2: viii+420pp, V 3: ix+448pp, (**X** 0805) Cambridge Univ Pr: Cambridge, GB V 4: viii+282pp
• DAT&PL 1965, Jul; London, GB • ISBN 0-521-07826-1 (V4), LC-No 67-20007 (V1), LC-No 67-28648 (V2), LC-No 67-28649 (V3), LC-No 78-105496 (V4)
• REM Vol 1: Problems in the Philosophy of Mathematics. Vol 2: The Problem of Inductive Logic. Vol 3: Problems in the Philosophy of Science. Vol 4: Criticism and the Growth of Knowledge.

Source Index

P 2312 Contemp Res in Found & Philos of Quantum Th;1973 London ON • CDN
[1973] *Contemporary Research in the Foundations and Philosophy of Quantum Theory* ED: HOOKER, C.A. • SER (**S** 3308) Univ Western Ontario Ser in Philos of Sci 2 • PUBL (**X** 0835) Reidel: Dordrecht xx+385pp
• DAT&PL 1973 -?-; London, ON, CDN • ISBN 90-277-0271-3, LC-No 73-178864

P 2326 Action Th;1975 Winnipeg • CDN
[1976] *Action Theory. Proceedings of the Winnipeg Conference on Human Action* ED: BRAND, M. & WALTON, D. • SER (**S** 3307) Synth Libr 97 • PUBL (**X** 0835) Reidel: Dordrecht vi+345pp
• DAT&PL 1975 May; Winnipeg MB, CDN • LC-No 76-6882

P 2410 Measure Th & Appl;1982 Sherbrooke • CDN
[1983] *Measure Theory and Its Applications* ED: BELLEY, J.M. & DUBIOS, J. & MORALES, P. • SER (**S** 3301) Lect Notes Math 1033 • PUBL (**X** 0811) Springer: Heidelberg & New York xv+317pp
• DAT&PL 1982 Jun; Sherbrooke, Que, CDN • ISBN 3-540-12703-8, LC-No 83-20209

P 2411 Found Probab Th, Stat Inf & Stat Th Sci;1973 London ON • CDN
[1976] *Foundations of Probability Theory, Statistical Inference and Statistical Theories of Science* ED: HARPER, W.L. & HOOKER, C.A. • SER (**S** 3308) Univ Western Ontario Ser in Philos of Sci 6 • PUBL (**X** 0835) Reidel: Dordrecht 3 Vols: x+308pp,x+455pp,xii+241pp
• DAT&PL 1973 May; London,ON, CDN • ISBN 90-277-0616-6, ISBN 90-277-0617-4 (V1), ISBN 90-277-0618-2, ISBN 90-277-0619-0 (V2), ISBN 90-277-0620-4, ISBN 90-277-0621-2 (V3), LC-No 75-34354 (V1), LC-No 75-38667 (V2), LC-No 75-33879 (V3), ISBN 90-277-0614-X (V1-V3)
• REM Vol.1: Foundations and Philosophy of Epistemic Applications of Probability Theory. Vol.2: Foundations and Philosophy of Statistical Inference. Vol.3: Foundations and Philosophy of Statistical Theories in the Physical Sciences.

P 2539 Frege Konferenz (1);1979 Jena • DDR
[1979] *"Begriffsschrift". Jenaer Frege-Konferenz* ED: BOLCK, F. • PUBL (**X** 2211) Schiller Univ: Jena iii+548pp
• DAT&PL 1979 May; Jena, DDR

P 2552 Conf Finite Algeb & Multi-Val Log;1979 Szeged • H
[1981] *Proceedings of the Conference on Finite Algebra and Multiple-Valued Logic* ED: CSAKANY, B. & ROSENBERG, J.G. • SER (**S** 3312) Coll Math Soc Janos Bolyai 28 • PUBL (**X** 0809) North Holland: Amsterdam 880pp • ALT PUBL (**X** 3725) Bolyai Janos Mat Tars: Budapest
• DAT&PL 1979 Aug; Szeged, H • ISBN 0-444-85439-8, LC-No 81-214217

P 2554 All-Union Symp Th Log Infer;1974 Moskva • SU
[1979] *Logicheskij Vyvod. (Logical Inference. Proceedings of the All-Union Symposium on the Theory of Logical Inference.)* ED: SMIRNOV, V.A. • PUBL (**X** 2027) Nauka: Moskva 312pp
• DAT&PL 1974 Mar; Moskva, SU • LC-No 79-380844

P 2566 All-Union Symp Log & Method of Sci (7);1976 Kiev • SU
[1976] *(7th All-Union Symposium on Logic and Methodology of Science)*
• DAT&PL 1976 -?-; Kiev, SU

P 2570 Intens Log & Log Analiz Estestv Yazyk;1979 Moskva • SU
[1979] *2. Sovetsko-Finskij Kollokvium "Intensional'nye Logiki i Logicheskij Analiz Estestvennykh Yazykov" (Intensional Logics and Logic Analysis of Natural Languages)* PUBL (**X** 3353) Akad Nauk SSSR Inst Istor Filol & Filos: Novosibirsk
• DAT&PL 1979 -?-; Moskva, SU

P 2585 All-Union Conf Math Log (2);1972 Moskva • SU
[1972] *2 Vsesoyuznaya Konferentsiya po Matematicheskoj Logike. Tezitsy Doklady i Soobshcheniya (On Relative Recursiveness and Computability. 2nd All-Union Conference on Mathematical Logic)*
• DAT&PL 1972 -?-; Moskva, SU

P 2615 Scand Logic Symp (5);1979 Aalborg • DK
[1979] *Proceedings of the 5th Scandinavian Logic Symposium* ED: JENSEN, F.V. & MAYOH, B.H. & MOLLER, K.K. • PUBL (**X** 2646) Aalborg Univ Pr: Aalborg vii+361pp
• DAT&PL 1979 Jan; Aalborg, DK • ISBN 87-7307-037-8, LC-No 80-464603

P 2633 Autom Deduct (7);1984 Napa • USA
[1984] *7th International Conference on Automated Deduction* ED: SHOSTAK, R.E. • SER (**S** 3302) Lect Notes Comput Sci 170 • PUBL (**X** 0811) Springer: Heidelberg & New York vi+508pp
• DAT&PL 1984 May; Napa, CA, USA • ISBN 3-540-96022-8, LC-No 84-5441

P 2896 Algeb Method & Anwendgn in Automatenth;1976 Weissig • DDR
[1977] *Algebraische Methoden und ihre Anwendungen in der Automatentheorie* SER Vortraege zur Automatentheorie 21 • PUBL Weiterbildungszentrum fuer Mathematische Kybernetik und Rechentechnik, Technische Universitaet Dresden; 55pp
• DAT&PL 1976 Apr; Weissig, DDR • LC-No 78-379953

P 2901 Appl Sheaves;1977 Durham • GB
[1979] *Applications of Sheaves. Proceedings of the Research Symposium on Applications of Sheaf Theory to Logic, Algebra and Analysis* ED: FOURMAN, M.P. & MULVEY, C.J. & SCOTT, D.S. • SER (**S** 3301) Lect Notes Math 753 • PUBL (**X** 0811) Springer: Heidelberg & New York xiv+779pp
• DAT&PL 1977 Jul; Durham, GB • ISBN 3-540-09564-0, LC-No 79-23219

P 2903 Automata, Lang & Progr (8);1981 Akko • IL
[1981] *Automata, Languages and Programming. 8th Colloquium* ED: EVEN, S. & KARIV, O. • SER (**S** 3302) Lect Notes Comput Sci 115 • PUBL (**X** 0811) Springer: Heidelberg & New York viii+552pp
• DAT&PL 1981 Jul; Akko, IL • ISBN 3-540-10843-2, LC-No 81-9053
• REM Also Abbreviated as ICALP 81

P 2904 Automata, Lang & Progr (7);1980 Noordwijkerhout • NL
[1980] *Automata, Languages and Programming. 7th Colloquium* ED: BAKKER DE, J.W. & LEEUWEN VAN, J. • SER (**S** 3302) Lect Notes Comput Sci 85 • PUBL (**X** 0811) Springer: Heidelberg & New York viii+671pp
• DAT&PL 1980 Jul; Noordwijkerhout, NL • ISBN 3-540-10003-2, LC-No 81-156919
• REM Also Abbreviated as ICALP 80

P 2918 ACM Symp Princ Progr Lang (6);1979 San Antonio • USA
[1979] *Conference Record of the 6th Annual ACM Symposium on Principles of Programming Languages (Association for Computing Machinery)* PUBL (**X** 2205) ACM: New York iv+290pp
• DAT&PL 1979 Jan; San Antonio, TX, USA • LC-No 84-643240

P 2923 CAAP'81 Arbres en Algeb & Progr (6);1981 Genova • I
[1981] *CAAP '81. Les Arbres en Algebre et en Programmation. 6eme Colloque* ED: ASTESIANO, E. & BOEHM, C. • SER (S 3302) Lect Notes Comput Sci 112 • PUBL (**X** 0811) Springer: Heidelberg & New York vi+364pp
• DAT&PL 1981 Mar; Genova, I • ISBN 3-540-10828-9, LC-No 81-8959

P 2930 Formal of Progr Concepts;1981 Peniscola • E
[1981] *Formalization of Programming Concepts.* ED: DIAZ, J. & RAMOS, I. • SER (S 3302) Lect Notes Comput Sci 107 • PUBL (**X** 0811) Springer: Heidelberg & New York vii+478pp
• DAT&PL 1981 Apr; Peniscola, E • ISBN 3-540-10699-5, LC-No 81-5715

P 2931 Found & Appl of Decision Th;1975 London ON • CDN
[1978] *Foundations and Applications of Decision Theory. Papers from an International Workshop held at the University of Western Ontario* ED: HOOKER, C.A. & LEACH, J.J. & MCCLENNEN, E.F. • SER (S 3308) Univ Western Ontario Ser in Philos of Sci 13 • PUBL (**X** 0835) Reidel: Dordrecht 2 Vols: xxiii+446pp,xxiii+208pp
• DAT&PL 1975 May; London, ON, CDN • ISBN 90-277-0842-8 (V1), ISBN 90-277-0844-4 (V2), LC-No 77-25329
• REM Vol 1: Theoretical Foundations. Vol 2: Epistemic and Social Applications

P 2935 FCT'79 Fund of Comput Th;1979 Berlin/Wendisch-Rietz • DDR
[1979] *Fundamentals of Computation Theory - FCT '79. Proceedings of the Conference on Algebraic, Arithmetic, and Categorical Methods in Computation Theory* ED: BUDACH, L. • SER Mathematical Research - Mathematische Forschung 2 • PUBL (**X** 0911) Akademie Verlag: Berlin 576pp
• DAT&PL 1979 Sep; Berlin/Wendisch-Rietz, DDR • LC-No 82-460828

P 2936 Fuzzy Sets;1980 Durham • USA
[1980] *Fuzzy Sets. Theory and Applications to Policy Analysis and Information Systems* ED: WANG, PAUL P. & CHANG, S.K. • PUBL (**X** 1332) Plenum Publ: New York ix+413pp
• DAT&PL 1980 Jun; Durham, NC, USA • ISBN 0-306-40557-1, LC-No 80-19934

P 2943 Symp Math Found Comp Sci; 1982 Diedrichshagen • DDR
[1983] *Symposium on Mathematical Foundations of Computer Science* SER (S 3382) Sem-ber, Humboldt-Univ Berlin, Sekt Math 52 • PUBL (**X** 2219) Humboldt-Univ Berlin: Berlin iv+173pp
• DAT&PL 1982 Dec; Diedrichshagen, DDR • LC-No 84-214921

P 2949 Int Wittgenstein Symp (4);1979 Kirchberg • A
[1980] *Language, Logic, and Philosophy. Proceedings of the 4th International Wittgenstein Symposium* ED: HALLER, R. & GRASSL, W. • SER Schriftenreihe der Wittgensteingesellschaft 2 • PUBL (**X** 2728) Hoelder-Pichler-Tempsky: Wien 617pp
• DAT&PL 1979 Aug; Kirchberg am Wechsel, A • ISBN 3-209-00249-5, LC-No 81-107419

P 2958 Latin Amer Symp Math Log (4);1978 Santiago • RCH
[1980] *Mathematical Logic in Latin America. Proceedings of the 4th Latin American Symposium on Mathematical Logic* ED: ARRUDA, A.I. & CHUAQUI, R. & COSTA DA, N.C.A. • SER (S 3303) Stud Logic Found Math 99 • PUBL (**X** 0809) North Holland: Amsterdam xii+392pp
• DAT&PL 1978 Dec; Santiago, RCH • ISBN 0-444-85402-9, LC-No 79-20797

P 2964 Max Entropy Formalism;1978 Cambridge MA • USA
[1979] *Maximum Entropy Formalism. A Conference held at the Massachusetts Institute of Technology* ED: LEVINE, R.D. & TRIBUS, M. • PUBL (**X** 0865) MIT Pr: Cambridge, MA xiii+498pp
• DAT&PL 1978 May; Cambridge, MA, USA • ISBN 0-262-12080-1, LC-No 80-10799

P 2975 Asilomar Conf Circ, Syst & Comput (9);1975 Pacific Grove • USA
[1976] *9th Asilomar Conference on Circuits, Systems, and Computers* ED: CHAN, SHU PARK • PUBL (**X** 1777) Western Periodicals: Hollywood xiii+638pp
• DAT&PL 1975 Nov; Pacific Grove, CA, USA • LC-No 86-643237

P 2989 Log of Progr; 1983 Pittsburgh • USA
[1984] *Logics of Programs* ED: CLARKE, E. & KOZEN, D. • SER (S 3302) Lect Notes Comput Sci 164 • PUBL (**X** 0811) Springer: Heidelberg & New York vi+527pp
• DAT&PL 1983 Jan; Pittsburgh, PA, USA • ISBN 3-540-12896-4, LC-No 84-3123

P 2995 Manitoba Conf Num Math (5);1975 Winnipeg • CDN
[1976] *Proceedings of the 5th Manitoba Conference on Numerical Mathematics* ED: HARTNELL, B.L. & WILLIAMS, H.C. • SER (S 3726) Congressus Numerantium 16 • PUBL (**X** 2420) Utilitas Mathematica Publ: Winnipeg vi+658pp
• DAT&PL 1975 Oct; Winnipeg, MB, CDN • LC-No 77-378411

P 2999 Proc Conf Databasis (Calzone);1985 Heidelberg • D
[1985] *Proceedings of a Conference on Databases* ED: GSCHNITZER, W. & SPRENGER, H. & STEIN, H. & THURN, K. • PUBL Apl & Wemding Co: Mosbach 294pp
• DAT&PL 1985 Dec; Heidelberg, D

P 3001 Proc Inst Math;1974 Iasi • RO
[1976] *Proceedings of the Institute of Mathematics* ED: CARAMAN, P. • PUBL (**X** 0871) Acad Rep Soc Romania: Bucharest 306pp
• DAT&PL 1974 Jun; Iasi, RO

P 3003 Int Symp Multi-Val Log (9);1979 Bath • GB
[1979] *Proceedings of the 9th International Symposium on Multiple-Valued Logic* PUBL (**X** 2179) IEEE: New York v+303pp
• DAT&PL 1979 May; Bath, GB • LC-No 79-641110

P 3004 Manitoba Conf Num Math (6);1976 Winnipeg • CDN
[1977] *Proceedings of the 6th Manitoba Conference on Numerical Mathematics* ED: HARTNELL, B.L. & WILLIAMS, H.C. • SER (S 3726) Congressus Numerantium 18 • PUBL (**X** 2420) Utilitas Mathematica Publ: Winnipeg iii+504pp
• DAT&PL 1976 Oct; Winnipeg, MB, CDN • ISBN 0-919628-18-4

P 3006 Brazil Conf Math Log (3);1979 Recife • BR
[1980] *Proceedings of the 3rd Brazilian Conference on Mathematical Logic* ED: ARRUDA, A.I. & COSTA DA, N.C.A. & SETTE, A.M. • PUBL (**X** 2836) Soc Brasil Log: Sao Paulo vi+336pp
• DAT&PL 1979 Dec; Recife, BR

P 3016 Quant Th & Struct of Time & Space;1974 Feldafing • D
[1975] *Quantum Theory and the Structure of Time and Space* ED: CASTELL, L. & DRIESCHNER, M. & WEIZSAECKER VON, C.F. • PUBL (**X** 3223) Hanser: Muenchen 252pp
• DAT&PL 1974 Jul; Feldafing, D • ISBN 3-446-12075-0, LC-No 76-450053

P 3059 Algeb Method Automatenth;1978 Altenberg • DDR
[1978] *Vortraege aus dem Problemseminar: Algebraische Methoden der Automatentheorie* SER Vortraege zur Theorie der Automaten und Sprachen. Weiterbildungszentrum fuer Mathematische Kybernetik und Rechentechnik, Informationsverarbeitung 35 • PUBL Technische Universitaet Dresden:Dresden 106pp
• DAT&PL 1978 Apr; Altenberg, DDR

P 3062 IEEE Symp Switch & Automata Th (14);1973 Iowa City • USA
[1973] *14th Annual IEEE Symposium on Switching and Automata Theory* PUBL (**X** 2179) IEEE: New York v+213pp
• DAT&PL 1973 Oct; Iowa City, IA, USA • LC-No 80-646635

P 3081 IFAC Symp Fuzzy Inf, Knowl Repr & Decis. Anal;1983 Marseille • F
[1984] *Fuzzy Information, Knowledge Representation and Decision Analysis. Proceedings of the IFAC Symposium* ED: SANCHEZ, E. • SER IFAC Proc Ser 6 • PUBL (**X** 0869) Pergamon Pr: Oxford xiii+480pp
• DAT&PL 1983 Jul; Marseille, F • LC-No 83-10994

P 3092 Congr Naz Logica;1979 Montecatini Terme • I
[1981] *Atti del Congresso Nazionale di Logica* ED: BERNINI, S. • PUBL (**X** 1732) Bibliopolis: Napoli 735pp
• DAT&PL 1979 Oct; Montecatini Terme, I • LC-No 81-198713

P 3096 Conf Math Service of Man (2,Pap);1982 Las Palmas • E
[1984] *Aspects of Vagueness. Papers presented at the 2nd World Conference on Mathematics at the Service of Man* ED: SKALA, H.J. & TERMINI, S. & TRILLAS, E. • SER Theory & Decision Library 39 • PUBL (**X** 0835) Reidel: Dordrecht vii+304pp
• DAT&PL 1982 Jun; Las Palmas, E • ISBN 90-277-1692-7, LC-No 83-26994

P 3103 Adv Cybern Syst;1972 Oxford • GB
[1974] *Advances in Cybernetics and Systems* ED: ROSE, J. • PUBL (**X** 0836) Gordon & Breach: New York 3 Vols: xx+1730pp
• DAT&PL 1972 -?-; Oxford, GB • LC-No 75-316901

P 3108 AISB/GI Conf Artif Intel;1978 Hamburg • D
[1979] *Proceedings of the AISB/GI Conference on Artificial Intelligence* ED: SLEEMAN, D. & NAGEL, H.-H. • PUBL SSAISB and GI: Hamburg x+379pp
• DAT&PL 1978 Jul; Hamburg, D
• REM The Society for the Study of Artificial Intelligence and Simulation of Behaviour (SSAISB) and Gesellschaft fuer Informatik (GI)

P 3113 Aspects Philos Logic;1977 Tuebingen • D
[1981] *Aspects of Philosophical Logic* ED: MOENNICH, U. • SER (S 3307) Synth Libr 147 • PUBL (**X** 0835) Reidel: Dordrecht vii+283pp
• DAT&PL 1977 Dec; Tuebingen, D • ISBN 90-277-1201-8, LC-No 81-7358

P 3114 Conv Geom Combin & Appl;1970 Perugia • I
[1971] *Atti del Convegno di Geometria Combinatoria e sue Applicazioni* PUBL Universita degli Studi di Perugia, Istituto di Matematica:Perugia; iii+432pp
• DAT&PL 1970 Sep; Perugia, I

P 3139 Classific Autom & Percept par Ordin;1978/79 Rocquencourt • F
[1980] *Classification Automatique et Perception par Ordinateur* ED: DIDAY, E. & LECHEVALLIER, Y. • SER Seminaires IRIA • PUBL (**X** 2732) INRIA: Le Chesnay Cedex 345pp
• DAT&PL 1978 Oct - 1979 Jun; Rocquencourt, F • ISBN 2-7261-0238-7

P 3143 Comput Orient Learn Process;1974 Bonas • F
[1976] *Computer Oriented Learning Processes* ED: SIMON, J.C. • SER NATO Advanced Study Institutes Series, Series E: Applied Science; 14 • PUBL (**X** 1317) Noordhoff: Groningen vi+595pp
• DAT&PL 1974 Aug; Bonas, F • LC-NO 76-378425

P 3146 Constr Math;1980 Las Cruces • USA
[1981] *Constructive Mathematics. Proceedings of the New Mexico State University Conference* ED: RICHMAN, F. • SER (S 3301) Lect Notes Math 873 • PUBL (**X** 0811) Springer: Heidelberg & New York vii+347pp
• DAT&PL 1980 Aug; Las Cruces, NM, USA • ISBN 3-540-10850-5, LC-No 81-9345

P 3165 FCT'81 Fund of Comput Th;1981 Szeged • H
[1981] *Fundamentals of Computation Theory. Proceedings of the 1981 International FCT-Conference* ED: GECSEG, F. • SER (S 3302) Lect Notes Comput Sci 117 • PUBL (**X** 0811) Springer: Heidelberg & New York xi+471pp
• DAT&PL 1981 Aug; Szeged, H • ISBN 3-540-10854-8, LC-No 81-13533

P 3167 Fuzzy Set Theory (2);1980 Linz • A
[1981] *Fuzzy Set Theory. Proceedings of the 2nd International Seminar* ED: KLEMENT, E.P. • PUBL (**X** 2767) J Kepler Univ: Linz i+188pp
• DAT&PL 1980 Sep; Linz, A

P 3168 Fuzzy Set Theory (3);1981 Linz • A
[1981] *Fuzzy Set Theory. Proceedings of the 3rd International Seminar* ED: KLEMENT, E.P. • PUBL (**X** 2767) J Kepler Univ: Linz i+251pp
• DAT&PL 1981 Sep; Linz, A

P 3185 Interpr & Found of Quantum Th;1979 Marburg • D
[1981] *Interpretations and Foundations of Quantum Theory* ED: NEUMANN, H. • SER Grundlagen der Exakten Naturwissenschaften 5 • PUBL (**X** 0876) Bibl Inst: Mannheim 144pp
• DAT&PL 1979 May; Marburg, D • ISBN 3-411-01601-9, LC-No 81-150650

P 3210 Math Founds of Comput Sci (9);1980 Rydzyna • PL
[1980] *Mathematical Foundations of Computer Science. Proceedings of the 9th Symposium* ED: DEMBINSKI, P. • SER (S 3302) Lect Notes Comput Sci 88 • PUBL (**X** 0811) Springer: Heidelberg & New York viii + 723pp
• DAT&PL 1980 Sep; Rydzyna, PL • ISBN 3-540-10027-X, LC-No 80-20087

P 3234 Probl Founds of Physics;1977 Varenna • I
[1979] *Problems in the Foundations of Physics. Proceedings of the International School of Physics "Enrico Fermi". Course 72* ED: TORALDO DI FRANCIA, G. • PUBL (**X** 0809) North Holland: Amsterdam xii + 497pp
• DAT&PL 1977 Jul; Varenna, I • ISBN 0-444-85285-9, LC-No 79-13069

P 3237 Colloq Universal Algeb;1977 Esztergom • H
[1982] *Proceedings of the Colloque on Universal Algebra* ED: CSAKANY, B. & FRIED, E. & SCHMIDT, E.T. • SER (S 3312) Coll Math Soc Janos Bolyai 29 • PUBL (**X** 0809) North Holland: Amsterdam 804pp
• DAT&PL 1977 Jun; Esztergom, H • ISBN 0-444-85405-3, LC-No 79-350398

P 3244 Hawaii Int Conf Syst Sci (6);1973 Honolulu • USA
[1973] *Proceedings of the 6th Hawaii International Conference on System Sciences* ED: LEW, A. • PUBL (**X** 1777) Western Periodicals: Hollywood xx + 533pp
• DAT&PL 1973 Jan; Honolulu, HI, USA • LC-No 72-180444

P 3246 Progr Constr;1978 Marktoberdorf • D
[1979] *Program Construction. International Summer School* ED: BAUER, F.L. & BROY, M. • SER (S 3302) Lect Notes Comput Sci 69 • PUBL (**X** 0811) Springer: Heidelberg & New York vii + 651pp
• DAT&PL 1978 Jul; Marktoberdorf, D • ISBN 3-540-09251-X, LC-No 79-13704

P 3269 Set Th Found Math (Kurepa);1977 Beograd • YU
[1977] *Set Theory. Foundations of Mathematics* SER (**J** 2887) Zbor Radova, NS 2(10) • PUBL (**X** 3727) Beograd Mat Inst: Belgrade 152pp
• DAT&PL 1977 Aug; Beograd, YU • LC-No 79-373865

P 3299 Progr Kiso Riron, Algor Okeru Shomei Ron;1973/74 Kyoto • J
[1975] *"Program no Kiso Riron" Kenkyu Shugo Oyobi Tanki Kyodo Kenkyu "Algorithm ni Okeru Shomei Ron". Hokoku Shu (The 3rd Symposium on Basic Theory of Programs and the Workshop for Proof Theory about Algorithms. Proceedings)* ED: IGARASHI, S. • SER (S 2308) Symp Kyoto Univ Res Inst Math Sci (RIMS) 236 • PUBL (**X** 2441) Kyoto Univ Res Inst Math Sci: Kyoto 215pp
• DAT&PL 1973 May; Kyoto, J, 1974 Nov; Kyoto, J

P 3355 Algeb Conf (1);1980 Skopje • YU
[1980] *Algebraic Conference '80* ED: CELAKOSKI, N. • PUBL (**X** 3760) Univ Kiril et Metodij, Mat Fak: Skopje vi + 152pp
• DAT&PL 1980 Feb; Skopje, YU • LC-No 82-205527

P 3367 Proc Comput Linguistics;1980 Tokyo • J
[1980] *COLING 80. Computational Linguistics. Proceedings of the 8th International Conference* ED: WADA, H. • PUBL Internat Comittee on Computational Linguistics ix + 616pp
• DAT&PL 1980 Sep; Tokyo, J • LC-No 81-176861

P 3373 Form Meth in Stud of Lang;1980 Amsterdam • NL
[1981] *Formal Methods in the Study of Language. Part 1, 2. Proceedings of the 3rd Amsterdam Colloquium* ED: GROENENDIJK, J.A.G. & JANSSEN, T.M.V. & STOKHOF, M.B.J. • SER (S 1605) Math Centr Tracts 135, 136 • PUBL (**X** 1121) Math Centr: Amsterdam 2 Vols: v + 599pp
• DAT&PL 1980 Mar; Amsterdam, NL • ISBN 90-6196-211-0 (V1), ISBN 90-6196-213-7 (V2)

P 3375 Found of Comput Sci (3);1978 Amsterdam • NL
[1979] *Foundations of Computer Science. III. Part 1, 2* ED: BAKKER DE, J.W. & LEEUWEN VAN, J. • SER (S 1605) Math Centr Tracts 108,109 • PUBL (**X** 1121) Math Centr: Amsterdam 2 Vol: iii + 112pp, i + 164pp
• DAT&PL 1978 Aug; Amsterdam, NL • ISBN 90-6196-176-9 (V 1), ISBN 90-6196-177-7 (V 2)
• REM Vol 1: Automata, Data Structures, Complexity. Vol 2: Languages, Logic, Semantics.

P 3406 Congr Cybern & Syst (3);1975 Bucharest • RO
[1977] *Modern Trends in Cybernetics and Systems. Vol 1-3. Proceeedings of the 3rd International Congress of Cybernetics and Systems* ED: ROSE, J. & BILCIU, C. • PUBL (**X** 0811) Springer: Heidelberg & New York 3 Vols: 3326pp
• DAT&PL 1975 Aug; Bucharest, RO • ISBN 3-540-08199-2
• REM Vol 1: Proceedings of Official Meetings, Symposia and Section 1

P 3422 Oper Res DGOR Jahrestag (8);1978 Berlin • D
[1979] *Proceedings in Operations Research 8. Vortraege der Jahrestagung 1978 DGOR (= Deutsche Gesellschaft fuer Operartions Research)* ED: GAEDE, K.-W. & PRESSMAR, D.B. & SCHNEEWEISS, C. & SCHUSTER, K.P. & SEIFERT, O. • PUBL (**X** 0870) Physica: Wuerzburg 700pp
• DAT&PL 1978 Oct; Berlin, D • ISBN 3-7908-0212-3, LC-No 78-641068

P 3428 Prog in Cybern & Syst Res;1978 Wien • A
[1978] *Progress in Cybernetics and Systems Research. Vol. 3: General Systems Methodology, Fuzzy Mathematics and Fuzzy Systems, Biocybernetics and Theoretical Neurobiology.* ED: TRAPPL, R. & KLIR, G.J. & RICCIARDI, L. • PUBL (**X** 0827) Wiley & Sons: New York xiv + 674pp • ALT PUBL (**X** 2437) Hemisphere Publ: Washington
• DAT&PL 1978 -?-; Wien, A • LC-No 75-6641
• REM At least 4 Vols

P 3429 Math Founds of Comput Sci (10);1981 Strbske Pleso • CS
[1981] *Mathematical Foundations of Computer Science 1981. Proceedings of the 10th Symposium* ED: GRUSKA, J. & CHYTIL, M. • SER (S 3302) Lect Notes Comput Sci 118 • PUBL (**X** 0811) Springer: Heidelberg & New York xi + 589pp
• DAT&PL 1981 Aug; Strbske Pleso, CS • ISBN 3-540-10856-4, LC-No 81-9302

P 3432 Prog in Cybern & Syst Res;1972 Wien • A
[1979] *Progress in Cybernetics and Systems Research. Vol. 5: Organization and Management, Organic Problem-Solving in Management, System Approach in Urban and Regional Planning, Computer Performance, Control and Evaluation, Computer Linguistics.* ED: TRAPPL, R. & HANIKA, F. DE P. & PICHLER, F.R. • PUBL (**X** 0827) Wiley & Sons: New York xv+683pp • ALT PUBL (**X** 2437) Hemisphere Publ: Washington
• DAT&PL 1972 -?-; Wien, A • ISBN 0-470-26553-1, LC-No 75-6641

P 3474 ACM Symp Th of Comput (13);1981 • USA
[1981] *Proceedings of the 13th ACM Symposium on Theory of Computing (Association for Computing Machinery)* PUBL (**X** 2205) ACM: New York
• LC-No 82-645581

P 3487 IFAC World Congr (7);1978 Helsinki • SF
[1979] *A Link between Science and Applications of Automatic Control. Proceedings of the 7th Triennial World Congress of the International Federation of Automatic Control (IFAC)* ED: NIEMI, A. & WAHLSTROEM, B. & VIRKKUNEN, J. • PUBL (**X** 0869) Pergamon Pr: Oxford 4 Vols: xlix+2667pp
• DAT&PL 1978 Jun; Helsinki, SF

P 3497 Log of Progr;1979 Zuerich • CH
[1981] *Logic of Programs* ED: ENGELER, E. • SER (**S** 3302) Lect Notes Comput Sci 125 • PUBL (**X** 0811) Springer: Heidelberg & New York v+245pp
• DAT&PL 1979 May; Zuerich, CH • ISBN 3-540-11160-3, LC-No 82-137449

P 3512 Jorn Mat Luso-Espanol (7);1980 St Feliu de Guixois • E
[1980] *Actas de las Septimas Jornadas Luso-Espanolas de Matematica* SER (**J** 2562) Publ Sec Mat Univ Autonoma Barcelona 20,21,22 • PUBL Univ. Autonoma de Barcelona: Barcelona, E 276pp,263pp,329pp
• DAT&PL 1980 May; Sant Feliu de Guixois, E

P 3516 Proc Bienn Meet Phil of Sci Ass;1974 Notre Dame • USA
[1976] *PSA 1974. Proceedings of the 1974 Biennial Meeting of the Philosophy of Science Association* ED: COHEN, R.S. & HOOKER, C.A. & MICHALOS, A.C. & EVRA VAN, J.W. • SER (**S** 3311) Boston St Philos Sci 32, (**S** 3307) Synth Libr 101
• PUBL (**X** 0835) Reidel: Dordrecht xiii+739pp
• DAT&PL 1974 -?-; Notre Dame, IN, USA • LC-No 72-624169

P 3535 IEEE Symp Founds of Comput Sci (20);1979 San Juan • PRI
[1979] *20th Annual IEEE Symposium on Foundations of Computer Science* PUBL (**X** 2179) IEEE: New York vii+431pp
• DAT&PL 1979 Oct; San Juan, PRI • LC-No 80-646634
• REM IEEE Publication No. 79CH1471-26

P 3538 Artif Intell & Inf-Control Syst Robot (3);1984 Smolenice • CS
[1984] *Artificial Intelligence and Information-Control Systems of Robots. Proceedings of the 3rd International Conference* ED: PLANDER, I. • PUBL (**X** 0809) North Holland: Amsterdam xvi+413pp
• DAT&PL 1984 Jun; Smolenice, CS • LC-No 84-8139

P 3539 ACM Symp Princ Progr Lang (5);1978 Tucson • USA
[1978] *Conference Record of the 5th Annual ACM Symposium on Principles of Programming Languages (Association for Computing Machinery)* PUBL (**X** 2205) ACM: New York iv+264pp
• DAT&PL 1978 Jan; Tucson, AZ, USA

P 3542 ACM Symp Th of Comput (11);1979 Atlanta • USA
[1979] *Conference Record of the 11th Annual ACM Symposium on Theory of Computing (Association for Computing Machinery)* PUBL (**X** 2205) ACM: New York vii+368pp
• DAT&PL 1979 Apr; Atlanta, GA, USA • ISBN 0-89791-003-6, LC-No 82-642181

P 3565 Symp of Th Aspects of Comput Sci (1);1984 Paris • F
[1984] *STACS 84. Symposium of Theoretical Aspects of Computer Science by AFCET (Association Francaise pour la Cybernetique Economique et Technique) and GI (Gesellschaft fuer Informatik)* ED: FONTET, M. & MEHLHORN, K. • SER (**S** 3302) Lect Notes Comput Sci 166 • PUBL (**X** 0811) Springer: Heidelberg & New York vi+338pp
• DAT&PL 1984 Apr; Paris, F • ISBN 3-540-12920-0, LC-No 84-5299

P 3572 IEEE Symp Found of Comput Sci (18);1977 Providence • USA
[1977] *18th Annual IEEE Symposium on Foundations of Computer Science* PUBL (**X** 2179) IEEE: New York v+269pp
• DAT&PL 1977 Oct; Providence, RI, USA • LC-No 80-646634

P 3580 Repr des Conn & Raison dans Sci Homme;1979 St Maximin • F
[1980] *Representation des Connaissances et Raisonnement dans les Sciences de l'Homme. Papers from the IRIA-LISH Colloquium* ED: BORILLO, M. • PUBL (**X** 2732) INRIA: Le Chesnay Cedex iii+607pp
• DAT&PL 1979 Sep; Saint Maximin, F • ISBN 2-7261-0234-4

P 3584 Joint Sess Aristot Soc & Mind Assoc;1975 Canterbury • GB
[1975] *Symposia Read at the Joint Session of the Aristotelian Society and Mind Association* SER (**J** 2119) Suppl Aristotelian Soc 49 • PUBL (**X** 0816) Methuen: London & New York v+247pp
• DAT&PL 1975 Jul; Canterbury, GB • LC-No 76-373816

P 3621 Frege Konferenz (2);1984 Schwerin • DDR
[1984] *Frege Konferenz* ED: WECHSUNG, G. • SER Mathematical Research - Mathematische Forschungen 20
• PUBL (**X** 0911) Akademie Verlag: Berlin 408pp
• DAT&PL 1984 Sep; Schwerin, DDR • LC-No 84-248695

P 3622 Int Congr Log, Meth & Phil of Sci (6,Proc);1979 Hannover • D
[1982] *Proceedings of the 6th International Congress for Logic, Methodology and Philosophy of Science* ED: COHEN, L.J. & LOS, J. & PFEIFFER, H. & PODEWSKI, K.-P. • SER (**S** 3303) Stud Logic Found Math 104 • PUBL (**X** 0809) North Holland: Amsterdam xiv+842pp
• DAT&PL 1979 Aug; Hannover, D • LC-No 80-12713

P 3642 Colloq Math Log in Computer Sci;1978 Salgotarjan • H
[1981] *Proceedings of the Colloquium on Mathematical Logic in Computer Science* ED: DOEMALKI, B. & GERGELY, T. • SER (S 3312) Coll Math Soc Janos Bolyai 26 • PUBL (X 0809) North Holland: Amsterdam 758pp • ALT PUBL (X 3725) Bolyai Janos Mat Tars: Budapest
• DAT&PL 1978 Sep; Salgotarjan, H • ISBN 0-444-85440-1, LC-No 82-101557

P 3653 Modal du Jugement Chez Aristote & Log Moderne;1969 Brasov • RO
[1969] *La Modalite du Jugement chez Aristote et dans la Logique Moderne. Compte-Rendu des Travaux du Colloque* SER (J 0147) An Univ Bucuresti, Acta Logica 12 • PUBL Univ Bucuresti: Bucharest
• DAT&PL 1969 Aug; Brasov, RO

P 3654 Mat Model Slonykh Sist;1969 Brasov • RO
[1973] *Matematicheskie Modeli Slonykh Sistem (Mathematical Models of Complex Systems)* PUBL (X 2522) Akad Nauk Inst Kibernet: Kiev 214pp
• DAT&PL 1969 Aug; Brasov, RO

P 3658 Math Founds of Comput Sci (11);1984 Prague • CS
[1984] *Mathematical Foundations of Computer Science 1984. Proceedings of the 11th Symposium* ED: CHYTIL, M.P. & KOUBEK, V. • SER (S 3302) Lect Notes Comput Sci 176
• PUBL (X 0811) Springer: Heidelberg & New York 581pp
• DAT&PL 1984 Sep; Prague, CS • ISBN 3-540-13372-0, LC-No 83-13980

P 3669 SE Asian Conf on Log;1981 Singapore • SGP
[1983] *Southeast Asian Conference on Logic* ED: CHONG, CHI TAT & WICKS, M.J. • SER (S 3303) Stud Logic Found Math 111 • PUBL (X 0809) North Holland: Amsterdam xiv+210pp
• DAT&PL 1981 Nov; Singapore, SGP • ISBN 0-444-86706-6, LC-No 83-11458

P 3673 Int Symp Multi-Val Log (10);1980 Evanston • USA
[1980] *Proceeding of the 10th International Symposium on Multiple-Valued Logic* PUBL (X 2179) IEEE: New York iv+281pp
• DAT&PL 1980 May; Evanston, IL, USA • LC-No 79-641110

P 3693 Int Wittgenstein Symp (2);1977 Kirchberg • A
[1978] *Wittgenstein and His Impact on Contemporary Thought. Proceeedings of the 2nd International Wittgenstein Symposium in Kirchberg* PUBL (X 2728) Hoelder-Pichler-Tempsky: Wien 550pp
• DAT&PL 1977 Aug; Kirchberg, A • LC-No 79-303201

P 3705 Int Symp Multi-Val Log (11);1981 Oklahoma City & Norman • USA
[1981] *Proceedings of the 11th International Symposium on Multiple-Valued Logic.* PUBL (X 2179) IEEE: New York vi+298pp
• DAT&PL 1981 May; Oklahoma City, OK, USA & Norman, OK, USA • LC-No 79-641110

P 3715 Symp Princ of Distrib Computing • USA
[1982] *Symposium on Principles of Distributed Computing* PUBL (X 2205) ACM: New York

P 3738 Log of Progr;1981 Yorktown Heights • USA
[1982] *Logics of Programs. Papers Presented at the Workshop* ED: KOZEN, D. • SER (S 3302) Lect Notes Comput Sci 131
• PUBL (X 0811) Springer: Heidelberg & New York vi+429pp
• DAT&PL 1981 May; Yorktown Heights, NY, USA • ISBN 3-540-11212-X, LC-No 82-3219

P 3739 Phil & Grammar;1978 Uppsala • S
[1981] *Philosophy and Grammar. Papers from the Symposium on the Occasion of the Quincentennial of Uppsala University* ED: KANGER, S. & OEHMAN, S. • SER (S 3307) Synth Libr 143 • PUBL (X 0835) Reidel: Dordrecht ii+158pp
• DAT&PL 1978 -?-; Uppsala, S • ISBN 90-277-1091-0, LC-No 80-15692

P 3754 Argumentation;1978 Groningen • NL
[1982] *Argumentation. Approaches to Theory Formation. Papers from the Groningen Conference on the Theory of Argumentation* ED: BARTH, E.M. & MARTENS, J.L. • SER Studies in Language Companion Series (SLCS) 8 • PUBL (X 2257) Benjamins: Amsterdam xvii+333pp
• DAT&PL 1978 Oct; Groningen, NL • ISBN 90-272-3007-2

P 3759 Phil of Econ;1981 Muenchen • D
[1982] *Philosophy of Economics* ED: STEGMUELLER, W. & BALZER, W. & SPOHN, W. • SER Studies in Contemporary Economics 2 • PUBL (X 0811) Springer: Heidelberg & New York viii+306pp
• DAT&PL 1981 Jul; Muenchen, D • ISBN 3-540-11927-2, LC-No 82-16947

P 3764 Sci Discover, Log & Ration (Leonard);1978 Reno • USA
[1980] *Scientific Discovery, Logic, and Rationality. Paper Presented at the 1st Guy L. Leonard Memorial Conference in Philosophy* ED: NICKLES, T. • SER (S 3311) Boston St Philos Sci 56 • PUBL (X 0835) Reidel: Dordrecht xii+385pp
• DAT&PL 1978 Oct; Reno, NV, USA • ISBN 90-277-1069-4, ISBN 90-277-1070-8, LC-No 80-11548

P 3787 Discr Math;1977 Warsaw • PL
[1982] *Discrete Mathematics* ED: KULIKOWSKI, J.L. & YABLONSKIJ, S.V. & ZHURAVLEV, JU.I. & MICHALEWICZ, M.
• SER (S 1926) Banach Cent Publ 7 • PUBL (X 1034) PWN: Warsaw 224pp
• DAT&PL 1977 -?-; Warsaw, PL • ISBN 83-01-01494-6, LC-No 82-207902

P 3800 Intens Log: Th & Appl;1979 Moskva • SU
[1982] *Intensional Logic: Theory and Applications. Proceedings of the 2nd Soviet-Finnish Logic Conference* ED: NIINILUOTO, I. & SAARINEN, E. • SER (J 0096) Acta Philos Fenn 35 • PUBL (X 0809) North Holland: Amsterdam 301pp • ALT PUBL (X 2812) Philos Soc Finland : Helsinki
• DAT&PL 1979 Dec; Moskva, SU • ISBN 951-95054-8-2

P 3814 Int Conf Collective Phenomena (4);1981 Moskva • SU
[1981] *4th International Conference on Collective Phenomena* ED: LEBOWITZ, J.L. • SER (J 1377) Ann New York Acad Sci 373 • PUBL (X 2241) New York Acad Sci: New York ix+233pp
• DAT&PL 1981 Apr; Moskva, SU • ISBN 0-89766-135-4, LC-No 81-16792

P 3820 Curr Iss in Quantum Log;1979 Erice • I
[1981] *Current Issues in Quantum Logic* ED: BELTRAMETTI, E.G. & FRAASSEN VAN, B.C. • SER Ettore Majorana Internat. Sci. Ser.: Physical Sci. 8 • PUBL (X 1332) Plenum Publ: New York ix+492pp
• DAT&PL 1979 Dec; Erice, I • ISBN 0-306-40652-7, LC-No 80-29505

P 3821 Mat & Mat Obrazov (12);1983 Sl"nchev Bryag • BG
[1983] *Matematika i Matematichesko Obrazovanie (Mathematics and Mathematical Education. Proceedings of the 12th Spring Conference of the Union of Bulgarian Mathematicians)* ED: GEROV, G. • PUBL (**X** 2237) Publ Bulg Acad Sci: Sofia 393pp
• DAT&PL 1983 Apr:Sl"nchev Bryag (Sunny Beach), BG
• LC-No 80-643123

P 3829 Atti Incontri Log Mat (1);1982 Siena • I
[1983] *Atti Degli Incontri di Logica Matematica* ED: BERNARDI, C. • PUBL (**X** 3812) Univ Siena, Dip Mat: Siena 398pp
• DAT&PL 1982 Jan; Siena, I, 1982 Apr; Siena, I, 1982 Jun; Siena, I

P 3830 Logics of Progr & Appl;1980 Poznan • PL
[1983] *Logics of Programs and Their Applications* ED: SALWICKI, A. • SER (**S** 3302) Lect Notes Comput Sci 148
• PUBL (**X** 0811) Springer: Heidelberg & New York vi + 324pp
• DAT&PL 1980 Aug; Poznan, PL • ISBN 3-540-11981-7, LC-No 83-158934

P 3831 Universal Algeb & Appl;1978 Warsaw • PL
[1982] *Universal Algebra and Applications* ED: TRACZYK, T.
• SER (**S** 1926) Banach Cent Publ 9 • PUBL (**X** 1034) PWN: Warsaw 454pp
• DAT&PL 1978 Feb-Sep; Warsaw, PL • ISBN 83-01-02145-4, LC-No 83-167533

P 3833 Symp Semigroups (6);1982 Kyoto • J
[1982] *Semigroup Theory and Its Related Fields. Proceedings of the 6th Symposium on Semigroups* ED: YOSHIDA, R. • PUBL Osaka College of Pharmacy ii + 79pp • ALT PUBL Ritsumeikan Univ: Kyoto
• DAT&PL 1982 Oct; Kyoto, J

P 3836 Automata, Lang & Progr (9);1982 Aarhus • DK
[1982] *Automata, Languages and Programming. 9th Colloquium* ED: NIELSEN, M. & SCHMIDT, E.M. • SER (**S** 3302) Lect Notes Comput Sci 140 • PUBL (**X** 0811) Springer: Heidelberg & New York vii + 614pp
• DAT&PL 1982 Jul; Aarhus, DK • ISBN 3-540-11576-5, LC-No 83-10430
• REM Also Abbreviated as ICALP '82

P 3840 Autom Deduct (6);1982 New York • USA
[1982] *6th Conference on Automated Deduction* ED: LOVELAND, D.W. • SER (**S** 3302) Lect Notes Comput Sci 138
• PUBL (**X** 0811) Springer: Heidelberg & New York vii + 389pp
• DAT&PL 1982 Jun; New York, NY, USA • ISBN 3-540-11558-7, LC-No 82-5948

P 3844 IMA Conf Control Th (3);1980 Sheffield • GB
[1981] *3rd IMA Conference on Control Theory. The Institute of Mathematics and Its Applications (IMA)* ED: MARSHALL, J.E. & COLLINS, W.D. & HARRIS, C.J. & OWENS, D.H. • PUBL (**X** 0801) Academic Pr: New York xii + 922pp
• DAT&PL 1980 Sep; Sheffield, GB • ISBN 0-12-473960-1, LC-No 81-67923

P 3845 Conf Math Service of Man (2,Proc)(Feriet);1982 Las Palmas • E
[1982] *2nd World Conference on Mathematics at the Service of Man. Dedicated to the Remembrance of Joseph Kampe de Feriet (1893-1982)* ED: BALLESTER, A. & CARDUS, D. & TRILLAS, E. • PUBL Universidad Politecnica: Las Palmas xxx + 696pp
• DAT&PL 1982 Jun; Las Palmas, E

P 3846 Cybern & Systems Research (6);1982 Wien • A
[1982] *Cybernetics and Systems Research. Proceedings of the 6th European Meeting on Cybernetics and Systems Research* ED: TRAPPL, R. • PUBL (**X** 0809) North Holland: Amsterdam xviii + 986pp
• DAT&PL 1982 Apr; Wien, A • LC-No 82-14288

P 3851 Automata, Lang & Progr (10);1983 Barcelona • E
[1983] *Automata, Languages and Programming. 10th Colloquium* ED: DIAZ, J. • SER (**S** 3302) Lect Notes Comput Sci 154 • PUBL (**X** 0811) Springer: Heidelberg & New York viii + 734pp
• DAT&PL 1983 Jul; Barcelona, E • ISBN 3-540-12317-2, LC-No 83-10435
• REM Also Abbreviated as ICALP 83

P 3858 Adequate Modeling of Syst;1982 Bad Honnef • D
[1983] *Adequate Modeling of Systems. Proceedings of the International Working Conference on Model Realism* ED: WEDDE, H. • PUBL (**X** 0811) Springer: Heidelberg & New York xi + 336pp
• DAT&PL 1982 Apr; Bad Honnef, D • ISBN 3-540-12567-1, LC-No 83-10335

P 3859 Th & Pract Multiple Criteria Decis Making;1981 Moskva • SU
[1983] *Theory and Practice of Multiple Criteria Decision Making* ED: CARLSSON, C. & KOCHETKOV, Y. • PUBL (**X** 0809) North Holland: Amsterdam ix + 170pp
• DAT&PL 1981 May; Moskva, SU • LC-No 82-24873

P 3860 Coll Papers to Farah on Retirement;1981 Sao Paulo • BR
[1982] *Collected Papers Dedicated to Professor Edison Farah on the Occasion of His Retirement* ED: ALAS, O.T. & COSTA DA, N.C.A. & HOENIG, C.S. • PUBL Inst Mat Estat Univ Sao Paulo: Sao Paulo v + 134pp
• DAT&PL 1981 May; Sao Paulo, BR

P 3864 FCT'83 Found of Comput Th;1983 Borgholm • S
[1983] *Foundation of Computation Theory. 1983 FCT-Conference* ED: KARPINSKI, M. • SER (**S** 3302) Lect Notes Comput Sci 158 • PUBL (**X** 0811) Springer: Heidelberg & New York xi + 517pp
• DAT&PL 1983 Aug; Borgholm, S • ISBN 3-540-12689-9, LC-No 83-16736

P 3870 Congr Catala de Log Mat (1);1982 Barcelona • E
[1982] *1er Congres Catala de Logica Matematica. Actes* PUBL Univ Politecnica & Univ Barcelona: Barcelona 130pp
• DAT&PL 1982 -?-; Barcelona, E

P 3876 Mat & Mat Obrazov (11);1982 Sl"nchev Bryag • BG
[1982] *Matematika i Matematichesko Obrazovanie (Mathematics and Mathematical Education. Proceedings of the 11th Spring Conference 1979-81)* PUBL (**X** 2237) Publ Bulg Acad Sci: Sofia 596pp
• DAT&PL 1982 Apr; Sl"nchev Bryag (Sunny Beach), BG
• LC-No 80-643123, LC-No 82-211699

P 3906 Th Found of Progr Methodol;1981 Marktoberdorf • D
[1982] *Theoretical Foundations of Programming Methodology* ED: BROY, M. & SCHMIDT, G. • SER NATO Adv Study Inst Ser C 91 • PUBL (**X** 0835) Reidel: Dordrecht xiii + 658pp
• DAT&PL 1981 -?-; Marktoberdorf, D • LC-No 82-12347

P 3908 Graph-Gram & Appl to Comput Sci (2);1982 Neunkirchen • D
[1983] *Graph-Grammars and Their Application to Computer Science.* ED: EHRIG, H. & NAGL, M. & ROZENBERG, G. • SER (**S** 3302) Lect Notes Comput Sci 153 • PUBL (**X** 0811) Springer: Heidelberg & New York vii + 452pp
• DAT&PL 1982 Oct; Neunkirchen, D • ISBN 3-540-12310-5, LC-No 83-6677

P 4001 Int Conf Collective Phenomena (5);1981 Moskva • SU
[1983] *5th International Conference on Collective Phenomena* ED: LEBOWITZ, J.L. • SER (**J** 1377) Ann New York Acad Sci 410 • PUBL (**X** 2241) New York Acad Sci: New York x + 365pp
• DAT&PL 1981 -?-; Moskva, SU • LC-No 83-13175;, ISBN 0-89766-213-X

P 4002 Int Symp Multi-Val Log (12);1982 Paris • F
[1982] *Proceedings of the 12th International Symposium on Multiple-Valued Logic* PUBL (**X** 2179) IEEE: New York x + 341pp
• DAT&PL 1982 May; Paris, F • LC-No 79-641110

P 4008 Cybern & Systems Research (7);1984 Wien • A
[1984] *Cybernetics and Systems Research. Proceedings of the Seventh European Meeting on Cybernetics and Systems Research* PUBL (**X** 0809) North Holland: Amsterdam xiv + 840pp
• DAT&PL 1984 Apr; Wien, A • LC-No 84-4073

P 4010 Found of Softw Tech & Th Comput Sci (4);1984 Bangalore • IND
[1984] *Foundations of Software Technology and Theoretical Computer Science. Proceedings of the 4th Conference* ED: JOSEPH, M. & SHYAMASUNDAR, R. • SER (**S** 3302) Lect Notes Comput Sci 181 • PUBL (**X** 0811) Springer: Heidelberg & New York viii + 468pp
• DAT&PL 1984 Dec; Bangalore, IND • ISBN 3-540-13883-8, LC-No 84-23528

P 4041 Log, Inform, Law;1981 Firenze • I
[1982] *Selected Papers from the International Conference on Logic, Informatics, Law.* ED: CIAMPI, C. (VOL 1) & MARTINO, A.A. (VOL 2) • PUBL (**X** 0809) North Holland: Amsterdam 2 Vols: xii + 476pp, xii + 518pp
• DAT&PL 1981 Apr; Firenze, I • ISBN 0-444-86414-8 (V1), ISBN 0-444-86415-6 (V2), LC-No 82-6465
• REM Vol 1: Artificial Intelligence and Legal Information Systems. Vol 2: Deontic Logic, Computational Linguistics, and Legal Information Systems.

P 4043 Winter School on Abstract Anal (11);1983 Zelezna Ruda • CS
[1984] *Proceedings of the 11th Winter School on Abstract Analysis* SER (**J** 3522) Rend Circ Mat Palermo, Ser 2 Suppl 3 • PUBL (**X** 2528) Circolo Mat: Palermo 411pp
• DAT&PL 1983 Jan; Zelezna Ruda, CS

P 4047 Allerton Conf Commun, Control & Comput (15);1977 Monticello • USA
[1977] *15th Annual Allerton Conference on Communication, Control and Computing* PUBL (**X** 1285) Univ Ill Pr: Urbana xii + 761pp
• DAT&PL 1977 Sep; Monticello, IL, USA

P 4050 Appl Syst & Cybern;1980 Acapulco • MEX
[1981] *Applied Systems and Cybernetics, Vol. I, II, III, IV, V, VI* ED: LASKER, G.E. • PUBL (**X** 0869) Pergamon Pr: Oxford 6 Vols: xxxii + 3294pp
• DAT&PL 1980 Dec; Acapulco, MEX • ISBN 0-08-027196-0, LC-No 81-13765

P 4051 Fuzzy Set & Possibility Th;1980 Acapulco • MEX
[1982] *Fuzzy Set and Possibility Theory. Recent Developments* ED: YAGER, R.R. • PUBL (**X** 0869) Pergamon Pr: Oxford xiv + 633pp
• DAT&PL 1980 Dec; Acapulco, MEX • ISBN 0-08-026294-5, LC-No 81-10582

P 4053 IFAC Symp Large Scale Syst: Th & Appl;1983 Warsaw • PL
[1984] *Large Scale Systems: Theory and Applications 1983. Proceedings of the IFAC/IFORS Symposium* ED: STRASZAK, A. • SER IFAC Proc Ser 10 • PUBL (**X** 0869) Pergamon Pr: Oxford xviii + 710pp
• DAT&PL 1983 Jul; Warsaw, PL • LC-No 84-14822

P 4054 Int Symp Multi-Val Log (13);1983 Kyoto • J
[1983] *Proceedings of the 13th International Symposium on Multiple-Valued Logic* PUBL (**X** 2179) IEEE: New York xii + 431pp
• DAT&PL 1983 May; Kyoto, J • LC-No 79-641110

P 4081 Many-Val Log & Appl;1982 Kyoto • J
[1982] *Many-Valued Logic and Its Applications* ED: HASEGAWA, T. • SER (**S** 2308) Symp Kyoto Univ Res Inst Math Sci (RIMS) 455 • PUBL (**X** 2441) Kyoto Univ Res Inst Math Sci: Kyoto iii + 280pp
• DAT&PL 1982 Jan; Kyoto, J

P 4096 Jorn Mat Luso-Espanol (9);1982 Salamanca • E
[1982] *Actas 9 Jornadas Luso-Espanolas de Matematica* PUBL (**X** 2239) Univ Sec Publ: Salamanca 488pp
• DAT&PL 1982 Apr; Salamanca, E • ISBN 84-7481-230-5

P 4097 Jorn Mat Luso-Espanol (6);1979 Santander • E
[1979] *Actas 6 Jornadas Luso-Espanolas de Matematica* SER Rev Univ Santander 2 • PUBL Univ. Santander: Santander 2 Vols
• DAT&PL 1979 Apr; Santander, E

P 4117 Conf Math Log (Markov);1980 Sofia • BG
[1984] *Matematicheskaya Logika: Trudy Konferentsij po Matematicheskoj Logike, Posvyashchennoj Pamiati A.A.Markova 1903-1979 (Proceedings of a Conference on Mathematical Logic)* ED: SKORDEV, D. • PUBL Bulgar Akad Nauk 172pp
• DAT&PL 1980 Sep; Sofia, BG • LC-No 84-215513

P 4136 Proc Symp Circuit Anal;1955 Urbana • USA
[1955] *Proceedings of the Symposium on Circuit Analysis* PUBL (**X** 1285) Univ Ill Pr: Urbana
• DAT&PL 1955 -?-; Urbana, IL, USA • LC-No 79-642703

P 4153 B-Val Anal & Nonstand Anal;1981 Kyoto • J
[1981] *Boolean-Algebra-Valued Analysis and Nonstandard Analysis* ED: NANBA, K. • SER (**S** 2308) Symp Kyoto Univ Res Inst Math Sci (RIMS) 441 • PUBL (**X** 2441) Kyoto Univ Res Inst Math Sci: Kyoto ii + 158pp
• DAT&PL 1981 May; Kyoto, J

P 4180 Int Congr Log, Meth & Phil of Sci (7,Pap);1983 Salzburg • A
[1985] *Foundations of Logic and Linguistics. Problems and Their Solutions. Papers from the 7th International Congress of Logic, Methodology and Philosophy of Science* ED: DORN, G. & WEINGARTNER, P. • PUBL (**X** 1332) Plenum Publ: New York xi+715pp
• DAT&PL 1983 Jul; Salzburg, A • ISBN 0-306-41916-5, LC-No 84-26518

P 4235 IEEE Symp Found of Comp Sci (22);1981 Nashville • USA
[1981] *22nd Annual IEEE Symposium on Foundations of Computer Science* PUBL (**X** 2179) IEEE: New York x+429pp
• DAT&PL 1981 Oct; Nashville, TN, USA • LC-No 80-646634

P 4244 Rewriting Techn & Appl (1);1985 Dijon • F
[1985] *Rewriting Techniques and Applications. Papers from the 1st Conference* ED: JOUANNAUD, J.-P. • SER (**S** 3302) Lect Notes Comput Sci 202 • PUBL (**X** 0811) Springer: Heidelberg & New York vi+441pp
• DAT&PL 1985 May; Dijon, F • ISBN 3-540-15976-2, LC-No 85-22164

P 4254 Joint Autom Control Conf;1980 San Francisco • USA
[1980] *Proceedings of the 1980 Joint Automatic Control Conference. Vol 1, 2. An ASME Century 2 Emerging Technology Conference* PUBL (**X** 2179) IEEE: New York 2 Vols
• DAT&PL 1980 Aug; San Francisco, CA, USA • LC-No 80-67739

P 4288 Predst Znan Chelov-Mash & Robot Sist • SU
[1984] *Predstavlenie Znanij v Cheloveko-Mash. i Robototekhn. Sistemakh T.A.RG-18 KNVVT*

P 4289 IEEE Symp Found of Comput Sci (23);1982 Chicago • USA
[1982] *23rd Annual IEEE Symposium on Foundations of Computer Science* PUBL (**X** 2179) IEEE: New York vii+387pp
• DAT&PL 1982 -?-; Chicago, IL, USA • LC-No 80-646634

P 4301 Parallel Comp, Parallel Math;1977 Muenchen • D
[1977] *Parallel Computers – Parallel Mathematics. Proceedings of the IMACS-GI Symposium* ED: FEILMEIER, M. • PUBL (**X** 0809) North Holland: Amsterdam xii+331pp
• DAT&PL 1977 Mar; Muenchen, D • ISBN 0-444-85042-2, LC-No 77-10345

P 4302 Symp on Progr (6);1984 Toulouse • F
[1984] *International Symposium on Programming. 6th Colloquium* SER (**S** 3302) Lect Notes Comput Sci 167 • PUBL (**X** 0811) Springer: Heidelberg & New York vi+262pp
• DAT&PL 1984 Apr; Toulouse, F • ISBN 3-540-12925-1, LC-No 84-174810

P 4309 Space, Time & Causality;1981 Kiel • D
[1983] *Space, Time and Causality* ED: SWINBURNE, R. • SER (**S** 3307) Synth Libr 157 • PUBL (**X** 0835) Reidel: Dordrecht xvi+211pp
• DAT&PL 1981 Sep; Kiel, D • LC-No 82-18123, ISBN 90-277-1437-1

P 4314 Canad Math Congr;1969 Montreal • CDN
[1969] *Canadian Math Congress. Reports of the Summer Research Institute*
• DAT&PL 1969 -?-; Montreal, Que, CDN

P 4317 Anal, Geom & Probab;1981 Valparaiso • RCH
[1985] *Analysis, Geometry and Probability: Proceedings of the 1st Chilean Symposium on Mathematics* ED: CHUAQUI, R. • SER (**S** 3310) Lect Notes Pure Appl Math 96 • PUBL (**X** 1684) Dekker: New York viii+274pp
• DAT&PL 1981 Dec; Valparaiso, RCH • ISBN 0-8247-7419-1, LC-No 85-1630

P 4381 Int Wittgenstein Symp (7);1982 Kirchberg • A
[1983] *Akten des 7ten Internationalen Wittgenstein Symposium* ED: CZERMAK, H. & WEINGARTNER, P. • PUBL (**X** 2728) Hoelder-Pichler-Tempsky: Wien 573pp
• DAT&PL 1982 Aug; Kirchberg, A • LC-No 83-227181, ISBN 3-209-00499-4

P 4383 Found Quant Mech;1983 Kokubunji-shi • J
[1984] *Proceedings of the International Symposium on Foundations of Quantum Mechanics in the Light of New Technology* ED: KAMEFUCHI, S. • PUBL Physical Soc of Japan: Tokyo 377pp
• DAT&PL 1983 Aug; Kokubunji-shi, J • LC-No 84-256146, ISBN 4-89027-001-9

P 4391 Mat & Mat Obrazov (14);1985 Sl"nchev Bryag • BG
[1985] *Matematika i Matematichesko Obrazovanie. Doklady na 14a Proletna Konferentsiya na S"juza na Matematicite v B"lgariya (Mathematics and Mathematical Education. Proceedings of the 14th Spring Conference of the Union of Bulgarian Mathematicians)* PUBL (**X** 2237) Publ Bulg Acad Sci: Sofia
• DAT&PL 1985 Apr; Sl"nchev Bryag (Sunny Beach), BG
• LC-No 80-643123

P 4392 Mat Logika (Markov);1980 Sofia • BG
identical with P 4117

P 4400 Hist of Log (24); Cracow • PL
[1980] *Proceedings 24-Conference on History of Logic*
• DAT&PL -??- -?-; Cracow, PL

P 4401 Rect Devel in Quant Log;1984 Koeln • D
[1985] *Recent Development in Quantum Logic* ED: MITTELSTAEDT, P. & STACHOW, E.-W. • PUBL (**X** 0876) Bibl Inst: Mannheim 352pp
• DAT&PL 1984 Jun; Koeln, D • ISBN 3-411-01695-7

P 4423 Semigroups (8);1984 Matsue • J
[1985] *Proceedings of the 8th Symposium on Semigroups* ED: YAMADA, U. & FUJIWARA, T. • PUBL Shimane Univ: Matsue ii+83pp
• DAT&PL 1984 Oct; Matsue

P 4460 Stoch Diff Syst;1984 Marseille • F
[1985] *Stochastic Differential Systems.Proceedings of the IFIP-WG 7/1* SER Lect Notes in Control & Information Sci 69 • PUBL (**X** 0811) Springer: Heidelberg & New York viii+322pp
• DAT&PL 1984 Mar; Marseilles-Luminy, F • ISBN 3-540-15176-1

P 4508 Congr Int Phil (8); • CS
[1936] *Actes du 8eme Congres International de Philosophie* PUBL (**X** 1325) Orbis: Prague lxxii+1103pp

P 4564 Math Logic;1963 Xi-An • TJ
[1965] *1963 Nian Cueanguo Shuli Luoji Zhuanye Xueshu Huiyi Lunwen Xuanji (Mathematical Logic. Proceedings of the National Symposium)* PUBL Defence Industry Press: Beijing
• DAT&PL 1963 Oct; Xi-An, TJ

P 4571 Log of Progr;1985 Brooklyn • USA
Logic of Programs ED: PARIKH, R. • SER (S 3302) Lect Notes Comput Sci 193 • PUBL (**X** 0811) Springer: Heidelberg & New York vi+424pp
• DAT&PL 1985 Jun; Brooklyn, NY, USA • ISBN 3-540-15648-8

P 4580 Semin Concurrency;1984 Pittsburgh • USA
[1985] *Seminar of Concurrency* ED: BROOKES, S.D. & ROSCOE, A.W. & WINSKEL, G. • SER (S 3302) Lect Notes Comput Sci 197 • PUBL (**X** 0811) Springer: Heidelberg & New York x+523pp
• DAT&PL 1984 Jul; Pittsburgh, PA, USA • ISBN 3-540-15670-4

P 4588 Logic in China;1978 Beijing • TJ
(Issue of the Symposium on Logic in China) PUBL Jilin People Press
• DAT&PL 1978 May; Beijing, TJ

P 4605 Inform Syst Th & Formal Aspects;1985 Sitges • E
[1985] *Information Systems: Theoretical and Formal Aspects. Proceedings of the IFIP Workshop* ED: SERNADAS, A. & BUBENKO JR., J. & OLIVE, A. • PUBL (**X** 0809) North Holland: Amsterdam x+236pp
• DAT&PL 1985 Apr; Sitges, E • ISBN 0-444-87706-1

P 4620 Model & Simulation (13);1982 Pittsburgh • USA
[1982] *Modeling and Simulation, Part I, II, III, IV. Proceedings of the 13th Pittsburgh Conference* ED: VOGT, W.C. & MICKLE, M.H. • PUBL Instrument Soc Amer: Research Triangle Park, NC 4 Vols: 1614pp
• DAT&PL 1982 Apr; Pittsburgh, PN, USA • ISBN 0-87664-712-3 (V1), ISBN 0-87664-713-1 (V2), ISBN 0-87664-714-X (V3), ISBN 0-87664-715-8 (V4)
• REM Part 1: Control, Communications, and Biomedical. Part 2: Computers and Computer Modeling. Part 3: Socioeconomic Systems. Part 4: Energy and environmental Systems.

P 4621 Log & Models of Concurrent Syst;1984 La Colle-sur-Loup • F
[1985] *Logics and Models of Concurrent Systems. Proceedings of the NATO Advanced Study Institute* ED: APT, K.R. • SER NATO Adv Sci Inst, Ser F 13 • PUBL (**X** 0811) Springer: Heidelberg & New York viii+498pp
• DAT&PL 1984 Oct; La Colle-sur-Loup, F • ISBN 3-540-15181-8

P 4627 CAAP'85 Arbres en Algeb & Progr (10);1985 Berlin • D
[1985] *Mathematical Foundations of Software Development, Vol 1. Colloquium on Trees in Algebra and Programming (CAAP '85)* ED: EHRIG, H. & FLOYD, C. & NIVAT, M. & THATCHER, J. • SER (S 3302) Lect Notes Comput Sci 185 • PUBL (**X** 0811) Springer: Heidelberg & New York xiii+418pp
• DAT&PL 1985 Mar; Berlin, D • ISBN 3-540-15198-2

P 4628 Automata, Lang & Progr (12);1985 Nafplion • GR
[1985] *Automata, Languages and Programming.* ED: BRAUER, W. • SER (S 3302) Lect Notes Comput Sci 194 • PUBL (**X** 0811) Springer: Heidelberg & New York viii+520pp
• DAT&PL 1985 Jul; Nafplion, GR • ISBN 3-540-15650-X

P 4629 Dynamical Syst & Cellular Autom;1983 Luminy • F
[1985] *Dynamical Systems and Cellular Automata* ED: DEMONGEOT, J. & GOLES, E. & TCHUENTE, M. • PUBL (**X** 0801) Academic Pr: New York xv+399pp
• DAT&PL 1983 Sep; Luminy, F • ISBN 0-12-209060-8

P 4646 Atti Incontri Log Mat (2);1983/84 Siena • I
[1985] *Atti degli Incontri di Logica Matematica* ED: BERNARDI, C. & PAGLI, P. • PUBL (**X** 3812) Univ Siena, Dip Mat: Siena 648pp
• DAT&PL 1983 Jan; Siena, I, 1983 Apr; Siena, I, 1984 Jan; Siena, I, 1984 Apr; Siena, I

P 4647 FCT'85 Fund of Comput Th;1985 Cottbus • DDR
[1985] *Fundamentals of Computation Theory. FCT '85* ED: BUDACH, L. • SER (S 3302) Lect Notes Comput Sci 199 • PUBL (**X** 0811) Springer: Heidelberg & New York xii+542pp
• DAT&PL 1985 Sep; Cottbus, DDR • ISBN 3-540-15689-5

P 4651 Algeb Methods in Semantics;1982 Fontainebleau • F
[1985] *Algebraic Methods in Semantics. Seminar on 'The Applications of Algebra to Language Definition and Compilation'* ED: NIVAT, M. & REYNOLDS, J.C. • PUBL (**X** 0805) Cambridge Univ Pr: Cambridge, GB xv+634pp
• DAT&PL 1982 Jun; Fontainebleau, F

P 4661 Algeb & Log;1984 Zagreb • YU
[1985] *Proceedings of the Conference 'Algebra and Logic'* ED: STOJAKOVIC, Z. • PUBL (**X** 4030) Univ Novom Sadu, Inst Mat: Novi Sad vi+193pp
• DAT&PL 1984 Jun; Zagreb, YU

P 4670 Comput Th (5);1984 Zaborow • SU
[1985] *Computation Theory. Proceedings of the 5th Symposium* ED: SKOWRON, A. • SER (S 3302) Lect Notes Comput Sci 208 • PUBL (**X** 0811) Springer: Heidelberg & New York vii+397pp
• DAT&PL 1984 Dec; Zaborow, SU • ISBN 3-540-16066-3

P 4672 Found of Softw Tech & Th Comput Sci (5);1985 New Delhi • IND
[1985] *Foundations of Software Technology and Theoretical Computer Science. Proceedings of the 5th Conference* ED: MAHESHWARI, S.N. • SER (S 3302) Lect Notes Comput Sci 206 • PUBL (**X** 0811) Springer: Heidelberg & New York x+522pp
• DAT&PL 1985 Dec; New Delhi, IND • ISBN 3-540-16042-6

P 4677 Math Syst Th;1975 Udine • I
[1976] *Mathematical Systems Theory* ED: MARCHESINI, G. & MITTER, S.K. • SER (S 3314) Lect Notes Econ & Math Syst 131 • PUBL (**X** 0811) Springer: Heidelberg & New York x+408pp
• DAT&PL 1975 Jun; Udine, I • ISBN 3-540-07798-7

P 4724 Winter School on Abstract Anal (10);1982 Srni • CS
[1982] *Proceedings of the 10th Winter School on Abstract Analysis* SER (J 3522) Rend Circ Mat Palermo, Ser 2 Suppl 2 • PUBL (**X** 2528) Circolo Mat: Palermo 320pp
• DAT&PL 1982 Jan; Srni, CS

Collection volumes

C 0552 Phil Contemp - Chroniques • I
[1968] *La Philosophie Contemporaine: Chroniques (Contemporary Philosophy: A Survey)* ED: KLIBANSKY, R. • PUBL (**X** 1319) La Nuova Italia: Firenze 4 Vols
• LC-No 68-55649

C 0555 Topic Philos Logic • NL
[1969] *Topics in Philosophical Logic* ED: RESCHER, N. • SER (**S** 3307) Synth Libr • PUBL (**X** 0835) Reidel: Dordrecht xiv+350pp • ALT PUBL (**X** 2696) Humanities Pr: Atlantic Highlands
• LC-No 74-141565

C 0556 Words & Objections (Quine) • NL
[1969] *Words and Objections. Essays on the Work of W.V.Quine* ED: DAVIDSON, D. & HINTIKKA, J. • SER (**S** 3307) Synth Libr 19 • PUBL (**X** 0835) Reidel: Dordrecht vii+366pp • ALT PUBL (**X** 2696) Humanities Pr: Atlantic Highlands
• LC-No 79-495176
• REM Reprint from J0154 19/1-2

C 0558 Contemp Phil Scand • USA
[1972] *Contemporary Philosophy in Scandinavia* ED: OLSON, R.E. & PAUL, A.M. • PUBL (**X** 1291) Johns Hopkins Univ Pr: Baltimore x+508pp
• ISBN 0-8018-1315-8, LC-No 70-148242

C 0567 Form & Strategy in Sci (Woodger) • NL
[1964] *Form and Strategy in Science: Studies Dedicated to Joseph Henry Woodger on the Occasion of His 70th Birthday* ED: GREGG, J.R. & HARRIS, F.T.C. • PUBL (**X** 0835) Reidel: Dordrecht vii+476pp
• LC-No 65-87332

C 0569 Phil of Math Oxford Readings • GB
[1969] *The Philosophy of Mathematics.* ED: HINTIKKA, K.J.J. • SER Oxford Readings in Philosophy • PUBL (**X** 0894) Oxford Univ Pr: Oxford v+186pp
• ISBN 0-19-875011-0, LC-No 71-441791

C 0571 Basic Issues Philos of Time • USA
[1971] *Basic Issues in the Philosophy of Time* ED: FREEMAN, E. & SELLARS, W. • SER Monist Library of Philosophy 53/3 • PUBL (**X** 1324) Open Court: LaSalle 241pp
• LC-No 73-128197

C 0574 Semantics of Nat Lang • NL
[1972] *Semantics of Natural Language* ED: DAVIDSON, D. & HARMANN, G. • SER (**S** 3307) Synth Libr • PUBL (**X** 0835) Reidel: Dordrecht x+769pp
• ISBN 90-277-0304-3, ISBN 90-277-0310-8, LC-No 74-159655
• REM 2nd ed. 1973

C 0582 Log Semant & Modal Logika • SU
[1967] *Logicheskaya Semantika i Modal'naya Logika (Logical Semantics and Modal Logic)* ED: TAVANETS, P.V. • PUBL (**X** 2027) Nauka: Moskva 280pp
• LC-No 68-53796

C 0585 Struct, Meth & Meaning (Sheffer) • USA
[1951] *Structure, Method and Meaning. Essays in Honor of Henry M.Sheffer* ED: HENLE, P. & KALLEN, H.M. & LANGER, S.K. • PUBL New York: Liberal Arts Press xvi+306pp
• LC-No 51-2957

C 0601 Encycl of Philos • USA
[1967] *Encyclopedia of Philosophy* ED: EDWARDS, P. • PUBL (**X** 0843) Macmillan : New York & London 8 Vols
• LC-No 67-10059

C 0609 Lukasiewicz: Log & Philos Papers • PL
[1961] *Logical and Philosophical Papers by Jan Lukasiewicz* ED: SLUPECKI, J. • PUBL (**X** 1034) PWN: Warsaw 249pp
• REM Translations and Reprints

C 0615 Polish Logic 1920-39 • GB
[1967] *Polish Logic 1920-39.* ED: MCCALL, S. • PUBL (**X** 0815) Clarendon Pr: Oxford viii+406pp
• LC-No 67-106639
• REM Translations of Several Articles

C 0647 Encycl Britannica • GB
[1950ff] *Encyclopaedia Britannica: A New Survey of Universal Knowledge* ED: YUST, W. • PUBL Encyclopaedia Britannica: Chicago-London-Toronto 24 Vols
• LC-No 50-4840

C 0668 Neklass Log • SU
[1970] *Neklassucheskaya Logika (Nonclassical Logic)* ED: TAVANETS, P.V. • PUBL (**X** 2027) Nauka: Moskva 384pp
• LC-No 70-592611

C 0673 Phil of Carnap • GB
[1963] *The Philosophy of Rudolf Carnap.* ED: SCHILPP, P.A. • SER The Library of Living Philosophers 11 • PUBL (**X** 1324) Open Court: LaSalle xvi+1088pp • ALT PUBL (**X** 0805) Cambridge Univ Pr: Cambridge, GB
• LC-No 62-9577

C 0675 From Frege to Goedel • USA
[1967] *From Frege to Goedel: A Source Book in Mathematical Logic 1879-1931* ED: HEIJENOORT VAN, J. • PUBL (**X** 0858) Harvard Univ Pr: Cambridge x+660pp • ALT PUBL (**X** 0894) Oxford Univ Pr: Oxford
• LC-No 67-10905
• REM 2nd ed. 1971; xi+660pp. Some Articles have been Reprinted 1970; iv+117pp

C 0712 Logik & Logikkalkuel • D
[1962] *Logik und Logikkalkuel* ED: KAESBAUER, M. & KUTSCHERA VON, F. • PUBL (**X** 0826) Alber: Freiburg 249pp

C 0717 Automata Studies • USA
[1956] *Automata Studies* ED: SHANNON, C.E. & MCCARTHY, J.E. • SER (**S** 3513) Ann Math Stud 34 • PUBL (**X** 0857) Princeton Univ Pr: Princeton viii+285pp
• LC-No 56-7637
• TRANSL IN [1974] (**C** 1902) Stud Th Automaten
• REM 4th ed. 1965

C 0721 Contemp Readings in Log Th • USA
[1967] *Contemporary Readings in Logical Theory* ED: COPI, I.M. & GOULD, J.A. • PUBL (**X** 0843) Macmillan : New York & London 342pp
• LC-No 67-15535

C 0727 Logic Found of Math (Heyting) • NL
[1968] *Logic and Foundations of Mathematics. Papers Dedicated to A.Heyting on the Occasion of His 70th Birthday* ED: DALEN VAN, D. & DYKMAN, J.G. & KLEENE, S.C. & TROELSTRA, A.S. • PUBL (**X** 0812) Wolters-Noordhoff : Groningen 249pp
• LC-No 70-408259
• REM Also published as J0020 vol. 20

C 0735 Logic & Value (Dahlquist) • S
[1970] *Logic and Value. Essays Dedicated to Thorild Dahlquist on His 50th Birthday* ED: PAULI, T. • SER Filosofiska Studier 9 • PUBL (**X** 0882) Univ Filos Foeren: Uppsala vi+247pp
• LC-No 72-186683

C 0736 Aristotle: Coll Critical Essays • GB
[1967] *Aristotle. A Collection of Critical Essays* ED: MORAVCSIK, J.M.E. • SER Modern Studies in Philosophy • PUBL (**X** 0878) Doubleday: London 341pp
• LC-No 66-24304

C 0742 Phil Mid-Century • I
[1958-1959] *Philosophy in the Mid-Century. A Survey * La Philosophie au Milieu du Vingtieme Siecle. Chroniques* ED: KLIBANSKY, R. • PUBL (**X** 1319) La Nuova Italia: Firenze 4 Vols

C 0749 Contrib Logic & Methodol (Bochenski) • NL
[1965] *Contributions to Logic and Methodology in Honour of J.M.Bochenski* ED: TYMIENIECKA, A.T. & PARSONS, C. • PUBL (**X** 0809) North Holland: Amsterdam xviii+326pp
• LC-No 66-1566

C 0758 Logic Colloq Boston 1972-73 • D
[1975] *Logic Colloquium. Symposium on Logic Held at Boston, MA, USA, 1972-1973* ED: PARIKH, R.J. • SER (S 3301) Lect Notes Math 453 • PUBL (**X** 0811) Springer: Heidelberg & New York iv+251pp
• ISBN 3-540-07155-5, LC-No 75-11528

C 0766 Probl Log & Teor Poznan • SU
[1968] *Problemy Logiki i Teorii Poznaniya (Problems in Logic and Epistemology)* ED: NARSKIJ, I.S. • PUBL (**X** 0898) Moskov Gos Univ: Moskva 319pp
• LC-No 73-399696

C 0768 Stud Algeb Logic • USA
[1974] *Studies in Algebraic Logic* ED: DAIGNEAULT, A. • SER Studies in Mathematics 9 • PUBL (**X** 1298) Math Ass Amer: Washington vii+207pp
• ISBN 0-88385-109-1, LC-No 74-84580

C 0799 Stud Induct Logic & Probab, Vol 1 • USA
[1971] *Studies in Inductive Logic and Probability. Vol 1* ED: CARNAP, R. & JEFFREY, R.C. • PUBL (**X** 0926) Univ Calif Pr: Berkeley iv+264pp
• ISBN 0-520-01866-4, LC-No 77-136025 • REL PUBL (**C** 3537) Stud Induct Logic & Probab, Vol 2

C 0800 New Catholic Encycl • USA
[1967ff] *The New Catholic Encyclopedia* PUBL (**X** 0822) McGraw-Hill: New York 15 Vols
• LC-No 66-22292

C 1009 Wang: Survey Math Logic • TJ
[1963] *Wang,H.: Survey of Mathematical Logic* PUBL (**X** 1876) Kexue Chubanshe: Beijing • ALT PUBL (**X** 0809) North Holland: Amsterdam 1963

C 1013 Phil Essays (Husserl) • USA
[1940] *Philosophical Essays in Memory of Edmund Husserl.* ED: FARBER, M. • PUBL (**X** 0858) Harvard Univ Pr: Cambridge viii+332pp
• LC-No 40-35634

C 1017 Logical Studies • USA
[1957] *Logical Studies* ED: WRIGHT VON, G.H. • SER International Library of Psychology, Philosophy and Scientific Method • PUBL (**X** 2696) Humanities Pr: Atlantic Highlands xi+195pp
• LC-No 58-732

C 1029 Readings Phil of Lang • USA
[1971] *Readings in Philosophy of Language* ED: ROSENBERG, J.F. & TRAVIS, C. • PUBL (**X** 0819) Prentice Hall: Englewood Cliffs viii+645pp
• ISBN 0-13-759332-5, LC-No 70-132170

C 1098 Skolem: Select Works in Logic • N
[1970] *Skolem,T.A.: Selected Works in Logic* ED: FENSTAD, J.E. • SER Scandinavian University Books • PUBL (**X** 1554) Universitesforlaget: Oslo 732pp
• LC-No 74-485971

C 1101 Phil Logic • NL
[1969] *Philosophical Logic.* ED: DAVIS, J.W. & HOCKNEY, D.J. & WILSON, W.K. • SER (S 3307) Synth Libr • PUBL (**X** 0835) Reidel: Dordrecht viii+277pp
• LC-No 78-437276

C 1105 Phil of Math. Sel Readings • USA
[1964] *Philosophy of Mathematics. Selected Readings.* ED: BENACERRAF, P. & PUTNAM, H. • SER Prentice Hall Philosophy Series • PUBL (**X** 0819) Prentice Hall: Englewood Cliffs vii+563pp • ALT PUBL (**X** 1096) Blackwell: Oxford
• LC-No 64-13252

C 1107 Aspects Inductive Log • NL
[1966] *Aspects of Inductive Logic* ED: HINTIKKA, J. & SUPPES, P. • SER (S 3303) Stud Logic Found Math • PUBL (**X** 0809) North Holland: Amsterdam vi+322pp
• LC-No 67-84072

C 1134 Log Way of Doing Things • USA
[1969] *The Logical Way of Doing Things* ED: LAMBERT, K. • PUBL (**X** 0875) Yale Univ Pr: New Haven xii+325pp
• LC-No 69-15450

C 1190 Measurement • USA
[1959] *Measurement: Definitions and Theories* ED: CHURCHMAN, C.W. & RATOOSH, P. • PUBL (**X** 0827) Wiley & Sons: New York 274pp
• LC-No 59-11796

C 1222 Existence & Possible Worlds • USA
[1973] *Existence and Possible Worlds* ED: MUNITZ, M. • SER Studies in Contemporary Philosophy 2 • PUBL -?-: New York

C 1389 Modality, Morality, Probl of Sense & Nonsense (Hallden) • S
[1973] *Modality, Morality and other Problems of Sense and Nonsense. Essays Dedicated to Soeren Hallden.* ED: HANSSON, B. ET AL. • PUBL (**X** 1493) Gleerup: Lund vi+221pp
• ISBN 91-40-02780-5, LC-No 73-180115

Source Index

C 1413 Tekh Kibernetika • SU
[1965] *Tekhnicheskaya Kibernetika (Technical Cybernetics)* ED: TSIPKIN, YA.Z. ET AL • PUBL (**X** 2027) Nauka: Moskva 429pp
• LC-No 67-115871

C 1491 Stud in Logic (J. Hopkins Univ) • USA
[1883] *Studies in Logic: by Members of the Johns Hopkins University* ED: PEIRCE, C. • PUBL (**X** 0960) Little, Brown & Co: Boston & Toronto vi+203pp
• LC-No 6-39867

C 1495 Encycl Universalis • F
[1968-1976] *Encyclopaedia Universalis* PUBL (**X** 2524) Encyclopaedia Universalis: Paris 20 Vols
• ISBN 2-85229-281-5, LC-No 75-516014

C 1499 Raboty Tekhn Kibern • SU
[1967-1971] *Raboty po Tekhnicheskoj Kibernetike (Papers on Engineering Cybernetics)* ED: SMIRYAGIN, V.P. • PUBL (**X** 2265) Akad Nauk Vychis Tsentr: Moskva 3 Vols
• LC-No 68-35977

C 1530 Issl Log Sist (Yanovskaya) • SU
[1970] *Issledovaniya Logicheskikh Sistem (Investigations of Logical Systems)* ED: TAVANETS, P.V. • PUBL (**X** 2027) Nauka: Moskva 336pp
• LC-No 70-554105

C 1533 Quantoren, Modal, Paradox • DDR
[1972] *Quantoren, Modalitaeten, Paradoxien. Beitraege zur Logik* ED: WESSEL, H. • PUBL (**X** 0806) Dt Verlag Wiss: Berlin 524pp
• LC-No 72-3662278

C 1547 Deontic Log Readings • NL
[1971] *Deontic Logic. Introductory and Systematic Readings* ED: HILPINEN, R. • SER (**S** 3307) Synth Libr • PUBL (**X** 0835) Reidel: Dordrecht vi+184pp
• ISBN 90-277-0167-9, LC-No 72-135103

C 1577 Induct, Accept & Rat Belief • NL
[1970] *Induction, Acceptance, and Rational Belief* ED: SWAIN, M. • SER (**S** 3307) Synth Libr • PUBL (**X** 0835) Reidel: Dordrecht 240pp
• LC-No 70-469001

C 1587 Quine: Ways of Paradox & Essays • USA
[1966] *Quine, W.V.C.: The Ways of Paradox and Other Essays* PUBL (**X** 0981) Random House: New York x+258pp • ALT PUBL (**X** 0858) Harvard Univ Pr: Cambridge x+335pp
• ISBN 0-674-94835-1 (0858), LC-No 66-11146, LC-No 76-4200,
• REM 2nd revised and enlarged ed. 1976

C 1602 Lect on Modern Math • USA
[1963-1965] *Lectures on Modern Mathematics* ED: SAATY, T.L. • PUBL (**X** 0827) Wiley & Sons: New York 3 Vols
• LC-No 63-20369
• TRANSL IN [1965] (**C** 4652) Lect on Modern Math (Japanese)

C 1654 Probl Logiki • SU
[1963] *Problemy Logiki (Problems of Logic)* ED: YANOVSKAYA, S.A. ET AL. • PUBL (**X** 0899) Akad Nauk SSSR : Moskva 152pp

C 1662 Teor Log Vyvoda • SU
[1973] *Teoriya Logicheskogo Vyvoda (Theory of Logical Deductions)* ED: TAVANETS, P.V. • PUBL (**X** 2027) Nauka: Moskva 272pp
• LC-No 73-353812

C 1701 Lang in Focus (Bar-Hillel) • NL
[1976] *Language in Focus. Foundations, Methods & Systems: Essays in Memory of Y.Bar-Hillel* ED: KASHER, A. • SER (**S** 3307) Synth Libr 89, (**S** 3311) Boston St Philos Sci 43 • PUBL (**X** 0835) Reidel: Dordrecht xxviii+679pp
• ISBN 90-277-0644-1, ISBN 90-277-0645-X, LC-No 75-33775

C 1702 Local Induction • NL
[1976] *Local Induction* ED: BOGDAN, R.J. • SER (**S** 3307) Synth Libr 93 • PUBL (**X** 0835) Reidel: Dordrecht xiv+340pp
• ISBN 90-277-0649-2, LC-No 75-34922

C 1706 Essays Honour J. Hintikka • NL
[1979] *Essays in Honour of Jaakko Hintikka: On the Occasion of His 50th Birthday on January 12, 1979* ED: SAARINEN, E. & HILPINEN, R. & NIINILUOTO, I. & PROVENCE, M. • SER (**S** 3307) Synth Libr 124 • PUBL (**X** 0835) Reidel: Dordrecht x+386pp
• ISBN 90-277-0916-5, LC-No 78-11364

C 1786 Lukasiewicz: Select Works • NL
[1970] *Selected Works by Jan Lukasiewicz* ED: BORKOWSKI, L. • SER (**S** 3303) Stud Logic Found Math • PUBL (**X** 0809) North Holland: Amsterdam xii+405pp • ALT PUBL (**X** 1034) PWN: Warsaw xii+405pp
• ISBN 0-7204-2252-3, LC-No 74-96734

C 1808 Logica del Tempo • I
[1974] *La Logica del Tempo* ED: PIZZI, C. • PUBL (**X** 0905) Boringhieri: Torino

C 1810 Semant Filos, Probl & Disc • RA
[1973] *Semantica Filosofica, Problemas y Discussiones* ED: SIMPSON, T.M. • PUBL Siglo XXI Argentina Editores: Buenos Aires xvii+476pp

C 1811 Phil & Analysis • GB
[1954] *Philosophy and Analysis. A Selection of Articles Published in Analysis Between 1933-40 and 1947-53* ED: MACDONALD, M. • PUBL (**X** 1096) Blackwell: Oxford viii+296pp • ALT PUBL (**X** 0920) Philos Lib: New York 1974
• LC-No 55-14290

C 1812 Refer & Modal • GB
[1971] *Reference and Modality* ED: LINSKY, L. • SER Oxford Readings in Philosophy • PUBL (**X** 0894) Oxford Univ Pr: Oxford v+177pp
• ISBN 0-19-875017-X, LC-No 72-595855
• TRANSL IN [1974] (**C** 1853) Refer & Modal (Italian)

C 1826 Boston Colloq Philos Sci 1960-64 • USA
[1965] *Proceedings of the Boston Colloquium for the Philosophy of Science* ED: COHEN, R.S. & WARTOFSKY, M.W. • SER (**S** 3311) Boston St Philos Sci • PUBL (**X** 0835) Reidel: Dordrecht viii+212pp
• DAT&PL 1960-1964; Boston, MA, USA • LC-No 73-20858

C 1828 Readings Semantics • USA
[1974] *Readings in Semantics* ED: ZABECH, F. & KLEMKE, E.D. & JACOBSON, A. • PUBL (**X** 1285) Univ Ill Pr: Urbana v+853pp
• ISBN 0-252-00196-6, LC-No 74-639

C 1852 Entailment - Log of Relev & Nec • USA
[1975-1977] *Entailment: The Logic of Relevance and Necessity* ED: ANDERSON, A.R. & BELNAP JR., N.D. • PUBL (**X** 0857) Princeton Univ Pr: Princeton 2 Vols
• ISBN 0-691-07192-6, LC-No 72-14016

C 1853 Refer & Modal (Italian) • I
[1974] *(Reference and Modality (Italian))* ED: LINSKY, L. • TRANSL OF [1971] (**C** 1812) Refer & Modal

C 1856 Log Enterprise • USA
[1975] *The Logical Enterprise* ED: MARCUS, R.B. & ANDERSON, A.R. & MARTIN, R.M. • PUBL (**X** 0875) Yale Univ Pr: New Haven x+261pp
• ISBN 0-300-01790-1, LC-No 74-20084

C 1902 Stud Th Automaten • D
[1974] *Studien zur Theorie der Automaten* ED: SHANNON, C.E. & MCCARTHY, J. • SER R & B Studien 1 • PUBL Rogner & Bernhard: Muenchen xxxii+452pp
• TRANSL OF [1956] (**C** 0717) Automata Studies

C 1936 Log Th & Semant Anal (Kanger) • NL
[1974] *Logical Theory and Semantic Analysis. Essays Dedicated to Stig Kanger on His 50th Birthday* ED: STENLUND, S. & HENSCHEN-DAHLQUIST, A.-M. & LINDAHL, L. & NORDENFELT, L. & ODELSTAD, J. • SER (**S** 3307) Synth Libr 63 • PUBL (**X** 0835) Reidel: Dordrecht v+217pp
• ISBN 90-277-0438-4, LC-No 73-94456

C 1963 Probl in Th of Electr Digital Comp, No 3 (Kiev) • SU
[1967] *(Problems in the Theory of Electronic Digital Computers No. 3)* PUBL (**X** 2199) Naukova Dumka: Kiev 111pp

C 1993 Form Logik & Dialektik • DDR
[1952] *Ueber Formale Logik und Dialektik, Diskussionsbeitraege* PUBL Kultur und Fortschritt: Berlin 216pp

C 2041 Formal della Dialettica • I
[1979] *La Formalizzazione della Dialettica* ED: MARCONI, D. • PUBL (**X** 1416) Rosenberg & Sellier: Torino
• LC-No 80-495224

C 2049 Leggi di Nat, Modalita, Ipotesi • I
[1978] *Leggi di Natura, Modalita, Ipotesi. La Logica del Ragionamento Controfattuale* ED: PIZZI, C. • PUBL (**X** 0844) Feltrinelli: Milano
• LC-No 80-471451

C 2078 Contemp Philos • NL
[1981] *Contemporary Philosophy* ED: FLOISTAD, G. & WRIGHT VON, G.H. • PUBL (**X** 1316) Nijhoff: Leiden & s'Gravenhage
• LC-No 81-3972

C 2084 Stud Hist of Math Log • PL
[1973] *Studies in the History of Mathematical Logic* ED: SURMA, S.J. • PUBL (**X** 2212) Ossolineum: Wroclaw 288pp
• LC-No 73-170578

C 2098 Dict Log Appl Stud of Language • NL
[1981] *Diction Log Appl Study of Language.* ED: MARANEWSKI, W. • PUBL (**X** 1316) Nijhoff: Leiden & s'Gravenhage

C 2103 Develop of Log Probab (Lakatos) • NL
[1976] *The Development of Logical Probability. Essays in Honour of Imre Lakatos* ED: COHEN, R.S. & FEYERABEND, P.K. & WARTOFSKY, M.W. • SER (**S** 3311) Boston St Philos Sci 39, (**S** 3307) Synth Libr 99 • PUBL (**X** 0835) Reidel: Dordrecht xi+767pp
• LC-No 76-16770

C 2141 Filos Matematica • I
[1967] *La Filosofia della Matematica* ED: CELLUCCI, C. • PUBL Laterza: Bari 320pp
• LC-No 68-120024

C 2253 Log & Philos for Linguist Readings • F
[1974] *Logic and Philosophy for Linguistics: A Book of Readings* ED: MORAVCSIK, J.M.E. • PUBL (**X** 0873) Mouton: Paris 347pp • ALT PUBL (**X** 0846) Schoeningh: Paderborn
• ISBN 0-391-00399-2, LC-No 75-23015

C 2318 Entwicklung Math in DDR • DDR
[1974] *Entwicklung der Mathematik in der DDR: Zum 25. Jahrestag der Gruendung der Deutschen Demokratischen Republik* ED: SACHS, H. & AHRENS, H. ET AL. • PUBL (**X** 0806) Dt Verlag Wiss: Berlin xx+756pp
• LC-No 75-569431

C 2557 Teor & Priklad Zad Mat & Mekh • SU
[1977] *Teoreticheskie i Prikladnye Zadachi Matematiki i Mekhaniki. Sbornik Trudov (Theoretical and Applied Problems in Mathematics and Mechanics. Work Collection)* ED: AMANOV, T.I. • PUBL Nauka Kazakh SSR:Alma-Ata 322pp

C 2567 Modal & Intentsional Log • SU
[1978] *Modal'nye i Intentsional'nye Logiki (Modal and Intentional Logics)* PUBL -?-: Moskva

C 2568 Sb Log, Semant, Metodol • SU
[1978] *Sbornik Logika, Semantika, Metodologiya (Logic, Semantics, Methodology)* PUBL (**X** 3790) Metsniereba: Tbilisi

C 2577 Issl Formaliz Yazyk & Neklass Log • SU
[1974] *Issledovaniya po Formalizovannym Yazykam i Neklassicheskim Logikam (Investigations on Formalized Languages and Non-Classical Logics)* ED: BOCHVAR, D.A. • PUBL (**X** 2027) Nauka: Moskva 275pp
• LC-No 75-554105

C 2578 Filos & Logika • SU
[1974] *Filosofiya i Logika. Filosofiya v Sovremennom Mire (Philosophy and Logic. Philosophy in the Modern World)* ED: TAVANETS, P.V. & SMIRNOV, V.A. • SER Filosofiya v Sovremennom Mire • PUBL (**X** 2027) Nauka: Moskva 479pp
• LC-No 74-358872

C 2581 Issl Neklass Log & Teor Mnozh • SU
[1979] *Issledovaniya po Neklassicheskim Logikam i Teorii Mnozhestv (Investigations on Non-Classical Logics and Set Theory)* ED: MIKHAJLOV, A.I. ET AL. • PUBL (**X** 2027) Nauka: Moskva 374pp
• LC-No 80-475529

C 2583 Aktual Probl Log & Metodol Nauki • SU
[1980] *Aktual'nye Problemy Logiki i Metodologii Nauki (Current Problems of Logic and the Methodology of Science. Collection of Scientific Works)* ED: POPOVICH, M.V. • PUBL (**X** 2199) Naukova Dumka: Kiev 336pp
• LC-No 80-506762

C 2617 Modern Log Survey • NL
[1981] *Modern Logic - A Survey. Historical, Philosophical and Mathematical Aspects of Modern Logic and Its Applications.* ED: AGAZZI, E. • SER (**S** 3307) Synth Libr 149 • PUBL (**X** 0835) Reidel: Dordrecht viii+475pp
• ISBN 90-277-1137-2, LC-No 80-22027

C 2620 Teor Model & Primen • SU
[1980] *Teoriya Modelej i Ee Primeneniya (Theory of Models and Its Applications)* ED: BAJZHANOV, B.S. & TAJTSLIN, M.A. • PUBL (**X** 2769) Kazakh Gos Univ: Alma-Ata 83pp

Source Index

C 2621 Mal'tsev: Metamath of Algeb Syst • NL
[1971] *The Metamathematics of Algebraic Systems. Mal'tsev,A.I: Collected Papers, 1936-1967* ED: WELLS III, B.F. • SER (S 3303) Stud Logic Found Math 66 • PUBL (X 0809) North Holland: Amsterdam xviii+494pp
• LC-No 73-157020

C 2638 Log-Gnoseolog Issl Kategorial Strukt Myshleniya • SU
[1980] *Logiko-Gnoseologicheskie Issledovaniya Kategorial'noj Struktury Myshleniya (Logical-Gnoseological Studies of the Categorical Structure of Thought)* ED: POPOVICH, M.V. • PUBL (X 2199) Naukova Dumka: Kiev 339pp
• LC-No 81-127134

C 2953 Log & Probab in Quant Mech • NL
[1976] *Logic and Probability in Quantum Mechanics* ED: SUPPES, P. • SER (S 3307) Synth Libr 78 • PUBL (X 0835) Reidel: Dordrecht xv+541pp

C 3019 Appreciation (Fisher) • D
[1980] *R.A. Fisher: An Appreciation.* ED: FIENBERG, E. & HINKLEY, D.V. • SER Lecture Notes in Statistics 1 • PUBL (X 0811) Springer: Heidelberg & New York x+208pp
• ISBN 0-387-90476-X, LC-No 80-255

C 3026 Lukasiewicz Sel Pap on Sent Calc • PL
[1977] *Selected Papers on Lukasiewicz Sentential Calculi* ED: MALINOWSKI, G. & WOJCICKI, R. • PUBL (X 2885) Zakl Narod Wyd Pol Ak: Wroclaw 199pp
• LC-No 80-515606

C 3036 Vopr Teor Grupp & Kolets • SU
[1973] *Nekotorye Voprosy Teorii Grupp i Kolets (Some Questions on the Theory of Groups and Rings)* PUBL Inst. Fiz. im. Kirenskogo Sibirsk. Otdel. Akad. Nauk SSSR: Krasnoyarsk; 178pp
• LC-No 74-331837

C 3038 Log-Metodol Issl • SU
[1980] *Logiko-Metodologicheskie Issledovaniya (Studies in Logical Methodology)* ED: STARCHENKO, A.A. • PUBL (X 0898) Moskov Gos Univ: Moskva 375pp
• LC-No 81-482681

C 3045 Log-Algeb Appr to Quant Mech • NL
[1975,1979] *The Logico-Algebraic Approach to Quantum Mechanics. Vol. I: Historical Evolution. Vol.II: Contemporary Consolidation* ED: HOOKER, C.A. • SER (S 3308) Univ Western Ontario Ser in Philos of Sci 5 • PUBL (X 0835) Reidel: Dordrecht 2 Vols
• ISBN 90-277-0567-4 (V1), ISBN 90-277-0709-X (V2), LC-No 75-8737

C 3050 Essays Combin Log, Lambda Calc & Formalism (Curry) • USA
[1980] *To H.B. Curry: Essays on Combinatory Logic, Lambda Calculus and Formalism* ED: SELDIN, J.P. & HINDLEY, J.R.
• PUBL (X 0801) Academic Pr: New York xxv+606pp
• ISBN 0-12-349050-2, LC-No 80-40139

C 3094 Moisil: Essais Logiques Non Chrysippiennes • RO
[1972] *Moisil,G.C.: Essais sur les Logiques Non Chrysippiennes* PUBL (X 0871) Acad Rep Soc Romania: Bucharest 820pp
• LC-No 72-345318

C 3095 Approaching Vagueness • NL
[1983] *Approaching Vagueness* ED: BALLMAR, T.T. & PINKAL, M. • SER North-Holland Linguistic Ser. 50 • PUBL (X 0809) North Holland: Amsterdam x+429pp
• ISBN 0-444-86745-7, LC-No 83-14079

C 3174 Twenty-five Years Log Meth Poland • NL
[1977] *Twenty-five Years of Logical Methodology in Poland* ED: PRZELECKI, M. & WOJCICKI, R. • SER (S 3307) Synth Libr 87 • PUBL (X 0835) Reidel: Dordrecht viii+735pp
• LC-No 76-7064
• REM Translated from the Polish

C 3189 Issl Oper Mod, Sist, Reshen, Vyp 4 • SU
[1974] *Issledovanie Operatsij. Modeli, Sistemy, Resheniya. Vypusk 4 (Operations Research. Models, Systems, Solutions. No. 4)* ED: GERMEJER, YU.B. & EVTUSHENKO, YU.G. & IVANILOV, YU.P. & PAVLOVSKIJ, YU.N. & PETROV, A.A. & SOLODOV, V.M. • PUBL (X 2265) Akad Nauk Vychis Tsentr: Moskva 240pp

C 3233 Probl & Ergebn Math 1974/75 • DDR
[1977] *Probleme und Ergebnisse aus der Klasse Mathematik 1974/1975.* ED: SCHEEL, H. • SER Sitzungsberichte der Akademie der Wissenschaften der DDR. Mathematik - Naturwissenschaften - Technik. Jahrgang 1976. Nr. 18/N
• PUBL (X 0911) Akademie Verlag: Berlin 72pp
• LC-No 79-640513

C 3263 Sint Diskr Avtom Upravl Ustr • SU
[1968] *Sintez Diskretnykh Avtomatov i Upravlyayushchikh Ustrojstv (Synthese Diskreter Automaten und Steuervorrichtungen)* PUBL (X 2027) Nauka: Moskva 92pp
• LC-No 70-379696

C 3271 Issl Teor Mnozh & Neklass Logik • SU
[1976] *Issledovaniya po Teorij Mnozhestv i Neklassicheskim Logikam. Sbornik Trudov (Studies in Set Theory and Nonclassical Logics. Collection of Papers)* ED: BOCHVAR, D.A. & GRISHIN, V.N. • PUBL (X 2027) Nauka: Moskva 328pp
• LC-No 77-501571

C 3389 Fuzzy Sets & Decision Anal • NL
[1984] *Fuzzy Sets and Decision Analysis* ED: ZIMMERMANN, H.-J. & ZADCH, L.A. & GAINES, B.R. • SER TIMS Studies in the Management Sciences 20 • PUBL (X 0809) North Holland: Amsterdam vii+522pp
• ISBN 0-444-86593-4, LC-No 83-8199

C 3433 Prog in Cybern & Syst Res, Vol 6 • A
[1982] *Progress in Cybernetics and Systems Research. Vol. 6 Cybernetics in Biology and Medicine, Systems Analysis, Systems Engineering Methodology, Mathematical Systems Theory.* ED: PICHLER, F.R. & TRAPPL, R. • PUBL (X 0822) McGraw-Hill: New York xv+398pp • ALT PUBL (X 2437) Hemisphere Publ: Washington
• ISBN 0-89116-194-5, LC-No 75-6641

C 3484 Diff, Beskoal, Koop & Stat Igry • SU
[1979] *Differentsial'nye, Beskoalitsionnye, Kooperativnye i Statisticheskie Igry. Mezhvuzovskij Tematicheskij Sbornik (Differential, Non-Coalitional, Cooperative and Statistical Games. Interuniversity Thematic Work Collection)* ED: ASKENASY, V.O. • PUBL (X 1434) Kalinin Gos Univ: Kalinin 109pp

C 3514 Adv Fuzzy Sets & Appl • NL
[1979] *Advances in Fuzzy Set Theory and Applications* ED: GUPTA, M.M. & RAGADE, R.K. & YAGER, R.R. • PUBL (X 0809) North Holland: Amsterdam xv+753pp
• ISBN 0-444-85372-3, LC-No 79-17151

C 3515 Ital Studies in Phil of Sci • NL
[1981] *Italian Studies in the Philosophy of Science* ED: CHIARA SCABIA DALLA, M.L. • SER (S 3311) Boston St Philos Sci 47 • PUBL (X 0835) Reidel: Dordrecht xi+525pp
• ISBN 90-277-0735-9, LC-No 80-16665

C 3537 Stud Induct Logic & Probab, Vol 2 • USA
[1980] *Studies in Inductive Logic and Probability. Vol 2* ED: JEFFREY, R.C. • PUBL (**X** 0926) Univ Calif Pr: Berkeley i+305pp
• ISBN 0-520-03826-6, LC-No 77-136025 • REL PUBL (**C** 0799) Stud Induct Logic & Probab, Vol 1

C 3566 Phys Theor as Log-Operat Struct • NL
[1979] *Physical Theory as Logico-Operational Structure* ED: HOOKER, C.A. • SER (**S** 3308) Univ Western Ontario Ser in Philos of Sci 7 • PUBL (**X** 0835) Reidel: Dordrecht xvii+334pp
• ISBN 90-277-0711-1, LC-No 78-12481

C 3573 Log-Metodol Probl Estestv & Odshchestbennykh Nauk • SU
[1977] *Logiko-Metodologicheskie Problemy Estestvennykh i Odshchestbennykh Nauk (Logico-Methodological Problems of the Natural Science and the Social Science)* ED: TSELISHCHEV, V.V. • PUBL (**X** 2642) Nauka: Novosibirsk 192pp
• LC-No 77-517318

C 3582 Adv Fuzzy Sets, Possibility Th & Appl • USA
[1983] *Advances in Fuzzy Sets, Possibility Theory, and Applications* ED: WANG, P.P. • PUBL (**X** 1332) Plenum Publ: New York xii+421pp
• ISBN 0-306-41390-6, LC-No 83-11077

C 3585 Sb Rabot Mat Kibernetike, Vyp 1 • SU
[1976] *Sbornik Rabot po Matematicheskoj Kibernetike. Vyp 1 (A Collection of Papers on Mathematical Cybernetics. No. 1)* ED: ZHURAVLEV, YU.I. • PUBL (**X** 2265) Akad Nauk Vychis Tsentr: Moskva 247pp

C 3611 Semantics from Diff Points of View • D
[1979] *Semantics from Different Points of View* ED: BAEUERLE, R. & EGLI, U. & STECHOW VON, A. • PUBL (**X** 0811) Springer: Heidelberg & New York viii+419pp
• ISBN 3-540-09676-0, LC-No 80-509679

C 3613 Literary Semant & Possible Worlds • H
[1980] *Literary Semantics and Possible Worlds* ED: CSURI, K. • PUBL J.Attila Tudomanyegyetem: Szeged 344pp
• LC-No 81-112079

C 3619 Log Anal of Tense & Aspect • D
[1977] *On the Logical Analysis of Tense and Aspect* ED: ROHRER, C. • PUBL Narr: Tuebingen
• LC-No 77-574989

C 3659 Intens Math • NL
[1985] *Intensional Mathematics* ED: SHAPIRO, S. • SER (**S** 3303) Stud Logic Found Math 113 • PUBL (**X** 0809) North Holland: Amsterdam
• ISBN 0-444-87632-4, LC-No 84-18856

C 3670 Intention & Intentionalities • USA
[1979] *Intention and Intentionalities* ED: DIAMOND, C. & TEICHMAN, J. • PUBL (**X** 3719) Harvester Pr: Sussex 265pp
• LC-No 79-2478

C 3680 Platts: Refer Truth & Reality • GB
[1980] *Platts,M.: Reference Truth and Reality* PUBL (**X** 0866) Routledge & Kegan Paul: Henley on Thames 350pp
• LC-No 79-42859

C 3731 New Studies Deontic Log • NL
[1981] *New Studies in Deontic Logic. Norms, Actions, and the Foundations of Ethics* ED: HILPINEN, R. • SER (**S** 3307) Synth Libr 152 • PUBL (**X** 0835) Reidel: Dordrecht ix+256pp
• ISBN 90-277-1278-6, ISBN 90-277-1346-4, LC-No 81-12079

C 3737 Grundl-probl Modern Physik (Mittelstaedt) • D
[1981] *Grundlagenprobleme der Modernen Physik. Festschrift for Peter Mittelstaedt on His 50th Birthday* ED: NITSCH, J. & PFARR, J. & STACHOW, E.W. • PUBL (**X** 0876) Bibl Inst: Mannheim 317pp
• ISBN 3-411-01600-0, LC-No 82-179246

C 3743 Probl Log & Metodol Nauk • SU
[1982] *Problemy Logiki i Metodologii Nauki (Problems of Logic and the Methodology of Science)* ED: TSELISHCHEV, V.V. • PUBL (**X** 2027) Nauka: Moskva 336pp
• LC-No 82-197677

C 3747 Mat Log & Mat Lingvistika • SU
[1981] *Matematicheskaya Logika i Matematicheskaya Lingvistika (Mathematical Logic and Mathematical Linguistics)* ED: GLADKIJ, M. • PUBL (**X** 1434) Kalinin Gos Univ: Kalinin 172pp

C 3778 Fuzzy Inform & Decis Processes • NL
[1982] *Fuzzy Information and Decision Processes* ED: GUPTA, M.M. & SANCHEZ, E. • PUBL (**X** 0809) North Holland: Amsterdam xxi+451pp
• ISBN 0-444-86491-1, LC-No 82-14367

C 3786 Approx Reason in Decis Anal • NL
[1982] *Approximative Reasoning in Decision Analysis* ED: GUPTA, M.M. & SANCHEZ, E. • PUBL (**X** 0809) North Holland: Amsterdam xix+455pp
• ISBN 0-444-86492-X, LC-No 82-14308

C 3792 Log & Filos Kategorii • SU
[1982] *Logika i Filosofskie Kategorij (Logic and Philosophical Categories)* ED: PLOTNIKOV, A.M. & CHUPAKHIN, I.YA. • PUBL (**X** 0938) Leningrad Univ: Leningrad 152pp
• LC-No 82-229178

C 3798 Mat Log, Mat Ling & Teor Algor • SU
[1983] *Matematicheskaya Logika, Matematicheskaya Lingvistike i Teoriya Algoritmov (Mathematical Logic, Mathematical Linguistics and Theory of Algorithms)* ED: GLADKIJ, A.V. • PUBL (**X** 1434) Kalinin Gos Univ: Kalinin 116pp
• LC-No 84-157035

C 3799 Logic & Religion • B
[1982] *Logic and Religion* ED: HUBBELING, H. & APOSTEL, L. & VANDAMME, F. • SER (**J** 2705) Comm Cognition Monograph 15/2 • PUBL (**X** 2301) Commun & Cognition: Ghent 73pp
• LC-No 84-206716

C 3807 Issl Neklass Log & Formal Sist • SU
[1983] *Issledovaniya po Neklassicheskim Logikam i Formal'nym Sistemam (Studies in Nonclassical Logics and Formal Systems)* ED: MIKHAJLOV, A.I. • PUBL (**X** 2027) Nauka: Moskva 360pp
• LC-No 83-181942

C 3810 Log-Semat Issl (Tbilisi) • SU
[1981] *Logiko-Semanticheskie Issledovaniya (Studies in Logic and Semantics)* ED: MIKELADZE, Z.N. • PUBL (**X** 3790) Metsniereba: Tbilisi 144pp
• LC-No 82-159969

C 3811 Logic 20th Century • I
[1983] *Logic in the 20th Century. A Series of Papers on the Present State and Tendencies of Studies* PUBL Scientia: Milano 246pp
• LC-No 83-212666

C 3813 Ponim Kak Log-Gnoseolog • SU
[1982] *Ponimanie Kak Logiko-Gnoseologicheskaya (Understanding as a Logical-Epistemological Problem)* ED: POPOVICH, M.V. • PUBL (**X** 2199) Naukova Dumka: Kiev 272pp
• LC-No 83-124729

C 3832 Avtom Issl Mat • SU
[1982] *Avtomatizatsiya Issledovanij v Matematike. Sbornik Nauchnykh Trudov (Automatization of Investigations in Mathematics. Collection of Scientific Works)* ED: KAPITONOVA, YU.V. • PUBL (**X** 2522) Akad Nauk Inst Kibernet: Kiev 98pp
• LC-No 83-165025

C 3834 Lang, Logic and Method • NL
[1983] *Language, Logic, and Method* ED: COHEN, R.S. & WARTOFSKY, M.W. • SER (**S** 3311) Boston St Philos Sci 31
• PUBL (**X** 0835) Reidel: Dordrecht viii+464pp
• LC-No 82-7558

C 3849 Modal & Relevant Log, Vyp 1 • SU
[1982] *Modal'nye i Relevantnye Logiki. Trudy Nauchno-Issledovatel'skogo Seminara po Logike Institute Filosofii AN SSSR Vyp.1 (Modal and Relevant Logics. Vol.1)* ED: SMIRNOV, V.A. & KARPENKO, A.S. • PUBL (**X** 0899) Akad Nauk SSSR : Moskva 107pp

C 3902 Met & Modeli Upravleniya & Kontrolya • SU
[1979] *Metody i Modeli Upravleniya i Kontrolya. Mezhvuzovskij Nauchno-Tekhnicheskij Sbornik (Methods and Models of Control and Monitoring. Interuniversity Scientific-Technical Collection)* ED: AUZIN', P.K. • PUBL Rizhskij Politekh Inst: Riga 180pp

C 3953 Mat Log & Teor Algor • SU
[1982] *Matematicheskaya Logika i Teoriya Algoritmov (Mathematical Logic and the Theory of Algorithms)* ED: SOBOLEV, S.L. • SER Trudy Inst Mat 2 • PUBL (**X** 2642) Nauka: Novosibirsk 176pp

C 4026 Unters Log & Methodol, Vol 1 • DDR
[1984] *Untersuchung zur Logik und zur Methodologie, Vol 1* ED: KREISER, L. & STREHLE, P. • PUBL (**X** 2373) Karl-Marx-Univ: Leipzig ii+93pp

C 4055 Wajsberg: Logical Works • PL
[1977] *Wajsberg,M.: Logical Works* ED: SURMA, S.J. • PUBL (**X** 2882) Pol Akad Nauk: Wroclaw 216pp
• LC-No 84-246718

C 4062 Montague: Formal Philos • USA
[1974] *Formal Philosophy. Selected Papers of Richard Montague* ED: THOMASON, R.H. • PUBL (**X** 0875) Yale Univ Pr: New Haven 369pp
• ISBN 0-300-01527-5, LC-No 73-77159

C 4065 Theorie der Normen • D
[1984] *Theorie der Normen* ED: KRAWIETZ, W. & SCHELSKY, H. & SCHRAMM, A. & WINKLER, G. • PUBL (**X** 1258) Duncker & Humblot: Berlin xiii+627pp

C 4085 Handb Philos Log • NL
[1983 & 1984] *Handbook of Philosophical Logic* ED: GABBAY, D. & GUENTHNER, F. • SER (**S** 3307) Synth Libr 164,165
• PUBL (**X** 0835) Reidel: Dordrecht 2 Vols: xi+493pp,xi+776pp
• ISBN 90-277-1542-4 (V1), ISBN 90-277-1604-8 (V2), LC-No 83-4277

C 4091 Algeb & Diskret Mat (Riga) • SU
[1984] *Algebra i Diskretnaya Matematika (Algebra and Discrete Mathematics)* ED: IKAUNIEKS, EH.A. • PUBL (**X** 0895) Latv Valsts (Gos) Univ : Riga 164pp

C 4106 Phil of Russell • USA
[1944] *The Philosophy of Bertrand Russell* ED: SCHILPP, P.A.
• PUBL (**X** 1318) Northwestern Univ Pr: Evanston 829pp
• LC-No 44-6786

C 4138 Feys: Modal Logika • SU
[1974] *Feys,R.: Modal'naya Logika (Modal Logic)* ED: MINTS, G.E. • PUBL (**X** 2027) Nauka: Moskva 520pp
• TRANSL OF Modal Logic
• REM Contains translations of Kripke,S.A. 1959 ID 07554, 1962 ID 07555, 1963 ID 07558, 1965 ID 07559 and of Schuette,K. 1968 ID 24822

C 4139 Reforging Great Chain of Being • NL
[1981] *Reforging the Great Chain of Being. Studies of the History of Modal Theories* ED: KNUUTTILA, S. • SER Synthese Historical Library 20 • PUBL (**X** 0835) Reidel: Dordrecht xiv+320pp
• ISBN 90-277-1125-9, LC-No 80-19869

C 4140 Ifs • NL
[1981] *Ifs. Conditionals, Belief, Decision, Chance, and Time* ED: HARPER, W.L. & STALNAKER, R. & PEARCE, G. • SER (**S** 3308) Univ Western Ontario Ser in Philos of Sci 15
• PUBL (**X** 0835) Reidel: Dordrecht x+345pp
• ISBN 90-277-1184-4, ISBN 90-277-1220-4, LC-No 80-21638

C 4181 Math Log & Formal Syst (Costa da) • USA
[1985] *Mathematical Logic and Formal Systems. A Collection of Papers in Honor of Newton C.A. da Costa* ED: ALCANTARA DE, L.P. • SER (**S** 3310) Lect Notes Pure Appl Math 94 • PUBL (**X** 1684) Dekker: New York xv+297pp
• ISBN 0-8247-7330-6, LC-No 84-25984

C 4270 Stud in Probab & Rel Topics (Onicescu) • I
[1983] *Studies in Probabilistic and Related Topics. Papers in Honour of Octav Onicescu on His 90th Birthday* ED: DEMETRESCU, M.C. & IOSIFESCU, M. • PUBL Ed Nagard Srl: Roma 479pp

C 4283 Reduct, Time & Reality • GB
[1981] *Reduction, Time and Reality. Studies in the Philosophy of the Natural Science* ED: HEALEY, R. • PUBL (**X** 0805) Cambridge Univ Pr: Cambridge, GB xi+202pp
• ISBN 0-521-23708-4, LC-No 80-41535

C 4290 Log & Epistem Stud Contemp Phys • NL
[1974] *Logical and Epistemological Studies in Contemporary Physics* ED: COHEN, R.S. & WARTOFSKY, M.W. • PUBL (**X** 0835) Reidel: Dordrecht
• LC-No 73-83557

C 4318 Quant, Essay in Th Phys (Wentzel) • USA
[1970] *Quanta, Essays in Theoretical Physics. Dedicated to Gregor Wentzel* ED: FREUND, P.G.O. & GOEBEL, C.J. & NAMBU, Y. • PUBL (**X** 0862) Univ Chicago Pr: Chicago xiv+414pp
• LC-No 70-108268, ISBN 0-226-26280-4

C 4327 Problemsemin.Algeb & Log Grundl Inform • DDR
Problemseminar. Algebraische und Logische Grundlagen der Informatik ED: WEISSIG • PUBL Technische Univ: Dresden

C 4366 Modal & Intens Log & Primen Probl Metodol Nauk • SU
Modalnaya i Intensionalnaya Logiki i Ikh Primeneniek Problemam Metodologii (Modal- und Intensional-Logik und ihre Anwendung auf die Probleme der Methodologie)

C 4403 Logika, Pozn, Otrazh • SU
[1984] *Logika, Poznanie, Otrazhenie - Sverdlovsk (Logik, Erkenntnis, Reflektion - Sverdlovsk)*

C 4451 Sci Phil Today (Bunge) • NL
[1981] *Scientific Philosophy Today. Essays in Honour of M. Bunge* PUBL (**X** 0835) Reidel: Dordrecht

C 4454 Seman Inf Process • USA
[1964] *Semantic Information Processing* ED: MINSKY, M. • PUBL (**X** 2360) MIT Alumni Ass: Cambridge vii + 440pp

C 4528 Handlg-Th Interdiszipl, Vol 1 • D
[1980] *Handlungstheorien Interdisziplinaer I* ED: LENK, H. • PUBL (**X** 2162) Fink: Muenchen

C 4638 Stud in Modelth Seman • NL
[1983] *Studies in Modeltheoretic Semantics* ED: MEULEN TER, A.G.B. • SER (**S** 4560) Groningen-Amsterdam Stud Semant 1 • PUBL (**X** 4217) Foris: Dordrecht x + 206pp

C 4652 Lect on Modern Math (Japanese) • J
[1965] *(Lectures on Modern Mathematics)* ED: SAATY, T.L. • PUBL (**X** 3552) Iwanami Shoten: Tokyo
• TRANSL OF [1963] (**C** 1602) Lect on Modern Math

C 4675 Hintikka: Logiko-Epist Issled • SU
[1980] *Hintikka, K.J.J.: Logiko-Epistemologicheskie Issledovanniya (Hintikka, K.J.J.: Logical-Epistemological Investigations.)* ED: SADOVSKIJ, V.N. & SMIRNOV, V.A. • PUBL (**X** 2055) Progress: Moskva 448pp

C 4678 Enzykl Geisteswiss Arbeitsmethoden • D
[1968] *Enzyklopaedie der Geisteswissenschaftlichen Arbeitsmethoden. 3. Lieferung: Methoden der Logik und Mathematik. Statistische Methoden* PUBL (**X** 0814) Oldenbourg: Muenchen

C 4679 Stud & Essays to Courant • USA
[1948] *Studies and Essays Presented to R. Courant on His 60th Birthday, January 8, 1948* PUBL (**X** 0820) Intersci Publ: New York

C 4684 Probab Method in Appl Math, Vol 2 • USA
[1970] *Probabilistic Methods in Applied Mathematics, Vol 2* ED: BHARUCHA-REID, A.T. • PUBL (**X** 0801) Academic Pr: New York x + 220pp

C 4688 Prog in Math, Vol 12 • USA
[1972] *Progress in Mathematics. Vol 12: Algebra and Geometry* ED: GAMKRELIDZE, R.V. • PUBL (**X** 1332) Plenum Publ: New York ix + 254pp
• TRANSL OF [1972] (**J** 1501) Itogi Nauki Tekh, Ser Algeb, Topol, Geom 1968

C 4691 Dialektika & Logika, Formy Mysl • SU
[1962] *Dialektika i Logika, Formy Myshleniya (Dialectic and Logic, The Forms of Thought)* PUBL (**X** 0899) Akad Nauk SSSR : Moskva

C 4722 Inform, Infer & Decis • NL
[1974] *Information, Inference and Decision* ED: MENGES, G. • SER Theory & Decis Library 1 • PUBL (**X** 0835) Reidel: Dordrecht viii + 195pp
• ISBN 90-277-0422-8

C 4723 Math Founds of Comput Sci • PL
[1977] *Mathematical Foundations of Computer Science* ED: MAZURKIEWICZ, A. & PAWLAK, Z. • SER (**S** 1926) Banach Cent Publ 2 • PUBL (**X** 1034) PWN: Warsaw 259pp

Publishers

X 0221 *Universidad de la Habana* (Havanna, C)

X 0272 *Editions du Griffon* (Neuchatel, CH) ISBN 2-88006

X 0670 *Inst Problem Upravlen* (Moskva, SU)

X 0740 *Shanghai Kexue Jishu Chubanshe (Scientific and Technical Press)* (Shanghai, TJ)

X 0801 *Academic Press* (New York, NY, USA & London, GB) ISBN 0-12

X 0802 *Allyn & Bacon* (London, GB & Boston, MA, USA & Spit Junction, NSW, AUS) ISBN 0-205, ISBN 0-695

X 0803 *American Mathematical Society* (Providence, RI, USA) ISBN 0-8218

X 0804 *Birkhaeuser Verlag* (Basel, CH & Stuttgart, D & Cambridge, MA, USA) ISBN 3-7643

X 0805 *The Cambridge University Press.* (Cambridge, GB & New York, NY, USA & Melbourne, Vic, AUS) ISBN 0-521

X 0806 *VEB Deutscher Verlag der Wissenschaften* (Berlin, DDR) ISBN 3-326

X 0807 *Duke University Press* (Durham, NC, USA) ISBN 0-8223

X 0808 *W. Kohlhammer* (Stuttgart, D & Koeln, D & Berlin, D & Mainz, D) ISBN 3-17

X 0809 *North-Holland Publishing Company.* (Amsterdam, NL & Oxford, GB) ISBN 0-7204 • REL PUBL (**X** 0838) Amer Elsevier: New York

X 0810 *W.B.Saunders Company* (Philadelphia, PA, USA & Eastbourne, GB & Toronto, ON, CDN) ISBN 0-7216 • REL PUBL (**X** 0818) Holt Rinehart & Winston: New York • REM In United Kingdom: Holt-Saunders: Eastbourne, GB

X 0811 *Springer-Verlag* (Heidelberg, D & Berlin, D & New York, NY, USA & Tokyo, J) ISBN 3-540, ISBN 0-387 • REL PUBL (**X** 1231) Barth: Leipzig & (**X** 0902) Springer: Wien

X 0812 *Wolters-Noordhoff* (Groningen, NL) ISBN 90-01 • REL PUBL (**X** 1317) Noordhoff: Groningen

X 0813 *Dover Publications* (New York, NY, USA) ISBN 0-486

X 0814 *R.Oldenbourg Verlag* (Muenchen, D & Wien, A) ISBN 3-486

X 0815 *The Clarendon Press* (Oxford, GB) ISBN 0-19 • REL PUBL (**X** 0894) Oxford Univ Pr: Oxford • REM This Imprint is Used for Academic Books Published by X0894.

X 0816 *Methuen & Company* (London, GB & New York, NY, USA & Agincourt, M, CDN & North Ryde, AUS) ISBN 0-416

X 0818 *Holt Rinehart & Winston* (New York, NY, USA & Toronto, ON, CDN & Artarmon, NSW, AUS & & Eastbourne,GB) ISBN 0-03 • REM In Australia & United Kingdom: Holt-Saunders: Eastbourne, GB & Artarmon, NSW, AUS

X 0819 *Prentice Hall* (Englewood Cliffs, NJ, USA & Brookvale, NSW, AUS & Scarborough, ON, CDN) ISBN 0-13 • REL PUBL (**X** 2040) Winthrop: Cambridge

X 0820 *Interscience Publishers* (New York, NY, USA & Chichester, GB) ISBN 0-470 • REL PUBL (**X** 0827) Wiley & Sons: New York

X 0821 *Wadsworth Publishing Co.* (Belmont, CA, USA & Crows Nest, NSW, AUS & Artamon, NSW, AUS) ISBN 0-534

X 0822 *McGraw-Hill Book Company* (New York, NY, USA & Roseville, NSW, AUS & Isando, ZA & Maidenhead, GB & Singapore, SGP & Scarborough, CDN & Sao Paulo, BR) ISBN 0-07 • REM Member Firms: 1) CRM/McGraw-Hill, Del Mar, CA, USA. 2) CTB/McGraw-Hill, Monterey, CA, USA. 3) Edutronics/McGraw-Hill, Los Angeles, CA, USA. 4) Instruto/McGraw-Hill, Paoli, PA, USA. 5) McGraw-Hill Continuing Education Center, Washington, DC, USA. 6) McGraw-Hill International Book Company, Singapore, SGP. 7) Sheperd's/McGraw-Hill, Colorado Springs, CO, USA. 8) McGraw-Hill do Brasil, Sao Paulo, BR.

X 0823 *B.G.Teubner* (Stuttgart, D) ISBN 3-519 • REM See also X1079

X 0824 *The Free Press* (New York, NY, USA) ISBN 0-02 • REL PUBL (**X** 0843) Macmillan : New York & London

X 0825 *Westkulturverlag Anton Hain* (Meisenheim am Glan, D & Koenigstein, D) ISBN 3-445

X 0826 *Verlag Karl Alber* (Freiburg, D & Muenchen, D) ISBN 3-495 • REL PUBL (**X** 1279) Herder: Freiburg

X 0827 *J.Wiley & Sons* (New York, NY, USA & Chichester, GB & Rexdale, ON, CDN & Auckland, NZ) ISBN 0-471 • REL PUBL (**X** 0942) Norton: New York & (**X** 0820) Intersci Publ: New York & (**X** 0880) Ronald Press: New York & (**X** 2737) Israel Progr Sci Transl: Jerusalem

X 0828 *Edwards Brothers* (Ann Arbor, MI, USA) ISBN 0-910546

X 0832 *Addison-Wesley Publishing Co.* (Reading, MA, USA & London, GB & Don Mills, ON, CDN & North Ryde, NSW, AUS) ISBN 0-201 • REL PUBL (**X** 0867) Benjamin: Reading

X 0833 *Jerusalem Academic Press* (Jerusalem, IL)

X 0834 *Gauthier-Villars Editeur* (Paris, F) ISBN 2-04

X 0835 *D.Reidel Publishing Company* (Dordrecht, NL & Hingham, MA, USA) ISBN 90-277

X 0836 *Gordon & Breach, Science Publishers* (New York, NY, USA & London, GB & Paris, F) ISBN 0-677

X 0837 *Harper & Row, Publishers* (New York, NY, USA & London, GB & Artamon, NSW, AUS & Petone, NZ) ISBN 0-06 • REM Member Firm: Lippincott, J.B.,Company: New York, NY, USA

X 0838 *American Elsevier Publishing Co.* (New York, NY, USA & Amsterdam, NL & London, GB) ISBN 0-444, ISBN 0-525, ISBN 0-87690 • REL PUBL (**X** 0809) North Holland: Amsterdam

X 0840 *Editions Presses Universitaires de France* (Paris, F) ISBN 2-13 • REL PUBL (**X** 2435) Pr Univ France Period: Paris

X 0841 *Blaisdell Publishing Company* (New York, NY, USA & London, GB & Toronto, CDN)

X 0842 *University of Wisconsin Press* (Madison, WI, USA & London, GB) ISBN 0-299 • REL PUBL (**X** 3828) Amer Univ Publ Group: London

X 0843 *Macmillan Publishing Company* (New York, NY, USA & Melbourne, Vic, AUS & London, GB & Toronto, Ont, CDN) ISBN 0-02 • REL PUBL (**X** 2375) Macmillan Journal: London & (**X** 0824) Free Press: New York

X 0844 *Giangiacomo Feltrinelli Editore* (Milano, I) ISBN 88-07

X 0845 *University of Notre Dame Press* (Notre Dame, IN, USA & London, GB) ISBN 0-268

X 0846 *Ferdinand Schoeningh* (Paderborn, D & Muenchen, D & Wien, A) ISBN 3-506

X 0847 *Houghton Mifflin Company* (Boston, MA, USA & Markham, CDN) ISBN 0-395, ISBN 0-89289

X 0848 *Chelsea Publishing Company* (New York, NY, USA) ISBN 0-8284

X 0849 *F.G.Kroonder, Uitgeverij* (Bussum, NL)

X 0850 *Armand Colin, Editeur* (Paris, F) ISBN 2-200

X 0851 *Edmund C. Berkeley and Associates* (New York, NY, USA)

X 0854 *Littlefield, Adams & Co.* (Totowa, NJ, USA) ISBN 0-8226

X 0856 *Dunod, Editeur* (Paris, F) ISBN 2-04

X 0857 *Princeton University Press* (Princeton, NJ, USA & Guildford, GB) ISBN 0-691

X 0858 *Harvard University Press* (Cambridge, MA, USA & London, GB) ISBN 0-674

X 0859 *Hermann, Editeurs des Sciences et des Arts* (Paris, F) ISBN 2-7056

X 0862 *University of Chicago Press* (Chicago, IL, USA & London, GB) ISBN 0-226

X 0863 *Harcourt Brace Jovanovich* (New York, NY, USA & London, GB & Boston, MA, USA & North Ryde, NSW, AUS) ISBN 0-15, ISBN 0-515 • REM Also: HBJ Press: Boston, MA, USA

X 0864 *Van Nostrand, Reinhold* (New York, NY, USA & Scarborough, ON, CDN & Mitcham, Vic, AUS & & Wokingham, GB & Florence, KY, USA) ISBN 0-442

X 0865 *The MIT Press* (Cambridge, MA, USA & London, GB) ISBN 0-262

X 0866 *Routledge & Kegan Paul* (Henley on Thames, GB & Boston, MA, USA) ISBN 0-7100, ISBN 0-7102

X 0867 *W.A. Benjamin* (Reading, MA, USA) ISBN 0-8053 • REL PUBL (**X** 0832) Addison-Wesley: Reading

X 0868 *Penguin Books* (Harmondsworth, GB & New York, NY, USA & Ringwood, Vic, AUS & & Auckland, NZ & Markham, CDN) ISBN 0-14

X 0869 *Pergamon Press* (Oxford, GB & Elmsford, NY, USA & Rushcutters Bay, NSW, AUS & & Willowdale, ON, CDN & Paris, F) ISBN 0-08 • REL PUBL (**X** 0900) Vieweg: Wiesbaden

X 0870 *Physica-Verlag Rudolf Liebing* (Wuerzburg, D & Wien, A) ISBN 3-7908

X 0871 *Academiei Republicii Socialiste Romania Editura (RSR)* (Bucharest, RO)

X 0872 *Albert Blanchard* (Paris, F)

X 0873 *Mouton et Cie.* (Paris, F) ISBN 2-7193

X 0875 *Yale University Press* (New Haven, CT, USA & London, GB) ISBN 0-300

X 0876 *Bibliographisches Institut* (Mannheim, D & Wien, A & Zuerich, CH) ISBN 3-411

X 0877 *Max Niemeyer Verlag* (Tuebingen, D & Halle, DDR) ISBN 3-484

X 0878 *Doubleday & Co.* (London, GB & New York, NY, USA & Garden City, NY, USA & Auckland, NZ & Toronto, ON, CDN & Sydney, AUS) ISBN 0-385 • REM In Australia: Anchor Books: Sydney, AUS

X 0879 *Institut Superieur de Philosophie* (Louvain, B)

X 0880 *Ronald Press & Co.* (New York, NY, USA) ISBN 0-8260 • REL PUBL (**X** 0827) Wiley & Sons: New York

X 0881 *Societa Editrice Il Mulino S.R.I.* (Bologna, I)

X 0882 *Uppsala Universitet, Filosofiska Foereningen och Filosofiska Institutionen* (Uppsala, S)

X 0885 *Izdatel'stvo Mir* (Moskva, SU)

X 0886 *Leicester University Press* (Leicester, GB) ISBN 0-7185

X 0888 *New York University, Institute of Mathematical Sciences.* (New York, NY, USA)

X 0890 *Wissenschaftliche Buchgesellschaft* (Darmstadt, D) ISBN 3-534

X 0891 *Larousse International* (Paris, F) ISBN 2-03

X 0892 *Georg Olms Verlag* (Hildesheim, D & New York, NY, USA & Zuerich, CH) ISBN 3-487

X 0893 *Les Presses de l'Universite de Montreal* (Montreal, PQ, CDN) ISBN 2-7606

X 0894 *Oxford University Press* (Oxford, GB & London, GB & Melbourne, Vic, AUS & Don Mills, ON, CDN & Nairobi, EAK & Auckland, NZ & Petaling Jaya, MAL & New York, NY, USA & Karachi, PAK & Harare, ZW) ISBN 0-19 • REL PUBL (**X** 0815) Clarendon Pr: Oxford

X 0895 *Latvijas Valsts Universitate* (Riga, SU)

X 0898 *Izdatel'stvo Moskovskogo Gosudarstvennogo Universiteta* (Moskva, SU)

X 0899 *Izdatel'stvo Akademii Nauk SSSR* (Moskva, SU)

X 0900 *Vieweg, Friedrich & Sohn Verlagsgesellschaft* (Wiesbaden, D) ISBN 3-528 • REL PUBL (**X** 0869) Pergamon Pr: Oxford

X 0901 *C.F. Winter'sche Verlagshandlung* (Braunschweig, D) ISBN 3-8065

X 0902 *Springer-Verlag* (Wien, A) ISBN 3-211 • REL PUBL (**X** 0811) Springer: Heidelberg & New York

X 0903 *Vandenhoeck & Ruprecht* (Goettingen, D) ISBN 3-525 • REM Member Firms: 1) E. Klotz Verlag, Goettingen, D. 2) Verlag der Medizinischen Psychologie Goettingen, Dr D.& Dr A. Ruprecht, Goettingen, D

X 0904 *Australian National University, Research School of Social Sciences, Department of Philosophy.* (Canberra, ACT, AUS) ISBN 0-909596

X 0905 *Boringhieri Editore* (Torino, I) ISBN 88-339

X 0908 *Universitaet Bonn, Mathematisches Institut* (Bonn, D)

X 0909 *Cedam* (Padova, I)

X 0910 *Aschendorffsche Verlagsbuchhandlung* (Muenster, D) ISBN 3-402

X 0911 *Akademie Verlag* (Berlin, DDR)

X 0913 *Novosibirskij Gosudarstvennyj Universitet* (Novosibirsk, SU)

X 0915 *Galois Institute of Mathematics and Art* (Brooklyn, NY, USA)

X 0917 *Nymphenburger Verlagshandlung* (Muenchen, D) ISBN 3-485

X 0918 *Ernst Klett Verlag* (Stuttgart, D) ISBN 3-12 • REL PUBL (**X** 1255) Deuticke: Wien

X 0920 *Philosophical Library* (New York, NY, USA) ISBN 0-8022

X 0922 *Editura Stiintificha si Enciclopedica* (Bucharest, RO)

X 0923 *Universidade Federal de Pernambuco, Instituto de Matematica* (Recife, BR)

X 0924 *New York University Press* (New York, NY, USA) ISBN 0-8147

X 0926 *University of California Press* (Berkeley, CA, USA & London, GB) ISBN 0-520 • REL PUBL (**X** 1291) Johns Hopkins Univ Pr: Baltimore

X 0927 *University Tutorial Press* (Slough, GB) ISBN 0-7231

X 0928 *Akademiai Kiado, Publishing House of the Hungarian Academy of Sciences.* (Budapest, H) ISBN 963-05

X 0929 *Weizmann Science Press of Israel* (Jerusalem, IL) ISBN 965-270

X 0935 *Southern Illinois University Press* (Carbondale, IL, USA) ISBN 0-8093

X 0938 *Izdatel'stvo Leningradskogo Universiteta* (Leningrad, SU)

X 0939 *Hutchinson Publishing Group* (London, GB & Willowdale, ON, CDN & Salem, NH, USA & Auckland, NZ & Richmond, Vic, AUS) ISBN 0-09

X 0942 *W.W.Norton & Co.* (New York, NY, USA & London, GB) ISBN 0-393 • REL PUBL (**X** 0827) Wiley & Sons: New York

X 0943 *Ginn and Company* (Boston, MA, USA & Aylesbury, GB & Scarborough, ON, CDN) ISBN 0-663

X 0944 *Harvard Cooperative Society* (Cambridge, MA, USA)

X 0949 *St. Martin's Press* (New York, NY, USA) ISBN 0-312, ISBN 0-320

X 0958 *Izdatel'stvo Saratovskogo Universiteta* (Saratov, SU)

X 0959 *George Allen & Unwin* (London, GB & North Sydney, NSW, AUS & Bombay, IND) ISBN 0-04

X 0960 *Little, Brown and Co.* (Boston, MA, USA & Toronto, ON, CDN) ISBN 0-316

X 0981 *Random House* (New York, NY, USA & Mississauga, ON, CDN) ISBN 0-394

X 0992 *Cornell University Press* (Ithaca, NY, USA & London, GB) ISBN 0-8014

X 0994 *W.H.Freeman and Co.* (San Francisco, CA, USA & Oxford, GB) ISBN 0-7167

X 0995 *Beck'sche C.H. Verlagsbuchhandlung Oscar Beck* (Muenchen, D) ISBN 3-406

X 0999 *Centre National de la Recherche Scientifique (CNRS), Institut Blaise Pascal* (Paris, F)

X 1000 *Barnes & Noble Books* (Totowa, NJ, USA & New York, NY, USA) ISBN 0-389

X 1006 *Irkutskij Gosudarstvennyj Universitet* (Irkutsk, SU)

X 1007 *Gerold* (Wien, A) ISBN 3-900190

X 1010 *Loffredo Editore* (Napoli, I)

X 1014 *Universidad Nacional Autonoma de Mexico* (Mexico City, MEX)

X 1026 *Humboldt Verlag* (Frankfurt, D & Muenchen, D & Wien, A) ISBN 3-581

X 1034 *Panstwowe Wydawnictwo Naukowe (PWN)* (Warsaw, PL) ISBN 83-01

X 1035 *Thomas Nelson and Sons* (Walton on Thames, GB & Melbourne, Vic, AUS & Scarborough, ON, CDN & South Nashville, TN, USA) ISBN 0-17

X 1036 *Volk und Wissen, Volkseigener Verlag Berlin* (Berlin, DDR) ISBN 3-353

X 1039 *Moritz Diesterweg* (Frankfurt am Main & Berlin & Duesseldorf & Hamburg & Muenchen) ISBN 3-425

X 1042 *Espasa-Calpe* (Madrid, E & Buenos Aires, RA) ISBN 84-239

X 1052 *Izdatel'stvo Tbilisskogo Universiteta* (Tbilisi, SU)

X 1054 *Verlag Harri Deutsch* (Frankfurt, D & Thun, CH) ISBN 3-87144

X 1056 *Valention Bompiani* (Milano, I)

X 1071 *F. Alcan* (Paris, F)

X 1073 *Adam Hilger* (Bristol, GB) ISBN 0-85274

X 1079 *B.G.Teubner Verlagsgesellschaft* (Leipzig, DDR) ISBN 3-519 • REM See also X0823

X 1080 *Scottish Academic Press* (Edinburgh, GB) ISBN 0-7073

X 1088 *Felix Meiner Verlag* (Hamburg, D) ISBN 3-7873 • REL PUBL (**X** 4718) Meiner: Leipzig

X 1096 *Basil Blackwell* (Oxford, GB) ISBN 0-631, ISBN 0-632, ISBN 0-86286, ISBN 0-86793 • REM Also: Blackwell Scientific Publications: Oxford, GB & Carlton, Vic, AUS

X 1121 *Mathematisch Centrum* (Amsterdam, NL) ISBN 90-6196

X 1163 *Almqvist & Wiksell Foerlag* (Stockholm, S & Bromma, S & Goeteborg, S & Malmoe, S) ISBN 91-20

X 1164 *Charles E. Merrill Publishing Co.* (Columbus, OH, USA & Wembley, GB & Pyrmont, NSW, AUS & Sydney, NSW, AUS) ISBN 0-675 • REM In Canada: Division by Bell & Howell

X 1165 *Hodder & Stroughton Educational* (London, GB & Sevenoaks, GB & Glenfield, NZ & Don Mills, ON, CDN & Lane Cove, NSW, AUS & Ashwood, Vic, AUS & Brisbane, Qld, AUS & Adelaide, SA, AUS & West Perth, WA, AUS) ISBN 0-340 • REM Formerly: The English University Press: London, GB

X 1166 *PZW (= Panstwowe Zaklady Wydawnictwo Szkolnych)* (Warsaw, PL)

X 1167 *Holden-Day* (San Francisco, CA, USA) ISBN 0-8162

X 1168 *Kyoritsu Shuppan Company* (Tokyo, J) ISBN 4-320

X 1169 *Akademische Verlagsgesellschaft* (Wiesbaden, D & Leipzig, DDR) ISBN 3-400

X 1172 *Les Editions de l'Office Central de Librairie S.A.R.L. (O.C.D.L.)* (Paris, F) ISBN 2-7043

X 1174 *Walter de Gruyter* (Berlin, D) ISBN 3-11 • REM Member Firms: 1) de Gruyter: Hawthorne, NY, USA. 2) Mouton Publishers: Berlin, D

X 1187 *Sankibo* (Tokyo, J)

X 1201 *Dickenson Publishing Company* (Belmont, CA, USA) ISBN 0-8221

X 1205 *Junker und Duennhaupt Verlag* (Berlin, D)

X 1212 *Izdatel'stvo Belorusskogo Gosudarstvennogo Universiteta* (Minsk, SU)

X 1214 *New York University, Courant Institute of Mathematical Science.* (New York, NY, USA)

X 1218 *Uzhgorodskij Gosudarstvennyj Universitet* (Uzhgorod, SU)

X 1221 *Frederick Ungar Publishing* (New York, NY, USA & London, GB) ISBN 0-8044

X 1225 *Drei Masken Verlag* (Muenchen, D & Berlin, D)

X 1226 *Academia* (Prague, CS) • REM Publishing House of the Czechoslovak Academy of Science

X 1228 *Appleton-Century-Crofts* (New York, NY, USA) ISBN 0-8385

X 1229 *Edizione della Libreria L'Ateneo* (Napoli, I)

X 1230 *Aufbau-Verlag* (Berlin, DDR & Weimar, DDR) ISBN 3-351

X 1231 *Johann Ambrosius Barth* (Leipzig, DDR) ISBN 3-335 • REL PUBL (**X** 0811) Springer: Heidelberg & New York

X 1233 *Bedminster Press* (Totowa, NJ, USA) ISBN 0-87087

X 1234 *G.Bell & Sons* (London, GB) ISBN 0-7135

X 1236 *Ernest Benn Ltd.-Benn Bros plc* (Tunbridge, GB) ISBN 0-510, ISBN 0-85314, ISBN 0-85459 • REM Also: Benn Business Information Services: Tunbridge Wells, GB

X 1238 *The Bobbs-Merril Company* (Indianapolis, IN, USA) ISBN 0-672 • REL PUBL (**X** 1320) Odyssey Pr: Indianapolis

X 1243 *William C.Brown Co., Publishers* (Dubuque, IA, USA) ISBN 0-697

X 1248 *Chandler Publishing Company* (San Francisco, CA, USA & Clifton, NJ, USA & London, GB) ISBN 0-8102 • REL PUBL (**X** 1288) Intext Pr: New York

X 1253 *Crowell, Collier & Macmillan* (New York, NY, USA & West Drayton, GB) ISBN 0-02 • REM Also: 1) Macmillan Publishing: New York, NY, USA. 2) Collier-Macmillan: West Drayton, GB

X 1255 *Franz Deuticke, Verlagsgesellschaft* (Wien, A) ISBN 3-7005 • REL PUBL (**X** 0918) Klett: Stuttgart

X 1257 *Dryden Press* (Hinsdale, IL, USA & Darlinghurst, NSW, AUS) ISBN 0-8498

X 1258 *Duncker & Humblot* (Berlin, D) ISBN 3-428

X 1261 *Edinburgh University Press* (Edinburgh, GB) ISBN 0-85224

X 1262 *Giuliu Einaudi, Editore* (Torino, I) ISBN 88-06

X 1263 *N.G. Elwert Verlag* (Marburg, D) ISBN 3-7708

X 1265 *Gustav Fischer Verlag* (Stuttgart, D) ISBN 3-437

X 1272 *Gutenberg-Gesellschaft* (Mainz, D) ISBN 3-7755

X 1274 *Hafner Press* (New York, NY, USA) ISBN 0-02 • REL PUBL (**X** 1253) Crowell Collier & Macmillan: New York & (**X** 0843) Macmillan : New York & London

X 1275 *Anton Hain K.G. Meisenheim Verlag* (Koenigstein, D & Meisenheim, D) ISBN 3-445 • REL PUBL (**X** 1588) Athenaeum/Hain/Hanstein: Koenigstein

X 1277 *Hayden Book Company* (Rochelle Park, NY, USA) ISBN 0-8104 • REL PUBL (**X** 1354) Spartan Books : Sutton

X 1279 *Herder & Co.* (Freiburg, D & Roma, I) ISBN 3-451

X 1283 *Ulrico Hoepli, Casa Editrice Libreria* (Milano, I) ISBN 88-203

X 1285 *University of Illinois Press* (Urbana, IL, USA & London, GB) ISBN 0-252 • REL PUBL (**X** 3828) Amer Univ Publ Group: London

X 1286 *Indiana University Press* (Bloomington, IN, USA & London, GB) ISBN 0-253 • REL PUBL (**X** 3828) Amer Univ Publ Group: London

X 1288 *Intext Press* (New York, NY, USA) ISBN 0-88444 • REL PUBL (**X** 1248) Chandler: San Francisco

X 1290 *Richard D.Irwin* (Homewood, IL, USA & London, GB) ISBN 0-256 • REL PUBL Pandemic: London, GB

X 1291 *Johns Hopkins University Press* (Baltimore, MD, USA) ISBN 0-8018

X 1292 *Alfred Kroener Verlag* (Stuttgart, D) ISBN 3-520

Source Index

X 1296 *Longman Group* (Harlow, GB & New York, NY, USA & Melbourne, Vic, AUS & Surry Hills, NSW, AUS & Brisbane, Qld, AUS & Adelaide, SA, AUS & Perth, WA, AUS & Quarry Bay, HK & Auckland, NZ & Jurong Town, SGP & & Kampala, EAU & Dar es Salaam, EAT & Cape Town, ZA & Kingston,JA) ISBN 0-582 • REM Member Firm: Keesing's Reference Publications: Harlow, GB

X 1297 *J.B.Lippincott* (Philadelphia, PA, USA) ISBN 0-397, ISBN 0-06 • REL PUBL (**X** 0837) Harper & Row: New York & (**X** 1096) Blackwell: Oxford

X 1298 *Mathematical Association of America* (Washington, DC, USA) ISBN 0-88385

X 1306 *University of Miami Press* (Austin, TX, USA) ISBN 0-87024

X 1307 *University of Minnesota Press* (Minneapolis, MN, USA & London, GB) ISBN 0-8166

X 1308 *E.S.Mittler & Sohn* (Herford, D & Bonn, D) ISBN 3-87547, ISBN 3-8132

X 1309 *J.C.B.Mohr (Paul Siebeck)* (Tuebingen, D) ISBN 3-16

X 1310 *Monarch Press* (New York, NY, USA) ISBN 0-671

X 1312 *Editions Aubier-Montaigne* (Paris, F) ISBN 2-7007

X 1313 *Nauwelaerts, Beatrice* (Louvain, B) ISBN 2-900014

X 1316 *Martinus Nijhoff Social Sciences Devision* (Leiden, NL & s'Gravenhage, NL) ISBN 90-207, ISBN 90-247 • REL PUBL (**X** 4624) Kluwer-Nijhoff: Boston

X 1317 *P.Noordhoff International Publishing* (Groningen, NL & Leiden, NL) ISBN 90-01 • REL PUBL (**X** 0812) Wolters-Noordhoff : Groningen & (**X** 1352) Sijthoff: Leiden • REM Now: Sijthoff & Noordhoff International Publishers: Leiden, NL

X 1318 *Northwestern University Press* (Evanston, IL, USA) ISBN 0-8101

X 1319 *La Nuova Italia Editrice* (Firenze, I) ISBN 88-221

X 1320 *Odyssey Press* (Indianapolis, IN, USA) ISBN 0-8399 • REL PUBL (**X** 1238) Bobbs-Merril: Indianapolis

X 1321 *Oeffentliches Leben* (Frankfurt, D)

X 1323 *Oliver & Boyd* (Edinburgh, GB) ISBN 0-05

X 1324 *Open Court Publishing Co.* (LaSalle, IL, USA) ISBN 0-87548, ISBN 0-89688

X 1325 *Orbis Nakladatelstvi* (Prague, CS)

X 1328 *Pennsylvania State University, Department of Mathematics.* (University Park, PA, USA)

X 1330 *Pitman Publishing* (Belmont, CA, USA & London, GB & Toronto, ON, CDN & & Carlton, Vic, AUS & Johannesburg, ZA & Wellington, NZ & & Auckland, NZ) ISBN 0-273, ISBN 0-8224, ISBN 0-915092, ISBN 0-272

X 1331 *University of Pittsburgh Press* (Pittsburgh, PA, USA) ISBN 0-8229

X 1332 *Plenum Publishing Corporation* (New York, NY, USA & London, GB) ISBN 0-306 • REM Also Called Plenum Press

X 1334 *Praeger Publishers* (New York, NY, USA) ISBN 0-03, ISBN 0-275

X 1337 *Prindle, Weber & Schmidt* (Boston, MA, USA) ISBN 0-87150 • REM Also Called: PWS Publishers

X 1340 *Henry Regnery Co.* (Chicago, IL, USA) ISBN 0-8092

X 1342 *Royal Irish Academy* (Dublin, IRL) ISBN 0-901714

X 1343 *Russell & Russell Publishers* (New York, NY, USA) ISBN 0-8462

X 1346 *G.C.Sansoni, Editrice* (Firenze, I) ISBN 88-383

X 1347 *Charles Scribner's Sons Publishers* (New York, NY, USA) ISBN 0-684

X 1348 *La Scuola Editrice* (Brescia, I)

X 1349 *Editions du Seuil* (Paris, F) ISBN 2-02

X 1352 *A.W.Sijthoff International Publishing Co.* (Leiden, NL) ISBN 90-218, ISBN 90-286 • REL PUBL (**X** 1317) Noordhoff: Groningen

X 1354 *Spartan Books* (Sutton, GB & Rochelle Park, NJ, USA) ISBN 0-905532 • REL PUBL (**X** 1277) Hayden: Rochelle Park

X 1355 *Stanford University Press* (Stanford, CA, USA) ISBN 0-8047

X 1358 *University of Texas Press* (Austin, TX, USA & London, GB) ISBN 0-292 • REL PUBL (**X** 3828) Amer Univ Publ Group: London

X 1367 *University of Washington Press* (Seattle, WA, USA & London, GB) ISBN 0-295 • REL PUBL (**X** 3828) Amer Univ Publ Group: London

X 1368 *C.A.Watts & Co.* (London, GB) ISBN 0-296

X 1370 *Wesleyan University Press* (Middletown, CT, USA) ISBN 0-8195

X 1375 *Nicola Zanichelli Editore* (Bologna, I) ISBN 88-08

X 1407 *Publicaciones del Instituto de Matematicas "Jorge Juan"* (Madrid, E) ISBN 84-00

X 1408 *Deerven F. Bohn* (Amsterdam, NL & Haarlem, Nl) ISBN 90-6051

X 1416 *Rosenberg & Sellier* (Torino, I) ISBN 88-7011

X 1425 *Wydawnictwo Uniwersitetu Slaskiego* (Katowice, PL)

X 1434 *Kalininskij Gosudarstvennyj Universitet* (Kalinin, SU)

X 1453 *Moskovskij Gosudarstvennyj Istoriko-Arkh. Institut* (Moskva, SU)

X 1465 *Gondolat Koenyvkiado* (Budapest, H) ISBN 963-280, ISBN 963-281

X 1466 *Mueszaki Koenyvkiado (Institut of Technology)* (Budapest, H) ISBN 963-10

X 1493 *C.W.K.Gleerup Bokfoerlag a.B.* (Lund, S) ISBN 91-40

X 1494 *Munksgaard International Publishers* (Copenhagen, DK) ISBN 87-16

X 1553 *Verlag Dokumentation Saur* (Muenchen, D & London, GB) ISBN 3-7940, ISBN 3-598 • REL PUBL (**X** 2797) Minerva Publ: Muenchen • REM Member Firm: Zell Publishers: Oxford, GB

X 1554 *Universitetsforlaget* (Oslo, N & Bergen, N & Tromsoe, N & New York, NY, USA) ISBN 82-00 • REM Also: UB-Forlaget: Oslo, N

X 1574 *Vyssheyshaya Shkola* (Minsk, SU)

X 1588 *Verlagsgruppe Athenaeum/ Hain/ Hanstein* (Koenigstein/Ts, D) • REL PUBL (**X** 1275) Hain: Koenigstein & (**X** 2665) Athenaeum: Frankfurt

X 1599 *Aarhus Universitet, Matematisk Institut* (Aarhus, DK) ISBN 87-87436

X 1649 *Rusvozizdat* (Leningrad, SU)

X 1656 *Izdatel'stvo Inostr. Lit.* (Moskva, SU)

X 1684 *Marcel Dekker* (New York, NY, USA & Basel, CH) ISBN 0-8247 • REL PUBL (**X** 2442) Dekker Journal: New York

X 1715 *Edunsa Editorial Universitaria* (Santiago, RCH)

X 1732 *Bibliopolis* (Napoli, I) ISBN 88-7088

X 1744 *Editorial Ariel* (Esplugas de Llobregat, E & Barcelona, E) ISBN 84-344

X 1752 *Masson, Editeur* (Paris, F) ISBN 2-225

X 1758 *Ars Polona* (Warsaw, PL)

X 1761 *Hekdosis Hellinikis Mathematikes Hetaireias-Greek Mathematical Society* (Athens, GR)

X 1763 *Royal Society* (London, GB) ISBN 0-85403

X 1764 *Ksiaznica Atlas T.N.S.W.* (Warsaw, PL & Lwow, PL)

X 1773 *VEDA - Vydavatelstvo Slovenskej Akademie Vied* (Bratislava, CS)

X 1775 *Societe Royale des Sciences de Liege* (Liege, B)

X 1777 *Western Periodicals Company* (North Hollywood, CA, USA)

X 1781 *Tecnos* (Madrid, E) ISBN 84-309

X 1876 *Kexue Chubanshe (Science Press)* (Beijing, TJ)

X 1925 *Nauka i Izkustwo* (Sofia, BG)

X 1941 *Nanyang University Publications* (Singapore, MAL)

X 1946 *University Press of America* (Lanham, MD, USA & Washington, DC, USA) ISBN 0-8191

X 1987 *A. Francke Verlag* (Muenchen, D) ISBN 3-7720

X 2027 *Izdatel'stvo Nauka* (Moskva, SU & Alma-Ata, SU & Leningrad, SU & Novosibirsk, SU)

X 2036 *Franz Ehrenwirth Verlag* (Muenchen, D) ISBN 3-431

X 2039 *Polytechnic Institute of New York* (Brooklyn, NY, USA)

X 2040 *Winthrop* (Cambridge, MA, USA) ISBN 0-87626 • REL PUBL (**X** 0819) Prentice Hall: Englewood Cliffs

X 2045 *Perry Lane Press* (Stanford, CA, USA)

X 2055 *Progress* (Moskva, SU)

X 2075 *E.Flammarion & Cie, Librairie* (Paris, F) ISBN 2-08

X 2091 *Consejo Superior de Investigaciones Cientificas.* (Madrid, E) ISBN 84-00

X 2109 *Angelicum* (Roma, I)

X 2110 *N.V. Dekker & van de Vegt* (Nijmegen, NL) ISBN 90-255

X 2113 *University of Uppsala* (Uppsala, S) ISBN 91-506

X 2121 *Accademia Nazionale dei Lincei* (Roma, I) ISBN 88-218

X 2162 *Wilhelm Fink Verlag* (Muenchen, D) ISBN 3-7705

X 2178 *Gruner* (Amsterdam, NL) ISBN 90-6032

X 2179 *IEEE (Institute of Electrical and Electronics Engineers* (New York, NY, USA & Long Beach, CA, USA & Piscataway, CA, USA) ISBN 0-87942 • REM Section: IEEE Computer Society: Long Beach, CA, USA. Section: IEEE United States Activities Commitee: Piscataway, CA, USA

X 2183 *Electrical Engineering Department, State University of New York (SUNY) at Buffalo* (Buffalo, NY, USA)

X 2188 *Princeton University, Department of Electrical Engineering and Computer Science* (Princeton, NJ, USA)

X 2192 *Government College, Research Council* (Lahore, PAK)

X 2194 *Sociedad Matematica Mexicana* (Mexico City, MEX)

X 2195 *Peeters S.P.R.L.* (Louvain, B) ISBN 2-8017

X 2196 *History of Science Society* (Philadelphia, PA, USA) ISBN 0-934235

X 2197 *Tohoku University, Mathematical Institute* (Sendai, J)

X 2199 *Izdatel'stvo Naukova Dumka* (Kiev, SU)

X 2201 *Australasian Association of Philosophy.* (Bundoora, Vic, AUS) ISBN 0-9592545

X 2203 *Philosophy of Science Association* (East Lansing, MI, USA) ISBN 0-917586

X 2204 *American Institute of Physics* (New York, NY, USA) ISBN 0-88318

X 2205 *(ACM) Association for Computing Machinery* (New York, NY, USA) ISBN 0-89791

X 2206 *Accademia delle Science* (Torino, I)

X 2207 *Universitatea "Al. I. Cuza" din Iasi* (Jassy (Iasi), RO)

X 2208 *Taylor & Francis* (London, GB) ISBN 0-85066

X 2209 *Izdatel'stvo Akademii Nauk Gruzinskoj SSR* (Tbilisi, SU)

X 2211 *Friedrich Schiller Universitaet* (Jena, DDR)

X 2212 *Ossolineum, Publishing House of the Polish Academy of Sciences* (Wroclaw, PL & Warsaw, PL) • REL PUBL (**X** 2885) Zakl Narod Wyd Pol Ak: Wroclaw & (**X** 2882) Pol Akad Nauk: Wroclaw

X 2213 *State University of New York at Buffalo (SUNY)* (Buffalo, NY, USA)

X 2214 *Dialectica* (Bienne/Biel, CH)

X 2215 *Indiana University, Department of Philosophy* (Bloomington, IN, USA)

X 2216 *Cornell University, Sage School of Philosophy* (Ithaca, NY, USA)

X 2217 *Academia Brasileira de Ciencias.* (Rio de Janeiro, BR)

X 2218 *University of Poona* (Poona, IND)

X 2219 *Humboldt-Universitaet zu Berlin* (Berlin, DDR)

X 2220 *Buchdruckerei und Verlag Leemen* (Zuerich, CH)

X 2221 *Philosophy Education Society* (Washington, DC, USA)

X 2224 *Matematisk Institut* (Bergen, N)

X 2225 *Izdatel'stvo Akademii Nauk Armyanskoj SSR* (Erevan, SU)

X 2227 *Scuola Normale Superiore di Pisa* (Pisa, I) ISBN 88-7642

X 2228 *Hegler Institute* (La Salle, IL, USA)

X 2229 *Universita J.E.Purkyne* (Brno, CS) • REM Formerly Named Masarykova Universita

X 2230 *Isdevnieciba Zinatne* (Riga, SU)

X 2231 *Charles S. Pierce Society. State University of New York (SUNY) at Buffalo, Department of Philosophy* (Buffalo, NY, USA)

X 2234 *B.Pellerano & S. del Gaudia Editrice Libraria* (Napoli, I)

X 2235 *Vsesoyuznyj Institut Nauchnoj i Tekhnicheskoj Informatsii (VINITI), Gosudarstvennyj Komitet SSSR po Nauke: Tekhnike (GKNT SSSR), Akademiya Nauk (AN) SSSR* (Moskva, SU)

X 2237 *Izdatelstvo na Bulgarskata Akademia na Naukite (Publishing House of the Bulgarian Academy of Sciences.)* (Sofia, BG)

X 2239 *Universidad de Salamanca Ediciones* (Salamanca, E) ISBN 84-7481

X 2241 *New York Academy of Sciences* (New York, NY, USA) ISBN 0-89072, ISBN 0-89766

X 2242 *Osaka University, Department of Mathematics* (Osaka, J)

X 2243 *University of Osaka Prefecture, Department of Mathematics* (Osaka, J)

X 2244 *Societe Mathematique de France* (Paris, F) ISBN 2-85629

X 2248 *Drushtvo Matematichara i Fizichara Sr Hrvatske.* (Zagreb, YU)

X 2256 *Casa Editrice Felice le Monnier* (Firenze, I)

X 2257 *John Benjamins B.V.* (Amsterdam, NL & Philadelphia, PA, USA) ISBN 90-272

X 2258 *Periodika* (Tallinn, SU)

X 2265 *Vychislitel'nyj Tsentr Akademii Nauk SSSR* (Moskva, SU)

X 2299 *Universidade de Campinas "EDUC" (State University of Campinas)* (Sao Paulo, BR)

X 2301 *Communication and Cognition* (Ghent, B)

X 2360 *Massachussetts Institute of Technology, Alumni Association* (Cambridge, MA, USA)

X 2370 *British Computer Society* (London, GB) ISBN 0-901865

X 2373 *Karl-Marx-Universitaet, Direktorat fuer Forschung, Abteilung Wissenschaftliche Publikationen* (Leipzig, DDR)

X 2374 *University of Oklahoma Press* (Norman, OK, USA) ISBN 0-8061

X 2375 *MacMillan Journals* (London, GB) ISBN 0-333 • REL PUBL (**X** 0843) Macmillan : New York & London

X 2378 *Society for Industrial and Applied Mathematics (SIAM)* (Philadelphia, PA, USA) ISBN 0-89871

X 2392 *University of Electro-Communications* (Tokyo, J)

X 2393 *Universidad de Chile, Facultad de Filosofia y Letras.* (Santiago, RCH)

X 2395 *Turun Ylioposto* (Turku, SF) ISBN 951-641

X 2396 *Hitotsubashi University* (Tokyo, J)

X 2397 *Bar Ilan University Press* (Ramat Gan, IL) ISBN 965-226

X 2399 *Sigma Xi, Scientific Research Society of North America* (New Haven, CT, USA) ISBN 0-914446

X 2406 *Bibliotheque de l'Universite* (Sofia, BG)

X 2409 *Suomalainen Tiedakatemia.* (Helsinki, SF) ISBN 951-41

X 2420 *Utilitas Mathematica Publications, University of Manitoba* (Winnipeg, MB, CDN) ISBN 0-919628

X 2421 *University of Tokyo, Faculty of Science* (Tokyo, J)

X 2422 *Canadian Philosophical Association* (Montreal, PQ, CDN) ISBN 0-9690153

X 2423 *Scripta Publishing Co.* (Silver Spring, MT, USA & Washington, DC, USA) ISBN 0-88380

X 2426 *Universa, P.V.B.A.* (Wetteren, B) ISBN 90-6281

X 2434 *University of Tokyo College of General Education* (Tokyo, J)

X 2435 *Presses Universitaires de France, Service Periodiques* (Paris, F) ISBN 2-13 • REL PUBL (**X** 0840) Pr Univ France: Paris

X 2436 *Indian Academy of Sciences, Publications Department* (Bangalore, IND)

X 2437 *Hemisphere Publishing* (Washington, DC, USA & New York, NY, USA & London, GB) ISBN 0-89116

X 2438 *Edizioni Scientifiche Inglesi Americane* (Roma, I)

X 2441 *Kyoto University, Research Institute for Mathematical Sciences* (Kyoto, J)

X 2442 *Marcel Dekker Journals.* (New York, NY, USA) ISBN 0-8247 • REL PUBL (**X** 1684) Dekker: New York

X 2443 *Izdatel'stvo Nauka, Otdelenie v Kazakhstane SSR* (Alma-Ata, SU)

X 2445 *Academia Sinica, Institute of Mathematics* (Taipei, RC)

X 2446 *Martin-Luther-Universitaet Halle-Wittenberg* (Halle, DDR)

X 2447 *Technische Hochschule Ilmenau* (Ilmenau, DDR)

X 2448 *Akademiya Navuk Belorusskaj SSR* (Minsk, SU)

X 2451 *Uniwersitet Jagiellonskie, Instytut Matematiyczny* (Krakow, PL)

X 2455 *Yokohama City University, Department of Mathematics* (Yokohama, J)

X 2456 *Tamkang College of Arts and Sciences, Research Institute of Mathematics* (Taipei, RC)

X 2457 *Allerton Press* (New York, NY, USA) ISBN 0-89864

X 2461 *Jugoslovenska Akademija Zananosti i Umjetnosti Rad.* (Zagreb, YU) ISBN 86-407

X 2463 *Tartuskij Gosudarstvennyj Universitet* (Tartu, SU)

X 2464 *Mathematisches Seminar, Selbstverlag* (Giessen, D)

X 2465 *Kinokuniya Company* (Tokyo, J) ISBN 4-314

X 2467 *Societatea de Stiinte Matematice din Republica Socialista Romania (RSR)* (Bucharest, RO)

X 2469 *Data, A/S af 2. april 1971* (Copenhagen, DK) ISBN 87-980512

X 2472 *Centro Superiore di Logica e Scienze Comporate, Editrice Franco Spisani* (Bologna, I)

X 2473 *Belfort Graduate School of Science; Yeshiva University* (New York, NY, USA)

X 2476 *Journal of Philosophy Inc.* (New York, NY, USA) ISBN 0-931206

X 2479 *Rocky Mountain Mathematics Consortium* (Tempe, AZ, USA)

X 2480 *Operations Research Society of Japan* (Tokyo, J)

X 2492 *Japan Association for the Philosophy of Science* (Tokyo, J)

X 2494 *Leo S.Olschki Editore* (Firenze, I) ISBN 88-222

X 2496 *Suomen Fyysikkoseura* (Helsinki, SF)

X 2505 *Savez Drushtava Matematichara, Fizichara, Astronoma. Jugoslavije-Nacionali Matematichki Komitet* (Belgrade, YU)

X 2510 *Elm (Izdatel'stvo Akademiya Nauk Azerbajdzhanskoj SSR)* (Baku, SU)

X 2511 *Universita Karlova, Matematicky Ustav* (Prague, CS)

X 2512 *Institut Matematiki Akademii Nauk SSSR* (Moskva, SU)

X 2513 *Mathematical Association* (Leicester, GB) ISBN 0-906588

X 2516 *Union Matematica Argentina* (Buenos Aires, RA)

X 2518 *Aberdeen University Press* (Aberdeen, GB) ISBN 0-900015, ISBN 0-08

X 2519 *Izdatel'stvo Akademii Nauk Uzbekskoj SSR* (Tashkent, SU)

X 2520 *National Academy of Sciences, Transportation Research Board* (Washington, DC, USA) ISBN 0-309

X 2522 *Akademiya Nauk USSR, Nauchnyj Sovet po Kibernetike, Institut Kibernetiki.* (Kiev, SU)

X 2524 *Encyclopaedia Universalis* (Paris, F) ISBN 2-85229

X 2526 *Gazeta de Matematica* (Lisboa, P)

X 2528 *Circolo Matematico di Palermo* (Palermo, I)

X 2529 *Consultants Bureau* (New York, NY, USA) • REL PUBL (**X** 1332) Plenum Publ: New York

X 2530 *Societe Mathematique de Belgique* (Bruxelles, B)

X 2531 *Japan Academy* (Tokyo, J)

X 2596 *Akademische Druck- und Verlagsanstalt Dr P. Struzl* (Graz, A) ISBN 3-201, ISBN 3-900144

X 2600 *DEB - Verlag. Das Europaeische Buch* (Berlin, DDR)

X 2637 *Taylor & Walton* (London, GB)

X 2641 *Izdatel'stvo Nauka Leningradskoe Otdelenie* (Leningrad, SU)

X 2642 *Izdatel'stvo Nauka Sibirskoe Otdelenie* (Novosibirsk, SU)

X 2643 *Izdatel'stvo Sovetskoe Radio* (Moskva, SU)

X 2644 *Izdatel'skoe Ob"edineniya "Vishcha Shkola". Izdatel'stvo pri Khar'kovskom Gosudarskvennom Universitete* (Khar'kov, SU)

X 2645 *Izdatel'skoe Ob"edineniye "Vishcha Shkola". Izdatel'stvo pri Kievskom Gosudarstvennom Universitete* (Kiev, SU)

X 2646 *Aalborg University Press= Aalborg Universitetsforlag* (Aalborg, DK) ISBN 87-7307

X 2652 *Institut Matematiki Sibirskogo Otdeleniya Akademii Nauk SSSR (SOAN SSSR)* (Novosibirsk, SU)

X 2665 *Athenaeum Verlag* (Frankfurt, D) ISBN 3-7610 • REL PUBL (**X** 1588) Athenaeum/Hain/Hanstein: Koenigstein

X 2671 *Basic Books* (New York, NY, USA) ISBN 0-465

X 2673 *Biblioteka Akademiya Nauk SSSR* (Leningrad, SU)

X 2674 *Bouvier Verlag Herbert Grundmann* (Bonn, D) ISBN 3-416

X 2686 *Compania Editorial Continental (CECSA)* (Mexico City, MEX)

X 2692 *University of Queensland Press* (St. Lucia, Qld, AUS) ISBN 0-7022

X 2696 *Humanities Press* (Atlantic Highlands, NY, USA) ISBN 0-391

X 2698 *Departamento de Logica y Filosofia de la Ciencia, Universidad de Valencia* (Valencia, E)

X 2709 *Editura Tehnica* (Bucharest, RO)

X 2710 *Eidgenoessische Technische Hochschule Zuerich* (Zuerich, CH)

X 2717 *Franz Steiner Verlag* (Wiesbaden, D & Stuttgart, D) ISBN 3-515

X 2725 *Hackett Publishing Co.* (Indianapolis, IN, USA) ISBN 0-915144

X 2728 *Hoelder-Pichler-Tempsky* (Wien, A) ISBN 3-209

X 2732 *Institut National de Recherche en Informatique et en Automatique (INRIA)* (Le Chesnay Cedex, F) ISBN 2-7261 • REM Also Called: Institut de Recherche en Informatique et en Automatique (IRIA)

X 2733 *Polish Academy of Sciences, Institute of Philosophy and Sociology* (Wroclaw, PL)

X 2737 *Israel Program for Scientific Translations* (Jerusalem, IL) ISBN 0-7065 • REL PUBL (**X** 0827) Wiley & Sons: New York

X 2741 *Shtiintsa* (Kishinev, SU)

X 2765 *Japan Association of Automatic Control Engineers* (Kyoto, J)

X 2766 *Yaroslavskij Gosudarstvennyj Universitet* (Yaroslavl', SU)

X 2767 *Johannes Kepler Universitaet* (Linz, A)

X 2769 *Kazakhskij Gosudarstvennyj Universitet* (Alma-Ata, SU)

X 2776 *Les Presses de l'Universite Laval* (Ste-Foy, PQ, CDN) ISBN 0-7746, ISBN 2-7637

X 2786 *Universitet i Oslo, Matematisk Institut* (Oslo, N) ISBN 82-553

X 2791 *Universiteit van Amsterdam, Mathematisch Instituut* (Amsterdam, NL)

X 2797 *Minerva Publikation Saur* (Muenchen, D) ISBN 3-597 • REL PUBL (**X** 1553) Dokumentation Saur: Muenchen

X 2799 *Mladinska Knjiga* (Ljubljana, YU) ISBN 86-11

X 2812 *The Philosophical Society of Finland* (Helsinki, SF) ISBN 951-95053

X 2828 *Robert E. Krieger Publishing Co.* (Melbourne, FL, USA) ISBN 0-88275, ISBN 0-89874

X 2836 *Sociedade Brasileira Logica* (Sao Paulo, BR)

X 2859 *Universite de Toulouse - Le Mirail* (Toulouse, F)

X 2865 *Verlag Peter Lang* (Frankfurt, D & Bern, CH, Las Vegas, USA) ISBN 3-8204, ISBN 3-261

X 2866 *Verlag Ullstein* (Berlin, D) ISBN 3-550

X 2880 *Wydawnictwa Naukowo-Techniczne* (Warsaw, PL)

X 2881 *Wydawnictwa Szkolne i Pedagogiczne* (Warsaw, PL) ISBN 83-02

X 2882 *Wydawnictwo Polskiej Akademii Nauk (Publisher of the Polish Academy of Science)* (Wroclaw, PL) • REL PUBL (**X** 2885) Zakl Narod Wyd Pol Ak: Wroclaw & (**X** 2212) Ossolineum: Wroclaw

X 2885 *Zaklad Narodowy imienia Ossolinskich, Wydawnictwo Polskiej Akademii Nauk* (Wroclaw, PL) ISBN 83-04 • REL PUBL (**X** 2882) Pol Akad Nauk: Wroclaw & (**X** 2212) Ossolineum: Wroclaw

X 3111 *Philosophia Verlag* (Muenchen, D & Wien, A) ISBN 3-88405

X 3119 *Intertext Books* (Aylesbury, GB)

X 3123 *Beijing Shifan Daxue (Beijing Normal University)* (Beijing, TJ)

X 3151 *Marx Karoly Koezgazdasagi Egyetem Matematikai Intezet (Karl Marx University of Economics, Department of Mathematics)* (Budapest, H)

X 3176 *Fernuniversitaet Hagen* (Hagen, D)

X 3215 *Mathematisch-Naturwissenschaftliche Fakultaet der Rheinisch-Westfaelischen Technischen Hochschule Aachen.* (Aachen, D)

X 3219 *Leningradskij Ordena Trudovogo Krasnogo Znameni Institut Tochnoj Mekhaniki i Optiki* (Leningrad, SU)

X 3223 *Carl Hanser Verlag* (Muenchen, D & Wien, A) ISBN 3-446

X 3249 *The Weizmann Institute of Science* (Rehovot, IL) ISBN 965-281

X 3333 *Unknown Publisher*

X 3353 *Akademiya Nauk SSSR. Institut Istor Filol Filos* (Novosibirsk, SU)

X 3357 *Institute of Education. University of Keele* (Keele, GB)

X 3407 *"Vysshaya Shkola"* (Moskva, SU)

X 3552 *Iwanami Shoten Publishers* (Tokyo, J) ISBN 4-00

X 3555 *Clarkson N. Potter, Publishers* (New York, NY, USA) ISBN 0-8257

X 3560 *Paraninfo* (Madrid, E) ISBN 84-283

X 3562 *Ecole Normale Superieure de Jeunes Filles* (Paris, F) ISBN 2-85929

X 3581 *Computer Science Press* (Rockville, MD, USA) ISBN 0-914894

X 3602 *Kraus Reprint* (Vaduz, FL & Nendeln, FL) ISBN 3-262 • REM Parent Firm: Kraus Thomson Organisation: Vaduz, FL

X 3604 *Istituto Nazionale di Alta Matematica (INDAM)* (Roma, I)

X 3605 *Izdatel'stvo Kazanskogo Gosudarstvennogo Universiteta* (Kazan', SU)

X 3606 *Izdadel'stvo Tomskogo Universiteta* (Tomsk, SU)

X 3615 *Il Saggiatore* (Milano, I)

X 3636 *Tokyo Tosho* (Tokyo, J) ISBN 4-489

X 3709 *Gosudarstvennoye Izdatel'stvo Fiziko-Matematicheskoj Literatury* (Moskva, SU)

X 3711 *Koseisha Koseikaku* (Tokyo, J) ISBN 4-7699

X 3718 *L'Enseignement Mathematique, Universite de Geneve* (Geneve, CH)

X 3719 *The Harvester Press* (Sussex, GB) ISBN 0-85527, ISBN 0-901759, ISBN 0-7108

X 3725 *Bolyai Janos Matematikai Tarsulat (Janos Bolyai Mathematical Society)* (Budapest, H)

X 3727 *Beograd. Matematicki Institut* (Belgrade, YU)

X 3734 *Cabay Libraire-Editeur* (Louvain-la-Neuve, B) ISBN 2-87077

X 3760 *Universite "Kiril et Metodij", Matematicki Fakultet* (Skopje, YU)

X 3775 *Radio i Svyaz* (Moskva, SU)

X 3777 *F. Angeli Editore* (Milano, I) ISBN 88-204

X 3790 *Metsniereba* (Tbilisi, SU)

X 3794 *B. Edsall & Co.* (London, GB) ISBN 0-902623

X 3805 *The Blindern Theoretic Research Team* (Oslo, N)

X 3812 *Universita di Siena, Dipartimento di Matematica, Scuola di Specializzazione in Logica Matematica* (Siena, I)

X 3828 *American University Publishers Group* (London, GB) • REL PUBL (**X** 0842) Univ Wisconsin Pr: Madison & (**X** 1285) Univ Ill Pr: Urbana & (**X** 1286) Indiana Univ Pr: Bloomington & (**X** 1358) Univ Texas Pr: Austin & (**X** 1367) Univ Washington: Seattle

X 3890 *Verlag TUEV Rheinland* (Koeln, D) ISBN 3-921059, ISBN 3-88585

X 3905 *Ridgeview Publishing Company* (Atascadero, CA, USA) ISBN 0-917930

X 4030 *Univerzitet u Novom Sadu, Institut za Matematiku* (Novi Sad, YU)

X 4056 *Editorial Labor* (Barcelona, E) ISBN 84-335

X 4217 *Foris Publications* (Dordrecht, NL) ISBN 90-70176

X 4315 *Haven Publishing Corporation* (New York, NY, USA)

X 4321 *Sterling Publishers Pvt.* (New Delhi, IND)

X 4322 *National Publishing House* (New Delhi, IND & Bihar, IND & Calcutta, IND & Ambala, IND)

X 4328 *Editora Pedagogica Universitaria* (Sao Paulo, BR) ISBN 85-202

X 4374 *Huazhong Gongxueyuan Chubanshe (Huazhong Institute of Technology Press)* (Wuhan, TJ)

X 4380 *Posebna Izdanja Matematichok Instituta* (Belgrade, YU)

X 4450 *Willem A. Meeuws* (Oxford, GB) ISBN 0-902672

X 4457 *Presses Polytechniques* (Lausanne, CH)

X 4465 *Obshchestvennaya Pol'za i Provintsiya* (Leningrad, SU)

X 4477 *System Develpment Corporation* (Santa Monica, CA, USA)

X 4484 *Licenciatura Universidad de Concepcion* (Concepcion, RCH)

X 4499 *Ehlins Handboeker* (Stockholm, S)

X 4502 *University of Tehran* (Tehran, IR)

X 4506 *Institut d'Etudes Polonaises en Belgique* (Bruxelles, B)

X 4512 *Alfred A. Knopf* (New York, NY, USA) ISBN 0-394

X 4562 *Udmurtskij Gosudarstvennij Universiteta* (Izhevsk, SU)

X 4563 *Liberal Arts Press* (New York, NY, USA)

X 4569 *(Commercial Press)* (Beijing)

X 4579 *Gnodeng Jiaoyu Chubanshe (Higher Education Press)* (Beijing, TJ)

X 4589 *Shanghai Jiaoyu Chubanshe (Shanghai Education Press)* (Shanghai, TJ)

X 4624 *Kluwer - Nijhoff Publishing* (Boston, MA, USA) ISBN 0-89838 • REL PUBL (**X** 1316) Nijhoff: Leiden & s'Gravenhage

X 4718 *Felix Meiner Verlag* (Leipzig, DDR) • REL PUBL (**X** 1088) Meiner: Hamburg

X 4721 *Libreria Fratelli Bocca & C. Clausen* (Torino, I & Roma, I)

Miscellaneous Indexes

External classifications

This index complements the Subject Index at the beginning of this volume; it lists the items which, in addition to classifications in the present volume, have classifications *external to this volume*. These items are ordered by external classification code and within each code by author (the first alphabetically in the case of multi-author items), year and identification number (thus an item with, for example, two external classifications occurs twice in this listing). This index provides another way to search the bibliography. With it, the user can easily identify those items in this volume classified also in some area external to this volume.

B05

Arnold, B.H. [1962] 23577
Baginski, M. [1973] 75585
Devide, V. [1964] 43409
Freudenthal, H. [1958] 04610
Goodstein, R.L. [1958] 05182
Khasin, L.S. [1969] 05733
Leblanc, H. [1960] 07913
Leone, M. [1985] 49912
Lukasiewicz, J. [1921] 28632
MacColl, H. [1906] 22595
Nieland, J.J.F. [1967] 09951
Pospesel, H. [1974] 44546
Post, E.L. [1921] 10681
Prior, A.N. [1961] 10782
Quine, W.V.O. [1953] 10892
Schmidt, H.A. [1960] 12892
Smith, Henry Bradford [1936] 41315
Strazdins, I.E. [1981] 55620
Wasserman, H.C. [1976] 65990

B10

Dreben, B. [1957] 29366
Glubrecht, J.-M. [1983] 36976
Goddard, L. [1966] 15234
Hermes, H. [1957] 21083
Hermes, H. [1965] 06014
Hintikka, K.J.J. [1973] 44480
Hintikka, K.J.J. [1981] 42772
Hunter, G. [1971] 22356
Los, J. [1963] 31415
Malinowski, G. [1985] 49715
Markwald, W. [1971] 08782
Markwald, W. [1974] 08784
Novikov, P.S. [1949] 10018
Paulos, J.A. [1981] 56466
Pliuskevicius, R. [1968] 10543
Pospesel, H. [1976] 44547
Quine, W.V.O. [1954] 10895
Rasiowa, H. [1953] 11004
Reichbach, J. [1962] 11065
Reichbach, J. [1963] 11067
Reichbach, J. [1964] 11068
Reichbach, J. [1965] 11070
Robinson, A. [1959] 19551
Robinson, John Alan [1979] 53769
Winnie, J.A. [1970] 14236

B15

Asser, G. [1981] 55197
Cirulis, J. [1982] 35570
Cresswell, M.J. [1972] 02551
Hintikka, K.J.J. [1981] 42772
Lehmann, S.K. [1976] 14765
Prawitz, D. [1965] 22164
Quine, W.V.O. [1934] 19580
Rosser, J.B. [1953] 11565
Scedrov, A. [1985] 44771
Surma, S.J. [1977] 51405

B20

Dodgson, C.L. [1895] 02237
Feys, R. [1937] 41048
Khomich, V.I. [1979] 53468
Komori, Y. [1982] 37775
Lambert, K. [1967] 07823
Marsden, E.L. [1974] 63732
Mo, Shaokui [1950] 12028
Post, E.L. [1921] 10681
Prior, A.N. [1964] 10787
Rasiowa, H. [1965] 29826
Renardel de Lavalette, G.R. [1981] 55207
Rose, A. [1951] 11403
Sanmartin Esplugues, J. [1973] 78125
Schroeder, F.W.K.E. [1890] 19702
Schroeder, F.W.K.E. [1891] 19703
Schroeder, F.W.K.E. [1895] 19704
Sobocinski, B. [1964] 21264
Trew, A. [1970] 13673
Venn, J. [1881] 21755
Wozniakowska, B. [1978] 31565

B22

Gabbay, D.M. [1973] 30381
Gabbay, D.M. [1974] 17555
Grant, J. [1975] 21774
Los, J. [1949] 19296
Magari, R. [1966] 49174
Makinson, D. [1974] 08573
Maksimova, L.L. [1977] 52433
Murskij, V.L. [1965] 09654
Shekhtman, V.B. [1982] 38116
Stepien, T. [1985] 48971
Wojcicki, R. [1982] 35669

B25

Adamson, A. [1979] 53245
Alves, E.H. [1978] 52943
Anderson, A.R. [1954] 00309
Baudisch, A. [1980] 56368
Boolos, G. [1976] 14578
Boolos, G. [1977] 26442
Boolos, G. [1979] 56312
Boricic, B.R. [1983] 44659
Bozzi, S. [1980] 54221
Bryll, G. [1973] 24728
Burgess, J.P. [1980] 55004
Burgess, J.P. [1985] 42591
Cavalli, A.R. [1984] 43192
Chlebus, B.S. [1979] 54219
Chlebus, B.S. [1982] 34361
Costa da, N.C.A. [1974] 22926
Cresswell, M.J. [1985] 42657
Danko, W. [1983] 39522
Davidson, B. [1977] 23697
Deutsch, H. [1985] 47508
Drugush, Ya.M. [1984] 45347
Emerson, E.A. [1984] 45349
Engeler, E. [1975] 21738
Ferenczi, M. [1977] 55783
Fidel, M.M. [1977] 24188
Fidel, M.M. [1980] 54154
Gabbay, D.M. [1970] 04721
Gabbay, D.M. [1971] 04723
Gabbay, D.M. [1971] 30000
Gabbay, D.M. [1973] 17911
Gabbay, D.M. [1974] 17555
Gabbay, D.M. [1975] 04738
Gabbay, D.M. [1976] 32148
Gaerdenfors, P. [1975] 21522
Giambrone, S. [1985] 42658
Goad, C.A. [1978] 73424
Goldblatt, R.I. [1973] 05104
Goldblatt, R.I. [1974] 05106
Goldblatt, R.I. [1981] 55422
Gurevich, Y. [1985] 47372
Hailperin, T. [1984] 41842
Hajek, P. [1977] 31126
Hajek, P. [1981] 54605
Hallden, S. [1949] 05586
Hallden, S. [1949] 05587
Hanson, W.H. [1966] 05648
Ackermann, W. [1928] 00107
Hodes, H.T. [1984] 41142

Ito, Makoto [1965] 27530
Iwanus, B. [1973] 17553
Kramosil, I. [1978] 54777
Kron, A. [1978] 53971
Kuznetsov, A.V. [1975] 63246
Langford, C.H. [1932] 22592
Lehmann, D.J. [1983] 41143
Lewis, D. [1971] 08107
Marconi, D. [1980] 56614
Matsumoto, K. [1957] 16393
Matsumoto, K. [1959] 16395
Matsumoto, K. [1960] 08893
McKay, C.G. [1968] 08986
McKay, C.G. [1971] 08988
McKinsey, J.C.C. [1941] 09020
Meredith, D. [1956] 09147
Montgomery, H. [1976] 18366
Nakamura, A. [1980] 54126
Nieland, J.J.F. [1967] 09951
Ohnishi, M. [1961] 10085
Poliferno, M.J. [1961] 10611
Rasiowa, H. [1953] 11004
Reichbach, J. [1974] 31999
Rennie, M.K. [1968] 11098
Rogava, M.G. [1979] 77850
Routley, R. [1968] 11609
Routley, R. [1968] 11610
Routley, R. [1970] 11613
Routley, R. [1974] 15192
Rybakov, V.V. [1978] 53114
Rybakov, V.V. [1978] 53244
Rybakov, V.V. [1984] 39654
Saeli, D. [1975] 78070
Sawamura, H. [1985] 47625
Segerberg, K. [1968] 11935
Segerberg, K. [1968] 41673
Segerberg, K. [1976] 65185
Shukla, A. [1970] 19678
Skolem, T.A. [1928] 12385
Swart de, H.C.M. [1980] 54220
Thomason, R.H. [1967] 13579
Thomason, R.H. [1975] 51538
Ursini, A. [1985] 47525
Valiev, M.K. [1980] 54018
Vojshvillo, E.K. [1983] 48317
Wasserman, H.C. [1976] 18423
Weispfenning, V. [1985] 47574
Wright von, G.H. [1951] 14305
Wright von, G.H. [1951] 14313
Zeman, J.J. [1969] 14399

B28

Baginski, M. [1973] 75585
Beth, E.W. [1962] 01176
Boolos, G. [1979] 56312
Hajek, P. [1981] 35379
Hajek, P. [1983] 34630

B30

Carnap, R. [1954] 01852
Carnap, R. [1971] 25461
Dalla Chiara Scabia, M.L. [1979] 56529
Evans, H.P. [1939] 30933
Howson, C. [1976] 62614
Leblanc, H. [1960] 07913
Muchnik, A.A. [1959] 00382
Mueller, Gert H. [1957] 33928
Munoz, V. [1961] 28705
Rasiowa, H. [1971] 22289
Reichenbach, H. [1932] 11080
Reichenbach, H. [1939] 43787
Roeper, P. [1980] 54022
Tarski, A. [1937] 19757

B35

Case, J. [1983] 37600
Chang, C.L. [1973] 03980
Farinas del Cerro, L. [1982] 36159
Foxley, E. [1962] 04474
Kalman, J.A. [1976] 62846
Lee, R.C.T. [1972] 28860
Ohlbach, H.-J. [1984] 43199
Pliuskeviciene, A. [1985] 49019
Pnueli, A. [1985] 49386
Pnueli, A. [1985] 49524
Richter, M.M. [1978] 51870
Robinson, John Alan [1979] 53769
Schwartz, J.T. [1967] 23572
Schwind, C.B. [1980] 78311
Silvestrini, D. [1981] 48725
Venkatesh, G. [1985] 49739

B40

Barendregt, H.P. [1981] 55008
Berkling, K.J. [1982] 43431
Bunder, M.W. [1979] 56104
Bunder, M.W. [1980] 53770
Bunder, M.W. [1980] 54073
Church, A. [1936] 23466
Courvoisier, M. [1976] 51687
Curry, H.B. [1957] 29324
Curry, H.B. [1958] 02627
Curry, H.B. [1967] 02637
Fitch, F.B. [1952] 04335
Fitch, F.B. [1974] 24019
Hindley, J.R. [1972] 23471
Kanda, A. [1985] 47568
Mirkowska, G. [1977] 31914
Scott, D.S. [1975] 27594
Scott, D.S. [1982] 38593
Vanderveken, D.R. [1975] 22976

B80

Bressan, A. [1974] 60704

B96

Radu, E. [1978] 51580
Suppes, P. [1969] 12737

C05

Bloom, S.L. [1982] 36525
Chang, C.C. [1966] 02410
Denecke, K. [1982] 37824
Maksimova, L.L. [1977] 52433
Maksimova, L.L. [1977] 54760
Maksimova, L.L. [1979] 55851
Rosenberg, I.G. [1973] 29025
Selman, A.L. [1972] 11964
Ursini, A. [1985] 47525
Zygmunt, J. [1974] 14462

C07

Bloom, S.L. [1982] 36525
D'Ottaviano, I.M.L. [1985] 41799
Fine, K. [1975] 22897
Fraassen van, B.C. [1968] 04480
Fraisse, R. [1982] 43874
Grant, J. [1978] 28232
Halkowska, K. [1976] 52421
Krabbe, E.C.W. [1978] 51481
Manaster, A.B. [1975] 23120
Manin, Yu.I. [1977] 51984
Mostowski, Andrzej [1962] 09574
Paulos, J.A. [1976] 18327
Rao, A.P. [1977] 56420
Robinson, A. [1958] 25442
Robinson, A. [1959] 19551
Selman, A.L. [1972] 11964
Skolem, T.A. [1928] 12385
Smirnov, V.A. [1972] 30359
Suszko, R. [1971] 12777

C10

Baudisch, A. [1980] 56368
Langford, C.H. [1932] 22592
Skolem, T.A. [1928] 12385
Weispfenning, V. [1985] 47574

C13

Bellissima, F. [1983] 42314
Denecke, K. [1982] 37824
Gao, Hengshan [1976] 61914
Hajek, P. [1976] 31127
Hajek, P. [1977] 31126
Halpern, M. [1983] 45211
Smorynski, C.A. [1982] 43687
Yankov, V.A. [1967] 06513

C20

Chang, C.C. [1962] 01946
Chang, C.C. [1966] 02410
Gabbay, D.M. [1972] 04730
Gabbay, D.M. [1975] 04739
Goldblatt, R.I. [1975] 05108
Grandy, R.E. [1976] 18176
Mangani, P. [1976] 51187
Nishimura, H. [1980] 54084
Ono, H. [1983] 34652
Rasiowa, H. [1970] 11031
Rauszer, C. [1980] 56381
Vojvodic, G. [1978] 52815
Waszkiewicz, J. [1969] 14081
Zygmunt, J. [1974] 14462

C25

Gerla, G. [1983] 44834
Gerla, G. [1984] 39889

C30

Benthem van, J.F.A.K. [1983] 37277
Waszkiewicz, J. [1969] 14081

C35

Danko, W. [1983] 39522
Skolem, T.A. [1928] 12385
Vojvodic, G. [1978] 52815

C40

Alves, E.H. [1984] 39893
Barwise, J. [1975] 60316
Bowen, K.A. [1980] 54337
Cantini, A. [1980] 54349
Dahn, B.I. [1974] 02746
Dahn, B.I. [1975] 61216
Fajardo, S. [1985] 39901
Fine, K. [1978] 30694
Fitting, M. [1984] 42443
Gabbay, D.M. [1971] 04726
Gabbay, D.M. [1972] 04733
Gabbay, D.M. [1977] 26456
Goranko, V.F. [1985] 47504
Goranko, V.F. [1985] 49398
Gumb, R.D. [1979] 62230
Hajek, P. [1977] 31126
Kawai, H. [1982] 36043
Komori, Y. [1978] 52844
Krivine, J.-L. [1974] 07566
Krzystek, P.S. [1977] 52053
Lopez-Escobar, E.G.K. [1981] 55143
Lopez-Escobar, E.G.K. [1983] 39786
Lue, Qici [1983] 40907
Maksimova, L.L. [1977] 52433
Maksimova, L.L. [1977] 54760
Maksimova, L.L. [1979] 55851
Maksimova, L.L. [1982] 37067
Maksimova, L.L. [1982] 40416
Maksimova, L.L. [1982] 40502
Malinowski, G. [1981] 54348
Malinowski, G. [1982] 42305
Miyama, T. [1974] 29672
Mostowski, Andrzej [1962] 09574
Nishimura, H. [1980] 54084
Nishimura, H. [1980] 56424
Nishimura, H. [1981] 54611
Nishimura, H. [1981] 55147
Ono, H. [1983] 34652
Porebska, M. [1985] 48967
Przymusinska, H. [1980] 54915
Radev, S.R. [1984] 44072
Rasiowa, H. [1972] 11032
Rauszer, C. [1985] 41793
Rautenberg, W. [1983] 39823
Renardel de Lavalette, G.R. [1981] 55207
Rockingham Gill, R.R. [1970] 22231
Smirnov, V.A. [1972] 30359
Smorynski, C.A. [1978] 54148
Vincenzi, A. [1985] 49797
Vojvodic, G. [1980] 36055
Wasilewska, A. [1979] 56380
Zachorowski, S. [1977] 54806
Zachorowski, S. [1978] 56212

C50

Chang, C.C. [1962] 01946
Chang, C.C. [1966] 02410
Nishimura, H. [1980] 56424
Nishimura, H. [1981] 55147
Rosenberg, I.G. [1973] 29025

C52

Alves, E.H. [1984] 39893
D'Ottaviano, I.M.L. [1985] 41799
Hauschild, K. [1982] 34362

C55

Przymusinska, H. [1980] 54915
Tiuryn, J. [1981] 34010

C57

Kanda, A. [1985] 47568
Mostowski, Andrzej [1962] 09574
Shanin, N.A. [1977] 53812
Vuckovic, V. [1967] 27478

C60

Baudisch, A. [1980] 56368
Krivine, J.-L. [1974] 07566
Lorenzen, P. [1962] 08303
Robinson, A. [1963] 28804
Saeli, D. [1975] 78070

C62

Hajek, P. [1977] 26853
Montagna, F. [1983] 34635
Montague, R. [1963] 14478
Schultz, Konrad [1970] 13035
Visser, A. [1984] 42643

C65

Bugajski, S. [1983] 42320
Fajardo, S. [1985] 41798
Goldblatt, R.I. [1984] 40768
Grzegorczyk, A. [1967] 05412
Gurevich, Y. [1985] 47372
Johnstone, P.T. [1979] 56505
Keisler, H.J. [1977] 16614
Kreisel, G. [1954] 16973
McKinsey, J.C.C. [1941] 09020
Weispfenning, V. [1985] 47574

C70

Barwise, J. [1975] 60316
Ershov, Yu.L. [1983] 40496

C75

Costa da, N.C.A. [1976] 14829
Gerla, G. [1983] 44834
Gerla, G. [1984] 39889
Goad, C.A. [1978] 73424
Hickin, K.K. [1975] 28362
Jongh de, D.H.J. [1980] 55677
Kalicki, C. [1980] 53776
Kawai, H. [1982] 36043
Keisler, H.J. [1977] 16614
Krauss, P.H. [1966] 27709
Kreisel, G. [1975] 75092
Morais, R. [1976] 54742
Morais, R. [1977] 16605
Nguyen Cat Ho [1972] 09939
Nguyen Cat Ho [1973] 29577
Przymusinska, H. [1980] 54914
Przymusinska, H. [1980] 54915
Radev, S.R. [1984] 44072
Rasiowa, H. [1982] 45056
Sundholm, G. [1977] 50941
Tauts, A. [1975] 52459
Tiuryn, J. [1981] 34010

C80

Altham, J.E.J. [1971] 00285
Barwise, J. [1981] 55362
Baudisch, A. [1980] 56368
Canty, J.T. [1970] 27980
Carlstrom, I.F. [1975] 60828
Hajek, P. [1976] 31127
Hajek, P. [1977] 31126
Havranek, T. [1975] 31144
Hintikka, K.J.J. [1982] 55406
Hoover, D.N. [1978] 29154
Keisler, H.J. [1977] 16614
Montagna, F. [1980] 54076
Mostowski, Andrzej [1962] 09574
Peterson, P.L. [1979] 52390
Shanin, N.A. [1977] 53812
Shrejder, Yu.A. [1977] 79715
Sobolev, S.K. [1977] 51014
Thompson, B. [1982] 55079

C85

Burgess, J.P. [1980] 55004
Gallin, D. [1975] 29739
Girard, J.-Y. [1976] 50375
Hajek, P. [1981] 35692
Mostowski, Andrzej [1962] 09574

C90

Alves, E.H. [1984] 39893
Anderson, J.G. [1969] 00344
Anshakov, O.M. [1984] 44158
Anshakov, O.M. [1984] 47425
Artemov, S.N. [1980] 54546
Bell, J.L. [1983] 43055
Bellissima, F. [1983] 42314
Benthem van, J.F.A.K. [1979] 28157
Benthem van, J.F.A.K. [1983] 37277
Bernhardt, K. [1977] 53552
Bigelow, J.C. [1976] 60510
Blok, W.J. [1983] 40651
Bowen, K.A. [1975] 29636
Bowen, K.A. [1978] 30659
Bozic, M. [1984] 40033
Bozic, M. [1985] 48301
Bozzi, S. [1980] 54221
Brady, R.T. [1984] 45522
Bugajski, S. [1983] 42320
Chang, C.C. [1962] 01946
Chang, C.C. [1966] 02410
Chang, C.C. [1973] 03978
Chuaqui, R.B. [1980] 55151
Chudnovsky, D.V. [1968] 02732
Chudnovsky, D.V. [1970] 26322
Dahn, B.I. [1973] 04057
Dahn, B.I. [1974] 02746
Dahn, B.I. [1974] 04058
Dahn, B.I. [1974] 22893
Dahn, B.I. [1975] 61216

Dziobiak, W. [1978] 53122
Ebbinghaus, H.-D. [1969] 03219
Fajardo, S. [1985] 39901
Fajardo, S. [1985] 41798
Ferenczi, M. [1977] 55783
Fine, K. [1975] 22897
Fine, K. [1978] 30694
Fine, K. [1981] 54607
Fitting, M. [1969] 04349
Fitting, M. [1972] 31074
Follesdal, D. [1966] 32328
Gabbay, D.M. [1971] 04723
Gabbay, D.M. [1971] 04726
Gabbay, D.M. [1971] 30000
Gabbay, D.M. [1972] 04730
Gabbay, D.M. [1972] 04732
Gabbay, D.M. [1972] 04733
Gabbay, D.M. [1975] 04738
Gabbay, D.M. [1975] 04739
Gabbay, D.M. [1976] 32148
Gabbay, D.M. [1977] 26456
Gabbay, D.M. [1982] 38464
Gabbay, D.M. [1984] 41831
Gaerdenfors, P. [1975] 21522
Gaifman, H. [1960] 16922
Gaifman, H. [1964] 33352
Gallin, D. [1975] 29739
Garson, J.W. [1984] 41834
Gerla, G. [1983] 44834
Gerla, G. [1984] 39889
Girard, J.-Y. [1976] 50375
Goble, L.F. [1973] 05058
Goernemann, S. [1971] 39863
Goldblatt, R.I. [1973] 05104
Goldblatt, R.I. [1975] 05108
Goldblatt, R.I. [1981] 55422
Grant, J. [1974] 05305
Hailperin, T. [1984] 41842
Hajek, P. [1976] 31127
Havranek, T. [1975] 31144
Heijenoort van, J. [1979] 79641
Hickin, K.K. [1975] 28362
Hoogewijs, A. [1979] 53121
Hoogewijs, A. [1983] 37062
Hoover, D.N. [1978] 29154
Johnstone, P.T. [1979] 56505
Jongh de, D.H.J. [1980] 55677
Kawai, H. [1982] 36043
Keisler, H.J. [1977] 16614
Kirk, R.E. [1979] 53554
Kirk, R.E. [1980] 56546
Klemke, D. [1971] 07229
Koppelberg, S. [1985] 39673
Kozen, D. [1981] 36840
Krauss, P.H. [1966] 27709
Kripke, S.A. [1965] 07557
Krivine, J.-L. [1974] 07566
Leblanc, H. [1983] 37685
Leone, M. [1985] 49912
Lopez-Escobar, E.G.K. [1981] 55143
Lopez-Escobar, E.G.K. [1983] 39786
Makowsky, J.A. [1977] 31849
Maksimova, L.L. [1977] 52433
Maksimova, L.L. [1977] 54760
Maksimova, L.L. [1979] 55851
Maksimova, L.L. [1982] 37067
Maksimova, L.L. [1982] 40416
Maksimova, L.L. [1982] 40502
Mangani, P. [1976] 51187
Minari, P. [1983] 42336
Miura, S. [1972] 09336
Montagna, F. [1980] 54076
Montagna, F. [1984] 42567
Morgan, C.G. [1982] 36507
Morgan, C.G. [1982] 37860
Mortimer, M. [1974] 09506
Mostowski, Andrzej [1962] 09574
Mundici, D. [1984] 44636
Muravitskij, A.Yu. [1981] 36087
Muravitskij, A.Yu. [1983] 40399
Muravitskij, A.Yu. [1984] 45360
Nat vander, A. [1979] 56121
Nepejvoda, N.N. [1982] 46228
Nishimura, H. [1980] 54084
Nishimura, H. [1980] 56424
Nishimura, H. [1981] 54611
Nishimura, H. [1981] 55147
Ono, H. [1983] 34652
Paulos, J.A. [1976] 18327
Pollock, J.L. [1966] 10622
Rantala, V. [1982] 37647
Rasiowa, H. [1953] 11004
Rasiowa, H. [1955] 11011
Rasiowa, H. [1955] 11014
Rasiowa, H. [1970] 11031
Rasiowa, H. [1972] 11032
Rasiowa, H. [1985] 48433
Rauszer, C. [1977] 50668
Rauszer, C. [1977] 50669
Rauszer, C. [1980] 56381
Rauszer, C. [1985] 41793
Rautenberg, W. [1979] 77698
Reichbach, J. [1974] 31999
Reiterman, J. [1980] 54019
Rescher, N. [1969] 14969
Rescher, N. [1969] 25168
Routley, R. [1974] 15192
Routley, R. [1978] 53083
Ruzsa, I. [1969] 11692
Ruzsa, I. [1969] 19613
Ruzsa, I. [1971] 11693
Ruzsa, I. [1971] 19614
Rybakov, V.V. [1977] 56122
Sanmartin Esplugues, J. [1973] 78125
Sato, M. [1977] 30800
Schuette, K. [1968] 24822
Schultz, Konrad [1970] 13035
Schumm, G.F. [1978] 29254
Schumm, G.F. [1978] 52126
Segerberg, K. [1970] 11938
Segerberg, K. [1972] 11939
Segerberg, K. [1976] 65185
Segerberg, K. [1982] 41144
Shekhtman, V.B. [1978] 52052
Shekhtman, V.B. [1978] 52941
Shekhtman, V.B. [1979] 78450
Shekhtman, V.B. [1980] 54274
Skvortsov, D.P. [1979] 55940
Smorynski, C.A. [1973] 27058
Sobolev, S.K. [1977] 51014
Sonobe, O. [1975] 51586
Suszko, R. [1967] 26275
Szatkowski, M. [1984] 44572
Takeuti, G. [1978] 52447
Takeuti, G. [1984] 39581
Tauts, A. [1974] 17694
Tauts, A. [1975] 52459
Thomason, S.K. [1975] 29635
Ursini, A. [1979] 52568
Vojvodic, G. [1978] 52815
Vojvodic, G. [1980] 36055
Vuckovic, V. [1967] 27478
Waszkiewicz, J. [1969] 14081
Waszkiewicz, J. [1971] 14084
Yashin, A.L. [1977] 36830
Zachorowski, S. [1978] 56212
Zakhar'yashchev, M.V. [1983] 40499
Zlatos, P. [1981] 55634

C95

Ebbinghaus, H.-D. [1978] 28201
Makowsky, J.A. [1980] 54945
Mundici, D. [1984] 44636
Nepejvoda, N.N. [1982] 46228

C97

Addison, J.W. [1974] 70206
Arruda, A.I. [1977] 16589
Bernardi, C. [1983] 34945
Harrington, L.A. [1985] 49810
Hodges, W. [1972] 70226
Loeb, M.H. [1968] 37293
Macintyre, A. [1978] 53596
Prisco di, C.A. [1985] 41792
Sobolev, S.L. [1982] 34492

C98

Barwise, J. [1975] 60316
Barwise, J. [1977] 70117
Baudisch, A. [1980] 56368
Chang, C.C. [1966] 02410
Dalen van, D. [1980] 55705
Ebbinghaus, H.-D. [1978] 28201
Enderton, H.B. [1972] 03355
Ershov, Yu.L. [1973] 32036
Fitting, M. [1969] 04349
Fraisse, R. [1967] 04529
Fraisse, R. [1972] 04535
Friedman, H.M. [1975] 04296
Gabbay, D.M. [1973] 17911
Gabbay, D.M. [1983] 41457
Gabbay, D.M. [1984] 41831
Garson, J.W. [1984] 41834
Goldblatt, R.I. [1979] 55754
Halkowska, K. [1976] 52421
Koppelberg, S. [1985] 39673
Kreisel, G. [1967] 22411
Langford, C.H. [1932] 22592
Lightstone, A.H. [1978] 51926
Malitz, J. [1979] 56167
Marek, W. [1972] 34216
Mostowski, Andrzej [1965] 09578
Rasiowa, H. [1963] 11028
Rasiowa, H. [1982] 45056
Robinson, A. [1959] 19551
Robinson, A. [1963] 28804
Rogers, R. [1971] 11365
Shoenfield, J.R. [1967] 22384
Smirnov, V.A. [1972] 30359
Stillwell, J.C. [1977] 52422

Tajtslin, M.A. [1970] 28675
Wang, Hao [1981] 34116

D03

Gabbay, D.M. [1976] 50854
Gorlov, V.V. [1973] 05201
Maslov, S.Yu. [1972] 19328
Romov, B.A. [1971] 11387
Skolem, T.A. [1954] 21133
Smullyan, R.M. [1961] 12542
Smullyan, R.M. [1961] 21025

D05

Ajzerman, M.A. [1963] 00220
Arbib, M.A. [1975] 23098
Chandra, A.K. [1985] 49198
Chomsky, N. [1972] 25314
Danko, W. [1979] 54156
Dassow, J. [1976] 61236
Denning, P.J. [1978] 37736
Emerson, E.A. [1985] 49191
Engeler, E. [1971] 23569
Gaines, B.R. [1975] 35823
Gurevich, Y. [1985] 47372
Heringer, H.J. [1972] 21764
Hilton, A.M. [1963] 23532
Isard, S.D. [1977] 52338
Kalmar, L. [1965] 48630
Kloetzer, G. [1972] 30031
Kudryavtsev, V.B. [1980] 36855
Levin, V.A. [1972] 63410
Levin, V.A. [1974] 63412
Lindner, R. [1974] 24790
McCulloch, W.S. [1943] 08977
Michel, M. [1984] 43712
Muchnik, A.A. [1983] 43519
Rasiowa, H. [1974] 29624
Shu, Yongchang [1981] 47233
Sistla, A.P. [1985] 49529
Smullyan, R.M. [1961] 21025
Sobolev, S.K. [1977] 51279
Turquette, A.R. [1968] 13749
Wechler, W. [1978] 54647

D10

Ajzerman, M.A. [1963] 00220
Blum, L. [1975] 51605
Chlebus, B.S. [1979] 54219
Denning, P.J. [1978] 37736
Havranek, T. [1978] 31145
Kalmar, L. [1965] 48630
Lindner, R. [1974] 24790
Moshchenskij, V.A. [1973] 21271
Muchnik, A.A. [1983] 43519
Rose, A. [1962] 11465
Santos, E.S. [1976] 65059
Schwartz, J.T. [1967] 23572
Stillwell, J.C. [1977] 52422
Yasuhara, A. [1971] 18028
Zaslavskij, I.D. [1979] 55622

D15

Adamson, A. [1979] 53245
Blum, L. [1973] 71122
Blum, L. [1975] 51605
Cherniavsky, J.C. [1973] 71692
Chlebus, B.S. [1979] 54219
Chlebus, B.S. [1981] 38156
Chlebus, B.S. [1982] 37748
Daley, R.P. [1982] 33516
Emerson, E.A. [1984] 45349
Engeler, E. [1975] 21738
Fischer, Michael J. [1979] 56253
Frejvald, R.V. [1968] 04599
Genrich, H.J. [1976] 61945
Harel, D. [1984] 44121
Ladner, R.E. [1977] 51489
Muchnik, A.A. [1983] 43519
Nakamura, A. [1980] 54959
Pogosyan, Eh.M. [1975] 77403
Rasiowa, H. [1973] 19572
Rennie, M.K. [1968] 11098
Tkachev, G.A. [1977] 69960
Valiev, M.K. [1980] 54018

D20

Aron, E. [1975] 35833
Botusharov, O.I. [1983] 45089
Botusharov, O.I. [1985] 48979
Case, J. [1983] 37600
Church, A. [1936] 23466
Crossley, J.N. [1965] 31683
Danko, W. [1979] 54156
Fisher, A. [1982] 36967
Frejvald, R.V. [1982] 43888
Frejvald, R.V. [1984] 42396
Hermes, H. [1958] 14545
Ito, Makoto [1965] 27530
Kleene, S.C. [1967] 45895
Kosovskij, N.K. [1981] 55663
Lorenzen, P. [1962] 08303
Moshchenskij, V.A. [1973] 21271
Prank, R.K. [1979] 53437
Smullyan, R.M. [1961] 21025
Sorensen, R.A. [1984] 43297
Thomason, S.K. [1975] 13597
Wette, E. [1976] 31554
Yablonskij, S.V. [1980] 55350
Yasuhara, A. [1971] 18028
Zaslavskij, I.D. [1979] 55622

D25

Cresswell, M.J. [1985] 42657
Gnani, G. [1974] 62056
Gurevich, Y. [1985] 47372
Kleene, S.C. [1967] 45895
Meyer, A.R. [1981] 54996
Prank, R.K. [1980] 77510
Prank, R.K. [1981] 35552
Smullyan, R.M. [1961] 12542
Smullyan, R.M. [1961] 21025
Stephan, B.J. [1975] 13174

D30

Kreczmar, A. [1972] 07466

D35

Artemov, S.N. [1985] 48972
Bazhanov, V.A. [1984] 46738
Cresswell, M.J. [1985] 42657
Gao, Hengshan [1973] 46707
Gao, Hengshan [1976] 61914
Goddard, L. [1965] 05062
Hauschild, K. [1982] 34362
Ackermann, W. [1928] 00107
Hodes, H.T. [1984] 41142
Isard, S.D. [1977] 52338
Kilmister, C.W. [1967] 22338
Kreczmar, A. [1971] 28106
Kreczmar, A. [1972] 07466
Kripke, S.A. [1962] 07555
Lorenzen, P. [1962] 08303
Manaster, A.B. [1975] 23120
Meyer, A.R. [1981] 54996
Meyer, R.K. [1968] 09181
Meyer, R.K. [1973] 09201
Nakamura, A. [1970] 09789
Orevkov, V.P. [1967] 10149
Popov, S.V. [1977] 77439
Popov, S.V. [1981] 34293
Ragaz, M. [1983] 36532
Scarpellini, B. [1962] 11877
Shekhtman, V.B. [1978] 78453
Shekhtman, V.B. [1982] 38116
Smorynski, C.A. [1982] 43687
Sobolev, S.K. [1977] 51014
Stillwell, J.C. [1977] 52422
Urquhart, A.I.F. [1984] 40056
Ursini, A. [1985] 47525
Yablon, P. [1975] 17472

D40

Kreisel, G. [1975] 75092
Meyer, R.K. [1973] 09201

D45

Frejvald, R.V. [1982] 43888
Kanda, A. [1985] 47568
Mostowski, Andrzej [1962] 09574

D55

Artemov, S.N. [1985] 48972
Chauvin, A. [1960] 01998
Gnani, G. [1974] 62056
Hajek, P. [1977] 26853
Jeroslow, R. [1975] 28273
Kleene, S.C. [1967] 45895
Kreczmar, A. [1971] 28106
Marek, W. [1976] 24424
Ragaz, M. [1983] 36532
Thomason, S.K. [1975] 13597

D60

Barwise, J. [1975] 60316

D70

Boyer, R.S. [1979] 56665

D75

Scott, D.S. [1982] 38593
Tsichritzis, D.C. [1969] 65798

D80

Baez, J.C. [1983] 40205
Blum, L. [1973] 71122
Cresswell, M.J. [1985] 42657
Hermes, H. [1958] 14545
Peters, F.E. [1974] 64464
Schwartz, J.T. [1967] 23572

D97

Barzdins, J. [1974] 46742
Barzdins, J. [1975] 46743
Crossley, J.N. [1965] 31683
Ershov, A.P. [1979] 45742
Harrington, L.A. [1985] 49810
Kalmar, L. [1965] 48630
Khomich, V.I. [1979] 70052
Marek, W. [1976] 24424
Schwartz, J.T. [1967] 23572
Sobolev, S.L. [1982] 34492

D98

Ajzerman, M.A. [1963] 00220
Barwise, J. [1975] 60316
Barwise, J. [1977] 70117
Boolos, G. [1974] 03933
Cohen, L.J. [1982] 36588
Davis, Martin D. [1974] 21268
Denning, P.J. [1978] 37736
Friedman, H.M. [1975] 04296
Kleene, S.C. [1952] 07173
Kleene, S.C. [1960] 07186
Kleene, S.C. [1967] 45895
Kloetzer, G. [1972] 30031
Korfhage, R. [1966] 07376
Kosovskij, N.K. [1981] 55663
Lavrov, I.A. [1970] 07889
Lavrov, I.A. [1975] 21765
Lorenzen, P. [1962] 08303
Malitz, J. [1979] 56167
Manaster, A.B. [1975] 23120
Manin, Yu.I. [1977] 51984
Marek, W. [1972] 34216
Mostowski, Andrzej [1965] 09578
Peters, F.E. [1974] 64464
Porte, J. [1972] 17471
Shoenfield, J.R. [1967] 22384
Smullyan, R.M. [1961] 21025
Yasuhara, A. [1971] 18028

D99

Hofstadter, D.R. [1979] 74228

E07

Gumb, R.D. [1982] 35230
Kaluzhnin, L.A. [1979] 53329
Menger, K. [1951] 31954
Moisil, G.C. [1941] 09362
Poeschel, R. [1975] 21515
Salij, V.N. [1965] 16216

E10

Christian, C.C. [1976] 71770

E15

Morais, R. [1977] 16605

E20

Kudryavtsev, V.B. [1981] 35548
Rosenberg, I.G. [1976] 26563
Schmidt, J. [1971] 12921
Shrejder, Yu.A. [1977] 79715
Yablonskij, S.V. [1964] 00078

E25

Davis, Charles C. [1976] 18114
Fitting, M. [1969] 04349

E30

Arruda, A.I. [1971] 00494
Bar-Hillel, Y. [1958] 00805
Barwise, J. [1975] 60316
Christian, C.C. [1976] 71770
Costa da, N.C.A. [1967] 02476
Dalla Chiara Scabia, M.L. [1977] 31066
Glubrecht, J.-M. [1983] 36976
Grishin, V.N. [1981] 54617
Hatcher, W.S. [1968] 22644
Montagna, F. [1983] 34635
Petersen, U. [1983] 46146
Prior, A.N. [1955] 22325
Quine, W.V.O. [1963] 23295
Rockingham Gill, R.R. [1977] 77833
Skolem, T.A. [1963] 12431
Solovay, R.M. [1976] 26088
Sublet, J. [1954] 28005
Wiredu, J.E. [1974] 18430

E35

Arruda, A.I. [1971] 00494
Brady, R.T. [1971] 01576
Brady, R.T. [1972] 01577
Brady, R.T. [1980] 53774
Costa da, N.C.A. [1974] 02487
Fitting, M. [1969] 04349
Fitting, M. [1972] 31074
Flagg, R.C. [1985] 49605
Gallin, D. [1975] 29739
Koppelberg, S. [1985] 39673
Krajicek, J. [1984] 43351
Manin, Yu.I. [1977] 51984
Manin, Yu.I. [1979] 54716
Rose, A. [1978] 52835
Schultz, Konrad [1970] 13035
White, R.B. [1979] 53322
Zachorowski, S. [1975] 66185

E40

Takeuti, G. [1978] 52447
Takeuti, G. [1981] 39565
Takeuti, G. [1984] 39581
Zhang, Jinwen [1980] 54107

E45

Fitting, M. [1969] 04349
Koppelberg, S. [1985] 39673

E47

Reinhardt, W.N. [1980] 54025

E50

Fitting, M. [1969] 04349
Gallin, D. [1975] 29739
Goedel, K. [1947] 05073
Koppelberg, S. [1985] 39673
Manin, Yu.I. [1977] 51984
Manin, Yu.I. [1979] 54716

E55

Costa da, N.C.A. [1967] 02476
Reinhardt, W.N. [1980] 54025

E65

Chilin, V.I. [1978] 56688
Koppelberg, S. [1985] 39673

E70

Aczel, P. [1980] 70302
Arruda, A.I. [1970] 00490
Arruda, A.I. [1970] 00492
Arruda, A.I. [1970] 00493
Arruda, A.I. [1971] 00494
Arruda, A.I. [1974] 16566
Arruda, A.I. [1975] 27257
Arruda, A.I. [1977] 31721
Arruda, A.I. [1980] 53627
Arruda, A.I. [1980] 54036
Arruda, A.I. [1982] 42801
Asenjo, F.G. [1975] 00521
Asenjo, F.G. [1977] 52236
Bar-Hillel, Y. [1958] 00805
Bencivenga, E. [1976] 29770
Bencivenga, E. [1977] 51280
Brady, R.T. [1971] 01576
Brady, R.T. [1972] 01577
Brady, R.T. [1980] 53774
Brady, R.T. [1980] 56394
Bressan, A. [1973] 17268
Bressan, A. [1973] 60703
Bressan, A. [1974] 17269
Bressan, A. [1974] 29947
Canty, J.T. [1970] 27980
Chang, C.C. [1963] 01961
Chang, C.C. [1965] 01968
Chwistek, L.B. [1938] 02672
Clay, R.E. [1975] 17538
Costa da, N.C.A. [1964] 02466
Costa da, N.C.A. [1967] 02475
Costa da, N.C.A. [1967] 02476
Costa da, N.C.A. [1970] 02481
Costa da, N.C.A. [1971] 02484
Costa da, N.C.A. [1974] 02487
Davis, Charles C. [1976] 18114
Ehsakia, L.L. [1979] 55773
Fadini, A. [1962] 17180
Fadini, A. [1978] 72779
Fenstad, J.E. [1964] 03740
Fine, K. [1981] 72893
Finkelstein, D. [1976] 61694
Finkelstein, D. [1981] 34713
Finkelstein, D. [1985] 49064
Fitch, F.B. [1967] 04347

Flagg, R.C. [1985] 49605
Gallin, D. [1975] 29739
Goe, G. [1968] 05067
Goodman, Nicolas D. [1981] 55017
Gottwald, S. [1977] 50788
Grishin, V.N. [1974] 73648
Grishin, V.N. [1981] 54617
Hatcher, W.S. [1968] 22644
Iwanus, B. [1973] 17553
Jahn, K.-U. [1975] 21673
Jahn, K.-U. [1975] 62724
Kaufmann, A. [1973] 62912
Klaua, D. [1965] 07147
Krajicek, J. [1984] 43351
Lejewski, C. [1955] 27074
Lejewski, C. [1974] 29952
Moisil, G.C. [1972] 42022
Negoita, C.V. [1974] 64172
Petersen, U. [1983] 46146
Pirog-Rzepecka, K. [1973] 12500
Prade, H. [1982] 40989
Quine, W.V.O. [1940] 10867
Rasiowa, H. [1965] 29826
Reinhardt, W.N. [1980] 54025
Rickey, V.F. [1977] 53290
Rose, A. [1978] 52835
Rosser, J.B. [1953] 11565
Salij, V.N. [1965] 16216
Scedrov, A. [1985] 44771
Schock, R. [1971] 28602
Schultz, Konrad [1970] 13034
Schultz, Konrad [1970] 13035
Scott, D.S. [1975] 27594
Skolem, T.A. [1957] 12418
Skolem, T.A. [1957] 12420
Skolem, T.A. [1960] 12425
Skolem, T.A. [1963] 12431
Stout, L.N. [1979] 56341
Takano, M. [1985] 47523
Takeuti, G. [1981] 39565
Wang, Hao [1961] 14036
Welsh Jr., P.J. [1978] 27072
White, R.B. [1979] 53322
Wiredu, J.E. [1974] 18430
Yasugi, M. [1981] 55376
Zachorowski, S. [1975] 66185

E72

Albert, P. [1978] 56197
Almog, J. [1976] 50468
Almog, J. [1976] 50469
Alsina, C. [1980] 44653
Atanassov, K. [1984] 43217
Baldwin, J.F. [1979] 53009
Baldwin, J.F. [1979] 70620
Baldwin, J.F. [1979] 80615
Bandler, W. [1980] 70651
Bellman, R.E. [1973] 24614
Botta, O. [1982] 35770
Brown, J.G. [1971] 04263
Cerny, M. [1984] 44291
Cools, M. [1975] 62910
Czogala, E. [1983] 34693
Dishkant, G.P. [1981] 55165
Domingo, X. [1980] 44665
Domingo, X. [1981] 43204
Dubois, Didier [1979] 55724
Dubois, Didier [1983] 39264
Dubois, Didier [1985] 48203
Dubois, Didier [1985] 49478
Dyckhoff, H. [1984] 43917
Edmonds, E.A. [1980] 72572
Eytan, M. [1977] 32319
Gaines, B.R. [1976] 61897
Gaines, B.R. [1977] 50094
Giles, R. [1975] 73364
Gol'dman, R.S. [1980] 46184
Gottwald, S. [1979] 53061
Gottwald, S. [1979] 73554
Grattan-Guinness, I. [1976] 18482
Grattan-Guinness, I. [1979] 53654
Hajnal, M. [1977] 50471
Hajnal, M. [1977] 53810
Hamacher, H.C. [1979] 55713
He, Zhongxiong [1984] 49002
Hirota, K. [1979] 81567
Hisdal, E. [1979] 74198
Hisdal, E. [1981] 55208
Iwai, S. [1982] 46116
Iwai, S. [1982] 46458
Kampe de Feriet, J. [1980] 81766
Kaufmann, A. [1973] 62912
Klement, E.P. [1981] 56587
Koczy, L.T. [1979] 56339
Li, Bangrong [1980] 82094
Liao, Qun [1985] 49012
Liu, Yunfeng [1982] 34899
Moisil, G.C. [1972] 42022
Nalimov, V.V. [1979] 76753
Negoita, C.V. [1974] 64172
Negoita, C.V. [1975] 50043
Nekola, J. [1984] 46735
Novak, Vilem [1983] 38250
Prade, H. [1982] 46444
Pultr, A. [1982] 35536
Sanchez, E. [1976] 65045
Skala, H.J. [1978] 52634
Spigler, R. [1980] 78882
Togai, M. [1983] 46900
Trillas, E. [1979] 45484
Trillas, E. [1979] 53378
Wang, Jincheng [1982] 49625
Wechler, W. [1978] 54647
Willmott, R. [1980] 79990
Wu, Xuemou [1978] 83180
Wu, Xuemou [1979] 83179
Wu, Xuemou [1980] 83177
Wu, Xuemou [1981] 36019
Wu, Xuemou [1981] 83178
Yager, R.R. [1980] 53809
Yager, R.R. [1980] 54211
Yager, R.R. [1983] 35027
Yanase, M.M. [1985] 49435
Yu, Liansheng [1982] 35782
Zadeh, L.A. [1965] 24934
Zadeh, L.A. [1971] 26291
Zadeh, L.A. [1976] 83203
Zhang, Jinwen [1980] 54107
Zhang, Jinwen [1981] 34874
Zimmermann, H.-J. [1985] 49392

E75

Cooper, W.S. [1974] 21633
Good, I.J. [1962] 16283
Jankowski, A.W. [1985] 47520
Lungarzo, C.A. [1978] 52163
Sikorski, R. [1958] 12300
Takeuti, G. [1978] 52447
Takeuti, G. [1984] 39581

E97

Bochvar, D.A. [1976] 52367
Harrington, L.A. [1985] 49810
Marek, W. [1976] 24424

E98

Avenoso, F.J. [1970] 22631
Baginski, M. [1973] 75585
Bar-Hillel, Y. [1958] 00805
Barwise, J. [1975] 60316
Barwise, J. [1977] 70117
Bernays, P. [1958] 16969
Borkowski, L. [1963] 19509
Chenique, F. [1974] 23053
Christian, R.R. [1958] 22478
Combes, M. [1971] 02334
Cools, M. [1975] 62910
Costa da, N.C.A. [1963] 42829
Detlovs, V.K. [1970] 17027
Dinkines, F. [1964] 22482
Dodge, C.W. [1969] 22633
Drake, F.R. [1985] 47704
Durnev, V.G. [1978] 72511
Edwards, R.E. [1979] 52995
Ehlers, F. [1968] 22634
Fiorentini, M. [1970] 17901
Fraenkel, A.A. [1959] 28712
Friedman, H.M. [1975] 04296
Gladkij, A.V. [1974] 21763
Goedel, K. [1947] 05073
Goldblatt, R.I. [1979] 55754
Grzegorczyk, A. [1955] 31447
Halkowska, K. [1976] 52421
Halkowska, K. [1978] 54789
Hanna, S.C. [1971] 22641
Hasse, M. [1966] 22643
Hatcher, W.S. [1968] 22644
Karpov, V.G. [1977] 51362
Kaufmann, A. [1973] 62912
Krempa, J. [1977] 51983
Lavrov, I.A. [1970] 07888
Lavrov, I.A. [1975] 21765
Marek, W. [1972] 34216
Mostowski, Andrzej [1965] 09578
Oberschelp, A. [1970] 31198
Oberschelp, A. [1974] 31195
Oberschelp, A. [1978] 31196
Otepanov, V.I. [1972] 13931
Penzov, Yu.E. [1968] 16550
Quine, W.V.O. [1963] 23295
Rasiowa, H. [1971] 22289
Rasiowa, H. [1972] 31450
Rosser, J.B. [1953] 11565
Shoenfield, J.R. [1967] 22384
Stoll, R.R. [1961] 19683
Stoll, R.R. [1963] 13197
Vasilache, S. [1977] 51636

Wang, Hao [1981] 34116
Wojciechowska, A. [1979] 80045

F05

Avron, A. [1984] 42415
Borga, M. [1983] 42318
Dopp, J. [1962] 03085
Girard, J.-Y. [1976] 50375
Goodman, Nicolas D. [1984] 42452
Bernays, P. [1934] 01098
Bernays, P. [1939] 01082
Leivant, D. [1981] 54609
Lopez-Escobar, E.G.K. [1983] 39786
Malmnaes, P.E. [1968] 24815
Meyer, R.K. [1976] 27122
Petkov, P.P. [1984] 46576
Prawitz, D. [1965] 22164
Richter, M.M. [1978] 51870
Rogava, M.G. [1975] 18361
Rogava, M.G. [1979] 77850
Sambin, G. [1980] 54341
Sambin, G. [1982] 35627
Schuette, K. [1960] 13010
Smirnov, V.A. [1972] 30359
Takeuti, G. [1978] 52447
Tennant, N. [1980] 53989
Umezawa, T. [1980] 56395
Valentini, S. [1983] 38327
White, R.B. [1979] 53322
Zaslavskij, I.D. [1978] 80231

F07

Anderson, J.M. [1962] 00341
Corcoran, J. [1969] 02388
Dopp, J. [1962] 03085
Fitch, F.B. [1966] 17115
Fitting, M. [1983] 34628
Giambrone, S. [1985] 47499
Grishin, V.N. [1974] 73648
Grishin, V.N. [1981] 54617
Lopez-Escobar, E.G.K. [1985] 47264
Meloni, G.C. [1979] 56742
Nieland, J.J.F. [1967] 09951
Pliuskevicius, R. [1983] 43884
Prawitz, D. [1965] 22164
Sikic, Z. [1984] 42266
Smirnov, V.A. [1972] 30359
Smullyan, R.M. [1963] 28562

F10

Kreisel, G. [1965] 27577

F15

Bernays, P. [1939] 01082
Schuette, K. [1960] 13010
Schuette, K. [1962] 22382
Schuette, K. [1963] 22383

F20

Cherniavsky, J.C. [1973] 71692

F25

Danko, W. [1983] 39522
Hajek, P. [1981] 35379
Lambert, K. [1967] 07823
Malmnaes, P.E. [1968] 24815
Svejdar, V. [1983] 40906

F30

Artemov, S.N. [1982] 38128
Artemov, S.N. [1984] 45832
Artemov, S.N. [1985] 48972
Artemov, S.N. [1985] 49322
Avron, A. [1984] 42415
Bendova, K. [1982] 34402
Bernardi, C. [1975] 31011
Bernardi, C. [1976] 60478
Bernardi, C. [1984] 42581
Beth, E.W. [1962] 01176
Boolos, G. [1976] 14578
Boolos, G. [1977] 26442
Boolos, G. [1979] 56312
Boolos, G. [1979] 56313
Boolos, G. [1980] 55341
Boolos, G. [1982] 35340
Boolos, G. [1984] 43982
Boolos, G. [1985] 47529
Borga, M. [1983] 42318
Cantini, A. [1980] 54349
Crossley, J.N. [1975] 31689
Enderton, H.B. [1972] 03355
Fisher, A. [1982] 36967
Gargov, G.K. [1984] 46969
Goodman, Nicolas D. [1984] 42452
Goodstein, R.L. [1951] 25669
Goodstein, R.L. [1958] 05182
Guaspari, D. [1979] 53658
Hajek, P. [1977] 26853
Hajek, P. [1981] 35379
Bernays, P. [1934] 01098
Bernays, P. [1939] 01082
Kirov, K.A. [1984] 46974
Kleene, S.C. [1952] 07173
Kleene, S.C. [1967] 45895
Kosovskij, N.K. [1981] 55663
Ladriere, J. [1957] 07777
Leivant, D. [1981] 54609
Lopez-Escobar, E.G.K. [1985] 47264
Magari, R. [1974] 08539
Magari, R. [1982] 35470
Manin, Yu.I. [1977] 51984
Manin, Yu.I. [1979] 54716
Meyer, R.K. [1976] 27121
Meyer, R.K. [1985] 48205
Montagna, F. [1983] 34635
Montagna, F. [1983] 37078
Montagna, F. [1984] 42567
Montague, R. [1963] 14478
Muravitskij, A.Yu. [1983] 34636
Rose, A. [1955] 24903
Sambin, G. [1978] 52783
Sambin, G. [1980] 54341
Sambin, G. [1982] 35627
Schuette, K. [1960] 13010
Simmons, H. [1975] 18388
Skolem, T.A. [1952] 21071
Skolem, T.A. [1953] 21072
Smorynski, C.A. [1978] 54148
Smorynski, C.A. [1979] 52968
Smorynski, C.A. [1980] 56734
Smorynski, C.A. [1982] 36167
Smorynski, C.A. [1982] 43687
Smorynski, C.A. [1984] 44891
Smorynski, C.A. [1985] 44663
Smullyan, R.M. [1985] 44678
Smullyan, R.M. [1985] 44783
Solitro, U. [1983] 35696
Solovay, R.M. [1976] 26088
Steen, S.W.P. [1972] 22386
Stralberg, A.H. [1965] 13218
Svejdar, V. [1983] 40906
Takeuti, G. [1978] 52447
Valentini, S. [1983] 34778
Valentini, S. [1983] 38327
Visser, A. [1984] 42643
Vuckovic, V. [1967] 27478
Vuckovic, V. [1967] 27479
Wang, Hao [1981] 34116
Zaslavskij, I.D. [1978] 80231
Zinov'ev, A.A. [1974] 17312

F35

Girard, J.-Y. [1976] 50375
Goldblatt, R.I. [1979] 55754
Kreisel, G. [1965] 27577
Meyer, R.K. [1976] 27122
Prank, R.K. [1980] 77510
Scedrov, A. [1985] 44771
Schuette, K. [1960] 13010
Takeuti, G. [1978] 52447
Troelstra, A.S. [1981] 55352
Wendel, N. [1978] 53217

F40

Bernays, P. [1939] 01082

F50

Akimov, A.P. [1979] 70347
Almukdad, A. [1984] 42414
Arruda, A.I. [1978] 31617
Artemov, S.N. [1980] 54546
Asanidze, G.Z. [1981] 34653
Bertolini, F. [1971] 01565
Bessonov, A.V. [1977] 53417
Beth, E.W. [1962] 01176
Blanche, R. [1957] 01538
Boricic, B.R. [1981] 35211
Bowen, K.A. [1971] 01504
Bozic, M. [1983] 46894
Bozzi, S. [1980] 54221
Burgess, J.P. [1981] 56494
Cherniavsky, J.C. [1973] 71692
Dardzhaniya, G.K. [1977] 53196
Dardzhaniya, G.K. [1979] 53268
Destouches-Fevrier, P. [1948] 02965
Dopp, J. [1962] 03085
Dosen, K. [1981] 35691
Dosen, K. [1981] 55339
Dosen, K. [1984] 44539
Dosen, K. [1984] 44540
Dosen, K. [1985] 47541
Drugush, Ya.M. [1978] 52635
Epstein, G. [1974] 22895
Epstein, G. [1976] 18456
Fidel, M.M. [1978] 52314
Fischer Servi, G. [1976] 55901
Fischer Servi, G. [1977] 50945

Fischer Servi, G. [1978] 52127
Fischer Servi, G. [1981] 54078
Fitch, F.B. [1948] 04332
Fitting, M. [1969] 04349
Fitting, M. [1973] 04355
Fitting, M. [1983] 34628
Flagg, R.C. [1985] 49605
Frink, O. [1938] 04671
Gabbay, D.M. [1970] 04721
Gabbay, D.M. [1971] 04726
Gabbay, D.M. [1972] 04730
Gabbay, D.M. [1973] 17911
Gabbay, D.M. [1975] 04738
Gabbay, D.M. [1976] 14748
Gabbay, D.M. [1976] 50854
Gabbay, D.M. [1977] 26456
Gabbay, D.M. [1977] 50853
Gabbay, D.M. [1981] 54131
Gabbay, D.M. [1982] 38464
Gargov, G.K. [1975] 21677
Georgacarakos, G.N. [1982] 43512
Gimon, V.V. [1979] 53494
Girard, J.-Y. [1976] 50375
Goad, C.A. [1978] 73424
Goble, L.F. [1974] 05060
Goedel, K. [1931] 22120
Goernemann, S. [1971] 39863
Goldblatt, R.I. [1978] 56314
Goldblatt, R.I. [1979] 55754
Goldblatt, R.I. [1981] 55422
Golota, Ya.Ya. [1969] 05123
Goodman, Nicolas D. [1984] 42452
Goodstein, R.L. [1958] 05182
Gorgy, F.W. [1979] 53991
Grzegorczyk, A. [1967] 62216
Hart, A.M. [1978] 51664
Heijenoort van, J. [1979] 79641
Hermes, H. [1976] 51102
Heyting, A. [1958] 06062
Ibragimov, S.G. [1973] 74359
Jaskowski, S. [1936] 06544
Johnstone, P.T. [1979] 56505
Jongh de, D.H.J. [1980] 55677
Kalicki, C. [1980] 53776
Kirk, R.E. [1979] 53554
Kirk, R.E. [1980] 56546
Kleene, S.C. [1952] 07173
Klemke, D. [1971] 07229
Kondo, M. [1984] 44181
Kreisel, G. [1957] 07492
Kreisel, G. [1965] 27577
Kripke, S.A. [1965] 07557
Lavendhomme, R. [1984] 40558
Leblanc, H. [1983] 37685
Leblanc, H. [1983] 54992
Levin, L.A. [1969] 08046
Levy, P. [1927] 08092
Lopez-Escobar, E.G.K. [1981] 55142
Lopez-Escobar, E.G.K. [1981] 55143
Lopez-Escobar, E.G.K. [1983] 39786
Lorenz, K. [1968] 28288
Lorenz, K. [1973] 27473
Lorenzen, P. [1955] 08290
Lorenzen, P. [1972] 08310
Lorenzen, P. [1972] 08311
Malmnaes, P.E. [1968] 24815
Martin, J.N. [1984] 40627
Matsumoto, K. [1950] 08888
McKay, C.G. [1967] 08983
Meyer, R.K. [1973] 09200
Meyer, R.K. [1976] 21939
Meyer, R.K. [1977] 21964
Mints, G.E. [1972] 09288
Morgan, C.G. [1976] 76542
Muchnik, A.A. [1983] 43519
Muravitskij, A.Yu. [1981] 36087
Nepejvoda, N.N. [1982] 46228
Nepejvoda, N.N. [1983] 43233
Petkov, P.P. [1971] 10430
Porebska, M. [1975] 17277
Prank, R.K. [1979] 53437
Prank, R.K. [1980] 77510
Prank, R.K. [1981] 35552
Prawitz, D. [1965] 22164
Previale, F. [1975] 31421
Rasiowa, H. [1951] 11001
Rasiowa, H. [1953] 11004
Rasiowa, H. [1955] 11011
Rasiowa, H. [1963] 11028
Rasiowa, H. [1974] 31451
Rauszer, C. [1974] 15188
Rauszer, C. [1976] 18356
Rauszer, C. [1977] 50668
Rauszer, C. [1977] 50669
Rauszer, C. [1980] 56381
Renardel de Lavalette, G.R. [1981] 55207
Rescher, N. [1966] 11118
Rousseau, G. [1970] 11601
Routley, R. [1974] 15192
Routley, R. [1978] 53083
Scedrov, A. [1984] 44085
Scedrov, A. [1985] 44771
Schmidt, H.A. [1960] 12892
Schuette, K. [1968] 24822
Segerberg, K. [1968] 11934
Segerberg, K. [1974] 11941
Semenenko, M.I. [1970] 11972
Shanin, N.A. [1977] 53812
Sikorski, R. [1956] 12298
Sikorski, R. [1958] 12300
Skolem, T.A. [1952] 21071
Skolem, T.A. [1953] 21072
Skvortsov, D.P. [1979] 55940
Skvortsov, D.P. [1980] 55941
Smorynski, C.A. [1973] 27058
Smorynski, C.A. [1978] 29151
Smullyan, R.M. [1973] 14740
Sobolev, S.K. [1977] 51014
Sobolev, S.K. [1977] 51279
Sonobe, O. [1975] 51586
Stralberg, A.H. [1965] 13218
Surma, S.J. [1973] 42072
Surma, S.J. [1974] 16561
Surma, S.J. [1975] 23141
Takeuti, G. [1981] 39565
Tauts, A. [1974] 17694
Tauts, A. [1975] 52459
Tolstova, Yu.N. [1971] 13633
Tragesser, R.S. [1978] 30597
Troelstra, A.S. [1981] 55352
Tsitkin, A.I. [1977] 50214
Umezawa, T. [1980] 56395
Vakarelov, D. [1977] 31280
Varpakhovskij, F.L. [1965] 27855
Vorob'ev, N.N. [1964] 16331
Wendel, N. [1978] 53217
Wojcicki, R. [1970] 14254
Wronski, A. [1971] 14337
Wronski, A. [1974] 18433
Yankov, V.A. [1963] 16539
Yankov, V.A. [1963] 19006
Yankov, V.A. [1967] 06513
Yankov, V.A. [1968] 06514
Yankov, V.A. [1969] 06518
Zachorowski, S. [1975] 23947
Zarnecka-Bialy, E. [1973] 14364
Zaslavskij, I.D. [1978] 80231
Zeman, J.J. [1968] 14395

F55

Arruda, A.I. [1978] 31617
Belding, W.R. [1971] 00957
Beth, E.W. [1956] 01166
Heyting, A. [1958] 06062
Skolem, T.A. [1954] 21133

F60

Goodstein, R.L. [1951] 25669
Kosovskij, N.K. [1981] 55663

F65

Asanidze, G.Z. [1981] 34653
Ehrenfeucht, A. [1974] 03261
Lorenzen, P. [1955] 08290
Miglioli, P.A. [1983] 37781

F97

Harrington, L.A. [1985] 49810
Loeb, M.H. [1968] 37293
Orevkov, V.P. [1968] 37197

F98

Anderson, J.M. [1962] 00341
Bar-Hillel, Y. [1958] 00805
Barwise, J. [1977] 70117
Beth, E.W. [1962] 01176
Boolos, G. [1979] 56312
Dopp, J. [1962] 03085
Fisher, A. [1982] 36967
Fitting, M. [1969] 04349
Fitting, M. [1983] 34628
Friedman, H.M. [1975] 04296
Gabbay, D.M. [1981] 54131
Goldblatt, R.I. [1979] 55754
Goodstein, R.L. [1951] 25669
Haas, G. [1984] 44527
Heyting, A. [1958] 06062
Bernays, P. [1934] 01098
Bernays, P. [1939] 01082
Hindley, J.R. [1972] 23471
Kleene, S.C. [1952] 07173
Kleene, S.C. [1967] 45895
Kosovskij, N.K. [1981] 55663
Kreisel, G. [1965] 27577
Ladriere, J. [1957] 07777
Lorenzen, P. [1955] 08290
Lorenzen, P. [1962] 08303
Manin, Yu.I. [1979] 54716

Mostowski, Andrzej [1965] 09578
Prawitz, D. [1965] 22164
Rasiowa, H. [1963] 11028
Schmidt, H.A. [1960] 12892
Schuette, K. [1960] 13010
Schuette, K. [1968] 24822
Shoenfield, J.R. [1967] 22384
Smirnov, V.A. [1972] 30359
Smorynski, C.A. [1985] 44663
Smullyan, R.M. [1963] 28562
Steen, S.W.P. [1972] 22386
Wang, Hao [1981] 34116

F99

Krabbe, E.C.W. [1985] 42654
Leivant, D. [1981] 55013
Lindner, R. [1975] 51606
Lorenz, K. [1978] 28287
Smullyan, R.M. [1983] 41200
Tirnoveanu, M. [1970] 30384

G30

Ehsakia, L.L. [1974] 04119
Goldblatt, R.I. [1979] 55754
Goldblatt, R.I. [1981] 55422
Hatcher, W.S. [1968] 22644
Johnstone, P.T. [1979] 56505
Stout, L.N. [1979] 56341
Vasilache, S. [1977] 51636

H05

Goze, M. [1981] 34857
Hirschfeld, J. [1969] 08459
Lightstone, A.H. [1978] 51926
Lindstroem, T.L. [1982] 36617
Robert, A. [1985] 48149
Robinson, A. [1966] 11263

H10

Robinson, A. [1966] 11263
Sain, I. [1983] 40674

H15

Robinson, A. [1963] 28804
Robinson, A. [1966] 11263

Alphabetization and alternative spellings of author names

The purpose of this index is to help the user find an author in whom he is interested. We begin by outlining both the general principles of alphabetization followed in the Author Index and the systems of transliteration used. The second half of this index addresses the problems which arise with author names for which there may be many variants in the literature. How do you find the primary form of a name used in the Bibliography? The ideal would be to have a table linking all the 'imaginable' versions of an author name to the unique primary form used here, but the obstacles to realizing this are obvious: one 'imaginable' form may correspond to two different authors and, worse, 'imaginable' itself depends on the linguistic background of the user. We have instead suggested some guidelines for identifying the primary form of a name from one of its variants. Finally, there is a list of alternative forms of names for those cases in which the difference between the alternative and the primary forms is particularly striking. For an author whose name has changed, each publication is listed under the name form used on that publication. Pointers to the other name form are given in the Author Index.

The Roman alphabet is as usual alphabetized in the following form:

A B C D E F G H I J K L M N O P Q R S T U V W X Y Z

Within this general framework, the ordering for hyphenated and double names is illustrated by the following example:

Ab,G. ; Ab-Aa,G. ; Ab Aa,G. ; Aba,G.

Apostrophes in a name are disregarded: Mal'tsev, for example, is treated as Maltsev for alphabetical purposes.

Titular prefixes such as von, du, de la, etc. come immediately after the surname (family name), and before the given name (or initials); so, e. g., J. von Neumann appears as Neumann von, J. Similarly J. Smith, Jr., and C.F. Miller III are given as Smith Jr, J. and Miller III, C.F., respectively.

In general, initials are used for given names. The full given name(s) are used only where necessary or helpful to distinguish between authors with the same surnames and initials.

As has been mentioned in the Preface, diacritical marks have, for practical reasons, mostly been disregarded. The following lists those diacritical marks of Scandinavian and German languages that have been transliterated:

æ to ae,
ø to oe,
å to aa,
ä to ae,
ö to oe,
ü to ue.

By the way one cannot infer that every ae, oe, or ue in German comes from ä, ö, or ü; e.g. Gloede is the correct spelling, not Glöde!

Note that the hacek in languages written in the Roman alphabet (e. g. Serbo-Croatian) has not been transliterated (so, for example, Šešelja appears here as Seselja).

The transliteration used for Cyrillic is explained in another index. (For a Russian author who has emigrated to the West the primary name is usually the form used by the author in Western publications. This form does not always agree with the transliteration of the Cyrillic name.) For Chinese names, the Pinyin system of transliteration has been used as far as possible, and commas have been added to separate the surname and given names (which are not abbreviated to initials) to accord with Western style. However, for Korean names no commas are used.

Over the last hundred years there have been in general use several different systems for transliterating Cyrillic into the Roman alphabet. This has given rise to many variants for author names originally written in Cyrillic. We list here our transliteration of those Cyrillic letters for which there have been several variants and give the most common alternative transliterations. If you are searching for an author name you suspect may be of Slavic origin this list will help you to find the most likely form used here: simply replace each (block of) letter(s) on the right occurring in your version by the appropriate letters given on the left.

Our transliteration	Possible alternatives
ya	ja a
yu	ju
eh	e
e	je ye
ts	c
ch	c tsh tch tsch
sh	s sch
zh	z

Our transliteration	Possible alternatives
z	s
j	i y
kh	h
v	w ff
”	y
ks	x
u	ou

The following is a selective listing of alternative forms of author names. It contains only those alternative forms from which the primary may not be guessed by using the guidelines above.

for	see
Abellanas Cebollero, P.	Abellanas, P.F.
Adams, M.M.	McCord Adams, M.
Albuquerque, J.	Ribeiro de Albuquerque, J.
Angulin, D.	Angluin, D.
Artalejo, R.M.	Rodriguez Artalejo, M.
Asjwiniekoemaar	Ashvinikumar
Avraham, U.	Abraham, U.
Barzdin', Ya.M.	Barzdins, J.
Benlahcen, D.	Benhalcen, D.
Bhaskara Rao, K.P.S.	Rao, K.P.S.Bhaskara
Bhaskara Rao, M.	Rao, M.Bhaskara
Bloch, A.S.	Blokh, A.Sh.
Blochina, G.N.	Blokhina, G.N.
Carroll, L.	Dodgson, C.L.
Chakan, B.	Csakany, B.
Char-Tung, R.	Lee, R.C.-T.
Chen, T.T.	Tang, Caozhen
Choodnovsky, D.V.	Chudnovsky, D.V.
Chu, W.J.	Zhu, Wujia
Cohen, E.L.	Longini Cohen, E.
Colburn, C.J.	Colbourn, C.J.
Colburn, M.J.	Colbourn, M.J.
Coppola, L.G.	Gonzalez Coppola, L.
Costa, A.A.	Almeida Costa, A.
Cresswell, M.M.	Meyerhoff Cresswell, Mary
Dao, D.H.	Dang Huu Dao
Decew, J.W.	Wagner Decew, J.
Dieu, P.D.	Phan Dinh Dieu
Duncan Luce, R.	Luce, R.D.
Dyson, V.H.	Huber-Dyson, V.
Fan Din' Zieu'	Phan Dinh Dieu
Foellesdal, D.	Follesdal, D.
Frejvald, R.V.	Freivalds, R.
Gegalkine, I.	Zhegalkin, I.I.
Gibbelato Valabrega, E.	Valabrega, E.G.
Greendlinger, M.	Grindlinger, M.
Hoo, T.-H.	Hu, Shihua
Hsu, L.C.	Xu, Lizhi
Hsueh, Yuang Cheh	Xueh, Yuangche
Jutting, L.S.B.	Benthem Jutting van, L.S.
Kao, H.	Gao, Hengshan
Kapinska, E.	Capinska, E.
Keldych, L.	Keldysh, L.V.
Khunyadvari, L.	Hunyadvari, L.
Kister, J.E.	Bridge, J.
Klein, F.	Klein-Barmen, F.
Kroonenberg, A.V.	Verbeek-Kroonenberg, A.
Kurepa, G.	Kurepa, D.
Kurkova-Pohlova, V.	Pohlova, V.
Kwei, M.S.	Mo, Shaokui
Lifshits, V.	Lifschitz, V.
Lo, Li Bo	Luo, Libo
Loewenthal, F.	Lowenthal, F.
Macdonald, S.O.	Oates MacDonald, S.
Malyaukene, L.K.	Maliaukiene, L.
Markus, S.	Marcus, S.
Moenting, J.S.	Schulte-Moenting, J.
Moh, S.-K.	Mo, Shaokui
Moura, J.E.A.	Almeida Moura de, J.E.
Nardzewski, C.R.	Ryll-Nardzewski, C.
Nash, W.C.S.J.A.	Nash-Williams, C.St.J.A.
Oates-Williams, S.	Oates MacDonald, S.
Plattner, A.	Pieczkowski, A.
Plyushkevichus, R.A.	Pliuskevicius, R.
Plyushkevichene, A.Yu.	Pliuskeviciene, A.
Poprougenko, G.	Popruzenko, J.
Puzio-Pol, E.	Pol, E.
R.-Salinas, B.	Rodriguez-Salinas, B.
Reymond, A.	Virieux-Reymond, A.
Riccioli, B.V.	Veit Riccioli, B.
Rucker, R.	Bitter-Rucker von, R.
Russi, G.Z.	Zubieta Russi, G.
Salinas, B.	Rodriguez-Salinas, B.
Schmir-Hay, L.	Hay, L.
Shain, B.M.	Schein, B.M.
Shaw, M.K.	Mo, Shaokui
Shih-Hua, H.	Hu, Shihua
Shlyakhovaya, N.I.	Slyakhova, N.I.
Solans, V.	Verdu i Solans, V.
Strazdin', I.Eh.	Strazdins, I.E.
Themaat, W.A.v.	Verloren van Themaat, W.A.
Toa van, T.	Tran van Toan
Toth, P.	Ecsedi-Toth, P.
Tsao-Chen, T.	Tang, Caozhen
Tseng, Y.X.	Zheng, Yuxin
Tsirulis, Ya.P.	Cirulis, J.
Tulcea, C.	Ionescu Tulcea, C.
Turksen, I.B.	Tuerksen, I.B.
Tzeng, O.C.	Tseng, O.C.
Vinter, H.	Winter, H.
Williams, C.St.J.A.N.	Nash-Williams, C.St.J.A.
Wou, Shou Zhi	Wu, Shouzhi
Wu, K.J.	Johnson Wu, K.
Yukna, S.P.	Jukna, S.
Yuting, S.	Shen, Y.-T.
Zhay, B.	Zhang, Bosheng
Zhen, Z.	Zhao, Zhen
Zilli, M.V.	Venturini Zilli, M.
Zou, Juan	Zhou, Juan

International vehicle codes

The following abbreviations are used as *codes for the country* in which a conference took place or in which a publishing company is located. (These abbreviations are those used internationally for vehicles.)

Code	Country
A	Austria
ADN	People's Dem. Rep. Yemen (South Yemen)
AFG	Afghanistan
AL	Albania
AND	Andorra
AUS	Australia
B	Belgium
BD	Bangladesh
BDS	Barbados
BG	Bulgaria
BH	Belize
BOL	Bolivia
BR	Brazil
BRN	Bahrain
BRU	Brunei
BS	Bahamas
BU	Burundi
BUR	Burma
C	Cuba
CDN	Canada
CH	Switzerland
CI	Ivory Coast
CL	Sri Lanka
CO	Columbia
CR	Costa Rica
CS	Czechoslovakia
CY	Cyprus
D	Fed. Rep. Germany (West Germany)
DDR	German Dem. Rep. (East Germany)
DK	Denmark
DOM	Dominican Republic
DZ	Algeria
E	Spain
EAK	Kenya
EAT	Tanzania
EAU	Uganda
EC	Ecuador
ES	El Salvador
ET	Egypt
ETH	Ethiopia
F	France
FJL	Fiji Islands
FL	Liechtenstein
FR	Faeroes
GB	Great Britain and Northern Ireland
GBA	Alderney
GBG	Guernsey
GBJ	Jersey
GBM	Isle of Man

Code	Country
GBZ	Gibraltar
GCA	Guatemala
GH	Ghana
GR	Greece
GUY	Guyana
H	Hungary
HK	Hong Kong
HV	Upper Volta
I	Italy
IL	Israel
IND	India
IR	Iran
IRL	Ireland (Eire)
IRQ	Iraq
IS	Iceland
J	Japan
JA	Jamaica
JOR	Jordan
K	Cambodia
KWT	Kuwait
L	Luxembourg
LAO	Laos
LAR	Libya
LB	Liberia
LS	Lesotho
M	Malta
MA	Morocco
MAL	Malaysia
MC	Monaco
MEX	Mexico
MS	Mauritius
MW	Malawi
N	Norway
NA	Netherlands Antilles
NIC	Nicaragua
NL	Netherlands
NZ	New Zealand
P	Portugal
PA	Panama
PAK	Pakistan
PE	Peru
PL	Poland
PNG	Papua-New Guinea
PRI	Puerto Rico
PRK	People's Rep. Korea (North Korea)
PY	Paraguay
Q	Qatar
RA	Argentina
RB	Botswana
RC	Taiwan
RCA	Central African Republic
RCB	Congo

Code	Country
RCH	Chile
RFC	Cameroon
RH	Haiti
RI	Indonesia
RIM	Mauritania
RL	Lebanon
RM	Madagascar
RMM	Mali
RN	Niger
RO	Romania
ROK	South Korea
ROU	Uruguay
RP	Philippines
RPB	Benin
RSM	San Marino
RWA	Ruanda
S	Sweden
SA	Saudi Arabia
SCV	Vatican
SD	Swaziland
SF	Finland
SGP	Singapore
SME	Surinam
SN	Senegal
SP	Somalia
STL	Windward Islands St. Lucia
SU	Soviet Union
SY	Seychelles
SYR	Syria
TG	Togo
THA	Thailand
TJ	People's Rep. China
TN	Tunisia
TR	Turkey
TT	Trinidad and Tobago
USA	United States of America
VN	Vietnam
WAG	Gambia
WAL	Sierra Leone
WAN	Nigeria
WD	Dominica
WG	Grenada
WS	Samoa
WV	Windward Islands St. Vincent
Y	Arabic Rep. Yemen (North Yemen)
YU	Yugoslavia
YV	Venezuela
Z	Zambia
ZA	South Africa
ZRE	Zaire
ZW	Zimbabwe

Transliteration scheme for Cyrillic

Author names and titles originally in *Cyrillic* have been transliterated into the Roman alphabet using the following scheme. (It is the same as the scheme curently used by Zbl and differs only slightly from that used by MR.)

Cyrillic		Roman
а	А	a
б	Б	b
в	В	v
г	Г	g
д	Д	d
е(ё)	Е(Ё)	e
ж	Ж	zh
з	З	z
и	И	i
й	Й	j
к	К	k

Cyrillic		Roman
л	Л	l
м	М	m
н	Н	n
о	О	o
п	П	p
р	Р	r
с	С	s
т	Т	t
у	У	u
ф	Ф	f
х	Х	kh

Cyrillic		Roman
ц	Ц	ts
ч	Ч	ch
ш	Ш	sh
щ	Щ	shch
ъ	Ъ	”
ы	Ы	y
ь	Ь	’
э	Э	eh
ю	Ю	yu
я	Я	ya